T0205719

# Lecture Notes in Computer Science 11506

More information about this series at http://www.springer.com/series/7407

Ignacio Rojas · Gonzalo Joya · Andreu Catala (Eds.)

# Advances in Computational Intelligence

## 15th International Work-Conference on Artificial Neural Networks, IWANN 2019 Gran Canaria, Spain, June 12–14, 2019 Proceedings, Part I

*Editors*
Ignacio Rojas
University of Granada
Granada, Spain

Andreu Catala
Polytechnic University of Catalonia
Barcelona, Spain

Gonzalo Joya
University of Malaga
Malaga, Spain

ISSN 0302-9743 ISSN 1611-3349 (electronic)
Lecture Notes in Computer Science
ISBN 978-3-030-20520-1 ISBN 978-3-030-20521-8 (eBook)
https://doi.org/10.1007/978-3-030-20521-8

LNCS Sublibrary: SL1 – Theoretical Computer Science and General Issues



This Springer imprint is published by the registered company Springer Nature Switzerland AG
The registered company address is: Gewerbestrasse 11, 6330 Cham, Switzerland

# Preface

We are proud to present the set of final accepted papers for the 13th edition of the IWANN conference—the International Work-Conference on Artificial Neural Networks—held in Gran Canaria, (Spain) during June 12–14, 2019.

IWANN is a biennial conference that seeks to provide a discussion forum for scientists, engineers, educators, and students about the latest ideas and realizations in the foundations, theory, models, and applications of hybrid systems inspired by nature (neural networks, fuzzy logic and evolutionary systems) as well as in emerging areas related to these topics. As in previous editions of IWANN, it also aims to create a friendly environment that could lead to the establishment of scientific collaborations and exchanges among attendees. The proceedings will include all the communications presented at the conference. A publication of an extended version of selected papers in a special issue of several specialized journals (such as *Neural Computing and Applications, PLOS One,* and *Neural Proccesing Letters*) is also foreseen.

Since the first edition in Granada (LNCS 540, 1991), the conference has evolved and matured. The list of topics in the successive Call for Papers has also evolved, resulting in the following list for the present edition:

1. **Mathematical and theoretical methods in computational intelligence**. Mathematics for neural networks. RBF structures. Self-organizing networks and methods. Support vector machines and kernel methods. Fuzzy logic. Evolutionary and genetic algorithms.
2. **Neurocomputational formulations**. Single-neuron modeling. Perceptual modeling. System-level neural modeling. Spiking neurons. Models of biological learning.
3. **Learning and adaptation**. Adaptive systems. Imitation learning. Reconfigurable systems. Supervised, non-supervised, reinforcement, and statistical algorithms.
4. **Emulation of cognitive functions**. Decision-making. Multi-agent systems. Sensor mesh. Natural language. Pattern recognition. Perceptual and motor functions (visual, auditory, tactile, virtual reality, etc.). Robotics. Planning motor control.
5. **Bio-inspired systems and neuro-engineering**. Embedded intelligent systems. Evolvable computing. Evolving hardware. Microelectronics for neural, fuzzy and bioinspired systems. Neural prostheses. Retinomorphic systems. Brain–computer interfaces (BCI) Nanosystems. Nanocognitive systems.
6. **Advanced topics in computational intelligence**. Intelligent networks. Knowledge-intensive problem-solving techniques. Multi-sensor data fusion using computational intelligence. Search and meta-heuristics. Soft computing. Neuro-fuzzy systems. Neuro-evolutionary systems. Neuro-swarm. Hybridization with novel computing paradigms.
7. **Applications**. Expert systems. Image and signal processing. Ambient intelligence. Biomimetic applications. System identification, process control, and manufacturing. Computational biology and bioinformatics. Parallel and distributed computing. Human-computer interaction, Internet modeling, communication and networking.

Intelligent systems in education. Human–robot interaction. Multi-agent systems. Time series analysis and prediction. Data mining and knowledge discovery.

At the end of the submission process, and after a careful peer review and evaluation process (each submission was reviewed by at least 2, and on average 2.9, Program Committee members or additional reviewers), 150 papers were accepted for oral or poster presentation, according to the reviewers' recommendations and the authors' preferences.

In this edition of IWANN 2019, a workshop entitled "Artificial Intelligence in Nanophotonics" was presented, organized by Dr. Nikolay Zheludev, University of Southampton, UK, and NTU Singapore and Dr. Cesare Soci, NTU, Singapore.

During IWANN 2019, several special sessions were held. Special sessions are a very useful tool for complementing the regular program with new and emerging topics of particular interest for the participating community. Special sessions that emphasize multi-disciplinary and transversal aspects, as well as cutting-edge topics, are especially encouraged and welcome, and in this edition of IWANN 2019 comprised the following:

- **SS01: Artificial Neural Network for Biomedical Image Processing**
  Organized by: Dr. Yu-Dong Zhan
- **SS02: Deep Learning Models in Health Care and Biomedicine**
  Organized by: Dr. Leonardo Franco, Dr. Ruxandra Stoean and Dr. Francisco Veredas
- **SS03: Deep Learning Beyond Convolution**
  Organized by: Dr. Miguel Atencia
- **SS04: Machine Learning in Vision and Robotics**
  Organized by: Dr. José García-Rodríguez, Dr. Enrique Domínguez and Dr. Ramón Moreno
- **SS05: Data-Driven Intelligent Transportation Systems**
  Organized by: Dr. Ignacio J. Turías Domínguez, Dr. David Elizondo and Dr. Francisco Ortega Zamorano
- **SS06: Software Testing and Intelligent Systems**
  Organized by: Dr. Juan Boubeta, Dr. Pablo C. Cañizares and Dr. Gregorio Díaz
- **SS07: Deep Learning and Natural Language Processing**
  Organized by: Dr. Leonor Becerra-Bonache, Dr. M. Dolores Jiménez-López and Dr. Benoit Favre
- **SS08: Random-Weights Neural Networks**
  Organized by: Dr. Claudio Gallicchio
- **SS09: New and Future Tendencies in Brain–Computer Interface Systems**
  Organized by: Dr. Ricardo Ron and Dr. Ivan Volosyak
- **SS10: Human Activity Recognition**
  Organized by: Dr.-Ing. habil. Matthias Pätzold
- **SS11: Computational Intelligence Methods for Time Series**
  Organized by: Dr. Héctor Pomares
- **SS12: Advanced Methods for Personalized/Precision Medicine**
  Organized by: Dr. Luis Javier Herrera and Dr. Fernando Rojas

- **SS13: Exploring Document Information to Improve Neural Summarization Models**
  Organized by: Dr. Luigi Di Caro
- **SS15: Machine Learning in Weather Observation and Forecasting**
  Organized by: Dr. Juan Luis Navarro-Mesa, Dr. Antonio Ravelo-García and Dr. Carmen Paz Suárez Araujo

In this edition of IWANN, we were honored to have the presence of the following invited speakers:

1. Dr. Nuria Oliver, Director of Research in Data Science, Vodafone Chief Data Scientist, Data-Pop Alliance
2. Dr. Aureli Soria-Frisch, Director of Neuroscience, Starlab Consulting Division
3. Dr. Jose C. Principe, Distinguished Professor ECE, Eckis Professor of ECE, Director Computational NeuroEngineering Lab, University of Florida
4. Dr. Marin Soljacic, Professor of Physics at MIT

It is important to note that for the sake of consistency and readability of the book the presented papers are not organized as they were presented in the IWANN 2019 sessions, but classified under 22 chapters. The papers are organized in two volumes arranged basically following the topics list included in the call for papers. The first volume (LNCS 11506), entitled *Advances in Computational Intelligence. IWANN 2019. Part I*, is divided into ten main parts and includes contributions on:

1. Machine learning in weather observation and forecasting
2. Computational intelligence methods for time series
3. Human activity recognition
4. New and future tendencies in brain–computer interface systems
5. Random-weights neural networks
6. Pattern recognition
7. Deep learning and natural language processing
8. Software testing and intelligent systems
9. Data-driven intelligent transportation systems
10. Deep learning models in health care and biomedicine

In the second volume (LNCS 11507), entitled *Advances in Computational Intelligence. IWANN 2019. Part II*, is divided into 12 main parts and includes contributions on:

1. Deep learning beyond convolution
2. Artificial neural network for biomedical image processing
3. Machine learning in vision and robotics
4. System identification, process control, and manufacturing
5. Image and signal processing
6. Soft computing
7. Mathematics for neural networks
8. Internet modeling, communication, and networking
9. Expert systems

10. Evolutionary and genetic algorithms
11. Advances in computational intelligence
12. Computational biology and bioinformatics

The 14th edition of the IWANN conference was organized by the University of Granada, University of Malaga, and Polytechnical University of Catalonia. We wish to thank to the University of Gran Canaria for their support and grants.

We would also like to express our gratitude to the members of the different committees for their support, collaboration, and good work. We especially thank our honorary chairs (Prof. Joan Cabestany, Prof. Alberto Prieto and Prof. Francisco Sandoval), the technical program chairs (Prof. Miguel Atencia, Prof. Francisco García-Lagos, Prof. Luis Javier Herrera and Prof. Fernando Rojas), the local Organizing Committee (Prof. Domingo J. Benítez Díaz, Prof. Carmen Paz Suárez Araujo and Prof. Juan Luis Navarro Mesa), the Program Committee, the reviewers, invited speaker, and special session organizers. Finally, we want to thank Springer and especially Alfred Hofmann and Anna Kramer for their continuous support and cooperation.

June 2019

Ignacio Rojas
Gonzalo Joya
Andreu Catala

# Organization

## Program Committee

| | |
|---|---|
| Kouzou Abdellah | Djelfa University, Algeria |
| Vanessa Aguiar-Pulido | Weill Cornell Medicine, Cornell University, USA |
| Arnulfo Alanis Garza | Instituto Tecnologico de Tijuana, Mexico |
| Ali Alkaya | Marmara University, Turkey |
| Amparo Alonso-Betanzos | University of A Coruña, Spain |
| Jhon Edgar Amaya | University of Tachira, Venezuela |
| Gabriela Andrejkova | Slovakia |
| Davide Anguita | University of Genoa, Italy |
| Javier Antich Tobaruela | University of the Balearic Islands, Spain |
| Alfonso Ariza | University of Málaga, Spain |
| Angelo Arleo | CNRS - University Pierre and Marie Curie Paris VI, France |
| Corneliu Arsene | SC IPA SA, Romania |
| Miguel Atencia | University of Málaga, Spain |
| Jorge Azorín-López | University of Alicante, Spain |
| Antonio Bahamonde | University of Oviedo at Gijón, Asturias, Spain |
| Halima Bahi | University of Annaba, Algeria |
| Javier Bajo | Polytechnic University of Madrid, Spain |
| Juan Pedro Bandera Rubio | ISIS Group, University of Malaga, Spain |
| Oresti Banos | University of Granada, Spain |
| Bruno Baruque | University of Burgos, Spain |
| Leonor Becerra Bonache | Laboratoire Hubert Curien, France |
| Lluís Belanche | Universitat Politècnica de Catalunya, Spain |
| Sergio Bermejo | Universitat Politècnica de Catalunya, Spain |
| Francisco Bonin-Font | University of the Balearic Islands, Spain |
| Juan Boubeta-Puig | University of Cádiz, Spain |
| Antoni Burguera | Universitat de les Illes Balears, Spain |
| Pablo C. Cañizares | Complutense University of Madrid, Spain |
| Tomasa Calvo | University of Alcala, Spain |
| Azahara Camacho | Carbures Defense, Spain |
| David Camacho | Autonomous University of Madrid, Spain |
| Francesco Camastra | University of Naples Parthenope, Italy |
| Hoang-Long Cao | Vrije Universiteit Brussel, Belgium |
| Carlos Carrascosa | GTI-IA DSIC University Politecnica de Valencia, Spain |
| Pedro Castillo | University of Granada, Spain |
| Andreu Catala | Universitat Politècnica de Catalunya, Spain |
| Ana Cavalli | Institut Mines-Telecom/Telecom SudParis, France |

| | |
|---|---|
| Miguel Cazorla | University of Alicante, Spain |
| Wei Chen | Fudan University, China |
| Valentina Colla | Scuola Superiore S. Anna, Italy |
| Francesco Corona | Aalto University, Finland |
| Ulises Cortés | Universitat Politècnica de Catalunya, Spain |
| Marie Cottrell | SAMM Université Paris 1 Panthéon-Sorbonne, France |
| Raúl Cruz-Barbosa | University Tecnológica de la Mixteca, Mexico |
| Erzsébet Csuhaj-Varjú | Eötvös Loránd University, Hungary |
| Daniela Danciu | University of Craiova, Romania |
| Angel Pascual Del Pobil | University of Jaume I, Spain |
| Enrique Dominguez | University of Malaga, Spain |
| Richard Duro | Universidade da Coruna, Spain |
| Gregorio Díaz | University of Castilla - La Mancha, Spain |
| David Elizondo | Centre for Computational Intelligence, UK |
| Enrique Fernandez-Blanco | University of A Coruña, Spain |
| Carlos Fernandez-Lozano | University of A Coruña, Spain |
| Jose Manuel Ferrandez | P. University of Cartagena, Spain |
| Oscar Fontenla-Romero | University of A Coruña, Spain |
| Leonardo Franco | University of Málaga, Spain |
| Claudio Gallicchio | University of Pisa, Italy |
| Esther Garcia Garaluz | Eneso Tecnología de Adaptación SL, Spain |
| Francisco Garcia-Lagos | University of Malaga, Spain |
| Jose Garcia-Rodriguez | University of Alicante, Spain |
| Pablo García Sánchez | University of Granada, Spain |
| Rodolfo García-Bermúdez | University Técnica de Manabí, Ecuador |
| Angelo Genovese | University of Milan, Italy |
| Peter Gloesekoetter | Münster University of Applied Sciences, Germany |
| Juan Gomez Romero | University of Granada, Spain |
| Karl Goser | Technical University Dortmund, Germany |
| Manuel Graña | UPV/EHU, Spain |
| Jose Guerrero | Universitat de les Illes Balears, Spain |
| Bertha Guijarro-Berdiñas | University of A Coruña, Spain |
| Nicolás Guil Mata | University of Málaga, Spain |
| Alberto Guillen | University of Granada, Spain |
| Pedro Antonio Gutierrez | University of Cordoba, Spain |
| F. Luis Gutiérrez Vela | University of Granada, Spain |
| Marco A. Gómez-Martín | Complutense University of Madrid, Spain |
| Luis Herrera | University of Granada, Spain |
| Cesar Hervas | University of Cordoba, Spain |
| Mercedes Hidalgo-Herrero | Complutense University of Madrid, Spain |
| Wei-Chiang Hong | Jiangsu Normal University, China |
| Petr Hurtik | IRAFM, Czechia |
| Jose M. Jerez | University of Málaga, Spain |
| M. Dolores Jimenez-Lopez | Rovira i Virgili University, Spain |
| Juan Luis Jiménez Laredo | Université du Havre Normandie, France |
| Gonzalo Joya | University of Málaga, Spain |

| | |
|---|---|
| Vicente Julian | Universitat Politècnica de València, Spain |
| Raul Lara-Cabrera | Polytechnic University of Madrid, Spain |
| Nuno Lau | University of Aveiro, Portugal |
| Amaury Lendasse | University of Houston, USA |
| Otoniel Lopez Granado | Miguel Hernandez University, Spain |
| Rafael M. Luque-Baena | University of Extremadura, Spain |
| Fernando López Pelayo | University of Castilla-La Mancha, Spain |
| Ezequiel López-Rubio | University of Málaga, Spain |
| Kurosh Madani | LISSI/Université Paris-EST Creteil, France |
| Mario Martin | Universitat Politècnica de Catalunya, Spain |
| Bonifacio Martin Del Brio | University of Zaragoza, Spain |
| Luis Martí | University Federal Fluminense, Brazil |
| Montserrat Mateos | Pontifical University of Salamanca, Spain |
| Jesús Medina | University of Cádiz, Spain |
| Mercedes Merayo | Complutense University of Madrid, Spain |
| Gustavo Meschino | National University of Mar del Plata, Argentina |
| Salem Mohammed | Mustapha Stambouli University, Algeria |
| Jose M. Molina | University Carlos III de Madrid, Spain |
| Augusto Montisci | University of Cagliari, Italy |
| Antonio Mora | University of Granada, Spain |
| Jose A. Moral-Munoz | University of Cadiz, Spain |
| Gines Moreno | University of Castilla-La Mancha, Spain |
| Ramón Moreno | IK4-LOREK, Spain |
| Juan Moreno Garcia | University of Castilla-La Mancha, Spain |
| Juan L. Navarro-Mesa | University of Las Palmas de Gran Canaria, Spain |
| Nadia Nedjah | State University of Rio de Janeiro, Brazil |
| Bizdoaca Nicu | University of Craiova, Romania |
| Alberto Núñez | Complutense University of Madrid, Spain |
| Manuel Núñez | Complutense University of Madrid, Spain |
| Ivan Olier | Liverpool John Moores University, UK |
| Madalina Olteanu | SAMM, Université Paris 1, France |
| Alberto Ortiz | Universitat de les Illes Balears, Spain |
| Osvaldo Pacheco | University of Aveiro, Portugal |
| Patricia Paderewski | University of Granada, Spain |
| Esteban José Palomo | University of Malaga, Spain |
| Miguel Angel Patricio | University Carlos III de Madrid, Spain |
| Ricardo Pinto Ferreira | University Nove de Julho, Brazil |
| Vincenzo Piuri | University of Milan, Italy |
| Hector Pomares | University of Granada, Spain |
| Alberto Prieto | University of Granada, Spain |
| Alexandra Psarrou | University of Westminster, UK |
| Francisco A. Pujol | University of Alicante, Spain |
| Matthias Pätzold | University of Agder, Norway |
| María Pérez Ortiz | University of Córdoba, Spain |
| Pablo Rabanal | Complutense University of Madrid, Spain |
| Juan Rabuñal | University of A Coruña, Spain |

| | |
|---|---|
| Vladimir Rasvan | University of Craiova, Romania |
| Antonio Ravelo-García | University of Las Palmas de Gran Canaria, Spain |
| Ismael Rodriguez | Complutense University of Madrid, Spain |
| Fernando Rojas | University of Granada, Spain |
| Ignacio Rojas | University of Granada, Spain |
| Ricardo Ron-Angevin | University of Málaga, Spain |
| Francesc Rossello | University of the Balearic Islands, Spain |
| Fabrice Rossi | SAMM - Université Paris 1, France |
| Peter M. Roth | Graz University of Technology, Austria |
| Fernando Rubio | Complutense University of Madrid, Spain |
| Ulrich Rueckert | Bielefeld University, Germany |
| Addisson Salazar | Universitat Politècnica de València, Spain |
| Francisco Sandoval | University of Málaga, Spain |
| Jorge Santos | ISEP, Portugal |
| Jose Santos | University of A Coruña, Spain |
| Jose A. Seoane | Stanford Cancer Institute, Stanford University, USA |
| Cesare Soci | Nanyang Technological University, Singapore |
| Jordi Solé-Casals | University of Vic - Central University of Catalonia, Spain |
| Catalin Stoean | University of Craiova, Romania |
| Ruxandra Stoean | University of Craiova, Romania |
| Carmen Paz Suárez-Araujo | University Las Palmas de Gran Canaria, Spain |
| Peter Szolgay | Pazmany Peter Catholic University, Hungary |
| Claude Touzet | Aix-Marseille University, France |
| Ignacio Turias | University of Cádiz, Spain |
| Daniel Urda | University of Cádiz, Spain |
| Olga Valenzuela | University of Granada, Spain |
| Oscar Valero | University of las Islas Baleares, Spain |
| Francisco Velasco-Alvarez | University of Málaga, Spain |
| Marley Vellasco | Pontifical Catholic University of Rio de Janeiro, Brazil |
| Alfredo Vellido | Universitat Politècnica de Catalunya, Spain |
| Francisco J. Veredas | University of Málaga, Spain |
| Ivan Volosyak | Rhine-Waal University of Applied Sciences, Germany |
| Yudong Zhang | Nanjing Normal University, China |
| Nikolay I. Zheludev | University of Southampton, UK |
| Igor Zubrycki | Lodz University of Technology, Poland |
| Juan Antonio Álvarez García | University of Seville, Spain |

## Additional Reviewers

Abdelgawwad, Ahmed
Almendros-Jimenez, Jesus M.
Azorín-López, Jorge
Basterrech, Sebastian
Benito-Picazo, Jesus
Bermejo, Sergio
Borhani, Alireza
Brazalez-Segovia, Enrique

Cazorla, Miguel
Cuartero, Fernando
Dapena, Adriana
Delecraz, Sebastien
Duro, Richard
Escalona, Félix
Fuster-Guillo, Andres
Garcia-Garcia, Alberto
García-González, Jorge
Gomez-Donoso, Francisco
Gorostegui, Eider
Graña, Manuel
Hicheri, Rym
Hinaut, Xavier
Hoermann, Timm
Korthals, Timo
Kouzou, Abdellah
Lachmair, Jan
Luque-Baena, Rafael M.
López-García, Guillermo
López-Rubio, Ezequiel
Macià Soler, Hermenegilda
Mattos, César Lincoln
McCabe, Philippa Grace
Medina-Bulo, Inmaculada
Molina-Cabello, Miguel A.
Muaaz, Muhammad
Muniategui, Ander
Nguyen, Huu Nghia
Oneto, Luca
Oprea, Sergiu
Ortiz-De-Lazcano-Lobato, Juan Miguel
Orts-Escolano, Sergio
Palomo, Esteban José
Pedrelli, Luca
Riaza Valverde, José Antonio
Riley, Patrick
Rincon, Jaime A.
Ruiz Delgado, M. Carmen
Safont, Gonzalo
Saval-Calvo, Marcelo
Scardapane, Simone
Segovia, Mariana
Thurnhofer-Hemsi, Karl

# Contents – Part I

**New and Future Tendencies in Brain-Computer Interface Systems**



**Random-Weights Neural Networks**

**Software Testing and Intelligent Systems**

**Data-Driven Intelligent Transportation Systems**

## Deep Learning Models in Healthcare and Biomedicine

# Contents – Part II

## Machine Learning in Vision and Robotics

**Image and Signal Processing**



**Soft Computing**

**Expert Systems**



**Evolutionary and Genetic Algorithms**

**Advances in Computational Intelligence**

## Computational Biology and Bioinformatics

# Machine Learning in Weather Observation and Forecasting

# A Deeper Look into 'Deep Learning of Aftershock Patterns Following Large Earthquakes': Illustrating First Principles in Neural Network Physical Interpretability

Arnaud Mignan(✉) and Marco Broccardo

Swiss Federal Institute of Technology, 8092 Zurich, Switzerland
arnaud.mignan@sed.ethz.ch

**Abstract.** In the last years, deep learning has solved seemingly intractable problems, boosting the hope to find (approximate) solutions to problems that now are considered unsolvable. Earthquake prediction - a recognized moonshot challenge - is obviously worthwhile exploring with deep learning. Although encouraging results have been obtained recently, deep neural networks (DNN) may sometimes create the illusion that patterns hidden in data are complex when this is not necessarily the case. We investigate the results of De Vries et al. [*Nature,* vol. 560, 2018] who defined a DNN of 6 hidden layers with 50 nodes each, and with an input layer of 12 stress features, to predict aftershock patterns in space. The performance of their DNN was assessed using ROC with AUC = 0.85 obtained. We first show that a simple artificial neural network (ANN) of 1 hidden layer yields a similar performance, suggesting that aftershock patterns are not necessarily highly abstract objects. Following first principle guidance, we then bypass the elastic stress change tensor computation, making profit of the tensorial nature of neural networks. AUC = 0.85 is again reached with an ANN, now with only two geometric and kinematic features. Not only seems deep learning to be "excessive" in the present case, the simpler ANN streamlines the process of aftershock forecasting, limits model bias, and provides better insights into aftershock physics and possible model improvement. Complexification is a controversial trend in all of Science and first principles should be applied wherever possible to gain physical interpretations of neural networks.

**Keywords:** Aftershock modelling · Pattern recognition · Applied deep learning

## 1 Introduction

Deep learning is rapidly rising as the go-to technique not only in data science [1, 2] but also for solving hard problems of Physics [3–5]. This is justified by the superior performance of deep learning in discovering complex patterns in very large datasets. One of the major advantages of deep neural networks (DNN) is that, generally, there is no need for feature extraction and engineering, as data can be used directly to train the network with great results. It comes as no surprise that DNNs are also becoming


I. Rojas et al. (Eds.): IWANN 2019, LNCS 11506, pp. 3–14, 2019.
https://doi.org/10.1007/978-3-030-20521-8_1

popular in statistical seismology [6] and give fresh hope for earthquake prediction [7], a challenge which has long been considered impossible [8]. The black-box nature of neural networks (NN) was considered an advantage in early attempts of predicting earthquakes from apparently complex patterns [9, 10]. Such a view is only valid in the context of pragmatic model applicability, not physical interpretability.

Recently, De Vries et al. [11], hereafter referred to as DeVries18, proposed a DNN to study the spatial distribution of aftershocks in the aftermath of a main seismic event. The goal of the authors was to design a stress-change based classifier for determining the spatial likelihood of aftershocks. Once the DNN was trained and tested (see Sect. 2.2 for details on the dataset and model), they provided a physical interpretation of the aftershock pattern. In particular, they analyzed three stress metrics and showed that each alone could lead to similar results as their DNN classifier. In this regard, the DeVries18 study is an improvement over the back-box approach [9, 10].

Designing a suitable DNN is a highly iterative process, based on hyper-parameterization tuning and, sometimes (even if not highly recommendable in this context), feature engineering. How do such choices affect, not only model performance, but physical interpretability? In view of the flexibility of deep learning, can we miss first-principles in the modelling process? We aim at answering these questions in the context of aftershock spatial pattern prediction, taking the DeVries18 study as baseline model.

## 2 Artificial Neural Networks in Statistical Seismology

### 2.1 Literature Survey

**The Earthquake Prediction Challenge.** '*The subject of Statistical Seismology aims to bridge the gap between physics-based models without statistics, and statistics-based models without physics*' [12] - This scientific domain can be divided roughly into two categories, with earthquakes as point sources (i.e. seismicity) in stochastic non-stationary processes or earthquakes as seismic waves radiating from finite sources.

Earthquake prediction remains the Grail of Seismology, as well as a moonshot challenge for all of society that is under the threat of large earthquakes. Earthquake predictability research has already gone through several cycles of enchantments and disillusions [13] with earthquake physics remaining derivative: It is still in a cataloguing phase of seismicity patterns, akin the naturalists collecting animals and plants in past centuries before modern biology emerged. It remains to be identified whether earthquake patterns are complex, in the holistic sense of the Critical Point Theory [14] which relates to Thermodynamics and Chaos, or complicated, in the reductionist sense of the recent Solid Seismicity Theory [15–17] which relates to Geometry. In this context, machine learning should help improving earthquake pattern recognition. While those algorithms are theoretically agnostic, they may give some insights into the theoretical directions to follow. For a review of pattern recognition algorithms used specifically for earthquake forecasting, see [18].

It should be mentioned that aftershock prediction (or forecasting) is different and much easier than mainshock prediction since aftershocks follow well known statistics in time (see review by [19]), space [17, 20, 21] and productivity [17, 22]. Operational aftershock forecasting is thus already possible [23].

**Neural Network Models for Earthquake Pattern Prediction.** The earliest attempts to apply ANNs to earthquake predictions dates back, to the best of our knowledge, to 1994 [24, 25]. Few studies followed in the next decade [9, 26, 27]. The next milestone was in 2007 with the comprehensive work of Panakkat and Adeli [28], one of the first to use deep learning (2 hidden layers) as well as radial basis function (RBF) networks, and the first to test recurrent neural networks (RNN). Neural network research for earthquake prediction took off from there, but mostly from a computer science perspective and from a limited number of teams [7, 29, 30].

All the aforementioned works used structured seismic data as input, i.e. earthquake catalogues in tabular format with location, occurrence time, and magnitude of events (except [24] who used seismic electric signals). For two different approaches to feature engineering on structured data, compare [27] for a financial market approach to e.g. [30] for a 'seismicity law' approach. The most common outputs are the predicted magnitude, occurrence time and location of large earthquakes. Overall, mixed results were obtained so far, with performance decreasing with increasing mainshock magnitude [7, 10]. The gain of using neural networks instead of simpler methods remains unclear since performance was rarely compared to a baseline model.

The most recent DNN model was proposed by De Vries et al. [11], which was different from previous studies in various ways. First, it did not try to predict mainshock characteristics but the spatial patterns of aftershocks (an early attempt at predicting aftershock spatial distribution had already been done by [31] but for one sequence only). DeVries18 used a global earthquake catalogue for aftershock binary classification (aftershocks present or not in geographic cells) but with features engineered from stress computed from mainshock rupture models. Their DNN model will be fully described in Sect. 2.2, and critically assessed and improved upon in Sect. 3 in terms of physical interpretability.

Let us mention that raw earthquake data is unstructured, in the form of seismic wave timeseries. To the best of our knowledge, neural networks have not yet been applied to seismic wave data to predict earthquakes. However Random Forests have recently been used to predict lab earthquakes [32]. We should also mention earthquake early warning (EEW), which consists in predicting the arrival time of S waves and surface waves based on P wave data (which is however not to be confused with earthquake prediction). This work was pioneered with NNs in 1996 [33] but NN applications in EEW still remain at an early stage. The best-known example is the MyShake project where a neural network classifies the amplitude and frequency content of smartphone movements to discriminate early earthquake shaking from human activities [34].

With the recent advances in applying CNNs [35, 36] and RNNs [37] to automatically pick seismic waves, one could easily imagine applying those techniques to predict a mainshock based on foreshock seismic waves [38], as already tested with Random Forests [32]. Moreover, those techniques tend to improve the quality of earthquake catalogues by increasing the number of events ten-folds [35]. which can in turn be used as features in earthquake prediction based on structured data. A recent meta-analysis indeed showed that an increase in the amount of micro-seismicity improves anomaly detection [39]. Hence, CNN-based earthquake catalogues would likely help improving model performances. A semi-supervised mixture model has also recently been proposed towards that goal [40]. Finally, autoencoder networks [41] and generative adversarial networks (GAN) [42] of seismic wave data could help reducing data dimensions, which could further facilitate feature engineering.

## 2.2 The DeVries18 Study

The DeVries18 study [11] defined a DNN for binary classification of spatial patterns of aftershocks. It was made of 6 hidden layers, each composed of 50 nodes with hyperbolic tangent activation and a 0.5 dropout rate, for a total of 13,451 weights and biases. Their input layer contained 12 nodes, representing features engineered from the elastic stress change tensor of the related mainshocks. Their workflow is illustrated in Fig. 1.

DeVries18 defined aftershocks as all events - as catalogued in the International Seismological Center (ISC) global catalogue [43] - which occurred between one second and one year after a mainshock, and within 100 km horizontally of the mainshock rupture (50 km vertically). They repeated the operation for 199 mainshocks which occurred between 1968 and 2012. Then they gridded aftershocks in 5 km-wide XYZ cells, each cell labelled 1 if it contained *at least* one aftershock, 0 otherwise. They retrieved mainshock rupture data (geometry, mechanism, slip) from the Finite-Source Rupture Model Database (SRCMOD) [44]. They computed the elastic stress tensor $\Delta\boldsymbol{\sigma}$ [45] at the centroid of each cell and finally defined their DNN input layer based on the 12 following features (normalized by $10^{-6}$): the absolute values of the six independent components of $\Delta\boldsymbol{\sigma}$, which are $|\Delta\sigma_{xx}|$, $|\Delta\sigma_{xy}|$, $|\Delta\sigma_{xz}|$, $|\Delta\sigma_{yy}|$, $|\Delta\sigma_{yz}|$, $|\Delta\sigma_{zz}|$, and their opposites $-|\Delta\sigma_{xx}|$, $-|\Delta\sigma_{xy}|$, $-|\Delta\sigma_{xz}|$, $-|\Delta\sigma_{yy}|$, $-|\Delta\sigma_{yz}|$, $-|\Delta\sigma_{zz}|$.

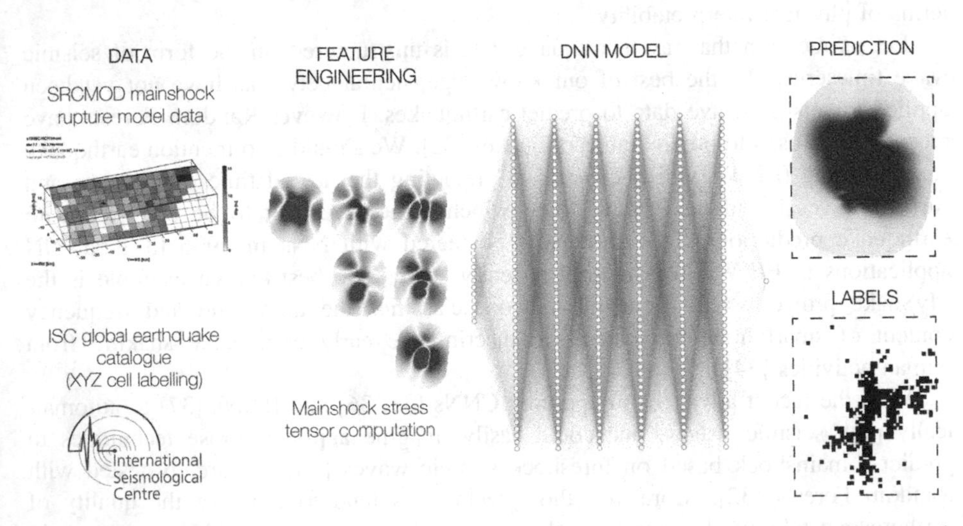


**Fig. 1.** The DeVries18 workflow to predict aftershock patterns with deep learning (illustrated with SRCMOD eventTag 1999CHICHI01MAxx).

Their model was trained on c. 75% of the data, with each XYZ cell considered one sample. They tested the performance of their DNN on the remaining 25% of the data, by accessing model accuracy via the Receiver Operating Characteristic (ROC) curve, calculating the Area Under the Curve (AUC). Based on their model input and topology, DeVries18 obtained AUC = 0.849, which appears impressive compared to AUC = 0.583 (near-random performance) obtained for the classical Coulomb failure criterion [46], which represents the main earthquake-triggering model of the current paradigm. It remains unclear why such low performance was obtained for Coulomb stress. It is possible that cherry-picking or overfitting had been previously achieved or it is maybe DeVries18, investigating Coulomb stress on a global dataset, that made assumptions which are too generic (e.g. by not investigating possible changes in the regional stress field [17]). This debate is worth mentioning but it is however outside the scope of the present paper.

Previous NN studies on earthquake prediction (Sect. 2.1) did not relate their findings to any physical process (black-box approach) although they assumed, explicitly or implicitly, that highly non-linear patterns were due to complex processes which can only be investigated holistically (i.e. mainshock as a critical point). To the best of our knowledge, only DeVries18 used their DNN results to seek for interpretable and meaningful physical patterns.

# 3 Applying First Principles to Neural Network Interpretability

## 3.1 Was High Abstraction Required to Predict Aftershock Patterns?

Deep learning has dramatically improved the state-of-the-art in computer vision, among other fields [1], and it is often accepted that defining a larger and deeper DNN does not hurt model performance. However, a deeper NN can be interpreted as a model of higher abstraction. In computer vision, for instance, a first layer may represent simple shapes, a second layer parts of a face (such as eye, nose, ear), and a third layer, different faces. When aftershock patterns are predicted by a 6-hidden-layer DNN [11], it captivates the collective imagination as to the degree of abstraction that seismicity patterns carry. This can explain a certain media euphoria about artificial intelligence predicting earthquakes [47–50]. This is unfortunately misleading. As proven below, a shallow neural network can predict aftershock patterns with a similar performance as the DNN of DeVries18.

We should first emphasize that DeVries18 did not flatten aftershock maps as in computer vision made with fully-connected feedforward networks, but used each XYZ cell as sample. This turns the study case from unstructured to structured data, with features defined in tabular form (X, Y, Z, stress features). Standard machine learning algorithms are performant on such data. This especially means that we only deal with a handful of features instead of the hundreds or thousands that usually call for deep learning.

We simplified the topology of the DeVries18 model and found, for example, that both a DNN with topology 12-8-8-1 (with dropout rate reduced to 0.2) or just an ANN with topology 12-30-1 yield a similar AUC of 0.85 (Fig. 2). The newly predicted aftershock patterns remained similar to the ones of DeVries18, with spatial lobes related to the spatial distribution of the stress features (similar as in Fig. 1). This demonstrates that deep learning is so far unnecessary for aftershock pattern prediction. We will show how simple aftershock patterns are in the next subsection.

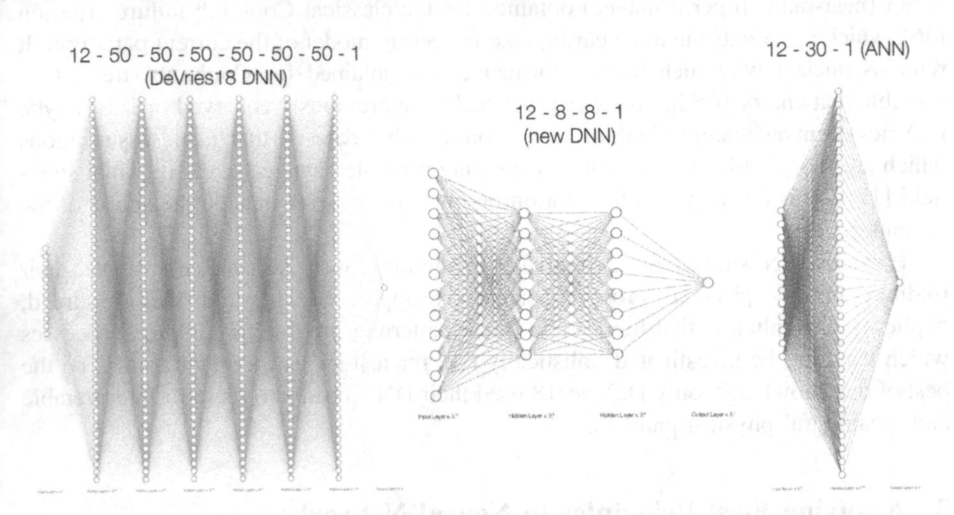


**Fig. 2.** Different topologies with similar AUC performances and similar predicted aftershock patterns. Network topology plots generated with alexlenail.me/NN-SVG/.

### 3.2 Were Stress Metrics the Most Pertinent Physical Parameters?

**Taking Advantage of the Tensorial Definition of Neural Networks.** NNs consist of chains of tensor operations, and following also the Universal Approximation Theorem, we can intuitively deduct that any stress tensor could be mimicked by a certain NN topology and weight combination.

Assuming the linearized theory of elasticity, the stress-change tensor can be generally written as

$$\Delta\boldsymbol{\sigma} = \boldsymbol{C}\boldsymbol{\varepsilon} \tag{1}$$

where $\boldsymbol{C}$ is the 4th order elasticity tensor (which, in the case of isotropic elasticity, has two independent constants, i.e. the Lamé parameters) and where $\boldsymbol{\varepsilon}$ is the linear strain tensor defined as the symmetric part of the displacement gradient

$$\boldsymbol{\varepsilon}(\boldsymbol{r},\boldsymbol{d}) = \frac{1}{2}\left(\nabla \boldsymbol{u}(\boldsymbol{r},\boldsymbol{d}) + \nabla \boldsymbol{u}^{T}(\boldsymbol{r},\boldsymbol{d})\right) \tag{2}$$

where $\boldsymbol{u}(\boldsymbol{r},\boldsymbol{d})$ is the displacement field at a distance $\boldsymbol{r}$ from the rupture, and $\boldsymbol{d}$ is the finite rupture displacement. Note the possible parallel between Eq. (1) and a neuron linear function $z = \mathbf{w}^{\mathrm{T}}x + \mathrm{b}$ where the activation $g(z)$ (tanh, relu, or else) could also relate to triggering of an earthquake above a threshold $\Delta\sigma$.

Following first principles, one shall define the NN input layer from displacement data directly, avoiding any model assumption. Having a neural network doing (eventually) a mapping from deformation (mainshock geometry and kinematics) back to deformation (simplified to aftershocks occurring or not) avoids making any assumption on stress (elasticity versus poroelasticity theory, plasticity, *etc.*), material properties (Lamé parameters), and other unknowns. In particular, this avoids having large uncertainties potentially affecting the quality of the classifier, or in other words, this avoids theoretical model bias. Recall that deformation is measurable and should be used as input layer while stress is derivative, representing subjective feature engineering.

**New Neural Network with Displacement-Based Input Layer.** The two variables in Eq. (2) are the distance to the rupture $r$ and the rupture displacement $d$. We here define $r$ as the minimum distance between cell XYZ and the mainshock rupture plane, and $d$ as the mean slip on the rupture obtained from the SRCMOD database [44] (observe that both $r$ and $d$ are scalars in this version and orthogonal features). $r$ represents a geometric feature and $d$ a kinematic feature. Such a simple parameterization of aftershock patterns is compatible with the observation that aftershocks occur closest to the mainshock rupture with their likelihood decreasing as a power-law with increasing distance [17, 20, 21]. Considering the mean slip provides a physical link to deformation and a way to scale aftershock patterns to the size of the mainshock [17].

Those two features are found sufficient to get a similar model performance as the DeVries18 model, with both a simple DNN 2-6-6-1 or with an ANN 2-30-1 yielding AUC = 0.84−0.86 (Fig. 3; note that we here changed tanh activation to rectified linear unit; a dropout rate of 0.2 was again used for the DNN). The predicted aftershock patterns are simpler and blurrier than the ones produced by DeVries18 but generalize well (Figs. 3 and 4).

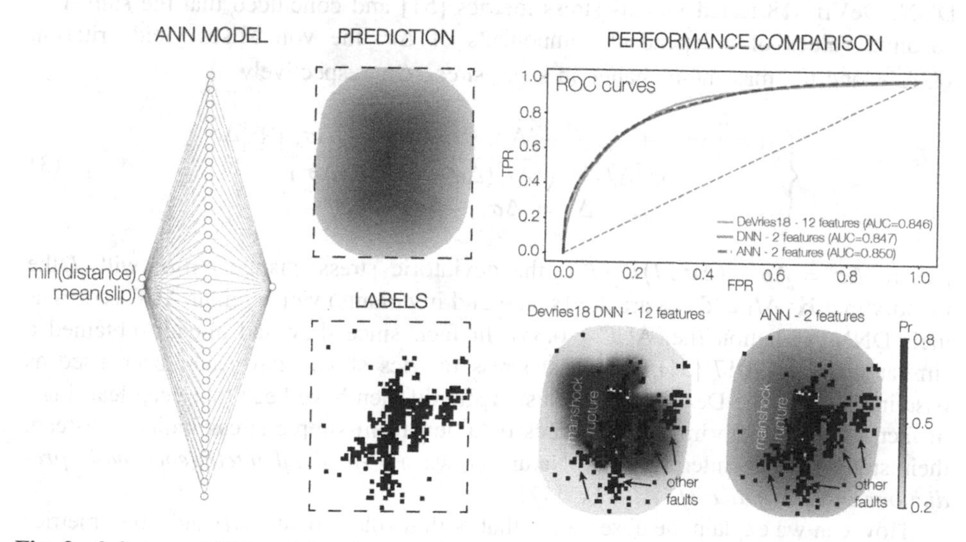


**Fig. 3.** 2-features ANN model and comparison with the DeVries18 deep learning performance. Aftershock maps for SRCMOD eventTag 1999CHICHI01MAxx; ROC curves for the full test set.

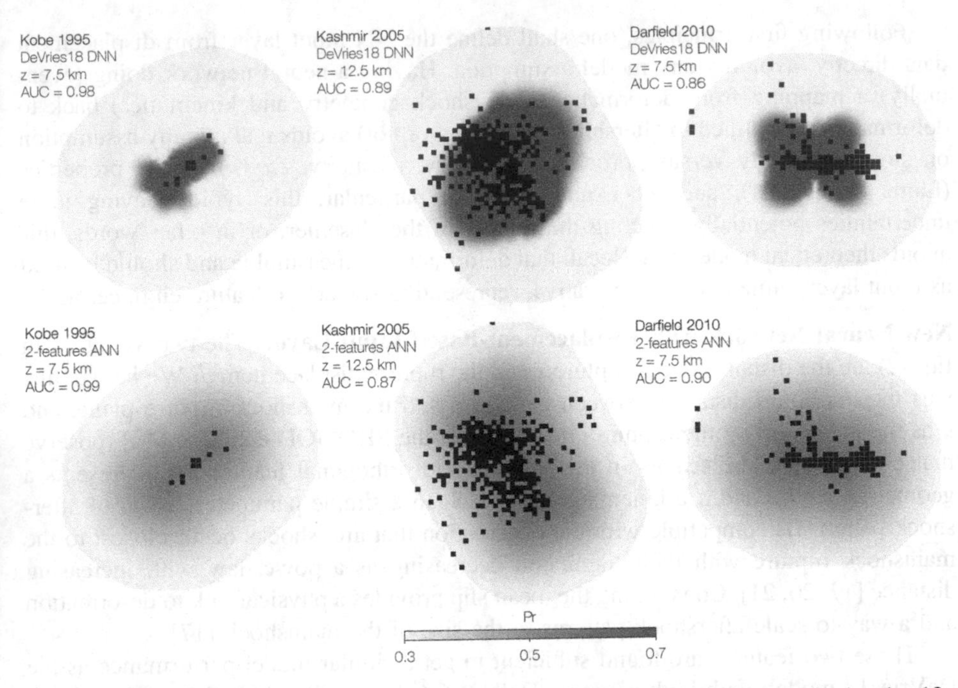


**Fig. 4.** Other examples from the test set of aftershock pattern predictions made by the 12-features DeVries18 DNN and the new 2-features ANN model (SRCMOD eventTags s1995KOBEJA01YOSH, s2005KASHMI01SHAO, and s2010DARFIE01ATZO).

**Interpreting the DeVries18 Stress-Based Input Layer.** In order to interpret their DNN, DeVries18 tested various stress metrics [51] and concluded that the sum $A$ of absolute values of independent components of $\Delta\boldsymbol{\sigma}$, the von Mises yield criterion $\sqrt{3\Delta J_2}$, and the maximum change in shear stress $\Delta\tau$, respectively

$$\begin{cases} A = |\Delta\sigma_{xx}| + |\Delta\sigma_{yy}| + |\Delta\sigma_{zz}| + |\Delta\sigma_{xy}| + |\Delta\sigma_{xz}| + |\Delta\sigma_{yz}| \\ \sqrt{3\Delta J_2} = \sqrt{\Delta I_1^2(\Delta\boldsymbol{\sigma}') - 3\Delta I_2(\Delta\boldsymbol{\sigma}')} \\ \Delta\tau = |\Delta\sigma_1 - \Delta\sigma_3|/2 \end{cases} \tag{3}$$

(where $\Delta\boldsymbol{\sigma}' = \Delta\boldsymbol{\sigma} - (\Delta\boldsymbol{\sigma} : I)/3 \cdot I$ is the deviatoric stress change tensor with $I$ the identity matrix; $\Delta I_1$ and $\Delta I_2$ are the 1st and 2nd invariants) yield similar AUC scores as their DNN prediction (i.e. AUC = 0.85). In fact, since they had already obtained a similar result in 2017 [51], their best stress metrics should have later been used as baseline model. The DeVries18 conclusion should then have been that deep learning - in fact - did not help improving scores obtained from simple stress indices. Instead, their study was misinterpreted by Nature News as *'Artificial intelligence nails predictions of earthquake aftershocks'* [47].

How can we explain the observation that both displacement ($r$, $d$) and stress metrics lead to similar performances? We see that in both the DeVries18 feature engineering and the three metrics of Eq. (3), only absolute values of the stress components are considered, which means that all dipolar information is lost. What remains at first order is the distance $r$ from mainshock rupture (subject to some rotations at second order) and a spatial scaling, which can be calibrated by mainshock rupture displacement $d$ (see e.g. [17] for an analytical expression of static stress changes along $r$ for different mainshock magnitudes). The top row of Fig. 4 indeed shows that aftershock patterns do not seem to be correlated to the stress lobe geometry but simply with distance from the mainshock rupture (bottom row).

Interestingly, the AUC seems to plateau at c. 0.85 for the full test set for any mainshock-based feature and all the tested hyper-parameterizations. This suggests that valuable information is currently missing from the models. Comparing stress-based and displacement-based pattern maps shows that some aftershock clusters occur on well-defined lineaments which are likely representative of other fault segments (Fig. 3). Deviations from seismicity spatial laws have already been related to the presence of faults at proximity of the mainshock rupture [16, 17, 21]. We suggest that model performance could further improve, once features based on fault network data are added.

## 4 Conclusions

The present work is not a purely intellectual exercise on Occam's Razor, nor an incremental work on neural network topology simplification and feature engineering. It has important implications in operational aftershock forecasting, physical research, and communication of AI results in general:

(1) By entirely bypassing stress computation, aftershock forecasting could be streamlined with the proposed geometric and kinematic features possibly retrievable in real-time. It also avoids making physical assumptions that may lead to some model bias (compare Fig. 3 to Fig. 1). For this, we took advantage of the tensorial nature of a neural network. We also used an ANN instead of a DNN in view of the very small number of features.

(2) By using first principles, we were able to show that aftershock patterns are simple after all, in agreement with the literature on the statistical properties of the spatial distribution of aftershock [17, 20, 21]. While this does not tell us what the physics is behind the mapping of mainshock deformation to aftershock deformation, it already clarifies the importance of distance to rupture and rupture slip as main physical measurable parameters.

(3) Deep learning branding helps to captivate the popular imagination. This, in itself, is fine for model applications and to revive potentially moribund topics. However, it has dangerous consequences in physical interpretability of neural networks, as demonstrated above. Whenever possible, first principles should be applied even if this means using shallower rather than deeper learning. This can only be decided on a case-to-case basis.

## References

1. LeCun, Y., Bengio, Y., Hinton, G.: Deep learning. Nature **521**, 436–444 (2015)
2. Jordan, M.I., Mitchell, T.M.: Machine learning: trends, perspectives, and prospects. Science **349**(6245), 255–260 (2015)
3. Carleo, G., Troyer, M.: Solving the quantum many-body problem with artificial neural networks. Science **355**, 602–606 (2017)
4. Han, J., Jentzen, A., Weinan, E.: Solving high-dimensional partial differential equations using deep learning. PNAS **115**(34), 8505–8510 (2018)
5. Pathak, J., Hunt, B., Girvan, M., Lu, Z., Ott, E.: Model-free prediction of large spatiotemporally chaotic systems from data: a reservoir computing approach. Phys. Rev. Lett. **120**, 024102 (2018)
6. Kong, Q., Trugman, D.T., Ross, Z.E., Bianco, M.J., Meade, B.J., Gerstoft, P.: Machine learning in seismology: turning data into insights. Seismol. Res. Lett. **90**(1), 3–14 (2019)
7. Panakkat, A., Adeli, H.: Recurrent neural network for approximate earthquake time and location prediction using multiple seismicity indicators. Comput.-Aided Civ. Infrastruct. Eng. **24**, 280–292 (2009)
8. Geller, R.J., Jackson, D.D., Kagan, Y.Y., Mulargia, F.: Earthquakes cannot be predicted. Science **275**(5306), 1616–1617 (1997)
9. Brodi, B.: A neural-network model for earthquake occurrence. J. Geodyn. **32**, 289–310 (2001)
10. Moustra, M., Avraamides, M., Christodoulou, C.: Artificial neural networks for earthquake prediction using time series magnitude data or seismic electric signals. Expert Syst. Appl. **38**, 15032–15039 (2011)
11. DeVries, P.M.R., Viégas, F., Wattenberg, M., Meade, B.J.: Deep learning of aftershock patterns following large earthquakes. Nature **560**, 632–634 (2018)
12. Vere-Jones, D., Ben-Zion, Y., Zuniga, R.: Statistical seismology. Pure Appl. Geophys. **162**, 1023–1026 (2005)
13. Mignan, A.: Retrospective on the Accelerating Seismic Release (ASR) hypothesis: controversy and new horizons. Tectonophysics **505**, 1–16 (2011)
14. Sornette, D.: Critical Phenomena in Natural Sciences, Chaos, Fractals, Selforganization and Disorder: Concepts and Tools. Springer, New York (2009). https://doi.org/10.1007/3-540-33182-4
15. Mignan, A.: Seismicity precursors to large earthquakes unified in a stress accumulation framework. Geophys. Res. Lett. **39**, L21308 (2012)
16. Mignan, A.: Static behaviour of induced seismicity. Nonlin. Process. Geophys. **23**, 107–113 (2016)
17. Mignan, A.: Utsu aftershock productivity law explained from geometric operations on the permanent static stress field of mainshocks. Nonlin. Process. Geophys. **25**, 241–250 (2018)
18. Tiampo, K.F., Shcherbakov, R.: Seismicity-based earthquake forecasting techniques: ten years of progress. Tectonophysics **522–523**, 89–121 (2012)
19. Mignan, A.: Modeling aftershocks as a stretched exponential relaxation. Geophys. Res. Lett. **42**, 9726–9732 (2015)
20. Richards-Dinger, K., Stein, R.S., Toda, S.: Decay of aftershock density with distance does not indicate triggering by dynamic stress. Nature **467**, 583–586 (2010)
21. Hainzl, S., Brietzke, G.B., Zöller, G.: Quantitative earthquake forecasts resulting from static stress triggering. J. Geophys. Res. **115**, B11311 (2010)
22. Båth, M.: Lateral inhomogeneities of the upper mantle. Tectonophysics **2**(6), 483–514 (1965)

23. Gerstenberger, M.C., Wiemer, S., Jones, L.M., Reasenberg, P.A.: Real-time forecasts of tomorrow's earthquakes in California. Nature **435**, 328–331 (2005)
24. Lakkos, S., Hadjiprocopis, A., Compley, R., Smith, P.: A neural network scheme for earthquake prediction based on the seismic electric signals. In: Proceedings of the IEEE Conference on Neural Networks and Signal Processing, pp. 681–689. IEEE, Ermioni (1994)
25. Alves, E.I.: Notice on the predictability of earthquake occurrences. Memórias e Notícias **117**, 51–61 (1994)
26. Liu, Y., Wang, Y., Li, Y., Zhang, B., Wu, G.: Earthquake prediction by RBF neural network ensemble. In: Yin, F.-L., Wang, J., Guo, C. (eds.) ISNN 2004. LNCS, vol. 3174, pp. 962–969. Springer, Heidelberg (2004). https://doi.org/10.1007/978-3-540-28648-6_153
27. Alves, E.I.: Earthquake forecasting using neural networks: results and future work. Nonlin. Dyn. **44**, 341–349 (2006)
28. Panakkat, A., Adeli, H.: Neural network models for earthquake magnitude prediction using multiple seismicity indicators. Int. J. Neural Syst. **17**(1), 13–33 (2007)
29. Martínez-Álvarez, F., Reyes, J., Morales-Esteban, A., Rubio-Escudero, C.: Determining the best set of seismicity indicators to predict earthquakes. Two case studies Chile and the Iberian Peninsula. Knowl.-Based Syst. **50**, 198–210 (2013)
30. Asencio-Cortés, G., Martínez-Álvarez, F., Morales-Esteban, A., Reyes, J.: A sensitivity study of seismicity indicators in supervised learning to improve earthquake prediction. Knowl.-Based Syst. **101**, 15–30 (2016)
31. Madahizadeh, R., Allamehzadeh, M.: Prediction of aftershocks distribution using artificial neural networks and its application on the May 12, 2008 Sichuan earthquake. JSEE **11**(3), 111–120 (2009)
32. Rouet-Leduc, B., Hulbert, C., Lubbers, N., Barros, K., Humphreys, C.J., Johnson, P.A.: Machine learning predicts laboratory earthquakes. Geophys. Res. Lett. **44**, 9276–9282 (2017)
33. Leach, R., Dowla, F.: Earthquake early warning system using real-time signal processing. In: Proceedings of the 1996 IEEE Signal Processing Society Workshop, pp. 463–472. IEEE, Kyoto (1996)
34. Kong, Q., Allen, R.M., Schreier, L., Kwon, Y.-W.: MyShake: a smartphone seismic network for earthquake early warning and beyond. Sci. Adv. **2**, e1501055 (2016)
35. Perol, T., Gharbi, M., Denolle, M.: Convolutional neural network for earthquake detection and location. Sci. Adv. **4**, e1700578 (2018)
36. Ross, Z.E., Meier, M.-A., Hauksson, E.: P wave arrival picking and first-motion polarity determination with deep learning. J. Geophys. Res. Solid Earth **123**, 5120–5129 (2018)
37. Ross, Z.E., Yue, Y., Meier, M.-A., Hauksson, E.: Phaselink: a deep learning approach to seismic phase association. J. Geophys. Res. Solid Earth (2019). https://doi.org/10.1029/2018jb016674
38. Bouchon, M., Karabulut, H., Aktar, M., Özalaybey, S., Schmittbuhl, J., Bouin, M.P.: Extended nucleation of the 1999 Mw 7.6 Izmit earthquake. Science **331**(6019), 877–880 (2011)
39. Mignan, A.: The debate on the prognostic value of earthquake foreshocks: a meta-analysis. Sci. Rep. **4**, 4099 (2014)
40. Mignan, A.: Asymmetric Laplace mixture modelling of incomplete power-law distributions: application to 'seismicity vision'. In: Arai, K., Kapoor, S. (eds.) CVC 2019. AISC, vol. 944, pp. 30–43. Springer, Cham (2019). https://doi.org/10.1007/978-3-030-17798-0_4
41. Valentine, A.P., Trampert, J.: Data space reduction, quality assessment and searching of seismograms: autoencoder networks for waveform data. Geophys. J. Int. **189**, 1183–1202 (2012)

42. Li, Z., Meier, M.-A., Hauksson, E., Zhan, Z., Andrews, J.: Machine learning seismic wave discrimination: application to earthquake early warning. Geophys. Res. Lett. **45**, 4773–4779 (2018)
43. International Seismological Center. http://www.isc.ac.uk/. Accessed 29 Jan 2019
44. Finite-Source Rupture Model Database. http://equake-rc.info/SRCMOD/. Accessed 29 Jan 2019
45. Okada, Y.: Surface deformation due to shear and tensile faults in a half-space. Bull. Seismol. Soc. Am. **75**(4), 1135–1154 (1985)
46. King, G.C.P.: Fault interaction, earthquake stress changes, and the evolution of seismicity. Treatise Geophys. **4**, 225–255 (2007)
47. Nature News: Artificial intelligence nails predictions of earthquake aftershocks. https://www.nature.com/articles/d41586-018-06091-z. Accessed 29 Jan 2019
48. The New York Times: A.I. is Helping Scientists Predict When and Where the Next Big Earthquake Will Be. https://www.nytimes.com/2018/10/26/technology/earthquake-predictions-artificial-intelligence.html. Accessed 29 Jan 2019
49. Futurism: Google's AI can help predict where earthquake aftershocks are most likely. https://futurism.com/the-byte/aftershocks-earthquake-prediction. Accessed 29 Jan 2019
50. The Verge: Google and Harvard team up to use deep learning to predict earthquake aftershocks. https://www.theverge.com/2018/8/30/17799356/ai-predict-earthquake-aftershocks-google-harvard. Accessed 29 Jan 2019
51. Meade, B.J., DeVries, P.M.R., Faller, J., Viegas, F., Wattenberg, M.: What is better than coulomb failure stress? A ranking of scalar static stress triggering mechanisms from $10^5$ mainshock-aftershock pairs. Geophys. Res. Lett. **44**, 11409–11416 (2017)

# Boosting Wavelet Neural Networks Using Evolutionary Algorithms for Short-Term Wind Speed Time Series Forecasting

Hua-Liang Wei(✉)

Automatic Control and Systems Engineering, University of Sheffield,
Sheffield S1 3JD, UK
w.hualiang@sheffield.ac.uk

**Abstract.** This paper addresses nonlinear time series modelling and prediction problem using a type of wavelet neural networks. The basic building block of the neural network models is a ridge type function. The training of such a network is a nonlinear optimization problem. Evolutionary algorithms (EAs), including genetic algorithm (GA) and particle swarm optimization (PSO), together with a new gradient-free algorithm (called coordinate dictionary search optimization – CDSO), are used to train network models. An example for real speed wind data modelling and prediction is provided to show the performance of the proposed networks trained by these three optimization algorithms.

**Keywords:** Neural network · Wavelet · Boosting · Optimization · Evolutionary algorithms · Time series · Wind speed · Forecasting · Data-Driven modelling

## 1 Introduction

Many practical time series modelling problems can be described as follows. There is a response variable $y$ (also known as output or dependent variable) that depends on a set of independent variables $\mathbf{x} = \{x_1, x_2, \ldots, x_n\}$ (also known as input or explanatory variables). Usually, a number of observations of both the output and input variables are available, which are denoted by $\{y_k, \mathbf{x}_k\}(k = 1, 2, \ldots, N)$. The true quantitative representation of the relationship between the output $y$ and the input $\mathbf{x}$ is in general not known. The central task of data modelling is to establish quantitative representations, e.g. mathematical models such as $y = f(\mathbf{x}) + e$ (where $e$ is model error), to approximate the input-output relationship as close as possible.

There are a variety of methods and algorithms in the literature for dealing with different types of nonlinear data based modelling problems, such as system identification [1–4], data mining [5, 6], pattern recognition and classification [7], supervised statistical learning [8, 9]. Among these methods, system identification techniques provide a tool for deducing mathematical models from measured input and output data for dynamic processes. In general, the output signal $y$ at time instant $t$ depends on previous values of the input and output signals, such as $u_r(t-1)$, $u_r(t-2)$, …, $u_r(t-n_u)$, $y(t-1)$, …, $y(t-n_y)$, where $r$ is the number of exogenous input variables, $n_y$ is the time


I. Rojas et al. (Eds.): IWANN 2019, LNCS 11506, pp. 15–26, 2019.
https://doi.org/10.1007/978-3-030-20521-8_2

lag in the output, $n_u$ is the time lag in the inputs. For a time series without any external input, $y(t)$ only depends on the previous output values such as $y(t-1)$, …, $y(t-n_y)$.

There is a diversity of methods and approaches for building a good function to approximate the function $f$ or $F$ for a given problem, including machine learning and neural networks, among others. In recent years, boosted regression has attracted extensive attention due to the work of [9–11], which connects boosting to general regression models such as Gaussian, logistic and generalized linear models. In [12], a new form of boosted trees called aggregated boosted trees was proposed for ecological system modelling and prediction. In [13], a boosting ridge regression was proposed for solving a medical image processing problem. A boosted L1 regularized projection pursuit for additive model learning was proposed and applied to face caricature generation and gender classification in [14]. An image based regression algorithm using boosting method has been proposed for image detection and feature selection in [15]. In [16–18], a boosting method was integrated to projection pursuit regression [19] to construct neural networks for spatio-temporal system identification. In order to improve the accuracy of flood forecasting, boosting approaches were proposed and incorporated to kernel based modelling and forecasting systems in [20] and [21], respectively. Most recently, comparative studies have been conducted on random forest regression, gradient boosted regression and extreme gradient boosting to tackle wind energy prediction and solar radiation problem [22]. It has been shown that ensemble methods can improve the performance of support vector regression for individual wind farm energy prediction [22].

It is known that wavelet basis functions have the property of localization in both time and frequency [23]. Due to this inherent property, wavelet approximations provide the foundation for representing arbitrary functions economically using just a small number of basis functions, and this makes wavelet representations more adaptive compared with other basis functions [24–29]. This motivates us to develop a new family of neural networks where wavelet are used as the building blocks.

The training of such networks is a nonlinear optimization problem, which can be solved by using either a classical gradient based algorithm or a modern meta-heuristic search algorithm. In this study, two population based algorithms, namely genetic algorithm (GA) [30] and particle swarm optimization (PSOs) [31], together with a new derivative-free algorithm (called coordinate dictionary search optimization – CDSO), are applied to estimate the hyper-parameters of the wavelet network models.

## 2 Structure of the Wavelet Neural Network

### 2.1 The Framework of the Network

Following the commonly used notation, it is assumed that the system $y$ is related to the input vector $\mathbf{x} = [x_1, x_2, \ldots, x_n]^T$ as below:

$$y(t) = F[\mathbf{x}(t); \boldsymbol{\theta}] + e(t) \tag{1}$$

where $F[\bullet]$ is a function which is normally unknown or unavailable but can be approximated by a set of functions estimated through machine learning, system identification or other data modelling techniques, $\boldsymbol{\theta}$ is a model parameter vector which can be estimated from data, and $e(t)$ is unmeasurable noise sequence.

This study considers to use a one-hidden-layer neural network to approximate the unknown function $F$ as:

$$F[\mathbf{x}(t);\boldsymbol{\theta}] = \sum_{k=1}^{K} w_k g_k(\mathbf{x}(t);\boldsymbol{\theta}_k) + r_K(t) \tag{2}$$

where $g_k$ ($k$ = 1, 2, …, $K$) are basis functions whose structure and property are known, $\boldsymbol{\theta}_k$ are parameter vectors, $w_k$ are weight coefficients (connection coefficients), $K$ is the number of basis functions, and $r_K(t)$ is model error (residual).

## 2.2 Ridge Type Wavelet Basis Function

In this study, each of the functions $g_k$($k$ = 1, 2 …, $K$) in (2) is chosen to be the ridge type wavelet, which is of the form:

$$h(x_1,\ldots,x_n) = \psi(a_0 + a_1x_1 + \ldots + a_nx_n) = \psi(\boldsymbol{\theta}^T\mathbf{x}) \tag{3}$$

where $\psi$ is a scalar function, $a_0$, $a_1$, …, $a_n$ are called direction parameters, $\boldsymbol{\theta} = [a_0, a_1, \ldots, a_n]^T$, and $\mathbf{x} = [x_1, x_2, \ldots, x_n]^T$. The function $\psi$ in (3) can be any functions with good representation properties including wavelet basis functions. Such a function is used as the elementary building block for model construction.

## 2.3 Training of the Network

Let $\mathbf{y} = [y(1), y(2), \ldots, y(N)]^T$ be the observation vector of the output signal and $\mathbf{x}_k(t) = [x_k(t), x_k(t), \ldots, x_k(t)]^T$ be the observation vector ($t$ = 1, 2, …, $N$) at instant $t$. Let $\boldsymbol{\psi}(t) = \psi(\boldsymbol{\theta}^T\mathbf{x}(t))$, $\mathbf{g}(\Phi;\boldsymbol{\theta}) = [\boldsymbol{\psi}(1), \boldsymbol{\psi}(2), \ldots, \boldsymbol{\psi}(N)]^T$, with $\Phi = [\mathbf{x}_1, \mathbf{x}_2, \ldots, \mathbf{x}_n]$.

The boosting procedure of the network is carried out in a stepwise manner; at each step a function that minimizes the projection error is determined. Starting with $\mathbf{r}_0 = \mathbf{y}$, the first step is to find an element vector $\mathbf{g}_1 = \mathbf{g}(\mathbf{X};\boldsymbol{\theta}_1)$ such that $(\boldsymbol{\theta}_1, w_1) = \arg\min_{(\boldsymbol{\theta},w)}\{||\mathbf{r}_0 - w\mathbf{g}(\Phi;\boldsymbol{\theta})||^2\}$. The resulting residual vector is defined as $\mathbf{r}_1 = \mathbf{r}_0 - w_1\mathbf{g}_1$, which can be used as the "pseudo-reference" signal to find the second element vector $\mathbf{g}_2$, and so on. This procedure may repeat many times. At the $k$th step, we use the sum of squared errors, $||\mathbf{r}_k||^2$, to define a measure called the *error-to-signal ratio*: $\mathrm{ESR}_k = ||\mathbf{r}_k||^2 \big/ ||\mathbf{r}_0||^2$ and use this to monitor the iterative procedure - when ESR becomes smaller than a pre-specified threshold value. This measure can be used to define a criterion called the *penalized error-to-signal ratio:* $\mathrm{PESR}_k = [N/(N - \lambda k)]^2\mathrm{ESR}_k$ [16–18], where $\lambda$ is a small positive number which is normally in the range $1 \leq \lambda \leq 0.005\,N$. The maximum number of basis functions included in the network can be determined as the number of iterations where PESR reaches its minimum.

The cost function in the algorithm is defined as

$$J_{k-1}(\boldsymbol{\theta}) = ||\mathbf{r}_{k-1} - w\mathbf{g}(\Phi;\boldsymbol{\theta})||^2 = \sum_{t=1}^{N} [r_{k-1}(k) - wg(\mathbf{x}(t);\boldsymbol{\theta})]^2 \tag{4}$$

which can be solved through a boosted regression algorithm which is briefly summarized below.

---

**The Boosted Projection Pursuit Regression algorithm:**

Initialization: $\mathbf{r}_0 = \mathbf{y}$; $k$=1; ESR($k$) = 0; PESR($k$) = 0;

while $\{k \le K\}$; //$\{K$ is the maximum number of iterations$\}$//

$$[\boldsymbol{\theta}_k; w_k] = \arg\min_{\boldsymbol{\theta},w} \{ \, || \mathbf{r}_{k-1} - w\mathbf{g}(\Phi;\boldsymbol{\theta}) ||^2 \} \; ;$$

$$\mathbf{g}_k = \mathbf{g}(\Phi;\boldsymbol{\theta}_k);$$

$$\mathbf{r}_k = \mathbf{r}_{k-1} - w_k \mathbf{g}_k;$$

$$\text{ESR}(k) = || \mathbf{r}_k ||^2 / || \mathbf{r}_0 ||^2;$$

$$\text{PESR}_k = [N/(N - \lambda k)]^2 \text{ESR}_k;$$

$$k = k + 1;$$

end while

---

## 3 Network Training

The optimization of the parameters in the cost function (4) can be achieved by means of either a classical gradient based algorithm or a modern meta-heuristic search algorithm. Once the estimates of the required parameters are available, the function $F[\bullet]$ in (2) can then be represented as a linear combination of the estimated functions $g_k$ ($k$ = 1, 2 …,).

### 3.1 Evolutionary Algorithms

Two evolutionary algorithms, namely, genetic algorithm (GA) [30] and particle swarm optimization (PSO) [31] are considered in this study. Matlab toolbox for GA and PSO is available in Mathworks products (Matlab 2018b). A large amount of information on evolutionary algorithms is readily and easily available publically in the literature, descriptions for these algorithms are therefore omitted here to save space.

### 3.2 Coordinate Dictionary Search Optimization (CDSO) Algorithm

For comparison purpose, a new coordinate dictionary search optimization (CDSO) algorithm is presented in this section.

**2D Case**

Let $f(x_1, x_2)$ be a function defined in 2D, and the objective is to find its global minimum point, with a box constraint $a_i \leq x_i \leq b_i (i = 1, 2)$. We first define the basic unit search directions as $D_B = \{d_1, d_2, \ldots, d_8\}$, where the $d_i$'s are:

$$\overset{d_1}{\begin{bmatrix}1\\0\end{bmatrix}}, \overset{d_2}{\begin{bmatrix}0\\1\end{bmatrix}}, \overset{d_3}{\begin{bmatrix}-1\\0\end{bmatrix}}, \overset{d_4}{\begin{bmatrix}0\\-1\end{bmatrix}}, \overset{d_5}{\frac{1}{\sqrt{2}}\begin{bmatrix}1\\1\end{bmatrix}}, \overset{d_6}{\frac{1}{\sqrt{2}}\begin{bmatrix}-1\\1\end{bmatrix}}, \overset{d_7}{\frac{1}{\sqrt{2}}\begin{bmatrix}-1\\-1\end{bmatrix}}, \overset{d_8}{\frac{1}{\sqrt{2}}\begin{bmatrix}1\\-1\end{bmatrix}}$$

The eight unit directions are shown in Fig. 1, where the first four directions are in the first, third, fifth and seventh quadrant lines, while the other four are in the second, fourth, sixth and eighth quadrant lines, respectively.

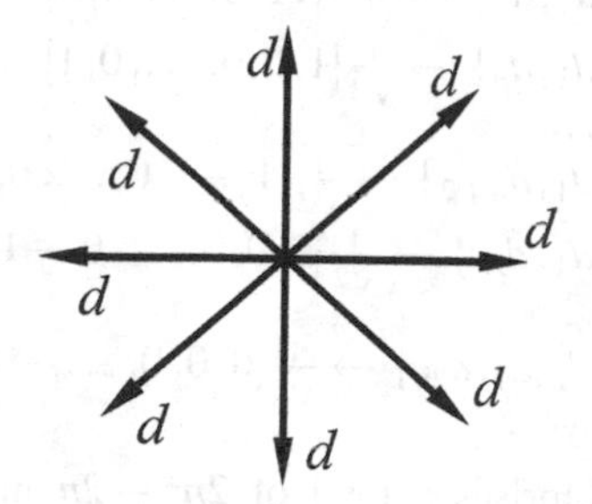


**Fig. 1.** An illustration of the basic unit search directions for 2D case functions.

Then we define the scaled search directions as below:

$$\begin{cases} D_{s_0} = s_0 D_B = \{s_0 d_1, s_0 d_2, \ldots, s_0 d_8\} \\ D_{s_1} = s_1 D_B = \{s_1 d_1, s_1 d_2, \ldots, s_1 d_8\} \\ \cdots \\ D_{s_{10}} = s_{10} D_B = \{s_{10} d_1, s_{10} d_2, \ldots, s_{10} d_8\} \end{cases} \tag{5}$$

where $s_m (m = 0, 1, \ldots, 10)$ are scale coefficients which are defined as:

$$s_m = 10^{-m} r \tag{6}$$

The parameter $r$ in (6) is adjustable, it determines the maximum step-size (learning rate) for network training. The dictionary used for parameter optimization is a combination of all these scaled dictionaries, that is, $D = D_{s_0} + D_{s_1} + \ldots + D_{s_{10}}$.

**n-Dimensional Case**

The 2D case can easily be extended to a general $n$ dimensional case. Assume that a box constraint is given as $a_i \leq x_i \leq b_i$ ($i$ = 1, 2, …, $n$). We define the basic coordinate search directions $D_1 = \{d_1, d_2, \ldots, d_{2n}\}$ as below:

$$\overset{d_1}{\begin{bmatrix} 1 \\ 0 \\ \cdots \\ 0 \end{bmatrix}}, \overset{d_2}{\begin{bmatrix} 0 \\ 1 \\ \cdots \\ 0 \end{bmatrix}}, \ldots, \overset{d_n}{\begin{bmatrix} 0 \\ 0 \\ \cdots \\ 1 \end{bmatrix}}, \overset{d_{n+1}}{\begin{bmatrix} -1 \\ 0 \\ \cdots \\ 0 \end{bmatrix}}, \overset{d_{n+1}}{\begin{bmatrix} 0 \\ -1 \\ \cdots \\ 0 \end{bmatrix}}, \ldots, \overset{d_{2n}}{\begin{bmatrix} 0 \\ 0 \\ \cdots \\ -1 \end{bmatrix}}$$

We then use the $2n$ unit vectors to generate new unit vectors as:

$$D_2: \begin{cases} \{d_1, d_2\} \rightarrow \frac{1}{\sqrt{2}}[1, 1, 0, \ldots, 0, 0]^T \\ \{d_1, d_n\} \rightarrow \frac{1}{\sqrt{2}}[1, 0, 0, \ldots, 0, 1]^T \\ \cdots \\ \{d_1, d_{n+2}\} \rightarrow \frac{1}{\sqrt{2}}[1, -1, 0, \ldots, 0, 1]^T \\ \{d_1, d_{2n}\} \rightarrow \frac{1}{\sqrt{2}}[1, 0, 0, \ldots, 0, -1]^T \\ \cdots \\ \{d_{2n-1}, d_{2n}\} \rightarrow \frac{1}{\sqrt{2}}[0, 0, 0, \ldots, -1, -1]^T \end{cases} \tag{7}$$

Note that group (7) comprises a total of $2n^2 - 2n$ unit vectors. The basic unit dictionary $D_B$ is made up of all the elements of basic coordinate dictionary $D_1$ and all the elements of the group $D_2$. The basic dictionary therefore contains a total of $2n^2$ unit vectors, which are denoted by: $D_B = \{d_1, d_2, \ldots, d_M\}$, with $M = 2n^2$.

Similar to the 2D case, we define the scaled search directions as:

$$D_{s_m} = s_m D_B = \{s_m d_1, s_m d_2, \ldots, s_m d_{2n}\} \tag{8}$$

where $s_m$($m$ = 0, 1, …, 10) are scaling coefficients which are the same as in (6). The dictionary used for optimization is a combination of all these scaled dictionaries, that is, $D = \bigcup_{m=0}^{10} D_{s_m}$.

**Outline of the CDSO Algorithm**

The implementation of the proposed CDSO algorithm briefly summarised below.

---

Initialization: Number of decision variables (dimension);
Constraint boundary [lb, ub];
Maximum search distance (parameter R);
Maximum iteration (itMax);
Tolerate threshold (Tol )
Guessed initial condition (x0);

1. Generate the dictionary D
2. t =1;
3. x(t) = x0;
4. Find the best direction (denoted by dbest);
5. Record the current best solution xbest(t) = x(t) + dbest;
6. Check the stop criterion;
7. t = t+1; x0 = xbest(t);

Repeat 3 to 7 until the specified stop criterion is met.

---

# 4 Case Study – Wind Speed Forecasting

The proposed method is applied to real wind speed data, which were acquired from our research collaborators. The hourly wind speed data were collected at the Berkhout wind station, Netherlands, for the period of January-December 2004. The data were measured by the Royal Netherlands Meteorological Institute. For demonstration purpose, we use the data of November 2004 to train the network model and use the data of December of 2004 to test the model prediction performance. The training data (1–30 November 2004) and test data (1–31 December 2004) are shown in Fig. 2.

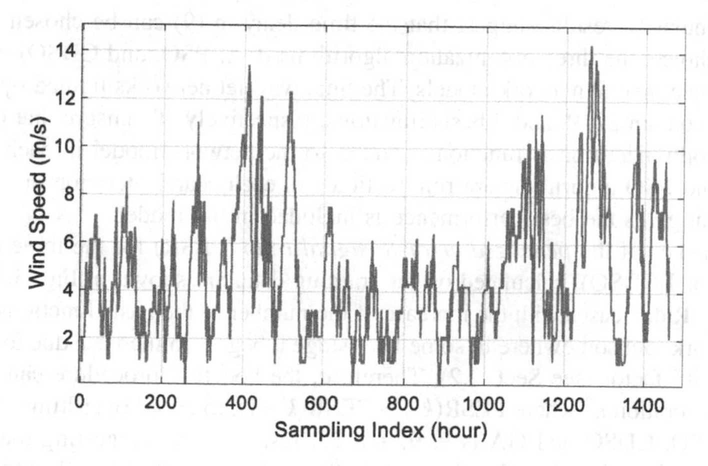


**Fig. 2.** Graphical illustration of the hourly wind speed data (1 November–31 December 2004).

### 4.1 The Model

Let the value of wind speed at time instant $t$ be designated by $y(t)$ ($t = 1, 2, \ldots, N$). We are interested in predicting $y(t)$ using the previous values at the time instants $t-1$, $t-2$, ..., $t-n$. We consider the following model:

$$y(t) = F[y(t-1), y(t-2), \ldots, y(t-n)] + e(t) \tag{9}$$

For convenience of description, we use $x_j(t)$ to denote $y(t-j)$, with $j = 1, 2, \ldots, n$. So model (9) can be written as:

$$y(t) = F[x_1(t), x_2(t), \ldots, x_n(t)] + e(t) \tag{10}$$

We then use the training data to train a wavelet neural network model by the three algorithms: GA, PSO and CDSO. The following well-known *sinc* function (also known as the Shannon wavelet scaling function) [32] is used as the basis function to build the network model:

$$\varphi(x) = \frac{\sin \pi x}{\pi x} \tag{11}$$

With the above *sinc* function, the ridge type function $\psi(\boldsymbol{\theta}^T \mathbf{x})$ is:

$$\psi(\boldsymbol{\theta}^T \mathbf{x}) = \varphi(\boldsymbol{\theta}^T \mathbf{x}) = \frac{\sin(\pi[a_0 + a_1 x_1(t) + \ldots + a_n x_n(t)])}{\pi[a_0 + a_1 x_1(t) + \ldots + a_n x_n(t)]} \tag{12}$$

where $\boldsymbol{\theta} = [a_0, \ldots, a_n]^T$ and $\mathbf{x} = [x_1(t), \ldots, x_n(t)]^T$. Note that the Shannon wavelet scaling function (11) is not differentiable, meaning that the conventional gradient descent type algorithms cannot be directly used to train the associated wavelet neural network.

### 4.2 Model Performance

Primary simulation results suggest that the time delay in (9) can be chosen as $n = 4$. With this choice, the three optimization algorithms (GA, PSO, and CDSO) were used to train wavelet neural network models. The final wavelet networks trained by the three algorithms contain 22, 9, and 9 basis functions, respectively. To ensure that the "best" (i.e. better optimized) basis function is added to the network model in each iteration, both GA and PSO algorithms are run 10 times in each search iteration and the basis function that gives the best performance is included in the model.

The changes of the *penalized error-to-signal ratio* (PESR) for the three algorithm (GA, PSO and CDSO), calculated on the training data, are shown in Fig. 3. Note that initially PESR decreases with the increase of the number of the basis functions included in the network, but somewhere in some later stage it begins to increase due to the effect of the penalty factor (see Sect. 2.2). Therefore, the boosting procedure can be terminated at an iteration $k^*$ where PESR($k^*$) > PESR($k^*-1$), to avoid overfitting. The value of $k^*$ for PSO, CDSO and GA is 9, 9, and 22, respectively, suggesting that the best network models trained by the three algorithms should include 9, 9, and 22 basis functions, respectively.

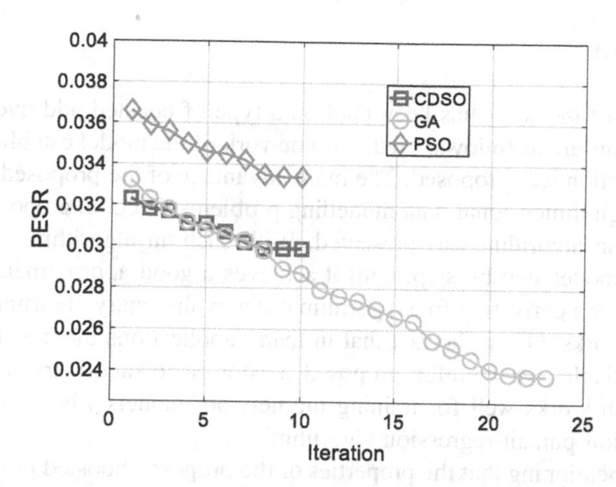


**Fig. 3.** The change plot of PESR on the training data.

The PESR values of the best models generated by PSO (with 9 basis functions), CDSO (with 9 basis functions) and GA (with 22 basis functions) are 0.0334, 0.0296, and 0.0238, using respectively. In terms of time complexity, it turns out that the CPU time for PSO, CDSO and GA to achieve the three best models is 48.23 s, 61.33 s and 436.70 s, respectively.

For an illustration, the model predicted values, produced by the network trained using the CDSO algorithm over the test data (1–31 December 2004), are plotted in Fig. 4, where the corresponding measurements are also displayed for a comparison purpose. It can be seen that the obtained wavelet neural network model shows excellent prediction performance.

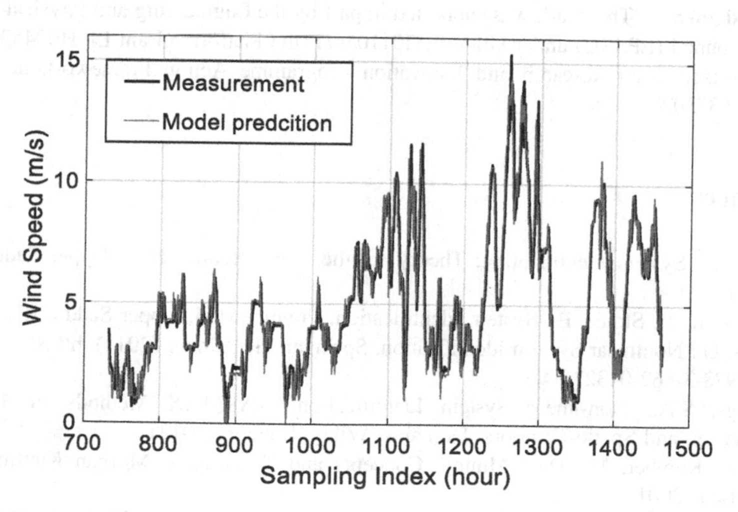


**Fig. 4.** A comparison of the model predicted values with the measurements on the test data (1–31 December 2004).

## 5 Conclusion

The main focus of the work has been paid on a type of boosted additive models. The main contributions are as follows. First, a framework of the model established based on a ridge type function was proposed. The main advantage of the proposed framework is that it allows high dimensional data modelling problems. Second, a boosted projection pursuit regression algorithm was presented. With such an algorithm, we can conveniently build a model step by step, until it achicves a good approximation. Third, we showed that either a derivative-free algorithm or an evolutionary algorithm can be used to train the networks. Given the fact that in many applications the cost functions may not be differentiable, we therefor proposed a coordinate dictionary search (CDSO) algorithm, which works well for training the network models when integrated to the boosted projection pursuit regression algorithm.

It is worth mentioning that the properties of the proposed boosted projection pursuit regression algorithm and the coordinate dictionary search algorithm have not been fully investigated. There are still several open questions that remain to be explored and answered. For example, in addition to Shannon wavelet scaling function, there perhaps exist many better alternative choices (e.g. Gaussian wavelet, radial basis function could be one of them); we will explore these in future work. We would also carry out further assessments on the performance of the proposed method and compare with traditional feedforward neural networks and state-of-the-art approaches.

While CDSO, GA and PSO algorithms all provide a zeroth-order optimization approach, meaning that they do not need gradient information, it does not necessarily mean that these methods would always be superior to gradient based algorithms. In this respect, it would be interesting to integrate the gradient boosting machine (GBM) to the boosted projection pursuit regression algorithm, to explore the advantage of GBM and investigate the potential to improve the performance of gradient-free algorithms.

**Acknowledgments.** This work was supported in part by the Engineering and Physical Sciences Research Council (EPSRC) under Grant EP/I011056/1, the Platform Grant EP/H00453X/1, and the EU Horizon 2020 Research and Innovation Programme Action Framework under grant agreement 637302.

## References

1. Ljung, L.: System Identification: Theory for the User. Prentice-Hall, Upper Saddle River (1987)
2. Sodestrom, T., Stoica, P.: System Identification. Prentice Hall, Upper Saddle River (1988)
3. Nelles, O.: Nonlinear System Identification. Springer, Heidelberg (2011). https://doi.org/10.1007/978-3-662-04323-3
4. Billings, S.A.: Non-linear System Identification: NARMAX Methods in the Time, Frequency, and Spatio-Temporal Domains. Wiley, London (2013)
5. Han, J., Kamber, M.: Data Mining: Concepts and Techniques. Morgan Kaufmann, San Francisco (2001)
6. Witten, I.H., Frank, E.: Data Mining: Practical Machine Learning Tools and Techniques. Morgan Kaufmann, San Francisco (2005)

7. Bishop, C.M.: Neural Networks for Pattern Recognition. Oxford University Press, New York (1995)
8. Vapnik, V.: The Nature of Statistical Learning Theory, 2nd edn. Springer, New York (1999). https://doi.org/10.1007/978-1-4757-3264-1
9. Friedman, J., Hastie, T., Tibshirani, R.: The Elements of Statistical Learning. Springer, New York (2001). https://doi.org/10.1007/978-0-387-84858-7
10. Friedman, J.: Greedy function approximation: a gradient boosting machine. Ann. Stat. **29**, 1189–1232 (2001)
11. Friedman, J., Hastie, T., Tibshirani, R.: Additive logistic regression: a statistical view of boosting. Ann. Stat. **28**, 337–407 (2000)
12. Zhou, S.K., Georgescu, B., Zhou, X.S., Comaniciu, D.: Image based regression using boosting method. In: Proceedings 10th IEEE International Conference on Computer Vision (ICCV 2005), pp. 541–548. IEEE, Beijing (2005)
13. De'ath, G.: Boosted trees for ecological modeling and prediction. Ecology **88**(1), 243–251 (2007)
14. Zhou, S., Zhou, J., Comaniciu, D.: A boosting regression approach to medical anatomy detection. In: Proceedings IEEE Conference on Computer Vision and Pattern Recognition, Minneapolis, MN (2007)
15. Zhang, X., Liang, L., Tang, X., Shum, H.: L1 regularized projection pursuit for additive model learning. In: IEEE Conference on Computer Vision and Pattern Recognition (CVPR 2008), Anchorage, AK, USA (2008)
16. Wei, H.-L., Billings, S.A.: Generalized cellular neural networks (GCNNs) constructed using particle swarm optimization for spatio-temporal evolutionary pattern identification. Int. J. Bifurcat. Chaos **18**(12), 3611–3624 (2008)
17. Wei, H.-L., Billings, S.A., Zhao, Y., Guo, L.: Lattice dynamical wavelet neural networks implemented using particle swarm optimization for spatio–temporal system identification. IEEE Trans. Neural Netw. **20**(1), 181–185 (2009)
18. Wei, H.-L., Billings, S.A., Zhao, Y., Guo, L.: An adaptive wavelet neural network for spatio-temporal system identification. Neural Netw. **23**(10), 1286–1299 (2010)
19. Friedman, J.H., Stuetzle, W.: Projection pursuit regression. J. Amer. Statist. Assoc. **76**(376), 817–823 (1981)
20. Li, S., Ma, K., Jin, Z., Zhu, Y.: A new flood forecasting model based on SVM and boosting learning algorithms. In Proceedings IEEE Congress on Evolutionary Computation (CEC 2016), pp. 1343–1348, Vancouver, BC, Canada (2016)
21. Zhang, D., Zhang, Y., Niu, Q., Qiu, X.: Rolling forecasting forward by boosting heterogeneous kernels. In: Phung, D., Tseng, V.S., Webb, G.I., Ho, B., Ganji, M., Rashidi, L. (eds.) PAKDD 2018. LNCS (LNAI), vol. 10937, pp. 248–260. Springer, Cham (2018). https://doi.org/10.1007/978-3-319-93034-3_20
22. Torres-Barrána, A., Alonsoa, A., Dorronsoroa, J.R.: Regression tree ensembles for wind energy and solar radiation prediction. Neurocomputing **326–327**, 151–160 (2019)
23. Mallat, S.: A Wavelet Tour of Signal Processing. Academic Press, San Diego (1998)
24. Wei, H.-L., Billings, S.A.: A unified wavelet-based modelling framework for non-linear system identification: the WANARX model structure. Int. J. Control **77**(4), 351–366 (2004)
25. Billings, S.A., Wei, H.-L.: The wavelet-NARMAX representation: a hybrid model structure combining polynomial models with multiresolution wavelet decompositions. Int. J. Syst. Sci. **36**(3), 137–152 (2005)
26. Wei, H.-L., Billings, S.A.: Long term prediction of non-linear time series using multiresolution wavelet models. Int. J. Control **79**(6), 569–580 (2006)

27. Li, Y., Cui, W., Guo, Y.Z., et al.: Time-varying system identification using an ultra-orthogonal forward regression and multiwavelet basis functions with applications to EEG. IEEE Trans. Neural Netw. Learn. Syst. **29**(7), 2960–2972 (2018)
28. Li, Y., Lei, M., Guo, Y., Hu, Z., Wei, H.-L.: Time-varying nonlinear causality detection using regularized orthogonal least squares and multi-wavelets with applications to EEG. IEEE Access **6**, 17826–17840 (2018)
29. Li, Y., Lei, M., Cui, W., et al.: A parametric time frequency-conditional Granger causality method using ultra-regularized orthogonal least squares and multiwavelets for dynamic connectivity analysis in EEGs. IEEE Trans. Biomed. Eng. (in press)
30. Back, T.: Evolutionary Algorithms in Theory and Practice: Evolution Strategies, Evolutionary Programming, Genetic Algorithms. Oxford University Press, Oxford (1996)
31. Kennedy, J., Eberhart, R.: Particle swarm optimization. In: Proceedings IEEE Conference Neural Networks, vol. 4, pp. 1942–1948, Piscataway, NJ (1995)
32. Cattani, C.: Shannon wavelets theory. Math. Prob. Eng. (2008). Art. no. 164808

# An Approach to Rain Detection Using Sobel Image Pre-processing and Convolutional Neuronal Networks

José A. Godoy-Rosario(✉), Antonio G. Ravelo-García(✉), Pedro J. Quintana-Morales, and Juan L. Navarro-Mesa

Institute for Technological Development and Innovation in Communications-IDeTIC, University of Las Palmas de Gran Canaria (ULPGC), 25017 Las Palmas de Gran Canaria, Spain
{jose.godoy,antonio.ravelo}@ulpgc.es

**Abstract.** Rain fall detection has been an important factor under study in a multitude of applications: estimation of floods in order to minimize damage before an environmental risk situation, rain removal from images, agriculture field, etc. Actually, there are numerous methods implemented in order to try to solve this issue. For example, some of them are based on the traditional weather station or in the use of radar technology. In this work, we propose an approach to rain detection using image processing techniques and Convolutional Neuronal Networks (CNN). In order to improve the results of classification, images in rain and no rain conditions are pre-processed using the Sobel algorithm to detect edges. The architecture that defines the CNN is LeNet and it is carried out with three convolutional layers, three pooling layers and a soft max layer. With the proposed method, it is possible to detect the presence of rain in certain region of the image with an accuracy of 89%. The purpose of the proposed system is just to complete with a different added value, other traditional methods for detection of rain.

**Keywords:** Rain fall detection · Sobel image processing · Convolutional Neuronal Network (CNN)

## 1 Introduction

The determination of weather conditions has supposed an important research factor during years. Sectors which are interested in this issue are numerous: farmers, automotive industry, emergency situations in order to forecast actions before water floods, video processing, etc. In addition, the concept of climate change has acquired a growing interest with the consequent concern of people, governments and in the economic sector [1]. A result of this climatic change is the increase in the risk of storms which could have terrible effects on urban and rural areas. The research work we present here is intended to be part of a monitoring system in order to detect meteorological phenomena. Traditionally, methods based on weather station are used to detect rain precipitation, however, it is interesting to provide other meteorological measurement sources in order to complete the information provided by traditional weather stations that not always are automatized, thus avoiding real-time rain monitoring.


I. Rojas et al. (Eds.): IWANN 2019, LNCS 11506, pp. 27–38, 2019.
https://doi.org/10.1007/978-3-030-20521-8_3

The presence of rain over images or videos has been widely studied in order to detect the effect of this and to remove this information. In [2], a study to detect and remove rain from images and videos was carried out. Furthermore, they developed a model which describes rain intensities due to rain streaks and a dynamic model that captures the spatial-temporal properties of rain. A similar work is presented in [3] where a system based on computer is showed to detect the presence of rain or snow. However, that work was focused in the detection of the presence of rain or snow streaks. The method consisted on the separation of the foreground from the background in images. It is in the foreground plane where it was possible to detect rain or snow conditions. Then, they applied a Histogram of Orientations of rain or snow Streaks (HOS) to detect the presence of these phenomena. In [4], an image-based disdrometer is developed with the objective of detecting rainfalls due to the movement of raindrops. Other authors have been working in the detection of raindrop using slow motion cameras. Thus, in [5] a method for counting the number of drops and measure their size in a video frame is presented. In [6], a method for real-time raindrops detection in a car windscreen is proposed using cellular neural networks and support Vector Machines. However, the method presented in [6] is quite different to our proposal. We are centered in rainfall detection in certain distant regions in an image, so the raindrop estimation is not really useful for our objective. Although in the state of the art few works related to rain detection from images are described, there are not enough material to make a comparison analysis in similar conditions.

Other interesting works are related to the possibility of detecting fog through image processing techniques as is the case of [7] where a real-time fog detection system for vehicles is proposed for driving applications. This system is based on the estimation of the visibility distance and blurring due to the fog. In this case, the Sobel algorithm was used as sunny/foggy detector. This method consists on detecting edges on the images. When an image is in presence of fog, it appears a reduction of high frequency components, that can be measured using edge detection methods as Sobel. A similar work to [7] it is presented in [8]. However, in this case, they focused on analyzing the properties of local objects of the roads: lane markings, traffic signs, back lights of vehicles, etc.

Our work is based on a similar principle as studied in [7], however, we used a preprocessing based on Sobel algorithm to make an emphasis on the high frequency components of the image and, then, we train a Convolutional Neuronal Networks (CNN) to classify between rain and no rain images. The use of Neuronal Networks for rain detection has been approached using other sources of information: received signal strength [9] and the carrier-to-noise power ratio (C/N) [10].

An alternative for identification of adverse weather phenomena is the use of neuronal networks tools. These systems are able to learn in a wide range of situations, or meteorological event classes, from a set of multivariate data or images, in order to identify characteristic patterns with complex network models. In recent years, the introduction of Deep Neuronal Networks (DNN), particularly Convolutional Neuronal Networks, which allow automatic learning of feature from images, has obtained great acceptance. The convolutional layers allow the convolution of the input image with a filter, whose weight can be initialized randomly. These weights will be those that are update during the learning process. More recent models of neuronal networks are being development with Convolutional Neuronal Networks architectures which use a variable and numerous sets of filters and convolutional layers that are able to deepen in the learning of very complex models.

The purpose of this paper is to propose an approach to provide general aspects for the detection of distant rain falls using images. For this purpose, we build a CNN based on LeNet architecture. This architecture is able to recognize patterns in an image characterized by high variability and with robustness to distortions [11, 12]. It has been applied in a great number of applications like robot vision, image classification, handwriting recognition, etc.

The network will be trained to the identification of rain. The images will be taken from fixed cameras and they will be pre-processed with an algorithm for edge detection to remark the variability of the phenomena. Further experiments about the application of a transformation matrix based on the Discrete Cosine Transform (DCT) have been conducted to study the possibility to improve the classification scores.

This paper is organized as follow: first, in Sect. 2, a brief explaining about the methods used in this work, we introduced the image weather dataset, image pre-processing techniques and the classification system used in this work. In Sect. 3, we explain the performed experiments and obtained results. Finally, some future works and discussions about the study realized are summarized in Sect. 4.

# 2 Methods

In this section the methods and materials needed to carry out this research are presented. Firstly, it is detailed the dataset used to train and test the CNN. Secondly, it is presented the methods to determinate a Region of Interest (ROI) and the application of the Sobel algorithm for detecting edges in the ROI. Thirdly, the details of the CNN architecture are presented as the basis of the classification system.

## 2.1 Dataset

In this section, we explain the dataset used in this work. The dataset is divided in two different sections. First of all, we analyze the images which constitutes the dataset. Secondly, we introduce the weather data used to classify the images in rain or no rain conditions.

**Dataset of Images**

The dataset of images is composed by 2.098 images. These images are distributed as follow:

- 552 images with rainy conditions.
- 1.544 images where no raining conditions are presented in the image.

The images were taken by a video camera [13] located in Campus of Tafira at University of Las Palmas de Gran Canaria, Spain (see point 1 in Fig. 1). The images were obtained in different time intervals from January to December of 2018. All images were captured in JPG format with a resolution of 1920 $\times$ 1080 pixels. An example of no rain conditions and rain conditions images are shown in Fig. 2.

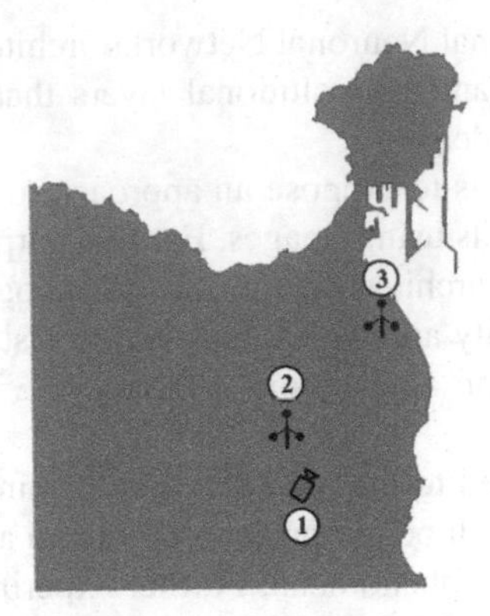


**Fig. 1.** Location map of the camera and weather stations in the city of Las Palmas de Gran Canaria

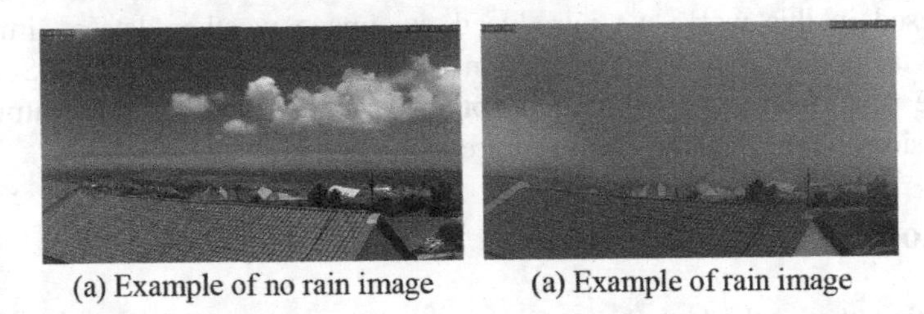

(a) Example of no rain image (a) Example of rain image

**Fig. 2.** Examples of images of rain and no rain conditions used in the image dataset

**Weather Dataset**

The method of classification used in this work consisted of crossing data of AEMET in order to know when precipitation occurred in the weather stations located in "Zurbarán, Tafira" (see point 2 in Fig. 1) and "Plaza de la Feria" (see point 3 in Fig. 1), both located in the city of Las Palmas de Gran Canaria, because they are the stations which are closed to the position that the camera was filming. Thus, differences between rain or no rain images is possible to obtain. In Table 1, it is shown the geographical coordinates of each weather station and camera.

**Table 1.** Location table of camera and weather stations

| Id | Type | Location | Latitude | Longitude | Altitude |
|---|---|---|---|---|---|
| 1 | Camera | University Campus of Tafira | 28.07125 | −15.45342 | 317 m |
| 2 | Weather station | Zurbarán, Tafira | 28.07829 | −15.45304 | 265 m |
| 3 | Weather station | Plaza de la Feria | 28,11317 | −15.42116 | 6 m |

## 2.2 Preprocessing

In order to extract relevant information of the image, a Sobel algorithm is applied over a region of interest to detect edges and thus the high frequency components of the image than can be correlated with the presence of rain.

**Determination of Regions of Interest**

A ROI is extracted from the images where we want to detect the presence or absence of rain conditions. We have decided to choose a ROI characterized by heterogeneous information due to the presence of buildings. This ROI presents acute high frequency components so we hypothesize that the application of Sobel algorithm can detect with good performance the edges of the image. The selected ROI is specified by coordinates x and y of the pixels that define a rectangle portion of the image. The ROI selected is defined by the points specified in Table 2.

**Table 2.** Pixel coordinates used to extract the ROI in images

| Parameter | Value |
|---|---|
| Origin (x) | 726 |
| Origin (y) | 675 |
| Destination (x) | 844 |
| Destination (y) | 724 |

This ROI was strategically selected due to that part of the image is static, in other words, not others factors such as vehicles movement, bird flies, etc. modifies the characteristics of the image in this zone. The unique factors that can produced changes in the image are weather conditions: rain, airborne dust or fog. Thus, it is possible to apply images techniques in order to detect changes in the ROI image which corresponds with alterations due to weather phenomena. In the topic of this paper, the most relevant factor that can affect the image is the rain. The ROI that is considered in this approach is marked by a green box and shown in Fig. 3.

(a) Studied ROI for a rain conditions image

(b) Studied ROI for a no-rain conditions image

**Fig. 3.** Selected ROIs for the rain/no rain study

**Sobel Edge Detection Algorithm**

The main purpose of Sobel algorithm is to detect edges in an image, thus it is a very suitable option to the needs presented in this work. An edge can be defined as a sector of an image where the variability between two consecutive pixels is maximum.

According to [14], most edge detection system consider that an edge appears when there is a discontinuity in the intensity function or a very steep intensity gradient in the image. The use of Sobel algorithm as edge detector has two main advantages [15]: First, Sobel introduces an average factor so it has a smoothing effect of the image noise. The other advantage is that Sobel is the result of differentiating two rows or two columns so the edge detected is clearer than the borders detected by other methods.

Sobel is a kind of orthogonal gradient operator. Gradient corresponds to the first derivative and gradient operator corresponds to the derivative factor. For a continuous function or image g(x, y), in a certain position, its gradient can be represented as a vector as follow [15]:

$$\nabla g(x,y) = \left[G_x G_y\right]^T = \left[\frac{\partial g}{\partial x}\frac{\partial g}{\partial y}\right] \tag{1}$$

The magnitude and direction angle of the vector can be calculated as follows:

$$mag\left(\nabla g\right) = \left|\nabla g_{(2)}\right| = \sqrt{G_x^2 G_y^2} \tag{2}$$

$$\varnothing(x,y) = arctang\left(G_x/G_y\right) \tag{3}$$

The partial derivatives of mathematical expressions should be calculated for each image pixel. $G_x$ and $G_y$ are defined by two matrixes which they are showed in (4) and (5). Each image pixel is convoluted with these two matrixes. The first matrix offers the maximum response in the vertical edge and the second matrix gives the maximum response to the level edge. The maximum value of these two convolutions is used as the output bit of the point, and the result is an image of edge amplitude.

$$G_x = \begin{pmatrix} -1 & -2 & -1 \\ 0 & 0 & 0 \\ 1 & 2 & 1 \end{pmatrix} \tag{4}$$

$$G_y = \begin{pmatrix} -1 & 0 & 1 \\ -2 & 0 & 2 \\ -1 & 0 & 1 \end{pmatrix} \tag{5}$$

For a better application of Sobel algorithm, first it is necessary to change de map color of the image from RGB to grayscale.

**Classification System**

The method used to classify a ROI consists on the use of a CNN which is able to label the images in rain or no rain conditions. A CNN can be defined as a set of convolutional layers which conform a network totally connected. The convolutional layers perform as a wide and varied set of convolutional filters which try to detect patterns in the image. Each convolutional layer has associated an activation layer and a pooling layer which introduces non-linearity and it reduces the characteristics of each layer which enhance the learning and the abstraction of the CNN. The input of the network is composed by images and the output corresponds with a vector of 2 units which are related to the probability that an input image belongs to rain or no rain type.

Filters used in each layer are fixed sized (I × J × C), where I is the image width, J is image height and C corresponds with the number of channels. Each layer can have a different number of filters and different sizes. Each convolutional filter works with the image regardless of the layer which it is processing. The number of resultant characteristic maps of the convolutions corresponds with the number of filters which are working in the layer. Once, the images pass the activation layer, a pooling layer serves to the next convolutional layer or to the end of the layer. All filters will be adjusted as adaptive form along the training process. These filters will extract the most outstanding features, not only of the original image, also the features of the characteristics of the successive convolutional layers. The pooling layer reduces the number of characteristics taking the maximum or other representative value of a kernel of size A × B (where A is the width and B is the height of the filter mask) which segment the image by steps. The kernel and the steps could be different from one layer to the next. As we are advancing through the convolutional layers, the characteristics will be easier to detect. This factor indicates an increase in the abstraction level of the extracted features. Usually, as the number of characteristics extracted by pooling layer is reduced, the number of characteristics extracted by the filters increase in each convolutional layer. Finally, the last layer feeds the completely connected layer which ends to extract the more relevant characteristics to identify the evaluated event and it is possible to provide a probability measurement in the last layer.

The training of the CNN will adjust the parameters of the filters and the weights completely connected, spreading the error from the input to the output of the CNN. The system will minimize a loss function based on crossed entropy.

The designed CNN is compound by 3 convolutional layers, the first has 20 filters, the second layer has 40 filters and the third layer has 60 filters. Each filter has a size of 5 × 5 × 3 and with adjustable parameters. Each convolutional layer has an activation layer type ReLu that implements the next function:

$$f(x) = \max(x, 0) \tag{6}$$

Where x is the neuron input. Then of each activation layer there is a pooling layer of size 2 × 2 and a step of 2 pixels. Finally, a completely connected network of 2 layers is performed, which corresponds with the classes we want to identify, rain vs no rain.

## 3 Experiments and Results

In order to train and test the CNN and measure the performance of the system a set of tests were performed. First, it is necessary to select those images used to test or train the CNN. The classification of these images was realized according to data of the Spanish Meteorology Agency (AEMET).

Once the images were classified in rain or no rain conditions, the next step is to select the number of images used to train the CNN. Typically, the number of images of each type should be balanced. Thus, we use the same number of images of rain and no rain to train: 280 images. This number of images corresponds with, approximately 50% of rain images of our dataset due to the little number of samples of rain condition images that we dispose. The rest of images will be used to test the network. Subsequently, the pre-processing explained in Sect. 2 is applied to all the images (training and testing). First of all, the same ROI is extracted to each image and later the Sobel algorithm to detect edges is also applied. As a result, an example for rain and no rain images after applying this process is shown in Fig. 4. With these resultant images, the training of the CNN is performed. The result of this training is a model which was used to test the CNN.

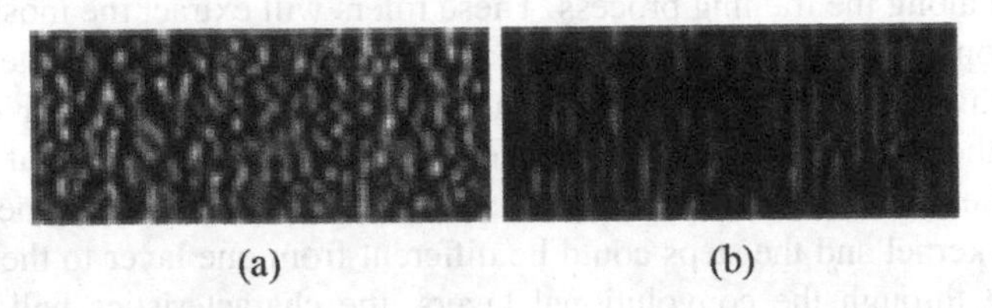

**Fig. 4.** Comparison between no rain (a) or rain (b) images after Sobel algorithm was applied to the selected image ROIs.

With the resultant images from applying the pre-processing techniques explained in Sect. 2 the training of the network is performed with the objective of create a model of CNN.

We train three different models of CNN: the first CNN was trained with images not previously filtered, the second of them was trained with the images where a transformation matrix based on the Discrete Cosine Transform (DCT) was applied, and finally we train the CNN with the images filtered by Sobel algorithm. Some interesting parameters for the evaluation of the CNN are collected in Table 3. The result of the training was a generated model for each one.

As this work represents an approach, and the size of the database has a limited number of examples, the design of the experiments considers aspects related to generalization. This, we repeat the test 20 times and the number of epochs is 30. In Table 4, it is shown a comparison of percentage of success (accuracy) and standard deviation of each model of neuronal network used in each test.

**Table 3.** Interesting parameters for CNN training.

| Parameter | Value |
|---|---|
| Epochs | 30 |
| Initial Learning Rate (LR) | $1e^{-5}$ |
| Batch Size (BS) | 20 |

**Table 4.** Comparison between accuracy and standard deviation of each CNN model

| Model | Accuracy | Standard deviation |
|---|---|---|
| Original ROI images | 84.47% | 1.74 |
| Applying Sobel filter | 91.97% | 1.79 |
| Applying DCT | 79.89% | 1.29 |

As can be seen in Table 4, the percentage of mean accuracy increases approximately in 8% in the detection of rain/no rain images when the CNN is trained with images filtered with Sobel. On the other hand, the application of DCT worsen the accuracy results of the proposed system in this approach. As it can be seen in Table 4, the results in each iteration are quite stable as standard deviation indicates.

Furthermore, to continue analyzing the performance of the proposed system, a set of tests was realized with the goal to determine accuracy, sensitivity and specificity of the use of CNN trained with images in which the Sobel algorithm was applied. This test consists in making an analysis of true positives, false positives, true negatives and false negatives to the results obtained from the test of the CNN. In this work, a true positive is considered as the determination of rain realized by the CNN of an image with raining conditions. Consequently, a true negative can be considered as the performance of the CNN to classify a no rain image as no rain. Thus, we analyzed a total of 400 images in where rain conditions occur and other 400 images with no rain conditions. The images were randomly selected from the total of images that make up our dataset.

In Fig. 5, each column corresponds with the real value that the CNN should predict and each row indicates the value offered by CNN as output. The figure specifies the values of truth positives (rain asserts), truth negatives (no rain asserts), false positives (rain fails), false negative (no rain fails).

| Values obtained by CNN \ Reference values | Rain | No rain |
|---|---|---|
| Rain | 385 | 73 |
| No rain | 15 | 327 |

**Fig. 5.** CNN results for rain and no rain images.

According to data of Fig. 5, it is possible to determinate the accuracy, sensitivity and specificity of the CNN for raining detection in images making use of the following expressions:

$$Accuracy = \frac{T_p + T_n}{T_p + T_N + F_p + F_n} \tag{7}$$

$$Sensitivity = \frac{T_p}{T_p + F_p} \tag{8}$$

$$Specificity = \frac{T_n}{T_n + F_p} \tag{9}$$

Where $T_p$ is the true positive values, $T_n$ is the true negative, $F_p$ is the false positive and $F_n$ is the false negative results of the CNN output. If data shown in Fig. 5 is applied in (7), (8) and (9) it is possible to calculate the performance of the proposed system:

$$\text{Accuracy} = 0.89\,(89\%) \tag{10}$$

$$\text{Sensitivity} = 0.84\,(84\%) \tag{11}$$

$$\text{Specificity} = 0.817\,(81.7\%) \tag{12}$$

## 4 Discussion and Future Work

The research work we present here is a contribution to the design of a monitoring system to detect meteorological phenomena in risk situations. Particularly, we make an approach to provide video systems a complementary value in weather observation services.

The determination of a ROI is not a trivial task. As mentioned previously, the ROI should be selected in such a way that a meteorological phenomenon produces changes in the image. However, other weather conditions similar to rain could produce misleading results, for example, fog or airborne dust. Thus, the results of detection through the proposed method can be improved correlating other types of data as, for example, weather stations measures. Furthermore, other methods to detect edges can be used [16], some based on gradient operators as: Canny operator [17], Prewitt operator [18] and Roberts operator [19]. Other methods can be based on the use of Laplacian operators: Marr-Hildreth [20] or Zero Cross [21]. As a last limitation we should expose the lack of rain images used in the dataset that could reduce the performance of the model as a classification system.

In any case, we can conclude that it is possible to detect meteorological phenomena in the images making use of difference techniques. In this case, the suggested system is composed by pre-processing tasks of the images in order to extract the characteristic parameters. The Sobel algorithm has shown a good performance due to the capacity the algorithm has to detect edges in rain conditions images. The main advantage of performing a pre-processing and training based on ROI is that we can segment the images in different zones. The use of a CNN has reached good performance as the results indicate. In this work, we obtained an accuracy of about 89% and a sensitivity of 84%. These values give us an idea of the reliability of the implemented system for detecting rain conditions in the image.

As part of our future work, we are also working on the generation of a database with more examples. This objective can be achieved with longer recordings and a higher number of cameras. Thus, it will be possible to test the proposed system in other heterogeneous scenarios.

Additionally to detection, rainfall quantification can also be proposed using the powerful of CNN and other machine learning techniques. Another open issues are: (1) the selection of several ROIs and the integration of the detection over them in order to increase the quality of our detector; and (2) analyze and compare different techniques of detecting edges can use different neuronal network architectures.

**Acknowledgements.** The authors acknowledge the involvement of the Spanish Meteorological Agency (AEMET) [22] for the support in the acquisition of meteorological data. The research work presented in this paper has been financed by the Programa de Cooperación Territorial. INTERREG V A España-Portugal. MAC 2014-2020, Eje 3 [23]. Project VIMETRI-MAC "Sistema de vigilancia meteorológica para el seguimiento de riesgos medioambientales/Meteorological monitoring system for tracking environmental risks", código MAC/3.5b/065 [24].

## References

1. United Nations Website. http://www.un.org/en/sections/issues-depth/climate-change/index.html. Accessed 4 Mar 2019
2. Garg, K., Nayar, S.K.: Vision and rain. Int. J. Comput. Vis. **75**(1), 3–27 (2007)
3. Bossu, J., Hautiere, N., Tarel, J.P.: Rain or snow detection in image sequences through use of a histogram of orientation of streaks. Int. J. Comput. Vis. **93**(3), 348–367 (2011)
4. Sawant, S., Ghonge, P.A.: Estimation of rain drop analysis using image processing. Int. J. Sci. Res. **4**(1), 1981–1986 (2015)
5. Al Machot, F., Ali, M., Haj Mosa, A., Schwarzlmüller, C., Gutmann, M., Kyamakya, K.: Real-time raindrop detection based on cellular neuronal networks for ADAS. J. Real-Time Image Process., 1–13 (2016)
6. Chen, C.Y., Hsieh, C.W., Chi, P.W., Lin, C.F., Weng, C.J., Hwang, C.H.: High-speed image velocimetry system for rainfall measurement. IEEE Access **6**, 20929–20936 (2018)
7. Bronte, S., Bergasa, L.M., Alcantarilla, P.F.: Fog detection system based on computer vision techniques. In: 12th International IEEE Conference on Intelligent Transportation Systems (2009)
8. Pavlic, M., Belzner, H., Rigoll, G., Ilic, S.: Image based fog detection in vehicles. In: IEEE Intelligent Vehicles Symposium (2012)
9. Beritelli, F., Capizze, G., Lo Sciuto, G., Napoli, C., Scaglione, F.: Rainfall estimation based on the intensity of received signal in a LET/4G mobile terminal by using a probabilistic neuronal network. IEEE Access **6**, 30865–30873 (2018)
10. Gharanjik, A., Bhavami Shankar, M.R., Zimmer, F., Ottersten, B.: Centralized rainfall estimation using carrier to noise of satellite communication links. IEEE J. Sel. Areas Commun. **36**(5), 1065–1073 (2018)
11. Joshi, A.M., Thakar, K.: A survey on digit recognition using deep learning. Int. J. Novel Res. Dev. **3**(4), 112–118 (2018)
12. Dung, P.V.: Multiple convolution neural networks for an online handwriting recognition system. In: 6th International Conference on Advances in System Simulation – SIMUL (2014)
13. Oneway devices Homepage. http://www.onewaydevices.com/product/owipcam25.php. Accessed 27 Feb 2019
14. Vicent, O.R., Folorunso, O.: A descriptive algorithm for Sobel image edge detection. In: Proceedings of Informing Science & IT Education Conference (2009)
15. Gao, W., Yan, L., Zhan, X., Liu, H.: An improved Sobel edge detection. In: 3rd International Conference on Computer Science and Information Technology (2010)
16. Anphy, J., Deepa, K., Naiji, J., Silpa George, E., Anjitha, V.: Performance study of edge operators. In: International Conference on Embedded Systems (2014)

17. Qiang, S., Guoying, L., Jingqi, M., Hongmei, Z.: An edged-detection method based on adaptive canny algorithm and iterative segmentation threshold. In: 2nd International Conference on Control Science and System Engineering (ICCSSE) (2016)
18. Dong, W., Shinsheng, Z.: Color image recognition method based on Prewitt operator. In: International Conference on Computer Science and Software Engineering (2008)
19. Rosenfeld, A.: The Max Roberts operator is a Hueckel-type edge detector. IEEE Trans. Pattern Anal. Mach. Intell. **PAMI-3**(1), 101–103 (1981)
20. Öztürk, C.N., Albayrak, S.: Edge detection on MR images with Marr-Hildreth method extended to third dimension. In: 23rd Signal Processing and Communications Applications Conference (2015)
21. Haralick, R.M.: Zero crossing of directional derivative edge operator. In: Proceedings of SPIE, vol. 0336, Robot Vision (1982)
22. AEMET Homepage. http://www.aemet.es/en/portada. Accessed 27 Feb 2019
23. Interreg-Mac Homepage, approved projects. https://www.mac-interreg.org. Accessed 6 Feb 2019
24. ViMetRiMAC Homepage. http://www.vimetrimac.ulpgc.es/. Accessed 6 Feb 2019

# On the Application of a Recurrent Neural Network for Rainfall Quantification Based on the Received Signal from Microwave Links

Ivan Guerra-Moreno[1](✉), Juan L. Navarro-Mesa[1], Antonio G. Ravelo-García[1], and Carmen Paz Suarez-Araujo[2]

[1] Institute for Technological Development and Innovation in Communications (IDeTIC), University de Las Palmas de Gran Canaria, Las Palmas de Gran Canaria, Spain
{ivan.guerramoreno, juanluis.navarro}@ulpgc.es

[2] Institute of Cybernetic Science and Technology, University de Las Palmas de Gran Canaria, Las Palmas de Gran Canaria, Spain

**Abstract.** The detection and quantification of rainfall is of paramount importance in many application contexts. The research work we present here is devoted to design a system to detect meteorological phenomena in situations of risk. Particularly, we extend the usage of systems designed for other specific purposes incorporating them weather observation as a new service. We investigate how machine learning techniques can be used to design rain detection and quantification algorithms that learn directly from data and become robust enough to perform detection under changing conditions over time. We show that Recurrent Neural Networks are well suited for rainfall quantification, and no precise knowledge of the underlying propagation model is necessary. We propose a recurrent neural architecture with two layers. A first layer acts as a detector-quantifier and it is trained using known precipitations from near rain gauges. The second layer plays the role of calibration that transforms previous levels into rainfall quantitation values. A very important aspect of our proposal is the feature extraction module, which allow a reliable detection and accurate quantification. We study several options for extracting the most discriminative features associated to rain and no rain events. We show that our system can detect and quantify rain-no rain events with promising results in terms of sensitivity (76,6%), specificity (97,0%) and accuracy (96,5%). The histograms of error and the accumulated rain rates show good performance, as well.

**Keywords:** Recurrent Neural Networks · Pattern recognition · Rain detection · Rainfall quantification · Received signal strength · Microwave links

## 1 Introduction

In many application contexts, the detection and quantification of rainfall is of paramount importance. Examples of such contexts are planning and management of water resources, agriculture, hydraulic structure design, home and business supplies, etc. In addition, the concept of climate change has acquired a growing interest with the


I. Rojas et al. (Eds.): IWANN 2019, LNCS 11506, pp. 39–51, 2019.
https://doi.org/10.1007/978-3-030-20521-8_4

consequent concern of people, governments and various economic sectors [1]. One of the results of this change is an increased risk of storms, while drying up other areas of the planet. In parallel, the interest on specialized systems for rainfall observation and prediction has also grown. Traditionally, methods based on weather station are used to detect rain precipitation. The research work we present here is intended to be part of a monitoring system in order to detect meteorological phenomena, e.g., in situations of risk. Particularly, we extend the usage of systems intended for other specific purposes giving to them added value by incorporating weather observation as a new service.

Rain gauges are probably the most common devices for measuring the rainfall, and can provide accurate estimations with a fine temporal resolution. Despite rain gauges can provide good measurements in specific geographic locations, the information in the whole area under observation is an open problem.

There are several solutions to improve the monitoring. Among them, we can mention weather radars and infrared satellite imagery. Weather radars can monitor much larger area than rain gauges and determining the areal distribution of precipitation [2], but it is more expensive. Satellite rainfall estimations assure greater spatial and temporal resolution [3], but the rain intensity estimation is less accurate [4] and is also expensive.

In recent years, the idea of giving added value to systems designed for other specific uses has gained acceptance thus opening the possibility of improving the observation network without incurring excessive expense. In this context, it has become interesting to use the attenuations of microwave signals due to meteorological phenomena, for example, rain or snow. Since 2006, several studies have shown that it is possible to calculate precipitation by observing the reception power of the microwave links of telecommunications networks [5]. This method is applicable for both quality control of rain gauge measurements using telecommunication microwave links [6] and for precipitation observation using microwave backhaul links [7].

In this paper, we investigate machine learning techniques to design rain detection and quantification algorithms. We show that these algorithms are robust enough to perform detection under changing propagation conditions over time. This approach is particularly effective in emerging weather observation technologies where appropriate models may not exist or are difficult to derive analytically. Our approach is different from prior works since we assume that the mathematical models for the propagation channel are incomplete or unknown. This is motivated by the recent success using neural networks (NNs) for rainfall estimation. For instance, in [8] and [9] the authors use the impact of the rain on the cellular system performance and on the carrier-to-noise power ratio (C/N) levels, respectively.

Recurrent neural network (RNN) [10] process the input sequences one element at a time, maintaining in their hidden units a 'state vector' that implicitly contains information about the history of all the past elements of the sequence. We show that the philosophy inherent in RNN is well suited for rainfall quantification [11]. In our approach, the quantification goes through an architecture with two layers. A first layer acts as a detector-quantifier and it is trained using known precipitations from near rain gauges. A second layer plays the role of calibration, that is, the output of the first layer is transformed from detection levels into calibrated rainfall quantitation values. In our network, we pay attention to the feature extraction module as it is of major importance in order to make reliable detection and accurate quantification. We study several options for extracting the most discriminative features associated to rain and no rain events.

The rest of the paper is organized as follows. In Sect. 2 we present the experimental setup and how to process data in order to make appropriate design and experiments. Section 3 we show our design for the feature extraction module, and we highlight its importance as first block of the system. In Sect. 4 is devoted to explain the rationale that led us to use Recurrent Neural Networks in a layered architecture. In Sect. 5 we present the set of experiments and results obtained. We show that our system can detect and quantify rain-no rain events, study the optimum number of filter masks for the radio link under study, and show good results by means of histograms of absolute error, and accumulated rain. Finally, Sect. 6 is devoted to the conclusions and future work.

## 2 Experimental Setup and Data Processing

The experimental setup starts from a signal set corresponding to mean received signal level (RSLmean) values measured from 106 radio communication links placed in several locations of the Canary Islands. The transmission frequencies range from 6 to 38 GHz with some antenna polarizations being vertical and the rest are horizontal. The distances between antennas range between 1 to 12 km. The antennas are installed at heights of more than a thousand meters to antennas installed a few meters above the sea level.

The sampling period, T, to capture the RSL signal is a parameter conditioned by the radiofrequency system. In our case, the RSL values represent the average reception power in the previous 15 min (over 9000 samples, one every 100 ms.) [12].

The Agencia Estatal de Meteorología (AEMET) of Spain has provided the reference pluviometric measures. The sampling period of the measurements is 60 min. Therefore, we have a problem of asynchrony between the sampling of the RSL and the precipitation signals. We solve this problem by using the largest period. That is, the RSL signal will be related to the hourly precipitation data. This means that every hour 't' we have a short sequence of four RSL values, St, that form our basic observation vector.

An important issue is to take the decision of which is the most appropriate reference rain gauge to the radio link. For this purpose, we take into consideration the distance (e.g., the nearest one) and orography so as there is as much correlation as possible between the measurements from the stations and the links. In Fig. 1 we show the radio link and the rain gauge used in this work. The length of the link is 1,1384 km and the rain gauge is 300 m from the path, the frequency is 23.086 GHz and the polarization is horizontal. Regarding the measures, it is important to note that there are often discontinuities in them. That is, although the sampling period is constant, some data is missing because of system failures. This may happen in rain gauges as well as in radio links, although it is more common in the former.

**Fig. 1.** Map with the radio link and rain gauge used in this work

In Fig. 2a we can see an architecture composed of four links in spatial diversity belonging to the radio link of Fig. 1. The four RSL signals shown in Fig. 2b are continuous, that is, there is no loss of data. However, on the 7th, 8th and 9th of January 2016 there are missing precipitation data (blue). For convenience, we represent this lack of data with negative values. Since our interest is to make supervised learning, throughout this work, we exclude sampling instances in which RSL or precipitation data are lacking.

In Fig. 2b we can see that in January 5th there were several rain events with precipitation rates of 0'1, 0'4 and 3'8 mm/h. As expected, [5–8], these events are manifested in the four RSL signals and, in turn, occur in the four links. In this Figure, 'X' is the antenna on the left and 'Y' is the one on the right. 'A' is the upper antenna and 'B' is the lower one.

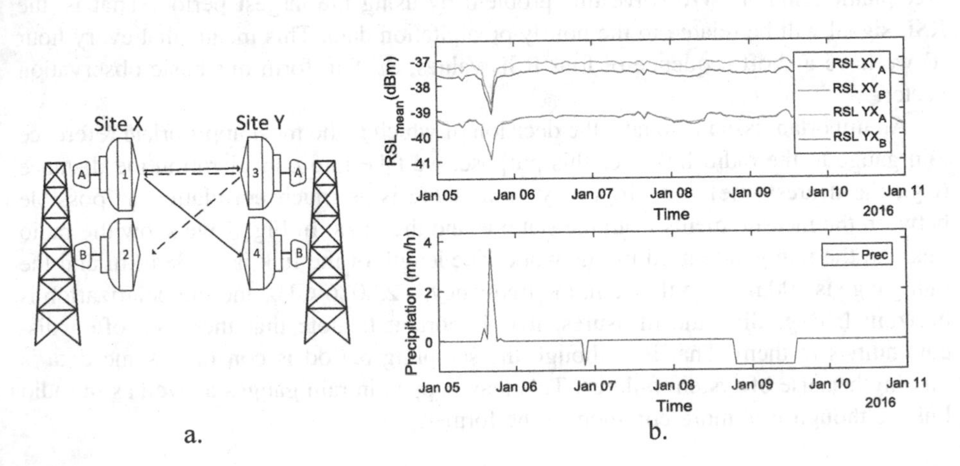


**Fig. 2.** (a) Receiver diversity of the radio link; (b) RSL measurements and precipitation (Color figure online)

## 3 Feature Extraction

The idea is to perform a preliminary processing of the data to facilitate the extraction of the relevant information with the minimum loss of information. The module for feature extraction must be seen as part of the network since the vectors of characteristics feed it in order to be able to discern the different effects of rain in the signal.

### 3.1 Variability of the RSL with Time

The nature of the RSL signal is illustrated in Fig. 3, which shows the evolution over 6 days of June 2016. Both the RSL and the Temperature signals exhibit a dry-air baseline with some degree of periodicity with a period of 24 h, which is due to the daily cycle of temperature and humidity, also air pressure (not present in the Figure). The deepest fading in the Figure happens by the end of June 10th, and it is due to rainfall. Without rain, the attenuation fluctuates (more or less) smoothly within some fixed range. In turn, the RSL follows a temporal evolution quite parallel to the temperature thus indicating dependency.

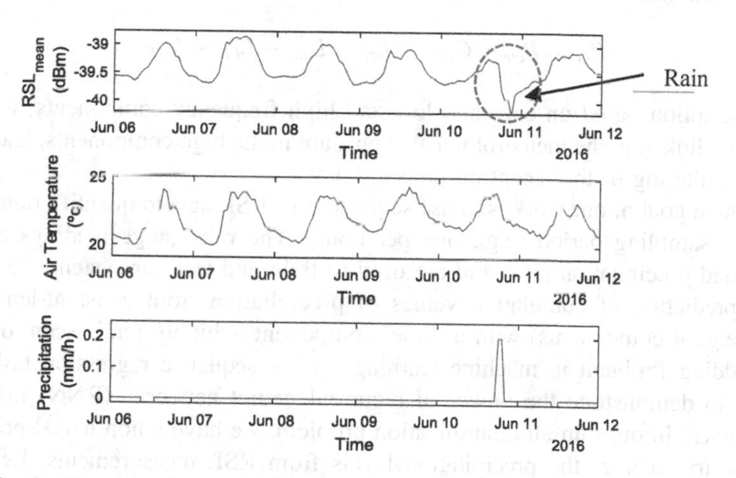


**Fig. 3.** Example of typical RSL, Temperature and precipitation in daily cycle

Paying attention to the RSL signal, the basic principle about detecting and estimating precipitation by using microwave radio links is based on the ability of rain to attenuate electromagnetic signals, and the relation between the attenuation and the rain rate is given by the following expression:

$$Att_n = k\,R^{\alpha}\ (dB/km),$$

where R(mm/h) is the precipitation, k and $\alpha$ are given in the UIT-R 838 [16] recommendation and are determined as a function of frequency and antenna polarization. The potentiality of using $Att_n$ is that it can be estimated from the received power, seen as a time series ($Att_n(t)$), and can serve as a basis for quantifying R. We do not take this approach in our work, we will use machine learning techniques. Let's now look at some aspects of the temporal dynamics inherent in the RSL signal, how we can take advantage of it. So, we can use of these aspects in an RNN-based architecture.

Each link is characterized as follows by its transmission power, its reception power at the destination, a set of gains and a set of events that cause transmission losses (Friis equation) [17]:

$$P_{rx}(dBm) = P_{tx} + G_{tx} + G_{rx} - FSPL - L_{Att} - L_{Rain} - L_{fog}$$

The terms are: $P_{rx}$ reception power, $P_{tx}$ transmission power, $G_{cte}$ reception and transmission antenna gains, FSPL Free-space path loss, $L_{Att}$ atmospheric loss, $L_{Rain}$ loss due to precipitation and $L_{fog}$ loss due to clouds or mists. Losses in the transmission can be classified as: stable losses once the radio link is installed, and the variant losses in time such as those due to atmospheric conditions, rainfall or other hydrometeors, etc. Therefore, it can be determined that the reception power $P_{rx}$ is equivalent to the result of the calculation of a set of constant values and the previously described losses, some of which are constant, others of low frequency (LF) and others of high frequency and impulsive-like (HF). Thus, the previous equation can be rewritten as:

$$P_{rx} = P_{cte} + G_{cte} + G_{cte} - L_{cte} - L_{LF} - L_{HF}$$

A description based on constant, low and high frequency components, where the information linked to the meteorological events are in the high components, leads to the need for a filtering of the reception power signal.

The main goal of our work is to use sequences of RSL data to quantify rainfall with a constant sampling period (e.g., one per hour). The rain gauge readings represent accumulated precipitation rates and are used to train and test our system.

The prediction of cumulative values of precipitation from constant-length RSL vectors (e.g., 4 components) with a 'time' component is highly reminiscent of the so-called Adding Problem in machine learning—a toy sequence regression task that is designed to demonstrate the power of recurrent neural networks (RNN) in learning dependencies. In our rainfall quantification problem, we have a non-trivial problem in which we try to infer the precipitation levels from RSL measurements. Let us pay attention, for example, to the effect of rain on the signal in Fig. 2b.

One can do the abstraction of the RSL signal carrying rain events, and we are designing a detector that also quantifies its levels. On the other hand, it is easy to understand that the rain at a given moment can be related to the one that occurred in previous moments. Let us see this more in detail.

The attenuations of a radio link at a given moment, 't', correspond to the integration of all the individual attenuations. Particularly, we can consider that the attenuation due to rain at the receiving antenna, $A'_{Rain}(t)$ (dB), is equivalent to the contribution of a sequence of differential segments of attenuations, $A_{Rain}(x, t)$ (dB/Km). Thus, we can mathematically express this total contribution in the following way:

$$A'_{Rain}(t) = \int_{x=1}^{L} A_{Rain}(\mathrm{x}, \mathrm{t})\, dx\, (dB),$$

where 'x' expresses the path. The amount of attenuation on a radio link during a period, T, is the average of attenuations in that period (e.g., every 15 min in our database):

$$A_{Rain}(t) = \frac{1}{T} \int_{\mathrm{t}-T}^{t} A'_{Rain}(t')dt'\, (dB).$$

In order to make a proper matching of the precipitation causing $\underline{A'_{Rain}(t)}$ with the real one, we must take a good reference. We do this by considering the precipitation signal obtained by a reference pluviometer over time, Prec(t), and it can be expressed as follows:

$$Prec(t) = \int_{\mathrm{t}-Tp}^{\mathrm{t}} Prec(s)ds\, (mm/h),$$

where $T_p$ is the period under consideration. Regarding this period, we would desire that T and $T_p$ be equal. Unfortunately, this is not usually the case since radio links and pluviometer usually work with different sampling time as they belong to different systems. We have solved this problem in Sect. 2.

As it is our interest to track of the precipitation by measuring of the attenuation, one would expect that $A_{Rain}(t)$ and $P_{rec}(t)$ be synchronized, and both should change in time guided by the evolution of $P_{rec}(t)$. In any case, both precipitation and the subsequent attenuation evolve in time.

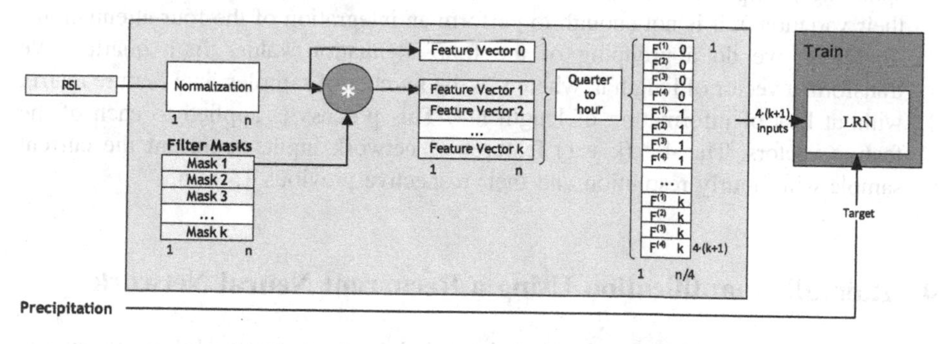


**Fig. 4.** Module for feature extraction and its connection to the LRN

### 3.2 Feature Extraction Module

In this module, vectors of characteristics are generated in order to offer discriminative information to discern the different effects of the rainfall in the signal. For this purpose, we make the following steps (Fig. 4):

- **Data normalization.** Data is normalized by calculating the median of the RSL signal, from the current time t to t-N where N (= 672) is the length of the normalization window, N samples corresponding to 7 days, 4 samples per hour. The subtraction of this median to the reception power eliminates the attenuations due to constant losses.
- **Filtering**. Considering that the attenuation due to precipitation is present in the high frequencies, a special processing is necessary to eliminate most of the constant components of the signal and low frequencies. We proceed to perform a filtering by convolving the signal under study against a set of 'k' convolution masks (Fig. 4, Filter Masks). These convolution masks have been devised based on several days without precipitation in the rain gauge data and with absence of attenuations in the RSL signal. We generate 'k' convolution masks that represent different typical dry days. The convolution of each of the masks with the input signal generates 'k' new filtered signals (Feature Vectors 1 to k). To this set of signals that characterize the input signal, we add the original input signal without the continuous component (Feature Vector 0).
- **Time scale adaptation.** The third step is the adaptation of the time scale of the feature vectors against the time scale of the precipitation (target). Knowing that the precipitation given at a certain time is the integration of rainfall produced throughout the hour, and that the RSL is a vector every 15 min we can affirm that there will be four sections of RSL whose integration will be related to the corresponding precipitation. Since we are interested in the waveform (frequencies and their variations), it is not enough to perform an integration of the four attenuations. Therefore, we do a grouping of the four attenuation values in a quartet. We transform a vector of length n, where n is the number of samples, in a feature matrix without loss of information of length n/4. This process is applied to each of the feature vectors. The 4 * (k + 1) features of network inputs represent the current sample with hourly resolution and their respective previous 15 min.

# 4 Rainfall Quantification Using a Recurrent Neural Network

With the previous considerations in mind, the underlying time structural characteristics of the $A_{Rain}(t)$ and $P_{rec}(t)$ signals are compelling enough to suggest that RNN are well suited for the problem. The main characteristic of an architecture based on RNN is the representation of time implicitly by its effects on processing. It involves the use of recurrent connections to provide networks with a dynamic memory. In this approach, we have opted for an architecture based on a Layered Recurrent Network (LRN), where hidden layer patterns feedback to themselves. In our LRN (Fig. 5), there is a feedback loop, with a single delay, around each layer of the network, except for the last stage. Because we are interested in knowing past events in the detection of precipitation, we add a block of one delay in the design that returns the output of the first layer of the network to the network itself to relate it to the input of the characteristics. The amount of delay depends on several aspects like the sampling period, how fast the precipitation may change, etc. From our experience with the data we are using, on delay is appropriate. The final number of neurons in Layer 1 was the result of trial and error process.

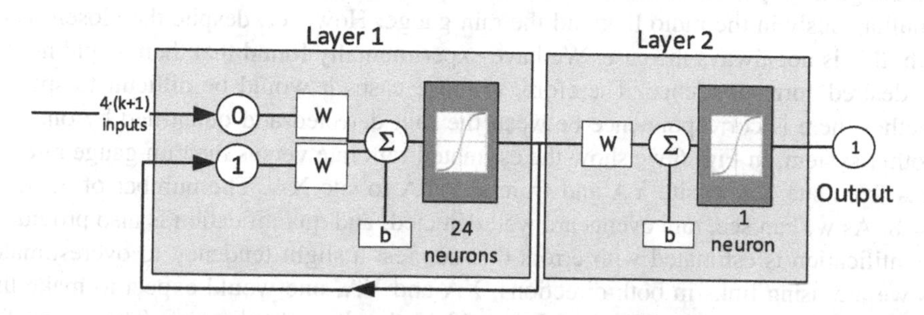


**Fig. 5.** Architecture of the layered recurrent neural network

In our experiments, we started by one neuron and increased the number successively being the classification error minimization the stopping criterion. We have experimentally found that, as the number of neurons exceeded 24, the error began to increase again. Therefore, we have taken 24 neurons, which coincides with the 24-h period representing a full day. We make the conjecture that this result is because the RSL signal follows a day-night dynamic and this is transferred to the temporal representation that the RNN makes of the process. This one corresponds mainly to the daily cycle effect on the RSL signal. A second layer of a single level adjustment neuron allows the network to be 'calibrated', that is, we add a function that transforms the output of the RNN in Layer 1 into precipitation values at the output of Layer 2. With such a relation between input vectors and the output of Layer 2 an output of '0' represents no-rain. Finally, we perform train and tests experiments using the Neural Network toolbox of Matlab [13]. The training function updates the weight and bias values according to Levenberg-Marquardt optimization [14]. It minimizes a combination of squared errors and weights, and then determines the correct combination to produce a network with a good generalize by Bayesian regularization [15].

## 5 Experiments and Results

For the experiments, we have used the database introduced in Sect. 2. Between the months of March 2014 and March 2016 we have selected 13057 samples of precipitation (one per hour) and the quadruple of RSL (four per hour, St) in each of the four links (see Fig. 2). There are 304 (2'33%) rain events of different intensities as collected by the rain gauge. Before proceeding, we must make the following consideration. The rain gauge is very close to the link, one can expect that if there is rain it will be reflected simultaneously in the radio link and the rain gauge. However, despite the closeness of both, this is not always the case. We have experimentally found that there could not be the desired correspondence. Therefore, in those cases it would be difficult to specify whether there is correspondence between the rain detected and quantified by one and another system. In Fig. 6 we show the estimated rain rate versus the rain gauge rate for links from site XA to site YA and from site YA to site XA. The number of filters is k = 8. As we can see, rain events are well detected, and quantification is also provided. Quantification is estimated with errors that suggest a slight tendency to overestimate. As we are using links in both directions, XA and YA, one would expect to make the same detections and quantification. It is evident that it is not the case. The reason for this is that the RSL signals are not the same thus showing different behavior.

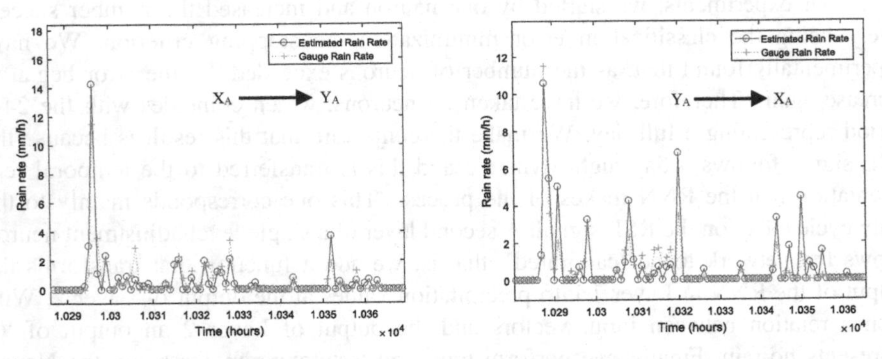


**Fig. 6.** Estimated and Rain rates (left) from site $X_A$ to $Y_A$; (right) from $Y_A$ to $X_A$

In Fig. 7, we show a different experiment. In this case there are several gaps (missed samples) from the rain gauge (see orange + under the 0 level). These gaps are due to failures, exhausted battery energy, etc. Since these failures may happen and it could take a long time until they are solved, we can wonder whether the quantification of our system is reliable for filling in the measurements in the meteorological station. As we do not have the exact precipitation references, we can only make conjectures about whether our system is able to fill the gaps from the rain gauges. The reliability in the detection, and the quality of the quantification suggest that it is viable to use our system. Even, the estimates obtained from both links can complement each other.

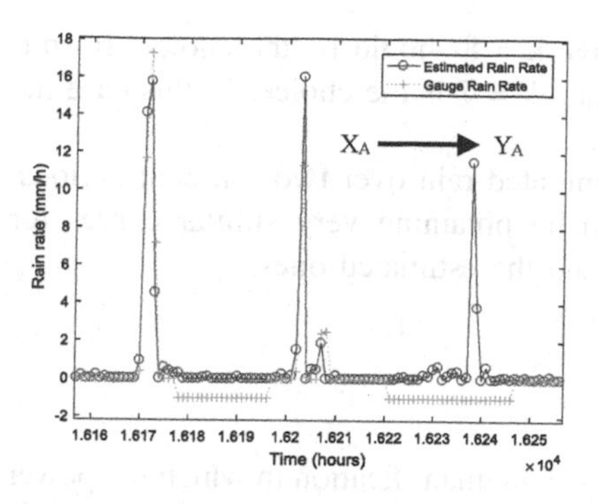


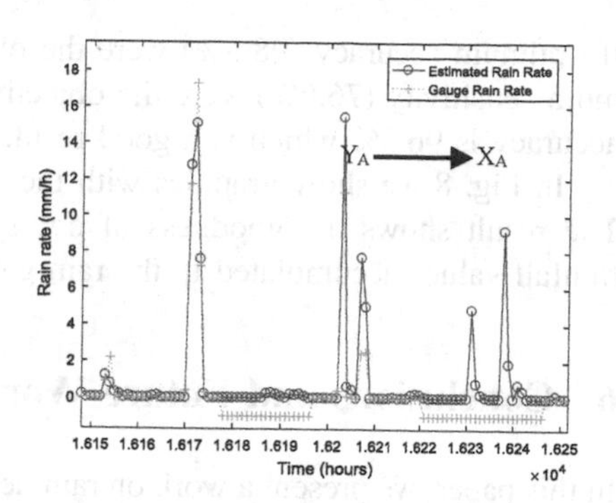


**Fig. 7.** Estimated rates during gaps in the Rain Gauge (left) from site $X_A$ to $Y_A$; (right) from site $Y_A$ to $X_A$

We must recall from Sect. 2, that the training function updates the weight and bias values according to Levenberg-Marquardt optimization. It works in such a way that produce a network that generalizes well by means of a process called Bayesian regularization. We have performed a set of experiments. In each one, the data were randomly divided in three data groups of each radio link, in 70% for training (9139 no rain and 213 rain events), 15% validation and 15% test (1913 no rain and 46 rain events each). The experiments were repeated 100 epochs.

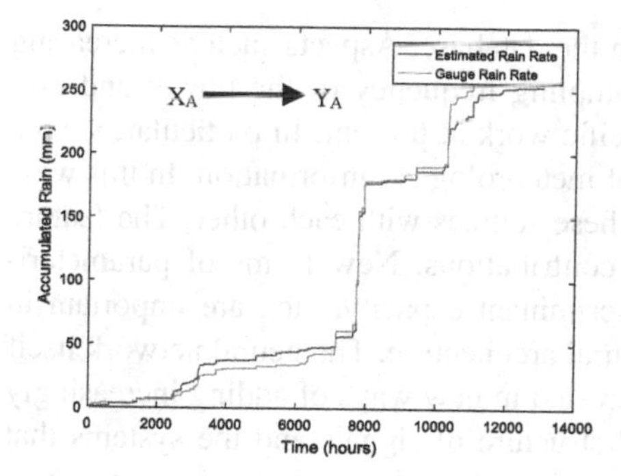


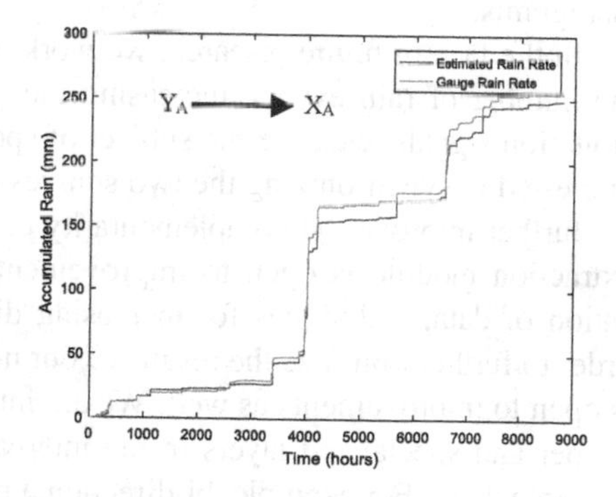


**Fig. 8.** Accumulated rain in two different periods of time

In the horizontal axis k = 1 is for Feature Vector 0, that is, no mask is used, and k = [2, …, 11] represent Feature Vector 1 to Feature Vector 10 plus Feature Vector 0. The specificity (Sp), sensitivity (Se) and accuracy (Ac) are measured. FP and FN are the false positives and false negatives, respectively. As we can see, specificity and accuracy show very similar behavior. This is because there are many negatives and the detection (TN) is very high. The sensitivity to rain events (TP) is quite stable among the number of filter masks, with a maximum for k = 7. The maximum accuracy and sensitivity are achieved for different but close number of masks, K = 7 and 8.

If optimum accuracy (98,3%) were the objective k = 8 should be the choice. If optimum sensitivity (76,6%) were the objective, then k = 7 is the choice. In this case the accuracy is 96,5% which is a good result, too.

In Fig. 8 we show graphics with the accumulated rain over two different periods. The result shows the goodness of our system by obtaining very similar curves for rainfall values accumulated in the rain gauge and the estimated ones.

## 6 Conclusions and Future Works

In this paper, we present a work on rain detection and quantification in which the power levels received in microwaves radio links are used as base information. This work is part of a larger one in which we look for new technologies to expand the network of weather monitoring stations. Starting from the well-known sensitivity of the signals to meteorological phenomena, we propose a module to extract characteristics that conveniently feed the neural network that we have designed. Up to our knowledge, the module we present is novel for feature extraction, and it a key element in the system. We have justified that given the temporal structure of rainfall precipitation, a recurrent neural network is appropriate to follow the dynamics of the signals. At the same time, this is achieved without needing an accurate description of the propagation.

Our experiments show truly promising results. We successfully detected and quantified the huge number of events, especially not rain. Rain events have a lower but equally good sensitivity, which results in good error and cumulative rainfall histograms.

In the face of future research, we work on the database. Aspects such as increasing the number of rain events, increasing the sampling frequency of the power and precipitation signals, etc., are the subject of specific work at present. In particular, we are interested in synchronizing the two sources of meteorological information. In this way, we further improve the complementarity of these sources with each other. The feature extraction module is open to improvement contributions. New forms of parameterization of data, techniques for increasing discriminant capacity, etc., are important in order to further optimize the results of our neural architecture. The neural network itself is open to improvements as well. We are interested in new ways of adding increasingly deeper and specialized layers in the internal structure of signals and the systems that generate them. For example, bi-directional recurring networks, convolutional networks, Long Short-Term Memory (LSTM) networks etc., receive our attention. We have worked with the signals of each link individually. Although the results obtained are good, it would be convenient to integrate the four sources in the feature extraction module, in such way that is individualized in several LRNs and integrate the results, or a hybrid formula.

**Acknowledgments.** INTERREG VA España-Portugal has financed the research work presented in this paper. MAC 2014-2020, Eje 3. Project VIMETRI-, code MAC/3.5b/065.

We are thankful to the Dirección General de Telecomunicaciones y Nuevas Tecnologías and the Dirección General de Seguridad y Emergencias, Autonomous Government of the Canary Islands, and the State Meteorological Agency-AEMET (Government of Spain), for providing the data and their valuable comments.

## References

1. http://www.un.org/en/sections/issues-depth/climate-change/index.html. Accessed 4 Mar 2019
2. Yeary, M., Cheong, B.L., Kurdzo, J.M., Yu, T.-Y., Palmer, R.: A brief overview of weather radar technologies and instrumentation. IEEE Instrum. Meas. Mag. (2014)
3. Levizzani, V.: Precipitation measurements from space. In: Proceedings EuCAP 2006, (ESA SP-626), Nice, France, 6–10 November 2006, October 2006
4. Geerts, B.: Satellite-Based Rainfall Estimation. http://www.das.uwyo.edu/~geerts/cwx/notes/chap10/satellite_rain.html. Accessed 3 Mar 2019
5. Messer, H., Zinevich, A., Alpert, P.: Environmental monitoring by wireless communication networks. Science **312**(5774), 713 (2006)
6. Bianchi, B., Rieckermann, J., Berne, A.: Quality control of rain gauge measurements using telecommunication microwave links. J. Hydrol. **492**, 15–23 (2013). ISSN 0022-1694
7. Chwala, C., et al.: Precipitation observation using microwave backhaul links in the alpine and pre-alpine region of Southern Germany, Hydrology and Earth System Sciences, vol. 16, no. 8, pp. 2647–2661 (2012)
8. Beritelli, F., Capizzi, G., Sciuto, G.L., Napoli, C., Scaglione, F.: Rainfall estimation based on the intensity of the received signal in a LTE/4G mobile terminal by using a probabilistic neural network. IEEE Access **6**, 30865–30873 (2018). https://doi.org/10.1109/access.2018.2839699
9. Gharanjik, A., Bhavani Shankar, M.R., Zimmer, F., Ottersten, B.: Centralized rainfall estimation using carrier to noise of satellite communication links. IEEE J. Sel. Areas Commun. **36**(5) (2018)
10. LeCun, Y., et al.: Deep learning. Nature **521**(7553), 436–444 (2015). https://doi.org/10.1038/nature14539. Accessed 1 Mar 2019
11. Luk, K.C., Ball, J.E., Sharma, A.: An application of artificial neural networks for rainfall forecasting. Math. Comput. Model. **33**(6–7), 683–693 (2001)
12. AVIAT networks. http://aviatnetworks.com/. Accessed 1 Mar 2019
13. Matlab, Campus Licence. https://es.mathworks.com/academia/tahportal/universidad-de-las-palmas-de-gran-canaria-40760023.html. Accessed 3 Mar 2019
14. Yu, H., Wilamowski, B.M.: Levenberg–marquardt training. In: Industrial Electronics Handbook, vol. 5, no. 12, p. 1 (2011)
15. MacKay, D.J.C.: Bayesian interpolation. Neural Comput. **4**(3), 415–447 (1992)
16. Specific Attenuation Model for Rain for Use in Prediction Methods, document ITU-R P.838-3, ITU-R Recommendation (2005)
17. Friis, H.T.: A note on a simple transmission formula. Proc. IRE **34**(5), 254–256 (1946)

# Ambient Temperature Estimation Using WSN Links and Gaussian Process Regression

Sofia I. Inácio(✉) and Joaquim A. R. Azevedo

Faculty of Exact and Engineering Sciences, University of Madeira,
Colégio dos Jesuítas - Rua dos Ferreiros, 9000-039 Funchal, Portugal
sofia.inacio@staff.uma.pt

**Abstract.** After several outdoor installations of wireless sensor networks (WSN) with low-cost, low-power and short-range radios, it was found that the quality of the connections between nodes was negatively influenced by the increase of temperature. Through several experiments with radios, it is believed that this is due to the degradation of electronic circuits with increasing temperature. However, this degradation can be used to monitor the ambient temperature when there are no temperature sensors. In this work, a Gaussian Process Regression (GPR) is applied for probabilistic temperature estimation, having as prediction variables the signal level of the WSN connections. Using five links and one month of training data, the temperature estimation on one week of test data was very satisfactory.

**Keywords:** Wireless sensor networks · Connections · Temperature · Machine learning

## 1 Introduction

In 1959, Arthur Samuel defined the concept of machine learning as a "field of study that gives computers the ability to learn without being explicitly programmed" [1]. Since then new and different approaches have been gradually emerging, but all with the same goal, learning through a machine. Thanks to the evolution of computational power and given the large amounts of data available, the concept is now well defined, with several machine learning algorithms becoming more reliable. Artificial intelligence is already in people's daily lives even without them realizing it. As examples, we can mention predictive maintenance in several areas, face, motion and object recognition, tumor detection through images, among others.

The main goal of artificial intelligence is to learn through experience. To do a good training of the model, that is, for it to gain experience, it is necessary that the data be representative of the space of occurrence, in order to be able to construct a general model on that space, allowing the creation of accurate predictions in new cases.


I. Rojas et al. (Eds.): IWANN 2019, LNCS 11506, pp. 52–62, 2019.
https://doi.org/10.1007/978-3-030-20521-8_5

Machine learning techniques can be divided into two groups, supervised learning and unsupervised learning. In the first, the model is trained on known input and output data, in order to predict future outputs. In the second group, the model is trained to find patterns in the input data. No machine learning algorithm is the best for all problems. It is usually through testing that the best algorithm for a certain problem is determined. The performance of each model depends on the nature of the input data and its size.

With the evolution of technology, small radio modules have been developed. These modules are often low-cost, low-power and short-range, and therefore suitable for innumerable applications in remote areas and without access to the power grid. However, there are many more applications, from automation of a house to environmental monitoring, among others. One advantage is that, unlike structured networks, communication between two network nodes can be direct without having to go through an access point. In order to evaluate the performance of the connections in the network, some devices already provide quality parameters of the received signal, such as the received signal strength indication (RSSI).

In most cases wireless sensor networks are deployed outdoors, subject to weather conditions, which can be severe and degrade the network nodes. For example, radios are specified for a range of operating temperature, which is usually between $-40\,^{\circ}\mathrm{C}$ and $+85\,^{\circ}\mathrm{C}$. Although this range is wide it is known that there is a degradation of the performance of the radio when the temperature is high, which may be due to a change in the oscillator frequency, to an increase of the noise level in the receiver or even to the saturation of the amplifiers [2–5]. It is therefore important to understand the influence of the various meteorological parameters in order to implement a robust wireless sensor network, taking into account that reducing signal quality results in a reduction in range. However, this degradation can be considered from another point of view, such as the possibility of extracting data about the meteorological parameters that degrade the network and try to quantify them.

At the University of Madeira there has been a network of wireless sensors installed on the terrace for a long time. The radios used in the network are the XBee S2C model, and it has been shown in previous work [6], that the quality of the signal level is degraded with increasing temperature and with precipitation. In this work it is shown that there is a relationship between the ambient temperature and the signal level of the connections in a wireless sensor network and that it is possible to predict the ambient temperature value through the RSSI value received from the connections (no precipitation days were analysed). For this, a supervised machine learning algorithm, a Gaussian Process with an exponential covariance function, was used. However, there are always errors in the predicted data, which are minimized by using multiple connections.

The rest of this paper is organized as follows: Sect. 2 presents a brief reference to related work found in literature. In Sect. 3 is shown the methodology and in Sect. 4 are presented the experimental results obtained in this work. The main conclusions are addressed in Sect. 5.

## 2 Related Work

With the appearance of small radio modules for the Internet of things, many networks are being deployed for the most diverse applications. With this, several studies evaluating the performance of the networks in the most varied situations are also appearing. The deterioration of the electronics behavior with temperature is known, with the datasheets of the different electronic components usually highlighting that relationship. Radios are electronic components and their performance is, naturally, influenced by temperature. It should further be noted that when the nodes are placed in boxes exposed to the sun, the temperature inside the boxes may be much higher than the ambient temperature. In the literature there are already some published works showing exactly this, that the temperature negatively influences the quality of the connections in a network of wireless sensors, and, therefore, their performance.

In [3], author C. A. Boano states that temperature must be considered in the planning and installation phase of a WSN and that sensor nodes should not be placed near the limits of the communication range, especially if the installation is carried out in colder periods of the year. This author also concluded that when the temperature is lower (for example at night), it is possible to reduce the transmission power and thus save energy [4].

In some applications it is necessary, through the signal level, to determine the location of the nodes. Obviously it is necessary to know the effect of temperature on such a system. In [5], the authors report that node localization errors can grow up to 150%. In [2] a quality analysis of the connections in a WSN was performed, correlating that data with meteorological data. It was concluded that it was the increase in temperature that most negatively influenced the quality of the network. In [6] a radio module was heated and it was observed that the emitter and the transmitter are affected differently by the temperature. It has also been found that when the temperature increases, the transmission power of the radio drops considerably.

As far as we know, in literature no work has been presented that uses machine learning techniques to obtain the ambient temperature through the analysis of the quality of wireless sensor network connections. However, there are many works that use these techniques for environmental monitoring, such as precipitation forecasting [7–9], creation of an early warning system for very short-term heavy rainfall events [10], forecast of river flow and runoff models for flood risk management [11–15], prediction of solar radiation [16–18], estimation of soil temperature [19], water temperature in a river forecast [20], wind velocity forecast [21] and estimation of daily crop evapotranspiration for efficient irrigation management [22]. Most of these studies use meteorological data as prediction variables. In the previously mentioned works several models are used, from machine learning algorithms such as decision trees, gaussian process, k-nearest neighbors, autoregressive models, artificial neural networks, support vector machine, among others.

As already mentioned, machine learning algorithms can be divided in two groups: supervised learning and unsupervised learning. The algorithm used in this work is a supervised learning algorithm, but this group can be, in turn, divided into regression or classification problems. In the first, the prediction is related to a continuous quantity, whereas the output of classification problems consist of discrete class labels. In any case, the algorithms consist of mathematical models which, through input data and learning, can make predictions and decisions.

The process of building a model requires at least three datasets in different stages. The build of the model is initiated by the training, in this phase the model is adjusted to the data, and therefore, the training data dataset is used. Subsequently, the validation phase begins, where the validation dataset allows the adjustment the hyperparameters of the model, providing an unbiased evaluation. Finally, the test dataset is used and a final evaluation of the model is performed.

# 3 Methodology

## 3.1 Gaussian Processes for Machine Learning

In order to solve a problem through machine learning techniques there are several questions that have to be answered before starting the training. The most important question is probably which model will be the best for the type of available data. For this work the Gaussian Process Regression (GPR), an effective kernel-based machine learning algorithm, was chosen with a exponential covariance function. This model belongs to the group of supervised learning models, being a regression model and therefore suitable for the data that will be treated, since the target variable is continuous and this value is available as predictor variable for training.

The main objective in a machine learning model is to determine the functional relationship between the $d$ variables of the predictor group $\mathbf{x}$ and the target variable:

$$y = f(x) \tag{1}$$

being $f$ the function to be determined by regression.

Due to measurement errors, problems with data communication, among others, the data is often noisy. Therefore, we can consider each observation $y$ as related to an underlying function $f(x)$ through a Gaussian noise model [23]:

$$y = f(x) + \mathcal{N}(0, \sigma_n^2) \tag{2}$$

A Gaussian Process by definition is a collection of random variables of which any finite subset has a joint Gaussian distribution. This is completely specified by its second-order statistics, such as:

$$f(x) \sim \mathrm{GP}(m(x), k(x, x')) \tag{3}$$

where $m(x)$ is the mean and $k(x, x')$ is the covariance function of $f$. Often it is assumed that the mean is zero, and therefore it is only the covariance function that relates one observation to another. Thus, the covariance function is the crucial ingredient in a Gaussian process predictor, since it encodes the assumptions about the function to be learned. The most commonly-used covariance functions are shown in Table 1 [24].

**Table 1.** Most commonly-used covariance functions.

| Covariance function | Expression |
|---|---|
| Constant | $\sigma_0^2$ |
| Linear | $\sum_{d=1}^{D} \sigma_d^2 x_d x'_d$ |
| Polynomial | $(\mathbf{x} \cdot \mathbf{x}' + \sigma_0^2)^p$ |
| Squared exponential | $\exp(-\frac{r^2}{2l^2})$ |
| Matérn | $\frac{1}{2^{\nu-1}\Gamma(\nu)}(\frac{\sqrt{2\nu}}{l}r)^{\nu} K_{\nu}(\frac{\sqrt{2\nu}}{l}r)$ |
| Exponential | $\exp(-\frac{r}{l})$ |
| $\gamma$-exponential | $\exp(-(\frac{r}{l})^{\gamma})$ |
| Rational quadratic | $(1 + \frac{r^2}{2\alpha l^2})^{-\alpha}$ |
| Neural network | $\sin^{-1}(\frac{2\tilde{\mathbf{x}}^{\top} \sum \tilde{\mathbf{x}}'}{\sqrt{(1+2\tilde{\mathbf{x}}^{\top} \sum \tilde{\mathbf{x}})(1+2\tilde{\mathbf{x}}'^{\top} \sum \tilde{\mathbf{x}}')}})$ |

The regression process is based on calculating the covariance function with the combination of prediction variables using the following three matrices [24]:

$$K = \begin{bmatrix} k(x_1, x_1) & k(x_1, x_2) & \dots & k(x_1, x_n) \\ k(x_2, x_1) & k(x_2, x_2) & \dots & k(x_2, x_n) \\ \vdots & \vdots & \ddots & \vdots \\ k(x_n, x_1) & k(x_n, x_2) & \dots & k(x_n, x_n) \end{bmatrix}$$

$$K_* = [k(x_*, x_1) \;\; k(x_*, x_2) \;\; \cdots \;\; k(x_*, x_n)] \tag{4}$$

$$K_{**} = k(x_*, x_*) \tag{5}$$

As the main assumption in Gaussian process modelling is that the data can be represented as a sample of a multivariate Gaussian distribution with zero mean, then it is possible to write:

$$\begin{bmatrix} \mathbf{y} \\ y_* \end{bmatrix} = \mathcal{N}\left(0, \begin{bmatrix} K & K_*^{\top} \\ K_* & K_{**} \end{bmatrix}\right)$$

where $\top$ indicates the transpose of a matrix. The conditional distribution of $y_*$ given $\mathbf{y}$ is itself Gaussian-distributed and it can be written as follows:

$$y_* | \mathbf{y} \sim \mathcal{N}(K_* K^{-1} \mathbf{y}, K_{**} - K_* K^{-1} K_*^{\top}) \tag{6}$$

The mean of this distribution is the matrix of the regression coefficients, and it is the best estimate for $y_*$:

$$\overline{y}_* = K_* K^{-1} \mathbf{y} \tag{7}$$

The uncertainty of the estimate is determined by its variance, as follows:

$$\text{var}(y_*) = K_{**} - K_* K^{-1} K_*^{\top} \tag{8}$$

### 3.2 Experimental Setup

A few years ago a wireless sensor network was installed on the terrace of the University of Madeira. The main objective was to collect meteorological data and to send it to a database for later viewing and analysis. Over time, this network has served as a basis for students work, and so nodes were often added, moved, or removed from the network. This fact makes it difficult to analyse signal level over time, so in this work only one month of data was used, ensuring that in that month there was no modification of the network that could change the arrangement of nodes.

The radios used were XBee S2C and they are at a maximum distance between them of 10 m. The region where the network was deployed is relatively open, with no obstacles between the nodes of the network. The setup is shown in Fig. 1, where the connections used are shown. The coordinator is responsible for receiving the data from the network and inserting it into an SQL database.

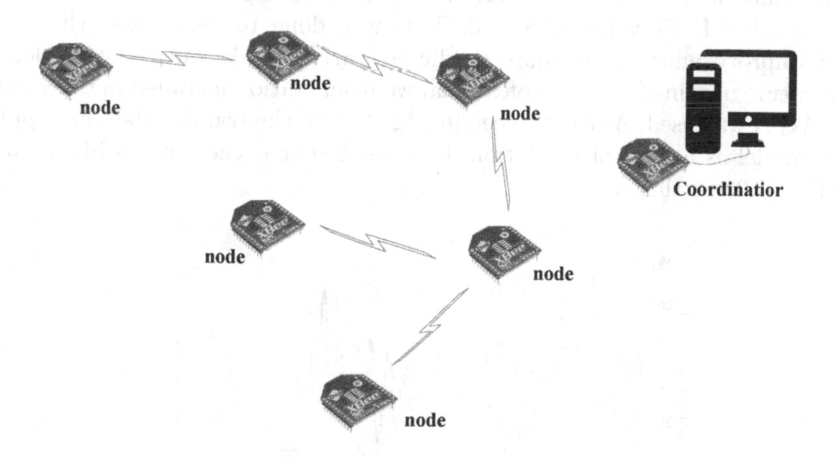


**Fig. 1.** Setup of the outdoor sensor network.

One limitation of these radios is the resolution of their RSSI signal, being only 1 dB. All antennas used were commercial quarter-wavelength antennas (W1030). For this work data from five network nodes was used. It was necessary to ensure that the connection between two radios did not change, and that the RSSI was always measured between the same two radios.

A professional weather station, Davis Vantage Pro 2, was installed to correlate the RSSI with meteorological data. This station has sensors for temperature, humidity, wind direction and speed and a rain gauge.

The data obtained from the RSSI of the chosen connections and the temperature were obtained during the month of August 2018. The test data was from the first week of September of the same year. Before entering the training phase, a pre-processing step was performed, consisting of filter in order to remove any high-frequency component. With this data the gaussian process was trained using Matlab software.

To determine the performance of the predictive model, the mean square error (RMSE) was used in this work. This measures the deviation between the predicted values and the observed values, and smaller values of this index indicate a better predictive performance. This is defined as:

$$\mathrm{RMSE} = \sqrt{\frac{\sum_{n=1}^{N}(y_i - \hat{y}_i)^2}{N}} \tag{9}$$

where $y_i$ and $\hat{y}_i$ are the $i$-index of the observed and predicted variables, respectively and $N$ is the number of predicted data.

## 4 Experimental Results

The training of the model was started only with an input variable, in this case only with the RSSI value of a link. This was done to verify how the model would improve when using more predicted variables. As output variable, the temperature obtained by the professional weather station installed in the vicinity of the WSN was used. As can be seen in Fig. 2 when the training data are applied to the model as a form of validation. It is verified that the error is high, with a RMSE of 1.23 in this case.

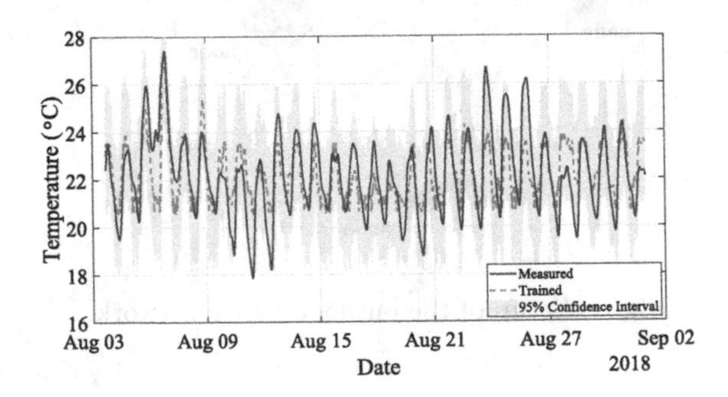


**Fig. 2.** Validation of the model with only one predicted variable.

In Fig. 3 the training result is shown when using the RSSI of five connections in conjunction with the time of day value at which the measurement was obtained. It is possible to verify that in this case the two curves are identical, as also shows the obtained RMSE of just 0,1.

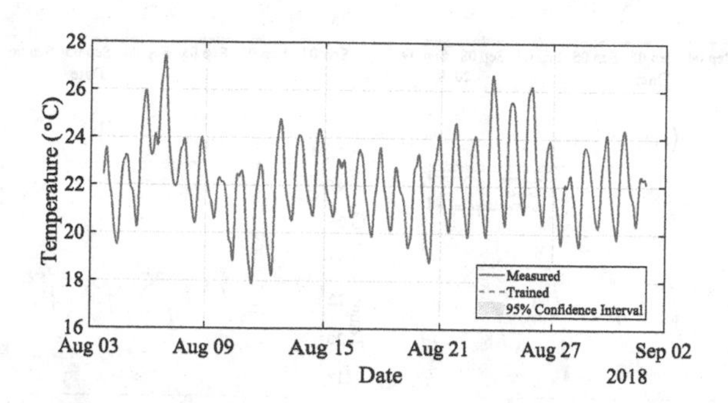


**Fig. 3.** Validation of the model with six predicted variables.

Table 2 shows the RMSE value for the training performed. It turns out that the more input variables the better the RMSE and therefore the better the model fits into the expected data as output.

As previously refered, the test data is only from one week, the week following the training data. In this case, the graphs were also obtained, and the corresponding RMSE value for each set of input variables, presented in Table 2. In Fig. 4 the results obtained with the different input variables are shown. It can be seen that the best case is clearly the last, where more prediction variables are used and the trained model best fits the output data.

**Table 2.** Performance evaluation.

| Prediction variables | RMSE (training) | RMSE (testing) |
|---|---|---|
| 1 connection | 1.23 | 1.75 |
| 2 connection | 1.08 | 1.67 |
| 3 connection | 0.49 | 0.78 |
| 4 connection | 0.29 | 0.99 |
| 5 connection | 0.15 | 0.69 |
| 5 connection and time | 0.10 | 0.56 |

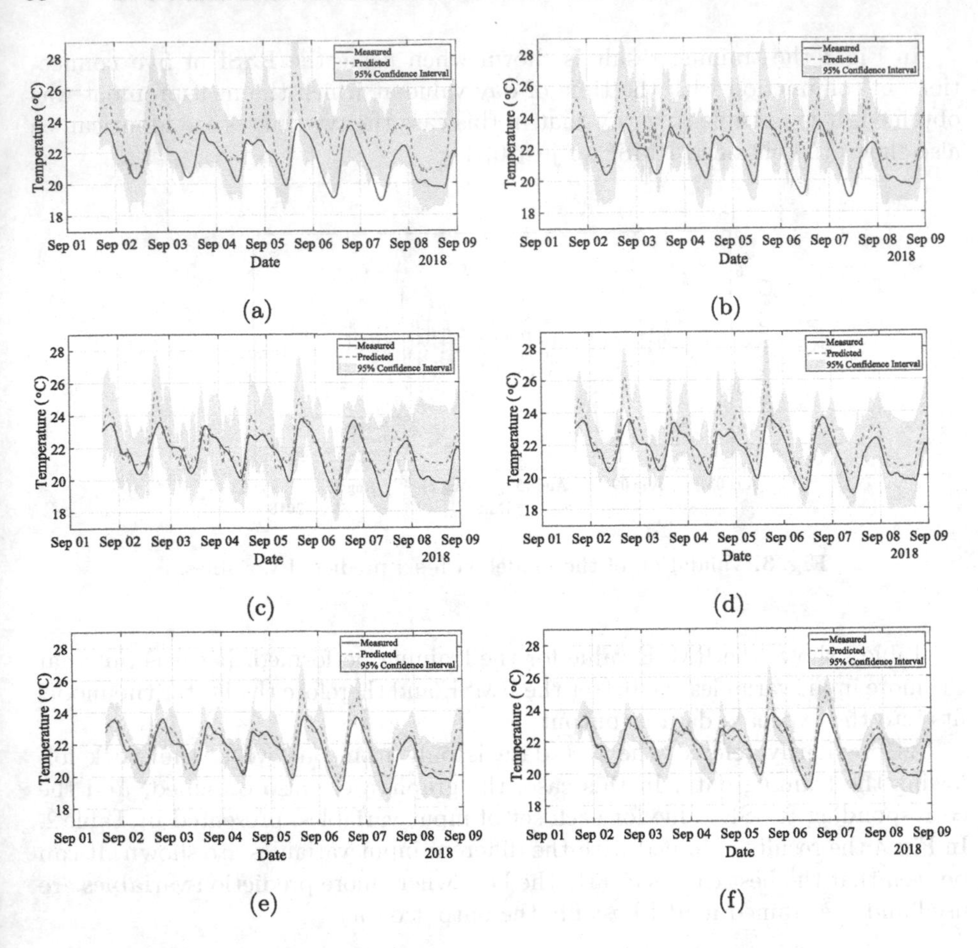


**Fig. 4.** Results of the temperature prediction with model training for (a) one connection, (b) two connections, (c) three connections, (d) four connections, (e) five connections and (f) five connections and time of measurement.

## 5 Conclusions

In this work a Gaussian process regression approach was presented in order to model an ambient temperature prediction problem. This model has as input variables the quality of the connections of a wireless sensor network. In this case, the XBee S2C radio nodes already provide the RSSI value which is a good link performance indication.

In order to verify the performance of the model with different quantities of input variables, we started by modelling with only one XBee connection. Then the number of connections was successively increased and finally the time of day when the value was obtained was also added. It was concluded that the more input variables the better the model fits the expected data as output, in training. This also occurs when the models were tested with the test data set, obtaining very acceptable predictions of the ambient temperature only from the quality of the connections. The RMSE, which provides a measure of the deviation between predicted and expected value, was used as the metric for evaluating the performance of the model.

Thus, a wireless sensor network can be seen as a virtual weather station, and the data, in this case temperature, can serve as support/confirmation to the data obtained by the network itself through temperature sensors, for example. This adds value to the wireless sensor network, which provides a meteorological parameter indirectly, while performing its main task, which is communication between nodes.

In this case the data was limited to one month of summer, therefore the model is not guaranteed to make good predictions of the temperature in the winter. More work has to be done in this regard if it is found that it may be important and necessary to obtain temperature data via the quality of the connections in wireless sensor networks.

Models based on Gaussian processes are interesting since they provide confidence intervals, while other models do not. Gaussian processes have been widely used to model several and very different problems, such as prediction of daily global solar irradiation, robotics and control, facial expression recognition, streamflow forecasting, wind speed prediction, among others.

**Acknowledgements.** This work was developed in the research unit CIMA, of the University of Madeira, supported by the INTERREG MAC program through the VIMETRI-MAC/3.5/065 project.

## References

1. Awad, M., Khanna, R.: Efficient Learning Machines: Theories, Concepts, and Applications for Engineers and System Designers. Apress, New York (2015)
2. Wennerström, H., Hermans, F., Rensfelt, O., Rohner, C., Nordén, L. Å.: A long-term study of correlations between meteorological conditions and 802.15. 4 link performance. In: 2013 IEEE International Conference on Sensing, Communications and Networking (SECON), pp. 221–229. IEEE (2013)
3. Boano, C.A., Brown, J., He, Z., Roedig, U., Voigt, T.: Low-power radio communication in industrial outdoor deployments: the impact of weather conditions and ATEX-compliance. In: Komninos, N. (ed.) Sensappeal 2009. LNICST, vol. 29, pp. 159–176. Springer, Heidelberg (2010). https://doi.org/10.1007/978-3-642-11870-8_11
4. Boano, C.A., Tsiftes, N., Voigt, T., Brown, J., Roedig, U.: The impact of temperature on outdoor industrial sensornet applications. IEEE Trans. Ind. Inform. **6**(3), 451–459 (2010)
5. Bannister, K., Giorgetti, G., Gupta, S.: Wireless sensor networking for hot applications: effects of temperature on signal strength, data collection and localization. In: Proceedings of the 5th Workshop on Embedded Networked Sensors (HotEmNets 2008). Citeseer (2008)
6. Inácio, S., Azevedo, J.: Influence of meteorological parameters in the received signal of a 2.4 GHz wireless sensor network. In: Proceedings of The Loughborough Antennas and Propagation Conference (2018)

7. Ingsrisawang, L., Ingsriswang, S., Somchit, S., Aungsuratana, P., Khantiyanan, W.: Machine learning techniques for short-term rain forecasting system in the northeastern part of Thailand. In: Proceedings of World Academy of Science, Engineering and Technology, vol. 31 (2008)
8. Chen, S.-T., Yu, P.-S., Tang, Y.-H.: Statistical downscaling of daily precipitation using support vector machines and multivariate analysis. J. Hydrol. **385**(1–4), 13–22 (2010)
9. Cramer, S., Kampouridis, M., Freitas, A.A., Alexandridis, A.K.: An extensive evaluation of seven machine learning methods for rainfall prediction in weather derivatives. Expert. Syst. Appl. **85**, 169–181 (2017)
10. Moon, S.-H., Kim, Y.-H., Lee, Y.H., Moon, B.-R.: Application of machine learning to an early warning system for very short-term heavy rainfall. J. Hydrol. **568**, 1042–1054 (2019)
11. Rasouli, K., Hsieh, W.W., Cannon, A.J.: Daily streamflow forecasting by machine learning methods with weather and climate inputs. J. Hydrol. **414**, 284–293 (2012)
12. Badrzadeh, H., Sarukkalige, R., Jayawardena, A.: Hourly runoff forecasting for flood risk management: application of various computational intelligence models. J. Hydrol. **529**, 1633–1643 (2015)
13. Humphrey, G.B., Gibbs, M.S., Dandy, G.C., Maier, H.R.: A hybrid approach to monthly streamflow forecasting: integrating hydrological model outputs into a Bayesian artificial neural network. J. Hydrol. **540**, 623–640 (2016)
14. Jiang, Z., Wang, H.-Y., Song, W.-W.: Discharge estimation based on machine learning. Water Sci. Eng. **6**(2), 145–152 (2013)
15. Sun, A.Y., Wang, D., Xu, X.: Monthly streamflow forecasting using Gaussian process regression. J. Hydrol. **511**, 72–81 (2014)
16. Voyant, C., Motte, F., Notton, G., Fouilloy, A., Nivet, M.-L., Duchaud, J.-L.: Prediction intervals for global solar irradiation forecasting using regression trees methods. Renew. Energy **126**, 332–340 (2018)
17. Alfadda, A., Rahman, S., Pipattanasomporn, M.: Solar irradiance forecast using aerosols measurements: a data driven approach. Sol. Energy **170**, 924–939 (2018)
18. Kaba, K., Sarıgül, M., Avcı, M., Kandırmaz, H.M.: Estimation of daily global solar radiation using deep learning model. Energy **162**, 126–135 (2018)
19. Feng, Y., Cui, N., Hao, W., Gao, L., Gong, D.: Estimation of soil temperature from meteorological data using different machine learning models. Geoderma **338**, 67–77 (2019)
20. Grbić, R., Kurtagić, D., Slišković, D.: Stream water temperature prediction based on Gaussian process regression. Expert. Syst. Appl. **40**(18), 7407–7414 (2013)
21. Zhang, C., Wei, H., Zhao, X., Liu, T., Zhang, K.: A Gaussian process regression based hybrid approach for short-term wind speed prediction. Energy Convers. Manag. **126**, 1084–1092 (2016)
22. Holman, D., et al.: Gaussian process models for reference et estimation from alternative meteorological data sources. J. Hydrol. **517**, 28–35 (2014)
23. Ebden, M.: Gaussian processes for regression: a quick introduction. The website of robotics research group in department on engineering science, University of Oxford (2008)
24. Williams, C.K., Rasmussen, C.E.: Gaussian Processes for Machine Learning, vol. 2. MIT Press Cambridge, MA (2006)

# Computational Intelligence Methods for Time Series

# Voice Command Recognition Using Statistical Signal Processing and SVM

Aleksandra Osowska[1] and Stanislaw Osowski[1,2(✉)]

[1] Warsaw University of Technology, Warsaw, Poland
olaosows@gmail.com, sto@iem.pw.edu.pl
[2] Military University of Technology, Warsaw, Poland

**Abstract.** The paper presents automatic system for recognition of the voice commands. The neural based system for recognition of spoken commands, registered using smartphone is proposed. It applies the statistical processing of data leading to diagnostic feature generation and application of support vector machine as the final classifier. The recognized words are typical commands that might be used in automatic controlling of the wheelchair or represent the passwords used in speaker identification. The results of numerical experiments will be presented and discussed. They show the applicability of the presented method in simplified solution of voice command recognition.

**Keywords:** Voice command recognition · Speaker identification · Support Vector Machine · Classification

## 1 Introduction

Voice recognition is a very popular subject of research, since it covers large field of applications in different areas of our life, especially in telecommunication, health care, military, system protection, etc. [1]. In general case very sophisticated systems have been developed. They include hidden Markov models combined with artificial neural networks [2], dynamic time warping [3, 4] or nowadays the deep learning method called long short-term memory [5–7]. The results presented in these papers depend on the applied method and type of speech under recognition. For example the survey paper on voice command recognition technique [4] has compared different advanced techniques, leading to the accuracy changing from 80% up to 98% in different applications.

However, there are still some applications, where much simpler approaches are possible to get the acceptable results. This paper will be concerned with the investigation of the special types of voice command recognition problem. Two tasks can be formulated here. The first: individual person pronounces some particular set of commands and the task is to recognize them. Vocal tract is not important in such task, since each command is spoken by the same person. The second problem: many persons speak the same command treated as a password. In such case the contents of the password is not important and only vocal tract of each speaker is under recognition. Such commands might find application in automatic controlling of the some vehicles (for example wheelchairs) or may be used in speaker identification in protection techniques of the special objects.


I. Rojas et al. (Eds.): IWANN 2019, LNCS 11506, pp. 65–73, 2019.
https://doi.org/10.1007/978-3-030-20521-8_6

The solutions presented in the paper will use the direct smartphone recordings of the commands spoken in different ways by the person. The spoken statements may differ significantly from case to case with the duration, way of speaking, loudness, noisy background, etc. Therefore, different stages of signal preprocessing must be applied. They include specific filtering of signals using wavelet transformation, statistical processing of signals to develop the set of numerical descriptors, selection of diagnostic features from these descriptors and finally application of the support vector machine (SVM) as the final classifier.

Two types of applications will be studied. The first one is recognition of commands typically used in controlling the wheelchair. The second one will show its application in speaker identification of the basis of spoken password.

## 2 Command Recognition System

### 2.1 Data Base of Commands

In the numerical experiments we have used the set of 8 spoken commands, typically used in wheelchair controlling.

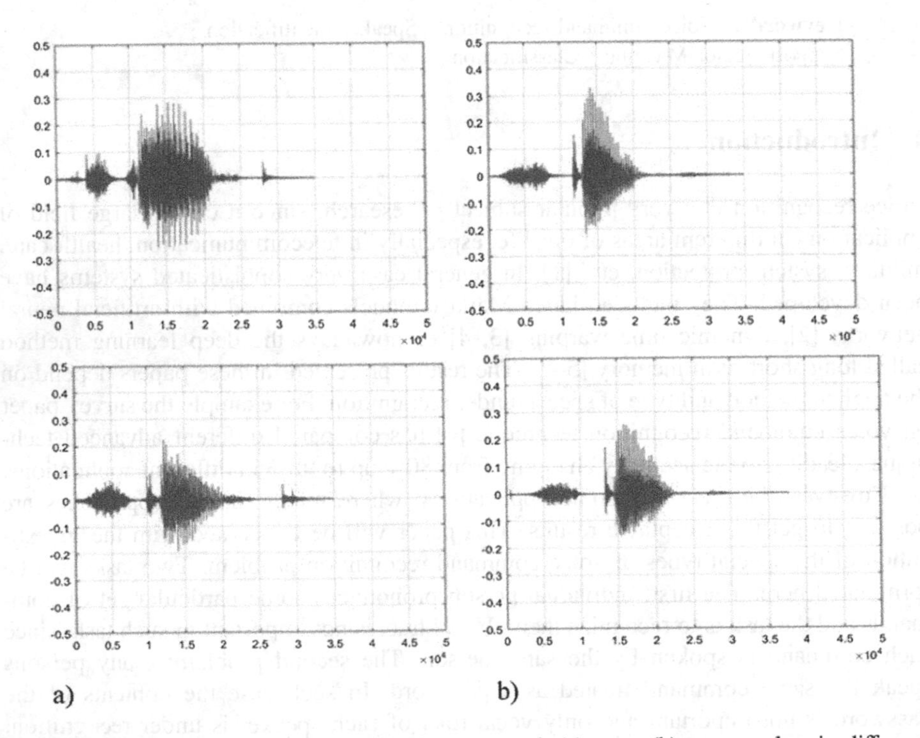


**Fig. 1.** The pairs of signals representing the command: (a) *start,* (b) *stop*, spoken in different time by the same person.

They include the following Polish words: naprzod (forward), w tyl (backward), w lewo (left), w prawo (right), szybciej (quicker), wolniej (slower), start (start) and stop (stop).

The recordings have been acquired at different days at various environmental conditions, in the presence of the surrounding noise. The individual recordings representing the same word and spoken by the same person differ from each other. This is well seen on the example of two recordings of the commands *start* and *stop* spoken by one person as presented in Fig. 1. The signals differ in some details of shape and length of particular segments.

Many recordings of the same commands, spoken by the person, are needed in preparation of system. The population of recordings, made for 8 commands, that will be used in experiments is shown in Table 1. They have been gathered in different days at different environmental conditions. The commands will represent the classes and denoted further by the numbers from C1 to C8.

**Table 1.** The population of recordings of different commands used in experiments.

| Class notation | Command | No. of recordings |
|---|---|---|
| C1 | Forward | 131 |
| C2 | Backward | 113 |
| C3 | Left | 107 |
| C4 | Right | 112 |
| C5 | Quicker | 104 |
| C6 | Slower | 101 |
| C7 | Start | 108 |
| C8 | Stop | 117 |
| Total | | **893** |

## 2.2 Signal Preprocessing for Feature Extraction

Signal preprocessing, leading to creation of the proper set of numerical descriptors of the command, is a very important stage in classification. Such descriptors should be insensitive to the possible variations (length of signal, magnitude, changes of intonation, presence of surrounding noise, fluctuation of the speaking environment, etc.) of the spoken word [8].

In the first step of preprocessing the normalization of signal has been applied by dividing the real signal by its maximum absolute value. Thanks to this all signals are in the range from −1 to 1. In the next step the signals were subject to wavelet transformation [9]. Its aim is to reduce the noise present in the recordings. The approximated signal $x(t)$ on $4^{th}$ level has been found as the most appropriate for further preprocessing stages.

In the next stage the numerical descriptors, based on statistical parametrization of the signal, have been applied [10]. They include such parameters as: mean of absolute values $mean(abs(x))$, standard deviation $std(x)$, $skewness(x)$, $kurtosis(x)$, $mean(x^2)$, interquartile range $iqr(x)$, $median(x)$, the ratio $iqr(x)/median(x)$ and cumulants of the $3^{rd}$ ($K_3$), $4^{th}$ ($K_4$), and the $5^{th}$ ($K_5$) orders. Additional three descriptors have been defined on the basis of histogram of the signals. They include the area under histogram $S_h$, the ratio of area under histogram related to the number of samples $S_h$/sum($n$) and the absolute value of area under histogram $|S_h|$. This way the total number of descriptors was equal 14. In further presentation they will be denoted in the following way: *f*1 - *mean*(*abs*(*x*)), *f*2 - *std*(*x*), *f*3 - *skewness*(*x*), *f*4 - *kurtosis*(*x*), *f*5 - *mean*($x^2$), *f*6 - *median*(*x*), *f*7 - *iqr*(*x*), *f*8 - *iqr*(*x*)/*median*(*x*), *f*9 - $K_3$, *f*10 - $K_4$, *f*11 - $K_5$, *f*12 - $S_h$, *f*13 - $S_h$/sum($n$), *f*14 - $|S_h|$.

In the next step the generated descriptors have been subject to selection. We have applied the Fisher criteria of significance [8]. In Fisher test the descriptors *f* were selected according to its 2-class discrimination measure $S_{12}(f)$, where

$$S_{12}(f) = \frac{|c_1 - c_2|}{\sigma_1 + \sigma_2} \tag{1}$$

in which $c_1$ and $c_2$ are the mean values of feature *f* for classes 1 and 2, respectively, while $\sigma_1$ and $\sigma_2$ - the appropriate standard deviations. A large value of this measure indicates good individual feature ability for recognition between class 1 and 2.

**Table 2.** Fisher measures of class discrimination ability of the defined descriptors.

| | f1 | f2 | f3 | f4 | f5 | f6 | f7 | f8 | f9 | f10 | f11 | f12 | f13 | f14 |
|---|---|---|---|---|---|---|---|---|---|---|---|---|---|---|
| $S_{12}$ | 6.92 | 8.70 | 7.73 | 4.15 | 3.48 | 3.31 | 1.94 | 0.09 | 0.27 | 1.86 | 0.69 | 0.02 | 5.20 | 0.01 |
| $S_{13}$ | 6.37 | 8.47 | 4.06 | 5.02 | 3.59 | 1.74 | 3.65 | 0.09 | 2.09 | 1.71 | 0.73 | 1.52 | 6.65 | 1.68 |
| $S_{14}$ | 4.21 | 4.80 | 3.58 | 3.08 | 2.27 | 2.86 | 2.04 | 18.87 | 0.24 | 0.84 | 0.08 | 0.31 | 4.05 | 0.29 |
| $S_{15}$ | 5.64 | 6.92 | 230.39 | 1.52 | 3.07 | 6.58 | 2.57 | 0.08 | 1.51 | 1.40 | 0.88 | 0.68 | 4.00 | 0.67 |
| $S_{16}$ | 4.92 | 5.25 | 2.84 | 4.07 | 2.56 | 2.59 | 3.72 | 0.09 | 1.57 | 1.27 | 0.85 | 1.39 | 3.94 | 2.06 |
| $S_{17}$ | 4.43 | 5.03 | 6.68 | 16.41 | 2.03 | 6.21 | 0.33 | 0.08 | 1.16 | 0.90 | 0.64 | 0.53 | 3.98 | 0.57 |
| $S_{18}$ | 9.92 | 7.61 | 258.34 | 2.89 | 2.35 | 3.27 | 2.98 | 0.10 | 0.91 | 0.83 | 0.48 | 0.08 | 3.30 | 0.07 |
| $S_{23}$ | 5.34 | 6.37 | 3.48 | 7.16 | 4.69 | 1.55 | 4.37 | 0.10 | 91.62 | 2.27 | 0.29 | 0.40 | 9.24 | 0.49 |
| $S_{24}$ | 18.70 | 19.76 | 0.43 | 0.30 | 4.26 | 4.63 | 2.25 | 0.12 | 0.28 | 2.06 | 0.86 | 0.12 | 6.94 | 0.13 |
| $S_{25}$ | 10.23 | 18.31 | 29.69 | 2.09 | 4.10 | 16.52 | 1.52 | 0.06 | 1.11 | 9.93 | 0.96 | 2.90 | 91.63 | 2.57 |
| $S_{26}$ | 2.55 | 1.33 | 2.30 | 4.00 | 1.82 | 2.46 | 4.43 | 0.08 | 123.06 | 2.07 | 2.01 | 0.36 | 2.28 | 0.28 |
| $S_{27}$ | 7.74 | 10.71 | 6.41 | 5.93 | 3.78 | 0.54 | 2.17 | 0.07 | 0.17 | 1.94 | 0.70 | 0.28 | 5.51 | 0.47 |
| $S_{28}$ | 6.78 | 8.89 | 39.45 | 3.26 | 3.58 | 3.41 | 2.00 | 0.08 | 0.20 | 1.92 | 0.71 | 0.08 | 5.49 | 0.10 |
| $S_{34}$ | 8.77 | 15.86 | 4.25 | 0.72 | 4.32 | 1.64 | 4.92 | 0.10 | 2.41 | 1.97 | 0.92 | 2.19 | 11.75 | 2.65 |
| $S_{35}$ | 7.01 | 13.46 | 0.50 | 2.51 | 4.22 | 1.26 | 3.79 | 1.21 | 2.72 | 0.65 | 4.98 | 0.51 | 19.20 | 0.46 |
| $S_{36}$ | 2.44 | 3.31 | 0.29 | 2.49 | 7.82 | 0.13 | 5.05 | 0.26 | 56.73 | 1.96 | 2.30 | 12.69 | 1.05 | 1.24 |
| $S_{37}$ | 6.74 | 10.09 | 1.58 | 6.25 | 3.88 | 1.45 | 3.78 | 1.38 | 2.20 | 1.82 | 0.73 | 0.90 | 7.43 | 0.56 |
| $S_{38}$ | 6.26 | 8.61 | 1.59 | 3.68 | 3.69 | 1.62 | 3.63 | 0.09 | 2.23 | 1.78 | 0.74 | 2.45 | 7.22 | 1.72 |
| $S_{45}$ | 6.99 | 21.91 | 13.65 | 5.18 | 4.55 | 31.20 | 1.84 | 0.11 | 2.01 | 1.51 | 1.13 | 0.80 | 3.93 | 0.80 |
| $S_{46}$ | 5.48 | 5.67 | 2.65 | 1.11 | 2.70 | 2.56 | 4.93 | 0.10 | 1.80 | 2.00 | 1.67 | 1.83 | 3.88 | 0.73 |
| $S_{47}$ | 4.12 | 4.59 | 8.82 | 5.40 | 2.46 | 8.04 | 2.20 | 0.11 | 1.69 | 0.79 | 0.14 | 2.02 | 4.09 | 1.35 |
| $S_{48}$ | 3.88 | 4.09 | 17.90 | 14.78 | 2.25 | 2.68 | 2.08 | 0.20 | 1.20 | 0.85 | 0.07 | 3.08 | 4.26 | 0.98 |
| $S_{56}$ | 4.46 | 3.17 | 0.44 | 2.50 | 2.06 | 1.96 | 3.87 | 2.20 | 1.64 | 1.45 | 2.70 | 0.54 | 3.85 | 0.53 |
| $S_{57}$ | 6.27 | 8.29 | 0.02 | 0.79 | 3.48 | 7.31 | 3.62 | 2.36 | 1.59 | 1.45 | 0.90 | 0.74 | 4.01 | 0.73 |
| $S_{58}$ | 5.35 | 6.77 | 535.02 | 1.17 | 3.19 | 20.65 | 2.63 | 0.04 | 1.65 | 1.44 | 0.91 | 0.98 | 4.11 | 1.09 |
| $S_{67}$ | 5.00 | 5.31 | 0.87 | 4.92 | 2.65 | 2.23 | 3.85 | 2.66 | 1.62 | 1.44 | 1.08 | 0.67 | 3.93 | 0.01 |
| $S_{68}$ | 4.80 | 4.99 | 1.22 | 3.46 | 2.58 | 2.51 | 3.71 | 0.08 | 1.65 | 1.39 | 1.05 | 2.18 | 3.99 | 0.83 |
| $S_{78}$ | 3.19 | 3.13 | 1.59 | 1.86 | 1.73 | 10.24 | 1.27 | 0.04 | 24.87 | 1.05 | 6.12 | 51.01 | 5.20 | 15.78 |

Table 2 presents the Fisher measures of all 14 descriptors in recognition between all pairs of classes. As we can see they take various values for different combinations of two classes. The least important as the diagnostic features seem to be descriptors *f*9 and *f*14. Only the features above some selected threshold for each pair of classes have been applied as the diagnostic features in classification process using SVM.

The detailed analysis of the distribution of descriptors for all classes has revealed some outliers (the samples very far from the mean). They are the results of some disturbances in speech recordings. To avoid such problems these recordings have been eliminated from further experiments.

## 2.3 Support Vector Machine Classifier

The support vector machine of Gaussian kernel has been found the best in the role of final classifier [11]. Since SVM is able to recognize only 2 classes at a time we have used many SVM networks for recognition of two pairs of classes. At 8 classes we have built 28 classifiers working in one-against-one mode.

The important parameters for efficiency of operation is proper choice of the regularization constant $C$ and the parameter $\gamma$ of the Gaussian function $K(\boldsymbol{x},\boldsymbol{x}_i) = \exp\left(-\gamma\|\boldsymbol{x}-\boldsymbol{x}_i\|^2\right)$. These parameters have been selected after additional introductory experiments performed on a small set of data using the set of predefined values for $C$ and $\gamma$. The values leading to the highest efficiency in recognition process have been selected for further experiments.

To get the most reliable assessment of the developed classification system the experiments of learning and testing have been repeated 1000 times at random choice of learning and testing sets. The applied proportion for these two sets was 70% (learning) and 30% (testing) samples. The results of all trials have been averaged and the mean value is assumed as the final quality measure.

## 2.4 Numerical Results of Command Recognition

The detailed results of command recognition will be presented for one person on the basis of 893 recordings of 8 commands, as presented in Table 1. Symlet *sym2* was selected in wavelet analysis. The hyperparameters of SVM $= 0.7$ and $C = 1000$ were found as the most optimal. The experiments have been performed using Matlab [12]. The average error of command recognition on the testing data not taking part in learning, obtained in 1000 trials was equal 6.91% ± 2.05%.

In the next experiments we have changed the number of recognized commands and observed the quality of system performance. In the case of 6 class commands (*forward, backward, left, right, quicker, slower*) the average error of recognition was 4.05% ± 1.6%. The next reduction of the classes to only four (*forward, backward, left, right*) has resulted in further decrease of error to the value 3.90% ± 1.72%.

The experiments performed for the next 5 persons have shown similar results. The command recognition error was changing in the same range and was dependent on the number of commands subject to recognition.

These quality values indicate high potential of the proposed approach in this simplified solution of command recognition. The accuracy of results might be further improved by increasing the number of recordings, taking part in learning. Additional possibility is to use more technically advanced system of recordings (substitution of smartphone by professional recorder, etc.).

## 3 Speaker Identification Using Developed System

In the previous section we have shown the possibility of applying the recorded speech of one person to recognize the specific commands. The opposite application of the developed system is to identify person among the analyzed group of people on the basis of one spoken password.

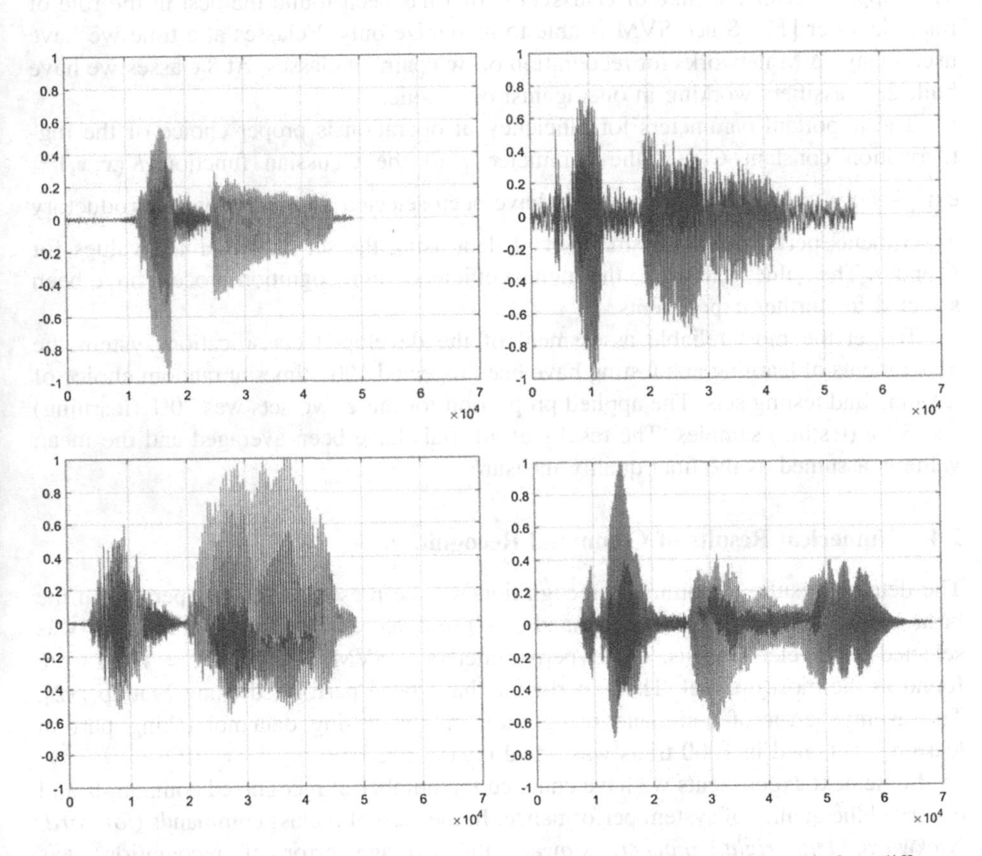

**Fig. 2.** The examples of signals representing the password "*Rosjanin*" spoken by four different speakers.

This time the system should react to the vocal tract of the particular speaker and not to the spoken word. Figure 2 shows four examples of the password "*Rosjanin*" spoken by four different persons after magnitude normalization. Large differences in the shape, time duration and frequency spectrum can be observed.

### 3.1 Data Base

The recordings of the password "*Rosjanin*" have been acquired from 20 persons using smartphone. Each person has spoken this password in different days in changing environmental conditions. The age of speakers was changing from 23 to 70 years. The way of speaking, even of the same person, was also changing (slower or quicker pronunciation, more or less quietly, etc.). Table 3 presents the population of recordings from 20 speakers, denoted by S1 to S20. Each speaker represents the separate class under recognition.

**Table 3.** The population of password recordings of different speakers

| Speaker S (class) | 1 | 2 | 3 | 4 | 5 | 6 | 7 | 8 | 9 | 10 | 11 | 12 | 13 | 14 | 15 | 16 | 17 | 18 | 19 | 20 |
|---|---|---|---|---|---|---|---|---|---|---|---|---|---|---|---|---|---|---|---|---|
| No of recordings | 68 | 64 | 90 | 96 | 71 | 58 | 50 | 54 | 82 | 63 | 69 | 73 | 54 | 82 | 59 | 63 | 47 | 62 | 67 | 69 |

The total number of recordings taking part in this stage of experiments was equal 1341. However, their distribution among different speakers, representing individual classes, was not equal (minimum number of recordings 47 and the maximum number 96).

### 3.2 Details of System

The signals representing all recordings have been preprocessed in the similar way as in the first part of research. After magnitude normalization they were filtered using biorthogonal wavelet *bior*1.3. The approximated signal on 4$^{th}$ level and detailed signal on the first level have been selected as the most representative for this problem. These two signals have been converted to the statistical descriptors in the way described in Sect. 2.2. In this way the individual set of descriptors from *f*1 to *f*14 for the approximated signal and for the selected detailed signal have been generated (total number of potential diagnostic features 28).

Fisher analysis of class discriminative properties of these features has found the proper set of input attributes for the classifier. On the basis of the approximated signals the following features have been selected: *f*1 to *f*7, *f*9 and *f*13 (9 features). The detailed signals on the first level has delivered the following set of features: *f*1 to *f*5, *f*7, *f*8 and *f*13 (8 features). In this way the total number of features used as the input attributes to the final SVM classifier was 17.

The SVM with Gaussian kernel function was used as the basic classifier. The regularization constant adjusted to this case was 100 and the Gaussian parameter $\gamma = 0.6$. These values have been found after some introductory experiments using the set of predefined prototypes. The classification system, based on SVM networks at application of one against one mode, was used. At 20 classes 190 two-class SVM units were trained.

### 3.3 Results of Experiments

The numerical experiments of speaker recognition have been performed using the set of data samples presented in Table 3. To obtain the most objective results of testing all experiments have been repeated 1000 times at random choice of learning and testing parts. In each run the whole data set was split in proportion: 75% for learning and 25% for testing. This large number of experiments allows getting the credible statistics.

The average error of speaker recognition achieved in this series of experiments was **1.85%** at standard deviation equal 0.82%. Seven speakers (out of 20) were identified in all trials without any error. Table 4 presents the typical confusion matrix corresponding to the testing data not taking part in experiments. The last column (denoted by **Sens**) represents sensitivity of system in recognition of the particular class. Only in four cases the sensitivity value was below 90%. All of them correspond to the small number of samples taking part in testing in this particular run of system.

**Table 4.** The confusion matrix in speaker recognition problem

| | S*1* | S2 | S3 | S4 | S5 | S6 | S7 | S8 | S9 | S10 | S11 | S12 | S13 | S14 | S15 | S16 | S17 | S18 | S19 | S20 | Sens |
|---|---|---|---|---|---|---|---|---|---|---|---|---|---|---|---|---|---|---|---|---|---|
| **S1** | *11* | 0 | 0 | 0 | 0 | 0 | 0 | 0 | 0 | 0 | 0 | 0 | 0 | 0 | 0 | 0 | 0 | 0 | 0 | 0 | 100% |
| **S2** | *0* | 19 | 0 | 0 | 0 | 0 | 0 | 0 | 0 | 0 | 0 | 0 | 0 | 0 | 0 | 0 | 0 | 0 | 0 | 0 | 100% |
| **S3** | *0* | 0 | 26 | 0 | 0 | 0 | 0 | 0 | 0 | 0 | 0 | 0 | 0 | 0 | 0 | 0 | 0 | 0 | 0 | 0 | 100% |
| **S4** | *0* | 0 | 0 | 21 | 0 | 0 | 0 | 1 | 0 | 0 | 0 | 0 | 0 | 0 | 0 | 1 | 0 | 0 | 0 | 0 | 91% |
| **S5** | *0* | 0 | 0 | 0 | 21 | 0 | 0 | 0 | 0 | 0 | 0 | 0 | 0 | 0 | 0 | 0 | 0 | 0 | 0 | 0 | 100% |
| **S6** | *0* | 0 | 0 | 0 | 0 | 17 | 0 | 0 | 1 | 0 | 0 | 0 | 0 | 0 | 0 | 0 | 0 | 0 | 0 | 0 | 94% |
| **S7** | *0* | 0 | 0 | 0 | 0 | 0 | 12 | 1 | 0 | 0 | 0 | 1 | 0 | 0 | 0 | 0 | 0 | 0 | 0 | 0 | 86% |
| **S8** | *0* | 0 | 0 | 0 | 0 | 0 | 0 | 11 | 1 | 0 | 0 | 2 | 1 | 0 | 0 | 0 | 0 | 0 | 0 | 1 | 69% |
| **S9** | *0* | 0 | 0 | 0 | 1 | 1 | 0 | 0 | 24 | 0 | 0 | 0 | 0 | 0 | 0 | 0 | 0 | 0 | 0 | 0 | 92% |
| **S10** | *0* | 0 | 0 | 0 | 0 | 0 | 0 | 0 | 0 | 14 | 0 | 0 | 0 | 0 | 0 | 0 | 0 | 0 | 0 | 0 | 100% |
| **S11** | *0* | 0 | 0 | 0 | 0 | 0 | 0 | 0 | 0 | 0 | 16 | 0 | 0 | 0 | 0 | 0 | 0 | 0 | 0 | 0 | 100% |
| **S12** | *0* | 0 | 0 | 0 | 0 | 0 | 0 | 0 | 1 | 0 | 0 | 17 | 0 | 0 | 0 | 0 | 0 | 0 | 0 | 0 | 94% |
| **S13** | *0* | 0 | 0 | 1 | 0 | 0 | 0 | 2 | 0 | 0 | 0 | 0 | 10 | 0 | 0 | 0 | 0 | 0 | 0 | 0 | 77% |
| **S14** | *0* | 0 | 0 | 0 | 0 | 0 | 0 | 0 | 0 | 0 | 0 | 0 | 0 | 14 | 0 | 0 | 1 | 0 | 0 | 0 | 93% |
| **S15** | *0* | 0 | 0 | 0 | 0 | 0 | 1 | 0 | 0 | 0 | 0 | 0 | 0 | 0 | 12 | 0 | 0 | 0 | 0 | 0 | 92% |
| **S16** | *0* | 0 | 0 | 0 | 0 | 0 | 0 | 0 | 0 | 0 | 0 | 0 | 0 | 1 | 0 | 8 | 0 | 1 | 0 | 0 | 80% |
| **S17** | *0* | 0 | 0 | 0 | 0 | 0 | 0 | 0 | 0 | 0 | 0 | 0 | 0 | 0 | 0 | 0 | 12 | 0 | 0 | 0 | 100% |
| **S18** | *0* | 0 | 0 | 0 | 0 | 0 | 0 | 0 | 0 | 0 | 0 | 1 | 0 | 0 | 0 | 0 | 0 | 16 | 0 | 0 | 94% |
| **S19** | *0* | 0 | 0 | 0 | 0 | 0 | 0 | 0 | 0 | 0 | 0 | 0 | 0 | 0 | 0 | 1 | 0 | 0 | 16 | 0 | 94% |
| **S20** | *0* | 0 | 0 | 0 | 0 | 0 | 0 | 0 | 0 | 1 | 0 | 0 | 0 | 0 | 0 | 0 | 0 | 0 | 0 | 18 | 95% |

## 4 Conclusions

The paper has presented the simplified system for voice command recognition. Two aspects of the problem have been considered. The first task was to recognize the spoken commands by particular person. In such case the vocal tract of the speaker is the same and only the contents of the spoken word is important. This contents is under analysis. In the second task the spoken word (password) is the same for all speakers, and only the vocal tract of speakers is under recognition.

The important point in this solution is to define and select the proper descriptors of the signals. The statistical parameters describing the signals have been applied. They refer to such basic parameters of time series as mean, median, skewness, kurtosis, interquartile range and more advanced extension in the form of cumulants and chosen descriptors of the histogram.

The obtained results confirmed good performance of such a system. The average classification accuracy in command recognition was changing from 93.09% (8 commands) to 96.1% (4 commands). In the case of speaker identification the average accuracy was 98.15%. Further increase of accuracy can be achieved by including higher number of recordings.

## References


1. Beigi, H.: Fundamentals of Speaker Recognition. Springer, New York (2011). https://doi.org/10.1007/978-0-387-77592-0
2. Juang, B.-H.: On the hidden Markov model and dynamic time warping for speech recognition—a unified view. AT T Tech. J. **63**(7), 1213–1243 (1984)
3. Kate, R.J.: Using dynamic time warping distances as features for improved time series classification. Data Min. Knowl. Disc. **30**(2), 283–312 (2015)
4. Prabhakar, O.P., Sahu, N.K.: A survey on voice command recognition technique. Int. J. Adv. Res. Comput. Sci. Softw. Eng. **3**(5), 576–585 (2013)
5. Dong, Y., Li, D.: Automatic Speech Recognition: A Deep Learning Approach. Springer, New York (2015). https://doi.org/10.1007/978-1-4471-5779-3
6. LeCun, Y., Bengio, Y.: Convolutional networks for images, speech, and time-series. In: Arbib, M.A. (ed.) The Handbook of Brain Theory and Neural Networks. MIT Press, Cambridge (1995)
7. Schmidhuber, J.: Deep learning in neural networks: an overview. Neural Netw. **61**, 85–117 (2015)
8. Tan, P.N., Steinbach, M., Kumar, V.: Introduction to Data Mining. Pearson Education Inc., Boston (2006)
9. Daubechies, I.: Ten Lectures on Wavelets. SIAM, Philadelphia (1992)
10. Sprent, P., Smeeton, N.C.: Applied Nonparametric Statistical Method. Chapman & Hall/CRC, Boca Raton (2007)
11. Schölkopf, B., Smola, A.: Learning With Kernels. MIT Press, Cambridge (2002)
12. Matlab user manual. MathWorks, Natick (2018)

# Enterprise System Response Time Prediction Using Non-stationary Function Approximations

K. Ravikumar[1(✉)], Kriti Kumar[2(✉)], Naveen Thokala[2(✉)], and M. Girish Chandra[2(✉)]

[1] Bangalore, India
[2] TCS Research and Innovation, Bangalore, India
{kriti.kumar,naveen.thokala,m.gchandra}@tcs.com

**Abstract.** We consider the problem of predicting response time of a large scale enterprise system using causal forecasting models. Specifically, the problem pertains to predicting potential system failure well in advance so that preventive actions can be initiated. Various influential factors are identified and their relationship with the system response time is estimated from data using non-stationary (time dependent) functional approximations. Experimental results on the prediction performance of different methods are presented and their discriminative characteristics with regard to error distribution are used to suggest a recommendation for practical implementation.

**Keywords:** Multivariate time series forecasting · Machine learning · Enterprise systems · Predictive analytics

## 1 Introduction

An Enterprise System is a large-scale system of integrated applications that will help an organization manage its business and automate many back office functions. It integrates all facets of an operation, including product planning, development, manufacturing, sales and marketing and thus, acts as the backbone for the enterprise. Unplanned downtimes of such enterprise system due to unforeseen failures in hardware or software can be extremely costly to the organization. The source of unplanned downtime can be in any of the layers that make up the complete software and hardware environment and is hard to trace due to the size of the system and scale of its usage. For the same reason, it is very hard to build physics-based dynamical system models that can analyze system performance. *Average dialog (on-line transaction) response time* is defined as the average time between the time at which a dialog process sends a request to a dispatcher work process, and the time at which the dialog is

K. Ravikumar was with TCS Research and Innovation till recently.


I. Rojas et al. (Eds.): IWANN 2019, LNCS 11506, pp. 74–87, 2019.
https://doi.org/10.1007/978-3-030-20521-8_7

complete and the data is transferred to the presentation layer. Average dialog response time (to be referred to as *response time* hereafter) is an important indicator of the health of the system and is affected by many factors associated with the operating system, or databases or application servers. In view of the complexity of enterprise systems, only periodic and concurrent measurements of response time (*output*) and associate factors (*inputs*) can be carried out. Data-driven and machine learning methodologies need to be adapted to glean functional relationship between the output and inputs. Advanced prediction of response time obtained from the functional map can be used to design appropriate predictive maintenance schedules to take preventive actions against system outages. In this paper, we consider a production order management system of an enterprise and develop data-driven models to learn the relationship between the response time and its associated factors. These factors have been identified in consultation with subject matter experts in Information Systems Management team. Please refer to Table 1 for a partial list of factors that affect dialog response time. Under normal working conditions, the average response time will be less than 1 s (1000 ms) and response times falling in the range 1500–2200 ms can potentially lead to system downtime and failure. In the following, we use the terms *output* and *response time* or *response variable*, or simply *response* interchangeably and similarly, the terms *inputs*, *influence factors*, or simply, *factors*. We highlight few challenges in modeling the foregoing functional relationships between the output and the inputs. Primarily all the interacting variables- output (response time) and inputs (influencing factors)- exhibit temporal patterns and hence, constitute longitudinal time series. While highly advanced function approximation schemes for cross-sectional data exist in machine learning literature, building causal relationship models in time series, particularly in high dimension settings, is still an active area of research [3,4] and yet to mature. We shun using causal graph models as their primary focus is on dimensionality reduction trying to elicit important causal relationship across time series and in our setting, the causal factors are already identified, if not their interrelationship. Majority of the studies on multivariate time series modeling focus on Auto Regressive Integrated Moving Average with eXternal factors (ARIMAX), or on state space models such as transfer functions. ARIMAX and transfer functions are linear models in nature and can deal only with stationary time series (under weak-sense stationarity) or more precisely, on *stationarized* time series of inputs and outputs which facilitate invocation of linear models. Residuals of the linear fit are assumed to behave like ARIMA errors and an ARIMA model of right order is fit on the residual time series. While these are plausible models for our analysis, linearity and low-dimensional modeling restricts their applicability to analysis of complex systems. Readers are urged to refer to [1] and [2] for a detailed treatment of these models. In fact, our experiments for the best ARIMAX model selection failed to converge, so are not discussed here. The second challenge, in our particular problem, arises from the fact that prediction model needs to work on aggregated data with aggregation as explained below. Data on inputs and output are gathered periodically at hourly time intervals. Measured values of inputs and outputs within each interval are aggregated to give average (or an appropriate statistic) values of the parameters. As a result, reported input as well as output series do not exhibit smooth continuous variation but rather intermittent bursts. This poses problem in learning as no indication of trend can be captured at consecutive lags making it difficult for multi-variate time series models to work properly with high accuracy. The third, and importantly the unique challenge of this problem comes from the fact the system under study is a critical enterprise system supporting various operational functions, so any preventive action against a possible outage would need a lead-time of 24 h to put various applications and servers to halt. This entails, at any time, $t$, next 24 h (or next 24 steps) forecast of response time. That is, any predictive model has to make 24-step ahead forecast of input factors along with temporally modulated function approximation. Predictive accuracy is seriously affected by the large number of inputs (>50) that need to be forecast, and also by the length of prediction (24 steps ahead). While forecasting for input series, it has been assumed that no interventions were executed in the past to control the response time.

As the focus of the problem is on predicting well in advance the system downtime, any predictive model to be developed needs to achieve high accuracy on peak values of response time. To this end, in this paper, by analyzing temporal patterns observed in various time series, we develop different prediction models based on machine learning. Machine learning techniques have been applied on univariate time series for prediction [7] and on multivariate time series for time series clustering [5]. We are not aware of any work on causal forecasting applications of machine learning. In our work, we focus on Weighted Least Squares Linear Regression (WLR) with appropriate weights assigned to control training errors at peaks and two non-linear models: Support Vector Regression (SVR) and Artificial Neural Networks (NN) with an additional set of temporal features. Given the complexity of data with dynamically varying range, these models are appropriately designed to predict the 'troughs' (or low values) also with good accuracy to keep the overall prediction error low.

Our analysis shows that in our particular setting, all the algorithms offer similar performance on Mean Absolute Percent Error (MAPE) but show significant difference percentage error (PE) distribution. Depending on the discriminative performance of various algorithms on weekends, weekdays and peak prediction accuracy, appropriate recommendations have been made for implementation. A working prototype of this methodology has been developed in R and has been integrated with Microsoft Azure platform for cloud based Analytics -as- a- Service/offering. See Fig. 1.

The overall methodology is summarized as follows:

- Data pre-processing for missing value treatment and dimensionality reduction
- Feature extraction from historical data to capture temporal and seasonal patterns in the data which are later appended to the original inputs our models

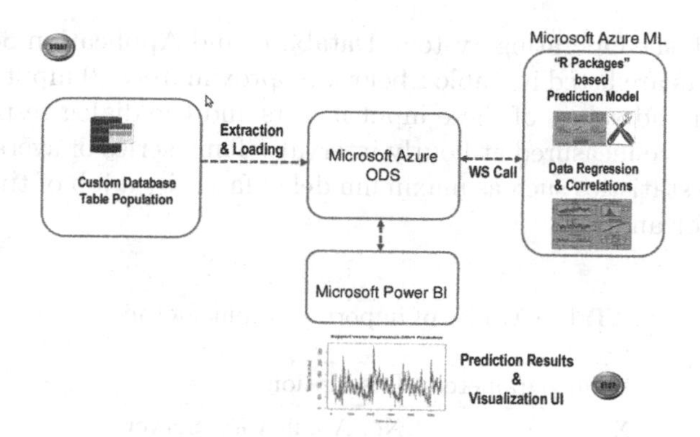


**Fig. 1.** Predictive analytics service offered over Microsoft Azure platform.

- Developing different prediction models, namely, Weighted Least Squares Linear Regression, Support Vector Regression, Artificial Neural Network with temporal features
- Recommendations on a "fused" methodology based on Percentage Error (PE) distribution of various predictive models

The remaining part of the paper is structured as follows. In Sect. 2, we describe the prediction model in detail. Section 3 highlights various data preprocessing and new features creation and develops the prediction models that includes forecasting of input series for given prediction horizon. Section 4 gives experimental results. We conclude with future directions in Sect. 5.

## 2 Problem Description

We consider a large scale production order system of an enterprise where multiple types of jobs are generated by users with requests for processing. As the complete description of the system is beyond the scope this paper, we highlight some salient operational features in the sequel.

Responsiveness of the system at any time is dependent on the types of jobs and operational workload at any time and can degrade depending on various factors, eventually heading to failure if unattended. Three important system failure modes are identified:

- Users are able to login but business process is at halt
- Users are able to login but system response is slow
- Users are unable to login

An important common metric, or a Key Performance Indicator (KPI), for the above three modes of failure is the dialog (on-line transaction) response time which is the elapsed time between the initiation of dialog dispatch to the completion of the dialog. The dialog response time is dependent on the various factors associated with the Operating System, Database, and Application Servers. Sample input factors are listed in Table 1 below. Approximately 50 input factors have been identified and values of these input factors and the dialog response for different requests are measured at hourly intervals. Time series of average values or of appropriate statistics such as maximum delay faced by a job of these variables are collected for analysis.

**Table 1.** List of important input factors.

| Input parameters | Description |
|---|---|
| $X_5$ | No. Application Server |
| $X_{10}$ | No. Dia St |
| $X_{21}$ | No. Dumps |
| $X_{25}$ | Inst. Dia |
| $X_{27}$ | tRFC Failure |
| $X_{30}$ | Extd. Memory Utilization |
| $X_{31}$ | Private Memory |
| $X_{34}$ | User Count |
| $X_{37}$ | Roll Area |
| $X_{38}$ | Page Area |
| $X_{41}$ | Memory Usage |
| $X_{42}$ | Swap Usage |
| $X_{44}$ | % File |

The system is considered to work under normal operational conditions if the average response time is below 1000 ms (or 1 s), and to enter a warning zone if the average falls between 1500 to 2200 ms. The system with measured response time crossing the warning zone is likely to head towards failure or an outage. Since these downtimes could be prohibitively costly for the organization, advanced prediction of potential failures can help in timely actuation of failure preventive interventions. A causal-factor based response prediction model is needed for identifying right interventions. To this end, we devise various time dependent function approximations on data and model input-output relationships.

## 3 Model Development

Data preprocessing, dimensionality reduction, new feature creation, designing predictive models, forecasting of inputs and predicting the output are the different tasks involved in developing a prediction system model. Figure 2 gives the flowchart of the proposed methodology.

### 3.1 Data Preprocessing

**Dimensionality Reduction:** As the original input time series are large in number, to work in reduced dimensional space and thus to develop a parsimonious model, we employed stepwise linear regression ignoring the temporal relationship among the input variables. In fact, it has been observed that majority of the input factors have exactly similar temporal pattern as that of the response variable supporting the use of such regression (under the first-cut assumption of linear relationship). See Fig. 3.

The stepwise regression starts off by choosing a regression equation containing the single best input variable and then attempts to build up with subsequent additions of inputs one at a time as long as these additions are worthwhile. Partial F-test has been used to determine the order of addition. After iterative executions, the number of significant variables has been reduced to 13 from 50.

**Missing Value Treatment:** The data sets have been observed to have many incomplete values with a sizable portion of them missing due to the possible disparate origins.

Missing values are imputed for all the original variables under study based on two methods: (i) Linear interpolation - where the value of a missing value at $t$ is a linear interpolation of the latest value available at a time earlier than $t$ and the earliest value available at a time later than $t$. (ii) Mean-based method - where missing value at time $t$ is the average of values corresponding to same-day same-time of all previous weeks. For example, for the output variable $Y(t)$,

$$Y_{miss}(t) = avg(Y[t - (7 * 1 * 24)], Y[t - (7 * 2 * 24)], ... \\ ...Y[t - (7 * (N - 1) * 24)], Y[t - (7 * N * 24)]) \quad (1)$$

where $N$ is a natural number satisfying $t - (7 * N * 24) > 0$.

### 3.2 New Feature Identification

New input features are derived from historical data to assist the models in (i) predicting 'peaks' in system response time with high accuracy; (ii) predicting 'troughs' with good accuracy and (iii) capturing the temporal/seasonal patterns in the data. Due to the rare occurrence of 'peaks' and 'troughs' in the data, a threshold based method is used to derive new features and train the predictive models. These threshold values are computed automatically by studying the historical data.

The new input features required the computation of four thresholds. Two thresholds $H_1$ and $H_2$ to cater for high values and two thresholds $L_1$, and $L_2$ for low values of $Y(t)$. These thresholds were calculated after removing the outliers from $Y(t)$. The outlier computation is based on one sample standard deviation on either side. A new sample mean and sample standard deviation are computed from $Y(t)$ after the removal of all outliers. Then, the threshold values are set at one and two sample standard deviations on either side of the new sample mean

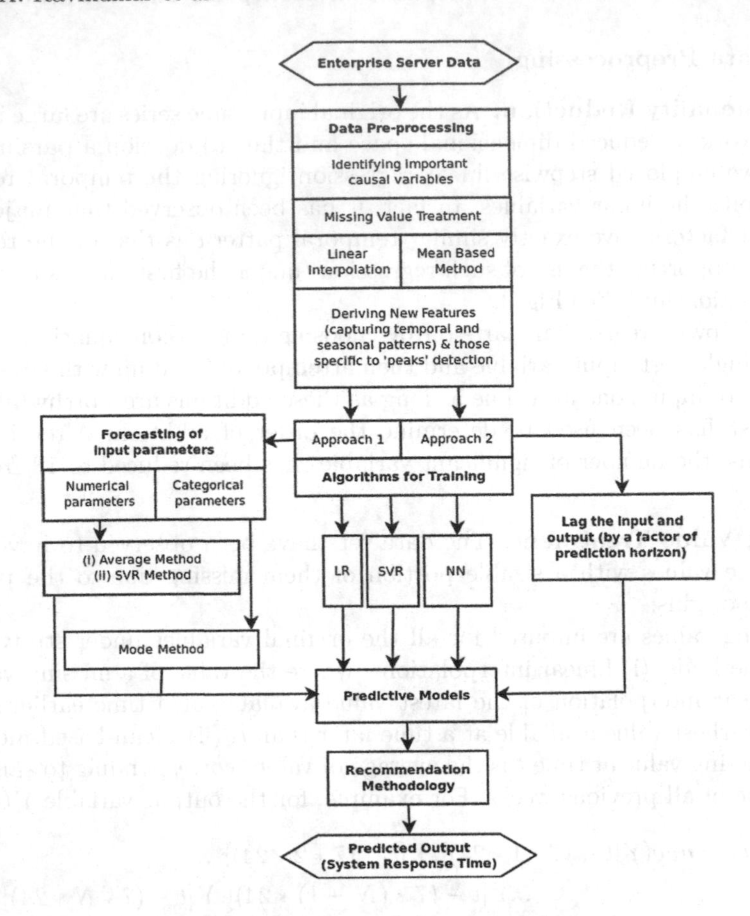


**Fig. 2.** Flowchart of the methodology.

(obtained after outlier removal). $H_1$ = *sample mean* + $2*$ (*sample std dev*); $H_2$ = *sample mean*+*sample std dev*; $L_1$ = *sample mean*−$2*$(*sample std dev*); $L_2$ = *sample mean* − *sample std dev*.

It is important to note that these thresholds are adaptive in nature will vary with time and hence, with sample size used for training. Occurrence hours of these threshold values in training data define new feature vectors (dummy variables) which are later assigned weights appropriately from improved peak or trough prediction accuracy. As the sample size increases these thresholds vary, and hence, the associated variables described below also evolve with time building a non-stationary approximation. Formally, we construct the new features as follows:

(i) Deriving $X_{d1}$ by extracting the time instants $t$ from historical data at which system response time $Y(t) > H_1$.

$$X_{d1}(t) = \begin{cases} 1, \forall t \quad s.t\, Y(t) > H_1 \\ 0, \text{otherwise} \end{cases} \tag{2}$$

(ii) Deriving $X_{d2}$ by extracting the time instants $t$ from historical data at which system response time $H_2 < Y(t) \leq H_1$.

$$X_{d2}(t) = \begin{cases} 1, \forall t \;\; s.t\, H_2 < Y(t) \leq H_1 \\ 0, \text{otherwise} \end{cases} \quad (3)$$

These two vectors define the 'peaks' in the response time.

(iii) Deriving $X_{d3}$, by extracting the time instants $t$ from historical data at which system response time $Y(t) \leq L_1$.

$$X_{d3}(t) = \begin{cases} 1, \forall t \;\; s.t\, Y(t) \leq L_1 \\ 0, \text{otherwise} \end{cases} \quad (4)$$

(iv) Deriving $X_{d4}$, by extracting the time instants $t$ from historical data at which system response time $L_1 < Y(t) \leq L_2$.

$$X_{d4}(t) = \begin{cases} 1, \forall t \;\; s.t\, L_1 < Y(t) \leq L_2 \\ 0, \text{otherwise} \end{cases} \quad (5)$$

These two vectors define the 'troughs' in the response time.

(v) Deriving $X_{d5}$, based on historical data – applying higher weights to time instants in the past where system response time was identified as 'peaks' or 'troughs' by the above four vectors and is represented as:

$$X_{d5}(t) = \begin{cases} 10, \forall t \;\; s.t\, X_{d1} \; or \; X_{d2} \; or \; X_{d3} \; or \; X_{d4} = 1 \\ 0, \text{otherwise} \end{cases} \quad (6)$$

In addition to the above, 23 dummy variables based on time in hours (2 to 24) and 6 based on day of the week (Saturday to Thursday) are included. This again is motivated by the strong correlation in temporal patterns. In the case of neural network model, temporal features like previous-day same-time output (system response time) and previous week same-day same-time output, maximum of last three weeks same-day same-time output are used to capture the temporal and seasonal variations in the data. Three features are derived of this type and are formally represented as:

(i) Deriving $X_{d6}$, based on previous day same-time $Y$ and is represented as:

$$X_{d6}(t) = Y(t - 24) \quad (7)$$

(ii) Deriving $X_{d7}$, based on previous week same-day same-time $Y$ and represented as:

$$X_{d7}(t) = Y(t - 168) \quad (8)$$

(iii) Deriving $X_{d8}$, based on the maximum of the last three weeks same-day same-time system response time $Y$ and represented as:

$$X_{d8}(t) = max(Y(t - 168), Y(t - 336), Y(t - 504)) \quad (9)$$

These additional derived input features help in increasing the prediction accuracy of 'peaks' in the system response time which is critical to system performance.

### 3.3 Modeling Input Forecasts

Two different approaches were followed for designing predictive models based on WLR, SVR and NN depending the type of input time series forecast methodology.

The **first approach** uses a direct relationship between input $X(t)$ and output $Y(t)$. Here the input $X(t)$ includes the original inputs $X_i(t)$ and the derived inputs $X_{di}(t)$. So,

$$Y_{predicted}(t) = f(X(t)) \tag{10}$$

To predict the output in advance (prediction horizon at least up to 24 h) it is required to forecast the inputs for the same prediction horizon. Since the input vector has both numerical and categorical values, appropriate time-series forecasting method is used for the prediction of individual input vector. To predict the numerical inputs, SVR model is used. This model predicts the individual input vectors by using a combination of one day (24 h), two days (48 h) and one week (168 h) lagged version of the same vector as inputs.

$$X_{pred\ numerical\ i}(t) = f_{SVR}(X_i(t-24), X_i(t-48), X_i(t-168)) \tag{11}$$

Another method is based on taking the mean of the input parameter value at the corresponding time instants is used.

$$X_{pred\,numerical\,i}(t_1) = \frac{1}{N}\sum_{k=1}^{N} X_{numerical\,i\,k}(t_1) \tag{12}$$

To predict the categorical inputs, mode of the input parameter value at the corresponding time instants is used.

$$X_{pred\,categorical\,i}(t_1) = mode(X_{categorical\,i}(t_1)) \tag{13}$$

where, $t_1$ is the time instant in hours which takes values from 1 to 24.

The **second approach** uses a modified naive method for predicting the output $Y(t)$ using a one day (24 h) lagged version of the inputs and output. This approach does not require the inputs to be forecast. Usually the lag is the same as the prediction horizon. Since here we are focusing on the prediction horizon of at least 24 h, the lag factor is taken as 24 h.

$$Y_{predicted}(t) = f(X(t-24), Y(t-24)) \tag{14}$$

### 3.4 Prediction Models

Three different models are trained for both the approaches by applying historical data to form a basket of solutions for the prediction of system response time of the enterprise.

**Weighted Least Squares Linear Regression:** The linear model given below is optimized for weighted error described below.

$$\overline{Y}(t+n) = \phi_1 Y(t) + \beta_0 + \beta_1 X_1(t) + ... + \beta_n X_n(t) + \gamma_1 X_{d1}(t) + ............ + \gamma_m X_{dm}(t) \tag{15}$$

where $X_i$ are the significant input variables (influential factors) obtained from dimensionality reduction and $X_{di}$ are the derived input variables. The coefficients $\phi_1$, $\beta_i$'s and $\gamma_i$'s above are obtained by optimizing the weighted error given below on training data. $n$ above represents the lag and equals 1 if the SVR based input forecasts model is used and equals 24 if modified naive method is used instead.

$$\min \sum_t w_t(\overline{Y}(t) - Y(t))^2 \tag{16}$$

where $w_t$ are the weights assigned based on the importance of observation at time $t$.

**Support Vector Regression (SVR):** It is an extension of Support Vector Machines-based classification to real-valued functions [8]. As in any regression scheme, $\epsilon-$ insensitive loss function is used to measure deviations, and the goal is to find a function $f_{SVR}(X(t))$ that has at most $\epsilon$ deviation from the actual $Y(t)$ for all training data, and at the same time as flat as possible. In our SVR model, the input variables are appended with weekday and time of the day dummy variables to model temporal affects and hence, model non-stationarity indirectly in the function approximation. Since the temporal variables do not have any preference connotation, dummy variable representation allows for unbiased representation of time variables in Radial Basis Function (RBF) kernels. We use RBF kernels to estimate the regression function.

$$\overline{Y} = f_{SVR}(Y_{lag}, X_1, .., X_n; X_{d1}, ..., X_{dm}) \tag{17}$$

**Artificial Neural Network (ANN):** We considered one hidden-layer feed-forward network with 15 hidden neurons having sigmoidal activation function. The novelty is in arriving at the appropriate inputs for the ANN, where apart from using system inputs ($X_i$), historical inputs $X_{d6}$, $X_{d7}$ and $X_{d8}$ are also considered. Additionally, the maximum of the response time out of the previous three weeks (on the same day) and which week day, are also injected as inputs. The former helps in capturing the information relevant to peaks. For ANN models in time series, refer [6] and [7].

$$\overline{Y} = f_{NN}(Y_{lag}, X_1, .., X_n, X_{d6}, X_{d7}, X_{d8}) \tag{18}$$

where $X_i$ are the significant input variables (influential factors) obtained from dimensionality reduction and $X_{di}$ are the derived input variables as above. Note that for NN only the derived features $X_{d6}$, $X_{d7}$, $X_{d8}$ are used in prediction.

## 4 Experimental Results and Discussion

A prototype of the proposed methodology was developed in R to demonstrate its capability in predicting the system response time with good accuracy. The enterprise system data consisted of more than 50 influential input features $X_i$ affecting the system response time, $Y(t)$. Dimensionality reduction using step-wise regression identified 13 of them to be most significant. Figure 3 shows three such inputs ($X_{21}(t)$, $X_{31}(t)$, $X_{34}(t)$) that have a strong influence on the system response time, $Y(t)$.

The prediction model for $\epsilon$-SVR was tuned with parameter values of penalty cost, $C = 1$ and loss parameter $\epsilon = 0.01$ using the RBF kernel for prediction analysis.

The prediction models were developed on ≈26 weeks of enterprise system data and tested for a span of 8 weeks (56 days) using a sliding window of 1 week to arrive at a detailed performance evaluation. Figures 4, 5, 6 and 7 show the actual system response time with the predicted ones using the above three algorithms with their average MAPE values over a span of one week. Figure 8 shows the performance of the algorithms in tracking 'peaks'. The 'peaks' here are defined by those values of $Y(t)$ which go beyond the user defined 'critical threshold' (taken as 1500). WLR model outperforms the rest of the algorithms in tracking these 'peaks' as is obvious from the plot.

**Table 2.** Overall prediction performance on 24 h rolling horizon.

| Algorithm error | <5% | <10% | <15% | <20% | <25% | <30% |
|---|---|---|---|---|---|---|
| WLR | 15% | 35% | 50% | 65% | 75% | 84% |
| SVR | 21% | 42% | 60% | 70% | 79% | 85% |
| NN | 19% | 38% | 53% | 65% | 73% | 80% |

**Table 3.** Performance at peaks.

| Algorithm error | <5% | <10% | <15% | <20% | <25% | <30% |
|---|---|---|---|---|---|---|
| WLR | 8% | 57% | 77% | 91% | 94% | 94% |
| SVR | 25% | 53% | 62% | 68% | 81% | 81% |
| NN | 7.7% | 18% | 46% | 56% | 59% | 62% |

**Table 4.** Performance over the weekends.

| Algorithm error | <5% | <10% | <15% | <20% | <25% | <30% |
|---|---|---|---|---|---|---|
| WLR | 13% | 31% | 45% | 59% | 69% | 83% |
| SVR | 17% | 37% | 53% | 62.5% | 73% | 79% |
| NN | 11% | 25% | 37% | 47% | 55% | 61% |

As the MAPE values are comparable across all the models, we analyzed Percentage Error (PE) distribution across models to understand their discriminative capability. Tables 2, 3 and 4 present the results on PE distribution. Overall SVR has high concentration of low individual errors compared to other models. However, it is interesting to observe that Weighted Linear Regression offers good prediction accuracy for peaks predicting with 90% of the cases having less than 20% error.

Based on the above results, a recommendation methodology is suggested to choose an appropriate predictive model or a mix of predictive model depending on the service on demand. Since the methods have been tested thoroughly for their robustness by examining their performance over 8 weeks with 24 h rolling horizon, their discriminative performance can be ascertained to be robust. For good overall accuracy, SVR is recommended and if the service requires good accuracy for peaks, WLR prediction is recommended. In general, the predictions

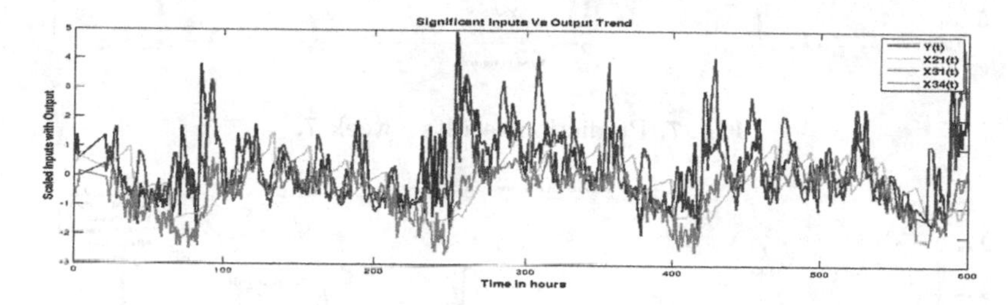


**Fig. 3.** Significant inputs vs output.

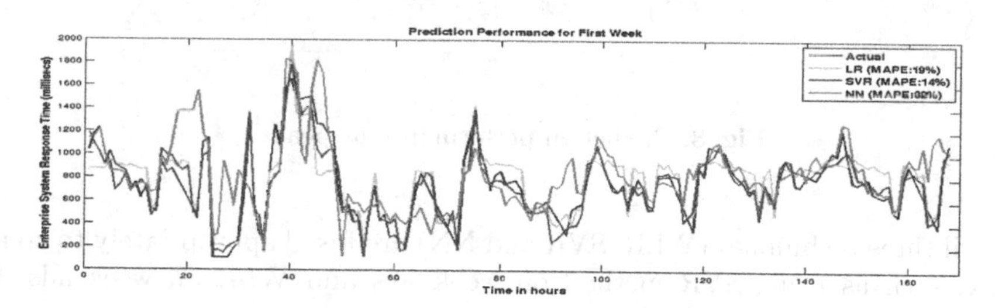


**Fig. 4.** Prediction results - week 1.

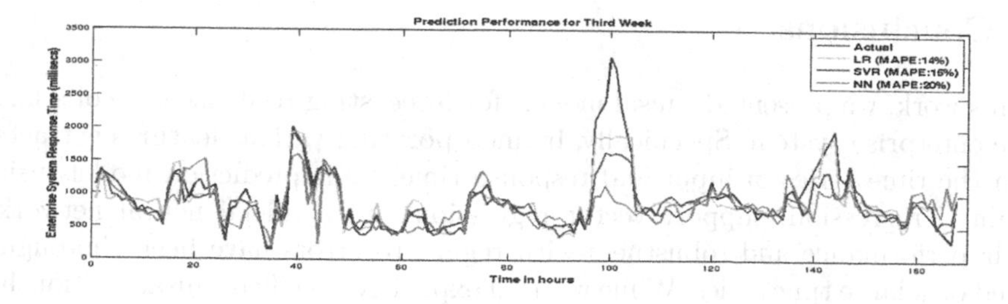


**Fig. 5.** Prediction results - week 3.

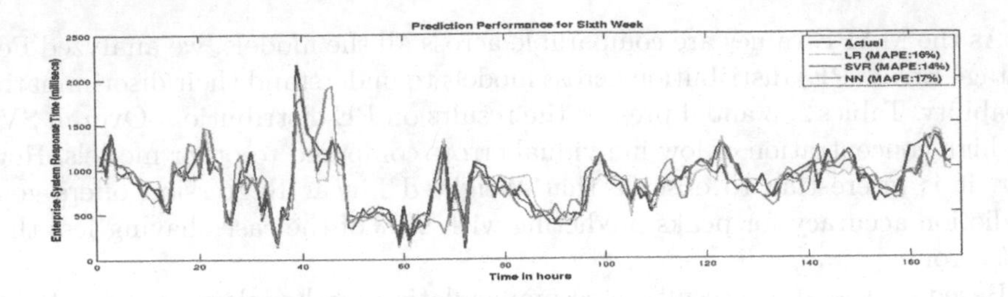


Fig. 6. Prediction results - week 6.

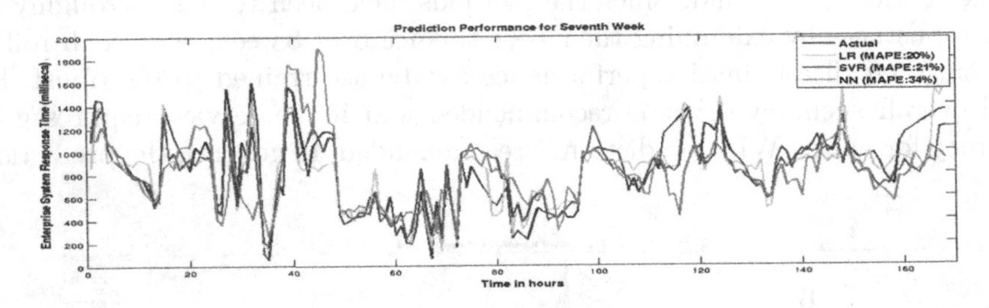


Fig. 7. Prediction results - week 7.

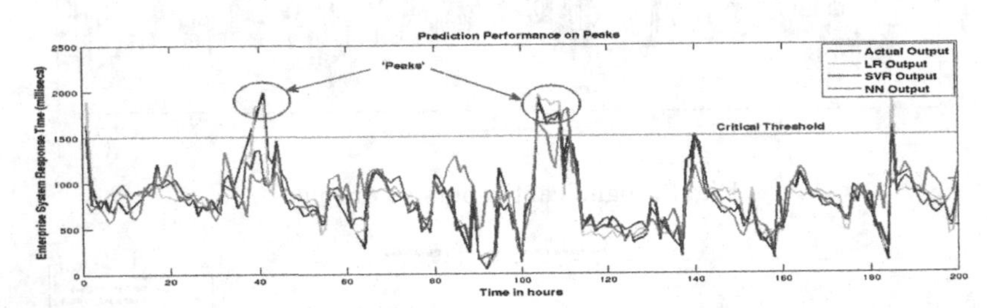


Fig. 8. Prediction performance at peaks.

from all three techniques (WLR, SVR and NN) are fused appropriately to create a mix, such as using SVR method on weekdays and WLR on weekends, for example.

## 5 Conclusions

In this work, we presented causal models for forecasting response time of a large scale enterprise system. Specifically, by incorporating various features extracted from the time series of input and response time, built prediction models using weighted regression, support vector regression and artificial neural networks. Their performance and robustness with regard to errors have been thoroughly tested on a large time series. While we used explicitly specified causal relationship in our model building, it would be interesting to obtain these causal relationships in general from data and analyze performance of causal relationship based low rank representation of multivariate time series.

## References

1. Montgomery, C.D., Jennings, C.L., Kulahci, M.: Introduction to Time Series Analysis and Forecasting. Wiley, Hoboken (2008)
2. Box, G.E.P., Jenkins, G.M.: Time Series Analysis: Forecasting and Control. Holden-Day, San Francisco (1976)
3. Moneta, A., Spirtes, P.: Graphical models for the identification of causal structures in multivariate time series models. In: Proceeding of Fifth International Conference on Computational Intelligence in Economics and Finance (2006)
4. Dahlhaus, R.: Graphical interaction models for multivariate time series. Metrika **51**(2), 157–172 (2000)
5. Spiegel, S., Gaebler, J., Lommatzsch, A., Luca, E., Albayrak, S.: Pattern recognition and classification for multivariate time series. In: SensorKDD 2011, San Diego (2011)
6. Cheng, H., Tan, P., Gao, J., Scripps, J.: Multistep-ahead time series prediction. In: PAKDD, pp. 765–774 (2006)
7. Kline, D.M.: Methods for multi-step time series forecasting with neural networks. In: Peter Zhang, G. (ed.) Neural Networks in Business Forecasting, pp. 226–250. Information Science Publishing, Hershey (2004)
8. Smola, A., Scholkopf, B.: A tutorial on support vector regression. J. Stat. Comput. **14**(3), 199–222 (2004)

# Using Artificial Neural Networks for Recovering the Value-of-Travel-Time Distribution

Sander van Cranenburgh[1(✉)] and Marco Kouwenhoven[1,2]

[1] Delft University of Technology, Jaffalaan 5, 2286 BX Delft, The Netherlands
{s.vancranenburgh,m.l.a.kouwenhoven}@tudelft.nl
[2] Significance, Grote Marktstraat 47, 2511 BH Den Haag, The Netherlands

**Abstract.** The Value-of-Travel-Time (VTT) expresses travel time gains into monetary benefits. In the field of transport, this measure plays a decisive role in the Cost-Benefit Analyses of transport policies and infrastructure projects as well as in travel demand modelling. Traditionally, theory-driven discrete choice models are used to infer the VTT distribution from choice data. This study proposes an alternative data–driven method to infer the VTT distribution based on Artificial Neural Networks (ANNs). The strength of the proposed method is that it is possible to uncover the VTT distribution (and its moments) without making strong assumptions about the shape of the distribution or the error terms, while being able to incorporate covariates and account for panel effects. We apply our method to data from the 2009 Norwegian VTT study. Finally, we cross-validate our method by comparing it with a series of state-of-the-art discrete choice models and other nonparametric methods used in the VTT literature. Based on the very encouraging results we have obtained, we believe that there is a place for ANN-based methods in future VTT studies.

**Keywords:** Artificial Neural Network · Value of Travel Time · Random Valuation · Nonparametric methods · Discrete choice modelling

## 1 Introduction

The Value-of-Travel Time (VTT) expresses travel time gains into monetary benefits [1] and plays a decisive role in the Cost-Benefit Analyses (CBA) of transport policies and infrastructure projects as well as in travel demand modelling. Not surprisingly in this regard, the VTT is one of the most researched notions in transport economics [2]. Most Western societies conduct studies to determine VTTs on a regular basis. But, despite decades of experience with data collection and VTT inference, the best way to obtain the VTT is still under debate. Early studies predominantly used Revealed Preference (RP) data in combination with Multinomial Logit (MNL) models [3]. However, despite the well-known advantages of RP data over data collected via Stated Choice (SC) experiments, nowadays RP data are seldom used for VTT studies. The main reason is that while the travellers' choices are observable (in a real-life setting), their actual trade-offs across alternatives are not – which hampers estimation of the VTT using RP data. More recent VTT studies therefore favour using SC data in combination with sophisticated


I. Rojas et al. (Eds.): IWANN 2019, LNCS 11506, pp. 88–102, 2019.
https://doi.org/10.1007/978-3-030-20521-8_8

discrete choice models that account for (some of the) potential artefacts of SC experiments (notably so-called size and sign effects) [4–8].

Besides discrete choice models, nowadays nonparametric methods are increasingly used in VTT studies [9, 10]. These methods are methodologically appealing as they do not make assumptions regarding the shape of the VTT distribution and the structure of the error terms. However, despite their methodological elegance they are typically not used to derive VTTs for appraisal. Rather, they are used as a first, complementary, step to learn about the shape of the distribution of the VTT, after which parametric discrete choice models are estimated to derive VTTs for appraisal. Börjesson and Eliasson [4] argue that nonparametric methods are not suitable to compute VTTs for appraisal for three reasons. First, they (often) cannot incorporate covariates. Second, they (often) cannot account for panel effects. Third, they do not recover the VTT distribution over its entire domain. That is, the distribution right of the highest VTT bid is not recovered, which hinders computation of the mean VTT.

Very recently, Artificial Neural Networks (ANNs) are gaining ground in the travel behaviour research arena [e.g. 11, 12–20]. A fundamental difference between discrete choice models and ANNs is the modelling paradigm to which they belong. Discrete choice models are theory-driven, while ANNs are data-driven. Theory-driven models work from the principle that the true Data Generating Process (DGP) is a (stochastic) function, which can be uncovered. To do so, the analyst imposes structure on the model. In the context of discrete choice models this is done by prescribing the utility function, the decision rule, the error term structure, etc. Then, the analyst estimates the model's parameters, usually compares competing models, and interprets the results. A drawback of this approach is that it heavily relies on potentially erroneous assumptions regarding choice behaviour, i.e. the assumptions may not accurately describe the true underlying DGP – potentially leading to erroneous inferences. Data-driven methods work from the principle that the true underlying process is complex and inherently unknown. In a data-driven modelling paradigm the aim is not to uncover the DGP, but rather to learn a function that accurately approximates the underlying DGP. The typical outcome in a data-driven modelling paradigm is a network which has very good prediction performance [18]. A major drawback of data-driven methods is that – without further intervention – they provide very limited (behavioural) insights on the underlying DGP, such as the relative importance of attributes, Willingness-to-Pay, or VTT. Yet, these behavioural insights are typically most valuable to travel behaviour researchers and for transport policy-making.

There is a general sense that ANNs (and other data-driven models), could complement existing (predominantly) theory-driven research efforts. In light of that spirit, this paper develops an ANN-based method to investigate the VTT distribution. Specifically, we develop a novel pattern recognition ANN which is able to estimate travellers' individual underlying VTTs. Our method capitalises on the strong prediction performance of ANNs (see [21] for a comprehensive review of articles that involve a comparative study of ANNs and statistical techniques). The strength of this method is that it is possible to uncover the VTT distribution (and its moments) without making strong assumptions on the underlying behaviour. For instance, it does not prescribe the utility function, the shape of the VTT distribution, or the structure of the error terms. Moreover, the method can incorporate covariates, account for panel effects and does yield a distribution right of the maximum VTT bid. Thereby, it overcomes important limitations associated with other nonparametric methods. As such, this method can be used to derive VTTs for appraisal. Finally, the method does not require extensive software coding on the side of the analyst as the method is built on a standard MultiLayer Perceptron (MLP) architecture. Hence, the method can be applied using off-the-shelf (open-source) software.

The remainder of this paper is organised as follows. Section 2 develops the ANN-based method for uncovering the VTT distribution. Section 3 applies the method to an empirical VTT data set from a recent VTT study. Section 4 cross-validates the method by comparing its results with those obtained using a series of state-of-the-art discrete choice models and other nonparametric methods. Finally, Sect. 5 draws conclusions and provides a discussion.

# 2 Methodology

## 2.1 Preliminary

**Panel Data Format (time series)**

Throughout this paper we suppose that we deal with data from a classic binary SC experiment, consisting of $T + 1$ choice tasks per individual, in which within-mode trade-offs between travel cost $TC$ and travel time $TT$ are embedded. This data format is in line with standard VTT practice in many Western European countries, including the UK [22], The Netherlands [7, 23], Denmark [8], Norway [5] and Sweden [4]. Figure 1 shows a choice task from such a SC experiment. Choice tasks are pivoted around the respondents current travel time and travel cost, which are typically elicited prior to the SC experiment. In the SC experiment respondents are confronted with $T + 1$ choice tasks consisting of two alternatives, in each choice task one trip being their current one and the other one being either a faster and more expensive, or a slower and cheaper trip. In each choice task there is an implicit price of time which is commonly referred to as the Boundary VTT (BVTT). The BVTT is defined as:

$$BVTT = -\frac{\Delta TC}{\Delta TT} = \frac{-(TC_2 - TC_1)}{(TT_2 - TT_1)} \tag{1}$$

where alternative 1 denotes the fast and expensive alternative and the alternative 2 denotes the slow and cheap alternative. The BVTT can be perceived as a valuation threshold as a respondent choosing the fast and expensive alternative signals a VTT which is (most likely) above the BVTT, while a respondent choosing the slow and cheap alternative signals a VTT which is (most likely) below the BVTT.

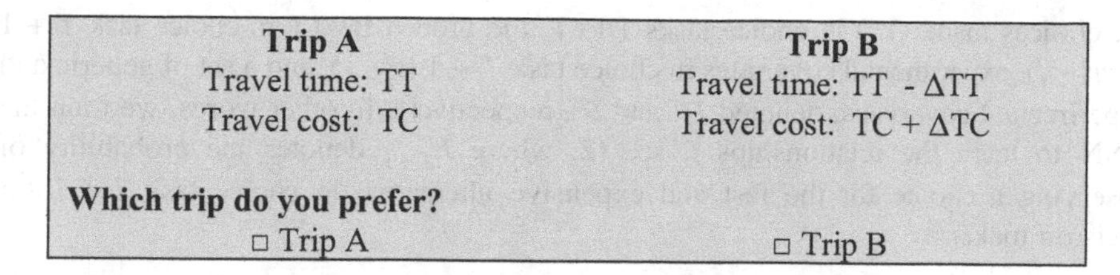


**Fig. 1.** Example choice task

**Covariates in VTT Studies**

It is important to incorporate covariates in models that aim to infer the VTT. Börjesson and Eliasson [4] provide four reasons for this. Firstly, accounting for covariates in VTT models allows better extrapolating the VTT to new situations. Secondly, accounting for covariates in VTT models allows better understanding what trip characteristics influence the VTT. Thirdly, accounting for covariates in VTT models allows the analyst to remove the influence of undesirable factors, such as income or urbanisation level from the VTT used for appraisal. Fourthly, accounting for covariates in VTT models allows accounting for so-called size and sign effect stemming from the experimental design [24]. Size effects are due to the behavioural notion that the VTT is dependent on the size of the difference in travel time and travel cost across alternatives in the choice task [25]. Sign effects are due to the behavioural notion that losses (e.g. higher travel cost and longer travel time) loom larger than equivalently sized gains (e.g. lower travel cost and shorter travel time) [24, 26].

## 2.2 Uncovering Individual VTTs Using ANNs

The ANN-based method is based on three observations. The first observation is that ANNs are very good at making predictions [21]. Their good prediction performance stems from the versatile structure of ANNs, which allow them to capture non-linearity, interactions between variables, and other peculiarities in the DGP, for instance in this context relating to the set-up of the experimental design. The second observation is that we can use the ANN to find the BVTT where it is maximally uncertain on the choice of the decision maker. The third observation is that we can give a behavioural interpretation to this BVTT and recover the individual's VTT from it. That is, we can interpret the BVTT where the ANN is maximally uncertain as the point where the individual is indifferent between choosing the fast and expensive alternative and the slow and cheap alternative. From this behavioural perspective, this BVTT reflects the VTT of the individual. Taking these three observations together, we can develop an ANN-based method that recovers individual level VTTs and can be used to derive VTTs for appraisal.

To do so, we take the following 5 steps:

(1) **Data preparation and training**

The aim of this step is to train an ANN to (probabilistically) predict, for decision maker $n$ the choice in the hold-out choice task $T + 1$, based on the BVTTs ($BVTT^n$) and the choices made ($Y^n$) in choice tasks 1 to $T$, the probed BVTT in choice task $T + 1$ ($bvtt^n_{T+1}$), experimental covariates in choice task $T + 1$ ($s^n_{T+1}$), and a set of generic and experimental covariates, denoted $D^n$ and $S^n$, respectively. In other words, we train the ANN to learn the relationships $f$, see (2, where $P^n_{T+1}$ denotes the probability of observing a choice for the fast and expensive alternative in choice task $T + 1$ for decision maker $n$.

$$P^n_{T+1} = f\left(BVTT^n, Y^n, bvtt^n_{T+1}, s^n_{T+1}, D^n, S^n\right) \tag{2}$$

$$\begin{aligned} where\, BVTT^n &= \left\{bvtt^n_1, bvtt^n_2, \ldots, bvtt^n_T\right\} \\ Y^n &= \left\{y^n_1, y^n_2, \ldots, y^n_T\right\} \\ S^n &= \left\{s^n_1, s^n_2, \ldots, s^n_T\right\} \end{aligned}$$

Figure 2 shows the proposed architecture of the ANN. At the input layer, the independent variables enter the network. At the top, there are the generic covariates (green). Typical generic covariates encountered in VTT studies are mode, purpose, age, income, distance, etc. Below the generic covariates are the variables associated with choice tasks 1 to $T$ (red). These include the BVTTs, the choices $y$ and experimental covariates $s$ (e.g. sizes and signs). Below the variables for choice tasks 1 to $T$ is an extra set of input nodes for choice task $R$ (blue). Choice task $R$ is a replication of one choice task, randomly picked from the set choice tasks 1 to $T$. These input nodes come in handy later when the ANN is used for simulation (they make it possible to use all $T + 1$ observations instead of only $T$ observations in the simulation). Finally, at the bottom are the variables associated with hold-out choice task $T + 1$ (yellow). These are essentially the 'knobs' of the model that can be used for simulation. The output layer consists of the dependent variable, which is the probability for choosing the fast and expensive alternative in choice task $T + 1$. One or multiple hidden layers can be used. In our analyses we find two layers to work well. However, the optimal number of hidden layer and the number of nodes depends on the complexity of the DGP that needs to be learned from the data, and hence may vary across applications.

To train the network in Fig. 2, we need to prepare the data. To do so, for each decision maker in the data we randomly draw $T$ explanatory choice tasks from the $T + 1$ choice tasks that are available in the data for each decision maker. These $T$ choice tasks are used as independent variables to predict the remaining choice. To avoid that the network undesirably learns a particular structure in the data, rather than the explanatory power of the variables it is crucial that the order in the set of $T$ explanatory choice tasks is randomised.[1] We randomise the order in the set of explanatory choice tasks $K$ times, each time creating a 'new' observation. The idea behind this is that the weights associated with the choice tasks attain (roughly) similar sizes. By doing so, we create a network that produces stable predictions, which is insensitive to the order of the explanatory choice tasks. In each manifestation of the

[1] Unless the order of the choice tasks is randomised during the data collection. Note that by doing so the network becomes blind to potential learning effects on the side of the respondent when conducting the survey. We come back to this point in the discussion.

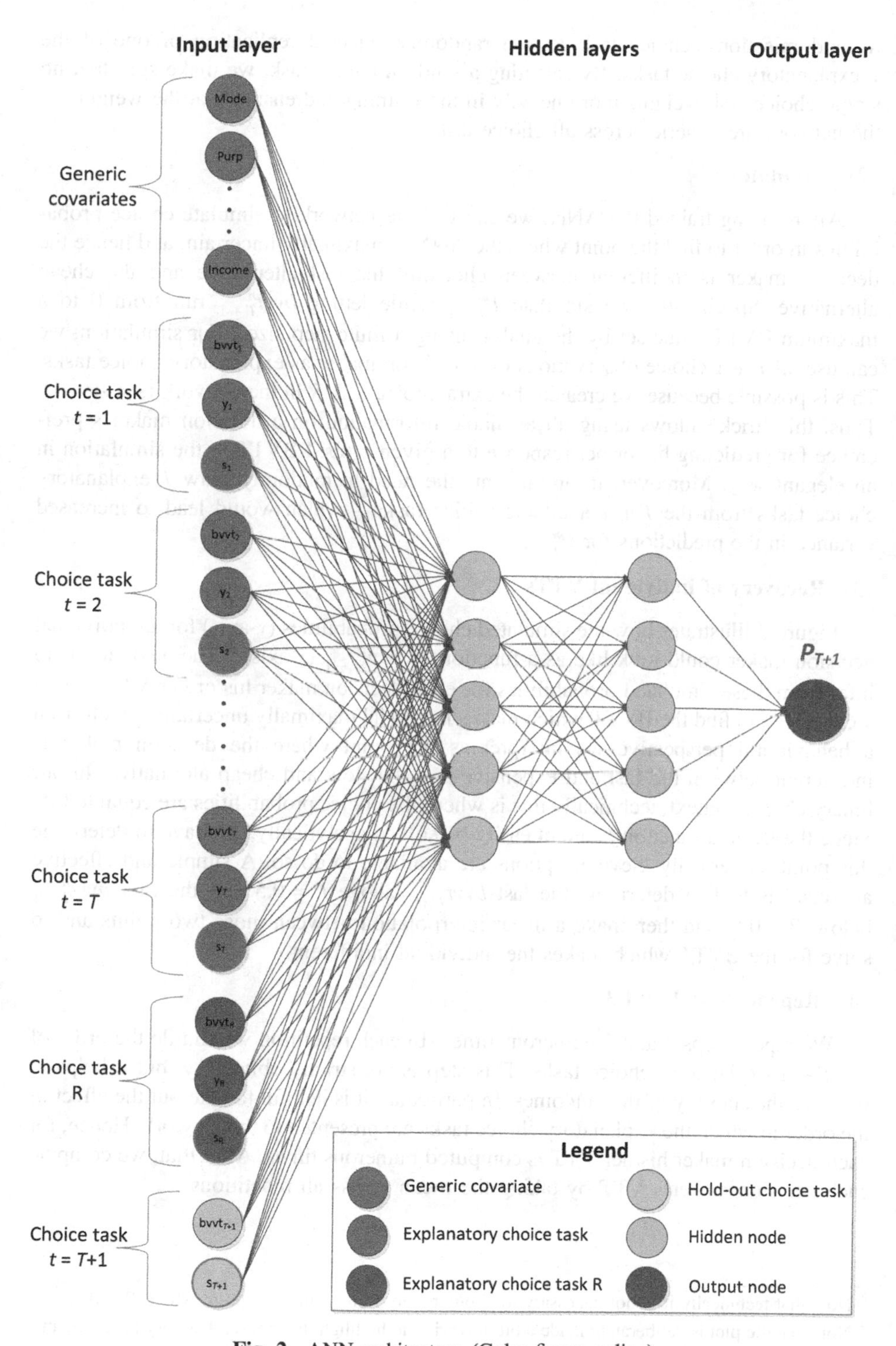


**Fig. 2.** ANN architecture (Color figure online)

$K$ randomisations, choice task $R$ is a randomly selected replication of one of the $T$ explanatory choice tasks. By selecting a random choice task, we make sure that no single choice task weights more heavily in the training and ensure that the weights of the network are generic across all choice tasks.

(2) **Simulate**

After having trained the ANN, we can use the network to simulate choice probabilities in order to find the point where the ANN is maximally uncertain, and hence the decision maker is indifferent between choosing the fast alternative and the cheap alternative. Specifically, we simulate $P^n_{T+1}$ while letting $bvvt^n_{T+1}$ run from 0 to a maximum BVTT value set by the analyst using a finite step size.[2] For simulation, we can use all $T + 1$ choice observations of a decision maker as explanatory choice tasks. This is possible because we created the extra choice task $R$ in the network (see step 1). Thus, this 'trick' allows using all available information on a decision maker's preference for predicting his or her response to a given probed BVTT in the simulation in an elegant way. Moreover, it circumvents the need to randomly draw $T$ explanatory choice tasks from the $T + 1$ available choice tasks – which would lead to increased variance in the predictions for $P^n_{T+1}$.

(3) **Recovery of individual VTTs**

Figure 3 illustrates how the simulated choice probabilities ($y$-axis) for an individual decision maker could look like as a function of $bvvt^n_{T+1}$ ($x$-axis).[3] The next step is to infer from these simulated probabilities for each decision maker his or her VTT. To do so, we need to find the BVTT which makes the ANN maximally uncertain, which from a behavioural perspective we interpret as the point where the decision maker is indifferent between the fast and expensive and the slow and cheap alternative. In our binary choice context, technically this is where the choice probabilities are equal to 0.5. Since the learned function $f$ cannot easily be solved analytically, we have to determine this point numerically. Several options are available to do so. A simple and effective approach is to first determine the last $bvvt^n_{T+1}$ above $P = 0.5$ and the first $bvvt^n_{T+1}$ below $P = 0.5$, and then make a linear interpolation between those two points and to solve for the BVTT which makes the individual indifferent.

(4) **Repeat steps 2 and 3**

We repeat steps 2 and 3 numerous times. In each repetition we shuffle the order of the $T + 1$ explanatory choice tasks. This step is not strictly obligatory, but it helps to improve the stability of the outcomes. In particular, it is helpful to take out the effect of the order in which the explanatory choice tasks are presented to the network. Hence, for each decision maker his/her VTT is computed numerous times. After that, we compute each decision maker's VTT by taking the mean across all repetitions.

---

[2] Note that technically it is not necessary to simulate any further than the point where $P < 0.5$.

[3] Note that the plot is deliberately made a bit quivering to highlight the notion that very little structure is imposed by the ANN on the functional form.

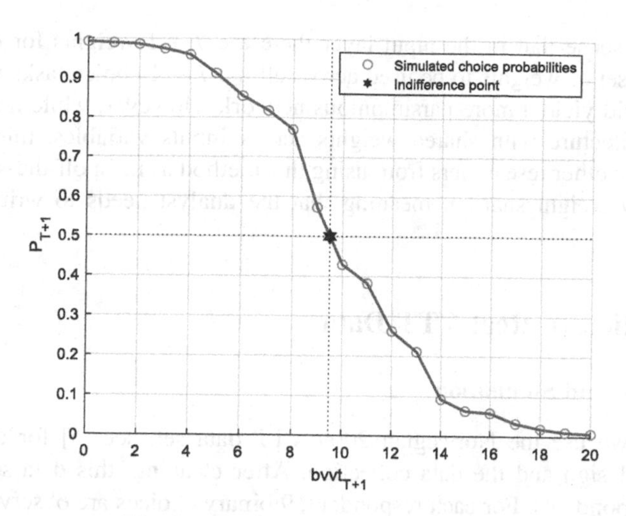


**Fig. 3.** Simulated choice probabilities

**(5) Construct the VTT distribution**

Having an estimate of the VTT for each decision maker, we can construct an empirical distribution of the VTT. Also, from the constructed empirical distribution we can readily compute the mean and standard deviation of the VTT.

## 2.3 ANN Development

In Sect. 2.2 we presented the ANN without going into much detail on its architecture or on underlying design choices. In this subsection we discuss these in more detail. To develop an ANN capable of learning function (2), we have tested numerous different architectures, including fully and semi-connected networks, different numbers of hidden layers, the presence or absence of bias nodes, and we have tried several different activation functions. The two-hidden layer architecture presented in Fig. 2 with ten nodes at each hidden layer is found to work particularly well for our data.[4] The proposed architecture is a so-called Multilayer Perceptron (MLP). This is one of the most widely used ANNs architectures and is implemented in virtually all off-the-shelf machine learning software packages. For the activation functions in the network we find good results using a softmax function both at the nodes of the hidden layers as well as at the nodes of the output layer. Using a softmax function at the output layer ensures that the sum of the predicted choice probabilities across the two alternatives add up to 1. Finally, note that while no bias nodes are depicted in Fig. 2 bias nodes are present as they are found to improve the classification performance.

The fact that off-the-shelf software can be used is a desirable feature of this method, as it makes the method accessible for a wide research community. Admittedly, from a methodological perspective our network consumes more weights than is strictly

[4] The network consumes 491 weights in total.

needed, in the sense that in the input layer there are $T + 1$ weights for *bvtt*, *y* and *s*, while just one set of weights to be used across all the $T + 1$ choice tasks would suffice and hence would yield a more parsimonious network. However, while it is possible to create an architecture with shared weights across inputs variables, this would substantially hinder other researchers from using this method as most off-the-shelf software does not allow weight sharing, meaning that the analyst needs to write customised codes.

# 3 Application to Real VTT Data

## 3.1 Training and Simulation

In this study we use the Norwegian 2009 VTT data set, see [5] for details on the experimental design and the data collection. After cleaning, this data set consists of 5832 valid respondents. For each respondent, 9 binary choices are observed. While the currency in the SC experiment was Norwegian Kronor, for reasons of communication we converted all costs into euros. To train the network on these empirical data, 70% of the data were used for training, 15% for validation and 15% for testing. The observations were randomly allocated to these subsets. We use $K = 20$ randomisations (see Sect. 2.2). The trained ANN acquires a cross-entropy of 0.36. Table 1 shows the confusion plot. It shows that overall about 85% of the choices are correctly predicted (based on highest probability). To obtain the VTT distribution, we use the network to simulate choice probabilities and search for the BVTTs where the ANN is maximally uncertain. We do this 20 times[5] for each respondent (i.e., steps 2 to 4, see Sect. 2.2).

**Table 1.** Confusion plot (based on validation and test data)

| | Target 1 (fast and expensive) | Target 2 (slow and cheap) | Σ |
|---|---|---|---|
| Output class 1 (fast and expensive) | **26.7%** | 6.9% | 79.4% *(Positive predictive value)* |
| Output class 2 (slow and cheap) | 8.3% | **58.1%** | 87.5% *(Negative predictive value)* |
| Σ | 76.3% *(Sensitivity)* | 89.4% *(Specificity)* | **84.8%** *(Overall accuracy)* |

## 3.2 Results

Figure 4 shows the recovered distribution of the VTT (step 5). For eight respondents, it has not been possible to obtain a VTT estimate. For these respondents, the ANN predicts choice probabilities below 0.5, even for very small BVTTs, suggesting a zero or even a negative VTT. While this seems behaviourally unrealistic, from a data-driven

[5] We find that after 20 times the results are stable.

modelling perspective it can be well understood why it is not possible to obtain a VTT for each and every respondent, especially considering that over 13% of the respondents in the data always choose the slow and cheap alternative. The ANN may have learned that some respondents just never choose the fast and expensive alternative, even if it is just a fraction more expensive than the slow and cheap alternative. Close inspection of the eight respondents for which we have not been able to obtain a VTT estimate, shows that indeed these respondents never chose the expensive and fast alternative and that they all had low income levels. In the remainder of our analyses these eight respondents are given a VTT of zero. About 2% of the respondents always chose the fast and expensive alternative in each choice task. For all these respondents a VTT has been recovered, in between €20 and €123 per hour, with a median VTT of €85 per hour.

Figure 4 shows that the shape of the VTT distribution is positively skewed. The lognormal-like shape is behaviourally intuitive and has occasionally been found in previous VTT studies. However, when fitting the lognormal distribution onto these data (not shown), we find that it does not fit the data very well: in particular, it cannot accommodate for the spike at around VTT = €2/h and the drop at VTT = €16/h. Close inspection of the bins around VTT = €2/h reveal that they are predominantly populated with those respondents that always choose the slow and cheap alternative (for clarity, non-traders are depicted in red in the right-hand side plot). The bimodal shape of this distribution essentially emphasises the need for flexible methods to uncover the distribution of the VTT.

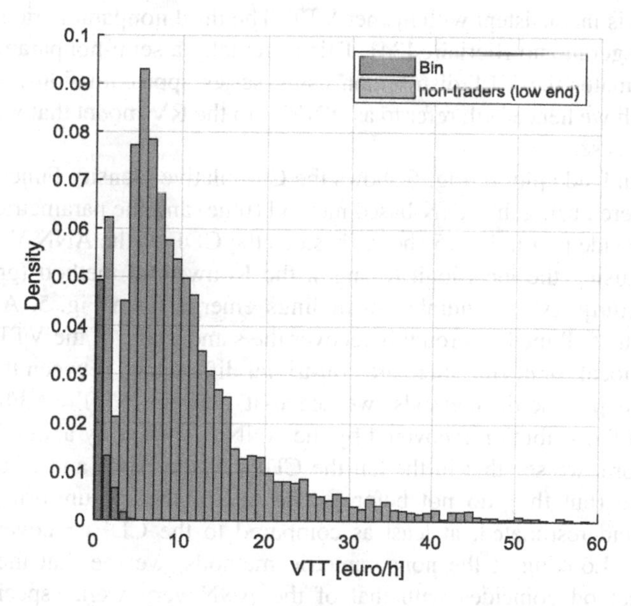


**Fig. 4.** VTT distribution

## 4 Cross-validation

To cross-validate the shape and mean of the recovered VTT distribution by the ANN-based method, we compare with state-of-the-art (parametric) choice models as well as with three (semi) nonparametric methods that have been used in recent VTT studies. The parametric models that we use in this cross-validation study are Random Valuation (RV) models [27, 28], with two types of distributions, namely the lognormal and the log-uniform distributions. The log-normal distribution has been used in the most recent Swedish VTT study; the log-uniform has been used in the most recent UK VTT study. Note that we also have estimated more conventional Random Utility Maximisation (RUM) models [29], but the RV models are found to outperform their random utility counterparts [30]. Therefore, we report only on the RV models. Regarding the non-parametric methods, the first nonparametric method that we consider is called local-logit. This method is developed by Fan, Heckman and Wand [31], pioneered in the VTT research literature by Fosgerau [10] and further extended by [32]. The local-logit method essentially involves estimation of logit models at 'each' value of the BVTT using a kernel with some shape and bandwidth. In our application we use a triangular shaped kernel with a bandwidth of 10 euro. The second nonparametric method is developed by Rouwendal, de Blaeij, Rietveld and Verhoef [33]. Henceforth, we refer this method as 'The Rouwendal method'. This method assumes that everybody has a unique VTT and makes consistent choices accordingly. But, at each choice there is a fixed probability that the decision maker makes a mistake and hence chooses the alternative that is inconsistent with his/her VTT. The third nonparametric method is put forward by Fosgerau and Bierlaire [34]. This is actually a semi-nonparametric method which approximates the VTT distribution using series approximations. We apply the method – which we henceforth refer to as 'SNP' – to the RV model that we also used in the parametric case.

The left-hand side plot in Fig. 5 shows the Cumulative Density Function (CDF) of the VTT recovered using the ANN-based method (blue) and the parametric RV models. The right-hand side plot in Fig. 5 shows, besides the CDF of the ANN VTT (blue), the CDFs created using the local-logit (orange), the Rouwendal method (green) and the SNP method (turquoise). A number of findings emerge from Fig. 5. A first general observation is that all methods roughly recover the same shape of the VTT distribution, except for the local-logit. But, there are non-trivial differences between the shapes too. Looking at the parametric methods, we see that between VTT = €3/h and VTT = €10/h, the VTT distribution recovered by the ANN is shifted by about 2 euros to the left. Furthermore, we see that in the tail the CDFs of the ANN and of the lognormal neatly coincide (but they do not before). The tail of the log-uniform seems to be substantially underestimated, at least as compared to the CDFs recovered using the other methods. Looking at the nonparametric methods, we see that the CDF of the Rouwendal method coincides with that of the ANN very well, especially up until VTT = €14/h and in the tail above €55/h. The CDF of the SNP method coincides well with that of the ANN for VTTs of €5/h and higher. The local-logit CDF deviates most from the other CDFs, in particular below VTT = €30/h. Possibly, this is caused by its inability to account for the panel nature of the data and its inability to disentangle unobserved heterogeneity from irreducible noise in the data. After all, the local-logit method only considers choices from several respondents around the same BVTT, without considering the other choices made by these (or other) respondents.

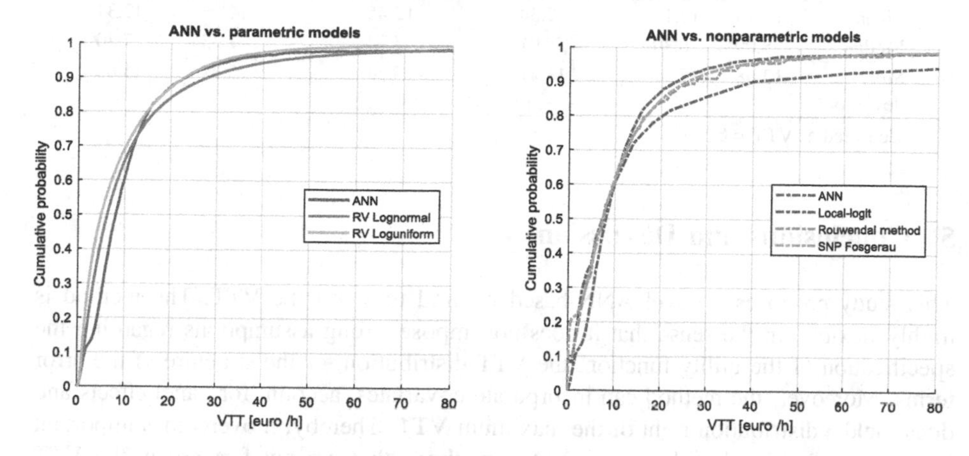


**Fig. 5.** Cross-validation of shape (Color figure online)

Table 2 summarises key statistics of the recovered VTT distributions for the methods that we have used. Since the nonparametric methods do not recover the VTT distribution beyond the maximum VTT bid, the presented statistics for these methods can be considered as lower bounds. However, in these data the maximum VTT is set very high (see [5]) and only 2% of the respondents in these data always choose for the fast and expensive alternative, many of whom did not receive a very high VTT bid. As such, the unrecovered tail for the Rouwendal method is very small, representing less than 0.05% of the density. But, the unrecovered tail for the local-logit still represents about 5% of the density (rendering computation of its moments unreliable). The statics for the SNP method are computed from the CDF. In line with previous practice using this method, we censored the right-hand side tail above VTT = €200/h. Not doing so, would substantially inflate the recovered standard deviation of this distribution. The overview shows that the mean recovered by the ANN-based method is close to those of all other methods, except the RV-log-uniform. The median VTT recovered by the ANN is higher than those of the parametric methods. This is presumably due to the limited flexibility of the latter methods to account for the substantial number of respondents having a very low VTT (13% of the respondents always choose the slow and cheap alternative), while still covering the VTT distribution over a large range. Altogether, it can be concluded that the shape, mean and median recovered by the ANN seem very plausible.

**Table 2.** Mean, median and standard deviations of recovered VTT distributions

| | ANN | RV Lognormal | RV Log-uniform | Rouwendal method | Local-logit | SNP[a] |
|---|---|---|---|---|---|---|
| Mean | 11.75 | 12.13 | 9.34 | 12.45 | 12.16 | 12.34 |
| Median | 8.09 | 6.30 | 5.01 | 7.74 | 7.33 | 7.40 |
| Std deviation | 13.68 | 17.57 | 11.41 | 15.54 | 15.24 | 15.64 |

[a]Censored at VTT = €200/h

## 5 Conclusions and Discussion

This study proposes a novel ANN-based method to study the VTT. The method is highly flexible, in the sense that it does not impose strong assumptions regarding the specification of the utility function, the VTT distribution, or the structure of the error terms. Moreover, the method can incorporate covariates, account for panel effects and does yield a distribution right of the maximum VTT. Thereby, it overcomes important limitations associated with nonparametric methods that are put forward in the VTT literature. In this study we have cross-validated the proposed method by comparing it with a series of state-of-the-art discrete choice models and nonparametric methods. Based on the encouraging results of this study, we believe that there is a place for ANN-based methods in future VTT studies.

The method proposed in this study provides ample scope for further research. A first direction for further research involves acquiring a good understanding regarding the data requirements for this method to work well. For instance, how many respondents are at least needed for this method? A commonly used rule-of-thumb in the Machine Learning field is that the number of observations needs to be at least ten times more than the number of estimable weights. However, a recent study on this topic in the context of choice data suggest a more conservative factor of 50 times more observations than weights [35]. Likewise, what is the 'minimum' number of choice tasks per respondents that is needed? In our study we found good results with nine choice tasks per respondents. But, will the method also work with just five choice tasks per respondent, or will it work even better with fifteen choice tasks? A second, related, direction for further research concerns the design of the SC experiment. Current SC experiments are optimised for estimation of discrete choice models. However, data from these experiments may actually be suboptimal for the ANN-based method. A question that remains to be answered therefore is how to design experiments optimised for this method? A third direction for further research concerns the generalisation of the method to work with choice tasks having three or more attributes. While it is clear that it becomes more difficult to recover a VTT from a choice task consisting of three or more alternatives using this method, there are – as far as we can see tell – no fundamental reasons why the method would be confined to data from two-attribute experiments only. A fourth interesting direction is investigating whether it is possible to also capture and incorporate learning and ordering effects. Some empirical studies suggest that respondents are subject to learning effects and ordering anomalies [36].

A fifth research direction for further research is application of this method to other VTT data sets, as well as applying the method to other areas of application, such as inference of the distribution of the value of reliability. Finally, a drawback of the ANN-based method relates to the opaque nature of ANNs: they cannot easily be diagnosed, e.g. by looking at its weights. Future research may be directed to illuminate the black boxes of ANNs, especially in contexts where they are used for behavioural analysis [37].

## References


1. Small, K.A.: Valuation of travel time. Econ. Transp. **1**, 2–14 (2012)
2. Abrantes, P.A.L., Wardman, M.R.: Meta-analysis of UK values of travel time: an update. Transp. Res. Part A Policy Pract. **45**, 1–17 (2011)
3. Wardman, M., Chintakayala, V.P.K., de Jong, G.: Values of travel time in Europe: review and meta-analysis. Transp. Res. Part A Policy Pract. **94**, 93–111 (2016)
4. Börjesson, M., Eliasson, J.: Experiences from the Swedish value of time study. Transp. Res. Part A Policy Pract. **59**, 144–158 (2014)
5. Ramjerdi, F., Flügel, S., Samstad, H., Killi, M.: Value of time, safety and environment in passenger transport–Time. TØI report 1053-B/2010. Institute of Transport Economics (TØI) (2010)
6. Hess, S., Daly, A., Dekker, T., Cabral, M.O., Batley, R.: A framework for capturing heterogeneity, heteroskedasticity, non-linearity, reference dependence and design artefacts in value of time research. Transp. Res. Part B Methodol. **96**, 126–149 (2017)
7. Kouwenhoven, M., et al.: New values of time and reliability in passenger transport in The Netherlands. Res. Transp. Econ. **47**, 37–49 (2014)
8. Fosgerau, M., Hjorth, K., Lyk-Jensen, S.V.: The Danish value of time study: Final Report (2007)
9. Fosgerau, M.: Investigating the distribution of the value of travel time savings. Transp. Res. Part B Methodol. **40**, 688–707 (2006)
10. Fosgerau, M.: Using nonparametrics to specify a model to measure the value of travel time. Transp. Res. Part A Policy Pract. **41**, 842–856 (2007)
11. Alwosheel, A., Van Cranenburgh, S., Chorus, C.G.: Artificial neural networks as a means to accommodate decision rules in choice models. ICMC2017, Cape Town (2017)
12. Mohammadian, A., Miller, E.: Nested logit models and artificial neural networks for predicting household automobile choices: comparison of performance. Transp. Res. Rec.: J. Transp. Res. Board **1807**, 92–100 (2002)
13. Omrani, H., Charif, O., Gerber, P., Awasthi, A., Trigano, P.: Prediction of individual travel mode with evidential neural network model. Transp. Res. Rec.: J. Transp. Res. Board **2399**, 1–8 (2013)
14. Wong, M., Farooq, B., Bilodeau, G.-A.: Discriminative conditional restricted Boltzmann machine for discrete choice and latent variable modelling. J. Choice Model. **29**, 152–168 (2017)
15. Sifringer, B., Lurkin, V., Alahi, A.: Enhancing discrete choice models with neural networks. In: hEART 2018–7th Symposium of the European Association for Research in Transportation Conference (2018)
16. Cantarella, G.E., de Luca, S.: Multilayer feedforward networks for transportation mode choice analysis: an analysis and a comparison with random utility models. Transp. Res. Part C Emerg. Technol. **13**, 121–155 (2005)

17. Golshani, N., Shabanpour, R., Mahmoudifard, S.M., Derrible, S., Mohammadian, A.: Modeling travel mode and timing decisions: comparison of artificial neural networks and copula-based joint model. Travel. Behav. Soc. **10**, 21–32 (2018)
18. Karlaftis, M.G., Vlahogianni, E.I.: Statistical methods versus neural networks in transportation research: differences, similarities and some insights. Transp. Res. Part C Emerg. Technol. **19**, 387–399 (2011)
19. Lee, D., Derrible, S., Pereira, F.C.: Comparison of four types of artificial neural network and a multinomial logit model for travel mode choice modeling. Transp. Res. Rec. **2672**, 101–112 (2018). 0361198118796971
20. Van Cranenburgh, S., Alwosheel, A.: An artificial neural network based approach to investigate travellers' decision rules. Transp. Res. Part C Emerg. Technol. **98**, 152–166 (2019)
21. Paliwal, M., Kumar, U.A.: Neural networks and statistical techniques: a review of applications. Expert Syst. Appl. **36**, 2–17 (2009)
22. Batley, R., et al.: New appraisal values of travel time saving and reliability in Great Britain. Transportation, 1–39 (2017). https://link.springer.com/article/10.1007/s11116-017-9798-7
23. HCG: The second Netherlands' value of time study - final report (1998)
24. De Borger, B., Fosgerau, M.: The trade-off between money and travel time: a test of the theory of reference-dependent preferences. J. Urban Econ. **64**, 101–115 (2008)
25. Daly, A., Tsang, F., Rohr, C.: The value of small time savings for non-business travel. J. Transp. Econ. Policy (JTEP) **48**, 205–218 (2014)
26. Ramjerdi, F., Lindqvist Dillén, J.: Gap between willingness-to-pay (WTP) and willingness-to-accept (WTA) measures of value of travel time: evidence from Norway and Sweden. Transp. Rev. **27**, 637–651 (2007)
27. Cameron, T.A., James, M.D.: Efficient estimation methods for "closed-ended" contingent valuation surveys. Rev. Econ. Stat. **69**, 269–276 (1987)
28. Fosgerau, M., Bierlaire, M.: Discrete choice models with multiplicative error terms. Transp. Res. Part B Methodol. **43**, 494–505 (2009)
29. McFadden, D.L.: Conditional logic analysis of qualitative choice behavior. In: Zarembka, P. (ed.) Frontiers in Econometrics, pp. 105–142. Academic Press, New York (1974)
30. Ojeda-Cabral, M., Hess, S., Batley, R.: Understanding valuation of travel time changes: are preferences different under different stated choice design settings? Transportation **45**, 1–21 (2018)
31. Fan, J., Heckman, N.E., Wand, M.P.: Local polynomial kernel regression for generalized linear models and quasi-likelihood functions. J. Am. Stat. Assoc. **90**, 141–150 (1995)
32. Koster, P.R., Koster, H.R.A.: Commuters' preferences for fast and reliable travel: a semi-parametric estimation approach. Transp. Res. Part B Methodol. **81**, 289–301 (2015)
33. Rouwendal, J., de Blaeij, A., Rietveld, P., Verhoef, E.: The information content of a stated choice experiment: a new method and its application to the value of a statistical life. Transp. Res. Part B Methodol. **44**, 136–151 (2010)
34. Fosgerau, M., Bierlaire, M.: A practical test for the choice of mixing distribution in discrete choice models. Transp. Res. Part B Methodol. **41**, 784–794 (2007)
35. Alwosheel, A., van Cranenburgh, S., Chorus, C.G.: Is your dataset big enough? sample size requirements when using artificial neural networks for discrete choice analysis. J. Choice Model. **28**, 167–182 (2018)
36. Day, B., Pinto Prades, J.-L.: Ordering anomalies in choice experiments. J. Environ. Econ. Manag. **59**, 271–285 (2010)
37. Castelvecchi, D.: Can we open the black box of AI? Nat. News **538**, 20 (2016)

# Sparse, Interpretable and Transparent Predictive Model Identification for Healthcare Data Analysis

Hua-Liang Wei[1,2](✉)

[1] Automatic Control and Systems Engineering, University of Sheffield, Sheffield S1 3JD, UK
w.hualiang@sheffield.ac.uk

[2] INSIGNEO Institute for in Silico Medicine, University of Sheffield, Sheffield, UK

**Abstract.** Data-driven modelling approaches play an indispensable role in analyzing and understanding complex processes. This study proposes a type of sparse, interpretable and transparent (SIT) machine learning model, which can be used to understand the dependent relationship of a response variable on a set of potential explanatory variables. An ideal candidate for such a SIT representation is the well-known NARMAX (nonlinear autoregressive moving average with exogenous inputs) model, which can be established from measured input and output data of the system of interest, and the final refined model is usually simple, parsimonious and easy to interpret. The performance of the proposed SIT models is evaluated through two real healthcare datasets.

**Keywords:** System identification · Data-driven modelling · Prediction · Healthcare · Machine learning · NARMAX

## 1 Introduction

Data analysis and data modelling are perhaps the most commonly used approaches to acquiring insightful understanding and characterization of complex systems or phenomena where the changes of relevant factors or variables can be quantitatively measured and recorded but the inherent mechanisms or first principle models are not available. Traditionally speaking, data analysis is a technique to achieve insight into data (e.g. organization's data, business data, or whatever other data of interest). Tasks of data analysis range from data management, pre-processing and evaluation to data mining and data modelling. Data modelling provides necessary techniques for understanding and analyze data; one important aspect of data modelling is to understand the relationships between different features and factors of interest through mathematical, statistical and/or other quantitative analysis approaches. System identification and predictive modelling are two important classes of data-driven modelling techniques, the former concerns the development of mathematical models using data observed from dynamical systems, whilst the latter concerns the revealing of relationships of features of interest from any collected data.


I. Rojas et al. (Eds.): IWANN 2019, LNCS 11506, pp. 103–114, 2019.
https://doi.org/10.1007/978-3-030-20521-8_9

The past decades have witnessed tremendous developments and applications of system identification and predictive modelling techniques [1–4], which have been applied in diverse areas including space weather [6–13], climate and geophysics [14–18], medicine and healthcare [19–22], environments [23–26], societal wellbeing studies [27], and engineering [28, 29]. In concept, there are some subtle differences in system identification and predictive modelling. System identification concerns how to find a good model, from measured input and output data of a system of interest, that is as closely as possible to represent the input-output behavior. In doing so, it requires that the identified model should be as accurate as possible to characterize the underlying dynamics hidden in the data. Predictive modelling concerns the detection of dependence relationships among a group of variables by analyzing and modelling the relevant data; the goal is to determine if the change in some variables would affect the other variables, or if the attribute of some specific variables of interest (e.g. response, dependent or output variables) can be characterized by other variables (commonly known as explanatory, predictor, independent or input variables). Classification problem solving with either parametric or non-parametric data modelling methods is a typical example of the application of predictive modelling.

In more detail, system identification [1–4] is different from the conventional concept of predictive modelling [5] in that the former pays more attention on system dynamics. In system identification, it is usually assumed that the measurements or recorded data come from dynamic systems, whose current behaviour (often referred to as the system output) depends on previous or historical states of both the inputs (stimuli or driving signals) and the output itself. In system identification or dynamic modelling, all the input and output signals should strictly be recorded chronologically. Unlike in predictive modelling where the order of data records can be altered and normally the change will not affect the overall modelling performance, in system identification altering data record order is not allowed as the data records virtually reflect the change of the system behaviour with 'time' (this is an implicit independent variable in all time-invariant dynamic systems).

Despite the difference between system identification and predictive modelling, they share many similarities in model construction and algorithm implementations. For example, the commonly used methods of generalized linear models in predictive modelling, including model variable/term selection, model structure detection, model validation and so on, can be borrowed to deal with nonlinear dynamical models e.g. NARMAX (nonlinear autoregressive moving average with exogenous input) model with an appropriate modification, and vice versa. As highlighted in [4], NARMAX model can be considered a dynamically driven single hidden-layer recurrent neural network, which include many neural network structures e.g. radial basis function neural networks (RBFNs) as a special case.

While NARMAX model has been extensively applied in many interdisciplinary fields, its application potential in healthcare and related area has not yet been well explored. So, this study aims to introduce a type of sparse, interpretable and transparent (SIT) model for healthcare and related data modelling problems. We propose to use the NARMAX model, which possesses a number of attractive 'smart' properties, namely, simple and simulatable, meaningful, accountable, reproducible, and transparent. Two examples are provided to show the performance of the proposed approach.

## 2 Model Representation

A wide range of dynamic systems or processes can be represented using NARMAX model [4]. Taking the case of multiple inputs (designated by $u_1, u_2, \ldots, u_r$) and one output (designated by $y$) problem as an example, the NARMAX model that links the output $y$ to the inputs $u_1, u_2, \ldots, u_r$ is of the form:

$$\begin{aligned} y(t) = & f[y(t-1), \cdots, y(t-n_y), u_1(t-1), \cdots, u_1(t-n_u), u_2(t-1), \ldots, u_2(t-n_u), \ldots, \\ & u_r(t-1), \ldots, u_r(t-n_u), e(t-1), \cdots, e(t-n_e)] + e(t) \end{aligned} \tag{1}$$

where $y(t)$, $u(t)$ and $e(t)$ are the measured system output, input and noise sequences respectively at time instant $t$; $n_y$, $n_u$, and $n_e$ are the maximum lags for the system output, input and noise; $f[\cdot]$ is some non-linear function to be estimated from data. Note that the noise $e(t)$ is unmeasurable but can be replaced by the model prediction error in system identification procedure. The noise terms are included to accommodate the effects of measurement noise, modelling errors, and/or unmeasured disturbances.

Now define a group of new variables (i.e., lagged versions of the original input and output variables) as

$$x_m(t) = \begin{cases} y(t-m), & 1 \le m \le n_y \\ u(t-m+n_y), & n_y+1 \le m \le n_y+n_u \\ e(t-m+n_y+n_u), & n_y+n_u+1 \le m \le n \end{cases} \tag{2}$$

where $n = n_y + n_u + n_e$. Model (1) can then be written as

$$y(t) = f[x_1(t), x_2(t), \cdots, x_n(t)] + e(t) \tag{3}$$

In practice, many types of functions are available to approximate the unknown function $f[\cdot]$ in (1), including power-form polynomial models and rational models [28, 30], radial basis function (RBF) [8, 31, 32], and wavelet expansions [33]. In this study, power-form polynomial basis is considered. Expanding model (1) by defining the function $f[\cdot]$ to be a polynomial of degree $\ell$ gives the representation:

$$\begin{aligned} y(t) = & \theta_0 + \sum_{i_i=1}^{n} \theta_{i_1} x_{i_1}(t) + \sum_{i_1=1}^{n} \sum_{i_2=i_1}^{n} \theta_{i_1 i_2} x_{i_1}(t) x_{i_2}(t) + \cdots \\ & + \sum_{i_1=1}^{n} \cdots \sum_{i_\ell=i_{\ell-1}}^{n} \theta_{i_1 i_2 \cdots i_\ell} x_{i_1}(t) x_{i_2}(t) \cdots x_{i_\ell}(t) + e(t) \end{aligned} \tag{4}$$

where $\theta_{i_1 i_2 \cdots i_m}$ are parameters. The degree of a multivariate polynomial is defined as the highest order among the terms. For example, the degree of the polynomial $h(x_1, x_2) = a_1 x_1 + a_2 x_1 x_2 + a_3 x_1^2 + a_4 x_1 x_2^2$ is $\ell = 1 + 2 = 3$, which is determined by the last term, $a_4 x_1 x_2^2$. Similarly, a polynomial model with degree $\ell$ means that the order of each term in the model is not higher than $\ell$. Note that the polynomial representation (4) belongs to the family of linear-in-the-parameters (LIP) but nonlinear-in-the-variables (NIV) models.

In many applications, the noise signal $e(t)$ in the NARMAX model (1) can be reasonably assumed to be an i.i.d. or white noise. In this case, model (1) can be reduced to a NARX model which only involves lagged input and output variables as below:

$$x_m(t) = \begin{cases} y(t-m), & 1 \leq m \leq n_y \\ u(t-m+n_y), & n_y+1 \leq m \leq n = n_y + n_u \end{cases} \tag{5}$$

With the above definition, (1) can easily be re-arranged to a LIP-NIV form.

Note that in nonlinear dynamical system identification, attention is always focused on building mathematical models, from experimental data, that can represent the inherent dynamics or system input-output relationship as accurate as possible. From a neural network perspective, a NARX model can be considered a dynamically driven 1-hidden-layer neural network, which is referred to as recurrent NARX network (R-NARX-NN) [4]. For example, for a NARX model of nonlinear degree $\ell = 3$, define

$$F_1[\mathbf{x}(t);\theta] = \sum_{i=1}^{n} \theta_i x_i(t)$$

$$F_2[\mathbf{x}(t);\theta] = \sum_{i=1}^{n}\sum_{j=i}^{n} \theta_{ij} x_i(t) x_j(t)$$

$$F_3[\mathbf{x}(t);\theta] = \sum_{i=1}^{n}\sum_{j=i}^{n}\sum_{k=j}^{n} \theta_{ijk} x_i(t) x_j(t) x_k(t)$$

The recurrent neural network structure of the NARX model is shown in Fig. 1.

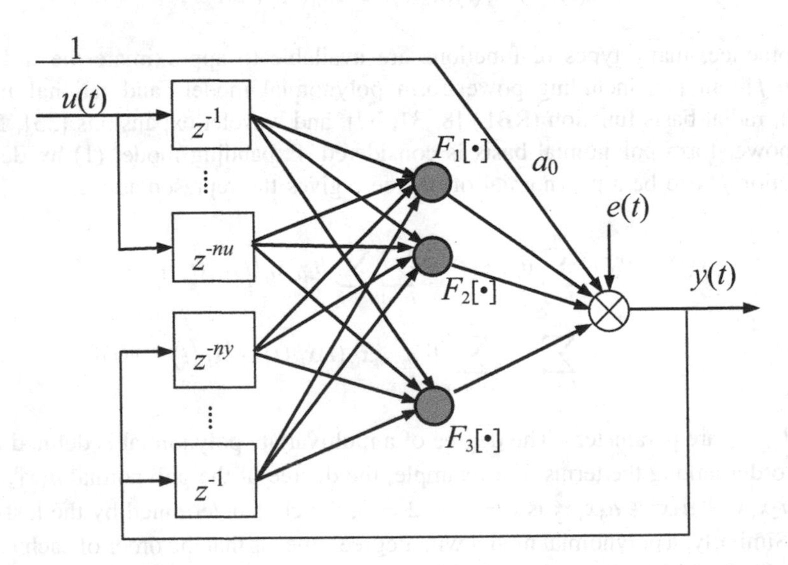


**Fig. 1.** The structure of the NARX model, which is a typical recurrent neural network.

## 3 Sparse Dictionary Learning and NARMAX Model Estimation

For convenience of description, we first focus on the NARX model estimation, for which the procedure starts with setting up a few parameters, namely, the maximum lags $n_y$ and $n_u$, and the nonlinearity degree $\ell$. Now, take a simple case as an example, where a system only has one input and one output signal, and assume $n_y = 1$, $n_u = 1$ and $\ell = 3$. We define the following distributed lag sub-dictionaries:

$$D_0 = \{1\} \text{ (constant term)}$$
$$D_1 = \{y(t-1), u(t-1)\},$$
$$D_2 = \left\{ \begin{array}{l} y(t-1)y(t-1) \\ y(t-1)u(t-1) \\ u(t-1)u(t-1) \end{array} \right\},$$
$$D_3 = \left\{ \begin{array}{l} y(t-1)y(t-1)y(t-1) \\ y(t-1)y(t-1)u(t-1) \\ y(t-1)u(t-1)u(t-1) \\ u(t-1)u(t-1)u(t-1) \end{array} \right\}$$

The four sub-dictionaries will then be used to form a dictionary:

$$D = D_0 + D_1 + D_2 + D_3 \tag{6}$$

The task of finding a good model is equivalent to selecting important model terms from the dictionary $D$, which well represents the input-output relation of the system.

For complex cases (e.g. with many inputs, and with large time lags), we can define $D_0$, $D_1$, $D_2$, $D_3$, etc. in the same way. Note that the total number of potential model terms in a polynomial NARX model is $M = (n + \ell)!/[n!\ell!]$. For example, if $\ell = 3$, $n_y = 10$, $n_u = 5$, then $M = (15 + 3)!/(15!3!) = 153$. For large $n_y$ and $n_u$, the number of initial candidate model terms included in the initial full model can be very large. However, for a given system, the $M$ candidate model terms in $D$ are not necessarily equally important for representing the system. Some terms may be irrelevant or only make very tiny contribution to explaining the system input-output behavior, thus should not be included in the model, because an inclusion of irrelevant model terms can generally lead to model overfitting, and may adversely make it more difficulty to reveal the true system dynamics. The forward regression orthogonal least squares (FROLS) algorithm [4, 31, 32] and its variants [34–38] provide an efficient, powerful tool for nonlinear significant model term selection and model structure detection. Detailed discussions on the FROLS algorithm can be found in [4, 39, 40].

This study uses the FROLS algorithm with ridge regularization to select significant model terms and determine the model structure. Once a NARX model structure is determined, the noise variables $e(t-1), \ldots, e(t-n_e)$ and the model terms involving these noise variables are accommodated to the NARX model, to develop a NARMAX model structure. Note that the noise signal $e(t)$ is not observable but can only be estimated from the prediction errors: $\xi(t) = y(t) - \hat{y}(t)$, where $\hat{y}(t)$ is the model prediction at time instant $t$. Detailed discussions may be found in [4].

# 4 Case Studies and Real Applications

## 4.1 The Relation Between Influenza-Like Illness Incidence Rate and Deaths

The weekly influenza-like illness (ILI) incidence rate and deaths data were acquired from the Office for National Statistics (ONS), The Royal College of General Practitioners Research and Surveillance Centre and Public Health Wales. The dataset contains a total of 991 weekly records starting in week 31 of 1999 and ending in week 30 of 2018. The raw data are plotted in Fig. 2.

The objective here is twofold. One is to quantify the relation between the week mortality and the ILI incidence rate, and another is to do a week ahead prediction of the death mortality. We consider two types of models: one using autoregressive variables and another one without using autoregressive variables. For both cases, the 991 data points are split into two parts: the first 600 samples are used for model training and the remaining 391 are used for model testing.

**The Model Without Including Autoregressive Variables**

Volterra model is special case of NARMAX model, without including autoregressive variables. The best Volterra model identified by the FROLS algorithm with Ridge regularization is:

$$\left.\begin{aligned} y(t) = {} & 8636.0572 + 64.6550u(t-1) + 55.6953u(t-4) \\ & -0.5110u^2(t-1) + 0.0015u^2(t-1) - 0.8304u^2(t-4) - 0.0026u^3(t-4) \end{aligned}\right\} \tag{7}$$

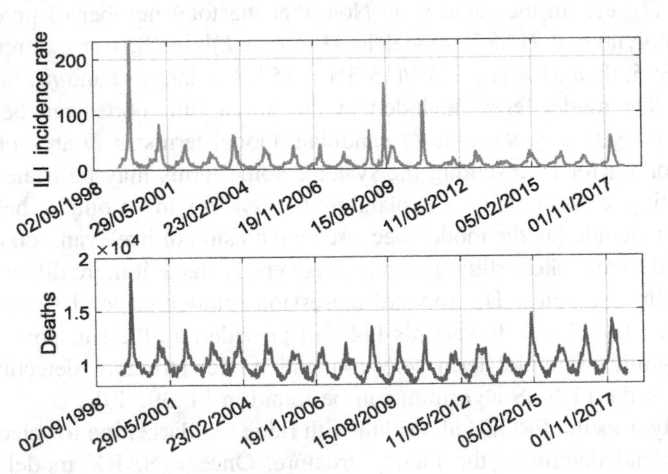


**Fig. 2.** Weekly influenza-like illness (ILI) incidence rate and deaths, England and Wales, between week 31 of 1999 and week 30 of 2018.

where $u(t)$ represents the weekly ILI incidence rate and $y(t)$ represents the number of weekly deaths. A comparison of the model predicted deaths and the corresponding true values, on the training and test data sets, are shown in Figs. 3 and 4, respectively. It can be seen that the simple NARMAX model (7) shows an overall very good prediction performance. Note that the model predictions of the short period centred on 15 August 2009 are quite bad, this is because there is some extremely odd behaviour in then ILI incidence rate as shown in Fig. 2.

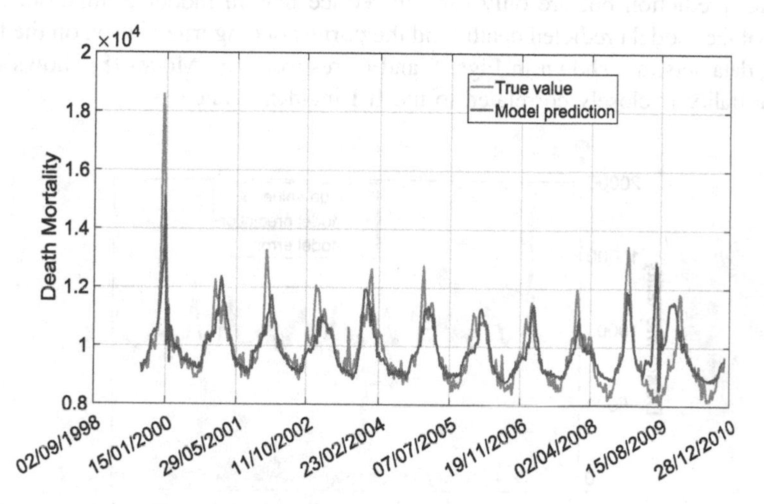


**Fig. 3.** A comparison of the model prediction with the corresponding true number of deaths, on the training dataset of the period between week 31 of 1993 and week 47 of 2010.

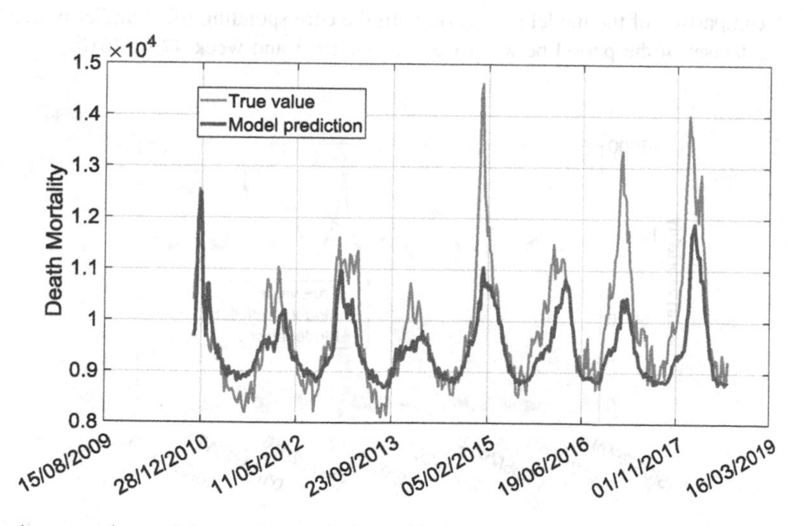


**Fig. 4.** A comparison of the model prediction with the corresponding true number of deaths, on the test dataset of the period between week 48 of 2010 and week 30 of 2018.

**The Model with Autoregressive Variables**

With the same training data, the FROLS algorithm produces the following NARMAX model:

$$y(t) = 616.435147 + 0.927840y(t-1) - 0.114871u(t-1)u(t-3) + 10.535455u(t-1) \quad (8)$$

Note that all the model terms involving noise variables such as $u(t-1)e(t-1)$ are omitted and not included in the final model, because all these noise terms are not useful for model prediction but are only used to reduce bias in model estimation. A comparison of the model predicted deaths and the corresponding true values, on the training and test data sets, are shown in Figs. 5 and 6, respectively. Model (8) shows that the death mortality is closely correlated to the ILI incidence rate.

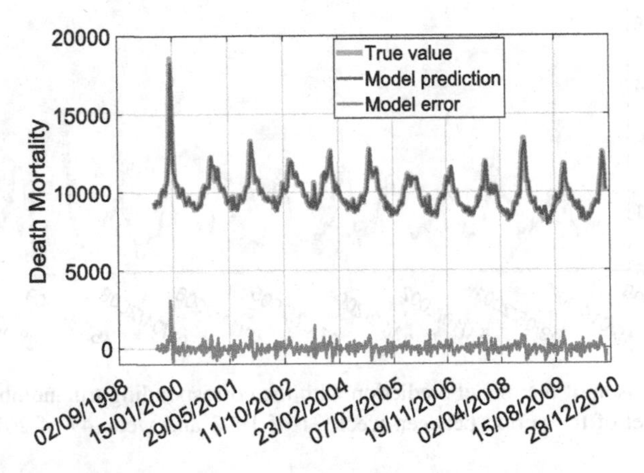


**Fig. 5.** A comparison of the model prediction with the corresponding true number of deaths, on the training dataset of the period between week 31 of 1993 and week 47 of 2010.

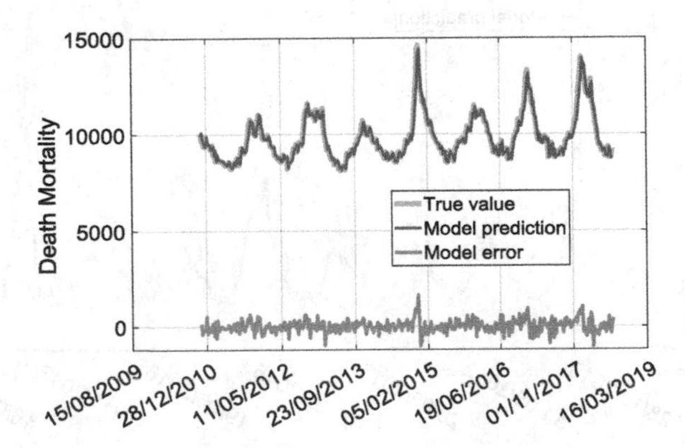


**Fig. 6.** A comparison of the model prediction with the corresponding true number of deaths, on the test dataset of the period between week 48 of 2010 and week 30 of 2018.

### 4.2 Analysis of Beijing Air Quality

A dataset of Beijing air quality is obtained from http://www.tianqihoubao.com/aqi/. The dataset contains six variables, namely, PM2.5 (μg/m$^3$), PM10 (μg/m$^3$), SO2 (μg/m$^3$), NO2 (μg/m$^3$), CO (mg/m$^3$), O3 (μg/m$^3$), all of which were measured daily. Here, in this study we are interested in understanding how PM2.5 depends on or is delated to the other five variables. We therefore treat PM2.5 as an output variable and the other five variables are treated to be the inputs.

We use 732 sample data of the period from 1 January 2016 to 31 December 2017 to train the model, and use the data of the period of 1 January 2018–31 January 2019 to test the model performance. The identified model is:

$$\left.\begin{aligned} PM_{2.5}(t) =& 0.43615PM_{10}(t) + 38.07453CO(t) - 17.90321 \\ &+ 0.31627CO(t)O_3(t) - 0.00119[O_3(t)]^2 - 7.48672[CO(t)]^2 \\ &+ 0.11175PM_{10}(t)CO(t) + 0.00033[PM_{10}(t)]^2 \\ &+ 0.01899NO_2(t)NO_2(t-1) - 0.007294PM_{10}(t)NO_2(t-1) \\ &+ 0.06749CO(t-1)O_3(t-1) + 0.00100SO_2(t)O_3(t-2) \\ &+ 0.01151PM_{10}(t)SO_2(t-1) - 0.02525SO_2(t-1)NO_2(t) \\ &+ 0.03123PM_{10}(t)CO(t-2) - 0.39556NO_2(t) \\ &+ 0.26760SO_2(t-2)CO(t) - 0.72491SO_2(t) \end{aligned}\right\} \quad (9)$$

The values of RMSE (root mean squared error), MAE (mean absolute error), Correlation (between the measurement and model prediction), and R$^2$ (coefficient of determination) of model (9) over the training data are 15.4160, 10.8202, 0.9656, and

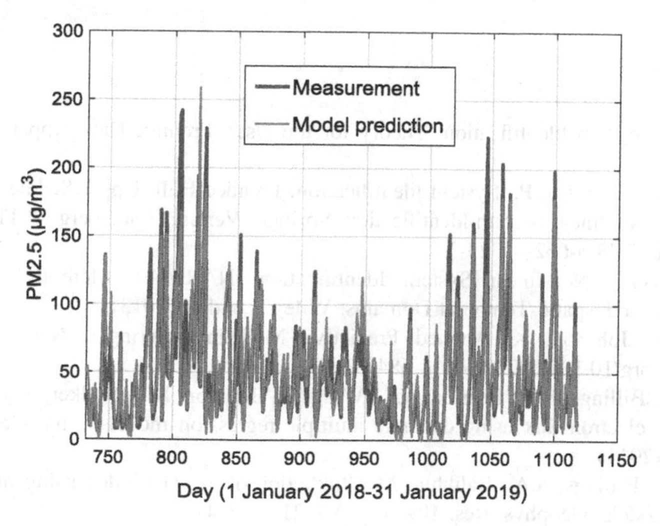


**Fig. 7.** A comparison between the model predicted values and the corresponding measurements of Beijing air quality (PM2.5) over the test dataset (RMSE = 16.9764, MAE = 11.2995, r = 0.9193 and R$^2$ = 0.8450).

0.9924, respectively, and 16.9764, 11.2995, 0.9193 and 0.8450, respectively, over the test data. These statistics show that PM2.5 has a very strong relation or dependence with other five variables. A comparison between the model predictions and the corresponding true measured values over test datasets is shown in Fig. 7.

# 5 Conclusion

Dara-driven modelling and data based quantitative analysis play a key instrumental role in knowledge discovery from healthcare data. In many application scenarios, it is interested in knowing how a variable is explicitly related to other variables or factors. To answer this question, this study proposed a type of SIT (sparse, interpretable and transparent) approach, called NARMAX model, which possess 'smart' properties: simple/sparse/simulatable, meaningful, accountable, reproducible, and transparent. SIT-NARMAX model can be written as a LIP-NIV (linear-in-the-parameters and nonlinear-in-the-variables) form, which can easily be estimated using the state-of-the-art linear regression methods. By applying the forward regression orthogonal least squares (FROLS) algorithms to this type of model, it usually leads to parsimonious representations for most real data modelling problems. The main advantage of the proposed modelling approach is that the resulting model is not only parsimonious but also can be written down and easily interpreted. As illustrated in the case studies, the proposed approach provides a powerful and effective tool for dealing with real healthcare and related data modelling problems.

**Acknowledgments.** This work was supported in part by the Engineering and Physical Sciences Research Council (EPSRC) under Grant EP/I011056/1, the Platform Grant EP/H00453X/1, and the EU Horizon 2020 Research and Innovation Programme Action Framework under grant agreement 637302.

# References

1. Ljung, L.: System Identification: Theory for the User. Prentice-Hall, Upper Saddle River (1987)
2. Sodestrom, T., Stoica, P.: System Identification. Prentice Hall, Upper Saddle River (1988)
3. Nelles, O.: Nonlinear System Identification. Springer-Verlag, Heidelberg (2011). https://doi.org/10.1007/978-3-662-04323-3
4. Billings, S.A.: Non-linear System Identification: NARMAX Methods in the Time, Frequency, and Spatio-Temporal Domains. Wiley, London (2013)
5. Kuhn, M.: Johnson, K: Applied Predictive Modeling. Springer, New York (2013). https://doi.org/10.1007/978-1-4614-6849-3
6. Wei, H.L., Billings, S.A., Sharma, A.S., Wing, S., Boynton, R.J., Walker, S.N.: Forecasting relativistic electron flux using dynamic multiple regression models. Ann. Geophys. **29**(2), 415–420 (2011)
7. Wei, H.L., Billings, S.A., Balikhin, M.: Prediction of the Dst index using multiresolution wavelet models. Geophys. Res. **109**(A7), A07212 (2004)

8. Wei, H.L., Zhu, D.Q., Billings, S.A., Balikhin, M.A.: Forecasting the geomagnetic activity of the Dst index using multiscale radial basis function networks. Adv. Space Res. **40**(12), 1863–1870 (2007)
9. Balikhin, M.A., Boynton, R.J., Walker, S.N., et al.: Using the NARMAX approach to model the evolution of energetic electrons fluxes at geostationary orbit. Geophys. Res. Lett. **38**(18), L18105 (2011)
10. Boynton, R.J., Balikhin, M.A., Billings, S.A.: Using the NARMAX OLS-ERR algorithm to obtain the most influential coupling functions that affect the evolution of the magnetosphere. Geophys. Res. Space Phys. **116**, A05218 (2011)
11. Gu, Y., Wei, H.L., Boynton, R.J., Walker, S.N., Balikhin, M.A.: System identification and data-driven forecasting of AE index and prediction uncertainty analysis using a new cloud-NARX model. J. Geophys. Res. Space Phys. **124**(1), 248–263 (2019)
12. Boynton, R., Balikhin, M., Wei, H.-L., Lang, Z.-Q.: Applications of NARMAX in space weather. In: Machine Learning Techniques for Space Weather, pp. 203–236 (2018)
13. Camporeale, E.: The challenge of machine learning in space weather nowcasting and forecasting. arXiv preprint arXiv:1903.05192 (2019)
14. Wei, H.-L., Billings, S.A.: An efficient nonlinear cardinal B-spline model for high tide forecasts at the Venice Lagoon. Nonlinear Process. Geophys. **13**(5), 577–584 (2006)
15. Karsten, S., Nitesh, V.C., Auroop, R.G.: Complex networks as a unified framework for descriptive analysis and predictive modeling in climate science. Stat. Anal. Data Min. **4**(5), 497–511 (2011)
16. Bigg, G.R., Wei, H.L., Wilton, D.J., et al.: A century of variation in the dependence of Greenland iceberg calving on ice sheet surface mass balance and regional climate change. Proc. R. Soc. A: Math. Phys. Eng. Sci. **470**(2166), 2013066 (2014)
17. Pearson, R.G., Dawson, T.P.: Predicting the impacts of climate change on the distribution of species: are bioclimate envelope models useful? Glob. Ecol. Biogeogr. **12**(5), 361–371 (2003)
18. Helmuth, B.: From cells to coastlines: how can we use physiology to forecast the impact of climate change? J. Exp. Biol. **212**, 753–760 (2009)
19. Billings, C.G., Wei, H.-L., Thomas, P., Linnane, S.J., Hope-Gill, B.D.M.: The prediction of in-flight hypoxaemia using non-linear equations. Respir. Med. **107**(6), 841–847 (2013)
20. Shamanand, J., Karspeck, A.: Forecasting seasonal outbreaks of influenza. Proc. Nat. Acad. Sci. USA **109**(50), 20425–20430 (2012)
21. Zhang, Y., Bambrick, H., Mengersen, K., Tong, S., Hu, W.: Using Google trends and ambient temperature to predict seasonal influenza outbreaks. Environ. Int. **117**, 284–291 (2018)
22. Osthus, D., Gattiker, J., Priedhorsky, R., Del Valle, S.Y.: Dynamic Bayesian influenza forecasting in the United States with hierarchical discrepancy. Bayesian Analysis (2019, in press)
23. Pisoni, E., Farina, M., Carnevale, C., Piroddi, L.: Forecasting peak air pollution levels using NARX models. Eng. Appl. Artif. Intell. **22**(4–5), 593–602 (2009)
24. Feng, X., Li, Q., Zhu, Y., Hou, J., Jin, L., Wang, J.: Artificial neural networks forecasting of PM2.5 pollution using air mass trajectory based geographic model and wavelet transformation. Atmos. Environ. **107**, 118–128 (2015)
25. Bai, Y., Li, Y., Wang, X., Xie, J., Li, C.: Air pollutants concentrations forecasting using back propagation neural network based on wavelet decomposition with meteorological conditions. Atmos. Pollut. Res. **7**, 557–566 (2016)
26. Sun, W., Sun, J.: Daily PM2.5 concentration prediction based on principal component analysis and LSSVM optimized by cuckoo search algorithm. J. Environ. Manag. **188**, 144–152 (2017)

27. Gu, Y., Wei, H.-L.: Significant indicators and determinants of happiness: evidence from a UK survey and revealed by a data-driven systems modelling approach. Soc. Sci. **7**(4), 53 (2018)
28. Zhang, W., Zhu, J., Gu, D.: Identification of robotic systems with hysteresis using nonlinear autoregressive exogenous input models. Int. J. Adv. Robot. Syst. **14**(3), 1729881417705845 (2017)
29. Santos, R.F., Pereira, G.A.S., Aguirre, L.A.: Learning robot reaching motions by demonstration using nonlinear autoregressive models. Robot. Auton. Syst. **107**, 182–195 (2018)
30. Billings, S.A., Zhu, Q.M.: Rational model identification using an extended least-squares algorithm. Int. J. Control **54**(3), 529–546 (1991)
31. Chen, S., Billings, S.A., Luo, W.: Orthogonal least squares methods and their application to non-linear system identification. Int. J. Control **50**(5), 1873–1896 (1989)
32. Chen, S., Cowan, C., Grant, P.: Orthogonal least squares learning algorithm for radial basis function networks. IEEE Trans. Neural Netw. **2**(2), 302–309 (1991)
33. Billings, S.A., Wei, H.-L.: The wavelet-NARMAX representation: a hybrid model structure combining polynomial models with multiresolution wavelet decompositions. Int. J. Syst. Sci. **36**(3), 137–152 (2005)
34. Chen, S., Hong, X., Luk, B.L., Harris, C.J.: Orthogonal-least-squares regression: a unified approach for data modelling. Neural Comput. **72**(10), 2670–2681 (2009)
35. Zhang, L., Li, K., Bai, E.-W., Irwin, G.W.: Two-stage orthogonal least squares methods for neural network construction. IEEE Trans. Neural Netw. Learn. Syst. **26**(8), 1608–1621 (2014)
36. Guo, Y., Guo, L.Z., Billings, S.A., Wei, H.-L.: Identification of nonlinear systems with non-persistent excitation using an iterative forward orthogonal least squares regression algorithm. Int. J. Model. Ident. Control **23**, 1–7 (2015)
37. Yaghoobi, M., Davies, M. E.: Fast non-negative orthogonal least squares. In: Proceedings of European Signal Processing Conference, pp. 479–483, Nice, France (2015)
38. Li, Y., Cui, W.G., Guo, Y.Z., et al.: Time-varying system identification using an ultra-orthogonal forward regression and multiwavelet basis functions with applications to EEG. IEEE Trans. Neural Netw. Learn. Syst. **29**(7), 2960–2972 (2018)
39. Wei, H.-L., Billings, S.A., Liu, J.: Term and variable selection for nonlinear system identification. Int. J. Control **77**, 86–110 (2004)
40. Wei, H.-L., Billings, S.A.: Model structure selection using an integrated forward orthogonal search algorithm assisted by squared correlation and mutual information. Int. J. Model. Ident. Control **3**(4), 341–356 (2008)

# Use of Complex Networks for the Automatic Detection and the Diagnosis of Alzheimer's Disease

Aruane Mello Pineda[1], Fernando M. Ramos[2], Luiz Eduardo Betting[3], and Andriana S. L. O. Campanharo[1](✉)

[1] Institute of Biosciences, Department of Biostatistics, São Paulo State University (UNESP), Botucatu, São Paulo, Brazil
{aruane.pineda,andriana.campanharo}@unesp.br

[2] Laboratory for Computing and Applied Mathematics, National Institute for Space Research (INPE), São José dos Campos, São Paulo, Brazil
fernando.ramos@inpe.br

[3] Institute of Biosciences, Department of Neurology, Psychology and Psychiatry, Botucatu Medical School, São Paulo State University (UNESP), Botucatu, São Paulo, Brazil
luiz.betting@unesp.br

**Abstract.** Alzheimer's disease (AD) is classified as a chronic neurological disorder of the brain and affects approximately 25 million elderly individuals worldwide. This disorder leads to a reduction in people's productivity and imposes restrictions on their daily lives. Studies of AD often rely on electroencephalogram (EEG) signals to provide information on the behavior of the brain. Recently, a map from a time series to a network has been proposed and that is based on the concept of transition probabilities; the series results in a so-called "quantile graph" (QG). Here, this map, which is also called the QG method, is applied for the automatic detection of healthy patients and patients with AD from recorded EEG signals. Our main goal is to illustrate how the differences in dynamics in the EEG signals are reflected in the topology of the corresponding QGs. Based on various network metrics, namely, the clustering coefficient, the mean jump length and the betweenness centrality, our results show that the QG method can be used as an effective tool for automated diagnosis of Alzheimer's disease.

## 1 Introduction

Alzheimer's disease (AD) can be understood as a degenerative and progressive dementia of the Central Nervous System. It is irreversible and with unknown cause. This disease is mainly characterized by a progressive intellectual deterioration, loss of memory and disorientation in time and space [5]. AD is the leading dementia among older people over 65 and affects approximately 25 million individuals worldwide [19]. Currently, accurate diagnosis of AD can be only made through the examination of the brain tissue obtained by biopsy or necropsy [12].


I. Rojas et al. (Eds.): IWANN 2019, LNCS 11506, pp. 115–126, 2019.
https://doi.org/10.1007/978-3-030-20521-8_10

Since only after the patient's death it is possible to be sure that he or she had AD, the approximate diagnosis is made by excluding other causes of dementia. In parallel, studies have been developed for the study of AD and other diseases with the use of Electroencephalography (EEG), which is a technique for the recording of electrical activity arising from the human brain with high temporal resolution [1,3,6,7,24]. However, the study of AD using EEG data is recent and still a challenge. In this sense, various techniques have been proposed for the AD detection based on Fast Fourier Transform (FFT) [2,16], Wavelet Transform (WT) [2,20,27], Phase-Space Reconstruction [2,15,23], Eigenvector Methods (EMs) [25,26], Time Frequency Distributions (TFDs) [21] and the Auto-Regressive Method (ARM) [29]. Since these techniques require assumptions about stationarity, high time and frequency resolution, length of the signal and/or noise level, there is considerable research being performed on the development of novel methods for the analysis of EEG time series.

In recent past, we have proposed a new and simple method to convert a time series to a graph (also called quantile graph or network) and showed that such a graph inherits some of the properties of the time series such as periodic time series resulting in regular networks, random time series resulting in random graphs and pseudo-periodic time series resulting in small-world networks [10]. We also showed that this method can be used to estimate the Hurst exponent of fractional motions and noises [9] and to detect differences in the data structures of physiological signals of healthy and unhealthy subjects [6,8,10]. Finally, we have shown that this method can not only differentiate epileptic from normal data but also distinguish the different abnormal stages/patterns of a seizure, such as pre-ictal (EEG changes that precede a seizure) and ictal (EEG changes during a seizure) [7].

In this paper, we investigate whether the proposed method can be used as a tool for diagnosing AD from recorded EEG signals. The remainder of this paper is organized as follows: after this introduction, Sect. 2 describes the QG method for mapping a time series into a network. Network measures for the characterization, analysis and discrimination of complex networks are presented in Sect. 3. Section 4 describes the data set that is used in this study. The results are presented and discussed in Sect. 5, and the conclusions are presented in Sect. 6.

## 2 Methods

Let $\mathcal{M}$ be a map to convert a time series $X = \{x(t)|t \in \mathbb{N}, x(t) \in \mathbb{R}\}$ into a network $g = \{\mathcal{N}, \mathcal{A}\} \in \mathcal{G}$ with $\mathcal{N}$ nodes (or vertices) and $\mathcal{A}$ arcs (or edges). The $Q$ quantiles of a time series, which are defined by the cutting points that divide $X$ into $Q-1$ equally sized intervals, are identified by ranking $X$ and splitting this ranking through the $Q-1$ intervals [28]. Once the $Q$ quantiles have been identified, each value in $X$ is mapped to its corresponding $q_i$ for $i = 1, 2, \ldots, Q$. After that, $\mathcal{M}$ assigns each quantile $q_i$ to a node $n_i \in \mathcal{N}$ in the corresponding network. Two nodes $n_i$ and $n_j$ are connected in the network with a weighted arc $(n_i, n_j, w_{ij}^k) \in \mathcal{A}$ whenever two values $x(t)$ and $x(t+k)$ belong to quantiles $q_i$ and $q_j$, respectively, with $t = 1, 2, \ldots, T$ and the time differences $k = 1, \ldots, k_{max} < T$. Each weight $w_{ij}^k$ in the weighted directed adjacency matrix, which is denoted as $A_k$, is equal to the number of times a value in quantile $q_i$ at time $t$ is followed by a point in quantile $q_j$ at time $t+k$. Therefore, repeated transitions through the same arc increase the value of the corresponding weight [7]. With proper normalization, the weighted directed adjacency matrix $A_k$ becomes a Markov transition matrix $W_k$, with $\sum_j^Q w_{ij}^k = 1$ [9]. In this work, the number of quantiles has been defined as $Q \approx 2T^{1/3}$ [17] (Fig. 1).

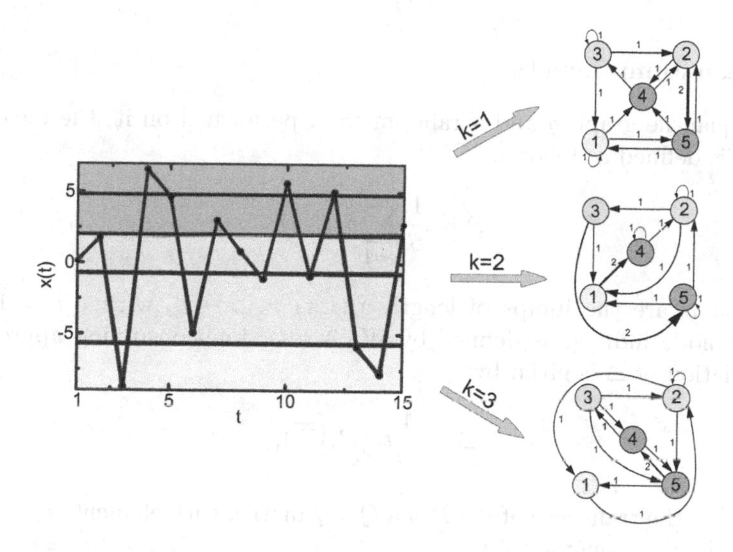


**Fig. 1.** Illustration of the QG method for a time series $X$ with $T = 15$ time points, $Q = 5$ quantiles and $k = 1, 2$ and 3. The quantile intervals are given by $[x(0), x(3)]$, $[x(3), x(6)]$, $[x(6), x(9)]$, $[x(9), x(12)]$ and $[x(12), x(15)]$ for the ordered data, i.e., $[-8.740, -4.912]$, $[-4.912, 0.128]$, $[0.128, 2.743]$, $[2.743, 5.000]$ and $[5.000, 6.571]$.

## 3 Network Measures

The characterization, analysis and discrimination of complex networks rely on the use of measures that are capable of expressing their most relevant topological features. Based on the values of $A$ and $W$, we describe the network measures that are used in Sect. 5, that are, the clustering coefficient ($CC$), the mean jump length ($\Delta$) and the betweenness centrality ($BC$).

### 3.1 Clustering Coefficient

In graph theory, the clustering coefficient is a measure of the degree to which nodes in a graph tend to cluster together. The literature presents various attempts at developing or implementing a clustering coefficient for weighted networks. In the approach used here [22], the clustering coefficient of a given node $n_i$ is given by:

$$CC_i = \frac{1}{s_i(d_i - 1)} \frac{\sum_{j,d}(w_{ij} + w_{id})}{2}(a_{ij}a_{jd}a_{id}), \tag{1}$$

where $w_{ij}$ is the element from the weighted matrix $W$ and $a_{ij} = 1$ if there is an edge between nodes $n_i$ and $n_j$, and 0 otherwise. $d_i$ is the total degree of node $n_i$, and $s_i$ is the strength of connectivity of node $n_i$. The value of $CC$ is given by the average of the local clustering coefficients of all the nodes.

### 3.2 Mean Jump Length

Given a quantile graph $g$ and a random walk performed on it, the mean jump length $\Delta$ is defined as follows:

$$\Delta = \frac{1}{S}\sum_{s=1} \delta_s(i,j), \tag{2}$$

where $s = S$ are the jumps of length $\delta_s(i,j) = |i - j|$, with $i, j = 1, \ldots, Q$ being the node indices, as defined by $W$. A less time-consuming approach for the calculation of $\Delta$ is given by:

$$\Delta = \frac{1}{Q} tr(PW^T), \tag{3}$$

where $W^T$ is the transpose of $W$, $P$ is a $Q \times Q$ matrix with elements $p_{i,j} = |i-j|$, and $tr$ is the trace operation [6].

### 3.3 Betweenness Centrality

The betweenness centrality of a given node $n_u$ is given by:

$$BC_{n_u} = \sum_{ij} \frac{\sigma(n_i, n_u, nj)}{\sigma(n_i, n_j)}, \tag{4}$$

where $\sigma(n_i, n_u, n_j)$ is the number of shortest paths between nodes $n_i$ and $n_j$ that pass through node $n_u$, $\sigma(n_i, n_j)$ is the total number of shortest paths between $n_i$ and $n_j$, and the sum is calculated over all pairs $n_i$, $n_j$ of distinct nodes [11,13]. The value of $BC$ is given by the average of the local betweenness centralities of all the nodes.

## 4 Data

In this study, we use an artifact-free EEG database that was obtained using a strict protocol from Florida Hospital, Orlando, US and provided by Florida

State University. The database contains 19-channel EEGs recorded at a sampling frequency of 128 Hz and includes four sets (denoted as A, B, C and D) under two rest states: eyes open (sets A and C) and eyes closed (sets B and D). Sets A and B consist of 10 healthy elderly (control group; aged 72 ± 11 years) with no history of neurological or psychiatric disorder. Sets C and D consist of 10 probable AD patients (aged 69 ± 16 years) diagnosed through National Institute of Neurological and Communicative Disorders and Stroke and the Alzheimer's Disease and Related Disorders Association (NINCDS-ADRDA) and Diagnostic and Statistical Manual of Mental Disorders (DSM)-III-R criteria (Pritchard et al. 1991). Eight-second EEG (1,024 time samples) segments free from eye blink, motion, and myogenic artifacts are extracted from the EEG recordings (one segment for each subject). The EEGs are band-limited to the range of 1–30 Hz during the EEG recording and preprocessing stages and were recorded using the traditional 10–20 standard system. Exemplary EEG's are depicted in Fig. 2. EEG time series recordings of healthy patients during the relaxed state and with eyes closed (Fig. 2(B)) show a predominant physiological rhythm, namely, the "alpha rhythm", in a frequency range of 8–13 Hz. In contrast, broader frequency characteristics are obtained for patients with eyes open (Fig. 2(A)). An increase of delta (range of 1–4 Hz) and theta (range of 4–8 Hz) power, and a decrease of alpha and beta power can be found at EEG time series recordings of AD patients (Figs. 2(C) and (D)) [4].

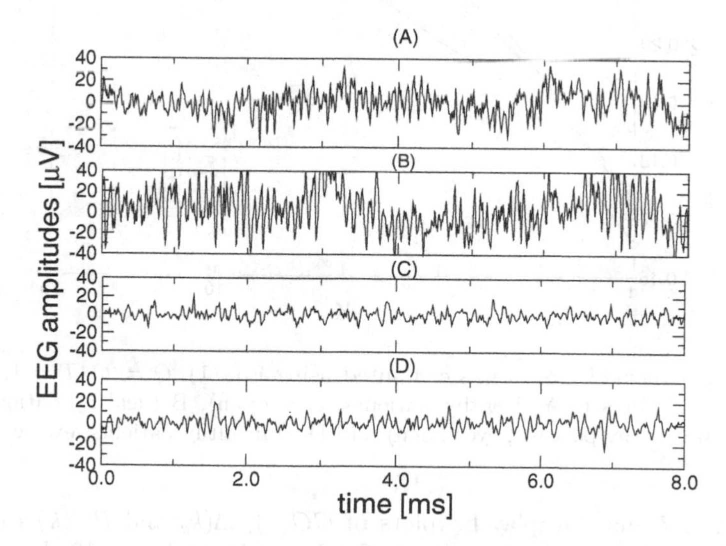


**Fig. 2.** Exemplary EEG signals from each of the four sets. From top to bottom: set A (healthy patient with eyes open), B (healthy patient with eyes closed), C (patient with AD and eyes open), D (patient with AD and eyes closed).

## 5 Results

We apply the QG method to the problem of differentiating patients with AD from normal data. Since all time series are of equal length ($T = 1{,}024$), we used $Q = 2(1{,}204)^{1/3} \approx 20$ and $k = 1, 2, \ldots, 25$ in all computations. Thus, we mapped 20 signals into 500 quantile graphs (or 500 $A_k$ matrices), and therefore, we obtained 500 $W_k$ matrices with $Q^2 = 400$ elements each. After that, for each set and for a specified $k$, we calculated the median over all $A_k$ matrices and obtained the weighted directed adjacency matrix of medians and the Markov transition matrix of medians.

For all sets, we computed $CC(k)$, $\Delta(k)$ and $BC(k)$ versus $k$ using Eqs. (1), (3) and (4), respectively (Figs. 3, 4 and 5). Observe in all cases that the curves for healthy patients (A and B) and patients with AD (C and D) form two distinct clusters at approximately $k = 12$ for $CC(k)$ (Fig. 3), $k = 12$ for $\Delta(k)$ (Fig. 4) and $k = 13$ for $BC(k)$ (Fig. 5). For all metrics, for $k > 20$, as correlations between QG nodes disappear, all curves tend to merge into one.

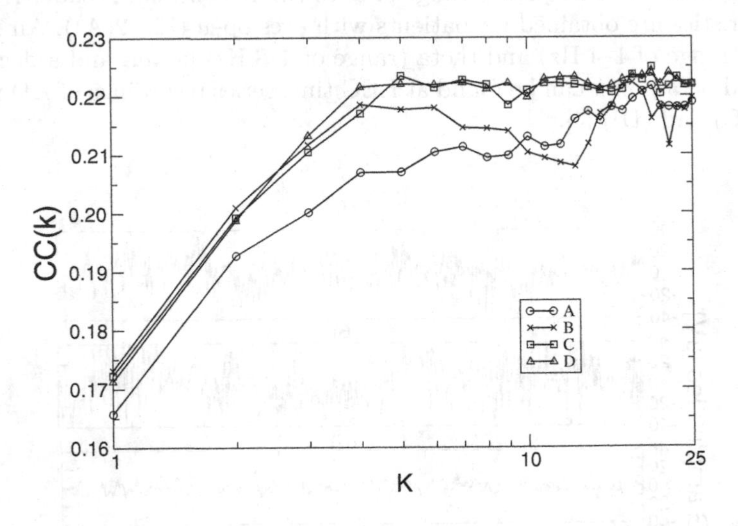


**Fig. 3.** $CC(k)$ versus $k$, which was computed using Eq. (1), $Q = 20$, $T = 1{,}024$ and $k = 1, 2, \ldots, 25$ for sets A (healthy patients, eyes open), B (healthy patients, eyes closed), C (unhealthy patients, eyes open) and D (unhealthy patients, eyes closed).

Figures 6, 7 and 8 display boxplots of $CC(k)$, $\Delta(k)$ and $BC(k)$ that were computed over 5 segments for sets A, B, C and D, and $k = 12$, $k = 12$ and $k = 13$, respectively. Regardless of the network measure that was used, the QG method enables robust discrimination between: (i) healthy patients with eyes open (set A) and closed (set B); (ii) patients with AD with eyes open (set C) and closed (set D), and more important, (iii) healthy patients (with eyes open or closed) and patients with AD (with eyes open or closed). Moreover, in all cases there is a statistically significant difference for a 95% confidence interval (CI)

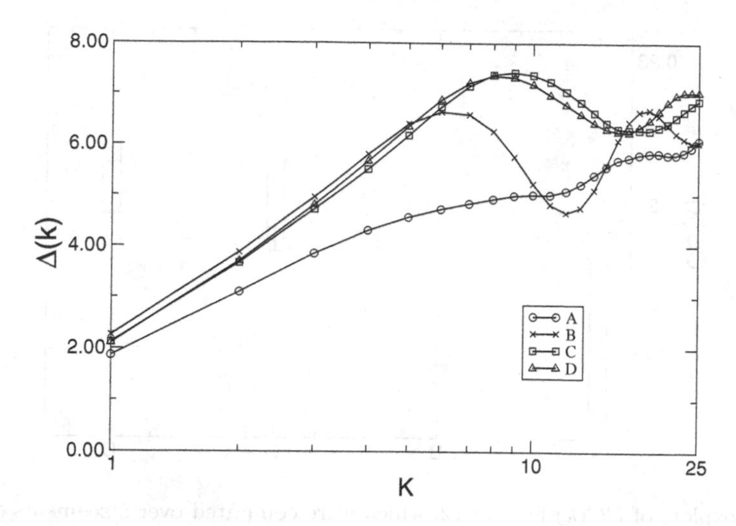


**Fig. 4.** $\Delta(k)$ versus $k$, which was computed using Eq. (3), $Q = 20$, $T = 1{,}024$ and $k = 1, 2, \ldots, 25$ for sets A (healthy patients, eyes open), B (healthy patients, eyes closed), C (unhealthy patients, eyes open) and D (unhealthy patients, eyes closed).

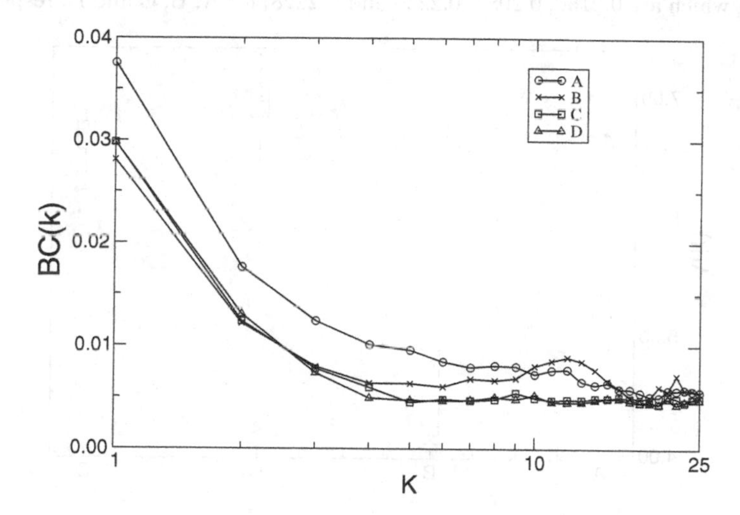


**Fig. 5.** $BC(k)$ versus $k$ using Eq. (4), $Q = 20$, $T = 1{,}024$ and $k = 1, 2, \ldots, 25$ for sets A (healthy patients, eyes open), B (healthy patients, eyes closed), C (unhealthy patients, eyes open) and D (unhealthy patients, eyes closed).

and a $p$-value of less than 0.05 between the corresponding sample means, which is more favorable between sets B and D and less favorable between sets A and C (Table 1). Receiver Operating Characteristic (ROC) analysis was performed, which is used here to quantify how accurately our map can discriminate between patients in two groups [14,18]. Table 2 shows the areas under the ROC curves (AUCs) of the network measures $CC$, $\Delta$ and $BC$ between patients in sets B

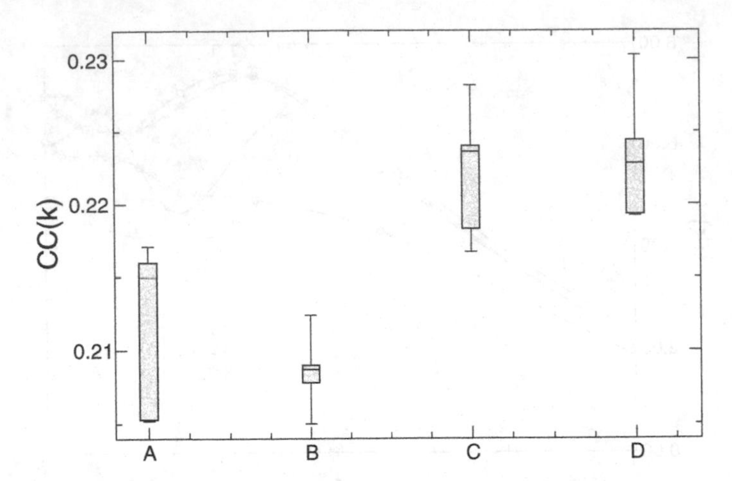


**Fig. 6.** Boxplots of $CC(k)$ for $k = 12$, which were computed over 5 segments each, for sets A (healthy patients, eyes open), B (healthy patients, eyes closed), C (unhealthy patients, eyes open) and D (unhealthy patients, eyes closed). Boxplots from patients with different health conditions have different medians (shown as a line in the center of each box), which are 0.2150, 0.2087, 0.2236 and 0.2228, for A, B, C and D, respectively.

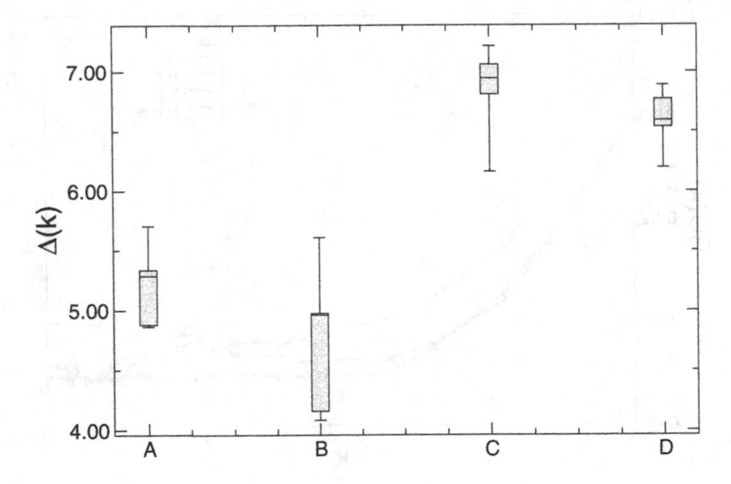


**Fig. 7.** Boxplots of $\Delta(k)$ for $k = 12$, which were computed over 5 segments each, for sets A (healthy patients, eyes open), B (healthy patients, eyes closed), C (unhealthy patients, eyes open) and D (unhealthy patients, eyes closed). Boxplots from patients with different health conditions have different medians (shown as a line in the center of each box), which are 5.2900, 4.9600, 6.9420, and 6.5900, for A, B, C and D, respectively.

and D and between patients in sets A and C. In all cases, the QG method performs very well in the differentiation between individuals with different health conditions. Overall, comparing the metrics that are used to discriminate the sets, $\Delta$ and $BC$ display the best results.

**Table 1.** Statistical comparison between the sample means of the network measures $CC$, $\Delta$ and $BC$ for the sets A (healthy patients, eyes open) and C (patients with AD, eyes open), the sets B (healthy patients, eyes closed) and D (patients with AD, eyes closed), the sets A and D and the sets B and C.

| | CC | $\Delta$ | BC |
|---|---|---|---|
| $\mu_A - \mu_C$ | −0.010 | −1.620 | 0.002 |
| $CI_{AC}$ | [−0.018, −0.002] | [−2.177, −1.064] | [0.001, 0.004] |
| t statistic | −3.104 | −6.741 | 4.085 |
| $\mu_B - \mu_D$ | −0.014 | −1.838 | 0.004 |
| $CI_{BD}$ | [−0.020, −0.008] | [−2.616, −1.060] | [0.002, 0.005] |
| t statistic | −6.214 | −5.964 | 6.377 |
| $\mu_A - \mu_D$ | −0.011 | −1.379 | 0.003 |
| $CI_{AD}$ | [−0.019, −0.003] | [−1.839, −0.919] | [0.001, 0.004] |
| t statistic | −3.437 | −7.013 | 4.452 |
| $\mu_B - \mu_C$ | −0.013 | −2.080 | 0.004 |
| $CI_{BC}$ | [−0.019, −0.007] | [−2.884, −1.276] | [0.002, 0.005] |
| t statistic | −5.690 | −6.156 | 6.020 |

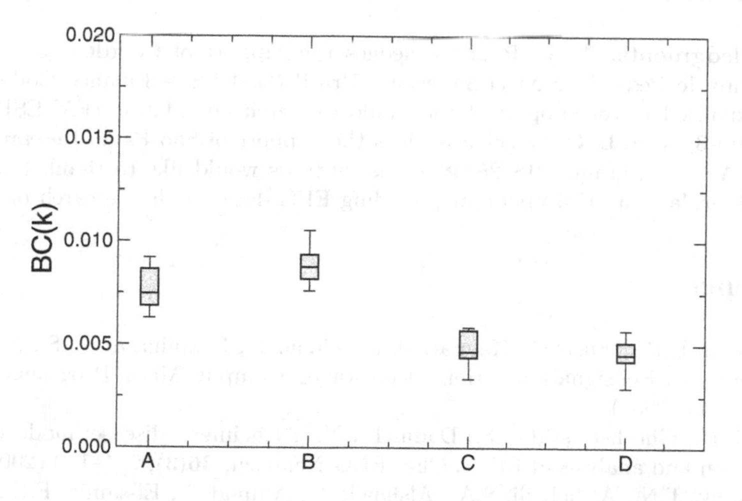


**Fig. 8.** Boxplots of $BC(k)$ for $k = 13$, which were computed over 5 segments each, for sets A (healthy, eyes open), B (healthy, eyes closed), C (unhealthy, eyes open) and D (unhealthy, eyes closed). Boxplots from patients with different health conditions have different medians (shown as a line in the center of each box), which are 0.0074, 0.0087, 0.0046 and 0.0045, for A, B, C and D, respectively.

**Table 2.** Areas under the ROC curves (AUCs) of the network measures $CC$, $\Delta$ and $BC$ between patients in sets A (healthy patients, eyes open) and C (patients with AD, eyes open) and between patients in sets B (healthy patients, eyes closed) and D (patients with AD, eyes closed) for $k = 12, 12$ and 13, respectively.

| | CC | Δ | BC |
|---|---|---|---|
| $AUC_{AC}$ | 0.960 | 1.000 | 1.000 |
| $AUC_{BD}$ | 1.000 | 1.000 | 1.000 |

## 6 Conclusions

Alzheimer's disease is a highly incapacitating disease of difficult precise diagnostics, which affects millions of people worldwide. A robust, non-invasive automatic diagnostics technique, based solely on the analysis of EEG data would thus potentially have a wide application. In this study we presented an original approach based on the use of quantile graphs for the classification of EEG data. The three network topological measures used here - clustering coefficient, mean jump length and betweenness centrality - showed that the QG method is able to differentiate healthy patients (with eyes open or closed) from patients with AD (with eyes open or closed) over a range of values of the time difference parameter $k$. Results are statistically significant and attest the QG method is a useful tool for the analysis of complex, nonlinear signals like those generated by EEGs from human patients.

**Acknowledgments.** A. M. P. acknowledges the support of Coordenação de Aperfeiçoamento de Pessoal de Nível Superior - Brasil (CAPES) - Finance Code 001. L. E. B. acknowledges the support of São Paulo Research Foundation (FAPESP), grant 2016/17914-3. A. S. L. O. C. acknowledges the support of São Paulo Research Foundation (FAPESP), grant 2018/25358-9. The authors would like to thank Dr. Dennis Duke of Florida State University for providing EEG data for this research project.

## References

1. Acharya, U.R., Faust, O., Kannathal, N., Chua, T., Laxminarayan, S.: Non-linear analysis of EEG signals at various sleep stages. Comput. Meth. Programs Biomed. **80**, 37–45 (2005)
2. Adeli, H., Ghosh-Dastidar, S., Dadmehr, N.: Alzheimer's disease: models of computation and analysis of EEGs. Clin. EEG Neurosci. **36**(3), 131–140 (2005)
3. Alotaiby, T.N., Alshebeili, S.A., Alshawi, T., Ahmad, I., El-samie, F.E.A.: EEG seizure detection and prediction algorithms: a survey. EURASIP J. Adv. Signal Process, p. 183 (2014)
4. Besthorn, C., et al.: Discrimination of Alzheimer's disease and normal aging by EEG data. Electroencephalogr. Clin. Neurophysiol. **103**(2), 241–248 (1997)
5. Budson, A., Solomon, P.: Memory Loss, Alzheimer's Disease, and Dementia. Elsevier, New York (2015)

6. Campanharo, A.S.L.O., Doescher, E., Ramos, F.M.: Automated EEG signals analysis using quantile graphs. In: Rojas, I., Joya, G., Catala, A. (eds.) IWANN 2017. LNCS, vol. 10306, pp. 95–103. Springer, Cham (2017). https://doi.org/10.1007/978-3-319-59147-6_9
7. Campanharo, A.S.L.O., Doescher, E., Ramos, F.M.: Application of quantile graphs to the automated analysis of EEG signals. Neural Process. Lett., 1–16 (2018)
8. Campanharo, A.S.L.O., Ramos, F.M.: Distinguishing different dynamics in electroencephalographic time series through a complex network approach. In: Proceeding Series of the Brazilian Society of Applied and Computational Mathematics. vol. 5. SBMAC (2017)
9. Campanharo, A.S.L.O., Ramos, F.M.: Hurst exponent estimation of self-affine time series using quantile graphs. Physica A **444**, 43–48 (2016)
10. Campanharo, A.S.L.O., Sirer, M.I., Malmgren, R.D., Ramos, F.M., Amaral, L.A.N.: Duality between time series and networks. PLoS ONE **6**, e23378 (2011)
11. Costa, L.F., Rodrigues, F.A., Travieso, G., Villas, P.R.: Characterization of complex networks. Adv. Phys. **56**, 167–242 (2007)
12. Feldman, H.H., Woodward, M.: The staging and assessment of moderate to severe Alzheimer disease. Neurology **65**, S10–S17 (2005)
13. Freeman, L.C.: A set of measures of centrality based on betweenness. Sociometry **40**, 35–41 (1977)
14. Hajian-Tilaki, K.: Receiver operating characteristic (ROC) curve analysis for medical diagnostic test evaluation. Caspian J. Intern. Med. **4**, 627 (2013)
15. Kantz, H., Schreiber, T.: Nonlinear Time Series Analysis. Cambridge University Press, Cambridge (2003)
16. Korner, T.W.: Fourier Analysis. Cambridge University Press, Cambridge (1988)
17. Morris, A.S., Langari, R.: Measurement and Instrumentation. Academic Press, San Diego (2012)
18. Obuchowski, N.A., Bullen, J.: Receiver operating characteristic (ROC) curves: review of methods with applications in diagnostic medicine. Phys. Med. Biol. **63**, 07TR01 (2018)
19. Organization, W.H.: Dementia: World Health Organization, Switzerland (2012)
20. Percival, D.B., Walden, A.: Wavelet Methods for Time Series Analysis. Cambridge University Press, Cambridge (2000)
21. Ridouh, A., Boutana, D., Bourennane, S.: EEG signals classification based on time frequency analysis. J. Circ. Syst. Comput. **26**, 1750198 (2017)
22. Saramäki, J., Kivelä, M., Onnela, J.P., Kaski, K., Kertesz, J.: Generalizations of the clustering coefficient to weighted complex networks. Phys. Rev. E **75**, 027105 (2007)
23. Stam, C., Jelles, B., Achtereekte, H., Van Birgelen, J., Slaets, J.: Diagnostic usefulness of linear and nonlinear quantitative EEG analysis in Alzheimer's disease. Clin. Electroencephalogr. **27**(2), 69–77 (1996)
24. Tsolaki, A., Kazis, D., Kompatsiaris, I., Kosmidou, V., Tsolaki, M.: Electroencephalogram and Alzheimer's disease: clinical and research approaches. Int. J. Alzheimer's Dis. **2014**, 10 (2014)
25. Ubeyli, E.D.: Analysis of EEG signals by combining eigenvector methods and multiclass support vector machines. Comput. Biol. Med. **38**, 14–22 (2011)
26. Ubeyli, E.D., Guler, I.: Features extracted by eigenvector methods for detecting variability of EEG signals. Comput. Biol. Med. **28**, 592–603 (2007)

27. Yagneswaran, S., Baker, M., Petrosian, A.: Power frequency and wavelet characteristics in differentiating between normal and alzheimer EEG. In: Engineering in Medicine and Biology, 2002 24th Annual Conference and the Annual Fall Meeting of the Biomedical Engineering Society EMBS/BMES Conference, 2002 Proceedings of the Second Joint, vol. 1. IEEE (2002)
28. Zar, J.H.: Biostatistical Analysis. Prentice Hall, New Jersey (2010)
29. Zhang, Y., Liu, B., Ji, X., Huang, D.: Classification of EEG signals based on autoregressive model and wavelet packet decomposition. Neural Process. Lett. **45**, 365–378 (2016)

# The Generalized Sleep Spindles Detector: A Generative Model Approach on Single-Channel EEGs

Carlos A. Loza[1(✉)] and Jose C. Principe[2]

[1] Department of Mathematics, Universidad San Francisco de Quito, Quito, Ecuador
cloza@usfq.edu.ec

[2] Computational NeuroEngineering Laboratory (CNEL), University of Florida, Gainesville, USA
principe@cnel.ufl.edu

**Abstract.** We propose a data-driven, unsupervised learning framework for one of the hallmarks of stage 2 sleep in the electroencephalogram (EEG)—sleep spindles. Neurophysiological principles and clustering of time series subsequences constitute the underpinnings of methods fully based on a generative latent variable model for single-channel EEG. Learning on the model results in representations that characterize families of sleep spindles. The discriminative embedding transform separates potential micro-events from ongoing background activity. Then, a hierarchical clustering framework exploits Minimum Description Length (MDL) encoding principles to effectively partition the time series into patterns belonging to clusters of different dimensions. The proposed algorithm has only one main hyperparameter due to online model selection and the flexibility provided by cross-correlation operators. Methods are validated on the DREAMS Sleep Spindles database with results that echo previous approaches and clinical findings. Moreover, the learned representations provide a rich parameter space for further applications such as sparse encoding, inference, detection, diagnosis, and modeling.

**Keywords:** EEG · Generative model · Representation learning · Sleep spindles

## 1 Introduction

Sleep spindles constitute the hallmark of stage 2 non-REM sleep. Their generation is attributed to the mutual interaction between GABAergic reticular neurons and excitatory thalamic cells [24] while long-range cortical projections are believed to regulate their temporal synchronization [5]. In terms of behavioral and functional correlates, sleep spindles have been associated to memory consolidation processes [4,22], cortical development [14], sleep deprivation [7], and are even regarded as potential biomarkers for psychiatric disorders, such as schizophrenia [8,17]. Therefore, principled detection and modeling are crucial.


I. Rojas et al. (Eds.): IWANN 2019, LNCS 11506, pp. 127–138, 2019.
https://doi.org/10.1007/978-3-030-20521-8_11

In the clinical field, EEGers usually utilize scoring rules and norms well documented in the literature [18,21]; yet, the ever-increasing amount of data and the advent of machine learning have bolstered the use of automatic sleep spindle detectors as an additional tool for clinicians and neuroscientists [6,12,19].

A sleep spindle is defined as a burst in the 11–15 Hz range (sigma band) with duration between 0.5 and 2 s. and a distinctive waxing-wanning envelope. Moreover, as a type of transient events in the EEG, sleep spindles require particular constraints for their detection. Such conditions are derived from both neurophysiology [18] and empirical clinical findings [21]. Automatic detectors explicitly incorporate such constraints into their framework, e.g. amplitude-based thresholds that are cross-validated to ground truth [6,8]. Classic detectors focus only on timing and amplitude features; yet, as neurophysiological micro-events, sleep spindles are also characterized by duration, frequency, modulation, and, from a generative model stance, by their encoding indexes. All these features can be collectively deemed as representations. A generalized detector should, then, be able to learn such representations in a data-driven manner.

By leveraging the sparse nature of the micro-events, we pose them as samples from a Temporal Marked Point Process (TMPP) that activate elements from a set of vectors (i.e. a dictionary) over time—a generative latent variable model for sleep spindles. A fully unsupervised framework aims to estimate the conditional densities of the latent variables given a set of neuromodulations: generating dictionary (centers of mass in vectors spaces of different dimensionalities), intensity function of the TMPP timings and density of the TMPP marks (amplitudes and indexes). We propose a novel learning algorithm that incorporates neurophysiological constraints and principled techniques for clustering of time series subsequences. We exploit Freeman's theories [9] to restrict the search space of relevant events. Then, a hierarchical clustering algorithm creates a partitioning of the search space by means of Minimum Description Length (MDL) encoding [1]. The result is twofold: learned representations suitable for modeling, and sets of latent variables appropriate for inference.

One of the major advantages of the proposed method is its data-driven nature. Durations of relevant neuromodulations are not limited to a set of user-defined inputs; they are learned on an unsupervised fashion as a result of cross-correlation operators that guarantee flexibility and fine temporal resolution. Also, MDL encoding performs online model selection in a fast, greedy manner. Thus, the proposed algorithm virtually requires only one hyperparameter that is closely related to amplitude-based thresholds of classic sleep spindle detectors.

We validate the methods on the DREAMS Sleep Spindles database [25] and obtain estimates of the representation densities. We compare them with their counterparts from visual scorers and quantify their similarity via the Kullback-Leibler (KL) divergence. We also analyze the effect of the only hyperparameter in terms of receiver operating characteristics (ROC) curves and KL divergences. The results highlight the potential of the proposed method when dealing with principled detection and modeling of sleep spindles. The rest of the paper is organized as follows: Sect. 2 details the problem to solve while Sect. 3 presents the methods and rationales behind their choice, Sect. 4 showcases the results, and, lastly, Sect. 5 concludes the paper.

## 2 A Problem Beyond Detection

Let $\tilde{y}[n]$ be a bandpassed single-channel EEG trace that can be decomposed into two time series according to the dynamical regimes of the generating network:

$$\tilde{y}[n] = \begin{cases} y[n] & \text{if Network is Active (Y State)} \\ z[n] & \text{if Network is at Rest (Z State)} \end{cases} \tag{1}$$

where $y[n]$ is the ideal, noiseless component with scale-specific micro-events (sleep spindles), and $z[n]$ is the ongoing, background activity. A mixture model characterizes the probability density of $\tilde{y}[n]$ as:

$$P(\tilde{y}[n]) = p_Y P(\tilde{y}[n]|Y, \Theta_Y) + p_Z P(\tilde{y}[n]|Z, \Theta_Z) \tag{2}$$

where $p_Y$ and $p_Z$ represent the probabilities of states $Y$ and $Z$ parameterized by $\Theta_Y$ and $\Theta_Z$, respectively ($p_Y = 1 - p_Z$).

A linear model posits $y[n]$ as the weighted sum of $N$ ideal patterns, $\mathbf{d} \in \mathbf{D}$, shifted over time:

$$y[n] = \sum_{i=1}^{N} \sum_{m=-\infty}^{\infty} \alpha_i \delta[n - \tau_i - m] \mathbf{d}_{\omega_i}[m] \tag{3}$$

where $\delta[n]$ is the Dirac delta function. The elements of the dictionary, $\mathbf{D} = \{\mathbf{d}_{\omega_j}\}_{j=1}^{K}$, do not necessarily have the same dimensionality, i.e. they represent templates with different durations or centers of mass in vector spaces of different dimensions; let such dimensions be the set $\{\phi_i\}_{i=1}^{N}$. Similar generative models with sparsity constraints have been proposed for the auditory nerve [23].

$\tilde{y}[n]$ can be either modeled as the noisy superposition of samples from a TMPP or as the observable variable from a generative latent variable model (Fig. 1) with two distinctive modes: $Z$, a background component that encodes the spontaneous, disorganized activity of the generating network during rest, and $Y$, an active component represented as reoccurring transient micro-events that reflect the spatiotemporal synchronization of neuronal assemblies [3,18]. Freeman posited that $Z$ can be modeled as a Gaussian distribution [10] ($\Theta_Z \triangleq \{\mu_Z, \sigma_Z\}$). $Y$ is the result of joint contributions from latent variables in the form of timings ($\tau$), amplitudes ($\alpha$), encoding indices ($\omega$), and generating dictionary $\mathbf{D}$, i.e. $\Theta_Y \triangleq \{\tau, \alpha, \omega, \mathbf{D}\}$. The shallow nature of the graph admits the equivalence between latent variables and features or representations. Moreover, $\mathbf{D}$ carries its own features, e.g. duration, frequency, and Q-factor. Consequently, learning on the model can be posed as a type of unsupervised representation learning [2].

Classic sleep spindles detectors usually estimate $\phi$ and $\tau$ leaving $\mathbf{D}$ unaddressed [6,12,19]. Given the dictionary $\mathbf{D}$ (where usually $K \geq N$), the goal in analysis, inference, encoding or detection is to estimate the shifts $\tau$, indices $\omega$, and a surrogate of the weights $\alpha$, for a constrained optimization problem, e.g. Matching Pursuit with the overcomplete Gabor basis as $\mathbf{D}$ [26]. Estimating $\mathbf{D}$ is challenging due to the inherent dynamics of the EEG—unlike classic blind source separation problems, the relevant sources of $y[n]$ are shift-invariant, non-overlapping and transient. Also, a principled decomposition should discriminate between $y[n]$ and $z[n]$ while preserving the original micro-events. The generalized detector should, then, estimate parameters beyond shifts and durations, i.e. given $\tilde{y}[n]$, the goal in an unsupervised framework is to learn $\omega$, $\tau$, $\alpha$, $\phi$, and $\mathbf{D}$.

## 3 Methods

The solution to the problem of the previous section is combinatorial in nature. In [15], we propose a solution based on shift-invariant k-means that results in $\phi \in \mathbb{R}^M$, i.e. dictionary elements with predefined duration. Now, we generalize the implementation by, first, restricting the search space and, then, exploiting MDL-based hierarchical clustering to yield prototypical patterns of different durations corresponding to reoccurring sleep spindles in the EEG. Densities of the remaining latent variables naturally arise from the learning process (Fig. 2).

### 3.1 The Discriminative Embedding Transform

The first step is to isolate $y[n]$ by exploiting the dynamical properties of the EEG. According to Freeman's experimental results, the EEG amplitudes during rest periods resemble a Gaussian distribution, while transitions to work states result in deviations from Gaussianity according to higher-order moments [9]. Both stages, rest and work, alternate in the EEG traces and give rise to transient

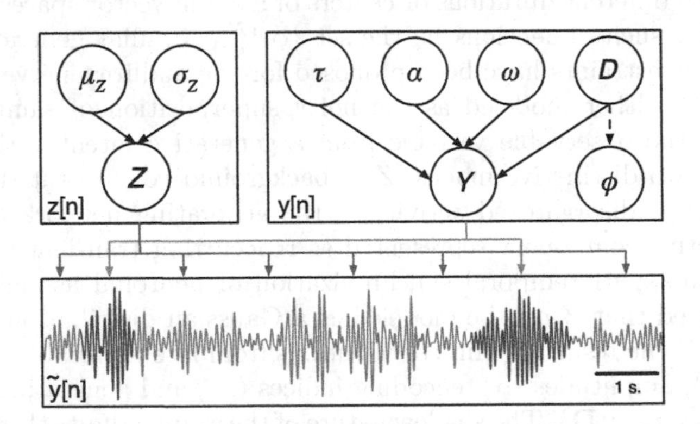


**Fig. 1.** Generative model of bandpassed single-channel EEG. $Z$ is characterized by the background EEG mean and standard deviation (blue). $Y$ consists of latent variables (red) in the form of timings, weights, indices, and generating dictionary with elements from vector spaces of different dimensions. Durations, $\phi$ (green), are features from $\mathbf{D}$. Mixing between regimes over time results in $\tilde{y}[n]$. (Color figure online)

neuromodulations as a result of the spatiotemporal synchronization of neuronal assemblies [18]. If $\tilde{y}[n]$ is the result of linear filtering, the Gaussian/Non-Gaussian properties during rest/active regimes are preserved for the bandpassed traces.

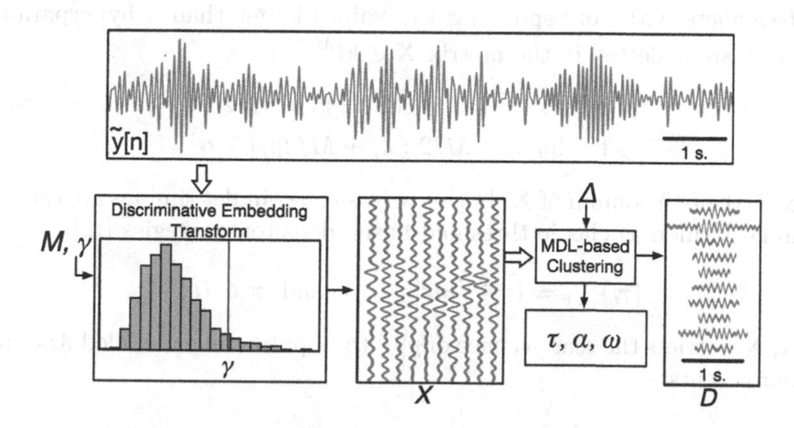


**Fig. 2.** Proposed generalized sleep spindles detector. Input: Bandpassed single-channel EEG trace, $\tilde{y}[n]$ (top row). Discriminative embedding transform restricts the search space and MDL-based clustering creates a hierarchical structure of patterns from vector spaces of different dimensions. Output: Representations and set of sleep spindles prototypes (**D**), i.e. clusters, of different durations.

**Definition 1.** *The M-sample-long subsequence from $\tilde{y}[n]$ centered at the time instance $t = i$ is known as M-snippet:*

$$\tilde{y}_i = \tilde{y}[i - M/2 : i + M/2] \quad \text{s.t} \quad i = M/2, M/2+1, \ldots, \eta - M/2 \tag{4}$$

where $\eta$ is the number of sampled values in $\tilde{y}[n]$. One of the goals of the generalized detector is to discriminate between $M$-snippets generated by $Z$ (background subsequences) and $M$-snippets with embedded micro-events generated by $Y$.

In [16], the embedding transform was introduced as a novel tool to assess stationary of bandpassed single-channel EEG recordings. In particular, the input time series is non-linearly decomposed into two components based on a surrogate distribution of constrained $\ell_2$-norms, $\beta_M$:

$$\begin{aligned}\beta_M &= \beta_M(\tilde{y}[n]) \\ &= ||\tilde{y}[\pi_i - M/2 : \pi_i + M/2]||_2 \quad \text{s.t.} \quad \pi_i \in \Pi\end{aligned} \tag{5}$$

What differentiates this approach from a regular embedding is the way the set $\Pi$ is built: the algorithm starts by isolating the indices where relevant modulatory activity is present (peak detection via moving averages or instantaneous amplitudes). Then, the indices corresponding to the remaining unmodulated patterns complete the set $\Pi$. For further details of the algorithm, refer to [16].

After $\beta_M$ is built, we posit that the $M$-snippets generated by $Z$ are mapped to a chi-distribution with $M$ degrees of freedom in the $\beta_M$ space; this density, for large $M$, results in a Gaussian by the Central Limit Theorem. Conversely, potential relevant sleep spindles, i.e. $y[n]$, are mapped to a second mode in $\beta_M$. $M$-snippets with corresponding $\beta_M$ values larger than a hyperparameter threshold $\gamma$ are collected in the matrix $\mathbf{X} \in \mathbb{R}^{M \times \hat{N}}$:

$$\begin{aligned} \mathbf{x}_i &= \tilde{y}[\pi_i - M/2 : \pi_i + M/2]^T \\ \text{s.t.} \quad & ||\tilde{y}[\pi_i - M/2 : \pi_i + M/2]||_2 \geq \gamma \end{aligned} \tag{6}$$

where $\mathbf{x}_i$ is the $i$-th column of $\mathbf{X}$. In this way, sleep spindles timings are estimated in a similar fashion as classic threshold-based detector strategies [12]:

$$\{\tau_i\}_{i=1}^{\hat{N}} = \{\pi \mid \; ||\tilde{y}_\pi||_2 \geq \gamma \;\text{ and }\; \pi \in \Pi\} \tag{7}$$

In short, $\mathbf{X}$ restricts the search space of $\tilde{y}[n]$ to $\hat{N}$ potential embedded $M$-sample-long micro-events.

### 3.2 MDL-Based Clustering

After the search space is efficiently restricted, it is necessary to find reoccurring patterns in $\mathbf{X}$. A naive solution would exploit classic clustering algorithms, such as k-means, in the $M$-dimensional space of the inputs; yet, the solution would include clusters deemed as meaningless [13] due to two main reasons: variable time offsets of patterns from the same cluster, i.e. shift-invariance, and the presence of micro-events of different durations embedded in $M$-snippets. A plausible solution must address both problems in a principled manner. The former problem is managed via template matching (nearest neighbor search) based on cross-correlations while the latter exploits principles of MDL encoding.

A hierarchical clustering framework greedily selects the number of clusters and estimates reoccurring patterns of different durations by exploiting principles of MDL compression. Let $DL(T)$ be the length of the bit level representation of time series $T$ with length $m$, i.e. the entropy of $T$ times $m$.

$$DL(T) = -m \sum_t P(T = t) \log_2 P(T = t) \tag{8}$$

Similarly, the conditional description length of a sequence $A$ after being encoded with a hypothesis $H$ is given by $DL(A|H) = DL(A - H)$, e.g. the cost of the encoding. This principle was applied to hierarchical clustering of time series in [20] under the connotation of *time series epenthesis*. Essentially, the $DL(\cdot)$ operator is a parameter-free tool to evaluate the 3 basic operations in hierarchical clustering: creation of a cluster, assignment of input to existing cluster, and merging of clusters. The cost function $BS$ represents the bits saved after performing the three basic operations, for instance:

$BS$ after creating cluster $C$ from subsequences $A$ and $B$:

$$BS = DL(A) + DL(B) - DLC(C) \tag{9}$$

where $DLC(C) = DL(H) + \sum_{A \in C} DL(A|H) - \max_{A \in C} DL(A|H)$ is the number of bits needed to represent all subsequences belonging to cluster $C$ and $H$ is the center subsequence of the cluster under consideration.

$BS$ after adding subsequence $A$ to cluster $C$:

$$BS = DL(A) + DLC(C) - DLC(C') \tag{10}$$

where $C'$ is the new cluster after adding $A$ to $C$.

$BS$ for merging clusters $C_1$ and $C_2$ into new cluster $C'$:

$$BS = DLC(C_1) + DLC(C_2) - DLC(C') \tag{11}$$

The algorithm exposed in [20] utilizes motif discovery algorithms to initialize novel clusters. We propose cross-correlation operations as suitable alternatives for discovering such motifs from $\mathbf{X}$, estimating distances between subsequences and corresponding clusters ($\alpha$), updating membership vectors ($\omega$), and, ultimately, learning shift-invariant prototypical patterns from vectors spaces of different dimensions. MDL-based encoding basically allows comparison of costs involving clusters of different dimensionalities, which would be prohibitive and unprincipled for the Euclidean distance. The final version of the algorithm evaluates at each step which operation results in the maximal $BS$ and proceeds with such option. Iterations continue until the set $\mathbf{X}$ is exhausted.

The major advantage of this hierarchical clustering framework is twofold: model selection is performed in a greedy manner ($K$ is learned from data) and the resulting sleep spindles prototypes are not restricted to fixed-length patterns. The first hyperparameter, $\gamma$, plays the role of an $\ell_2$-norm-based threshold of the generalized detector, while the second hyperparameter, $\Delta$, is a set of approximate durations of prospective sleep spindles. Yet, thanks to cross-correlation operators, the final clusters are not necessarily restricted to the elements of $\Delta$; this emphasizes the inherent adaptive nature of the proposed framework. Moreover, $\Delta$ can be chosen in a principled manner according to the rhythm under consideration [18]. Therefore, the proposed generalized detector is a virtually one-hyperparameter learning mechanism that sequentially selects in a greedy manner the operation that maximizes the bits saved among the possible $\delta$-dimensional inputs, where $\delta \in \Delta$ and $\delta \leq M$.

## 4 Results

Sleep spindles latent variables are estimated on the DREAMS Sleep Spindles database [25] for the available 8 subjects. Single-channel (either CZ-A1 or C3-A1), 30-min-long EEG traces were made available with their corresponding visual scorings of sleep spindles (sampling frequencies ranging from 50 to 200 Hz). The sigma band is isolated using Butterworth filters with quality factor (ratio of central frequency and bandwidth) $Q \approx 2$. $M$ is set equal to the sample equivalent of 1.5 s. $\Delta$ is set to [0.5:0.1:1.5] sec. according to scoring criteria of sleep spindles [18,19,21]. For proper validation and estimation of density similarities,

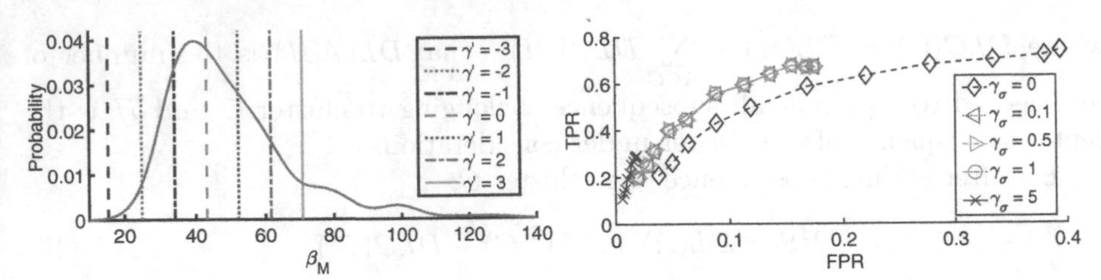


(a) Embedding Transform (Subject 2). Vertical lines indicate estimated $\gamma$ thresholds according to $\gamma = \mu_{Z_M} + \gamma' \times \sigma_{Z_M}$.

(b) ROC curves over $\gamma'$ for several values of $\gamma_\sigma$. For any ROC curve, points correspond to $\gamma'$ values in the interval [-3:0.5:3].

**Fig. 3.** Effects of the Embedding Transform-based threshold, $\gamma$, over the $M$-snippet domain and performance in terms of sleep spindles detection.

we only utilize the scores from one (visual scorer 1) of the two experts due to the strong bias in durations from visual scorer 2—zero standard deviations and mean durations of 1 s.

ROC curves are a good starting point to validate estimated timings and durations in terms of True Positive Rates (TPR) and False Positive Rates (FPR). Figure 3 shows ROC curves (grand average) over $\gamma = \mu_{Z_M} + \gamma' \times \sigma_{Z_M}$ where $\mu_{Z_M}$ and $\sigma_{Z_M}$ are the mean and standard deviation of the set of $M$-snippets generated from $Z$, respectively. We test $\gamma'$ in the range $[-3{:}0.5{:}3]$. Also to reduce the FPR, the sigma index [11,12] is exploited to reject alpha intrusions and EMG interference. In particular, the sigma index threshold, $\gamma_\sigma$, is a lower bound for the ratio between powers in the sigma band and neighboring rhythms. Figure 3b compares ROC curves for several values of $\gamma_\sigma$ while Fig. 3a indicates the estimated $\gamma$ thresholds in $\beta_M$ for a sample subject. In general, the ROC curves are very robust for a wide range of $\gamma_\sigma$ and quickly saturate for $2 \leq \gamma' \leq 3$. Best cases correspond to a global sensitivity of 67.7% and FPR = 0.154 compared to 70.2% and 0.264 from the original report [6], respectively.

Next, we validate the estimated amplitudes and durations. Here it is worth noting that classic sleep spindles detectors (and visual scorers) do not share the generative nature of the generalized detector and, hence, define amplitude as the absolute value of the peak amplitude during a micro-event. Duration, as previously noted, is not a latent variable, but rather a representation inferred from the learned dictionary; hence, durations parameterize—and act as surrogate features of—**D**. Similarity between scored and estimated representations is assessed via the KL divergence; in this way, we generalize the concepts classic detectors gauge with TPRs and FPRs. Amplitude representations display a local minimum KL divergence at $\gamma' = 0.5$, while durations are relatively unaffected (Fig. 4a). This implies that the generalized detector robustly learns the duration density, and therefore, is able to robustly learn the generating dictionary (at least in terms of surrogate features). Conversely, KL divergences of the amplitudes provide a novel criterion for threshold selection based on representation

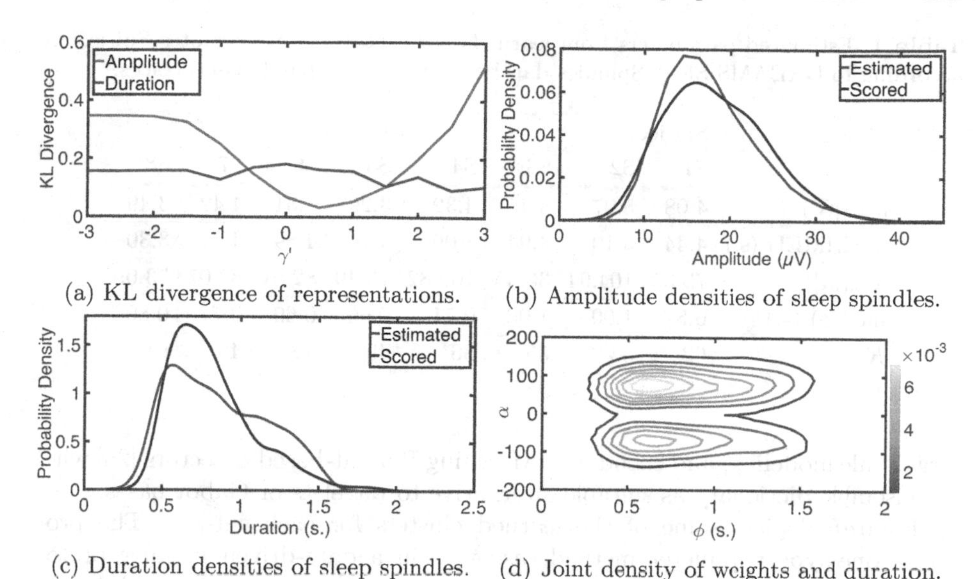


(a) KL divergence of representations. (b) Amplitude densities of sleep spindles. (c) Duration densities of sleep spindles. (d) Joint density of weights and duration.

**Fig. 4.** Learned densities from single-channel EEGs and associated measures. (a) KL divergence between learned and scored representations. (b) and (c) Estimated densities for sleep spindles amplitude and duration ($\gamma' = 0.5$). (d) Contour plot of bivariate joint density of weights and duration from generative model ($\gamma' = 0.5$). Color bar represents probability density—a surrogate of the membership index latent variable, $\omega$.

densities from a generative model instead of classic performance measures from ROC curves. Learned densities (Fig. 4b and c) echo the experimental results of Purcell et al. [19] in a massive study of sleep spindles characterization.

The remaining latent variables and representations are analyzed next. The weights, $\alpha$, represent the distance between detected micro-events and their corresponding clusters or prototypes, whereas $\mathbf{D}$ can be partially characterized by the duration $\phi$. Their joint bivariate density clearly indicates two main modes corresponding to sleep spindles in the range [0.5, 1] seconds and symmetric marginal weight densities (Fig. 4d). Membership indices, $\omega$, shape the probability density over the weight-duration space and, hence, define the bimodal density. The latent variable model can easily sample from this distribution to generate micro-events, and hence, simulate single-channel EEG traces with embedded sleep spindles.

Table 1 summarizes some statistics from the learned latent variables for each subject. $\sigma_Z$ is the estimated standard deviation of the background component (a measure of the rest RMS of this rhythm), the median of the inter micro-event interval (IMEI) characterizes the shifts $\tau$ in a similar manner as interspike intervals for units. Averages of weight magnitudes are also reported ($\alpha$ densities were bimodal and symmetric around zero). Lastly, median durations and number of clusters, $K$, parameterize the learned dictionaries succinctly. The measures of Table 1 can be further exploited for inference, e.g. sleep disorder diagnosis;

**Table 1.** Estimated parameters from learned representations. SX denotes Subject X according to DREAMS Sleep Spindles database. $\mu_Z = 0$ for bandpassed traces.

| | Subject | | | | | | | |
|---|---|---|---|---|---|---|---|---|
| | S1 | S2 | S3 | S4 | S5 | S6 | S7 | S8 |
| $\sigma_Z$ (μV) | 4.08 | 5.07 | 3.45 | 4.32 | 3.29 | 3.51 | 4.42 | 3.49 |
| med(IMEI) (s.) | 4.44 | 5.10 | 4.94 | 8.09 | 4.16 | 4.89 | 4.07 | 3.30 |
| avg($\|\alpha\|$) | 73.35 | 104.04 | 36.45 | 100.82 | 73.49 | 82.91 | 87.07 | 73.09 |
| med($\phi$) (s.) | 0.87 | 1.00 | 1.06 | 0.71 | 0.66 | 0.69 | 0.77 | 0.80 |
| $K$ | 62 | 53 | 55 | 36 | 49 | 42 | 48 | 61 |

large-scale modeling; and encoding—Matching Pursuit-based detectors [26] with an ensemble dictionary as suitable alternative to wavelets or Gabor bases.

Figure 5 depicts some of the learned clusters for each subject. The proposed one-hyperparameter method discovers in a data-driven manner prototypical sleep spindles with a wide range of temporal supports and modulatory patterns. Moreover, **D** can be characterized by its own features or representations beyond duration, e.g. frequency, number of oscillations, symmetry, and $Q$-factor. This opens the door to rich parameter spaces where inference and modeling are appealing. For instance, Fig. 6 characterizes the dictionaries via their power spectral density (PSD). The smoothing effect in the cluster estimation resembles ensemble averages in spectral estimation, which helps mitigate the bias.

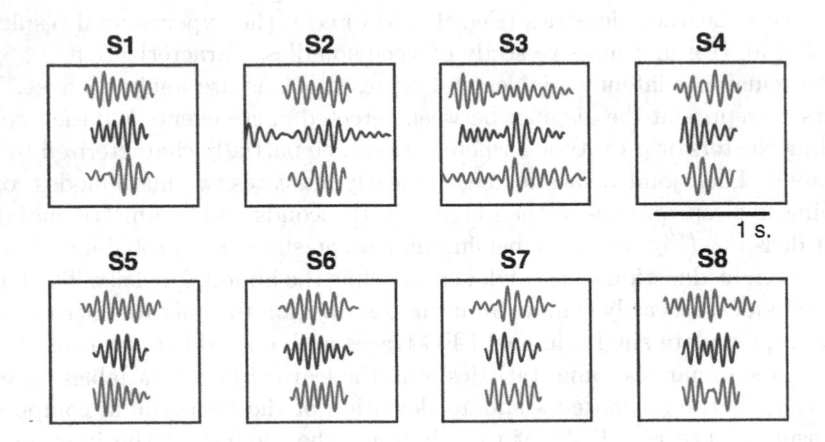


**Fig. 5.** Sample prototypical sleep spindles learned from single-channel EEGs. Clusters with highest number of micro-events assignments are shown. ($\gamma' = 0.5$).

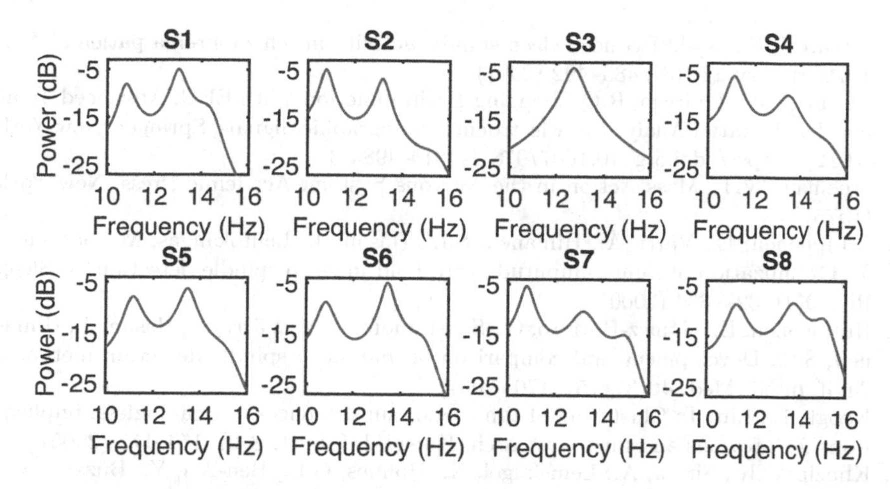


**Fig. 6.** Estimated power spectral densities of sleep spindles prototypes (Autoregressive model-based estimation with AIC model selection, $\gamma' = 0.5$).

## 5 Conclusion

A generative latent variable model for sleep spindles generalizes classic detectors to be able to learn representations from two physiological regimes in an unsupervised manner. The proposed methodology discovers bases from vector spaces of different dimensions, i.e. clusters of time series subsequences of different durations. The one-hyperparameter algorithm efficiently learns features that can be further exploited by clinicians as tools in encoding, detection, inference, and modeling. Future work includes iterative estimation of the latent variables in a Bayesian framework where the EM algorithm results advantageous.

## References

1. Barron, A., Rissanen, J., Yu, B.: The minimum description length principle in coding and modeling. IEEE Trans. Inf. Theory **44**(6), 2743–2760 (1998)
2. Bengio, Y., Courville, A., Vincent, P.: Representation learning: a review and new perspectives. IEEE Trans. Pattern Anal. Mach. Intell. **35**(8), 1798–1828 (2013)
3. Buzsáki, G., Anastassiou, C.A., Koch, C.: The origin of extracellular fields and currents–EEG, ECoG, LFP and spikes. Nat. Rev. Neurosci. **13**(6), 407 (2012)
4. Clemens, Z., Fabo, D., Halasz, P.: Overnight verbal memory retention correlates with the number of sleep spindles. Neuroscience **132**(2), 529–535 (2005)
5. Contreras, D., Destexhe, A., Sejnowski, T.J., Steriade, M.: Control of spatiotemporal coherence of a thalamic oscillation by corticothalamic feedback. Science **274**(5288), 771–774 (1996)
6. Devuyst, S., Dutoit, T., Stenuit, P., Kerkhofs, M.: Automatic sleep spindles detection–overview and development of a standard proposal assessment method. In: 2011 Annual International Conference of the IEEE Engineering in Medicine and Biology Society, EMBC, pp. 1713–1716. IEEE (2011)
7. Dijk, D.J., Hayes, B., Czeisler, C.A.: Dynamics of electroencephalographic sleep spindles and slow wave activity in men: effect of sleep deprivation. Brain Res. **626**(1–2), 190–199 (1993)

8. Ferrarelli, F., et al.: Reduced sleep spindle activity in schizophrenia patients. Am. J. Psychiatry **164**(3), 483–492 (2007)
9. Freeman, W., Quiroga, R.Q.: Imaging Brain Function With EEG: Advanced Temporal and Spatial Analysis of Electroencephalographic Signals. Springer, New York (2012). https://doi.org/10.1007/978-1-4614-4984-3
10. Freeman, W.J.: Mass Action in the Nervous System. Academic Press, New York (1975)
11. Huupponen, E., Värri, A., Himanen, S.L., Hasan, J., Lehtokangas, M., Saarinen, J.: Optimization of sigma amplitude threshold in sleep spindle detection. J. Sleep Res. **9**(4), 327–334 (2000)
12. Huupponen, E., Gómez-Herrero, G., Saastamoinen, A., Värri, A., Hasan, J., Himanen, S.L.: Development and comparison of four sleep spindle detection methods. Artif. Intell. Med. **40**(3), 157–170 (2007)
13. Keogh, E., Lin, J.: Clustering of time-series subsequences is meaningless: implications for previous and future research. Knowl. Inf. Syst. **8**(2), 154–177 (2005)
14. Khazipov, R., Sirota, A., Leinekugel, X., Holmes, G.L., Ben-Ari, Y., Buzsáki, G.: Early motor activity drives spindle bursts in the developing somatosensory cortex. Nature **432**(7018), 758 (2004)
15. Loza, C.A., Okun, M.S., Príncipe, J.C.: A marked point process framework for extracellular electrical potentials. Front. Syst. Neurosci. **11**, 95 (2017)
16. Loza, C.A., Principe, J.C.: The embedding transform. a novel analysis of non-stationarity in the EEG. In: 2018 IEEE 40th Annual International Conference of the Engineering in Medicine and Biology Society (EMBC). IEEE (2018)
17. Manoach, D.S., Pan, J.Q., Purcell, S.M., Stickgold, R.: Reduced sleep spindles in schizophrenia: a treatable endophenotype that links risk genes to impaired cognition? Biol. Psychiatry **80**(8), 599–608 (2016)
18. Niedermeyer, E., da Silva, F.L.: Electroencephalography: Basic Principles, Clinical Applications, and Related Fields. Lippincott Williams & Wilkins, Philadelphia (2005)
19. Purcell, S., et al.: Characterizing sleep spindles in 11,630 individuals from the national sleep research resource. Nat. Commun. **8**, 15930 (2017)
20. Rakthanmanon, T., Keogh, E.J., Lonardi, S., Evans, S.: Time series epenthesis: clustering time series streams requires ignoring some data. In: 2011 IEEE 11th International Conference on Data Mining (ICDM), pp. 547–556. IEEE (2011)
21. Rechtschaffen, A., Kales, A., University of California Los Angeles Brain Information Service, NINDB Neurological Information Network (US).: A Manual of Standardized Terminology, Techniques and Scoring System for Sleep Stages of Human Subjects. Publication, Brain Information Service/Brain Research Institute, University of California (1968)
22. Schabus, M., et al.: Sleep spindles and their significance for declarative memory consolidation. Sleep **27**(8), 1479–1485 (2004)
23. Smith, E.C., Lewicki, M.S.: Learning efficient auditory codes using spikes predicts cochlear filters. In: Advances in Neural Information Processing Systems, pp. 1289–1296 (2005)
24. Steriade, M., McCormick, D.A., Sejnowski, T.J.: Thalamocortical oscillations in the sleeping and aroused brain. Science **262**(5134), 679–685 (1993)
25. TCTS Lab: The DREAMS sleep spindles database (2011). http://www.tcts.fpms.ac.be/~devuyst/Databases/DatabaseSpindles/
26. Żygierewicz, J., Blinowska, K.J., Durka, P.J., Szelenberger, W., Niemcewicz, S., Androsiuk, W.: High resolution study of sleep spindles. Clin. Neurophysiol. **110**(12), 2136–2147 (1999)

# DeepTrace: A Generic Framework for Time Series Forecasting

Nithish B. Moudhgalya[1(✉)], Siddharth Divi[1], V. Adithya Ganesan[1], S. Sharan Sundar[1], and Vineeth Vijayaraghavan[2]

[1] Sri Sivasubramaniya Nadar College of Engineering, Chennai, India
{nithish.moudhgalya,siddharthdivi,v.adithyaganesan, sharan.sundar}@ieee.org
[2] Solarillion Foundation, Chennai, India
vineethv@ieee.org

**Abstract.** We propose a generic framework for time-series forecasting called DeepTrace, which comprises of 5 model variants. These variants are constructed using two or more of three task specific components, namely, Convolutional Block, Recurrent Block and Linear Block, combined in a specific order. We also introduce a novel training methodology by using future contextual frames. However, these frames are dropped during the testing phase to verify the robustness of DeepTrace in real-world scenarios. We use an optimizer to offset the loss incurred due to the non-provision of future contextual frames. The genericness of the framework is tested by evaluating the performance on real-world time series datasets across diverse domains. We conducted substantial experiments that show the proposed framework outperforms the existing state-of-art methods.

**Keywords:** Time Series · Deep framework · Bidirectional

## 1 Introduction

Time Series forecasting has a wide range of applications across domains like finance (stocks prediction), economics (interest rates of central bank, monthly changes to macro-economic indicators like GDP), engineering (speech and audio signals, video analysis, image recognition), natural sciences (temperature, seismic signals) and neuroscience (brain activity measurements via EEG and fMRI), to name a few. Several approaches have been used in time series modeling. Traditionally, statistical models include linear forecasting methods like moving average, exponential smoothing, and autoregressive integrated moving average (ARIMA). Because of their relative simplicity, linear models have been the main research focus during the past few decades. Lately, to overcome the limitations of the linear models and to account for the nonlinear patterns observed in real-world problems, several classes of nonlinear models like the bilinear model [1], the autoregressive conditional heteroscedastic (ARCH) model [2] and the threshold


I. Rojas et al. (Eds.): IWANN 2019, LNCS 11506, pp. 139–151, 2019.
https://doi.org/10.1007/978-3-030-20521-8_12

autoregressive (TAR) model [3] have been proposed. Although non-linear models show significant improvement, using them as general forecasting solutions is still limited [4]. With the advent of artificial neural networks and deep learning concepts that showcase flexible nonlinear modeling capability, time series forecasting has seen a massive enhancement.

Currently, recurrent neural networks (RNNs), and in particular the long-short term memory unit (LSTM) are referred to as the state-of-art solutions for time series forecasting [5,6]. The efficiency of these networks can be explained by the recurrent connections that allow the network to access previous time series values. Recently, convolutional neural networks (CNN) have also been widely used for time series forecasting because of their relative computational efficiency and ability to effectively capture local patterns. Convolutions with auto regressive properties [7], multiple layers of dilated convolutions [8] were used for time series forecast and classification. Multi-scale convolutional neural network (MCNN) [9] and fully convolutional network (FCN) [10] are deep learning approaches that take advantage of CNN for end-to-end classification of univariate time series. However, in all these works, the proposed algorithms and models are specific to the use case and dataset considered.

In this paper, a generic framework is proposed for time series forecasting, using a novel training methodology that involves providing future contextual frames (refer Sect. 2.3) during training. The framework achieves state-of-art results across diverse domains of time series datasets (mentioned in Sect. 6). We begin by defining some basic concepts, definitions and formulae. We then describe the components used in the framework, followed by the various architectural variants. Following this, we describe the novel training methodology used and the results of the proposed framework across 3 diverse datasets. In the end, we conclude with some inferences drawn from the results of the architectural variants.

## 2 Preliminary

### 2.1 Autocorrelation

Autocorrelation is the Pearson correlation coefficient calculated between two values of the same variable at times $X_t$ and $X_{t+k}$, where $k$ denotes the delay value. It is calculated as

$$R(k) = \frac{E[(X_t - \mu)(X_{t+k} - \mu)]}{\sigma^2} \tag{1}$$

Where $\mu$ is the mean of the time series, $\sigma$ is its standard deviation and $E$ refers to expectation. According to Eq. (1), a value close to +1 signifies a positively correlated data, a value close to −1 signifies a negatively correlated data and that with 0 has no correlation. Autocorrelation at k = 0 is always +1 as a time series is always perfectly correlated with itself.

### 2.2 1-D Convolutions

Convolution is mathematically defined as an integral that expresses the amount of overlap of one function $g(x)$ as it is shifted over another function $f(x)$ i.e, common properties and characteristics of both the functions.

$$\begin{aligned} h(x) &= f(x) * g(x) = g(x) * f(x) \\ h(x) &= \int_{-\infty}^{\infty} f(u).g(u-x)du \end{aligned} \tag{2}$$

The Fig. 1d below depicts the outcome of applying a convolution function on $g(x)$ using $f(x)$ shown in Fig. 1b and a respectively. In Convolutional Neural Networks (CNN), inputs are analogous to the function $g(x)$ and kernels are analogous to $f(x)$. Each layer in a CNN operates with $n$ number of kernels $\{f_1(x), f_2(x), ... f_n(x)\}$ on the input data $g(x)$.

### 2.3 Data Preparation

The data fed into the model is a 3-dimensional structure with dimensions as number of samples, time steps and the features. The data cube thus created can be visualized as a temporally increasing order from left to right as shown in Fig. 2a. If $x_t$ is the current data frame, the data would be segmented into 2 sets of frames - $x_{t-k}$ to $x_{t-1}$ and $x_{t+1}$ to $x_{t+k}$, see Fig. 2a. The former segment is referred as the past and the latter as the future frames. The variable $k$ refers to the span of the contextual vectors, i.e. the parameter that controls the amount of contextual information fed to the model.

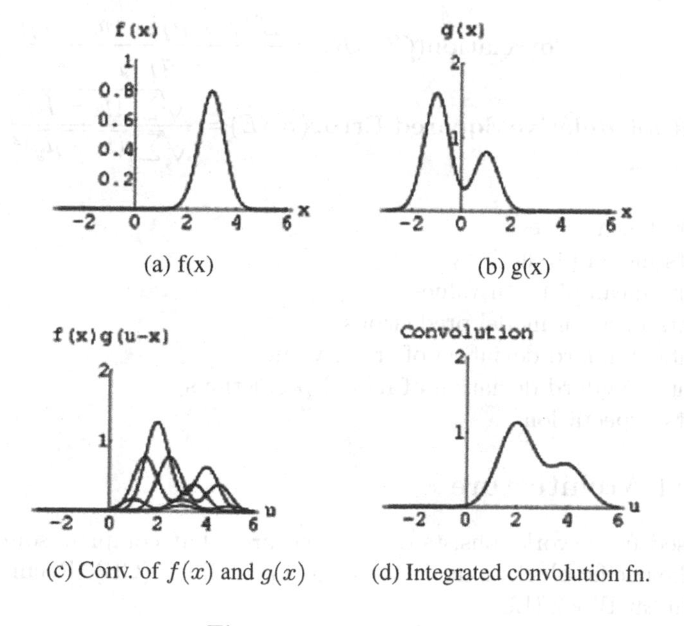


(a) f(x) (b) g(x)

(c) Conv. of $f(x)$ and $g(x)$ (d) Integrated convolution fn.

**Fig. 1.** 1D - Convolution

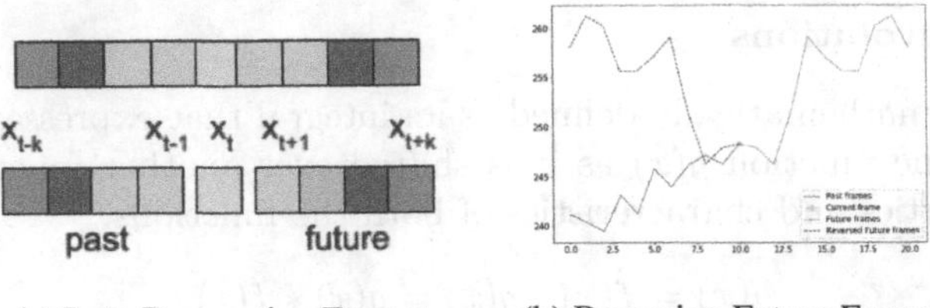


(a) Data Preparation Format (b) Reversing Future Frames

**Fig. 2.** Data preparation

It is conclusive from the works of [11], that providing reversed sequences provides further context to the model. From Fig. 2b, it can be seen that both the past and reversed future data frames converge from $t-k$ and $t+k$ to $t$ respectively. Hence the future frames were flipped along the temporal axis providing data frames ranging from $x_{t+k}$ to $x_{t+1}$. The past frames are henceforth referred to as $P$ and the future frames as $F$. By convention, $F'$ represents the reversed future frames.

### 2.4 Metrics

We use the following metrics to measure and compare the performance of DeepTrace across different domains of datasets.

$$\textbf{Coefficient of Determination}(R^2) = 1 - \frac{\sum_i (y_i - f_i)^2}{\sum_i (y_i - \mu_y)^2}$$

$$\textbf{Mean Squared Error}(MSE) = \sum_i (f_i - y_i)^2$$

$$\textbf{Correlation}(CORR) = \frac{E|(f_i - \mu_f) * (y_i - \mu_y)|}{\sigma_f \sigma_y}$$

$$\textbf{Root Relative Squared Error}(RSE) = \frac{\sqrt{\sum_i (y_i - f_i)^2}}{\sqrt{\sum_i (y_i - \mu_y)^2}}$$

Where,
$y_i$ represents truth values
$f_i$ represents model predictions
$\mu_y$ represents mean of truth values
$\mu_f$ represents mean of model predictions
$\sigma_y$ represents standard deviation of truth values
$\sigma_f$ represents standard deviation of model predictions
$E$ represents expectation

## 3 Model Architecture

The proposed framework consists of architectures that comprise some or all of 3 main task-specific blocks namely Convolutional Block (CB), Recurrent Block (RB) and linear Block (LB).

### 3.1 Convolutional Block (CB)

This block makes use of 1 dimensional convolutional layers as mentioned in Sect. 2.2 to detect and extract characteristic patterns present in the data. Several works have leveraged the temporal application of convolutions for time series classification problems [7,8,12,13]. In time series forecasting, the kernels of the CNN, as mentioned in Sect. 2.2, convolve over the data along the temporal dimension, in other words, it explores how different patterns vary with time in the given data. Also instead of integrating, the results of the convolutions are stacked one behind the other as shown in Fig. 1c, giving multiple non-linear higher dimensional representations of the input data that can better disentangle the factors of variation among the input data [14]. The convolutions operate only along the temporal dimension of the data thus also maintaining its sequential context as shown in Fig. 3a. In essence, the convolutional layers work towards detecting short-term patterns present in the data and the use of stacked convolutions provide higher level representations of these patterns [15].

### 3.2 Recurrent Block (RB)

Recurrent Neural Networks (RNNs) are known for their wide ranges of applications in sequential learning tasks that spread across various fields such as natural language, speech recognition, audio and video processing [5,16]. Long Short-Term Memory Networks (LSTMs), which were originally described in [17], are a special kind of recurrent units capable of learning long-term dependencies present in the sequence data. The 3 gates (input, forget and output) along with the cell state enhances their capabilities in processing longer sequences when compared to simple RNN units. To further capture the temporal dependencies among the localized features generated by the CB, we introduce the Recurrent Block (RB) in the framework shown in Fig. 3b. However RNNs lack the capability to convert the raw input sequence to higher dimensions that can better distinguish the data points if used directly for sequence prediction [18]. While both CNN and the fully connected layers help combat the above mentioned issue, the latter can provide only 1 such representation whereas the former can provide multiple higher dimensional representation. Also, the localized features generated by CNN are temporally consistent giving the LSTMs a latent view of the original sequence. Hence the CB precedes the RB in the framework.

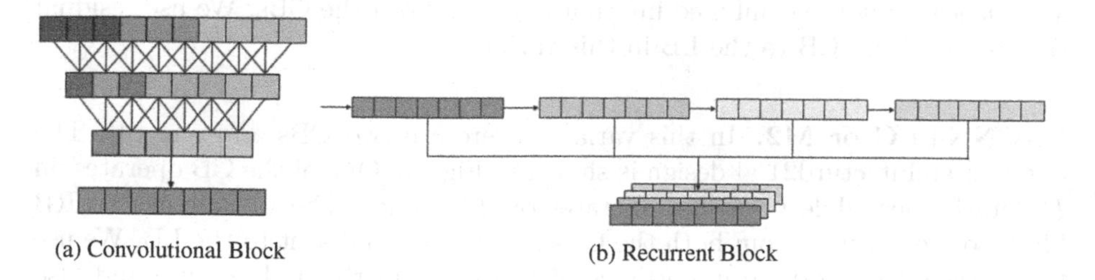

(a) Convolutional Block (b) Recurrent Block

**Fig. 3.** Block components

### 3.3 Linear Block (LB)

The last component is the Linear Block (LB). The outputs of the recurrent layers have a lot of variance among their hidden state values [18]. Also, the sequences and features from the RB and CB have to be merged to predict the target value in the current time frame. We make use of some fully connected layers in the LB to solve them. The fully connected layers learn to predict the target value using the information provided by both the CB and the RB through residual connections described in Sect. 3.4.

### 3.4 Residual Connections

Residual connections in the framework overcome vanishing gradients problem that might occur in the RB. Moreover, the CB learns directly from the errors of the LB, thus giving multiple learning sources during back propagation. The residual connections are added from the CB to the LB thus merging the features from the CB with the subsequences from the LB. These indirectly help in modeling the short-term and the long-term features of the data.

### 3.5 Model Variants

We set the following assumptions before generating the model variants.

- The training methodology of using future contextual frames is applied only to CB or RB.
- Variants with both CB and RB must have the CB preceding RB.
- Every variant comprising of CB has a residual link from CB to LB.
- Every variant must have either CB or RB before LB.

All architectural variants that abide by the above said assumptions were considered and are listed below.

**Bi-CNN+LSTM+FC or M1.** In this variant, there are two CBs, one RB and one LB. The abstract architectural design is shown in Fig. 4a. One of the CB takes only $P$ data frames and the other takes only $F'$ data frames. The output features from both the CBs are merged and sent into the RB. The RB produces a sequence using the combined information from both the CBs. We use residual connections from CB to the LB in this variant.

**Bi-CNN+FC or M2.** In this variant, there are two CBs and one LB. The abstract architecturd21 al design is shown in Fig. 4b. One of the CB operates on $P$ data frames while the other operates on $F'$ frames. The variant has no RB block so the features from both the CBs are merged and sent to the LB. We use this variant to check the importance of RB block in the architecture and also use it as a baseline model for time series forecast using CNNs.

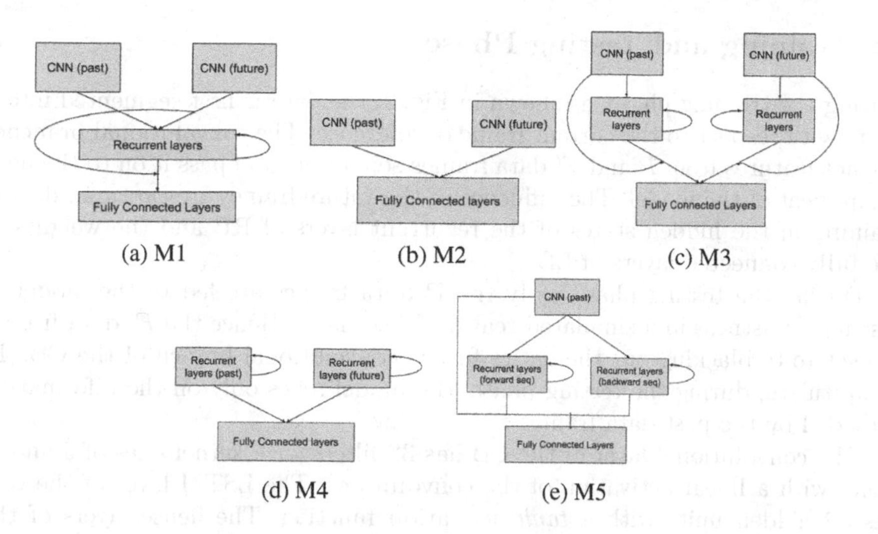


**Fig. 4.** Model variants

**Bi-(CNN+LSTM)+FC or M3.** In this variant, there are two CBs, two RBs and one LB. The abstract architectural design is shown in Fig. 4c. One of the CBs and one of the RBs are used to operate on the $P$ data frames while the other CB and RB operate on $F'$ data frames. The features from the respective CBs are fed into the respective RBs. The features and sequences generated by the combined CB+RB blocks are sent together to the LB. This variant is used to understand to which block should the future context be explicitly given and how to merge the block outputs.

**Bi-LSTM or M4.** In this variant, there are 2 RBs and one LB. The abstract architectural design is shown in Fig. 4d. One of the RBs takes $P$ data frames as input while the other takes the $F'$ data frames. The predicted sequences from both the RBs were merged and sent to the LB. We use this variant to understand the importance of CB in the architecture and to use it as a baseline model for time series forecast using RNNs.

**CNN+Bi-LSTM+FC or M5.** In this variant, there is one CB, two RBs and one LB. The abstract architectural design is shown in Fig. 4e. The CB operates only on $P$ data frames. This variant is an adapted version of the CLDNN model [14] with bidirectional property added to RB. The feature outputs from CB are sent to one RB while the reversed outputs are sent to the other one. The combined sequences from both these RBs along with the features from CB are merged and sent to the LB.

## 4 Training and Testing Phase

During the training phase, as shown in Fig. 2, the data is first segmented into 2 parts with respect to the current frame of reference. The convolutional branches extract features from $P$ and $F'$ data frames separately and pass it on to the next component of the model. The influence of the future frames are captured during training in the hidden states of the recurrent layers of RB and the weights of the fully connected layers of LB.

During the testing phase, only the $P$ data frames are fed to the model to test its robustness in a simulated real-world scenario. Hence the $F'$ data frames are set to 0, blacking out the entire future convolutional branch of the CB. To recapitulate, during the testing phase, the model relies only on the information provided by the past data frames.

The convolutional layer of the CB has 32 filters with kernel sizes of 5 and 2, along with a linear activation of the convolutions. The LSTM layer of the RB, has 32 hidden units with a *tanh* activation function. The dense layers of the LB have 256 neurons with a *LeakyRelu* activation. All the model variants are trained to minimize the *huber* loss function. The optimizers used are Stochastic Gradient Descent (SGD) and Adam, with custom learning rates.

## 5 Optimization

During testing, as the model is deprived of the future contextual frames, they do not contribute to the activations of the downstream neurons. Consequently, through the experiments conducted, it was observed that during the testing phase, the model predictions were always lower than the actual value despite not setting the weights of the LB to be positive. An optimizer is applied to offset the loss incurred during the testing phase, caused by blacking out the neurons corresponding to $F'$ in the CB. In essence, an optimizer is to be used on top of all those model variants that use future context during the training phase. Figure 5 shows the scatter plot of the difference between model predictions and actual values against the actual values during the testing phase. From the plots it is evident that the model predictions and the actual values are linearly dependent for all domains of the data considered. Hence we use a linear regressor to optimize the model predictions.

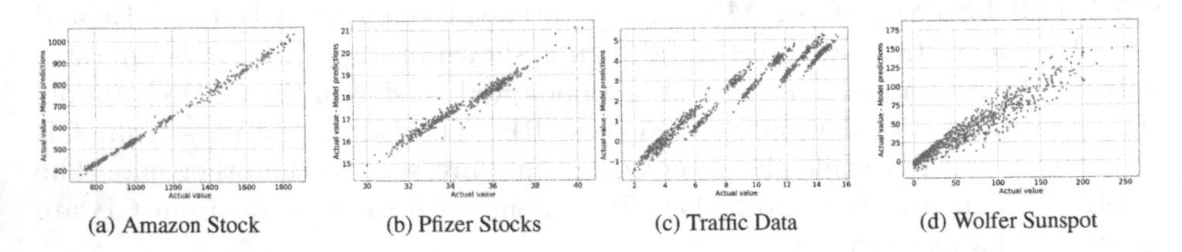

(a) Amazon Stock (b) Pfizer Stocks (c) Traffic Data (d) Wolfer Sunspot

**Fig. 5.** Linear difference of model predictions and actual values

## 6 Experimentation

To test the genericness and robustness of the framework, 3 specific classes of datasets were chosen that cut across diverse domains and have different statistical characteristics.

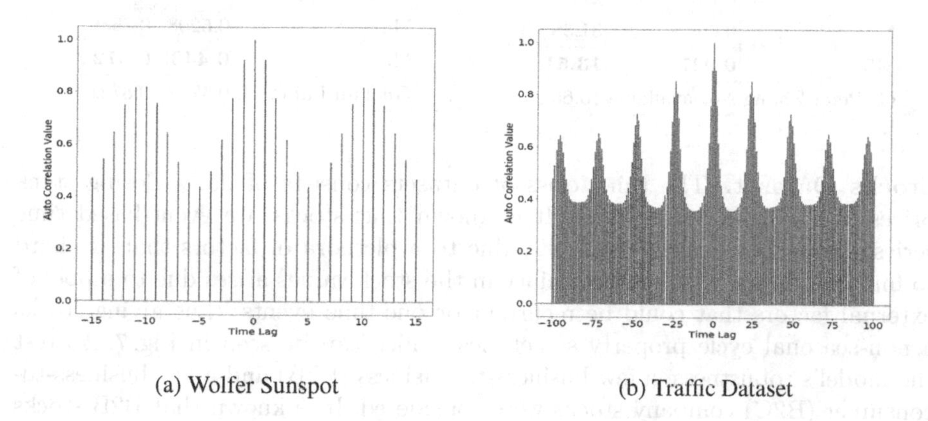


(a) Wolfer Sunspot (b) Traffic Dataset

**Fig. 6.** Autocorrelation plots

**Wolf's Sunspot Dataset.** The second dataset considered is the Wolf's Sunspot dataset, which is one of the well-known time series data sets. The data consists of sunspot numbers from 1700 to 1987 collected annually, giving a total of 288 observations. The data has an average seasonal cyclic pattern repeating itself over a mean period of about 10 years as observed from the autocorrelation plot in Fig. 6a. But experimentally it was observed that the framework performed best with $k$ (refer Sect. 2.3) set as 7. The data is non-linear and non-Gaussian in nature that makes it harder to model than conventional time series datasets. The benchmark uses a hybrid of conventional ARIMA and ANN models [19] resulting in MSE of 186.27. From Table 1, it can be observed that M5 outperforms the baseline and also all the other variants.

**Traffic Dataset.** The first datatset is traffic data that contains the hourly road occupancy rates (between 0 and 1) measured over 185 sensors on San Francisco Bay area freeways during the years 2015–2016, provided by the California Department of Transportation. The autocorrelation plot in Fig. 6b shows the existence of a strong correlation at a periodicity of 24 h. Hence, the models were trained with $k$ (refer Sect. 2.3) set as 24. Metrics for comparison are Correlation and Root Relative Squared Error (refer Sect. 2.4). The benchmark results are an RSE of 0.47 and a CORR of 0.87, achieved by LSTNet [20]. LSTNet combines the strengths of convolutional and recurrent neural networks with an autoregressive component. Results for 3 h ahead predictions are shown in Table 2. From Table 2 we can observe that M1 and M5 outperform the benchmark models by a great margin.

**Table 1.** Wolf's Sunspot results

| Model variants | R2 | RMSE |
|---|---|---|
| M1 | 0.855 | 18.32 |
| M2 | 0.759 | 24.61 |
| M3 | 0.707 | 27.12 |
| M4 | 0.608 | 31.38 |
| M5 | **0.927** | **13.51** |
| G. Peter Zhang | Not available | 13.66 |

**Table 2.** Traffic dataset results

| Model variants | RSE | CORR |
|---|---|---|
| M1 | **0.4552** | **0.9242** |
| M2 | 0.5204 | 0.8795 |
| M3 | 0.5450 | 0.8390 |
| M4 | 0.5208 | 0.8539 |
| M5 | **0.4432** | **0.9724** |
| Guokun Lai et al. | 0.4777 | 0.8721 |

**Stocks Dataset.** The third class of datasets consists of daily closing stock prices of 5 different companies. It is known that stocks are event-based time series values that change erratically due to a plethora of factors that are hard to find and factor in. The uncertainty in the stock values arises due to a pool of external factors that could be recurrent or one-time events, thus giving stocks a non-seasonal cycle property sometimes, which can be seen in Fig. 7. To test the model's robustness, a few business-to-business (B2B) and a few business-to-consumer (B2C) company stocks were considered. It is known that B2B stocks depend on factors like inter-company relations, trust factors, etc. and are harder to predict than B2C stock values. The B2C stocks considered are Amazon, Walmart and Facebook. The B2B stocks considered are Dun & Bradstreet Corp and Pfizer Inc. The framework was trained across all stocks with $k$ (refer Sect. 2.3) set as 10. The efficiencies of variants of framework have been tabulated below.

- It can be observed from Fig. 7a that the stock prices of Amazon have a global increasing trend. Steps, defined as a sudden permanent shift in values, exist throughout the data. The step values keep varying frequently, and hence there exists a difference between the current and future values within the contextual frames considered.
- In the case of Facebook stocks, Fig. 7b shows that the trend of the curve is progressive and smoother than Amazon. The step also changes frequently but there aren't any seasonal and non-seasonal cycles that might affect the performance of the model.
- In the case of Walmart stocks, Fig. 7c shows how significantly different it is from Facebook and Amazon, even though all these companies fall under the ambit of B2C. One reason could be because Walmart is a chain of retail stores, whereas the other two are technological ventures. The factors affecting them would be very different from the ones that influence the retail supermarket.
- In the case of Dun & Bradstreet and Pfizer stocks, there exists some seasonality which can be attributed to the fact that they are B2B. Compared to other stocks, these exhibit more irregularities which can be seen from Fig. 7d and e.

It can be observed from the Tables 3 and 4 that M1 performs the best across all the 5 stock datasets considered. Also, from these tables, we can observe that the framework performs better at univariate than multivariate forecasting.

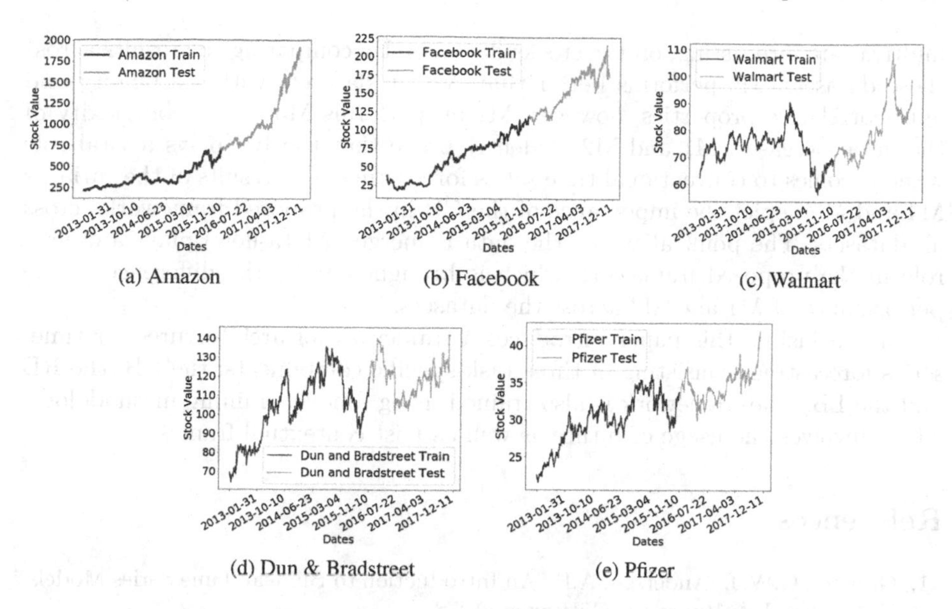


(a) Amazon (b) Facebook (c) Walmart

(d) Dun & Bradstreet (e) Pfizer

**Fig. 7.** Stock curves

**Table 3.** Stocks dataset $R^2$ values

| Model variants | PFE | | WMT | | AMZN | | DNB | | FB | |
|---|---|---|---|---|---|---|---|---|---|---|
| | Uni | Multi | Uni | Multi | Uni | Multi | Uni | Multi | Uni | Multi |
| M1 | **0.988** | 0.957 | 0.9878 | 0.983 | **0.998** | 0.995 | **0.956** | 0.943 | **0.991** | 0.971 |
| M2 | 0.966 | 0.960 | 0.985 | 0.982 | 0.996 | 0.995 | 0.946 | 0.937 | 0.983 | 0.981 |
| M3 | 0.965 | 0.949 | 0.987 | 0.986 | 0.994 | 0.993 | 0.947 | 0.936 | 0.984 | 0.983 |
| M4 | 0.915 | 0.933 | 0.806 | 0.783 | −4.35 | −6.534 | 0.930 | 0.519 | −2.07 | −2.167 |
| M5 | 0.963 | 0.961 | **0.9882** | 0.9881 | 0.996 | 0.996 | 0.952 | 0.948 | 0.982 | 0.982 |

**Table 4.** Stocks dataset *RMSE* values

| Model variants | PFE | | WMT | | AMZN | | DNB | | FB | |
|---|---|---|---|---|---|---|---|---|---|---|
| | Uni | Multi | Uni | Multi | Uni | Multi | Uni | Multi | Uni | Multi |
| M1 | **0.20** | 0.39 | **0.72** | 1.36 | **8.22** | 21.82 | **1.09** | 1.85 | **1.37** | 4.14 |
| M2 | 0.34 | 0.37 | 0.82 | 1.40 | 13.67 | 23.58 | 1.18 | 1.94 | 1.86 | 3.36 |
| M3 | 0.34 | 0.42 | 0.72 | 1.18 | 16.00 | 21.96 | 1.17 | 1.97 | 1.82 | 3.21 |
| M4 | 0.54 | 0.48 | 2.20 | 4.80 | 68.53 | 78.34 | 1.35 | 5.39 | 37.24 | 43.39 |
| M5 | 0.35 | 0.36 | 0.89 | 1.12 | 13.46 | 19.78 | 3.19 | 1.77 | 2.08 | 3.24 |

## 7 Conclusions

The proposed framework outperforms the baseline and the benchmark results across all the datasets considered. From Tables 3 and 4, we can observe that the framework performs better for univariate forecasting compared to multivariate forecasting, on the stocks datasets. By comparing the results across these datasets, M5 performs better than M1 on datasets with seasonality and auto-correlative properties, however, M1 outperforms M5 as the complexity of the data increases. M1 and M2 evidently prove that the RB plays a vital role when it comes to conventional time series forecasting. The results of the variants M4 and M5 depict the importance of the CB in the proposed framework across all datasets. The point at which the data is merged (data-flow) plays a crucial role in the proposed framework which is brought out by the difference in the performance of M1 and M3 across the datasets.

In conclusion, this paper introduces a framework of architectures for time-series forecasting, consisting of three task-specific components: the CB, the RB and the LB. The framework is also trained using a novel training methodology, which involves the usage of future as well as past contextual frames.

## References


1. Granger, C.W.J., Andersen, A.P.: An Introduction to Bilinear Time Series Models. Vandenhoeck & Ruprecht, Göttingen (1978)
2. Engle, R.F.: Autoregressive conditional heteroscedasticity with estimates of the variance of United Kingdom inflation. Econom. J. Econom. Soc. **50**, 987–1007 (1982)
3. Tong, H.: Threshold Models in Non-linear Time Series Analysis, vol. 21. Springer, New York (2012)
4. De Gooijer, J.G., Kumar, K.: Some recent developments in non-linear time series modelling, testing, and forecasting. Int. J. Forecast. **8**(2), 135–156 (1992)
5. Krizhevsky, A., Sutskever, I., Hinton, G.E.: Imagenet classification with deep convolutional neural networks. In: Advances in Neural Information Processing Systems, pp. 1097–1105 (2012)
6. Bao, W., Yue, J., Rao, Y.: A deep learning framework for financial time series using stacked autoencoders and long-short term memory. PloS One **12**(7), e0180944 (2017)
7. Bińkowski, M., Marti, G., Donnat, P.: Autoregressive convolutional neural networks for asynchronous time series (2017). arXiv preprint: arXiv:1703.04122
8. Borovykh, A., Bohte, S., Oosterlee, C.W.: Dilated convolutional neural networks for time series forecasting, March 2017
9. Cui, Z., Chen, W., Chen, Y.: Multi-scale convolutional neural networks for time series classification (2016). arXiv preprint: arXiv:1603.06995
10. Long, J., Shelhamer, E., Darrell, T.: Fully convolutional networks for semantic segmentation. CoRR, abs/1411.4038 (2014)
11. Schuster, M., Paliwal, K.K.: Bidirectional recurrent neural networks. Trans. Signal Process. **45**(11), 2673–2681 (1997)
12. Zheng, Y., Liu, Q., Chen, E., Ge, Y., Zhao, J.L.: Time series classification using multi-channels deep convolutional neural networks. In: Li, F., Li, G., Hwang, S., Yao, B., Zhang, Z. (eds.) WAIM 2014. LNCS, vol. 8485, pp. 298–310. Springer, Cham (2014). https://doi.org/10.1007/978-3-319-08010-9_33
13. Yang, J., Nguyen, M.N., San, P.P., Li, X., Krishnaswamy, S.: Deep convolutional neural networks on multichannel time series for human activity recognition. In: IJCAI, vol. 15, pp. 3995–4001 (2015)

14. Sainath, T., Vinyals, O., Senior, A., Sak, H.: Convolutional, long short-term memory, fully connected deep neural networks. In: ICASSP (2015)
15. Lin, T., Guo, T., Aberer, K.: Hybrid neural networks for learning the trend in time series. In: Proceedings of the Twenty-Sixth International Joint Conference on Artificial Intelligence, IJCAI 2017, pp. 2273–2279 (2017)
16. Graves, A., Mohamed, A.-R., Hinton, G.: Speech recognition with deep recurrent neural networks. In: 2013 IEEE International Conference on Acoustics, Speech and Signal Processing (ICASSP), pp. 6645–6649. IEEE (2013)
17. Hochreiter, S., Schmidhuber, J.: Long short-term memory. Neural Comput. **9**(8), 1735–1780 (1997)
18. Pascanu, R., Gulcehre, C., Cho, K., Bengio, Y.: How to construct deep recurrent neural networks. In: Proceedings of the Second International Conference on Learning Representations (ICLR 2014) (2014)
19. Peter Zhang, G.: Time series forecasting using a hybrid ARIMA and neural network model. Neurocomputing **50**, 159–175 (2003)
20. Lai, G., Chang, W.-C., Yang, Y., Liu, H.: Modeling long- and short-term temporal patterns with deep neural networks. CoRR, abs/1703.07015 (2017)

# Automatic Identification of Interictal Epileptiform Discharges with the Use of Complex Networks

Gustavo H. Tomanik[1], Luiz E. Betting[2], and Andriana S. L. O. Campanharo[1(✉)]

[1] Institute of Biosciences, Departament of Biostatistics, São Paulo State University (UNESP), Botucatu, São Paulo, Brazil
{gustavo.tomanik,andriana.campanharo}@unesp.br

[2] Institute of Biosciences, Departament of Neurology, Psychology and Psychiatry, Botucatu Medical School, São Paulo State University (UNESP), Botucatu, São Paulo, Brazil
luiz.betting@unesp.br

**Abstract.** The identification of Interictal Epileptiform Discharges (IEDs), which are characterized by spikes and waves in electroencephalographic (EEG) data, is highly beneficial to the automated detection and prediction of epileptic seizures. In this paper, a novel single-step approach for IEDs detection based on the complex network theory is proposed. Our main goal is to illustrate how the differences in dynamics in EEG signals from patients diagnosed with idiopathic generalized epilepsy are reflected in the topology of the corresponding networks. Based on various network metrics, namely, the strongly connected component, the shortest path length and the mean jump length, our results show that this method enables the discrimination between IEDs and free IEDs events. A decision about the presence of epileptiform activity in EEG signals was made based on the confusion matrix. An overall detection accuracy of 98.2% was achieved.

**Keywords:** Electroencephalographic time series · Interictal Epileptiform Discharges · Complex networks · Network measures

## 1 Introduction

Epilepsy is as a brain disorder predominantly characterized by recurrent and unpredictable interruptions of the normal brain function. Epileptic seizures are episodes characterized by synchronous neuronal activity in the brain that vary from brief and nearly undetectable focal seizures to long periods of vigorous shaking (generalized tonic-clonic seizures) [13]. Nearly 50 million people worldwide have epilepsy [1]. Approximately 90% of these people live in developing


I. Rojas et al. (Eds.): IWANN 2019, LNCS 11506, pp. 152–161, 2019.
https://doi.org/10.1007/978-3-030-20521-8_13

countries, and three-fourths of them do not have access to the necessary treatment. Sudden and abrupt seizures can have significant impacts on the daily lives of sufferers. Epilepsy can be divided by their clinical manifestation into two main classes, partial and generalized [39]. Partial or focal epilepsies involve only a bounded region of the brain (epileptic focus) and remain restricted to this region while generalized epilepsies involve almost the entire brain [16].

Electroencephalogram is one of the most important diagnostic test for investigation of patients with seizures disorders and epilepsies [38]. Through the EEG analysis is possible to identify pathological patterns of activity between seizures called *Interictal Epileptiform Discharges* (IEDs). The IEDs in EEG signals, clearly distinguished from the activity observed during the epileptic seizure itself, are characterized by spikes followed by slow waves [10]. Visual inspection of EEG recordings for those patterns remains the most common approach, though it is exhausting, time-consuming and can cover hidden characters. Furthermore, it requires a team of experts to analyze the entire length of the EEG recordings in order to detect the IEDs, which may cause disagreement among neurophysiologists concerning the same data due to subjective differences [16]. Therefore, a reliable technique for IEDs detection in EEG signals can support the automated detection and prediction of epileptic seizures in real time and improve quality of life of people suffering from the disease.

In recent years, single-step algorithms have been proposed for spike detection in IEDs based on the following techniques: Fast Fourier Transform, Wavelet Transform, Artificial Neural Networks, Fuzzy Clustering, among others [21,23,24,32,35,40,41]. Moreover, multi-step algorithms have been proposed for spike and slow waves in IEDs based on the combination of at least two of these techniques [3,5,12,16–18,20,22,26,33,37]. Considering that them require assumptions about stationary, high time and frequency resolution, length of the signal and/or noise level and multiple choice parameters, there is still substantial research being performed on the development of a simple single-step algorithm for spikes and slow waves detection in IEDs [7,26].

In the last two decades, research on complex networks became the focus of widespread attention, with developments and applications spanning different scientific areas, from sociology and biology to physics [2,4,30]. One of the reasons behind the growing popularity of complex networks is that almost any discrete structure can be suitably represented as special cases of graphs, whose features may be characterized, analyzed and, eventually, related to its respective dynamics [7,11]. Therefore, several investigations into complex networks involve the representation of the structure of interest as a network, followed by the analysis of the topological features of the obtained representation in terms of a set of informative measures, such as the Strongly Connected Component ($SCC$), the Average Shortest Path Length ($L$) and the Mean Jump Length ($\Delta$) [7,14,31].

Recently, an algorithm to convert a time series to a network has been proposed, and previous works have shown that such a network inherits some of the time series properties [6–9]. In particular, this algorithm was used to distinguish the different abnormal stages of a seizure, such as pre-ictal and ictal [7]. Here, we propose the use of a subtle modification on this algorithm in a novel application: in the automated detection of IEDs in EEG signals of patients diagnosed with idiopathic generalized epilepsy. The remainder of this paper is organized as follows: after this introduction, Sect. 2 describes the proposed method and the methodology used to identify the IEDs. Section 3 presents the topological measures used to characterize and analyze the resulting complex networks. Section 4 describes the data set that is used in this study. The results are presented and discussed in Sect. 5, and the conclusions are presented in Sect. 6.

## 2 Methods

Let $\mathcal{M}$ be a map from a time series $X \in \mathcal{T}$ to a network $g \in \mathcal{G}$, with $X = \{x(t)|t \in \mathbb{N}, x(t) \in \mathbb{R}\}$ and $g = \{\mathcal{N}, \mathcal{E}\}$ being a set of nodes (or vertices) $\mathcal{N}$ and edges (or arcs) $\mathcal{E}$. Let $B$ be the number of bins [25] of $X$, with $B \approx 2*(T)^{(1/3)}$ [27,29], being $T$ the number of points in the time series. The time series $X$ is first normalized by its highest absolute value. Once the $B$ bins have been identified, each value in $X$ is mapped to its corresponding $b_i$, with $i = 1, 2, \ldots, B$. Each bin $b_i$ is assigned to a node $n_i \in \mathcal{N}$ in the corresponding network, and two nodes, $n_i$ and $n_j$ are connected with a weighted edge whenever two values $x(t)$ and $x(t+1)$ belong to $b_i$ and $b_j$, respectively, with $t = 1, 2, \ldots, T-1$. Each weight $w_{ij}$ in the weighted directed adjacency matrix, which is denoted as $A$, is equal to the number of times a value in bin $b_i$ at time $t$ is followed by a value in bin $b_j$ at time $t+1$. Therefore, repeated transitions through the same edge increase the value of the corresponding weight. With proper normalization, the weighted directed adjacency matrix $A$ becomes a Markov transition matrix $W$, with $\sum_j^B w_{ij} = 1$ [8]. In our method, empty bins which correspond to disconnected nodes are removed from the network.

The resulting network is directed, weighted and admits at most $B$ nodes. It is worth mentioning that this method does not require assumptions about stationarity, length of the signal, or noise level [8,9]. It is a numerically simple method and has only one free parameter, namely, $B$, which represents the number of bins/nodes and is typically much smaller than $T$. Figure 1 shows an illustration of our method. An EEG signal with $T = 7,500$ time points is split into 5 segments and all segments are mapped in networks with $B = 11, 23, 20, 11$ and 8 nodes, respectively. IEDs events characterized by spikes and waves (depicted by red colors) are mapped in networks with higher number of nodes and edges.

## 3 Network Measures

Network measures typically quantify connectivity profiles associated with nodes or links and thus reflect the way in which these elements are embedded in the networks. We describe the network measures used to characterize and discriminate the complex networks in Sect. 5, that are, $SCC$, $L$ and $\Delta$.

### 3.1 Strongly Connected Component ($SCC$)

Given an unweighted and directed network, a path from node $n_1$ to node $n_r$ in $g$ is an alternating sequence $(n_1, (n_1, n_2), n_2, (n_2, n_3), \ldots, n_{r-1}, (n_{r-1}, n_r), n_r)$ of

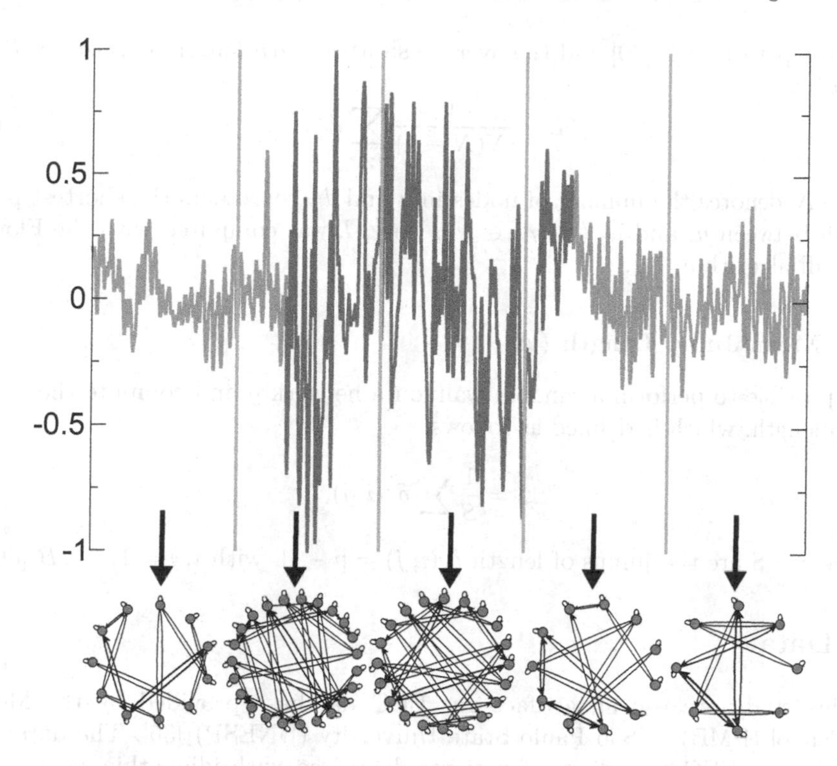


**Fig. 1.** Illustration of the proposed method for a time series with $T = 7,500$ time points split into 5 segments with IEDs events (depicted by red colors) and free IEDs events (depicted by blue colors). The segments are mapped in networks with $B = 11, 23, 20, 11$ and 8 nodes, respectively. IEDs events, characterized by spikes and waves, are mapped in networks with higher number of nodes and edges. (Color figure online)

nodes and edges that belong to $\mathcal{N}$ and $\mathcal{E}$, respectively. Two nodes $n_i$ and $n_j$ in $g$ are path equivalent if there is a path from $n_i$ to $n_j$ and a path from $n_j$ to $n_i$. Path equivalence partitions $\mathcal{N}$ into maximal disjoint sets of path equivalent nodes. These sets are called the strongly connected components of the graph [31]. In order to calculate the strongly connected component of a weighed network, the threshold operation [11] must first be applied to produce the unweighted counterpart.

### 3.2 Average Shortest Path Length ($L$)

A shortest path between two nodes in an unweighted network is a path with the minimum number of edges. If the network is weighted, it is a path with the minimum sum of all edges weights. The length of a shortest path is called shortest path length [30] and the average shortest path length, defined by $L$, is given by:

$$L = \frac{1}{N(N-1)} \sum_{i \neq j} d_{ij}, \quad (1)$$

where $N$ denotes the number of nodes in $g$ and $d_{ij}$ represents the shortest path length between $n_i$ and $n_j$ ($n_i$, $n_j \in N$). Here, $L$ was computed with the Floyd-Washall algorithm [14].

### 3.3 Mean Jump Length ($\varDelta$)

It is possible to perform a random walk on a network $g$ and compute the mean jump length, which is defined as follows:

$$\varDelta = \frac{1}{S} \sum \delta_s(i,j), \quad (2)$$

where s = S are the jumps of length $\delta_s(i,j) = |i - j|$, with $i, j = 1, \ldots, B$ [6].

## 4 Data

In this study, we use an artifact-free EEG database provided by the Medical School (FMB) of São Paulo State University (UNESP) [36]. The database includes ten EEG recordings of patients diagnosed with idiopathic generalized epilepsy. EEGs were performed with a 32 channel recorder with electrodes positioned according to the international 10–20 system of electrode placement [28]. Approximately, twenty-one minute EEGs were recorded at a sampling rate of 1,000 Hz. All electrode impedances were kept below 10 Ω. Exemplary EEG's are depicted in Fig. 2.

## 5 Results

We apply our method to the problem of identifying IEDs events. Each time series with $T = 1,304,000$ was divided in 326 segments with 4,000 points each. After, we mapped 326 time segments into 326 networks with at most $B = 2 * (4,000)^{1/3} \approx 30$ nodes each. In total, 3,260 signals were mapped into 3,260 networks and we obtained 3,260 Markov matrices with at most $B^2 = 900$ elements each. After, we computed $SCC$, $L$ and $\varDelta$ for all networks.

Figure 3 shows $SCC$, $L$ and $\varDelta$ versus the network index $k$, with $k = 1, \ldots, 326$, for the time series in Fig. 2(A). Observe that the proposed method enables the discrimination between IEDs and free IEDs events. More specifically, the spikes and slow waves found in the IEDs are mapped in networks with the highest values of $SCC$, $L$ and $\varDelta$, called here as outliers [15,19]. Figure 4 shows $SCC$, $L$ and $\varDelta$ versus $k$ for the time series in Fig. 2(B). Note that in this case, the highest values of the network measures can either correspond to IEDs or free IEDs events.

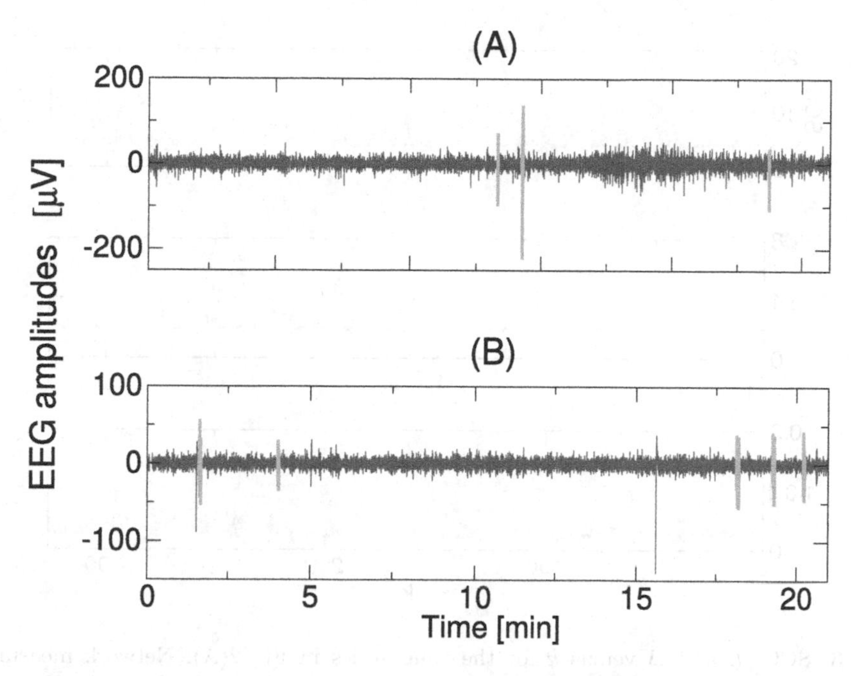


**Fig. 2.** Exemplary of two EEG signals with $T = 1,304,000$ time points each. In all cases, IEDs events are depicted with red colors and free IEDs events with blue colors. IEDs events in (A) have duration of 1, 3 and 1 s, respectively. IEDs events in (B) have duration of 4, 2, 5, 4 and 3 s, respectively. (Color figure online)

The performance of our method in the proper identification of the IEDs was evaluated based on the confusion matrix [34], with $TP$, $FP$, $TN$ and $FN$ being the number of true positives, false positives, true negatives and false negatives, respectively. Here, networks related to IEDs events which are classified as outliers are the true positives and IEDs events which are not classified as outliers are the false negatives. Free IEDs events which are classified as outliers are the false positives and free IEDs events which are not classified as outliers are the true negatives. Sensitivity ($S_e$), Specificity ($S_p$) and Accuracy ($A_c$) were computed. Sensitivity reflected the ability of detecting spikes and waves, while specificity evaluated the ability of discriminating non-spikes and non-waves. These quantities were defined as $S_e = TP/(TP + FN)$ and $S_e = TN/(TN + FP)$, respectively. The accuracy of the classification was calculated by $A_c = (TP+TN)/(TP+TN+FP+FN)$. Table 1 shows the values of $S_e$, $S_p$ and $A_c$ for all EEG signals. Considering all networks measures and all EEG signals, the sensitivity average of 82.3% shows that the proposed method was able to identify most of the networks related to IEDs events. Note that the corresponding values of $\Delta$ for the 3-rd and 8-th EEGs signals are approximately the same for IEDs and free IEDs events. These values are not considered as outliers and therefore present the lowest values of $S_c$, given by 0.25 and 0.2, respectively. The specificity average of 98.9% shows that the method was also

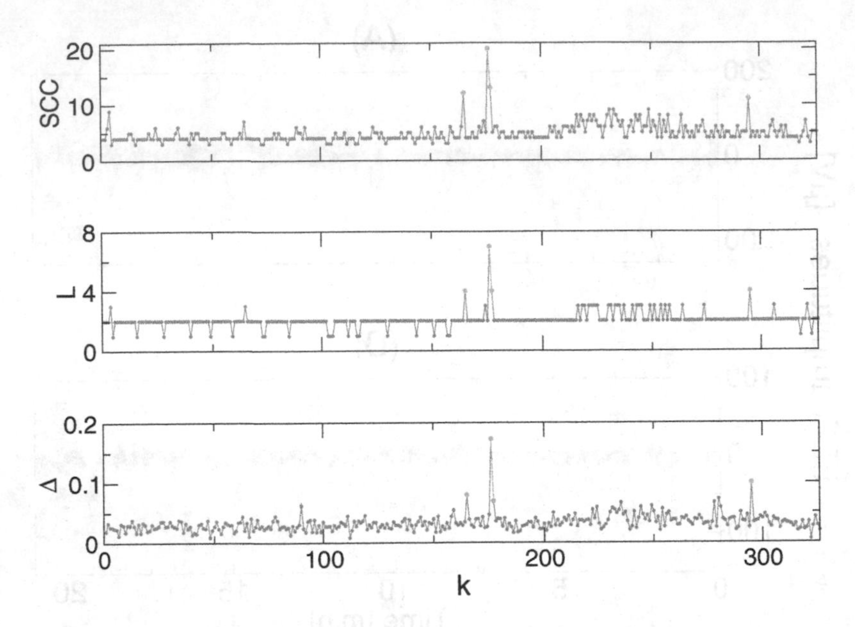


**Fig. 3.** SCC, $L$ and $\Delta$ versus $k$ for the time series in Fig. 2(A). Network measures represented with red points correspond to IEDs events and with blue points to free IEDs events. (Color figure online)

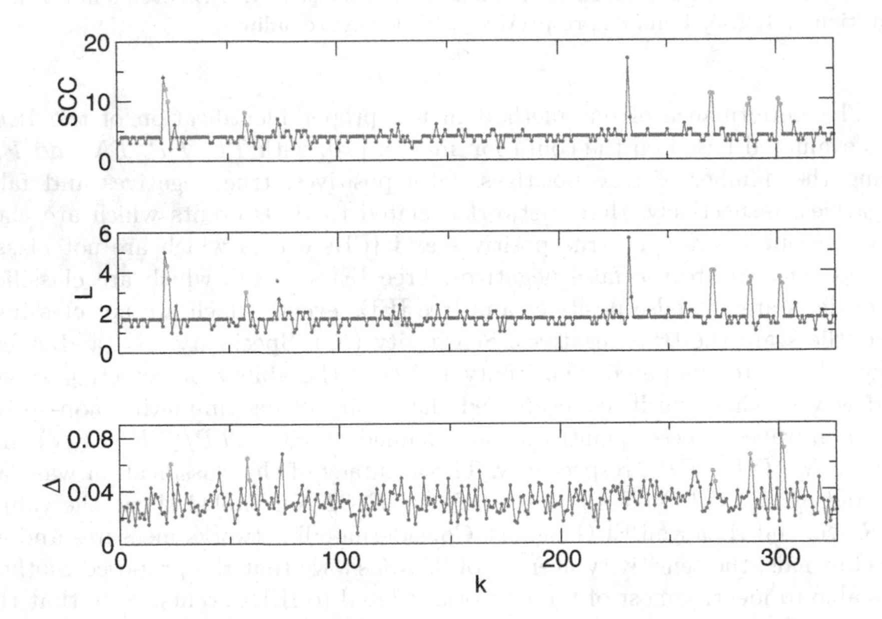


**Fig. 4.** SCC, $L$ and $\Delta$ versus $k$ for the time series in Fig. 2(B). Network measures represented with red points correspond to IEDs events and with blue points to free IEDs events. (Color figure online)

able to properly identify networks related to free IEDs. Moreover, the accuracy average of 98.2% shows that the method properly classified most of the networks. Comparing all the metrics used to identify the IEDs, $L$, followed by $SCC$ and $\Delta$, display the best results.

**Table 1.** Sensitivity ($S_e$), Specificity ($S_p$) and Accuracy ($A_c$) for $SCC$, $L$ and $\Delta$ of ten EEG recordings of patients diagnosed with idiopathic generalized epilepsy.

| EEG signals | SCC | | | L | | | Δ | | |
|---|---|---|---|---|---|---|---|---|---|
| | $S_e$ | $S_p$ | $A_c$ | $S_e$ | $S_p$ | $A_c$ | $S_e$ | $S_p$ | $A_c$ |
| 1 | 1 | 0.954 | 0.954 | 1 | 0.954 | 0.954 | 0.75 | 0.997 | 0.993 |
| 2 | 0.5 | 1 | 0.993 | 0.75 | 0.997 | 0.993 | 0.75 | 0.997 | 0.993 |
| 3 | 0.5 | 1 | 0.993 | 0.5 | 1 | 0.993 | 0.25 | 0.997 | 0.987 |
| 4 | 1 | 0.997 | 0.996 | 1 | 0.997 | 0.996 | 0.75 | 0.985 | 0.981 |
| 5 | 0.75 | 0.994 | 0.991 | 0.75 | 0.994 | 0.991 | 0.75 | 0.994 | 0.991 |
| 6 | 1 | 0.987 | 0.987 | 1 | 0.987 | 0.987 | 0.9 | 1 | 0.996 |
| 7 | 1 | 0.997 | 0.996 | 1 | 0.997 | 0.996 | 0.8 | 1 | 0.994 |
| 8 | 1 | 0.883 | 0.886 | 1 | 0.883 | 0.886 | 0.2 | 1 | 0.976 |
| 9 | 1 | 0.985 | 0.984 | 1 | 0.985 | 0.984 | 0.9 | 1 | 0.996 |
| 10 | 1 | 0.991 | 0.991 | 1 | 0.991 | 0.991 | 0.9 | 1 | 0.997 |

## 6 Conclusions

In this paper, a new single-step approach for spikes and slow waves detection in IEDs was proposed. It is a simple method based on the complex network theory that has only one free parameter and does not require assumptions about stationarity, length of the signal, or noise level. Computational simulations performed here show that this method enables the discrimination between IEDs and free IEDs events, with spikes and slow waves (found in the IEDs) mapped in networks with the highest values of $SCC$, $L$ and $\Delta$. An overall detection accuracy of 98.2% was achieved. In conclusion, the proposed method has good performance in detecting epileptiform activities and it reflects potential for clinical applications in EEG epileptic diagnosis and prediction of epileptic seizures.

**Acknowledgements.** We thank the members of the Amaral Lab at Northwestern University (USA) for insightful comments and suggestions. G. H. T. acknowledges the support of São Paulo Research Foundation (FAPESP), grants 2015/22293-5, 2017/09216-7 and 2018/02014-2 and Coordenação de Aperfeiçoamento de Pessoal de Nível Superior - Brasil (CAPES) - Finance Code 001. L. E. B. acknowledges the support of São Paulo Research Foundation (FAPESP), grant 2016/17914-3. A. S. L. O. C. acknowledges the support of São Paulo Research Foundation (FAPESP), grant 2018/25358-9.

## References


1. Seizures and epilepsy: Hope through research. www page (2004). http://www.ninds.nih.gov/disorders/epilepsy/detail_epilepsy.htm
2. Albert, R., Barabási, A.L.: Statistical mechanics of complex networks. Rev. Mod. Phys. **74**(1), 47 (2002)
3. Azami, H., Sanei, S.: Spike detection approaches for noisy neuronal data: assessment and comparison. Neurocomputing **133**, 491–506 (2014)
4. Boccaletti, S., Latora, V., Moreno, Y., Chavez, M., Hwang, D.U.: Complex networks: structure and dynamics. Phys. Rep. **424**(4–5), 175–308 (2006)
5. Calvagno, G., Ermani, M., Rinaldo, R., Sartoretto, F.: A multiresolution approach to spike detection in EEG. In: 2000 IEEE International Conference on Acoustics, Speech, and Signal Processing, vol. 6 (2000)
6. Campanharo, A.S.L.O., Doescher, E., Ramos, F.M.: Automated EEG signals analysis using quantile graphs. In: Rojas, I., Joya, G., Catala, A. (eds.) IWANN 2017. LNCS, vol. 10306, pp. 95–103. Springer, Cham (2017). https://doi.org/10.1007/978-3-319-59147-6_9
7. Campanharo, A.S.L.O., Doescher, E., Ramos, F.M.: Application of quantile graphs to the automated analysis of EEG signals. Neural Process. Lett. **48**, 1–16 (2018)
8. Campanharo, A.S.L.O., Ramos, F.M.: Hurst exponent estimation of self-affine time series using quantile graphs. Phys. A Stat. Mech. Appl. **444**, 43–48 (2016)
9. Campanharo, A.S.L.O., Sirer, M.I., Malmgren, R.D., Ramos, F.M., Amaral, L.A.N.: Duality between time series and networks. PloS ONE **6**(8), e23378 (2011)
10. Chatrian, G.: A glossary of terms most commonly used by clinical electroencephalographers. Electroencephalogr. Clin. Neurophysiol. **37**, 538–548 (1974)
11. Costa, L.F., Rodrigues, F.A., Travieso, G., Villas, P.R.: Characterization of complex networks. Adv. Phys. **56**(1), 167–242 (2007)
12. El-Samie, F.E.A., Alotaiby, T.N., Khalid, M.I., Alshebeili, S.A., Aldosari, S.A.: A review of EEG and MEG epileptic spike detection algorithms. IEEE Access **6**, 60673–60688 (2018)
13. Fisher, R.S., et al.: ILAE official report: a practical clinical definition of epilepsy. Epilepsia **55**(4), 475–482 (2014)
14. Floyd, R.W.: Algorithm 97: shortest path. Commun. ACM **5**, 345 (1962)
15. Frigge, M., Hoaglin, D.C., Iglewicz, B.: Some implementations of the boxplot. Am. Stat. **43**(1), 50–54 (1989)
16. Gajic, D., Djurovic, Z., Gligorijevic, J., Di Gennaro, S., Savic-Gajic, I.: Detection of epileptiform activity in EEG signals based on time-frequency and non-linear analysis. Front. Comput. Neurosci. **9**, 38 (2015)
17. Geva, A.B., Kerem, D.H.: Forecasting generalized epileptic seizures from the EEG signal by wavelet analysis and dynamic unsupervised fuzzy clustering. IEEE Trans. Biomed. Eng. **45**(10), 1205–1216 (1998)
18. Harner, R.: Automatic EEG spike detection. Clin. EEG Neurosci. **40**(4), 262–270 (2009)
19. Hawkins, D.M.: Identification of Outliers. MSAP, vol. 11. Springer, Dordrecht (1980). https://doi.org/10.1007/978-94-015-3994-4
20. İnan, Z.H., Kuntalp, M.: A study on fuzzy C-means clustering-based systems in automatic spike detection. Comput. Biol. Med. **37**(8), 1160–1166 (2007)
21. Indiradevi, K., Elias, E., Sathidevi, P., Nayak, S.D., Radhakrishnan, K.: A multilevel wavelet approach for automatic detection of epileptic spikes in the electroencephalogram. Comput. Biol. Med. **38**(7), 805–816 (2008)

22. Khalid, M.I., et al.: Epileptic MEG spikes detection using common spatial patterns and linear discriminant analysis. IEEE Access **4**, 4629–4634 (2016)
23. Ko, C.W., Chung, H.W.: Automatic spike detection via an artificial neural network using raw EEG data: effects of data preparation and implications in the limitations of online recognition. Clin. Neurophysiol. **111**(3), 477–481 (2000)
24. Latka, M., Was, Z., Kozik, A., West, B.J.: Wavelet analysis of epileptic spikes. Phys. Rev. E **67**(5), 052902 (2003)
25. Legg, P.A., Rosin, P.L., Marshall, D., Morgan, J.E.: Improving accuracy and efficiency of mutual information for multi-modal retinal image registration using adaptive probability density estimation. Comput. Med. Imaging Graph. **37**(7–8), 597–606 (2013)
26. Liu, H.S., Zhang, T., Yang, F.S.: A multistage, multimethod approach for automatic detection and classification of epileptiform EEG. IEEE Trans. Biomed. Eng. **49**(12), 1557–1566 (2002)
27. Lohaka, H.O.: Making a grouped-data frequency table: development and examination of the iteration algorithm. Ph.D. thesis, Ohio University (2007)
28. Malmivuo, J., Plonsey, R., et al.: Bioelectromagnetism: Principles and Applications of Bioelectric and Biomagnetic Fields. Oxford University Press, Oxford (1995)
29. Morris, A.S., Langari, R.: Measurement and Instrumentation: Theory and Application. Academic Press, San Diego (2012)
30. Newman, M.E.: The structure and function of complex networks. SIAM Rev. **45**(2), 167–256 (2003)
31. Nuutila, E., Soisalon-Soininen, E.: On finding the strongly connected components in a directed graph. Inf. Process. Lett. **49**(1), 9–14 (1994)
32. Özdamar, Ö., Kalayci, T.: Detection of spikes with artificial neural networks using raw EEG. Comput. Biomed. Res. **31**(2), 122–142 (1998)
33. Pang, C.C., Upton, A.R., Shine, G., Kamath, M.V.: A comparison of algorithms for detection of spikes in the electroencephalogram. IEEE Trans. Biomed. Eng. **50**(4), 521–526 (2003)
34. Powers, D.M.: Evaluation: from precision, recall and f-measure to ROC, informedness, markedness and correlation. J. Mach. Learn. Technol. **2**(1), 37–63 (2011)
35. Sartoretto, F., Ermani, M.: Automatic detection of epileptiform activity by single-level wavelet analysis. Clin. Neurophysiol. **110**(2), 239–249 (1999)
36. da Silva Braga, A.M., Fujisao, E.K., Betting, L.E.: Analysis of generalized interictal discharges using quantitative EEG. Epilepsy Res. **108**(10), 1740–1747 (2014)
37. Sitnikova, E., Hramov, A.E., Koronovsky, A.A., van Luijtelaar, G.: Sleep spindles and spike-wave discharges in EEG: their generic features, similarities and distinctions disclosed with fourier transform and continuous wavelet analysis. J. Neurosci. Methods **180**(2), 304–316 (2009)
38. Smith, S.: EEG in the diagnosis, classification, and management of patients with epilepsy. J. Neurol. Neurosurg. Psychiatry **76**(suppl. 2), ii2–ii7 (2005)
39. Tzallas, A.T., Tsipouras, M.G., Fotiadis, D.I.: Automatic seizure detection based on time-frequency analysis and artificial neural networks. Comput. Intell. Neurosci. **2007**(18), (2007)
40. Valenti, P., Cazamajou, E., Scarpettini, M., Aizemberg, A., Silva, W., Kochen, S.: Automatic detection of interictal spikes using data mining models. J. Neurosci. Methods **150**(1), 105–110 (2006)
41. Webber, W., Lesser, R.P., Richardson, R.T., Wilson, K.: An approach to seizure detection using an artificial neural network (ANN). Electroencephalogr. Clin. Neurophysiol. **98**(4), 250–272 (1996)

# Anomaly Detection for Bivariate Signals

Marie Cottrell[1(✉)], Cynthia Faure[1,3], Jérôme Lacaille[2], and Madalina Olteanu[1,4]

[1] SAMM, EA 4543, Panthéon-Sorbonne University, 90 rue de Tolbiac, 75013 Paris, France
cottrell@univ-paris1.fr
[2] Safran Aircraft Engines, Rond Point René Ravaud, Réau, 77550 Moissy Cramayel, France
[3] Aosis Consulting, 20 impasse Camille Langlade, 31100 Toulouse, France
[4] MaIAGE, INRA, Paris-Saclay University, Domaine de Vilvert, 78352 Jouy en Josas, France
http://samm.univ-paris1.fr
https://www.safran-aircraft-engines.com
http://www.aosis.net/
http://maiage.jouy.inra.fr

**Abstract.** The anomaly detection problem for univariate or multivariate time series is a critical question in many practical applications as industrial processes control, biological measures, engine monitoring, supervision of all kinds of behavior. In this paper we propose an empirical approach to detect anomalies in the behavior of multivariate time series. The approach is based on the empirical estimation of conditional quantiles. The method is tested on artificial data and its effectiveness is proven in the real framework of aircraft-engines monitoring.

## 1 Introduction

Detecting anomalies in univariate and multivariate time series is a critical question in many practical applications, such as fault or damage detection, medical informatics, intrusion or fraud detection, and industrial processes control. The present contribution stems from a joint work with the Health Monitoring Department of Safran Aircraft Engines Company. The motivation behind this collaboration was to find a judicious framework for mining the multivariate high-frequency data recorded on board computers during flights, and isolate unusual patterns, abnormal behaviors of the engine, and possibly anomalies. The issue of anomaly detection on flight data is not new, and some previous results of the joint work with Safran may be found in [2] and [18].

In a broader context, the literature on anomaly detection is quite abundant and was developed for decades in various fields: machine learning, statistics, signal processing. The techniques used for addressing the matter use supervised or unsupervised learning, model-based algorithms, information theory or spectral decomposition. For quite an exhaustive review, the reader may refer to [3].


I. Rojas et al. (Eds.): IWANN 2019, LNCS 11506, pp. 162–173, 2019.
https://doi.org/10.1007/978-3-030-20521-8_14

More particularly, we focus here on the issue of detecting collective anomalies or discords – an unusual subsequence of a time-series, in contrast to local anomalies which consist in unique abnormal time-instants – in a multivariate context. Detecting collective anomalies or unusual patterns in univariate time-series has been extensively studied, and we may cite, for instance, algorithms based on piecewise aggregated approximation [14], nearest-neighbor distances [11], Fourier or wavelet transforms [5,16], Kalman filters [12], ... In the multivariate case, anomaly detection has to take into account both the multivariate aspect of the data, and the temporal span, the possibly existing correlations and dependencies. Whereas the initial approaches used time series projection [10] and independent component analysis [1] to convert the multivariate time series into a univariate one, or performed separate anomaly detection for each variable [13], global approaches have been developed only recently. Among these recent works, one may cite, for instance [6], who use a kernel-based method for capturing dependencies among variables in the time series, and [15] who use neural networks for isolating anomalous regions in a multivariate time-series.

This paper addresses the issue of detecting anomalies in a multivariate time series context. Unlike some of the cited literature above, we suppose the data is a set of multivariate time-series, which have already been segmented into patterns of unequal lengths, using some change-point detection technique. Our goal is to find the most unusual of them, and for doing so we take the unsupervised learning approach *(as defined by the AI)*. No hypothesis whatsoever is made on some underlying model, the only constraint is to suppose that one component of the time series, called *key variable* in the sequel, exists, may be distinguished, and its behavior strongly influences the behavior of the rest. The approach we introduce here may be briefly described as follows: first, the initial patterns of the *key variable* are summarized by a fixed number of numerical features, second, they are clustered into an optimal number of clusters, third, the multi-variate patterns are realigned and synchronized within each cluster, fourth, unusual patterns are extracted after computing confidence tubes from empirical quantiles in each cluster. This approach was introduced in [7], and the contribution of the present paper relies in the use of conditional first order quantiles for computing the confidence tubes, instead of quantiles computed at each time instant, independently of the past. As will be illustrated in the Experiments section, this conditional approach greatly improves the ratio of false positives detection.

For simplicity purposes, the bivariate case only is presented here, but the algorithm can be easily extended to higher dimensional data.

The rest of the paper is organized as follows: Sect. 2 describes the main steps of the proposed methodology, Sect. 3 contains results on simulated examples and a comparison between the previous version of the method and the modified one based on conditional quantiles, while Sect. 4 illustrates the method on real-life dataset stemming from flight data.

## 2 Methodology

Let $S_a = (X_a, Y_a)$ be a set of bivariate $\mathbb{R}^2$-valued time series, $a = 1, \ldots A$. For each $a$, $X_a$ and $Y_a$ are of equal length $l_a$. Note that the lengths $l_a$ can be different from one time series to another one. We assume that one of the two variables is a *key variable* (easier to observe, with a limited number of different behaviors, which influences the behavior of the other). This hypothesis is not very restrictive since in many processes, there is a measure which gives a first information on all the others variables (water temperature, blood composition, temperature of the core, etc. according to the application field.) This hypothesis leads us to define two successive levels of analysis: the first one deals with the key variable (say $X_a$) and the second one will further take into account the second variable ($Y_a$).

When assessing the possible existence of abnormal elements, a straightforward approach would consist in mixing together all signals $(X_a)_a$, compute an *average signal*, and say that all signals *far* from this average may be labelled as anomalies. However, there is one major issue with this approach, coming from the fact that the lengths $l_a$ are different, so how does one actually compute an *average signal*? Furthermore, even if one was to find a way to define the *average signal*, there is no reason to summarize all signals behaviors by the average one. Hence, in order to have a better representation of the data, we choose to cluster signals $X_a$. Clustering will provide a limited number of homogeneous groups, and within each of them, one may define a representative signal.

### 2.1 Clustering

The difficulty of dealing with signals of different lengths is overcome as suggested in [9]: each signal $X_a$ is replaced by a fixed-length vector composed of its relevant numerical features (length, midpoint value, median, variance, variances on the two halves, means of the two halves, ...). Let $M$ be the number of relevant features for the set of $(X_a)_a$ time series. Any clustering algorithm may then be used on the feature-vectors data. In the following, let $C_1, C_2, \ldots, C_I$ denote the clusters obtained on the feature vectors, where $I$ is the number of clusters.

In the following, the clustering procedure consisted into first training a self-organizing map (SOM) with a large number of clusters, and second computing an optimal partitioning through an hierarchical agglomerative clustering (HAC) applied to the code-vectors computed by SOM. The optimal number of clusters is selected using an empirical criterion based on the percentage of explained variance. It is worth mentioning at this point that one may avoid summarizing time-series by a fixed number of features, and use some time-series dissimilarity measure (dynamic time warping, the distance defined in Eq. (1), ...) instead. In this case, relational or kernel SOM [17] may be used for clustering.

Once the clustering is trained, each cluster $C_i$ contains a set of time-series $X_a$, say $X_a^i$ for simplicity. They are grouped together based on the similarities of their extracted features, but may have different lengths. Hence, the next step of our methodology is to summarize each cluster by a *reference curve*, $RC_i$, which will serve hereafter for computing quantiles and for visualisation purposes.

### 2.2 Introducing Reference Curves for Summarizing the Clusters

The notion of *reference curve* for a set on univariate unequal time-series as defined hereafter was introduced in [8]. Let us briefly recall here the main steps of how does one compute it.

First, one has to define the dissimilarity between two curves with different lengths, say $X_{a_1}$ and $X_{a_2}$, with lengths $l_1$ and $l_2$, and $l_1 < l_2$. If $X_{a_1} = (x_1, x_2, \ldots, x_{l_1})$, its extended version $\tilde{X}_{a_1}$ of length $l_2 + l_1 + l_2$ is

$$\tilde{X}_{a_1} = (\underbrace{x_1, x_1, \ldots, x_1}_{l_2} \,|\, \underbrace{x_1, x_2, \ldots, x_{l_1}}_{l_1} \,|\, \underbrace{x_{l_1}, x_{l_1}, \ldots, x_{l_1}}_{l_2}).$$

Note that if $X_{a_1}$ is extracted from a longer time series, the extensions at left and at right may be done using the true values in the complete series. The dissimilarity between $X_{a_1}$ and $X_{a_2}$ is the defined as:

$$diss(X_{a_1}, X_{a_2}) = \min_{q \in 1, \ldots, l_1 + l_2 - 1} \frac{\|I_q(\tilde{X}_{a_1}) - X_{a_2}\|}{2 l_2}, \tag{1}$$

where $I_q(\tilde{X}_{a_1}) = \tilde{X}_{a_1}[q, q + l_2 - 1]$ is a $l_2$-long section of $\tilde{X}_{a_1}$ taken between indexes $q$ and $q + l_2 - 1$, for $q = 1, \ldots, l_1 + l_2 + 1$.

Next, one computes the reference curve $RC_i$ of a cluster $C_i$ described by the curves $X_a^i$ as being the one curve among the $|C_i|$ available which minimizes the sum of dissimilarities with respect to all curves in the cluster. Let $L_i$ be the length of $RC_i$. Once $RC_i$ is computed, all the curves in the cluster are realigned with respect to it. This step is achieved by applying a transformation which combines translation, completion and truncation, as described in the next section.

### 2.3 Time-Series Realignment Within Clusters

For realigning curves within a cluster, the idea is to use a similar approach to that in the previous section. We briefly describe it here, using the notations and approach in [8]. Consider a curve $X_a^i$ in $C_i$ with length $l_a^i$. $X_a^i$ is then extended at its left by $L_i$ constant values equal to its first value $X_a^i(1)$, and at its right by $L_i$ constant values equal to its last value $X_a^i(l_a^i)$. The resulting curve is denoted $\hat{X}_a^i$. One may then compute

$$diss(X_a^i, RC_i) = \min_{q \in 1, \ldots, l_a^i + L_i + 1} \frac{\|I_q(\hat{X}_a^i) - RC_i\|}{2 L_i} \tag{2}$$

and

$$q_a^i = \arg \min_{q \in 1, \ldots, l_a^i + L_i + 1} \frac{\|I_q(\hat{X}_a^i) - RC_i\|}{2 L_i}, \tag{3}$$

where $I_q(\hat{X}_a^i) = \hat{X}_a^i[q, q + L_i - 1]$ is a $L_i$-long section of $\hat{X}_a^i$, computed between instants $q$ and $q + L_i - 1$, for $q = 1, \ldots, l_a^i + L_i + 1$.

Each curve $X_a^i$ of cluster $C_i$ is thus replaced by $I_{q_a^i}(\hat{X}_a^i)$, denoted hereafter $\breve{X}_a^i$. In practice, this means replacing the initial unequal-length curves by a set of new curves, similar to the initial ones, and having all length $L_i$. The same *synchronization-transformation* is applied to the second signal $Y_a^i$: it is extended, translated, cut at the same indexes as $X_a^i$. The transformed signal is denoted by $\breve{Y}_a^i$ and has also the same length $L_i$. For the sake of simplicity, we denote by $C_i$ the set of signals $X_a^i$ as well as that of the transformed signals $\breve{X}_a^i$. The corresponding second components $Y_a^i$ or $\breve{Y}_a^i$ define a set $D_i$. We denote by $E_i$ the set of all the couples of transformed signals $(\breve{X}_a^i, \breve{Y}_a^i)$.

### 2.4 Anomaly Detection

Let us recall at this point that our main goal is to detect possibly atypical curves in $E_i$. The anomalies can be related to the first component $X$, to the second one $Y$, or to both. Our approach consists in building quantile-based confidence tubes in each set $C_i$ and $D_i$, for a given confidence level. The simplest way to do this is to compute point-by-point empirical quantiles, which is equivalent to supposing there is no time dependency in the data. Another solution is to take the past instants into account and consider rather conditional quantiles, as suggested in [4] and [19]. We describe next both approaches.

**Point-by-Point Confidence Tubes (CT Method).** Confidence tubes are computed in each cluster, for each set of realigned curves $(\breve{X}_a^i(t), \breve{Y}_a^i(t))$, where $t = 1, \ldots, L_i$. For a given confidence level $1-\alpha$ (typically $\alpha = 5\%$), one denotes by $q_{t,\frac{\alpha}{2}}^X$ (resp. $q_{t,\frac{\alpha}{2}}^Y$) and $q_{t,1-\frac{\alpha}{2}}^X$ (resp. $q_{t,1-\frac{\alpha}{2}}^Y$) the $\alpha$-quantiles computed for each time instant $t$.

The $1-\alpha$ confidence tube of the set $\breve{X}_a^i$ in $C_i$ is defined with a lower bound curve given by $(q_{t,\frac{\alpha}{2}}^X)_{t=1,\ldots,L_i}$ and an upper bound curve given by $(q_{t,1-\frac{\alpha}{2}}^X)_{t=1,\ldots,L_i}$. The $1-\alpha$ confidence tube of the set $\breve{Y}_a^i$ in $D_i$ are similarly computed.

With the previous definition of confidence tubes, we may now introduce the notion of *anomaly.* In the subsequent, a curve in $C_i$ or $D_i$ is considered as *anomalous* if at least $P\%$ consecutive instants are outside the corresponding confidence tube. $P$ is generally to be tuned by the user; for the examples in this manuscript, its value was fixed to 10%.

**Conditional Quantiles (CQ Method).** Since data are time series and since one has strong reasons to suppose a dependency structure in time, another approach for computing empirical quantiles consists in taking the past values of the series into account. Conditional quantiles are also computed on the realigned curves $(\breve{X}_a^i(t), \breve{Y}_a^i(t))$, where $t = 1, \ldots, L_i$. For a given confidence level $1-\alpha$ (typically $\alpha = 5\%$), one denotes by $\tilde{q}_{t,\frac{\alpha}{2}}^X$ (resp. $\tilde{q}_{t,\frac{\alpha}{2}}^Y$) and $\tilde{q}_{t,1-\frac{\alpha}{2}}^X$ (resp. $\tilde{q}_{t,1-\frac{\alpha}{2}}^Y$) the $\alpha$-quantiles computed for each time instant $t$, conditionally on the very recent past, $t-1$. We only consider here a dependency structure of order 1, but more sophisticated ones could be used, with an optimal selection of the number of lags.

The lower conditional quantiles will be computed by solving

$$\mathbb{P}\left(\breve{X}_a^i(t) \leq \tilde{q}_{t,\frac{\alpha}{2}}^X(x) / \breve{X}_a^i(t-1) = x\right) = \frac{\alpha}{2},$$

while the upper ones by solving

$$\mathbb{P}\left(\breve{X}_a^i(t) \geq \tilde{q}_{t,1-\frac{\alpha}{2}}^X(x) / \breve{X}_a^i(t-1) = x\right) = \frac{\alpha}{2}.$$

As the conditional distribution of $\breve{X}_a^i(t)$ conditionally to $\breve{X}_a^i(t-1)$ is generally unknown, the values in $t-1$, $\breve{X}_a^i(t-1)$, have to be discretized in order to have a sufficient number of values of $\breve{X}_a^i(t)$, conditionally to one given value of $\breve{X}_a^i(t-1)$.

Once the conditional quantiles are computed for each discretized value ${}^d\breve{X}_a^i(t-1)$ of $\breve{X}_a^i(t-1)$, a curve in $C_i$ is detected as an *anomaly* if and only if the number of couples $(\breve{X}_a^i(t-1), \breve{X}_a^i(t))$ such that

$$\breve{X}_a^i(t) \notin \left[\tilde{q}_{t,\frac{\alpha}{2}}^X({}^d\breve{X}_a^i(t-1)), \tilde{q}_{t,1-\frac{\alpha}{2}}^X({}^d\breve{X}_a^i(t-1))\right]$$

is greater than a certain threshold $P$, fixed by the user (usually 10 %). The $1-\alpha$ conditional quantiles of the set $\breve{Y}_a^i$ in $D_i$ are similarly computed.

## 3 An Experimental Example on Simulated Data

We first illustrate the proposed methodology and the interest of using conditional quantiles instead of point-by-point ones on a simulated example. 2,000 artificial bivariate time series were built, with an $X$-variable having one of the four shapes described in Fig. 1. These shapes were inspired by the real-data from aircraft flights that will be described in the next section. For each time series, its length is randomly generated (between 500 and 3,000 time instants), as well as the instant where the slope changes (between the first and the second third of the series), the slope values. A Gaussian centered noise with a variance varying between 10 and 100 is also added for supplementary noise, and eventually the resulting curves are smoothed using a 5-degree polynomial.

The $Y$-variables are simulated starting from $X$: for a given time series $X_a$, a change-point $z$ is randomly selected, and a new time series is simulated with two

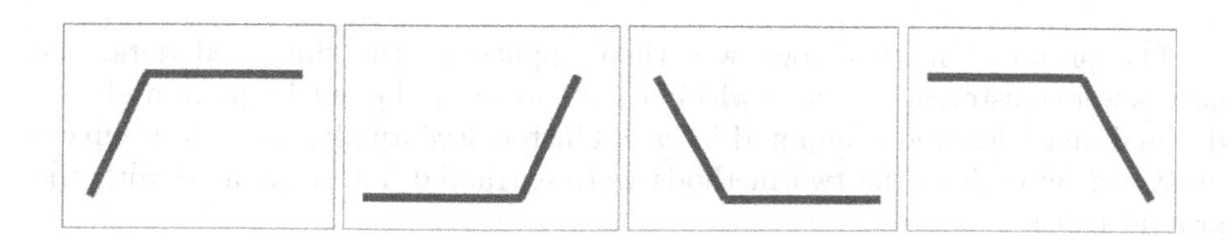

**Fig. 1.** Four shapes used for the simulated $X$-signals

random slopes before and after the change-point. The slopes are selected more or less in the same range as the $X_a$'s. If necessary, $Y_a$ is extended or cut so as to have the same length as $X_a$. Eventually, the resulting curves are also smoothed using a 5-degree polynomial. The resulting $X$-curves, $Y$-curves, and a bivariate example $(X_a, Y_a)$ of simulated time series are illustrated in Fig. 2.

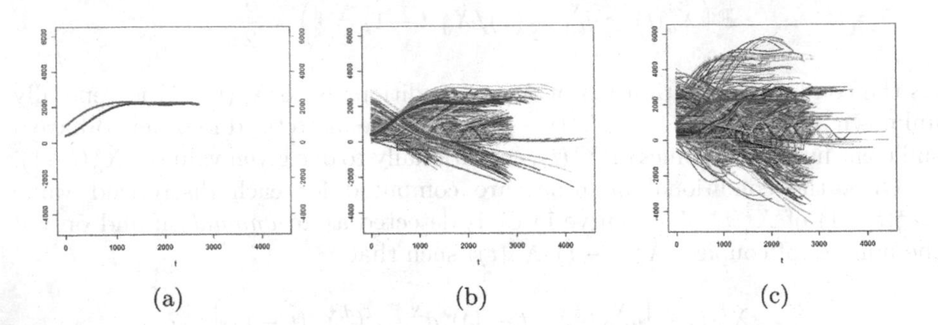

**Fig. 2.** (a): Example of bivariate signal, $X_a$ in blue and $Y_a$ in red; (b): All signals $X_a$; (c): All signals $Y_a$ (Color figure online)

Furthermore, we added 50 anomalies to the simulated data. Four types of anomalies were introduced, as shown in Fig. 3: sinusoidal, "hat"-shaped, and linear. A couple of variables $(X, Y)$ can be anomalous in $X$ only, in $Y$ only or in both components.

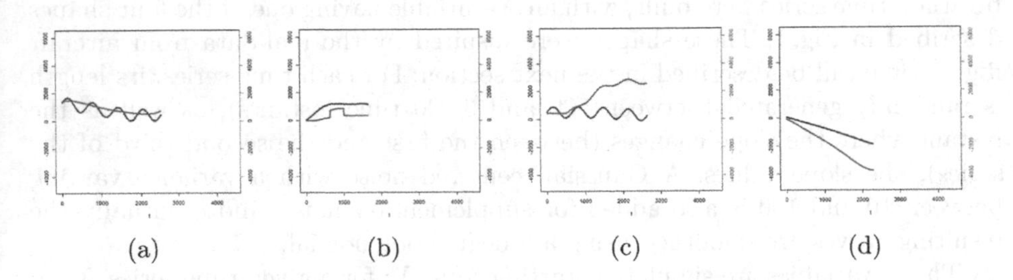

**Fig. 3.** Examples of atypical curves ($X$ blue, $Y$ red) (Color figure online)

The proposed methodology was then applied to the simulated data: the curves were clustered, realigned within each cluster, and point-by-point and conditional quantiles were computed in each cluster. Eventually, anomalous curves identified by each of the two methods were extracted, and compared with the ground truth.

Clustering with SOM followed by HAC yielded five final clusters, denoted $C_1, C_2, \ldots, C_5$ for the $X$ curves, and $D_1, D_2, \ldots, D_5$ for the $Y$ curves. We recall here that $D_i$ are defined by $D_i = \{Y_a / X_a \in C_i\}$, $i = 1, \ldots, 5$. The resulting clusters, which are globally homogeneous, are illustrated in Fig. 4. The reference curves $RC_1, \ldots, RC_5$ computed according to the definition in Sect. 2.2, are drawn with solid red lines in Fig. 4.

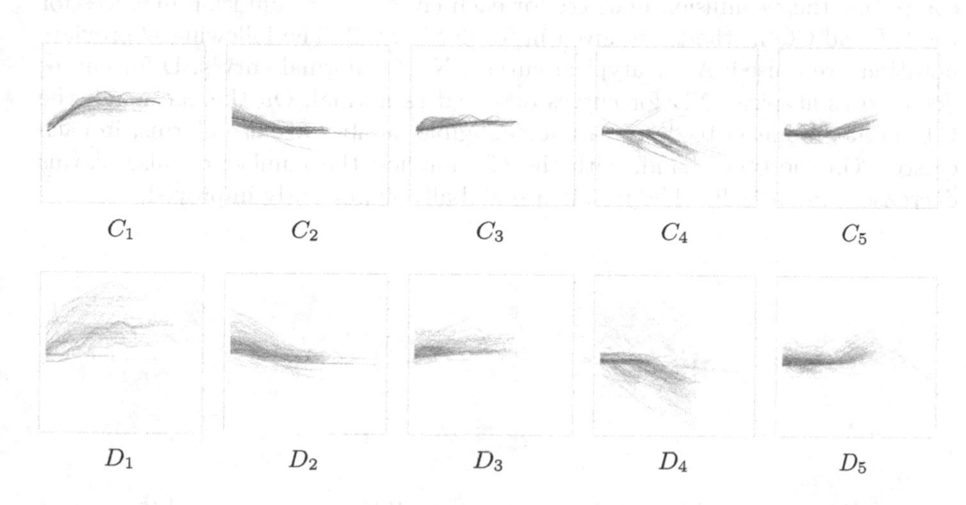


**Fig. 4.** Clustering ($X$-curves on top and $Y$-curves below) (Color figure online)

Next, $X$ and $Y$ time-series within each cluster are realigned using the transformation in Sect. 2.3. We only illustrate here the results for the first cluster. Figure 5a and b contains the initial curves $X_a^1$ of cluster $C_1$ and their transformed $\breve{X}_a^1$. Similarly, Fig. 5c and d displays the initial curves $Y_a^1$ of $D_1$ and their transformed $\breve{Y}_a^1$.

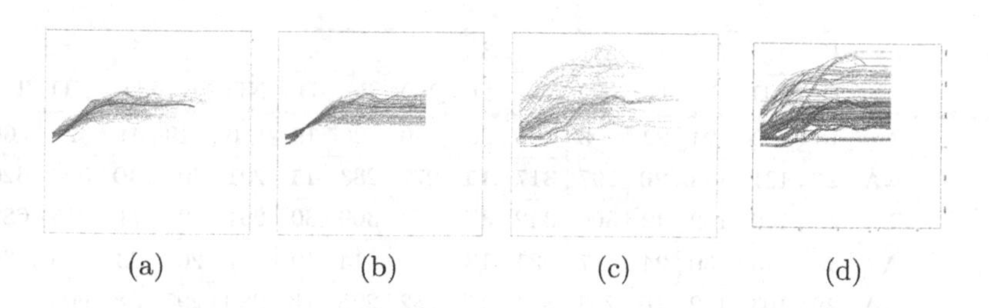


**Fig. 5.** Realignment and transformation of the curves in $C_1$ (a, b) and $D_1$ (c, d), initial curves in (a) and (c), transformed ones in (b) and (d)

Eventually, the last step consists in detecting the atypical curves in each cluster after having computed the confidence tubes (CT) with point-by-point empirical quantiles, and empirical conditional quantiles (CQ) (see Sect. 2.4). Figure 6a presents the results of the CT detection method for cluster $C_1$: all the $X$- curves in $C_1$ are drawn and the detected atypical ones are highlighted in red. The $Y$-curves and the detected atypical ones are displayed in Fig. 6b. In the same way,

Fig. 6c and d presents the result of the CQ detection method. At first glance, both methods seem to detect the same atypical curves.

The two approaches for identifying anomalous curves are then compared by computing their confusion matrices for each cluster. The confusion matrices for the CT and CQ methods are given in Tables 1 and 2. The following abbreviate notations were used: A for atypical curves, NA for normal curves, D for curves detected as atypical, ND for curves detected as normal. On the one hand, the CT method appears to detect a not negligible number of false alarms, in each cluster. On the other hand, with the CT method the number of false alarms decreases dramatically. The results are globally significantly improved.

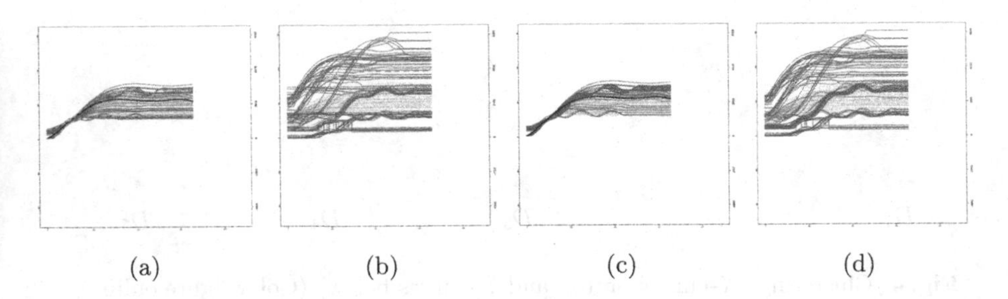

(a) (b) (c) (d)

**Fig. 6.** Anomalies detected in cluster 1 with the CT method: (a) - on $X$; (b) - on $Y$. Anomalies detected in cluster 1 with the CQ method: (c) - on $X$; (d) - on $Y$. Normal curves are in blue, abnormal in red. (Color figure online)

**Table 1.** Confusion matrices for the CT method, in bold the number of false alarms

| CT | $i$ | 1 | | | 2 | | | 3 | | | 4 | | | 5 | | |
|---|---|---|---|---|---|---|---|---|---|---|---|---|---|---|---|---|
| | | D | ND | T | D | ND | T | D | ND | T | D | ND | T | D | ND | T |
| $C_i$ | A | 15 | 9 | 24 | 22 | 3 | 25 | 26 | 0 | 26 | 19 | 0 | 19 | 41 | 25 | 66 |
| | NA | **27** | 131 | 158 | **20** | 297 | 317 | **11** | 272 | 283 | **11** | 291 | 302 | **30** | 590 | 620 |
| | T | 42 | 140 | 182 | 42 | 300 | 342 | 37 | 272 | 309 | 30 | 291 | 321 | 71 | 615 | 686 |
| $D_i$ | A | 35 | 15 | 50 | 24 | 7 | 31 | 13 | 1 | 14 | 19 | 7 | 26 | 23 | 16 | 39 |
| | NA | **29** | 103 | 132 | **18** | 293 | 311 | **13** | 282 | 295 | **15** | 280 | 295 | **38** | 609 | 647 |
| | T | 64 | 118 | 182 | 42 | 300 | 342 | 26 | 283 | 309 | 34 | 287 | 321 | 61 | 625 | 686 |
| $C_i$ & $D_i$ | A | 6 | 5 | 11 | 9 | 2 | 11 | 12 | 1 | 13 | 9 | 0 | 9 | 20 | 16 | 36 |
| | NA | **11** | 160 | 171 | **10** | 321 | 331 | **10** | 286 | 296 | **10** | 302 | 312 | **16** | 634 | 650 |
| | T | 17 | 165 | 182 | 19 | 323 | 342 | 22 | 287 | 309 | 19 | 302 | 321 | 36 | 650 | 686 |

**Table 2.** Confusion matrices for the CQ method, in bold the number of false alarms

| CQ | $i$ | 1 | | | 2 | | | 3 | | | 4 | | | 5 | | |
|---|---|---|---|---|---|---|---|---|---|---|---|---|---|---|---|---|
| | | D | ND | T | D | ND | T | D | ND | T | D | ND | T | D | ND | T |
| $C_i$ | A | 23 | 1 | 24 | 24 | 1 | 25 | 25 | 1 | 26 | 17 | 2 | 19 | 45 | 21 | 66 |
| | NA | **0** | 158 | 158 | **1** | 316 | 317 | **5** | 278 | 283 | **5** | 297 | 302 | **25** | 595 | 620 |
| | T | 23 | 159 | 182 | 25 | 317 | 342 | 30 | 279 | 309 | 22 | 299 | 321 | 70 | 616 | 686 |
| $D_i$ | A | 37 | 13 | 50 | 26 | 5 | 31 | 14 | 0 | 14 | 23 | 3 | 26 | 26 | 13 | 39 |
| | NA | **13** | 119 | 132 | **13** | 298 | 311 | **12** | 283 | 295 | **15** | 280 | 295 | **35** | 612 | 647 |
| | T | 50 | 132 | 182 | 39 | 303 | 342 | 26 | 283 | 309 | 38 | 283 | 321 | 61 | 625 | 686 |
| $C_i$ & $D_i$ | A | 8 | 3 | 11 | 9 | 2 | 11 | 12 | 1 | 13 | 7 | 2 | 9 | 19 | 17 | 36 |
| | NA | **7** | 164 | 171 | **8** | 323 | 331 | **8** | 288 | 296 | **10** | 302 | 312 | **19** | 631 | 650 |
| | T | 15 | 167 | 182 | 17 | 325 | 342 | 20 | 289 | 309 | 17 | 304 | 321 | 38 | 648 | 686 |

## 4 An Application to Real-World Data

Let us now illustrate the method on a real dataset, containing bivariate time-series recorded during aircraft flights. Some of the following results are excerpted from Faure's PhD thesis [7], completed in collaboration with the Health Monitoring Department of Safran Aircraft Engines Company. In actuality, the real data contained much higher dimensional time series, since the sensors placed on the engines register more than 50 different signals. Here, for illustration purposes, we only considered the fan speed and the temperature inside the engine. The fan speed is the *key variable* $X$ and the temperature is $Y$.

The data was initially made of 549 flights and 8 different engines, with a mean duration of 2.8 h per flight. After having partitioned the flight data using some change-point detection algorithm, 4500 transient ascending phases (time-series with an ascending behavior) were extracted, clustered, and the rest of the methodology described in Sect. 2 was applied to them. Their lengths are comprised between 200 and 10,000 time units (8 Hz).

We describe next the results obtained in one cluster only, that mainly contains take-offs. Figure 7 contains the atypical curves (red) with respect to the normal ones (blue), detected with the CT ((a) and (b) for the $X$ and $Y$ curves) and the CQ methods ((c) and (d) for the $X$ and $Y$ curves). The CT method detects 12 atypical $X$-curves, while the CQ detects 14 (6 common ones). As for the $Y$-curves, CT detects 24 anomalies, CQ 18, of which 14 anomalies are common. Let us stress also the case of bi-dimensional curves which are detected as atypical for $X$ **and** for $Y$ by both methods (see Fig. 8). The CT method detects three couples which are atypical for $X$ and for $Y$, while the CQ method finds ten couples, that include the first three. Again, the CQ method has better performances than the CT method, even though it is not possible to compute the confusion matrices here, since we have no a priori knowledge about the existence of the anomalies. In this real-world study, the experts have been able to bring a validation to our findings. The detected cases corresponded to some events that they could identify.

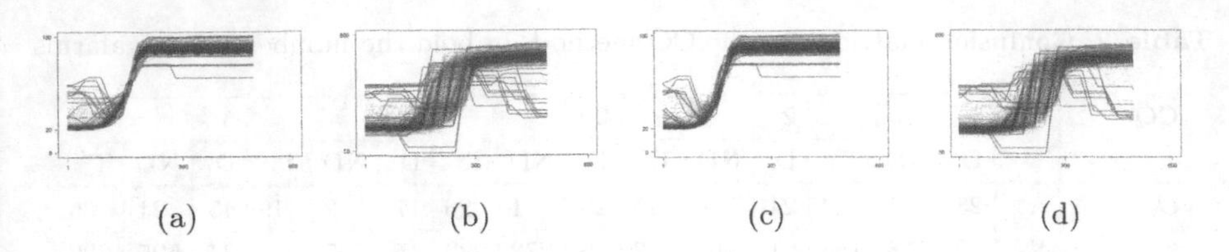

(a) (b) (c) (d)

**Fig. 7.** Anomalies detected in cluster 1 with the CT method: (a) - on $X$; (b) - on $Y$. Anomalies detected in cluster 1 with the CQ method: (c) - on $X$; (d) - on $Y$. Normal curves are in blue, abnormal in red. (Color figure online)

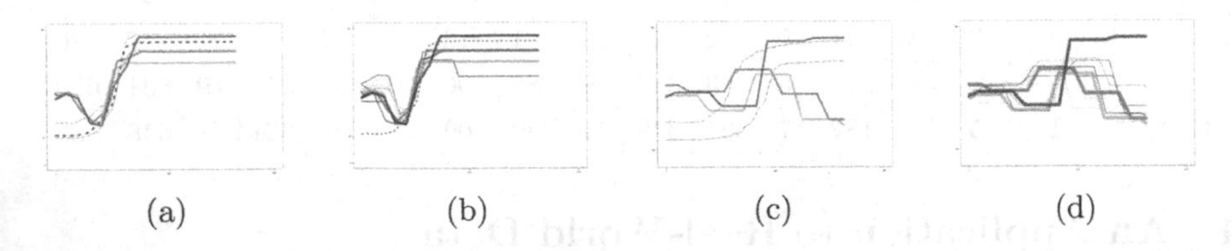

(a) (b) (c) (d)

**Fig. 8.** Three atypical couples are detected by the CT method ($X$-curves in (a) and $Y$-curves in (b)). The CQ method detects seven more couples drawn in (c) and (d). The reference curve of the cluster is represented in red dots in (a) and (c). The confidence tubes defined by the CT method are in green dots. (Color figure online)

## 5 Conclusion

We describe a complete methodology, based upon clustering, curve realignment and empirical quantiles computation, that allows one to detect abnormal elements in a large sample of time series with unequal lengths. When using conditional quantiles, the results are dramatically improved and the number of false alarms significantly reduced. We strongly believe that these results could be further improved by an optimal selection of the time lag in the conditional quantiles computation. From a practical point of view, the proposed methodology may be very useful in helping experts and engineers identify abnormal behaviors in the signals recorded during aircraft engines utilization. In the case of aircraft engine real data, companies may use this technique to increase the probability of detecting any kind of atypical, abnormal behavior of some recorded variable, in order to prevent any incident and to plan the maintenance events.

## References


1. Baragona, R., Battaglia, F.: Outliers detection in multivariate time series by independent component analysis. Neural Comput. **19**(7), 1962–1984 (2007)
2. Bellas, A., Bouveyron, C., Cottrell, M., Lacaille, J.: Anomaly detection based on confidence intervals using SOM with an application to health monitoring. In: Villmann, T., Schleif, F.-M., Kaden, M., Lange, M. (eds.) Advances in Self-Organizing Maps and Learning Vector Quantization. AISC, vol. 295, pp. 145–155. Springer, Cham (2014). https://doi.org/10.1007/978-3-319-07695-9_14
3. Chandola, V., Banerjee, A., Kumar, V.: Anomaly detection: a survey. ACM Comput. Surv. **41**(3), 15:1–15:58 (2009)

4. Charlier, I., Paindaveine, D., Saracco, J.: Conditional quantile estimation based on optimal quantization: from theory to practice. Comput. Stat. Data Anal. **91**, 20–39 (2015). https://hal.inria.fr/hal-01108504
5. Chen, X.y., Zhan, Y.y.: Multi-scale anomaly detection algorithm based on infrequent pattern of time series. J. Comput. Appl. Math. **214**(1), 227–237 (2008)
6. Cheng, H., Tan, P., Potter, C., Klooster, S.: A robust graph-based algorithm for detection and characterization of anomalies in noisy multivariate time series. In: 2008 IEEE International Conference on Data Mining Workshops, pp. 349–358 (2008)
7. Faure, C.: Détection de ruptures et identification des causes ou des symptômes dans le fonctionnement des turboréacteurs durant les vols et les essais. Ph.D. thesis, Université Paris 1 Panthéon-Sorbonne (2018)
8. Faure, C., Bardet, J.M., Olteanu, M., Lacaille, J.: Design aircraft engine bivariate data phases using change-point detection method and self-organizing maps. In: Conference: ITISE - International Work-Conference on Time Series. University of Granada, Granada, Spain, September 2017
9. Faure, C., Bardet, J.M., Olteanu, M., Lacaille, J.: Using self-organizing maps for clustering anc labelling aircraft engine data phases. In: Lamirel, J.C., Cottrell, M., Olteanu, M. (eds.) 12th International Workshop on Self-Organizing Maps and Learning Vector Quantization, Clustering and Data Visualization (WSOM+ 2017), pp. 1–8. IEEE, Piscataway (2017)
10. Galeano, P., Peña, D., Tsay, R.S.: Outlier detection in multivariate time series by projection pursuit. J. Am. Stat. Assoc. **101**(474), 654–669 (2006)
11. Keogh, E., Lin, J., Fu, A.: HOT SAX: efficiently finding the most unusual time series subsequence. In: Fifth IEEE International Conference on Data Mining (ICDM 2005), pp. 226–233 (2005)
12. Knorn, F., Leith, D.J.: Adaptive Kalman filtering for anomaly detection in software appliances. In: IEEE INFOCOM Workshops 2008, pp. 1–6 (2008)
13. Lakhina, A., Crovella, M., Diot, C.: Characterization of network-wide anomalies in traffic flows. In: Proceedings of the 4th ACM SIGCOMM Conference on Internet Measurement, IMC 2004, pp. 201–206. ACM, New York (2004)
14. Lin, J., Keogh, E., Lonardi, S., Chiu, B.: A symbolic representation of time series, with implications for streaming algorithms. In: Proceedings of the 8th ACM SIGMOD Workshop on Research Issues in Data Mining and Knowledge Discovery, DMKD 2003, pp. 2–11. ACM, New York (2003)
15. Malhotra, P., Vig, L., Shroff, G., Agarwal, P.: Long short term memory networks for anomaly detection in time series. In: Verleysen, M. (ed.) European Symposium on Artificial Neural Networks, Computational Intelligence and Machine Learning (ESANN 2015), pp. 89–94. Presses universitaires de Louvain, Louvain-la-Neuve (2015)
16. Michael, C.C., Ghosh, A.: Two state-based approaches to program-based anomaly detection. In: Proceedings of the 16th Annual Computer Security Applications Conference, pp. 21–30. IEEE Computer Society (2000)
17. Olteanu, M., Villa-Vialaneix, N.: On-line relational and multiple relational SOM. Neurocomputing **147**(1), 15–30 (2015)
18. Rabenoro, T., Lacaille, J., Cottrell, M., Rossi, F.: Anomaly detection based on indicators aggregation. In: International Joint Conference on Neural Networks (IJCNN 2014), Beijing, China, pp. 2548–2555, July 2014
19. Samanta, T.: Non-parametric estimation of conditional quantiles. Stat. Probab. Lett. **7**, 407–412 (1989)

# A Scalable Long-Horizon Forecasting of Building Electricity Consumption

Naveen Kumar Thokala(✉), S. Spoorthy(✉), and M. Girish Chandra(✉)

TCS Research and Innovation, Bangalore 66, Karnataka, India
{naveen.thokala,s.spoorthy,m.gchandra}@tcs.com

**Abstract.** Load Forecasting plays a key role in the efficient operation of the building energy management systems. In this work, a framework is proposed for effective scalable implementation of long-term (month and quarter ahead) building load forecasting. It comprises of techniques to deal with outliers and missing values, dynamic input feature selection as well as a hybrid algorithm combining direct and recursive strategies for forecasting. The solution is successfully validated using the real consumption data of six office buildings and further the average accuracies of 92–95% and 88–92% for month and quarter ahead respectively, corroborates its usefulness.

## 1 Introduction

Continuous power supply is vital for the effective functioning of commercial buildings. Electrical load forecasting solutions help commercial building managers assess their energy demands, and at the same time, help electrical utilities in planning their supply operations. These aspects help to avoid revenue losses due to power supply disruptions and align supply with the demand and vice-versa. Needless to say, this is vital for present-day demand-supply conditions. To help in such energy management, a range of load forecasting solutions have been developed of late for short term, medium term and long term, depending on the horizon of the forecast. These horizons may cater to the requirements of hours ahead, day-ahead, quarter-ahead and month-ahead forecasting of power consumption. Usually, *Short-term Load Forecasts*, such as hours-ahead and day-ahead forecasts, help the building manager to streamline the power consumption by adopting peak-load shaving, time-of-use pricing/demand response and energy bidding approach [1,3,4]. The *Medium-term* or *Long-term* building energy forecasts, i.e month and quarter-ahead forecasts, respectively, are useful in assessing fuel resources required for the continuous operation of the building, budgeting etc. Medium-term and Long-term forecasts also can be used at the distribution level so that electrical utilities can plan their operations of the electrical power system efficiently.

In the literature, various solutions were proposed for *Short-term load forecasting* of building-level power consumption, based on statistical and machine


I. Rojas et al. (Eds.): IWANN 2019, LNCS 11506, pp. 174–185, 2019.
https://doi.org/10.1007/978-3-030-20521-8_15

learning approaches [2,3,5,6,10]. However limited attention was given to *Long-term forecasting.* Recently, Naveen et al. [4] have proposed Non-linear AutoRegressive with eXogenous (NARX) Neural Network and SVR based month ahead forecasting. One of the major challenges in long-term building load forecasting is the time horizon for which independent (explanatory) variables also need to be forecasted. Therefore, in the current work, we focus mainly on the strategy to be used for long-term forecasting rather than resolve the function to be used to model the time-series. There are two well-known strategies in the literature for forecasting (i) Direct Forecast (DF) and (ii) Recursive Forecast (RF). Direct Forecast uses the data until the current instant and maps all the future values as a function of the past values of the time series. On the other hand, RF forecasts one value in future and the forecasted value is augmented to the input while forecasting the second value and so on until the last forecast value. Both of these strategies have their own strengths and weaknesses [11]. RF performs better when the model is correctly specified [9], else the forecasts will be largely biased. With DF, it is robust to model misspecification but the approach may lead to too much variance in the forecast with respect to the time-series data input. The selection of any one of these methods is a compromise between bias and variance. In order to take advantage of strengths in both the methods, a Hybridized Direct-Recursive multi-step ahead strategy is proposed in this paper. The whole proposed solution consists of many stages to implement long-term building load forecasting.

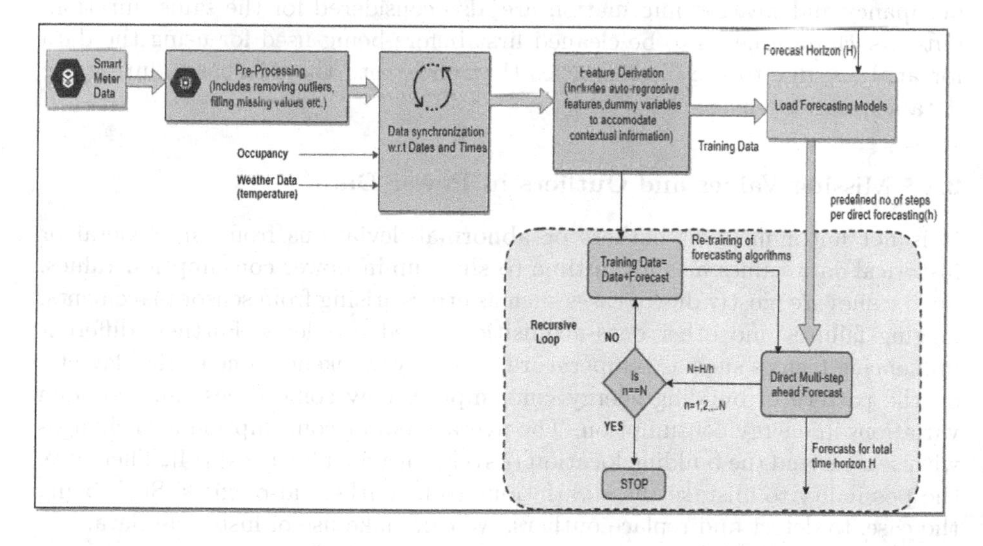


**Fig. 1.** Block diagram for building load forecasting

The block diagram in Fig. 1 depicts the framework for the proposed hybrid recursive-direct forecasting strategy. The solution includes a pre-processing stage to deal with outliers/missing values, followed by synchronization of smart meter data with other sensory data. In feature derivation stage additional features necessary for the forecast are derived. The algorithm or the method to be employed for modelling the building load consumption depends on the time horizon of forecasting and the granularity of the data. Figure 1 captures this aspect as well.

The major contributions of this paper are

- An effective algorithm for detecting the outliers and treatment of missing values.
- An useful strategy for long-horizon forecasting using hybridized direct and recursive methods.

The organization of the paper is as follows. Data pre-processing and Synchronization steps are detailed in Sect. 2. The proposed hybridized algorithm is explained in Sect. 3.3. The month ahead load forecasting algorithms and the associated results are explained in Sect. 4. Corresponding details for quarter-ahead are provided in Sect. 5. There is a discussion on the results in Sect. 6 and conclusions are presented in Sect. 7.

## 2 Data Pre-processing

While developing the models for long-horizon forecasting, we have considered the past data of buildings' power consumption for a period of 1.5 years. The other sensory information useful in modelling the building's power consumption like occupancy and weather information are also considered for the same duration. Often, such data needs to be cleaned first before being used for using the data for analysis, due to the discrepancies that enter into the databases during the data acquisition phase.

### 2.1 Missing Values and Outliers in Power Data

It is not uncommon for outliers or abnormal deviations from the general or historical data values at a given time to show up in power consumption values. Such values are mostly due to cases such as errors arising from sensor placements, logging failures and other data-acquisition based problems. Further, different influencing factors such as temperature, weekday/weekend, time of the day etc. on the pattern of building energy consumption may some times cause sudden variations in energy consumption. The average power consumption also changes with seasons and the building location (psychrometric influences) [14]. Therefore, the possibility to mistake these variations to be outliers also exists. Such being the case, to detect and replace outliers, we can make use of historical data.

We propose the following process to detect and replace outliers. Let us consider detecting and replacing a value at an instant $i$ ($Y(i)$) for a particular building. We prepare a block of data values from the historical data by picking the values existing during the same season, same day of the week and at the same time of the day. This block can be further refined by considering whether the instant $i$ belongs to either a working day or holiday, and correspondingly choosing the values from the historical data; this block can be represented by the vector $\mathbf{d}(i)$. The entries in this constructed block are averaged and the resulting average is compared with the value in question ($Y(i)$). If the latter deviates from the average by a large amount, we can declare the value as an outlier and is replaced with the average. The deviation can be ascertained by considering a suitable threshold; denoting the average by $\mu_i$ and the threshold by $\tau$, $Y(i)$ is an outlier if $Y(i) > \tau * \mu_i$. Similarly, we can handle the missing values; for this purpose, the missing value at the instant $i$ is estimated as $\mu_i$, that is, we would be using $Y(i) = \mu_i$. It is useful to note that this simple technique is effective in handling long duration of missing values or a burst of no data, which is often encountered in the realistic scenarios.

With the good amount of historical data at our disposal, the aforementioned strategy is adopted in the present work. In fact, when experimented with the other existing sophisticated techniques, the proposal outscored them; comparisons are not captured here. When the prior data is limited, other techniques (for example, using interpolation/filtering in the graph signal domain) can be used to negotiate the missing values and outliers to some extent. These are again not covered in this paper.

### 2.2 Data Synchronization

The next step, *data synchronization*, is carried out to have an appropriate mapping of the power consumption data with other sensory information like occupancy and temperature. The occupancy and temperature data available with us are at one-hour granularity. To synchronize, the temperature is interpolated every 15 min using the hourly values (as temperature varies slowly), occupancy is maintained constant and replicated every 15 min during the one-hour time-period.

## 3 Hybridized Recursive-Direct (HRD) Multi-step Ahead Forecast

Time-series forecasting is defined as an extrapolation of the time series for the future dates or times and it requires modelling the time-series in terms of its components like the trend, seasonality, cyclic patterns, and exogenous variables if any. Forecasting involves developing the models using the historical data and forecasting the future values of the time-series [12].

Let $Y$ be a stationarized time series and $Y(t+1)$ the value of the time series $Y$ at $(t+1)$. Then $Y(t+1)$ can be modeled as follows:

$$\hat{Y}(t+1)|_t = f(Y, X) \tag{1}$$

where $f$ is a function and its properties are decided by the learning algorithm considered for modelling the time series. In the linear case, $Y(t+1)$ would be the linear function of lag values of $Y$ and other independent variables $x_i$, then the function $f$ takes the following form

$$\hat{Y}(t+1)|_t = \sum_{i=1}^{n} \phi_i * Z_i + \epsilon_t \tag{2}$$

where, $\phi$ represents the parameters of the function learnt and $Z$ represents the vector consisting of both lagged values of $Y$ and independent variables $X$ that impact $Y$. In general, all the past data might not be used for modelling, only appropriate lags that are useful are used. There are many ways suggested in the literature to decide upon the lag-length, such as Akaike information criterion (AIC), Bayesian Information Criteria (BIC) and auto-correlation function or correlogram.

In multi-step ahead forecasting, it requires to forecast values at multiple steps ahead, for example forecasting $Y_{(t+1)-(t+h)}$ at $t$. There are two methods that are used in the state-of-art to do multi-step ahead forecasting namely (i) Direct forecast and (ii) Recursive forecast.

### 3.1 Direct Forecast

Direct forecast uses a static mapping to forecast future values. Direct forecasts are made using a horizon specific estimated model. For example, multi-step ahead forecast for $h$ steps at $t$ for the series $y$ is done in the following way using direct forecast methodology.

$$\begin{aligned} \hat{Y}(t+1)|_t &= f_1(Y, X) \\ \hat{Y}(t+2)|_t &= f_2(Y, X) \\ &. \\ &. \\ \hat{Y}(t+h)|_t &= f_h(Y, X) \end{aligned} \tag{3}$$

where $Y(t), Y(t-1), ..Y(1)$ is the time series $Y$ at time $t$. $X$ represents the eXogenous variables. $\hat{Y}(t+1), \hat{Y}(t+2), ..., \hat{Y}(t+h)$ are the forecasted values given the time series $Y$ at $t$. $f_1, f_2, ..., f_h$ are the different functions trained to forecast the values of the series at different instants of time. For every step in the forecast horizon, separate function is trained using the past data of $Y$.

### 3.2 Recursive Forecast

On the other hand, recursive forecast involves forecasting the multiple values of the series each at a time in a recursive fashion. In recursive forecasting, a single function is trained and parameters of the function are re-estimated at every time step with the new sample adding to the time series. The recursive forecasting for multi-step ahead forecast is of the form

$$\begin{aligned} \hat{Y}(t+1)|_t &= f(Y, X) \\ \hat{Y}(t+2)|_t &= f(\hat{Y}(t+1)|_t, Y, X) \\ &. \\ &. \\ \hat{Y}(t+h)|_t &= f(\hat{Y}(t+h-1)|_t, \hat{Y}(t+h-2)|_t, ..Y, X) \end{aligned} \tag{4}$$

where $\hat{Y}(t+1)$ is forecasted using the data from the beginning to the current instant $t$, while forecasting $\hat{Y}(t+2)$, the forecasted value at instant $t+1$ i.e $\hat{Y}(t+1)$ is added to the input series to re-train the function $f$, similarly $\hat{Y}(t+h-1)$, $\hat{Y}(t+h-2), ..., \hat{Y}(t+1)$ are used while forecasting $\hat{Y}(t+h)$ in a recursive fashion.

### 3.3 Hybridized Direct-Recursive (HDR) Forecast

In theory *recursive forecasts* are more accurate than *direct forecasts*, if models are specified correctly [9]. As the direct forecast uses the separate models for each and every step in the forecast horizon, information between the consecutive points of the time-series is not considered resulting in high variance in the forecast. Recursive forecasts suffer from biases as the forecasted values are iteratively used as inputs to forecast future values in the time series. The error in forecasted values propagates as these are used as inputs for further forecasts. Choosing between these two forecasts is a tradeoff between the bias and estimation variance. Therefore a **Hybrid Strategy** using both direct and recursive methods is devised to address the weaknesses of both the methods. The proposed hybridized direct-recursive (HDR) strategy is to have a total forecast horizon $H$ divided into $n$ slots each of length $h$. Direct multi-step forecast strategy is used to forecast the first $h$ values and then these forecasted values of the initial slot ($h$ steps) are used as an input for forecasting the second slot. This continues until the last slot.

$$
\begin{aligned}
&\hat{Y}(t+1,.,t+h)|_t = f(Y,X)\\
&\hat{Y}(t+h+1,.,t+2h)|_t = f(\hat{Y}(t+h),.,\hat{Y}(t+1),Y,X)\\
&.\\
&.\\
&\hat{Y}(t+(n-1)h+1,.,t+nh)|_t = f(\hat{Y}(t+(n-1)h),.,\hat{Y}(t+1),Y,X)
\end{aligned}
\tag{5}
$$

**Estimation Variance and Bias:** In Direct-Forecasting method, multiple functions are used to obtain forecasts at multiple time-instants due to which the total variance in the forecast is the addition of all the individual variances at each time-step. In Recursive-Forecasting, a certain bias, $b_i$, is induced at every time-instant $i$ due to the recursive nature of forecasting. This bias is additive in nature, in that it adds up to the previous bias at every recursive step.

In HDR, we observe that the sum total variance is restricted to $\Sigma_1^h \sigma_i$, as against $\Sigma_1^n \sigma_i$ in Direct-Forecasting ($h << n$). Similarly, the total bias in HDR is $\Sigma_h^n b_i$ as against $\Sigma_i^n b_i$ for Recursive-Forecasting. This observation re-affirms the robustness of HDR against Direct-Forecasting and Recursive-Forecasting methods.

The proposed HDR forecast strategy is used to forecast month and quarter ahead buildings' total power consumption. The actual office buildings' data is used for demonstrating the performance of the proposed approach.

# 4 Buildings' Month Ahead Load Forecasting

Smart meter measures total buildings' power consumption once in every 15 min. The buildings' power consumption is influenced by many factors like temperature (HVAC loads), occupancy and other factors like the working day or holiday etc. making it difficult for the linear algorithms like linear regression (LR), ARIMA etc., to forecast accurately. Artificial Neural Networks (ANN) and the Support Vector Regression (SVR) are the two well-known techniques for modelling non-linear and complex time series as mentioned in [4,7,8]. The modelling of the time series for the month ahead forecasting includes (i) Data Pre-processing as explained in Sect. 2. (ii) Feature Derivation and Selection (iii) Modelling of the time series using ANN/SVR.

## 4.1 Feature Derivation and Selection

New features derivation and selection are carried out for time-series modelling of buildings' power consumption. Dummy variables are created to capture the contextual information like Day of the week, Time of the Day, Holidays etc. The lags of buildings' power consumption are selected using the partial autocorrelation function (pacf). The power consumption lags (order selected using the pacf function) together with dummy variables used to capture contextual information forms the *Predictor Matrix.* Predictor Matrix is the input to the learning model and the buildings' power consumption is the output.

## 4.2 HDR Based Month Ahead Forecasting Using ANN and SVR

The modelling of the month ahead forecasting is carried out as mentioned in Sect. 3.3. The total forecast horizon for the month ahead forecasting at 15 min granularity is $H = 2880$ (30 days with 96 samples every day). The total forecasting horizon is divided into 5 slots, i.e. each slot will have 576-time steps. The direct forecast is used to forecast 576-time steps (6 days) and these forecasted values are added to input to forecast for the next 6 days of buildings power consumption. This is repeated five times to get a month ahead forecasting. The implementation is as explained in the below set of equations (Eq. 6).

$$
\begin{aligned}
&\hat{Y}(t+1,.,t+576)|_t = f(Y,X)\\
&\hat{Y}(t+577,.,t+1152)|_t = f(\hat{Y}(t+576),.,\hat{Y}(t+1),Y,X)\\
&.\\
&.\\
&\hat{Y}(t+2305,.,t+2880)|_t = f(\hat{Y}(t+2304),.,\hat{Y}(t+1),Y,X)
\end{aligned}
\tag{6}
$$

ANN is used as the function $f$ in the above equations for performance comparison. The performance of the proposed approach is demonstrated using the actual buildings' power consumption data.

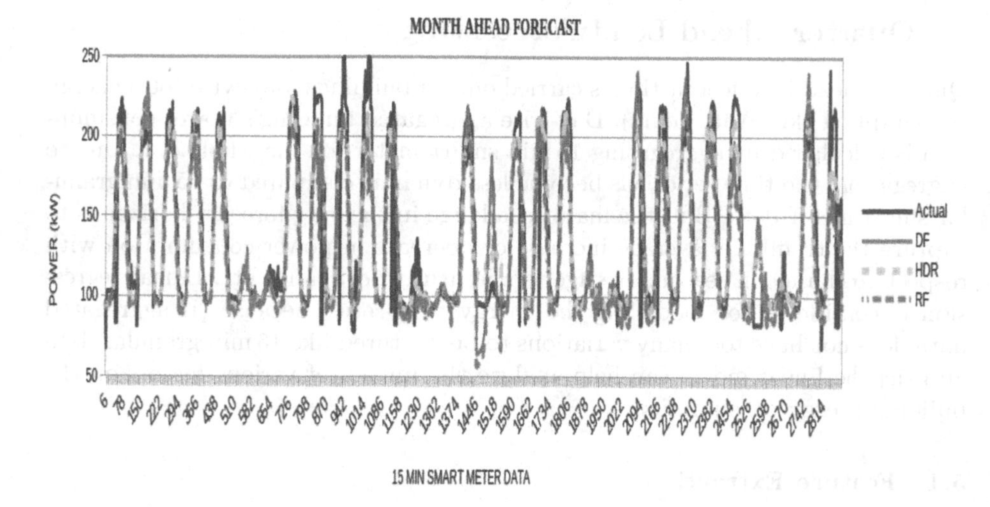


**Fig. 2.** Month ahead forecast comparison

### 4.3 Results

The proposed strategy is tested on six buildings for the month ahead forecasting. From Fig. 2, It is clear that the performance of the proposed strategy HDR is either better than the other two approaches DF and RF or matched with the best of the two approaches. For performance evaluation, Symmetric Mean Absolute Percentage Error (sMAPE) and Normalized Root Mean Squared Error (NRMSE) are considered as the error terms as these are scale independent making them applicable for comparing algorithms' performance across buildings' of different capacities. Symmetric MAPE (sMAPE) is considered over MAPE to avoid over penalty to the negative errors. Figures 3 and 4 clearly indicate that the proposed approach has improved forecasts for most of the buildings.

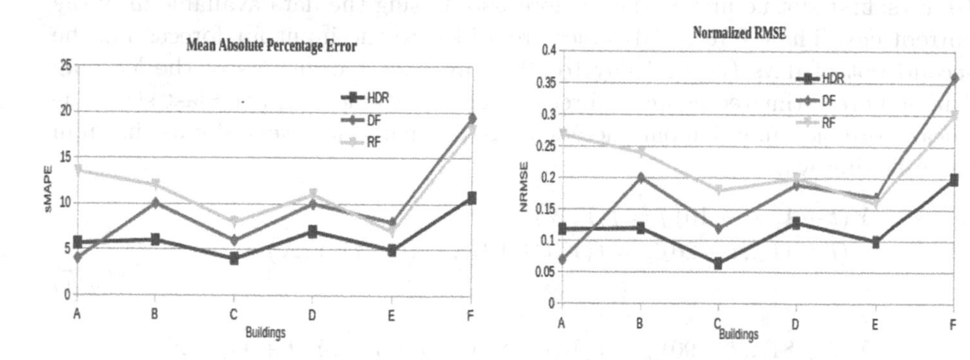


**Fig. 3.** Symmetric MAPE comparison

**Fig. 4.** Normalized RMSE comparison

# 5 Quarter Ahead Load Forecasting

Quarter ahead load forecasting is carried out on buildings' day-wise total energy consumption (kiloWatt hours). Day-wise aggregated building's power consumption is calculated by aggregating 15 min smart meter for the whole day. Due to aggregation, the time series has become less dynamic compared to 15 min granular smart meter data. But, the major challenge in the long forecast horizon is to capture the trend, i.e, average increase or decrease in power consumption with respect to change in season, average temperature, occupancy etc. Linear regression is considered for modelling the energy time-series because (i) aggregated data does not have too many variations to be captured like 15 min granular data and (ii) the linear model can help analyze the impact of various factors on the building energy consumption.

## 5.1 Feature Extraction

Similar to day-wise energy consumption, day-wise maximum and minimum temperatures, as well as occupancy, are considered as time-series. For the forecast horizon, maximum and minimum temperature forecasts are taken from the weather websites like [13], and the occupancy is extrapolated using ARIMA. Contextual information such as working day, month of the year is captured in the form of dummy variables.

## 5.2 HDR Based Quarter Ahead Forecasting Using Linear Regression

Multivariate Linear Regression (LR) is used to model energy time-series. The input feature vector for the LR function consists of contextual information (day of the week, working day), minimum and maximum temperatures for a day, day-wise average occupancy and auto-regressive terms (lagged values). The output of the function is buildings day-wise energy consumption. As explained in the Sect. 4.2, the total forecast horizon 90 days is divided into 9 slots, each having 10 days; first slot i.e first 10 days is forecasted using the data available until the current day. These forecasted values are added to the input for forecasting the second slot of days (i.e $11^{th}$ day to $20^{th}$ day), the parameters of the learning function are estimated again to forecast. This continues until the last slot. The linear regression function parameters are re-estimated for every slot as shown in the following way.

$$\begin{aligned} &\hat{Y}(t+1,.,t+10)|_t = f(Y,X) \\ &\hat{Y}(t+11,.,t+20)|_t = f(\hat{Y}(t+10),.,\hat{Y}(t+1),Y,X) \\ &. \\ &. \\ &\hat{Y}(t+81,.,t+90)|_t = f(\hat{Y}(t+80),\hat{Y}(t+79).,\hat{Y}(t+1),Y,X) \end{aligned} \tag{7}$$

In the above equation, the learning function $f$ is of the form,

$$\hat{Y}(t+1,.,t+10)|_t = \phi_1 * X_1 + \phi_2 * X_2 .. + \phi_n * X_n + \epsilon_t \tag{8}$$

where, $X_1, X_2, ...X_n$ are the inputs for the function, $\phi_i$ represents the weights/parameters trained which could signify the impact of the input features on the output i.e buildings' overall consumption. This model could be of help in understanding and taking control measures for building energy management.

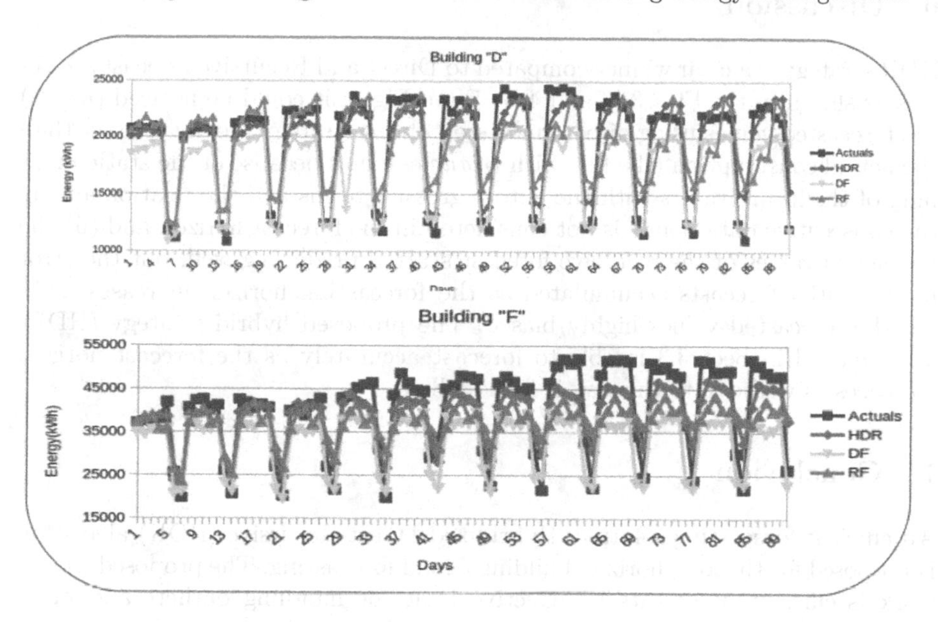


**Fig. 5.** Quarter ahead forecast comparison

### 5.3 Results

The proposed approach is used to forecast the six office buildings future energy consumption, the forecast is compared over actual consumption data in real-time. The performance of the proposed algorithms is given in Fig. 5. It could be clearly noticed that HDR forecast strategy has improved performance compared to both the techniques (Direct Forecast and Recursive Forecast). Figures 6 and 7 show that HDR forecast strategy has out-performed the other two strategies

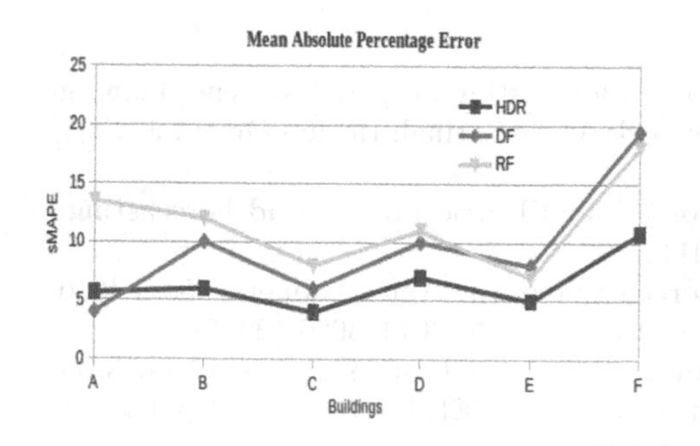


**Fig. 6.** sMAPE comparison

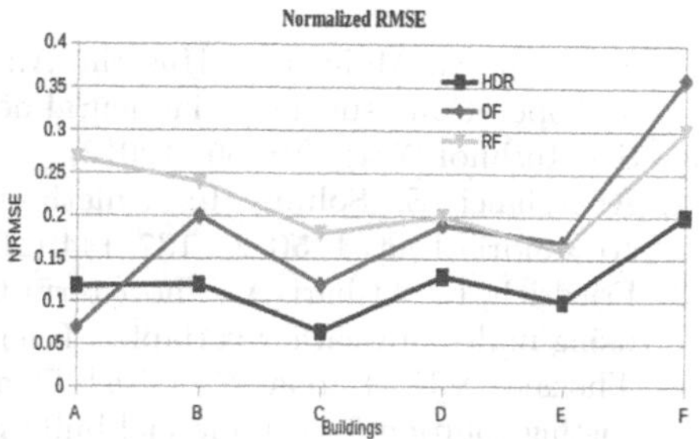


**Fig. 7.** NRMSE comparison

for all the buildings ($A$-$F$). Building $A$ is exceptional, where direct forecast (DF) has superior performance compared to the HDR approach.

# 6 Discussion

HDR strategy is a clear winner compared to Direct and Recursive forecast strategies as shown in the Figs. 3, 6 and 4, 7. From Fig. 5, it could be noticed that (i) the forecasted values using direct multi-step ahead strategy are much lower than the actual consumption (the line with *triangles*); it is because of the static mapping of the future values with the auto-regressive terms and the relation among the consecutive data points is not considered in the forecast horizon and (ii) the forecasted values of the recursive forecast started predicting well, but the error in the initial forecasts accumulated as the forecasting horizon increases making the forecasted values highly biased. The proposed hybrid strategy (HDR) as explained in Sect. 3.3 is able to forecast accurately as the forecast horizon increases as well due to the way it is modelled.

# 7 Conclusion

An efficient forecasting solution, Hybridized Direct-Recursive (HDR) algorithm is proposed for the long-horizon buildings' load forecasting. The proposed framework is efficient due to its (i) effective logic for handling outliers and missing values (ii) additional contextual features derived to capture the dynamics of the buildings' energy consumption and (iii) the algorithms capability in re-estimating the functional parameters for every new slot in forecast horizon making it more efficient in forecasting accurately even as the horizon increases. We have been able to test the performance of the proposed solution on actual buildings' energy consumption in real-time and the efficacy of the proposed algorithm is demonstrated in the results Sect. 5.3. The proposed framework covers all the steps required for the real-time implementation of the algorithm. Additionally, it scales well in terms of using it across a large number of buildings. Further, the framework can be adapted for different applications.

# References


1. Swaroop, R., Abdulqader Hussein, A.: Load forecasting for power system planning and operation using artificial neural network At Al Batinah Region Oman. J. Eng. Sci. Technol. **7**(4), 498–504 (2012)
2. Almeshaiei, E., Soltan, H.: A methodology for Electric Power Load Forecasting. Alexandria Eng. J. **50**(2), 137–144 (2011)
3. Friedrich, L., Afshari, A.: Short-term forecasting of the Abu Dhabi electricity load using multiple weather variables. Energy Procedia **75**, 3014–3026 (2015)
4. Thokala, N.K., Bapna, A., Girish Chandra, M.: A deployable electrical load forecasting solution for commercial buildings. In: 2018 IEEE International Conference on Industrial Technology (ICIT). IEEE (2018)

5. Fernández, I., Borges, C.E., Penya, Y.K.: Efficient building load forecasting. In: 2011 IEEE 16th Conference on Emerging Technologies and Factory Automation (ETFA). IEEE (2011)
6. Mai, W., Chung, C.Y., Wu, T., Huang, H.: Electric load forecasting for large office building based on radial basis function neural network. In: 2014 IEEE PES General Meeting — Conference and Exposition, National Harbor, MD, pp. 1–5 (2014)
7. Ojemakinde, B.T.: Support vector regression for non-stationary time series (2006)
8. Tealab, A., Hefny, H., Badr, A.: Forecasting of nonlinear time series using ANN. Future Comput. Inf. J. **2**(1), 39–47 (2017)
9. Marcellino, M., Stock, J.H., Watson, M.W.: A comparison of direct and iterated multistep AR methods for forecasting macroeconomic time series. J. Econometrics **135**(1–2), 499–526 (2006)
10. Kaur, N., Kaur, A.: Electricity demand prediction using artificial neural networks in data mining. Int. J. Technol. Res. Eng. **2**(8) (2015)
11. Hamzaçebi, C., Akay, D., Kutay, F.: Comparison of direct and iterative artificial neural network forecast approaches in multi-periodic time series forecasting. Expert Syst. Appl. **36**(2), 3839–3844 (2009)
12. Brockwell, P.J., Davis, R.A., Calder, M.V.: Introduction to Time Series and Forecasting, vol. 2. Springer, New York (2002). https://doi.org/10.1007/978-3-319-29854-2
13. http://www.accuweather.com/en/in/india-weather
14. Paresh, S., Thokala, N., Girish Chandra, M.: Facility energy management based on adaptive thermal comfort protocols: a case-study. In: 5th IEEE International Conference on Signal Processing and Integrated Networks (SPIN) (2018)

# Long-Term Forecasting of Heterogenous Variables with Automatic Algorithm Selection

Naveen Kumar Thokala[1(✉)], Kriti Kumar[1(✉)], M. Girish Chandra[1(✉)], and Karumanchi Ravikumar[2(✉)]

[1] TCS Research and Innovation, Bangalore, India
{naveen.thokala,kriti.kumar,m.gchandra}@tcs.com
[2] Bangalore, India

**Abstract.** An Enterprise System Bus (ESB) is a software which is used to communicate between various mutually interacting software applications in manufacturing plants. ESB performance is very important for the smooth functioning of the system. Any degradation or failure of the ESB results in huge revenue loss due to production discontinuity. Therefore, maintaining ESB in a healthy state is very essential and there are multiple factors related to *resource utilization, workload and number of interfaces etc.*, which influences the performance of the ESB. Forecasting these variables at least a day ahead (24 h ahead) is required to take appropriate actions by the business team to maintain the ESB performance under control. But, these variables are heterogeneous (continuous, discrete and percentages) in nature, highly non-linear and non-stationary. The challenges associated with forecasting of these variables are (i) long horizon (24 h ahead forecast at 5 min granularity requires to forecast 288 steps) (ii) data generated from these kinds of systems makes it very difficult to use any linear statistical methods like state-space models, ARIMA etc. To address these challenges, the paper presents a framework where a basket of learning algorithms based on Artificial Neural Network (ANN), Support Vector Regression (SVR) and Random Forests (RF) were used to model the chaotic behavior of the time series with a real-time automatic algorithm selection mechanism which enables appropriate forecasting algorithm to be chosen dynamically based on the performance over a time window, resulting in different algorithms being used for forecasting the same target variable on different days. Importance of the proposed strategy was demonstrated with suitable forecasting results for different variables/parameters impacting the performance of the critical Enterprise System Bus of an automotive manufacturing setup.

## 1 Introduction

In many situations, there is a need to forecast many variables of heterogeneous nature for a long time horizon (window) with high granularity. Especially, if

K. Ravikumar was with TCS R&I till recently.


I. Rojas et al. (Eds.): IWANN 2019, LNCS 11506, pp. 186–197, 2019.
https://doi.org/10.1007/978-3-030-20521-8_16

the variables under consideration are chaotic in nature, then the application of standard methods for forecasting is not easy. This requires different strategies or models to be adapted for forecasting these variables. These kinds of situations, for instance, can arise when we try to forecast different variables of interest which affects the performance of **Enterprise System Bus (ESB)** in an automotive manufacturing setup. An ESB is a software which is used to communicate between various mutually interacting software applications [1]. The functionality of an ESB is to act as an interface between heterogeneous and complex services. Figure 1 shows a typical ESB in any automotive manufacturing setup. The primary roles of ESB are (i) message routing between different interfaces (ii) monitor and control routing of messages between applications (iii) Improving regulatory compliance and risk management by using a single version of shared information (iv) message event queuing and sequencing etc. In the automotive industry, ESB is a highly business critical system as it acts as a common communication interface between the engineers, vendors, producers and consumers. Any delay or failure in transmission of the messages can potentially cause line stoppage and result in a considerable loss of business in the form of automobiles being produced. Thus, it is very important to monitor and hence predict the performance of the ESB for the effective operation of the manufacturing plant so that timely alerts can be given and potential failures are avoided before they become a bottleneck. The potential factors [3], that could impact ESB performance are (i) Resource Utilization (CPU load, memory usage etc.) (ii) Workload (data volume, scheduled batches, interface file processing time, concurrent interfaces, number of EGs and flows, number of active interfaces in ESB etc.) and (iii) Frame Resources and Performance (frame level CPU, frame level memory consumption, frame level i/o). All these factors need to be tracked for predicting the ESB performance. Most of these factors vary in a random fashion with too many fluctuations making it very challenging for any single model to be used for forecasting. Further, even for a single variable there would be a necessity or need to change the model for forecasting as we progress in time. This requires different strategies or models to be adapted for forecasting the variables/parameters.

In this paper, we solved the problem of heterogeneous variable forecasting using automatic algorithm selection. The problem considered has around 20 variables, all of which needs to be forecasted for ESB performance prediction. The business requirement is to have the forecasts at least 24 h ahead to proactively work for a remedy for any variables going out of control. The major challenges encountered with this requirement are

- Forecasting variables for a long horizon - 24 h advance prediction with inputs coming at 5 min interval; amount to 288-time steps ahead prediction which is unusual in forecasting.
- Majority of the variables, being percentages, are constrained to lie between 0 and 100. No meaningful aggregation (at hourly or daily level) is possible from 5 min interval samples.
- Variables behaviour being highly non-linear and non-stationary making it very difficult to establish any absolute relationships between the variables.

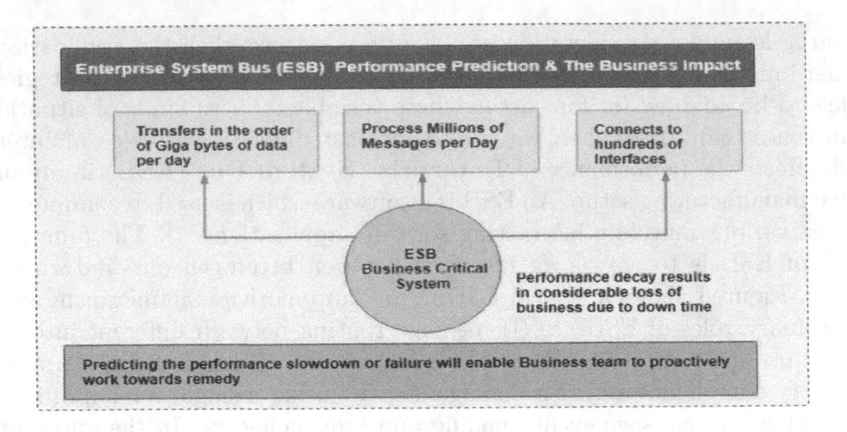


**Fig. 1.** Typical ESB in a manufacturing setup

Though there are many works on time-series forecasting, most of them are based on Auto-Regressive Integrated Moving Average (ARIMA) which can deal with only stationary time series [2]. One of its variant ARIMA with exogenous variables (ARIMAX) can be used if any causal relationships exist with the other variables, but again it cannot capture the non-linear relationship between the target variable and the causal variables, as it uses linear functions to map the inputs to the output.

In order to address all these challenges, we propose a framework for forecasting the key variables (parameters) of the ESB for its performance evaluation as shown in Fig. 2. It contains the following blocks (i) Granger's causality test to identify the inter-relationships between variables (ii) Feature generation by augmenting the causal variables obtained earlier with contextual information like time of the day, the day of the week etc. Additional lagged features derived for the target variable to capture the periodic patterns (iii) Feature transformation using log transformation and Principal Component Analysis (PCA) in the case of too many causal variables (iv) Model building for all the time series using different machine learning algorithms like, Artificial Neural Networks (ANNs), Support Vector Regression (SVR) and Random Forests (RF) which are well-known for their ability to model the chaotic and non-stationary time series [14]. (v) Automatic switching between the three algorithms designed in real-time for forecasting the time series of interest.

The major contributions of this paper are

- Using appropriate learning algorithms based on the target variable behavioural pattern.
- Modeling the time-series using the advanced machine learning algorithms like Artificial Neural Networks, Support Vector Regression and Random Forests.
- Designing a real-time switch selects an appropriate forecasting algorithm dynamically based on the performance, resulting in using different algorithms for forecasting the same variable on different days.

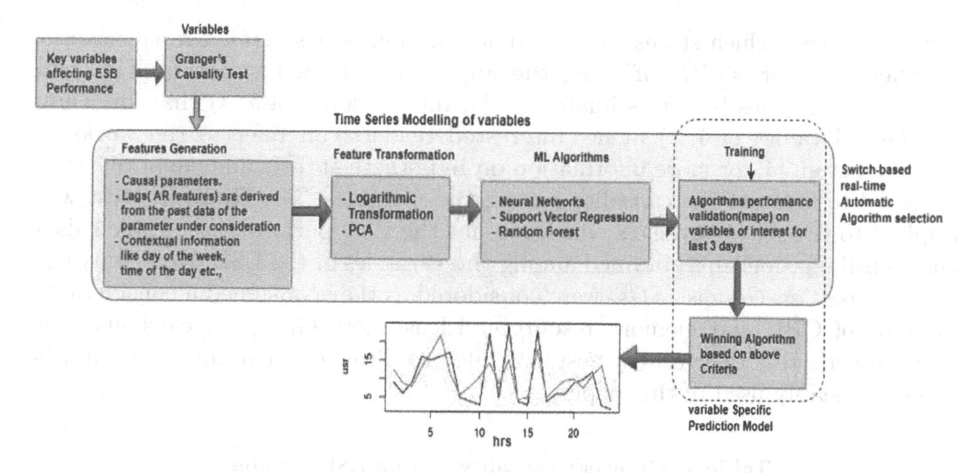


**Fig. 2.** Block diagram for ESB performance prediction

Towards demonstrating the effectiveness and usefulness of the proposed framework for time series forecasting for the said application, the paper is organized as follows. Section 2 presents the data pre-processing stage which includes the Grangers causality test and feature transformation. Section 3 presents the different algorithms used to model the time series. This is followed by Sect. 4 which presents the design of the real-time switch for automatic algorithm selection for time series forecasting. Section 5 summarizes the results obtained for different key parameters of ESB with different models. Finally, Sect. 6 concludes the work.

## 2 Data Pre-processing

Data pre-processing involves the following two tasks.

- identifying the causality among the variables using Granger's causality test and
- transforming the data using log transformation to reduce the dynamic range (of the target variable) and using PCA for input transformation if there are too many causal parameters for the target variable. In addition, deriving relevant inputs features for modelling the time series.

### 2.1 Granger's Causality Test

This is one of the most popular tests carried out to determine if any causal relationships exist between different times series which need to be predicted. Since the effect is always preceded by a cause, unlike computing correlations, this method exploits this cause and effect relationship to identify the time series that can be used to predict other time series. By definition it is a statistical hypothesis test which states that a particular time series $X(t)$, Granger-causes another time series $Y(t)$, if using the appropriate lagged values of $X(t)$ and $Y(t)$ both provides better estimates of the future values of $Y(t)$ than by using the lagged values of $Y(t)$ alone. Interested readers can refer to the works of Freeman et al. [4] for more information on hypothetical tests and equations that are used to find out the causality among the variables. The Granger's test was applied to the set of variables which impact the ESB performance. Table 1 lists the causal relationships obtained among the variables of the ESB. Here, only the top 7 *Execution Groups (EGs)* were considered as they consumed a considerable amount of CPU and memory resources. Please note Granger's causality test or Granger's test or causality test, all refer to Granger's causality test and is interchangeably used in this paper.

**Table 1.** Granger's causality test on ESB variables

| ESB variables | Causal variables |
|---|---|
| Page Space | Top Execution Groups (EGs) |
| CPU Utilization | Top Execution Groups (EGs) |
| Frame Level CPU Utilization | Disk Space Utilization, Data Size |
| Frame Level Memory Utilization | Data Size |
| Frame Level I/O | Data Size |
| Disk Space utilization | CPU Utilization, EG1 |
| Memory Usage | Page Space, Disk Space utilization |
| Data Volume | CPU Utilization, Data Size, EG4 |
| Data Size | Disk Space utilization, Data Volume, EG4 |
| Number of Interfaces | Frame Level I/O, Data Volume, Data Size, EG3 |
| EG1 | EG3 |
| EG2 | Page Space |
| EG3 | CPU Utilization, EG7 |
| EG4 | Page Space, EG3, EG7 |
| EG5 | EG1, EG7 |
| EG6 | CPU Utilization, EG3, EG5 |
| EG7 | Page Space, EG6 |

From the causality test, it was observed that the *EGs* were the independent variables causing/affecting the other variables to vary like *Page Space, CPU Utilization* etc. Also, it was seen that few *EGs* affect other *EGs*, this might be due to the competition among them to consume the available resources. Some other interesting relationships were revealed like, *Data Size* affected *Frame level CPU, memory utilization and I/O* along with *Data Volume and Number of Interfaces. Data Volume* affected *Data Size and Number of interfaces.* The causal relationships picked up by this test helped in getting a preliminary understanding of the different variables impacting the ESB performance.

### 2.2 Data Transformation and Predictor Matrix

The "target variable" was transformed using the natural logarithm to reduce the dynamic range for the variables having large deviations from the mean value. From Granger's causality test, all the variables that could impact the target variable were identified (also referred as causal variables). Although, appropriate lags of all these causal variables could be used as the inputs while forecasting the target variable; but, in practice only selected variables were used depending on their variational patterns. The variables with abrupt jumps and random patterns were avoided as causal inputs while modelling as it could adversely affect the performance of the forecasting algorithm. Apart from computing the causal variables lags, lagged features were derived from the target variable using partial auto-correlation function (pacf) to capture the target time series behaviour. Contextual information like time of the day, the day of the week, working day/holiday etc., were also used as the input for modelling the target variable time series.

Using all this information, "Predictor Matrix" was constructed for all the target variables containing the following three factors (i) Causal variables (ii) lagged features derived from the target variable using its past data and (iii) contextual information containing the calendar information to capture the user behavioural patterns.

## 3 Time Series Modelling Using Machine Learning Algorithms

The ever-increasing demand for operational efficiency demands all the performance metrics to be forecasted in order to take the appropriate action to keep all the key performance indicators (variables impacting ESB performance) under control. This requires sophisticated algorithms like machine learning strategies to be used for accurate forecasting of the various performance metrics. For ESB performance prediction, the potential factors like CPU load, memory usage, number of active interfaces etc., as mentioned in Table 1 required to be forecasted. Time series forecasting involves estimating the future values of the target variable as a function of the "Predictor Matrix" presented in Sect. 2.2.

The potential factors that could impact ESB performance are (i) Resource Utilization (CPU load, memory usage etc.) (ii) Workload (data volume, scheduled batches, interface file processing time, concurrent interfaces, number of Execution Groups (EG) and flows, number of active interfaces in ESB etc.) and (iii) Frame Resources and Performance (frame level CPU, frame level memory consumption, frame level I/O).

All these factors are highly varying and non-linear in nature that no single model can capture/learn the model. Therefore, multiple algorithms like Artificial Neural Networks (ANN), Support Vector Regression (SVR) and Random Forests (RF) were explored to model all the target variables time-series. These algorithms are well known in the literature for their strength in modelling the chaotic and highly non-stationary time-series [5,6,10]. The subsequent sections give details on the models learnt using these techniques.

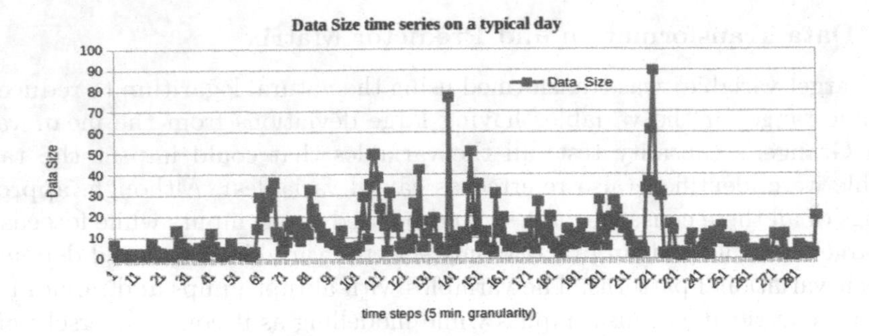


**Fig. 3.** Data Size variations on a typical day

### 3.1 Artificial Neural Networks

Artificial Neural Networks (ANNs) are mathematical models inspired by biological neurons. These are inherently non-linear and can estimate non-linear functions well. In principle, ANNs can approximate any order of non-linearity and are usually called as universal approximators. There are many works [6,8] showing that ANNs can approximate ordinary least squares, non-linear regression etc. Therefore, ANNs can learn the functional form that best describes the time series. The neural network with one hidden layer and sigmoid activation function was used in this work. The number of neurons in the hidden layer was decided based on the size of the "Predictor Matrix" which formed the input to the neural network for a particular target variable.

All the critical factors considered for ESB performance prediction were highly random and chaotic as shown in Fig. 3. ANNs are suited for modelling of these kinds of variables because of its capability of modelling non-linear behaviour. All the Neural Network based predictions are made by using the following equation.

$$\bar{Y}(t) = f_{ANN}(X(t - l_n), Y(t - l_p), Z(t)) \tag{1}$$

where $Y(t)$ is the output variable forecasted at time instant $t$, $Y(t-l_p)$ represents the significant lag values of $Y$ as given by the pacf function, $X(t - l_n)$ is the vector of causal variables and different causal variables can have different lags based on the causality test and $Z(t)$ represents the contextual information like *Day of the week, Time of the day, weekend* etc.

### 3.2 Support Vector Regression

Support Vector Machines (SVM) can be applied to both classification and regression tasks depending on whether the response variable is categorical or real-valued. The regression counterpart is referred to as Support Vector Regression (SVR) and is also well known in the literature for modelling the complex time series [7]. For input $X(t)$, let $Y(t)$ be the real-valued output, a non-linear function $f_{SVR}(X(t))$ is derived by projecting $X(t)$ to higher $m$-dimensional feature space (with the help of kernel) where linear machine mapping is learnt using

$\epsilon$-insensitive loss function. The non-linear function is learnt in such a way that it has at most $\epsilon$ deviations from the actual observed $Y(t)$ for all the training data and is as flat as possible at the same time [5]. The generalization ability of the SVR majorly depends on the hyper-parameters $C$, $\epsilon$ and the kernel type. The constant $C$ acts as a trade-off between the flatness of the function (model simplicity) and the amount up to which the deviations above $\epsilon$ are permitted. The choice of $\epsilon$ and the kernel along with the associated kernel function parameters are usually based on the knowledge of the application domain.

In this work, the time series $Y(t)$ was predicted using the appropriate lag vector of the time series $Y(t-l_p)$, $X(t-l_n)$ as inputs along with contextual variables like the day of the week and time of the day $Z(t)$. The lag vector $Y(t-l_p)$, comprised of top lag values of $Y(t)$ which turned out to be significant using the partial autocorrelation function (pacf).

$$\bar{Y}(t) = f_{SVR}(X(t-l_n), Y(t-l_p), Z(t)) \tag{2}$$

The model considered 'RBF' kernel with $C = 1$ and $\epsilon = 0.1$.

### 3.3 Random Forests

Random Forests (RFs) is a supervised learning algorithm. It is an ensemble of multiple decision trees that are built on different subsets of the data/features not necessarily exclusive [9] called as "bagging" method. The idea behind this approach is that multiple learning models increase the overall inference. Random Forests can be used for both regression and classification. RFs are extensively used in many prediction applications as they are very simple to apply and robust to missing values and any discontinuities that exist in the data [11,12]. The performance of the RF algorithm depends on the input variables selection and parameter tuning. In general, the random forest builts multiple decision trees using the randomly selected subset of input features and data. The final result is the ensemble of the outputs from the multiple decision trees making it robust to any noise in the data. In this work, RFs for regression is used to model the time-series of all the ESB variables.

The target variable is forecasted by learning the function $f$ using the Random Forest; the function is learnt using the lagged values of target variable $Y(t-l_p)$, causal variables, $X(t-l_n)$ and contextual information, $Z(t)$. The forecasted value $Y(t)$ is given by

$$\bar{Y}(t) = f_{RF}(X(t-l_n), Y(t-l_p), Z(t)) \tag{3}$$

## 4 Real-Time Switch for Automatic Algorithm Selection

Artificial Neural Networks (ANNs) and Support Vector Regression (SVR) have performed equally well for many variables. Random Forests were better for variables that had highly unpredictable behaviour. But, most of these variables were changing over time due to various factors that are unknown or not captured.

Due to this, no single algorithm performed well consistently for all days. Also, taking the average output from all the algorithms did not help as they further reduced the forecasting accuracy. All the variables were forecasted for 24 h ahead and at 5 min granularity, i.e 288-time steps ahead except for *Number of Interfaces, Data Volume* and *Data Size*. These three variables were hourly aggregated as they had random and abrupt patterns at 5 min granularity. This did not affect the actual goal of prediction and resulted in better performance.

**Table 2.** Day ahead forecasting accuracies for ESB variables

| ESB variables | Week 1 | | | Week 2 | | |
|---|---|---|---|---|---|---|
| | ANN | SVR | RF | ANN | SVR | RF |
| Page Space | 95% | **99%** | 90% | 90% | 90% | **95%** |
| CPU Utilization | 85% | 85% | **90%** | **95%** | 85% | 90% |
| Frame Level CPU Utilization | 90% | 85% | **92%** | **88%** | 85% | 85% |
| Frame Level Memory Utilization | **98%** | 95% | 97% | 96% | 95% | **98%** |
| Frame Level I/O | 55% | 60% | **65%** | 79% | 65% | **80%** |
| Disk Space utilization | **99%** | 90% | 95% | 85% | **95%** | 90% |
| Memory Usage | **90%** | 80% | 83% | **90%** | 85% | **90%** |
| Data Volume | **90%** | 85% | 80% | 80% | **90%** | 85% |
| Data Size | 80% | **85%** | 75% | 90% | **95%** | 80% |
| Number of Interfaces | **90%** | 85% | 80% | 86% | 88% | **90%** |
| EG1 | 72% | **95%** | 90% | 78% | **90%** | **90%** |
| EG2 | 80% | **94%** | 85% | 84% | **94%** | 90% |
| EG3 | 90% | **95%** | **95%** | **95%** | 92% | 90% |
| EG4 | **95%** | **95%** | 90% | **92%** | 86% | 88% |
| EG5 | 55% | **62%** | 60% | 74% | 71% | **75%** |
| EG6 | 95% | 90% | **97%** | 94% | 91% | **96%** |
| EG7 | **67%** | **55%** | 65% | 82% | 85% | **90%** |

Therefore, a "real-time switch" was developed which automatically selects one of the algorithms for forecasting a particular target variable for a particular day based on its performance on the previous days. With this in place, it was possible to use different algorithms for the same variable on different days.

### 4.1 Real-Time Switch

The real-time switch was based on the following logic. All the three algorithms were trained for modelling the time series leaving aside the last 3 days for testing. The trained algorithms were tested on the 3 days of data, previous to the day to be forecasted. The algorithm having the highest average accuracy for these three days was selected for forecasting the target variable for that particular day. This procedure was repeated for all the variables to be predicted so that algorithms could be randomly chosen based on the set criterion.

## 5 Results and Discussion

The forecasting results of different algorithms for all the ESB variables are given in Table 2 in terms of percentages (100-Mean Average Percentage Error (MAPE)). They were obtained for all the algorithms when testing was carried out on a rolling window basis for two weeks. The best performing algorithm's accuracy was highlighted in bold and it was observed that different algorithms performed better on different variables. Again, the best performing algorithm for a particular variable in week 2 was different from week 1. This re-emphasizes the fact that no single algorithm was able to perform consistently well for any variable. Therefore, the proposed framework of selecting the best algorithm in a dynamic way was necessary to have a very accurate forecast. The accuracy is very important as it affects the decisions taken by the business team in taking any remedial actions. Most of the variables have forecasting accuracies around 90%, though for few variables it is more than 95%. The forecasting accuracy of the variables related to *Frame Level CPU Utilization, Frame Level Memory Utilization, Disk Space Utilization* and *EG4* was greater than or equal to 90%. Forecasts of *CPU Utilization, Data Volume, Data Size* and *Number of Interfaces* was around 80% to 85%. Few *EGs* were predicted with high accuracy and few other *EGs* were predicted with pretty low accuracy. But, the situation would vary and could be completely opposite for other days as *EGs* were independent and had more abrupt and random changes. The forecasting results obtained through **real-time switch based automatic selection of algorithms** for few ESB variables are shown in Figs. 4, 5, 6, 7, 8, 9.

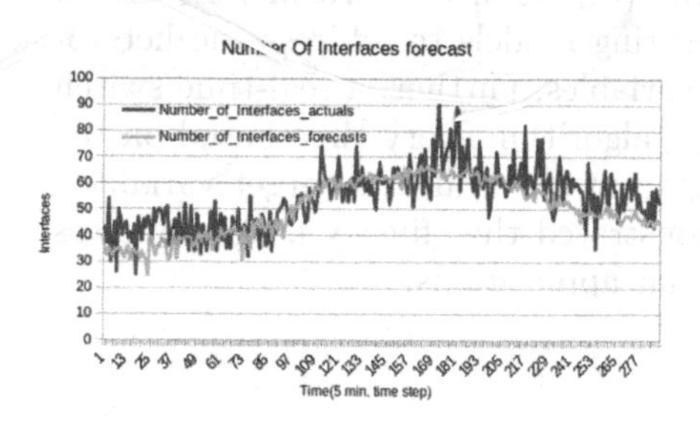


**Fig. 4.** Interfaces forecast

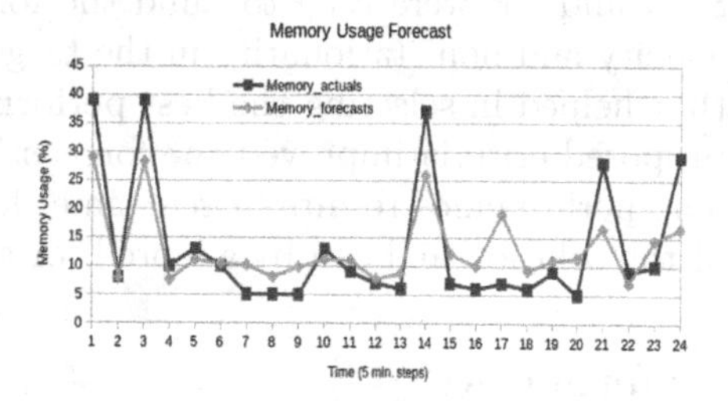


**Fig. 5.** Memory usage forecast

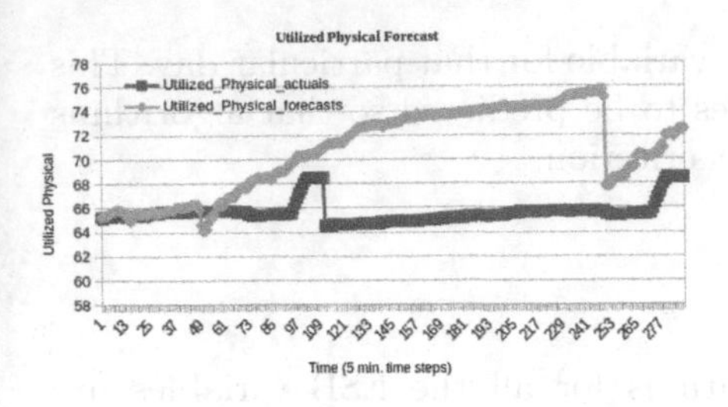


Fig. 6. Utilized physical forecast

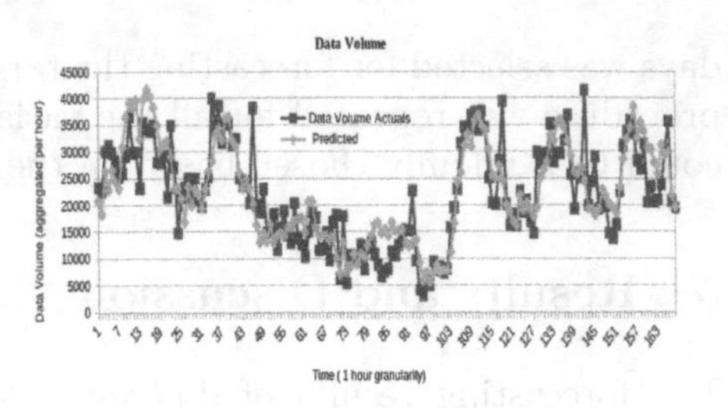


Fig. 7. Data volume forecast

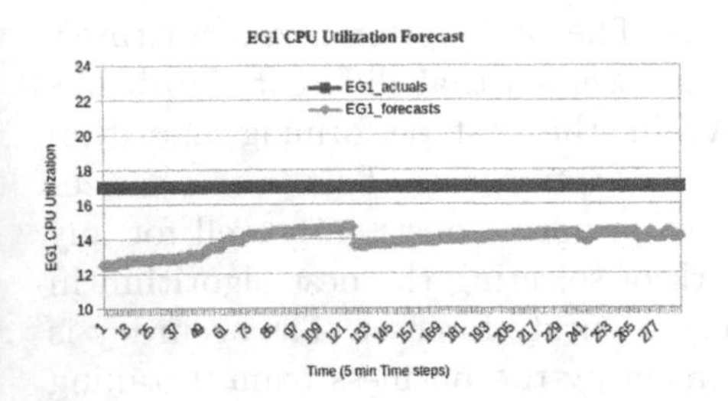


Fig. 8. EG1 forecast

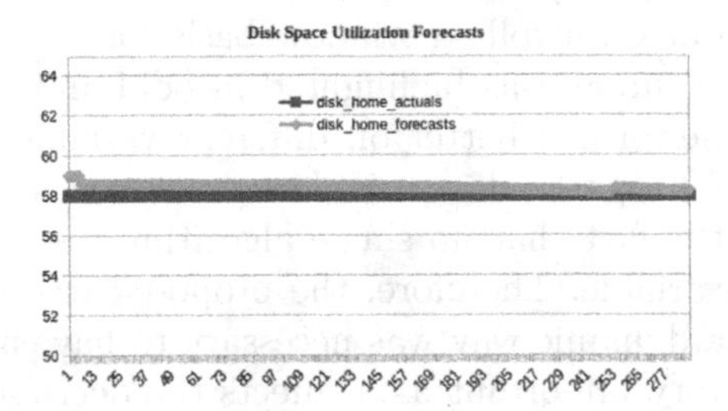


Fig. 9. Disk space forecast

## 6 Conclusion

The framework for Enterprise System Bus (ESB) performance prediction has been worked out using a real-time automatic selection of algorithms. The heterogeneous variables long-term forecasting at a high granularity has been successfully modelled, by addressing (i) identification of the causal variables for every target variable which helped in improving the target variables' predictions (ii) selection of appropriate causal variables and contextual information to be used in target variables time series modelling (iii) Robust algorithms like ANN, SVR and RF were used to build the forecasting models to address the heterogeneity and non-stationarity in the target variables. Further, a real-time switch that helped in selecting the best performing algorithm every time based on the proposed criteria improved the forecasting accuracy of all the target variables. The performance results in real-time demonstrated the efficacy and usefulness of the solution and can be adopted for similar applications.

## References

1. Bhadoria, R.S., Chaudhari, N.S., Tomar, G.S.: The performance metric for enterprise system bus (ESB) in SOA system: theoretical underpinnings and empirical illustrations for information processing. Inf. Syst. **65**, 158–171 (2017)
2. Brockwell, P.J., Davis, R.A., Calder, M.V.: Introduction to Time Series and Forecasting, vol. 2. Springer, New York (2002)
3. Schmidt, M.-T., et al.: The enterprise service bus: making service-oriented architecture real. IBM Syst. J. **44**(4), 781–797 (2005)

4. Freeman, J.R.: Granger causality and the times series analysis of political relationships. Am. J. Polit. Sci. **27**(2), 327–358 (1983)
5. Smola, A.J., Schölkopf, B.: A tutorial on support vector regression. Stat. Comput. **14**(3), 199–222 (2004)
6. Hill, T., O'Connor, M., Remus, W.: Neural network models for time series forecasts. Manage. Sci. **42**(7), 1082–1092 (1996)
7. Ojemakinde, B.T.: Support vector regression for non-stationary time series (2006)
8. Tealab, A., Hefny, H., Badr, A.: Forecasting of nonlinear time series using ANN. Future Comput. Inform. J. **2**(1), 39–47 (2017)
9. Breiman, L.: Random forests. Mach. Learn. **45**(1), 5–32 (2001)
10. Dudek, G.: Short-term load forecasting using random forests. In: Filev, D., et al. (eds.) Intelligent Systems'2014. AISC, vol. 323, pp. 821–828. Springer, Cham (2015). https://doi.org/10.1007/978-3-319-11310-4_71
11. Tyralis, H., Papacharalampous, G.: Variable selection in time series forecasting using random forests. Algorithms **10**(4), 114 (2017)
12. Verikas, A., Gelzinis, A., Bacauskiene, M.: Mining data with random forests: a survey and results of new tests. Pattern Recogn. **44**, 330–349 (2011)
13. Marcellino, M., Stock, J.H., Watson, M.W.: A comparison of direct and iterated multistep AR methods for forecasting macroeconomic time series. J. Econom. **135**(1–2), 499–526 (2006)
14. Ahmed, N.K., et al.: An empirical comparison of machine learning models for time series forecasting. Econom. Rev. **29**(5–6), 594–621 (2010)

# Automatic Time Series Forecasting with GRNN: A Comparison with Other Models

Francisco Martínez[1(✉)], Francisco Charte[1], Antonio J. Rivera[1], and María P. Frías[2]

[1] Andalusian Research Institute in Data Science and Computational Intelligence, Computer Science Dept., Universidad de Jaén, Jaén, Spain
{fmartin,fcharte,arivera}@ujaen.es

[2] Statistics and Operations Research Dept., Universidad de Jaén, Jaén, Spain
mpfrias@ujaen.es

**Abstract.** In this paper a methodology based on general regression neural networks for forecasting time series in an automatic way is presented. The methodology is aimed at achieving an efficient and fast tool so that a large amount of time series can be automatically predicted. In this sense, general regression neural networks present some interesting features, they have a fast single-pass learning and produce deterministic results. The methodology has been implemented in the R environment. A study of packages in R for automatic time series forecasting, including well-known statistical and computational intelligence models such as exponential smoothing, ARIMA or multilayer perceptron, is also done, together with an experimentation on running time and forecast accuracy based on data from the NN3 forecasting competition.

**Keywords:** Time series forecasting · General regression neural networks · Automatic forecasting

## 1 Introduction

Automatic time series forecasting allows to predict the future behavior of a time series with a minimal human intervention. This can be very useful when either the user is not an expert on a forecasting methodology or the volume of time series to be forecast is high enough to prevent the use of a human assisted procedure. In this later case, it would also be desirable an efficient forecasting methodology.

In this paper we present an automatic time series forecasting scheme based on generalized regression neural networks [16], which are a variation to radial basis function networks (RBFN). Generalized regression neural networks (GRNNs)

This work is supported by the Spanish National Research Project TIN2015-68454-R.



I. Rojas et al. (Eds.): IWANN 2019, LNCS 11506, pp. 198–209, 2019.
https://doi.org/10.1007/978-3-030-20521-8_17

have a single-pass learning so they can learn quickly. Furthermore, they produce deterministic results, avoiding the need to train several neural networks to achieve accurate and stable predictions.

Our scheme has been implemented in the R environment. We have considered interesting to compare this new method with other automatic methodologies for time series forecasting currently found in R as packages. In this comparison we have used both statistical and computational intelligence based techniques such as exponential smoothing, ARIMA, k-nearest neighbors (KNN) or multilayer perceptron. This comparison can shed some light on the controversial subject of whether statistical methodologies are better than computational intelligence ones [6].

The rest of the paper is structured as follows. In Sect. 2 generalized regression neural networks are analyzed. Section 3 explains our methodology to apply GRNNs to time series forecasting. In Sect. 4 R packages for automatic time series forecasting based on computational intelligence and statistical techniques are described. Furthermore, an R package that can be applied to combine the forecasts of several models is also discussed. In Sect. 5 a comparison among the different analyzed methods is done using data from the NN3 forecasting competition. Finally, Sect. 6 draws some conclusions.

## 2 Generalized Regression Neural Networks

A general regression neural network is a variant of a RBF network [15] characterized by a fast single-pass learning. A GRNN consists of a hidden layer with RBF neurons. Normally, the hidden layer has so many neurons as training examples. The center of a neuron is its associated training example and so, its output gives a measure of the closeness of the input vector to the training example. Commonly, a neuron will use the multivariate Gaussian function:

$$G(x, x_i) = \exp\left(-\frac{\|x - x_i\|^2}{2\sigma^2}\right) \tag{1}$$

where $x_i$ and $\sigma$ are the center and the smoothing parameter respectively—$x$ is the input vector.

Given a training set consisting of $n$ training patterns—vectors $\{x_1, x_2, \ldots x_n\}$—and their associated $n$ targets, normally scalars—$\{y_1, y_2, \ldots y_n\}$—, the GRNN output for an input pattern $x$ is computed in two steps. First, the hidden layer produces a set of weights representing the closeness of $x$ to the training patterns:

$$w_i = \frac{\exp\left(-\frac{\|x-x_i\|^2}{2\sigma^2}\right)}{\sum_{j=1}^{n} \exp\left(-\frac{\|x-x_j\|^2}{2\sigma^2}\right)} \tag{2}$$

Note that the weights decay exponentially with distance to the training pattern. The weights sum to one and represent the contribution of every training pattern

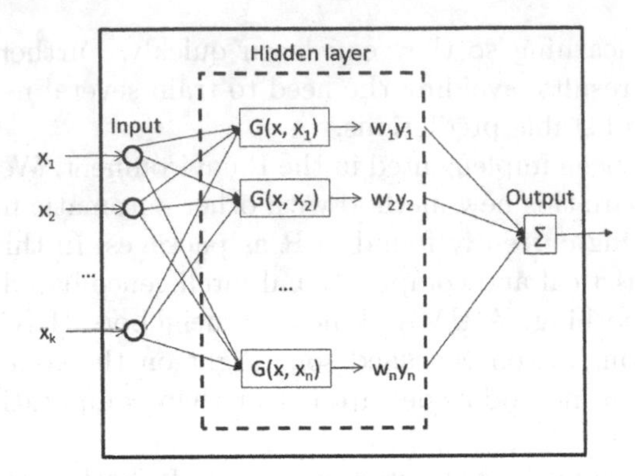


**Fig. 1.** Topology of a GRNN.

to the final result. The GRNN output layer computes the output as follows:

$$\hat{y} = \sum_{i=1}^{n} w_i y_i \tag{3}$$

so a weighted average of the training targets is obtained, where the weights decay exponentially with distance to the training patterns—see Fig. 1. The smoothing parameter controls how many targets are important in the weighted average. When $\sigma$ is very large the result is close to the mean of the training targets because all of them have a similar weight. When $\sigma$ is small only the closest training targets to the input vector have significant weights. In Fig. 2(a) the function $x^3$ over the interval $[-2, 2]$ is shown, together with a sample of 50 $(x, y)$-values of the function with an added random noise. In Fig. 2(b) the values of the function are predicted from the sample of 50 pairs of $(x, y)$-values with noise using GRNN regression and a small smoothing parameter. In Fig. 3(a) $\sigma$ is larger and the regression seems better. Finally, in Fig. 3(b) $\sigma$ seems too high. As can be seen in the example, GRNN regression is very sensitive to the smoothing parameter.

## 3 Time Series Forecasting with GRNN

In this section the methodology that has been developed to forecast a time series with GRNN is explained. The goal is to obtain a fast and automatic tool. In the following subsections the different design choices are explained.

### 3.1 Preprocessing

In our methodology the time series has been scaled to the range $[0, 1]$. However, no other preprocessing, including detendring or deseasonalizing—that is, transforming the time series to remove trend or seasonality—, is done.

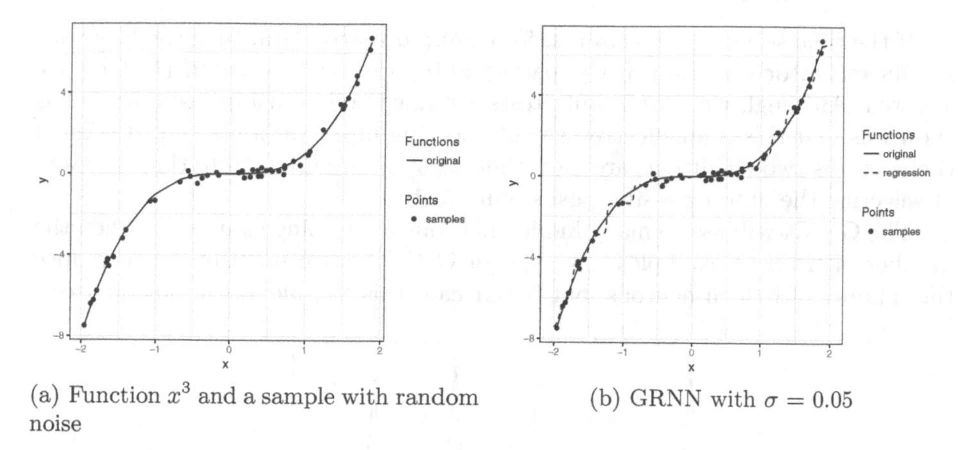


(a) Function $x^3$ and a sample with random noise

(b) GRNN with $\sigma = 0.05$

**Fig. 2.** A sample of a function with random noise and its regression with GRNN

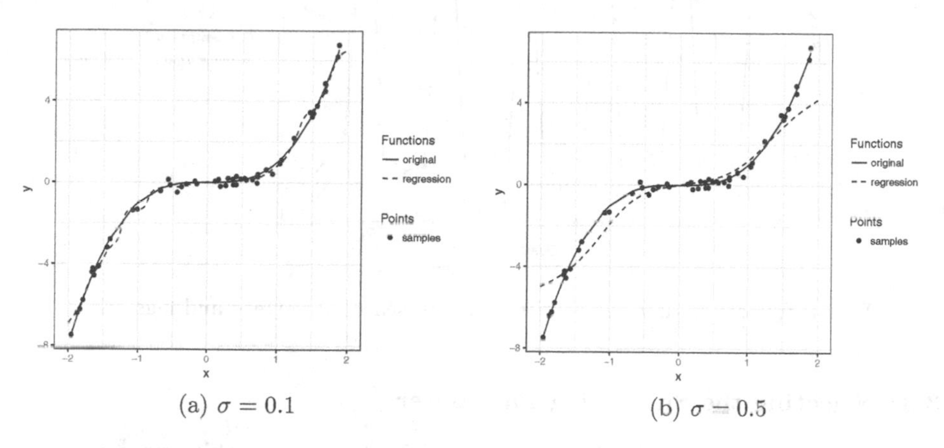


(a) $\sigma = 0.1$

(b) $\sigma = 0.5$

**Fig. 3.** Regression with GRNN and different smoothing parameters

### 3.2 Autoregressive Lags and Number of Neurons

In order to select the autoregressive lags the following strategy will be used. If the time series is seasonal and the length of the seasonal period is $m$, then $m$ consecutive lags, starting from lag 1, are used. For example, for quarterly data lags 1:4 are used and for monthly data lags 1:12. This way, seasonal patterns can be captured more easily. Let us see why. In Fig. 4 an artificial quarterly time series with a strong seasonal pattern is shown. The first quarter has a mean level higher than the other quarters that have a similar level. The autoregressive lags are lags 1:2 and we want to generate a one-step ahead forecast. These autoregressive lags can lead to an unsuitable forecast. As can be seen in the figure, the closest training pattern to the input pattern has a fourth quarter as target, when a first quarter value is going to be predicted. In Fig. 5 lags 1:4 are used and the target associated to the closest training pattern is a first quarter value.

If the time series is not seasonal, for example yearly data, then the lags with significant autocorrelation in the partial autocorrelation function (PACF) are selected. Although the PACF only tests for linear relationship, experience has shown us that this is an effective way of selecting input variables [14]. If none of the previous two conditions are met, then lags 1:5 are used. Note that this way of selecting the autoregressive lags is quite fast.

The GRNN will use so many hidden neurons as training examples. When the number of training examples is very high GRNN can use clustering to reduce the number of hidden neurons, but in our case this technique will not be used.

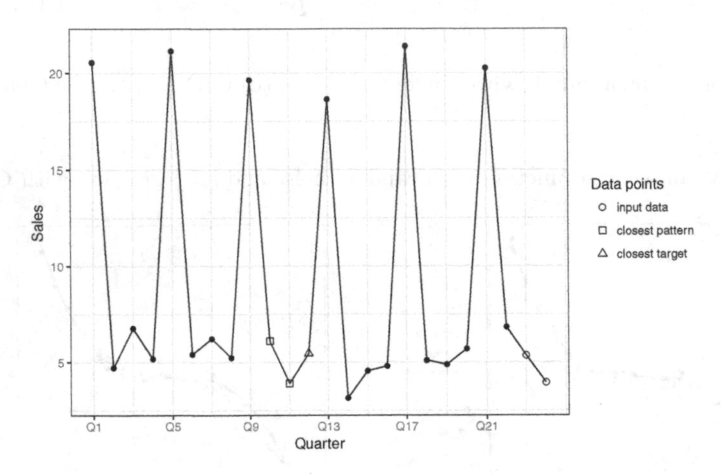


**Fig. 4.** Quarterly time series with a strong seasonal pattern and lags 1:2.

### 3.3 Selecting the Smoothing Parameter

As previously mentioned, GRNN is very sensitive to the smoothing parameter so it is vital to select a suitable value for it. In order to make a good choice we have applied an optimization tool for finding $\sigma$ using the rolling origin technique. The historical data is divided into a training and a validation set and $\sigma$ is selected so that it minimizes a forecast accuracy measure for the validation data using the training data.

### 3.4 Multi-step Ahead Strategy

When more than one future value of a time series has to be predicted a multi-step ahead strategy must be applied [2]. The classical strategies are direct, Multiple-Input Multiple-Output (MIMO) and iterative. We rule out MIMO because normally produces less accurate forecasts. There is no clear evidence that the direct strategy outperforms the iterative strategy so we choose the last one because it is straightforward and faster.

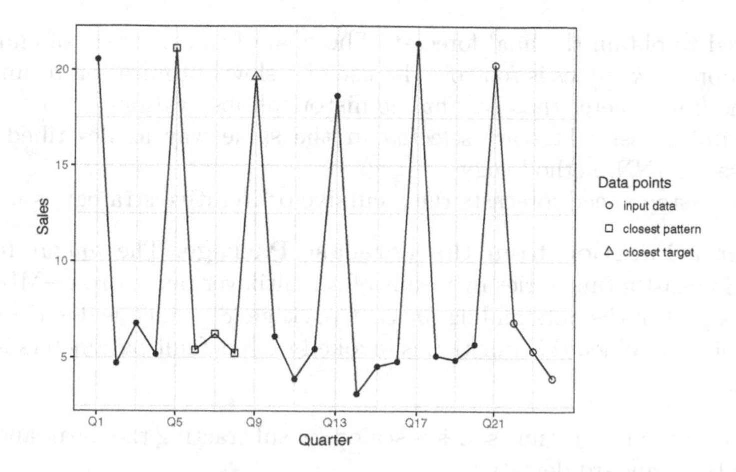


**Fig. 5.** Quarterly time series with a strong seasonal pattern and lags 1:4.

# 4 Automatic Time Series Forecasting in R

In this section we briefly describe other automatic time series forecasting methods that can be found in the R environment. We have classified them according to whether they are based on computational intelligence or statistical techniques. In the next subsections they are described. For every method, it is also explained how it works when it is used with default parameters. In general, when you call a forecasting function with default parameters the function tries to automatically select the best model to predict the time series.

## 4.1 Computational Intelligence Methods

This section analyzes the computational intelligence methods related to time series forecasting. We first have searched in the CRAN task view Time Series Analysis [5] to see what R packages are related to time series forecasting. Although CRAN—the Comprehensive R Archive Network—contains a lot of packages for regression based on computational intelligence techniques [3], just a few are specially devoted to time series forecasting.

**The *tsfknn* Package.** This package allows to forecast a time series using KNN regression, the methodology used is partially derived from this study [14]. When forecasts are generated with default parameters the package employs the following strategies:

- No preprocessing is done for detendring, deseasonalizing or scaling the time series.
- Instead of using a unique $k$ parameter, three preset $k$ parameters are used. Three diverse KNN models are built, each one with a different value of the preset $k$ values. Forecasts are generated using the three KNN models and are

averaged to obtain the final forecast. The goal of this strategy of employing several preset $k$ values is to avoid the use of a slow optimization technique to find the $k$ parameter that best fit the historical observations.
- The autoregressive lags are selected in the same way as described in the proposed GRNN methodology.
- For multi-step ahead forecasts the recursive or iterative strategy is used.

**The *nnetar* Function from the Forecast Package.** The *nnetar* function allows to forecast a time series by means of a multilayer perceptron—MLP. This function is part of the outstanding *forecast* package [8], which is described in an online book [11]. When this function is invoked with default parameters it works as follows:

- As preprocessing, the time series is scaled by subtracting the mean and dividing by the standard deviation.
- The autoregressive lags are selected as follows. For non-seasonal time series, $p$ consecutive lags are selected, starting from lag 1. The number of lags is the optimal number of lags of an autoregressive ARIMA model—i.e., an ARIMA model with 0 differences and no moving average terms—using the AIC to select the model. For seasonal time series, the $p$ consecutive lags are chosen from the optimal AR linear model fitted to the seasonally adjusted data. Furthermore, lag $m$, where $m$ is the seasonal period, is also used. For example, for monthly data lag 12 is also used.
- Only a hidden layer is used and its number of neurons is half of the number of input nodes—autoregressive lags—plus 1.
- In order to achieve more stable and accurate forecasts, 20 neural networks are fitted with different random starting weights. To produce the final forecast, the forecasts of these 20 networks are averaged.
- For multi-step ahead forecasts the recursive or iterative strategy is used.

**The *mlp* Function from the nnfor Package.** The *mlp* function from the *nnfor* package also implements multilayer perceptron for time series forecasting. Let us see how it automatically predicts a time series:

- As preprocessing, the time series is linearly scaled to $[-.8, .8]$. A test for finding trend is done and first differences are applied if necessary. Then the series is tested for seasonality, if the test successes seasonal differences are taken.
- The autoregressive lags are selected as described here [7]. The model can also include seasonal dummy variables.
- By default only a hidden layer with 5 nodes is used. Setting the appropriate parameters the number of neurons can be selected using an optimization algorithm.
- 20 neural networks are fitted with different random starting weights. The forecasts of these 20 networks are combined using the median operator to obtain the final forecast.
- For multi-step ahead forecasts the recursive or iterative strategy is used.

### 4.2 Statistical Models

In this section we analyze the two workhorses of statistical models for time series forecasting: exponential smoothing [10] and ARIMA [4]. The previously mentioned *forecast* package includes implementations of both methods with automatic selection of models and model parameters.

**Exponential Smoothing.** Exponential smoothing encompasses a set of models—e.g., SES, Holt or Holt-Winters—to predict a time series taking into account its trend and seasonal components. The function *ets* uses the AICc criterion to select an appropriate exponential smoothing model—for example, a model with damped trend and multiplicative seasonality—among all the possible exponential smoothing models.

**ARIMA.** Normally, an ARIMA model is selected by an expert after making the series stationary, analyzing the autocorrelation and partial autocorrelation functions and trying several models. The *auto.arima* function of the *forecast* package uses a variation of the Hyndman-Khandakar algorithm [8] to select an ARIMA model. Basically, it uses unit root tests to check stationarity and possibly take differences. Several ARIMA models are fitted and the AICc criterion is used to select the best one. The search for the ARIMA model is not exhaustive, so the selected model might not be the optimal one—according to AICc—but a good model is found.

### 4.3 A Combination of Methods

Since Bates and Granger [1] proposed to combine the forecasts of several methods, forecasting by combining different techniques has not stopped growing. As an example, the recent M4 forecasting competition [13], wherein 100,000 time series had to be forecast, has been dominated by combinations of models.

Taking into account the success of the ensembles of methods we have considered interesting to include a combination of models in our comparison. The *forecastHybrid* package combines, by default, seven models from the *forecast* package averaging their forecasts. Three of the models used in the combination have been described here: *nnetar*, *auto.arima* and *ets*—exponential smoothing. The package can be applied in an automatic way in which the user only sets the time series and the forecast horizon.

## 5 Experimentation

In this section a comparison among the proposed time series forecasting methodology based on GRNN and the other techniques explained in the previous section is carried out. To compare the methods the data from the NN3 competition [6] has been used. In this competition 111 time series drawn from the M3 monthly industry data [12] were used. The data set contains a balanced mix of 25 short seasonal series, 25 short non-seasonal series, 25 long seasonal series, 25 long non-seasonal series and a collection of 11 experimental series. Short series have about 52 observations per series and long series more than 120 observations.

In the NN3 competition the next 18 future months for all the time series should be predicted. To assess the accuracy of a forecast the NN3 organizers decided to use the sMAPE—symmetric absolute percentage error. Given the forecast $F$ for a NN3 time series with actual values $X$:

$$\text{sMAPE} = \frac{1}{18}\sum_{t=1}^{18} \frac{|X_t - F_t|}{(|X_t| + |F_t|)/2} 100$$

The sMAPE of each series will then be averaged over all the 111 time series for a global mean sMAPE. Although some experts discourage the use of sMAPE [9], it was the main measure for assessing forecast accuracy in the NN3 competition, so we decided to use it to compare our results with the NN3 contenders.

In Table 1 the results of our experiments are shown. The methods are listed in the first column, sorted by forecast accuracy according to the global mean sMAPE—second column—on the 111 time series, computed as described previously. The third column shows the average rank of every method over the 111 time series. The next four columns include the average sMAPE of the methods on the different data conditions evaluated in the NN3 competition: S.S. (short seasonal), S.N. (short non-seasonal), L.S. (long seasonal) and L.N. (long non-seasonal). The last column of the table shows the time in minutes needed by the method to forecast the 111 time series. For benchmarking purposes, the sMAPE of the 8 top contenders of the NN3 competition is shown in Table 2. In the ID field of this table, B stands for statistical benchmark and C for computational intelligence method.

After these results several conclusions can be drawn:

- The combination method is the winner, being its accuracy similar to the winner of the NN3 competition. This is an outstanding result, because the methods we have compared are applied automatically, in the sense that no study of the characteristics of the time series has been taken into account in order to apply the methods. The contenders of the NN3 competition knew the historical data and they could analyze their features to improve their models. The fact that the winner is a combination methodology is consistent with previous studies [12]. Its performance is very robust, beating the other methods over all data conditions.
- The statistical methods, exponential smoothing and ARIMA, have obtained a similar global sMAPE. Their results clearly outweigh the computational intelligence techniques. This superiority also corroborates other comparisons [13]. Furthermore, the *ets* function is remarkably fast. In spite of having a better mean rank, *auto.arima* has a worse global sMAPE due to its poor performance in the short seasonal series.

- The GRNN methodology developed in this paper has achieved slightly better performance than the KNN and *nnetar*. No wonder that KNN and GRNN get similar results because both are based on combining the targets of the patterns that are similar to the input pattern. GRNN is a bit slow because the smoothing parameter is selected by optimization.
- The *nnetar* function, based on multilayer perceptron, has achieved modest results, being beaten by a simple method such as a KNN. However, its mean rank is lower than GRNN or KNN and its results on long series are acceptable. Maybe, a proper training is not possible with short series.
- The result of the *mlp* function from the *nnfor* package is a bit disappointing. It is nearly the worst method over all the conditions. Its default configuration can possibly be improved.

**Table 1.** Comparison of the different methods.

| Method | sMAPE | Rank | S.S. | S.N. | L.S. | L.N. | Time |
|---|---|---|---|---|---|---|---|
| forecastHybrid | 14.85 | 2.79 | 13.66 | 19.21 | 9.38 | 16.92 | 16:11 |
| ets | 15.52 | 3.67 | 14.69 | 19.57 | 10.50 | 17.53 | 01:29 |
| auto.arima | 15.64 | 3.47 | 16.66 | 19.36 | 10.14 | 17.01 | 09:12 |
| GRNN | 16.71 | 4.29 | 16.11 | 19.57 | 13.80 | 18.35 | 06:45 |
| tsfknn | 16.96 | 4.31 | 15.38 | 21.90 | 12.85 | 19.18 | 00:03 |
| nnetar | 17.05 | 4.18 | 16.00 | 24.73 | 11.94 | 18.26 | 00:14 |
| mlp | 21.22 | 5.30 | 22.28 | 24.31 | 16.09 | 20.23 | 11:10 |

**Table 2.** The top NN3 competition contenders.

| ID | Method | sMAPE |
|---|---|---|
| B09 | Wildi | 14.84 |
| B07 | Theta | 14.89 |
| C27 | Echo state networks | 15.18 |
| B03 | ForecastPro | 15.44 |
| B16 | DES | 15.90 |
| B17 | Comb S-H-D | 15.93 |
| B05 | Autobox | 15.95 |
| C03 | Linear model + GA | 16.31 |

Figure 6 shows the boxplots of the accuracy according to sMAPE for the different methods over the 111 time series. We would have liked to add some recurrent network to the comparison, but to our knowledge R lacks of a package of that kind for time series forecasting.

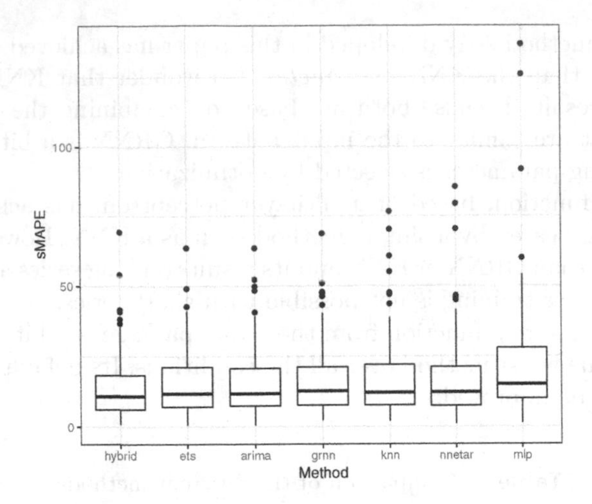


**Fig. 6.** Boxplots of the different methods.

## 6 Conclusions

In this paper a methodology based on GRNN regression for forecasting time series in an automatic way has been presented. The methodology uses straightforward strategies in order to obtain a fast forecasting tool. This goal has been facilitated by the intrinsic features of GRNN regression, such a single-pass learning or deterministic predictions. Currently, the bottleneck of the tool is the selection of the smoothing parameter. Packages for automatic time series forecasting in the R environment have also been described, together with a comparison among the studied tools in terms of forecasting accuracy and running time. In this comparison, based on monthly time series, statistical models seem to be more accurate and the proposed methodology has achieved good result among the computational intelligence models.

## References

1. Bates, J.M., Granger, C.W.J.: The combination of forecasts. Oper. Res. Q. **20**, 451–468 (1969)
2. Ben Taieb, S., Bontempi, G., Atiya, A.F., Sorjamaa, A.: A review and comparison of strategies for multi-step ahead time series forecasting based on the NN5 forecasting competition. Expert Syst. Appl. **39**(8), 7067–7083 (2012)
3. Bhatia, A., Chiu, Y.: Machine Learning with R Cookbook: Analyze Data and Build Predictive Models. Packt Publishing, Birmingham (2017)
4. Box, G.E.P., Jenkins, G.M., Reinsel, G.C.: Time Series Analysis: Forecasting and Control, 4th edn. Wiley, Hoboken (2008)
5. CRAN Task View: Time Series Analysis. https://cran.r-project.org/view=TimeSeries. Accessed 26 Feb 2019

6. Crone, S.F., Hibon, M., Nikolopoulos, K.: Advances in forecasting with neural networks? Empirical evidence from the NN3 competition on time series prediction. Int. J. Forecast. **27**(3), 635–660 (2011)
7. Crone, S.F., Kourentzes, N.: Feature selection for time series prediction - a combined filter and wrapper approach for neural networks. Neurocomputing **73**(10), 1923–1936 (2010)
8. Hyndman, R., Khandakar, Y.: Automatic time series forecasting: the forecast package for R. J. Stat. Softw. **27**(1), 1–22 (2008)
9. Hyndman, R.J., Koehler, A.B.: Another look at measures of forecast accuracy. Int. J. Forecast. **22**(4), 679–688 (2006)
10. Hyndman, R.J., Koehler, A.B., Ord, J.K., Snyder, R.D.: Forecasting with Exponential Smoothing: The State Space Approach. SSS. Springer, Heidelberg (2008). https://doi.org/10.1007/978-3-540-71918-2
11. Hyndman, R.J., Athanasopoulos, G.: Forecasting: Principles and Practice, 2nd edn. OTexts, Melbourne (2018). OTexts.com/fpp2. Accessed 26 Feb 2019
12. Makridakis, S., Hibon, M.: The M3-competition: results, conclusions and implications. Int. J. Forecast. **16**(4), 451–476 (2000)
13. Makridakis, S., Spiliotis, E., Assimakopoulos, V.: The M4 competition: results, findings, conclusion and way forward. Int. J. Forecast. **34**(4), 802–808 (2018)
14. Martínez, F., Frías, M.P., Pérez, M.D., Rivera, A.J.: A methodology for applying k-nearest neighbor to time series forecasting. Artif. Intell. Rev. (2017)
15. Moody, J., Darken, C.J.: Fast learning in networks of locally-tuned processing units. Neural Comput. **1**(2), 281–294 (1989)
16. Specht, D.F.: A general regression neural network. Trans. Neural Netw. **2**(6), 568–576 (1991)

# Improving Online Handwriting Text/Non-text Classification Accuracy Under Condition of Stroke Context Absence

Serhii Polotskyi(✉), Ivan Deriuga, Tetiana Ignatova, Volodymyr Melnyk, and Hennadii Azarov

Samsung R&D Institute Ukraine (SRK), 57 L'va Tolstogo Str., Kyiv 01032, Ukraine
{s.polotskyi,i.deriuga,te.ignatova,v.melnyk,gen.azarov}@samsung.com

**Abstract.** In this paper, an approach to text/non-text stroke classification for on-line handwriting recognition is proposed. This approach allows to improve classification accuracy when the stroke's context is absent. Having a label for the input stroke, we set this label for a feature vector which corresponds to each timestamp of the given input stroke. A trained neural network classifies each timestamp for the given input sequence. Finally, a decoder assigns a class label to the whole stroke. This approach was tested on the online handwritten dataset IAMonDO in online mode and has shown 98.5% accuracy. This approach could be used for other time-series classification tasks when the context information is not available.

**Keywords:** Recurrent neural network · Time series classification · Text/Non-text separation

## 1 Introduction

There are a lot of interactive devices nowadays starting from smartphones and tablets and finishing by interactive large-format boards. The user can input information on touch screen using a variety of devices like stylus, their own finger, etc. There is a question: do we need handwriting in modern digitized world? Our analysis has shown that handwriting input type is much faster than keyboard-based one especially for CJK (Chinese, Japanese, Korean) languages. As it was mentioned above, there are large interactive wide-boards that are used both in academia and in business. As a result, information of many types like text, formulas, diagrams, charts, tables, graphics, sketches, shapes, etc could be inputted. For accurate recognition it is important to know which type of content the user is inputting. Mode detection serves exactly this purpose. In this work we will discuss text/non-text mode detection. It's the most common and important use case. It can be used as a separate system for text/non-text recognition and as a starting point for further mode detection like formula, tables, etc. In the


I. Rojas et al. (Eds.): IWANN 2019, LNCS 11506, pp. 210–221, 2019.
https://doi.org/10.1007/978-3-030-20521-8_18

mode detection task we start to classify stroke immediately after its input has been finished.

A common approach to mode detection can be found in [9–11]. These works are among the first dedicated to this problem. They analyzed single strokes and worked with their own datasets. The accuracy of text/non-text mode detection was higher than 98%. Indermühle et al. [14] checked their approaches on the IAMonDO benchmark dataset [15] and got an accuracy of 91%. Weber et al. [13] got 97% accuracy using SVM, MLP, KNN. 97.01% accuracy was obtained in [16] with a bidirectional RNN with LSTM cells. It was one of the first applications of RNN to solving this task. Taking into account the powerful properties of RNN, Otte et al. obtained 98.47% accuracy in [12]. The state of the art result of 98.5% accuracy was gained in [2]. All these results were obtained when groups were pre-assigned. Also in [2], instead of pre-assigned grouping a few grouping algorithms were applied and classification was done for detected groups. The best result had the accuracy of 91.3%.

## 2 Proposed Solution

In all the papers cited above, it is supposed that there is a group of strokes and then these strokes are classified together. More precisely, a group is classified as a whole putting all its strokes into the same class. We have investigated and analyzed 14 features used in handwriting recognition according to [1]. The main contributions of this paper is the following:

- we use features ($\Delta x, \Delta y, penup/pendown$) from a handwriting recognition problem instead of complicated ones that are used in previous works;
- we can classify strokes in one group to different classes.

These properties helped us to achieve 92.3% and 99.4% accuracy of stroke classification using the same grouping algorithm as before on IAMonDO and our dataset respectively. As it was mentioned, the previous best result was 91.3% achieved by [2]. Let us briefly consider a classification task with an RNN. RNN is a state-of-the-art solution for time-series tasks. We need this for better understanding of the proposed approach. We have objects that belong to one class and our task is to find the class to which each object belongs. Let our classes be $C_1, C_2, \ldots, C_k$. For each label we do a one-hot encoding. For RNN training we need a three dimensional input tensor (batch size, timestamp sequence length, feature amount) and one-hot encoded labels. Taking into account the above information, we can imagine the training RNN for classification problem as following. Assume that the timestamp sequence length is $M$ and amount of features for each timestamp is $D$. For the $i$th element from a train set of size $N$ denote by $f_i^{(m)} \in \mathbb{R}^D$ a feature vector for each timestamp $m = \overline{1, M}$. We schematically displayed our train dataset in Table 1.

So $f_i^{(m)}, m = \overline{1, M}$ denotes feature vectors for all timestamp sequences for the $i$-th element from the train set.

We propose to rebuild our train data in the following manner. Without loss of generality, consider the first row of Table 1. Here we see feature vectors for all timestamps for the first element with the same label. Our idea is to assign the same label for each feature vector calculated for each timestamp. We can be see that the label for the first $M$ feature vectors in Table 2 is the same as the label for feature vectors in the first row of Table 1.

**Table 1.** Features for RNN in common case

| Features $\in \mathbb{R}^D$ | Label $\in \mathbb{R}^k$ | Position of one in label |
|---|---|---|
| $f_1^{(m)}, m = \overline{1, M}$ | $(0, 0, \ldots, 0, 1, 0, \ldots, 0)$ | $i_1$ |
| $f_2^{(m)}, m = \overline{1, M}$ | $(0, 1, \ldots, 0, 0, 0, \ldots, 0)$ | $i_2$ |
| ... | ... | ... |
| $f_N^{(m)}, m = \overline{1, M}$ | $(0, 0, \ldots, 0, 0, 1, \ldots, 0)$ | $i_N$ |

**Table 2.** Features for RNN in our case

| Features $\in \mathbb{R}^D$ | Label $\in \mathbb{R}^k$ | Position of one in label |
|---|---|---|
| $f_1^{(1)}$ | $(0, 0, \ldots, 0, 1, 0, \ldots, 0)$ | $i_1$ |
| $f_1^{(2)}$ | $(0, 0, \ldots, 0, 1, 0, \ldots, 0)$ | $i_1$ |
| ... | ... | ... |
| $f_1^{(M)}$ | $(0, 0, \ldots, 0, 1, 0, \ldots, 0)$ | $i_1$ |
| $f_2^{(1)}$ | $(0, 1, \ldots, 0, 0, 0, \ldots, 0)$ | $i_2$ |
| $f_2^{(2)}$ | $(0, 1, \ldots, 0, 0, 0, \ldots, 0)$ | $i_2$ |
| ... | ... | ... |
| $f_2^{(M)}$ | $(0, 1, \ldots, 0, 0, 0, \ldots, 0)$ | $i_2$ |
| ... | ... | ... |
| $f_N^{(1)}$ | $(0, 0, \ldots, 0, 0, 1, \ldots, 0)$ | $i_N$ |
| $f_N^{(2)}$ | $(0, 0, \ldots, 0, 0, 1, \ldots, 0)$ | $i_N$ |
| ... | ... | ... |
| $f_N^{(M)}$ | $(0, 0, \ldots, 0, 0, 1, \ldots, 0)$ | $i_N$ |

In such case, our model will classify the feature vector for each timestamp for the given element (group of strokes). We need a so called decoder that will take classification results for each timestamp and classify the input stroke. Let us define the decoder more precisely. As we mention above, each element of our set belongs to the one of the $k$ classes. For a given element, our trained model predicts the vector $p = (p^{(1)}, p^{(2)}, \ldots, p^{(M)})$ where each component $p^{(i)}$ has the form

$$p^{(i)} = (p_1^i, p_2^i, \ldots, p_k^i), \sum_{j=1}^{k} p_j^i = 1, \tag{1}$$

where $p_j^i$ is the probability that the feature vector calculated for the $i$-th timestamp $i = \overline{1, M}$ belongs to the $j$-th class $j = \overline{1, k}$. With these notations, we can define our decoder *Decoder* as following

$$Decoder : \mathbb{R}^{M*D} \rightarrow [C_1, C_2, \ldots, C_k]. \quad (2)$$

In simple terms, our decoder is a mapping of the vector of vector probabilities to the one of the $k$ classes. Let us write $p$ in the extended form in order to better understand the decoders form:

$$p = (p^{(1)}, p^{(2)}, \ldots, p^{(M)}) =$$
$$((p_1^1, p_2^1, \ldots, p_k^1), (p_1^2, p_2^2, \ldots, p_k^2), \ldots, (p_1^M, p_2^M, \ldots, p_k^M)).$$

Different *Decoder* variants are possible. You can find more information about decoding in handwriting recognition in the [1]. In our case, the model predicts probabilities of being text or non-text ($k = 2$ in the above notations) for each stroke's point. With these probabilities we can find cumulative (stroke) probability as their product. For example, assume that $k = 2$ in the above equation. Then probability that stroke belongs to first class could be calculated as

$$p_{first} = \prod_{i=1}^{M} p_1^i.$$

It is common to take logarithm in the last equation and continue to work with sum of log-probabilities. We could do this but we wanted to compare two inference strategies for the whole stroke classification probability. In the first strategy we calculate the contribution of all points together and in the second one we calculate the sum of contribution of each stroke's point. Taking into account, we consider two form of decoders.

$$Decoder_1 = argmax(\sum_{i=1}^{M} p_1^i, \sum_{i=1}^{M} p_2^i, \ldots, \sum_{i=1}^{M} p_k^i), \quad (3)$$

$$Decoder_2 = argmax(\sum_{i=1}^{M} g(p_1^i, p_2^i, \ldots, p_k^i)), \quad (4)$$

where $g(q_1, q_2, \ldots, q_k) = (0, 0, \ldots, 0, 1, 0, \ldots, 0)$ is a zero vector except one coordinate on position $j = argmax(q_1, q_2, \ldots, q_k)$ equalling 1.

Consider example for better understanding. Assume we have stroke with three points and predicted probabilities

$$p = ((0.51, 0.49), (0.51, 0.49), (0.1, 0.9)).$$

Using decoders (3) and (4) we get the classification probabilities $P_1$, $P_2$ respectively

$$P_1 = argmax((0.51 + 0.51 + 0.1), (0.49 + 0.49 + 0.9)) = argmax(1.12, 1.88) = 2,$$
$$P_2 = argmax((1, 0) + (1, 0) + (0, 1)) = argmax(2, 1) = 1.$$

From this small example you can see that it is possible to get different stroke classification with both decoders.

## 3 Evaluation of Our Solution

We use the Theano deep learning framework with Lasagne backend to train our models. We consider a classification task with a GRU neural network. We have also used an LSTM during evaluation of our approach and got the same accuracy, but training took more time. More precisely, we trained 100 epochs on the GeForce GTX 1080. The average training time for one epoch was 7500 s and 6000 s for an LSTM and GRU cell respectively. All results were obtained with a bidirectional GRU neural network with two layers for each direction and 40 neurons in each layer. Softmax was used in the output layer.

As we mentioned above, we are working with the on-line case where we try to detect stroke type immediately after its drawing was finished. It's a much harder problem than stroke classification when we have the whole document, because we have no context and cannot use global features like in [4].

**Table 3.** Datasets information

| Characteristic | IAMonDO | MHWD_M | MHWD |
|---|---|---|---|
| File amount, pcs | 1000 | 65000 | 10000 |
| Strokes per file, pcs | 378 | 30 | 297 |
| Text strokes, % | 81 | 81 | 91 |
| Amount of points in text strokes (mean $\pm$ std), pcs | $7 \pm 22$ | $18 \pm 16$ | $22 \pm 77$ |
| Amount of points in non-text strokes (mean $\pm$ std), pcs | $10 \pm 13$ | $82 \pm 218$ | $58 \pm 145$ |
| Points belong to text strokes, % | 79 | 49 | 71 |

The proposed solution for mode detection was evaluated on the IAMonDO dataset [15] and a specially collected Samsung Mobile HandWriting Document (MHWD, MHWD_M) datasets. Let us briefly describe MHWD and MHWD_M. MHWD contains free-form documents that combine unconstrained handwriting in seven different languages with different heterogeneous elements. MHWD_M is constructed from MHWD entities such as formulas, shapes, drawings, text blocks, tables, diagrams, etc. Its documents contain a smaller number of strokes. You can see the detailed statistics in Table 3. Documents forming both datasets contain text strokes in text blocks, lists, tables, formulas, diagrams and non-text ones as part of diagrams and drawings. These datasets have detailed element annotations. This allows us to derive the ground truth as the following: for strokes that are part of text blocks, labels in diagrams, lists, formulas and table content, the text class is assigned. The remaining strokes are considered non-text.

Together strokes form ink documents. Each stroke is a sequence of two-dimensional points with coordinates $x$ and $y$. Assume that we have a group of strokes. The concept of a "group" was taken from the labeling structure of the IAMonDO dataset. Strokes of the same group form some entity such as a word, a formula, a drawing, a chart, a diagram, etc. That is why the document can be seen as a set of groups. Consider a group $G$ with strokes $s_1, s_2, \ldots, s_n$ and their coordinates given below

$$\begin{aligned} s_1 &= [(x_1^{(1)}, y_1^{(1)}), (x_2^{(1)}, y_2^{(1)}), \ldots, (x_{c_1}^{(1)}, y_{c_1}^{(1)})] \\ s_2 &= [(x_1^{(2)}, y_1^{(2)}), (x_2^{(2)}, y_2^{(2)}), \ldots, (x_{c_2}^{(2)}, y_{c_2}^{(2)})] \\ &\ldots \\ s_n &= [(x_1^{(n)}, y_1^{(n)}), (x_2^{(n)}, y_2^{(n)}), \ldots, (x_{c_n}^{(n)}, y_{c_n}^{(n)})], \end{aligned}$$

where $|s_k| = c_k, k = \overline{1, n}$. For such group we calculate features by the following rule

$$\begin{aligned} f = [&(x_2^{(1)} - x_1^{(1)}, y_2^{(1)} - y_1^{(1)}, 0), (x_3^{(1)} - x_2^{(1)}, y_3^{(1)} - y_2^{(1)}, 0), \\ &\ldots, (x_{c_1}^{(1)} - x_{c_1-1}^{(1)}, y_{c_1}^{(1)} - x_{c_1-1}^{(1)}, 1), \\ &(x_2^{(2)} - x_1^{(2)}, y_2^{(2)} - y_1^{(2)}, 0), (x_3^{(2)} - x_2^{(2)}, y_3^{(2)} - y_2^{(2)}, 0), \\ &\ldots, (x_{c_2}^{(2)} - x_{c_2-1}^{(2)}, y_{c_2}^{(2)} - x_{c_2-1}^{(2)}, 1), \\ &\ldots, \\ &(x_2^{(n)} - x_1^{(n)}, y_2^{(n)} - y_1^{(n)}, 0), (x_3^{(n)} - x_2^{(n)}, y_3^{(n)} - y_2^{(n)}, 0), \\ &\ldots, (x_{c_n}^{(n)} - x_{c_n-1}^{(n)}, y_{c_n}^{(n)} - x_{c_n-1}^{(n)}, 1)]. \end{aligned}$$

Having done the same calculation for all groups of all documents, we get our train set. Below are the approaches we have checked.

**Approach 1**

We have one label for the whole object (group of strokes) as shown in Table 1. Assume that a group contains only strokes of one type. We consider a task of classification into two classes: non-text (first class) and text (second class). We set one label for the feature vector, calculated for the whole group. Applying this to all groups, we get the train set as shown in Table 1. Training a GRU neural network on such set, we got state of the art 98.5% accuracy on the IAMonDO test set and more than 99% on our datasets. Let us make some remarks. We classified group of strokes of the same type. Accuracy that we have mentioned above was calculated for strokes, not for groups.

**Approach 2**

We have a label for each timestamp as in Table 2. Without loss of generality, assume that we have a group of text strokes. Then, for feature vector $f$, we will have labels given in Table 4.

If a group contained non-text strokes, all features would have $(0, 0)$ labels. After training a GRU neural network on such features, we got the model and evaluated it using the two decoders defined in (3) and (4) and have achieved the state of the art 98.5% accuracy in both cases.

**Table 4.** Features

| Features $\in \mathbb{R}^3$ | Label $\in \mathbb{R}^2$ |
|---|---|
| $(x_2^{(1)} - x_1^{(1)}, y_2^{(1)} - y_1^{(1)}, 0)$ | $(0, 1)$ |
| $(x_3^{(1)} - x_2^{(1)}, y_3^{(1)} - y_2^{(1)}, 0)$ | $(0, 1)$ |
| ... | ... |
| $(x_{c_1}^{(1)} - x_{c_1-1}^{(1)}, y_{c_1}^{(1)} - x_{c_1-1}^{(1)}, 1)$ | $(0, 1)$ |
| $(x_2^{(2)} - x_1^{(2)}, y_2^{(2)} - y_1^{(2)}, 0)$ | $(0, 1)$ |
| $(x_3^{(2)} - x_2^{(2)}, y_3^{(2)} - y_2^{(2)}, 0)$ | $(0, 1)$ |
| ... | ... |
| $(x_{c_2}^{(2)} - x_{c_2-1}^{(2)}, y_{c_2}^{(2)} - x_{c_2-1}^{(2)}, 1)$ | $(0, 1)$ |
| ... | ... |
| $(x_2^{(n)} - x_1^{(n)}, y_2^{(n)} - y_1^{(n)}, 0)$ | $(0, 1)$ |
| $(x_3^{(n)} - x_2^{(n)}, y_3^{(n)} - y_2^{(n)}, 0)$ | $(0, 1)$ |
| ... | ... |
| $(x_{c_n}^{(n)} - x_{c_n-1}^{(n)}, y_{c_n}^{(n)} - x_{c_n-1}^{(n)}, 1)$ | $(0, 1)$ |

**Approach 3**

Similarly to approach 2, we have a label for each timestamp. The difference is that we classify the whole group instead of individual strokes. Using decoders (3) and (4), 98.53% and 98.55% accuracy was obtained respectively on the IAMonDO dataset.

As we mention above, now we are going to explain how decoder (4) works. We have a text/non-text probabilities vector for each segment of the stroke. Then for each stroke we calculate the amount of segments where the probability of it being text is greater than non-text. If the amount of such segments is no less than half of the amount of stroke segments, we classify this stroke as text. Otherwise, we classify it as non-text.

You could mention that we've worked with a group of strokes. Our approach could be applied to the whole document. As we mention above, the problem of detection what type of drawing (text, non-text) the user is doing now is very popular. From cited works it follows that for such a classification problem better accuracy is achieved when we classify the whole group of strokes instead of each stroke separately. Splitting documents into groups and classifying each group was done in [2] for the first time. Let us remind what it means. We process document's strokes one by one. After processing each new stroke, we check if a new group has been created. There are several algorithms for solving this issue. We use a GRU neural network for this task similarly to [5]. More precisely, our GRU receives two consequent strokes as input and predicts the probability that these strokes belong to the same or to the different groups. After forming a new group we classify all strokes from this newly created group. Table 5 contains all results that we have got.

Also we have trained and evaluated approaches similar to ones described in [3,5] which work with the whole document's strokes. The reported accuracy in these works is more than 98%. As we mentioned above, this task is much easier because we use global features and the whole document's context. RNNs were used in the above works. We have repeated their workflows and achieved the reported accuracies on the IAMonDO and MHWD datasets.

Then we did the following. We have trained a solution for whole document stroke classification on the MHWD_M dataset. From Table 3 you can see that its documents contain small number of strokes. We got 90% accuracy. It was a very surprising result. Why do we have such a considerable accuracy drop from more than 98% to 90%? We have started investigating this issue. In all cited works that were using RNNs for stroke classification, the feature vector for one time stamp was calculated using the whole stroke. In this case, the number of document's strokes defines the length of input sequence data. It's a well known fact that RNN's ability to remember input data is its main powerful tool. That's why we suggest that this fact is crucial in dramatic accuracy drop. More precisely, documents with small number of strokes do not provide enough information for the RNN.

## 4 Conclusion

Let us come back to the mode detection problem. As we mentioned before, our approach achieved the state of the art accuracy shown in [2]. All papers concerning this problem that achieved good results worked with sets of strokes, which are called "groups" in our terminology. But while checking their approaches on the whole documents with some grouping algorithms, we also experienced a dramatic accuracy drop. For example, 91.3% accuracy was achieved in [2] using a grouping algorithm and subsequent group classification. The grouping algorithm that was used had accuracy more than 94%. This again leads to the question: why do we have such accuracy drop from 98.5% to 91.3%? As we have already mentioned, the mode detection problem was very sensitive to accuracy of the grouping algorithm. It could be explained as follows. Assume that the grouping algorithm has predicted the creation of a new group incorrectly. For example, we have a group with three non-text and four text strokes and this group is classified as text. In this case we have three incorrectly classified strokes. So you can see the obvious lack of similar approaches: if we have stokes of different classes in one group, then our error will increase with any classification outcome.

In Figs. 1 and 2 we demonstrate how our solution works. There is a number of bounding boxes (dashed rectangles) surrounding strokes belonging to the same group. Let us remind that classification was done for each group separately. You can observe an excellent quality of classification for such a difficult case as in Fig. 1. In Fig. 2 you can see a document with misclassified strokes. For example, some strokes that belong to a formula were classified as non-text. We think this happened because of the formula's font size. It is much larger than for all other text. Also, you can see that the grouping algorithm made mistakes here.

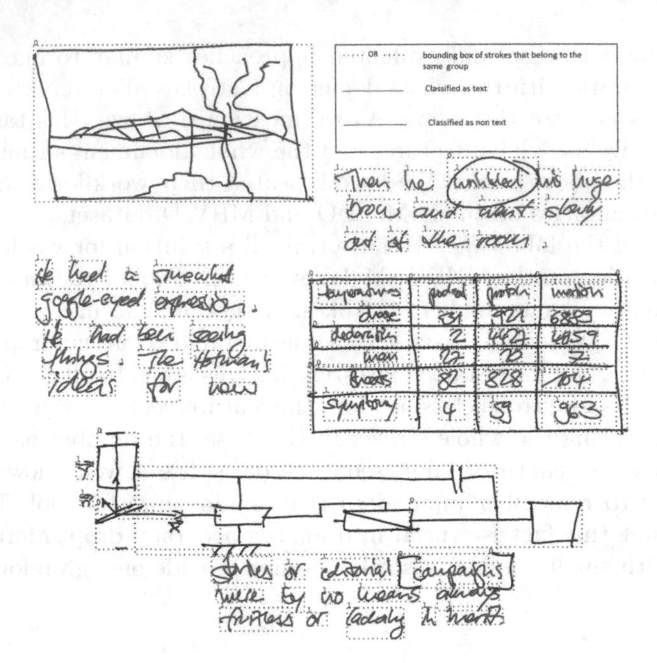


**Fig. 1.** Example of correct classification

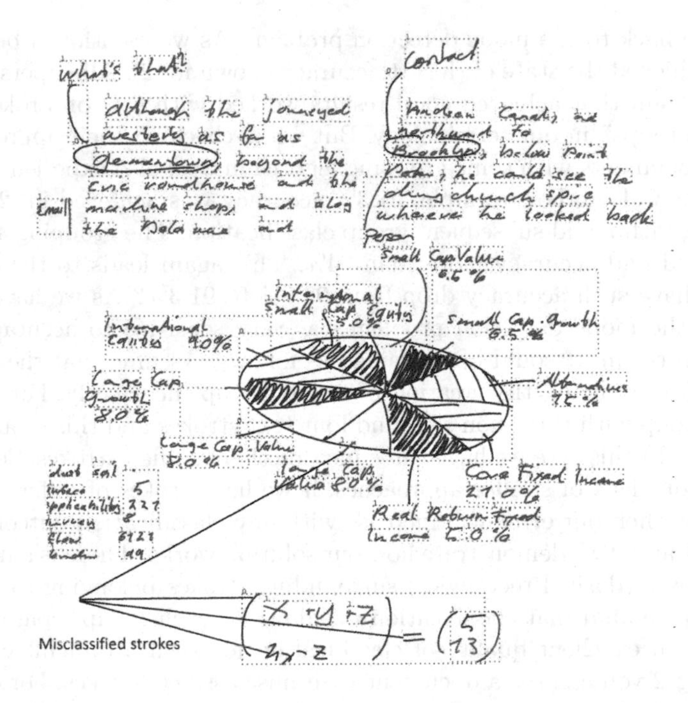


**Fig. 2.** Illustration of misclassification

**Advantages**

We use grouping just for dividing document's strokes into sets in order to decrease sequence data length fed to the RNN. In our approach, we classify each stroke of a group separately. This means that we could make correct classification for a group that contains text and non-text strokes at the same time. Looking at Table 5, you can see 98.55% accuracy for Approach 3 (reference grouping) that is better than 98.5% for Approach 2, but with a grouping algorithm Approach 2 shows better results. Additionally our solution shows similar results on both long and small documents because it uses information only about several last strokes instead of the whole document's strokes.

**Table 5.** Comparative table of the classification accuracy results, %

| Solution | Dataset | | | | | |
|---|---|---|---|---|---|---|
| | IAMonDO | | MHWD_M | | MHWD | |
| | RG* | AG* | RG | AG | RG | AG |
| Proposed in [13] | 97.0 | - | - | - | - | - |
| Proposed in [12] | 98.47 | - | - | - | - | - |
| Proposed in [2] | 98.5 | 91.3 | - | - | - | - |
| Approach 1 using decoder (3) | 98.5 | 91.4 | 99.93 | 97.0 | 99.14 | 94.0 |
| Approach 1 using decoder (4) | 98.5 | 91.4 | 99.93 | 97.0 | 99.14 | 94.0 |
| Approach 2 using decoder (3) | 98.5 | **92.3** | 99.93 | 99.4 | 99.14 | 96.0 |
| Approach 2 using decoder (4) | 98.5 | **92.3** | 99.93 | 99.4 | 99.14 | 96.0 |
| Approach 3 using decoder (3) | **98.53** | 92.2 | 99.95 | 98.5 | 99.20 | 95.1 |
| Approach 3 using decoder (4) | **98.55** | 92.2 | 99.95 | 98.5 | 99.20 | 95.1 |

RG* - reference(ideal) grouping
AG* - adaptive grouping (using grouping algorithm)

**Disadvantages**

The number of strokes defines the length of the input sequence data fed to RNN in the referenced papers. In our approach, the number of points in the group defines such length. This makes our solution slower. As we mentioned above, we also have implemented solution similar to ones that were considered in referenced papers. Below you can see comparative test results for stroke based and point based (proposed) solutions. The test was done on the our special test set that we use for speed measure and the same PC with i7 CPU and 8 GB RAM. This test set contains approximately 100 000 strokes and more than 3 million points. From these results it can be seen that our solution is on average 7 times slower.

**Table 6.** Speed test

| Metric | Stroke based | Point based |
|---|---|---|
| Average time per document, sec | 0.14 | 0.92 |
| Maximum time per document, sec | 0.42 | 2.84 |

This is one of the challenges we are planning to address in the future.

### Further Research

We are planning to do further research in the following directions:

- considering new, more sophisticated decoders;
- modifying the training process. More precisely, in our training set we had groups only with strokes from one class as you can see from Table 4. We are going to consider a case when we have strokes of different classes in one group during training. We hope that this will help to improve classification rate when we have groups of strokes from different classes during evaluation;
- considering evaluation of our approach for more than two classes. For example for working with sketches and shapes similarly to [6–8];
- finding new areas for application of our approach.

## References

1. Graves, A., Liwicki, M., Bunke, H., Schimdhuber, J., Fernández, S., Bertolami, R.: A novel connectionist system for unconstrained handwriting recognition. IEEE Trans. Pattern Anal. Mach. Intell. **31**(5), 855–868 (2009)
2. Khomenko, V., Degtyarenko, I., Radyvonenko, O., Bokhan, K.: Text/shape classifier for mobile applications with handwriting input. Int. J. Doc. Anal. Recognit. **19**(4), 369–379 (2016)
3. Van Phan, T., Nagakawa, M.: Text/non-text classification in online handwritten documents with recurrent neural networks. In: Proceedings of the ICFHR, pp. 23–28 (2014)
4. Van Phan, T., Nagakawa, M.: Combination of global and local contexts for text/non-text classification in heterogeneous online handwritten documents. Pattern Recognit. **51**, 112–124 (2016)
5. Khomenko, V., Degtyarenko, I., Radyvonenko, O., Bokhan, K., Volkoviy, A.: Document structure analysis for online handwriting recognition in mobile applications. In: Proceedings of ICAPR 2017. https://www.isical.ac.in/icapr17/apl.php. (in publish)
6. Bresler, M., Průša, D., Hlaváč, V.: Online recognition of sketched arrow-connected diagrams. Int. J. Doc. Anal. Recognit. **19**(3), 253–267 (2016)
7. Ghodrati, A., Blagojevic, R., Guesgen, H., Marsland, S., Plimmer, B.: The role of grouping in sketched diagram recognition. ACM Press (2018). https://doi.org/10.1145/3229147.3229160

8. Fahmy, A., Abdelhamid, W., Atiya, A.: Interactive sketch recognition framework for geometric shapes. In: Cheng, L., Leung, A.C.S., Ozawa, S. (eds.) ICONIP 2018, Part V. LNCS, vol. 11305, pp. 323–334. Springer, Cham (2018). https://doi.org/10.1007/978-3-030-04221-9_29
9. Jain, A., Namboodiri, A., Subrahmonia, J.: Structure in online documents. In: Proceedings of the the ICDAR, pp. 844–848 (2001)
10. Rossignol, S., Willems, D., Neumann, A., Vuurpijl, L.: Mode detection and incremental recognition. In: Proceedings of the ICFHR, pp. 597–602 (2004)
11. Willems, D., Rossignol, S., Vuurpijl, L.: Features for mode detection in natural online pen input. In: Proceedings of 12th Biennial Conference of the International Graphonomics Society, pp. 113–117 (2005)
12. Otte, S., Krechel, D., Liwicki, M., Dengel, A.: Local feature based online mode detection with recurrent neural networks. In: Proceedings of the ICFHR, pp. 533–537 (2012)
13. Weber, M., Liwicki, M., Schelske, Y., Schoelzel, C., Strauß, F., Dengel, A.: A MCS for online mode detection: evaluation on penabled multi-touch interfaces. In: Proceedings of the ICDAR, pp. 957–961 (2011)
14. Indermühle, E., Bunke, H., Shafait, F., Breuel, T.: Text vs. non-text distinction in online handwritten documents. In: Proceedings of the 25th Annual ACM Symposium on Applied Computing, vol. 1, pp. 3–7 (2010)
15. Indermühle, E., Liwicki, M., Bunke, H.: IAMonDo database: an online handwritten document database with non-uniform contents. In: Proceedings of 9th International Workshop on Document Analysis Systems, pp. 97–104 (2010)
16. Indermühle, E., Frinken, V., Bunke, H.: Mode detection in online handwritten documents using BLSTM neural networks. In: Proceedings of the ICFHR, pp. 302–307 (2012)

# Improving Classification of Ultra-High Energy Cosmic Rays Using Spacial Locality by Means of a Convolutional DNN

Francisco Carrillo-Perez[1(✉)], Luis Javier Herrera[1], Juan Miguel Carceller[2], and Alberto Guillén[1]

[1] Computer Architecture and Technology Department, University of Granada, Granada, Spain
franciscocp@correo.ugr.es
[2] Theoretical and Cosmos Physics Department, University of Granada, Granada, Spain

**Abstract.** Machine learning algorithms have shown their usefulness in a countless variety of fields. Specifically in the astrophysics field, these algorithms have helped the acceleration of our understanding of the Universe and the interaction between particles in recent years. Deep learning algorithms, enclosed in machine learning field, are showing outstanding performance in problems where spatial information is crucial, such as images or data with time dependency. Cosmic rays are high-energy radiation, mainly originated outside the Solar System and even from distant galaxies that constitute a fascinating problem in Physics today. When a Ultra-High Energy Cosmic Ray enters the Earth's atmosphere an extensive air shower is generated. An air shower is a cascade of particles and can be recorded with surface detectors. This work develops a supervised learning algorithm to classify the signals recorded by surface detectors with the aim of identifying the primary particle giving rise to the extensive air shower. Convolutional Neural Networks along with Feed Forward Neural Networks will be compared. Also, the aggregation of information from different surface detectors recording the same phenomenon will be studied against using the information of a single surface detector.

**Keywords:** Cosmic rays · Ultra high energy · Mass composition · Convolutional Neural Network · Deep learning

## 1 Introduction and Problem Description

The Pierre Auger Observatory [1] is an international collaboration that is carrying out the biggest experiment on earth to date. Its aim is understand the ultimate nature of UHECRs. To do so, a large array of Water-Cherenkov Detectors (WCD) occupying an area of 30000 $\mathrm{Km}^2$ operates with a duty cycle close to 100%.


I. Rojas et al. (Eds.): IWANN 2019, LNCS 11506, pp. 222–232, 2019.
https://doi.org/10.1007/978-3-030-20521-8_19

Once the data are analyzed and interpreted by experts, it may shine some light on many essential questions that still remain unanswered. Where do this radiation come from? What is the mechanism that accelerates them? What type of object creates them? are just a few of many unknowns that should be answered in the next years. Recent research has been focused on how Machine Learning (ML) techniques and methods could be applied to astroparticle physics problems like [4,5] and [8] with interesting results. In [5] authors were able to reconstruct cosmic air showers signals using CNNs and, recently, in [7] the muon number was determined using a Deep Neural Network. In [4] authors used Graph Neural Networks to classify IceCube events, with remarkable results. Also in [8] authors applied CNNs to IceCube pulses in order to reconstruct muon-neutrino events for further study.

The idea proposed in this paper is to apply Deep Learning algorithms to the problem of particle identification. As a first approach, Convolutional Neural Networks (CNN) models, which exploit the spatial locality of data, which correspond to temporal traces of particles showers, will be compared with Feedforward Neural Network classification models. Thus the work is structured as follows: Sect. 2 will describe with some detail the data available for the experiments and discuss some considerations on their use. Section 3 describes the different alternatives for DL models used. Section 4 presents and discuss the results of the experiments carried out. Finally, in Sect. 5, conclusions are drawn.

## 2 Data Description

In order to train a supervised learning model, labeled data is needed. In our case, these data are provided by the QGSJET-II package [13]. Events are simulated using the software described in [2]. Each surface detector records a signal each 25 ns being encapsulated in traces of 200 values. The simulator output is the total signal measured in Vertical Equivalent Muons (VEMS), $S_{total}$, which is the sum of the muonic component of the signal $S\mu$ and the electromagnetic part, $S_{em}$ for each surface detector at each time stamp. These two signals are also provided by the simulator so the models can use these components isolatedly or as a whole.

Simulations are quite expensive in terms of computing and storage needs, therefore, a reduced subset of possible types of particles is simulated. This work considers five types: photon, proton, helium, nitrogen and iron. Classifying between these types of particles is arduous, because there are complex non-linear relationships in the variables measured. Therefore blurry frontiers among these types of particles was observed in the first test performed using supervised learning. An interesting problem is to distinguish air showers generated by photons, from those originated by nuclei.

Examples of muonic and total signals of particles used in this work can be observed in Fig. 1. In the plots, it is possible to see how the muonic component is concentrated at the beginning of the event including high peaks. Samples considered in the dataset were recorded by surface detectors specifically placed between 500 m and 1000 m distance to the core of the shower. Typically, depending on the energy and zenith angle, events could be recorded by one, two or three surface detectors (or in few occasions even more). In this work, the comparison of using the information of one surface detector alone against the aggregation of information from three surface detectors hit by the same event for a supervised classification task is performed.

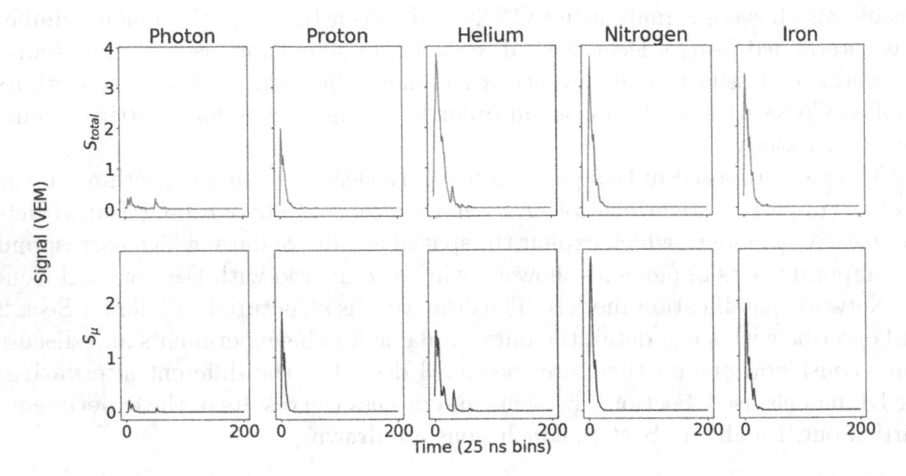


**Fig. 1.** Example traces with total and muonic signal in the same event for photon, proton, helium, nitrogen and iron particles. The five upper plots present examples of total signal. The inferior five plots represent examples of the muonic component of the total signal. The heavier the particle is, the more energy should be recorded, however, the tank might be far from the core.

## 2.1 Data Preprocessing

The data available are extremely imbalanced since there are 5 types of particles but the classifier considers only Photon vs. Hadron. In order to tackle this problem an under-sampling of the majority class is performed. When using one trace only 10000 samples per particle in the Hadrons group are available after under-sampling, to reach the total number of 39195 photon events available. Similarly, when taking into account three traces per phenomenon, 900 samples per hadron particle are used. This leads to the sample distribution shown in Table 1.

**Table 1.** Number of samples per class.

| | Photon | Hadrons | Total |
|---|---|---|---|
| Using 1 trace/event | 39195 | 40000 | 79195 |
| Using 3 traces/event | 3312 | 3600 | 6912 |

When dealing with the problem of using three traces to perform the classification, two different strategies were followed to aggregate the data, depending on the type of neural network used. For the Feed Forward Neural Network (FFNN) option, the three traces were concatenated along the row axis. Therefore the input for FFNNs will be of shape 600. Regarding the CNN architecture, as will be discussed in detail in the next subsection, they were concatenated along column axis, and thus input will be of shape (3,200). No particular order was used for the concatenation of the traces of a given event; experiments were performed to test if ordering the traces based on the proximity of the surface detector to the core of the shower affected the results, and no significant differences in accuracy were found. This was observed for both types of networks.

# 3 Methodology: Feed-Forward Neural Network (FFNN) and Convolutional Neural Network (CNN)

Artificial Neural Networks (ANNs) are loosely based on biological neural networks. In the latter, neurons are interconnected and the information is transmitted through these connections by electrical pulses known as synapses. In Artificial or Deep Neural Networks (DNNs), each neuron computes a linear combination of its inputs: the activation function. Neurons are grouped in layers, in which there is always an input layer to which we feed the data and an output layer which represents the prediction. In the middle of both of these layers we find the so-called hidden layers. The input of the activation function is the data that we feed the NN with (if it is the first layer) or the output of other neurons (in the case of hidden and output layers).

In this work we use supervised learning. Therefore, in the first stage, both types of neural nets are trained on labelled data: the predictions of the neural net and the true values are compared, and the error is minimized. This is done by changing the weights in the linear combinations taking place within each neuron [6,11].

Convolutional Neural Networks [12] are an special type of ANNs in which the convolution operation is applied over the input data. The neural net learns filters that previously needed to be hand-engineered. This independence from prior knowledge and human effort in feature design is a major advantage. CNNs have been successfully applied both to image and signal processing [3,5,10].

Usually in CNN architectures for image classification, three channels are used. This means that for each pixel there are three values, one for each primary color. Trying to simulate this behaviour in images, signals will be treated in this work in the same way when using three traces. For each position in time there is an array of three values. One value for each surface detector trace from the same phenomenon.

### 3.1 FFNN

For this type of neural network, different architectures with several layers and number of neurons were assessed but finally, two different topologies were chosen for this problem. One topology was formed by three layers, where the number of neurons for each layer was fifty, twenty and ten respectively. The second topology was formed only by two layers of one hundred neurons each. They were chosen due to experimentation as well. The selected FFNN topologies are depicted in Fig. 3. First topology will be denoted as FFNN1 and second one as FFNN2 for further use in this work.

### 3.2 CNN

Regarding the CNN architecture only two convolutional layers where chosen as a sufficient number to encapsulate the inner complexity of the data. As seen in Fig. 1, there is a main point where maximum energy is found, but most of the trace is equal to zero; therefore, more layers were expected not to be required. Also, filters of size 5 were chosen in accordance with the length of the peak of maximum energy. Since the maximum peak doesn't maintain long in time, with small filters is easier to spatially locate them. Two dense layers were chosen due to experimentation as well conformed by one hundred neurons each. We recall here that CNN architectures have been applied previously in literature for reconstruction of cosmic air showers traces with remarkable results [5]. The CNN topology is depicted in Fig. 4.

## 4 Results

Both the FFNN and the CNN were coded using **Pytorch** [14] framework in Python3.6. Both networks were trained using Adam Optimizer [9] with the following parameters, **learning rate** = 0.001, **betas** = (0.09, 0.999) during 40 epochs with a batch size of 64. These parameters were chosen due to experimentation. The best accuracy in validation during this 40 epochs was saved, and the respective test accuracy and confusion matrix were calculated. **Cross-Entropy Loss** was used in order to train the model. The loss can be expressed as:

$$loss(x, class) = -\log\left(\frac{exp(x[class])}{\sum_j exp(x[j])}\right) = -x[class] + \log\left(\sum_j exp(x[j])\right) \quad (1)$$

For a better assessment of the models in the problem tackled, five different random train-test subdivisions of the whole available dataset were performed, using a specific seed in order to maintain consistency across different experiments. The function used for the splitting is provided by the Python's library Scikit-Learn [15]. Accuracy represents the overall performance of the network

**Table 2.** Mean accuracy for each FFNN and CNN topologies using one and three traces. Results are presented for $S_{total}$ and $S_{\mu}$ both for test and validation sets.

| One trace | | |
|---|---|---|
| $S_{total}$ | **Test Acc.** | **Val. Acc.** |
| FFNN1 | 82.12(0.65) | 82.11(0.27) |
| FFNN2 | 84.14(0.43) | 84.18(0.17) |
| CNN | 87.23(0.22) | 88.74(0.14) |
| $S_{\mu}$ | **Test Acc.** | **Val. Acc.** |
| FFNN1 | 78.15(0.074) | 78.55(0.37) |
| FFNN12 | 77.92(0.074) | 78.61(0.37) |
| CNN | 79.05(0.12) | 79.05(0.34) |

| Three traces | | |
|---|---|---|
| $S_{total}$ | **Test Acc.** | **Val. Acc.** |
| FFNN1 | 84.03(1.40) | 86.20(0.68) |
| FFNN2 | 85.79(0.80) | 88.28(0.58) |
| CNN | 88.48(0.02) | 90.95(0.58) |
| $S_{\mu}$ | **Test Acc.** | **Val. Acc.** |
| FFNN1 | 87.08(0.41) | 86.98(0.65) |
| FFNN2 | 86.27(0.41) | 87.86(0.95) |
| CNN | 86.07(0.34) | 88.02(0.60) |

**Table 3.** Mean sensitivity and specificity for each FFNN and CNN topologies using one and three traces in the test set. Results are presented for the $S_{total}$ and for $S_{\mu}$.

| One trace | | |
|---|---|---|
| $S_{total}$ | **Sensitivy** | **Specificity** |
| FFNN1 | 81(0.9) | 83(0.8) |
| FFNN2 | 83(0.7) | 85(0.7) |
| CNN | 86(1.5) | 88(0.9) |
| $S_{\mu}$ | **Sensitivy** | **Specificity** |
| FFNN1 | 76(0.5) | 80(0.6) |
| FFNN2 | 77.5(0.7) | 78(1) |
| CNN | 76(0.9) | 80(1.2) |

| Three traces | | |
|---|---|---|
| $S_{total}$ | **Sensitivy** | **Specificity** |
| FFNN1 | 86(2) | 82(3) |
| FFNN2 | 89(2) | 83(3) |
| CNN | 88(6) | 90(4) |
| $S_{\mu}$ | **Sensitivy** | **Specificity** |
| FFNN1 | 86(0.9) | 88(1.7) |
| FFNN2 | 87(2) | 86(4) |
| CNN | 82(5) | 94(3) |

**Table 4.** Differences in accuracy, sensitivity and specificity for each FFNN and CNN topologies using one or three traces in the test set. Value for each metric obtained using three traces was subtract by the value obtained with one trace. Results are presented for the $S_{total}$ and for $S_{\mu}$.

| Differences | | | |
|---|---|---|---|
| $S_{total}$ | Test Acc. | Sensitivity | Specificity |
| FFNN1 | 1.91 | 5 | −1 |
| FFNN2 | 1.67 | 6 | −2 |
| CNN | 1.23 | 2 | 2 |
| $S_{\mu}$ | Test Acc. | Sensitivity | Specificity |
| FFNN1 | 8.93 | 10 | 8 |
| FFNN2 | 8.35 | 9.5 | 8 |
| CNN | 7.02 | 6 | 14 |

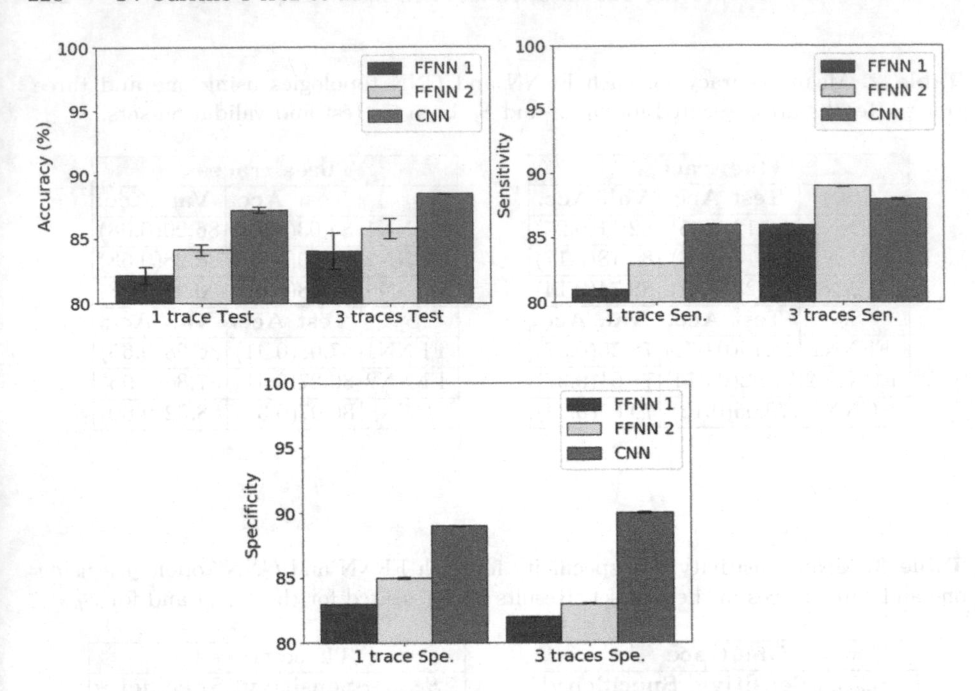


**Fig. 2.** In a mean accuracy for all architectures using total signal is presented. As observed, CNN outperforms FFNN architectures' results. A baseline model, where we could use a weak classifier which always predict the majority class, could obtain near 50% of accuracy, therefore all the three presented models here outperform this weak classifier. In b mean sensitivity in the test set using total signal is presented. In c mean specificity in the test set using total signal is presented. Y axis has been ranged between 80% and 100% in order to see the differences between different results.

over the test set. Sensitivity and specificity are computed based on the confusion matrix. Sensitivity represents how many true positives (photons) are correctly classified. Specificity represents how many true negatives (hadrons) are correctly classified.

Table 2 show the results obtained when comparing the accuracy obtained for each FFNN and CNN topologies. Accuracy using two alternatives for input data (muonic signal and total signal) is presented. Table 3 shows the results obtained when comparing the sensitivity and specificity for each topology. Sensitivities and specificities using muonic and total signal are presented (Table 4).

Mean accuracy, sensitivity and specificity for each architecture and each iteration has been described in the previous tables for total and muonic signals as inputs. Best accuracy is obtained using the aggregation of three traces per phenomenon for each architecture (see Table 2). The second architecture for the

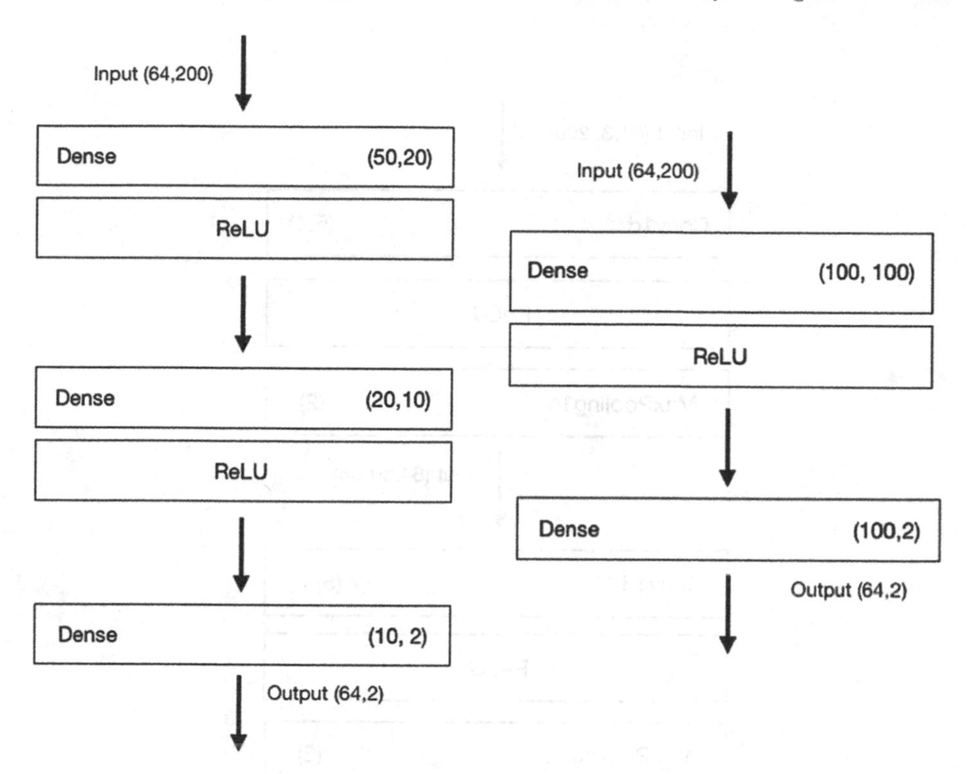


**Fig. 3.** The two selected Feed Forward Neural Network architectures (input layer-with no calculi- and output layer were omitted for simplicity).

FFNN, the one with two layers of one hundred neurons each, improves results over the other FFNN architecture. Even though none of them surpass the results obtained with the CNN using the information of three traces. Results obtained using total signal instead of muonic signal are superior for the CNN architecture. Nonetheless, for FFNN architectures superior results are obtained with muonic signal using three traces, but very close to results obtained with total signal. FFNNs using one trace have a superior performance when using total signal. Results for Sensitivity and Specificity (see Table 3) are superior in the CNN architecture. Again, using three traces improves results over using the information of one trace only. This was expected, since using three traces add more information of the phenomenon to classify. Also using total signal instead of muonic signal led to better results in both sensitivity and specificity. A graphical representation of the results for each architecture can be observed in Fig. 2.

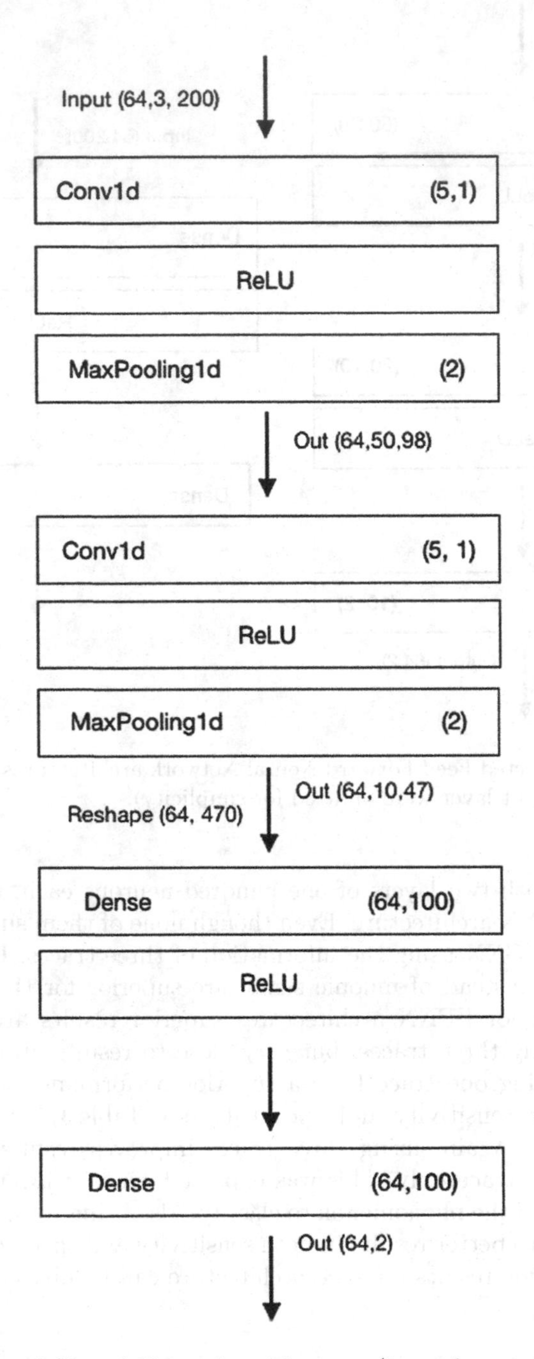


**Fig. 4.** Convolutional Neural Network architecture (input layer-with no calculi- and output layer were omitted for simplicity).

## 5 Conclusions

This work explores how machine learning algorithms can help solving open questions in the field of ultra-high energy cosmic rays. Specifically, in determining the type of the particle that generates the EAS recorded at the Piere Auger Observatory. The results obtained by the different approaches used are quite promising due to the high accuracy values obtained. Taking simulated events, it was shown how deep learning methods are a powerful tool to classify between photons and hadrons. This type of model show an improvement when processing the data and its spatial locality information by means of convolution operations in comparison with classical feedforward topologies. Two types of inputs were considered for the networks obtaining the best performances, those using the total signal recorded by the detector instead of the muonic part of the signal isolatedly. As future work, the roadmap consists on the inclusion of additional information, such as the zenith angle, the risetime, the falltime and other variables obtained for each event. Machine learning techniques seem the ideal tool to gain further insight into the mysteries of ultra-high energy cosmic rays and to increase the velocity of knowledge extraction from data.

**Acknowledgements.** This research has been possible thanks to the support of projects: FPA2015-70420-C2-2-R, FPA2017-85197-P and TIN2015-71873-R (Spanish Ministry of Economy and Competitiveness –MINECO– and the European Regional Development Fund. –ERDF). We thank the Pierre Auger Collaboration for letting us use the simulated event samples that are at the core of this study.

## References

1. The Pierre Auger Collaboration: The pierre auger cosmic ray observatory. Nucl. Instrum. Methods Phys. Res., Sect. A **798**, 172–213 (2015)
2. Argiro, S., et al.: The offline software framework of the pierre auger observatory. Nucl. Instrum. Methods Phys. Res., Sect. A **580**(3), 1485–1496 (2007)
3. Aznan, N.K.N., Bonner, S., Connolly, J., Al Moubayed, N., Breckon, T.: On the classification of ssvep-based dry-eeg signals via convolutional neural networks. In: 2018 IEEE International Conference on Systems, Man, and Cybernetics (SMC), pp. 3726–3731. IEEE (2018)
4. Choma, N., et al.: Graph neural networks for icecube signal classification. In: 2018 17th IEEE International Conference on Machine Learning and Applications (ICMLA), pp. 386–391. IEEE (2018)
5. Erdmann, M., Glombitza, J., Walz, D.: A deep learning-based reconstruction of cosmic ray-induced air showers. Astropart. Phys. **97**, 46–53 (2018)
6. Goodfellow, I., Bengio, Y., Courville, A.: Deep Learning. MIT Press (2016). http://www.deeplearningbook.org
7. Guillén, A., et al.: Deep learning techniques applied to the physics of extensive air showers. Astropart. Phys. (2019). https://doi.org/10.1016/j.astropartphys.2019.03.001, http://www.sciencedirect.com/science/article/pii/S0927650518302871
8. Huennefeld, M.: Deep learning in physics exemplified by the reconstruction of muon-neutrino events in icecube. PoS, p. 1057 (2017)

9. Kingma, D.P., Ba, J.: Adam: a method for stochastic optimization. CoRR abs/1412.6980 (2014). http://arxiv.org/abs/1412.6980
10. Krizhevsky, A., Sutskever, I., Hinton, G.E.: Imagenet classification with deep convolutional neural networks. In: Advances in Neural Information Processing Systems, pp. 1097–1105 (2012)
11. LeCun, Y., Bengio, Y., Hinton, G.: Deep learning. Nature **521**(7553), 436 (2015)
12. LeCun, Y., et al.: Backpropagation applied to handwritten zip code recognition. Neural Comput. **1**(4), 541–551 (1989)
13. Ostapchenko, S.: Monte carlo treatment of hadronic interactions in enhanced pomeron scheme: QGSJET-II model. Phys. Rev. D **83**(1), 014018 (2011)
14. Paszke, A., et al.: Automatic differentiation in pytorch (2017)
15. Pedregosa, F., et al.: Scikit-learn: machine learning in python. J. Mach. Learn. Res. **12**, 2825–2830 (2011)

# Model and Feature Aggregation Based Federated Learning for Multi-sensor Time Series Trend Following

Yao Hu, Xiaoyan Sun(✉), Yang Chen, and Zishuai Lu

School of Information and Control Engineering,
China University of Mining and Technology, Xuzhou 221116, China
xysun78@126.com

**Abstract.** In the industrial field, especially the work or environment condition monitoring, it is crucial but difficult to follow the trend of the time series monitoring data (TSD) when the TSD come from different kinds of sensors and are collected by different companies. The privacy of the multi-sensor TSD must be carefully treated. Few studies, however, have been devoted to solving such problems. Federated learning (FL) is a good structure developed by Google for well keeping the personal privacy. Motivated by this, we here present an improved FL structure for not only keeping the data privacy but also extracting and fusing the trends features of the multi-sensor TSD. In our work, the client models of FL are first designed and optimized for getting the initial parameters and features w.r.t. the corresponding sensor's TSD, and then both the model parameters and the extracted features of all the activated clients (sensors) are sent to the central server and aggregated. The fused parameters and features are returned to the clients and used to update the optimization of the model. Finally, the fused features of all multi-sensor TSD are put into an echo state network (ESN) to fulfill the trend following of the multi-sensor TSD. The proposed algorithm is applied to the multi-sensor electromagnetic radiation intensity TSD sampled from an actual coal mine, and its superiority in promoting the accuracy on every sensor is demonstrated.

**Keywords:** Feature aggregation · Federated learning · Multi-sensor data · Safety forecast · Trend following

## 1 Introduction

Time series data (TSD) is a series of data points indexed (or listed or graphed) in time order. Typical examples of time series are stock market transaction values, heights of ocean tides, and status data collected by sensors the like. In

This work is supported by the National Natural Science Foundation of China with Grant No. 61876184 and 61473298.


I. Rojas et al. (Eds.): IWANN 2019, LNCS 11506, pp. 233–246, 2019.
https://doi.org/10.1007/978-3-030-20521-8_20

some industrial applications, time series data can be taken as important indicators to judge incidents, such as the residual life prediction [14], equipment fault diagnoses [15]. But in terms of mining working environment safety pre-warning, Wang et al. [13] pointed out that the variation trends of sampled Electromagnetic Radiation Intensity time series data (ERI-TSD) within a certain period, rather than the specific values of time series at a certain moment, played a more important role. However, few studies about how to measure and follow such variation trends of TSD have been carried out. The concept of trend following of TSD was firstly introduced in the financial field, and there are less relevant research on the trend following of industrial TSD. With the formation of internet of things, large amount of diversified TSD are generated at high speeds from industrial equipments; the corresponding information about one object collected by only one single sensor can no longer meet practical requirements. Multi-sensor TSD are more and more general and important. In such cases, the trends measurement and following of TSD are becoming extremely difficult due to the TSD fusion.

Multi-sensor data fusion refers to develop a relatively integral description of the incomplete data information about the environmental feature provided by multiple sensors or information sources, and the formed integral description can be used for achieving more accurate recognitions and judgements. Compared with single source data processing, data fusion method can aggregate the information from multiple sensors at the same time, improve data detectability and credibility, enhance the system fault tolerance as well [11]. Common data fusion methods mainly contain some mathematical knowledge and technical tools [9], e.g., Kalman filtering [1] and neural networks [17]. Even the multi-sensor data fusion has been sufficiently concerned on, few studies consider the multi-sensor TSD fusion. Most important, in the multi-sensor TSD, the sensors can be made by different companies and the corresponding data can be managed separately and privately in the big data era.

In the process of multi-sensor data fusion, there exists lots of interaction operations in which sensitive data or confidential information might be revealed. Therefore, multi-sensor TSD fusion under the condition of privacy protection should be a great challenge. Traditional privacy protection manners adopted in multi-sensor data fusion methods mainly consider improving or modifying original data, but this kind of method may disturb the distribution of original data and then affect the obtained conclusions. Some methods aiming at solving these problems, e.g., [4] proposed a new technique which similar to an easy encryption algorithm, by fabricating of association rule using a stochastic standard map to maintain the confidentiality; another hybrid transformation which merges the entropy-based partition and combined distortion techniques was proposed in [10], but these proposed methods have some insufficients when it comes to the comparative complex problems or frequent interaction tasks.

From the viewpoint of privacy protection, Google proposed the framework of FL [8] to aggregate the information from every user without privacy leakage. In this setting, the entire process is divided into several communication rounds, based on different communication-status, facilities have different participation level. Local models are trained upon data that are stored in clients, and the server does not need training data containing privacy information. Instead of sharing local data, only local model parameters are sent to the server and are aggregated to obtain the central model. At the end of current communication round, clients receive the aggregated updates as initial parameters of the following training; then local models will be retrained on privacy data with local stochastic gradient descent (local SGD) [5]. The primary advantage of FL is that the security risks are greatly reduced, since only attacks on local devices might cause the privacy information leakage of these attacked clients. But this framework is not suitable for multi-sensor data fusion, because we are not able to utilize the central model which only aggregates the weights from local models to get the aggregated data. Motivated by this, under the FL framework, we decide to develop a data feature aggregation methods under the premise of avoiding local dataset exchange. As for the data features extraction, inspired by the successful application of deep learning on image classification [7] and machine translation [16], we can be fully convinced these networks can extract the reliable features of the relative datasets. FL is a good structure for privacy protection, yet has not been applied to multi-sensor TSD analysis, such a structure can be helpful for keeping data privacy of multi-sensor TSD trends following and features fusion.

As mentioned, the variation trend of ERI-TSD within a certain period can be taken as great indicators to judge the occurrence of dynamic disaster in the coal mine. Some research about trend following of time series data for safety forecast have proven efficient, e.g., Qiu et al. [12] defined a variable coefficient as the division between the standard deviation and mean value over the same time period, and took the coefficient as the trend variation over that period. Once the defined variation reached the threshold given by practice, the dynamic disaster was thought broke out. Based on that, three types of trend representations for TSD were defined, and an improved LSTM based trend following strategy was presented in our previous work [2]. Although the proposed strategy achieves high performance in the accuracy, there exists some deficiencies, for instance, we only consider the trends of dataset collected by one sensor which cause the final conclusion lake of reliability.

Thus, in order to aggregate the trends of multi-sensor data synchronously meanwhile avoid the privacy leakage, and perform an accurate trend following based on the aggregated features, we here propose an improved FL structure to aggregate not only the model parameters as traditional FL but also the features of multi-sensor TSD variation trends. The data from different sensors are stored and processed on local clients with long short term memory (LSTM) networks. The parameters of LSTM and its extracted features are uploaded to the central server. According to the aggregation of FL, the parameters and the features are fused and reused in the following optimization of each LSTM on every client. Finally, the aggregated features are used to analyze the variation trends of the multi-sensor TSD sets.

## 2 Multi-sensor TSD Trend Following with Model and Feature Aggregation in Federated Learning

### 2.1 Traditional Federated Learning

In the FL framework, the learning task is solved by a loose federation of participating devices which are coordinated by a central server. In every communication round, the structural weights will be computed in the clients and uploaded to the current global model maintained by the central server, the local dataset will never leave the client.

In the setting, ideal fundamental characteristics of data/clients are as follows [5,6]:

(1) Non-IID: The training data on clients are typically associated with the unique individual data feature, and hence the local datasets are not able to be representative of the overall distribution.
(2) Unbalanced: The unbalanced data are generated based on the sensors' different levels of participation.
(3) Massively distributed: The number of clients is large enough.
(4) Limited communication: Part of participated sensors may in bad communication quality, or in communication failure.

Due to the special characteristics of actual applications, Google proposed an algorithm named federated average algorithm. In this algorithm, only $C$-fraction or specified $K$ number of clients are randomly selected as the participating subset in every communication round, the participating clients subset are indexed by $k$. Stochastic gradient descent methods are carried out on the local train dataset, the obtained local trained model parameters $\omega^k$ are sent to the central server and aggregated by means of Eq. 1.

$$\omega_t \leftarrow \sum_{k=1}^{K} \frac{n_k}{n} \omega_t^k \quad where \quad n = \sum_{k=1}^{K} n_k \tag{1}$$

where $t$ represents the current round and $n_k$ means the amount of local examples of client $k$ while $n$ is sum number of selected participating subset examples. That is, each selected client takes several steps of gradient descent on the current model, and then the server takes a weighted average of the weights from the trained models. The selected clients will download the aggregated parameters from the central server as its initial parameters for local training process in next round. Clearly, the latent features of the data on each client are not concerned in the traditional FL. Here, we will modify this structure for getting the fused features of the TSD trend of each sensor to fulfill the multi-sensor TSD fusion and trend following.

### 2.2 Federated Learning Based Model and Feature Fusion for Multi-sensor TSD

The presented algorithm is visualized in Fig. 1. Similar to the traditional FL, it has the local clients training, central server aggregation and their upload/download. However, it is quite different from the traditional in two aspects: (1) In the clients, not only the model parameters, denoted as $\omega$, but also the corresponding features as $fea$ extracted by the learning model LSTM are separately treated. (2) For the aggregation, both the model and the features are fused. In every communication round, all selected clients are synchronously trained with local SGD on their local dataset $P_k$. Similar to the federated average algorithm, in each round the extracted features from selected individual client are aggregated in manner of Eq. 2.

$$Fea \leftarrow \sum_{t=1}^{R} \sum_{k}^{K} \frac{acc_k^t}{Acc_t} fea_t^k \quad where \quad Acc_t = \sum_{k=1}^{K} acc_k^t \tag{2}$$

where $acc_k$ refers to the local valid accuracy of $k$-th client, $fea_t^k$ means the extracted features from $k$-th client at $t$-th round.

This process will repeat $MAX\_NUM\_ROUNDS\_P$ rounds, the aggregated weights will be exchanged between clients and the central server in every round. But the aggregated features will only be updated in the central server until the iterations end, then they will be downloaded by every participation as the representation of their local data. The final outputs of this procedure are the global aggregated features.

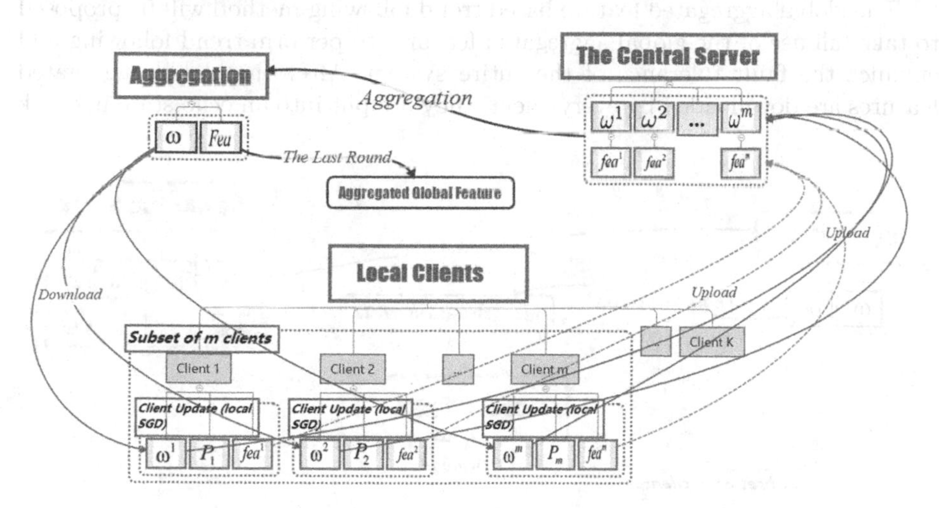


**Fig. 1.** Illustration of the Federated Learning based feature aggregated method

## 2.3 Global Aggregated Feature Based Trend Following of TSD

As for the trend following of TSD, once the trend of industrial TSD was defined, trend following became an essential new form of a TSD prediction. Therefore, we first calculate the trend representations of the multi-sensor TSD, and then get the aggregated feature of multi-source trends by the methods proposed in Sect. 2.2.

Experimental results from [2] show the average indicator ($A_{r,l}$) and the standard deviation method derived from the logarithmic ($\sigma_{r,l}$) within certain time spans can have a clear representation of the trends. With $T$ discrete-time index set and $M$ time points, the ERI-TSD are denoted as $Y = \{p_t, t \in T\}, T = \{1, 2, \cdots M\}$, consider a length $l$ for the time window of $Y$, the number of points in every time span which denoted as $N$ can be calculated as $N = \lfloor T/l \rfloor$, where $\lfloor . \rfloor$ represents a floor integer of an element. Equations 3 and 4 give details of these indicators.

$$A_{r,l} = \frac{\sum_{t=(k-1)l+1}^{rl} p_t}{l}, r = 1, 2, \cdots, N \tag{3}$$

$$\sigma_{r,l} = \sqrt{\frac{\sum_{t=(r-1)l+1}^{kl} (h_{t,l} - \bar{h}_{r,l})^2}{l-1}}, r = 1, 2, \cdots, N \tag{4}$$

$$where \quad h_{r,l} = \ln(\frac{p_{t+l-1}}{p_t}) = \ln(p_{t+l-1}) - \ln(p_t)$$

among that, $\bar{h}_{r,l}$ is the average of $h_{t,l}$ on the $r$-th time window. In each local client, $A_{r,l}$ and $\sigma_{r,l}$ are calculated as the trend representation, and given a general notation as $X_{r,l}$.

The global aggregated feature based trend following method will be proposed to take full use of the global aggregated features to perform trend following and enhance the fault tolerance of the entire system. After the global aggregated features are downloaded by every client, they are put into an echo state network

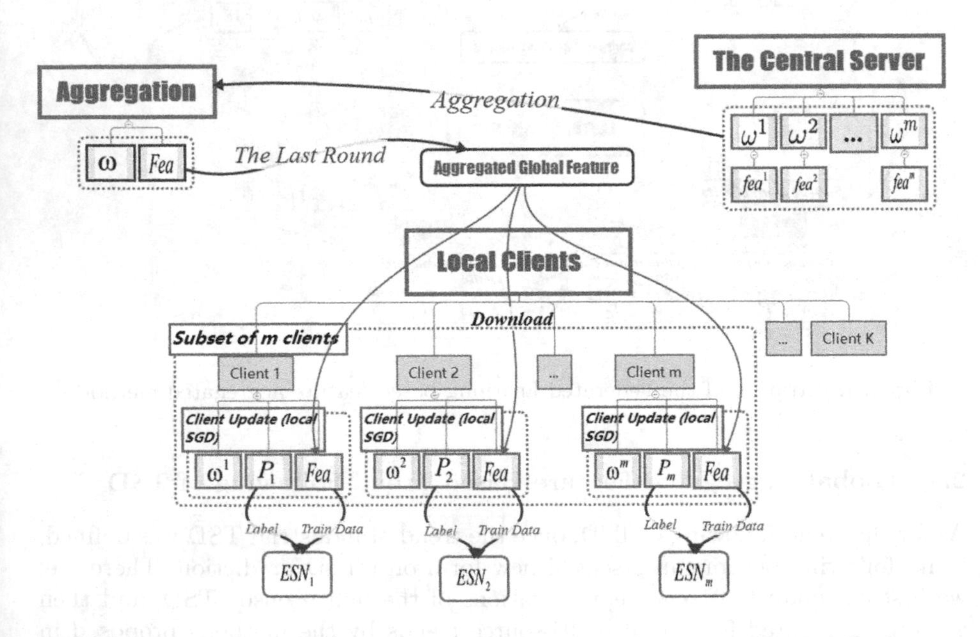


**Fig. 2.** Illustration of the process of multi-sensor feature based trend following

(ESN) [3] which is another variation of recurrent neural network with much lower computational cost. The aggregated features are taken as the training data, and the true values of local dataset in each client are set as the labels to get a well-trained ESN model, then training the ESN to perform the trend following. The entire processes are shown in Fig. 2, the specific experimental settings are given in the following section.

## 3 Experiments and Analysis

The proposed method for multi-sensor TSD trend following mainly contains two parts. First, we have to extract features of the dataset stored in the local clients and get the aggregated global feature; second, the aggregated feature will be downloaded by each client to realize the trend following. The effectiveness of these two parts are validated by applying the algorithm to the ERI-TSD in the coal mine. The corresponding settings of the experiment are shown as follows.

### 3.1 Settings for Local Feature Extraction

Since the sensors are set to different sampling frequencies, and we are unaware of the specific frequency values but sure that the dataset were collected in same period of time, thus the $X_{r,l}$ have to be calculated based on the number of data points in each time window. Specifically, the size of those local datasets stored in different clients and the corresponding number of data points in the time windows are shown in Table 1.

**Table 1.** Specific settings of the local clients

| Sensor# | Size of local dataset | Number of data points in each time window |
|---|---|---|
| Sensor_1 | 19064 | 1.914 |
| Sensor_2 | 17170 | 1.724 |
| Sensor_3 | 14723 | 1.478 |
| Sensor_4 | 9957 | 1 |
| Sensor_5 | 21709 | 2.180 |
| Sensor_6 | 14565 | 1.463 |

Numbers of data points in time windows of Client_4 are termed the baseline as 1, and the numbers in other clients are the relative magnifications of it. By this mean, the number of divided time windows in each client is similar, and the features extracted from $r$-th time window in different clients can represent the trend in same time period. In our experimental settings, we assign the data number in every time window of client_4 as 3, and obtain about 3310 time windows in each client, i.e., the trend of $i$-th client is represented as $Tr_i = \{X_{q,l}, q = 1, 2, \cdots, 3310\}$. In each communication round, $Tr$ will be firstly calculated according to the dataset stored in the selected clients then used to train LSTM, specifically, the first 8 points of $Tr$ are selected to predict the 9th value.

In terms of the structure of LSTM, we take the same model architecture as [2], the architecture is given in Fig. 3. There are one lstm_layer with $cell\_size = 5$, $time\_step = 8$ and a fully connected output layer without any activations. The common method of image feature extraction is to train a high-precision CNN model and take the output matrix of specific layer as the image feature. Combine this method with our special network structure, we separately build the LSTM model in every client and take outputs of lstm_layer as the representations of trend features.

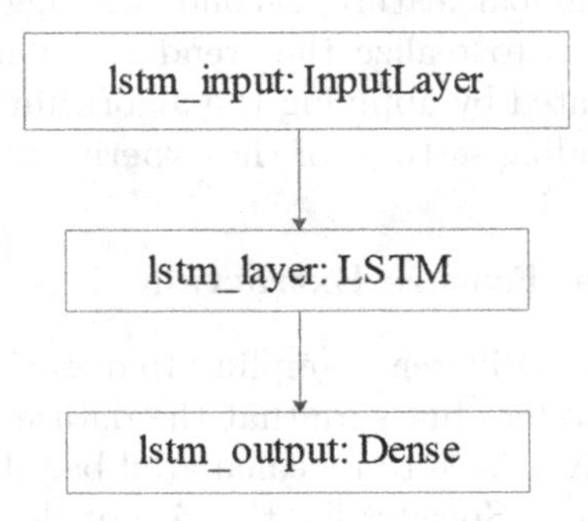


**Fig. 3.** The architecture of LSTM model

### 3.2 Settings for Feature Aggregation and Feature Based Trend Following

There are totally 6 sensors to synchronously sample the ERI-TSD in the coal mine, that is to say, under the framework of our algorithm, the local clients contain 6 components. The number of communication rounds are respectively assigned as $MAX\_NUM\_ROUNDS\_P$ = 10, 50, 100. In every communication round, we set the number of participated clients $K = 6$ and choose the Adam optimizer with the local client model training time equals 20, that means all clients are selected to be trained with Adam optimizer for 20 times and then participate the feature aggregating process. For the $k$-th client, 3310 five-dimensional vectors denoted as ${fea_t}^k$ are obtained as the feature representation of $Tr_k$.

Then according to the features aggregation manner, the global aggregated feature $Fea$ will be obtained on the final round and then they will be downloaded by every local client as the training set for the local ESN model and the true values of $X_{k,l}$ are labels. These ESNs are used to perform the trend following based on the new formed local dataset. All experiments are executed for 12 times, and their average values are calculated as experimental results.

### 3.3 Experimental Performance

With the default values of parameters, experiments are then implemented for testing the overall performance. As introduced, we judge the experimental results according to the specific trend following performance on each individual client. We compare the accuracy of global aggregated features based trend following method with that of individual local trained LSTM, ESN models. As an important contrast, the aggregated LSTM training method proposed in our previous work [2] is also applied to this problem, that method enhanced the performance of LSTM by combining the gradient descent (GD) method with PSO algorithm. According to the previous experimental results, the LSTM trained by a hybrid algorithm combining PSO algorithm and Adam method (abbreviated PSO-Adam-LSTM) is selected here. For the aggregated feature based methods, we set various communication rounds, e.g., 10, 50, 100, and the local data trained LSTM models follow the same structure as above mentioned will iterate 150 times in the training process. The hyper-parameters of PSO-Adam-LSTM can be referred to [2].

Five accuracy indicators, i.e., Root Relative Squared Error (RRSE) which is more sensitive to scale factors, Root Mean Square Error (RMSE), Mean Absolute Error (MAE), Mean Absolute Percent Error (MAPE) and the maximum error (Max_error) between the predicted values and corresponding true values, are used to measure the accuracy. Besides that, the time consumptions of different methods are also compared. The trend following results of $\sigma_{r,l}$ and $A_{k,l}$ by

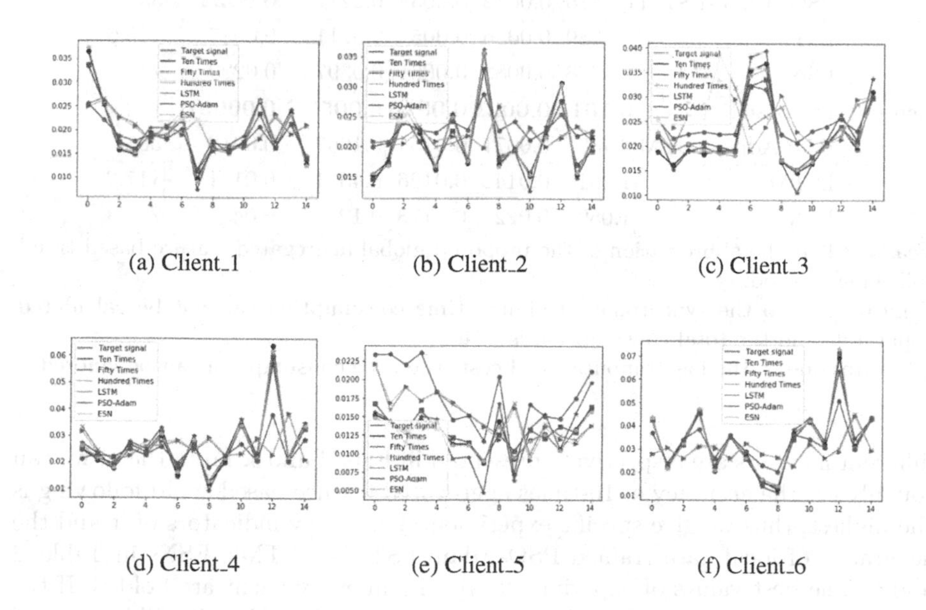


(a) Client_1 (b) Client_2 (c) Client_3

(d) Client_4 (e) Client_5 (f) Client_6

**Fig. 4.** Trend following results of $\sigma_{r,l}$ on each individual Client

**Table 2.** Specific experimental results of 10-times aggregated feature based and local data trained models on $\sigma_{r,l}$ trend following

| Sensor# | Methods | RRSE | RMSE | Mape | Mae | Max_error | Time_cost (s) |
|---|---|---|---|---|---|---|---|
| Sensor_1 | Fea_based* | **0.2322** | **0.0014** | **0.0012** | **0.0724** | **0.0007** | / |
| | PSO-Adam-LSTM | 0.5016 | 0.0033 | 0.0026 | 0.1471 | 0.0014 | 317 |
| | LSTM | 1.0233 | 0.0061 | 0.0049 | 0.3298 | 0.0073 | 116.5 |
| | ESN | 0.9977 | 0.0152 | 0.0123 | 0.036 | 0.032 | // |
| Sensor_2 | Fea_based* | **0.1783** | **0.0012** | **0.0009** | **0.0512** | **0.0011** | / |
| | PSO-Adam-LSTM | 0.7217 | 0.0047 | 0.0041 | 0.2251 | 0.0025 | 324 |
| | LSTM | 1.112 | 0.0072 | 0.0058 | 0.2936 | 0.0008 | 117.8 |
| | ESN | 1.383 | 0.025 | 0.0211 | 0.0528 | 0.0526 | // |
| Sensor_3 | Fea_based* | **0.2543** | **0.0021** | **0.0016** | **0.0785** | **0.0027** | / |
| | PSO-Adam-LSTM | 0.7389 | 0.0060 | 0.0053 | 0.296 | 0.0029 | 325 |
| | LSTM | 1.1853 | 0.0096 | 0.0081 | 0.4 | 0.004 | 121.8 |
| | ESN | 1.5032 | 0.021 | 0.0176 | 0.1734 | 0.0444 | // |
| Sensor_4 | Fea_based* | **0.2452** | **0.0026** | **0.0023** | **0.0912**** | **0.0021** | / |
| | PSO-Adam-LSTM | 0.5998 | 0.0065 | 0.0047 | 0.296 | 0.003 | 0314 |
| | LSTM | 1.2063 | 0.013 | 0.0097 | 0.4295 | 0.0141 | 121.1 |
| | ESN | 1.2116 | 0.0246 | 0.0208 | 0.1155 | 0.0472 | // |
| Sensor_5 | Fea_based* | **0.5301** | **0.0021** | **0.0017** | **0.1769** | 0.0035 | / |
| | PSO-Adam-LSTM | 0.9675 | 0.0042 | 0.0034 | 0.2211 | **0.0022** | 333 |
| | LSTM | 1.4389 | 0.0056 | 0.005 | 0.5211 | 0.0057 | 119.6 |
| | ESN | 1.0778 | 0.0088 | 0.006 | 0.0397 | 0.0229 | // |
| Sensor_6 | Fea_based* | **0.1614** | **0.0021** | **0.0016** | **0.0055**** | **0.0005** | / |
| | PSO-Adam-LSTM | 0.488 | 0.0071 | 0.0054 | 0.1697 | 0.0018 | 351 |
| | LSTM | 1.091 | 0.0142 | 0.0126 | 0.4331 | 0.0131 | 117.2 |
| | ESN | 1.059 | 0.022 | 0.0178 | 0.12 | 0.04 | // |

Fea_based* is the abbreviation of the proposed global aggregated feature based trend following method.
/ means due to the synchronous training, time consumption can not be calculated separately and the total consumption is 634.
// means due to the low computational cost, the time consumption can be ignored.

different methods are respectively presented in Figs. 4 and 5. From them, we can roughly get the accuracy of 10-times aggregated feature based trend following is the highest, thus we give specific experimental accuracy indicators of it and the accuracies of local data trained PSO-Adam-LSTMs, LSTMs, ESNs in Tables 2 and 3. The best values of experimental results in every client are bolded. If the value is significant better than the compared ones by T-testing, it will be marked with two stars.

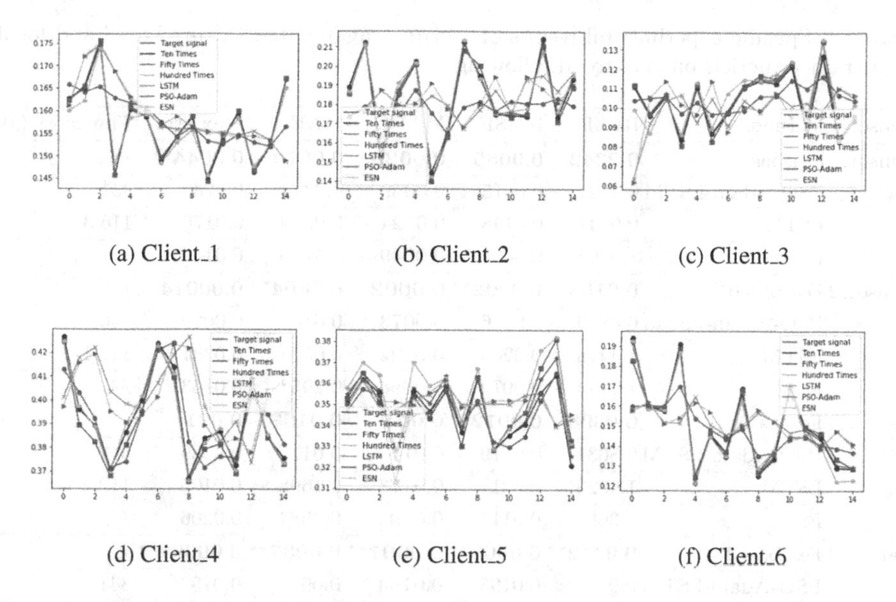


(a) Client_1 (b) Client_2 (c) Client_3

(d) Client_4 (e) Client_5 (f) Client_6

**Fig. 5.** Trend following results of $A_{r,l}$ on each individual Client

From Figs. 4, 5, Tables 2 and 3, the following observations can be reached: (1) When the $MAX\_NUM\_ROUNDS\_P = 10$, the results of the global aggregated feature based trend following are the highest, that is to say, we can obtain proper feature representations through few aggregation operations. (2) Compared with the local data trained models, the feature aggregation based trend following method achieve greater performance on both trend representations. Since the local data based models are trained separately, the sum of time costs is similar to the time consumption of our proposed method, the total time consumption of PSO-Adam-LSTMs is even higher. That's to say, with similar total time costs, our proposed methods can improve the trend following accuracy comprehensively. (3) In general, the trend following on average indicator $A_{r,l}$ can achieve higher accuracy with lower time consumption; among all data from sensors, data sampled from sensor_2 seem to be the most reliable for the trend following due to the higher predictive accuracy on both trend representations. (4) From the trend following results of ERI-TSD from different sensors, it seems the trends show comparative high fluctuations and even in some periods present a continuous rising tendency, that means we should pay great attention on the working face condition or activate some preventive measures.

**Table 3.** Specific experimental results of 10-times aggregated feature based and local data trained models on $A_{r,l}$ trend following

| Sensor# | Methods | RRSE | RMSE | MAPE | MAE | Max_error | Time_cost (s) |
|---|---|---|---|---|---|---|---|
| Sensor_1 | Fea_based* | **0.2342** | **0.0035** | **0.0026** | **0.0074** | **0.0044**** | / |
| | PSO-Adam-LSTM | 0.4821 | 0.0075 | 0.0059 | 0.0374 | 0.0286 | 331 |
| | LSTM | 0.9741 | 0.0148 | 0.0124 | 0.0359 | 0.0176 | 116.3 |
| | ESN | 1.0298 | 0.0062 | 0.0049 | 0.3285 | 0.0149 | // |
| Sensor_2 | Fea_based* | **0.0108**** | **0.0002**** | **0.0002** | **0.0004**** | **0.00014** | / |
| | PSO-Adam-LSTM | 0.4819 | 0.0086 | 0.0073 | 0.0187 | 0.0029 | 336 |
| | LSTM | 1.4478 | 0.0265 | 0.0222 | 0.0573 | 0.0331 | 112.1 |
| | ESN | 1.1184 | 0.0072 | 0.0058 | 0.295 | 0.0163 | // |
| Sensor_3 | Fea_based* | **0.0898**** | **0.0012**** | **0.001** | **0.0107** | **0.001** | / |
| | PSO-Adam-LSTM | 0.8631 | 0.0119 | 0.011 | 0.011 | 0.0029 | 342 |
| | LSTM | 1.4823 | 0.021 | 0.0173 | 0.1698 | 0.0105 | 112.2 |
| | ESN | 1.366 | 0.0111 | 0.0094 | 0.4584 | 0.0206 | // |
| Sensor_4 | Fea_based* | **0.0442**** | **0.0009**** | **0.0007**** | **0.0037**** | **0.00007**** | / |
| | PSO-Adam-LSTM | 0.9687 | 0.0196 | 0.0164 | 0.09 | 0.019 | 341 |
| | LSTM | 1.161 | 0.0236 | 0.0198 | 0.1096 | 0.0029 | 118.5 |
| | ESN | 1.22 | 0.0132 | 0.00989 | 0.444 | 0.03416 | // |
| Sensor_5 | Fea_based* | **0.03**** | **0.0002**** | **0.0021** | **0.0014**** | **0.00002**** | / |
| | PSO-Adam-LSTM | 0.841 | 0.0069 | 0.0055 | 0.0353 | 0.0129 | 333 |
| | LSTM | 1.0828 | 0.0089 | 0.006 | 0.0397 | 0.0182 | 111.2 |
| | ESN | 1.423 | 0.0055 | 0.0049 | 0.5159 | 0.0098 | // |
| Sensor_6 | Fea_based* | **0.0872**** | **0.0018** | **0.0016** | **0.011** | **0.00075** | / |
| | PSO-Adam-LSTM | 0.7404 | 0.0154 | 0.0112 | 0.0733 | 0.0065 | 340 |
| | LSTM | 1.019 | 0.0142 | 0.01686 | 0.1133 | 0.0106 | 115.7 |
| | ESN | 1.097 | 0.0143 | 0.0127 | 0.433 | 0.0339 | // |

Fea_based* is the abbreviation of the proposed global aggregated feature based trend following method.
/ means due to the synchronous training, time consumption can not be calculated separately and the total consumption is 624.
// means due to the low computational cost, the time consumption can be ignored.

## 4 Conclusions

In order to make full use of the information collected by multiple sensors with reducing privacy leakage, this paper considers a framework of conducting model and features aggregation under FL structure, and presents an aggregated feature based trend following method. Under this framework, deep neural networks are utilized to extract local features, then both the local trained weights and extracted features are sent to server to fuse. Then, we apply the proposed framework to an actual problem, i.e., the trend following for electromagnetic magnetic intensity data sampled by multi sensors. We first utilize LSTM to get the features and then carry out the feature aggregation operation under the proposed framework, finally we design a multi-sensor aggregated feature based trend following strategy by ESNs. As experimentally proven, under the proposed framework, we can aggregate the information from different sensors on the premise of privacy protection. With the similar time consumptions, the aggregated features based trend following strategy performs much better than the local dataset based method on the accuracy.

The present study follows the assumption that all clients adopt the same neural network. In the future research, new federated optimization allowing local clients utilize different hyper-parameters will be further studied.

## References


1. Chen, Y., Hu, Y., Liu, Y.n., Zhu, X.-d.: Processing and fusion for multi-sensor data. J. Jilin Univ. **56**(5), 1170–1178 (2018)
2. Hu, Y., Sun, X., Nie, X., Li, Y., Liu, L.: An enhanced LSTM for trend following of time series. IEEE Access **7**, 34020–34030 (2019)
3. Jaeger, H., Haas, H.: Harnessing nonlinearity: predicting chaotic systems and saving energy in wireless communication. Science **304**(5667), 78–80 (2004)
4. Kamakshi, P.: A survey on privacy issues and privacy preservation in spatial data mining. In: International Conference on Circuit, pp. 1759–1762 (2014)
5. Konecny, J., Mcmahan, H.B., Ramage, D.: Federated optimization: distributed optimization beyond the datacenter. arXiv: Learning, pp. 1–5
6. Konecny, J., Mcmahan, H.B., Ramage, D., Richtrik, P.: Federated optimization: distributed machine learning for on-device intelligence. arXiv preprint arXiv:1610.02527 (2016)
7. Kumar, A., Kim, J., Lyndon, D., Fulham, M.J., Feng, D.D.: An ensemble of fine-tuned convolutional neural networks for medical image classification. IEEE J. Biomed. Health Inform. **21**(1), 31–40 (2017)
8. Mcmahan, H.B., Moore, E., Ramage, D., Hampson, S., Arcas, B.A.Y.: Communication-efficient learning of deep networks from decentralized data. In: Artifical Intelligence and Statistics, pp. 1273–1282 (2017)
9. Pan, Q., Wang, Z.f., Liang, Y., Yang, F., Liu, Z.g.: Basic methods and progress of information fusion. Control Theory Appl. **29**(10), 1234–1244 (2012)
10. Putri, A.W., Hira, L.: Hybrid transformation in privacy-preserving data mining. In: International Conference on Data & Software Engineering, pp. 1–6 (2017)
11. Qi, Y.J., Wang, Q.: Review of multi-source data fusion algorithm. Aerosp. Electron. Warfare **6**(33), 37–41 (2017)
12. Qiu, L., et al.: Characteristics and precursor information of electromagnetic signals of mining-induced coal and gas outburst. J. Loss Prev. Process Ind. **54**, 206–215 (2018)
13. Wang, E.Y., Li, Z.H., He, X.Q., Liang, C.: Application and pre-warning technology of coal and gas outburst by electromagnetic radiation. Coal Sci. Technol. **42**(6), 53–57 (2014)
14. Wen, Y., Wu, J., Yuan, Y.: Multiple-phase modeling of degradation signal for condition monitoring and remaining useful life prediction. IEEE Trans. Reliab. **66**(3), 924–938 (2017)
15. Yao, Q., Tang, J., Zeng, F., Huang, X., Miao, Y., Pan, J.: Feature extraction of SF6 thermal decomposition characteristics to diagnose overheating fault. IET Sci. Meas. Technol. **9**(6), 751–757 (2015)

16. Ye, S., Wu, G.: Recursive annotations for attention-based neural machine translation. In: International Conference on Asian Language Processing (2018)
17. Zhang, B.C., Lin, J.Q., Chang, Z.C., Yin, X.J., Gao, Z.: The application of multi sensor data fusion based on the improved BP neural network algorithm. In: Control & Decision Conference, pp. 3842–3846 (2016)

# Robust Echo State Network for Recursive System Identification

Renan Bessa and Guilherme A. Barreto(✉)

Department of Teleinformatics Engineering, Federal University of Ceará, Center of Technology, Campus of Pici, Fortaleza, Ceará, Brazil
renanbessa.etec@gmail.com, gbarreto@ufc.br

**Abstract.** The use of recurrent neural networks in online system identification is very limited in real-world applications, mainly due to the propagation of errors caused by the iterative nature of the prediction task over multiple steps ahead. Bearing this in mind, in this paper, we revisit design issues regarding the robustness of the *echo state network* (ESN) model in such online learning scenarios using a recursive estimation algorithm and an outlier robust-variant of it. By means of a comprehensive set of experiments, we show that the performance of the ESN is dependent on the adequate choice of the feedback pathways and that the prediction instability is amplified by the norm of the output weight vector, an often neglected issue in related studies.

**Keywords:** Online system identification · Recurrent neural networks · Echo state network · Recursive estimation · Robustness to outliers

## 1 Introduction

The echo state network (ESN) [7] is a recurrent neural network (RNN) that has a large set of neurons, the so-called reservoir, with sparse interconnections and feedback pathways. The input weights of the reservoir neurons, the internal and the ones responsible for connecting the system input and output, are fixed and randomly assigned. Training of this network requires the estimation of the weights of the neurons in the output layer (aka, readout layer). This estimation is carried out via linear regression, usually by means of the well-known OLS method. The randomized nature of the ESN combined with the linear estimation of the output layer weights makes its design very simple if compared to the other RNNs. In Fig. 1 is depicted the standard architecture of the ESN.

The ESN has been used as a powerful tool for the prediction of chaotic series, attractor reconstruction, and for nonlinear system modelling in general [8]. In order to improve this network to handle real-world data, which commonly

This study was financed by the following Brazilian research funding agencies: CAPES (finance code 001), FUNCAP (BMD-008-01413.01.02-17) and CNPq (grant 309451/2015-9).


I. Rojas et al. (Eds.): IWANN 2019, LNCS 11506, pp. 247–258, 2019.
https://doi.org/10.1007/978-3-030-20521-8_21

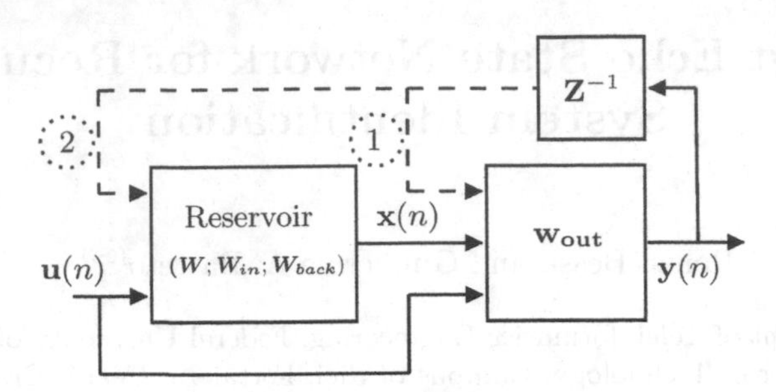


**Fig. 1.** The basic setup of ESN indicating feedforward (solid lines) and feedback (dashed lines) pathways. $\mathbf{W}$, $\mathbf{W}_{in}$, $\mathbf{W}_{back}$ and $\mathbf{W}_{out}$ are weight matrices.

contain non-Gaussian noise and inconsistent points (outliers), some works introduced robust estimation methods for the estimation of the readout layer weights. One of the simplest methods is the Tikhonov regularization, which penalizes weight vector with a high norm, reducing model overfitting to the corrupted data [10].

More complex approaches have also been proposed, including Bayesian inference, replacing the usual Gaussian likelihood function, which is very sensitive to the outliers, by a Laplacian [9] or mixed-Gaussian distribution [6]. Furthermore, the use of performance criteria based on the information theoretic learning, called Maximum Correntropy [5] and Generalized Correntropy [13], have also achieved good results in diminishing sensitivity to outliers.

Other robust ESN approaches, replace the usual linear output layer with a nonlinear framework, trying to take full advantage of the dynamics of the reservoir, whose output signals are inputted to the readout layer. One of such approaches applies SVM formulation with robust cost functions such as $\epsilon$-insensitive loss or Huber [11]. Other approaches use kernel adaptive filtering methods, such as the kernel recursive least squares (RLS) algorithm [14]. Although it has not been evaluated directly, there are references that in arrangements of the ESN with Laplacian Eigenmaps algorithm [6], mediating a decrease of the dimensionality of the reservoir states, and with Gaussian process regression models [3] may be able to decrease the sensitivity to outliers.

Despite the powerful modeling capabilities of the aforementioned ESN architectures, their application to nonlinear system identification is limited. This is particularly true for real-world application scenarios involving online long-term iterative predictions over multiple steps ahead. Since the chance to meet outliers in such scenarios is very high, the propagation of errors due to the iterative nature of the prediction task causes divergence (i.e. instability) in the predicted signals as time passes. Such instability is amplified by the norm of the output weight vector, an often neglected issue in previous studies.

Bearing the stability of long-term prediction in mind, in this work we revisit the RLS-ESN [8] model, ESN combined with the RLS estimation algorithm, for online system identification in the presence of outliers. By exploring variations in the feedback pathways of the ESN and replacing the RLS algorithm with a robust variant of it [2] we show that the ESN can be safely used for long-term multiple-step-ahead prediction tasks. The role of the norm of the weights in the propagation of errors is also evaluated. A comprehensive set of computer simulations using data from real-world and synthetic systems are carried out, which are contaminated by outliers in different proportions.

## 2 Fundamentals of the Echo State Network

The parameter estimation process in discrete-time system identification requires the availability of input and output observations from a system of interest: $(\mathbf{u}(n), \mathbf{t}(n))_{n=0}^{N_{all}}$, where $\mathbf{u}(n) = [u_1(n), u_2(n), \ldots, u_K(n)]^T$ is a $K$-dimensional input vector and $\mathbf{t}(n) = [t_1(n), t_2(n), \ldots, t_L(n)]^T$ is an $L$-dimensional output vector at a given instant $n$, $n = 1, \ldots, N_{all}$. This dataset is then divided into training and test subsets, $N_{all} = N_{train} + N_{test}$. Once adequate training is completed, the ESN is required to predict output vectors $\mathbf{y}(n) \in \mathbb{R}^{L\times 1}$ with the smallest possible deviation from $\mathbf{t}(n)$.

The activations of the $R$ reservoir neurons, also called state variables, are denoted by $\mathbf{x}(n) \in \mathbb{R}^{R\times 1}$, being computed as

$$\mathbf{x}(n) = \mathbf{f}(\mathbf{W}\mathbf{x}(n-1) + \mathbf{W}_{in}\mathbf{u}(n) + \mathbf{W}_{back}(\mathbf{t}(n-1) + \mathbf{v}_1(n)) + \mathbf{v}_2(n) - \mathbf{b}), \quad (1)$$

where $\mathbf{f}(\cdot)$ is a nonlinear activation function with element-wise operation, usually the logistic sigmoid or the tanh function, $\mathbf{W} \in \mathbb{R}^{R\times R}$ is the internal weight matrix of the reservoir, which is responsible for the reservoir feedback pathways, and $\mathbf{W}_{in} \in \mathbb{R}^{R\times K}$ and $\mathbf{W}_{back} \in \mathbb{R}^{R\times L}$ are the weight matrices that connect the input and output of the model to the reservoir, respectively. The terms $\mathbf{v}_1(n) \in \mathbb{R}^{L\times 1}$ and $\mathbf{v}_2(n) \in \mathbb{R}^{R\times 1}$ are optional white noise vectors, while $\mathbf{b} \in \mathbb{R}^{R\times 1}$ is the bias vector.

The output of the neural model is computed as

$$\mathbf{y}(n) = \mathbf{f}_{out}\left(\mathbf{W}_{out}\mathbf{h}(n)\right), \quad (2)$$

where $\mathbf{f}_{out}(\cdot)$ is the activation function of the output neurons (usually, the identity function), $\mathbf{W}_{out} \in \mathbb{R}^{L\times(1+K+R+L)}$ is the output weight matrix, $\mathbf{h}(n) = [-1, \mathbf{u}(n), \mathbf{x}(n), (\mathbf{t}(n-1) + \mathbf{v}_1(n))]^T \in \mathbb{R}^{H\times 1}$ is a concatenated vector, so that $H = 1 + K + R + L$.

The elements of the weight matrices $\mathbf{W}$, $\mathbf{W}_{in}$ and $\mathbf{W}_{back}$ are randomly assigned and kept fixed during training and testing of the model. However, $\mathbf{W}$ must be sparse, i.e. it must contain only a small percentage of nonzero elements. In order to guarantee the "echo state" property[1], the spectral radius of $\mathbf{W}$ must be chosen to be within the unit circle. The magnitudes of the noise vectors $\mathbf{v}_1(n)$ and $\mathbf{v}_2(n)$ must be chosen aiming at reaching a tradeoff between the stability and accuracy of the model. These vectors also help to regularize the solution. Details on the ESN configuration can be found in [7].

[1] That of relating asymptotic properties of the excited reservoir dynamics to the driving signal.

### 2.1 Recursive Algorithms for Parameter Estimation

In this paper we are dealing with SISO systems; thus, a single output is considered. The goal of training is to estimate recursively the weight vector $\mathbf{w}_{out} \in \mathbb{R}^{H\times 1}$. We aim at comparing two estimation methods, namely: the standard RLS algorithm, whose cost function is $J_{LS}(n) = \sum_{i=1}^{n} \gamma^{n-i} e^2(i)$, where $0 < \gamma \leq 1$ is the forgetting factor and $e(n) = t(n) - \mathbf{w}_{out}^T(n-1)\mathbf{h}(n)$ is the prediction error at instant $n$, and the *recursive Least M-estimate algorithm* (RLM), whose cost function is given by

$$J_\rho(n) = \sum_{i=1}^{n} \gamma^{n-i} \rho(e(i)), \tag{3}$$

where $\rho(e)$ is a function whose purpose is to limit the negative effect caused by very large errors (either caused by non-Gaussian noise or outliers). The RLM algorithm reduces to the standard RLS when $\rho(e(i)) = e^2(i)$.

The optimal output weights can be determined by differentiating $J_\rho(n)$ with respect to $\mathbf{w}_{out}$ and setting the derivatives to zero. After some algebric manipulation, the resulting RLM algorithm involves the following equations:

$$\mathbf{w}_{out}(n) = \mathbf{w}_{out}(n-1) + e(n)\mathbf{k}(n), \tag{4}$$

$$\mathbf{k}(n) = \frac{q(e(n))\mathbf{S}(n-1)\mathbf{h}(n)}{q(e(n))\mathbf{h}^T(n)\mathbf{S}(n-1)\mathbf{h}(n) + \gamma} \tag{5}$$

$$\mathbf{S}(n) = \gamma^{-1}\left(\mathbf{I} - \mathbf{k}(n)\mathbf{h}^T(n)\right)\mathbf{S}(n-1) \tag{6}$$

where $\mathbf{k}(n)$ is the gain vector and $q(e) = \frac{1}{e}\frac{\partial \rho(e)}{\partial e}$. The matrix $\mathbf{S}(n)$ is the online estimate of the inverse of the correlation matrix, i.e. $\mathbf{S}(n) = \mathbf{R}^{-1}(n)$.

In this paper, we use the classic Huber function

$$\rho(e) = \begin{cases} \frac{e^2}{2}, & |e| < \xi \\ \xi|e| - \frac{\xi^2}{2}, & |e| > \xi, \end{cases} \quad \Rightarrow \quad q(e) = \begin{cases} 1, & |e| < \xi \\ \frac{\xi}{|e|}, & |e| > \xi, \end{cases} \tag{7}$$

where $\xi$ is a threshold value. Errors high than this threshold are to be considered an outlier. It should be noted that $q(e) = 1$ for errors smaller than the threshold $\xi$; otherwise, $q(e) \to 0$ as $|e| \to \infty$.

It is possible to estimate $\xi$ continuously for each new input. For this purpose, one can use the following expression:

$$\xi(n) = 2.576\hat{\sigma}(n), \tag{8}$$

where $\hat{\sigma}(n)$ is a robust estimate of the standard deviation of the residuals:

$$\hat{\sigma}^2(n) = c_1 \text{med}\{e^2(n), ..., e^2(n - N_w + 1)\}, \tag{9}$$

so that $c_1 = 1.483(1 + 5/(N_w - 1))$, $\text{med}\{\cdot\}$ is the sample median and $N_w$ is the window length. The robustness and computational complexity of the method increases with increasing the value of $N_w$.

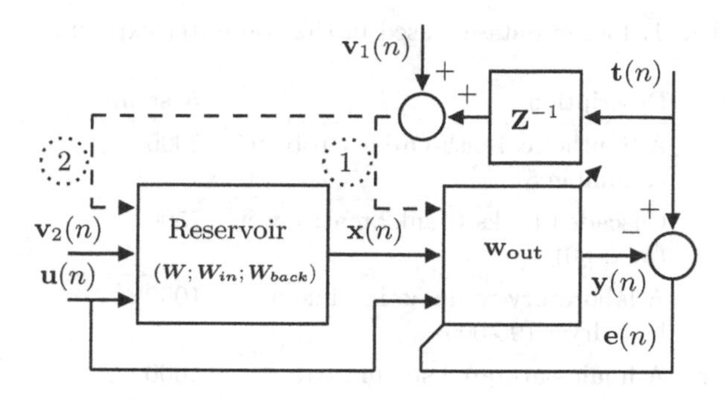


**Fig. 2.** ESN architecture during training (dashed lines are optional connections).

# 3 Methodology of Evaluation and Simulation

The ESN architecture diagram is outlined in Fig. 2. Our goal is to evaluate the following variants of the ESN architecture: (*i*) **Model 1** (M1) - connection pathways 1 and 2 disabled; (*ii*) **Model 2** (M2) - connection pathway 1 enabled, connection pathway 2 disabled; and (*iii*) **Model 3** (M3) - connection pathways 1 and 2 enabled.

Three approaches to estimating the output weight vector $\mathbf{w}_{out}$ are also evaluated. The standard RLS algorithm is used as a baseline of reference against which the plain RLM algorithm is contrasted. A third approach is specific to the task of iterated prediction, aiming at providing higher stability of the prediction over longer time horizons, when outliers are present in the data.

**RLM with Outlier Detection** (RLM-OD): At iteration $n$, the output prediction is computed as $y(n) = \mathbf{w}_{out}^T(n-1)\mathbf{h}(n)$. The prediction error is computed as $e(n) = t(n) - \mathbf{w}_{out}^T(n-1)\mathbf{h}(n)$ and used by the RLM algorithm as shown in Eqs. (4) and (5). However, if $|e(n)| > \xi$, this has happened because the target output $t(n)$ is probably an outlier. Then, in the next training iteration, we replace the actual observation $t(n)$ with its predicted value $y(n)$ in order to avoid filling in the input regression vector with outliers.

## 3.1 Evaluation and Simulation

All the datasets used in the experiments come from SISO systems. They are listed in Table 1, with some details added. For the sake of model building and validation, the datasets are divided as follows: the first half of the samples for training, from 0 to $N_{train} - 1$, and the second half for the test, from $N_{train}$ to $N_{all} - 1$. It is worth mentioning that once the systems of interest are SISO, the input vector $\mathbf{u} \in \mathbb{R}^K$, the target vector $\mathbf{t} \in \mathbb{R}^L$, and the random noise vector $\mathbf{v}_1 \in \mathbb{R}^K$ become scalars (i.e. $K = 1$ and $L = 1$). The random noise vector $\mathbf{v}_2 \in \mathbb{R}^R$ has the same dimension of the number of units in the reservoir.

**Table 1.** List of datasets used in the computer experiments.

| System | Description | # samples ($N_{all}$) | Ref. |
|---|---|---|---|
| TOP | A Synthetic Tenth-Order Problem (Example 5) | 2000 | [1] |
| Tank | Cascaded tanks (Tank2.mat; lower tank: $y0$) | 7500 | [12] |
| Dryer | A laboratory setup acting like a hair dryer (96-006) | 1000 | [4] |
| Exchanger | A liquid-saturated steam heat exchanger (97-002) | 4000 | [4] |
| Robot Arm | A flexible robot arm (96-009) | 1024 | [4] |

Two criteria were used for the evaluation of the ESN models. The root mean square error (RMSE) of the iterated predictions,

$$RMSE = \sqrt{\frac{\sum_{n=N_{train}}^{N_{all}-1}(t(n) - y(n))^2}{N_{test}}} \tag{10}$$

and the Euclidean norm ($l_2$) of the output weight vector, $||\mathbf{w}_{out}||$. These two figures of merit are important in assessing the model quality.

The default setting for all simulations of this work is shown in the Table 2. The number of reservoir neurons, $R$, is checked for each system, without the presence of outliers and with the RLS, seeking a balance between the RMSE, overfitting and stability, for $R \in \{10, 25, 50, 100, 150, ..., 500\}$. The RLM algorithm has an extra parameter, $N_w$, which defines the size of the sample window for calculating the threshold $\xi$. We decided to use all available samples up to the iteration $n$, that is, $N_w$ is increasing and ranges from 0 to $N_{train} - 1$.

The robustness of the methods are evaluated for outliers contamination scenarios of 5%, 10% e 15% of training samples. These are generated by $\sigma_{train}\mathcal{T}(0, 2)$, where $\sigma_{train}$ is the standard deviation of the original training data and $\mathcal{T}(0, 2)$ is a Student-T distribution with zero mean and two degrees of freedom. The values returned from the Student-T distribution are saturated at $\pm 20$. The systems' inputs and outputs, with and without outliers, are normalized by subtracting the mean and dividing by three times the standard deviation value of the training samples. The RMSE is calculated with the non-normalized data values. Twenty (20) independent training/testing runs are executed for each contamination scenario, for which all weights/biases and outliers are randomly initialized. In each runs, the same weights and the same contamination of the data are applied to each investigated model in order to carry out a fair performance comparison for the different ESM models (M1, M2 and M3) and the parameter estimation methods (RLS, RLM and RLM-OD).

## 4 Results

Before running the final simulations, the number of neurons in the reservoir was chosen by experimentation with outlier-free data. It was observed the convergence to a minimum value of the RMSE with the increase in the number of neurons for all datasets. In Table 3, it is shown the chosen number of neurons chosen and the results achieved by the RLS-ESN for the three evaluated architectures (M1, M2 and M3) using the RLS algorithm. These results are expressed in the following format: $mean \pm std(median)$, respectively, the mean, standard deviation and median of the RMSE values and the norm of $\mathbf{w}_{out}$ averaged along the 20 independent runs.

Keeping the same number of neurons in the reservoir, the models' performances are better illustrated in Figs. 3 and 4, where they can be compared with

**Table 2.** Default setting of several hyperparameters for the simulations.

| ESN CONFIGURATION | |
|---|---|
| Spectral Radius (**W**) | 0.98 |
| Sparsity (**W**) | 2% |
| Input Weights | $\mathbf{W}_{in} \sim U(-0.1, 0.1)$ |
| Feedback Weights | $\mathbf{W}_{back} \sim U(-0.1, 0.1)$ |
| Bias Weights | $\mathbf{b} \sim U(-0.1, 0.1)$ |
| Noise Vector 1 | $\mathbf{v}_1 \sim U(-0.0001, 0.0001)$ |
| Noise Vector 2 | $\mathbf{v}_2 \sim U(-0.0001, 0.0001)$ |
| RLS/RLM CONFIGURATION | |
| Forgetting Factor ($\gamma$) | 0.99998 |
| $\mathbf{S}(0)$ | $10^4 I$ |

**Table 3.** Models' performances using the RLS algorithm with outlier-free data.

| Datasets (value of $R$) | | Model M1 | Model M2 | Model M3 |
|---|---|---|---|---|
| TOP (500) | RMSE | $0.0312 \pm 0.0105(0.0279)$ | $0.0314 \pm 0.0110(0.0270)$ | $0.0344 \pm 0.0079(0.0316)$ |
| | Norm | $98.04 \pm 20.34(101.73)$ | $75.22 \pm 15.10(76.94)$ | $62.27 \pm 16.72(67.88)$ |
| Tank (500) | RMSE | $0.3227 \pm 0.0916(0.3202)$ | $0.1568 \pm 0.0244(0.1579)$ | $0.1056 \pm 0.0122(0.1033)$ |
| | Norm | $152.64 \pm 47.61(146.76)$ | $5.09 \pm 0.82(5.33)$ | $2.36 \pm 0.08(2.36)$ |
| Dryer (250) | RMSE | $0.1356 \pm 0.0377(0.1229)$ | $0.1238 \pm 0.0212(0.1165)$ | $0.1299 \pm 0.0170(0.1237)$ |
| | Norm | $18.79 \pm 2.33(19.01)$ | $8.59 \pm 0.85(8.43)$ | $8.34 \pm 1.24(8.17)$ |
| Exchanger (450) | RMSE | $0.2432 \pm 0.0305(0.2339)$ | $0.2507 \pm 0.0327(0.2400)$ | $0.2312 \pm 0.0348(0.2227)$ |
| | Norm | $37.05 \pm 4.28(37.24)$ | $33.449 \pm 2.80(33.23)$ | $24.03 \pm 2.86(24.74)$ |
| Robot Arm (500) | RMSE | $0.2840 \pm 0.0459(0.2785)$ | $0.2991 \pm 0.0552(0.2907)$ | $0.0238 \pm 0.0136(0.0198)$ |
| | Norm | $165.89 \pm 16.35(172.85)$ | $118.25 \pm 11.39(121.78)$ | $2.33 \pm 0.30(2.25)$ |

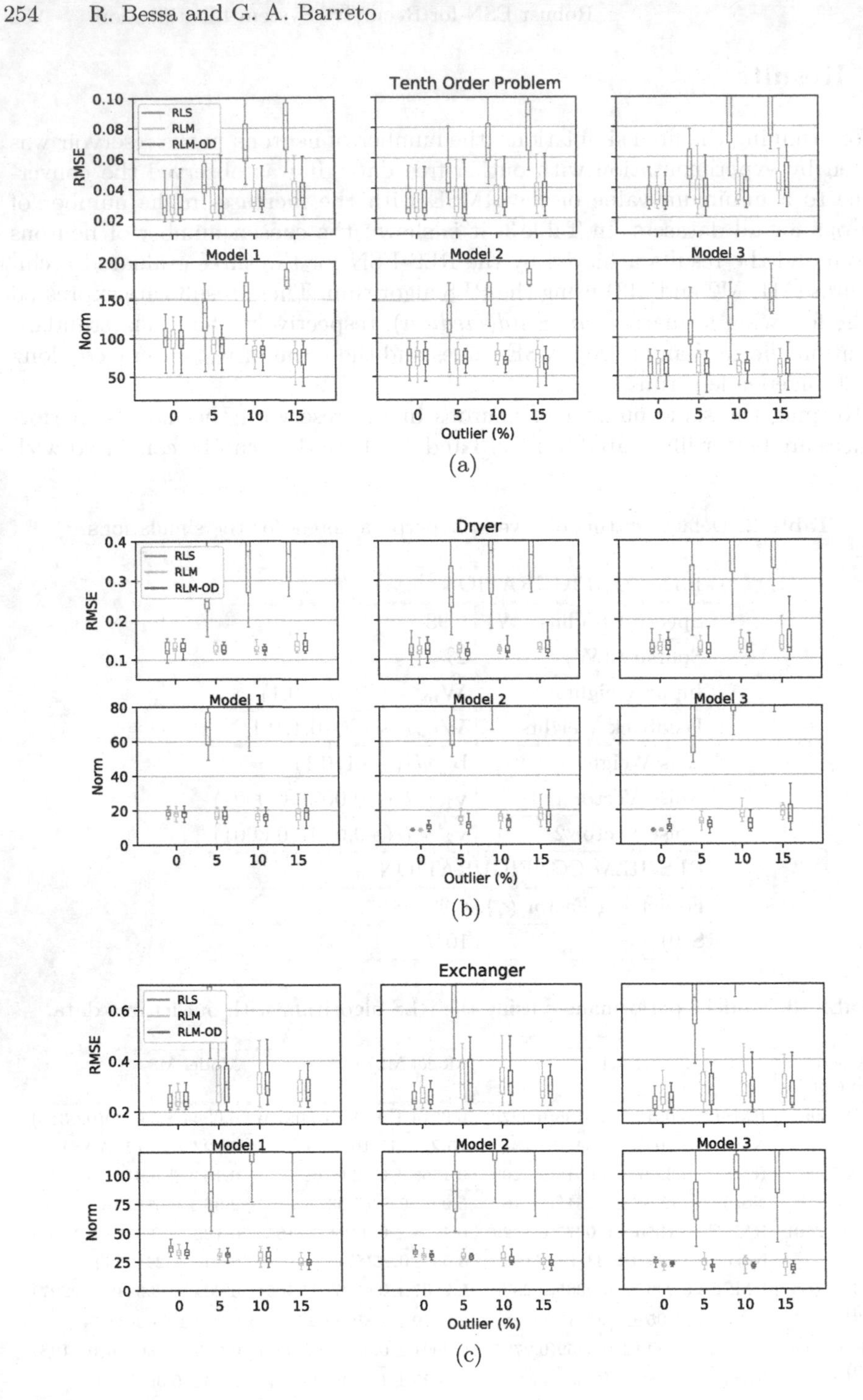


**Fig. 3.** Performances on the (a) 10-th order problem, (b) Dryer and (c) Exchanger datasets.

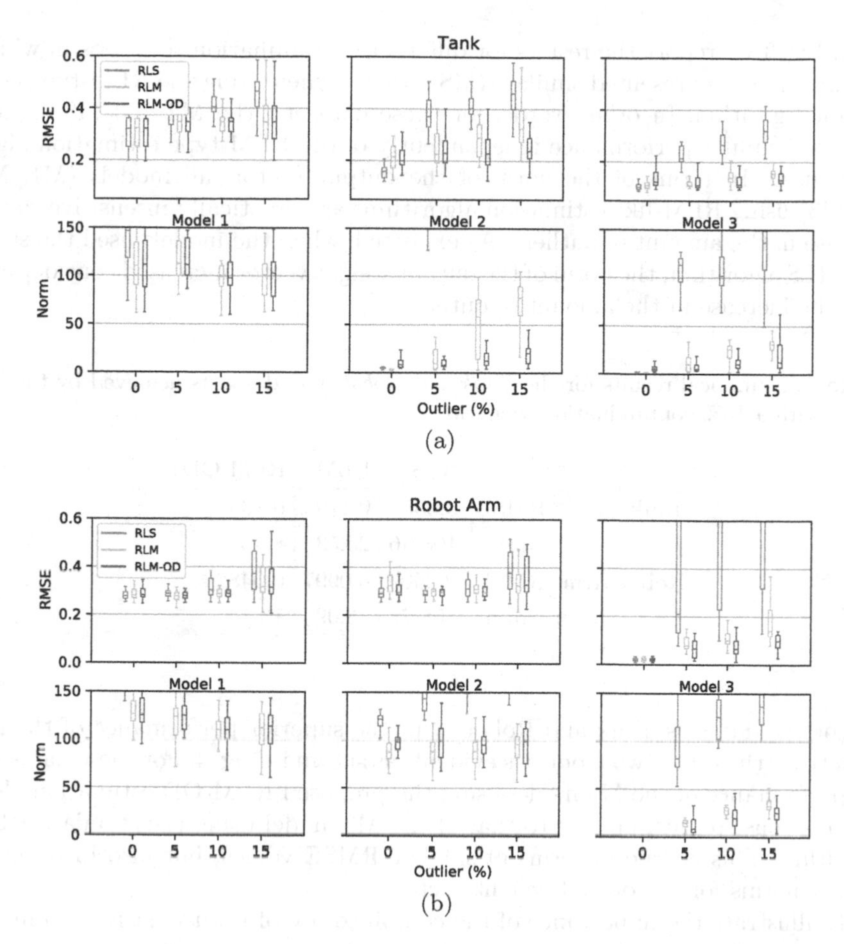


**Fig. 4.** Performances on the (a) Tank and (b) Robot arm datasets.

respect to the different contamination scenarios and to the choice of the robust estimator (RLM and RLM-OD). It should be noted that for the M1 model, the RLM-OD algorithm behaves exactly like the standard RLM, since the output is not fed back (i.e. connections pathways 1 and 2 are disabled).

As can be seen in Table 3, the presence of feedback pathways (Models M2 e M3) decreased the norm of the ESN output layer, which leads to an improvement in generalization and avoids overfitting. In other words, the feedback pathways acted as a model regularizer. The M3 model, which feeds the predicted output back to the readout layer and the reservoir, has the smallest norms and the corresponding RMSE values are smaller or very close compared to the other evaluated models.

In Fig. 3 we report the results for different contamination scenarios in which the three models presented similar RMSE values when using the RLM-type estimation algorithm. In other words, for these datasets, the M2 and M3 models presented similar performance independently of the RLM-type estimation algorithm used. In terms of the norm of the output vector, all models (M1, M2 and M3) using RLM-like estimation algorithms are practically insensitive to the increase in the amount of outliers. As expected, when the models used the standard RLS algorithm, the norm of the output weight vector increased considerably with the increase in the amount of outliers.

**Table 4.** Numerical results for the *Tank* and *Robot Arm* datasets achieved by the M3 model with a 15% contamination scenario.

| | | RLS | RLM | RLM-OD |
|---|---|---|---|---|
| Tank | RMSE | 0.3553 | 0.1796 | 0.1063 |
| | Norm | 108.46 | 25.53 | 8.40 |
| Robot Arm | RMSE | 0.2509 | 0.0997 | 0.0482 |
| | Norm | 108.75 | 46.09 | 13.14 |

For the datasets Tank and Robot Arm the superior performance of the M3 model over the other two models is evident, as shown in Fig. 4. For these datasets, the performance of the M3 model using the proposed RLM-OD estimation algorithm is consistently superior to that of the M3 model using the standard RLM algorithm. This is true in terms of smaller RMSE values, but also in terms of smaller norms for the output weight vector.

To illustrate the importance of the combined use of robust estimation methods and small norms of the output weight vector in the long-term performance of the M3 model, we report in Fig. 5 typical predicted time series for a single training/testing run and a scenario with 15% of outlier contamination. The vertical dotted lines indicate the end of training phase and beginning of the iterated long-term prediction task. The superior performance of the fully recurrent M3 model using the robust estimation algorithms (RLM and RLM-OD) is clear.

These models (M3-RLM and M3-RLM-OD) were both able to learn the underlying dynamics of the system of interest even in the presence of outliers during the training phase and capable of diminishing the effect of error propagation during the testing phase due to the smaller norms of the output weight vector. The corresponding numerical results for this example are reported in Table 4.

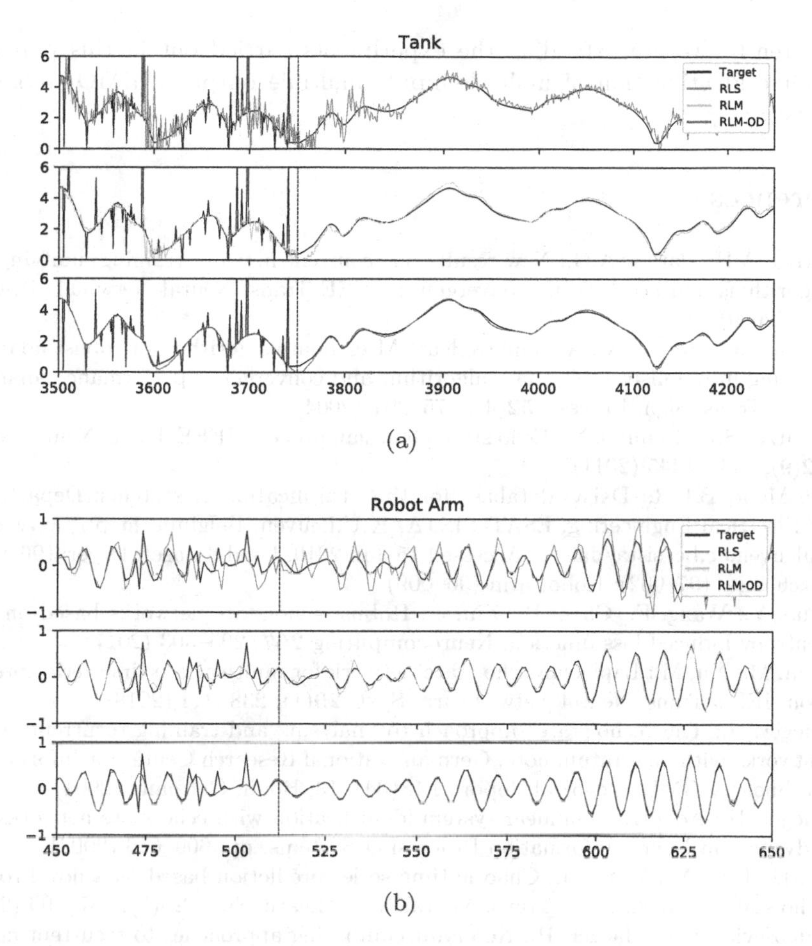


**Fig. 5.** Typical predicted time series for the (a) *Tank* and (b) *Robot Arm* datasets achieved by the M3 model with a 15% contamination scenario.

# 5 Conclusions and Further Work

In this paper, we evaluated the performance of the ESN model for recursive identification task under outlier-contaminated scenarios. Two outlier-robust variants of the RLS algorithm were used for estimating the output weight vector, namely, the RLM algorithm and the proposed RLM-OD algorithm.

It was verified by a comprehensive set of computer experiments using benchmarking datasets that the combined used of feedback pathways and robust estimation algorithms led to regularized recurrent neural network models. Among the evaluated models, the M3-RLM-OD model consistently presented very promising results, being able to keep the norm of the output weight vector small and, hence, reduce the negative influence of outliers in the long term iterated prediction performance of the ESN model.

Currently, we are extending the experiments carried out in this paper to the online identification of multiple-input, multiple output (MIMO) dynamic systems.

## References

1. Atiya, A.F., Parlos, A.G.: New results on recurrent network training: unifying the algorithms and accelerating convergence. IEEE Trans. Neural Netw. **11**(3), 697–709 (2000)
2. Chan, S.C., Zou, Y.X.: A recursive least M-estimate algorithm for robust adaptive filtering in impulsive noise: fast algorithm and convergence performance analysis. IEEE Trans. Sign. Process. **52**(4), 975–991 (2004)
3. Chatzis, S.P., Demiris, Y.: Echo state Gaussian process. IEEE Trans. Neural Netw. **22**(9), 1435–1445 (2011)
4. De Moor, B.L.R.: Daisy: database for the identification of systems. Department of Electrical Engineering, ESAT/SISTA, K.U.Leuven, Belgium. http://www.esat.kuleuven.ac.be/sista/daisy/. Accessed 15 Jan 2019. Used dataset: Dryer: (96–006); Exchanger (97–002); Robot arm (96–009)
5. Guo, Y., Wang, F., Chen, B., Xin, J.: Robust echo state networks based on correntropy induced loss function. Neurocomputing **267**, 295–303 (2017)
6. Han, M., Xu, M.: Laplacian echo state network for multivariate time series prediction. IEEE Trans. Neural Netw. Learn. Syst. **29**(1), 238–244 (2018)
7. Jaeger, H.: The "echo state" approach to analysing and training recurrent neural networks-with an erratum note. German National Research Center for Information Technology GMD Technical Report **148**(34), 13, Bonn, Germany (2001)
8. Jaeger, H.: Adaptive nonlinear system identification with echo state networks. In: Advances in Neural Information Processing Systems, pp. 609–616 (2003)
9. Li, D., Han, M., Wang, J.: Chaotic time series prediction based on a novel robust echo state network. IEEE Trans. Neural Netw. Learn. Syst. **23**(5), 787–799 (2012)
10. Lukoševičius, M., Jaeger, H.: Reservoir computing approaches to recurrent neural network training. Comput. Sci. Rev. **3**(3), 127–149 (2009)
11. Shi, Z., Han, M.: Support vector echo-state machine for chaotic time-series prediction. IEEE Trans. Neural Netw. **18**(2), 359–372 (2007)
12. Wigren, T.: Input-output data sets for development and benchmarking in nonlinear identification. Technical Reports from the Department of Information Technology 20, 2010–020 (2010)
13. Zhang, C., Guo, Y., Wang, F., Chen, B.: Generalized maximum correntropy-based echo state network for robust nonlinear system identification. In: 2018 International Joint Conference on Neural Networks (IJCNN), pp. 1–6. IEEE (2018)
14. Zhou, H., Huang, J., Lu, F., Thiyagalingam, J., Kirubarajan, T.: Echo state kernel recursive least squares algorithm for machine condition prediction. Mech. Syst. Sign. Process. **111**, 68–86 (2018)

# Random Hyper-parameter Search-Based Deep Neural Network for Power Consumption Forecasting

J. F. Torres(✉), D. Gutiérrez-Avilés, A. Troncoso(✉), and F. Martínez-Álvarez

Division of Computer Science, Pablo de Olavide University, Seville, Spain
{jftormal,dguvati,atrolor,fmaralv}@upo.es

**Abstract.** In this paper, we introduce a deep learning approach, based on feed-forward neural networks, for big data time series forecasting with arbitrary prediction horizons. We firstly propose a random search to tune the multiple hyper-parameters involved in the method performance. There is a twofold objective for this search: firstly, to improve the forecasts and, secondly, to decrease the learning time. Next, we propose a procedure based on moving averages to smooth the predictions obtained by the different models considered for each value of the prediction horizon. We conduct a comprehensive evaluation using a real-world dataset composed of electricity consumption in Spain, evaluating accuracy and comparing the performance of the proposed deep learning with a grid search and a random search without applying smoothing. Reported results show that a random search produces competitive accuracy results generating a smaller number of models, and the smoothing process reduces the forecasting error.

**Keywords:** Hyperparameters · Time series forecasting · Deep learning

## 1 Introduction

Deep learning is an emerging branch of machine learning that extends artificial neural networks. One of the main drawbacks that classical artificial neural networks exhibit is that, with many layers, its training typically becomes too complex. In this sense, deep learning consists of a set of learning algorithms to train artificial neural networks with a large number of hidden layers.

Deep learning models are also sensitive to a large numbers of hyper-parameters and much attention must be paid at this stage [6]. For Deep Feed Forward Neural Network (DFFNN), these hyper-parameters include the number of hidden layers, the number of neurons for hidden layers, the batch size and other parameters related to the optimization method used to compute the weights of the DFFNN in the training phase. There are many optimization methods such as gradient descend, gradient descend with momentum, RMSProp or Adam optimization algorithm, among others [14]. But the convergence of all of these algorithms depend on the learning rate, being one of the most important parameters.


I. Rojas et al. (Eds.): IWANN 2019, LNCS 11506, pp. 259–269, 2019.
https://doi.org/10.1007/978-3-030-20521-8_22

Therefore, the task of selecting an appropriate set of hyper-parameters is critical for the performance of the DFFNN.

In this context, we propose a DFFNN for time series forecasting that implements a random search to find the best values for the most relevant parameters related to the network structure and optimization method to compute the weights of the network. With this strategy, we aim at improving the performance of the DFFNN in terms of both learning time and accuracy. In addition, we propose a smoothing technique as last step of the proposed methodology, in order to minimize the prediction error. To evaluate the performance of the proposed approach, we use a real-world dataset composed of electricity consumption in Spain, and we compare the results with those generated by a grid search and a random search without smoothing.

The rest of the paper is organized as follows. Section 2 reviews relevant works related to time series forecasting based on deep learning and to the tuning of hyper-parameters in deep learning. Section 3 introduces the methodology proposed in this paper. The most relevant results obtained by the methodology are discussed in Sect. 4. Finally, the conclusions drawn from this research work are summarized in Sect. 5.

## 2 Related Work

In this section, we analyze recent and relevant state-of-the-art proposals in the fields of deep learning time series forecasting and the hyper-parameter tuning and optimization of deep neural networks.

Deep learning approaches for time series analysis have been widely applied during the last years and, indeed, several strategies to predict future values with deep neural networks models have been developed. The authors in [7] presented, in 2015, a novel deep learning-based solution to forecast event-driven stock market values. In particular, a deep convolutional neural network was used obtaining a remarkable performance.

A paradigmatic example of an effort for improving the predictions performance through the network architecture can be found in [10]. There, the authors designed a stacked auto-encoder model for feature extraction to predict air quality. In the proposal presented in [5], a full revision of the input variables was carried out to decrease the computational time related to the training of the proposed deep learning approach for time series forecasting.

Due to the nature of these neural networks architectures and the considerable length of the current time series, distributed computation and data storage approaches play a relevant role in this field of study. In this sense, the authors in [15] proposed a deep feed-forward solution deployed along with the Apache Spark [17] platform for distributed computing to predict electricity consumption in Spain.

The hyper-parameter tuning and optimization of the deep neural networks is a fundamental factor for obtaining a competitive performance of the results. In this regard, the authors in [9] introduced a Bayesian method for hyper-parameter optimization in which model the loss and the execution time in function of the dataset size. Random search and greedy methods for hyper-parameter tuning were applied in [1]. The authors concluded that the random search method can be useful in deep learning environments. The authors in [2] made a comparative study of three hyper-parameter optimization techniques: grid, experience-based, and random search methods. They concluded that the random one establishes a baseline to judge the performance of other hyper-parameter optimization algorithms.

Evolutionary strategies for optimization problems have been widely used, yielding competitive results. The authors in [16] addressed the hyper-parameter optimization problem with the approach mentioned above. Another specific approach for hyper-parameter optimization can be found in [8] where an efficient and deterministic method using radial functions was presented. Finally, in [11], the authors proposed a mixed strategy called Covariance Matrix Adaptation Evolution.

# 3 Methodology

This section describes the proposed methodology for time series forecasting using the DFFNN, which has been implemented in the H2O framework [3], under R language. It is also proposed an alternative method to the one implemented in H2O for the optimization of hyper-parameters and the use of a smoothing filter in order to minimize the impact of the time gap on each prediction. First, Sect. 3.1 describes a method for optimizing neural network hyper-parameters. After, Sect. 3.2 details the formulation that allows the multi-step forecast of a time series. Finally, the use of a smoothing filter to modulate the frequency of the prediction is introduced in Sect. 3.3. A complete workflow of the methodology proposed is illustrated in Fig. 1.

## 3.1 Hyper-parameters Tuning

It is well-known that the values of the hyper-parameters of the deep learning algorithm highly influence on the results. The algorithm implemented in H2O allows adjusting a large number of them, being worth highlighting some, such as the number of hidden layers or the number of neurons per layer or the learning rate.

In order to optimize the hyper-parameters described above, H2O implements two search options. One of them is a grid search, which performs an exhaustive search through the whole set of established hyper-parameters. The other one is a random search, which makes combinations of the defined hyper-parameters without a specific order or criteria. However, both search methods work with discrete values, which greatly limits the fine-tuning of the vast majority of hyper-parameters.

To avoid this problem, a random search is proposed in this article with continuous values. That is, given a set of hyper-parameters and a range for each

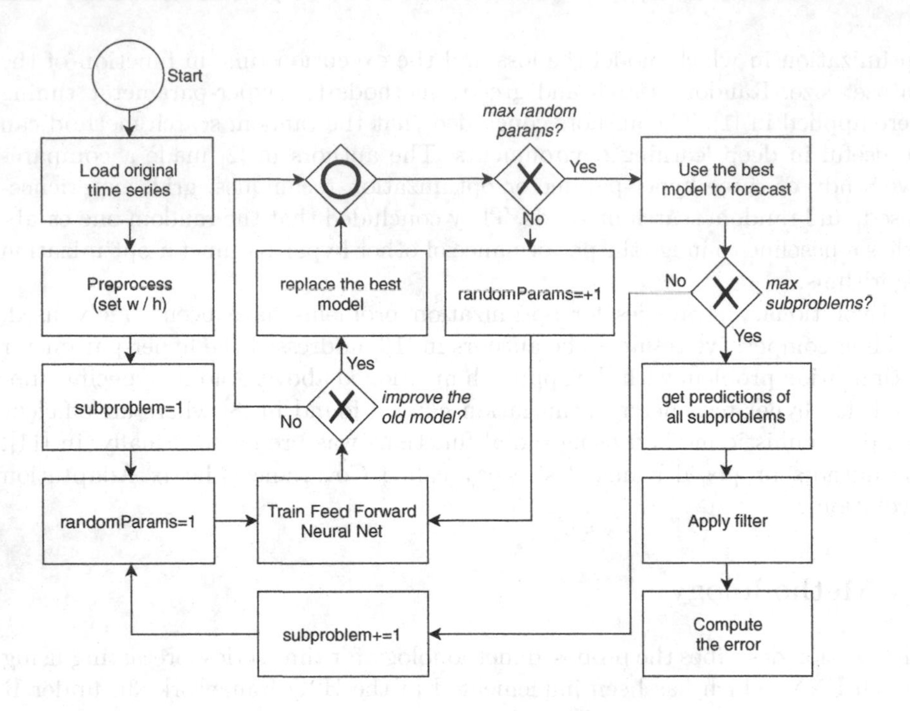


**Fig. 1.** Complete work-flow of the proposed methodology.

one, a random value is generated for each hyper-parameter and it is validated by computing the forecasting error using a validation set. This process is repeated during a certain number of iterations, storing the model that obtains the smallest error. Finally, a single model is stored for each sub-problem, corresponding to the one whose hyper-parameters offer the best results.

## 3.2 Multi-step to Single-step Regression

Given a time series expressed as $[x_1, x_2, \ldots, x_t]$, the main goal of this research is to forecast the future values of the time series. To do this, a predictive model is formed based on a historical window composed of $w$ past values that allow the prediction of the $h$ following values, also called the prediction horizon. This kind of problem is known as multi-step forecasting and can be formulated as:

$$[x_{t+1}, x_{t+2}, \ldots, x_{t+h}] = model(x_{t-(w-1)}, \ldots, x_{t-1}, x_t) \quad (1)$$

Regretfully, the deep learning algorithm included in the H2O framework does not support multi-step forecasting. To achieve this goal, a methodology has been developed. This methodology consists in focusing on the prediction of each instant of time individually, dividing the multi-step prediction into $h$ predictions of a single step. This methodology is formulated in Eqs. (2)–(5):

$$x_{t+1} = model_1(x_{t-(w-1)}, \ldots, x_{t-1}, x_t) \quad (2)$$

$$x_{t+2} = model_2(x_{t-(w-1)}, \ldots, x_{t-1}, x_t) \quad (3)$$

$$\ldots \quad (4)$$

$$x_{t+h} = model_h(x_{t-(w-1)}, \ldots, x_{t-1}, x_t) \quad (5)$$

As can be seen from these Equations, there is a gap in the data used in each prediction (e.g. the prediction of $x_{t+2}$ is not used to predict $x_{t+3}$). However, if these predictions were taken into account to forecast the next point of data, it would cause a propagation of the error, giving rise to a wrong prediction [4].

This formulation involves the training of $h$ different models instead of a single model, requiring a high computational cost. However, the implementation of the deep learning algorithm in H2O is optimized and parallelized, which minimizes this shortcoming.

### 3.3 Smoothing Filter

Once the hyper-parameters are calculated, the final task can be accomplished. The estimation of individual and independent models to forecast a set of values representing a prediction horizon has a consequence: the predicted values exhibit some significant ripple because the estimated values have no information about neither previous nor subsequent estimations. That is, sharp variations from one value to another may be generated.

For this reason, the application of a smoothing filter is also proposed, as the last step of the methodology. Different strategies can be chosen. For instance, filters based on Fourier transform are quite popular [12] but their quadratic cost function, $O(n^2)$, turn these filters into a not particularly suitable solution in the big data context.

Another much simpler, but powerful, filter has been selected: the one based on moving averages with linear cost function, $O(n)$, and, in particular, the one implemented in the *Stats* R package [13]. This low-pass filter is a common finite impulse response type that removes high frequencies, i.e. the sharp variations. It only needs to adjust the number of previous data that will be used to calculate the average, N.

Mathematically, the calculation of the first filtered value is formulated as follows:

$$x'(t) = \frac{1}{N} \sum_{i=1}^{N} x(t-i) \quad (6)$$

where $x(t)$ is the current smoothed value and x(t − i), for $i = 1$, are the N values preceding $x(t)$. Then, $x(t+i)$, for $i > 0$, are calculated by shifting forward x′(t) but excluding the first number of the time series and including its next value.

To adjust this parameter, N is trained using values from 1 to 12 (as it will explained in Sect. 4, $N = 12$ involves the two previous hours).

# 4 Results

This section presents the results obtained after applying the methodology described in Sect. 3 to the dataset detailed in Sect. 4.1. All the experiments have been executed into a Intel Core i7-5820K at 3.3 GHz with 15 MB of cache, 12 cores and 16 GB of RAM, working under an Ubuntu 18.04 operating system.

## 4.1 Dataset Description

The time series considered in this study is related to electrical electricity consumption in Spain, from January 2007 to June 2016. There is a total of 9 years and 6 months with a frequency of 10 min between each measure. This fact makes a time series with a total length of 497832 measures, stored into a 33 MB file in CSV format with a single column. For this reason, a preprocess has been applied to transform the time series into a supervised dataset with $w+h$ columns, where $w$ refers to the historical window of data used to predict the following $h$ values, called the prediction horizon. The whole dataset was split into 70% for the training set and 30% for the test set. In addition, a 30% from the training set has also been selected as the validation set in order to optimize the hyper-parameters of the deep learning algorithm as well as the smoothing filter.

## 4.2 Error Metrics

To measure the error of the methodology proposed in Sect. 3, the most used metrics in the literature for time series forecasting problems have been used. These metrics are the Mean Squared Error (MSE), Root Mean Squared Error (RMSE), Mean Absolute Error (MAE) and Mean Relative Error (MRE). The formulation of these error metrics is shown below:

$$MSE = \frac{1}{n}\sum_{i=1}^{n}(p_i - a_i)^2 \tag{7}$$

$$RMSE = \sqrt{\frac{1}{n}\sum_{i=1}^{n}(p_i - a_i)^2} \tag{8}$$

$$MAE = \frac{1}{n}\sum_{i=1}^{n}|p_i - a_i| \tag{9}$$

$$MRE = 100 \cdot \frac{1}{n}\sum_{i=1}^{n}\frac{|p_i - a_i|}{a_i} \tag{10}$$

where $n$, $p$ and $a$ mean the number of samples, predicted values and actual values, respectively.

### 4.3 Performance in Terms of Error

The experimental setting of the proposed methodology is as follows:

1. The historical window size used to predict the following four hours (24 values) has been set to 168, which represents a whole day and 4 h. This value has been chosen because the larger the historical window of data, the better results will be obtained, as demonstrated in [15].
2. The hyper-parameters that have been optimized are the number of layers, the number of neurons per layer and the value of learning ratio (rho). The hyper-parameters search ranges have been set to $[1, 5]$, $[10, 100]$ and $[0.9, 0.999]$, respectively.
3. A number of 50 epochs was established in the training phase of the deep learning algorithm. The rest of the deep learning hyper-parameters have default values.
4. To find the optimal hyper-parameters, a total of 50 iterations over each problem was carried out during the training and validation phase.
5. The possible values of $N$ for the smoothing filter have been set between 1 and 12. After the training phase of this parameter, the value has been set to 7.

The configuration of the experiments described above results in a total of 1200 calculated models. The best model for each sub-problem will be used to predict the test set. In order to have a reference point, the results obtained with the proposed methodology have been compared with the methodology proposed by the authors in [15]. This methodology applies an exhaustive search to optimize the size of the historical window, the value of the L1 penalty, distribution function, number of layers and number of neurons, calculating a total of 3120 different models. If only the optimized parameters proposed in this article are taken into account, the grid search calculates 1320 models, 120 more than the methodology proposed in this research.

After completing the training and validation step, 24 different network configurations were obtained, each corresponding to a sub-problem, as detailed in Table 1. It can be seen that the error increases when the timestamp to forecast increases. This fact is due to the time gap between the data to train the model and the timestamp to forecast.

Table 2 summarizes the errors reached by the different approaches. It can be seen how the use of the methodology proposed in this article improves by 20% the mean relative error obtained by the exhaustive search. This is because the exhaustive search only allowed the search for hyper-parameters in a discrete set of values. It is also observed how the application of the smoothed filter significantly improves the error.

A graphical comparison between the real data, non-smoothed predictions and smoothed predictions (described in Sect. 3.3) for an arbitrary day in the test set has been depicted in Fig. 2. It can be seen how the smoothed predictions remove the peaks of the non-smooth predictions, thus significantly decreasing the error.

**Table 1.** Best hyper-parameters for each subproblem (without smoothing).

| SP[1] | Hyper-parameters | | | Error in test phase | | | |
|---|---|---|---|---|---|---|---|
| | #hidden | #neurons | Rho | MSE | RMSE | MAE | MRE (%) |
| #1 | 4 | [66, 44, 99, 98] | 0.971 | 57099.12 | 238.95 | 186.20 | 0.69 |
| #2 | 3 | [91, 82, 11] | 0.922 | 90365.86 | 300.61 | 235.87 | 0.87 |
| #3 | 5 | [53, 59, 96, 29, 47] | 0.961 | 114441.50 | 338.29 | 265.31 | 0.96 |
| #4 | 5 | [79, 96, 94, 22, 44] | 0.937 | 121272.40 | 348.24 | 270.05 | 0.99 |
| #5 | 3 | [76, 86, 62] | 0.971 | 141457.60 | 376.11 | 288.54 | 1.07 |
| #6 | 5 | [3, 43, 27, 82, 53] | 0.928 | 157920.10 | 397.39 | 307.77 | 1.14 |
| #7 | 4 | [91, 48, 89, 83] | 0.988 | 178831.50 | 422.88 | 323.95 | 1.20 |
| #8 | 3 | [57, 99, 46] | 0.981 | 245192.60 | 495.17 | 383.04 | 1.43 |
| #9 | 4 | [41, 85, 46, 80] | 0.970 | 246930.00 | 496.92 | 383.03 | 1.42 |
| #10 | 5 | [49, 69, 62, 22, 27] | 0.917 | 245124.70 | 495.10 | 381.89 | 1.39 |
| #11 | 3 | [68, 47, 71] | 0.927 | 310147.90 | 556.91 | 430.91 | 1.59 |
| #12 | 4 | [89, 23, 96, 90] | 0.966 | 309112.60 | 555.98 | 432.56 | 1.60 |
| #13 | 4 | [36, 77, 45, 92] | 0.961 | 325379.70 | 570.42 | 438.93 | 1.64 |
| #14 | 3 | [77, 72, 81] | 0.969 | 336707.90 | 580.27 | 435.58 | 1.63 |
| #15 | 5 | [55, 61, 34, 91, 85] | 0.941 | 401978.60 | 634.02 | 478.17 | 1.77 |
| #16 | 5 | [45, 73, 38, 71, 61] | 0.963 | 373900.30 | 611.47 | 464.31 | 1.70 |
| #17 | 5 | [44, 41, 46, 98, 43] | 0.978 | 406642.40 | 637.69 | 489.32 | 1.80 |
| #18 | 2 | [88, 24] | 0.966 | 407873.10 | 638.65 | 482.05 | 1.79 |
| #19 | 5 | [91, 48, 89, 76, 46] | 0.907 | 395915.50 | 629.22 | 468.49 | 1.75 |
| #20 | 5 | [88, 37, 62, 78, 56] | 0.928 | 526235.70 | 725.42 | 541.03 | 2.01 |
| #21 | 3 | [53, 82, 33] | 0.962 | 657200.40 | 810.68 | 582.92 | 2.17 |
| #22 | 5 | [99, 89, 57, 27, 69] | 0.986 | 808235.20 | 899.02 | 648.59 | 2.43 |
| #23 | 4 | [75, 52, 88, 56] | 0.997 | 753634.70 | 868.12 | 622.51 | 2.33 |
| #24 | 3 | [82, 74, 63] | 0.941 | 689790.30 | 830.54 | 600.27 | 2.23 |

[1] Sub-problem

Figure 3 shows a comparison between actual and predicted data using the models obtained in Table 1. Figure 3(a) shows the prediction of the best day (144 values) for the entire test set. On the other contrary, Fig. 3(b) shows the forecast of the worst day.

**Table 2.** Comparison of the search metrics and the proposed methodology.

| | MSE | RMSE | MAE | MRE (%) |
|---|---|---|---|---|
| Grid | 380486.80 | 616.84 | 451.96 | 1.68 |
| Random | 345891.20 | 588.13 | 422.55 | 1.57 |
| Random + Filter | 251143.90 | 501.14 | 369.19 | 1.36 |

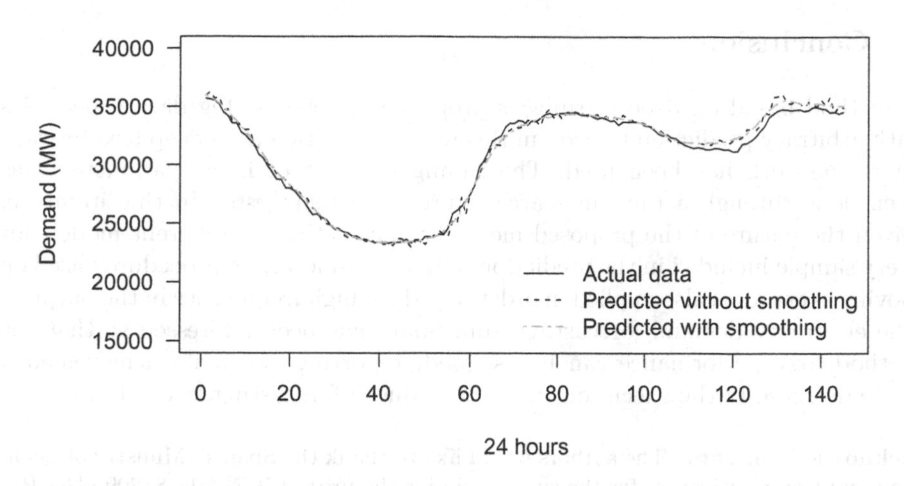


**Fig. 2.** Comparison between real data, non-smoothed and smoothed predictions.

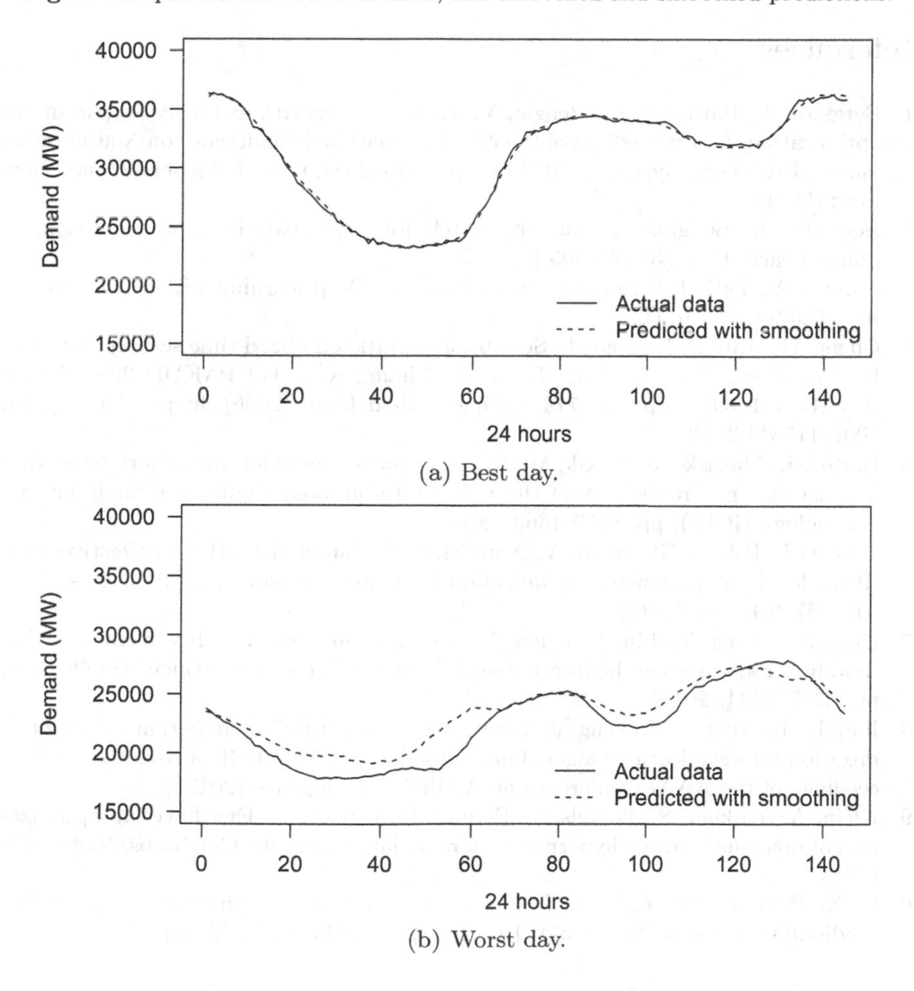


(a) Best day.

(b) Worst day.

**Fig. 3.** The best and worst day predicted by the proposed methodology.

## 5 Conclusions

A method based on deep learning is proposed to forecast big data time series with arbitrary prediction horizon in this work. In particular, a deep feed forward neural network has been used. The tuning of a set of hyper-parameters has been done through a random search approach, as suggested in the literature. Given the nature of the proposed method which estimates different models for every sample included in the prediction horizon, a smoothing procedure based on moving averages is also applied in order to reduce high frequencies in the outputs. The electricity demand forecasting from Spain has been addressed so that the methodology performance can be assessed, reporting two main achievements: acute decrease in the execution time and reduced forecasting error (1.36%).

**Acknowledgements.** The authors would like to thank the Spanish Ministry of Economy and Competitiveness for the support under the project TIN2017-88209-C2-1-R.

## References

1. Bergstra, J., Bardenet, R., Bengio, Y., Kégl, B.: Algorithms for hyper-parameter optimization. In: Proceedings of the 24th International Conference on Neural Information Processing Systems, NIPS'11, pp. 2546–2554. Curran Associates Inc., New York (2011)
2. Bergstra, J., Bengio, Y.: Random search for hyper-parameter optimization. J. Mach. Learn. Res. **13**, 281–305 (2012)
3. Candel, A., LeDell, E., Parmar, V., Arora, A.: Deep learning with H2O. H2O.ai, Inc., California (2017)
4. Cheng, H., Tan, P.-N., Gao, J., Scripps, J.: Multistep-ahead time series prediction. In: Ng, W.-K., Kitsuregawa, M., Li, J., Chang, K. (eds.) PAKDD 2006. LNCS (LNAI), vol. 3918, pp. 765–774. Springer, Heidelberg (2006). https://doi.org/10.1007/11731139_89
5. Dalto, M., Matusko, J., Vasak, M.: Deep neural networks for ultra-short-term wind forecasting. In: Proceedings of the IEEE International Conference on Industrial Technology (ICIT), pp. 1657–1663 (2015)
6. Diaz, G.I., Fokoue-Nkoutche, A., Nannicini, G., Samulowitz, H.: An effective algorithm for hyperparameter optimization of neural networks. IBM J. Res. Dev. **61**(4/5), 9:1–9:11 (2017)
7. Ding, X., Zhang, Y., Liu, T., Duan, J.: Deep learning for event-driven stock prediction. In: Proceedings of the International Joint Conference on Artificial Intelligence, pp. 2327–2334 (2015)
8. Ilievski, I., Akhtar, T., Feng, J., Shoemaker, C.A.: Efficient hyperparameter optimization for deep learning algorithms using deterministic RBF surrogates. In: Proceedings of the AAAI Conference on Artificial Intelligence (2017)
9. Klein, A., Falkner, S., Bartels, S., Hennig, P., Hutter, F.: Fast bayesian optimization of machine learning hyperparameters on large datasets. CoRR abs/1605.07079 (2016)
10. Li, X., Peng, L., Hu, Y., Shao, J., Chi, T.: Deep learning architecture for air quality predictions. Environ. Sci. Pollut. Res. Int. **23**, 22408–22417 (2016)

11. Loshchilov, I., Hutter, F.: CMA-ES for hyperparameter optimization of deep neural networks. arXiv preprint arXiv:1604.07269 (2016)
12. Manolakis, D.G., Ingle, V.K.: Applied Digital Signal Processing. Cambridge University Press, Cambridge (2011)
13. R Core Team: R: A Language and Environment for Statistical Computing. R Foundation for Statistical Computing, Vienna, Austria (2013). http://www.R-project.org/. ISBN 3-900051-07-0
14. Ruder, S.: An overview of gradient descent optimization algorithms. CoRR abs/1609.04747 (2016)
15. Torres, J., Galicia, A., Troncoso, A., Martínez-Álvarez, F.: A scalable approach based on deep learning for big data time series forecasting. Integr. Comput.-Aid. E. **25**(4), 335–348 (2018)
16. Young, S.R., Rose, D.C., Karnowski, T.P., Lim, S.H., Patton, R.M.: Optimizing deep learning hyper-parameters through an evolutionary algorithm. In: Proceedings of the Workshop on Machine Learning in High-Performance Computing Environments, p. 4. ACM, New York (2015)
17. Zaharia, M., Xin, R.S., Wendell, P., Das, T., Armbrust, M., Dave, A., Meng, X., Rosen, J., Venkataraman, S., Franklin, M.J., et al.: Apache spark: a unified engine for big data processing. Communications of the ACM **59**(11), 56–65 (2016)

# A First Approximation to the Effects of Classical Time Series Preprocessing Methods on LSTM Accuracy

Daniel Trujillo Viedma(✉), Antonio Jesús Rivera Rivas, Francisco Charte Ojeda, and María José del Jesus Díaz

Andalusian Research Institute on Data Science and Computational Intelligence (DaSCI), Computer Science Department, University of Jaén, 23071 Jaén, Spain
{dtviedma,arivera,fcharte,mjjesus}@ujaen.es

**Abstract.** A convenient data preprocessing has proven to be crucial in order to achieve high levels of accuracy, time series being no exception. For this kind of forecasting tasks, several specialized preprocessing methods have been described, being trend analysis and seasonal analysis some of them. Several have been formally grouped around a methodology that is always applied to state of the art time series forecasting models, like the well known ARIMA.

LSTM is a relatively novel architecture which has been specifically designed to get rid of the vanishing gradient problem. In these models, great results have been seen when applied for time series forecasting. Still, little is known about the impact of these traditional preprocessing methods on the accuracy of LSTM.

In this work an empirical analysis on how classical time series preprocessing methods influence LSTM results is conducted. That all considered ones can potentially improve LSTM performance is concluded, being the seasonal component removal the filter that achieves better, most robust accuracy gain.

**Keywords:** LSTM · Time series · Preprocessing

## 1 Introduction

Feeding not only clean, but also well-formatted, data to data mining models usually improves their overall performance. Many studies have highlighted the importance of a good preprocessing analysis [2, 10].

In the case of time series, this is not an exception, but given the peculiarity of this kind of datasets, specialized methods must be used along with several general purpose ones. Traditional time series modelling tools like ARIMA [9], which is considered the state of the art because of its good results and strong statistical foundations, are never applied without a set of preprocessing strategies called the Box-Jenkins methodology [1]. This demonstrate the great usefulness of applying these preprocessing techniques.


I. Rojas et al. (Eds.): IWANN 2019, LNCS 11506, pp. 270–280, 2019.
https://doi.org/10.1007/978-3-030-20521-8_23

Box-Jenkins methodology [1] can be divided into several simpler tasks to transform the original time series into an stationary one: trend and seasonal components detection and removal. They will be addressed in this work along with another useful one: an analysis of lags.

LSTM (*Long Short-Term Memory*) [7] are a kind of Deep Learning, recurrent neural networks that have been designed with an important limitation of other deep learning methods in mind: the vanishing gradient problem widely discussed in [6], in which several approaches to solve it are also introduced. LSTM includes several mechanisms to overcome this limitation, like the Constant Error Carrousel, that makes them able to remember filtered information. This model was later applied to time series with impressive results, mainly because of its capability to model long time dependencies.

This work highlights empirically the effect of the mentioned fundamental preprocessing methods on LSTM predictive performance, by predicting from time series that have been preprocessed in several ways.

This document is structured as follows. First, a briefly description of the LSTM is presented in Sect. 2, then the main preprocessing methods traditionally used for time series data mining is enumerated in Sect. 3. After that, the entire experimentation carried out and its results are detailed in Sect. 4. Lastly, a conclusion is given at Sect. 5.

## 2 LSTM

An LSTM is a recurrent artificial neural cell which has been specifically designed to solve the vanishing gradient problem. The proposal in [7] defines a complex neural cell in which information is filtered through several gates, and also transformed accordingly to a previously memorized value computed from the last forward propagation run. The LSTM's structure has been later successfully applied to time series to find that its internal mechanisms let it model very long time dependencies.

LSTM does so by minimizing the vanishing gradient problem, very common in deep learning models, like the recurrent neural networks. The vanishing gradient problem is detailed in [6]. It prevents a neural network from continuing to learn beyond a certain point due the lack of effect from layers which are furthest from the output.

An LSTM cell consist of a relatively complex data pipeline that makes it able to remember an internal state, while operating on input data to compute its output. The pipeline is composed of activation functions that filter and transform data; and additive and multiplicative operations that merge and also transform data.

In Fig. 1, a more detailed description of the contemporary pipeline, also referred to as *Vanilla LSTM* [5] and firstly described in [4], can be found. Comparing this LSTM with the original one, presented in [7], two main additions can be easily seen:

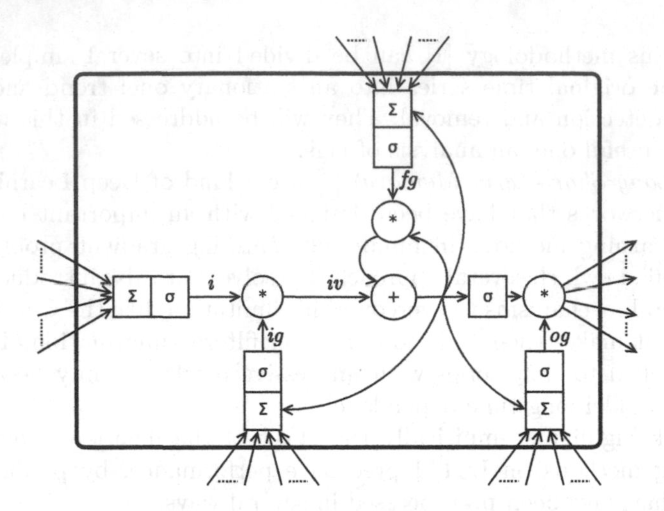


**Fig. 1.** Schematic of an LSTM cell.

**Forget gate** Introduced in [3], this gate demonstrated to be crucial to improve LSTM performance, at the cost of increasing the total number of parameters to train.

**Peephole connections** Data from inside the cell is made available through peephole connections to the gates, so data can be better refined before entering the cell.

Figure 1 also gives an easy way to understand how data flows inside the cell: The input data flows from the leftmost input gate to the output of the cell, located at the opposite side. In between, it is transformed by means of pointwise multiplications with values provided by additional gates, each one trained to filter data in a particular way: the input gate filter the input data to prevent noise coming from outside the cell disturb the memorized value, while the output gate performs a final filtering to protect next layer cells from bad memorized values.

Also, an internal loop can be appreciated, regulated by the forget gate. This is called CEC (*Constant Error Carousel*), and constitutes the mechanism by which the LSTM memorizes knowledge for managing long term dependencies. This memorized value can be influenced by new input data, but also filtered through the forget gate.

The data (both new and recurrent entries) flowing through the pipeline described in Fig. 1 first enter the cell through, performing the computations on the four input gates, then is merged by means of multiplicative and additive (pointwise or scalar) operations and then, feeded to a final activation function like an ordinary neural cell.

LSTM neural cells are commonly trained using the well-known BPTT (*Backpropagation Through Time*) algorithm [11], which models the contribution of each network parameter to the final error and greedily modifies it so the error is minimized.

# 3 Preprocessing Methods

Traditional time series data mining algorithms greatly improve their performance when are trained with a preprocessed dataset. ARIMA is one of these methods, whose prediction accuracy increases when the original time series is decomposed into trend, seasonal, and random components, and are analyzed separately. Typically, the preprocessing methods achieving better results have been:

**Significant lags detection** Adds additional values to every instance. More precisely, this new values are the very same time series, but lagged at certain times. As a drawback, the values corresponding to the most earliest time steps must be dropped or predicted, because the lack of real lagged data for the first values of the dataset.

**Trend component removal** A trend component in the time series is detected and removed. Trend removal may be useful to palliate the effect of the magnitude of the variable, but can be affected by abrupt variations on the time series. This preprocessing method feeds the network with values that are not present on the original time series (the actual real value of the series is never given to the network, but the result of the chosen transformation), so the network output prediction has to be rebuilt to revert this transformation.

**Seasonal component removal** A seasonal component in the time series is detected and removed. Like in the *trend* case, the input values are transformed when deleting the seasonal component, si the output of network trained with these versions of the original time series must be, as well, rebuilt.

# 4 Experiments and Results

Several experiments have been conducted to extract enough data to establish a comparison between the chosen preprocessing methods. The details regarding this experimentation are introduced below.

## 4.1 Error Metric

In order to compare the performance of the model, a quality measure of the prediction must be established. One broadly used method to quantify this is through an error measure, and then following the *less is best* criteria. RMSE (*Root Mean Squared Error*) is a well known error metric that emphasizes the deviations more than others. It is computed as follows: given T the true values of the time series, and P the predicted ones, RMSE can be described as:

$$RMSE(T, P) = \sqrt{MSE(T, P)} \tag{1}$$

$$MSE(T, P) = \frac{\sum_{i=1}^{n} (T_i - P_i)^2}{n} \tag{2}$$

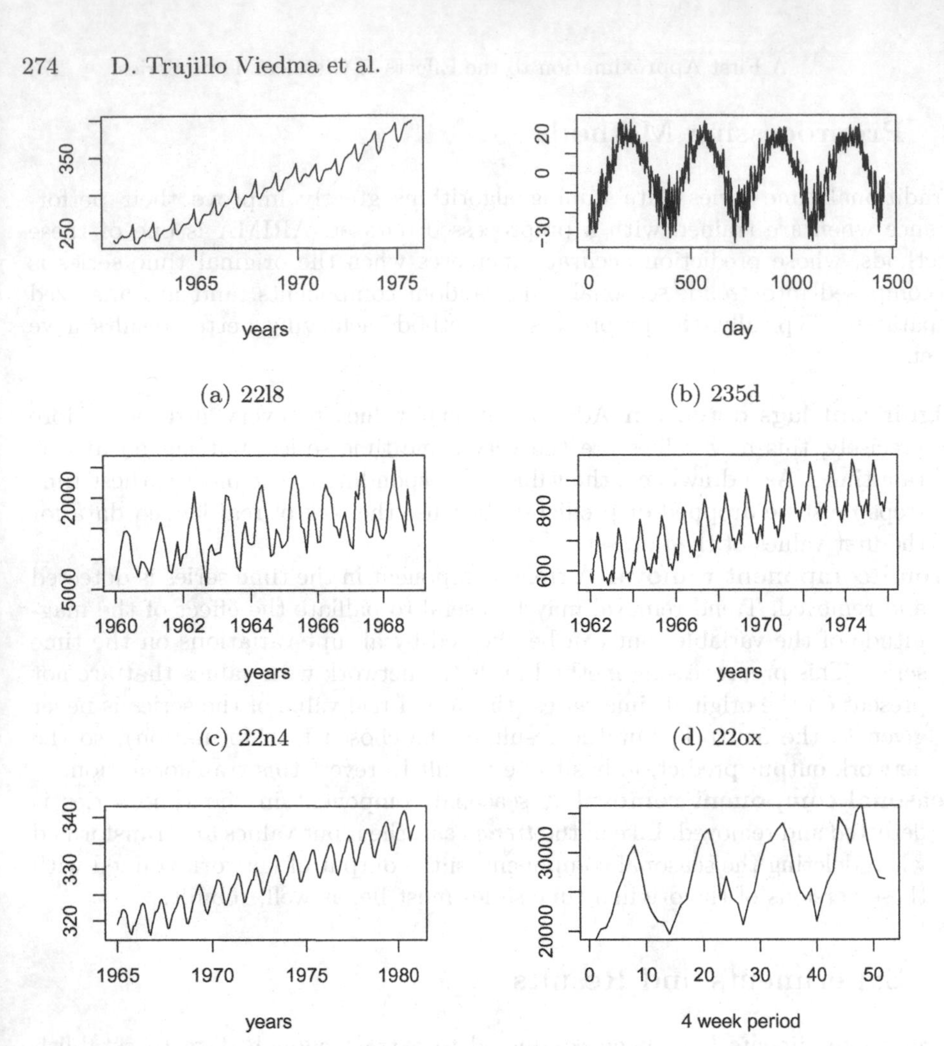


**Fig. 2.** Plots of the datasets considered

### 4.2 Datasets

In this work, 6 datasets from the Time Series Data Library [8] have been used. As it can be seen in Fig. 2, in general terms they expose a trending and seasonal behavior. This election of datasets has been made to maximize the effect of the preprocessing methods considered. In Table 1 descriptions of several aspects of them are included.

The size of these public, well-known time series, is small enough to let all the experimentation run in an acceptable time window while using relatively time-consuming models like the ones that are being trained.

**Table 1.** Datasets

| Code | Description | Resolution | Period length | Dates |
|---|---|---|---|---|
| 2218 | Wisconsin employment | Monthly | 12 | Jan. 1961 |
| | | | | Oct. 1975 |
| 22n4 | Car sales in Quebec | Monthly | 12 | Jan. 1960 |
| | | | | Dec. 1968 |
| 22ox | Milk production: pounds per cow | Monthly | 12 | Jan. 1962 |
| | | | | Dec. 1975 |
| 22v1 | $CO_2$ (ppm) mauna loa | Monthly | 12 | Jan. 1965 |
| | | | | Dec. 1980 |
| 235d | Mean temperature, Fisher River near Dallas | Daily | 365 | Jan. 1, 1988 |
| | | | | Dec. 31, 1991 |
| 2325 | Totals of beer shipments | Four-weekly | 14 | Week 2, 1970 |
| | | | | Week 49, 1973 |

**Table 2.** Hyperparameters

| Num. of LSTM | Epochs | Batch size | Topology | Repetitions |
|---|---|---|---|---|
| 10 | 200 | 1 | Dense | 10 |

For evaluation purposes, these time series are divided into training and test sets, the latter being formed by the last 24 observations of the series (the most recent ones), whilst the rest of the instances fall into the training dataset.

These datasets constitute the raw data from which a baseline for comparison has to be set. Several versions of each one will be computed accordingly to the preprocessing methods that have been selected. The creation procedure and the instance form for each of these variants are detailed in Subsect. 4.4.

### 4.3 Model

The experiments have been coded in *Python*, making use of the *Keras* library with the *TensorFlow* backend and the *Scikit-learn* library for error computing. A single set of hyperparameters has been heuristically selected, after performing several tests, and taking execution time concernings into account. These chosen hyperparameters can be seen in Table 2.

Preliminar experiments did not achieve better accuracy after increasing the number of LSTM cells. On the other hand, given the relatively small size of the time series being considered, it is possible to set a small value for the *batch size* parameter. Low values have a great negative impact on the execution time, but because of the size of the datasets and the available computational power, a value of 1 is possible.

Several repetitions of the experiments have to be made because of the random strategy the training algorithm follows to initialize the weights of the LSTM network. In order to minimize the effect of an accidentally advantageous initial weight configuration by luck, 10 repetitions have been made, and then conclusions are extracted from a summarization statistic. This also improves the generalization of our conclusions, which are not tied to a particular execution.

### 4.4 Preprocessing Methods

For each one of the distinct time series considered, several datasets have been created, as a result of applying a preprocessing method to the original time series. These versions are introduced below.

**Raw Time Series.** This is a trivial identity preprocessing. The original time series remains unchanged. It is interesting to test the LSTM network with raw, unpreprocessed data, in order to set a baseline to which compare the rest of the non-trivial preprocessing methods.

The definition of an instance is also trivial, but will serve as an introduction to the notation of the definition of the instances. The instances in these time series have the form:

$$I_t = (t, V_t) \tag{3}$$

Being:

$t$ Is the time period this instance refers to.
$I_t$ The instance of the dataset at time $t$.
$V_t$ The value of the dataset at time $t$.

**Lags.** In the experimentation carried out, the lags considered for each time depends on a very simple correlogram graph analysis consisting in selecting the lags whose partial autocorrelation function value exceeds the statistically significant threshold. Also, the first values have been dropped, in order to keep the lag window inside real data, instead of predicting those values and introducing noise. These datasets have enough values to be dramatically affected by this drop strategy.

The form of the lag-processed instances is defined as:

$$I_t = (t, V_t, V_{t-l_0}, \ldots, V_{t-l_n}) \tag{4}$$

Being:

$t$ Is the time period this instance refers to.
$I_t$ The instance of the dataset at time $t$.
$V_t$ The value of the dataset at time $t$.
$l_0, \ldots, l_n$ The selected lags.

**Trend.** The trend component is removed by applying differentiation to the dataset. This is done by subtracting a value from the time series to the following one, so the differences between two consecutive values are given to the neural network. Trend removal may be useful to palliate the effect of the magnitude of the variable, but can be affected by abrupt variations on the time series.

After differentiate, the instances have the following form:

$$I_t = (t, V_{t+1} - V_t) \tag{5}$$

In the experiments carried out, the rebuild procedure for a particular instance takes the value at the output of the neural network, and adds it to the true value from the original dataset, so errors are not accumulated, which is a usual problem when relying solely on the network output values, and applying cumulative sum.

**Seasonal.** For this preprocessing method, a very simple and intuitive seasonal component detection and deletion procedures have been adopted. Then, that de-seasonalized time series have been fed to the neural network.

The detection of the seasonal component in a given dataset is performed by calculating the average of the values at the same position relative to the start of a period (it is a prerequisite that the time series period length is known).

Then, once the seasonal component has been modelled, seasonality is removed by subtracting the corresponding value of the modelled seasonality to the value of the series:

$$D_t = V_t - s_{u-mod(t,l)} \tag{6}$$

Being:

$D_t$ Deseasonalized value at time $t$.
$s_u$ Detected seasonality component value at position $u$.
$l$ Period length, detected seasonality component length.

The instance for this particular preprocessing method is described as follows:

$$I_t = (t, D_t) \tag{7}$$

Just like in the *trend* case, transformed values are fed to the neural network, so another rebuilt has to be performed. This is done just by adding the previously detected seasonal component to each value on the prediction.

## 4.5 Results

In Table 3, average, standard deviation and variation of the error from each set of experiments can be found. These statistics are highlighted when the average of a preprocess improves the accuracy of the raw methodology. As pointed out in Subsect. 4.3, each value summarize 10 repetitions of the corresponding set of experiments that have been performed. The variation is calculated relative to the raw average statistic, so predictive performance can be compared across multiple datasets.

**Table 3.** LSTM RMSE

| Dataset | Statistic | Raw | Lags | Trend | Seasonal |
|---|---|---|---|---|---|
| 22l8 | Average | 8,982 | 14,453 | **7,528** | **3,517** |
| | Std. Dev. | 0,334 | 1,803 | **0,226** | **0,520** |
| | Var. | | 60,909% | **−16,181%** | **−60,844%** |
| 22n4 | Average | 3841,427 | **2944,574** | 4088,950 | **1705,087** |
| | Std. Dev. | 19,440 | **152,559** | 16,493 | **12,934** |
| | Var. | | **−23,347%** | 6,443% | **−55,613%** |
| 22ox | Average | 45,148 | **20,453** | **35,900** | **12,363** |
| | Std. Dev. | 0,194 | **2,102** | **1,074** | **0,756** |
| | Var. | | **−54,699%** | **−20,485%** | **−72,616%** |
| 22v1 | Average | 0,830 | **0,562** | 0,874 | **0,639** |
| | Std. Dev. | 0,031 | **0,060** | 0,003 | **0,059** |
| | Var. | | **−32,32%** | 5,26% | **−23,04%** |
| 235d | Average | 5,317 | 9,931 | 5,437 | **4,667** |
| | Std. Dev. | 0,019 | 0,205 | 0,021 | **0,037** |
| | Var. | | 86,794% | 2,263% | **−12,225%** |
| 2325 | Average | 2970,936 | **2221,160** | 2998,783 | **1778,490** |
| | Std. Dev. | 42,565 | **111,955** | 13,650 | **34,674** |
| | Var. | | **−25,237%** | 0,937% | **−40,137%** |

Analyzing average statistic values, we find out that the lagged preprocessing method makes the LSTM to predict better most of the cases. Something similar happens to the seasonal analysis, but with a more consistent behaviour: better LSTM performance is achieved in all cases studied. By contrast, the trend preprocessing method achieves almost the same performance.

In the case of datasets 22l8 and 235d, lagged preprocessing performs surprisingly bad, with almost double the error achieved without any data treatment. Taking a closer look at the plots (Figs. 2a and b) shows, for the dataset 235d, several differences between periods of a strong seasonal component, which may cause this preprocess to fail. On the other hand, on dataset 22l8 we found what probably is a slight concept drift at the end of the time series, falling most of it in the test partition. In the case of the trend analysis, two of the datasets with the strongest trend component, 22ox and 22l8, performs better than the raw training.

Looking at the standard deviation there is a large variability between the lagged preprocessing and the other ones, including the experiments made with raw time series. In a general sense, both the trend and the seasonal analysis have a standard deviation comparable with the raw time series experimentation. The unusual variability on the 22ox dataset may be explained by the stabilization of the overall trend observed at the last 5 years of the time series.

## 5 Conclusion

LSTM neural networks are being actively used recently, while being proposed time ago. In this work, a preliminar experimentation has been conducted to quantify how much this kind of networks are affected by the use of several preprocessing methods in the context of time series forecasting.

Performing a trend deletion (differentiation) on the input time series usually has no effect on the accuracy, compared with a raw time series. A lagged dataset, unlike a differentiated one, can make the LSTM predict better in some cases, while severely worsening it in other cases. Further work would be needed to determine theses cases. On the other hand, a seasonal component removal has achieved an important accuracy gain on all the datasets considered, dropping the error below a 50% of the raw time series error in many cases. Our results show that special care has to be taken with lags and differentiating preprocessing methods, in which a severe performance drop has been seen with some datasets. In the case of the seasonal component removal preprocessing method, a more robust, accurate predictive behavior has been found.

**Acknowledgement.** This work is partially supported by the Spanish Ministry of Science and Technology under project TIN2015-68454-R.

## References

1. Box, G.E., Jenkins, G.M.: Time Series Analysis: Forecasting and Control. Holden-Day Series in Time Series Analysis. Holden-Day, San Francisco (1976). Revised edition
2. Famili, A., Shen, W.M., Weber, R., Simoudis, E.: Data preprocessing and intelligent data analysis. Intell. Data Anal. **1**(1), 3–23 (1997). https://doi.org/10.3233/IDA-1997-1102
3. Gers, F.A., Schmidhuber, J., Cummins, F.: Learning to forget: continual prediction with LSTM. Neural Comput. **12**(10), 2451–2471 (2000). https://doi.org/10.1016/j.neunet.2014.09.003
4. Graves, A., Schmidhuber, J.: Framewise phoneme classification with bidirectional LSTM and other neural network architectures. Neural Netw. **18**(5–6), 602–610 (2005). https://doi.org/10.1016/j.neunet.2005.06.042
5. Greff, K., Srivastava, R.K., Koutník, J., Steunebrink, B.R., Schmidhuber, J.: LSTM: a search space odyssey. IEEE Trans. Neural Netw. Learn. Syst. **28**(10), 2222–2232 (2017). https://doi.org/10.1109/TNNLS.2016.2582924
6. Hochreiter, S., Bengio, Y., Frasconi, P., Schmidhuber, J., et al.: Gradient flow in recurrent nets: the difficulty of learning long-term dependencies (2001)
7. Hochreiter, S., Schmidhuber, J.: Long short-term memory. Neural Comput. **9**(8), 1735–1780 (1997)
8. Hyndman, R.J., Akram, M.: Time series data library (2010). http://robjhyndman.com/TSDL
9. Hyndman, R.J., Khandakar, Y., et al.: Automatic time series for forecasting: the forecast package for R. No. 6/07, Monash University, Department of Econometrics and Business Statistics (2007). https://doi.org/10.18637/jss.v027.i03

10. Kotsiantis, S., Kanellopoulos, D., Pintelas, P.: Data preprocessing for supervised leaning. Int. J. Comput. Sci. **1**(2), 111–117 (2006)
11. Werbos, P.J.: Generalization of backpropagation with application to a recurrent gas market model. Neural Netw. **1**(4), 339–356 (1988). https://doi.org/10.1016/0893-6080(88)90007-X

# Human Activity Recognition

# Detecting Driver Drowsiness in Real Time Through Deep Learning Based Object Detection

Muhammad Faique Shakeel, Nabit A. Bajwa, Ahmad Muhammad Anwaar, Anabia Sohail, Asifullah Khan, and Haroon-ur-Rashid(✉)

Pakistan Institute of Engineering and Applied Sciences (PIEAS), Nilore, Islamabad 45650, Pakistan
haroon@pieas.edu.pk

**Abstract.** Vehicle accidents due to drowsiness in drivers take thousands of lives each year worldwide. This fact clearly exhibits a need for a drowsiness detection application that can help prevent such accidents and ultimately save lives. In this work, we propose a novel deep learning methodology based on Convolutional Neural Networks (CNN) to tackle this problem. The proposed methodology treats drowsiness detection as an object detection task, and from an incoming video stream of a driver, detects and localizes open and closed eyes. MobileNet CNN architecture with Single Shot Multibox Detector (SSD) is used for this task of object detection. A separate algorithm is then used to detect driver drowsiness based on the output from the MobileNet-SSD architecture. In order to train the MobileNet-SSD Network a custom dataset of about 6000 images was compiled and labeled with the objects face, eye open and eye closed. Out of these, 350 images were randomly separated and used to test the trained model. The trained model was evaluated on the test dataset using the PASCAL VOC metric and achieved a Mean Average Precision (mAP) of 0.84 on these categories. The proposed methodology, while maintaining reasonable accuracy, is also computationally efficient and cost effective, as it can process an incoming video stream in real time on a standalone mobile device without the need of expensive hardware support. It can easily be deployed on cheap embedded devices in vehicles, such as the Raspberry Pi 3 or a mobile smartphone.

**Keywords:** Drowsiness detection · Deep learning · Object detection · MobileNets · Single Shot Multibox Detector (SSD) · Android

## 1 Introduction

Vehicle crashes and accidents due to drowsy driving are prevalent all over the world. Thousands of people die every year resulting from vehicle accidents due to drowsy driving [1, 2]. Finland, Australia, England and other European countries have consistent crash reporting procedures and in data analyzed from these countries drowsy driving represents 10 to 30% of all crashes [3]. In order to reduce such accidents and


I. Rojas et al. (Eds.): IWANN 2019, LNCS 11506, pp. 283–296, 2019.
https://doi.org/10.1007/978-3-030-20521-8_24

enhance the safety of the driver and the passengers, driver drowsiness detection systems have been worked on and developed by various researchers all across the world.

These drowsiness detection systems can be broadly categorized to depend on the following methods [4, 5]: Vehicle Based, Behavioral Based, and Physiological Based. Vehicle based drowsiness detection systems work by monitoring the vehicle's lane changes, steering wheel rotation, speed, pressure on accelerator pedal etc. Behavior based drowsiness detection systems on the other hand depend on the behavior of the driver. To be more specific eye closure, yawn, and head posture are monitored through a camera to detect drowsiness in such systems. Lastly, physiological based drowsiness detection systems rely on the correlation between physiological signals ECG (Electrocardiogram) and EOG (Electrooculogram) to detect driver drowsiness.

All these categories of drowsiness detection system have their respective advantages and limitations. Physiological based drowsiness detection systems such as [6] have the limitation that the driver has to wear electrodes on his body that could prove to be a hindrance and an annoyance to the driver. Vehicle based drowsiness detection systems such as [7] are not robust because they are subjected to constraints related to the kind of driver and vehicle, road conditions etc. Hence, it is most practical to develop drowsiness detection systems based on the visual assessment of the drivers face as these systems do not require the driver to wear anything, and they can be implemented in any type of vehicle without modifications. Moreover, the current computer vision techniques based on convolutional neural networks enable one to develop highly robust systems.

Broadly, there are two main categories of computer vision techniques [8], traditional vision and deep learning. The traditional vision approaches extract human engineered features like edges, colors, corners, texture and hence depend on traditional image processing techniques. Among the popular traditional vision approaches are The Viola-Jones detector, The SIFT (Scale-Invariant Feature Transform) [9] algorithm, Spatial Pyramid Matching [10] and Histogram of oriented gradients (HOG) [11]. Contrary to this, deep learning based methods can automatically learn excellent abstract features by exploiting the underlying relationships in image data on their own. Hence, they provide a better representation of raw data that can be used for prediction purposes. Tiresome efforts on hand crafting features and designing the right filters are not needed in the case of deep learning methods. Deep learning methods are also very good at generalizing and hence do not suffer from the limitations of traditional computer vision techniques. For example, the Viola Jones detector requires upright face images to detect faces otherwise it won't give the desired performance, whereas deep learning models overcome this limitation. Moreover, almost all the traditional approaches listed above are largely ineffective against illumination changes, occlusion, deformation, background clutter etc. It is for this reason that deep learning based methods are now being employed to solve numerous computer vision tasks. In this work, we also employ a CNN architecture to develop the drowsiness detector application.

The main contribution of this work can be categorized in the following two aspects: (1) A new, more resource efficient, and accurate drowsiness detection methodology based on object detection using Convolutional Neural Networks (2) A new, annotated, Drowsy dataset to support the presented drowsiness detection methodology.

## 2 Literature Review

In this section, we attempt to briefly review the approaches used previously by researchers for vision-based drowsiness detection, along with their limitations. Eyelid closure is considered to be the most reliable indicator of drowsiness [12], and hence a lot of the systems developed seem to depend on eyelid closure for driver drowsiness detection. However, the following visual characteristics are also an indicator of driver drowsiness: a longer blink duration, yawning, slow eyelid movement, frequent nodding, fixed gaze, sluggish facial expression, and drooping posture [13].

Different algorithms have been presented in the past to detect drowsiness. Some use standard cameras [14], and some IR and stereo cameras [13, 15]. Horng et al. [16] localize the eyes using edge information, and track them using dynamical template matching to detect driver fatigue. In [13], 6 parameters are measured: Blink frequency, nodding frequency, eye closure duration, percent eye closure (PERCLOS), fixed gaze, and face position. These measured parameters are then combined using a fuzzy classifier to detect driver's drowsiness. In [17], yawning detection is done to determine if the driver is drowsy or not. Yawn and mouth regions are detected using a modified version of the Viola Jones algorithm, and the face is tracked using Kalman filter motion tracking. Danisman et al. [18] measure the distance between eyelids. This distance is then used to measure the level of drowsiness. The level of drowsiness is distinguished by the blink frequency, where the frequency increases as the person becomes sleepier and vice versa. In [19] a drowsiness detector is presented based on PERCLOS measurement, which is more robust against strong illumination variations. To reiterate, many of these methods depend on eyelid closure to detect drowsiness of the driver.

All of the above mentioned algorithms to detect drowsiness suffer from typical limitations of traditional image processing techniques, and hence may fail in varying illumination conditions, varying user appearance and fast head movements. Recently, Convolutional Neural Networks have truly revolutionized the field of computer vision, outperforming every other algorithm/technique in many applications such as image classification, object detection, emotion recognition, scene segmentation [20–23] etc.

Hence, convolutional neural networks have been employed to develop drowsiness detection systems in latest research. Dwivedi et al. [24] is one of the first attempts to have used convolutional neural networks to address this problem. As stated earlier, convolutional neural networks are able to learn an automated and efficient set of features that provide a better representation of the raw data, and hence help us classify the driver as drowsy or non-drowsy very accurately. However, [24] focused only on increasing the accuracy of drowsiness detection, and in real applications speed of the system is also a major concern.

Reddy et al. [25] presented an accurate but fast real time driver drowsiness system for embedded systems. This feat was achieved by compressing the convolutional network model. The drowsiness detection system based on two successive compressed deep neural networks in [25] was deployed to a Jetson TK1 GPU kit that is reasonably expensive (It costs $199.99 [26]). Even though the neural network was compressed in [25], the algorithm was still not efficient enough to be deployed on a Raspberry Pi 3 or mobile platform. Jabbar et al. [27] presented a compressed light weight deep neural network based method for real-time drowsiness detection on an Android device. This method is efficient enough to be deployed to an Android device, however, it achieves a reported accuracy of slightly more than 80%, which can be improved. In [28], Lyu et al. proposed a Long-term Multi-granularity Deep framework to detect driver drowsiness. They were able to achieve an accuracy of 90.05%, but the framework was not capable of being deployed to mobile devices due to its complexity.

## 3 Proposed Methodology

This paper presents a method to design and develop an accurate, cost-effective, and robust drowsiness detection system for real driving conditions. The research objectives were to develop a robust algorithm based on an accurate and resource efficient deep learning architecture, which could be deployed to a development board like that of a Raspberry Pi 3 or to a mobile platform such as Android/iOS. In short, our main goal was to develop a system that benefits from the unmatchable accuracies of convolutional neural networks in computer vision tasks while at the same be computationally resource efficient so that it could be deployed to cheaper embedded devices.

In order to achieve the goal of developing an accurate, resource efficient and cost-effective drowsiness detection system, we treated drowsiness detection as an object detection task. In our work, we employ the lightweight but accurate convolutional neural network architecture MobileNets [29], along with the Single Shot Multibox Detector (SSD) [30] framework on top of the MobileNet architecture. MobileNets are specifically designed for mobile vision applications, and hence were particularly well-suited for our task. MobileNet-SSD [31] is an object detection framework capable of detecting and localizing multiple objects in an image, in a single forward pass of the network. For our task of drowsiness detection, we trained it to detect human face, eye open and eye closed. Eye open and eye closed were treated as two separate objects. The incoming video stream was taken from a standard camera, as we work on the assumption that this task will be completed in daylight conditions only, and the training dataset was compiled as such on this assumption, as detailed further along in this paper. Based on these detections in a given time duration, we use a separate algorithm to determine whether the driver is drowsy or not drowsy.

The proposed methodology is accurately summarized in the flow chart below.

The threshold in the flowchart in Fig. 1 is a function of the image frames processed per second (FPS), and the longest duration of a blink was found to be equal to 400 ms [32].

$$\text{Threshold} = 0.4 \times \text{FPS} \tag{1}$$

Hence, if the processing is being done at 24 FPS, the Threshold value could come out to be 9.6–10. This means that if both of the eyes of the driver are detected to be closed for 10 successive frames in the incoming video stream (or in other words if the eyes are found to be closed for a slightly longer time than the longest average blink duration) the proposed system would declare the driver to be drowsy, and an alarm would be generated to waken the driver. This is done to prevent accidents at high speeds because in such situations a few seconds of carelessness can result in a fatal accident.

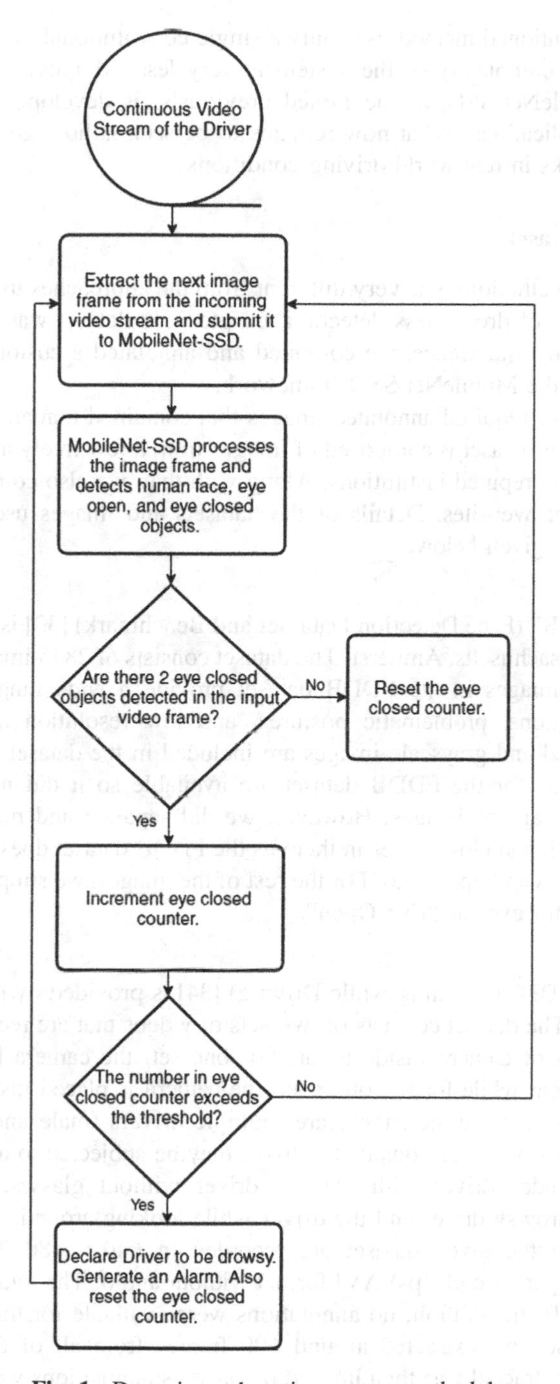


**Fig. 1.** Drowsiness detection system methodology

The above-mentioned method uses only a single convolutional neural network and hence the overall complexity of the system is very less. Moreover, the architecture being used, MobileNet-SSD, as mentioned previously, is developed specifically for mobile vision applications. What now remains to be seen is how good this proposed methodology works in real world driving conditions.

## 3.1 Drowsy Dataset

As our proposed methodology is very different from the approaches used previously to tackle the problem of drowsiness detection, no previous dataset was available which was tailored to our task. Hence, we compiled and annotated a custom dataset which was used to train the MobileNet-SSD framework.

For our task, we required annotated images that contained human faces, eye open, and eye closed. Our dataset is composed of images from a few freely available datasets online published by reputed institutions. Along with that, we also compiled data from online stock image websites. Details of the datasets and images used to create our custom dataset are given below.

**FDDB**

The dataset “FDDB” (Face Detection Data Set and Benchmark) [33] is provided by the University of Massachusetts, Amherst. The dataset consists of 2845 images that contain 5171 faces. The images in the FDDB dataset provide a wide range of difficulties including obstructions, problematic postures, and low resolution and out-of-focus faces. Both colored and grayscale images are included in the dataset.

The annotations for the FDDB dataset are available so it did not require us to manually annotate all the images. However, we did separate and manually annotate those images which had closed eyes in them as the FDDB dataset does not differentiate between closed eyes and open eyes. For the rest of the images, we simply re-labeled the given annotation for eyes as “Eye Open”.

**YawDD**

The dataset “YawDD” (Yawning while Driving) [34] is provided by the University of Ottawa, Canada. The dataset consists of two sets of videos that are recorded using two different locations of camera inside a car. For one set, the camera is placed on the dashboard of the car while for the other set, the camera is placed just underneath the front mirror of the car. Further, there are multiple drivers (male and female) in the videos. All the possible conditions that a driver may be subjected to are present in the videos. This includes driver with glasses, driver without glasses, smiling driver, yawning driver, drowsy driver and the driver while looking around.

The videos in the given dataset are recorded in $640 \times 480$ 24-bit true color (RGB) 30 frames per second (fps) AVI format without audio. The total data size of the data is about 5 GB. In addition, no annotations were available for this dataset.

For this dataset, we extracted around 600 frames from all of the videos. Each individual frame extracted was then labeled by hand as annotations were not provided.

**Closed Eyes in the Wild (CEW)**
The dataset closed eyes in the wild [35] was compiled by Department of Computer Science and Technology, Nanjing University of Aeronautics and Astronautics, China. The dataset Closed Eyes in the Wild contains all the images of humans having closed eyes.

The images are available in JPEG format. A total of 1192 images are available which is about 21 MBs. Out of the available images we used 137 images in our dataset. In addition to this, no annotations were available for this dataset. Since annotations for Closed Eyes in the Wild (CEW) dataset were not available so we hand-labeled each individual image.

**Custom Images Used**
In addition to the datasets listed above, our training dataset contained images acquired by us through web-search of different image databases. The images used were open-source and licensed for re-use. The purpose of using such images was to enhance our training dataset. We selected images such that our dataset had variety of illumination conditions, a variety of poses, and so that it was more diverse.

The images are available in JPEG format. A total of 2691 images are available which is about 235 MBs in size. As these were images downloaded through various web-sources, no annotations were available for them. For this dataset, we hand-labeled features in the images for training purposes.

It is to be noted that while the compiled custom dataset is labelled with the category "yawn", it is currently not being used in the proposed methodology to detect drowsiness.

The following Fig. 2 shows some images from the Drowsy Dataset.

It can be seen from the sample images above that the Drowsy dataset has been compiled with efforts to incorporate images from a wide variety of poses, angles and illumination conditions, so as to make a diverse dataset. This is done in order to achieve high accuracy and generalization ability of the object detection framework. This dataset, however, does not contain low light images as for now the task was limited to detecting drowsiness in daylight conditions.

### 3.2 Training Methodology

As the compiled dataset does not contain enough training images to train the object detection framework from scratch, the concept of transfer learning [36, 37] was utilized. A MobileNet-SSD model pre-trained on the MS COCO dataset [38] was taken from the TensorFlow Object Detection model zoo [31] and was first fine-tuned on the FDDB dataset to make it detect Face and Eyes in the image. After that, we further trained this model on our custom Drowsy dataset to develop the drowsiness detection system.

**Fig. 2.** Drowsy dataset images

### 3.3 Hardware and Software Environments

The model was trained on an NVIDIA GTX 1070 GPU. TensorFlow runtime version 1.6 was used for training and evaluation. A batch size of 7 was used during the training process.

## 4 Experimental Results

The trained MobileNet-SSD model was evaluated on the test dataset using the PASCAL VOC evaluation metric [39]. Average Precision (AP) was calculated at 0.5 Intersection over Union (IoU) ratio for each of the individual categories, and the AP are averaged to yield the Mean Average Precision (mAP). The results are summarized in Table 1 below.

**Table 1.** Results summary

| | AP @ 0.5 IoU eyes closed | AP @ 0.5 IoU eyes open | AP @ 0.5 IoU face | **mAP @ 0.5 IoU** |
|---|---|---|---|---|
| Drowsy trained MobileNet-v1-SSD | 0.776 | 0.763 | 0.971 | **0.837** |

Figure 3 below shows the trained model in action, with bounding boxes showing the confidence scores of the detections made by the model for a particular category at a particular instant in time.

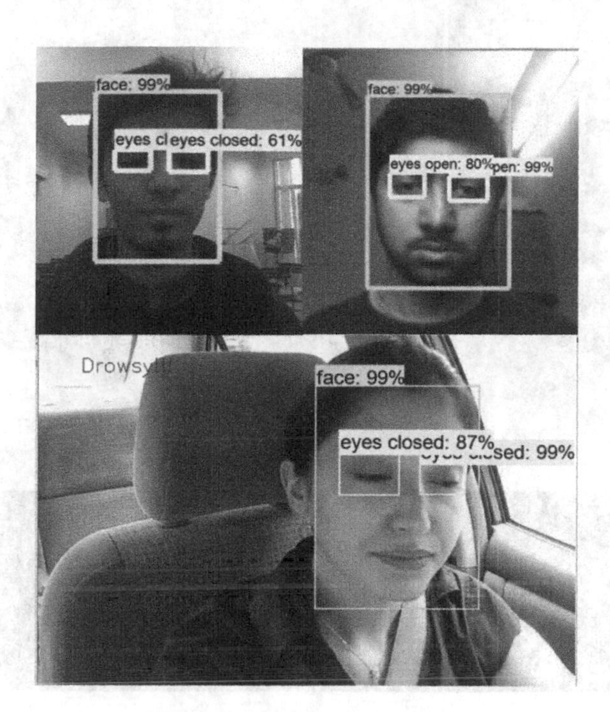


**Fig. 3.** Trained model in action

The trained model was also converted to a quantized TFLite model [40] for deployment to an Android mobile device. The deployment hardware used was a Sony Xperia Z smartphone [41], a somewhat outdated smartphone released in 2013. It is pertinent to mention that the smartphone used for testing was running Android OS 5.0, which does not support hardware acceleration using Android Neural Networks API. On the mentioned hardware, the trained and quantized TFLite model was able to process an incoming video frame in around 200 ms, which is deemed acceptable. On a high end phone like the Google Pixel 2, it is reported that 12–16 fps for inference can be achieved using TFLite quantized models for object detection and classification [42], but could not be independently tested on our given task.

Figures 4 and 5 illustrate the performance of the quantized TFLite model on the above mentioned Android smartphone.

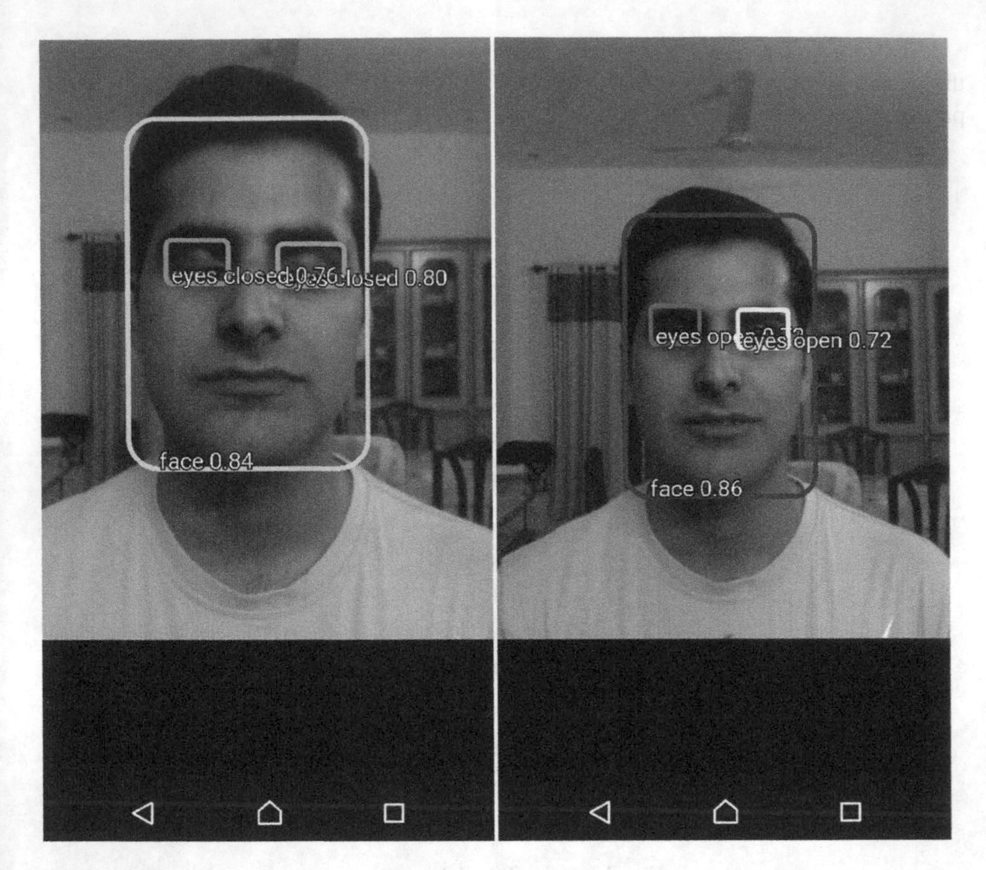


**Fig. 4.** Quantized model in action on an Android smartphone (Xperia Z)

## 5 Discussion

The custom Drowsy dataset trained object detection model was thoroughly tested in a wide variety of conditions, with varying illumination conditions, poses and occlusions, and it's performance in real world driving conditions was deemed outstanding. However, we found certain scenarios existed where its performance was not up to par. Situations where bright light was directed towards the camera lens in the background, or very low light conditions are examples of when the trained model did not perform well. However, this is deemed acceptable because the dataset compiled does not contain any images in low light conditions and hence, we do not expect it to work well in these conditions either.

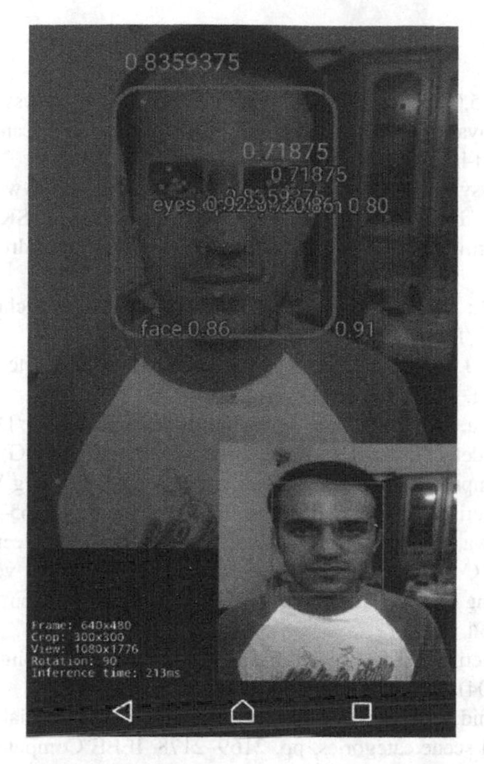


**Fig. 5.** Inference time for detections in the quantized TFlite model on the Sony Xperia Z

## 6 Conclusion and Future Works

Based on the results on the test dataset and real-world testing in a variety of illumination and occlusion conditions, we conclude that the proposed methodology of treating the task of drowsiness detection as an object detection task is practical and reliable. Moreover, this technique works in real time on an Android device, which makes it very accessible.

One major improvement that could be done in the future is that of enhancing the Drowsy dataset and adding low light images (in near infra-red light) to enable the model to detect drowsiness in low light conditions. This would be beneficial as drowsiness related accidents have a high chance of occurrence during night-time driving. Moreover, work needs to be done in incorporating yawning information (which is labeled in the dataset but not used in this methodology) in our algorithm to improve the accuracy and reliance of the drowsiness detection system.

Another area for improvement could be to modify the MobileNet-SSD architecture to better suit the drowsiness detection application.

Lastly, we would like to acknowledge the efforts of the FDDB, YawDD and CEW dataset creators for providing us with the data without which this work would not have been possible.

## References

1. Mohn, T.: Around 5,000 people were killed last year due to drowsy driving. https://www.forbes.com/sites/tanyamohn/2016/08/08/nearly-83-6-million-american-drivers-are-sleep-deprived-new-report-highlights-dangers-high-cost/#3bbc82664007
2. Rapaport, L.: Drowsy drivers often behind fatal crashes. https://www.reuters.com/article/us-health-driving-sleep/drowsy-drivers-often-behind-fatal-crashes-idUSKBN15P2PM
3. Facts and Stats: Drowsy driving – stay alert, arrive alive. http://drowsydriving.org/about/facts-and-stats/
4. Saini, V., Saini, R.: Driver drowsiness detection system and techniques: a review. Int. J. Comput. Sci. Inf. Technol. **5**, 4245–4249 (2014)
5. Bhatt, P.P., Trivedi, J.A.: Various methods for driver drowsiness detection: an overview. Int. J. Comput. Sci. Eng. **9**, 70–74 (2017)
6. Chieh, T.C., Mustafa, M.M., Hussain, A., Hendi, S.F., Majlis, B.Y.: Development of vehicle driver drowsiness detection system using Electrooculogram (EOG). In: 1st International Conference on Computers, Communications, and Signal Processing With Special Track on Biomedical Engineering (CCSP), Kuala Lumpur, Malaysia, pp. 165–168 (2005)
7. Takei, Y., Furukawa, Y.: Estimate of driver's fatigue through steering motion. In: 2005 IEEE International Conference on Systems, Man and Cybernetics, vol. 2, pp. 1–6 (2005)
8. Lee, A.: Comparing deep neural networks and traditional vision algorithms in mobile robotics. Swart. Coll. (2015)
9. Lowe, D.G.: Distinctive image features from scale-invariant keypoints. Int. J. Comput. Vis. **60**(2), 91–110 (2004)
10. Lazebnik, S., Schmid, C., Ponce, J.: Beyond bags of features: spatial pyramid matching for recognizing natural scene categories, pp. 2169–2178. IEEE Computer Society (2006)
11. Dalal, N., Triggs, B.: Histograms of oriented gradients for human detection. In: Conference on Computer Vision and Pattern Recognition (2005)
12. Bittner, R., Hána, K., Poušek, L., Smrka, P., Schreib, P., Vysoký, P.: Detecting of fatigue states of a car driver. In: Brause, R.W., Hanisch, E. (eds.) ISMDA 2000. LNCS, vol. 1933, pp. 260–273. Springer, Heidelberg (2000). https://doi.org/10.1007/3-540-39949-6_32
13. Bergasa, L.M., Nuevo, J.: Real-time system for monitoring driver vigilance. In: Proceedings of the IEEE International Symposium on Industrial Electronics, vol. III, pp. 1303–1308 (2005)
14. Suzuki, M., Yamamoto, N., Yamamoto, O., Nakano, T., Yamamoto, S.: Measurement of driver's consciousness by image processing - a method for presuming driver's drowsiness by eye-blinks coping with individual differences. In: Conference Proceedings - International Conference on Systems, Man and Cybernetics, vol. 4, pp. 2891–2896 (2007)
15. Technology | Smart Eye. http://smarteye.se/technology/
16. Fan, C.: Driver fatigue detection based. In: Proceedings of the 2004 IEEE International Conference on Networking, Sensing and Control, pp. 7–12 (2004)
17. Abtahi, S., Hariri, B., Shirmohammadi, S.: Driver drowsiness monitoring based on yawning detection. In: IEEE International Instrumentation and Measurement Technology Conference, Binjiang, Hangzhou, China (2011)
18. Danisman, T., Bilasco, I.M., Djeraba, C., Ihaddadene, N.: Drowsy driver detection system using eye blink patterns. In: Proceedings of the 2010 International Conference on Machine and Web Intelligence, ICMWI 2010, pp. 230–233 (2010)
19. Bronte, S., Bergasa, L.M., Almaz, J., Yebes, J.: Vision-based drowsiness detector for real driving conditions (2012)

20. Krizhevsky, A., Sutskever, I., Hinton, G.E.: ImageNet classification with deep convolutional neural networks. In: Neural Information Processing Systems (2012)
21. He, K., Zhang, X., Ren, S., Sun, J.: Deep residual learning for image recognition. In: Conference on Computer Vision and Pattern Recognition (2016)
22. Ren, S., He, K., Girshick, R., Sun, J.: Faster R-CNN: towards real-time object detection with region proposal networks. In: Neural Information Processing Systems, pp. 1–14 (2015)
23. Shelhamer, E., Long, J., Darrell, T.: Fully convolutional networks for semantic segmentation. IEEE Trans. Pattern Anal. Mach. Intell. **39**, 640–651 (2017)
24. Dwivedi, K., Biswaranjan, K., Sethi, A.: Drowsy driver detection using representation learning. In: Souvenir 2014 IEEE International Advanced Computing Conference, IACC 2014, pp. 995–999 (2014)
25. Reddy, B., Kim, Y.-H., Yun, S., Seo, C., Jang, J.: Real-time driver drowsiness detection for embedded system using model compression of deep neural networks. In: Conference on Computer Vision and Pattern Recognition Workshops (CVPRW), pp. 438–445 (2017)
26. Amazon.com: NVIDIA Jetson TK1 Development Kit: Computers & Accessories. https://www.amazon.com/NVIDIA-Jetson-TK1-Development-Kit/dp/B00L7AWOEC
27. Jabbar, R., Al-Khalifa, K., Kharbeche, M., Alhajyaseen, W., Jafari, M., Jiang, S.: Real-time driver drowsiness detection for android application using deep neural networks techniques. Procedia Comput. Sci. **130**, 400–407 (2018)
28. Lyu, J., Yuan, Z., Chen, D.: Long-term multi-granularity deep framework for driver drowsiness detection (2018)
29. Howard, A.G., et al.: MobileNets: efficient convolutional neural networks for mobile vision applications (2017)
30. Liu, W., et al.: SSD: single shot MultiBox detector. In: Leibe, B., Matas, J., Sebe, N., Welling, M. (eds.) ECCV 2016. LNCS, vol. 9905, pp. 21–37. Springer, Cham (2016). https://doi.org/10.1007/978-3-319-46448-0_2
31. Huang, J., et al.: Speed/accuracy trade-offs for modern convolutional object detectors. In: Proceedings - 30th IEEE Conference on Computer Vision and Pattern Recognition, CVPR 2017, pp. 3296–3305, January 2017
32. Schiffman, H.R.: Sensation and Perception: An Integrated Approach. Wiley, New York (2000)
33. Jain, V., Learned-Miller, E.: FDDB: a benchmark for face detection in unconstrained settings. UM-CS-2010-009 (2010)
34. Abtahi, S., Omidyeganeh, M., Shirmohammadi, S., Hariri, B.: YawDD. In: Proceedings of the 5th ACM Multimedia Systems Conference on - MMSys 2014, pp. 24–28. ACM Press, New York (2014)
35. Song, F., Tan, X., Liu, S.: Eyes closeness detection from still images with multi-scale histograms of principal oriented gradients (2014). http://parnec.nuaa.edu.cn/xtan/data/ClosedEyeDatabases.html
36. Pan, S.J., Fellow, Q.Y.: A survey on transfer learning. IEEE Trans. Knowl. Data Eng. **22**, 1345–1359 (2010)
37. Oquab, M., Bottou, L., Laptev, I., Sivic, J.: Learning and transferring mid-level image representations using convolutional neural networks. In: IEEE Conference on Computer Vision and Pattern Recognition, pp. 1717–1724 (2014)
38. Lin, T.-Y., et al.: Microsoft COCO: common objects in context. In: Fleet, D., Pajdla, T., Schiele, B., Tuytelaars, T. (eds.) ECCV 2014. LNCS, vol. 8693, pp. 740–755. Springer, Cham (2014). https://doi.org/10.1007/978-3-319-10602-1_48

39. Everingham, M., et al.: The PASCAL Visual Object Classes (VOC) challenge. Int. J. Comput. Vis. **88**, 303–338 (2010)
40. TensorFlow Lite | TensorFlow. https://www.tensorflow.org/lite/
41. Xperia TM Z (2013)
42. Krishnamoorthi, R.: Quantizing deep convolutional networks for efficient inference: a whitepaper (2018)

# The Influence of Human Walking Activities on the Doppler Characteristics of Non-stationary Indoor Channel Models

Muhammad Muaaz(✉), Ahmed Abdelgawwad, and Matthias Pätzold(✉)

Faculty of Engineering and Science, University of Agder, P.O. Box 509, 4898 Grimstad, Norway
{muhammad.muaaz,ahmed.abdel-gawwad,matthias.paetzold}@uia.no

**Abstract.** This paper analyzes the time-variant (TV) Doppler power spectral density of a 3D non-stationary fixed-to-fixed indoor channel simulator after feeding it with realistic trajectories of a walking person. The trajectories of the walking person are obtained by simulating a full body musculoskeletal model in OpenSim. We provide expressions of the TV Doppler frequencies caused by these trajectories. Then, we present the complex channel gain consisting of fixed scatterers and a cluster of moving scatterers. After that, we use the concept of the spectrogram to analyze the TV Doppler power spectral density of the complex channel gain. Finally, we present expressions of the TV mean Doppler shift and Doppler spread. The work of this paper is important for human activity recognition systems using radio-frequency (non-wearable) sensors as the demand for such systems has increased nowadays.

**Keywords:** Non-stationary channel model · OpenSim · Musculoskeletal model · Dynamic simulation · Spectrogram · Time-frequency distributions

## 1 Introduction

Human activity recognition (HAR) aims at inferring human activities from body motion and gesture data recorded by different types of wearable and non-wearable sensors. Systems with the ability to automatically recognize human activities can significantly improve and simplify our daily living in an increasingly complex society. Motivated by this, HAR is not only a well-researched, but still a very active research area mainly due to its dynamic nature, wide variety of applications, advancements in learning algorithms, and developments in sensing technologies. Depending on the type of sensors, HAR systems can be divided into vision-based [4,9], sensor-based [7,18], and device-free systems [14,15].

Vision-based systems record and process videos and still images of users to recognize their activities. They require a line-of-sight (LOS) link and may lead to possible violations of the user's privacy.


I. Rojas et al. (Eds.): IWANN 2019, LNCS 11506, pp. 297–309, 2019.
https://doi.org/10.1007/978-3-030-20521-8_25

Sensor-based systems use inexpensive sensors, such as inertial measurement units (IMUs) to capture human movement data. However, users need to carry these sensors all the time for continuous activity recognition. Furthermore, the performance of sensor-based systems is susceptible to the placement of the sensors on the human body. In the device-free approaches, such as radio-frequency (RF) sensing, the transmitter and receivers are placed in the environment and they continuously transmit and receive RF signals. These RF signals are very sensitive to the change in the domestic environment. Human movements in the environment introduce fluctuations in the RF signals. These variation-enriched RF signals are picked up by the receivers and used for activity recognition [15]. Since device-free HAR systems suffer less from privacy issues and do not require the users to carry sensors all the time, the device-free approach has become a preferred choice for applications such as health and indoor HAR systems for the elderly.

In recent years, various device-free HAR systems have been developed [14–17]. Although the existing device-free HAR systems have shown promising results, they face a major challenge, namely *limited capability to deal with changes in the environment* [6]. This is because RF signals picked up by the receivers usually carry information about the moving and non-moving scatterers[1] present in the environment. As a result, a HAR system trained by using the data collected from specific subjects in a specific environment may not perform well when applied to a different environment to recognize activities of other individuals. To overcome this challenge, researchers have proposed several ideas. For instance, Jiang et al. [17] used environment and subject-independent features to train HAR models. This approach requires training data to be collected from various subjects under different environmental settings. In [16], a semi-supervised learning approach is used to address this issue, where users are required to manually label the new instances of recorded channel state information (CSI) data upon detecting changes in the CSI-fingerprints of the activities. This approach requires manual interaction from users which is not feasible in various environments, such as public places and hospitals.

Radio communication researchers have proposed a three-dimensional (3D) indoor channel modeling[2] technique, which captures the Doppler effect caused by the body movements, for device-free HAR [1,2]. The basic idea is that fixed scatterers and different movements of moving scatterers affect the Doppler characteristics of the channel differently. After filtering out the effect of fixed scatterers on the Doppler characteristics of the channel, we can obtain Doppler signatures for different activities, which can be used for environment-independent activity recognition. The additional advantage of this approach is that the measurements from the simulated channel can be used to train the learning algorithms without additional efforts for the collection of training data.

[1] Here scatterers mean objects present in the environment. When a radio wave hits an object with a rough surface, then the wave will be redirected in many directions depending on the slope of the object.

[2] A channel model is a physical layer model of the mathematical representation of the effects of a communication channel through which RF signals propagate.

The contribution of this paper is the analysis of the Doppler power spectrum characteristics of a 3D non-stationary fixed-to-fixed (F2F) indoor channel simulator, which is fed with the trajectories of major body parts. For this purpose, we first simulate the human walking activity by using a full-body musculoskeletal model to obtain trajectories of different body parts, e.g., torso, pelvis, upper arms, lower arms, hands, upper legs, lower legs, and toes. Then, we model these moving body parts of the walking human by a cluster of synchronized moving scatterers. After that, we compute the time-variant (TV) Doppler frequencies caused by the trajectories of these major body parts. Furthermore, the expression of the instantaneous channel phase of each moving scatterer of the cluster is provided. Next, we present an expression of the complex channel gain of the 3D non-stationary F2F multipath fading channel with fixed scatterers and a cluster of synchronized moving scatterers. In addition, expressions of the TV mean Doppler shift and TV Doppler spread are presented. In this paper, we use the spectrogram approach to analyze the influence of a cluster of moving scatterers on the TV Doppler power spectrum. The spectrogram consists of two terms: the auto-term and the cross-term. The auto-term represents the desired Doppler power spectrum, while the cross-term represents an undesired interference component, which reduces the resolution of the spectrogram. Finally, we provide expressions for the TV mean Doppler shift and TV Doppler spread computed by using the spectrogram.

The rest of this paper is organized as follows. Section 2, succinctly explains the dynamic simulation of human walking activities. Section 3 presents the TV Doppler frequencies caused by the walking person, the complex channel gain, TV mean Doppler shift, TV Doppler spread, and the spectrogram of the complex channel gain. Section 4 discusses the numerical results of the spectrogram of the complex channel gain fed by the trajectories of the body parts of a walking human. Section 5 summarizes our work and proposes possible extensions for future work.

## 2 Simulation of Human Walking Activities

To study the influence of human walking activities on the Doppler characteristics of the indoor channel model, we need to obtain trajectories of human body parts during walking activities. Instead of defining and validating our own trajectory models for different human body parts, in this paper, we have used an OpenSim-based full-body musculoskeletal model to obtain realistic trajectories of human body parts.

### 2.1 Overview of OpenSim and Its General Workflow

OpenSim [5] is a publicly available open-source suit of tools for modeling musculoskeletal structures and analyzing dynamic simulation of a wide range of human movements [3] in rehabilitation science [8], sports science [10], and robotics research [11]. OpenSim provides a large repository of musculoskeletal models consisting of rigid bodies, joints, and specialized forces. The skeleton is modeled by rigid bodies which are interconnected by joints. Joints define the motion of a rigid body, with respect to its parent rigid body and specialized force elements are used to model the muscles of human bodies.

The first step in simulating any human movement in OpenSim is to formally define the dynamic musculoskeletal model and its interactions with the environment. Once the model is formulated, it is scaled by using the scaling tool of OpenSim. In the scaling process, the dimensions and mass properties of each body segment as well as the musculotendon properties (e.g., muscle fiber length) of the dynamic musculoskeletal model are scaled to match with the anthropometric data of real subjects [3,5].

In the next step, the inverse kinematics (IK) tool is used to find the generalized coordinates (including joint angels and positions) of the dynamic musculoskeletal model. At each time step, the IK tool computes the generalized coordinates that describes the model in a pose which "best matches" the experimental/motion-capture data at each time step [3,5]. At this stage, the measured ground reaction forces (GRFs) and joint moments are usually inconsistent with the model kinematics. To overcome this issue a residual reduction algorithm (RRA) is applied. The goal of the RRA is to make measured GRFs and moments more consistent with the model kinematics [3,5]. Thereafter, the computed muscle control (CMC) tool is used to compute muscle excitations that drive the generalized coordinated muscle-driven simulation of the movement [3,5].

### 2.2 Trajectories of Body Parts of a Moving Person

In this paper, we have employed a validated full-body musculoskeletal model and the experimental motion data from a previously published work by Rajagopal et al. [13]. This model consists of a bony geometry of the full body, 37 degrees of freedom (DoF), hill-type muscle models to model 80 musculotendon units actuating the lower body, and 17 ideal torque actuators for the upper body [13], as shown in Fig. 1.

The authors in this study [13] collected motion data using 41 retroreflective markers measured at a sampling rate of 100 Hz by using an eight-camera optical motion-capture system. GRFs and moments were measured with ground force plates at a sampling rate of 2000 Hz, and the EMG data were recorded by placing wireless surface electrodes on 10 different muscles. Further details about the dynamic model and data collection process can be found in [13]. First, the full-body musculoskeletal model is scaled according to the experimental data, then the general workflow of OpenSim (see Sect. 2.1) is followed to generate the dynamic simulation of the full gait cycle. Finally, the body kinematics analysis is performed to obtain the TV kinematics (including position and velocity) of the center of mass (CoM) of each body segment defined in the full body musculoskeletal model as shown in Fig. 2. The TV velocity vectors $\boldsymbol{v}_n(t)$ of CoM of each body segment are fed to the channel simulator described in Sect. 3 to compute the complex channel gain of the RF channel in presence of a walking human.

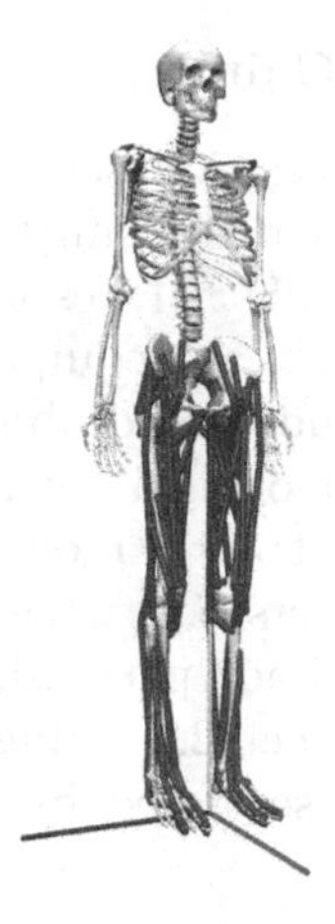

**Fig. 1.** A full body musculoskeletal model [x-axis (red), y-axis (green), and z-axis (blue)]. (Color figure online)

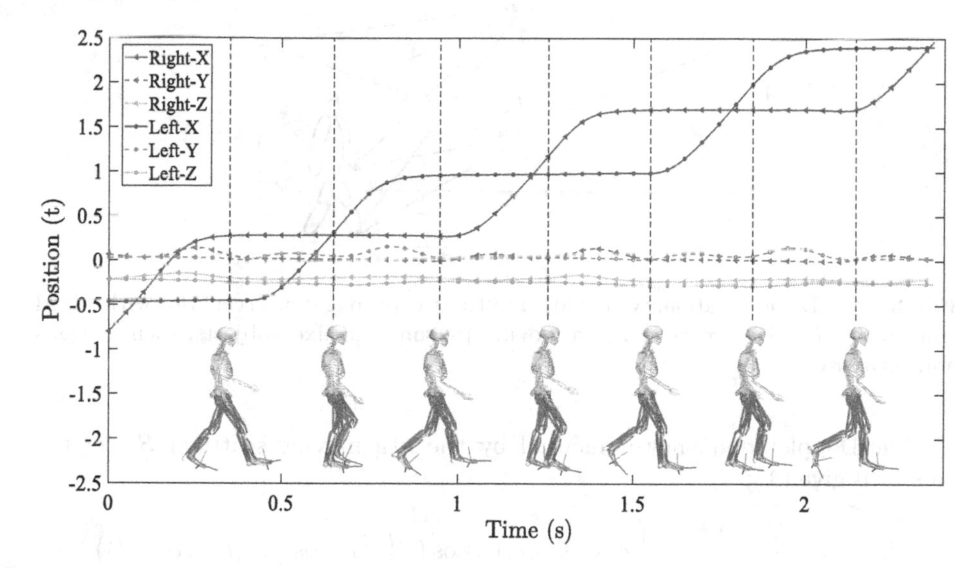


**Fig. 2.** 3D displacement of the CoM of the right and left toes during walking.

# 3 Monitoring Human Activity Using the Spectrogram

The trajectories of the major body parts discussed in Sect. 2 vary in time and space. In this section, we explore the impact of these temporal and spatial variations on the Doppler characteristics of the received signal. This is done by simulating the complex channel gain of F2F indoor channels with these trajectories generated in Sect. 2.2 as inputs. Then, the Doppler spectral characteristics of the complex channel gain are analyzed by using the concept of the spectrogram.

### 3.1 The Complex Channel Gain

Figure 3 shows a multipath propagation scenario of an F2F 3D indoor channel. In this scenario, we have a fixed transmitter ($T_x$) and a fixed receiver ($R_x$) located at $(x^T, y^T, z^T)$ and $(x^R, y^R, z^R)$, respectively. We assume that the LOS is obstructed and both $T_x$ and $R_x$ are equipped with omnidirectional antennas. The considered scenario includes fixed objects, such as walls and furniture. These fixed objects are simply modelled by $M$-fixed point scatterers $S_m^{\mathrm{F}}$ for $m = 1, 2, \ldots, \mathcal{M}$. Figure 3 also shows a moving person modelled by a cluster of $N$-synchronized moving scatterers $S_n^{\mathrm{M}}$, where $n = 1, 2, \ldots, \mathcal{N}$. Each moving scatterer $S_n^{\mathrm{M}}$ represents a moving body part, which has a trajectory described by its TV velocity vector $\boldsymbol{v}_n(t)$ (see Sect. 2.2). Single bounce scattering is assumed, i.e., each wave arriving at $R_x$ is scattered by either a fixed scatterer $S_m^{\mathrm{F}}$ or a moving scatterer $S_n^{\mathrm{M}}$.

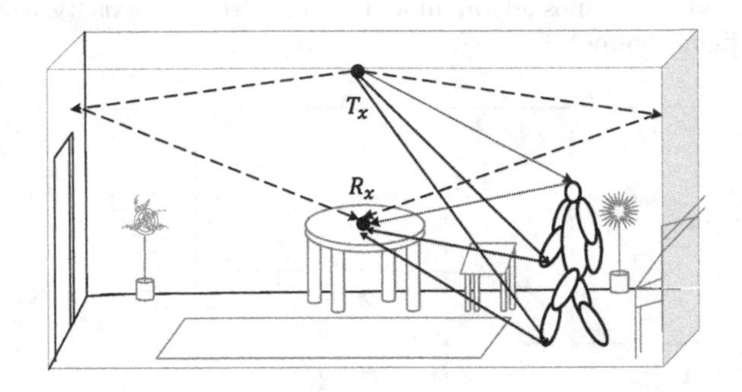


**Fig. 3.** A 3D non-stationary indoor multipath propagation scenario with fixed transmitter $T_x$, fixed receiver $R_x$, a moving person, and fixed objects, such as walls and furniture.

The Doppler frequency influenced by the $n$th moving scatterer $S_n^{\mathrm{M}}$ of the cluster is given by [1]

$$f_n(t) = -\frac{\mathrm{V}_n(t)\, f_0}{c_0} \times \Big\{ \cos(\beta_{v_n}(t)) \left[ \cos\left(\beta_n^T(t)\right) \cos\left(\alpha_n^T(t) - \alpha_{v_n}(t)\right) \right.$$
$$\left. + \cos\left(\beta_n^R(t)\right) \cos\left(\alpha_{v_n}(t) - \alpha_n^R(t)\right) \right]$$
$$+ \sin(\beta_{v_n}(t)) \times \left[ \sin\left(\beta_n^T(t)\right) + \sin\left(\beta_n^R(t)\right) \right] \Big\} \tag{1}$$

where the functions $\beta_{v_n}(t)$, $\alpha_{v_n}(t)$, $\beta_n^T(t)$, $\alpha_n^T(t)$, $\beta_n^R(t)$, and $\alpha_n^R(t)$ denote the TV horizontal angle of motion (HAOM), vertical angle of motion (VAOM), elevation angle of departure (EAOD), vertical angle of departure (AAOD), elevation angle of arrival (EAOA), and azimuth vertical angle of (AAOA) of the $n$th moving scatterer $S_n^{\mathrm{M}}$, respectively. The expressions of these functions are obtained by using the inverse trigonometric functions. Further details about these expressions are provided in [1]. The function $\mathrm{V}_n(t) = |\boldsymbol{v}_n(t)|$ designates the TV speed of the motion of the $n$th moving scatterer of the cluster. The parameters $f_0$ and $c_0$ are the carrier frequency of the signal and the speed of light, respectively. The instantaneous channel phase influenced by the $n$th moving scatterer of the cluster is expressed as [12]

$$\theta_{n,\mathrm{M}}(t) = 2\pi \int_{-\infty}^{t} f_n(t')dt' = 2\pi \int_{0}^{t} f_n(t')dt' + \theta_{n,\mathrm{M}} \tag{2}$$

where the first term of the right-hand side of (2) is the TV phase shift influenced by the motion of the $n$th moving scatterer of the cluster of the moving person. The second term $\theta_{n,\mathrm{M}}$ designates the initial phase shift which is modelled as a zero-mean random variable with uniform distribution with values from $-\pi$ to $\pi$, i.e., $\theta_{n,\mathrm{M}} \sim \mathcal{U}(-\pi, \pi]$. Thus, the instantaneous channel phase $\theta_{n,\mathrm{M}}(t)$ in (2) is a stochastic process. The complex channel gain $\mu(t)$ that consists of $\mathcal{N}+\mathcal{M}$ received multipath components can be expressed as [1]

$$\mu(t) = \sum_{n=1}^{\mathcal{N}} c_{n,\mathrm{M}}\, e^{j\theta_{n,\mathrm{M}}(t)} + \sum_{m=1}^{\mathcal{M}} c_{m,\mathrm{F}}\, e^{j\theta_{m,\mathrm{F}}}. \tag{3}$$

The first term in (3) represents the superposition of the received $\mathcal{N}$ waves corresponding to the moving scatterers. Each path in the first term of (3) is described by a constant path gain $c_{n,\mathrm{M}}$ and a stochastic phase process $\theta_{n,\mathrm{M}}(t)$ due to the motion of the moving scatterer. The second term represents the superposition of $\mathcal{M}$ received multipath components originating from the fixed scatterers. Each component in the second term is associated with a constant path gain $c_{m,\mathrm{F}}$ and a random phase $\theta_{m,\mathrm{F}}$ caused by the interaction with the fixed scatterer. The random variables $\theta_{n,\mathrm{M}}$ and $\theta_{m,\mathrm{F}}$ are independent and identically distributed (i.i.d.) with uniform distribution from $-\pi$ and $\pi$, i.e., $\theta_{m,\mathrm{F}}, \theta_{n,\mathrm{M}} \sim \mathcal{U}(-\pi, \pi]$. The expression of the complex channel gain $\mu(t)$ given by (3) is a stochastic model for a 3D non-stationary F2F multipath fading channel with fixed scatterers and a cluster of moving scatterers. The TV mean Doppler shift and TV Doppler spread of the model described by (3) are given by [12]

$$B_f^{(1)}(t) = \frac{\sum_{n=1}^{\mathcal{N}} c_{n,\mathrm{M}}^2 f_n(t)}{\sum_{n=1}^{\mathcal{N}} c_{n,\mathrm{M}}^2 + \sum_{m=1}^{\mathcal{M}} c_{m,\mathrm{F}}^2} \tag{4}$$

and

$$B_f^{(2)}(t) = \sqrt{\frac{\sum_{n=1}^{\mathcal{N}} c_{n,\mathrm{M}}^2 f_n^2(t)}{\sum_{n=1}^{\mathcal{N}} c_{n,\mathrm{M}}^2 + \sum_{m=1}^{\mathcal{N}} c_{m,\mathrm{F}}^2} - \left(B_f^{(1)}(t)\right)^2} \tag{5}$$

respectively.

### 3.2 Spectrogram of the Complex Channel Gain

To compute the TV Doppler power spectrum by using the spectrogram approach, a sliding window is required. In this paper, a Gaussian window described by

$$h(t) = \frac{1}{\sqrt{\sigma_w \sqrt{\pi}}} e^{-\frac{t^2}{2\sigma_w^2}} \quad (6)$$

is used, where the paramcter $\sigma_w$ is called the window spread. The window function $h(t)$ is positive, even, and has normalized energy, i.e., $\int_{-\infty}^{\infty} h^2(t) = 1$. Then, we compute the Fourier transform of the multiplication of the complex channel gain and the sliding window to obtain the short-time Fourier transform (STFT) $X(f,t)$. Finally, the STFT $X(f,t)$ is multiplied by its complex conjugate, which defines the spectrogram $S_\mu(f,t)$ as

$$S_\mu(f,t) \ = \ |X(f,t)|^2 \ = \ S_\mu^{(a)}(f,t) + S_\mu^{(c)}(f,t). \quad (7)$$

The spectrogram $S_\mu(f,t)$ in (7) consists of two terms, the auto-term $S_\mu^{(a)}(f,t)$ and the cross-term $S_\mu^{(c)}(f,t)$. The auto-term $S_\mu^{(a)}(f,t)$ represents the desired TV Doppler power spectral density. It is a real and positive function. The cross-term $S_\mu^{(c)}(f,t)$ represents an undesired spectral interference component, which is also real, but not necessarily positive. It reduces the resolution of the spectrogram. Further details about the expressions of $X(f,t)$, $S_\mu^{(a)}(f,t)$, and $S_\mu^{(c)}(f,t)$ can be found in [1]. It should be noted that the cross-term can be eliminated by taking the average of the spectrogram $S_\mu(f,t)$ over the random phase variables $\theta_{n,\mathrm{M}}$ and $\theta_{m,\mathrm{F}}$, i.e., $E\left\{S_\mu(f,t)\right\}|_{\theta_{n,\mathrm{M}},\theta_{m,\mathrm{F}}} = S_\mu^{(a)}(f,t)$. The TV mean Doppler shift and Doppler spread can be computed from the spectrogram using the following expressions [2]

$$B_\mu^{(1)}(t) = \frac{\int_{-\infty}^{\infty} f S_\mu(f,t)df}{\int_{-\infty}^{\infty} S_\mu(f,t)df} \quad (8)$$

and

$$B_\mu^{(2)}(t) = \sqrt{\frac{\int_{-\infty}^{\infty} f^2 S_\mu(f,t)df}{\int_{-\infty}^{\infty} S_\mu(f,t)df} - \left(B_\mu^{(1)}(t)\right)^2} \quad (9)$$

respectively.

# 4 Numerical Results

In this section, we will discuss the numerical results of the spectrogram $S_\mu(f,t)$, the auto-term $S_\mu^{(a)}(f,t)$ of the spectrogram, the TV mean Doppler shift, and TV Doppler spread. For the simulation scenario in Fig. 4, we have chosen a room with the dimensions $10\,\mathrm{m} \times 5\,\mathrm{m} \times 2.4\,\mathrm{m}$. For the locations of $T_x$ and $R_x$, we chose the coordinates $(5\,\mathrm{m},\ 2.5\,\mathrm{m},\ 2.3\,\mathrm{m})$ and $(5\,\mathrm{m},\ 3\,\mathrm{m},\ 2.3\,\mathrm{m})$, respectively. The number of the fixed scatterers $\mathcal{M}$ was chosen to be 6. With respect to Fig. 4, a person is walking parallel to the $x$-axis, towards the positive direction as a cluster of 18 moving scatterers, i.e., $\mathcal{N} = 18$. The path gains of each moving and fixed scatterer have been computed by

$$c_{n,\mathrm{M}} = \sqrt{\frac{2\eta}{\mathcal{N}}} \quad \text{and} \quad c_{m,\mathrm{F}} = \sqrt{\frac{2\,(1-\eta)}{\mathcal{M}}} \tag{10}$$

where the parameter $\eta$ is used to balance the contribution of the moving scatterers in the cluster and the fixed scatterers to the fading power. Its value was set to 0.6. The value for the carrier frequency $f_0$ was chosen to be 5.9 GHz. The indoor F2F channel simulator was monitoring the walking person for 6 s. The TV VAOMs $\beta_{v_n}(t)$ and TV HAOMs $\alpha_{v_n}(t)$ were computed from the TV velocities of the moving scatterers. The TV EAODs $\beta_n^T(t)$ and AAODs $\alpha_n^T(t)$ of the moving scatterers were calculated from their TV displacements and the location of $T_x$. Also, the TV EAOAs $\beta_n^R(t)$ and AAOAs $\alpha_n^R(t)$ of the moving scatterers were computed from their TV displacements and the location of $R_x$. Then, the Doppler frequencies of the moving scatterers were computed using (1). For computing the spectrogram, the window spread parameter $\sigma_w$ of the Gaussian window function $h(t)$ was set to 0.01 s. The values of the random channel phases $\theta_{n,\mathrm{M}}$ and $\theta_{m,\mathrm{F}}$ were obtained as outcomes from random generators with uniform distributions from $-\pi$ and $\pi$.

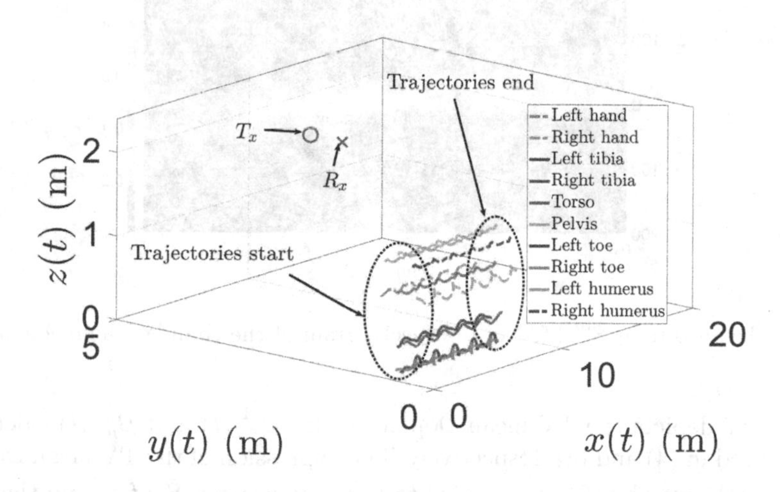


**Fig. 4.** 3D indoor simulation scenario showing the trajectories of the body parts (moving scatterers) of a walking person.

Figure 5 depicts the spectrogram $S_\mu(f,t)$ given by (7), showing the variations of the TV Doppler power spectrum. However, they are blurred due to the effect of the cross-term $S_\mu^{(c)}(f,t)$, which reduces the resolution of the spectrogram $S_\mu(f,t)$. Figure 6 shows the auto-term $S_\mu^{(a)}(f,t)$ of the spectrogram. The influence of the moving cluster on the TV Doppler power spectrum is clearly visible. The Doppler frequencies approach zero values at $t \approx 3.2$ s. The reason is that, at this time instant, the values of the TV EAODs $\beta_n^T(t)$, AAODs $\alpha_n^T(t)$, TV EAOAs $\beta_n^R(t)$, AAOAs $\alpha_n^R(t)$ of the moving scatterers tend to $-\pi/2$ rad.

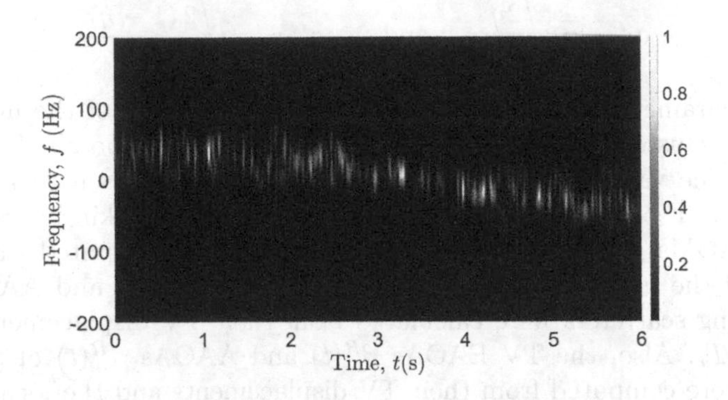


**Fig. 5.** Spectrogram $S_\mu(f,t)$ of the complex channel gain $\mu(t)$ given by (7).

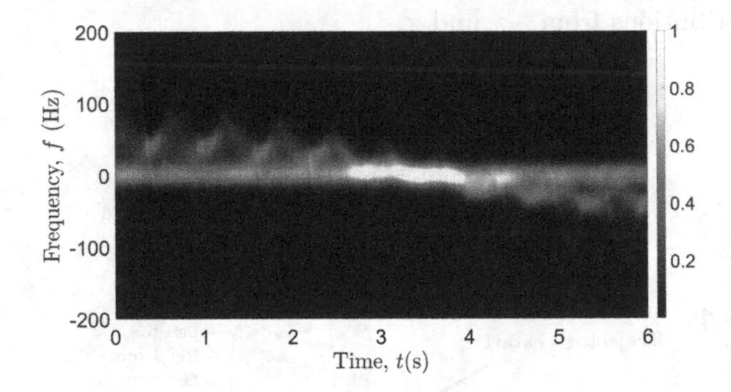


**Fig. 6.** The auto-term $S_\mu^{(a)}(f,t)$ of the spectrogram of the complex channel gain $\mu(t)$.

Figure 7 depicts the TV mean Doppler shifts $B_f^{(1)}(t)$ and $B_\mu^{(1)}(t)$ calculated as expressed in (4) and (8), respectively. The expression of the TV mean Doppler shift $B_\mu^{(1)}(t)$ given by (8) was applied to the spectrogram $S_\mu(f,t)$ and the auto-term $S_\mu^{(a)}(f,t)$. The variations of $B_\mu^{(1)}(t)$, calculated by using $S_\mu(f,t)$, are due to the impact of the cross-term $S_\mu^{(c)}(f,t)$. There is a good match between $B_\mu^{(1)}(t)$ and $B_f^{(1)}(t)$. The depicted functions in Fig. 7 approach zero-mean values at $t \approx 3.2$ s as the values of the TV EAODs $\beta_n^T(t)$, AAODs $\alpha_n^T(t)$, TV EAOAs $\beta_n^R(t)$, and AAOAs $\alpha_n^R(t)$ of the moving scatterers tend to $-\pi/2$ rad.

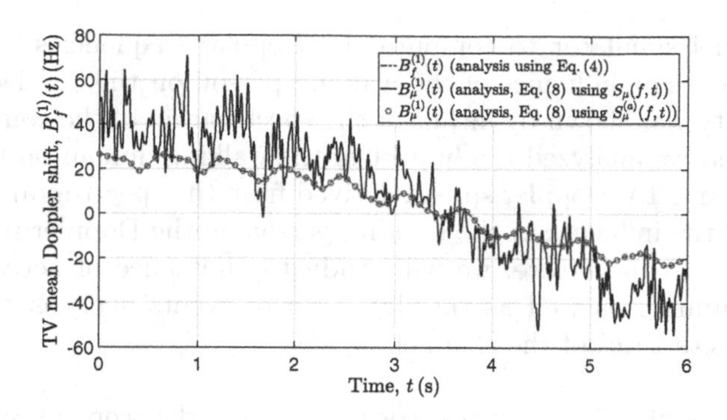


**Fig. 7.** TV mean Doppler shifts $B_f^{(1)}(t)$ and $B_\mu^{(1)}(t)$ given by (4) and (8), respectively.

Figure 8 depicts the TV Doppler spreads $B_f^{(2)}(t)$ and $B_\mu^{(2)}(t)$ calculated by using (5) and (9), respectively. The expression of TV Doppler spread $B_\mu^{(2)}(t)$ given by (8) was applied to the spectrogram $S_\mu(f,t)$ and the auto-term $S_\mu^{(a)}(f,t)$. The fluctuations of $B_\mu^{(2)}(t)$ calculated by using $S_\mu(f,t)$ are due to the impact of the cross-term $S_\mu^{(c)}(f,t)$. The functions $B_\mu^{(2)}(t)$ computed by utilizing the auto-term $S_\mu^{(a)}(f,t)$ and $B_f^{(2)}(t)$ do not match closely due to the influence of the Gaussian window spread $\sigma_w$ on $B_\mu^{(2)}(t)$.

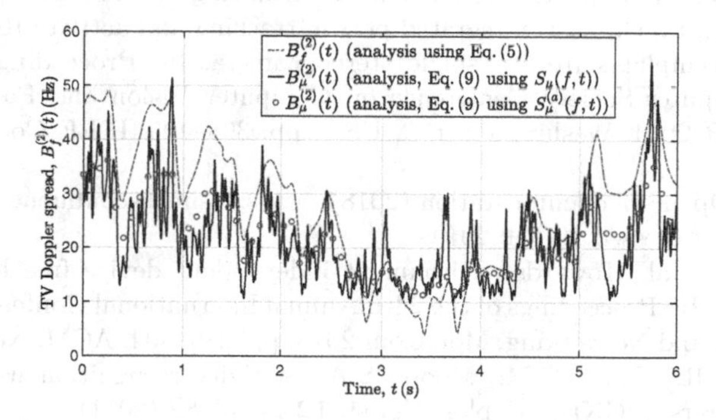


**Fig. 8.** TV Doppler spreads $B_f^{(2)}(t)$ and $B_\mu^{(2)}(t)$ computed by using (5) and (9), respectively.

## 5 Conclusion

In this paper, we studied the influence of human walking activity on the Doppler characteristics of an F2F non-stationary indoor channel model. At first, we used a full-body musculoskeletal model to simulate human walking activity to obtain trajectories of different body parts. We modelled the walking human as a cluster of synchronized moving scatterers. After that, we used the trajectories as inputs for this channel simulator to compute the Doppler frequencies caused by the walking human. The influence of the walking person on the TV Doppler power spectral density was shown by means of the spectrogram of the complex channel gain. Moreover, we analyzed the impact of the walking human on the TV mean Doppler shift and TV Doppler spread derived from the spectrogram. The results demonstrated the influence of the walking person on the Doppler power spectral characteristics. In the future, we will study the influence of activities such as sitting and running as well as the detection of events such as falling on the Doppler characteristics of the channel.

**Acknowledgement.** This work is carried out within the scope of WiCare project funded by the Research Council of Norway under the grant number 261895/F20.

## References

1. Abdelgawwad, A., Pätzold, M.: A framework for activity monitoring and fall detection based on the characteristics of indoor channels. In: IEEE 87th Vehicular Technology Conference (VTC Spring), Porto, Portugal, June 2018
2. Abdelgawwad, A., Pätzold, M.: On the influence of walking people on the Doppler spectral characteristics of indoor channels. In: 2017 IEEE 28th Annual International Symposium on Personal, Indoor, and Mobile Radio Communications (PIMRC), pp. 1–7. IEEE (2017)
3. Delp, S.L., et al.: OpenSim: open-source software to create and analyze dynamic simulations of movement. IEEE Trans. Biomed. Eng. **54**(11), 1940–1950 (2007)
4. Harville, M., Li, D.: Fast, integrated person tracking and activity recognition with plan-view templates from a single stereo camera. In: Proceedings of the 2004 IEEE Computer Society Conference on Computer Vision and Pattern Recognition, CVPR 2004, Washington, DC, USA, pp. 398–405. IEEE Computer Society (2004)
5. Hicks, J.: OpenSim documentation (2018). https://simtk-confluence.stanford.edu/x/pga9AQ. Accessed 19 Feb 2019
6. Jiang, W., et al.: Towards environment independent device free human activity recognition. In: Proceedings of the 24th Annual International Conference on Mobile Computing and Networking, MobiCom 2018, pp. 289–304. ACM. New York (2018)
7. Kwapisz, J.R., Weiss, G.M., Moore, S.A.: Activity recognition using cell phone accelerometers. SIGKDD Explor. Newsl. **12**(2), 74–82 (2011)
8. Lessard, S., Pansodtee, P., Robbins, A., Trombadore, J.M., Kurniawan, S., Teodorescu, M.: A soft exosuit for flexible upper-extremity rehabilitation. IEEE Trans. Neural Syst. Rehabil. Eng. **26**(8), 1604–1617 (2018)
9. Li, H., Cohen, I.: Inference of human postures by classification of 3D human body shape. In: 2003 IEEE International Workshop on Analysis and Modeling of Faces and Gestures (AMFG), p. 74, October 2003
10. Mahadas, S., Mahadas, K., Hung, G.K.: Biomechanics of the golf swing using OpenSim. Computers in Biology and Medicine **105**, 39–45 (2019)
11. Maldonado, G., Souères, P., Watier, B.: From biomechanics to robotics. In: Venture, G., Laumond, J.-P., Watier, B. (eds.) Biomechanics of Anthropomorphic Systems. STAR, vol. 124, pp. 35–63. Springer, Cham (2019). https://doi.org/10.1007/978-3-319-93870-7_3

12. Pätzold, M., Gutiérrez, C.A., Youssef, N.: On the consistency of non-stationary multipath fading channels with respect to the average Doppler shift and the Doppler spread. In: Proceedings of IEEE Wireless Communications and Networking Conference, WCNC 2017, San Francisco, CA, USA, March 2017
13. Rajagopal, A., Dembia, C.L., DeMers, M.S., Delp, D.D., Hicks, J.L., Delp, S.L.: Full-body musculoskeletal model for muscle-driven simulation of human gait. IEEE Trans. Biomed. Eng. **63**(10), 2068–2079 (2016)
14. Sigg, S., Scholz, M., Shi, S., Ji, Y., Beigl, M.: RF-sensing of activities from non-cooperative subjects in device-free recognition systems using ambient and local signals. IEEE Trans. Mob. Comput. **13**(4), 907–920 (2014)
15. Wang, W., Liu, A.X., Shahzad, M., Ling, K., Lu, S.: Device-free human activity recognition using commercial WiFi devices. IEEE J. Sel. Areas Commun. **35**(5), 1118–1131 (2017)
16. Wang, Y., Liu, J., Chen, Y., Gruteser, M., Yang, J., Liu, H.: E-eyes: device-free location-oriented activity identification using fine-grained WiFi signatures. In: Proceedings of the 20th Annual International Conference on Mobile Computing and Networking, MobiCom 2014, pp. 617–628. ACM, New York (2014)
17. Wei, B., Hu, W., Yang, M., Chou, C.T.: From real to complex: enhancing radio-based activity recognition using complex-valued CSI. CoRR abs/1804.09588 (2018)
18. Yatani, K., Truong, K.N.: BodyScope: a wearable acoustic sensor for activity recognition. In: Proceedings of the 2012 ACM Conference on Ubiquitous Computing, UbiComp 2012, pp. 341–350. ACM, New York (2012)

# A Neural Network for Stance Phase Detection in Smart Cane Users

Juan Rafael Caro-Romero[1], Joaquin Ballesteros[2(✉)], Francisco Garcia-Lagos[1], Cristina Urdiales[1], and Francisco Sandoval[1]

[1] Department of Electronic Technology, University of Malaga, Malaga, Spain
{jrcaro,fgl,acurdiales,fsandoval}@uma.es
[2] School of Innovation, Design and Engineering, Mälardalen University, Västerås, Sweden
joaquin.ballesteros@mdh.es

**Abstract.** Persons with disabilities often rely on assistive devices to carry on their Activities of Daily Living. Deploying sensors on these devices may provide continuous valuable knowledge on their state and condition. Canes are among the most frequently used assistive devices, regularly employed for ambulation by persons with pain on lower limbs and also for balance. Load on canes is reportedly a meaningful condition indicator. Ideally, it corresponds to the time cane users support weight on their lower limb (stance phase). However, in reality, this relationship is not straightforward. We present a Multilayer Perceptron to reliably predict the Stance Phase in cane users using a simple support detection module on commercial canes. The system has been successfully tested on five cane users in care facilities in Spain. It has been optimized to run on a low cost microcontroller.

**Keywords:** Multilayer Perceptron · Smart cane · Gait analysis · Phase detection

## 1 Introduction

Nowadays, new technologies are extensively accepted by elderly people. Hence, they have been extended to improve their social and cognitive daily routines [13]. For instance, social networking websites fight social isolation [1] and mobile devices and their applications have proven effective to reduce the effect of dementia [14]. However, regarding users' mobility, traditional assistive platforms are still the most frequent solution [3]. As these platforms are required by users for ambulation, recent research has focused on attaching technology to the device to continuously monitoring users [2,6].

Traditional assistive devices include rollators, walkers, wheelchairs, crutches or canes. The most appropriate platforms is selected for each user depending on their necessities [24], favoring the aid that provides the least amount of assistance to avoid loss of residual skills. Specifically, canes are the most popular solution


I. Rojas et al. (Eds.): IWANN 2019, LNCS 11506, pp. 310–321, 2019.
https://doi.org/10.1007/978-3-030-20521-8_26

to provide mobility for the elderly [9]. Canes distribute weight from an affected lower extremity and also improve balance.

Monitoring cane users during their Activities of Daily Living may provide valuable information to support diagnosis, evaluate degenerative process or obtain rehabilitation feedback. There are different approaches to monitoring cane's users. For instance, wearable sensors can be attached to their bodies to assessing gait [4]. These solutions provides accurate measurement, but users may forget to wear them for long-term monitoring and/or attach them incorrectly. This problem can be solved by attaching sensors directly to the cane that they require for mobility (cane on-board sensors) [2,6,25].

There are some approaches to obtain information on users condition using on-board cane sensors. For instance, supported weight can be estimated by using force sensors on the cane tip [2]. Other approaches rely on inertial measurement units to measure and classify user's activity (walking, standing, climbing stairs,...) [25] or estimate walking distance [6]. However, meaningful parameters directly related to user's rehabilitation have not been obtained yet using only on-board cane sensors.

One of these meaningful parameters is the Stance Phase on the affected leg [5], i.e. the amount of time that the affected leg is supporting the user's body weight. This parameter could be obtained from other ones like walking velocity. Unfortunately, although common devices like smart watches can approximate velocity for healthy users, they work with averages and assume that Stance Phases in both legs are symmetrical. This is typically not true for elderly users nor for persons with walking disabilities [20].

In this work we propose a system for long term Stance Phase Detection on the affected leg using only affordable on-board cane sensors. On-board sensors on a cane present two major limitations, mostly related to power consumption, ergonomics and cost. This work relies on a smart-cane released by the authors under an Open License that measures supported weight on the cane [2]. In order to predict the Stance Phase using only weight-bearing information, a Multilayer Perceptron (MLP) has been implemented and trained. The MPL has been designed keeping in mind all constraints imposed by the limited power performance of the Low Power Consumption microcontroller in the cane. The system has been validated with 5 elderly volunteer from 2 Senior Centers in Cordoba, Spain. Results prove that the proposed MLP achieves an accuracy above 90% in Stance Phase detection.

The rest of the paper is organized as follow. Section 2 describes our platform and also addresses the problem formulation. Next, Sect. 3 describes a Finite State Machine to process input data. This is the simplest approach to Stance Phase detection and its performance is later compared to the MLP. Section 4 presents the Neural Network design. Volunteer selection process, finite state machine and neural network configuration, and results are presented in Sect. 5. Finally, Sect. 6 shows our main conclusions and future work.

## 2 Methodology

### 2.1 Smart Cane Platform

The proposed cane is a commercial model modified with an Open License module that provides information when the cane is touching the ground. In this work, a simplified version of the smart-cane presented in [2] has been used. Specifically, this module relies on a single low cost force sensor (FSR 402, Interlink Electronics, Los Angeles, CA, USA) that decreases their resistance when applied force increases. This simplification means that the sensor saturates soon when users support weight on the cane. Hence, unlike the original cane, it simply provides temporal information about ground contact, i.e., readings under a threshold mean the cane is not touching the ground and viceversa. This modified module also fit standard canes. Its signal is processed by a microcontroller externally attached to the cane. Specifically, it works with the BLE Nano (NRF51822, Nordic Semiconductor, Norway) because is small, light and works with Bluetooth Low Energy. The BLE Nano reads the sensor every 0.02 s (50 Hz), and then it transmits a 20 Bytes package via Bluetooth to a paired Android device. Thus, all sensor data is stored into a text file in the mobile storage.

### 2.2 Problem Formulation

In a gait cycle, users only support weight on the cane during the stance phase of their affected leg. As commented, the cane provides information while it is in contact with the ground. Figure 1 shows the relationship between the Stance Phase and cane contact with the ground. The Stance Phase starts when a heel strikes the ground and finishes when the toe off. Ideally, the Stance Phase duration and the time the cane is in contact with the ground, i.e. the sensor is above a threshold, should be equal. However, this is typically not the case: while some users tend to hit the floor with the cane before they support any weight on their affected leg, others tend to do the contrary. Hence, the cane contact time with the ground is not equivalent to the Stance Phase [23].

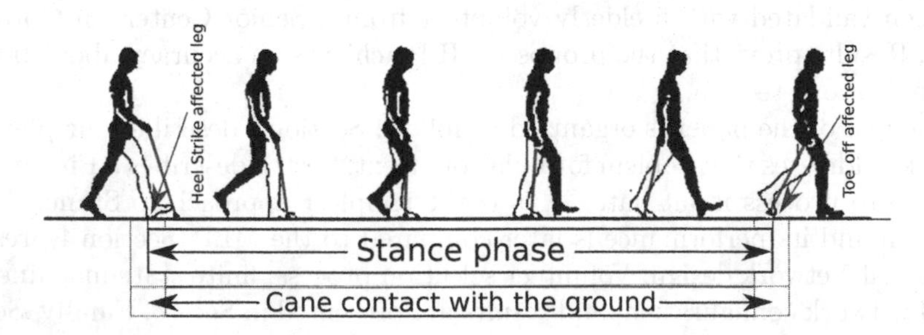


**Fig. 1.** Affected stance phase (white leg). The cane contact with ground start around the heel strike and finish around the toe off.

There are different ways to predict stance phase using the smart cane data. The most straightforwards one is to use a Finite State Machine (FSM), assuming that increases in cane data correspond to heel contact (affected leg) whereas decreases correspond to toe contact (Fig. 1). Cane data is thresholded to decide whether we have a positive or negative slope. Thresholds can be adapted to each person to improve state detection. However, once they are set, thresholds are static and do not adapt to variations. Alternatively, we can use a more complex model capable of capturing more subtle and/or hidden information. In this work, we propose to use a Neural Network to improve adaptability to users specific behaviors and, hence, detect stance phase more reliably. This NN will be compared to a FSM solution to compare their performances.

In order to test our system and also to create training patterns for the network, the stance phase for a set of volunteers that require a cane for ambulation has been manually measured using video capture (200 fps). The stance phase might vary while climbing a stair or walking in crowded areas. However, the commonest scenarios in cane's users are uncrowded areas with few stairs, due to their mobility limitations. For this reason, the testing area is an uncrowded area without stairs.

## 3 Finite State Machine to Process Data

We can model the cane state using a Finite State Machine (FSM). A FSM in this context is simply a behavior model which represents a number of finite situations, i.e. states, that a system could be in, depending on cane inputs and transitions. Let's define 5 states. Idle (I) is the state when the cane is not used to walk, i.e. it is not supported or it is in contact with the ground for a prolonged period of time. Support (S*) corresponds to the state where the cane is supporting weight during gait. Non-support (NS*) corresponds to the time when the cane is not touching the ground during gait. Additionally, we have two intermediate states to model transition from S* to NS*: Support Transition to NS* (SpTns) and Support Transition to S* (SpTs).

Transitions between states are ruled by two variables: sensor input at instant $t$ ($fsr_t$), which can be easily thresholded into TRUE or FALSE, and counter $i$, who controls time restrictions in each state. These restrictions are controlled by two time thresholds: $th_u$ and $th_l$.

The system is typically in idle state (I). If the sensor reading changes from FALSE to TRUE, a gait cycle starts on a support state (S*). On the other hand, if the sensor reading is TRUE (e.g. the user is sitting or standing while supporting the hand on the cane) and then changes to FALSE, a gait cycle starts with a non-support state (NS*). Afterwards, if the sensor value changes ($fsr_t \neq fsr_{t-1}$) during a valid period of time ($th_l < i < th_u$), the corresponding intermediate state is reached (SpTs or SpTns) and valid support or non-support is recorded. Then, the state changes automatically to the opposite one (NS* or S*). It is necessary to control time to discard errors. Standing or inactive periods of times are controlled by the maximum time interval $th_u$. If support or non-support states are prolonged ($i > th_u$), the state machine will go back to an idle state (I). On the other hand, the state machine will also return to an idle state (I) if support or non-support times are too short ($i < th_l$). These readings are usually due to spurious changes not related to gait cycles, i.e., accidental supports, cane falls, etc. (Fig. 2).

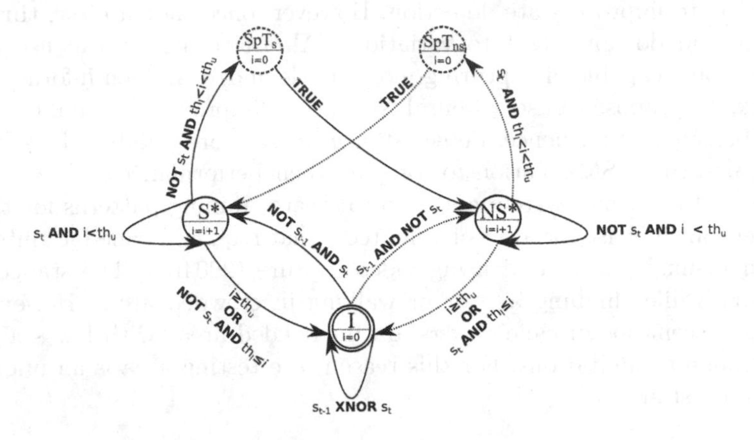


**Fig. 2.** Finite state machine to process sensor readings in instants $t$ and $t-1$ ($s_t$ and $s_{t-1}$). States are: (I) Idle, (S*) Support, (NS*) Non Support, Support Transition to NS* (SpTns) and Support Transition to S* (SpTs). Variable $i$ controls time in a state.

## 4 Neural Network Design

The presented FSM allows us to determine the stance phase of the affected leg but, as commented, does not adapt to the specifics of each user. Neural Networks (NNs) have proven particularly fit to classify complex patterns in different fields [11]. This is possible because NNs are capable of adjusting to specific data sets and to find non-linear relationships between input data and the desired outputs [26]. To do so, NNs require training, where the expected output of the input data set is known (target). Training can be done via different algorithms and its goal is usually to minimize the error between the network output and the target.

The required type of NN, its structure and the training algorithm depend on the nature of the processed data and the application goal. These choices usually have a direct impact on the resulting computational complexity of the system. Hence, design is of key importance if the NN is expected to run in a low cost microcontroller.

There are time dependencies on the data sets, so the best NN to solve the problem would be a Recurrent Neural Network (RNN), capable of recalling what happened in the past and to use it as context for the next input sequence (feedback). Context allows to solve time dependant problems, forecasting problems or speech recognition problems [8,16,18]. However, training RNNs is more computationally complex than training feedforward networks, plus it is and more difficult to guarantee the error function minimization. Hence, we propose to reorganize the inputs of the time series problem to solve it using a feedforward net, specifically a Multilayer Perceptron, as it is fit to run in a low cost microcontroller in our cane.

### 4.1 Multilayer Perceptron

The perceptron is the simplest unit in a Neural Network. It can be connected to others perceptrons to develop more complex networks with several layers, capable of solving more complex problems.

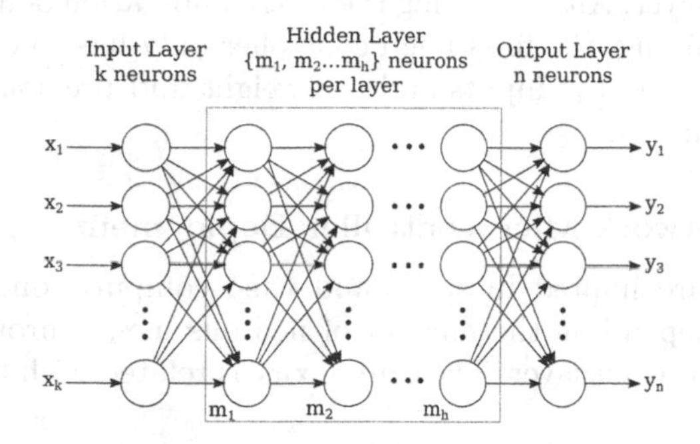


**Fig. 3.** Multilayer Perceptron with $h$ hidden layers.

Figure 3 shows the three distinct areas of a Multilayer Perceptron. The input layer propagates the input instance to the other layers. The hidden layer can be composed of more than one neuron layers. Hidden layers use non-linear activation functions to model the non-linear properties of the problem. Different layers can focus on different properties. The capacity to solve a complex problem will depend on the number and type of activation functions. Finally, the output layer for a classifier usually have as many neurons as classes are expected. The input of the net can be a time series vector, $\overline{x} = [x(n-1), x(n-2), ..., x(n-k)]$, where k is the prediction order (number of neurons in the input layer). In this way, the MLP model the time series prediction as a temporal pattern recognition problem [19]. Since we need to run our NN in the cane BLE nano microcontroller, we have implemented different versions of the MLP to choose the least computationally expensive that meets the required performance criteria (accuracy, sensitivity and specificity).

As commented, there are a number of configuration parameters (hyperparameters) to fix before training, namely the number of hidden layers, the number of neurons of each hidden layer, the learning rate, the loss function, the training algorithm, the transfer functions, etc. Hyperparameters adjustment can be heuristically performed [7,12].

In order to avoid computational constraints, training is performed off-line on a computer using MATLAB R2018b (The MathWorks, Inc., Natick, MA, USA).

The training function that we select is the Levenberg-Marquardt Backpropagation [15,17]. Its converge is stable and faster than Backpropagation algorithm [21] and avoid the possibility of the error valley during the gradient descent [22]. The other hyperparameters are set to default by MATLAB: maximum number of epochs to train equal to 500; performance goal equal to 0; maximum validations failures equal to 6; minimum performance gradient set equal to $1e^{-7}$; and initial $\mu$ factor 0.001 with increment (10), decrement (0.1) and maximum value of $1e^{10}$.

We have trained several MLPs with a single hidden layer to keep the NN complexity bounded. These NNs presented a different number of neurons in the input and hidden layer. After selecting the best combination of neurons per layers (Sect. 5.2) and training, the final microcontroller only has to compute a matrix multiplication between the inputs and the weight and previous calculated bias (Forward Propagation).

### 4.2 Neural Network Microcontroller Deployment

Microcontrollers are limited in space and time computation. As commented, NN calculations depend on the number of neurons, i.e., neurons in input layer, hidden layers and output layer. The complexity is related with the type of neural network.

The MLP is defined by the number of neurons in the input layer $k$, the number of hidden layers $h$ with $\{m_1, ..., m_h\}$ neurons per hidden layer, and the number of neurons in the output layer $n$ (Fig. 3). For simplicity, our NN will have the same activation function $g_1$ for hidden layers and an activation function for the output layer $g_2$. Given the number of additions $\alpha$, the number of products $\pi$, the number of divisions $\delta$ and their cost per unit $C_\alpha, C_\pi, C_\delta$, the following equation shows the computational cost of a given multilayer perceptron configuration:

$$\alpha C_\alpha + \pi C_\pi + \delta C_\delta \tag{1}$$

being

$$C_\alpha = m_1(k+g_1^s) + n(m_h + g_2^s) + \sum_{i=2}^{h} m_{i-1} * (m_i + g_1^s)$$

$$C_\pi = m_1(k+g_1^p) + n(m_h + g_2^p) + \sum_{i=2}^{h} m_{i-1} * (m_i + g_1^p)$$

$$C_\delta = n(g_2^d) + \sum_{i=1}^{h} m_i(g_1^d)$$

## 5 Results

### 5.1 Volunteer Selection

In order to check the performance of proposed system, we conducted tests in two senior centers in Cordoba, Spain. All volunteers signed an informed consent.

All our volunteers required a cane for everyday mobility, either for balance or to reduce pain in limbs, and supported (some) weight on the device. We selected 5 volunteers. Participants were on average 83.6 years old (range 74–91 years). Table 1 shows their ages, genders, average gait speeds, and their physical diseases.

**Table 1.** Condition and characteristics per users

| Id | Age | Gender | Gait speed | Physical issues |
|---|---|---|---|---|
| 1 | 80 | M | 0.615 m/s | Visual impairment; osteoarthritis; low back pain |
| 2 | 87 | M | 0.498 m/s | Osteoarthritis (left knee) |
| 3 | 86 | M | 0.687 m/s | Heart surgery; Lower limbs weakness |
| 4 | 91 | M | 0.597 m/s | Vestibular disorder |
| 5 | 74 | M | 0.792 m/s | Right knee prosthesis |

Volunteers were asked first to complete the 10 m Test. Then, they walked freely for at least 1 more minute. All data generated by users during tests was gathered via BT and processed off-line using MATLAB. During the tests, a high-speed camera recorded the feet and cane at 200 FPS.

### 5.2 FSM and MLP Configuration

Configuration of the proposed FSM is rather straightforwards. As commented, we only need to select time thresholds ($th_u$ and $th_l$) to determine when to revert to Idle State and a sensor threshold $s_{th}$ to determine whether is active or not. We have heuristically set $s_{th}$ to 15 due to the high sensibility of the sensor and its fast saturation. Regarding time thresholds, the cane support ideally is equal to the stance phase. In elderly users its average value is equal to 0.77 s with a variance of 0.09 s (men with plus than 85 years old) and 0.67 s with a variance of 0.08 s (women with 70–74 years old) [10]. Considering a normal distribution, the minimum Stance Phase is $th_l = 0.51$ s $(0.67 - 2 \cdot 0.08 =$ the and a maximum of $th_h = 0.95$ s $(0.77 + 2 \cdot 0.09)$ for 95% of the elderly.

In order to choose the best hyperparameters for the proposed MLP, we rely on a 5-fold validation process. The folders have been divided per user, i.e., first k subsample is: volunteer 1 in the test set whereas the others are in the training and validation sets. Initial testing proved that the best structures in all cases presented a single neuron in the input layer. Hence, we performed further testing for an increasing number of neurons in the hidden layer, keeping a single neuron both in the input and output layer. Figure 4 shows the best configurations in a set of 10 different 5-fold (50 tests) for a hidden layer ranging from 1 to 100 neurons. We can observe that configurations presenting 10 to 20 neurons in the hidden layer offered the best results 9 times out of the 50 tests (18%). Actually, there are no major differences in performance from 10 to 70 neurons in the hidden layer. From 70 neurons on, the performance of the resulting NN clearly decreases. However, the number of operations grows with the number of neurons in the hidden layer, so our final MLP was the best performing network among structures presenting 10 to 20 neurons. Specifically, our MLP has 17 neurons in the hidden layer (most selected solution in this interval) and it requires 767 operations.

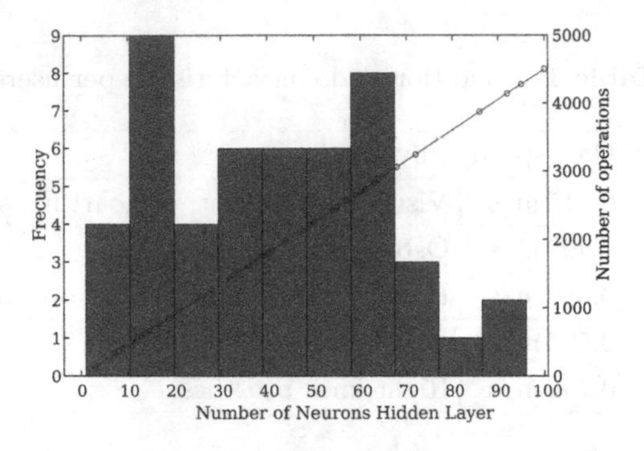


**Fig. 4.** Best MLP structures out of 50 tests

### 5.3 MLP vs FSM

After the best configurations for our MLP and FSM have been set, we can compare their performance using the previously described dataset. Once more, we have used a 5-fold validation, i.e. training data does not include testing data.

Figure 5(b) shows the results of all our tests for both techniques. We can observe that MLP outperforms FSM in all performance metrics, i.e. Accuracy, Sensitivity and Specificity. Average accuracy, for example, grows from 0.84 to 0.9. In both cases, Specificity is clearly larger than Accuracy and Sensitivity. This result makes sense, as it is the True Negative rate: the system is very good at deciding when the sensor is not touching the ground. However, it may miss some situations when the cane is touching the floor, but users are not really supporting any weight on it. It is also remarkable that MLP tests practically have no variation, meaning that their performance is more reliable for all volunteers, despite their condition.

We can evaluate the impact of user condition -which is related to how long they support weight on their affected leg- in prediction if we train the network with data from 4 volunteers and then test the MLP with the remaining one. Table 2 shows these results both for the proposed FSM and MLP. The best results for every patients are marked in blue. Accuracy and sensibility is significantly better in MLP than FSM. Wilcoxon Signed-Rank test using the 50 values from 10 tests returned a p-value lower than 0.01 for accuracy and sensibility. However, specificity in MLP is not significantly better than in FSM (p-value 0.48392). Indeed, it depends on the tested volunteer, e.g., volunteer 1 has a specificity value 6.04% higher in MLP than FSM while volunteer 4 has a specificity value 2.04% lower in MLP than FSM.

As commented, our microcontroller presents computational limitations. The selected NN has 1 neuron in the input layer, 17 in the hidden layer and 2 in

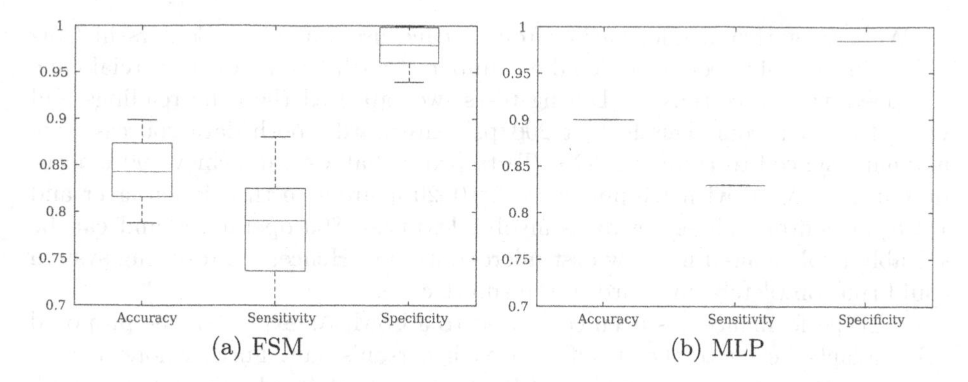


(a) FSM (b) MLP

**Fig. 5.** Accuracy, Sensitivity and Specificity using FSM in all volunteer (a). Average of those values in 10 repetitions for 5-Fold cross-validation using MLP (b). MPL is with the best configuration: 17 neurons hidden layer and 1 neuron in input layer.

**Table 2.** Prediction performance: MLP vs. FSM. MPL has been validated per volunteer and trained with the rest ones. Average value of ten repetitions is represented.

| | | Volunteers | | | | | |
|---|---|---|---|---|---|---|---|
| | | 1 | 2 | 3 | 4 | 5 | Average |
| FSM | Accuracy | 0.8987 | 0.8649 | 0.7878 | 0.8251 | 0.8426 | 0.8438 |
| | Specificity | 0.9392 | 0.9994 | 0.9668 | 0.998 | 0.9793 | 0.9766 |
| | Sensitivity | 0.8804 | 0.8061 | 0.6948 | 0.7503 | 0.7903 | 0.7844 |
| MLP | Accuracy | 0.9583 | 0.8802 | 0.8762 | 0.8996 | 0.8861 | 0.9001 |
| | Specificity | 0.9996 | 0.9922 | 0.9596 | 0.9776 | 0.9986 | 0.9855 |
| | Sensitivity | 0.9004 | 0.7921 | 0.8263 | 0.8385 | 0.7888 | 0.8292 |

the output layer. Additionally, the activation functions used for the hidden layer and output layer are Sigmoid and Softmax functions, which can be approximated using Taylor series. This implies a total of 767 operations per sensor readings. Our NN has been deployed in our BLE nano microcontroller and we have performed 1000 classification with random numbers. The average execution time has been 10.17 ms. This number is below the bound for our sampling rate of 50 Hz (20 ms). Hence, the proposed NN fits the requirements to be deployed in the selected microcontroller.

## 6 Conclusion and Future Work

This work has presented a MLP to predict how long a given cane user supports weight on his/her injured leg using input data from a cane equipped with a force sensor module. Ideally, this time should be equal to the time of support on the cane, but in reality there are differences due to the cane user's condition and habits. Additionally, some users do not support much weight on the cane or just do it occasionally for balance. Hence, Stance Phase detection is not straightforwards.

We have gathered information from 5 cane users at Care Centers in Cordoba, Spain. Volunteers completed a number of paths using a commercial cane equipped with force sensors. During tests, we captured the cane readings and video feedback from their feet at 200 fps. Afterwards, both data sources were manually synced to train the NNs. Tests prove that we can achieve 90% accuracy using a MLP with 1 input neuron, 10–20 neurons in the hidden layer and 1 output neuron. These networks involve less than 800 operations and can be suitably implemented in a low cost microcontroller. Hence, the resulting system could run completely on-board in the smart cane.

MLP performance has been compared to a FSM. As expected, the proposed MLP adapts better to the specifics of each person's gait. Furthermore, results present a lower variance for the MLP, meaning that it adapts better to each person's specifics.

Future work will focus on extracting more clinically relevant information from the processed data, like gait asymmetries, speed, load, etc. To do so, further tests with volunteers presenting different disabilities will be required. Additionally, the contact area with the sensor will be reduced to increase the weight-bearing range.

**Acknowledgements.** This work has been supported by: Proyectos Puente and programa operativo de empleo juvenil (UMAJI58) at Malaga University; and the Swedish Knowledge Foundation (KKS) through the research profile Embedded Sensor Systems for Health (ESS−H) at Mälardalen University, Sweden. Authors would like to acknowledge PONIENTE and LOS NARANJOS senior centers for their support during the tests.

## References

1. Ballantyne, A., Trenwith, L., Zubrinich, S., Corlis, M.: 'I feel less lonely': what older people say about participating in a social networking website. Qual. Ageing Older Adults **11**(3), 25–35 (2010)
2. Ballesteros, J., Tudela, A., Caro-Romero, J.R., Urdiales, C.: Weight-bearing estimation for cane users by using onboard sensors. Sensors **19**(3), 509 (2019)
3. Bradley, S., Hernandez, C.: Geriatric assistive devices. Am. Fam. Physician **84**(4), 405 (2011)
4. Brognara, L., Palumbo, P., Grimm, B., Palmerini, L.: Assessing gait in parkinson's disease using wearable motion sensors: a systematic review. Diseases **7**(1), 18 (2019)
5. Chen, C.L., Chen, H.C., Wong, M.K., Tang, F.T., Chen, R.S.: Temporal stride and force analysis of cane-assisted gait in people with hemiplegic stroke. Arch. Phys. Med. Rehabil. **82**(1), 43–48 (2001)
6. Dang, D.C., Suh, Y.S.: Walking distance estimation using walking canes with inertial sensors. Sensors **18**(1), 230 (2018)
7. Diaz, G.I., Fokoue-Nkoutche, A., Nannicini, G., Samulowitz, H.: An effective algorithm for hyperparameter optimization of neural networks. IBM J. Res. Dev. **61**(4/5), 9–11 (2017)
8. Elman, J.L.: Finding structure in time. Cogn. Sci. **14**(2), 179–211 (1990)

9. Gell, N.M., Wallace, R.B., Lacroix, A.Z., Mroz, T.M., Patel, K.V.: Mobility device use in older adults and incidence of falls and worry about falling: findings from the 2011–2012 national health and aging trends study. J. Am. Geriatr. Soc. **63**(5), 853–859 (2015)
10. Hollman, J.H., McDade, E.M., Petersen, R.C.: Normative spatiotemporal gait parameters in older adults. Gait Posture **34**(1), 111–118 (2011)
11. Huang, W.Y., Lippmann, R.P.: Neural net and traditional classifiers. In: Neural Information Processing Systems, pp. 387–396. Morgan Kaufman, San Mateo (1988)
12. Hunter, D., Yu, H., Pukish III, M.S., Kolbusz, J., Wilamowski, B.M.: Selection of proper neural network sizes and architectures-a comparative study. IEEE Trans. Industr. Inf. **8**(2), 228–240 (2012)
13. Khosravi, P., Ghapanchi, A.H.: Investigating the effectiveness of technologies applied to assist seniors: a systematic literature review. Int. J. Med. Informatics **85**(1), 17–26 (2016)
14. Leuty, V., Boger, J., Young, L., Hoey, J., Mihailidis, A.: Engaging older adults with dementia in creative occupations using artificially intelligent assistive technology. Assistive Technol. **25**(2), 72–79 (2013)
15. Levenberg, K.: A method for the solution of certain non-linear problems in least squares. Quart. Appl. Math. **12**, 164–168 (1944)
16. Lipton, Z.C.: A critical review of recurrent neural networks for sequence learning. CoRR abs/1506.00019 (2015). http://arxiv.org/abs/1506.00019
17. Marquardt, D.: An algorithm for least-squares estimation of nonlinear parameters. J. Soc. Indust. Appl. Math. **11**(2), 431–441 (1963)
18. Mikolov, T., Karafiát, M., Burget, L., Černocký, J., Khudanpur, S.: Recurrent neural network based language model. In: Eleventh Annual Conference of the International Speech Communication Association (2010)
19. Morariu, N., Iancu, E., Vlad, S., et al.: A neural network model for time series forecasting. Rom. J. Econ. Forecast. **12**(4), 213–223 (2009)
20. Patterson, K.K., et al.: Gait asymmetry in community-ambulating stroke survivors. Arch. Phys. Med. Rehabil. **89**(2), 304–310 (2008)
21. Rumelhart, D.E., Hinton, G.E., Williams, R.J.: Learning representations of back-propagation errors. Nature **323**, 533–536 (1986)
22. Sapna, S., Tamilarasi, A., Kumar, M.P., et al.: Backpropagation learning algorithm based on levenberg Marquardt algorithm. Comput. Sci. Inform. Technol. (CS IT) **2**, 393–398 (2012)
23. Sprint, G., Cook, D.J., Weeks, D.L.: Quantitative assessment of lower limb and cane movement with wearable inertial sensors. In: 2016 IEEE-EMBS International Conference on Biomedical and Health Informatics (BHI), pp. 418–421. IEEE (2016)
24. Van, F.H., Demonbreun, D., Weiss, B.D.: Ambulatory devices for chronic gait disorders in the elderly. Am. Fam. Physician **67**(8), 1717–1724 (2003)
25. Wade, J., et al.: Design and implementation of an instrumented cane for gait recognition. In: IEEE International Conference on Robotics and Automation (ICRA), pp. 5904–5909. IEEE (2015)
26. Zhang, G.P.: Neural networks for classification: a survey. IEEE Trans. Syst. Man Cybern. Part C (Appl. Rev.) **30**(4), 451–462 (2000)

# Closed-Eye Gaze Gestures: Detection and Recognition of Closed-Eye Movements with Cameras in Smart Glasses

Rainhard Dieter Findling(✉), Le Ngu Nguyen, and Stephan Sigg

Ambient Intelligence Group, Department of Communications and Networking, Aalto University, Maarintie 8, 02150 Espoo, Finland
rainhard.findling@aalto.fi
http://ambientintelligence.aalto.fi/

**Abstract.** Gaze gestures bear potential for user input with mobile devices, especially smart glasses, due to being always available and hands-free. So far, gaze gesture recognition approaches have utilized open-eye movements only and disregarded closed-eye movements. This paper is a first investigation of the feasibility of detecting and recognizing closed-eye gaze gestures from close-up optical sources, e.g. eye-facing cameras embedded in smart glasses. We propose four different closed-eye gaze gesture protocols, which extend the alphabet of existing open-eye gaze gesture approaches. We further propose a methodology for detecting and extracting the corresponding closed-eye movements with full optical flow, time series processing, and machine learning. In the evaluation of the four protocols we find closed-eye gaze gestures to be detected 82.8%–91.6% of the time, and extracted gestures to be recognized correctly with an accuracy of 92.9%–99.2%.

**Keywords:** Closed eyes · Gaze gestures · Machine learning · Mobile computing · Recognition · Smart glasses · Time series analysis

## 1 Introduction

Users interact with their mobile devices frequently and throughout their daily routines. Smart phones and tablets have an average of 60 and 23 interactions per day, respectively, with a total usage duration of on average 221 min per day [7]. For this reason, user input across mobile devices should be fast, easy, reliable, and convenient. Gaze gestures have been demonstrated feasible for input to mobile devices [4]. They bear potential as being conceptually both hands free and allow to perform quick input. However, gaze gestures with smart phones and tablets are usually done when users are holding them/looking at them. In contrast, smart glasses with embedded eye-facing cameras allow hands-free gaze gesture recognition, which does not require additional preparation time (i.e. taking devices out of a trousers pocket) or users to look at a device screen.


I. Rojas et al. (Eds.): IWANN 2019, LNCS 11506, pp. 322–334, 2019.
https://doi.org/10.1007/978-3-030-20521-8_27

So far, mobile gaze gesture sensing has utilized movements from opened eyes only and disregarded closed eyes. With gaze gesture alphabets in general being limited (e.g. 4 to 8 easily performable gaze gestures [4,6]), sensing and recognizing also closed-eye gaze gestures would allow for an extended alphabet that combines movements of both opened and closed eyes. Eyes being open or closed could thereby be distinguished by the presence or absence of pupils in recordings. The major challenge with recognizing closed-eye gaze gestures from optical sources is that prior work has mostly utilized pupil movements, which renders them inapplicable for detecting closed-eye movements. In this paper we therefore investigate whether detection and recognition of closed-eye gaze gestures from close-up optical sensors, e.g. from eye-facing cameras embedded in smart glasses, is feasible. Our contributions are:

- We propose a processing methodology to detect and recognize closed-eye gaze gestures from recordings of cameras embedded in smart glasses.
- We propose three basic closed-eye gaze gesture protocols for smart glasses, which contain different sets of gaze gestures, and which all effectively extend existing open-eye gaze gesture alphabets with closed-eye movements.
- We comparatively evaluate the proposed protocols for their gesture detection and confusion rates using different machine learning approaches.

## 2 Related Work

While there has been numerous prior work on open-eye gaze gestures (cf. [6]), closed-eye gaze gestures have so far not been investigated. EyeWrite [13] uses gaze gesture input for text composition. The concept is based on EdgeWrite [12], in which each character is replaced by a gesture. The alphabet therefore contains $26 + 10$ gestures for numeric characters and further gestures for punctuation and text control. The approach is designed to work with a stationary Tobii ET-1750 eye tracker in the form factor of a computer monitor. A set of 12 gaze gestures is used in [10]. Each gaze gesture in their alphabet consists of left, right, up, down, and optional diagonal movements. They evaluate their approach with a PC setup, and compare it to dwell-based gaze input in subsequent work [8]. In [5] an alphabet of 8 gaze gestures is used. Each gesture is an unidirectional stroke into a certain direction (left, right, up, down, and the four diagonals in between). The authors utilize a setup with a computer monitor and a camera with attached infrared light to perform pupil tracking and extract gaze gestures. The same alphabet is used in [4] for gaze gesture input to mobile devices. All those approaches have in common that they rely on optical pupil tracking, hence on opened eyes, for gaze gesture extraction. While cameras can be built into smart glasses frames [11], pupils are not visible with closed eyes, which prevents those approaches from being applied to closed-eye gaze gestures.

Electrooculography (EOG) recognition, while technically different in the sensing, is conceptually able to overcome this limitation, as it does not rely on pupil tracking. However, employing EOG gaze gestures with closed eyes too has not been investigated yet. Related work on EOG based gaze gestures with opened eyes [2] used a basic alphabet related to the one in [5], consisting of 8 gestures (left, right, up, down, and four diagonals). In subsequent work they expand user input to have either small or big eye movements in left, right, up, and down [1]. By combining two movements they thereby encode a total of 16 different characters for user input. When considering smart glasses, EOG sensors bring a significant drawback. While EOG should in general allow for sensing closed-eye movements, embedding those sensors into smart glasses impacts the usability of the devices as glasses. Firstly, in contrast optical sensors, EOG sensors need to touch the skin, which limits possible sensor positions to around the nose and close to the ears (similar e.g. to the Jins Meme device [9]). Those locations cause sensors to be positioned horizontally, but not vertically. While this allows for horizontal eye movement sensing, vertical eye movement sensing would require additional sensors above and below the eye (similar to the goggles utilized in [3]). Adding such sensor positions would make the device significantly more cumbersome due to increased size and weight. Secondly, for good EOG signal quality, sensors need to be connected well to the skin. This would require either wet electrodes, arguably reducing comfort and usability, or firm skin contact. The latter seems possible when utilizing the weight of common glasses, but only on positions where glasses abut on skin (nose, ears), which again disregards vertical sensing.

To summarize, while there is prior work on optical sensing for gaze gestures, the employed approaches are not applicable with closed eyes due to pupils being hidden. EOG sensors should conceptually allow for sensing closed-eye movements, but embedding them in smart glasses seems cumbersome. It would either limit vertical sensing capabilities or cause the device to become goggles instead of glasses, with drawbacks in size and weight. This paper addresses the remaining gap: it provides a methodology to detect and recognize closed-eye gaze gestures with close-up cameras, and thereby is a first step towards utilizing both open- and closed-eye gaze gestures with smart glasses, thereby allowing and extension of the restricted alphabet of gaze-interfaces.

## 3 Our Approach to Closed-Eye Gaze Gestures

Our approach enables recognition of closed-eye gaze gestures from cameras embedded in smart glasses. In this section we propose four closed-eye gaze gesture protocols, together with a technical methodology to detect and extract those from sensed data. From a technical perspective, our approach relies on machine learning for recognizing optically sensed closed-eye gaze gestures. Our recognition consists of a training part to enroll individual users, and a recognition part to utilize users' eye movements as input. Both training and recognition internally perform optical closed-eye gaze gesture detection and extraction from video data (Fig. 1). The details for the individual steps in those parts are discussed in the sections below.

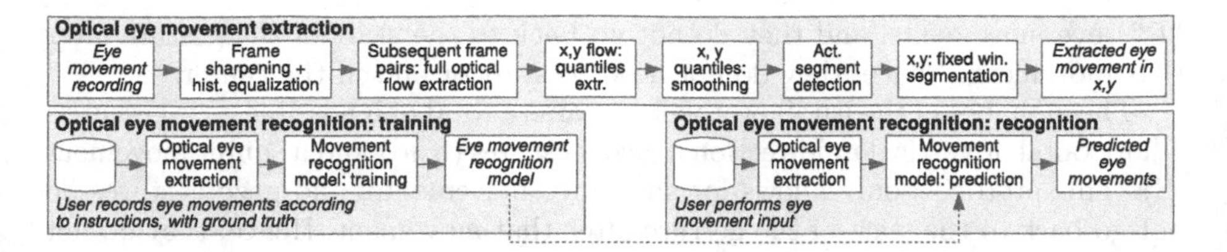


**Fig. 1.** Overview of processing in our approach to recognize closed-eye gaze gestures.

## 3.1 Closed-Eye Gaze Gesture Protocols

We propose 4 closed-eye gaze gesture protocols, which are related to protocols from prior work on open-eye gaze gestures (cf. [4,5,8,10]). With all protocols, users look straight on and close their eyes to begin user input, and they open their eyes to end user input. Each user input can contain multiple individual gaze gestures (Fig. 2). The alphabet of possible gaze gestures is defined by the respective protocol.

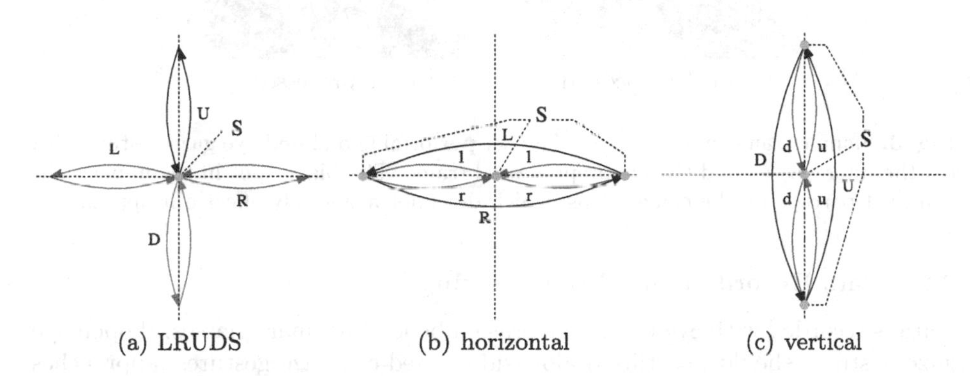


(a) LRUDS (b) horizontal (c) vertical

**Fig. 2.** Graphical depiction of possible gaze gestures with different protocols. While (a) the LRUDS protocol contains bidirectional gaze gestures only, all other protocols contain a set of small (e.g. "r") and/or big (e.g. "R") unidirectional gaze gestures, in both (b) horizontal and/or (c) vertical direction.

The alphabet of the LRUDS protocol contains a total of 5 possible gaze gestures: 4 horizontal or vertical bidirectional gaze gestures (LRUD) and a squint movement (S). All gaze gestures start and end in the center. The bidirectional gaze gestures go either left, right, up, or down, and backwards in the same gesture. The squint movement results from shortly squinting eyes. It therefore is equal to blinking except that users do not open their eyes during the process.

The alphabet of the LlRrUuDdS protocol contains a total of 9 possible gaze gestures: 8 horizontal or vertical unidirectional gaze gestures (LlRrUuDd) and a squint movement (S). The gaze gestures are either small (e.g. "r") or big (e.g.

"R") eye movements, and they do not go back to the start of the gesture after that movement. Hence, they do not need to start or end in the center.

The alphabet of the lrudS protocol contains a total of 5 possible gaze gestures: 4 horizontal or vertical unidirectional gaze gestures (lrud) and a squint movement (S). This protocols only allows for small eye movements, and gaze gestures do not go back to the start of the gesture after that movement. Hence, they do not need to start or end in the center.

The alphabet of the LlRrS protocol contains a total of 5 possible gaze gestures: 4 horizontal unidirectional gaze gestures (LlRr) and a squint movement (S). This protocol utilizes only horizontal eye movements. The gaze gestures are either small (e.g. "r") or big (e.g. "R") eye movements, and they do not go back to the start of the gesture after that movement. Hence, they do not need to start or end in the center.

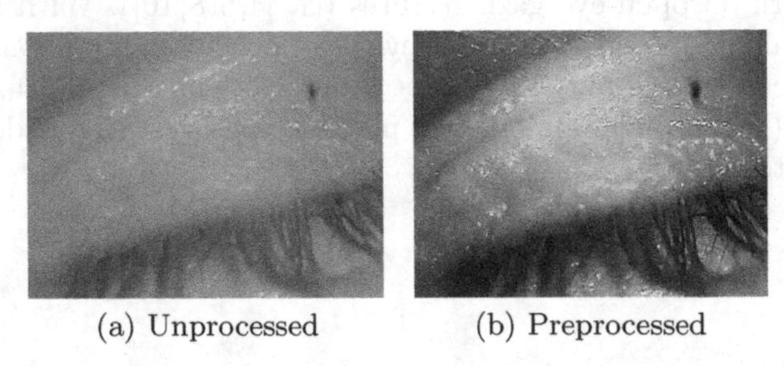

(a) Unprocessed (b) Preprocessed

**Fig. 3.** Sample frame of a recording of a user performing a closed-eye gaze gesture with (a) the unprocessed and (b) the preprocessed frame. The black spot in the upper right is a dust particle on the camera lens, which does not negatively effect our approach.

### 3.2 Data Recording and Preprocessing

Data is recorded with eye-facing cameras embedded in smart glasses. If open-eye gaze gestures should be utilized alongside closed-eye gaze gestures, approaches from related work can be used. We therefore declare open-eye gaze gestures out of scope for our approach. With all closed-eye gaze gesture protocols, users close their eyes to start closed-eye user input, and open them to end the input. The system detects that eyes are closed by pupils not being visible anymore. If the duration of eyes being closed is longer than a predefined threshold (1.5 s in our configuration) the system treats eye movements in between closing and opening eyes as potential closed-eye gaze gestures. The video captured from this input is then processed further once eyes are opened. For preprocessing, we apply frame-wise histogram equalization and image sharpening (Fig. 3). For the latter, with the $320 \times 240$ pixel resolution we employed in our evaluation, a $5 \times 5$ pixel luma matrix with a luma effect strength of 1.5 pixel was utilized.

### 3.3 Closed-Eye Gaze Gesture Detection and Extraction

On a preprocessed closed-eye movement video we employ frame-wise full optical flow to extract movements. For a pair of two subsequent $M \times N$ pixel frames, the optical flow yields two MxN matrices containing the horizontal and the vertical optical flow, respectively. Hence, a video consisting of F frames results in F-1 pairs of optical flow matrices. With that series of matrices, we compute the 10%, 25%, 50%, 75%, and 90% quantiles of optical flow per matrix. Those effectively capture both major positive and major negative closed-eye movements in optical flow. Each of them can therefore be understood as a one-dimensional time series, capturing closed-eye movements over time in either horizontal or vertical direction.

To reduce noise in each of those time series, we employ a Savitzky–Golay filter. For subsequent closed-eye gaze gesture detection, the filter configuration is a window length of 0.5 s and polynomial grade of 4. For feature extraction used in subsequent model training and prediction, we further utilize all permutations of window length $\{0.3\,\text{s}, 0.4\,\text{s}, 0.5\,\text{s}\}$ and polynomial grade $\{2, 3, 4\}$. For detecting segments of active eye movements in the filtered optical flow time series, we at first compute the sum of the filtered 10% and 90% quantile time series per axis, then the piecewise $L2$-norm over both axis. The variance of the resulting time series shows closed-eye movements in the video as non-zero periods. To automatically detect those periods we utilize a Schmitt-Trigger with a window length and a high and low trigger of $\frac{1}{2}$ s, 1.5, and 1 for the LRUDS protocol, and $\frac{1}{3}$ s, 2, and 1.5 for other protocols (Fig. 4). Each detected period of activity is segmented at its center with a fixed size window. The window length is 1.3 s for the LRDUS protocol and 1 s for other protocols. The reason for the LRUDS protocol utilizing different setting and longer windows is that its movements are bidirectional, hence take slightly longer. The segmented periods of activity become closed-eye gaze gestures (see example in Fig. 5), which we subsequently utilize for both training and user input recognition.

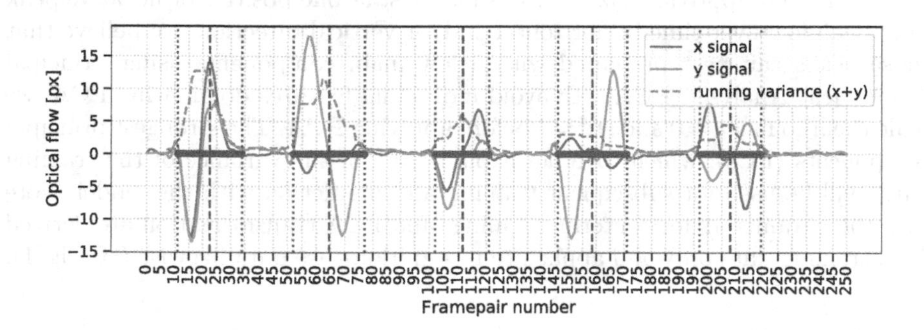


**Fig. 4.** Filtered optical flow in pixels between frame pairs, for an LRUDS video sample, containing "SDLRU" closed-eye gaze gestures, with detected gestures marked red. (Color figure online)

### 3.4 Closed-Eye Gaze Gesture Recognition: Training and Prediction

Users enroll by performing closed-eye gaze gestures according to instructions of the selected protocol. All possible gaze gestures of the corresponding protocol are thereby recorded multiple times together with ground truth. Gaze gestures are detected and segmented, and together with the ground truth form the basis for training a user specific closed-eye gaze gesture recognition model.

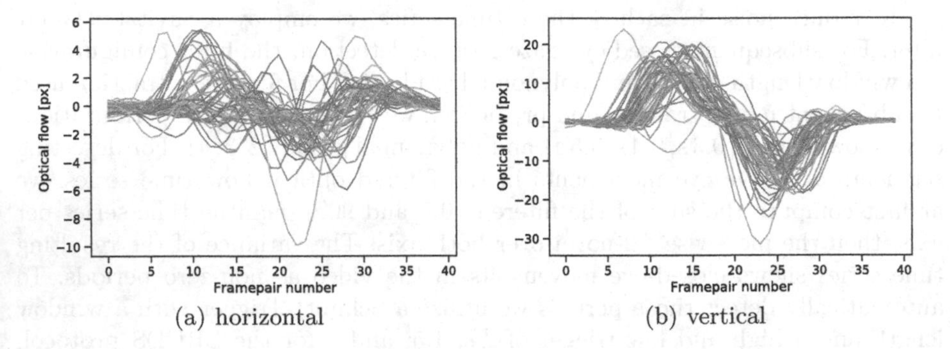


**Fig. 5.** Human understandable representation of multiple automatically extracted closed-eye gaze gesture samples, for the "U" movement from the LRUDS protocol. (a) and (b) show closed-eye movement in pixels in between frame pairs, in horizontal and vertical direction. While each sample contains multiple time series, this representation only shown the sum of the 10% and 90% quantile of the extracted 2D optical flow over time.

As with our approach gaze gestures manifest as one positive or negative peak in optical flow over time in horizontal and/or vertical direction, we believe that those peaks can be represented with a few main components using Principal Component Analysis (PCA). To avoid individual features dominating PCA, we scale them to mean = 0 and std = 1 before applying PCA. The strongest principal components (PCs), which together explain 80% of the variance of the training data, will be used for subsequent training and prediction with the model. Note that the parameters for centering, scaling, and PCA transformation are derived from training data only and applied to recognition-case data likewise (cf. Fig. 1).

## 4 Evaluation

To evaluate our approach we compare the gesture detection and classification rates of all proposed closed-eye gaze gesture protocols. For this we record test data, perform gaze gesture detection and extraction, and use extracted gaze gestures to comparatively evaluate the performance of different machine learning approaches on recognizing gaze gestures.

### 4.1 Evaluation Dataset

Data recording was done with a first generation Pupil Eye Tracker [11], which has the form factor of smart glasses. The device has a right-eye-facing camera to record videos of gaze gesture input with 60 Hz and a resolution of $320 \times 240$ pixels. We recorded data of one subject over a total of 8 sessions (2 per protocol), indoors, in office spaces. Over all protocols, the dataset thereby contains a total of 181 closed-eye gaze gesture video recordings, which together contain a total of 1024 closed-eye gaze gestures (Table 1).

**Table 1.** The utilized dataset: amount of recordings per protocol with their total contained closed-eye gaze gestures. The last two columns depict results of applying our gaze gesture extraction approach to this dataset (Sect. 5.1).

| Protocol | Recordings | Gaze gestures | Gaze gestures extracted | Extraction rate |
|---|---|---|---|---|
| LRUDS | 60 | 290 | 240 | 82.8% |
| LlRrUuDdS | 16 | 204 | 182 | 89.2% |
| lrudS | 16 | 148 | 126 | 85.1% |
| LlRrS | 89 | 382 | 350 | 91.6% |

### 4.2 Evaluation Procedure

Gaze gesture detection rates are quantified from the amount of correctly detected and extracted gaze gestures. With extracted gaze gestures, we then analyze components contained in gaze gestures. Further, we train and apply different machine learning models to perform gaze gesture classification, which will be quantified based on the amount of (in-)correctly classified gaze gestures.

For model tuning, selection, and reporting final performances, we utilize nested cross validation for data partitioning. We select models using the highest accuracies from the inner loop and report final results with gaze gesture confusion matrices for each protocol from the outer loop. The hyperparameter search for model tuning relies on logarithmic parameter grids. Data specific preprocessing (such as PCA) is done inside cross validation to not bias results.

## 5 Results

### 5.1 Closed-Eye Gaze Gesture Detection and Extraction Results

We found our closed-eye gaze gesture extraction to correctly detect and extract gestures for 82.8%–91.6% of all samples (Table 1). The threshold to accept a movement as gaze gesture has been configured to avoid false positives, which would be gaze gestures extracted from noise or small, unintended eye movements. This causes a trade-off, enabling low false positives by accepting certain false negatives. While with our data all intended gaze gestures, together with false positives, would initially have been recognized, the threshold was set too high for certain movements, which, as a consequence, were not recognized.

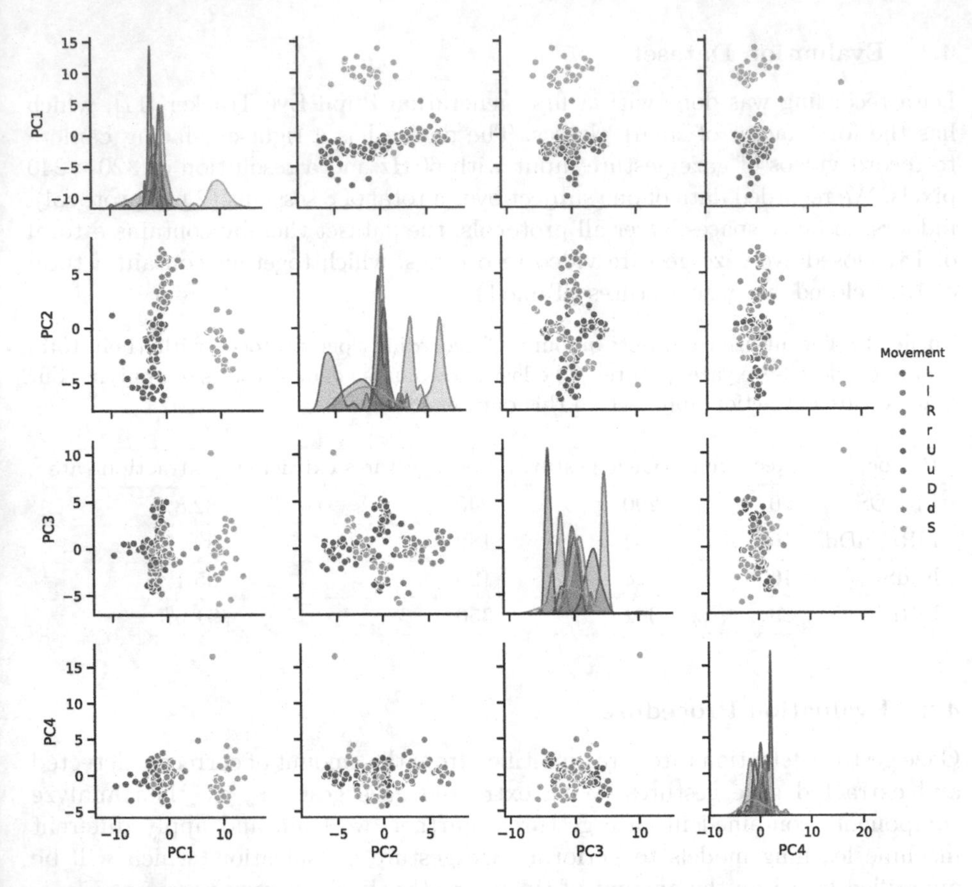


**Fig. 6.** Distribution of samples in PC space for the LlRrUuDdS protocol. While there is some overlap visible with certain classes, other classes seem clearly distinguishable.

## 5.2 Closed-Eye Gaze Gesture Decomposition Results

Applying PCA with extracted closed-eye gaze gestures indicates dominance of few PCs, which aligns with our expectations. To explain 80%, 90%, and 95% of the variance in the data, 9, 13, and 17 components are required for the LRUDS protocol, 7, 11, and 15 for the LlRrUuDdS protocol, 7, 11, and 14 for the lrudS protocol, and 5, 7, and 11 for the LlRrS protocol. An interesting detail thereby is that the protocol LRUDS, in which all gaze gestures contain bidirectional movements, requires the most components (9). In contrast, protocol LlRrUuDdS, which consists of more gaze gestures, but in which each gaze gesture only contains unidirectional movements, requires fewer components (7). Protocol LlRrS requiring the fewest components (5), as all of its gaze gestures only contain horizontal movements. Further, with all protocols, the distribution of samples in PC space indicates that some classes are easily separable, while others show stronger amounts of overlap (see LlRrUuDdS protocol example, Fig. 6).

### 5.3 Closed-Eye Gaze Gesture Recognition: Model Tuning Results

Evaluation results from tuning different models in the inner CV loop of our evaluation in general indicate good closed-eye gaze gesture recognition rates for all protocols (Table 2). Even simple models, such as a linear SVM, are able to achieve good results. Nevertheless, recognition accuracies over protocols vary in between 99.2%–93.7%, which indicates that gaze gestures of different protocols might be differently hard to distinguish.

### 5.4 Closed-Eye Gaze Gesture Recognition: Protocol Results

From applying the selected model types and hyperparameter configurations (Table 2) to the corresponding protocols in the outer CV loop of our evaluation, we obtain final gaze gesture confusion matrices (Fig. 7).

Gaze gestures in all except the LlRrS protocols seem well distinguishable. Overall closed-eye gaze gesture recognition accuracy is 99.2% with the LRUDS protocol, 95.6% with the LlRrUuDdS protocol, 97.6% with the lrudS protocol, and 92.9% with the LlRrS protocol. Most gaze gesture prediction errors confuse movements which are into the same direction but of different size. This is strongly visible with the LlRrS protocol, where 36% of "R" movements are wrongly predicted to be "r", and 19% of "L" as "l". It is also visible with the LlRrUuDdS protocol, which confuses 17% of "R" movements as "r", 6% of "D" as "d", 6% of "U" as "u", and 7% of "u" as "U".

To summarize: while the LlRrUuDdS and LlRrS protocols yield higher closed-eye gaze gesture detection and extraction rates (89.2% and 91.6%), gaze gesture recognition results seem to be in favor of the LRUDS and lrudS protocols (99.2% and 97.6%). While the LlRrU protocol does not require vertical and relies purely on horizontal eye movements, the LlRrUuDdS protocol provides the biggest alphabet of gaze gestures. In combination with the noticeably better

**Table 2.** Mean/standard deviation of gaze recognition accuracy per protocol and model, over all gaze gestures in that protocol and all inner CV repetitions. For each protocol, the result is emphasized for the model that was selected for the outer CV loop evaluation from those results, and its hyperparameters are stated in the bottom row.

| Model\Protocol | LRUDS | LlRrUuDdS | lrudS | LlRrS |
|---|---|---|---|---|
| KNN | 0.988/0.020 | 0.945/0.067 | 0.992/0.023 | 0.926/0.020 |
| LDA | 0.979/0.021 | 0.940/0.073 | 0.984/0.031 | 0.900/0.024 |
| CART | 0.958/0.037 | 0.857/0.089 | 0.944/0.049 | 0.906/0.044 |
| SVM Linear | 0.983/0.021 | 0.940/0.049 | **0.992/0.023** | **0.937/0.020** |
| SVM Radial | 0.988/0.020 | **0.961/0.049** | 0.992/0.023 | 0.937/0.034 |
| ANN | **0.992/0.017** | 0.912/0.064 | 0.992/0.023 | 0.920/0.033 |
| Hyperparam | Hidden = (50), $\alpha = 3^{-10}$ | $C = 3^2$, $\gamma = 3^{-6}$ | $C = 3^{-4}$ | $C = 3^4$ |

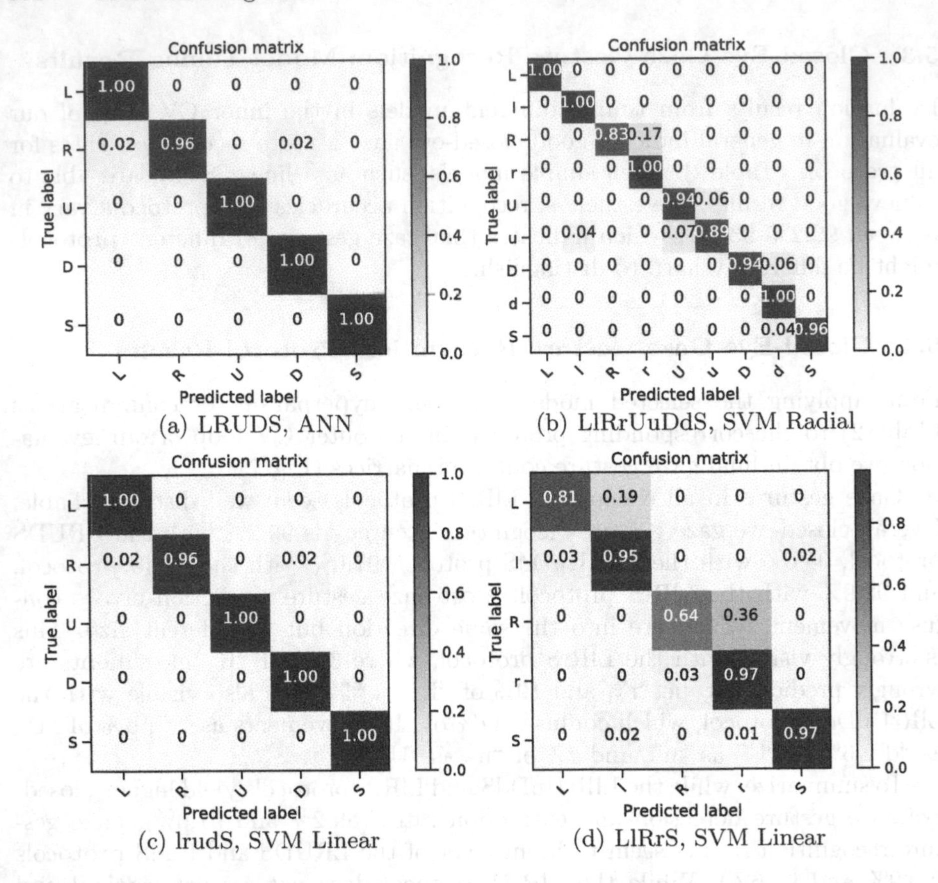


(a) LRUDS, ANN (b) LlRrUuDdS, SVM Radial

(c) lrudS, SVM Linear (d) LlRrS, SVM Linear

**Fig. 7.** Symbol confusion per protocol from the outer CV evaluation.

gaze gesture recognition results, the LlRrUuDdS protocol (89.2% gesture detection rate, 95.6% gesture classification accuracy) hence seems to be the most robust option within the evaluated protocols for closed-eye gaze gestures with cameras in smart glasses.

# 6 Conclusion

In this paper we investigated the feasibility of detecting and recognizing closed-eye gaze gestures from close-up optical sources, such as cameras embedded in smart glasses. We proposed four basic closed-eye gaze gesture protocols, consisting of different horizontal and vertical eye movements. We further proposed a methodology to detect, extract, and classify the corresponding gaze gestures, based on full optical flow extraction, time series processing, and machine learning. For our evaluation we utilized data from an eye tracker in the form factor of smart glasses, which records closed-eye movement videos with an embedded camera. For each closed-eye gaze gesture protocol, in the evaluation we investigated the detection rate of gestures, the amount of components required to represent extracted gestures, as well as gaze gesture confusion during classification. Results indicate gaze gesture detection rates of 82.8%–91.6% and average gaze gesture classification accuracies of 92.9%–99.2%. Further, with PCA, 80%, 90%, and 95% of variance in extracted closed-eye gaze gestures can be explained with a maximum of 9, 13, and 17 principal components, respectively, for all protocols. While those numeric results are based on limited amounts of data, our work has demonstrated the technical feasibility of detecting and recognizing closed-eye movements from optical sensors. It therefore is a first step towards utilizing closed-eye user input with cameras embedded in smart glasses, and extending gaze gesture alphabets with closed-eye gestures. Future work could investigate closed-eye gaze gestures across users with and without enrollment phases. Further, it could investigate the feasibility of combining open- and closed-eye gaze gestures in a single protocol. The feasibility and applicability of employing other sensors for closed-eye gaze gestures, such as EOG sensors embedded in the frame of smart glasses, would too be an interesting field of investigation for future work.

## References


1. Bulling, A., Ward, J.A., Gellersen, H., Troster, G.: Eye movement analysis for activity recognition using electrooculography. IEEE Trans. Pattern Anal. Mach. Intell. **33**(4), 741–753 (2011)
2. Bulling, A., Roggen, D., Tröster, G.: It's in your eyes - towards context-awareness and mobile HCI using wearable EOG goggles. In: Proceedings of the UbiComp 2008, vol. 344, pp. 84–93, September 2008
3. Bulling, A., Roggen, D., Tröster, G.: Wearable EOG goggles: eye-based interaction in everyday environments. In: CHI 2009 Extended Abstracts on Human Factors in Computing Systems, pp. 3259–3264. ACM, New York (2009)
4. Drewes, H., De Luca, A., Schmidt, A.: Eye-gaze interaction for mobile phones. In: Proceedings of the Mobility 2007, pp. 364–371. ACM, New York (2007)
5. Drewes, H., Schmidt, A.: Interacting with the computer using gaze gestures. In: Baranauskas, C., Palanque, P., Abascal, J., Barbosa, S.D.J. (eds.) INTERACT 2007. LNCS, vol. 4663, pp. 475–488. Springer, Heidelberg (2007). https://doi.org/10.1007/978-3-540-74800-7_43
6. Heikkilä, H., Räihä, K.J.: Speed and accuracy of gaze gestures. J. Eye Mov. Res. **3**(2), 1–14 (2009)
7. Hintze, D., Hintze, P., Findling, R.D., Mayrhofer, R.: A large-scale, long-term analysis of mobile device usage characteristics. In: Proceedings of the ACM on Interactive, Mobile, Wearable and Ubiquitous Technologies, vol. 1, no. 2, June 2017
8. Hyrskykari, A., Istance, H., Vickers, S.: Gaze gestures or dwell-based interaction? In: Proceedings of the ETRA 2012, pp. 229–232. ACM, New York (2012)
9. Ishimaru, S., Kunze, K., Tanaka, K., Uema, Y., Kise, K., Inami, M.: Smart eyewear for interaction and activity recognition. In: Proceedings of the Conference Extended Abstracts on Human Factors in Computing Systems, pp. 307–310. ACM (2015)
10. Istance, H., Hyrskykari, A., Immonen, L., Mansikkamaa, S., Vickers, S.: Designing gaze gestures for gaming: an investigation of performance. In: Proceedings of the ETRA 2010, pp. 323–330. ACM, New York (2010)

11. Kassner, M., Patera, W., Bulling, A.: Pupil: an open source platform for pervasive eye tracking and mobile gaze-based interaction. In: Proceedings of the UbiComp 2014, Adjunct Publication, pp. 1151–1160. ACM (2014)
12. Wobbrock, J.O., Myers, B.A., Kembel, J.A.: Edgewrite: A stylus-based text entry method designed for high accuracy and stability of motion. In: Proceedings of the UIST 2003, pp. 61–70. ACM, New York (2003)
13. Wobbrock, J.O., Rubinstein, J., Sawyer, M.W., Duchowski, A.T.: Longitudinal evaluation of discrete consecutive gaze gestures for text entry. In: Proceedings of the ETRA 2008, pp. 11–18. ACM, New York (2008)

# RF-Based Human Activity Recognition: A Non-stationary Channel Model Incorporating the Impact of Phase Distortions

Alireza Borhani(✉) and Matthias Pätzold

Faculty of Engineering and Science, University of Agder, P.O. Box 509, 4898 Grimstad, Norway
{alireza.borhani,matthias.paetzold}@uia.no
http://www.mcg.uia.no

**Abstract.** This paper proposes a non-stationary channel model that captures the impact of the time-variant (TV) phase distortion caused by hardware imperfections. The model allows for studying the spectrogram of in-home radio channels influenced by walking activities of the home user under realistic non-stationary propagation conditions. The resolution of the spectrogram is investigated for the von-Mises distribution of the phase distortion. It is shown that high-entropy distributions considerably mask fingerprints of the user activity on the spectrogram of the channel. For an orthogonal frequency-division multiplexing (OFDM) system, a computationally simple method for mitigating the undesired phase rotation is proposed. Both theoretical and simulation results confirm that the proposed method significantly reduces the impact of the phase distortion, allowing us to retrieve the desired spectrogram imprinted by the activity of the home user. The results of this paper are useful for the development of software-based radio frequency (RF)-based activity recognition systems.

**Keywords:** RF-based human activity recognition · Non-stationary channel modelling · Spectrogram analysis · Phase distortion

## 1 Introduction

The World Health Organization (WHO) states that the global average life expectancy increased by 5.5 years between 2000 and 2016, the fastest increase since the 1960s [1]. With such a trend, more and more seniors are expected to live a longer independent life. To assure the quality of independent living, in-home activity monitoring systems, such as video cameras, vision sensors, wearable

This work was supported by the WiCare Project through the Research Council of Norway under Grant 261895/F20.


I. Rojas et al. (Eds.): IWANN 2019, LNCS 11506, pp. 335–346, 2019.
https://doi.org/10.1007/978-3-030-20521-8_28

devices, and smart floors, are emerging. These systems send healthcare information, such as daily activity information and multimodal bio-sensors data, to remote caregivers, who act accordingly. These systems are often expensive, but more importantly, intrusive in terms of privacy and comfort of the home user. Such drawbacks have triggered a so-called radio frequency (RF)-based activity recognition approach, according to which no user involvement in the monitoring mechanism is required.

In this approach, a transmit antenna emits radio waves throughout the in-home propagation area, while a receive antenna receives the waves reflected off the body of the home occupant, and thus collecting fingerprints of the user activity in the environment. The new approach indeed allows for a passive indoor radar solution without the active participation of the users and without compromising their privacy. The collected radio waves are imprinted quite differently depending on the type of the activity. Therefore, sophisticated recognition algorithms and machine learning methods are required to distinguish activities from each other. The number of proposed activity recognition systems is increasing. Employing principles of radar systems [3,14], ultrawide band sensors [8], received signal strength indicator (RSSI) [7], and the channel state information (CSI) [4,9,18–20] are the main levers for the development of activity recognition systems. A comprehensive survey of the existing literature on the RF-based activity recognition can be found in, e.g., [5].

To the best knowledge of authors, all the proposed activity recognition systems have been developed using an experimental design approach, which is very time-consuming and costly. Often, numerous repetitions of the measurement are required in distinct experimental setups, such that machine learning algorithms can be sufficiently trained. However, a software-based design approach, in which channel models/simulators generate numerous datasets under different propagation conditions and different mobility patterns in a very short time, can save time and money compared to conducting a large number of real experiments. The data can then be used to train and to test detection algorithms. The software-based design approach has been very rarely, if at all, addressed in the literature. The channel model proposed in [2] is an exception that can generate experimentally verifiable complex channel gain data. However, the model in [2] does not capture the time-variant (TV) phase distortion, which in practice exists due to the hardware imperfection, such as carrier frequency offset (CFO) and sampling frequency offset (SFO). In fact, the undesired phase rotation caused by the imperfection of a device, e.g., Intel NIC 5300, significantly interferes with the desired phase rotation caused by the human activities, which ultimately reduces the performance of detection algorithms. Coping with the phase distortion has been a challenge for most of the literature listed above, as well as [6,16,17].

In this paper, we propose a non-stationary channel model that incorporates the effect of phase distortion in terms of a stochastic process. To study the time-frequency distribution of the channel, the spectrogram of the complex channel gain is computed. The spectrogram provides useful information about the variations of Doppler frequency components in time, allowing us to track the human walking pattern. The impact of different distributions of the phase distortion on the spectrogram of the complex channel gain is analyzed, showing that the expected observations are considerably spoiled when the undesired phase rotation is integrated into the model. To mitigate the impact of phase distortion, a computationally efficient method based on the principles of orthogonal frequency-division multiplexing (OFDM) systems is proposed. It is shown that the proposed method retrieves the desired spectrogram imprinted by the activity of the user.

The remainder of this paper is organized as follows. Sections 2 and 3 describe the in-home propagation scenario and present the phase distortions, respectively. The complex channel gain and its time-frequency distribution are presented in Sects. 4 and 5. Section 6 proposes the distortion mitigation technique. Simulation results are presented in Sect. 7. Finally, Sect. 8 concludes the paper.

## 2 Propagation Scenario

With reference to Fig. 1, a fixed-to-fixed propagation scenario is assumed in which an omnidirectional transmitter (Tx) antenna is placed on the floor and an omnidirectional receiver (Rx) antenna is mounted on the ceiling of the room. The position of the Tx (Rx) is denoted by $(x^T, y^T, z^T)$ $((x^R, y^R, z^R))$, while the positions of $N_F$ fixed scatterers (black stars in Fig. 1) are given by $(x^S_{n_F}, y^S_{n_F}, z^S_{n_F})$, where $n_F = 1, 2, ..., N_F$. It is assumed that the line-of-sight (LOS) between the Tx and the Rx is blocked.

A cluster of $N_M$ moving scatterers $S^M_n$ $(n_M = 1, 2, ..., N_M)$ accounts for the human body parts, such as the head, hands, and legs. The random trajectory approach in [2] is used to model a moving person, starting from $(x_s, y_s, z_s)$ and terminating at a predefined destination point $(x_d, y_d, z_d)$. In particular, a realization of the random trajectory (based on the first primitive of Brownian fields) gives a set of triples $(x_l, y_l, z_l)$, where $l$ denotes the position index (or equivalently $t$ represents the corresponding time). It is assumed that the random bridge is fully established, while the drift to the destination point exists. For simplicity reasons, the horizontal displacement of the path from the shortest path is assumed to be zero. A single realization of the random trajectory model generates a master trajectory $\mathcal{T}$, explaining the spatial behavior of each body part if it is shifted to its corresponding starting point $(x_s, y_s, z_s)$. The temporal features of the motion are added to the spatial trajectory model by employing horizontal and vertical speed vectors (similar to those used in [2]), describing the temporal behaviour of the body parts along the path (see Sect. 7 for details).

It is assumed that a plane wave emitted from the Tx with an azimuth angle-of-departure (AOD) $\alpha^T_n(t)$ and an elevation AOD $\beta^T_n(t)$ reaches the Rx with an azimuth angle-of-arrival (AOA) $\alpha^R_n(t)$ and an elevation AOA $\beta^R_n(t)$ after a single bounce with the $n$th scatterer $S^{(.)}_n$. Note that in case of $N_F$ fixed scatterers $S^F_n$, none of those angles change in time, but they change in time according to the trajectory of the $N_M$ moving scatterers $S^M_n$. It is also assumed that the reverberation effect between the fixed and moving scatterers does not exist.

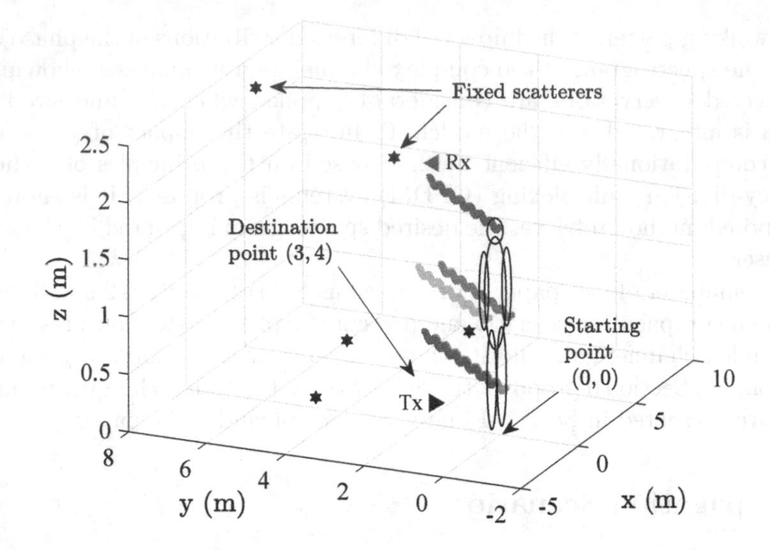


**Fig. 1.** The schema of an in-home propagation area, illustrating the trajectories of a person's head, arms, and legs, if the person walks from the starting point (0, 0) to the destination point (3,4).

Furthermore, it is assumed that the communication is established through $K$ subcarriers of an OFDM WiFi system. This corresponds to the principles of a typical commodity WiFi system, such as Intel NIC 5300, which has been the core tool of many experimental studies in this field, e.g., [4].

## 3 Phase Distortions

In practice, the phase of the received multipath components is often determined by two main factors, namely the phase shift caused by the propagation delay and the phase shift caused by the device characteristics (imperfection). The propagation delays and the corresponding phase shifts remain constant in time if the environment is stationary. However, if some objects (herein, the user) starts moving, the corresponding propagation delay and thus the associated phase vary in time. Such variations carry fingerprints of the user activities in the propagation area, which can be used for activity recognition purposes. With this token, the phase rotation caused by a change of the propagation delay is called the *desired phase rotation.*

In contrast, device imperfections result in an *undesired phase rotation* that can hardly be characterized, mainly because it is a superposition of several phase shifts originating from different sources [6]. The carrier frequency offset, sampling frequency offset, packet boundary offset, and the phase-locked loop offset are the main hardware-related sources of phase distortion in experimental studies [15,17,18,20]. The undesired phase distortion $\boldsymbol{\Phi}_1^{(k)}(t)$ changes across both the OFDM subcarrier index $k$ and the time variable $t$, making coping more challenging. Let us assume $H_1^{(k)}(f,t)$ denotes the channel transfer function (CTF) associated with the $k$th subcarrier of an OFDM WiFi system with phase distortions. This function can then be written as $H_1^{(k)}(f,t) = \tilde{H}^{(k)}(f,t)e^{j\boldsymbol{\Phi}_1^{(k)}(t)}$, in which $\tilde{H}^{(k)}(f,t)$ is the corresponding undistorted CTF.

In the literature, there exists a number of phase sanitization techniques[1], transforming the distorted CTF $H_1^{(k)}(f,t)$ to the calibrated CTF $H^{(k)}(f,t) = H_1^{(k)}(f,t)e^{-j\Phi_2{}^{(k)}(t)} = \tilde{H}^{(k)}(f,t)e^{j(\boldsymbol{\Phi}_1^{(k)}(t)-\Phi_2{}^{(k)}(t))}$, where $\Phi_2{}^{(k)}(t)$ accounts for the phase shift originated from the employed sanitization technique. However, these techniques cannot fully retrieve the desired phase rotation [6], meaning that the residual phase $\boldsymbol{\Phi}^{(k)}(t) = \boldsymbol{\Phi}_1^{(k)}(t) - \Phi_2{}^{(k)}(t)$ still distorts the time-frequency observations.

On the other hand, the randomness of the phase distortion adds a significant level of uncertainty to the residual phase rotation $\boldsymbol{\Phi}^{(k)}(t)$, spoiling the performance of activity recognition algorithms based on the time-frequency distribution of the channel. Therefore, it is of great importance to understand the impact of the undesired phase rotation $\boldsymbol{\Phi}^{(k)}(t)$ on the spectrogram of the complex channel gain.

In this paper, the undesired (residual) phase rotation $\boldsymbol{\Phi}^{(k)}(t)$ is modelled by a stochastic process following the von-Mises distribution, i.e.,

$$p_{\boldsymbol{\Phi}^{(k)}(t)}(\Phi|\,\Phi_0,\kappa) = \frac{e^{\kappa\cos(\Phi-\Phi_0)}}{2\pi I_0(\kappa)} \tag{1}$$

where $\Phi_0$ and $\kappa$ are the parameters of the circular distribution, while $I_0(.)$ denotes the zeroth-order modified Bessel function. The values of the undesired phase rotation across both subcarrier $k$ and time $t$ are independent outcomes of the distribution function in (1). The reason for this choice is the flexibility of the von-Mises distribution, ranging from the uniform distribution for $\kappa = 0$, over approximating the Gaussian distribution, and up to very concentrated distributions for large values of $\kappa$. Indeed, the von-Mises distribution allows us to assess the impact of different distortion entropies on the spectrogram of the channel.

## 4 Non-stationary Channel Model

Given the propagation scenario in Sect. 2, the complex channel gain $\boldsymbol{\mu}^{(k)}(t)$ associated with the $k$th subcarrier equals the CTF $H^{(k)}(f,t)$ of the same subcarrier frequency $f^{(k)}$. It follows $\boldsymbol{\mu}^{(k)}(t) = \tilde{\boldsymbol{\mu}}^{(k)}(t)e^{j\boldsymbol{\Phi}^{(k)}(t)}$, in which $\tilde{\boldsymbol{\mu}}^{(k)}(t)$ is the undistorted complex channel gain and $\boldsymbol{\Phi}^{(k)}(t)$ represents the TV residual phase distortion discussed in Sect. 3. Under NLOS propagation conditions, the non-distorted complex channel gain $\tilde{\boldsymbol{\mu}}^{(k)}(t)$ is modelled by a process representing

[1] A known example is to compensate the phase rotation by linearly removing the mean and the slope of the measured CSI phase [16,17]. This technique removes parts of the desired CSI, but more importantly has no physical justification [6].

the sum $\boldsymbol{\mu}_F$ of the scattered components due to the $N_F$ fixed scatterers and the sum $\boldsymbol{\mu}_M(t)$ of TV components due to the $N_M$ moving scatterers. It follows

$$\mu^{(k)}(t) = \tilde{\boldsymbol{\mu}}^{(k)}(t)e^{j\boldsymbol{\Phi}^{(k)}(t)} = \left(\tilde{\boldsymbol{\mu}}_F + \tilde{\boldsymbol{\mu}}_M^{(k)}(t)\right)e^{j\boldsymbol{\Phi}^{(k)}(t)}$$
$$= \left(\sum_{n_F=1}^{N_F} c_{n_F}e^{j\phi_{n_F}} + \sum_{n_M=1}^{N_M} c_{n_M}(t)e^{j\phi_{n_M}^{(k)}(t,\tau'_{n_M})}\right)e^{j\boldsymbol{\Phi}^{(k)}(t)} \quad (2)$$

in which $\phi_{n_M}^{(k)}(t,\tau'_{n_M}) = \phi_{n_M} - 2\pi f_0^{(k)}\tau'_{n_M}(t)$. In (2), the propagation path gain $c_{n_{(.)}}(t)$ is given by a negative path loss exponent $\gamma$ applied to the total travelling distance $D_{n_{(.)}}(t)$ of the $n_{(.)}$th plane wave, i.e., $c_{n_{(.)}}(t) = CD_{n_{(.)}}^{-\gamma}(t)$, where the constant $C$ accounts for the Tx(Rx) antenna gain, transmission power, and the wave length (see [11,12]). The total travelling distance $D_{n_{(.)}}(t)$ can be readily calculated by using the known coordinates of the trajectory $\mathcal{T}$ and those of the Tx/Rx. The constant phase shifts $\phi_{n_F}$ and $\phi_{n_M}$ account for the physical interaction of the emitted wave with the $n_F$th fixed scatterer and $n_M$th moving scatterer, respectively. These two shifts are assumed to be uniformly and independently distributed random variables ranging from $-\pi$ to $\pi$ [13, p. 47]. The propagation delay $\tau'_{(.)}(t)$ is proportional to the propagation path length $D_{n_{(.)}}(t)$, and is also related to the Doppler frequency $f_{(.)}^{(k)}(t)$ via (without proof) $f_{(.)}^{(k)}(t) = -f_0^{(k)}\dot{\tau}'_{(.)}(t)$, where $\dot{\tau}'_{(.)}(t)$ represents the derivation of the TV delay $\tau'_{(.)}(t)$ with respect to time $t$. From the latter relationship, it is straightforward to confirm that fixed scatterers contribute to the time-frequency distribution of the channel with zero Doppler shifts, as $\dot{\tau}'_{n_F}(t) = \dot{\tau}'_{n_F} = 0$.

## 5 Spectrogram Analysis

A practical approach to study the time-frequency distribution of the channel is to perform a spectrogram analysis on the complex channel gain process $\mu^{(k)}(t)$ (a sample function of the process $\boldsymbol{\mu}^{(k)}(t)$ in (2)). To this aim, one first needs to multiply $\mu^{(k)}(t)$ with a sliding window function $w(t'-t)$, i.e.,

$$x_\mu^{(k)}(t',t) = \mu^{(k)}(t)w(t'-t) \quad (3)$$

where $w(t)$ is a positive even function with normalized energy. In this paper, a Gaussian function is used to window the main signal $\mu^{(k)}(t)$. Applying the Fourier transform to the windowed signal $x_\mu^{(k)}(t',t)$ with respect to $t'$ gives then the short-time Fourier transform (STFT)

$$X_\mu^{(k)}(f,t) = \int_{-\infty}^{\infty} x_\mu^{(k)}(t',t)e^{-j2\pi ft'}dt' \quad (4)$$

of the original signal $\mu^{(k)}(t)$. The spectrogram $S_X^{(k)}(f,t)$ is then defined as the squared magnitude of the STFT $X_\mu^{(k)}(f,t)$, i.e., $S_{X_\mu}^{(k)}(f,t) = \left|X_\mu^{(k)}(f,t)\right|^2$.

## 6 Distortion Mitigation Method

The STFT $X^{(k)}(f,t)$ of the distorted complex channel gain $\mu^{(k)}(t)$ suffers from interfering terms, which can be analytically formulated as

$$\begin{aligned} X_{\mu}^{(k)}(f,t) &= \int_{-\infty}^{\infty} x_{\tilde{\mu}}^{(k)}(t',t)e^{j\Phi^{(k)}(t')}e^{-j2\pi ft'}dt' \\ &= \int_{-\infty}^{\infty} x_{\tilde{\mu}}^{(k)}(t',t)e^{-j2\pi ft'}\left(\sum_{l=0}^{\infty}\frac{(j\Phi^{(k)}(t'))^{l}}{l!}\right)dt' \\ &= X_{\tilde{\mu}}^{(k)}(f,t)+I_1(f,t)+jI_2(f,t) \end{aligned} \quad (5)$$

where $X_{\tilde{\mu}}^{(k)}(f,t)$ represents the desired STFT, while $I_1(f,t)$ and $I_2(f,t)$ stand for the interfering terms caused by the elements $l=1,2,...,\infty$ of the Maclaurin series of the phase distortion. Consequently, the spectrogram $S_{X_\mu}^{(k)}(f,t)$ of $\mu^{(k)}(t)$ includes complex-valued interfering terms that change not only in time $t$, but also in frequency $f$.

To mitigate the impact of such undesired components, a simple (computationally inexpressive) approach is to first average over $K$ distorted complex channel gain $\mu^{(k)}(t)$, and then to apply the spectrogram analysis on the average $\bar{\mu}(t)$. The theoretical justification of this technique is that $\bar{\mu}(t) = E\{\mu^{(k)}(t)\}$ equals to $E\{\tilde{\mu}^{(k)}(t)e^{j\Phi^{(k)}(t)}\}$, which after some mathematical manipulations can be approximated by $\tilde{\mu}^{(\cdot)}(t)\frac{I_1(\kappa)}{I_0(\kappa)}e^{j\Phi_0}$, where $\kappa$ and $\Phi_0$ are the parameters of the von-Mises distribution describing the phase distortion $\Phi^{(k)}(t)$ (see Sect. 3), and $I_1(.)$ denotes the modified Bessel function of the first kind. The STFT $X_{\bar{\mu}}(f,t)$ of the averaged complex channel gain $\bar{\mu}(t)$ can then be computed as follows

$$\begin{aligned} X_{\bar{\mu}}(f,t) &= \int_{-\infty}^{\infty} \bar{\mu}(t')w(t',t)e^{-j2\pi ft'}dt' \\ &\approx \frac{I_1(\kappa)}{I_0(\kappa)}e^{j\Phi_0}\int_{-\infty}^{\infty} |\tilde{\mu}^{(\cdot)}(t')|e^{j\angle\tilde{\mu}^{(\cdot)}(t)}w(t',t)e^{-j2\pi ft'}dt' \\ &= \frac{I_1(\kappa)}{I_0(\kappa)}e^{j\Phi_0}X_{\tilde{\mu}}^{(\cdot)}(f,t). \end{aligned} \quad (6)$$

From the equation above, it can be concluded that $X_{\bar{\mu}}(f,t)$ is a scaled version of the desired STFT $X_{\tilde{\mu}}^{(k)}(f,t)$, where the scaling factor $I_1(\kappa)e^{j\Phi_0}/I_0(\kappa)$ changes neither in time $t$, nor in frequency $f$. Therefore, the corresponding spectrogram $S_{X_{\bar{\mu}}}(f,t)$ is also a scaled, yet non-interfered, version of the desired spectrogram $S_{X_{\tilde{\mu}}}^{(\cdot)}(f,t)$. It can be shown that the proposed technique does not mitigate the impact of the phase distortion if the undesired phase follows an absolutely uniform distribution, i.e., if $\kappa = 0$.

## 7 Simulation Results

The central frequency $f_0 = 5.32\,\text{GHz}$ (associated with Channel 64) of an OFDM WiFi system is considered. The number of $K = 30$ equally spaced OFDM subcarriers around this central frequency has been generated. The Tx and Rx antennas are assumed to be placed on the floor and the ceiling of the room at a height of 0.1 m and 2.2 m, respectively (see Fig. 1). The free-space path loss exponent has been set to $\gamma = 2$, which matches our single-bounce scattering scenario, while the constant $C$ has been set to 1. A cluster of $N_M = 5$ moving scatterers, accounting for the head, hands, and legs of the person, as well as a number of $N_F = 5$ fixed scatterers are considered to be present in the propagation environment. The height $H$ of the person is assumed to be 178 cm. It is supposed that the user starts walking from the origin $(x_s, y_s) = (0\,\text{m}, 0\,\text{m})$ of the Cartesian coordinates to reach the preplanned destination point $(x_d, y_d) = (3\,\text{m}, 4\,\text{m})$ via a single realization of the 3D random trajectory $\mathcal{T}$. The user accelerates from a zero speed to a constant vertical (horizontal) walking speed of 0.1 m/s (1 m/s), followed by a deceleration to a zero speed when approaching the destination point. A Gaussian window of size $\sigma_\omega = 50$ ms has been used in the spectrogram analysis.

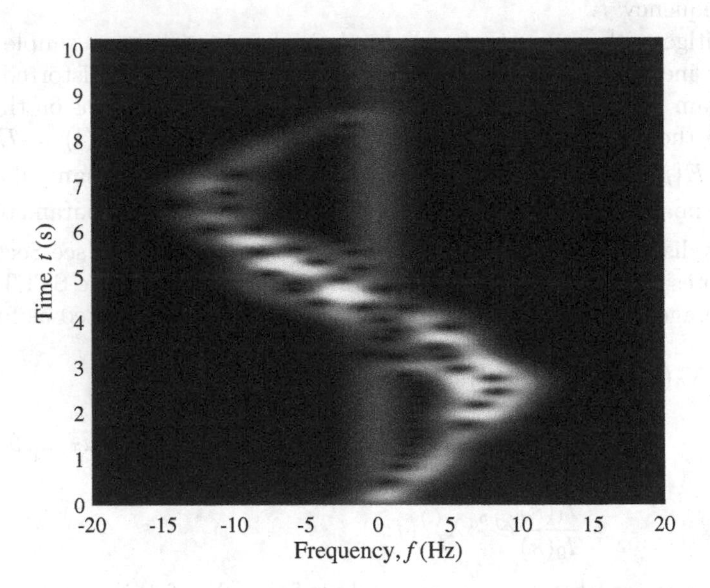


**Fig. 2.** The spectrogram $S^{(1)}_{X_{\tilde{\mu}}}(f,t)$ of the non-distorted complex channel gain $\tilde{\mu}^{(1)}(t)$ associated with the walking scenario.

Figure 2 displays the spectrogram $S^{(1)}_{X_{\tilde{\mu}}}(f,t)$ of the non-distorted complex channel gain $\tilde{\mu}^{(1)}(t)$ associated with the walking scenario above. The illustrated spectrogram $S^{(1)}_{X_{\tilde{\mu}}}(f,t)$ provides a non-distorted estimation of TV Doppler frequency components, representing our benchmark for the time-frequency observation associated with the first OFDM subcarrier. For the first three seconds, the Doppler shifts increase from zero (initial stop) to about 12 Hz, confirming the initial acceleration of the user for a normal walk. As the user approaches the vicinity of the RX/TX, the Doppler shifts decrease because the direction of motion becomes almost perpendicular to the direction of arrival. This trend continues almost to the middle of the path, from where the user starts leaving the transceiver, indicated by increasing Doppler frequency shifts towards negative values. For $t > 7$, the Doppler components start vanishing, as the person starts decelerating to a zero speed. The oscillatory behavior of the frequency components is caused by the sinusoidal variations of the user's height within a normal walk process. The time-invariant zero frequency components are due to the presence of fixed scatterers in the propagation environment, while the small frequency spread around $f = 0\,\text{Hz}$ is caused by interfering cross-terms as the artefact of the spectrogram analysis [10]. In a nutshell, the phase rotation caused by the walking activity of the user shapes the spectrogram into a clean $S$-pattern (see Fig. 2)[2].

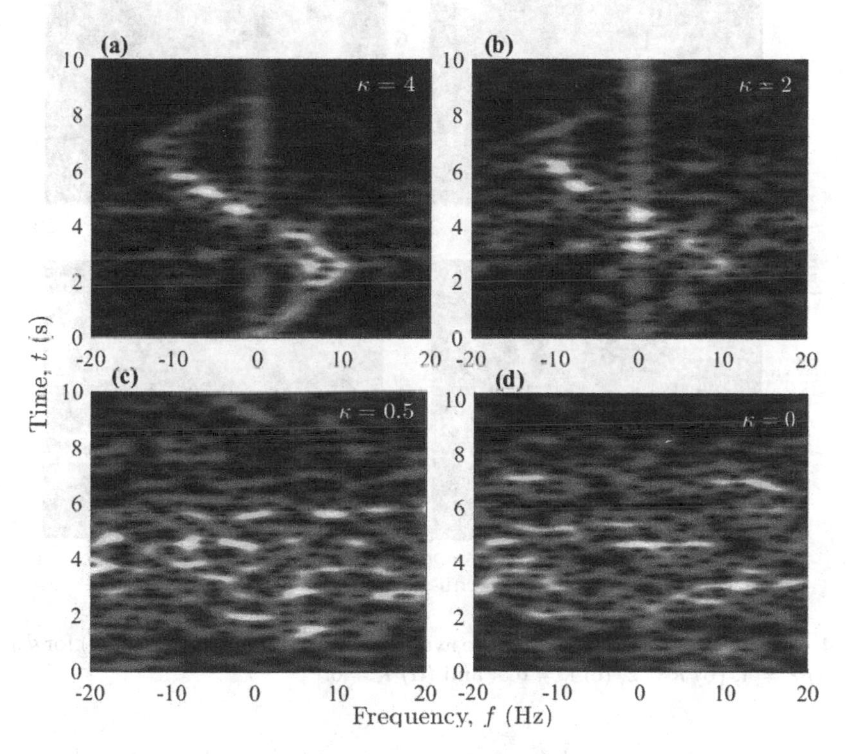


**Fig. 3.** The spectrogram $S^{(1)}_{X_\mu}(f,t)$ of the distorted complex channel gain $\mu^{(1)}(t)$ for $\Phi_0 = 0$ and (a) $\kappa = 4$, (b) $\kappa = 2$, (c) $\kappa = 0.5$, and (d) $\kappa = 0$.

Figure 3 exhibits the spectrogram $S^{(1)}_{X_\mu}(f,t)$ of the distorted complex channel gain $\mu^{(1)}(t)$ for the same walking scenario and for four different values of $\kappa$.

[2] The spectrogram associated with the other OFDM subcarriers is also imprinted with a similar $S-$pattern.

In all subfigures, one can observe that the expected $S$-pattern is interfered by the frequency contributions originating from the phase distortion. This interference increases if the value of $\kappa$ decreases. If the phase distortion approaches the uniform distribution, i.e., if $\kappa = 0$, no signature of the desired frequency components can be identified in the plot. For small values of $\kappa$, the $S-$pattern in the spectrogram of the channel is hardly distinguishable, while higher values of $\kappa$ allow for the appearance of the expected pattern. This can be attributed to the fact that higher values of $\kappa$ result in more concentrated phase distortions with more distinguishable fingerprints in the spectrogram.

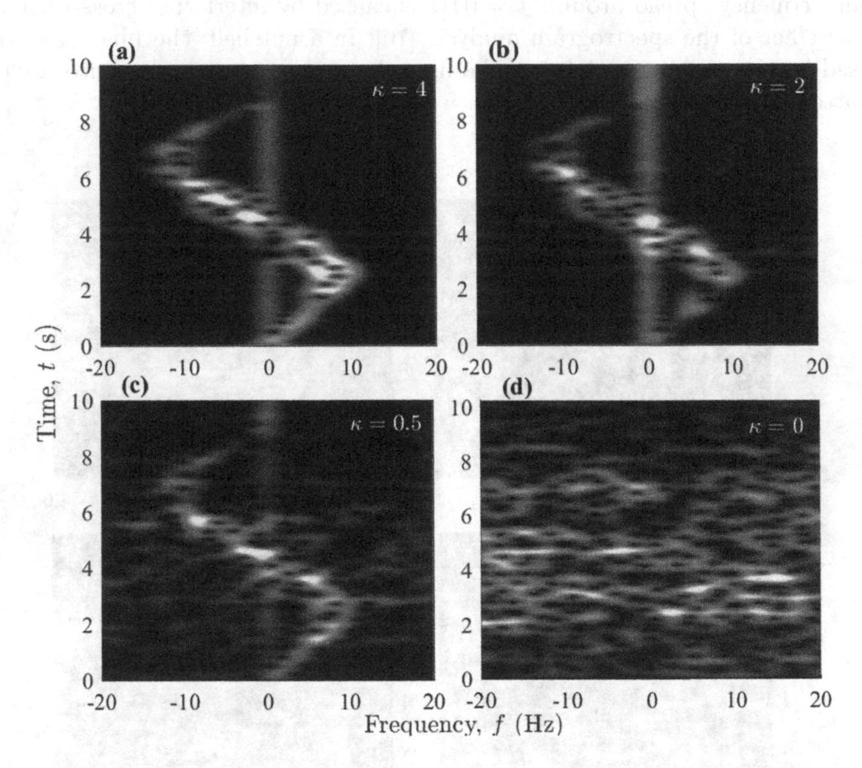


**Fig. 4.** The spectrogram $S_{X_{\bar{\mu}}}(f,t)$ of the averaged complex channel gain $\bar{\mu}(t)$ for $\Phi_0 = 0$ and (a) $\kappa = 4$, (b) $\kappa = 2$, (c) $\kappa = 0.5$, and (d) $\kappa = 0$.

Figure 4 demonstrates the spectrogram $S_{X_{\bar{\mu}}}(f,t)$ of the averaged complex channel gain $\bar{\mu}(t)$ for the same walking scenario and for four different values of $\kappa$. This set of figures is one-to-one comparable with those of the previous set shown in Fig. 3. As can be observed, the proposed mitigation technique can considerably reduce the interference caused by the phase distortion, so that the $S-$pattern can be restored as the original contribution of the walking person to the time-frequency distribution of the channel. The quality of the proposed sanitization technique decreases if $\kappa$ decreases. For $\kappa = 0$, the proposed algorithm cannot sanitize the spectrogram of the channel any longer. Theoretically, it can be shown that a pure uniform distribution of the phase distortion over time completely fades the desired spectrogram $S^{(.)}_{X_{\tilde{\mu}}}(f,t)$. However, an interesting feature of the proposed mitigation scheme is that even a slightly concentrated distribution of the phase distortion, e.g., if $\kappa = 0.5$, allows us to restore the expected $S-$pattern, as can be confirmed in Fig. 4(c).

## 8 Conclusion

A non-stationary channel model incorporating the impact of phase distortion caused by device imperfections has been developed in this paper. It has been shown that the phase distortion adds significant interference to the time-frequency distribution of the channel. A distortion mitigation technique based on the principles of OFDM systems was proposed. It has been shown that the proposal is robust with respect to the entropy of the distortion distribution. The experimental verification of the proposed method is a topic of future studies.

## References


1. Department of Ageing and Life Course: WHO global report on falls prevention in older age, Geneva, Switzerland (2007)
2. Borhani, A., Pätzold, M.: A non-stationary channel model for the development of non-wearable radio fall detection systems. IEEE Trans. Wirel. Commun. **17**(11), 7718–7730 (2018)
3. Erol, B., Amin, M.G., Boashash, B.: Range-Doppler radar sensor fusion for fall detection. In: 2017 IEEE Radar Conference (RadarConf), pp. 0819–0824, May 2017. https://doi.org/10.1109/RADAR.2017.7944316
4. Halperin, D., Hu, W., Sheth, A., Wetherall, D.: Tool release: gathering 802.11n traces with channel state information. ACM SIGCOMM Comput. Commun. Rev. **41**(1), 53 (2011)
5. Jiang, H., Cai, C., Ma, X., Yang, Y., Liu, J.: Smart Home Based on WiFi Sensing: A Survey. IEEE Access **6**, 13317–13325 (2018). https://doi.org/10.1109/ACCESS.2018.2812887
6. Keerativoranan, N., Haniz, A., Saito, K., Takada, J.: Mitigation of CSI temporal phase rotation with B2B calibration method for fine-grained motion detection analysis on commodity Wi-Fi devices. Sensors **18**(11) (2018). https://doi.org/10.3390/s18113795
7. Kianoush, S., Savazzi, S., Vicentini, F., Rampa, V., Giussani, M.: Device-free RF human body fall detection and localization in industrial workplaces. IEEE Internet Things J. **4**(2), 351–362 (2017). https://doi.org/10.1109/JIOT.2016.2624800
8. Mokhtari, G., Zhang, Q., Fazlollahi, A.: Non-wearable UWB sensor to detect falls in smart home environment. In: 2017 IEEE International Conference on Pervasive Computing and Communications Workshops (PerCom Workshops), pp. 274–278, March 2017. https://doi.org/10.1109/PERCOMW.2017.7917571
9. Palipana, S., Rojas, D., Agrawal, P., Pesch, D.: FallDeFi: ubiquitous fall detection using commodity Wi-Fi devices. Proc. ACM Interact. Mob. Wearable Ubiquitous Technol. **1**(4), 155:1–155:25 (2018). https://doi.org/10.1145/3161183

10. Pätzold, M., Gutiérrez, C.A.: Spectrogram analysis of multipath fading channels under variations of the mobile speed. In: 2016 IEEE 84th Vehicular Technology Conference (VTC-Fall), pp. 1–6, September 2016. https://doi.org/10.1109/VTCFall.2016.7881234
11. Phillips, C., Sicker, D., Grunwald, D.: A survey of wireless path loss prediction and coverage mapping methods. IEEE Commun. Surv. Tutorials **15**(1), 255–270 (2013)
12. Sarkar, T.K., Ji, Z., Kim, K., Medouri, A., Salazar-Palma, M.: A survey of various propagation models for mobile communication. IEEE Antennas Propag. Mag. **45**(3), 51–82 (2003)
13. Stüber, G.: Principles of Mobile Communications, 3rd edn. Springer, Heidelberg (2011)
14. Su, B.Y., Ho, K.C., Rantz, M.J., Skubic, M.: Doppler radar fall activity detection using the wavelet transform. IEEE Trans. Biomed. Eng. **62**(3), 865–875 (2015). https://doi.org/10.1109/TBME.2014.2367038
15. Vasisht, D., Kumar, S., Katabi, D.: Decimeter-level localization with a single WiFi access point. In: 13th USENIX Symposium on Networked Systems Design and Implementation (NSDI 16), pp. 165–178. USENIX Association, Santa Clara (2016). https://www.usenix.org/conference/nsdi16/technical-sessions/presentation/vasisht
16. Wang, W., Liu, A.X., Shahzad, M., Ling, K., Lu, S.: Device-free human activity recognition using commercial WiFi devices. IEEE J. Sel. Areas Commun. **35**(5), 1118–1131 (2017). https://doi.org/10.1109/JSAC.2017.2679658
17. Wang, X., Gao, L., Mao, S.: CSI phase fingerprinting for indoor localization with a deep learning approach. IEEE Internet Things J. **3**(6), 1113–1123 (2016). https://doi.org/10.1109/JIOT.2016.2558659
18. Wang, X., Yang, C., Mao, S.: PhaseBeat: exploiting CSI phase data for vital sign monitoring with commodity WiFi devices. In: 2017 IEEE 37th International Conference on Distributed Computing Systems (ICDCS), pp. 1230–1239, June 2017. https://doi.org/10.1109/ICDCS.2017.206
19. Wang, Y., Wu, K., Ni, L.M.: WiFall: device-free fall detection by wireless networks. IEEE Trans. Mob. Comput. **16**(2), 581–594 (2017). https://doi.org/10.1109/TMC.2016.2557792
20. Xie, Y., Li, Z., Li, M.: Precise power delay profiling with commodity WiFi. In: Proceedings of the 21st Annual International Conference on Mobile Computing and Networking, MobiCom 2015, pp. 53–64. ACM, New York (2015). https://doi.org/10.1145/2789168.2790124

# Workout Type Recognition and Repetition Counting with CNNs from 3D Acceleration Sensed on the Chest

Kacper Skawinski, Ferran Montraveta Roca, Rainhard Dieter Findling(✉), and Stephan Sigg

Department of Communications and Networking, Aalto University, Maarintie 8, 02150 Espoo, Finland
{kacper.skawinski,ferran.montravetaroca, rainhard.findling,stephan.sigg}@aalto.fi

**Abstract.** Sports and workout activities have become important parts of modern life. Nowadays, many people track characteristics about their sport activities with their mobile devices, which feature inertial measurement unit (IMU) sensors. In this paper we present a methodology to detect and recognize workout, as well as to count repetitions done in a recognized type of workout, from a single 3D accelerometer worn at the chest. We consider four different types of workout (pushups, situps, squats and jumping jacks). Our technical approach to workout type recognition and repetition counting is based on machine learning with a convolutional neural network. Our evaluation utilizes data of 10 subjects, which wear a Movesense sensors on their chest during their workout. We thereby find that workouts are recognized correctly on average 89.9% of the time, and the workout repetition counting yields an average detection accuracy of 97.9% over all types of workout.

**Keywords:** Acceleration · Activity recognition · CNN · Deep learning · Movesense · Neural Networks · Workout · Sensors

## 1 Introduction

In recent years the ability to track sports and workout activities with has broadly become available with off-the-shelf mobile devices. Those devices, including fitness trackers, smart phones, and alike, usually feature positioning sensors such as GPS as well as IMU sensors, such as accelerometers and gyroscopes. Tracking characteristics for sports thereby range from tracking distance with positioning sensors to step detection and counting when walking [4,5]. While extracting certain characteristics about a given sport by those sensors has previously been investigated, automatically extracting information for workout sessions with mixed workout types is more sparse [10]. It would be desirable for body-worn mobile devices to at first automatically recognize the type of workout done, and subsequently to analyze characteristics about the workout. Since nearly all


I. Rojas et al. (Eds.): IWANN 2019, LNCS 11506, pp. 347–359, 2019.
https://doi.org/10.1007/978-3-030-20521-8_29

mobile devices nowadays feature IMU sensors such as accelerometers, but due to their size and energy constraints not all devices feature position sensors such as GPS, the automatic recognition should ideally work with IMU sensor data only. Towards this goal, one aspect that has not yet been investigated is the suitability of using a *single body-worn 3D accelerometer* for automatically detecting workout, recognizing the workout type, and counting repetitions in a given workout type.

In this paper we therefore present a methodology to automatically detect and recognize workout in mixed workout type sessions, and to count repetitions once a workout type has been determined. Our methodology uses a single 3D accelerometer only, worn at the chest. The contributions of this paper are:

- We propose a deep learning based methodology to recognize exercise being performed and to distinguish between four different exercise types: pushups, situps, squats, and jumping jacks. For this we utilize only a single 3D accelerometer worn at the chest.
- Based on the predicted workout, we utilize PCA and peak detection to count repetitions within different workout types.
- We record a data set containing the four exercise types with 10 subjects and in between 2–3 exercise recordings per subject, which contains a total of 55 workouts and a total of 583 workout repetitions. We evaluate our approach with this data.

## 2 Related Work

Wearable mobile devices with IMU sensors have become prevalent, since they enable a wide range of applications for their users. Examples for such devices include the Nike FuelBand and FitBit Flex [10], which can be used to give activity and workout feedback. An exemplary study for workout feedback is the RecoFit study [10]. RecoFit aims to give real-time and post-work feedback to sport trainees. The system bases its information extraction on data gathered with embedded IMU sensors with 50 Hz sampling rate. RecoFit encompasses three stages: automatically segmenting exercise periods, recognizing the exercise type, and counting the repetitions done in a given exercise. They at first smooth data with a Butterworth Lowpass filter (−60 dB damping at 20 Hz). Then a sliding window (width of 5 s, overlapping between windows of 96%) is applied to achieve windows of uniform length for subsequent processing. Based on this data they extract 24 features (auto correlation, RMS, mean, standard deviation, variance, integrated RMS, and frequency power bands). To distinguish between workout and no-workout, they use those windows and features to train a L2 linear support vector machine (L2-SVM). To subsequently differentiate between the 26 different types of possible workout, they train a multiclass SVM classifier. Once the workout type is defined, they apply principal component analysis (PCA) on only acceleration data to reduce it to one dimension, then employ a repetition counting algorithm on that data. In their evaluation they use data of 114 participants and 146 sport sessions. Their results indicate precision and recall bigger than 95% in the detection phase. For exercise recognition, they used circuits of 4, 7, and 13 exercises and achieved an accuracy of 99%, 98%, and 96% respectively. The counting with ±1 accuracy reached a precision of 93%.

Javed [7] proposed a method for arm and elbow workout exercise recognition. They use the accelerometers embedded in a Samsung Galaxy S4 smartphone. The exercises they investigated were arm based, that means workouts such as Bicep Curl, Active Pronator, Active Supinator, Assisted Biceps, Isometric Biceps, and triceps workout. Their data recording uses raw accelerometer data, then apply a class conditional probabilities filter. The filtered data is classified with different classifiers from the Waikato Environment for Knowledge Analysis (WEKA) toolkit [13]. Their findings indicate that Random Forest and LMT classifiers yield better performance for their setup, with an accuracy of 99,5% and 99,83%, respectively.

Another study [3] also used accelerometers embedded in Android smartphones, but focused on activities such as walking (fast, slow, upstairs, downstairs), running, and aerobic dancing. They utilized mobile phones in two different postures: in-hand phone position and in-pocket phone position. They applied a digital low pass filter to separate the gravity component. Subsequently, data is handed to a robust supervised classifier. For their evaluation, they tested multiple classifiers and found Multilayer Perceptron (MLP) and SVM to yield best accuracies with 89.48% and 88.76% in the in-hand case, and MLP and RF to yield best accuracies with 89.72% and 72.27% in the in-pocket case. Finally, the combination of multiple classifier with fusion was evaluated. Accuracy thereby was enhanced to 91.15% with combining MLP, LogitBoost, and SVM for the in-hand case, and to 90.34% for in-pocket with combining MLP, RF, and SimpleLogistic.

In another study [6], several machine learning models, including KNN, SVM, and RF were utilized to classify transportation ways such as driving a car, riding a bicycle, riding a bus, walking, and running. The study utilizes data from accelerometers, gyroscope, and rotation sensors. To achieve uniform sampling rates across those sensors, upsampling with linear interpolation to a uniform sampling rate of 100 Hz was used, similar to [9]. A sliding window with length 1 s was applied to achieve uniform sample sizes. The best results were generally obtained with RF (overall accuracy 95,1%), although SVM (overall accuracy 94,41%) was better for walking and running.

Liang and Wang [8] utilized Convolutional Neural Networks (CNN) to enhance the accuracy of classification of transportation modes over traditional machine learning methods. In their approach, a sliding window is applied to acceleration sensor data recorded with 50 Hz using smartphones. Subsequently each window is processed by a CNN to differentiate between seven classes: stationary, walking, cycling, driving, taking a bus, subway or train. The data is smoothed with a Savitzky-Golay filter to reduce noise in mobile phone movements. From the 3D acceleration values they then computed the magnitude. In the sliding window they extracted a 512 value long window with window overlapping by 12.5%. Those windows are used as input for the CNN model, which after convolutional and pooling layers uses a fully connected layer to predict the transportation type target. Their system was able to obtain an accuracy of 94.48%.

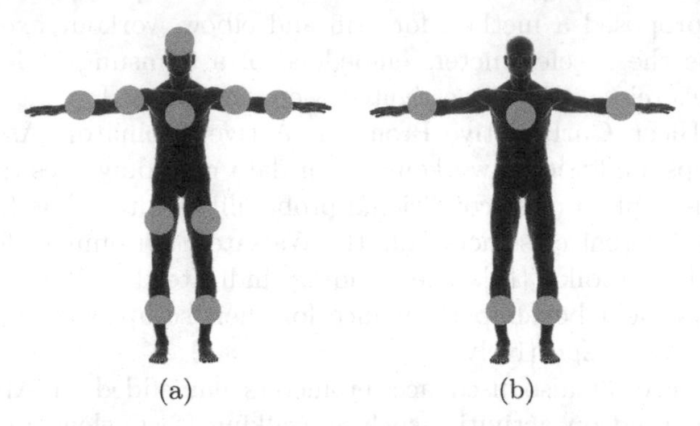


**Fig. 1.** Position of sensors for (a) most data gathering driven solution and (b) the setup evaluated to be best in literature [2,12]. Figure adapted from [1].

# 3 Method

Our goal is to automatically detect, recognize, and count repetitions in four different types of workout. In this section we describe the workout types, data source and sensor position, as well as technical details of the detection, recognition, and repetition counting.

## 3.1 Workout Types and Sensor Position

The workout types we consider are four-fold:

- Pushups
- Situps
- Squats
- Jumping jacks

As our approach automatically detects if one of those types of workout is done, we also consider a fifth type, named no-workout, which covers all other types of workout or no workout being done at all.

Our goal is to recognize workout from 3D accelerometer data of a single sensor only. The reason for using only one sensor is usability: using more than one sensor would be cumbersome for users who want their workout to automatically be recognized and counted. In a preliminary study we evaluated five different sensor positions on the human body that have been shown to be useful for human activity recognition in previous studies [2,12] (Fig. 1). This preliminary study yielded the chest to be the sensor position best suited to detect, recognize, and count repetitions from 3D acceleration sensor data only. For this reason we use an accelerometer on the chest in our approach.

### 3.2 Sensing Workout

We sense workouts with one 52 Hz 3D accelerometer on the chest. We use this sampling frequency due to related work with similar goals and good results utilizing similar frequencies [8,10]. On the continuous 3D time series we apply a sliding window to split the stream into fixed length samples. Our window length is 1 s, and windows overlap for $\frac{1}{4}$ of their lengths, which corresponds to a $\frac{13}{52}$ s overlap. Each window position thereby yields a sample that consists of $52 \cdot 3 = 156$ features.

### 3.3 Workout Detection and Recognition

From labeled samples, each having 156 features and representing 1 s of sensed workout, we train a model to distinguish between our five workout types (the four target workout types and the no-workout type). This model will be able to automatically distinguish for new samples if workout is done, and if yes, which type of workout it is. The chosen model type is a convolutional neural network (CNN) with 3 hidden layers (Fig. 2). The network thereby has an input layer (156 neurons), three hidden fully connected convolutional layers, and a Softmax classification output layer (5 neurons) after GlobalAveragePooling1D. The first convolution is performed with 2 filters and kernel size 15, the second one with 100 filters and size 10, and the third one with 8 filters with size 2.

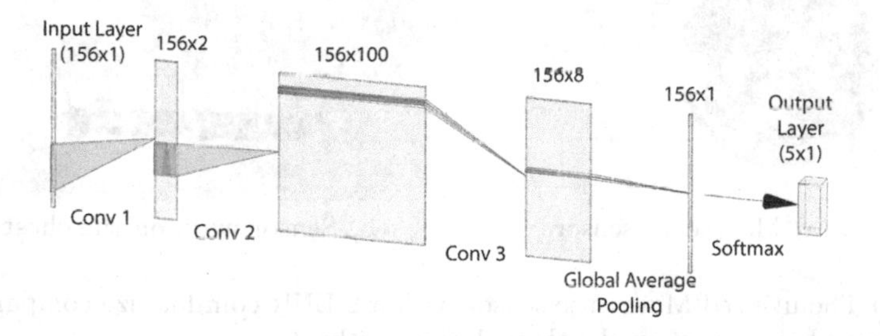


**Fig. 2.** The CNN architecture utilized for recognizing workout types.

### 3.4 Workout Repetition Counting

After the type of workout has been determined for a given workout recording, our approach automatically detects the amount of repetitions in this workout. In contrast to workout type recognition this requires a longer window. We therefore only count repetitions once the workout type recognition yields that workout for a certain workout type has finished (with either another or no workout being started afterwards). With the data of one such continuous workout that contains 3D accelerometer data, we at first apply PCA and extract only the strongest PC dimension. This transforms the 3D time series into a 1D time series. To the resulting time series we then apply a peak detection to detect repetitions. The peak detection has two parameters: the minimum distance $d_{min}$ between peaks, as well as the minimum height $h_{min}$ of the peak. For the latter we set an adaptive threshold $\alpha$ (Eq. 1). Once we surpass $\alpha$, the distance between this new candidate and the previous peak is calculated. If said distance is bigger than $d_{min}$, the candidate is counted as a peak.

$$\alpha = mean(\text{data}) + (max(\text{data}) - mean(\text{data})) \cdot h_{min} \quad (1)$$

We utilize different configuration of our peak detection for different workout types (Table 1). The amount of peaks corresponds to the amount of repetitions done in the workout, with exception of jumping jacks, for which the amount of repetitions is half the amount of peaks detected.

**Table 1.** Peak detection parameters for detecting repetitions in workout sessions.

| Workout | $d_{min}$ | $h_{min}$ |
|---|---|---|
| Pushups | $\frac{15}{52}$ s | 0.5 |
| Situps | $\frac{15}{52}$ s | 0.5 |
| Squats | $\frac{15}{52}$ s | 0.5 |
| Jumping jacks | $\frac{5}{52}$ s | 0.2 |

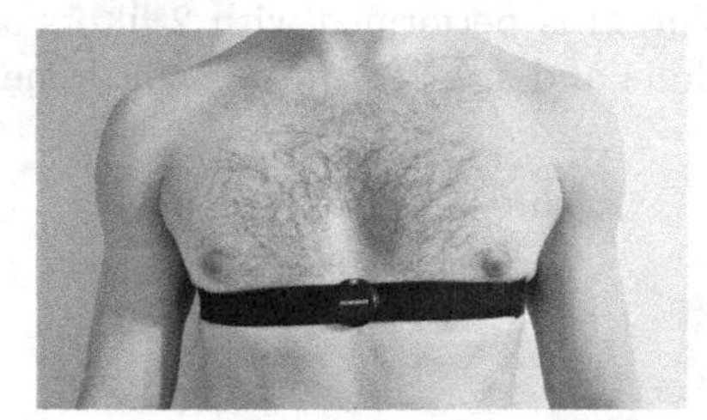

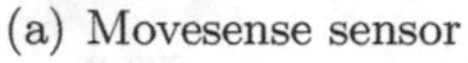
(a) Movesense sensor

(b) Sensor worn on the chest

**Fig. 3.** (a) The utilized Movesense sensor with a 2 EUR coin for size comparison, and (b) the sensor being worn on the chest during workout.

# 4 Evaluation

## 4.1 Utilized Sensor

For our study we utilize accelerometer values from a Movesense sensor [11] (Fig. 3). We selected the Movesense sensor due to its relatively compact hardware, with diameter of 36.6 mm and the thickness of 10.6 mm. The controlling unit is Nordic Semiconductor nRF52832 comprising a 32-bit ARM Cortex-M4 with 64 kB on-chip RAM and 512 kB on-chip FLASH. The communication of the sensor is based on Low Energy Bluetooth 4.0. Movesense is able to measure linear and angular acceleration, magnetic field intensity, temperature, heart rate and ECG. The sensors for this research were provided by the sensor manufacturer, MoveSensor, which is part of the Suunto Corporation. As our study aims to use accelerometer data only, we utilize only the accelerometer embedded in the movesense sensor. This accelerometer is a 3D accelerometer and provides a sampling frequency of 12.5/26/52/104/208 Hz. For our evaluation we configured the sensor to sample with 52 Hz as a trade-of between energy sampling accuracy and energy consumption, similar to sampling frequencies used in related work for similar purposes [8, 10]. Participants wear the sensor on their chest using an attachment strap which is part of Movesense Developer Kit.

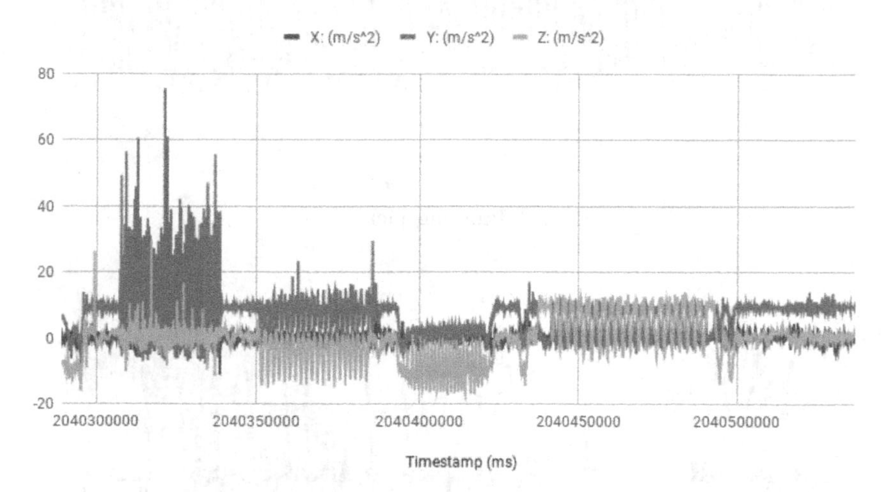


**Fig. 4.** Acceleration $[\frac{m}{s^2}]$ for an exercise set recorded on the chest. Contained exercises, from the left: jumping jacks, squats, pushups, and situps.

## 4.2 Evaluation Data

We recorded workout data of 10 different subjects with 2–3 workout sessions per subject, each performing the four workout types, and with no-workout phases in between workouts. Thereby, 11 workouts per workout type were recorded, resulting in a total of 44 workouts. Each workout thereby contains 10, 20 or 40 repetitions of the workout, depending on the workout type and the person. An example workout recording that contains jumping jacks, squats, pushups, and situps, in this order, is shown in Fig. 4. The no-workout parts of those samples contain diverse resting related activities (uncontrolled and individual for each participant), including sitting, standing, walking, drinking water, and the transition from one workout type to another one, like getting from push-ups. After data was recorded, the periods corresponding to the four target workout types, as well as all no-workout periods were annotated in the data. This resulted in a total of 55 workout samples (11 being no-workout). Examples for workout containing jumping jacks, situps, and squats, are shown in Fig. 5. Those extracted samples thereby form the basis for training and evaluating the workout type recognition and repetition counting.

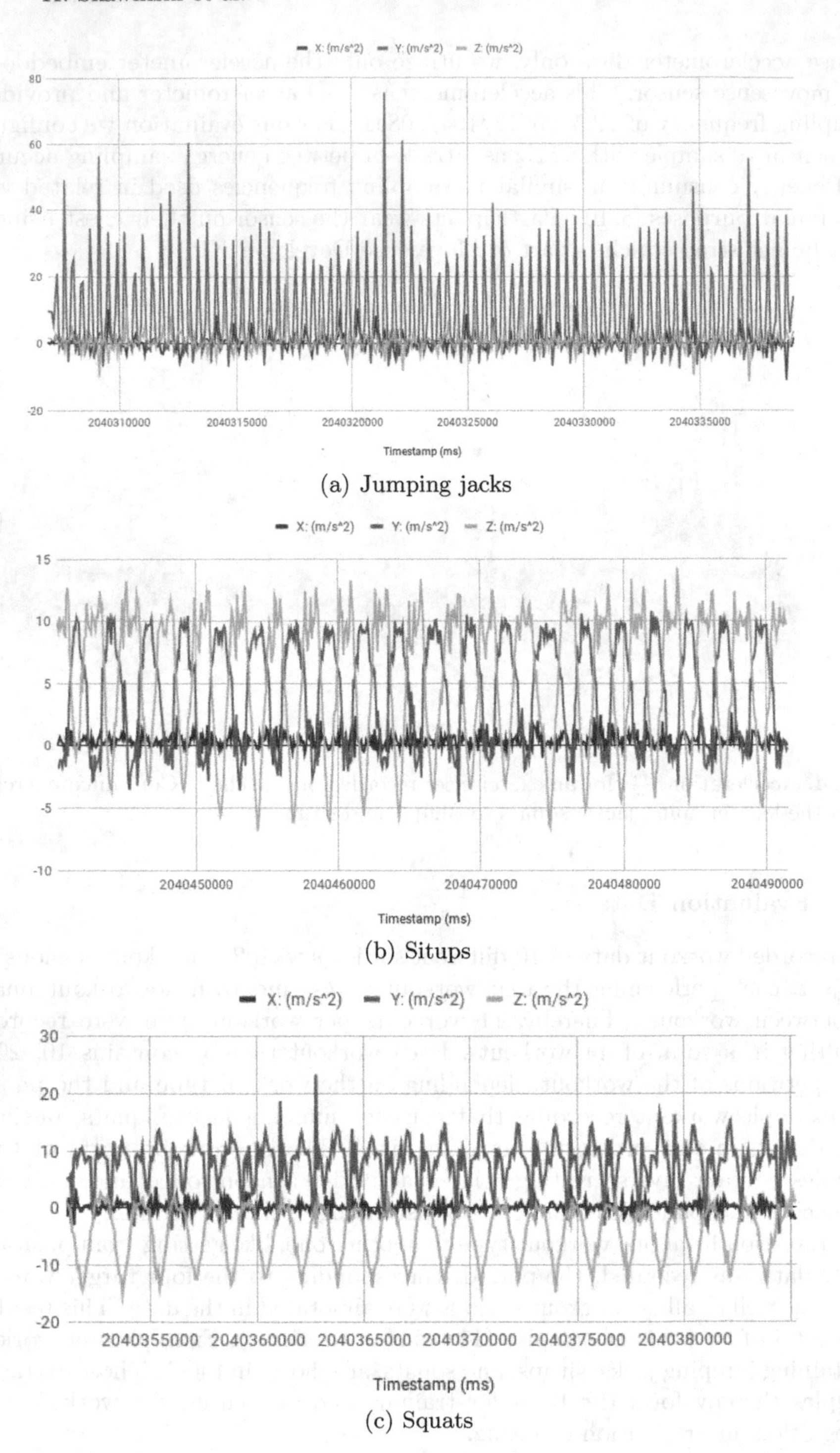


(a) Jumping jacks

(b) Situps

(c) Squats

**Fig. 5.** Acceleration $[\frac{m}{s^2}]$ for samples with different workout types recorded on the chest.

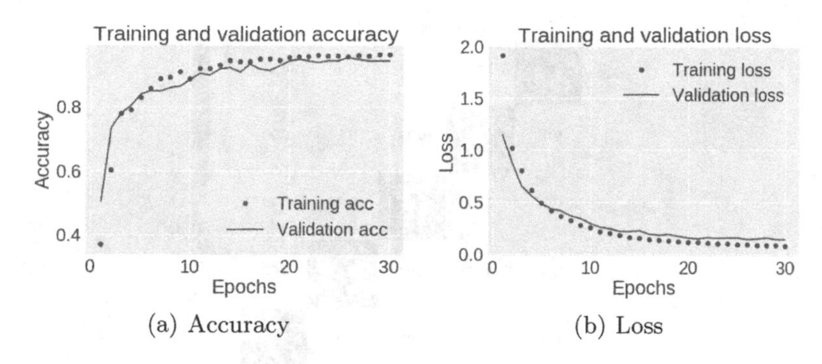


(a) Accuracy (b) Loss

**Fig. 6.** Accuracy (a) and loss (b) over training and validation data.

### 4.3 Evaluation Setup

To train and evaluate our approach, we utilize a training-validation test set data partitioning approach on the recorded evaluation data set. The training, validation, and test set contain 75%, 17.5%, and 7.5% of all samples, respectively. The CNN model is trained with the training data and a batch size of 512. The validation data is used to monitor the training progress and to stop training when the accuracy over the validation set does not improve anymore, within a small tolerance, for four consecutive epochs. The test data is used to report the final workout detection and type recognition rates once the model has been trained.

## 5 Results

### 5.1 Exercise Type Recognition Results

Results for workout type recognition in general indicate good recognition results. The overall accuracy is 90.6% for the validation set and 89.9% for the test set, with a final loss over test data of 0.206 (Fig. 6). The confusion matrix (Fig. 7) indicates only minor confusion between certain types of workout/no workout. Confusion is most frequent between no pushups and no exercise (10% of no pushups recognized as no exercise, 4% of no exercise recognized as pushups), and between situps and squats (10% of situps recognized as squats, and 7% of squats recognized as situps).

With the ready trained model, the time required to predict the workout type for one 1 s long workout sample was measured to be on average 39.74 µs, with a standard deviation of 6.79 µs (1000 predictions from test set samples). Those measurements were done on a Lenovo ThinkPad X1 Carbon with Intel Core i7-8550U 1.80 GHz × 8 processor and 16 GB of memory.

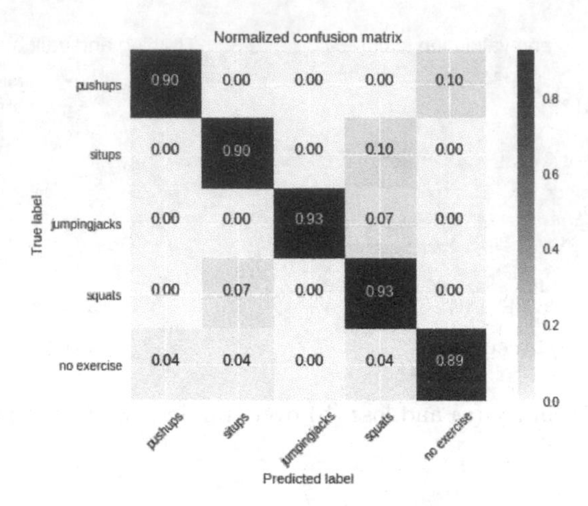


**Fig. 7.** Normalized confusion matrix.

## 5.2 Workout Repetition Counting Results

Once the workout type has been recognized, the workout repetition counting yields accuracies in between 97.4%–98.7% (Table 2). While errors with pushups and jumping jacks were caused by false negatives (repetitions not being detected), errors with situps and squats were caused by false positives (falsely detecting a repetition where there is none). The average detection accuracy of repetitions over the four workout types thereby is 97.9%. Examples for repetitions being detected in a continuous workout for doing pushups, situps, and jumping jacks are shown in Fig. 8.

**Table 2.** Repetition counting results per type of workout, with the total contained repetition, the repetitions detected by our approach, the amount of false positives and false negatives, and the accuracy resulting thereof.

| Workout | Contained | Detected | False negatives | False positives | Accuracy |
|---|---|---|---|---|---|
| Pushups | 116 | 113 | 3 | 0 | 97.4% |
| Situps | 159 | 161 | 0 | 2 | 98.7% |
| Squats | 136 | 139 | 0 | 3 | 97.8% |
| Jumping jacks | 172 | 168 | 4 | 0 | 97.7% |

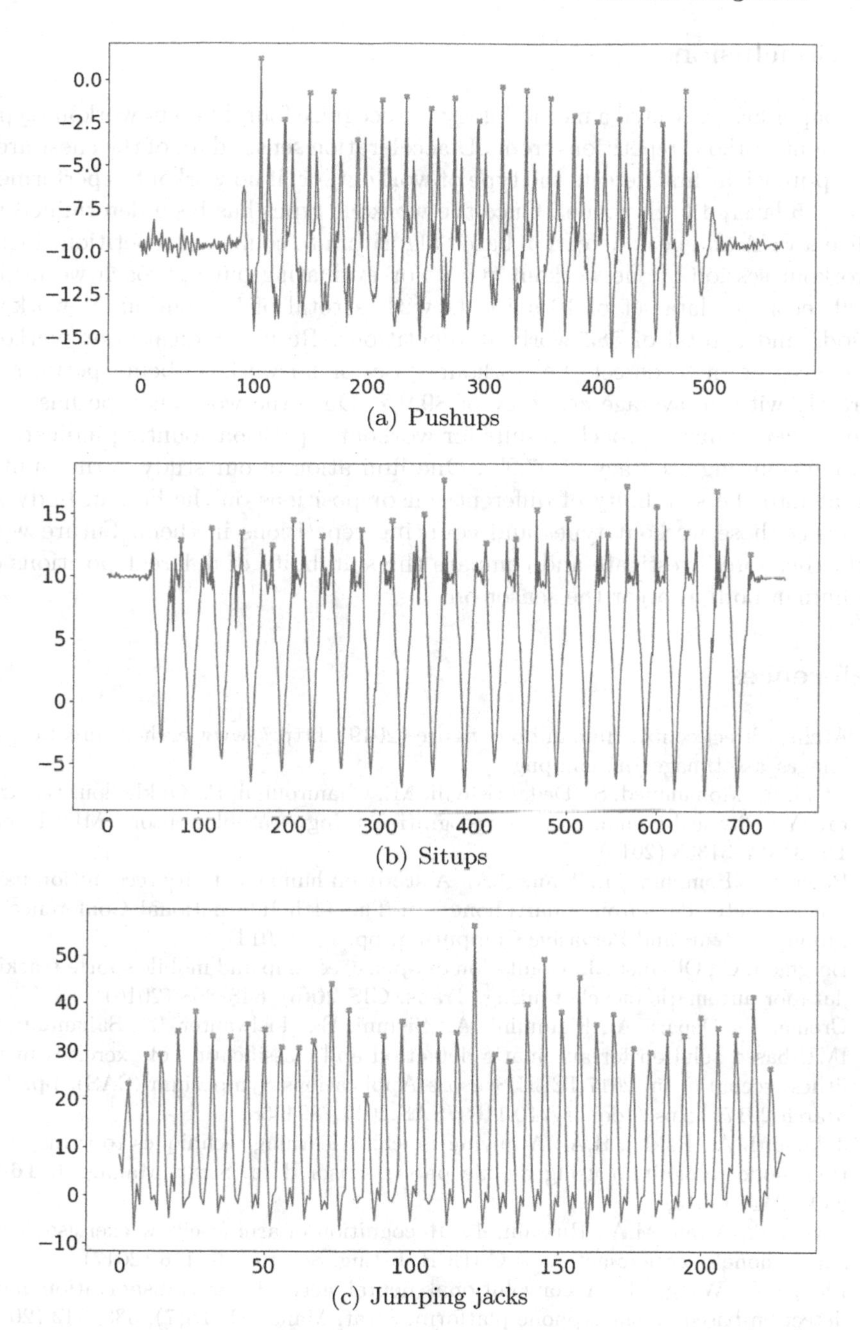


(a) Pushups

(b) Situps

(c) Jumping jacks

**Fig. 8.** Peak counting based on the acceleration $[\frac{m}{s^2}]$ sensed for samples with (a) pushups, (b) situps, and (c) jumping jacks.

## 6 Conclusions

This paper has presented a methodology to recognize four different workout types and count workout repetitions from 3D acceleration sensor data of the chest area. Our approach at first detects the type of workout, or, if no workout is performed, using a 5 layered CNN model. Once the workout types has been determined we utilize a PCA and peak detection based algorithm to count the repetitions inside a workout session of one workout type. For evaluating our approach we utilize a self-recorded data set of 10 subjects with a total of 55 continuous workout periods and a total of 583 workout repetitions. Results indicate our workout type recognition to detect the workout type, or no workout being performed, correctly with an average accuracy of 89.9%. Once the workout type has been determined by our approach, results for workout repetition counting indicate an average counting accuracy of 97.9%. One limitation of our study is the limited insight into the suitability of different sensor positions on the human body for detecting those workout types and counting repetitions in them. Future work could therefore investigate and compare the suitability of different positions on the human body to wear the sensor on.

## References

1. Archerydirect.co.nz.: Human body figure (2019). http://www.archerydirect.co.nz/images/assetimages/human.png
2. Attal, F., Mohammed, S., Dedabrishvili, M., Chamroukhi, F., Oukhellou, L., Amirat, Y.: Physical human activity recognition using wearable sensors. MDPI Sens. **15**, 31314–31338 (2015)
3. Bayat, A., Pomplun, M., Tran, D.A.: A study on human activity recognition using accelerometer data from smartphone. In: The 11th International Conference on Mobile Systems and Pervasive Computing, pp. 1–8 (2014)
4. Bergman, C., Oksanen, J.: Conflation of openstreetmap and mobile sports tracking data for automatic bicycle routing. Trans. GIS **20**(6), 848–868 (2016)
5. Crema, C., Depari, A., Flammini, A., Sisinni, E., Haslwanter, T., Salzmann, S.: IMU-based solution for automatic detection and classification of exercises in the fitness scenario. In: 2017 IEEE Sensors Applications Symposium (SAS), pp. 1–6, March 2017. https://doi.org/10.1109/SAS.2017.7894068
6. Jahangiri, A., Rakha, H.A.: Applying machine learning techniques to transportation mode recognition using mobile phone sensor data. Strat. Manag. J. **16**(5), 2406–2417 (2015)
7. Javed, T., Awan, M.A., Hussain, T.: Recognition of arm & elbow exercises using smartphone's accelerometer. NFC-IEFR J. Eng. Sci. Res. **5**, 1–6 (2017)
8. Liang, X., Wang, G.: A convolutional neural network for transportation mode detection based on smartphone platform. Strat. Manag. J. **18**(7), 338–342 (2017)
9. Manzoni, V., et al.: Transportation mode identification and real-time CO2 emission estimation using smartphones. Technical report, Massachusetts Inst. Technol., Cambridge, MA, USA (2010)
10. Morris, D., Saponas, T.S., Guillory, A., Kelner, I.: RecoFit: using a wearable sensor to find, recognize, and count repetitive exercises. In: Proceedings of the SIGCHI Conference on Human Factors in Computing Systems, pp. 3225–3234. ACM (2014)

11. Movesense: Movesense sensor description (2019). https://www.movesense.com/product/movesense-sensor/
12. Patel, S., Park, H., Bonato, P., Chan, L., Rodgers, M.: A review of wearable sensors and systems with application in rehabilitation. JNER **9**, 21 (2012)
13. Weka: Waikato environment for knowledge engineering. New Zealand Computer Science Research Students Conference, University of Waikato, Hamilton, New Zealand, pp. 57–64 (1995)

# Improving Wearable Activity Recognition via Fusion of Multiple Equally-Sized Data Subwindows

Oresti Banos[1](✉), Juan-Manuel Galvez[1], Miguel Damas[1], Alberto Guillen[1], Luis-Javier Herrera[1], Hector Pomares[1], Ignacio Rojas[1], and Claudia Villalonga[2]

[1] Research Center for Information and Communication Technologies, University of Granada, Granada, Spain
{oresti,jmgg,mdamas,aguillen,jherrera,hector,irojas}@ugr.es
[2] School of Engineering and Technology, Universidad Internacional de La Rioja, Logroño, Spain
claudia.villalonga@unir.net

**Abstract.** The automatic recognition of physical activities typically involves various signal processing and machine learning steps used to transform raw sensor data into activity labels. One crucial step has to do with the segmentation or windowing of the sensor data stream, as it has clear implications on the eventual accuracy level of the activity recogniser. While prior studies have proposed specific window sizes to generally achieve good recognition results, in this work we explore the potential of fusing multiple equally-sized subwindows to improve such recognition capabilities. We tested our approach for eight different subwindow sizes on a widely-used activity recognition dataset. The results show that the recognition performance can be increased up to 15% when using the fusion of equally-sized subwindows compared to using a classical single window.

**Keywords:** Activity recognition · Segmentation · Data window · Data fusion · Wearable sensors

## 1 Introduction

After some decades of research, the automatic recognition of human physical activity continues to attract an enormous interest from the scientific community. One major reason for such enduring interest is that activity recognition systems are used in a variety of application domains, ranging from industry [17,22], healthcare [11], transportation [12], gaming [13] or housekeeping tasks [16], to name a few. Beyond research prototypes, we can already find activity recognition-based systems in a handful of products, mostly oriented to wellness applications such as fitness trackers. However, various challenges remain


I. Rojas et al. (Eds.): IWANN 2019, LNCS 11506, pp. 360–367, 2019.
https://doi.org/10.1007/978-3-030-20521-8_30

for activity recognition systems to penetrate in more critical domains where a high level of accuracy and performance are required.

In the particular case of wearable activity recognition, i.e. the recognition of activities based on the automatic analysis of user's data collected through wearable sensors, some challenges have to do with the placement [19], orientation [20], robustness [4], battery [21] or number [8] of the sensors. Other important challenges relate to the design of the different phases of the so-called recognition chain, as to maximise the chances of recognising the activity carried out by the user. One of such phases is the segmentation, which refers to the partitioning of the continuous data stream produced by the sensors into discrete data windows. Each data window is sequentially processed to identify the activity performed by the user during that period of time. In this direction, we showed in a previous work [2] the great impact of the segmentation phase on the accuracy of the recognition models. Amongst other findings, this work proved the existing relation among activity categories and involved body parts with the window size utilized during the segmentation process. Driven by the goal of increasing the recognition performance, we also proposed in another work a multiwindow fusion mechanism that exploits the classification capabilities of several recognisers working on different window sizes [5]. This approach shows to generally improve the recognition accuracy at the expense of running as many recognition systems as window sizes are considered.

In this paper, we explore however the potential of fusing multiple subwindows but of a similar size. We hypothesize that while using a single window, especially of a relatively long duration, the system may not fully capture the dynamics of the performed activity, thus potentially leading to an erroneous recognition. To that end, we propose a methodology for splitting and characterising each data window with a number of subwindows (Sect. 2). Next, we evaluate the proposed approach on a well-known activity recognition dataset (Sect. 3) and discuss the results for each design choice (Sect. 4). Final conclusions are summarized at the end of the paper (Sect. 5).

## 2 Methods

The most common approach for recognising activities from sensor data involves a sequence of steps, a.k.a. activity recognition chain or pipeline, normally consisting of sensor data acquisition, segmentation, feature extraction and classification. Depending on the quality of the collected data, some level of preprocessing might be required (e.g., band-pass filtering to remove bias and high frequency noise). Such preprocessing is however most often available on the device as a hardware component; therefore, we assume here that the sensor data stream is ready for segmentation. Likewise, feature selection or dimensionality reduction techniques are sometimes used to determine the optimal set of features to be extracted from each data window. While the feature selection could be also considered as another step of the pipeline, we rely on the system designer's choice of features, and as such, this step is not explicitly addressed in the proposed method.

Let us consider a classical activity recognition system which segments the sensor data stream into fixed windows of $P$ seconds (i.e., recognition period). Such system will provide, after processing the considered data window, a label identifying the activity performed by the person during those $P$ seconds. The approach proposed here revolves around the idea of splitting such window into smaller windows, hereafter subwindows, processing each subwindow separately, and then fusing the classification decisions or labels yielded from each of these subwindows, delivering eventually a unique recognised activity. For practical reasons, the subwindow sizes should be divisors of the system's recognition period. The complete structure of the proposed model is shown in Fig. 1, while its mathematical foundation is described in the following.

Generally, let us consider a problem with $K$ classes or activities, $k = 1, ..., K$. Also, let us consider the system's data collection is composed by $M$ sensors, each one producing a separate data stream $u_m$, with $m = 1, ..., M$. Each sensor data stream $u_m$ is segmented by using $N$ different window sizes, $W_n$ with $W_{n-1} < W_n$ and $W_n$ divisor of $W_N$ for all $n = 1, ..., N$, and $W_N$ formally representing the system's recognition period. This leads to the creation of $M \times N$ segmentation pipelines, in which every data window of size $W_N$, i.e., $s^{W_N}$, is split into $W_N/W_n$ subwindows of size $W_n$, i.e., $\{s_{m1}^{W_n}, ..., s_{mi}^{W_n}, ..., s_{mW_N/W_n}^{W_n}\}$, for all $i = 1, ..., W_N/W_n$. All subwindows $s_{mi}^{W_n}$ are transformed into features, $f(s_{mi}^{W_n})$, which are then aggregated across sensors into feature vectors $\{f(s_{1i}^{W_n}), ..., f(s_{mi}^{W_n}, ..., f(s_{Mi}^{W_n}))\}$. These feature vectors are input to each corresponding classifier, yielding a recognised activity or label per subwindow, $c_i^{W_n}$.

As a result of the previous process, we obtain a vector with $W_N/W_n$ labels, $\{c_1^{W_n}, ..., c_i^{W_n}, ..., c_{W_N/W_n}^{W_n}\}$. These labels are then used to determine the most probable activity for each data window ($W_N$). Different approaches can be used, however, for the sake of simplicity, we propose to use a majority voting rule:

$$D_{ik}\left(c_i^{W_n}\right) = \begin{cases} 1, & c_i^{W_n} = k \\ 0, & c_i^{W_n} \neq k \end{cases} \tag{1}$$

The eventual recognised class is defined as the one obtaining the highest score:

$$c^{W_n} = \underset{k}{\text{argmax}} \left( \sum_{i=1}^{W_N/W_n} \cdot D_{ik}\left(c_i^{W_n}\right) \right) \tag{2}$$

## 3 Evaluation

### 3.1 Experimental Setup

For the evaluation of the proposed approach we use one of the most popular wearable activity recognition datasets publicly available [1,3]. This dataset comprises motion data, namely, acceleration, rate of turn and magnetic field orientation, recorded for 17 volunteers while carrying out 33 physical activities ($K = 33$).

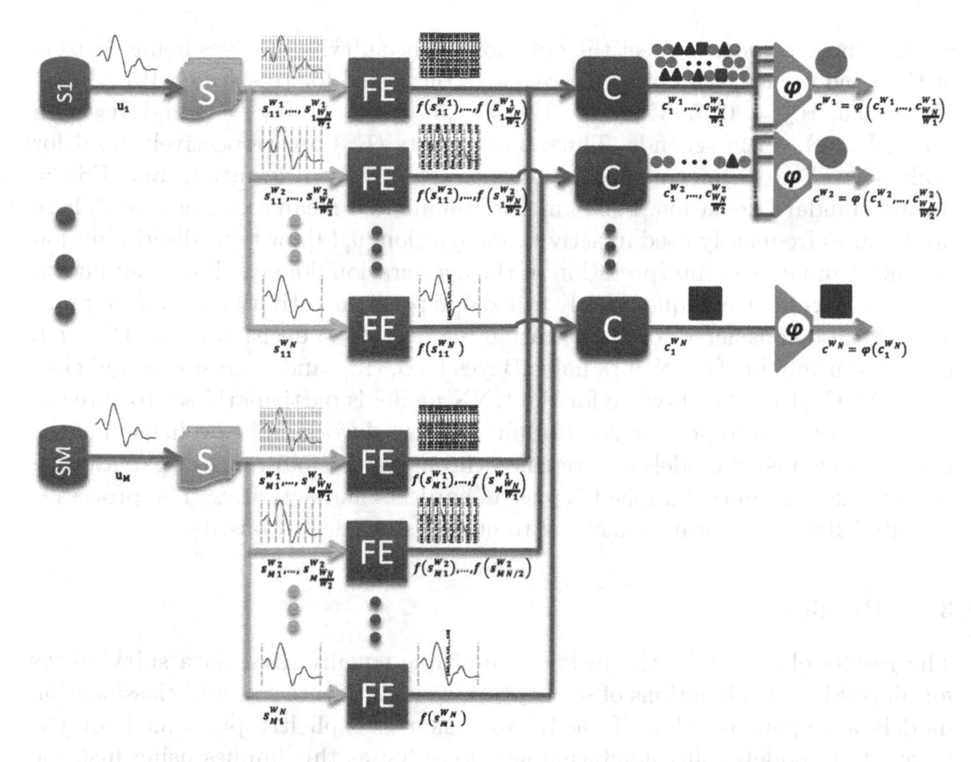


**Fig. 1.** Fusion of multiple equally-sized data subwindows. Given M raw sensor data streams, $u_m$ with $m \in [1, M]$, each one is segmented into $n$ data windows of size $W_n$, $s_{mi}^{W_n}$, with $n \in [1, N]$, $i \in [1, k]$ $\forall$ $k|N$. A set of features are extracted from each data subwindow, $f(s_{mi}^{W_n})$, which are then aggregated into a vector combining the computed features for the given subwindow for all sensors $\{f(s_{1i}^{W_n}), f(s_{2i}^{W_n}), ..., f(s_{Mi}^{W_n})\}$. Each feature vector is input to the n-*th* classifier (one per subwindow size), yielding a recognised activity label, $c_i^{W_n}$. The recognised labels are fused to determine, for each of the considered subwindow sizes, the eventually recognised activity $c^{W_n}$.

A set of nine inertial sensors ($M = 9$) attached to different parts of their bodies was used for the motion recording. From all measured magnitudes, only the acceleration data is considered here since this turns to be the most prevalent sensor modality in previous activity recognition works. The potential of this dataset stems from the number of considered activities, diversity of body parts involved, as well as the variety in intensity and dynamicity of the actions. Moreover, all the recordings were collected in the wild, with no constraints whatsoever on the way the activities must be executed.

The activity recognition models devised for evaluation are described next. As it was stated in Sect. 2, no preprocessing of the data is applied to avoid the removal of any relevant information. This is normal practice when the activities are of a diverse nature. A window size of 6 s is considered as the reference value for the recognition system, as this figure has been used in a number of previous works. Thus, in order to meet the criterion of the subwindow sizes being divisors of the window size, the following sizes are considered for this study, $W_1 = 0.25$, $W_2 = 0.5$, $W_3 = 0.75$, $W_4 = 1$, $W_5 = 1.5$, $W_6 = 2$, $W_7 = 3$ and $W_8 = 6$ (i.e., $N = 8$), all in seconds. Three feature sets (FS) are respectively used for evaluation: FS1 = 'mean', FS2 = 'mean and standard deviation' and FS3 = 'mean, standard deviation, maximum, minimum and mean crossing rate'. These are features frequently used in activity recognition [9,14] for their discrimination potential and ease of interpretation in the acceleration domain. Four well-known machine learning techniques widely utilized in previous activity recognition problems are also considered for classification, namely, C4.5 decision trees (DT, [7]), k-nearest neighbors (KNN, [6]), naive Bayes (NB, [18]) and nearest centroid classifier (NCC, [15]). The k-value for the KNN model is particularly set to three as it has been shown to provide good results in related works. The evaluation of the multiwindow fusion models is performed through a ten-fold random-partitioning cross validation process applied across all subjects and activities. The process is repeated 100 times for each method to ensure statistical robustness.

### 3.2 Results

The results obtained for the fusion of multiple equally-sized data subwindows for all possible combinations of segmentation sizes, feature sets and classification models are summarised in Table 1. No fusion is explicitly performed for the recognition models using a subwindow size of 6 s, as this implies using just one window. Thus, the results presented for this case coincide with the performance obtained at the classification level, i.e., before applying majority voting.

Generally speaking, the segmentation into subwindows has a great effect on the improvement of the recognition capabilities of the system. This result is observed across all feature sets and classification paradigms. Thus, for example, we can find a growth between 10% and 15% with respect to the baseline (i.e., 6 s) when using subwindows of 0.25 s and DT. For the same subwindow size, the results vary from 3% to 6% for KNN and up to 4% for NB, with nevertheless a notable decrease in performance of up to 6% for NCC. In the case of NB we obtain up to 8% improvement with respect to baseline when using subwindows of 0.75 s and 1 s. For subwindows of 1 and 1.5 s we can reach up to 10% improvement with respect to baseline for NCC. The already good-enough results for baseline in KNN are improved around 4% to 6% for subwindows of 0.25 and 0.5 s. It should be also noted that a perfect recognition capabilities (F-score = 1) are reached with this model.

**Table 1.** Multisubwindow fusion performance ($F - score$) for all subwindow sizes ($W_1 = 0.25$, $W_2 = 0.5$, $W_3 = 0.75$, $W_4 = 1$, $W_5 = 1.5$, $W_6 = 2$, $W_7 = 3$ and $W_8 = 6$), classification paradigms (DT, NB, NCC, KNN) and feature sets (FS1 = mean, FS2 = mean and standard deviation, FS3 = mean, standard deviation, maximum, minimum and mean crossing rate).

| Window size (s) | DT | | | NB | | | NCC | | | KNN | | |
|---|---|---|---|---|---|---|---|---|---|---|---|---|
| | FS1 | FS2 | FS3 | FS1 | FS2 | FS3 | FS1 | FS2 | FS3 | FS1 | FS2 | FS3 |
| 0.25 | **0.9995** | **0.9992** | **0.9994** | 0.9193 | 0.9366 | 0.9675 | 0.7340 | 0.8343 | 0.8842 | **1.0** | **1.0** | **1.0** |
| 0.5 | 0.9900 | 0.9987 | 0.9987 | 0.9345 | 0.9727 | 0.9851 | 0.8306 | 0.9164 | 0.9566 | **1.0** | **1.0** | **1.0** |
| 0.75 | 0.9827 | 0.9972 | 0.9980 | 0.9450 | 0.9813 | **0.9873** | 0.8626 | 0.9379 | 0.9707 | 0.9995 | 0.9999 | 0.9999 |
| 1 | 0.9817 | 0.9961 | 0.9970 | **0.9479** | **0.9824** | 0.9863 | 0.8843 | **0.9561** | **0.9765** | 0.9994 | 0.9999 | 0.9999 |
| 1.5 | 0.9702 | 0.9890 | 0.9908 | 0.9423 | 0.9792 | 0.9821 | **0.8906** | 0.9552 | 0.9750 | 0.9970 | 0.9997 | 0.9998 |
| 2 | 0.9559 | 0.9795 | 0.9782 | 0.9307 | 0.9800 | 0.9830 | 0.8751 | 0.9504 | 0.9703 | 0.9952 | 0.9988 | 0.9993 |
| 3 | 0.9128 | 0.9410 | 0.9324 | 0.9208 | 0.9675 | 0.9661 | 0.8565 | 0.9377 | 0.9604 | 0.9793 | 0.9885 | 0.9924 |
| 6 | 0.8506 | 0.8961 | 0.8929 | 0.8781 | 0.9337 | 0.9249 | 0.7956 | 0.8977 | 0.9011 | 0.9363 | 0.9643 | 0.9476 |

# 4 Discussion

Choosing a proper window size for the activity recognition model can often be the key to realising a system with good or poor recognition capabilities. While a simple solution may be to use those sizes that worked just fine for previous recognition systems, we could jeopardise by doing so the capabilities of our system. Hence, it is generally recommended to pay special attention to the segmentation phase and fine tune the window size value based on the considered sensors and targeted activity set.

In view of the results in this work, a valid alternative to selecting a single window size seems to be to use a combination of small windows, thus avoiding the need of engineering the optimal window size. In our earlier work [2] we demonstrated that very small windows (0.5 s, 0.75 s) normally lead to relatively poor recognition results, with some activity recognisers yielding F-scores below 0.5. However, in this paper we show that such small windows have a lot to offer when used in combination, outperforming even the performance reached with the optimal window sizes (1 to 2 s).

The above conclusion extends to any of the considered feature set and classification paradigms but for the NCC case. We do believe that such behaviour may have to do with the limitations of this classification model to capture the dynamics of the activities when considering such short windows, already reported in previous works when detecting short actions or gestures [10].

Despite the significant performance improvement, the proposed model is not without limitations. In this work we did not evaluate the complexity added from using a single window size to multiple subwindows. However, as it can be derived at a glance from Fig. 1, the smaller the subwindow size is the more subwindows have to be processed (i.e., features extracted and classifications made). The ever growing computational resources available on both edge and cloud computing devices make this however less of an issue compared to past years. Either way, this should be taken into consideration specially when it comes to embedded computing systems where battery optimisation is of much relevance.

With the latest breakthroughs in machine learning, the engineering of activity recognition systems has somehow shifted to the background. While deep learning models show splendid results within a number of application domains, surpassing in many cases expert-driven feature extraction and probabilistic classification, some genuine aspects of the activity recognition problem, like the sensor data stream segmentation, are still ahead of these models. By means of this work, we take the opportunity to highlight the importance of valuing the engineering of all phases of the activity recognition chain, as well as the existing need for defining new segmentation strategies.

## 5 Conclusions

Using a proper window size has been shown in prior works to play a major role in the activity recognition system capabilities, which becomes quite often more important than the choice of features or machine learning models. In this work we propose a new segmentation technique that exploits the potential of equally-sized subwindows to capture the dynamics of different activities. Our results prove this approach to perform better than relying on a single window size, at the expense of increasing computational costs. Future work will explore the validation of our preliminary findings for other sensor modalities, e.g., angular velocity, and activity types, e.g., activities of the daily living.

**Acknowledgments.** This work was partially supported by the Spanish Ministry of Economy and Competitiveness (MINECO) Projects TIN2015-71873-R and TIN2015-67020-P together with the European Fund for Regional Development (FEDER). This work was also partially funded by the "User Behaviour Sensing, Modelling and Analysis" contract OTRI-UGR-4071.

## References

1. Banos, O., Damas, M., Pomares, H., Rojas, I., Toth, M.A., Amft, O.: A benchmark dataset to evaluate sensor displacement in activity recognition. In: ACM International Conference on Ubiquitous Computing, pp. 1026–1035 (2012)
2. Banos, O., Galvez, J.M., Damas, M., Pomares, H., Rojas, I.: Window size impact in human activity recognition. Sensors **14**(4), 6474–6499 (2014)
3. Banos, O., Toth, M.A., Damas, M., Pomares, H., Rojas, I.: Dealing with the effects of sensor displacement in wearable activity recognition. Sensors **14**(6), 9995–10023 (2014)
4. Banos, O., et al.: Multi-sensor fusion based on asymmetric decision weighting for robust activity recognition. Neural Process. Lett. **42**(1), 5–26 (2015)
5. Banos, O., et al.: Multiwindow fusion for wearable activity recognition. In: Rojas, I., Joya, G., Catala, A. (eds.) IWANN 2015. LNCS, vol. 9095, pp. 290–297. Springer, Cham (2015). https://doi.org/10.1007/978-3-319-19222-2_24
6. Cover, T., Hart, P.: Nearest neighbor pattern classification. IEEE Trans. Inf. Theory **13**(1), 21–27 (1967)
7. Duda, R.O., Hart, P.E., Stork, D.G.: Pattern Classification, 2nd edn. Wiley, Hoboken (2000)

8. Ertugrul, O.F., Kaya, Y.: Determining the optimal number of body-worn sensors for human activity recognition. Soft Comput. **21**(17), 5053–5060 (2017)
9. Figo, D., Diniz, P.C., Ferreira, D.R., Cardoso, J.M.P.: Preprocessing techniques for context recognition from accelerometer data. Pers. Ubiquitous Comput. **14**(7), 645–662 (2010)
10. Forster, K., Roggen, D., Troster, G.: Unsupervised classifier self-calibration through repeated context occurences: is there robustness against sensor displacement to gain? In: International Symposium on Wearable Computers, pp. 77–84 (2009)
11. Guo, X., Liu, J., Chen, Y.: FitCoach: virtual fitness coach empowered by wearable mobile devices. In: IEEE Conference on Computer Communications, pp. 1–9. IEEE (2017)
12. Hur, T., Bang, J., Kim, D., Banos, O., Lee, S.: Smartphone location-independent physical activity recognition based on transportation natural vibration analysis. Sensors **17**(4), 1–21 (2017)
13. Jablonsky, N., McKenzie, S., Bangay, S., Wilkin, T.: Evaluating sensor placement and modality for activity recognition in active games. In: Australasian Computer Science Week Multiconference, p. 61 (2017)
14. Kwapisz, J.R., Weiss, G.M., Moore, S.A.: Activity recognition using cell phone accelerometers. Int. Conf. Knowl. Discov. Data Min. **12**, 74–82 (2011)
15. Lam, W., Keung, C.K., Ling, C.X.: Learning good prototypes for classification using filtering and abstraction of instances. Pattern Recognit. **35**(7), 1491–1506 (2002)
16. Liu, K.C., Yen, C.Y., Chang, L.H., Hsieh, C.Y., Chan, C.T.: Wearable sensor-based activity recognition for housekeeping task. In: International Conference on Wearable and Implantable Body Sensor Networks, pp. 67–70. IEEE (2017)
17. Malaisé, A., Maurice, P., Colas, F., Charpillet, F., Ivaldi, S.: Activity recognition with multiple wearable sensors for industrial applications. In: International Conference on Advances in Computer-Human Interactions (2018)
18. Theodoridis, S., Koutroumbas, K.: Pattern Recognition, 4th edn. Academic Press, Cambridge (2008)
19. Villalonga, C., Pomares, H., Rojas, I., Banos, O.: MIMU-Wear: ontology-based sensor selection for real-world wearable activity recognition. Neurocomputing **250**, 76–100 (2017)
20. Yurtman, A., Barshan, B.: Activity recognition invariant to sensor orientation with wearable motion sensors. Sensors **17**(8), 1–24 (2017)
21. Zappi, P., Roggen, D., Farella, E., Tröster, G., Benini, L.: Network-level power-performance trade-off in wearable activity recognition: a dynamic sensor selection approach. ACM Trans. Embed. Comput. Syst. **11**(3), 68:1–68:30 (2012)
22. Zhao, W., et al.: A human-centered activity tracking system: toward a healthier workplace. IEEE Trans. Hum.-Mach. Syst. **47**(3), 343–355 (2017)

# New and Future Tendencies in Brain-Computer Interface Systems

# Preliminary Results Using a P300 Brain-Computer Interface Speller: A Possible Interaction Effect Between Presentation Paradigm and Set of Stimuli

Álvaro Fernández-Rodríguez(✉), María Teresa Medina-Juliá, Francisco Velasco-Álvarez, and Ricardo Ron-Angevin

Departamento de Tecnología Electrónica, Universidad de Málaga, 29071 Malaga, Spain
{afernandezrguez,maytemed,rron}@uma.es, fvelasco@dte.uma.es

**Abstract.** Several proposals to improve the performance controlling a P300-based BCI speller have been studied using the standard row-column presentation (RCP) paradigm. However, this paradigm could not be suitable for those patients with lack of gaze control. To solve that, the rapid serial visual presentation (RSVP) paradigm, which presents the stimuli located in the same position, has been proposed in previous studies. Thus, the aim of the present work is to assess if a stimuli set of pictures that improves the performance in RCP, could also improve the performance in a RSVP paradigm. Six able-bodied participants have controlled four conditions in a calibration task: letters in RCP, pictures in RCP, letters in RSVP and pictures in RSVP. The results showed that pictures in RCP obtained the best accuracy and information transfer rate. The improvement effect given by pictures was greater in the RCP paradigm than in RSVP. Therefore, the improvements reached under RCP may not be directly transferred to the RSVP.

**Keywords:** Brain-computer interface (BCI) · P300 · Speller · Stimuli · Rapid serial visual presentation (RSVP)

## 1 Introduction

Several diseases may provoke deterioration in the motor abilities of affected patients. Some of these diseases, such as the amyotrophic lateral sclerosis (ALS), can cause the locked-in state, in which the patient has his whole body immobilized except for some ocular movements [1].

Numerous systems, that make use of the motor skills remaining in patients, have been developed to give those patients a communication channel. Some examples are push-buttons or eye-trackers [2]. However, more severe cases of the ALS disease lead to the so-called complete locked-in state, in which there is even no presence of gaze control [3, 4]. In the case of these patients, the communication system cannot be supported by any kind of muscular movements, including eye-gaze. This is when brain-computer


I. Rojas et al. (Eds.): IWANN 2019, LNCS 11506, pp. 371–381, 2019.
https://doi.org/10.1007/978-3-030-20521-8_31

interfaces (BCI) come into play. BCI is a type of technology that uses the patient's brain signal to establish a channel of communication between him/her and the device to be controlled [5]. Due to its relatively low cost and adequate temporal resolution, the technology most widely used for recording the brain signal is the electroencephalography (EEG).

Verbal communication is one of the main needs for patients, to the point of being able to take important decisions as receiving life-sustaining treatments [6]. Therefore, it is not surprising that the virtual speller is the application most developed by researchers in the field of BCI. The first BCI speller was proposed in 1988 by Farwell and Donchin [7]. It consisted of a matrix of letters whose rows and columns were flashed at random – it is called row-column presentation (RCP) – while the user had to focus his attention on the target letter. Each time that flashes the row or column which contains the target letter, the user perceives its flash and consequently it is evoked an event-related potential (ERP), known as P300, on the user's brain signal. Specifically, the P300 potential is a positive deflection in the voltage of the EEG signal, which is generally registered from the parietal lobe of the cortex and around 300 ms after the presentation of an uncommon target stimulus [8].

The P300 potential when used in a speller is influenced by numerous parameters, such as the number of elements of the matrix, the stimuli presentation duration or the use of some stimuli over others (see [9] for a detailed review of BCI spellers). In relation to this last parameter, the presented stimuli, it has been shown how the use of symbols of different colors or complex figures, instead of simple letters, can increase performance [10–12]. However, other paradigms have gone beyond the flashing of the symbol to be selected, and have opted for the superposition of the letters with certain images that increase the amplitude of the P300. To this day, the stimulus with the best results are familiar faces [13, 14]. However, a preliminary study carried out by the research group of the present study – the UMA BCI group – revealed the possibility that a set of varied different pictures (e.g., photographs of things, people or places) as flashing stimuli – a different one for each symbol – could improve the performance compared to the classical paradigm of letters [15].

It should not be forgotten that the main purpose of these applications is to offer an additional channel of communication to patients with severely impaired motor skills, including the ability of ocular mobility. Therefore, the presentation paradigm is a factor to consider in order to meet the aim of establishing a communication channel for those patients. Currently, the most used presentation paradigm for a speller is the RCP; however, this paradigm requires the user to focus on the target symbol. Consequently, it has been proved that the user's performance is drastically affected when the user cannot focus on the desired symbol [16, 17]. To solve this problem, another kind of presentation paradigm that does not require ocular mobility has been shown. The most developed paradigm for this purpose is the rapid serial visual presentation (RSVP) with the symbols located in the same position (e.g., [17–21]).

Despite the fact that the most suitable presentation paradigm for patients could be the RSVP, most of proposals have been developed under the RCP paradigm. Thus, it should be questioned if the findings made in RCP can be applied in RSVP with the stimuli presented in the same position. For example, in RSVP, the stimuli could compete with each other as they all appear in the same position in the screen. While in the RCP paradigm, in the position of the desired symbol, only the stimulus of the corresponding symbol appears. This implicates that the positive deflection of the EEG caused by the non-target stimuli can only occur due to the peripheral vision in RCP, but in RSVP it occurs due to the central vision. On the other hand, the positive deflection of the EEG caused by the target stimuli occurs due to the central vision in both paradigms. Therefore, those stimuli that increase the amplitude of the P300 signal could produce a clear improvement in RCP performance, but not in RSVP.

In short, the aim of the present study is to test two stimuli sets under two different presentation paradigms to assess the differences between conditions. Our hypothesis is that in RCP the benefit of pictures versus letters will be much greater than in the RSVP paradigm.

## 2 Method

### 2.1 Participants

The experiment involved six able-bodied participants (aged 27.17 ± 6.04, 3 females) who had normal or corrected-to-normal vision. Three subjects had previous experience controlling BCI systems and the other three did not. The study was approved by the Ethics Committee of the University of Malaga and met the ethical standards of the Helsinki Declaration. According to self-reports, none of the participants had any history of neurological or psychiatric illness or were taking any medication regularly.

### 2.2 Data Acquisition and Signal Processing

The EEG was recorded at a sample rate of 250 Hz using the electrode positions: Fz, Cz, Pz, Oz, P3, P4, PO7 and PO8, according to the 10/20 international system. All channels were referenced to TP8 and grounded to position AFz. Signals were amplified by an acti-CHamp amplifier (Brain Products GmbH, Munich, Germany). Neither online nor offline artifact detection techniques were employed. All aspects of EEG data collection and processing were controlled by the BCI2000 system [22]. A Stepwise Linear Discriminant Analysis (SWLDA) of the EEG data was performed to obtain the weights for the P300 classifier and calculate the accuracy. The software used to design the interface was developed by the UMA-BCI group from the University of Malaga. This software is named UMA-BCI Speller, and serves as a friendly front-end of BCI2000.

### 2.3 The Spelling Conditions

The present work employed four conditions, based on the combination of two paradigms (RCP and RSVP) and two sets of stimuli (letters versus pictures) (Fig. 1). The only difference between the compared conditions was the employed presentation paradigm and flashing stimuli for each condition. Thus, the four presented conditions were: (i) letters in RCP (L-RCP), (ii) pictures in RCP (P-RCP), (iii) letters in RSVP (L-RSVP), and iv) pictures in RSVP (P-RSVP) (Fig. 1). The font used for letters was arial bold in capital letters. On the one hand, the RCP conditions (2 × 3 matrix size) were initially based on the previously mentioned paradigm of [7]. On the other hand, for the RSVP, only the flashing stimuli were presented, placed in the middle of the screen. Letters had a size around 3 cm × 3.5 cm, depending on the letter. Regarding the pictures, their size was equal to 4.7 cm × 3.5 cm (187 px × 140 px). The conditions were displayed on a 15.6-in (39.6 cm) screen at a distance of 60 cm and at a refresh rate of 60 Hz. For all conditions, a stimulus onset asynchrony (SOA) of 288 ms and an inter-stimulus interval (ISI) of 96 ms were used, so each stimulus was presented for 192 ms. A 3968 ms pause was established between each selection.

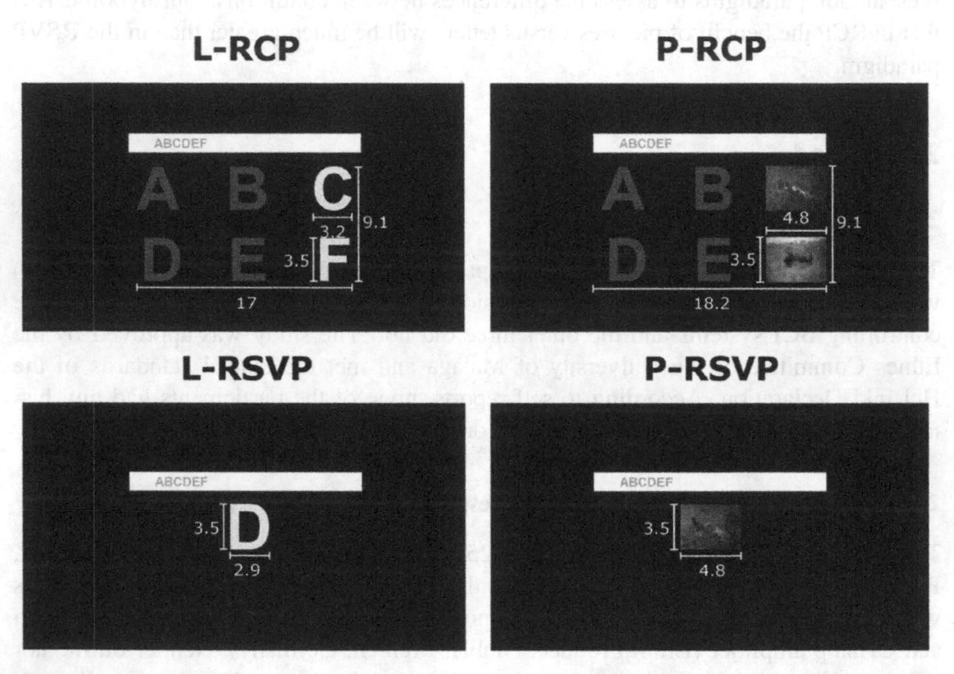


**Fig. 1.** Size metrics (cm) for all conditions: letters in row-column presentation (L-RCP), pictures in row-column presentation (P-RCP), letters in rapid serial visual presentation (L-RSVP), and pictures in rapid serial visual presentation (P-RSVP). Due to copyright reasons, the condition with pictures has been pixelated.

Regarding the pictures, they were obtained from the International Affective Picture System (IAPS) [23]. Those images with the lowest score in the level of excitation were chosen, and also that filled the proportion of the aforementioned size (i.e., those pictures that filled all the space and did not have black paddings). The specific selected pictures according to the IAPS codification for the matrix in major-row order were: 7175, 7010, 7004, 7031, 7020 and 7110.

## 2.4 Procedure

The experiment was carried out in an isolated room where only the participant was present at the time he/she was performing the task in order to concentrate on it without external distractions. It consisted of only one exercise: a calibration task to adapt the system to the user. There was no writing task in which the user actually controlled the interface. Consequently, the study was performed in one session.

An intrasubject, also called repeated measures, design was used, and so all the users went through all the experimental conditions.

The experiment was carried out in 16 blocks of six selections, so each condition was performed in four blocks (i.e., 24 selections per condition). The symbol selection order for each block was A, B, C, D, E and F. The conditions were presented pseudo-randomly to prevent any unwanted effect, such as learning or fatigue, and all conditions were equally distributed. There was a short break between blocks (variable at the request of the user). The number of sequences per run (when the user is selecting one symbol) was equal to 6, so each symbol flashed 12 times in RCP (6 flashes per row and 6 flashes per column) and 6 times in RSVP. The participant was asked to count these flashes in order to keep his/her attention on the task. The writing time for each symbol was 8.54 s in RCP and 10.27 s in RSVP.

### 2.5 Evaluation

Two parameters were used to evaluate the effect of the presentation paradigm and stimuli type on the performance: (i) the *accuracy* of the system classifying the selections (i.e., the number of correctly predicted selections divided by the total number of predicted selections) and (ii) the *information transfer rate* (*ITR*, bits/min) based on the formula presented in [24]. It should be advised that the pause between selections is not considered to calculate the *ITR*. In addition, the ERP signal was analyzed in order to observe how the brain signal activity is affected by the application of different sets of stimuli and paradigm. Specifically, the dependent variable employed was the *amplitude difference* in absolute value between target and no target stimuli signals (μV) from 200 to 600 ms.

Due to the small sample size, non-parametric analyses were carried out. Initially, a Freidman's test was performed to find differences between conditions. Then, multiple Wilcoxon signed-rank tests were carried out in order to find those specific differences between conditions. Due to the preliminary nature of the present study, no correction method was applied for multiple comparisons. The obtained conclusions should be considered carefully, so more tests will be necessary to replicate the results and avoid type I error.

## 3 Results and Discussion

### 3.1 Accuracy

Figure 2 visually shows how the *accuracy* of the different conditions become similar as the sequences increase. In general, all conditions offered good results in *accuracy*. However, it should be noted that, despite there being no significant differences, the P-RCP condition obtained 100% *accuracy* for all participants from sequence 2.

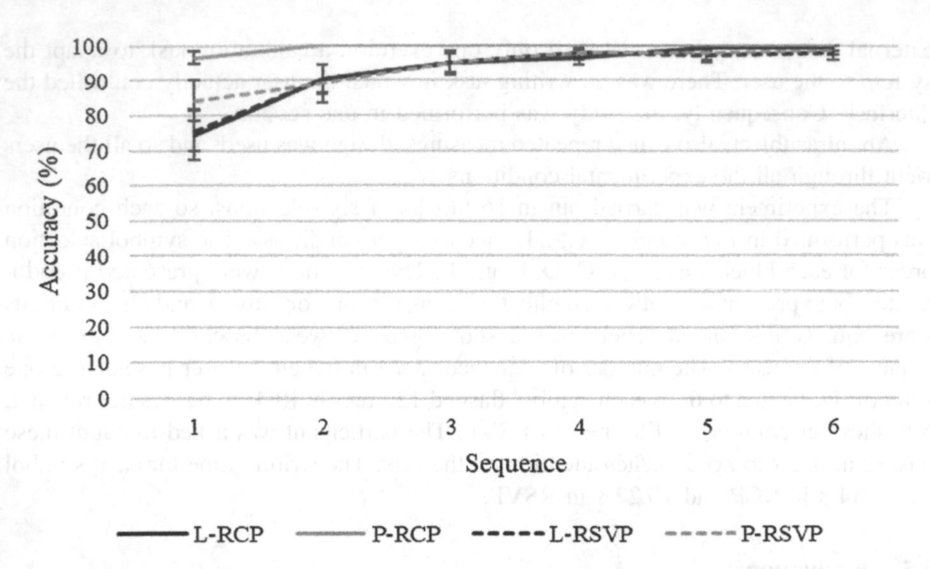


**Fig. 2.** Accuracy (mean ± standard error) of the different P300-speller conditions as a function of the number of sequences.

According to the Friedman's tests, significant differences between conditions were found only in the first sequence ($\chi^2$ (3) = 12.273; $p$ = .007). The specific differences were obtained between P-RCP versus L-RCP ($Z$ = 2.795; $p$ = .005) and versus L-RSVP ($Z$ = −2.795; $p$ = .005), i.e., between the P-RCP and the two conditions with letters. Meanwhile, the P-RSVP did not offer any significant difference, so the effect of pictures in RSVP is not as strong as in RCP. This could guide us in future studies to look for an interaction effect between the presentation paradigm and the stimulus employed.

Averaging the results of the picture conditions (P-RCP and P-RSVP) versus those of letters (L-RCP and L-RSVP), it was obtained that in sequence 1 the picture conditions obtained a better *accuracy* ($Z$ = 2.201; $p$ = .028). However, this effect was already lost from sequence 2. On the other hand, in reference to the averaging of conditions under RCP (L-RCP and P-RCP) and RSVP (L-RSVP and P-RSVP) paradigms, no significant differences between presentation paradigms were found, in any sequence. Nevertheless, in spite of not having found a significant effect between RCP and RSVP, in the visual analysis of Fig. 2 a trend can be observed in the first sequence according to which the benefit of applying pictures is greater in the RCP paradigm than in RSVP.

### 3.2 Information Transfer Rate (ITR)

Figure 3 shows the change of the *ITR* as the sequences increased. Due to the suitable results in accuracy, all conditions achieved a proper *ITR*. The superiority of the P-RCP paradigm was shown again, which in this case is further increased due to the trend of the RCP paradigms to offer a higher *ITR*. This superiority of the RCP paradigms in terms of *ITR* is observed in the fact that the L-RCP paradigm, although having a similar accuracy to the RSVP paradigms, obtains a higher *ITR*.

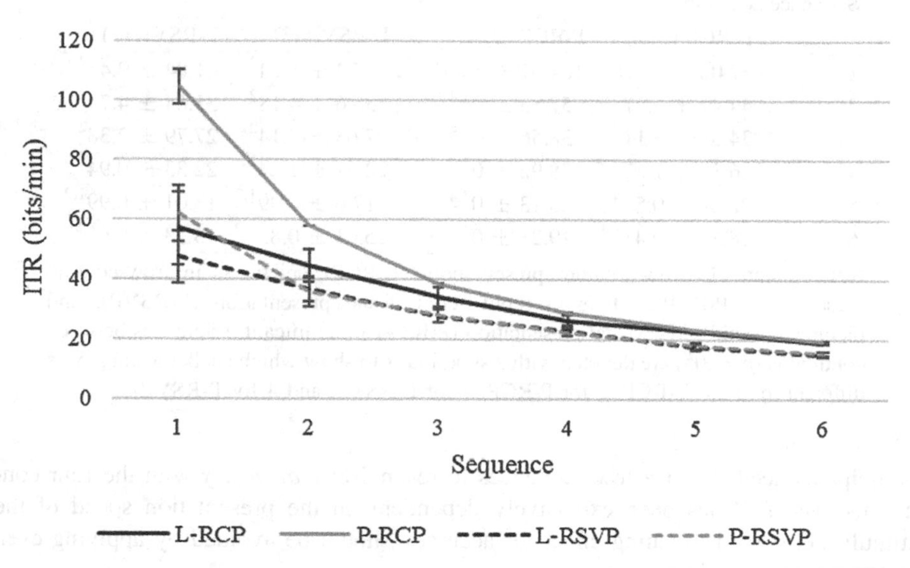


**Fig. 3.** Information transfer rate (mean ± standard error) of the different P300-speller conditions as a function of the number of sequences.

According to the Friedman's test, significant differences were found between conditions for every sequence: sequence 1 ($\chi^2$ (3) = 12.966; $p$ = .005), sequence 2 ($\chi^2$ (3) = 10.286; $p$ = .016), sequence 3 ($\chi^2$ (3) = 9.34; $p$ =.025), sequence 4 ($\chi^2$ (3) = 9.462; $p$ = .024), sequence 5 ($\chi^2$ (3) = 17.4; $p$ = .001) and sequence 6 ($\chi^2$ (3) = 17.4; $p$ = .001). Table 1 shows the multiple comparison test between conditions. In the first sequence, significant differences were obtained only for P-RCP compared to the rest of conditions (L-RCP, L-RSVP and P-RSVP). These results showed that the effect of improvement produced by pictures was only observed in RCP, while in RSVP the stimuli used do not influence significant differences in ITR. From sequence 2 both presentation paradigms are dissociated, leading to a complete dissociation in sequences 5 and 6 (Fig. 3).

In reference to the average of the conditions relative to the stimuli of letters and pictures, the picture conditions only offered a higher *ITR* versus the letters conditions in the first sequence (Z = 2.201; $p$ = .028). However, the effect of the presentation paradigm was found in all sequences: sequence 1 (Z = −1.992, $p$ = .046), sequence 2 (Z = −2.201; $p$ = .028), sequence 3 (Z = −2.201; $p$ = .028), sequence 4 (Z = −2.207; $p$ = .027), sequence 5 (Z = −2.226; $p$ = .026) and sequence 6 (Z = −2.264; $p$ = .024). These results corroborate what was previously declared in [9], according to which the paradigm RSVP trends to offer a lower *ITR* than RCP. In this case, this difference is especially notable since the *accuracy* obtained has been suitable, as most of

**Table 1.** Average information transfer rate (ITR; mean ± standard error) for each condition and sequence during the calibration task.

| Sequence | Condition | | | |
|---|---|---|---|---|
| | L-RCP (1) | P-RCP (2) | L-RSVP (3) | P-RSVP (4) |
| 1 | $57.06 \pm 12.18^{2}$ | $104.67 \pm 5.84^{1,3,4}$ | $47.4 \pm 8.74^{2}$ | $61.84 \pm 9.45^{2}$ |
| 2 | $44.47 \pm 5.27$ | $57.83 \pm 0^{3,4}$ | $36.6 \pm 4.18^{2}$ | $35.64 \pm 4.74^{2}$ |
| 3 | $34.04 \pm 3.11$ | $38.56 \pm 0^{3,4}$ | $27.63 \pm 2.14^{2}$ | $27.79 \pm 2.38^{2}$ |
| 4 | $26.32 \pm 1.67$ | $28.92 \pm 0^{3,4}$ | $22.51 \pm 1.24^{2}$ | $22.33 \pm 0.94^{2}$ |
| 5 | $22.64 \pm 0.5^{3,4}$ | $23.13 \pm 0^{3,4}$ | $17.6 \pm 0.99^{1,2}$ | $18.01 \pm 0.99^{1,2}$ |
| 6 | $18.86 \pm 0.41^{3,4}$ | $19.28 \pm 0^{3,4}$ | $15.01 \pm 0.82^{1,2}$ | $15.23 \pm 0.6^{1,2}$ |

Note: letters in row-column presentation (L-RCP), pictures in row-column presentation (P-RCP), letters in rapid serial visual presentation (L-RSVP), and pictures in rapid serial visual presentation (P-RSVP). Significant differences between conditions ($p < .05$) are denoted with a superindex to show which speller average was different to (1 for L-RCP, 2 for P-RCP, 3 for L-RSVP, and 4 for P-RSVP).

participants needed just a few sequences to reach 100% *accuracy* with the four conditions. The *ITR* has been exclusively dependent on the presentation speed of the stimuli. Perhaps, this ceiling effect for accuracy should be avoided by applying even faster presentation times.

### 3.3 Event-Related Potential (ERP) Waveform

Figure 4 shows the average amplitude of the EEG signal for each condition, electrode and stimuli (target or non-target). However, as it was previously explained, the dependent variable used for next analysis was the *amplitude difference* in absolute value between target and non-target stimuli (μV). According to the Friedman's test, significant differences were obtained in *amplitude differences* for the next electrodes: Cz ($\chi^2$ (3) = 8.6; $p$ = .035), Pz ($\chi^2$ (3) = 11; $p$ = .012), Oz ($\chi^2$ (3) = 10; $p$ = .019) and P4 ($\chi^2$ (3) = 15; $p$ = .002). Electrodes Fz, P3, PO7 and PO8 did not offer any significant differences. Table 2 shows the multiple comparison test between conditions.

As for the previous variables, it has been decided to perform the average *amplitude difference* in relation to the stimulus and presentation paradigm factors. On the one hand, for the averaging of the stimuli, it was obtained that the picture conditions presented a higher *amplitude difference* for the following positions: Oz (Z = 2.201, $p$ = .028), P4 (Z = 2.201, $p$ = .028), PO7 (Z = 1.992, $p$ = .046) and PO8 (Z = 1.992, $p$ = .046). On the other hand, contrary to what was expected, the different Wilcoxon signed-rank test did not offer significant differences by averaging the conditions according to the paradigm of the stimuli presentation.

An interesting point of Fig. 4 is that in the channels Fz and Cz for the conditions L-RCP, L-RSVP and P-RSVP, the peak of *amplitude difference* between target and non-target (around 200–400 ms) showed a higher amplitude for the target. However, for the P-RCP condition, the opposite effect is observed: clearly smaller amplitude of the target versus the non-target. This effect has not been previously reported in other work and, thus, it should be specifically studied in future works.

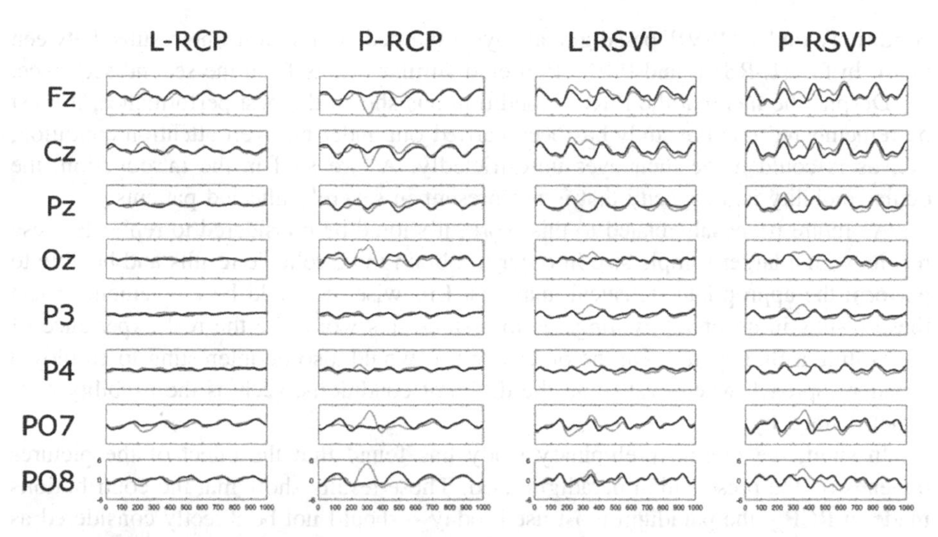


**Fig. 4.** Grand average event-related (ERP) potential waveforms (μV) for target – green – and non-target stimuli – black – for the eight used electrodes and spellers (letters in row-column presentation (L-RCP), pictures in row-column presentation (P-RCP), letters in rapid serial visual presentation (L-RSVP), and pictures in rapid serial visual presentation (P-RSVP)). (Color figure online)

**Table 2.** Average amplitude differences in absolute value (μV; mean ± standard error) for each condition and electrode position during the calibration task.

| Electrode | Condition | | | |
|---|---|---|---|---|
| | L-RCP (1) | P-RCP (2) | L-RSVP (3) | P-RSVP (4) |
| Fz | $1.39 \pm 0.18$ | $2.49 \pm 0.3$ | $1.93 \pm 0.33$ | $2.12 \pm 0.22$ |
| Cz | $1.32 \pm 0.17^{2,3,4}$ | $2.03 \pm 0.29^{1}$ | $2.67 \pm 0.41^{1}$ | $2.6 \pm 0.39^{1}$ |
| Pz | $0.9 \pm 0.1^{3,4}$ | $1.27 \pm 0.29^{4}$ | $2.38 \pm 0.37^{1}$ | $2.45 \pm 0.31^{1,2}$ |
| Oz | $0.71 \pm 0.14^{2}$ | $2.56 \pm 0.42^{1}$ | $1.23 \pm 0.26$ | $1.51 \pm 0.41$ |
| P3 | $0.87 \pm 0.14$ | $0.88 \pm 0.11$ | $1.82 \pm 0.41$ | $1.74 \pm 0.31$ |
| P4 | $0.56 \pm 0.07^{3,4}$ | $0.96 \pm 0.09^{4}$ | $1.65 \pm 0.27^{1}$ | $1.71 \pm 0.23^{1,2}$ |
| PO7 | $1.37 \pm 0.38$ | $2.31 \pm 0.4$ | $1.76 \pm 0.38$ | $1.78 \pm 0.49$ |
| PO8 | $1.02 \pm 0.21$ | $2.58 \pm 0.54$ | $1.03 \pm 0.17$ | $1.27 \pm 0.32$ |

Note: letters in row-column presentation (L-RCP), pictures in row-column presentation (P-RCP), letters in rapid serial visual presentation (L-RSVP), and pictures in rapid serial visual presentation (P-RSVP). Significant differences between conditions ($p < .05$) are denoted with a superindex to show which speller average was different to (1 for L-RCP, 2 for P-RCP, 3 for L-RSVP, and 4 for P-RSVP).

## 4 Conclusions

The general trend of the present work has been to show the P-RCP as the condition with the best performance. Otherwise, the other conditions were fairly even. While differences in performance between the RCP conditions have been found, the conditions under RSVP paradigm always obtained non-significant results between them. In fact, L-RSVP and P-RSVP offered similar results from the second sequence.

Despite the fact that the P-RCP condition has shown the best performance, it must be remembered that the study has been carried out under an overt attention condition, i.e., users could move their eyes unrestrictedly. As we said in the introduction, the ocular mobility is a capacity that is not present in severely affected patients.

As future proposals related to this work, it should be considered to replicate these results with a larger sample size in order to obtain more reliable results and be able to perform the appropriate statistical analyses. Likewise, it would be convenient to test these results in an online writing session, since this would be the real experience of controlling a BCI speller. On the other hand, it would also be interesting to employ a broader approach when evaluating the different conditions, such as the usability construct defined by [25].

In short, the present preliminary study has found that the effect of the pictures depends on the presentation paradigm used. These results show that the contributions made in RCP – the paradigm most used today – should not be directly considered as valid in RSVP without testing them.

**Acknowledgements.** This work was partially supported by the Spanish Ministry of Economy and Competitiveness through the projects LICOM (DPI2015-67064-R), by the European Regional Development Fund (ERDF) and by the University of Malaga. Moreover, the authors would like to thank all participants for their cooperation.

## References

1. Patterson, J.R., Grabois, M.: Locked-in syndrome: a review of 139 cases. Stroke **17**, 758–764 (1986)
2. Pal, S., Mangal, N.K., Khosla, A.: Development of assistive application for patients with communication disability. In: IEEE International Conference on Innovations in Green Energy and Healthcare Technologies - 2017, IGEHT 2017, pp. 1–4 (2017)
3. Bauer, G., Gerstenbrand, F., Rumpl, E.: Varieties of the locked-in syndrome. J. Neurol. **221**, 77–91 (1979)
4. Murguialday, A.R., et al.: Transition from the locked in to the completely locked-in state: a physiological analysis. Clin. Neurophysiol. **122**, 925–933 (2011)
5. Birbaumer, N.: Breaking the silence: brain-computer interfaces (BCI) for communication and motor control. Psychophysiology **43**, 517–532 (2006)
6. Lemoignan, J., Ells, C.: Amyotrophic lateral sclerosis and assisted ventilation: how patients decide. Palliat. Support. Care. **8**, 207–213 (2010)
7. Farwell, L.A., Donchin, E.: Talking off the top of your head: toward a mental prosthesis utilizing event-related brain potentials. Electroencephalogr. Clin. Neurophysiol. **70**, 510–523 (1988)
8. Nicolas-Alonso, L.F., Gomez-Gil, J.: Brain computer interfaces, a review. Sensors **12**, 1211–1279 (2012)
9. Rezeika, A., Benda, M., Stawicki, P., Gembler, F., Saboor, A., Volosyak, I.: Brain–computer interface spellers: a review. Brain Sci. **8** (2018)

10. Ryan, D.B., Townsend, G., Gates, N.A., Colwell, K., Sellers, E.W.: Evaluating brain-computer interface performance using color in the P300 checkerboard speller. Clin. Neurophysiol. **128**, 2050–2057 (2017)
11. Ryan, D.B., et al.: Evaluating brain-computer interface performance in an ALS population: checkerboard and color paradigms. Clin. EEG Neurosci. **49**, 114–121 (2018)
12. Ma, Z., Qiu, T.: Performance improvement of ERP-based brain–computer interface via varied geometric patterns. Med. Biol. Eng. Comput. **55**, 2245–2256 (2017)
13. Kaufmann, T., Schulz, S.M., Grünzinger, C., Kübler, A.: Flashing characters with famous faces improves ERP-based brain-computer interface performance (2011). http://stacks.iop.org/1741-2552/8/i=5/a=056016?key=crossref.04cfedc9b1db574d6b4a9c9ee9a759f9
14. Li, Q., Liu, S., Li, J., Bai, O.: Use of a green familiar faces paradigm improves P300-speller brain-computer interface performance. PLoS ONE **10**, 1–15 (2015)
15. Fernández-Rodríguez, Á., Velasco-Álvarez, F., Ron-Angevin, R.: Evaluation of a P300 brain-computer interface using different sets of flashing stimuli. In: Ron-angevin, R. (ed.) BRAININFO 2018: The Third International Conference on Neuroscience and Cognitive Brain Information, pp. 1–4. IARIA, Venice (2018)
16. Brunner, P., Joshi, S., Briskin, S., Wolpaw, J.R., Bischof, H., Schalk, G.: Does the "P300" speller depend on eye gaze? J. Neural Eng. **7**, 056013 (2010)
17. Treder, M.S., Blankertz, B.: (C)overt attention and visual speller design in an ERP-based brain-computer interface. Behav. Brain Funct. (2010)
18. Acqualagna, L., Treder, M.S., Schreuder, M., Blankertz, B.: A novel brain-computer interface based on the rapid serial visual presentation paradigm. In: 2010 Annual International Conference of the IEEE Engineering in Medicine and Biology Society, EMBC 2010 (2010)
19. Acqualagna, L., Treder, M.S., Blankertz, B.: Chroma Speller: isotropic visual stimuli for truly gaze-independent spelling. Int. IEEE/EMBS Conference on Neural Engineering, NER, pp. 1041–1044 (2013)
20. Sato, H., Washizawa, Y.: An N100-P300 spelling brain-computer interface with detection of intentional control. Computers **5**, 31 (2016)
21. Aricò, P., Aloise, F., Schettini, F., Riccio, A., Salinari, S.: GeoSpell: an alternative P300-based speller interface towards no eye gaze required. Int. J. Bioelectromagn. **13**, 152–153 (2011)
22. Schalk, G., McFarland, D.J., Hinterberger, T., Birbaumer, N., Wolpaw, J.R.: BCI2000: a general-purpose brain-computer interface (BCI) system (2004)
23. Lang, P.J., Bradley, M.M., Cuthbert, B.N.: International affective picture system (IAPS): affective ratings of pictures and instruction manual. Technical report A-8. University of Florida, Gainesville, FL (2008)
24. Wolpaw, J.R., Ramoser, H., McFarland, D.J., Pfurtscheller, G.: EEG-based communication: Improved accuracy by response verification. IEEE Trans. Rehabil. Eng. **6**, 326–333 (1998)
25. ISO: ISO9241-11: Ergonomic requirements for office work with visual display terminals (VDTs). Part11 Guid, usability. 11, 22 (1994)

# Custom-Made Monitor for Easy High-Frequency SSVEP Stimulation

Mihaly Benda[1], Felix Gembler[1], Piotr Stawicki[1], Sadok Ben-Salem[2], Zahidul Islam[2], Arne Vogelsang[2], and Ivan Volosyak[1](✉)

[1] Faculty of Technology and Bionics, Rhine-Waal University of Applied Sciences, 47533 Kleve, Germany
ivan.volosyak@hochschule-rhein-waal.de
[2] Polyoptics GmbH, 47533 Kleve, Germany
https://bci-lab.hochschule-rhein-waal.de

**Abstract.** In this paper, we present and evaluate a special Custom-Made Computer Display (CMCD) with additional background light, which is separately controlled in order to create visual stimuli for Brain-Computer Interfaces (BCIs). While the monitor itself is working with a 60 Hz refresh rate, twelve strips of LED lights that are placed in between the backlight allow for a higher frequency flickering than any flickering object on a conventional screen. The goal of this study is to evaluate the effectiveness of this CMCD, which is mostly based on a change in intensity rather than in contrast. Therefore, we compared the responses to both types of flickering at different frequency ranges, while also measuring the speed and accuracy of the BCI with short spelling tasks. The CMCD LED illumination yielded slightly superior performance in terms of offline ITR in comparison to the standard flickering.

**Keywords:** Brain-Computer Interface (BCI) · Human-Computer Interaction (HCI) · Steady-State Visual Evoked Potential (SSVEP) · Liquid-Crystal Display (LCD) · High-frequency stimuli

## 1 Introduction

Steady-State Visual Evoked Potentials (SSVEPs) are normal brain responses to visual stimuli, which are elicited in the visual cortex [5]. In SSVEP-based Brain-Computer Interfaces (BCIs), there are several stimuli flashing with different frequencies. By fixating on such a target, continuous brain responses are elicited in the brain, which correspond to the frequency of the stimulus. The frequency as well as its harmonics can be detected in the measured brain waves. The brain signals are usually recorded via Electroencephalography (EEG) and the BCI can interpret these brain responses online. This way, the attended target can be determined and the associated command is executed. Thus, the SSVEP paradigm allows hand-free communication, which can be useful in the development of various assistive technologies or as a method of control.


I. Rojas et al. (Eds.): IWANN 2019, LNCS 11506, pp. 382–393, 2019.
https://doi.org/10.1007/978-3-030-20521-8_32

The speed and accuracy of the system are essential for developing a practical and user-friendly BCI. Great improvements have been made in terms of speed due to new signal classification algorithms. For example Nakanishi et al. [7] achieved an Information Transfer Rate (ITR) of 325.33 bpm (bit per minute) by utilizing spatial filtering, filter-banks, and ensemble-based classification.

The highest ITR values have been achieved when stimulus presentation was used with multiple targets on standard screens [7]. As the frame-based display of a standard monitor does not easily allow for many different frequency stimuli [3], some form of frame-based frequency approximation method is typically utilized (see e.g. Wang et al. [10]), enabling the implementation of visual stimuli at flexible frequencies. Visual flickers with a frequency resolution as low as 0.05 Hz have been realized with this technique [9], still eliciting SSVEP responses the BCI could differentiate between. Therefore, a similar system was used for multi-target BCIs, yielding an overall higher ITR [4].

Regarding frequency choice, the best SSVEP responses are usually obtained using stimulation frequencies between 5 and 20 Hz [3]. However, in terms of user-friendliness, these frequencies are considered annoying by many participants, which is why many researchers developed BCIs with higher flickering rates [1, 2,8]. Higher stimulation frequencies, on the other hand, are harder to realize on a standard monitor due to the limitations of the vertical monitor refresh rate (typically 60 Hz, in this experiment: 30 Hz). They also evoke weaker neural responses, reducing the accuracy and speed of the whole BCI system.

In this paper, a Custom-Made Computer Display (CMCD) dedicated for the graphical presentation of VEP-evoking stimuli is presented and tested. The stimuli are presented by a change in illumination rather than in contrast. The monitor that was specially assembled for this purpose enables easy and exact high-frequency stimulation, without the need of approximation methods. The frequency of the flickering can be controlled freely, enabling easy setup and testing.

In order to investigate the effectiveness of this hardware, we tested three frequency sets utilizing both the typically used frequency approximation method, as well as the custom-made hardware to present the stimuli.

## 2 Methods and Materials

This section provides a description of the equipment, the setup of the study (Fig. 1) and the details of the used BCI system. First of all, information about the participants is given, which is followed by the description of the CMCD. Then, the recording hardware, the signal processing, the experimental procedure, and finally the Graphical User Interface (GUI) are discussed.

### 2.1 Participants

Nine healthy volunteers (four female) participated in this study. The mean age of the participants was 24.00 years with an SD (standard deviation) of 3.16 years.

**Fig. 1.** A participant during the experiment (the recording for the offline analysis). The flickering is realized by the CMCD's background illumination. In the picture, the segments 1, 4, and 5 are in an active state, while the other segments are not.

Written informed consent was given by each subject before the experiment, in accordance with the Declaration of Helsinki. This study was approved by the ethical committee of the medical faculty of the University Duisburg-Essen. All information was stored anonymously during the experiment and results cannot be traced back to the participants. Subjects had the opportunity to opt-out of the study at any time. Spectacles were worn when appropriate. The subjects received a small financial reward for participation in this study.

### 2.2 Stimulus Presentation

Two different methods of stimulus presentation where utilized, LED illumination (background light) and frequency approximation (stimuli shown by rendering on the screen) [6].

### 2.3 CMCD

The monitor was assembled by putting controllable LED strips into a Fujitsu ScenicView P22W-5 TFT-Monitor (Figs. 1 and 2). The whole screen was divided into six segments, in which the LED strips were separately controllable. To minimize their crosstalk, the intensity at the intersection of the segments was reduced (Fig. 3). This means that two types of LED lights were used with different brightness. For the most part, HD LED strips (LED-Emotion GmbH, 6500 K pure white, 2330 lm/m, 19.2 W/m, 140 LEDs/m, 24 V) were utilized, but along the middle horizontal line, less bright strips (Abrams & Mantler GmbH & Co.

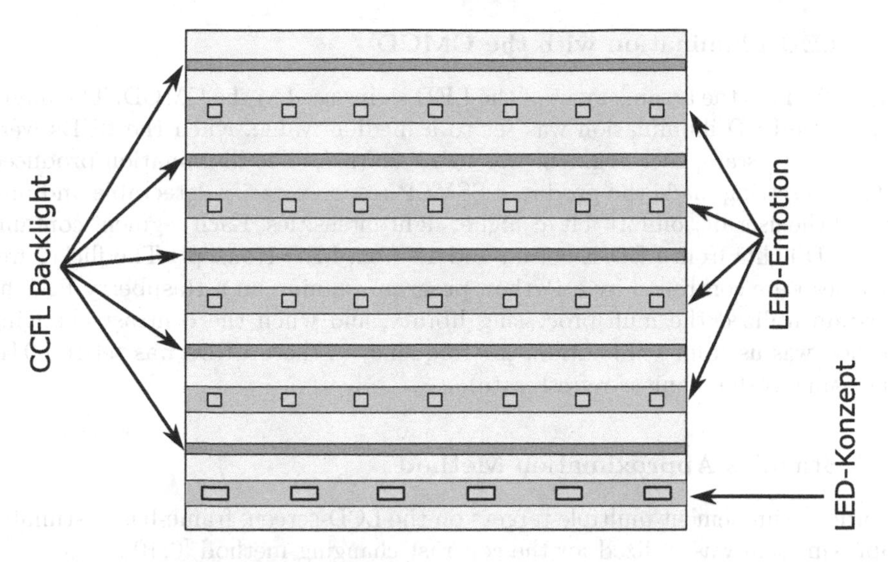


**Fig. 2.** The concept of the used LED structure inside one of the segments of the CMCD. The Cold Cathode Fluorescent Light (CCFL) backlight provides the standard display on the screen, the LED-Emotion and LED-Konzept lights provide the flickering by illumination. Six fields are separated, which can thus flicker at different frequencies without affecting the neighboring fields.

KG, 7000 K cold white, 950 lm/m, 9.6 W/m, 24 V) were used. The lights were controlled by a Raspberry Pi 3 B (Raspberry Pi Foundation), which controlled the six flickering fields of the monitor via a Gravity: MOSFET Power Controller SKU: DFR0457 (DFRobot). By providing power the LED strips turned on, and the changing of these on/off phases generated the flickering.

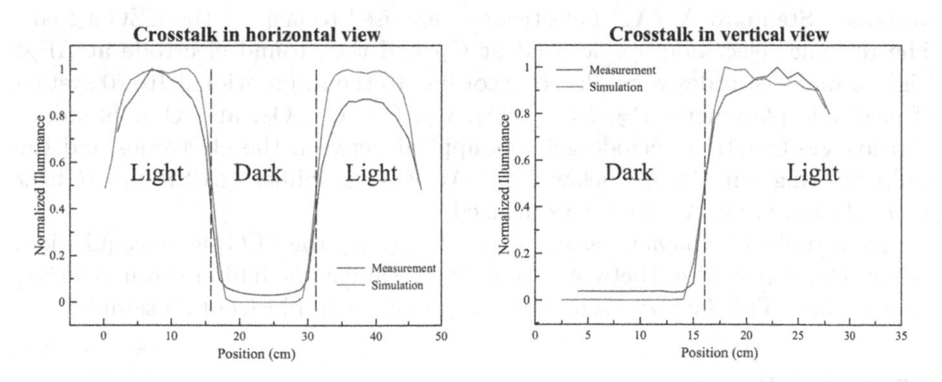


**Fig. 3.** On the left side the crosstalk between the segments of one row is shown, while on the right side the crosstalk between segments above each other is shown.

### 2.4 LED Illumination with the CMCD

Figure 2 shows the arrangement of the LED strips used in the CMCD. The intensity of the LED illumination was set to a medium value, when the LEDs were in the active state, each segment was using 260 mA. The illumination produced with this setting made the produced SSVEP response easily detectable and prevented the user-discomfort felt at higher light intensities. Each segment contains 105 LED lights from LED-Emotion and 18 from LED-Konzept. The flicker frequencies were controlled by a Python program running on a Raspberry Pi. The program utilized the multiprocessing library, and when the contrast changing method was used for stimulation, the frequency of these LEDs was set to 60 Hz (the same as the monitor refresh rate).

### 2.5 Stimulus Approximation Method

In order to implement multiple targets on the LCD-screen, frame-based stimulus approximation was utilized for the contrast changing method [6,10].

To realize the flashing pattern, a specific frequency $f$ was assigned to each target; the transparencies of the targets where sinusoidally modulated in accordance with the frequency [6].

For a monitor refresh rate of 60 Hz, the stimulus sequence for the $i$-th target is given by

$$c_i(t) = \frac{1}{2}\left(1 + \sin\left(2\pi f_i \frac{t}{60}\right)\right), \quad t = 0, 1, \ldots, \tag{1}$$

yielding values in the range of 0 to 1.

### 2.6 Signal Acquisition

The computer operated with Microsoft Windows 7 Enterprise running on an Intel processor. Standard Ag/AgCl electrodes were used to acquire the EEG signals. The reference electrode was located at $C_Z$ and the ground electrode at $AF_Z$. Eight signal electrodes were placed according to the international 10/20 system of electrode placement: $P_Z$, $PO_3$, $PO_4$, $O_1$, $O_Z$, $O_2$, $O_9$, and $O_{10}$. Standard abrasive electrolytic electrode gel was applied between the electrodes and the scalp to bring impedances below 5 k$\Omega$. An EEG amplifier, g.USBamp (Guger Technologies, Graz, Austria), was utilized.

The sampling frequency was set to 600 Hz. During the EEG signal acquisition, a digital band pass filter (between 2 and 100 Hz) and a notch filter (around 50 Hz) were applied. The data was sent from the amplifier in blocks of 30 samples.

### 2.7 Signal Processing

Canonical correlation analysis (CCA), was used for SSVEP signal classification [5]. In general, CCA is used to investigate the relationship between two sets of variables $\mathbf{X} \in \mathbb{R}^{p \times s}$ and $\mathbf{Y} \in \mathbb{R}^{q \times s}$. It computes two vectors $\mathbf{w}_X \in \mathbb{R}^p$

and $\mathbf{w}_Y \in \mathbb{R}^q$ that maximize the correlation $\rho$ between the linear combinations $\mathbf{x} = \mathbf{X}^T\mathbf{w}_X$ and $\mathbf{y} = \mathbf{Y}^T\mathbf{w}_Y$ by solving

$$\max_{\mathbf{w}_X,\mathbf{w}_Y} \rho(\mathbf{x},\mathbf{y}) = \frac{\mathbf{w}_X^T\mathbf{X}\mathbf{Y}^T\mathbf{w}_Y}{\sqrt{\mathbf{w}_X^T\mathbf{X}\mathbf{X}^T\mathbf{w}_X\ \mathbf{w}_Y^T\mathbf{Y}\mathbf{Y}^T\mathbf{w}_Y}}. \tag{2}$$

To classify the attended target with the CCA, cos and sin templates are compared to the EEG signal matrix $\mathbf{X} \in \mathbb{R}^{N\times M}$, which contains recorded EEG data; $M$ is the number of collected samples and $N$ is the number of signal channels. Let $N_h$ be the number of harmonics that are considered for frequency detection and $F_s$ the amplifier sampling rate. For each frequency $f_i$, $i = 1, \ldots, K$ the cos and sin templates $\mathbf{Y}_{f_i} \in \mathbb{R}^{2N\times M}$ are given by

$$\mathbf{Y}_{f_i} = \begin{bmatrix} \sin(2\pi f_i t) \\ \cos(2\pi f_i t) \\ \vdots \\ \sin(2\pi N_h f_i t) \\ \cos(2\pi N_h f_i t) \end{bmatrix}, \quad t = \frac{1}{F_s}, \frac{2}{F_s}, \ldots, \frac{M}{F_s}. \tag{3}$$

For each of these templates, CCA is used to compute the maximal canonical correlation to the signal matrix $X$. This yields correlations $\rho_i$, $i = 1, \ldots, K$. The classified target $T$ is then determined as

$$T = \max_i \rho_i \quad i = 1, \ldots, K. \tag{4}$$

Here, CCA-classification was performed on the basis of thresholds, i.e. if the difference between the highest and second highest $\rho_i$ did not surpass a specific threshold $\beta$, more data was collected, so that $M$ increased stepwise. More details about the threshold-based classification can be found in [5]. In this study, the minimal classification time window was set to 2 s, i.e. $M \geq 2F_s$. Moreover, the number of signal channels was set to $N = 8$, the amplifier sampling rate was $F_s = 600$, and $N_h = 3$ harmonics were considered to identify which of the $K = 6$ targets was attended.

### 2.8 Experimental Protocol

Participants sat on a chair facing the screen at a distance of approximately 80 cm. After the preparation for the EEG recording, they went through an offline recording phase, an online copy spelling phase, and a brief questionnaire. These steps were done with three different frequency sets in the $\alpha$ and $\beta$ bands (for details see Table 1).

For all participants the following order of sessions was utilized:

1. Alpha-band stimulation frequency approximation method
2. Alpha-band stimulation frequency CMCD LED illumination
3. Lower Beta-band stimulation CMCD LED illumination

4. Lower Beta-band stimulation frequency approximation method
5. Higher Beta-band stimulation frequency approximation method
6. Higher Beta-band stimulation CMCD LED illumination.

The order of the experiments was fixed in order to prevent user-frustration by possibly not working higher frequency stimuli (the experiment was stopped if the participants could not control the BCI with both types of flickering for at least a few minutes). This was necessary as the BCI scenarios which were not working caused such levels of user-frustration, that further tests would have been heavily biased. In order to avoid fatigue, participants took small breaks between the sessions. After each session, users were asked to rate the perceived level of annoyance of the stimuli on the Likert scale (1-5, 1 - not annoying at all, 5 - very annoying). This was done to compare the stimulus methods from this point as well. A disturbing stimulation would render the whole system not user-friendly, which opposes the goals of BCI development.

**Table 1.** Stimulation sets: Three sets of six frequencies ($f_1$–$f_6$) were tested (both offline and online) with both the frequency approximation method for standard LCD screens as well as the CMCD LED illumination presented in this paper.

| *Stimulus frequencies* (Hz) | $f_1$ | $f_2$ | $f_3$ | $f_4$ | $f_5$ | $f_6$ |
|---|---|---|---|---|---|---|
| Alpha-band | 7.0 | 7.5 | 8.0 | 8.5 | 9.0 | 9.5 |
| Lower Beta-band | 17.0 | 17.5 | 18.0 | 18.5 | 19.0 | 19.5 |
| Higher Beta-band | 27.0 | 27.5 | 28.0 | 28.5 | 29.0 | 29.5 |

**Offline Recording Phase.** In the offline recording phase, six boxes were presented to the user containing the numbers 1–6. Each of the boxes flickered with a specific frequency. Participants were asked to focus on each box three times, while EEG-data trials were recorded (Fig. 1 shows the offline recording phase). A green frame around the box indicated which target needed to be fixated on. After each trial, the next box was highlighted, the recording paused for one second, and the participant shifted his/her gaze to the next target.

**Online Copy Phase.** In the copy spelling phase, participants completed a spelling task by utilizing a six-target BCI spelling application. After each selection, the recording paused for two seconds, when the participant shifted his/her gaze to the next target (i.e. letter). The same classification threshold was used for each subject and task. After a brief familiarization run where participants spelled the word "BCI", participants were asked to spell the word "KLEVE". During the spelling task, accuracy and ITR were measured. As the interface provided a two-step selection mechanism for writing letters, a minimum of ten commands were necessary to write "KLEVE" (First the box "GHIJKL" had to be selected, followed by "K", etc.). Errors had to be corrected by fixating on the

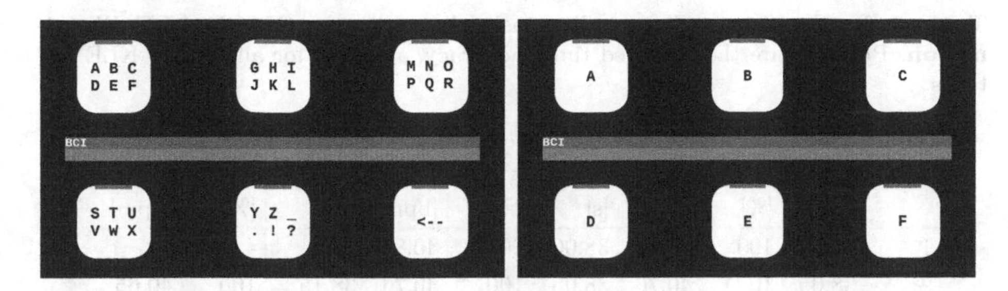


**Fig. 4.** Graphical user interface of the six-target speller. The participants wrote "BCI" in the familiarization session. In order to write the letter "B", two selections were required: First, the group containing the desired character (here: "ABCDEF") needed to be selected, then, the letter itself needed to be chosen.

**Table 2.** Results of the online spelling tasks of the **frequency approximation** test. Provided are the averaged time, accuracy, and ITR for all the tasks "KLEVE". For the higher Beta-band stimulation, participants who did not complete the task for both stimulation methods (marked with * or -) were excluded from the calculation of mean values.

| # | Alpha-band | | | Lower Beta-band | | | Higher Beta-band | | |
|---|---|---|---|---|---|---|---|---|---|
| | Time [s] | Acc [%] | ITR [bpm] | Time [s] | Acc [%] | ITR [bpm] | Time [s] | Acc [%] | ITR [bpm] |
| 1 | 38.4 | 100 | 40.44 | 38.25 | 100 | 40.55 | 57.10* | 100* | 27.16* |
| 2 | 38.1 | 100 | 40.71 | 38.00 | 100 | 40.82 | 39.10 | 100 | 39.67 |
| 3 | 38.4 | 100 | 40.44 | 38.50 | 100 | 40.29 | 117.80 | 82 | 16.57 |
| 4 | 65.6 | 92 | 23.98 | 41.75 | 100 | 37.15 | - | - | - |
| 5 | 38.0 | 100 | 40.82 | 38.00 | 100 | 40.82 | 38.00 | 100 | 40.82 |
| 6 | 38.2 | 100 | 40.6 | 38.20 | 100 | 40.60 | 231.00 | 65 | 5.74 |
| 7 | 53.5 | 85 | 23.45 | 38.05 | 100 | 40.76 | 239.60 | 71 | 7.42 |
| 8 | 40.9 | 100 | 37.92 | 38.00 | 100 | 40.82 | 87.00 | 100 | 17.83 |
| 9 | 38.0 | 100 | 40.82 | 38.00 | 100 | 40.82 | - | - | - |
| SD | 9.2 | 5.1 | 6.93 | 1.15 | 0 | 1.12 | 82.47 | 14.5 | 14.07 |
| Mean | 43.2 | 97.4 | 36.58 | 38.53 | 100 | 40.29 | 125.42 | 86.3 | 21.34 |

last target for the "UNDO" function of the interface. The goal of this phase was to provide information about the performance of the BCI with different types of stimuli generators (CMCD and approximation method).

**Table 3.** Results of the online spelling tasks done with the **CMCD LED illumination**. Provided are the averaged time, accuracy, and ITR for all of the "KLEVE" tasks.

| # | Alpha-band | | | Lower Beta-band | | | Higher Beta-band | | |
|---|---|---|---|---|---|---|---|---|---|
| | Time [s] | Acc [%] | ITR [bpm] | Time [s] | Acc [%] | ITR [bpm] | Time [s] | Acc [%] | ITR [bpm] |
| 1 | 38.00 | 100 | 40.82 | 38.00 | 100 | 40.82 | - | - | - |
| 2 | 38.05 | 100 | 40.76 | 38.05 | 100 | 40.76 | 38.15 | 100 | 40.65 |
| 3 | 38.45 | 100 | 40.34 | 38.00 | 100 | 40.82 | 189.50 | 62 | 7.94 |
| 4 | 38.10 | 100 | 40.71 | 54.20 | 92 | 29.00 | - | - | - |
| 5 | 38.00 | 100 | 40.82 | 38.00 | 100 | 40.82 | 98.60 | 74 | 13.21 |
| 6 | 38.00 | 100 | 40.82 | 38.05 | 100 | 40.76 | 79.20 | 75 | 14.46 |
| 7 | 44.15 | 100 | 35.13 | 38.00 | 100 | 40.82 | 182.40 | 65 | 8.39 |
| 8 | 39.95 | 100 | 38.82 | 51.15 | 85 | 24.53 | 109.80 | 68 | 9.88 |
| 9 | 38.05 | 100 | 40.76 | 38.00 | 100 | 40.82 | 76.80* | 81* | 18.17* |
| SD | 1.92 | 0 | 1.79 | 6.14 | 5.1 | 5.93 | 54.11 | 12.5 | 11.39 |
| Mean | 38.97 | 100 | 39.89 | 41.27 | 97.4 | 37.68 | 116.28 | 74.0 | 15.76 |

### 2.9 Graphical User Interface

For this experiment, a six-target BCI spelling application was designed. The GUI was arranged as a $2 \times 3$ stimulus matrix containing the letters of the alphabet as well as additional characters in five groups of six characters each (see Fig. 4). Each of these characters could be selected in two steps. The sixth box contained a correction option. Every command classification was followed by an audio feedback.

## 3 Results and Discussion

Tables 2 and 3 show results from the online experiment. ITRs were calculated according to [11]. For the CMCD LED illumination, mean ITRs of 39.89, 37.68, and 15.76 bpm were achieved with the Alpha-band, lower Beta-band and higher Beta-band stimulation, respectively. For the frequency approximation method, ITRs of 36.58, 40.29, and 21.34 bpm were achieved, respectively.

A paired t-test was performed to investigate differences in online performance between frequency approximation method and CMCD illumination. However, no statistically significant difference was found for either stimulation set. The reason for this result could be the length of the time window (2 s). This prevented faster classification, which would have otherwise occurred. If we calculate the ITR with maximum accuracy and the fastest possible classification (2 s) and include the gaze shifting time of 2 s (except before the first classification), we get the maximum achievable ITR with this window length: 40.82 bpm. There are

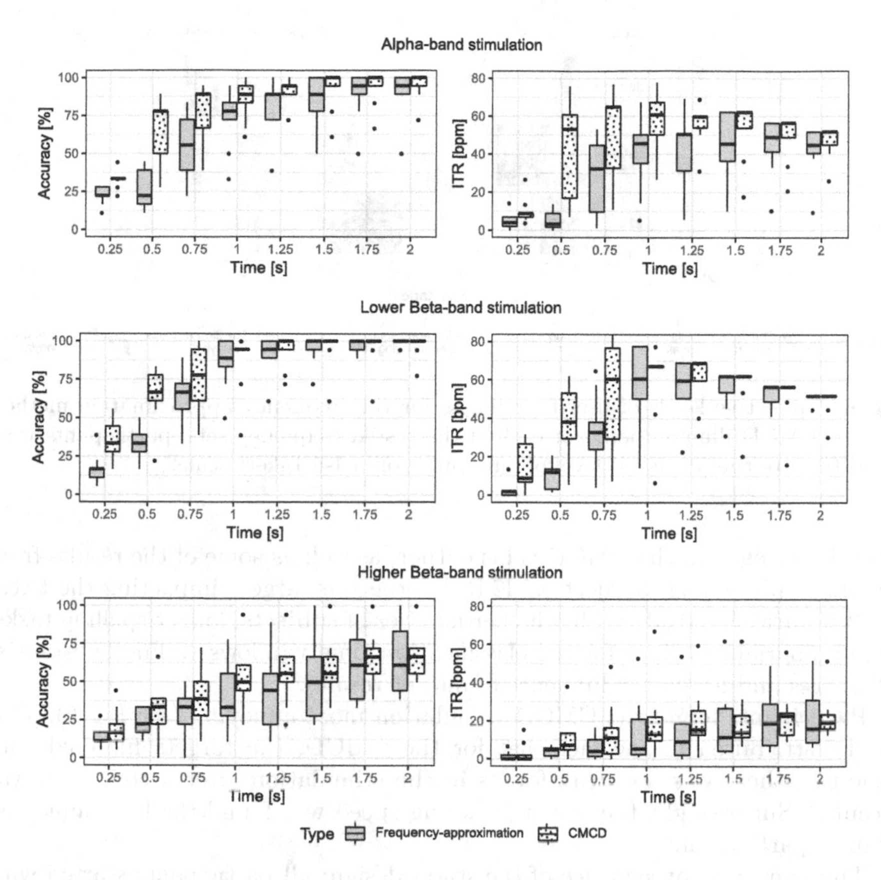


**Fig. 5.** Offline comparison of the CMCD LED illumination and the frequency approximation method. Provided are the accuracies and ITRs for classification windows between 0.25 and 2 s from the offline recording. Each of the six targets was attended three times. For classification, CCA was utilized.

no significant differences between the results most likely because most subjects reached an ITR close to the maximum, which suggests that using shorter time windows, which allow a higher maximum ITR could result in more dispersed (and higher) ITR results which could be significantly different for the two types of stimulation. To further investigate the difference in performance, offline analysis was conducted.

The results of the offline analysis are provided in Fig. 5. A clear trend can be observed, showing that the CMCD LED illumination yields faster system speeds for all tested frequency bands. The outlier results of the high Beta-band stimulation with CMCD originate from one subject, who had exceptionally good control over the BCI, achieving close to maximum ITRs in the online phase, and really high ITR and accuracy in the offline analysis. The same subject achieved also high ITR with the approximation method, however, as can be seen on Fig. 5 this result is not considered an outlier as the variance of ITR across subjects was

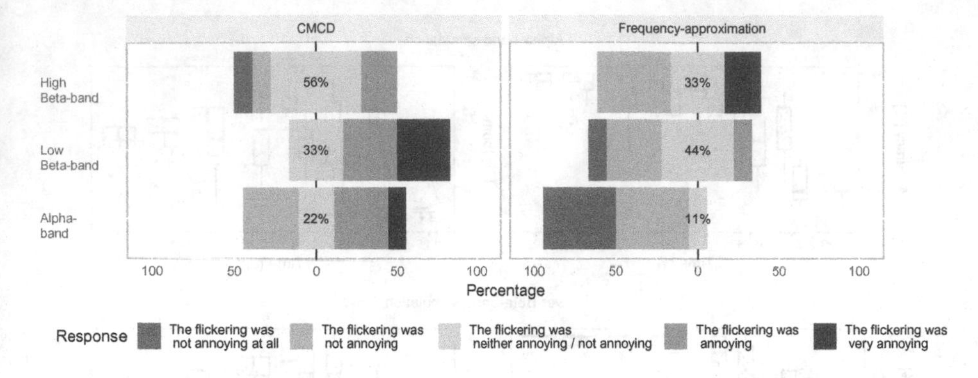


**Fig. 6.** Subjective levels of user-friendliness for the frequency approximation method and the CMCD illumination. For each of the tested frequency sets, participants were asked to state the perceived level of annoyance on a 1-5 Likert scale.

much larger as with the CMCD. This outlier, as well as some of the results from the online phase (e.g. Subject 5s' ITR difference is largely impacting the t-test results) warrant a study with a higher number of subjects, longer spelling tasks, and appropriate settings for the classification time windows to find substantial differences and lessen the influence of outlier results.

Participants found the CMCD stimulation more annoying (see Fig. 6). This can be attributed to the fact that for the CMCD, the targets flickered continuously; there was no break for 2 s in the stimulation after a command was executed. Surprisingly, the lowest flickering speed was found the least annoying by most participants.

This can be a consequence of the study design; all participants started with the low-frequency set and could have become more annoyed by the flickering during the experiment. A higher number of subjects are required for further investigation.

## 4 Conclusion

A custom-made monitor for VEP stimulation was presented. The system allows the setup of high stimulation frequencies without the typical limitations caused by the dependence on the monitor refresh rate. Though the system yielded superior performance in terms of offline ITR, some improvements and tests can still be made in the future, like the optimization of user-friendliness. In preliminary tests, we also tested flickering beyond the visible threshold (30 Hz). However, with the classification algorithms used here, only a few participants achieved reliable control over the system. We are planning further tests allowing for faster classification, and with more participants to fully assess the CMCD LED illumination. Higher stimulation frequencies, and the effect of the intensity of the LEDs are also planned to be examined in the future.

**Acknowledgment.** This research was funded by the European Fund for Regional Development (EFRD - or EFRE in German) under Grants GE-1-1-047 and IT-1-2-001. We thank all the participants of this research study as well as our student assistants.

## References

1. Aminaka, D., Makino, S., Rutkowski, T.M.: Chromatic and high-frequency cVEP-based BCI paradigm, pp. 1906–1909. IEEE, August 2015. https://doi.org/10.1109/EMBC.2015.7318755
2. Chabuda, A., Durka, P., Zygierewicz, J.: High frequency SSVEP-BCI with hardware stimuli control and phase-synchronized comb filter. IEEE Trans. Neural Syst. Rehabil. Eng. **26**(2), 344–352 (2017). https://doi.org/10.1109/TNSRE.2017.2734164
3. Gembler, F., Stawicki, P., Volosyak, I.: Autonomous parameter adjustment for SSVEP-based BCIs with a novel BCI wizard. Front. Neurosci. **9**, 474 (2015). https://doi.org/10.3389/fnins.2015.00474
4. Gembler, F., Stawicki, P., Volosyak, I.: Exploring the possibilities and limitations of multitarget SSVEP-based BCI applications. In: 2016 IEEE 38th Annual International Conference of the the IEEE Engineering in Medicine and Biology Society (EMBC), pp. 1488–1491. IEEE (2016). https://doi.org/10.1109/EMBC.2016.7590991
5. Lin, Z., Zhang, C., Wu, W., Gao, X.: Frequency recognition based on canonical correlation analysis for SSVEP-based BCIs. IEEE Trans. Biomed. Eng. **54**(6), 1172–1176 (2007). https://doi.org/10.1109/TBME.2006.889197
6. Manyakov, N.V., Chumerin, N., Robben, A., Combaz, A., van Vliet, M., Van Hulle, M.M.: Sampled sinusoidal stimulation profile and multichannel fuzzy logic classification for monitor-based phase-coded SSVEP brain–computer interfacing. J. Neural Eng. **10**(3), 036011 (2013). https://doi.org/10.1088/1741-2560/10/3/036011
7. Nakanishi, M., Wang, Y., Chen, X., Wang, Y.T., Gao, X., Jung, T.P.: Enhancing detection of SSVEPs for a high-speed brain speller using task-related component analysis. IEEE Trans. Biomed. Eng. **65**(1), 104–112 (2018). https://doi.org/10.1109/TBME.2017.2694818
8. Sakurada, T., Kawase, T., Komatsu, T., Kansaku, K.: Use of high-frequency visual stimuli above the critical flicker frequency in a SSVEP-based BMI. Clin. Neurophysiol. **126**(10), 1972–1978 (2015). https://doi.org/10.1016/j.clinph.2014.12.010
9. Stawicki, P., Gembler, F., Volosyak, I.: Evaluation of suitable frequency differences in SSVEP-based BCIs. In: Blankertz, B., Jacucci, G., Gamberini, L., Spagnolli, A., Freeman, J. (eds.) Symbiotic 2015. LNCS, vol. 9359, pp. 159–165. Springer, Cham (2015). https://doi.org/10.1007/978-3-319-24917-9_17
10. Wang, Y., Jung, T.P.: Visual stimulus design for high-rate SSVEP BCI. Electron. Lett. **46**(15), 1057–1058 (2010)
11. Wolpaw, J.R., Birbaumer, N., McFarland, D.J., Pfurtscheller, G., Vaughan, T.M.: Brain–computer interfaces for communication and control. Clin. Neurophysiol. **113**(6), 767–791 (2002). https://doi.org/10.1016/S1388-2457(02)00057-3

# A Comparison of cVEP-Based BCI-Performance Between Different Age Groups

Felix Gembler, Piotr Stawicki, Aya Rezeika, and Ivan Volosyak(✉)

Faculty of Technology and Bionics, Rhine-Waal University of Applied Sciences, 47533 Kleve, Germany
ivan.volosyak@hochschule-rhein-waal.de
https://bci-lab.hochschule-rhein-waal.de

**Abstract.** Persons who need assistive technologies to communicate with the relatives, for example amyotrophic lateral sclerosis (ALS) patients, could benefit from Brain-Computer Interface (BCI) technology if it is used alternatively or in addition to established communication tools. BCIs based on visual evoked potentials (VEPs) have shown high system speeds in many studies. However, some major issues need to be addressed: On the one hand, the strength of VEP responses varies across subjects. Especially age-related performance differences have been covered in several studies and need to be investigated further. On the other hand, the stimuli used to evoke the VEP response, are considered annoying for many people.

This paper investigates the subjective level of annoyance in different age groups for different flickering rates as well as age-related performance differences in cVEP based BCIs. In this regard, the cVEP-based eight target spelling interface was tested with two age groups (13 subjects each, ranging from 20 to 28 years and 62 to 83 years). Typically, 60 Hz monitor refresh rate is used to generate the cVEP stimuli. Here, three different flickering speeds were tested (m-sequences generated using refresh rates of 30, 60 and 120 Hz for stimuli presentation).

The mean ITR of the elderly age group was 42.03, 45.32 and 45.75 bits per minute (bpm) while the young group achieved an ITR of 53.09, 64.01 and 72.92 bpm for the 30, 60 and 120 Hz setups. The difference was significant for the faster flickering setups (60 and 120 Hz), respectively. Hence, our results show that elderly people have slightly worse BCI performance in terms of information transfer rate (ITR). Regarding the level of annoyance, subjects from both age groups preferred the 120 Hz setup which offered a more subtle visual stimulation.

**Keywords:** Brain-Computer Interface (BCI) · Human-Computer Interaction (HCI) · Steady-state visual evoked potentials (SSVEP) · Code-modulated visual evoked potentials (cVEP) · Monitor refresh rate


I. Rojas et al. (Eds.): IWANN 2019, LNCS 11506, pp. 394–405, 2019.
https://doi.org/10.1007/978-3-030-20521-8_33

## 1 Introduction

Brain-Computer Interfaces (BCIs) describe a field of technical systems that acquire and analyze brain activity patterns in real time to translate them into control commands for computers or external devices [19].

Visual evoked potentials (VEPs) are continuous responses that are elicited at the occipital and parietal cortical areas of the brain under visual stimulation; if a set of distinct stimuli is presented to the user, an attended stimuli can be identified by analyzing the brain data recorded via EEG [15,17]. VEP-BCIs are typically realized using brain responses to constant frequencies, so called steady state visual evoked potentials (SSVEPs) or brain responses to pseudorandom code patterns, which are referred to as code-modulated visual evoked potentials (cVEPs) [2]. For the latter, typically m-sequences are used because of their good autocorrelation properties. VEP-BCIs are described as reliable and responsive in the literature [2,12,15]. Indeed, the VEP approach utilized as a non-invasive communication tool yielded information transfer rates of more than 300 bits per minute (bpm) [12].

People who need assistive technologies to communicate with their relatives, such as ALS patients, could benefit from this technology if it is used alternatively or in addition to established communication tools. Several studies have investigated methods of data-fusion between eye tracking devices and VEP BCIs with promising results (see e.g. [14,16]). However, some major issues regarding this technology remain to be addressed.

On the one hand, the strength of VEP-responses vary across subjects. Although patients from all age groups could benefit from these technologies, the majority of studies are conducted with a relatively young subject group, typically students/research assistants. However, few articles indicate slightly worse BCI performance by older subjects. E.g. Dias et al. [3] found that elderly participants show smaller P300 amplitudes than younger ones. A correlation between age and BCI performance has also been found for the sensorimotor-rhythm (SMR) BCIs [8].

Research on SSVEP based BCIs frequently reported variations in performance between users. In two recent studies, we investigated age associated differences in SSVEP performance [7,17]. In the latter, a significant difference between young (a group of 10 participants between 19 and 27 years) and elderly (10 participants, age 64 to 76 years) was found. Users from both age groups were able to control the system, but the young age group performed much faster. The effects of age on BCI performance has also been addressed by Norton et al. [13], who showed that young children between 9–11 years can reliably control a SSVEP BCI.

Next to system responsiveness, overall user friendliness is another important factor in BCI research. The stimuli used to evoke the brain response are considered as annoying by many people. In a larger SSVEP study, it was observed that younger subjects were less annoyed by the flickering and tended to attain a higher information transfer rate (ITR) [1]. But only a few subjects were over 50 years old in this study.

In order to explore these differences further, we tested two equal sized groups of different age ranges with a spelling application. Here, the cVEP paradigm was used. Contrary to the SSVEP studies we conducted earlier, the cVEP BCI requires a training session were templates for the specific targets are recorded.

As the flickering rate plays a central role in the perceived level of annoyance as well as performance [4, 18], three flickering speeds were tested. In this respect, the cVEP stimuli were generated with monitor refresh rates set to 30, 60 and 120 Hz.

Although highest speeds are achieved with multi-target systems [12], here, only eight targets were displayed simultaneously. This ensures readability and simplicity of the graphical user interface and reduces cognitive load.

## 2 Methods and Materials

This section describes the hardware setup, stimulus presentation, experimental design, and presents details about the classification methods and spelling interface.

### 2.1 Participants

For the experiment, 26 healthy participants were recruited which were divided in two groups based on their age; one group, referred to as elderly group (average age 72.69 years, SD 6.27, range 62 to 83 years, eight female and five male), the other referred to as young group (average age 23.48 years, SD 2.61, range 20 to 28 years, 8 male and 5 female). All subjects had normal or corrected to normal vision. The research was approved by the ethical committee of the medical faculty of University Duisburg-Essen. Before the experiment started, the volunteers were informed about the involved risks and the experimental procedure. The subjects who agreed to participate signed a consent form in accordance with the Helsinki declaration. The subject information was stored anonymously. During the experiment, the participants had the opportunity to opt-out at any time. All participants received a financial reward for the participation in this study.

### 2.2 Hardware

The used computer (MSI GT 73VR with nVidia GTX1070 graphic card) operated on Microsoft Windows 10 Education running on an Intel processor (Intel Core i7, 2.70 GHz). A liquid crystal display screen (Asus ROG Swift PG258Q, 1920 × 1080 pixel, 240 Hz maximal refresh rate) was used for stimulus presentation.

An EEG amplifier (g.USBamp, Guger Technologies, Graz, Austria) was used, utilizing all 16 signal channels which were placed according to the international 10/5 system of electrode placement (see, e.g., [10] for more details): $P_Z$, $P_3$, $P_4$, $P_5$, $P_6$, $PO_3$, $PO_4$, $PO_7$, $PO_8$, $POO_1$, $POO_2$, $O_1$, $O_2$, $O_Z$, $O_9$, and $O_{10}$. Further, the reference electrode was placed at $C_Z$ and the ground electrode at $AF_Z$. Standard abrasive electrolytic electrode gel was applied between the electrodes and the scalp to bring impedances below 5 kΩ during the preparation phase.

### 2.3 Stimulus Presentation

Eight boxes (230 × 230 pixel) arranged as 2 × 4 stimulus matrix were utilized (see Fig. 1 for more details), which corresponded to $K = 8$ stimulus classes.

A distinct flashing pattern was assigned to each of these targets using the cVEP paradigm. For this, so-called m-sequences, non-periodic binary codes with good autocorrelation properties [2] were applied.

When flickering, the color of the target stimuli corresponding to the codes alternated between the background color 'black' (represented by '0') and 'white' (represented by '1').

The m-sequences used in this experiment, $c_i$, $i = 1, \dots, K$, were set recursively. The initial code, $c_1$ was set to

$$c_1 = 101011001101110110100100111000101111001010001100001000001111110.$$

The remaining $K - 1$ sequences were determined by employing a circular shift of 2 bits ($c_1$ had no shift, $c_2$ was shifted by 2 bits to the left, $c_3$ was shifted by 4 bits to the left, etc.).

The duration of the stimulation cycle was calculated by dividing the code length (here 63 bit) by the monitor refresh rate $r$ in Hz.

In this experiment, three setups were tested: $r$ was set to 30, 60 and 120 Hz yielding stimulus durations of 2.1, 1.05 and 0.525 s.

### 2.4 Experimental Protocol

For both age groups the same experimental design was used. All participants went through three sessions (30 Hz, 60 Hz and 120 Hz refresh rate setup), each consisting of a training phase and a copy spelling phase.

In the training phase, each of the eight stimuli was gazed at by the user several times. The recording was grouped in six training blocks, $n_b = 6$. In each block every stimulus was attended once, resulting in $6 * 8 = 48$ trials in total.

Each of these trials lasted for 4.2 s, i.e. the code pattern repeated for 2, 4 or 8 cycles depending on the used refresh rate (30, 60 and 120 Hz, respectively).

A green frame indicated which box the user needed to fixate on. The flickering was initiated by pressing the space bar. After each trial, the next box the user needed to focus on was highlighted, and the flickering paused for one second.

After each block of eight trials the user was allowed to rest. The next recording block was again initiated by pressing the space bar.

The copy spelling task consisted of a brief familiarization run were participants spelled the word BCI. If necessary, the automatically determined classification parameters were altered slightly. Thereafter, the German words BAUM, HAUS and WELT were spelled. Errors needed to be corrected using the UNDO function of the spelling interface. After each session, participants were asked to rate the level of annoyance caused by the flickering.

### 2.5 Signal Processing

Canonical-correlation analysis (CCA), a statistical method which investigates the relationship between two sets of variables $\mathbf{X} \in \mathbb{R}^{p \times s}$ and $\mathbf{Y} \in \mathbb{R}^{q \times s}$, can be used to create spatial filters for VEP-BCIs (see, e.g., [15]). In general, CCA determines weight vectors $\mathbf{w}_X \in \mathbb{R}^p$ and $\mathbf{w}_Y \in \mathbb{R}^q$ that maximize the correlation $\rho$ between the linear combinations $\mathbf{x} = \mathbf{X}^T \mathbf{w}_X$ and $\mathbf{y} = \mathbf{Y}^T \mathbf{w}_Y$ by solving

$$\max_{\mathbf{w}_X, \mathbf{w}_Y} \rho(\mathbf{x}, \mathbf{y}) = \frac{{\mathbf{w}_X}^T \mathbf{X} \mathbf{Y}^T \mathbf{w}_Y}{\sqrt{{\mathbf{w}_X}^T \mathbf{X} \mathbf{X}^T \mathbf{w}_X \mathbf{w}_Y^T \mathbf{Y} \mathbf{Y}^T \mathbf{w}_Y}}. \tag{1}$$

Here, templates and filters were determined for each stimulus individually. The training trials were stored in a $m \times n_t$ matrix, where $m$ is the number of electrode channels (here, $m = 16$) and $n_t$ is the number of sample points (here, $n_t = 2520$). Class specific templates $\mathbf{X} \in \mathbb{R}^{m \times n_t}$ were generated by averaging all $n_b$ trials corresponding to a specific target. Then, two matrices,

$$\hat{\mathbf{T}} = [\mathbf{T}_1 \mathbf{T}_2 \dots \mathbf{T}_{n_b}] \quad \text{and} \quad \hat{\mathbf{X}} = [\underbrace{\mathbf{X}\mathbf{X} \dots \mathbf{X}}_{n_b}] \tag{2}$$

were constructed and inserted into (1), yielding a weight vector $\mathbf{w} = \mathbf{w}_{\hat{\mathbf{X}}}$. Applying this procedure for all $K$ classes, a set of training templates $\mathbf{X}_i$ with corresponding filters $\mathbf{w}_i$, $i = 1, \dots, K$ was generated.

In this study, an ensemble based classification strategy [12] with sliding classification time windows of dynamic length [17] were used for signal classification. New EEG data blocks transferred from the amplifier were successively added to a EEG data buffer, $\mathbf{Y} \in \mathbb{R}^{m \times n_y}$.

For the implementation of the sliding window mechanism, the number of samples per EEG data block was selected as divider of the cycle length in samples. Here, the amplifier transferred the EEG data in blocks of 90 samples per channel. This ensured that data collection and stimulus presentation remained synchronized when shuffling out old data blocks.

The output command corresponding to a classification was only performed if a certain threshold was surpassed. In this regard, the data buffer $\mathbf{Y}$, storing the EEG changed dynamically, i.e. the length of the classification time window $n_y$ was extended incrementally as long as $n_y < n_t$.

The data buffer $\mathbf{Y}$ was compared to the reference signals $\mathbf{R}_i \in \mathbb{R}^{m \times n_y}$, $i = 1, \dots, K$ which were constructed as sub-matrix of the corresponding training template $\mathbf{X}_i$, from rows $1, \dots, m$ and columns $1, \dots, n_y$. In this regard, ensemble correlations, $\lambda_k$,

$$\lambda_k = \rho \left( \begin{bmatrix} \mathbf{Y}^T \mathbf{w}_1 \\ \vdots \\ \mathbf{Y}^T \mathbf{w}_K \end{bmatrix}, \begin{bmatrix} {\mathbf{R}_k}^T \mathbf{w}_1 \\ \vdots \\ {\mathbf{R}_k}^T \mathbf{w}_K \end{bmatrix} \right), \quad k = 1, \dots, K \tag{3}$$

were calculated. Further, a filter bank method, decomposing the EEG data in sub-band components was used as described e.g. in [12]. For this, an 8th order

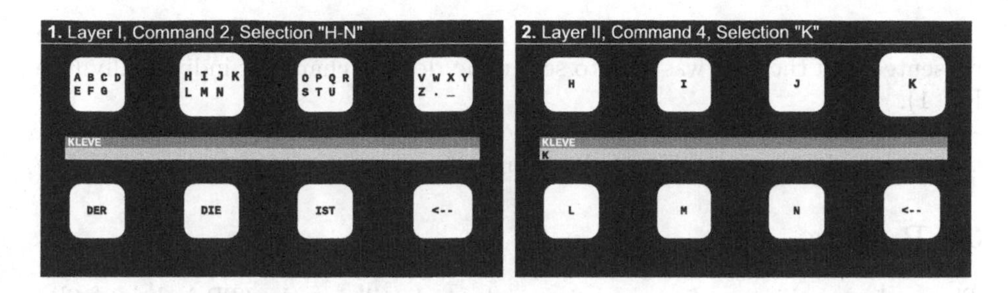


**Fig. 1. A participant is writing KLEVE with the eight-target spelling interface.** Letter selection required two steps: First, the character group was selected (here H-N), and then the character itself. The target word and the already spelled string were displayed in the center of the screen.

Butterworth filter was used, where the lower and upper cut-off frequencies of the $m$-th sub-band were set as $m * 8$ and 60 Hz.

The ensemble approach, Eq. (3), was then applied to each sub-band component individually, yielding a set of correlations $\lambda_k^{(1)}, \lambda_k^{(2)}, \ldots, \lambda_k^{(M)}$, $k = 1, \ldots, K$, where $M$ denotes the number of considered sub-bands (here $M = 5$). Then, the output command candidate $C$ was determined using linear combinations of the sub-band correlations,

$$C = \underset{k=1,\ldots,K}{\arg\max}\, \tilde{\lambda}_k, \text{ where } \tilde{\lambda}_k = \sum_{m=1}^{M} a_m \lambda_k^{(m)}. \tag{4}$$

Here, the weights were set to $a_1 = 0.4$, $a_2 = 0.2$, $a_3 = 0.15$, $a_4 = 0.13$ and $a_5 = 0.11$ (giving lower weights to the higher bands, which carry less relevant information, see also [12]). Noting that, optimal weight selection needs to be investigated further.

In the on-line application, the system output associated with a classified label $C$ was only performed if the decision certainty, $\Delta_C$, determined as the distance between the highest and second highest correlation exceeded a threshold value, $\beta$ which was determined individually on the basis of the training data.

If the threshold criterion was met, the data buffer $\mathbf{Y}$ was cleared and a two second gaze shifting period followed (data collection and flickering paused).

### 2.6 Spelling Interface

The graphical user interface presented eight selection options, see Fig. 1. Letter selection required two steps: Initially, the first row of the interface contained 28 characters (26 letters, underscore and full stop character) divided into four boxes (7 characters each). The second row contained a correction option and three dictionary suggestions based on word frequency lists (if no letter was typed, the most frequent words according to the list were suggested, see [5] for more details). By choosing the correction option, the last typed character was deleted, by choosing a box from the first row, the 7 characters contained in it were presented and the user was able to select the desired character individually (see Fig. 1).

Every selection was accompanied by audio and visual feedback (the size of the selected box increased for a short time).

## 3 Results

The on-line spelling performance was evaluated utilizing the ITR in bpm [19],

$$ITR = \frac{\log_2 K + p\log_2 p + (1-p)\log_2\left(\frac{1-p}{K-1}\right)}{t/60}, \tag{5}$$

with the target identification accuracy, $p$, calculated as the number of correctly classified commands divided by the total number of commands, and $t$ representing the average selection time (in s).

The results from all three spelling tasks are listed in Tables 1 and 2. The results from the three spelling tasks (BAUM, HAUS and WELT) were averaged. Young participants reached mean ITRs of 53.09, 64.01 and 72.92 bpm for the 30, 60 and 120 Hz system. The difference between refresh rate setups was analyzed using paired t-tests: For the young age group, the difference between 30 and 60 Hz was significant ($t = 3.25, p = 0.007$), the differences between 60 and 120 Hz ($t = 4.612, p = 0.0006$) as well as between 30 and 120 Hz ($t = 6.857, p = 0,00002$) were highly significant.

For the elderly participants differences between the refresh rates were not significant; this age group reached ITRs of 42.03, 45.32 and 45.74 bpm for 30, 60 and 120 Hz refresh rate. For all refresh rate settings and age groups average accuracies above 95% were achieved.

Regarding the performance difference between age groups, a Student's t-test (with unpooled variance) revealed highly significant differences between the mean ITRs of young and elderly subjects for the 120 Hz setup ($t = 5.2741, p = 0.00002$) as well as for the 60 Hz setup ($t = 3.9256, p = 0.0009$). For the 30 Hz setup however, no statistical difference between ITR of young and elderly ($t = 1.7502, p = 0.09287$) was found. An overview of the age related performance difference is presented in Fig. 2.

The results from the questionnaires are depicted in Fig. 3. The subjective level of annoyance was rated using a five-point Likert scale. The majority of the users did not find the flickering annoying. Overall, answers regarding the higher speed flickering confirmed that it was considerably less annoying for both age-groups.

## 4 Discussion

The presented results confirm that there is a significant difference between the performance of young participants and participants of advanced age. The differences were most noticeably for the 120 Hz refresh rate setting. While accuracies

**Table 1.** Results of the on-line spelling tasks of the elderly age group for the different refresh rate setups. Provided are the averaged time, accuracy and ITR for all three spelled words (BAUM, HAUS and WELT).

| # | 30 Hz | | | 60 Hz | | | 120 Hz | | |
|---|---|---|---|---|---|---|---|---|---|
| | Time [s] | acc [%] | ITR [bpm] | Time [s] | acc [%] | ITR [bpm] | Time [s] | acc [%] | ITR [bpm] |
| 1 | 34.00 | 100 | 42.94 | 31.47 | 100 | 47.11 | 35.98 | 93 | 37.08 |
| 2 | 29.12 | 100 | 50.59 | 23.47 | 100 | 63.32 | 24.12 | 100 | 59.90 |
| 3 | 17.85 | 100 | 80.76 | 24.00 | 100 | 62.90 | 18.03 | 100 | 79.91 |
| 4 | 37.53 | 96 | 36.64 | 24.60 | 100 | 51.66 | 25.82 | 95 | 43.89 |
| 5 | 44.43 | 93 | 29.72 | 48.15 | 97 | 31.21 | 45.52 | 94 | 31.88 |
| 6 | 46.52 | 95 | 35.28 | 27.20 | 100 | 53.49 | 29.22 | 95 | 53.93 |
| 7 | 37.73 | 100 | 38.19 | 51.30 | 100 | 29.74 | 27.83 | 100 | 52.43 |
| 8 | 25.25 | 100 | 57.11 | 33.45 | 100 | 45.31 | 42.45 | 97 | 34.06 |
| 9 | 28.08 | 93 | 50.06 | 19.80 | 100 | 72.91 | 27.83 | 97 | 53.92 |
| 10 | 48.28 | 97 | 32.62 | 38.30 | 91 | 36.74 | 32.90 | 100 | 44.18 |
| 11 | 55.87 | 93 | 23.98 | 42.12 | 100 | 34.25 | 60.10 | 89 | 24.11 |
| 12 | 52.08 | 92 | 28.95 | 42.33 | 91 | 36.41 | 48.33 | 97 | 33.43 |
| 13 | 36.37 | 93 | 39.58 | 58.62 | 94 | 24.13 | 34.82 | 100 | 45.90 |
| Mean | 37.93 | 96.4 | 42.03 | 35.75 | 97.9 | 45.32 | 34.84 | 96.6 | 45.74 |
| SD | 10.75 | 3.1 | 14.41 | 11.68 | 3.3 | 14.33 | 11.11 | 3.4 | 14.08 |

**Table 2.** Results of the on-line spelling tasks of the young age group for the different refresh rate setups. Provided are the averaged time, accuracy and ITR for all three spelled words (BAUM, HAUS and WELT).

| # | 30 Hz | | | 60 Hz | | | 120 Hz | | |
|---|---|---|---|---|---|---|---|---|---|
| | Time [s] | acc [%] | ITR [bpm] | Time [s] | acc [%] | ITR [bpm] | Time [s] | acc [%] | ITR [bpm] |
| 1 | 21.02 | 100 | 69.66 | 22.22 | 100 | 65.43 | 17.48 | 100 | 82.38 |
| 2 | 39.27 | 96 | 43.69 | 18.82 | 100 | 66.98 | 21.72 | 97 | 66.58 |
| 3 | 27.65 | 100 | 52.88 | 20.48 | 100 | 70.80 | 18.50 | 100 | 78.35 |
| 4 | 96.75 | 80 | 15.22 | 38.27 | 91 | 41.63 | 25.67 | 100 | 50.23 |
| 5 | 21.63 | 97 | 69.61 | 21.90 | 97 | 71.48 | 16.65 | 100 | 82.88 |
| 6 | 17.02 | 100 | 74.28 | 20.78 | 100 | 63.87 | 14.57 | 100 | 86.50 |
| 7 | 25.37 | 100 | 57.09 | 22.30 | 97 | 70.67 | 19.63 | 100 | 73.37 |
| 8 | 20.22 | 100 | 71.68 | 20.85 | 100 | 70.30 | 17.18 | 100 | 83.82 |
| 9 | 27.23 | 100 | 53.62 | 20.35 | 100 | 71.08 | 18.07 | 100 | 79.71 |
| 10 | 28.77 | 94 | 48.66 | 26.27 | 97 | 56.18 | 27.33 | 100 | 52.94 |
| 11 | 34.10 | 93 | 41.27 | 23.17 | 100 | 62.77 | 19.40 | 100 | 74.26 |
| 12 | 50.63 | 94 | 32.14 | 25.20 | 100 | 56.50 | 21.78 | 100 | 67.40 |
| 13 | 21.08 | 100 | 60.43 | 19.88 | 100 | 64.49 | 18.98 | 100 | 69.47 |
| Mean | 33.13 | 96.5 | 53.09 | 23.11 | 98.6 | 64.01 | 19.77 | 99.7 | 72.92 |
| SD | 20.36 | 5.4 | 16.49 | 4.81 | 2.5 | 8.17 | 3.44 | 0.9 | 10.98 |

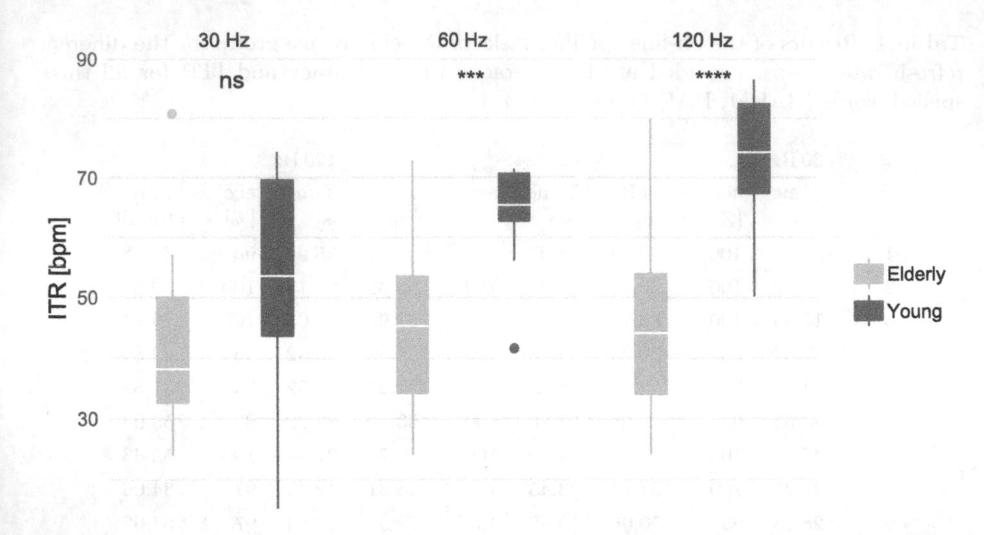


**Fig. 2.** Comparison of ITR performances. The results from young (20 to 28 years) and elderly (62 to 83 years) groups are compared. Displayed are the ITRs averaged over the three on-line spelling tasks (BAUM, HAUS, WELT) for three monitor refresh rates that were used to generate the stimuli. The asterisks mark statistical significance ($***p < 0.001$ and $****p <= 0.0001$).

were above 95% for all participants of both groups, the time needed to complete the spelling task was generally larger for the elderly resulting in overall lower ITR. This is in line with our findings from our previous studies with the SSVEP paradigm, where we also utilized threshold-based dynamic classification time windows; elderly participants typically needed to gaze at a stimulus for longer time to produce an output command as well.

According to the findings of Hsu et al. [9], elderly people show smaller VEP amplitudes, which could explain the performance gap. It should be noted, that the reaction time, which is typically shorter for younger participants, can also influence the ITR. To ensure that each participant understood the functioning of the interface, a familiarization run was performed before the spelling tasks started. Further, in this study, gaze-shifting phases of 2 s (providing enough time to find the next target on the GUI) were used to reduce these effects. The drawback of this approach is comparably low ITR as typically periods of one second or less are used for gaze shifting [6,12].

To our knowledge, a comparison between young and elderly users has not been made for the cVEP paradigm. The performance gap was investigated for different refresh rate setups and most noticeable for the 120 Hz setup (see Fig. 2).

As observed by Wittevrongel [18], the 120 Hz rate enabled faster communication speed in comparison to the typically used 60 Hz stimulus presentation. However, we found that the difference between refresh rates in lights of ITR is more relevant for the younger participants. The results indicate that faster

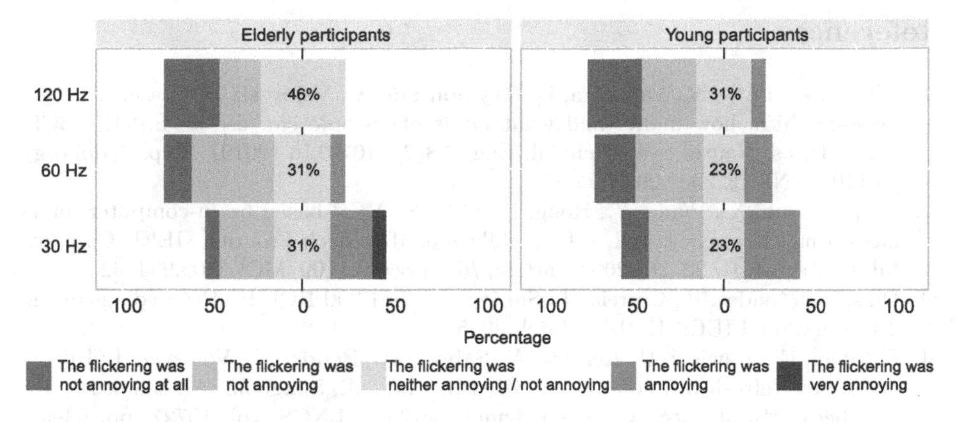


**Fig. 3.** Subjective level of user friendliness when using a cVEP based BCI system. Participants were asked to state the perceived level of annoyance on a 1 to 5 Likert scale.

flickering rates than tested here, would yield even better performance. However, in preliminary experiments we also tested the system utilizing 200 and 240 Hz refresh rates; while the stimulation was even more subtle, for most participants performance was significantly worse in comparison to the 120 Hz generated flickering (see also [4] for more details).

Aside from performance, the flickering speed has high impact on the user comfort of the BCI. To investigate user-friendliness, the participants were asked about the level of annoyance caused by flickering. As expected, the faster stimulation was preferred by most users. Hence, the 120 Hz stimulus presentation seems to be the better option in general. On the other hand, the temporal stability of higher frequency components might require the system to be recalibrated more regularly, i.e. for each session [11]. Future long term studies should address this question further.

# 5 Conclusion

Subject age is an important factor in VEP-BCI performance. BCI performance was accessed using an eight-target cVEP spelling interface. Three different refresh rate setups were used to generate the flickering. For the 120 Hz and 60 Hz flickering a highly significant difference between young and elderly users was found. The study confirms that elderly users yield poorer BCI performance in comparison to the younger users. As the 120 Hz setup was least annoying to users from both age groups, it seems to be a better choice than the typically used vertical refresh rate of 60 Hz.

**Acknowledgment.** This research was supported by the European Fund for Regional Development (EFRD - or EFRE in German) under Grants GE-1-1-047 and IT-1-2-001. We also thank all the participants of this research study and our student assistants.

## References

1. Allison, B., Luth, T., Valbuena, D., Teymourian, A., Volosyak, I., Graser, A.: BCI demographics: how many (and what kinds of) people can use an SSVEP BCI? IEEE Trans. Neural Syst. Rehabil. Eng. **18**(2), 107–116 (2010). https://doi.org/10.1109/TNSRE.2009.2039495
2. Bin, G., Gao, X., Wang, Y., Hong, B., Gao, S.: VEP-based brain-computer interfaces: time, frequency, and code modulations [Research Frontier]. IEEE Comput. Intell. Mag. **4**(4), 22–26 (2009). https://doi.org/10.1109/MCI.2009.934562
3. Dias, N., Mendes, P., Correia, J.: Subject age in P300 BCI. In: Proceedings of the 2 International IEEE EMBS, March 2005
4. Gembler, F., Stawicki, P., Rezeika, A., Saboor, A., Benda, M., Volosyak, I.: Effects of monitor refresh rates on c-VEP BCIs. In: Ham, J., Spagnolli, A., Blankertz, B., Gamberini, L., Jacucci, G. (eds.) Symbiotic 2017. LNCS, vol. 10727, pp. 53–62. Springer, Cham (2018). https://doi.org/10.1007/978-3-319-91593-7_6
5. Gembler, F., et al.: A dictionary driven mental typewriter based on code-modulated visual evoked potentials (cVEP). In: 2018 IEEE International Conference on Systems, Man, and Cybernetics (SMC), pp. 619–624. IEEE, Miyazaki, October 2018. https://doi.org/10.1109/SMC.2018.00114
6. Gembler, F., Stawicki, P., Volosyak, I.: Autonomous parameter adjustment for SSVEP-based BCIs with a novel BCI wizard. Front. Neurosci. **9** (2015). https://doi.org/10.3389/fnins.2015.00474
7. Gembler, F., Stawicki, P., Volosyak, I.: A comparison of SSVEP-based BCI-performance between different age groups. In: Rojas, I., Joya, G., Catala, A. (eds.) IWANN 2015, Part I. LNCS, vol. 9094, pp. 71–77. Springer, Cham (2015). https://doi.org/10.1007/978-3-319-19258-1_6
8. Grosse-Wentrup, M., Schölkopf, B.: A review of performance variations in SMR-based brain- computer interfaces (BCIs). In: Guger, C., Allison, B., Edlinger, G. (eds.) Brain-Computer Interface Research, pp. 39–51. Springer, Heidelberg (2013). https://doi.org/10.1007/978-3-642-36083-1_5
9. Hsu, H.T., et al.: Evaluate the feasibility of using frontal SSVEP to implement an SSVEP-based BCI in young, elderly and ALS groups. IEEE Trans. Neural Syst. Rehabil. Eng. **24**(5), 603–615 (2016). https://doi.org/10.1109/TNSRE.2015.2496184
10. Jurcak, V., Tsuzuki, D., Dan, I.: 10/20, 10/10, and 10/5 systems revisited: their validity as relative head-surface-based positioning systems. NeuroImage **34**(4), 1600–1611 (2007). https://doi.org/10.1016/j.neuroimage.2006.09.024
11. McCullagh, P., Lightbody, G., Zygierewicz, J., Kernohan, W.G.: Ethical challenges associated with the development and deployment of brain computer interface technology. Neuroethics **7**(2), 109–122 (2014). https://doi.org/10.1007/s12152-013-9188-6
12. Nakanishi, M., Wang, Y., Chen, X., Wang, Y.T., Gao, X., Jung, T.P.: Enhancing detection of SSVEPs for a high-speed brain speller using task-related component analysis. IEEE Trans. Biomed. Eng. **65**(1), 104–112 (2018). https://doi.org/10.1109/TBME.2017.2694818
13. Norton, J.J.S., Mullins, J., Alitz, B.E., Bretl, T.: The performance of 9–11-year-old children using an SSVEP-based BCI for target selection. J. Neural Eng. **15**(5), 056012 (2018). https://doi.org/10.1088/1741-2552/aacfdd

14. Saboor, A., et al.: Mesh of SSVEP-based BCI and eye-tracker for use of higher frequency stimuli and lower number of EEG channels. In: 2018 International Conference on Frontiers of Information Technology (FIT), pp. 99–104. IEEE, Islamabad, December 2018. https://doi.org/10.1109/FIT.2018.00025
15. Spüler, M., Rosenstiel, W., Bogdan, M.: Online adaptation of a c-VEP brain-computer interface (BCI) based on error-related potentials and unsupervised learning. PLoS ONE **7**(12), e51077 (2012). https://doi.org/10.1371/journal.pone.0051077
16. Stawicki, P., Gembler, F., Rezeika, A., Volosyak, I.: A novel hybrid mental spelling application based on eye tracking and SSVEP-based BCI. Brain Sci. **7**(4), 35 (2017). https://doi.org/10.3390/brainsci7040035
17. Volosyak, I., Gembler, F., Stawicki, P.: Age-related differences in SSVEP-based BCI performance. Neurocomputing **250**, 57–64 (2017). https://doi.org/10.1016/j.neucom.2016.08.121
18. Wittevrongel, B., Van Wolputte, E., Van Hulle, M.M.: Code-modulated visual evoked potentials using fast stimulus presentation and spatiotemporal beamformer decoding. Sci, Rep. **7**(1) (2017). https://doi.org/10.1038/s41598-017-15373-x
19. Wolpaw, J.R., Birbaumer, N., McFarland, D.J., Pfurtscheller, G., Vaughan, T.M.: Brain–computer interfaces for communication and control. Clin. Neurophysiol. **113**(6), 767–791 (2002). https://doi.org/10.1016/S1388-2457(02)00057-3

# Remote Steering of a Mobile Robotic Car by Means of VR-Based SSVEP BCI

Piotr Stawicki, Felix Gembler, Roland Grichnik, and Ivan Volosyak(✉)

Faculty of Technology and Bionics, Rhine-Waal University of Applied Sciences,
47533 Kleve, Germany
ivan.volosyak@hochschule-rhein-waal.de
https://bci-lab.hochschule-rhein-waal.de

**Abstract.** Brain-computer interface (BCI) technology, including applications based on the steady-state visual evoked potentials (SSVEPs) have proven to provide reliable and accurate control. In this paper, we present and evaluate remote steering of a previously developed and successfully tested mobile robotic car (MRC) utilizing the SSVEP-based BCI system. The visual stimulations were presented inside the head-mounted virtual reality (VR) glasses, here, the Oculus Go. The live video feedback from the MRCs point of view was displayed inside the custom made app of the VR environment. The three visual stimuli were located on both sides and above the video stream of the MRC camera.

The task of this study was to steer the MRC through a 8 m long path (in the real world) with 6 turns. Seven participants took part in the experiment reaching on average an accuracy of 98.1 (standard deviation: 5.04) %, an information transfer rate (ITR) of 10.71 (2.78) bits/min with an average command classification time of 3.95 (2.3) seconds. For classification, the minimum energy combination method (MEC) with 16 EEG electrodes as well as a filter bank decomposing method were utilized. All participants successfully completed the task, almost all subjects stated that the presented VR-based SSVEP-BCI was a highly immersive experience.

**Keywords:** Brain-Computer Interface (BCI) ·
Brain-Machine Interface (BMI) ·
Steady-State Visual Evoked Potentials (SSVEP) ·
Head-Mounted Display (HMD) · Virtual Reality (VR)

## 1 Introduction

Brain-computer interfaces are a fast developing branch of computer supported communication tools for patients with severe neuromuscular disorders, brainstem stroke, locked-in syndrome or amyotrophic lateral sclerosis (ALS) [14]. The BCI technology establishes an output channel between the brain and a computer system. The brain signals are collected via electroencephalography (EEG), and used for the one-way communication based on well defined patterns in the EEG.


I. Rojas et al. (Eds.): IWANN 2019, LNCS 11506, pp. 406–417, 2019.
https://doi.org/10.1007/978-3-030-20521-8_34

The possible combination of the BCI technology with simulated virtual environments and its' potentials has already been the subject of some studies. Kerous et al. reviewed the fusion of video games with BCI technology [5], where 14 of the 22 papers used the SSVEP-based BCI paradigm, demonstrating that SSVEP is a robust system and the overall convenience and popularity of the approach. In 2011, Ron-Angevin et al. demonstrated the control of a real robot and a simulated one inside a virtual environment using an SMR-based BCI [7]. Four participants took part in the experiment, three successfully steered the virtual robot (online) in the simulated environment.

The possible combination of BCI & mobile robots has been tested in recent years. Belraldo et al. used a 2-class Sensorimotor Rhythm (SMR) based BCI to control a telepresence robot (Pepper, SoftBank, Japan) with a live stream from the robots camera [1]. Wang et al. developed an SSVEP-based BCI for the navigation of a home-auxiliary robot (TurtleBot) [13]. Wu et al. tested a 4 class SSVEP-based BCI for remote steering of a mobile robot through a maze [16]. The classification was based on the power spectrum of the FFT and a fuzzy tracking system. On average, the subjects achieved an accuracy above 83%. Gergondet et al. presented a study with SSVEP-based BCI and live video feedback for steering a humanoid robot [4]. Five subjects tested the remote steering under static and dynamic background video conditions. Koo et al. compared a 3-class SSVEP-based BCI on a standard monitor (with a refresh rate of 60 Hz) and a virtual reality head mounted display (VRHMD, Oculus developers kit 1) [6]. In the used VR maze game (steering a ball), three participants reached a mean information transfer rate (ITR) of 24.58 and 22.17 bits/min with the VRHMD and the monitor setup, respectively. The minimal time for a target selection was set to 3 s. The participants voted in favor of the VRHMD setup, as the more immersive one.

With the variety of consumer grade market products the VR-experience can be a cost-effective VR headset for medical rehabilitation; yet the market aims at the entertainment sector.

In our recent studies we already explored the combination of SSVEP-based BCI and VR. In one of the tests, the participants had to navigate a virtual avatar through a VR-maze while the steering was realized with a visual flickering stimulation inside the VR [8]. In another study, the participants had to steer a VR vacuum cleaner robot to collect all obstacles [9]. The mobile robotic car (MRC) used in the current study has been previously tested with 61 participants that used an SSVEP-based BCI with live video feedback acquired from the robot's camera [11]. All participants steered the MRC successfully and effectively along a 15 m long path with 10 turns, achieving on average, an accuracy, total experiment time, and ITR of 93.03%, 207.08 s, and 14.07 bits/min, respectively.

In this paper, we extend our previous research in the VR area on remote steering of the MRC and provide a much more immersive user experience by using consumer grade VR glasses, the Oculus Go 64 GB.

## 2 Methods and Materials

In this section we describe the experimental design and conditions as well as the hard- and software setup, needed in order to reproduce the experiment.

### 2.1 Participants

Seven subjects (one female) with a mean age of 24.3 (SD: 5.1) years and a normal or corrected to normal vision participated in this study, all healthy adult volunteers (students of the Rhine-Waal University of Applied Sciences). This study was approved by the ethical committee of the medical faculty of the University Duisburg-Essen, all subjects signed a written consent form. Spectacles were worn when appropriate, the recording took place in a regular laboratory room, and the luminance was kept at an acceptable level (see e.g. Fig. 1). The participants received a financial reward for their participation.

**Fig. 1.** One of the participants immersed in controlling the MRC (visible in the lower left corner) during the online experiment. The end of the track is shown on the picture as well as the VRHMD and the recording hardware (gUSBamp and Alienware laptop).

### 2.2 Hardware

The data acquisition and analysis was running on a DELL (Round Rock, USA) Alienware laptop (Intel core i7-6820HK @2.7 GHz; Windows 10; build 1809).

The communication between the laptop, the VRHMD, and the robot was done with User Datagram Protocol (UDP) commands. To start the stimulation the BCI software sent a UDP command to the VRHMD headset, and when a command was classified, a UDP package was sent to the robot to execute the corresponding command.

The graphical user interface (GUI), along with the SSVEP stimulation and the live video view (from the robots point of view), was presented in the Oculus Go (Facebook Technologies Ireland Limited, Oculus VR, LLC, Menlo Park,

USA) headset. This VR headset is an Android (version 7.1) based stand alone head mounted display with a resolution of 2560 × 1440 pixels, refresh rate of 72 Hz, and allows a viewing angle of 100°.

**Mobile Robotic Car (MRC).** The Mobile robotic car (introduced in [11]) is a self-made line-tracking robot based on BeagleBone (ARM7, running Ubuntu for ARM), equipped with a WLAN access point and a live streaming camera software (for further information see [11]). Figure 2 shows the MRC used in this study.

**Fig. 2.** The mobile robotic car (MRC) used in this study. The camera and the line sensors are located in front of the robot [11]. The MRC has been previously tested with 61 participants that used an SSVEP-based BCI with live video feedback (from the robot's camera) and effectively steered it along a 15 m long path with 10 turns [11].

### 2.3 Software

**Graphical User Interface.** The graphical user interface (GUI) was developed in Unity (version 2017.4, Unity Technologies ApS, San Francisco, USA, unity3d.com, accessed on 7th January 2019). Inside the GUI app, the live video feedback stream from the robot was displayed in the center and the stimulation was located on both sides and above the video. Since the VR surrounding was not flat, the buttons were located closer to the user (on the Z axis) then the video screen and tilted 45° towards the user.

The VRHMD was connected to the access point located on the MRC, thus, the live video stream was continuously displayed. The stimulation was controlled through UDP commands sent from the BCI software running on the Alienware laptop. To increase the comfort of the BCI users, as the VR environment tends to be highly immersive (especially a fast moving environment can cause dizziness), the gaze shifting time (time between the selection of one command and the start of the flickering for next selection) was set to 5 s.

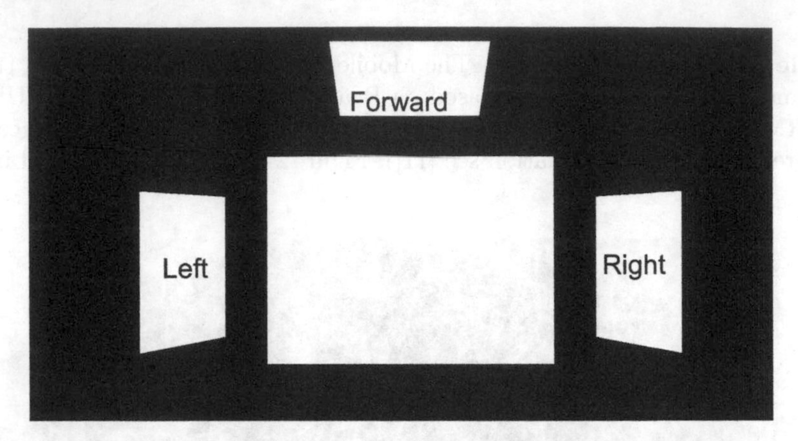


**Fig. 3.** GUI of the VR-based SSVEP-BCI. The aspect ratio of the live feedback screen (located in the center) was 1.33, and for the stimulation boxes 0.5 and 2, for the vertical and horizontal box, respectively. The horizontal stimulation box was located above the video screen (responsible for the "Forward" command) and the vertical stimulation boxes (for the "Left" and "Right" commands) were located on both sides of the video screen. The diagonal visual angle of the live feedback screen was 50(*) degrees, and for the boxes 20(*) degrees. *(Those values were calculated based on internal Unity units.)

**VR-Based Visual Stimulation.** The SSVEP stimulation inside the VR was located on the sides of the live video feedback. The *Forward* command was located above the upper edge of the live stream and the *Left* and *Right* commands on the left and right side of the screen with the video, respectively (see Fig. 3). Although the difference between the SSVEP stimulation frequencies can be as dense as 0.05 Hz [10], here we decided to select the divisors of the vertical refresh rate of the headset display (72 Hz) and used the following frequencies: 7.2, 9.0, and 12.0 Hz for the commands "Left", "Forward", and "Right", respectively. The flickering in the Unity Game Engine was produced by periodically alternating the colours of the GUI elements between black and white. Frame timing was governed by the Unity Engine running the VR software, thus, a time-based algorithm was used to determine for each frame whether a box/button should be white or black for this frame.

### 2.4 Signal Acquisition

For the EEG recording, a biomedical amplifier gUSBamp (Guger Technologies, Graz, Austria) was utilized. The sampling frequency was 256 Hz and data were collected at 64 samples/block. Digital filters: band-pass (2–100 Hz) and notch (around 50 Hz) were applied in the amplifier.

Sintered Ag/AgCl ring electrodes were used. The ground electrode was placed over $AF_Z$, the reference electrode over $C_Z$, and the sixteen signal electrodes were placed at predefined locations available in the g.GAMMAcap[2] (international 10 – 10 system, extended with intermediate positions), around the $P_Z$, $P_3$, $P_5$, $P_4$, $P_6$, $PO_3$, $PO_4$, $PO_7$, $PO_8$, $POO_1$, $POO_2$, $O_1$, $O_2$, $O_Z$, $O_9$, and $O_{10}$. Standard abrasive electrode gel was applied between the electrodes and the scalp to bring the impedances of the electrodes below 5 kΩ. The HMD stripes (for securing the position on the subjects' head) were placed on top of the electrodes (see Fig. 1).

### 2.5 Signal Processing

The acquired EEG data were processed online in a self developed (C++) software. During the experiment, a growing time window ranging from 4 blocks (1 s) till 16 blocks (4 s) was used for classification. The data was parallel band-pass filtered using the filter-bank (FB) method described by [2]. In those filter banks, the minimum energy combination (MEC) method [3] was used for classification (FBMEC).

The result of the FBMEC, the correlation coefficient for each class, was compared to a threshold value set for each subject individually after an introduction phase of the experiment (driving the path for the first time).

The MEC uses principal component analysis (PCA) to cancel out components that do not contribute to the response.

The response to a specific stimulus frequency $f$, as well as its $N_h$ harmonics can be defined as the voltage between the i-th electrode and a reference at time $t$, which is subject to a phase-shift and channel specific environmental nuisance and noise signal $E_i$,

$$y_i(t) \;=\; \sum_{k=1}^{N_h} a_{i,k}\sin(2\pi kft) + b_{i,k}\cos(2\pi kft) + E_i(t). \tag{1}$$

More generally, we considered $M$ samples of EEG data, recorded for each of $N$ signal electrodes at a sampling frequency of $F_s$ Hz. Let $N_h$ denote the number of harmonics that were considered for classification; for each stimulus frequency $f_i$, $i = 1, \ldots, K$ a reference matrix $\mathbf{R}_{f_i} \in \mathbb{R}^{2N_h \times M}$ is constructed as

$$\mathbf{R}_{f_i} = \begin{bmatrix} \sin(2\pi f_i t) \\ \cos(2\pi f_i t) \\ \vdots \\ \sin(2\pi N_h f_i t) \\ \cos(2\pi N_h f_i t) \end{bmatrix}, \quad t = \frac{1}{F_s}, \frac{2}{F_s}, \ldots, \frac{M}{F_s}. \tag{2}$$

For the reference $\mathbf{Y} \in \mathbf{R}_{f_1}, \ldots, \mathbf{R}_{f_K}$ corresponding to the fixed stimulation frequency $f$ and the EEG signal matrix $\mathbf{X} \in \mathbb{R}^{N \times M}$ holding the data for classification, we can generalize (1) to

$$\mathbf{X}^T = \mathbf{Y}^T\mathbf{A} + \mathbf{E}, \tag{3}$$

where the phases and amplitudes corresponding to $\mathbf{Y}$ are stored in the matrix $\mathbf{A} \in \mathbb{R}^{2N_h \times N}$ and the noise signals are stored in $\mathbf{E} \in \mathbb{R}^{M \times N}$.

The goal of the MEC is to amplify the SSVEP-amplitudes and filtering out the noise signal.

For this, the noise matrix $E$ is estimated using orthogonal projection

$$\tilde{\mathbf{E}} = \mathbf{X} - \mathbf{Y}(\mathbf{Y}^T\mathbf{Y})^{-1}\mathbf{Y}^T\mathbf{X}. \tag{4}$$

Now, a vector $\hat{w}$ that minimizes the energy of $\tilde{\mathbf{E}}$ is searched,

$$\min_{\hat{\mathbf{w}}} \left\| \tilde{\mathbf{E}}\hat{\mathbf{w}} \right\|^2 = \min_{\hat{\mathbf{w}}} \hat{\mathbf{w}}^T \tilde{\mathbf{E}}^T \tilde{\mathbf{E}} \hat{\mathbf{w}}. \tag{5}$$

This optimization problem can be solved by finding the eigenvector corresponding to the smallest eigenvalue. We calculated the eigenvalues $\lambda_1 \leq \lambda_2 \leq \ldots \leq \lambda_N$ and the corresponding eigenvectors $\mathbf{v}_1, \ldots, \mathbf{v}_N$ of $\tilde{\mathbf{E}}^T\tilde{\mathbf{E}}$. We then defined a set of weight vectors,

$$\mathbf{w}_i = \frac{\mathbf{v}_i}{\sqrt{\lambda_i}}, \quad i = 1, \ldots, N,$$

and further, a set of virtual channels

$$\mathbf{s}_i = \mathbf{X}^T \mathbf{w}_i, \quad i = 1, \ldots, N.$$

To detect the SSVEP response for the specific frequency $f$, the power of that frequency and its harmonics $N_h$ is estimated by

$$\hat{\mathbf{P}} = \frac{1}{NN_h} \sum_{l=1}^{N} \sum_{k=1}^{N_h} \| \mathbf{X}_k \mathbf{s}_l \|^2, \tag{6}$$

where $\mathbf{X}_k \in \mathbb{R}^{2 \times N}$ is defined as the sub-matrix of $\mathbf{X}$ constructed from the rows containing the sine and cosine data of the $k$-th harmonic [12]. These SSVEP power estimations are computed for all $K$ considered frequencies and then normalized, yielding the probabilities

$$p_i = \frac{\hat{P}_i}{\sum_{j=1}^{N_f} \hat{P}_j}, \quad i = 1, \ldots, K.$$

As a last step, a filter bank method utilizing an 8-th order Butterworth filter was applied (see e.g. in [2]). The lower and upper cut-off frequencies of the $m$-th sub-band were set as $m * 3 - 1$ and 90 Hz. Here, three sub-bands were analyzed, i.e., the MEC was used for each individual sub-band component, yielding a set of probabilities $p_k^1, \ldots, p_k^3$, $k = 1, \ldots, K$. The target identification label $C$ was determined as linear combinations of these sub-band probabilities,

$$C = \mathop{\arg\max}_{k=1,\ldots,K} \tilde{p}_k, \text{ where } \tilde{p}_k = \sum_{m=1}^{3} a_m p_k^m. \tag{7}$$

Here, the following weight set was used: $a_1 = 0.52$, $a_2 = 0.28$, and $a_3 = 0.20$.

The classifier output $C$ was ignored, if the distance between the highest and second highest probability did not exceed a threshold value ($C_{\text{Td}}$). This threshold value was set for each subject individually during an introduction phase of the experiment (selecting each command 2 times). In this experiment, the number of considered harmonics was set to $N_h = 3$, the number of signal channels was $N = 16$ (see signal acquisition) and the number of stimuli classes was $K = 3$.

BCI performance was evaluated using the ITR in bits/min (see e.g. [15]).

$$B_m = \left( \log_2 N + P \log_2 P + (1 - P) \log_2 \left[ \frac{1 - P}{N - 1} \right] \right) \times \frac{60}{T} \times C_N, \tag{8}$$

where $B_m$ represents the bits per minute, $P$ is the accuracy and $T$ is the total time of the experiment, and $N$ is the number of targets $C_N$ (here: three).

### 2.6 Procedure

The experiment took place in a typical PC laboratory room at the Rhine-Waal University of Applied Sciences. For user comfort, the blinds were pulled down so the luminance was kept at a comfortable level. After the experiment was explained, the participants signed the consent form and were prepared for the test. The task for each participant was to steer the MRC along the predefined path (see Fig. 4), which was marked on the floor and visible on the live streamed video inside the VRHMD headsets GUI). Each subject performed a familiarization run (one time driving of the whole track), where the individual threshold were adjusted (if necessary), after that all participants (expect P2, for whom the familiarization run results are presented, as she/he withdrew from the experiment after the first run) steered the MRC for a second time (main part of the experiment) through the track.

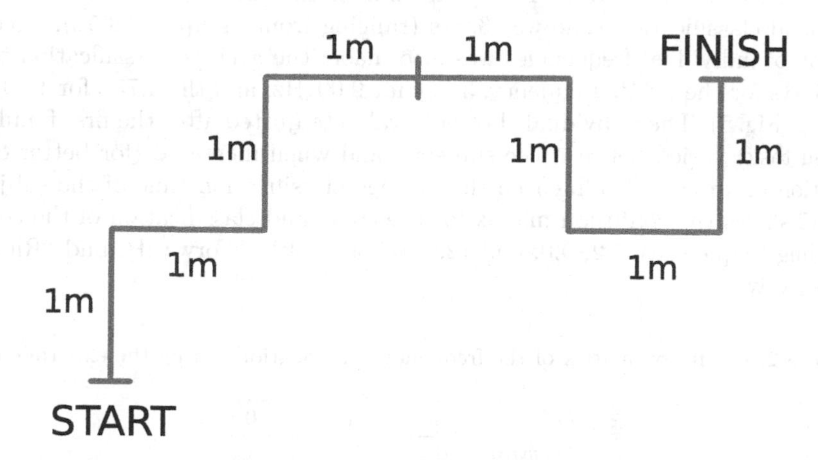


**Fig. 4.** MRC track from the experiment. It consists of 6 turns (3 left, 3 right, each aprox. 90°) and 8 forward commands that moved the MRC 1m/command along the path using the line sensors.

During the experiment, when a wrong command was classified (or selected by the user) the MRC was manually moved by the experimenter (this occurred only 3 times during the whole study).

## 3 Results

The results of the online experiment are presented in Table 1.

**Table 1.** Results of the online experiment. Presented are the total time of the experiment, the achieved accuracy (Acc), the information transfer rate (ITR) in bits/min (bpm), the average times for: a single command selection ($T_C$), selection of first frequency ($T_{7.2\text{Hz}}$), selection of the second frequency ($T_{9.0\text{Hz}}$), selection of the third frequency ($T_{12.0\text{Hz}}$). The last column provides the individual classification threshold value ($C_{\text{Td}}$).

| Subject | Time [s] | Acc. [%] | ITR [bpm] | $T_C$ | $T_{7.2\text{Hz}}$ | $T_{9.0\text{Hz}}$ | $T_{12.0\text{Hz}}$ | $C_{\text{Td}}$ |
|---|---|---|---|---|---|---|---|---|
| P1 | 108.75 | 87 | 7.33 | 2.59 | 2.75 | 2.13 | 4.63 | 0.15 |
| P2 | 139.25 | 100 | 9.56 | 5.31 | 5.26 | 5.47 | 4.91 | 0.20 |
| P3 | 93.75 | 100 | 14.20 | 2.06 | 1.59 | 1.94 | 2.83 | 0.15 |
| P4 | 118.5 | 100 | 11.24 | 3.38 | 3.59 | 2.69 | 7.08 | 0.17 |
| P5 | 185.25 | 100 | 7.19 | 8.59 | 10.58 | 8.32 | 7.33 | 0.19 |
| P6 | 108.75 | 100 | 12.24 | 3.13 | 1.08 | 3.82 | 3.34 | 0.20 |
| P7 | 100.75 | 100 | 13.21 | 2.56 | 2.00 | 2.53 | 3.16 | 0.18 |
| Mean | 122.14 | 98.10 | 10.71 | 3.95 | 3.83 | 3.84 | 4.75 | 0.18 |
| SD | 31.39 | 5.04 | 2.78 | 2.30 | 3.29 | 2.32 | 1.84 | 0.02 |

On average participants reached a total experiment time of 122.14 s, accuracy of 98.10% and ITR of 10.71 bits/min in the driving task. The average command classification time was 3.95 s (ranging from 1 s up to 18.5 s). A comparison of individual frequencies was also made: the average classification time was 3.83 s for the 7.2 Hz frequency, 3.84 s for 9.00 Hz, and the 4.75 s for 12.0 Hz (see e.g. Fig. 5). The individual threshold value (adjusted after the first familiarization task) varied between the subjects, and when increased (for better classification accuracy), it influenced the average classification time of the subject. Table 2 shows the confusion matrix for the command classification of the corresponding frequencies: 7.2, 9.0, and 12.0 Hz for "Left", "Forward", and "Right", respectively.

**Table 2.** Confusion matrix of the frequency classification during the experiment.

| Outcome \ Target | *Left* | *Forward* | *Right* |
|---|---|---|---|
| *Left* | 21 | 1 | 0 |
| *Forward* | 0 | 55 | 1 |
| *Right* | 0 | 0 | 20 |

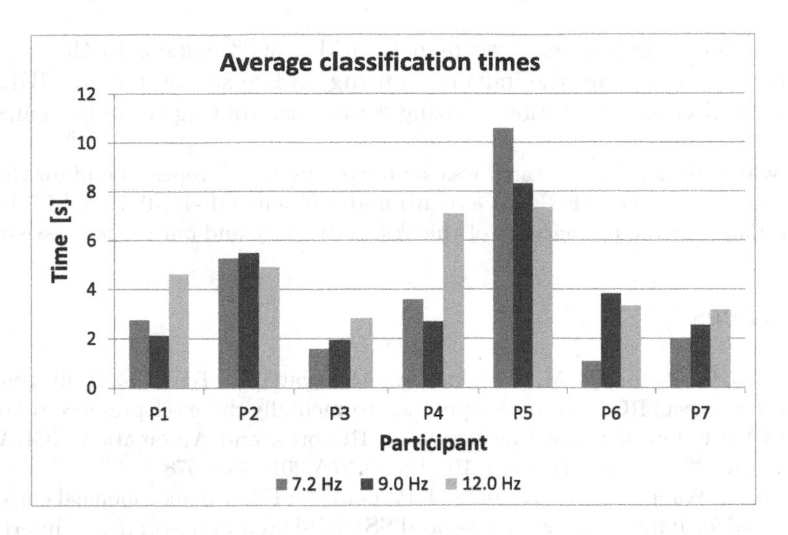


**Fig. 5.** Graphical representation of the differences in classification times for each frequency across participants.

# 4 Discussion and Conclusion

The high average accuracy of 98.1% achieved in this study proves the reliability of the proposed robot driving system. This result is comparable to our previous VR vacuum robot experiment [9], where eight participants with a mean age 27.6 years reached an average accuracy of 98.91% (for the HMD setup).

The ITR achieved here (10.71 bits/min) was lower than the previous experiments, 14.07 bits/min in the MRC experiment [11] and 24.54 bits/min in the simulated VRHMD experiment [9]. This was caused by the high gaze shifting time (here 5 s) compared to only 0.92 s in the previous ones [9,11]. If we compare the two studies with similar gaze shifting times (here we reduce it from 5 s to only 1 s), this recalculated ITR would reach on average 20.43 bits/min. The gaze shifting time of 5 s was set considering user safety, however, it increased the duration of the experiment and lowered the ITR significantly.

Almost all participants performed the driving task twice, P2 steered the MRC only one time (still reaching an accuracy of 100%) and did not want to drive for a second time, because of dizziness caused by the immersive VR experience and live video feedback. This effect is to be considered during design of VR-based experiments which include live video feedback.

The differences in average classification times for individual frequencies (see Fig. 5) could be influenced by the internal headset setup/hardware performance (Oculus Go, running on Android 7.1). Yet, paired t-tests revealed no statically significant differences between the classification times of the different frequencies. In contrast, the setup of our previous experiment [9] consisted of a much more powerful VR-gaming laptop, which enabled a more stable flickering and therefore shorter classification times. These differences and the frequency selection for this hardware should be further investigated.

In our future experiments we plan to add a 360° camera to the hardware setup to further improve the immersion using its live stream for the GUI. Also, a portable EEG-based solution utilizing remote computing could be realized.

**Acknowledgment.** This research was supported by the European Fund for Regional Development (EFRD or EFRE in German) under Grants GE-1-1-047, and IT-1-2-001. We also thank all the participants of this research study and our student assistants.

## References


1. Beraldo, G., Antonello, M., Cimolato, A., Menegatti, E., Tonin, L.: Brain-computer interface meets ROS: a robotic approach to mentally drive telepresence robots. In: 2018 IEEE International Conference on Robotics and Automation (ICRA), pp. 1–6, May 2018. https://doi.org/10.1109/ICRA.2018.8460578
2. Chen, X., Wang, Y., Gao, S., Jung, T.P., Gao, X.: Filter bank canonical correlation analysis for implementing a high-speed SSVEP-based brain-computer interface. J. Neural Eng. **12**(4), 046008 (2015)
3. Friman, O., Volosyak, I., Graser, A.: Multiple channel detection of steady-state visual evoked potentials for brain-computer interfaces. IEEE Trans. Biomed. Eng. **54**(4), 742–750 (2007)
4. Gergondet, P., Petit, D., Kheddar, A.: Steering a robot with a brain-computer interface: impact of video feedback on BCI performance. In: 2012 IEEE RO-MAN: The 21st IEEE International Symposium on Robot and Human Interactive Communication, pp. 271–276. IEEE (2012)
5. Kerous, B., Skola, F., Liarokapis, F.: EEG-based BCI and video games: a progress report. Virtual Reality **22**, 1–17 (2017)
6. Koo, B., Lee, H.G., Nam, Y., Choi, S.: Immersive BCI with SSVEP in VR head-mounted display. In: 2015 37th Annual International Conference of the IEEE Engineering in Medicine and Biology Society (EMBC), pp. 1103–1106. IEEE (2015)
7. Ron-Angevin, R., Velasco-Alvarez, F., Sancha-Ros, S., da Silva-Sauer, L.: A two-class self-paced BCI to control a robot in four directions. In: 2011 IEEE International Conference on Rehabilitation Robotics (ICORR), pp. 1–6. IEEE (2011)
8. Stawicki, P., et al.: Investigating spatial awareness within an SSVEP-based BCI in virtual reality. In: 2018 IEEE International Conference on Systems, Man, and Cybernetics (SMC), pp. 615–618. IEEE (2018)
9. Stawicki, P., et al.: SSVEP-based BCI in virtual reality-control of a vacuum cleaner robot. In: 2018 IEEE International Conference on Systems, Man, and Cybernetics (SMC), pp. 534–537. IEEE (2018)
10. Stawicki, P., Gembler, F., Volosyak, I.: Evaluation of suitable frequency differences in SSVEP-based BCIs. In: Blankertz, B., Jacucci, G., Gamberini, L., Spagnolli, A., Freeman, J. (eds.) Symbiotic 2015. LNCS, vol. 9359, pp. 159–165. Springer, Cham (2015). https://doi.org/10.1007/978-3-319-24917-9_17
11. Stawicki, P., Gembler, F., Volosyak, I.: Driving a semiautonomous mobile robotic car controlled by an SSVEP-based BCI. Comput. Intell. Neurosci. **2016**, 5 (2016)
12. Volosyak, I.: SSVEP-based bremen-BCI interface–boosting information transfer rates. J. Neural Eng. **8**(3), 036020 (2011)
13. Wang, F., Zhang, X., Fu, R., Sun, G.: Study of the home-auxiliary robot based on BCI. Sensors **18**(6), 1779 (2018)

14. Wolpaw, J.R., et al.: Independent home use of a brain-computer interface by people with amyotrophic lateral sclerosis. Neurology **91**(3), e258–e267 (2018)
15. Wolpaw, J., Birbaumer, N., McFarland, D., Pfurtscheller, G., Vaughan, T.: Brain-computer interfaces for communication and control. Clin. Neurophysiol. **113**, 767–791 (2002)
16. Wu, C.M., Chen, Y.J., Zaeni, I.A., Chen, S.C.: A new SSVEP based BCI application on the mobile robot in a maze game. In: 2016 International Conference on Advanced Materials for Science and Engineering (ICAMSE), pp. 550–553. IEEE (2016)

# A VR-Based Hybrid BCI Using SSVEP and Gesture Input

Roland Grichnik, Mihaly Benda, and Ivan Volosyak(✉)

Faculty of Technology and Bionics,
Rhine-Waal University of Applied Sciences, 47533 Kleve, Germany
ivan.volosyak@hochschule-rhein-waal.de
https://bci-lab.hochschule-rhein-waal.de

**Abstract.** Brain-Computer Interfaces (BCIs) using Steady-State Visual Evoked Potentials have been proven to work with many different display technologies for visual stimulation. The recent advent of consumer grade Virtual Reality (VR) Head-Mounted Devices (VR-HMDs) has made research in the area of VR-based BCIs more accessible than ever - yet the possibilities of such systems still have to be tested. In this paper, we present a BCI using a well-studied 3-step spelling interface converted into a Virtual Environment (VE). The Oculus Rift CV1 VR-HMD used in this study also provides motion tracking capability, which was used to implement a novel hybrid BCI utilizing gesture input. The interface consisted of three flickering boxes on a virtual screen in the VE for typing letters. Head shake gestures were used to intuitively trigger "Delete/Back" commands. A g.tec g.USBamp amplifier was used to record and filter the signal of eight electrodes mounted in an electroencephalography cap. The Minimum Energy Combination (MEC) method was used to classify commands in real time. Eighteen participants successfully performed seven spelling tasks each, reaching an accuracy of $91.11 \pm 10.26\%$ (mean $\pm$ Standard Deviation, SD) and an Information Transfer Rate of $23.56 \pm 7.54$ bit/minute (mean $\pm$ SD). Questionnaires filled out before and after the experiment show that most participants enjoyed the VR BCI experience and found the gesture input very natural. Future studies could expand the input mechanism by adding more head gestures, e.g. pecking, nodding or circling to control intuitively related software tasks.

**Keywords:** Virtual reality · Brain-Computer Interface · SSVEP · Gesture input

## 1 Introduction

Brain-Computer Interfaces (BCIs) allow control of computers or other machines using signals derived from brain activity. Such brain activity can arise from internal processes of the brain (e.g. thinking, imagination) or be created by stimulating brain areas with external stimuli (e.g. optical or acoustical stimuli) [15].


I. Rojas et al. (Eds.): IWANN 2019, LNCS 11506, pp. 418–429, 2019.
https://doi.org/10.1007/978-3-030-20521-8_35

Besides the nature of the brain signals used, BCIs also differ in the way those signals are measured. Most often, electroencephalography (EEG) [3] is used for measurement, while in some cases, functional Magnetic Resonance Imaging (fMRI) [6] or magnetoencephalography (MEG) [7] has been used as well.

Constant-frequency flickering light sources cause neural activity in the human visual cortex at the same frequency (and its harmonics) [5], known as Steady-State Visual Evoked Potentials (SSVEPs). These can be detected by measuring the electrical activity of the corresponding brain area, for example using EEG. SSVEPs are often used as control mechanisms of BCI systems, as they are easily detectable and no training is required for the users (they occur in every case, though the power of the neural response varies greatly).

Over the last two decades, a number of research groups has been working on BCIs in Virtual Reality (VR) head-mounted devices (VR-HMDs) (e.g. [1,14]). With the growing number of commercially available, consumer grade VR-HMDs, research in VR has become increasingly accessible.

Head-mounted VR is of interest to BCI research due to multiple reasons. First and foremost, research with users on the move and in changing environments can be bothersome. In VR, changes to the Virtual Environment (VE) can easily be made at the click of a button, facilitating research into attention levels and background distractions. Second, VR experiences tend to be highly immersive. Control mechanisms that work without obvious user interface (UI) elements, such as by using a BCI, could be favored over traditional methods in the future, as UI elements and the need to use a controller have a negative impact on immersion. Third, VR BCIs could play an important role for people with disabilities, be it for entertainment purposes or for communication via telepresence.

In recent years, hybrid solutions have been used to improve accuracy, Information Transfer Rate (ITR) and/or usability of BCIs (e.g.: increase of ITR in [8], increase of BCI literacy in [10]). Hybrid BCIs use other signals in conjunction with brain signals, such as surface myoelectric signals measured using surface electromyography (sEMG) [8] or gaze position on a screen measured using eye tracking technologies [10].

In VR-based BCI systems, very little use has been made of postural data so far. Wang et al. used rotational data of the VR-HMD to determine a user's gaze transitions between SSVEP invoking stimuli to improve system stability [14]. Postural data was used as a sort of switch for the BCI system, sending a hover command to the quadcopter controlled with the system whenever the user was transitioning between command stimuli. To our best knowledge, gestures, being chronological sequences of certain, defined movements, have not been used in VR-HMD-based BCI systems so far.

In the research presented, SSVEPs were used in an EEG-based BCI in conjunction with gesture input. A well-studied interface, a 3-step Speller (as in [13]), was tested in a VE presented on a VR-HMD. This type of speller was selected because in our prior, personal experience this system is highly accurate and every user is able to control it. However, its speed is limited as multiple classifications are necessary for a letter to be written.

To increase immersion of the user in the VR and improve the BCI system's performance, we propose a novel hybrid control mechanism. In addition to the SSVEP-based BCI, the motion tracking system of the VR-HMD is used to detect head gestures and add an additional way of control over the software. Intuitive gesture controls could provide a valuable addition to any future VR BCI system.

## 2 Methods and Materials

This chapter lists the materials needed to reproduce the experiment and describes the experimental procedure. Where applicable in this paper, mean values are given as mean ± Standard Deviation (SD).

### 2.1 Participants

Twenty adult students of the Rhine-Waal University of Applied Sciences tested the system. All experimental procedures involving human participants were approved by the Ethical Review Board of the medical faculty of the University Duisburg-Essen. Participants had the opportunity to opt-out of the study at any time. All participants gave written informed consent prior to the experiment in accordance with the Declaration of Helsinki. Information needed for the analysis of the experiments was stored pseudonomously during the experiment, so results and questionnaire data cannot be traced back to specific participants. The participants were financially compensated.

Out of all participants, two were excluded from the analysis due to technical reasons. One participant had problems with stereo vision and could not focus in VR at all, so the trial was canceled after VR setup. The other participant encountered graphical issues in VR during the trial, which led to unstable frequency generation. This trial was canceled and all previous data of this participant were excluded from analysis. The remaining 18 participants (seven male) had a mean age of 22.89 ± 2.97 years.

### 2.2 VR Hardware

For display of the VE, an Oculus Rift CV1 VR-HMD with a screen resolution per eye of 1080 × 1200 pixels was used. A laptop (MSI GE72MVR 7RG Apache Pro, Intel Core i7, 2.80 GHz) running Windows 10 Education was used for the experiment. The VR software ran on the dedicated Nvidia Geforce 1070 graphics processor, while the recording software was executed on the integrated Intel graphics processor to prevent resource conflicts.

### 2.3 VR Scene Set-Up

The VE for the spelling interface was designed using the Unity game engine, version 2018.2.13f1 [11]. The Unity software natively provides basic support for VR-HMDs, so no additional Software Development Kit was needed for VR-HMD.

*Virtual Environment.* In the VE, the user was placed in front of the 3-step spelling interface, which was constructed using the UI system of the Unity game engine. The stimulus size was chosen so as to resemble the visual angle of 12 cm × 12 cm stimuli on a physical 24" screen at a distance of 1 m. The head position was fixed in 3D space and could not be influenced by the user, so as to provide a mostly stable stimulus size (i.e.: The user could not lean in to get closer to the stimuli). Physical rotation of the head was translated to rotation of the camera in the VE though, so the user could look around freely and tilt his head left and right. The interface did not rotate with the user's head rotation, but stayed in place, emulating the user sitting in front of a virtual 24" screen.

*3-Step Speller Interface.* The 3-Step Speller Interface comprised of 3 square stimuli (size W × H: 6.8990° × 6.8990° visual angle) in a horizontal line, with a gap of 2.6250° visual angle between the stimuli (see Fig. 1). Underneath the stimulus boxes, there were two long text boxes with a gray background. The top text box in the interface contained the letters spelt by the participant, while the lower text box showed the spelling task at hand. The letters inside the stimulus boxes were 1.7506° high. During spelling only, the stimulus boxes, from left to right, flickered at 6.0 Hz, 7.5 Hz and 10 Hz, respectively, with a duty cycle of 50%. The boxes' fill color as well as the frame color alternated between black (RGBA 0-0-0-255) and white (RGBA 255-255-255-255), the text inside the stimuli was inversely colored to the box color at all times. The flickering was produced using a time-based, not a frame-based algorithm. Between spelling tasks, the user was surrounded by an ocean scene (free skybox "Skybox" from the Unity Asset Store, https://assetstore.unity.com, author: Clod). During spelling tasks, the surrounding skybox changed to a pure black background. In the speller's initial state, each box contained 9 letters and/or symbols (see Fig. 1). When a participant chose a box, its contents were equally split up into the 3 boxes, narrowing down the desired choice by a third with each step (see Fig. 1 for an exemplary typing cycle). When a selection was made in the third screen of the interface, the chosen letter was typed and the interface reverted to its initial state. In the initial screen, the user could delete the last letter by using the "Delete/Back" command. In any subsequent screen, usage of this command reverted to the previous screen.

For each selection made, there was an audio feedback: When the speller transitioned between screens, a short "beep" was played. When a final choice was made, the selected choice was read out to the user.

*Gesture Input.* The 3-step speller interface did not have a stimulus assigned for providing the "Delete/Back" command. Instead, the accelerometer built into the VR headset was utilized to provide a gesture-based input channel for these commands. The add-on "VR Head Gestures as Input" from the Unity Asset Store (author: Tom Farro, https://assetstore.unity.com) was used to detect head gestures and translate them into commands.

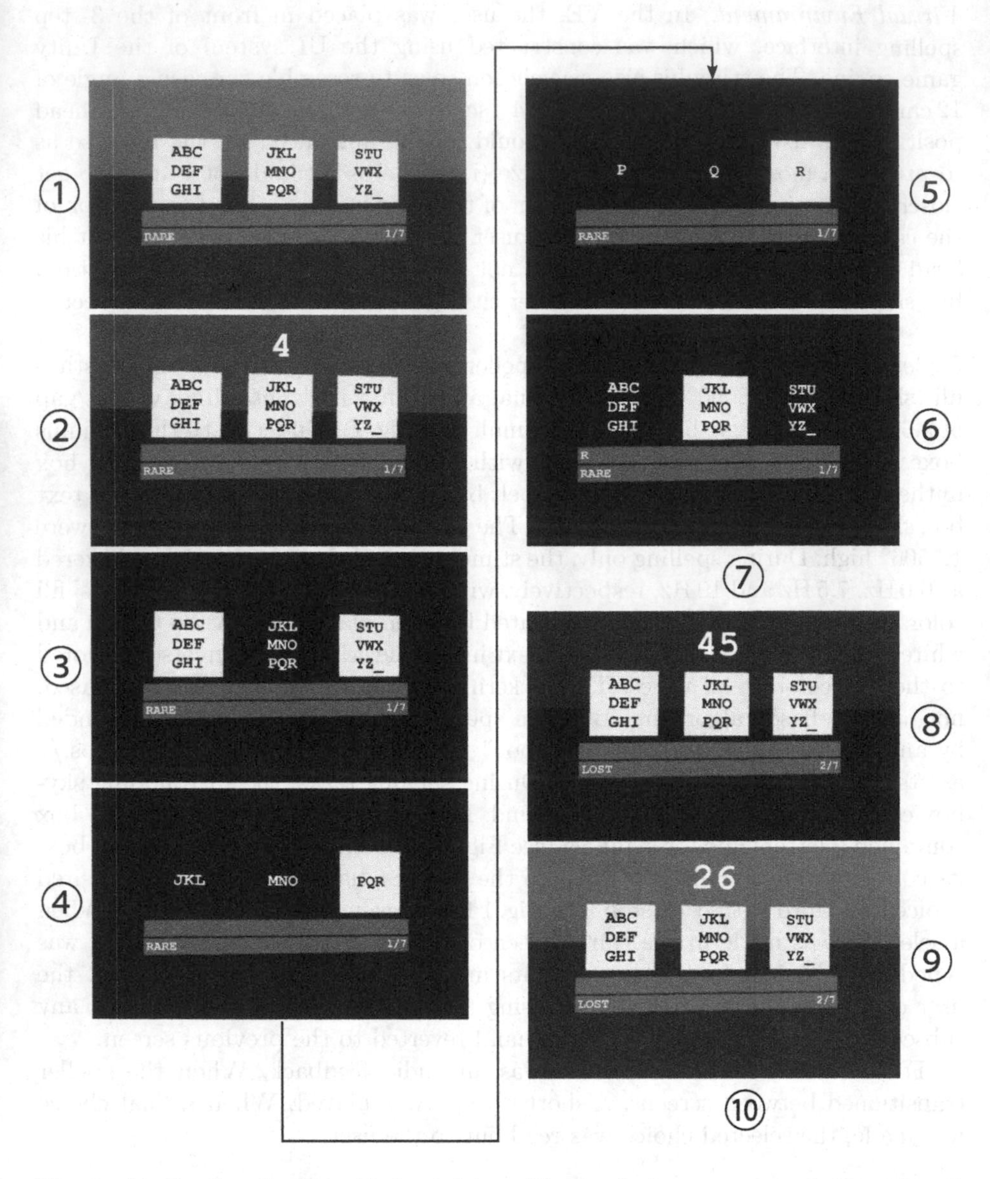


**Fig. 1.** Application Spelling Cycle and transitioning between screens. 1: Experiment starts, participant not typing yet, ocean scenery. 2: Fade to black over 5 s countdown. 3: Boxes flickering, participant typing, Box #2 is selected. 4: Box #2 contents are split up, Box #3 is selected. 5: Box #3 contents are split up, Box #3 is selected. 6: Single letter from Box #3 is spelt, interface reverts to first screen. 7: Steps 3–6 repeat until all letters are spelt. 8: Next target word shown, 45 s timer starts, fade to ocean scenery over 3 s. 9: Relaxation period. 10: Process repeats from step 2 onwards for all target words. In steps 2, 8, and 9, the countdown timer is visible at the top edge.

For usage of the script, a Gesture Manager object was attached to the main camera in the Unity scene, which was controlled by the physical rotation of the VR-HMD. The script checked the main camera's quaternion in regular, fixed intervals and recognized gestures when the rotation around a specific axis between two intervals exceeded the value `Delta Threshold`. The axis in question depends on whether head shaking (rotation around camera's local y axis) or head nodding (rotation around camera's local x axis) were to be detected. Figure 2A shows the gesture detected in relation to the VR-HMD. If at least as many gestures as defined in the value `Gesture Amount Threshold` were detected and the time gap between each subsequent gesture did not exceed the value `Time Threshold`, the final gesture was detected and the "Delete/Back" command was activated in the software. The add-on values used were: `Delta Threshold` = 0.015, `Time Threshold` = 0.5 s, `Gesture Amount Threshold` = 2.

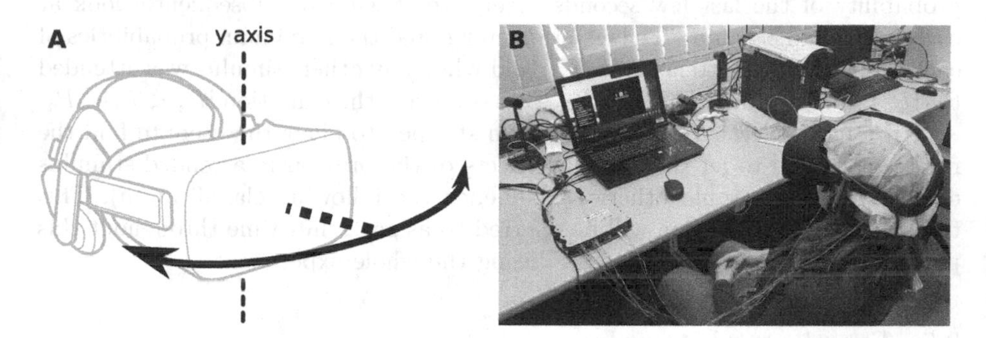


**Fig. 2.** A: Schematic of the gesture-based input mechanism. Rotation around the headset's vertical axis (game engine camera's local y axis) in form of head shaking is detected and a command triggered in the software. B: Participant during the experiment.

### 2.4 Classification Software

A custom classification software (written in C++) was used to communicate with the amplifier, allow the setup of classifier thresholds, store all measurements for further offline analysis and classify data into commands. Communication with the software providing the VE was achieved using Universal Datagram Protocol (UDP) communication on the localhost. All exchanged UDP datagrams were logged to assure no communication was lost during trials.

*Classification Method.* Minimum energy combination method (MEC) [12] was used for signal classification. Detecting the frequency of the attended stimulus was done in the following steps: The recorded signal was spatially filtered, the data from different electrodes were combined after weighting them according to the amount of noise in them, and a probability was calculated for each possible target frequency.

Several classifiers were working concurrently (differing only in the amount of EEG data processed: 2 s, 3 s, 4 s, and 5 s of recording, respectively), and the calculated probabilities were combined (as in [2]). The classifiers only started calculating if enough data were available. As soon as the probability of a frequency surpassed the corresponding threshold (which was set at the beginning of the experiment), a classification occurred. For more details please refer to [2]. If none of the thresholds were reached, the output was rejected and the recording continued with the newest data shuffling out the oldest data in the signals the classifiers were processing.

The thresholds were set so as to allow fast selections, and if possible eliminate the chance of misclassifications. They were set individually for each participant at the start of the session, then they were kept the same throughout the whole recording. For each target frequency, there was a row displaying the average probability of the last few seconds. Users were asked to subsequently look at each individual stimulus. The experimenter noted the maximum probabilities of each stimulus when attended to ($P_n$) and when any other stimulus was attended to ($!P_n$). The threshold ($T_n$) was set so as to satisfy the condition $!P_n < T_n < P_n$.

After each selection, the classification stopped to allow the users to find the next target, and to let the lingering effects of the previously attended stimulus disappear (which would otherwise influence the following classification). This time without classification will be referred to as gaze shift time throughout this paper, its duration was kept at 2 s during the whole experiment.

### 2.5 Experimental Procedure

The experiment took place in a standard, unshielded computer lab environment and was designed to take approximately 60 min. After filling out consent forms and pre-test questionnaires, participants were seated comfortably and an EEG cap was put on. The ground electrode was placed at $AF_Z$, the reference at $C_Z$, and eight electrodes were placed at predefined locations on the EEG-cap marked with $P_Z$, $PO_3$, $PO_4$, $O_1$, $O_2$, $O_Z$, $O_9$, and $O_{10}$ in accordance with the international 10/20 system of EEG electrode placement. Standard abrasive electrode gel was applied between the electrodes and the scalp to bring impedances below 5 kΩ. A shower cap was placed over the EEG cap to prevent soiling of the VR-HMD (see Fig. 2B for an exemplary setup).

A g.USBamp (Guger Technologies, Graz, Austria) amplifier was used for EEG recording. The sampling rate was 256 Hz. A digital band pass filter (between 2 and 60 Hz) and a notch filter (around 50 Hz) were applied to remove non-relevant signals and power line noise from the recording.

Participants were given electronic questionnaires before (pre) and after (post) the test run. The pre-test questionnaire asked for age and sex, a freely chosen, English five letter word as a word to be typed in the experiment and the tiredness state. The post-test questionnaire comprised of 12 questions. Table 3 in the Results part lists all questions asked and the resulting mean scores on a 1–5 Likert scale.

The VR-HMD was put onto the participant's head and straps were adjusted to provide a comfortable, yet firm seat. Participants were instructed to adjust the height of the headset as well as the lens distance to maximize the sharpness of the picture in VR as much as possible using the letters displayed in the speller as a guide. After a short familiarization with the software and set-up of the classification thresholds, participants had to complete seven spelling tasks: five predetermined ones (HUNG, LOST, RARE, STOP, RUBBER), one determined by the subject and another one derived from that. The target words were chosen so as to require all speller targets to be selected an equal amount of times.

*Actual, Maximum and Relative ITR Calculation.* The maximum ITR achievable $ITR_{max}$ was calculated using formula (1). It differed slightly by word length, as in our software implementation, there was no gaze shift phase for the first classification made (a gaze shift period was only granted between subsequent classifications). An average time needed for classification was calculated for each word length by dividing the total, optimal spelling time by the optimal number of commands used. This average classification time and a value of $n_{Commands} = 4$ were used in the formula to calculate the maximum achievable ITR per target word length. Table 1 shows the maximum ITRs achievable by word length. The actual ITR of the spelling tasks was calculated using the common ITR calculation formula, mentioned in [15]. Accuracy was calculated by dividing the number of correct commands by the total number of commands. Any command that resulted in getting closer to finishing the spelling task was correct, therefore, correcting an incorrect selection with the gesture input was considered correct. To evaluate the system performance in regards to its maximum, theoretical performance, a relative ITR was calculated by dividing measured ITR by the maximum ITR achievable.

$$ITR_{max} = \frac{60\ \mathrm{s}}{average\,Classification\,Time} * \log_2 n_{Commands} \quad (1)$$

**Table 1.** Maximum Information Transfer Rate achievable in the software presented for different target word lengths.

| Target Word Length [letters] | Best Case Commands | Best Case Average Time/Command [s] | max. ITR [bit/minute] |
|---|---|---|---|
| 4 | 12 | 3.83 | 31.30 |
| 5 | 15 | 3.87 | 31.03 |
| 6 | 18 | 3.89 | 30.86 |

## 3 Results

For all spelling tasks combined, the mean accuracy was $91.11 \pm 10.26\%$, with a mean ITR of $23.56 \pm 7.54$ bit/minute. Taking the maximum ITR achievable (see

Table 1) into account, the mean relative ITR was 75.60 ± 24.15%. Figure 3 shows ITR and accuracy by the target word as well as by the spelling task order (data mean values in Table 2). While target words did not differ in terms of accuracy and relative ITR, there was a trend towards gradually decreasing performance with each subsequent spelling task (see Fig. 3). In total, users needed to make use of the gesture input 295 times during the experiment (but every user also practiced the gesture five to ten times during familiarization).

For three participants, a different version of the software using fewer concurrent classifiers and shorter classifier lengths (1 s, 2 s and 4 s) was tested in addition to the general version (and always after the general run). In this version, the maximum ITR was higher (41.54–42.35 bit/minute, depending on the target word length) due to shorter classification times needed. In the general trial run, the performance of only these three participants resulted in a mean accuracy of 98.25%, a mean ITR of 29.51 bit/minute and mean relative ITR of 94.70%. In the second trial run, they achieved a mean accuracy of 94.70%, a mean ITR of 35.43 bits/minute and a mean relative ITR of 84.22%.

**Table 2.** Mean and median accuracy, Information Transfer Rate (ITR) and relative ITR (ITR/maximum ITR for the respective target word), grouped by the spelling target word and by the chronological order of the spelling tasks (Task #).

| | Accuracy [%] | | ITR [bit/minute] | | Relative ITR [%] | |
|---|---|---|---|---|---|---|
| | Mean | Median | Mean | Median | Mean | Median |
| **Target Word** | | | | | | |
| HUNG | 92.69 | 93.8 | 24.34 | 24.5 | 77.76 | 78.26 |
| LOST | 87.26 | 87.5 | 21.19 | 19.7 | 67.70 | 62.93 |
| RARE | 92.96 | 96.9 | 25.10 | 27.8 | 80.18 | 88.81 |
| STOP | 93.62 | 100.0 | 26.03 | 30.9 | 83.15 | 98.71 |
| RUBBER | 90.29 | 89.7 | 22.76 | 22.05 | 73.77 | 71.46 |
| *subject-chosen* | 90.66 | 90.5 | 23.17 | 22.5 | 74.67 | 72.5 |
| *complementary* | 90.32 | 92.3 | 22.32 | 23.25 | 71.93 | 74.92 |
| **Task Number** | | | | | | |
| Task #1 | 95.24 | 100.0 | 26.59 | 30.5 | 85.38 | 97.86 |
| Task #2 | 94.79 | 97.5 | 26.12 | 27.8 | 83.68 | 89.41 |
| Task #3 | 93.12 | 100.0 | 25.46 | 30.6 | 81.61 | 98.17 |
| Task #4 | 91.08 | 94.25 | 23.66 | 23.9 | 76.02 | 77.01 |
| Task #5 | 89.04 | 91.2 | 21.75 | 22.05 | 69.67 | 70.74 |
| Task #6 | 86.74 | 88.7 | 20.53 | 19.8 | 66.03 | 64.0 |
| Task #7 | 87.77 | 90.2 | 20.82 | 19.95 | 66.78 | 64.2 |
| **Total** | **91.11** | - | **23.56** | - | **75.60** | - |

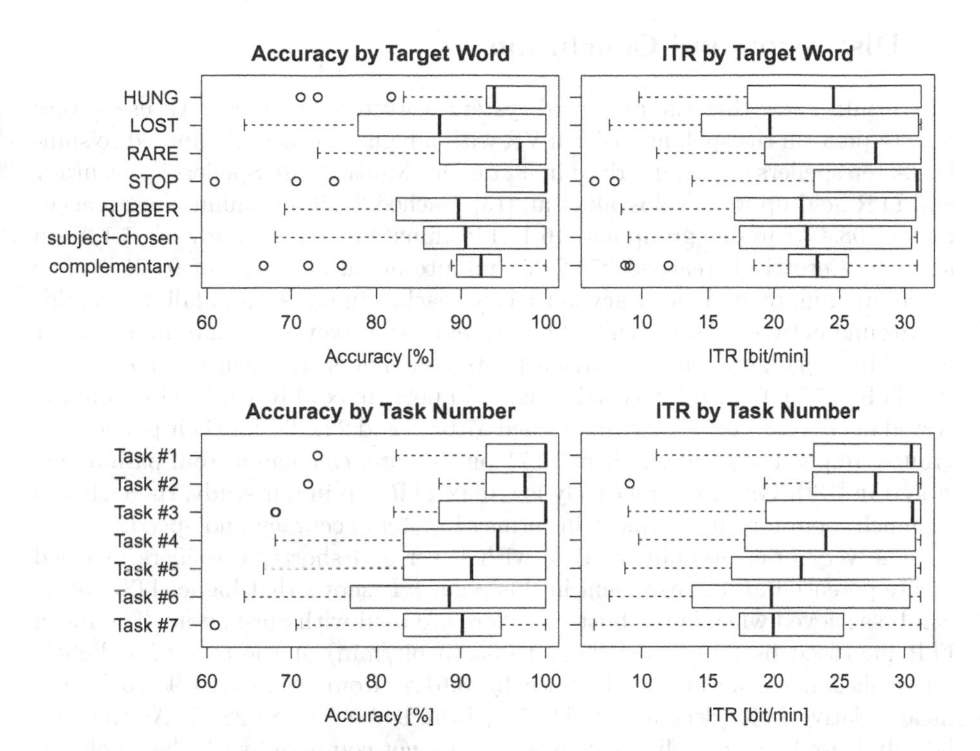


**Fig. 3.** Boxplots showing accuracy and ITR values achieved by 18 subjects in the spelling tasks. Data are shown grouped by the spelling target word and by the chronological order of the spelling tasks (Task #). **Legend:** *Median* values (solid, vertical lines), middle 50% of all data points (box outline), lowest/highest non-outlier data points (left/right whisker), outlier data points (open circles). Outliers were defined as values more than 1.5 × IQR lower than the 25th percentile.

**Table 3.** Questions asked in the pre- and post-test questionnaires and their mean score (mean ± Standard Deviation) over 18 participants. Participants had to state their agreement to the shown statements on the Likert scale, ranging from 1 (labeled "I don't agree at all!") to 5 (labeled "I absolutely agree!").

| Statement | | Score |
|---|---|---|
| I am tired right now. | *(pre)* | 2.33 ± 1.08 |
| I am tired right now. | *(post)* | 2.61 ± 1.04 |
| The system did work well for me. | *(post)* | 4.11 ± 0.90 |
| I would recommend the system. | *(post)* | 4.39 ± 0.61 |
| I found the stimuli (flickering boxes) annoying. | *(post)* | 3.17 ± 1.20 |
| Selecting the speller boxes was easy. | *(post)* | 3.89 ± 1.18 |
| The setup was comfortable on my head (in regards of pressure by the straps etc.). | *(post)* | 4.33 ± 0.69 |
| I found the picture inside Virtual Reality blurry. | *(post)* | 2.17 ± 1.15 |
| The audio feedback (beeps and letter readouts) were loud enough. | *(post)* | 4.59 ± 1.18 |
| The speller interface was big enough. | *(post)* | 4.61 ± 0.92 |
| I enjoyed resting my eyes between spelling tasks. | *(post)* | 4.50 ± 0.86 |
| The "Delete"/"Back" gesture (headshaking) worked well for me. | *(post)* | 4.06 ± 1.06 |
| The "Delete"/"Back" gesture (headshaking) felt natural to me. | *(post)* | 4.22 ± 0.88 |

## 4 Discussion and Conclusion

The results show that the presented system worked as intended. All users were able to perform the spelling tasks in VR with a high accuracy. Compared to similar 3-step spellers (as summarized in [9] under "Multi-Phase Spellers"), accuracy and ITR hold up well: Volosyak et al. [13] reached 27.36 bit/minute at an accuracy of 98.49% in one group and 16.10 bit/minute at an accuracy of 91.13% in another. Cecotti [4] reached 37.62 bit/minute at an accuracy of 92.25%. The 23.56 bit/minute at an accuracy of 91.11% reached in our system fall into a middle ground between those results. Our system did, however, seem to make better use of its capabilities in comparison to its contenders: It reached a mean relative ITR of 75.60%, while Cecotti reached 54.00% (max ITR of 69.66 bit/minute; stated in [4]) and Volosyak et al. reached 37.63% and 22.14% for their participant groups, respectively (max ITR of 72.71 bit/minute; calculated from parameters stated in [13]). Given the relatively low max ITR cap in our study, there should be much room for improvement in terms of system accuracy and speed.

The very brief preliminary trial with fewer and shorter classifiers revealed future potential for improvement in this regard: It shows that higher ITR values can be achieved when less robust classifiers are used with our system. The mean ITR increased by 20% (from 29.51 to 35.43 bit/min) at the cost of a slightly diminished mean accuracy (decrease by 3.61%, from 98.25% to 94.70%) and mean relative ITR (decrease by 11.07%, from 94.70% to 84.22%). As they are heavily biased, these preliminary results were not compared with the whole 18-participant group of the general trial. For one, the study duration in all cases was 60 min maximum. The additional trial was only conducted if participants had enough time left after their general trial to still finish the additional tasks in time. This means that participants performing the additional trial were already fast in the general trial and did not show any delaying problems in preparation or experiment. Also, the participants were already familiarized with the software when they performed the second run. Further trials with more participants would be needed to explore the speed boundaries of the system to full extent.

As the questionnaire results show, the system was well received by most users in terms of comfort, visual quality and design, sound feedback volume, and the feel of the gesture input. Most users had no troubles using the gesture-based input at all and seamlessly integrated the head shaking commands into their spelling routine when necessary. Users were neutral in their score regarding the annoyance by the flickering stimuli, yet the existence of an eye resting phase was highly appreciated. Most users also felt that the system worked well for them and would recommend its usage.

For future experiments, even shorter classifiers could be tested more thoroughly to test the boundaries of what is possible. More and different head gestures could be implemented for further improvement of system performance or to add other intuitive ways to access speller functions (e.g. nodding, pecking, tilting, circling, etc.). It would also be interesting to see whether the trend towards performance decrease over subsequent spelling tasks would continue when more spelling tasks had to be completed. Additionally, it would be interesting to investigate the effect of different lengths of resting phases between spelling tasks and whether they would influence the performance decrease over time. Finally, an extended experiment could be conducted with different age and social groups.

**Acknowledgments.** This work was supported by the European Fund for Regional Development (EFRD - or EFRE in German) under Grant IT-1-2-001. We thank all the participants of this research study as well as our student assistants.

## References

1. Bayliss, J.D., Ballard, D.H.: A virtual reality testbed for brain-computer interface research. IEEE Trans. Rehabil. Eng. **8**(2), 188–190 (2000)
2. Benda, M., Stawicki, P., Gembler, F., Grichnik, R., Rezeika, A., Saboor, A., Volosyak, I.: Different feedback methods for an SSVEP-based BCI. In: 2018 40th Annual International Conference of the IEEE Engineering in Medicine and Biology Society (EMBC), pp. 1939–1943. IEEE (2018). https://doi.org/10.1109/EMBC.2018.8512622
3. Birbaumer, N., et al.: A spelling device for the paralysed. Nature **398**(6725), 297–298 (1999)
4. Cecotti, H.: A self-paced and calibration-less SSVEP-based brain-computer interface speller. IEEE Trans. Neural Syst. Rehabil. Eng. **18**(2), 127–133 (2010)
5. Herrmann, C.S.: Human EEG responses to 1–100 Hz flicker: resonance phenomena in visual cortex and their potential correlation to cognitive phenomena. Exp. Brain Res. **137**(3–4), 346–353 (2001)
6. Lee, J.H., Ryu, J., Jolesz, F.A., Cho, Z.H., Yoo, S.S.: Brain-machine interface via real-time fMRI: preliminary study on thought-controlled robotic arm. Neurosci. Lett. **450**(1), 1–6 (2009)
7. Mellinger, J., Schalk, G., Braun, C., Preissl, H., Rosenstiel, W., Birbaumer, N., Kübler, A.: An MEG-based brain-computer interface (BCI). NeuroImage **36**(3), 581–593 (2007)
8. Rezeika, A., Benda, M., Stawicki, P., Gembler, F., Saboor, A., Volosyak, I.: 30-Targets hybrid BNCI speller based on SSVEP and EMG. In: 2018 IEEE International Conference on Systems, Man, and Cybernetics (SMC), pp. 153–158. IEEE (2018). https://doi.org/10.1109/SMC.2018.00037
9. Rezeika, A., Benda, M., Stawicki, P., Gembler, F., Saboor, A., Volosyak, I.: Brain-computer interface spellers: a review. Brain Sci. **8**(4), 57 (2018)
10. Stawicki, P., Gembler, F., Rezeika, A., Volosyak, I.: A novel hybrid mental spelling application based on eye tracking and SSVEP-based BCI. Brain Sci. **7**(4), 35 (2017)
11. Unity Technologies: Unity Game Engine version 2018.2.13f1 (2018). https://unity3d.com/de/unity
12. Volosyak, I.: SSVEP-based bremen-BCI interface - boosting information transfer rates. J. Neural Eng. **8**(3), 036020 (2011)
13. Volosyak, I., Gembler, F., Stawicki, P.: Age-related differences in SSVEP-based BCI performance. Neurocomputing **250**, 57–64 (2017)
14. Wang, M., Li, R., Zhang, R., Li, G., Zhang, D.: A wearable SSVEP-based BCI system for quadcopter control using head-mounted device. IEEE Access **6**, 26789–26798 (2018)
15. Wolpaw, J., Wolpaw, E.W.: Brain-Computer Interfaces: Principles and Practice. OUP, Oxford (2012)

# Word Prediction Support Model for SSVEP-Based BCI Web Speller

Abdul Saboor, Mihaly Benda, Felix Gembler, and Ivan Volosyak[(✉)]

Faculty of Technology and Bionics,
Rhine-Waal University of Applied Sciences, 47533 Kleve, Germany
ivan.volosyak@hochschule-rhein-waal.de
https://bci-lab.hochschule-rhein-waal.de

**Abstract.** Steady state visual evoked potential (SSVEP) based BCI-systems are dependent on the brain signals which are elicited in response to a visual stimuli presented to the user. The spelling systems are very popular applications for the SSVEP-based BCI. In this paper, we are presenting a web-based speller supported with word prediction. The emphasis of the study was on two main points: (1) provide a dictionary based web speller which could also be accessed through a widely available web browsers; (2) increase the accuracy and speed of the SSVEP-based BCI speller. Using the concept of three step speller, a web interface was developed which provided additional support of word predictions based on characters typed and the co-occurrence of the previously typed word. The architectural pattern for the word prediction support model was based on MVC (model-view-controller). The AJAX call was placed form the web speller interface to access the database using Java Servlet and Java Beans. The relational database for the word prediction was derived from the Leipzig corpora collection. The developed system was tested with eleven healthy subjects. An average accuracy of 92.5 % and ITR of 18.8 bits/min were achieved. The results showed that word suggestions can increase the typing speed and accuracy of the web speller.

**Keywords:** Brain-Computer Interface (BCI) · Steady-State Visual Evoked Potential (SSVEP) · Web speller · Word prediction · Dictionary supported speller

## 1 Introduction

A brain-computer interface (BCI) system detects the brain activity, analyze the signals in real time, and provide a command and/or control signal to interact with the external environment without using muscular activities of the BCI user [11]. BCI systems are not only helpful for people with disabilities, but can also be used by healthy persons to perform the spelling tasks [2], play video games [6], browse the web [12], and control the smart home environment [8].

The spelling systems are a key implementation area for BCI systems, thus a continuous effort is made to increase the performance and accuracy for the


I. Rojas et al. (Eds.): IWANN 2019, LNCS 11506, pp. 430–441, 2019.
https://doi.org/10.1007/978-3-030-20521-8_36

spellers. The performance of the speller can be increased by reducing the number of steps involved in writing a word, and the accuracy can be further increased by providing the word prediction support to the user. These features can be achieved by introducing the dictionary support into the BCI spellers.

In recent past, Volosyak et al. [10] presented one of the earliest SSVEP-based BCI spellers driven by a dictionary. In this study a custom-built dictionary, which consisted of 49,142 common English words, was used. At start, the occurrence frequency of every word in the dictionary was set to zero. During the experimental study, when a specific word was typed, the usage frequency value of that specific words was incremented. This feature provided the suggestions of most frequently used words, thus speeding up the whole process of the system. This study showed that the performance of the dictionary-driven speller was higher than the conventional Bremen-BCI speller. The solution was based on SQLite database and desktop application.

In an article by Stawicki et al. [9], a dictionary driven four class SSVEP-based speller was presented. It was a desktop application, which provided a dictionary mode consisting of 39,000 words derived from the list of the most frequently used words from spoken English. The word suggestion feature provided a list of six words, and the user had to shift from spelling mode to dictionary mode for the selection of the suggested word. The prediction model they relied on used the string matching method only for the currently typed word, while our study combined this with the suggestions based on the co-occurrence of the previously typed word.

A modified T9 interface was present in [1], using a 3 by 3 matrix of nine keys. The words suggestion list was based on the combinations of prefixes formed by the typed keys. The study was limited to dictionary support of 1000 most commonly used English words. The word suggestion list could only be used when the suggestion count reached a certain threshold value.

Few other studies were based on interaction with web the browsers, e.g. [12] provided the navigation control and limited speller interface, [4] parsed the web documents to provide the web page navigation and a speller driven by P300 event related potential (ERP) responses. But all these studies were limited to providing speller interfaces based on desktop applications or even providing the web spellers with limited or no access to the dictionary for the word predictions.

In this study, we introduce the SSVEP-based BCI web speller which was driven from the web browser, presented the CSS based flickering stimuli, and integrated the corpus collection based dictionary to provide better word prediction.

## 2 Methods and Materials

This section describes the web-speller software architecture, database structure, hardware used for signal acquisition, and the signal precessing application for the online processing of the EEG data. Moreover, it also describes the experimental study and provides the information about the participants of the study.

### 2.1 Subjects

In this study, eleven healthy volunteers (4 female, 7 male) participated. The mean age of the participants was 26.46 years (with the standard deviation of 3.92). This study was approved by the ethical committee of the medical faculty of the University Duisburg-Essen. All the information needed to conduct the analysis was stored anonymously, thus it cannot be traced back to the individual subjects. The participants had the opportunity to quit the study at any time. All participants gave the written informed consent in accordance with the Declaration of Helsinki. The subjects received the financial reward for conducting the tests.

### 2.2 Hardware

The signal classification application, the database server, and the web server were running on a MSI GE72MVR 7RG Apache Pro Gaming-Laptop (Intel Core i7-*7th* Generation CPU @ 2.8 GHz, 16 GB RAM). The laptop was running on Microsoft Windows 10 Education edition. An additional liquid crystal display (LCD) monitor was attached to the laptop to present the web speller interface using web browser. The LCD monitor was manufactured by DELL (Model Number: P2414Hb) and was running on 60 Hz refresh rate with the resolution of 1920 × 1080 pixels.

The subjects were seated in front of the LCD monitor at the distance of approximately 50 cm. Eight standard passive Ag/AgCl electrodes were used to get the brain signals non-invasively. The electrodes were placed in accordance to the international 10/20 system of EEG electrode placement. The locations of the electrodes were marked as $P_Z$, $PO_3$, $PO_4$, $O_1$, $O_2$, $O_Z$, $O_9$, and $O_{10}$. The ground electrode was placed at $AF_Z$, and the reference electrode was placed at $C_Z$. Standard abrasive electrolytic electrode gel was applied to bring the impedance below 5 kΩ. An EEG amplifier, g.USBamp (Guger Technologies, Graz, Austria) was used in the study. The sampling frequency of the amplifier was set to 128 Hz. A digital band pass filter between 2 and 60 Hz and a noch filter of around 50 Hz were applied in the amplifier.

### 2.3 Web Speller

This section describes the functionality of the web speller and it's dictionary assisted word prediction support model. The speller architecture (shown in Fig. 1) was based on MVC (Model–View–Controller) framework. The GlassFish server (version 4.1.1) was used to host the web speller and the Java Servlet. The web page request is initiated through HTTP request by calling the uniform resource locator (URL). The HTTP response consisting of HTML, CSS and JavaScript loads the web page which displays the graphical interface of the web speller.

An AJAX call is placed to retrieve the word suggestion based on the current state of the speller. The call to AJAX invokes a Java Servlet, which instantiate a Java Bean to retrieve the data from the relational database (explained later in the sub-section: *Word Prediction*). The Java Servlet then sends the three words list in JSON (JavaScript Object Notation) format. This words list is received by the AJAX engine, which process the information and updates the web page without refreshing the loaded web page.

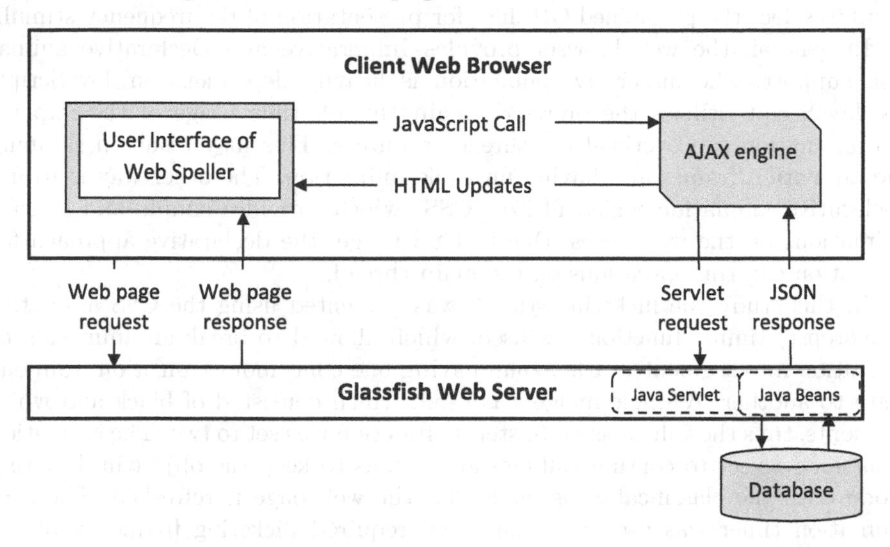


**Fig. 1.** An architecture of the web speller with word prediction

**Graphical Interface:** The web speller consisted of seven stimuli, flickering at the frequencies of 6 Hz, 6.6 Hz, 8 Hz, 9.5 Hz, 10 Hz, 11.11 Hz, and 12.5 Hz. The first three stimuli showed the text overly of twenty six characters of English alphabets and the underscore (to represent space).

At the first step of three step speller, twenty seven character (English alphabets + underscore, to represent space) were divided into the three sets of nine characters ({A,B,C,D,E,F,G,H,I}, {J,K,L,M,N,O,P,Q,R}, {S,T,U,V,W,X,Y,Z,_}) and overlaid on the first three stimulus. In second step, nine characters were divided into sets of three characters (e.g. the set {A,B,C,D,E,F,G,H,I} was divided into further three sets: {A,B,C}, {D,E,F}, {G,H,I}), and in the fourth step three characters set was divided into three individual characters (e.g. the set {A,B,C} will be divided to characters: A, B, C) and overlaid on the three stimuli. This behavior represented the working of the three step speller in the web browser. The fourth stimuli was used to delete the wrongly typed character/word, and was also used to move to previous steps of the three steps speller.

The last three flickering stimuli showed the word predictions based on the last word/character(s) typed. The prediction assisted the user with direct selection of complete word, instead of going through character by character to type the word. When the selection of the predicted word was made, a space is automatically added at the end of the word.

**Stimuli Presentation:** In a previous study [7] the flickering stimuli on the web page were presented using Graphics Interchange Format (GIF) files. It was found that GIF file animation sometimes skipped the frames during execution. Also to change the frequency of the target stimuli, the GIF files need to be replaced with the new files. This caused the serious limitation to the speller, and restricted the users to select the predefined GIF files for presentation of the frequency stimuli.

In general, the web browser provides Imperative and Declarative animation support. The imperative animation is heavily dependent on JavaScript. As JavaScript utilizes the browser's main thread, thus usage of the imperative animation can overload the thread resources. This may result in skipping the animation frames and having unstable animation. The other way is to use declarative animation which utilizes CSS, which provides simple and elegant animations on the web pages. Due to CSS usage, the declarative approach for animation can run operations off the main thread.

In this study the flickering stimuli was presented using the CSS animator. The steps() timing function was used, which allowed to break an animation or transition into segments, rather than having one continuous transition from one state to another. A single image was used, which consisted of black and white segments, thus the value passed to steps() function was set to two. The animation behavior was set to continue infinite loop, so as to keep the object in flickering mode until the classification is made and the web page is refreshed. The CSS animation timer was used to achieve the required flickering frequency of the animation. The timer value (T) for a specific frequency (Freq) was calculated using Eq. 1.

$$T(ms) = 1000/Freq \tag{1}$$

**Word Prediction:** The computational linguistic model called n-gram was used for the word prediction based on the text database. The n-gram model consisted of the contiguous sequence of n items for a given text. Thus, the n-gram is a set of co-occurrence of words within a given bracket, and they can be classified as uni-gram (size 1), bi-gram (size 2), tri-gram (size 3), etc. The number of n-grams ($g$) for a given sentence ($s$) can be calculated as

$$g_s = w - (m - 1) \tag{2}$$

where $m$ is gram number ($m = 1$ for uni-gram, $m = 2$ for bi-gram ... and so on), and $w$ is equal to number of words in a sentence. The n-gram model predicts $x_i$, based on

$$x_{i-(n-1)}, \dots, x_{i-1} \tag{3}$$

In terms of probability, we can represent it as

$$P(x_i \mid x_{i-(n-1)}, \dots, x_{i-1}) \tag{4}$$

In our study bi-gram (n-gram of size 2) was used, which anticipated the next letter(s)/word based on the probability P. When the typed text was null, then the n-gram was applied to find the combinations of adjacent letter(s), and when a word was typed then the n-gram model was also applied at the word level. The n-gram model provided simplicity and scalability for the implementation of prediction model.

The relational database for word prediction was set using MySql (version 8.0.12), and the data was derived from the monolingual dictionaries of the Leipzig Corpora Collection [5]. The corpus-based monolingual dictionary can be accessed at: http://corpora.uni-leipzig.de. The corpora collection of English news was used. The corpus data was extracted from around one million sentences. The data was provided in the text file, which was encoded in UTF-8 format.

The relational database consisted of two tables, named as *words* and *co_n*. The association between the entities is represented in entity relationship diagram (refer Fig. 2). The *words* table consisted of the word list of all word forms of the corpus (attribute name: word). The words were weighted by their frequency (attribute name: freq), and every word entry was identified by a unique identification number (attribute name: w_id). The *co_n* table consisted of the information that how often two words occurred in direct neighborhood. The table columns consisted of word id of first word (attribute name: w1_id) and it's immediate occurrence in the left neighborhood of second word (attribute name: w2_id). The frequency of such concurrences of two words was also stored in the table (attribute name: freq). The database consisted of large number of tuples, thus the database indexes were used to boost up the speed of operations (particularly SELECT statement) on the table.

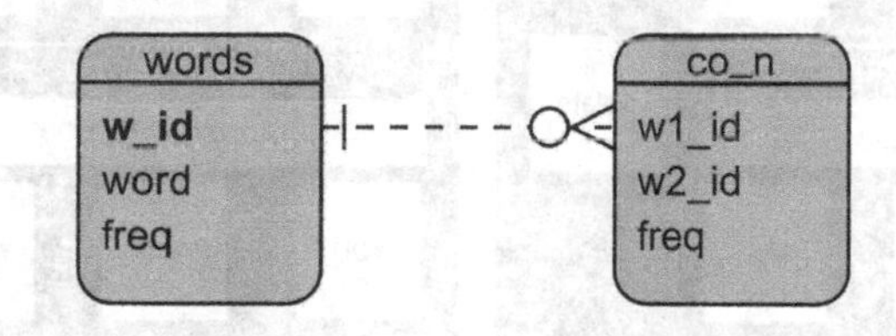


**Fig. 2.** Entity Relationship Diagram (ERD)

To understand the operational behavior of the web speller, the step by step visualization of the speller interface is shown in Fig. 3. The sentence typed is HOW_ARE_YOU, and the selection pattern for this sentence is: 132223563315. The initial layout of the proposed speller is shown in Fig. 3(a). The letter "H" was typed with the consecutive classification of index 1, 3, and 2. When the letter "H" was typed, the possible words starting with letter "H" were predicted. The top three words (HAVE, HAS, HE) were passed to the speller, sorted by the weighted frequency values, refer Fig. 3(b). The letter "O" was typed through consecutive classification of index 2, 2, and 3 stimuli. The word prediction was made based on the string "HO", which returned the word "HOW" as part of suggestion list, refer Fig. 3(c). Now the user made a direct selection of word "HOW" by the classification of index 5 stimuli, Fig. 3(d). This operational procedure continued (refer Fig. 3(e) and (f)) until the whole sentence was typed, refer Fig. 3(g).

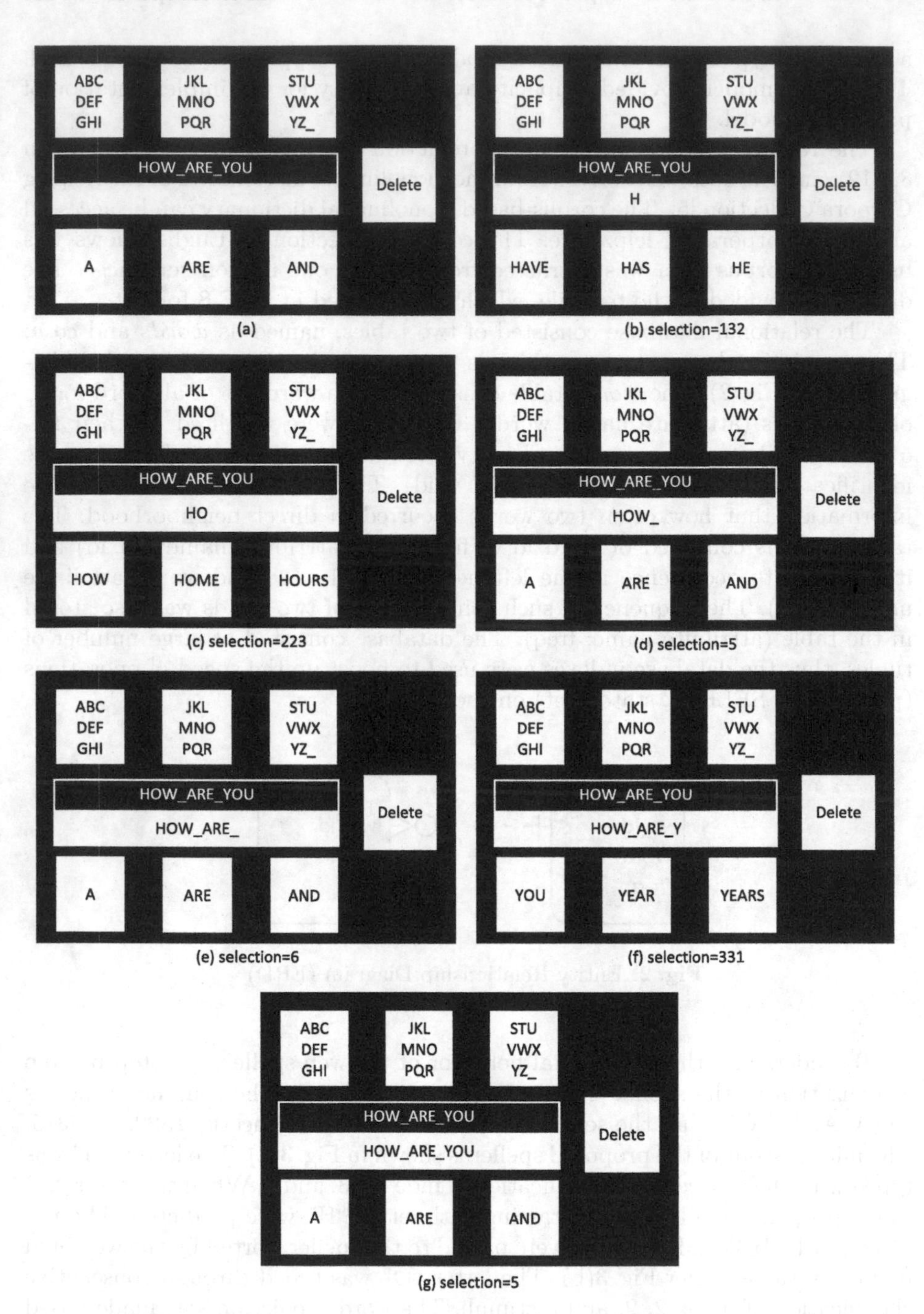


**Fig. 3.** Operational behavior of the web speller interface for typing the sentence HOW_ARE_YOU. (a) Initial web speller screen. (b) Typing character *H*. (c) Typing character *O*. (d) Selection of predicted word *HOW*. (e) Selection of predicted word *ARE*. (f) Typing character *Y*. (g) Selection of predicted word *YOU*.

The index of every classified frequency was passed as part of the query string. The JavaScript then processed the query string and updated the character list and word predictions accordingly. Please note that to simplify the illustration in Fig. 3, the steps to select individual letters were skipped.

### 2.4 Signal Processing

A well-known Minimum Energy Combination (MEC) method [3] was used for SSVEP signal classification. The sampling frequency was set to 256 Hz. The recorded EEG data were transferred to the processing unit in the blocks of 64 samples, and the classifiers used the fixed length of 2, 3, 4, and 5 s of EEG data. The acquired signals was processed online using a custom developed C++ application.

The SSVEP power assessment of seven frequencies were normalized into the corresponding probabilities, in order to detect a frequency in the spatially filtered signal. The classification was performed based on the combination of the individual classifier results. The classifier results were weighted based on their length, i.e. higher weight was given to the longer classifier. The classification was made when a certain threshold was reached and the classification ID was posted to the web browser. The classification threshold for each flickering frequency was set for every subject at the start of the experiment.

The classifier output was rejected for two seconds when a classification was made (either correct classification or miss-classification). This provided a time window to cancel out the effects of the previous stimuli in EEG data. First four stimuli (used for main speller interface) were not shown for the entire duration of the gaze shift period. During the gaze shift time, the database request was made to generate the list of word predictions. The word predictions were shown immediately on the last three flickering stimuli, so that the user could decide for a suitable dictionary option.

## 3 Experimental Setup

Before staring the experiment, the participants had to sign the consent form. Then the participants were seated on a chair facing the LCD screen (at a distance of approximately 50 cm). The participant was prepared for the EEG recording. The participant were explained about the working of the three step speller and how they can use the word prediction feature during the experiment. The participants were asked to look at each flickering stimuli and the threshold values for each stimuli was set individually. To familiarize with the web speller, the participants were asked to type random characters. Afterwards, the participants were given the opportunity to decide the sentence which they want to type, and the same sentence was typed by them during the test. While typing, the correction (character/word was typed or moved to wrong step) was made with help of Delete/Back option, which was counted as the correct command. The speller was stop manually, when the whole sentence was spelled. The spelling session took around two to nine minutes, depending upon the length of the sentence chosen by the participants.

# 4 Results

The performance of the word prediction model for SSVEP-based BCI web speller was evaluated in terms accuracy, information transfer rate (ITR), and time taken to complete a specific task. The accuracy of the web speller was determined by the number of correct command classifications divided by the total number of classified commands. The information transfer rate was calculated as follow (see [11]):

$$B = \log_2 N + P \log_2 P + (1 - P) \log_2 \left[ \frac{1 - P}{N - 1} \right] \quad (5)$$

where $B$ represents the information transfer rate in bits, $P$ is the classification accuracy, and $N$ is the number of targets (i.e. $N = 7$). The $ITR$ in bits/min was evaluated by Eq. 6

$$ITR = \frac{B}{T/60} \quad (6)$$

where $T$ is the total time of the experiment.

The experimental results are listed in Table 1. The subjects were given the chance to type the sentence of their own choice, which can be found in the *Sentence* column of the table.

**Table 1.** The results of the spelling tasks performed by the participants. The table shows the time, accuracy, ITR, and the sentence typed.

| Subject | Time (s) | Accuracy (%) | ITR (bits/min) | Sentence |
|---|---|---|---|---|
| S1 | 378.9 | 82.98 | 12.72 | I_LIVE_IN_GERMANY |
| S2 | 482.4 | 95 | 17.85 | PLEASE_FILL_THE_FORM |
| S3 | 173.4 | 100 | 27.19 | I_LIKE_TACOS |
| S4 | 342.4 | 90.91 | 12.33 | I_LIVE_IN_GERMANY |
| S5 | 323.6 | 100 | 19.77 | DOTA_IS_MY_FAVOURITE_GAME |
| S6 | - | - | - | - |
| S7 | 134.4 | 100 | 27.56 | ARE_YOU_FREE_TONIGHT |
| S8 | 473.1 | 85.48 | 14.42 | GOATS_ARE_BETTER_THAN_DOGS |
| S9 | 245.4 | 87.18 | 18.34 | ARE_YOU_FEELING_HUNGRY |
| S10 | 158.4 | 91.3 | 18.78 | IT_SNOW_TODAY |
| S11 | 267.4 | 92.5 | 20.01 | THIS_DOG_IS_CUTE |
| **Mean** | **297.94** | **92.53** | **18.8** | |

## 5 Discussion

The objective of this study was to develop and test the SSVEP-based BCI web speller which can provide word predictions based on the characters/words typed. A working model of a web speller was built using HTML, JavaScript, and CSS based animation objects. The word prediction was provided using the Leipzig corpora collection. The system was tested on eleven subjects, and the results achieved were quite satisfactory. For the experiment, the subjects had chosen the sentence of their choice, and the range of sentence length was between 12 to 26 character. In terms of number of words, the minimum length was 3 words, and maximum length was 5 words.

The mean ITR of 18.8 bits/min was achieved. In the previous study [7], the mean ITR of 12.7 was achieved without using the word prediction model. Thus the current study results showed that better ITR can be achieved using the given dictionary support. The usage of n-gram based prediction model proved to be effective and produced the in-time results for the web based Ajax requests. During the operation of the speller, the user of the system did not have to go through the typing procedure for each character, and instead can select the suggested word, which significantly improve the speed of the speller.

The results showed that mean accuracy of 92.53% was achieved, and three subjects were also able to complete the task without any error. Providing the suggested list of words was one of the major factors to achieve the better accuracy. During spelling task the users had to select the correctly spelled word from the suggestion list, and did not need to type each character of the word. Thus, cutting down the steps involved in typing significantly increased the accuracy. This also resulted in reducing the time required to complete the spelling task.

During the experimental study, only one of the subjects was not able to complete the task. The signal classification for subject $S6$ was taking long time and the subject decided to quit the experiment. It was found that the subject was tired before the start of experiment, and was also using the SSVEP-based BCI system for the very first time.

The flickering stimulus on the web was presented using CSS animator. The results showed that we can get a stimuli which can be configured easily for any flickering frequency. The CCS-based animation provided smooth operation and can be used as an alternative to GIF files.

The large pool of word forms of the Leipzig corpora collection provided a good basis for finding the words match based using n-gram model. The usage of the neighborhood concurrences, sorted on their usage frequency, provided the direct match of the upcoming word. This procedure improved the usability and the performance of the web speller.

In the future we can extend the system by providing multi-language support for the speller. Other techniques such as webGL, Flash, and Synchronized Multimedia Integration Language (SMIL) can be used for generating the flickering stimuli and the challenge will be to find the best possible option. Moreover, the sentence prediction model based on the typed word can also be introduced in the web speller.

## 6 Conclusion

We can conclude that the word prediction support model can significantly increase the ITR and accuracy of the SSVEP-based BCI web speller. The CSS based animation can provide better stimulation and easy configuration options. During the execution of the speller, the AJAX calls were placed to call the Java Servlet and Java Beans which provided the JSON format word list from the database. These calls proved to provide the in-time word prediction. Therefore, the web speller proposed in this study can be extended for the web based email interface, or to write a blog which require a user to type the text within the web browser.

**Acknowledgment.** This research was supported by the European Fund for Regional Development (EFRD - or EFRE in German) under Grants IT-1-2-001, and GE-1-1-047. We are thankful to the participants and student assistants who took part in this research study.

## References

1. Akram, F., Han, S.M., Kim, T.S.: An efficient word typing P300-BCI system using a modified T9 interface and random forest classifier. Comput. Biol. Med. **56**, 30–36 (2015)
2. Cao, T., et al.: A high rate online SSVEP based brain-computer interface speller. In: 2011 5th International IEEE/EMBS Conference on Neural Engineering (NER), pp. 465–468. IEEE (2011)
3. Friman, O., Volosyak, I., Gräser, A.: Multiple channel detection of steady-state visual evoked potentials for brain-computer interfaces. IEEE Trans. Biomed. Eng. **54**(4), 742–750 (2007)
4. Gannouni, S., Alangari, N., Mathkour, H., Aboalsamh, H., Belwafi, K.: BCWB: a P300 brain-controlled web browser. Int. J. Semant. Web Inf. Syst. (IJSWIS) **13**(2), 55–73 (2017)
5. Goldhahn, D., Eckart, T., Quasthoff, U.: Building large monolingual dictionaries at the leipzig corpora collection: from 100 to 200 languages. In: LREC, vol. 29, pp. 31–43 (2012)
6. Martišius, I., Damaševičius, R.: A prototype SSVEP based real time BCI gaming system. Comput. Intell. Neurosci. **2016**, 18 (2016)
7. Saboor, A., et al.: A browser-driven SSVEP-based BCI web speller. In: 2018 IEEE International Conference on Systems, Man, and Cybernetics (SMC), pp. 625–630. IEEE (2018)
8. Saboor, A., et al.: SSVEP-based BCI in a smart home scenario. In: Rojas, I., Joya, G., Catala, A. (eds.) IWANN 2017. LNCS, vol. 10306, pp. 474–485. Springer, Cham (2017). https://doi.org/10.1007/978-3-319-59147-6_41
9. Stawicki, P., Gembler, F., Volosyak, I.: A user-friendly dictionary-supported SSVEP-based BCI application. In: Gamberini, L., Spagnolli, A., Jacucci, G., Blankertz, B., Freeman, J. (eds.) Symbiotic 2016. LNCS, vol. 9961, pp. 168–180. Springer, Cham (2017). https://doi.org/10.1007/978-3-319-57753-1_15

10. Volosyak, I., Moor, A., Gräser, A.: A dictionary-driven SSVEP speller with a modified graphical user interface. In: Cabestany, J., Rojas, I., Joya, G. (eds.) IWANN 2011. LNCS, vol. 6691, pp. 353–361. Springer, Heidelberg (2011). https://doi.org/10.1007/978-3-642-21501-8_44
11. Wolpaw, J.R., Birbaumer, N., McFarland, D.J., Pfurtscheller, G., Vaughan, T.M.: Brain-computer interfaces for communication and control. Clin. Neurophysiol. **113**(6), 767–791 (2002)
12. Yehia, A.G., Eldawlatly, S., Taher, M.: WeBB: a brain-computer interface web browser based on steady-state visual evoked potentials. In: 2017 12th International Conference on Computer Engineering and Systems (ICCES), pp. 52–57. IEEE (2017)

# Is Stress State an Important Factor in the BCI-P300 Speller Performance?

Liliana Garcia[1](✉), Maud Zak[2], Celestin Grenier[2], Solene Hanrio[2], Dorine Henry[2], Romain Randriamanantena[2], Catherine Semal[3], Jean Marc Andre[1], Veronique Lespinet-Najib[1], and Ricardo Ron-Angevin[4]

[1] IMS Laboratory - CNRS, Bordeaux University, Bordeaux INP, Talence, France
liliana.audin@u-bordeaux.fr,
{jean-marc.andre, veronique.lespinet}@ensc.fr
[2] ENSC - Bordeaux INP, Talence, France
{mzak, cgrenier003, shanrio, dhenry002, rrandriamana}@ensc.fr
[3] INCIA Laboratory - CNRS UMR 5287, Bordeaux, France
catherine.semal@ensc.fr
[4] Dpto. Tecnología Electrónica, Universidad de Málaga, Málaga, Spain
rra@dte.uma.es

**Abstract.** Brain-Computer Interface (BCI) is an advanced human–machine interaction technology requiring higher-order cognitive functions for an efficient task execution. The relation between cognition, human performance and psychological state has been studied for many years. Nevertheless, the effect of acute stress on cognitive performance involving BCI systems has never been studied. Nowadays, people are more and more affected by stressful situations. Stress is an important human factor which can impact the ability to appropriately process cognitive information related to language, working memory, attention, or executive control.

Individuals are continuously interacting with technology to execute daily actions. BCI represent an alternative way to allow any individual, even with motor disabilities, to interact with that technology. BCI-P300 Speller is driven by EEG signals and enables communication without physical intervention. It is used in both clinical investigations and fundamental research.

In this work, we study the impact of acute stress effects on cognitive ability to control a BCI-P300 speller. Although we have observed a broad spectrum of response to stress, analyses show a correlation between BCI-speller performance and user's stress state. We have also noted that BCI performance seems to be improved if users have a good cognitive engagement in the task and if they showed an ability to develop efficient strategies, such as selective attention or increased effort, in order to cope with the stressful situations.

In conclusion, these preliminary results performed on a small sample (n = 7) show that BCI-P300 Speller is a robust and reliable tool and suggest that an optimal utilization of BCI systems could be assured despite the fluctuations of users' state. We assume that neural mechanisms involved in the BCI task,


I. Rojas et al. (Eds.): IWANN 2019, LNCS 11506, pp. 442–454, 2019.
https://doi.org/10.1007/978-3-030-20521-8_37

may set the brain in an adequate level of generalized arousal, which allows establishment of compensatory mechanisms in stressful states.

**Keywords:** Brain-Computer Interface (BCI) · P300-Speller · Acute stress · User's state

## 1 Introduction

Brain activation in response to visual stimuli can be used to reliably drive external devices through systems named Brain-Computer Interfaces (BCI). BCI requires a good cognitive state to be able to communicate with the external world. It is an excellent tool for medicine and research; especially regarding patients affected by Amyotrophic Lateral Sclerosis (ALS) who have impaired motor activity but intact cognitive functions, and thus are able to execute a spelling task with a BCI system [1]. The BCI-P300 system uses an Event Related Potential (ERP) which is a positive EEG signal triggered approximatively 300 ms after stimulus onset [2].

The BCI-P300 Speller used in this study relies on a Farwell-Donchin paradigm design, which consists of a 6 × 6 matrix of letters and numbers displayed as a virtual keyboard [3]. In this paradigm, subjects focus on a target character while rows and columns of characters flash randomly. The BCI's classifier algorithm detects the Event Related Potential (ERP-300) triggered by an attended visual stimulus (target) among many other irrelevant stimuli (non-targets) presented in rows and columns. It discriminates and classifies specific features of EEG signals, which are unique for each individual.

In fact, using BCI is a task requiring higher-order cognitive functions, which in turn are greatly influenced by psychological states. Attention and working memory are the main cognitive processes underlying the ERP-P300 signals [4]. Some authors reported correlations between P-300 signals (P-300 amplitude and P300 performance rate) or electroencephalographic (EEG) band activity, and emotional motivational or mental states [4–7].

Stress is a factor impacting normal homeostasis of nearly anyone once in their life [8]. Stress occurs when an individual perceives stimuli as being negative (stressors) representing a challenge, danger or threat and eliciting physiological, psychological and behavioral changes. This stress state appears when a demand exceeds the regulatory capacity of an organism [9]. But, subjects are able to develop coping strategies and adaptive homeostatic mechanisms to compensate the imbalance generated by the new state [10]. However, it is also known that not all individuals have similar capacities to cope with stress and some of them could trigger short or long periods of instability and even be the cause of the development of serious diseases [11].

Stress induces the parallel activation of two biological systems: the Sympathetic-Adrenal-Medullary (SAM) axis, and the Hypothalamic-Pituitary-Adrenal (HPA) system. The first triggers adrenaline and noradrenaline release some seconds after the stressful event. In a second time, a slower hormone cascade leads to glucocorticoid release from the adrenal cortex [9]. Both systems contribute to adapt to the stressor and to restore homeostasis [12]. Furthermore, in acute stress, sympathetic activation through mental stress is associated with the activation of a number of cerebral regions, some of them being task relevant (visual, motor, and premotor areas). It is known that stress impacts working memory (WM) [13–15], attention [16] and other functions involving the prefrontal cortex [17].

The aim of this study is to analyze how stress state induced by psychological methods could impinge on BCI effectiveness. Our hypothesis is that stress-control by the user is a determinant component of an appropriate state of mind for a good performance using a BCI. This would suggest that acute stress condition can alter selective attention and task engagement, which are necessary with BCI system.

## 2 Methods and Experimental Paradigm Design

### 2.1 Participants

Seven healthy subjects (4 women, 3 men, range age 20 to 22) participated in this experiment. Participants were volunteer students, all novices in BCI system. They were native French speakers, with normal or corrected vision, and they didn't suffer from Post-Traumatic Stress Disorder (PTSD). Drinking coffee or tea and smoking were not allowed 2 h prior to the experiment. Every experimental session was scheduled in the afternoon so that all participants had similar hormonal levels. They all gave informed consent through a protocol reviewed by the IMS Cognitive and the University of Malaga teams.

### 2.2 EEG Data Acquisition and Processing

EEG signals were recorded from 8 electrodes placed at positions Fz, Cz, Pz, Oz, P3, P4, PO7 and PO8, according to the 10/20 international system. All channels were referenced to the right earlobe, using FPz as ground. The EEG was amplified through a 16 channel biosignal amplifier gUSBamp (Guger Technologies). The amplifier settings were from 0.5 Hz to 100 Hz for the band-pass filter, the notch (50 Hz) was on, and the sensitivity was 500 μV. The EEG was then digitized at a rate of 256 Hz. EEG data collection and processing were controlled by the UMA-BCI Speller software) [18].

### 2.3 Experimental Setup

After signing the consent document and following recommendations of ethical committee for research human and data protection law (General Data Protection Regulation RGPD, France), participants sat 60 cm away from the screen and were given verbal instructions concerning the BCI-P300 Speller. They were asked to focus on the center of the screen where the spelling keyboard was presented. The matrix-based BCI Speller consists of a 6 $\times$ 6 matrix of symbols (letters and digits) arranged within rows and columns (see Fig. 1). It features the 26 characters of the Latin alphabet and digits from 1 to 9, 0 being replaced by a dash symbolizing a space.

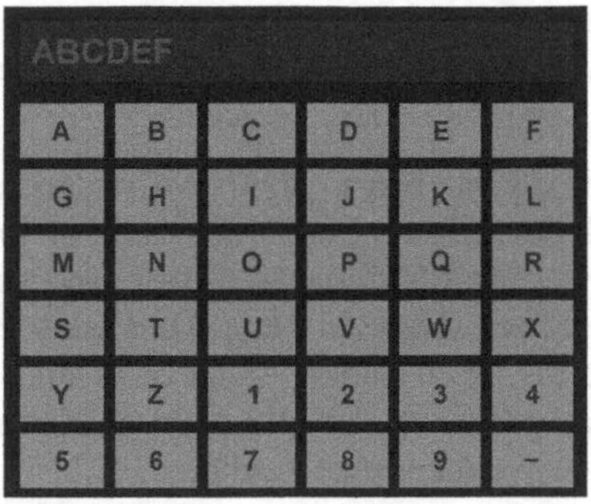

Fig. 1. Design of the input interface.

## 2.4 Protocol

Each participant took part in a 1 h session where they carried out two BCI tasks: a control condition task (BCI Control task) and then a stressful condition task (BCI test task). A description of the experimental design is shown in Fig. 2. A session began with a calibration phase where participants were asked to spell three different words in order to detect the specific features of their own P300 signal. Both BCI tasks constitute the second phase of the experiment, named copy-speller phase, where participants must spell predefined sentences. For the calibration and copy-speller phases, row and column of the matrix was randomly flashed 10 times by sequence (i.e. by letter to spell) and participants focused attention on desired character. Each character thus flashes 20 times by sequence. To spell a character, participants need to mentally count how many times the chosen character had flashed during a sequence.

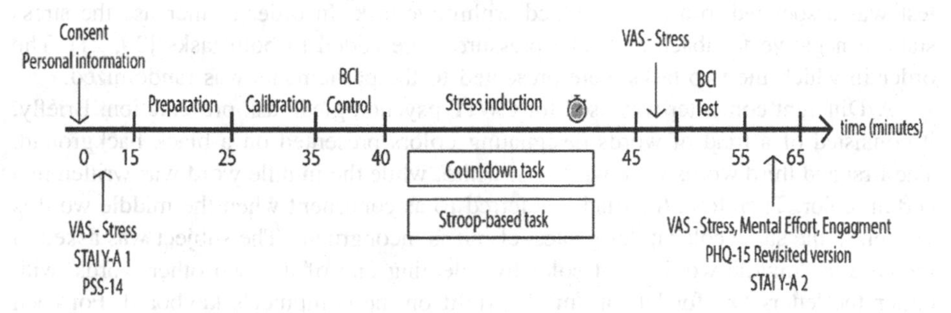


Fig. 2. Experimental design scheme showing the sequencing of stress-inducing tasks and collection of subjective metrics.

**Calibration Phase.** Three French words of 4 characters (TOUR, ZINC and LAVE) were chosen for this phase. Participants had to spell them without receiving any feedback (i.e. the spelled letters weren't displayed above the keyboard). Their EEG signals were recorded for analysis using a stepwise linear discriminant analysis (SWLDA) to obtain the weights for the on-line P300 classifier. This phase allows us to define the minimum number of flashes required for the BCI system to detect participant's P300. This minimum number corresponds to the number of flashes with which participants were able to spell with 100% of accuracy.

**Copy-Speller Phase.** Participants had to spell two sentences composed of 19 characters. The first BCI task (BCI Control task) consisted of spelling "IL COMPTE 18 NOYAUX" whereas the second BCI task (BCI Test task) consisted of spelling "TU LANCES 36 PHENIX". Spaces between each word were considered as characters. Participants were instructed not to correct themselves and to continue even if the classifier had chosen a wrong letter. In this phase, the characters typed by the user were displayed inside the box above the virtual keyboard.

### 2.5 BCI-P300 Classification Accuracy and Performance

During the calibration phase, brain signals were processed for feature extraction and classification using SWLDA algorithms. Then, those classified results were used for the BCI copy-spelling phase. BCI performance was measured by the percentage of spelling errors made during copy-spelling phase in control and stressful condition.

### 2.6 Stress Induction Protocol

Stress can be induced from physical, psychological or social tests, each one producing different effects on activation of physiological systems. Acute stress refers to episodes of limited duration (generally occurring within minutes or hours; e.g., a speech in public, academic exam) [19].

In this work, a validated Stroop-inspired color/word interference test (CWIT) was used to induce acute stress. This psychological task is used frequently in the academic environment to produce mental overstimulation as a result of cognitive conflict. This test was associated to a second added arithmetic task. In order to increase the stress state, a negative feedback and time pressure were added to both tasks [20, 21]. The order in which the two tasks were presented to the participants was randomized.

A Different computer was used for CWIT psychological task presentation. Briefly, it consisted of a triad of words designating colors presented on a black background. The first and third words were written in white, while the middle word was written in a certain color. Therefore, the triad is referred to as congruent when the middle word is written in the same color it designates, else it is incongruent. The subject was asked to indicate the middle word's font color by selecting one of the two other words, with either the letters "x" for left or "m" for right on the computer's keyboard. For each triad, participants had one second to answer. Otherwise, negative feedbacks were displayed on the screen (during 250 ms): "wrong" (faux!) for errors or, "faster" (plus vite!) when the time delays of responses were superiors to one second (Fig. 3).

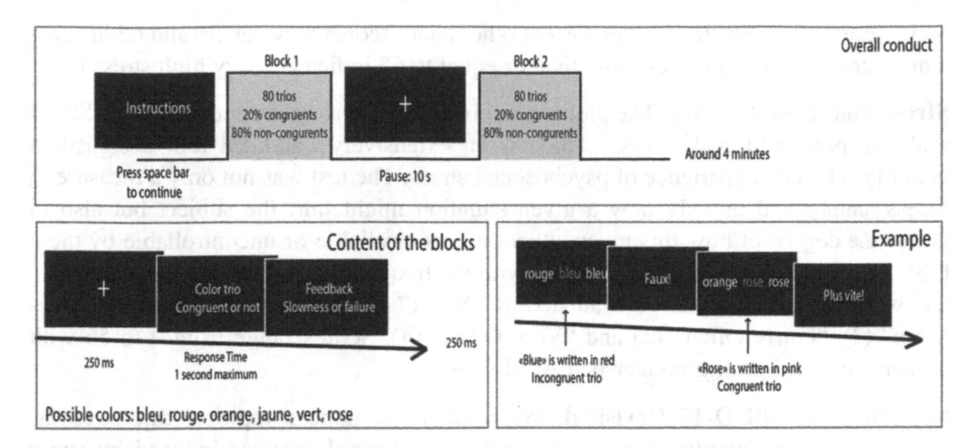


**Fig. 3.** Experimental design scheme of the Stroop-based task. Two blocks of congruent (matching between the word's font color and the color designated by the word), and incongruent (no matching between the word's font color and the color designated by the word) were presented to each participant and separated by a 10 s pause. Negative feedbacks were given for time delays > 1 s or wrong answers.

The entire task had a duration of about 4 min (250 s total) and was composed of two blocks of 80 trio of words (of 1 s duration and 0.5 interval between trios) separated by ten-seconds pause for a total of 160 trios (Fig. 3). To avoid habituation, we introduced 20% of congruent trios (the color and the color name of the middle word correspond). A central fixation cross was presented during 250 ms to maintain attention.

For the arithmetic task, participants were asked to count down out loud, 7 by 7, starting from 189 or 140 (x-7, Fig. 2). The starting number was changed each time between participants. An experimenter supervises the participant during the task to augment stressor.

## 2.7 Metrics to Stress Assessment

Stress reactions were assessed with four subjective measurements from tests used in biomedical and neuroergonomic research. Four types of scale were used to evaluate the stress level of the subjects at the time of the experiment. The execution of each test is indicated in Fig. 2.

**Current Stress (Anxiety-State) STAI Y-A Test.** We use the Scale of Anxiety-State A-form to measure perceived stress at the moment of experiment. The aim is to measure the anxiety arising from stress-inducing tasks. Participants had to evaluate each of the 20 items as "Not at all" (1), "Somewhat" (2), "Moderately so" (3), "Very much so" (4). Scores range from 20–80 with higher scores indicating a high level of stress [22]. On the one hand, scores less than or equal to 35 indicate a very low stress level, scores between 36 and 45 indicate a low stress level and scores between 46 and

55 indicate a medium stress level. On the other hand, scores between 56 and 65 indicate a high stress level and scores more than or equal to 65 indicate a very high stress level.

**Stress Tolerance PSS-14.** The global perceived stress was measured with the PSS-14 scale, or perceived scale stress, which is an extensively validated tool designed to quantify subjects' experience of psychosocial stress. The test was not only a measure to assess simply and quickly how a given situation might hurt the subject but also to assess the degree of how this given situation is controllable or uncontrollable by them [23]. It is composed of 14 items dealing with the frequency of some feelings during the last weeks. Each item can be evaluated as "Never" (0), "Almost Never" (1), "Sometimes" (2), "Fairly Often" (3) and "Very Often" (4). Scores range from 0 to 56 with higher scores indicating greater perceived stress.

**Somatic Stress PHQ-15 Revisited.** We used the somatic symptom scale stress as indicator for the severity of symptoms and physiological changes induced by stress task. Participants had to fill a 10 items somatic symptom form ("Stomach pain", "Back pain", "Pain in your arm, back, or joints", "Headaches", "Chest pain", "Dizziness", "Feeling your heart pound or race", "Shortness of breath", "Nausea or gas", "Feeling tired or having low energy") adapted from the Patient Health Questionnaire Physical Symptoms (PHQ-15) [24]. Each item asks the individual to rate the severity of his/her somatic symptom during the experiment and is rated on a 3-point scale (e.g., "Not bothered at all" = 0; "Bothered a little" = 1; "Bothered a lot" = 2). Total score can range from 0 to 20, with higher scores indicating greater severity of somatic symptoms.

**Visual Analogue Scale VAS Stress.** It is a psychometric response scale that measures subjective characteristics or attitudes that can't be measured directly. When the subject answers to a VAS element, he specifies the degree of agreement with a statement by indicating a position along a continuous line between two endpoints. In the case of our study, the subject had to evaluate her/his degree of stress on a scale of 0 to 10. The continuous scale is comprised of a vertical, 15 cm long line anchored by 3 verbal descriptors: "Not stressed at all" for the lower end of the scale, "Moderate stress" for middle scale, and "Very stressed" for the upper end. Score ranges from 0–10 with higher scores indicating a higher level of stress.

**Temporal Subjective Test Distribution.** PSS-14, STAI Y-A (STAI Y-A 1), and VAS stress were filled before the BCI-P300 speller task. PHQ-15 and VAS stress were filled after the stress induction. At the end of the session, participants had to fill another STAI Y-A (STAI Y-A 2), a VAS stress and two other VAS scale metrics (0 to 10) to assess task engagement and mental effort during the task (Fig. 2).

**Statistical Analyses.** Due to the low sample size, statistical analyses were performed using a non-parametric Wilcoxon test with a risk of error of 5%: p-value $< 0.05$ were accepted as significant.

# 3 Results

## 3.1 Measurements of Subjects' Stress State

The results of the stress induction test highlight a variety of effects on participants (see Fig. 4-left). However, as it is shown in Fig. 4-right, the combination of CWIT and arithmetic task produced an average increase in stress state.

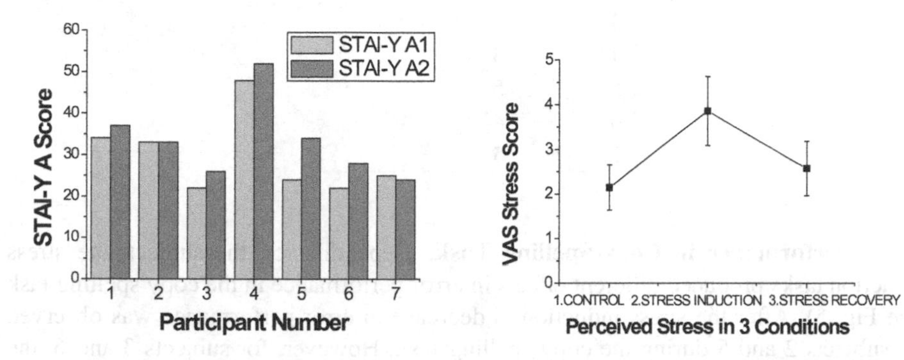


**Fig. 4.** Perceived stress using a STAI Y-A and VAS stress Score. Left: STAI Y-A1 and STAI Y- A2 measured, for each participant, at the beginning (basal state) and at the end of the session, respectively. Right: VAS Stress score averaged for the 3 conditions: 1. CONTROL, at the beginning of the session; 2. STRESS INDUCTION, after the stress induction and 3. RECOVERY, at the end of the session. Data are represented as averaged values + standard error mean (SEM) Concerning STAI Y-A scale, analyses show that the average score at the end of the session (STAI Y-A2) was significantly higher (33.43 ± 3.56 SEM) than the average score for the first STAI Y-A1 (21.79 ± 3.58 SEM) at the basal state (p-value = 0,029) (ranging score between 22 and 48).

Regarding the measured stress, we can notice how participants increase the VAS averaged score after the CWIT and arithmetical task (from 2.1 ± 0.5 SEM to 3.85 ± 0.7 SEM). In this sense, the post-stressful task was significant compared to VAS control stress score (p-value = 0.049). It is interesting to observe how, at the end of the session, the VAS stress score nearly drops back to initial values for 2.5 ± 0.6.

These results are in accord with the somatic symptoms measured by PHQ-10 questionnaire for which participants expressed principally: fatigability (5/7), headaches (4/7) and palpitations (3/7) symptoms, suggesting that stress induction produces attenuated physiological effects. However, objective metrics using electrophysiological techniques will be necessary for a closer study of these effects.

## 3.2 BCI-P300 Speller Performance

**Classifier Performance Accuracy.** All participants were able to control the BCI-speller, reaching, all of them, 100% classification accuracy during the calibration phase. However, each participant needed different number of flashes to get 100% classifier accuracy (see Table 1). On average, participants needed 4.4 ± 0.36 SEM flashes to ensure maximum reliability from the classifier.

**Table 1.** Number of flashes used by each participant in the copy-spelling phase

| Participant | Number of flashes |
|---|---|
| 1 | 4 |
| 2 | 6 |
| 3 | 3 |
| 4 | 4 |
| 5 | 5 |
| 6 | 4 |
| 7 | 5 |

**Error Performance in Copy-Spelling Task.** Depending on the subject, the stress induction tasks produced different effects in error performance in the copy spelling task (see Fig. 5). After the stress induction, a decrease in error performance was observed for subjects 2 and 5 during the copy spelling task. However, for subjects 3 and 6, the effect of stress induction was opposed, producing an increase in the error performance.

In this sense, from results we can deduce, that for some subjects, stressful conditions seem fail to modify performance indicating that these subjects were not sensitive to stressors.

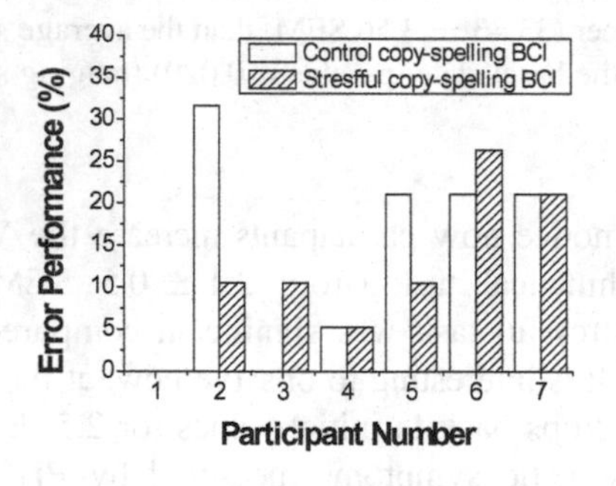


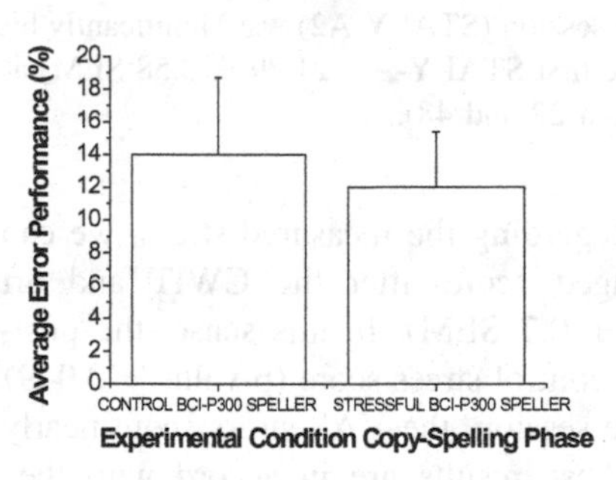


**Fig. 5.** BCI error performance of copy-spelling task for control and stressful conditions for each participant (left) and averaged values (right).

This variability suggests that performance could be influenced by a stress condition depending on the vulnerability of a subject to stress. To test this hypothesis we have put in relation performance versus perceived stress variables.

### 3.3 Correlation Between the VAS Stress Scale and Error Performance

As depicted in the Fig. 6 (left), no correlation was observed between error performances and VAS stress score during the copy-spelling task in control situation (r nearly equal to 0 in Pearson correlation). By contrast, there was a high correlation between error performance and VAS stress score in the stressful condition (r = –0.947, p-value = 0.001 in Pearson correlation). This last result suggests that moderate stressful situations have a tendency to reduce the number of mistakes, which is in accord with performance results showed in Fig. 5-right. However, complementary experiments are necessary to confirm or not this result.

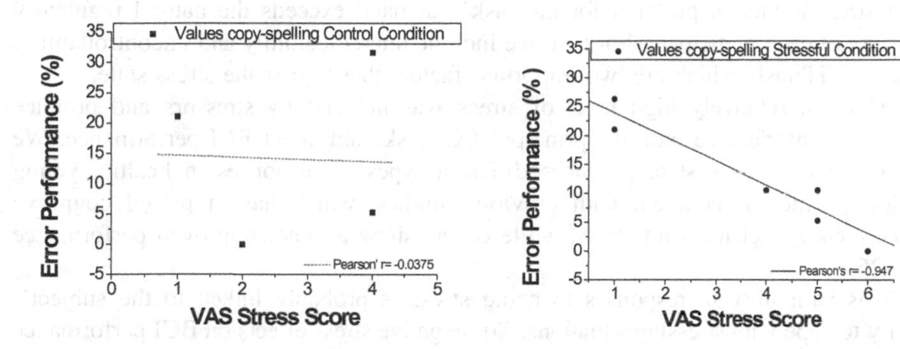


**Fig. 6.** Graphs representing the participant's BCI -Performances (errors %) as a function of perceived stress measured by VAS stress score. Left: In control condition. Right: In Stressful condition. Points can be coincident for some subjects.

### 3.4 Engagement and Effort Task Dependent

Engagement for the task was very high for control and test BCI task showing that a stressful situation doesn't altered motivation for BCI task (8.0 ± 0.3 SEM for control vs 8.3 ± 0.42 SEM for test BCI task, rating *scale from 0–10).* On the contrary, the participant seems to employ additional mental efforts for the second copy spelling task (Fig. 7), which was significant (8.5 ± 037 SEM), compared to the first BCI speller control (6.0 ± 0.64 SEM) p value = 0.05.

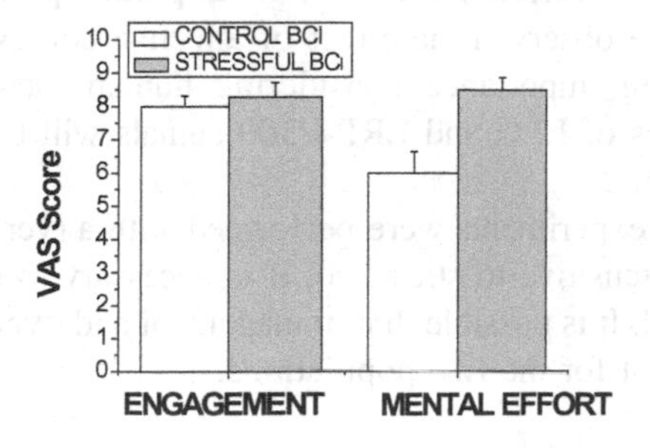


**Fig. 7.** Graph showing VAS score averaged for all participants for Engagement and Mental Effort during the two copy spelling tasks.

These two results are correlated with the task engagement of the participant and suggest that supplementary effort and commitment to the BCI task could improve performance.

## 4 Discussion

In this study, we have performed experiments to analyze the interaction between stress and BCI-P300 speller performance. Our hypothesis was that acute stress triggered by psychological tasks, could be correlated to ability to control BCI speller system. Starting from the idea that combined stressors increase probability to develop stress, we used stress induction protocol for the task's demand exceeds the natural regulatory capacity of each participant. For that, we include unpredictability and uncontrollability using CWIT task, which are two important factors that trigger the stress state.

Albeit, a relatively high level of stress was induced by stressors and produce behavioral interference with the principal BCI task, and affect BCI performance. We observe that a stress state produces different types of responses in healthy young subjects, which is coherent with previous studies, which have reported cognitive impairment associated with stress, while others show a rather improved performance [25, 26].

This variability in responses to acute stress is probably linked to the subject's ability to cope with stressing situations. So, negative stress effects on BCI performance observed for some subjects could be related to low task engagement or little to no effort invested in the task. Conversely, the improvement in performance observed for other subjects may be due to developed adaptation strategies, requiring additional resources, to manage stress.

It is possible, according to load theory [27], that people respond inappropriately to demands imposed by the task, particularly when demands are higher than the individual's cognitive capacity. In this work, we show that stress induces relatively high load only to subjects with vulnerability to stress. Moreover, some subjects apparently have difficulty for stress appraisal despite having low BCI performance.

In this study, the observed inter-individual variability could also be explained by multiple factors influencing individual responses such as, vulnerability, gender, or personality traits [28], which are known to affect cognitive performance [29]. Supporting these results, previous studies have described that personality factor, motivation or emotional state can be correlated with P300 amplitude performance [5, 7]. Moreover, other studies have observed the effect of affective contexts on classifier performance [30], reinforcing importance considering human factor on BCI utilization. Supplementary analyses of EEG and ERP-P300 signals will be necessary to advance conclusions on this topic.

Furthermore, these experiments were performed with a (very) young population of students who are less sensitive to stress, so, it is necessary to do similar experiments with older subjects [31]. It is possible that management and dynamics of recovery from stress would be different for the two populations.

## 5 Conclusion

This study shows that acute stress impacts differently BCI-P300 speller performance according to the subject's ability to manage stress. So, these results underline the importance to understand stress role on human-machine interaction. Although preliminary, this study emphasized the relevance of optimization and adaptation of interfaces to the user's mental state. Additional experiments are necessary to understand how psychological or emotional factors modulate BCI control capability, particularly for an end-users population. In the future, human factor analysis and its modulator effects represent a new challenge to BCI experiments.

**Acknowledgement.** This work was partially supported by the Spanish Ministry of Economy and Competitiveness through the projects LICOM (DPI2015-67064-R), by the European Regional Development Fund (ERDF) and by the University of Malaga. Moreover, the authors would like to thank all participants for their cooperation. This work has been carried out in a framework agreement between the University of Málaga and IMS Laboratory- CNRS, Bordeaux University, Bordeaux INP – France.

## References

1. Garcia, L., et al.: A comparison of a brain-computer interface and an eye tracker: is there a more appropriate technology for controlling a virtual keyboard in an ALS patient? In: Conference: International Work-Conference on Artificial Neural Networks (2015)
2. Polich, J.: Updating P300: an integrative theory of P3a and P3b. Clin. Neurophysiol. **118**(10), 2128–2148 (2007)
3. Farwell, L.A., Donchin, E.: Talking off the top of your head: toward a mental prosthesis utilizing event-related brain potentials. Electroencephalogr. Clin. Neurophysiol. **70**(6), 510–523 (1988)
4. Kleih, S.C., Kübler, A.: Empathy, motivation, and P300 BCI performance. Front. Hum. Neurosci. **17**(7), 642 (2013)
5. Kleih, S.C., Nijboer, F., Halder, S., Kübler, A.: Motivation modulates the P300 amplitude during brain-computer interface use. Clin. Neurophysiol. **121**(7), 1023–1031 (2010)
6. Käthner, I., Wriessnegger, S.C., Müller-Putz, G.R., Kübler, A., Halder, S.: Effects of mental workload and fatigue on the P300, alpha and theta band power during operation of an ERP (P300) brain-computer interface. Biol. Psychol. **102**, 118–129 (2014)
7. Hammer, E.M., Halder, S., Kleih, S.C., Kübler, A.: Prediction of auditory and visual p300 brain-computer interface aptitude. Front. Neurosci. **12**, 307 (2018)
8. Staal, M.A.: Stress, Cognition, and Human Performance: A Literature Review and Conceptual Framework. NASA/TM-2004-212824
9. Dickerson, S.S., Kemeny, M.E.: Acute stressors and cortisol responses: a theoretical integration and synthesis of laboratory research. Psychol. Bull. **130**, 355–391 (2004)
10. Hockey, G.R.: Compensatory control in the regulation of human performance under stress and high workload; a cognitive-energetical framework. Biol. Psychol. **45**(1–3), 73–93 (1997)
11. Kloet, E., Ron Joëls, M., Holsboer, F.: Stress and the brain: from adaptation to disease. Nat. Rev. Neurosci. **6**, 463–475 (2005)
12. Joëls, M., Baram, T.Z.: The neuro-symphony of stress. Nat. Rev. Neurosci. **10**(6), 459–466 (2009)

13. Elzinga, B.M., Roelofs, K.: Cortisol-induced impairments of working memory require acute sympathetic activation. Behav. Neurosci. **119**, 98–103 (2005)
14. Schwabe, L.: Memory under stress: from single systems to network changes. Eur. J. Neurosci. **45**(4), 478–489 (2017)
15. Lukasik, K.M., Waris, O., Soveri, A., Lehtonen, M., Laine, M.: The relationship of anxiety and stress with working memory performance in a large non-depressed sample. Front. Psychol. **10**(4) (2019)
16. Sänger, J., Bechtold, L., Schoofs, D., Blaszkewicz, M., Wascher, E.: The influence of acute stress on attention mechanisms and its electrophysiological correlates. Front. Behav. Neurosci. **8**, 353 (2014)
17. Fuster, J.M.: Cognitive functions of the prefrontal cortex. Front. Hum. Neurosci. **4**, 11–22 (2013)
18. Velasco-Álvarez, F., Sancha-Ros, S., García-Garaluz, E., Fernández-Rodríguez, A., Medina-Juliá, M.T., Ron-Angevin, R.: UMA-BCI speller: an easily configurable P300 speller tool for end users. Comput. Methods Programs Biomed. **172**, 127–138 (2019)
19. Owens, M., Stevenson, J., Hadwin, J.A.: Anxiety and depression in academic performance: an exploration of the mediating factors of worry and working memory. Sch. Psychol. Int. **33** (4), 433–449 (2012)
20. Fechir, M., et al.: Patterns of sympathetic responses induced by different stress tasks. Open Neurol. J. **2**, 25–31 (2008)
21. Van Oort, J., et al.: How the brain connects in response to acute stress: a review at the human brain systems level. Neurosci. Biobehav. Rev. **83**, 281–297 (2017)
22. Spielberger, C.D., Gorsuch, R.L., Lushene, R.E.: The State-Trait Anxiety Inventory. Consulting Psychologists Press Inc, Palo Alto (1970)
23. Cohen, S., Kamarck, T., Mermelstein, R.: A global measure of perceived stress. J. Health Soc. Behav. **24**(4), 385–396 (1983)
24. Kroenke, K., Spitzer, R.L., Williams, J.B.: The PHQ-15: validity of a new measure for evaluating the severity of somatic symptoms. Psychosom. Med. **64**(2), 258–266 (2002)
25. Eysenck, M., Derakshan, N.: New perspectives in attentional control theory. Personality Individ. Differ. **50**(7), 955–960 (2011)
26. Kudielka, B.M., Hellhammer, D.H., Wüst, S.: Why do we respond so differently? reviewing determinants of human salivary cortisol responses to challenge. Psychoneuroendocrinology **34**, 2–18 (2009)
27. Gerjets, P., Walter, C., Rosenstiel, W., Bogdan, M., Zander, T.O.: Cognitive state monitoring and the design of adaptive instruction in digital environments: lessons learned from cognitive workload assessment using a passive brain-computer interface approach. Front. Neurosci. **8**, 385 (2014)
28. Pesle, F., Lespinet-Najib, V., Garcia, L., Bougard, C., Diaz, E., Schneider, S.: Profils psychologiques et styles de conduite: Analyse exploratoire multidimensionnelle de la vulnérabilité au stress des conducteurs. ERGO'IA 2018, Bidart, France. Hal-01882632 (2018)
29. Mendl, M.: Performing under pressure: stress and cognitive function. Appl. Anim. Behav. Sci. **65**(3), 221–244 (1999)
30. Mühl, C., Jeunet, C., Lotte, F.: EEG-based workload estimation across affective contexts. Front. Neurosci. **8**, 114 (2014)
31. Zurrón, M., Lindín, M., Galdo-Alvarez, S., Díaz, F.: Age-related effects on event-related brain potentials in a congruence/incongruence judgment color-word Stroop task. Front. Aging Neurosci. **6**, 128 (2014)

# Random-Weights Neural Networks

# Echo State Networks with Artificial Astrocytes and Hebbian Connections

Peter Gergel'(✉) and Igor Farkaš

Faculty of Mathematics, Physics and Informatics, Comenius University in Bratislava, Mlynská dolina, 84248 Bratislava, Slovak Republic
{peter.gergel,farkas}@fmph.uniba.sk
http://cogsci.fmph.uniba.sk

**Abstract.** For the last few decades, the neuroscientific research has highlighted the importance of astrocytes, a type of glial cells, in the information processing capabilities. By dynamic bidirectional communication with neurons, astrocytes regulate their excitability through a variety of mechanisms. Traditional artificial neural networks (ANNs) are connectionist models that describe how information passes throughout layer of neurons abstracting from low-level mechanisms. However, very little research has addressed artificial astrocytes and their incorporation into ANNs. In this paper, we present an echo state network (ESN) extended with astrocytes which influence the neurons by fixed or Hebbian connections. By systematic analysis we investigate their role on five classification tasks and show that they can outperform the standard ESN without astrocytes. Although the model with fixed astrocytic weights yields from none to little improvement, the model with Hebbian weights from astrocytes to neurons is significantly superior.

**Keywords:** Glial cells · Astrocytes · ESN · Classification · Computational model

## 1 Introduction

Firstly identified in the 19th century, glial cells (often called glia) significantly contribute to the total brain mass with around 50% and their glia:neuron ratio in mammalian brains about 1:1 [3]. Although considered as non-functional and supportive units for over more than a decade, recently they gained lots of attention, as the emerging evidence indicates their role as active and equally important components compared to neurons. By interacting and cooperating with each other in nervous systems, they both take a significant part in various neurophysiological processes. Several functions of the glia are well characterized, including maintenance of homeostasis, being inevitable in the development of the central and peripheral nervous system, forming structural foundations that hold neurons, providing metabolic support and so on. However, the full characterization of their active roles and exact mechanisms still remain unresolved.


I. Rojas et al. (Eds.): IWANN 2019, LNCS 11506, pp. 457–466, 2019.
https://doi.org/10.1007/978-3-030-20521-8_38

The population of glia is commonly subdivided into four major groups: (1) astrocytes, (2) oligodendrocytes, (3) microglia and (4) their progenitors NG2-glia. According to recent evidence, astrocytes, the most abundant and probably the most complex group, play significant role in cognitive functions, traditionally attributed solely to neurons, such as learning and memory, information transfer and processing [2,8]. Although not being able to be excited electrically and generating action potentials as neurons do, they are incorporated in network *glial syncytium* where upon being excited they propagate $Ca^{2+}$ signals throughout gap junctions. They are characterized by having the resting membrane potential of $\sim -80\,\mathrm{mV}$ close to Nernst equilibrium for potassium ions ($K^+$) [5] and express both ionotropic and metabotropic glutamate receptors [19] allowing them to be highly sensitive to neuronal activity.

In order to better understand these low-level mechanisms, computational modelling is often employed which recently has become an essential part of neuroscience. Such models may provide testable predictions for processes that are built upon these mechanisms such as neuronal regulation, or synaptic plasticity. A better knowledge about astrocyte–neuron cooperation may also provide building blocks for studying the regulatory capability of glial syncytium on a larger scale. Computational models of ANNs extended with astrocytes may not only be used as an interesting novel concept, but mainly can provide space for hypotheses to explain the potential roles of glia in biological neuronal circuits and networks.

In the previous study, we investigated the role of astrocytes as neuronal regulators in a feedforward ANN [7]. The proposed models were superior to the traditional multi-layer perceptron on the same datasets, however, this was not the case for all of them. In addition, we detected unique astrocytic regimes in terms of output distributions that were different for each problem. In this paper, we transfer the same model of astrocytes to ESNs and systematically investigate their role using five classification tasks from UCR time series classification archive [4]. In addition, we incorporate and analyze Hebbian learning as a form of plasticity for astrocytic weights. The paper is organized as follows. Section 2 includes the related background and work. In Sect. 3, we describe proposed model in depth. In Sect. 4, we provide the experimental results. Section 5 concludes the paper.

## 2 Related Background

Computational neuroscience distinguishes two modeling paradigms: biophysical and connectionist. While the former focuses on physical and chemical properties of a biological system using various mathematical methods, the latter makes a significant reduction in the complexity of low-level mechanisms which in turn may lead to better comprehension of the system from a higher level. Despite the plethora of biophysical models of astrocytes per se and neuron-astrocyte coupling, connectionist modeling has so far been out of scientific interest. For an overview of biophysical models, we recommend reviews in [18,21,22].

As far as the connectionist models are concerned, we highlight the work by Ikuta et al. who initially introduced astrocytes[1] into ANN [10]. Their model of an astrocyte served as a chaotic noise generator which was being propagated to the neighboring units and impacted standard neuronal signalling. On the two-spirals classification problem, the performance in terms of mean absolute error and the rate of convergence was superior to the multilayer perceptron (MLP) without astrocytes. The architecture of the model can be described as an extension of MLP with astrocytes regulating neurons on the hidden layer. The authors in their later work investigated various activation functions for astrocytes and various architectures [11–13].

Instead of the neuronal regulation, Porto-Pazos et al. [20] and González et al. [1] focused on synaptic plasticity modulation. They presented an MLP with astrocytes regulating neural transmission on a larger temporal scale (hundreds of milliseconds and seconds) as opposed to fast neuronal and synaptic signaling (milliseconds). Astrocytes were activated by intense neural transmission and consequently regulated synaptic weights with a slow temporal time course. Each neuron was paired with a single astrocyte and each astrocyte only responded to the activity of the associated neuron by modulating its output synaptic weights. Since the model was dependent on various hyperparameters that needed to be fine-tuned, in their following work they presented a method for automatic search of these parameters based on cooperative coevolution [16].

## 3 Proposed Model and Training Methods

Here we present a novel neural network architecture based on an ESN with the reservoir extended with astrocytes. We first provide a brief overview of ESNs, the training procedure including weights initialization, and then describe the architecture of our model, parameter selection and incorporation of Hebbian learning.

### 3.1 Echo State Networks

Training traditional recurrent neural networks is considered to be difficult because of limitations of gradient descent methods which tends to be computationally expensive, to have slow convergence and to generally lead to poor local minima. Hence, the full adaption of all network weights is often omitted, yet still yielding excellent performance. This approach serves as a foundation for ESNs which were introduced by Jaeger for nonlinear system identification and time series modeling [14]. They are characterized by having randomly generated input weights and reservoirs with the training only on readout weights. However, in order to work

[1] Originally, authors use term *artificial glia* but we consider *artificial astrocytes* instead, since glia represent the vast majority of non-neuronal cells in the nervous system with multiple functions, whereas only astrocytes are currently considered to play a vital role in information processing tasks.

well, ESNs require delicate tuning of several hyperparameters including the size of the reservoir $N$, the spectral radius $\rho$, and input weight scaling $\tau$.

Reservoir activation vectors $\boldsymbol{x}(t) = [x_1(t), ..., x_N(t)]$ and output activations $\boldsymbol{y} = [y_1, ..., y_C]$ for given input pattern $\boldsymbol{u} = [u(1), ..., u(T)]$ are updated according to ESN dynamics given by the formulas

$$\boldsymbol{x}(t) = f(\boldsymbol{w}^{\text{in}}u(t) + \boldsymbol{W}^{\text{res}}\boldsymbol{x}(t-1)) \tag{1}$$

$$\boldsymbol{y}(t) = \boldsymbol{W}^{\text{out}}\boldsymbol{x}(T) \tag{2}$$

with the logistic sigmoid activation function $f(net) = 1/(1 + \exp(-net))$.

For dealing with classification problems, we consider the following training procedure:

1. Generate random input weights $\boldsymbol{w}^{\text{in}}$ and reservoir weights $\boldsymbol{W}^{\text{res}}$ scaled by $\rho/|\lambda_{\max}|$, where $\lambda_{\max}$ denotes the largest absolute eigenvalue of $\boldsymbol{W}^{\text{res}}$ and $\rho$ is manually selected.
2. Run ESN using the training inputs and for each $\boldsymbol{u}_{train}$ collect the last reservoir activation state $\boldsymbol{x}(T)$.
3. Compute the linear readout weights using formula
$$\boldsymbol{W}^{\text{out}} = \boldsymbol{Y}^{\text{tgt}}\boldsymbol{X}^{+} \tag{3}$$
where $\boldsymbol{Y}^{\text{tgt}}$ is a matrix of concatenated target vectors (in columns) with one-hot encoding and $\boldsymbol{X}^{+}$ is the pseudoinverse matrix of concatenated reservoir activation states from step 2.
4. Use the trained network on new input data $\boldsymbol{u}_{test}$ and decide the class by selecting output neuron with maximum activation
$$\text{class}(\boldsymbol{u}_{test}) = \arg\max_k y_k \tag{4}$$

### 3.2 Neuron–Astrocyte Coupling – A-ESN Model

Although astrocytes are interconnected within glial syncytium using gap junctions and communicate sharing slow $Ca^{2+}$ signals (as opposed to neuronal firing), we omit this concept for the sake of complexity and start exploring the simplest model of astrocytes. Upon investigating whether they work relatively well and produce favorable results, in the future research we plan to explore more complex and biologically plausible mechanisms. For now, we focus on this simple model.

In our model we consider an astrocyte to play a single role in neuronal regulation. Since it was discovered that mammalian cortices have glia:neuron ratio of about 1:1, as stated in the introduction, in the context of ESN we equip each reservoir neuron with one astrocyte as shown in Fig. 1. We call this model **A-ESN**.

Reservoir activation $x'_i(t)$ takes into account input pattern $u(t)$, previous time step activation vector $\boldsymbol{x}'(t-1)$ and astrocyte activation $\psi_i(t)$ weighted by a single shared weight $w^{\alpha}$, which is expressed in the vector form as

$$\boldsymbol{x}'(t) = f(\boldsymbol{w}^{\text{in}}u(t) + \boldsymbol{W}^{\text{res}}\boldsymbol{x}'(t-1) + w^{\alpha}\boldsymbol{\psi}(t)) \tag{5}$$

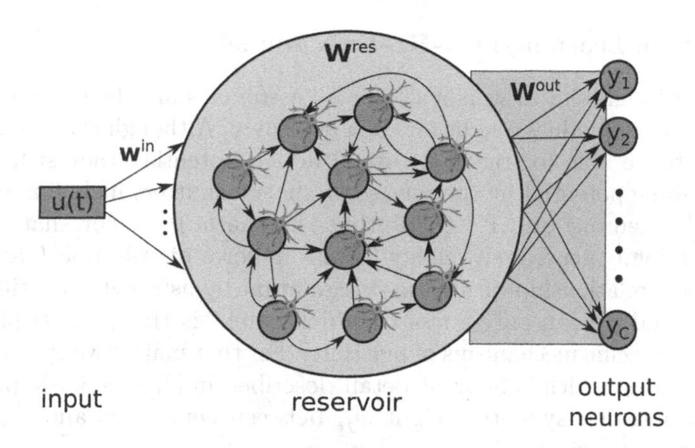


**Fig. 1.** The architecture of the proposed model, A-ESN, with a reservoir of neurons and astrocytes. Each neuron is paired with an astrocyte that listens to it and regulates its behaviour based on activity. Since we consider the single time series classification problems, we use a single input neuron, $N$ neurons and astrocytes within the reservoir, and $C$ output neurons representing the classes.

Astrocytes $\psi_i(t)$ listen to their associated neurons and when some of the neurons exceed the threshold $\theta$, astrocytes produce the activation value of 1. The rest of them decays by factor $\gamma$, as shown in Eq. 6.

$$\psi_i(t) = \begin{cases} 1, & \text{if } \theta < x'_i(t-1) \\ \gamma\psi_i(t-1), & \text{otherwise} \end{cases} \tag{6}$$

This ESN dynamics is graphically depicted in Fig. 2.

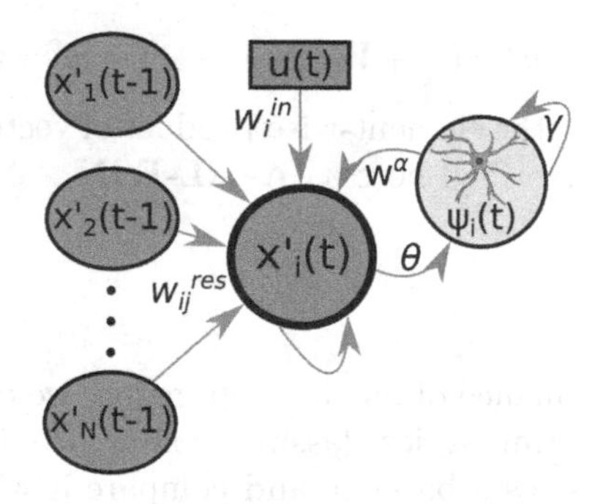


**Fig. 2.** Neuron–astrocyte coupling. The astrocyte, weighted by $w^{\alpha}$, regulates the associated neuron by contributing to its input. When the neuron surpasses the threshold $\theta$, the astrocyte outputs 1 and slowly decays by $\gamma$ in the next time steps. Blue arrows depict reservoir weights, the green arrow an input weight and orange arrows the astrocyte parameters. (Color figure online)

### 3.3 Hebbian Learning in A-HL-ESN Model

Since using a single shared weight $w^\alpha$ for all astrocytes may be too constraining, we consider an individual weight for each astrocyte. Although astrocytes are not considered to be able to trigger neuronal action potential, they still modulate their membrane potential by the release of gliotransmitters including glutamate (exciting the neuron) or ATP (inhibiting the neuron) [6]. For that reason we consider randomly generated weights from a uniform distribution $Uni(-1, 1)$.

The exact relationship of neuronal regulation by astrocytes is still not well understood and we can only guess to which extent is this process plastic and what are the specific mechanisms of plasticity. For that matter we speculate using Hebbian learning which is in great detail described in [9]. The basic principle is that the change of a synaptic weight $w_{ji}$ between neurons $x_i$ and $y_j$, with the learning rate $\eta$, is expressed as

$$\Delta w_{ji}(t) = \eta x_i(t) y_j(t) \tag{7}$$

In our case we apply this rule for the change of the weight $w^\alpha$ between a neuron $x'_i$ and an astrocyte $\psi_i$. Repeated application, however, may lead to an exponential change of the weight which is not biologically plausible, so this is solved by incorporating some form of stabilization. This is in many cases the normalization of the final weights. We consider Oja's rule [17] which introduces a nonlinear, forgetting factor for the weight change

$$\Delta w_i^\alpha(t+1) = \eta x'_i(t)[\psi_i(t) - x'_i(t) w_i^\alpha(t)] \tag{8}$$

To take into account this new dynamics, we split our training algorithm into two phases: (1) once the unsupervised learning of the weights $\boldsymbol{w}^\alpha$ (Eq. 8) in the reservoir is complete, (2) a supervised learning algorithm (from Sect. 3.1) is applied to the readout weights. Instead of using Eq. 5 for the reservoir update, we consider

$$\boldsymbol{x}'(t) = f(\boldsymbol{w}^{\text{in}} u(t) + \boldsymbol{W}^{\text{res}} \boldsymbol{x}'(t-1) + \boldsymbol{w}^\alpha * \boldsymbol{\psi}(t)) \tag{9}$$

with operator '$*$' denoting the element-wise product of vectors. We call the model with Hebbian learning described here as **A-HL-ESN**.

## 4 Experiments

For evaluating the performance of our new approach, we consider five classification problems from UCR time series classification archive [4]. We use a standard ESN (without astrocytes) as a baseline and compare it with proposed methods described in the previous section. Using grid search we systematically investigated each hyperparameter (averaged over several instances) and selected the values with the lowest error rate on the testing dataset. Regarding the ranges for each hyperparameter we chose the values presented in Table 1.

**Table 1.** Hyperparameter value ranges used in the grid search for each dataset.

| Parameter | Tested values |
|---|---|
| $N$ | 20 to 500 with step = 20 |
| $\tau$ | 10, 5, 1, 0.5, 0.1, 0.05, 0.01, 0.001, 0.0001 |
| $\rho$ | 0.8 to 1.4 with step = 0.05 |
| $w^{\alpha}$ | −1.0 to 1.0 with step = 0.1 |
| $\gamma$ | 0.0 to 1.0 with step = 0.1 |
| $\theta$ | 0.0 to 1.0 with step = 0.1 |

The UCR archive already provides train/test split of the datasets, but we found this rather problematic because of the high risk of overfitting the hyperparameters to a particular test dataset. In order to avoid this, we merged both train and test datasets into a single set and used 5-fold cross-validation instead. To eliminate the random fluctuation in performance, we executed training procedures with random weights, random permutations of datasets and averaged error rates over 100 instances.

Allowing for possibility of imbalanced datasets in which one class is over-represented with the respect to the others, we use *Matthews correlation coefficient* (MCC) as a metrics for performance evaluation score [15] rather than the mean-squared error, accuracy or F1-score which does not work relatively well on imbalanced datasets. The value MCC = 1 corresponds to a perfect match between model predictions and observations, whereas −1 indicates total disagreement between the two.

In all experiments, we used hyperparameters summarized in Table 2 resulting in the largest MCC on testing datasets.

**Table 2.** Optimal hyperparameters selected using the grid search for each dataset. Non-astrocytic hyperparameters ($N$, $\rho$, $\tau$) were shared in all models on a given dataset.

| Dataset | ESN | | | A-ESN | | | A-HL-ESN | |
|---|---|---|---|---|---|---|---|---|
| | $N$ | $\rho$ | $\tau$ | $w^{\alpha}$ | $\gamma$ | $\theta$ | $\gamma$ | $\theta$ |
| FaceFour | 20 | 0.95 | 0.05 | −0.4 | 0.1 | 0.2 | 0.2 | 0.2 |
| MoteStrain | 120 | 1.3 | 0.001 | −0.3 | 0.9 | 0.5 | 0.3 | 0.3 |
| OSULeaf | 60 | 0.95 | 0.01 | 0.6 | 0.6 | 0.8 | 0.2 | 0.1 |
| SwedishLeaf | 160 | 1.4 | 0.001 | −0.4 | 0.2 | 0.9 | 0.2 | 0.2 |
| ToeSegmentation1 | 60 | 1.3 | 1.0 | −0.5 | 1.0 | 0.9 | 0.3 | 0.1 |

Results in terms of MCC averaged over 100 simulations are presented in Table 3. It is clear that model with Hebbian connections, A-HL-ESN, significantly outperforms models ESN and A-ESN. Despite having more complex training procedure and thus higher time complexity, gain in terms of performance is clearly notable. Model with fixed connections, A-ESN, have yielded results equivalent to standard ESN (assuming correct settings of hyperparameters), although it is speculative why on the last dataset (ToeSegmentation1), the error rate is significantly better (MCC of $0.5 \pm 0.1$ vs $0.32 \pm 0.11$).

**Table 3.** MCC (mean + standard deviation) averaged over 100 simulations on each dataset. In each case, the model A-HL-ESN is superior regarding the performance.

| Dataset | ESN | A-ESN | A-HL-ESN |
|---|---|---|---|
| FaceFour | $0.44 \pm 0.12$ | $0.43 \pm 0.13$ | $0.56 \pm 0.14$ |
| MoteStrain | $0.65 \pm 0.04$ | $0.67 \pm 0.06$ | $0.85 \pm 0.03$ |
| OSULeaf | $0.41 \pm 0.06$ | $0.42 \pm 0.06$ | $0.57 \pm 0.06$ |
| SwedishLeaf | $0.64 \pm 0.03$ | $0.63 \pm 0.03$ | $0.84 \pm 0.03$ |
| ToeSegmentation1 | $0.32 \pm 0.11$ | $0.50 \pm 0.10$ | $0.59 \pm 0.11$ |

In order to better understand the role of astrocytes with Hebbian connections, we were interested to know how the astrocyte weights develop during learning. For the fully trained models (all 100 instances), we plotted final distributions of the weights $w_i^\alpha$ as depicted in Fig. 3. We can observe that the weight distributions are skewed in the interval (1,2), roughly independent of the dataset, with an exception being MoteStrain, where some of the weights are also between 0 and 1. We may conclude this implies excitatory nature of the astrocytes in terms of neuronal regulation.

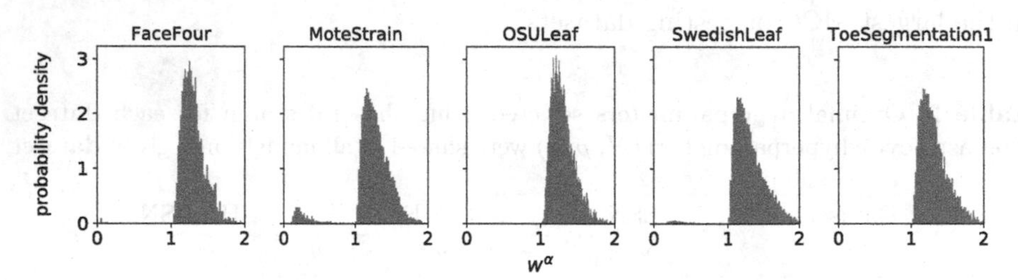


**Fig. 3.** Distribution of the weights $w_i^\alpha$ in the trained models A-HL-ESN reveals excitatory role of the astrocytes.

## 5 Conclusion

To advance the modeling of biological neuronal networks, which are inherently recurrent, it is inevitable to focus on models from the same domain. Since training recurrent neural networks is difficult for various problems, we considered ESNs instead. Moroever, the neuroscientific research for the last decades has highlighted the importance of glial cells in information processing context. Astrocytes regulate neuronal functionality in a variety of ways, particularly by maintaining the concentration of ions and neurotransmitters and by releasing gliotransmitters, and modulating both neuronal excitability and synaptic plasticity. However, limited amount of research has been done in the field of ANNs equipped with artificial astrocytes and basically none on the recurrent models, so the exact role of astrocytes remains speculative.

In our previous work [7], we investigated the role of astrocytes in feedforward models and inspired by positive results, we transferred the same model of astrocytes to ESNs and explored their influence. In addition, we incorporated Hebbian learning for weights between astrocytes and their associated neurons. By systematic analysis of this new dynamics on five classification tasks we found very little contribution of astrocytes with fixed weights, but in case of Hebbian learning the performance yielded significantly positive outcome. By analyzing the final distributions of astrocyte weights, we discovered that astrocytes operate as neuronal excitors by lowering the threshold required for firing. Out of curiosity we also examined various modifications including bipolar activation functions for neurons and astrocytes (with their output activation within the interval $(-1, 1)$) and swapping astrocytes with neurons in Eq. 8. However, these modifications did not perform that well.

Future research in this area may follow several directions. The activation function for the astrocyte, as formulated in Eq. 6, is definitely not the only one and there are several varieties to be considered. Since $Ca^{2+}$ signalling within glial syncytium operates on a much slower pace as opposed to neuronal firing, it may be beneficiary to incorporate this slow, temporal dynamics into astrocytic behaviour. Although our model of an artificial astrocyte includes slow decay, "firing", however, remained still instant. Despite focusing on the astrocytes as single separate units, it is possible to model glial syncytium and design an astrocytic network of astrocytes connected together, hence fulfilling the biologically plausible spatiotemporal dynamics. Last, but not least, instead of focusing on regulation of neuronal excitability, it is possible to design models that also incorporate rules for synaptic plasticity.

**Acknowledgments.** This work was supported by grant UK/250/2019 from Comenius University in Bratislava (P.G.) and Slovak Grant Agency for Science, project VEGA 1/0796/18 (I.F.).

## References

1. Alvarellos-González, A., Pazos, A., Porto-Pazos, A.B.: Computational models of neuron-astrocyte interactions lead to improved efficacy in the performance of neural networks. Comput. Math. Methods Med. 2012, 10 pages (2012)
2. Alvarez-Maubecin, V., García-Hernández, F., Williams, J.T., Van Bockstaele, E.J.: Functional coupling between neurons and GLIA. J. Neurosci. **20**(11), 4091–4098 (2000)

3. Azevedo, F.A., et al.: Equal numbers of neuronal and nonneuronal cells make the human brain an isometrically scaled-up primate brain. J. Comp. Neurol. **513**(5), 532–541 (2009)
4. Chen, Y., et al.: The UCR time series classification archive (2015). www.cs.ucr.edu/~eamonn/time_series_data/
5. Dallérac, G., Chever, O., Rouach, N.: How do astrocytes shape synaptic transmission? insights from electrophysiology. Front. Cell. Neurosci. **7**, 159 (2013)
6. Fellin, T., Pascual, O., Haydon, P.G.: Astrocytes coordinate synaptic networks: balanced excitation and inhibition. Physiology **21**(3), 208–215 (2006)
7. Gergel', P., Farkaŝ, I.: Investigating the role of astrocyte units in a feedforward neural network. In: Kurková, V., Manolopoulos, Y., Hammer, B., Iliadis, L., Maglogiannis, I. (eds.) ICANN 2018. LNCS, vol. 11141, pp. 73–83. Springer, Cham (2018). https://doi.org/10.1007/978-3-030-01424-7_8
8. Haydon, P.G.: Neuroglial networks: neurons and glia talk to each other. Curr. Biol. **10**(19), R712–R714 (2000)
9. Hebb, D.O.: The Organization of Behavior: A Neuropsychological Theory. Wiley, New York (1949)
10. Ikuta, C., Uwate, Y., Nishio, Y.: Chaos glial network connected to multi-layer perceptron for solving two-spiral problem. In: Proceedings of 2010 IEEE International Symposium on Circuits and Systems, pp. 1360–1363 (2010)
11. Ikuta, C., Uwate, Y., Nishio, Y.: Multi-layer perceptron with impulse glial network. In: IEEE Workshop on Nonlinear Circuit Networks, pp. 9–11 (2010)
12. Ikuta, C., Uwate, Y., Nishio, Y.: Performance and features of multi-layer perceptron with impulse glial network. In: International Joint Conference on Neural Networks, pp. 2536–2541 (2011)
13. Ikuta, C., Uwate, Y., Nishio, Y.: Multi-layer perceptron with positive and negative pulse glial chain for solving two-spirals problem. In: International Joint Conference on Neural Networks, pp. 1–6 (2012)
14. Jaeger, H.: The "echo state" approach to analysing and training recurrent neural networks-with an erratum note. Bonn, Ger.: Ger. Nat. Res. Cent. Inf. Technol. GMD Tech. Rep. **148**(34), 13 (2001)
15. Matthews, B.W.: Comparison of the predicted and observed secondary structure of t4 phage lysozyme. Biochimica et Biophysica Acta (BBA)-Protein Struct. **405**(2), 442–451 (1975)
16. Mesejo, P., Ibánez, O., Fernández-Blanco, E., Cedrón, F., Pazos, A., Porto-Pazos, A.B.: Artificial neuron-glia networks learning approach based on cooperative coevolution. Int. J. Neural Syst. **25**(04), 1550012 (2015)
17. Oja, E.: Simplified neuron model as a principal component analyzer. J. Math. Biol. **15**(3), 267–273 (1982)
18. Oschmann, F., Berry, H., Obermayer, K., Lenk, K.: From in silico astrocyte cell models to neuron-astrocyte network models: a review. Brain Res. Bull. **136**, 76–84 (2018)
19. Porter, J.T., McCarthy, K.D.: Astrocytic neurotransmitter receptors in situ and in vivo. Prog. Neurobiol. **51**(4), 439–455 (1997)
20. Porto-Pazos, A.B., et al.: Artificial astrocytes improve neural network performance. PloS ONE **6**(4), e19109 (2011)
21. Volman, V., Bazhenov, M., Sejnowski, T.J.: Computational models of neuron-astrocyte interaction in epilepsy. Front. Comput. Neurosci. **6**, 58 (2012)
22. Wade, J., Kelso, S., Crunelli, V., McDaid, L.J., Harkin, J.: Biophysically based computational models of astrocyte-neuron coupling and their functional significance. Frontiers E-books (2014)

# Multiple Linear Regression Based on Coefficients Identification Using Non-iterative SGTM Neural-like Structure

Ivan Izonin[1(✉)], Roman Tkachenko[1], Natalia Kryvinska[2,4], Pavlo Tkachenko[3], and Michal Greguš ml.[4]

[1] Lviv Polytechnic National University, Lviv, Ukraine
ivanizonin@gmail.com, roman.tkachenko@gmail.com

[2] University of Vienna, Vienna, Austria
natalia.kryvinska@univie.ac.at

[3] IT STEP University, Lviv, Ukraine
pavlo.tkachenko@gmail.com

[4] Comenius University in Bratislava, Bratislava, Slovakia
Michal.Gregusml@fm.uniba.sk

**Abstract.** In the paper, a new method for solving the multiple linear regression task via a linear polynomial as a constructive formula is proposed. It is based on the use of high-speed SGTM Neural-Like Structure. This linear non-iterative computational intelligence tool is used for identification of polynomial coefficients. As a result of the implementation of the learning algorithm and applied the matrix of test signals to the trained SGTM, the identification of the linear polynomial coefficients is carried out. A further solution of the task occurs by searching a dependent variable using the obtained polynomial. The results of the method have been tested on the task of the output of the electric power prediction of the combined-type factory. The method ensures the identification of five polynomial's coefficients at the high speed, which ensures high accuracy of the solution. Based on the comparison with known regression analysis methods, the highest accuracy of the work has been established. The transition from neural-like structure to the solution of the task in the form of a linear polynomial provides the possibility for the simple interpretation of the result of the regression or classification tasks. That does not require high qualifications from the user. In addition, the developed method, based on the repetition of training outcomes and the lack of debugging and parameter selection procedures, allows synthesizing linear polynomial for complex models that use various non-linear extensions of SGTM inputs while preserving the accuracy of their operation. The proposed approach can be used in the fields of medicine, economics, materials science, service sciences etc., for fast and accurate solution of regression or classification tasks with the possibility of easy interpretation of the result.

**Keywords:** Multiple regression · Linear precision model · Fast training · Coefficients identification · Neural-like structures · SGTM


I. Rojas et al. (Eds.): IWANN 2019, LNCS 11506, pp. 467–479, 2019.
https://doi.org/10.1007/978-3-030-20521-8_39

# 1 Introduction

An identification and mathematical description of the dependence of the dependent variable on several factors is a fairly widespread task in various branches of industry [1]. In most cases, almost all of the factorial attributes depend on each other, which greatly complicates the construction of the multiple regression equation [2]. In addition, existing methods, in particular, the method for selecting different values or the least squares method, are labor-intensive and time-consuming. The problem is deepened in the case of processing large volumes of data with a large number of attributes. In this case, the model is difficult to implement. Also, it requires significant time expenditures [3]. The application of computational intelligence methods, in particular, non-iterative [4], can provide a fast and accurate solution of this task.

# 2 Review and Problem Statement

Multiple regression based on the least squares method, as already pointed out above requires large time expenses to find the coefficients of the model. In [5] the hybrid scheme of the regression task solution with the use of nonparametric clustering method and multilayer perceptron is presented. The application of this approach increases the accuracy of the model, but significantly decrease the time of its training. Paper [6] describes the prediction method using GRNN. Despite the satisfactory results of solving the stated in [6] task, the model based on this type of neural networks may appear to be slow and large, especially in the case of the large volumes of data processing. Authors in [7] improved the properties of GRNN-based systems, in particular by applying an iterative learning algorithm to reduce the amount of memory that is characteristic of the GNNR. In [8] three different regression models based on SVM have been developed. Despite their satisfactory time characteristics, the SVR's does not always provide a fairly accurate solution. All described methods have several advantages and disadvantages, but the fast and accurate search of multiple regression coefficients is an actual task [2]. In addition, in particular, in the field of medicine, is an important interpretation of the results of one or another method by the user without high qualification in the intelligent data analysis. In this paper, we propose a new non-iterative algorithm for identifying multiple regression coefficients based on SGTM neural-like structure for the case of large volumes of data processing.

# 3 Linear Non-iterative SGTM Neural-like Structure

## 3.1 Topologies and Basic Provisions

The topology (Fig. 1) of the SGTM neural-like structure (Successive Geometric Transformations Model) reflects the algorithm of successive geometric transformations in the $n$-dimensionality space of implementations of vectors that realizes non-iterative training and the application modes of this model ($n$ is the number of variables) [4].

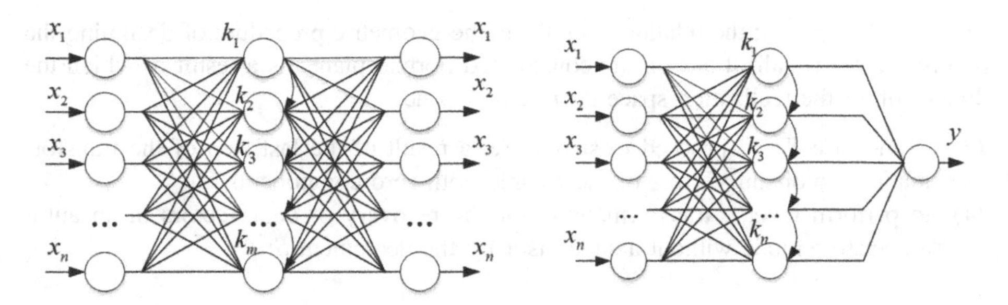


**Fig. 1.** SGTM neural-like structure topologies: (a) auto-associative mode, (b) supervised mode.

The sequence of step-by-step learning and the application procedures of the neural-like structure of the SGTM are described in [4, 9]. In this paper, we will demonstrate only the basic mathematical expressions of this procedure for giving it a detailed geometric interpretation.

Let us suppose, that $X_{i,j}^{(1)}$ is the an arbitrary input component of the vector that belonging to the training sample, $i = \overline{1,N}$ $j = \overline{1,n}$, where $i$ is the number of the observation and $j$ is the number of the attribute. After setting the value of the first step of the transformation S = 1, the further training process for Fig. 1a requires the execution of such steps [4, 9]:

(1) to choose $X_{b,j}^{(S)}$ - the base line-point that corresponds to the vector with the highest norm, where $j = \overline{1,n}$;
(2) to do the calculation:

$$K_i^{(S)} = \frac{\sum_{j=1}^{n} (X_{i,j}^{(S)} . X_{b,j}^{(S)})}{\sum_{j=1}^{n} (X_{b,j}^{(S)})^2} \tag{1}$$

for $i = \overline{1,N}$;

$$X_{b,j}^{(S)} = \frac{\sum_{i=1}^{N} (X_{j,i}^{T(S)} . K_i^{(S)})}{\sum_{i=1}^{N} (K_i^{(S)})^2} \tag{2}$$

for $j = \overline{1,n}$.

$$X_{i,j}^{(S+1)} = X_{i,j}^{(S)} - K_i^{(S)} . X_{b,j}^{(S)} \tag{3}$$

for $i = \overline{1, N}, j = \overline{1, n}$; the relation (3) reflects the geometric procedure of designing the point-vectors of realizations on the constructed normal plane, as a result of which the dimension of the realization space decreases by one.

(3) to increase $S$ and proceed to step 1. As a result of the last step of the transformation we obtain vectors of the sample with zero components.
(4) to perform inverse transformations for the recovery of data vectors in an auto-associative mode without a supervisor by the decrement $S$:

$$X_{i,j}^{(S)} = X_{i,j}^{S+1} + K_i^{(S)} \times X_{b,j}^{(S)} \tag{4}$$

for $S = \overline{S_{\max}, 1}$, where $X_{i,j}^{S\max+1} = \|0\|$, that is, non-iterative, fast SGTM algorithm is symmetric in training and application modes.

In supervised mode, (1) can be represented as an approximation of dependence [4, 10]:

$$K_i^{(S)} = f(\overline{K_i^{(S)}}). \tag{5}$$

where for the linear neural-like structure, the activation function is calculated by the formula [4]:

$$K_i^{(S)} = \alpha \times \overline{K_i^{(S)}}, \tag{6}$$

where coefficient $\alpha$ is determined by the least squares criterion.

### 3.2 The Geometric Interpretation of the SGTM Training Procedure

The geometric interpretation of the SGTM training procedure for a three-dimensional implementation space is demonstrated in Fig. 2. It can be extended to arbitrary dimensional space options. The training matrix is:

$$\left\|X_{i,j}^{(0)}\right\| = \begin{Vmatrix} x_{11} & x_{12} & x_{13} \\ x_{21} & x_{22} & x_{23} \\ \cdots & \cdots & \cdots \\ x_{i1} & x_{i2} & x_{i3} \\ x_{m1} & x_{m2} & x_{m3} \\ x_{p1} & x_{p2} & x_{p3} \\ x_{N1} & x_{N2} & x_{N3} \end{Vmatrix}. \tag{7}$$

The point of a three-dimensional implementation space corresponds to each line of the matrix $\left\|X_{i,j}^{(0)}\right\|$.

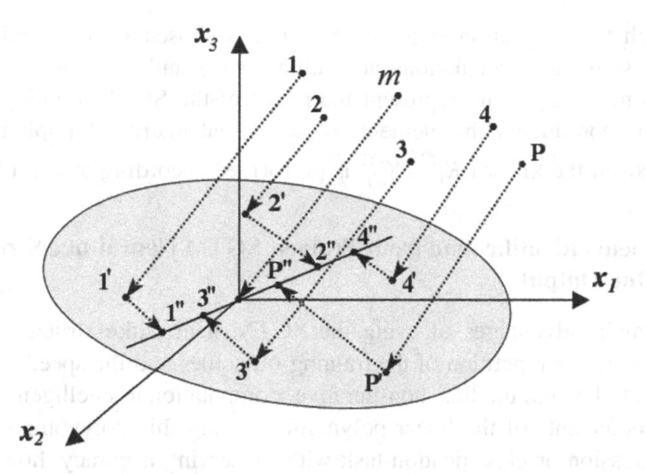


**Fig. 2.** Space of the implementation for the SGTM neural-like structure.

Let us suppose, that in the first training step, the base line with the number $m$ was chosen. In the first training step (S = 1) we make the plane through the origin of the coordinates, perpendicular to the normal vector, which connects this point with the start of coordinates (Fig. 2). We have gotten the equation of the plane:

$$x_{m1} \cdot x_1 + x_{m2} \cdot x_2 + x_{m3} \cdot x_3 = 0. \tag{8}$$

The distance from an arbitrary point in a given space of realizations to the constructed plane is equal to the absolute value $K$, where:

$$K = \frac{x_{m1} \cdot x_1 + x_{m2} \cdot x_2 + x_{m3} \cdot x_3}{\sqrt{x_{m1}^2 + x_{m2}^2 + x_{m3}^2}}. \tag{9}$$

Analyzing the obtained expression, we conclude that the value $K$ corresponds to the magnitude of the coefficient $K_i^{(S)}$, which is calculated by (1), divided by the norm of the base vector. By the value $K$, it is possible to determine the relative position of the base and current points. Obviously, after the first training step (transformations based on (2), the base point $m$ is projected to the origin, and the points $(1, 2, 3, 4, P)$ are projected to the plane that constructed by the method which was described above. This corresponds to the transformation of the base line of the matrix into a zero line. During the next training step $(S = 2)$, geometric transformations are carried out in space, which dimensionality is reduced by one (in this case, on a plane). Then the chosen base point $P'$ is projected to the origin of the coordinates, and the points $(1', 2', 3', 4')$ are straight, perpendicular to the normal vector that connects the base point P' with the start of the coordinates. The lengths of the distance of points to the straight line brought to the length of the normal vector are equal to the value of the corresponding coefficient $K_i^{(S)}$. Finally, as a result of the last transformations step $(S = 3)$, the points $(1'', 2'', 3'', 4'')$ are projected to the origin, corresponding to the transformation of the implementations matrix to zero matrices. To provide a minimal residual dispersion of the transformed matrix at each transformation steps the baseline is chosen by performing additional transposition steps and calculations according to (1) and (2). The elements of the baselines are memorized and represent the result of the SGTM training. The reverse procedure for reproducing the elements of the initial matrix of implementations for Fig. 1b, based on the known $K_i^{(S)}$, $X_{b,j}^{(S)}$ is performed according to Eq. (4).

### 3.3 Coefficients Identification from Trained SGTM Neural-like Structure with One Output

Among the main advantages of using the SGTM neural-like structure to solve the different tasks are the repetition of the training outcomes and the speed of its work [9]. In addition, at the output, this non-iterative computational intelligence tool allows getting the coefficients of the linear polynomial. Using this polynomial it possible to solve the regression or classification task with preserving accuracy, however, without using a neural-like structure.

An algorithmic implementation of the method for identifying coefficients via SGTM with one output (Fig. 1b) involves performing the following procedures:

1. *Data preprocessing and training of SGTM neural-like structure.*
2. An important step at this stage is the compulsory procedure for the normalization of inputs $x_1, x_2, x_3, \ldots, x_n$ of the entire data sample prior to the implementation of the training procedure.
3. *The formation of the matrix of test signals.*

The dimensionality of the matrix of test signals *B* depends on the number *n* of input parameters of the task (the independent variables that used for the learning process). *B* is a square identity matrix E(dim(E) = $n \times n$) with an extra row $\Omega$ ($\Omega = \left\{ \underbrace{0, \ldots, 0}_{n} \right\}$) whose elements will be zeros: $B = \begin{pmatrix} \Omega \\ E \end{pmatrix}$. In the case of solving the stated task, where the number of independent variables is equal to four, the matrix of test signals will have the following form:

$$B = \begin{pmatrix} 0 & 0 & 0 & 0 \\ 1 & 0 & 0 & 0 \\ 0 & 1 & 0 & 0 \\ 0 & 0 & 1 & 0 \\ 0 & 0 & 0 & 1 \end{pmatrix} \tag{10}$$

4. *The formation of coefficients of the surface of response.*

5. To implement this step, we use the matrix of the test signals in the application mode of the SGTM neural-like structure. As a result, we get a set of coefficients of the surface of response $\overline{\beta}_t, t = 1, \ldots, n+1$. For $n = 4$ we have obtained:

$$\begin{pmatrix} \beta_1 \\ \beta_2 \\ \beta_3 \\ \beta_4 \\ \beta_5 \end{pmatrix} \tag{11}$$

6. *The formation of coefficients of the linear polynomial.*

   The coefficients $\theta_t$ of a linear polynomial are formed according to:

$$\forall t \in \{1, n+1\} : \theta_t = \begin{cases} \beta_1, & t = 1; \\ \beta_t - \beta_1, & t \neq 1. \end{cases} \tag{12}$$

As a result of the execution (12), it can be received a linear polynomial. It can be written as follows:

$$y = \theta_1 + \theta_2 x_1 + \theta_3 x_2 + \theta_4 x_3 + \theta_5 x_4 \tag{13}$$

Among the obvious advantages of using the polynomial (13) for the performance increase of the intelligent systems built on SGTM, as well as the simplification of their architecture [11]. In addition, at the input of the SGTM neural-like structure, it can be performed any nonlinear transformations of the primary inputs, and at the output, it can be synthesized the linear polynomial more often with a greater number of members from the initial [9].

# 4 Simulation Results

## 4.1 The SGTM-Based Neurocomputer

The uniformity of training algorithms and the use of the SGTM neural-like structures simplifies the task of both software and hardware implementation of similar neurocomputers [4]. A window of one of the author's neurocomputer programs for solving prediction and classification tasks, which was used for modelling, is shown in Fig. 3.

The program provides the high speed of the training and in-depth analysis of available data, including the selection of the principal components as well as the trend and independent fluctuations in time sequences, the visualization of the body of the information object in the coordinates of the three chosen principal components, etc. There are software implementations of the other functions of data processing, in particular, a fundamentally new method of filling the spaces and recovering data in tables and time sequences based on the procedure of the projection of point-realizations on the hyperplane of the object. This method essentially outperforms for the accuracy the

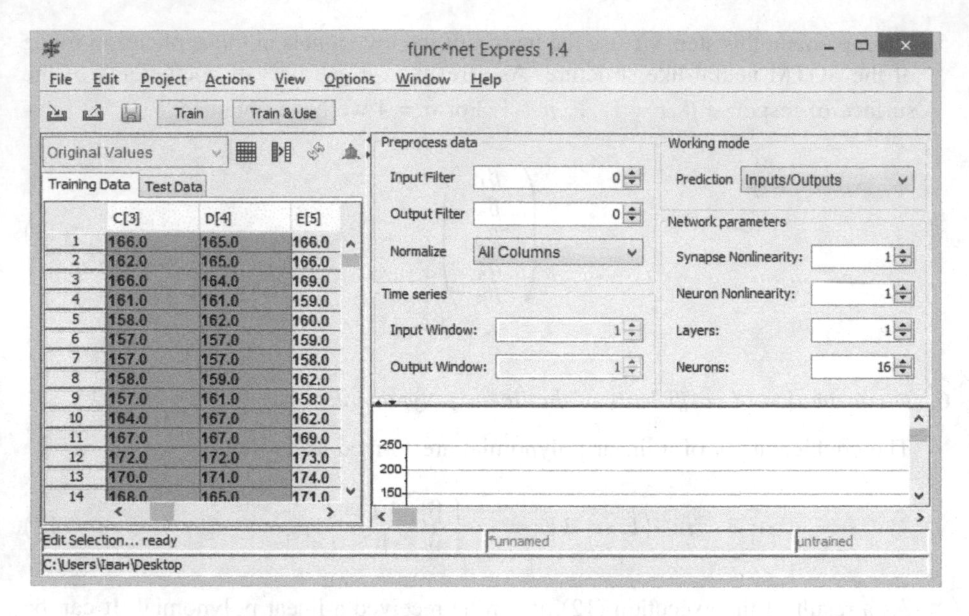


**Fig. 3.** Program window of the SGTM-based neurocomputer.

existing methods: substitutions by the average value, interpolation, and extrapolation, and can be the basis for designing the adaptive control models and autonomous intelligent agents. An important thing was the direction of constructing neuro-fuzzy structures of SGTM, which provide high accuracy and unambiguous of the dephasing function, the possibility of solving tasks of increased dimensionality [4].

### 4.2 Dataset Description

The simulation of the method was carried out using a dataset on the predicting of the electrical power outputs of the combining-type power factory. Dataset was taken from [12]. The authors of [13] collected the dataset from 1913 observations for the electrical power outputs of the combining-type power factory. Detailed characteristics of the dataset, as well as their statistical analysis, are given in [13]. Visualization of the dataset using [14] is shown in Fig. 4.

Dataset was randomly divided into training and test samples: 80% to 20% for the approbation of the method. It should be noted that in this paper authors used only one (first) of the five samples collected in [12, 13].

### 4.3 Obtained Results via SGTM Neural-like Structure

Experimental investigation on the application of the SGTM neural-like structure to the solution of the stated task in training and test modes shows the following results (Table 1). The main SGTM parameters are four neurons in the input and hidden layer, one output.

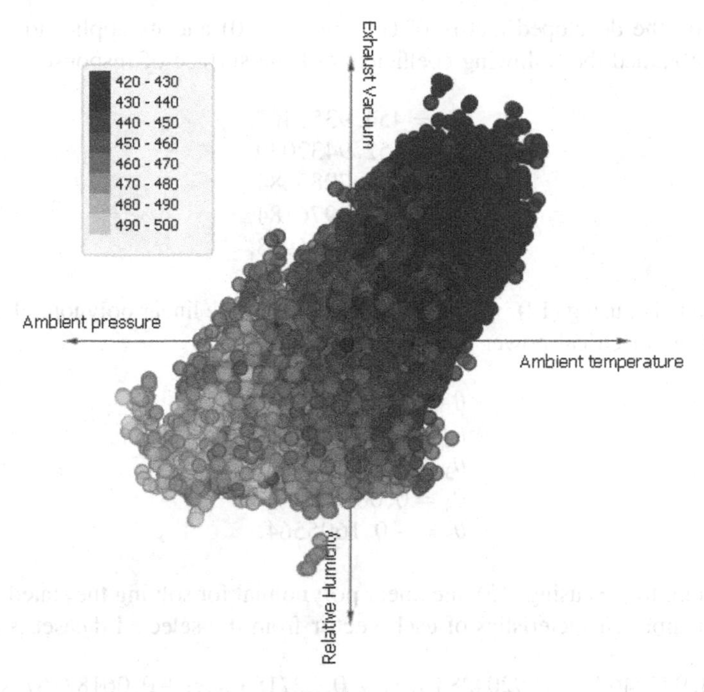


**Fig. 4.** Datasets visualization.

**Table 1.** Prediction results through the SGTM neural-like structure.

| MAPE, % | MAE | RMSE | Training time |
|---|---|---|---|
| Training mode | | | |
| 0,797253 | 3,6160809 | 4,5307881 | 0,0349068 |
| Test mode | | | |
| 0,808342 | 3,6597486 | 4,6623431 | |

As can be seen from Table 1, the errors in both modes of the SGTM neural-like structure are very small and close by values.

### 4.4 Obtained Polynomial Coefficients from Trained SGTM Neural-like Structure

To identify the coefficients of the linear polynomial, using the SGTM neural-like structure, we use the method described in Sect. 3.3.

Based on the developed matrix of test signals (10) and its application in the test mode, we obtained the following coefficients of the surface of response:

$$\begin{aligned} \beta_1 &= 454,9357467, \\ \beta_2 &= 452,9437039, \\ \beta_3 &= 454,7085684, \\ \beta_4 &= 454,9976184, \\ \beta_5 &= 454,7751903. \end{aligned} \quad (14)$$

Based on (14) using (12), we obtain coefficients of the linear polynomial for solving the task of the electrical power outputs prediction:

$$\begin{aligned} \theta_1 &= 454,9357467, \\ \theta_2 &= -1,99204284, \\ \theta_3 &= -0,2271783 \\ \theta_4 &= 0,06187167, \\ \theta_5 &= -0,1605564. \end{aligned} \quad (15)$$

According to (13) using (15), the linear polynomial for solving the stated task based on the four input characteristics of each vector from the selected dataset is as follow:

$$\begin{aligned} y = 454,9357467 - 1,99204284 \times x_1 - 0,227183 \times x_2 + 0,06187167 \times x_3 \\ -0,1605564 \times x_4 \end{aligned} \quad (16)$$

The further solution of the task occurs by searching for a dependent variable $y$ based on the received coefficients for each selected vector from the given dataset.

# 5 Comparisons and Discussions

## 5.1 The Comparison with Existing Methods

The solution of the task of the electrical power outputs prediction was carried out using both approaches: based on SGTM neural-like structure [4] and using polynomial (16). As expected, the results of both approaches based on all indicators were identical.

To compare our regression analysis method, some existing ones was chosen, in particular: GRNN (General Regression Neural Network), SVR (Support Vector regression) with *rbf*-kernel and linear regression based on SGD (Stochastic Gradient Descent) [15, 16]. We have used a software implementation of known methods from [17]. Parameters selected by default. The results of the conducted experiments concerning the errors (MAE, MAPE, RMSE) of all methods as well as the time of their training are shown in Figs. 5 and 6.

As can be seen from Figs. 5 and 6, SVR-based method with *rbf*-kernel and linear regression based on the SGD algorithm show great errors (MAE and RMSE from Fig. 5 and MAPE,% from Fig. 6). The GRNN-based regression method shows a fairly high accuracy, but less than the proposed approach (based on MAE, RMSE and MAPE). The developed approach provides the highest accuracy for all three indicators.

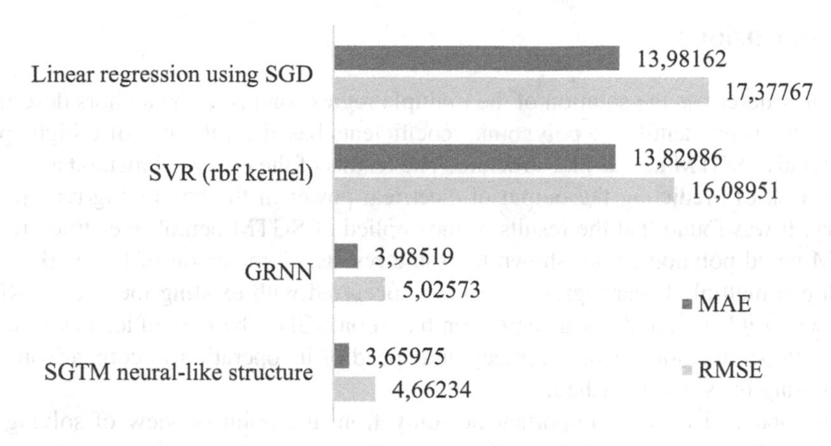


**Fig. 5.** The histogram of the comparison MAE errors (blue columns) and RMSE (yellow columns) for all methods. (Color figure online)

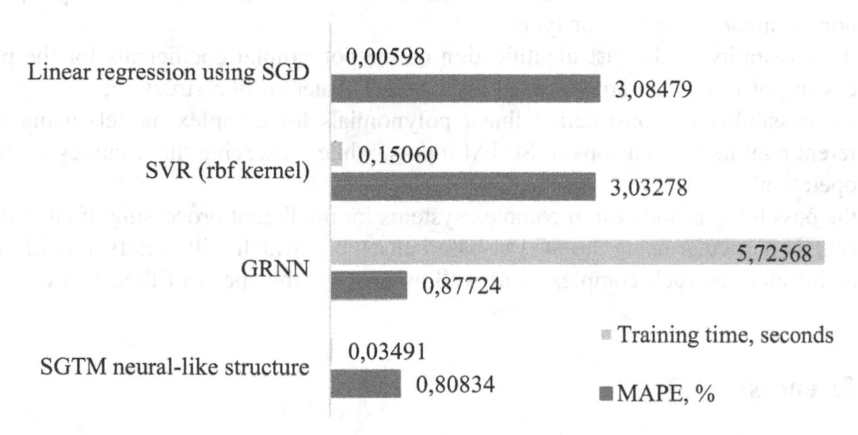


**Fig. 6.** The histogram of the comparison MAPE errors (blue columns) and training time (red columns) of all methods. (Color figure online)

In addition to the fact that the proposed method works by almost 8% better in a percentage ratio to GRNN, the duration of its training is 164 times faster than GRNN.

The highest speed in the training mode demonstrates linear regression based on SGD. It runs almost 6 times faster than the one developed. However, the percentage ratio between the errors of both methods is quite significant. The developed method demonstrates more than 26% better results than linear regression based on SGD.

## 6 Conclusions

The paper describes the solution of the multiple regression task. The authors describe a new method for identifying polynomial coefficients based on the use of a high-speed non-iterative SGTM neural-like structure. The results of the proposed method are tested on the task of predicting the output of electrical power in the combining-type power factory. It was found that the results of the applied of SGTM neural-like structure and SGTM-based polynomial are shown the same results. Comparison of the work of the developed multiple linear regression method occurred with existing methods: GRNN, SVR with *rbf* kernel and linear regression based on SGD. The high efficiency of using the developed method for the accuracy and speed of its operation in comparison with the existing ones is established.

The obtained result is important not only from the point of view of solving the stated task but also for the following main reasons:

- the possibility of a simple interpretation of the result of the regression or classification tasks through the transition from neural-like structures to the simple polynomial model of the linear type;
- the possibility of the fast identification of the polynomial coefficients for the processing of large data arrays based on the SGTM neural-like structure;
- the possibility of constructing linear polynomials for complex models using different nonlinear extensions of SGTM inputs, while preserving the accuracy of their operation;
- the possibility of the design complex systems for intelligent processing of large data sets, in particular using the SGTM-based ensemble, which will greatly simplify the architecture of such complexes, as well as increase the speed of their work.

## References

1. Berk, R.A.: Data mining within a regression framework. In: Maimon, O., Rokach, L. (eds.) Data Mining and Knowledge Discovery Handbook, pp. 231–255. Springer, Boston (2005). https://doi.org/10.1007/0-387-25465-X_11
2. Rathi, M.: Regression modeling technique on data mining for prediction of CRM. In: Das, V.V., Vijaykumar, R. (eds.) ICT 2010. CCIS, vol. 101, pp. 195–200. Springer, Heidelberg (2010). https://doi.org/10.1007/978-3-642-15766-0_28
3. Babichev, S., Lytvynenko, V., Gozhyj, A., Korobchynskyi, M., Voronenko, M.: A fuzzy model for gene expression profiles reducing based on the complex use of statistical criteria and Shannon entropy. In: Hu, Z., Petoukhov, S., Dychka, I., He, M. (eds.) ICCSEEA 2018. AISC, vol. 754, pp. 545–554. Springer, Cham (2019). https://doi.org/10.1007/978-3-319-91008-6_55
4. Tkachenko, R., Izonin, I.: Model and principles for the implementation of neural-like structures based on geometric data transformations. In: Hu, Z., Petoukhov, S., Dychka, I., He, M. (eds.) ICCSEEA 2018. AISC, vol. 754, pp. 578–587. Springer, Cham (2019). https://doi.org/10.1007/978-3-319-91008-6_58

5. Arouri, C., Nguifo, E.M., Aridhi, S., et al.: Towards a constructive multilayer perceptron for regression task using non-parametric clustering. A case study of Photo-Z redshift reconstruction (2019). https://arxiv.org/abs/1412.5513
6. Alomair, O.A., Garrouch, A.A.: A general regression neural network model offers reliable prediction of CO2 minimum miscibility pressure. J. Petrol. Explor. Prod. Technol. **6**(3), 351–365 (2016)
7. Song, J., Romero, C.E., Yao, Z., He, B.: A globally enhanced general regression neural network for online multiple emissions prediction of utility boiler. Knowl. Based Syst. **118**(C), 4–14 (2017)
8. Multi-target support vector regression via correlation regressor chains (2019). https://www.sciencedirect.com/science/article/pii/S0020025517307946
9. Tkachenko, R., Izonin, I., Vitynskyi, P., Lotoshynska, N., Pavlyuk, O.: Development of the non-iterative supervised learning predictor based on the ito decomposition and SGTM neural-like structure for managing medical insurance costs. Data **3**(4), 46 (2018)
10. Doroshenko, A.: Piecewise-linear approach to classification based on geometrical transformation model for imbalanced dataset. In: 2018 IEEE Second International Conference on Data Stream Mining Processing (DSMP), pp. 231–235 (2018)
11. Boyko, N., Sviridova, T., Shakhovska, N.: Use of machine learning in the forecast of clinical consequences of cancer diseases. In: 2018 7th Mediterranean Conference on Embedded Computing (MECO), Budva, 2018, pp. 1–6 (2018)
12. UCI Machine Learning Repository: Combined Cycle Power Plant Data Set (2019). https://archive.ics.uci.edu/ml/datasets/Combined+Cycle+Power+Plant
13. Hassan, A.H.A., Elfaki, E.: Prediction of electrical output power of combined cycle power plant using regression ANN model. Int. J. Comput. Sci. Control Eng. **6**(2), 9–21 (2018)
14. Demšar, J., et al.: Orange: data mining toolbox in Python. J. Mach. Learn. Res. **14**, 2349–2353 (2013)
15. Fedushko, S., Ustyianovych, T.: Predicting pupil's successfulness factors using machine learning algorithms and mathematical modelling methods. In: Hu, Z., Petoukhov, S., Dychka, I., He, M. (eds.) ICCSEEA 2019. AISC, vol. 938, pp. 625–636. Springer, Cham (2020). https://doi.org/10.1007/978-3-030-16621-2_58
16. Kazarian, A., Teslyuk, V., Tsmots, I., Mashevska, M.: Units and structure of automated 'smart' house control system using machine learning algorithms. In: 2017 14th International Conference The Experience of Designing and Application of CAD Systems in Microelectronics (CADSM), pp. 364–366 (2017)
17. Scikit-learn Machine Learning in Python (2019). http://scikit-learn.org/stable/

# Richness of Deep Echo State Network Dynamics

Claudio Gallicchio(✉) and Alessio Micheli

Department of Computer Science, University of Pisa,
Largo B. Pontecorvo, 3, 56127 Pisa, Italy
{gallicch,micheli}@di.unipi.it

**Abstract.** Reservoir Computing (RC) is a popular methodology for the efficient design of Recurrent Neural Networks (RNNs). Recently, the advantages of the RC approach have been extended to the context of multi-layered RNNs, with the introduction of the Deep Echo State Network (DeepESN) model. In this paper, we study the quality of state dynamics in progressively higher layers of DeepESNs, using tools from the areas of information theory and numerical analysis. Our experimental results on RC benchmark datasets reveal the fundamental role played by the strength of inter-reservoir connections to increasingly enrich the representations developed in higher layers. Our analysis also gives interesting insights into the possibility of effective exploitation of training algorithms based on stochastic gradient descent in the RC field.

**Keywords:** Deep Echo State Networks · Deep Reservoir Computing · Richness of RNN dynamics

## 1 Introduction

Randomized approaches to the design of neural networks are subject of considerable attention in the research community nowadays [2,21,29]. The idea of keeping the internal connections fixed is particularly intriguing when considering Recurrent Neural Networks (RNNs) [7]. In this case, indeed, the remarkable efficiency advantages of using untrained hidden weights are coupled with the need to control the resulting system dynamics, to make sure that they can be useful for learning. In this context, Reservoir Computing (RC) [17,24] represents the reference paradigm for the randomized design of RNNs. A promising research line in the current development of RC is given by the exploration of its extensions to deep learning [16,22], with the introduction of Deep Echo State Network (DeepESN) [5] providing a refreshing perspective to the study of hierarchically structured RNNs. On the one hand, results in [6,10] suggested that a proper architectural design of DeepESNs can have a tremendous impact on real-world applications. On the other hand, investigations on DeepESNs dynamics [4,5,11] revealed that a stacked composition of recurrent layers has the inherent ability to diversify the dynamical response to the driving input signals in successive


I. Rojas et al. (Eds.): IWANN 2019, LNCS 11506, pp. 480–491, 2019.
https://doi.org/10.1007/978-3-030-20521-8_40

levels of the architecture. However, the effects of network setup on the quality of state dynamics in deep RC models currently remain largely unexplored.

In this paper we study the richness of state dynamics developed in successive layers of DeepESNs. To do so, we make use of quantitative measures of different nature (including entropy, number of uncoupled state dynamics and condition number) and study how they vary in deeper network settings. Differently from the work in [5], here we do not consider any pre-training approach for the recurrent layers, and focus our analysis only on the intrinsic effects of layering. Besides, the experimental investigation reported in this paper also contributes to explore the viability of least mean squares (LMS)-based algorithms for training the output component of RC networks.

The rest of this paper is structured as follows. In Sect. 2 we present the basics of the DeepESN model, while the adopted quality measures of reservoir dynamics are introduced in Sect. 3. The experimental settings and the achieved results are reported in Sect. 4. Finally, Sect. 5 concludes the paper.

## 2 Deep Echo State Networks

The RC approach to RNN design is based on separating the dynamical recurrent (non-linear) component of the network, called *reservoir*, from the feed-forward (linear) output layer, called *readout*. While the application of training algorithms is limited to the readout, the reservoir is initialized under stability constraints and then is left untrained, making the overall approach extremely efficient in comparison to fully trained RNN models. The RC paradigm has several equivalent formulations, among which the Echo State Network (ESN) [13,14] is one of the most popular. In the rest of this section we deal with deep extensions of ESNs, referring the reader interested in basic aspects of RC to the extensive overviews available in literature [17,24].

A Deep Echo State Network (DeepESN) [5] is an RC model in which the reservoir part is structured into a stack of layers. The output of each reservoir layer constitutes the input for the next one in the deep architecture, and the external input signal is propagated only to the first reservoir in the stack. A comprehensive summary of properties and recent advancements in the study of DeepESNs can be found in [9].

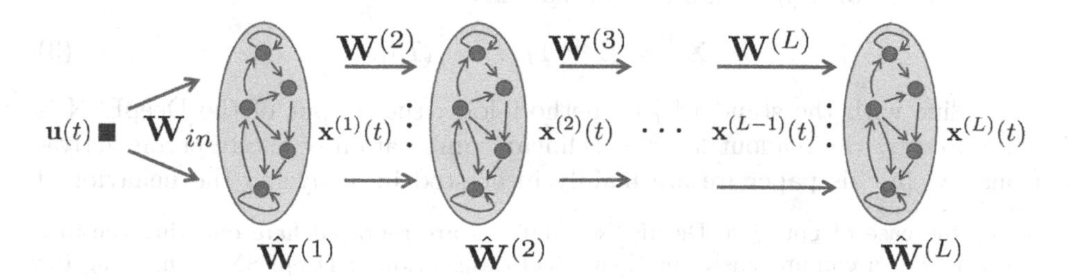


**Fig. 1.** The reservoir architecture of a DeepESN.

In what follows, we denote the input and the output sizes respectively by $N_U$ and $N_Y$, the number of layers is indicated by $L$, and we make the assumption that each layer contains the same number of $N_R$ recurrent units. An illustrative representation of the DeepESN reservoir architecture is given in Fig. 1. The state update equations of a DeepESN are described in terms of discrete-time iterated mappings[1]. At each time-step $t$, the state of the first layer, denoted by $\mathbf{x}^{(1)}(t)$, is computed as follows:

$$\mathbf{x}^{(1)}(t) = \tanh\left(\mathbf{W}_{in}\mathbf{u}(t) + \hat{\mathbf{W}}^{(1)}\mathbf{x}^{(1)}(t-1)\right), \tag{1}$$

while for successive layers $l > 1$, the state $\mathbf{x}^{(l)}(t)$ is updated according to the following equation:

$$\mathbf{x}^{(l)}(t) = \tanh\left(\mathbf{W}^{(l)}\mathbf{x}^{(l-1)}(t) + \hat{\mathbf{W}}^{(l)}\mathbf{x}^{(l)}(t-1)\right). \tag{2}$$

In the above Eqs. 1 and 2, $\mathbf{W}_{in}$ is the input weight matrix, $\mathbf{W}^{(l)}$ (for $l > 1$) is the inter-layer weight matrix connecting the $(l-1)$-th reservoir to the $l$-th reservoir, $\hat{\mathbf{W}}^{(l)}$ (for $l \geq 1$) is the recurrent weight matrix for layer $l$, and tanh denotes the element-wise application of the hyperbolic tangent non-linearity. Typically, a zero state is used as initial condition for each layer, i.e. $\mathbf{x}^{(l)}(0) = \mathbf{0}$. Note that whenever a single reservoir layer is considered in the architecture, i.e. if $L = 1$, the DeepESN reduces to a standard (shallow) ESN.

Following the RC paradigm, the values in all the above mentioned reservoir weight matrices (in Eqs. 1 and 2) are left untrained after initialization based on asymptotic stability criteria, commonly known under the name of Echo State Property (ESP) [17,27]. A detailed analysis of the ESP for the case of deep reservoirs is provided in one of our previous works [4], to which the reader is referred for a detailed description. Here we limit ourselves to recall that the analysis of stability of deep reservoir dynamics essentially suggests to constrain the magnitude of the involved weight matrices, which leads to a simple initialization procedure for a DeepESN. Accordingly, values in $\mathbf{W}_{in}$, $\{\mathbf{W}^{(l)}\}_{l=2}^{L}$ and $\{\hat{\mathbf{W}}^{(l)}\}_{l=1}^{L}$ are first chosen randomly from a uniform distribution in $[-1, 1]$, and then are re-scaled to control the values of the following hyper-parameters: *input scaling* $\omega_{in} = \|\mathbf{W}_{in}\|_2$, *inter-layer scaling* (for $l > 1$) $\omega_{il}^{(l)} = \|\mathbf{W}^{(l)}\|_2$, and *spectral radius*[2] (for $l \geq 1$) $\rho^{(l)} = \rho(\hat{\mathbf{W}}^{(l)})$. Given a driving input signal of length $T$, i.e. $\mathbf{u}(1), \ldots, \mathbf{u}(T)$, we find it useful to (column-wise) collect the states developed by each reservoir layer $l$ into a state matrix:

$$\mathbf{X}^{(l)} = \left[\mathbf{x}^{(l)}(1) \ldots \mathbf{x}^{(l)}(T)\right]. \tag{3}$$

In line with the standard RC methodology, the output of the DeepESN is computed by the readout layer as a linear combination of the reservoir activations. As in this paper we are mainly interested in analyzing the behavior of

[1] For the ease of notation, DeepESN equations are reported here omitting the bias terms. For a comprehensive mathematical description of DeepESN, comprising the case of leaky integrator reservoir units, the reader is referred to [5].

[2] The maximum among the eigenvalues in modulus.

the DeepESN locally to each layer, we study the output of the model when the readout is fed by the state developed individually by each reservoir. Accordingly, when layer $l$ is under focus, the output at time-step $t$, denoted by $\mathbf{y}^{(l)}(t)$, is computed as follows:

$$\mathbf{y}^{(l)}(t) = \mathbf{W}^{(l)}_{out}\mathbf{x}^{(l)}(t). \quad (4)$$

In previous Eq. 4, $\mathbf{W}^{(l)}_{out}$ denotes the readout weight matrix, whose values are adjusted on a training set to solve a linear regression problem given by:

$$\|\mathbf{W}^{(l)}_{out}\mathbf{X}^{(l)} - \mathbf{Y}_{tg}\|_2^2, \quad (5)$$

where $\mathbf{X}^{(l)}$ is as defined in Eq. 3, and $\mathbf{Y}_{tg} = \left[\mathbf{y}_{tg}(1)\ldots\mathbf{y}_{tg}(T)\right]$ is a matrix that collects the corresponding target signals (in a column-wise fashion). Due to a typically large condition number of the reservoir state matrices, readout training is commonly performed by means of non-iterative methods [17].

## 3 Richness of Deep Reservoir Dynamics

To quantify the richness of reservoir dynamics in DeepESNs, we make use of the following measures.

**Average State Entropy** (ASE) - From an information-theoretic perspective the richness of ESN dynamics can be effectively quantified by means of the entropy of instantaneous reservoir states [18]. Here we employ an efficient estimator of Renyi's quadratic entropy [19,20], which, for each time step $t$ and for each layer $l$ in the deep reservoir architecture, can be computed as follows:

$$H^{(l)}(t) = -log(\frac{1}{{N_R}^2}\sum_{j=1}^{N_R}(\sum_{i=1}^{N_R}\mathcal{K}(x_j^{(l)}(t) - x_i^{(l)}(t)))), \quad (6)$$

where $x_j^{(l)}(t)$ denotes the $j$-th component of $\mathbf{x}^{(l)}(t)$, and $\mathcal{K}$ is a gaussian kernel (whose size is obtained by shrinking the standard deviation of instantaneous reservoir activations by a factor of 0.3, in analogy to [18]). Given an input signal of length $T$, we compute the average state entropy (ASE) of layer $l$ as the time-averaged value of the Renyi's quadratic entropy in Eq. 6:

$$ASE^{(l)} = \frac{1}{T}\sum_{t=1}^{T} H^{(l)}(t). \quad (7)$$

**Uncoupled Dynamics** (UD) - It is a known fact in RC literature that reservoir units exhibit behaviors that are strongly coupled among each other [17,26]. In one of our previous works [8] we experimentally showed that this phenomenon can be understood in terms of the inherent Markovian characterization of ESN dynamics [3,23]. Essentially, the stronger is the contractive characterization of a reservoir (i.e. the smaller is the Lipschitz constant of its state transition function) the stronger is the observed redundancy of its units activations, and the poorer is the quality of reservoir dynamics provided to the readout learner. Following this spirit, we propose to evaluate the richness of reservoirs by measuring the actual dimensionality of (linearly) uncoupled state dynamics. To this aim, here we take a simple approach consisting in computing the number of the principal components (i.e. orthogonal directions of variability) that are able to explain the most of the variance in the reservoir state space. Specifically, given an input signal of length $T$, the $l$-th reservoir layer is driven into a set of states collected into the state matrix $\mathbf{X}^{(l)}$ (see Eq. 3). Denoting by $\sigma_1^{(l)}, \ldots, \sigma_{N_R}^{(l)}$ the singular values of $\mathbf{X}^{(l)}$ in decreasing order (i.e., the eigenvalues of the covariance matrix of reservoir units activations), the (normalized) relevance of the $i$-th principal component can be computed as follows:

$$R_i^{(l)} = \frac{\sigma_i^{(l)}}{\sum_{j=1}^{N_R} \sigma_j^{(l)}}. \tag{8}$$

Based on this, the uncoupled dynamics (UD) indicator of the $l$-th reservoir is given by:

$$UD^{(l)} = \arg\min_{d} \Big\{ \sum_{k=1}^{d} R_k^{(l)} \mid \sum_{k=1}^{d} R_k^{(l)} \geq \mathcal{A} \Big\}, \tag{9}$$

where $\mathcal{A}$ ranges in $(0, 1]$ and expresses the desired amount of explained variability. In this paper we considered $\mathcal{A} = 0.9$, meaning that $UD^{(l)}$ is the number of linearly uncoupled directions that explain the 90% of the state space variability.

**Condition Number** ($\kappa$) - Another well-known measure for reservoir quality is given by the conditioning of the resulting learning problem for the readout learner (see Eq. 5). Conventional ESNs are known to suffer from poor conditioning [15,17] (i.e., high eigenvalue spread), which (among the other downsides) prevents the use of efficient LMS-based learning algorithms employing stochastic gradient descent [12] in RC contexts. In this paper, we study the conditioning of the state representation developed in successive levels of DeepESNs. To this end, for each layer $l$ in the deep reservoir architecture, we compute the condition number of its reservoir state matrix $\mathbf{X}^{(l)}$, as follows:

$$\kappa^{(l)} = \frac{\sigma_1^{(l)}}{\sigma_{N_R}^{(l)}}, \tag{10}$$

where $\sigma_1^{(l)}$ and $\sigma_{N_R}^{(l)}$ are respectively the largest and the smallest singular values of $\mathbf{X}^{(l)}$, with smaller values of $\kappa^{(l)}$ indicating richer reservoirs.

## 4 Experiments

In this section we report the outcomes of our experimental analysis. Specifically, details on the considered datasets and experimental settings are given in Sect. 4.1, whereas numerical results are described in Sect. 4.2.

### 4.1 Datasets and Experimental Settings

We considered two well-known benchmark datasets in the RC area, both involving univariate time-series (i.e., $N_U = N_Y = 1$). Although the major focus of our analysis is on the evaluation of deep reservoir dynamics excited by the input (irrespective of the target output values), the datasets are also taken in into account for the definition of regression tasks.

The first dataset is related to the prediction of a 10th order non-linear auto-regressive moving average (NARMA) system, where at each time-step the input $u(t)$ is randomly sampled from a uniform distribution in $[0, 0.5]$, and the target $y_{tg}(t)$ is given by the following equation:

$$y_{tg}(t) = 0.3\, y_{tg}(t-1) + 0.05\, y_{tg}(t-1)\, \big(\sum_{i=1}^{10} y_{tg}(t-i)\big) + 1.5\, u(t-10)\, u(t-1) + 0.1. \tag{11}$$

The second dataset is the Santa Fe Laser dataset [25], consisting in a time-series of sampled intensities from a far-infrared laser in chaotic regime. The dataset enables the definition of a next-step prediction task where $y_{tg}(t) = u(t+1)$ for each time-step $t$. As a minimal pre-processing step, the original values present in the Laser dataset were scaled by a factor of 0.01.

For both the NARMA and the Laser datasets, we considered sequences of length $T = 5000$ to assess the richness of reservoir dynamics. The same data was used as training set in the regression experiments, where the continuation of the respective temporal sequences was considered as test set (of length 5000 for NARMA, and of length 5092 for Laser). In all the experiments the first 1000 time-steps were used as transient to washout the initial conditions from state dynamics.

In our experiments, we considered DeepESNs with a number of layers $L$ ranging from 1 to 5. All reservoir layers contained $N_R = 100$ fully connected recurrent units, and shared the same values of the spectral radius $\rho$ and inter-layer scaling $\omega_{il}$ (i.e., $\rho = \rho^{(1)} = \ldots = \rho^{(L)}$, $\omega_{il} = \omega_{il}^{(2)} = \ldots = \omega_{il}^{(L)}$). Table 1 summarizes the hyper-parameters of DeepESN and their values (or range of values) considered in our experiments. For every DeepESN hyper-parameterization, results were averaged (and standard deviations computed) over 15 networks realizations (with the same values of the hyper-parameters, but random initialization).

We focused our experimental analysis on the behavior of reservoir dynamics in progressively higher layers. As such, all the measures of state richness detailed in Sect. 3 (as well as the predictive performance) were computed on a layer-wise basis, i.e. individually for each layer in the deep architecture.

### 4.2 Results

Figure 2 shows the quality measures of DeepESN dynamics, computed in correspondence of progressively higher layers of the architecture. In particular, results correspond to DeepESNs with reservoir layers hyper-parameterized by input scaling $\omega_{in} = 1$ and spectral radius $\rho = 0.9$ (a value that is of common use in

**Table 1.** DeepESN hyper-parameters and corresponding values, or range of values, explored in this paper. In our experiments, all the reservoir layers share the same values of the hyper-parameters ($N_R$, $\rho$, $\omega_{il}$).

| Hyper-parameter | Symbol | Value(s) |
|---|---|---|
| Number of layers | $L$ | $1, 2, 3, 4, 5$ |
| Number of reservoir units | $N_R$ | 100 |
| Spectral radius | $\rho$ | 0.9 |
| Input scaling | $\omega_{in}$ | 1 |
| Inter-layer scaling | $\omega_{il}$ | $0.5, 1, 2$ |

ESN practice). Results are shown for values of $\omega_{il} = 2$, 1, and 0.5, as representatives for the cases of strong, medium and weak inter-layer connectivity strength, respectively.

Results indicate that when strong inter-layer connections are used, the quality of state representations in DeepESNs improves significantly in progressively higher layers. In this case ($\omega_{il} = 2$ in our experiments) we can appreciate a progressive increase of both the state entropy ($ASE$, first row in Fig. 2) and the number of relevant directions of state variability ($UD$, second row in Fig. 2), corresponding to a substantial decrease of the condition number ($\log_{10}(\kappa)$, third row in Fig. 2). With the increasing layer number we can also observe a saturation trend in the improvement of both entropy and condition number. Interestingly, from Fig. 2 we can note that the marked enrichment of reservoir dynamics is generally observed only for strong connections between consecutive layers. For medium values of $\omega_{il}$ we see that the reservoir quality improves only slightly or remains substantially unchanged ($\omega_{il} = 1$ in our experiments), and for small values of $\omega_{il}$ it eventually gets worse ($\omega_{il} = 0.5$ in our experiments). Overall, our results point out the major role played by the inter-layer scaling parameter $\omega_{il}$ in determining the trend of improvement/worsening of reservoir quality in deeper network settings. In this sense, the impact of other RC parameters, such as the spectral radius $\rho$ and the input scaling $\omega_{in}$, resulted to be much less relevant and is not shown here for brevity[3].

From the perspective of RC training algorithms, an interesting consequence of the possible decrease of the condition number in higher reservoir layers (third row in Fig. 2) is that it makes potentially more suitable the application of computationally cheap stochastic gradient descent algorithms. To start exploring this possibility, we performed a further set of experiments, training the readout component fed by the reservoir states at individual layers, applying a basic LMS algorithm with learning rate $\eta = 0.01$ for 5000 epochs. Note that our aim was not

[3] Changing the value of $\rho$ resulted in globally scaling (up/down for larger/smaller values, respectively) the achieved results for every layer. Changing the value of $\omega_{in}$ affected only the results in the first layer (scaling it up/down for larger/smaller values, respectively). In any case, changes in the values of $\rho$ and $\omega_{in}$ did not affect the quality of the results in Fig. 2 at the increase of network's depth.

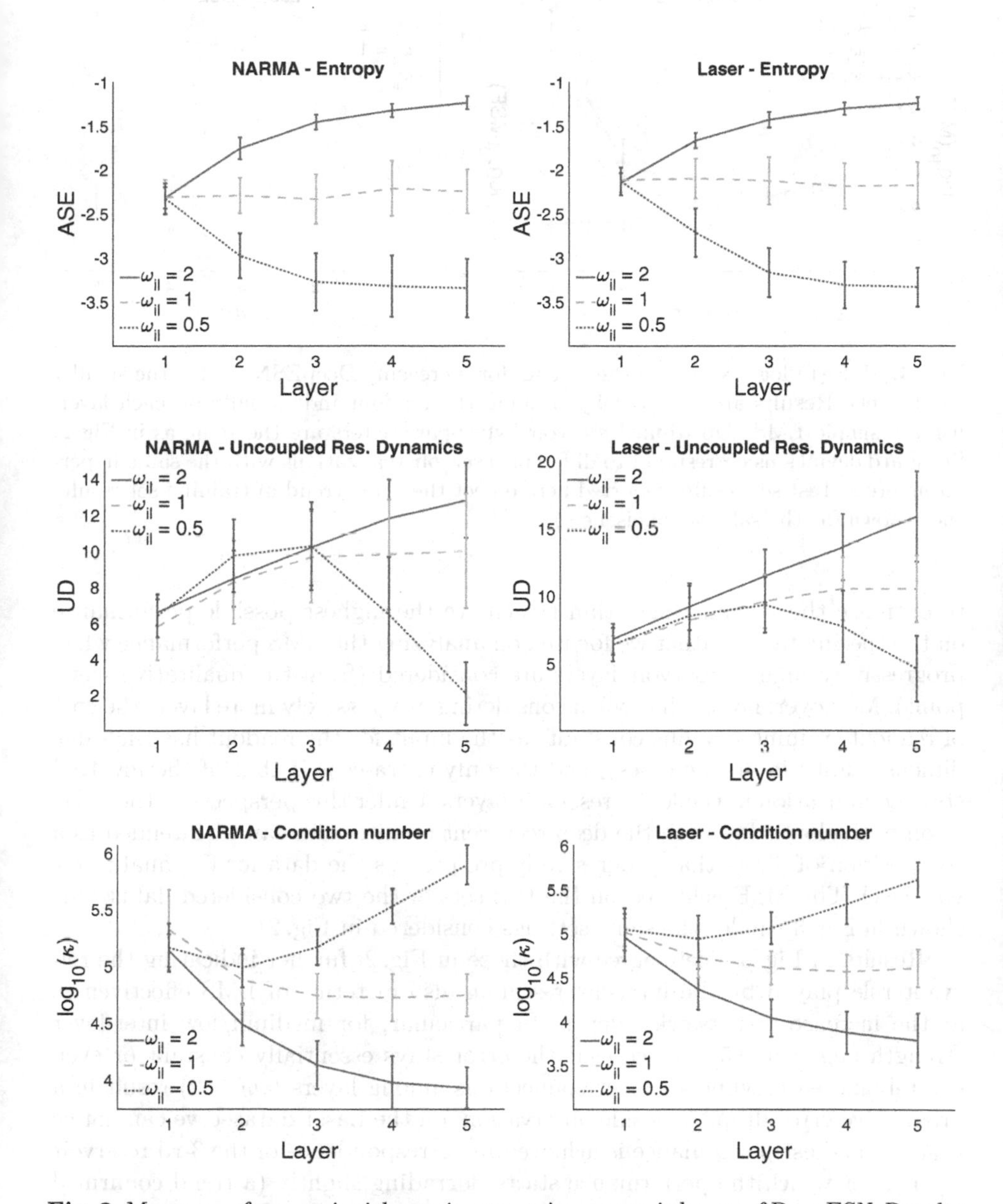


**Fig. 2.** Measures of reservoir richness in successive reservoir layers of DeepESN. Results are achieved using reservoir layers with 100 fully connected units, $\rho = 0.9$, $\omega_{in} = 1$, and varying $\omega_{il}$ in $\{0.5, 1, 2\}$. Standard deviations correspond to different reservoir realizations with the same hyper-parameters. *First row*: Average State Entropy (ASE, the higher the better). *Second row*: Uncoupled Reservoir Dynamics (UD, the higher the better). *Third row*: Condition number in log scale ($\log_{10}(\kappa)$, the lower the better).

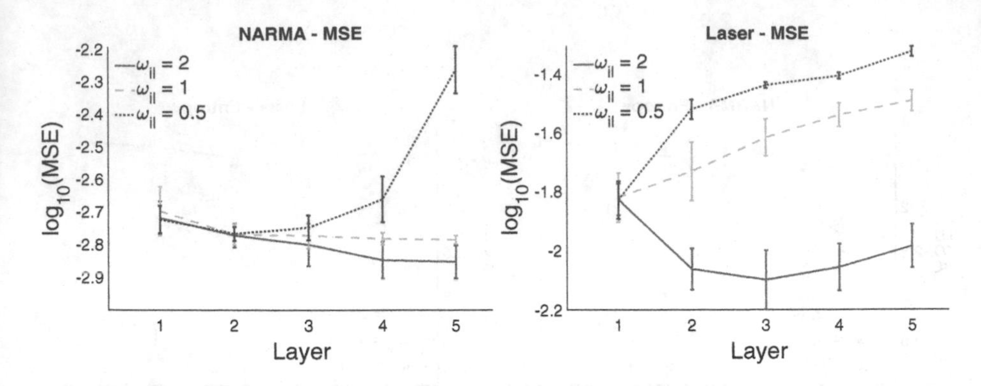


**Fig. 3.** MSE (in $\log_{10}$ scale) on the test set for increasing DeepESN depth (the smaller the better). Results are achieved by training the readout individually on each layer, using a simple LMS algorithm. Reservoir hyper-parameters are the same as in Fig. 2. Standard deviations correspond to different reservoir realizations with the same hyper-parameters. Test set results reported here reflect the same trend of training set results (not shown for the sake of conciseness).

to optimize the training algorithm to achieve the highest possible performance on the specific tasks, rather we focused on analyzing the LMS performance when progressively higher reservoir layers are considered (from the qualitative viewpoint). Moreover, notice that when considering progressively more layers the cost of readout training remains constant (as the input for the readout has the same dimensionality in all the cases), and the only extra-cost is that of the involved state computation in the lower reservoir layers. Under this perspective, the operation of the lower layers in the deep recurrent architecture can be intended as a composition of filters that progressively pre-process the data for the final reservoir level. The MSE achieved on the test sets of the two considered datasets is shown in Fig. 3 (under the same settings considered in Fig. 2).

Results in Fig. 3 nicely agree with those in Fig. 2, further indicating the relevant role played by the inter-layer scaling also in terms of LMS effectiveness at the increase of network's depth. In particular, for medium/low inter-layer strength ($\omega_{il} = 1, 0.5$) we see that the error stays essentially constant or even severely increase, while stronger connections among layers ($\omega_{il} = 2$) result in a progressive error drop. As a side observation, on the Laser dataset we can notice that the highest performance is achieved in correspondence of the 3-rd reservoir layer, after which the performance starts degrading slightly (a trend confirmed also on the training set, not reported in Fig. 3 for the sake of conciseness). This behavior reflects the fact that the performance on a supervised learning task clearly depends on the task characterizations in a broad sense (including the target properties), and it is in line with the observations already made in [1,11] in relation to the memorization skills of reservoirs.

## 5 Conclusions

In this paper we have studied the richness of reservoir representations in DeepESN architectures. Our empirical analysis, conducted on benchmark datasets in the RC area, pointed out the key role played by the strength of connections between the successive layers of recurrent units. Our major finding is that a strong inter-layer connectivity is required in order to determine a progressive enrichment of the state dynamics as the network's depth is increased. This outcome is interestingly related to recent results in the context of reservoirs composed of spiking neurons [28], showing the importance of inter-reservoirs connections to properly propagate information across the levels of a deep recurrent architecture.

Our empirical results indicate that in presence of strong inter-layer connections, reservoirs in higher layers (i.e., further away from the input) are able to develop internal representations featured by increasing entropy and with higher intrinsic (linearly uncoupled) dimensionality, at the same time leading to a decrease in the condition number that characterizes the resulting state matrices. Interestingly, this latter observation is paired with an improved effectiveness of LMS for training the readout, which mitigates the widely known issue with the applicability of gradient descent algorithms in the context of RC.

While already revealing on the potentialities of deep architectures in the RC context, the work presented in this paper allows us to pave the way for promising future research developments. In this regard, we find particularly interesting to extend the application of the analysis tools delineated here to drive the architectural design of DeepESNs in challenging real-world tasks. A starting point in this sense might be represented by a fruitful exploitation of the saturation effects shown for the quality of reservoirs in deeper networks settings. Besides, the insights on the improved amenability of LMS in DeepESNs give a further stimulus to investigate the use of efficient state-of-the-art training algorithms based on stochastic gradient descent in the RC field.

## References


1. Boedecker, J., Obst, O., Lizier, J.T., Mayer, N.M., Asada, M.: Information processing in echo state networks at the edge of chaos. Theory Biosci. **131**(3), 205–213 (2012)
2. Gallicchio, C., Martin-Guerrero, J., Micheli, A., Soria-Olivas, E.: Randomized machine learning approaches: recent developments and challenges. In: 25th European Symposium on Artificial Neural Networks, Computational Intelligence and Machine Learning (ESANN 2017), pp. 77–86. i6doc. com publication (2017)
3. Gallicchio, C., Micheli, A.: Architectural and markovian factors of echo state networks. Neural Netw. **24**(5), 440–456 (2011)
4. Gallicchio, C., Micheli, A.: Echo state property of deep reservoir computing networks. Cogn. Comput. **9**(3), 337–350 (2017)
5. Gallicchio, C., Micheli, A., Pedrelli, L.: Deep reservoir computing: a critical experimental analysis. Neurocomputing **268**, 87–99 (2017)

6. Gallicchio, C., Micheli, A., Pedrelli, L.: Comparison between DeepESNs and gated RNNs on multivariate time-series prediction. In: 27th European Symposium on Artificial Neural Networks, Computational Intelligence and Machine Learning (ESANN 2019). i6doc. com publication (2019)
7. Gallicchio, C., Micheli, A., Tiňo, P.: Randomized recurrent neural networks. In: 26th European Symposium on Artificial Neural Networks, Computational Intelligence and Machine Learning (ESANN 2018), pp. 415–424. i6doc. com publication (2018)
8. Gallicchio, C., Micheli, A.: A markovian characterization of redundancy in echo state networks by PCA. In: Proceedings of the 18th European Symposium on Artificial Neural Networks (ESANN). d-side publication (2010)
9. Gallicchio, C., Micheli, A.: Deep echo state network (deepesn): a brief survey. arXiv preprint arXiv:1712.04323 (2017)
10. Gallicchio, C., Micheli, A., Pedrelli, L.: Design of deep echo state networks. Neural Netw. **108**, 33–47 (2018)
11. Gallicchio, C., Micheli, A., Silvestri, L.: Local lyapunov exponents of deep echo state networks. Neurocomputing **298**, 34–45 (2018)
12. Haykin, S.: Neural Networks and Learning Machines, 3rd edn. Pearson, Upper Saddle River (2009)
13. Jaeger, H.: The "echo state" approach to analysing and training recurrent neural networks - with an erratum note. Tech. rep., GMD - German National Research Institute for Computer Science (2001)
14. Jaeger, H., Haas, H.: Harnessing nonlinearity: predicting chaotic systems and saving energy in wireless communication. Science **304**(5667), 78–80 (2004)
15. Jaeger, H.: Reservoir riddles: suggestions for echo state network research. In: Proceedings of the 2005 IEEE International Joint Conference on Neural Networks (IJCNN), vol. 3, pp. 1460–1462. IEEE (2005)
16. LeCun, Y., Bengio, Y., Hinton, G.: Deep learning. Nature **521**(7553), 436–444 (2015)
17. Lukoševičius, M., Jaeger, H.: Reservoir computing approaches to recurrent neural network training. Comput. Sci. Rev. **3**(3), 127–149 (2009)
18. Ozturk, M.C., Xu, D., Príncipe, J.C.: Analysis and design of echo state networks. Neural Comput. **19**(1), 111–138 (2007)
19. Principe, J.C.: Information Theoretic Learning: Renyi's Entropy and Kernel Perspectives. Springer, New York (2010). https://doi.org/10.1007/978-1-4419-1570-2
20. Principe, J.C., Xu, D., Fisher, J.: Information theoretic learning. Unsupervised Adapt. Filtering **1**, 265–319 (2000)
21. Scardapane, S., Wang, D.: Randomness in neural networks: an overview. Wiley Interdisc. Rev.: Data Min. Knowl. Discovery **7**(2), e1200 (2017)
22. Schmidhuber, J.: Deep learning in neural networks: an overview. Neural Netw. **61**, 85–117 (2015)
23. Tiňo, P., Hammer, B., Bodén, M.: Markovian bias of neural-based architectures with feedback connections. In: Hammer, B., Hitzler, P. (eds.) Perspectives of Neural-Symbolic Integration. Studies in Computational Intelligence, vol. 77, pp. 95–133. Springer, Heidelberg (2007). https://doi.org/10.1007/978-3-540-73954-8_5
24. Verstraeten, D., Schrauwen, B., d'Haene, M., Stroobandt, D.: An experimental unification of reservoir computing methods. Neural Netw. **20**(3), 391–403 (2007)
25. Weigend, A.S.: Time Series Prediction: Forecasting The Future and Understanding The Past. Routledge, Abingdon (2018)
26. Xue, Y., Yang, L., Haykin, S.: Decoupled echo state networks with lateral inhibition. Neural Netw. **20**(3), 365–376 (2007)

27. Yildiz, I., Jaeger, H., Kiebel, S.: Re-visiting the echo state property. Neural Netw. **35**, 1–9 (2012)
28. Zajzon, B., Duartel, R., Morrison, A.: Transferring state representations in hierarchical spiking neural networks. In: Proceedings of the 2018 International Joint Conference on Neural Networks (IJCNN), pp. 1785–1793. IEEE (2018)
29. Zhang, L., Suganthan, P.N.: A survey of randomized algorithms for training neural networks. Inf. Sci. **364**, 146–155 (2016)

# Image Classification and Retrieval with Random Depthwise Signed Convolutional Neural Networks

Yunzhe Xue and Usman Roshan(✉)

Department of Computer Science, New Jersey Institute of Technology, Newark, NJ 07090, USA
{yx277,usman}@njit.edu

**Abstract.** We propose a random convolutional neural network to generate a feature space in which we study image classification and retrieval performance. Put briefly we apply random convolutional blocks followed by global average pooling to generate a new feature, and we repeat this $k$ times to produce a $k$-dimensional feature space. This can be interpreted as partitioning the space of image patches with random hyperplanes which we formalize as a random depthwise convolutional neural network. In the network's final layer we perform image classification and retrieval with the linear support vector machine and $k$-nearest neighbor classifiers and study other empirical properties. We show that the ratio of image pixel distribution similarity across classes to within classes is higher in our network's final layer compared to the input space. When we apply the linear support vector machine for image classification we see that the accuracy is higher than if we were to train just the final layer of VGG16, ResNet18, and DenseNet40 with random weights. In the same setting we compare it to a recent unsupervised feature learning method and find our accuracy to be comparable on CIFAR10 but higher on CIFAR100 and STL10. We see that the accuracy is not far behind that of trained networks, particularly in the top-$k$ setting. For example the top-2 accuracy of our network is near 90% on both CIFAR10 and a 10-class mini ImageNet, and 85% on STL10. We find that $k$-nearest neighbor gives a comparable precision on the Corel Princeton Image Similarity Benchmark than if we were to use the final layer of trained networks. As with other networks we find that our network fails to a black box attack even though we lack a gradient and use the sign activation. We highlight sensitivity of our network to background as a potential pitfall and an advantage. Overall our work pushes the boundary of what can be achieved with random weights.

## 1 Introduction

Convolutional neural networks (CNNs) are the state of the art in image recognition benchmarks today [1]. Optimization methods such as stochastic gradient descent [2] combined with data augmentation [3], regularization [4], dropout and


I. Rojas et al. (Eds.): IWANN 2019, LNCS 11506, pp. 492–506, 2019.
https://doi.org/10.1007/978-3-030-20521-8_41

cutout [5,6] have made CNNs the de-facto approach for accurate image recognition [7]. Interestingly, several of these methods involve randomness.

For example the dropout method [5] ignores a random set of nodes during training. The cutout method [6] masks random square patches in the input training images. Stochastic gradient descent randomly uses a single training example at a time (or mini batches) to obtain the gradient as opposed to computing it from the entire dataset. This method has been shown to converge to the global optimum for convex functions as we increase the number of iterations [8]. All of these methods use randomness to avoid overfitting during training and thus give better model generalization.

Random weights have been explored previously in several studies [9–12] including generating images with random nets [13]. They show the importance of a network architecture in achieving high accuracies and connect unsupervised pre-training and discriminative fine tuning to architecture. We explore randomness from the perspective of random depthwise convolutional blocks.

Consider a series of convolutional blocks applied repeatedly to an image. The image representation in the final layer is then globally average pooled to obtain a single value. If we repeat this $k$ times we obtain a $k$ dimensional space. This can also be considered as a method for unsupervised feature learning since we make no use of labels in generating the new space. We show below that our method can be interpreted as applying random hyperplanes to all patches of all input images and average pooling the final image representations. We formalize this as a random depthwise convolutional neural network and study various aspects of image classification and retrieval in our network's final layer.

We present several experimental results on image classification and retrieval in our network's final layer. We start by showing that images across and within classes are better represented than the input feature space. We then compare to the final trained layer of other random networks and an unsupervised feature learning method. In both cases our network attains comparable or higher accuracies. Compared to trained networks our random weights are no match but are not far behind in accuracy, especially as we go into top-2 and top-3 accuracy.

Interestingly our network performs competitively to trained networks when it comes to the problem of image retrieval, particularly image retrieval by similarity such as the Corel-Princeton Image Similarity benchmark [14]. Finally we highlight some limitations sensitivity of our network's layer to background colors which can be a disadvantage but possibly also an advantage in some cases. Overall we push the envelope of random networks and show that accuracies better than expected can be achieved.

## 2 Methods

### 2.1 Motivation and Intuition Behind Depthwise Random Convolutions

We provide motivation and intuition with a simple toy example in Fig. 1. There we see four images $I_0, I_1, I_2$, and $I_3$ containing various objects. Clearly images

$I_0$ and $I_1$ are very similar to each other. Image $I_2$ is also similar since it contains common objects as $I_0$ and $I_1$ but they are in different positions. Image $I_3$ shares only one common object to the other images and thus is the most dissimilar. We seek a representation that would capture these similarities.

Suppose we divide each image into four equal quadrants (or patches) and consider four random hyperplanes $H_0, H_1, H_2$, and $H_3$ in the space of all patches as shown in Fig. 1. We place patches containing the same object nearby in the figure since they are likely to be similar in pixel values. For example the car in partitions $I_{02}, I_{12}, I_{23}$, and $I_{32}$ are clustered in the lower left in the middle figure near the origin.

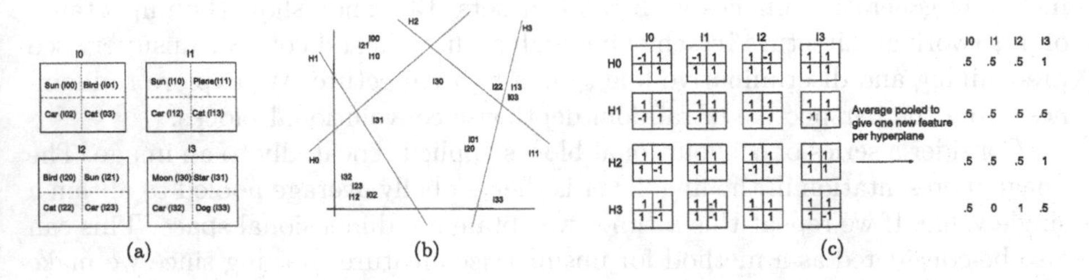


**Fig. 1.** Shown in (a) are four images each containing objects in different parts of the image and divided into four partitions. In (b) we show four random hyperplanes (in red) on the input space of features from all patches of the images. The outputs of the four random hyperplanes correspond to a convolutional kernel that considers just four partitions (part (c)). We average pool to obtain a feature value. (Color figure online)

We determine the sign of each patch according to each hyperplane and obtain a $2 \times 2$ matrix for each image. This is exactly the output of a convolution kernel that considers just four partitions of an image. We then average pool the matrix to obtain a single feature value for the image given by the hyperplane. By repeating this for many hyperplanes we obtain more features per image. We see in our toy example that the final feature representation for images $I_0, I_1$, and $I_2$ are more similar than image $I_3$.

We don't have a theoretical guarantee that applying random hyperplanes to image patches repeatedly (as we do in our network) would yield a linearly separable space. However, one can draw intuition from our toy example in Fig. 1. There we see that the four random hyperplanes partition the space and that patches in the same space will have exactly the same outputs for all four hyperplanes. Patches in adjacent partitions are likely to be less similar and will have exactly one of the four hyperplane outputs to be different.

### 2.2 Random Depthwise Convolutional Neural Networks

Before formalizing our notion into random depthwise networks we briefly review convolutional neural networks. Convolutional neural networks are typically composed of alternating convolution and pooling layers followed by a final flattened layer. A convolution layer is specified by a kernel size and the number of kernels in the layer. Briefly, the convolution layer performs a moving non-linearized dot product against pixels given by a fixed kernel size $k \times k$ (usually $3 \times 3$ or $5 \times 5$). The dot product is usually non-linearized with the sigmoid or hinge (relu) function since both are differentiable and fit into the gradient descent framework. The output of applying a $k \times k$ convolution against a $p \times p$ image is an image of size $(p - k + 1) \times (p - k + 1)$.

Consider applying random convolutional blocks repeatedly and then averaging all the values in the final representation of the image. If we repeat this $k$ times it gives us $k$ new features. This can be described as a random depthwise convolutional neural network (RDCNN). Each convolutional block in our network is a convolutional kernel followed by $2 \times 2$ average pooling with stride 2.

Our network is parameterized by the number of convolutional blocks $b$, the size of each kernel $k \times k$ and the number of kernels $m$ in each layer (this is the same in each layer). In Fig. 2 we show an example of our network with two layers ($l = 2$) and five $3 \times 3$ convolution blocks in each layer ($m = 5, k = 3$). We set the values in each convolutional kernel randomly from the Normal distribution with mean 0 and variance 1.

We non-linearize the output of each convolution with the sign function and our convolution is *depthwise*. This means the $i^{th}$ convolution is applied on the $i^{th}$ kernel only of the previous layer. In the input layer, however, the convolution is applied in the conventional way to account for RGB images that have three layers. After we are done with convolutions we globally average pool the final layer which gives us a flattened feature space. We then apply a linear support vector machine or stochastic gradient descent on the final feature space.

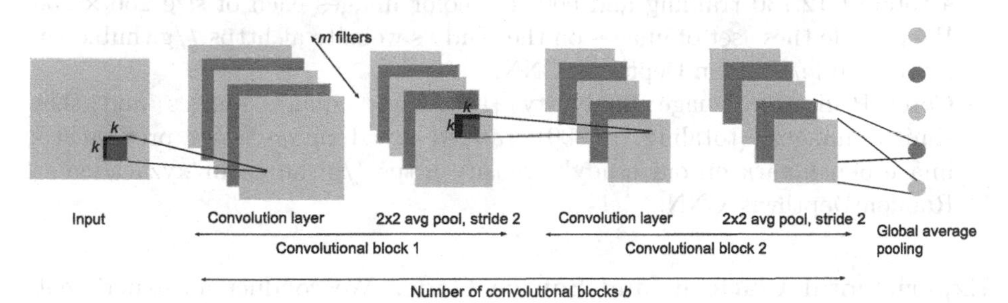


**Fig. 2.** A random depthwise convolutional neural network with two convolutional blocks, kernel size of $k$, and $m = 5$ kernels in each layer

## 2.3 Experimental Performance Study

In order to evaluate the empirical performance of our random network we compare it to three popular networks with random and trained weights and an unsupervised feature learning method on several image benchmarks.

**Deep Networks Compared in Our Study.** We compare our method to modern networks used in image recognition today. These are all convolutional neural networks designed to enable deeper architectures and are trained with stochastic gradient descent. We implement these networks ourselves with PyTorch and TensorFlow and make our implementations freely available from the study website https://github.com/xyzacademic/RandomDepthwiseCNN.

- ResNet18 [15]: Residual convolutional networks contain connections from previous layers and not just the last one.
- DensenNet40 [16]: Convolutional networks contain dense layers in between convolutions.
- VGG16 [17]: Deep convolutional neural network with layers of convolution and pooling.

**Datasets.** We collect several image benchmarks on which we evaluate our method.

- MNIST [18]: Handwritten digit recognition from 10 classes in $32 \times 32$ images, training size of 60,000 and test size of 10,000.
- CIFAR10 [19]: Object recognition from 10 classes in $32 \times 32$ color images, training size of 50,000, test size of 10,000.
- CIFAR100 [19]: As CIFAR10 except from 100 classes.
- STL10 [20]: Object recognition from 10 classes in $96 \times 96$ color images, training size of 5000, and test size of 8000.
- Mini-ImageNet [21]: We randomly select 10 classes from the benchmark giving a total of 12,730 training and 500 test color images each of size $256 \times 256$. We provide these set of images on the study's website at https://github.com/xyzacademic/RandomDepthwiseCNN.
- Corel Princeton Image Similarity [14]: Eight query images and their similar images (totaling 10,000) ranked by humans. We provide the image benchmark on our study's website https://github.com/xyzacademic/RandomDepthwiseCNN.

**Experimental Platform and Source Code.** We conduct all experiments on computing nodes equipped with Intel Xeon E5-2630-v4 CPUs and NVIDIA Tesla P100 16 GB Pascal GPUs. We implement our method to produce the final flattened layer with PyTorch and TensorFlow and make it available on this study's website https://github.com/xyzacademic/RandomDepthwiseCNN. We also provide there our implementations of other networks that we study in this paper including code to generate adversarial examples.

**Program Parameters and Training.** We use the libinear program [22] version 2.20 for determining a linear support vector machine on the final layer obtained by our model. Our liblinear parameters are -s 2 -B 1 which turns on primal optimization and a threshold value of non zero. Our $C$ (regularization) parameter values are 0.5 for CIFAR10, CIFAR100, and STL10, and 0.01 for MNIST and Mini-ImageNet. We optimize ResNet18, DenseNet40, and VGG16 networks with stochastic gradient descent. For CIFAR10 and CIFAR100 we use a batch size of 256 whereas for STL10 and Mini-ImageNet we use 32 and 16 respectively. On Mini-ImageNet we use 90 epochs and on the other benchmarks we use 300. We vary the learning rate across the number of epochs by starting with a large value of 0.1 and progressively reducing to 0.01 and 0.001 as the number of epochs increases.

# 3 Results

## 3.1 Effect of Number of Blocks and Kernel; Size

We start by exploring the effect of number of blocks and kernel size on our model test accuracy. In Table 1 we show the accuracy of our model with a fixed number of layers, fixed kernel size, and 100,000 kernels. We see that the number of layers and kernel size has a considerable effect on test accuracy

**Table 1.** Accuracy of our network RDCNN with different number of convolutional blocks and kernel size.

| Dataset | $k = 3, b = 1$ | $k = 3, b = 2$ | $k = 5, b = 1$ | $k = 5, b = 2$ |
|---|---|---|---|---|
| CIFAR10 | 75.8 | 74.8 | 76.4 | 70.1 |
| CIFAR100 | 51.8 | 51.9 | 53.3 | 47.5 |

In Table 2 we show the parameters used in RDCNN for each image benchmark. These parameters give us the best test accuracy over combinations of $k = 3, 5, 7$ and $b = 1, 2, 3$.

**Table 2.** Parameters used in RDCNN for each image benchmark

| Dataset | MNIST | CIFAR10 | CIFAR100 | STL10 | Mini ImageNet |
|---|---|---|---|---|---|
| Parameters | $k = 7, b = 1$ | $k = 5, b = 1$ | $k = 5, b = 1$ | $k = 3, b = 3$ | $k = 5, b = 2$ |

## 3.2 Effect of Number of Random Kernels on Test Accuracy

Having shown the effect of kernel size and number of convolutional blocks we proceed to determine the effect of number of kernels on the test accuracy. Each kernel gives rise to a new feature in the final flattened layer. We expect an improvement in test accuracy as we increase the number of kernels and indeed we see this is the case in Fig. 3. There we see that increasing the number of kernels improves the accuracy on the STL10 and CIFAR10 benchmarks. We also see that the train accuracy reaches 100% much faster and stays there while test accuracy continues to improve.

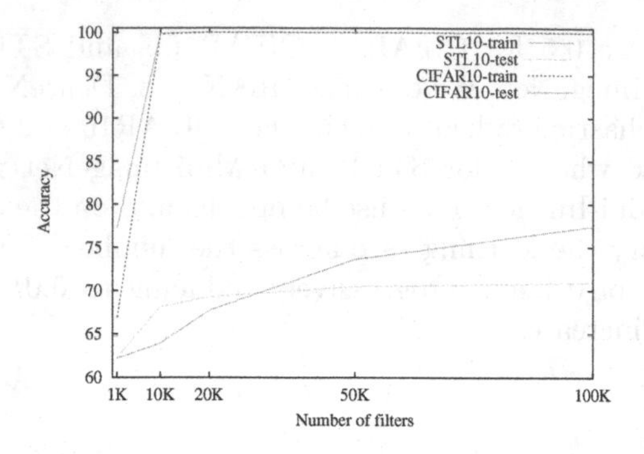


**Fig. 3.** Effect of number of kernels (final features) on the test accuracy

### 3.3 Image Pixel Distribution Similarity Across and Within Classes

We select 10 random images from STL10 (6 from class 0 and 4 from class 1) and show the distribution of the image pixel values before and after our network. In Fig. 4(a) we see the distribution of image pixel values in the original data and in (b) we show the distribution in our final layer before the SVM. We see that the random kernels smoothen the image distributions and appear to show a better separation of image distributions between the two classes.

In order to obtain a quantitative measure we report the average Jensen Shannon (avgJS) divergence [23] between image pixel distributions across classes and within classes. To obtain this we simply average the divergence across all pairs of image distributions across two classes (and within classes). We then measure the ratio of

$$\frac{avgJS(class0, class1)}{avgJS(class0, class0) + avgJS(class1, class1)}$$

For images in the original feature representation we find a ratio of $\frac{0.25}{0.3+0.18} = 0.52$ and in our RDCNN final layer the ratio increases to $\frac{0.02}{0.018+0.016} = 0.59$. This ratio varies across classes and the above values are for class 0 and 1 that we randomly chose. If we measure this ratio between classes 0 and 6 we see a larger difference in the ratio across the two feature representations. In the original representation this value is $\frac{0.24}{0.3+0.13} = 0.56$ and in our RDCNN final layer it increases to $\frac{0.025}{0.018+0.015} = 0.76$. Thus we see that our new feature representation gives a better *signal to noise* ratio.

### 3.4 Comparison to Random Weights in Other Networks and Unsupervised Feature Learning

Here we consider random weights in ResNet18 and DenseNet40 and except for the final layer that we train. We also consider a recent unsupervised feature learning method (Discriminative UFL) [24] that targets instance level discrimination with convolutional neural networks. In Fig. 5 we see that all the random networks have lower accuracy than ours (denoted as RDCNN) even after training the final layer. We obtain the same accuracy with Discriminative UFL on CIFAR10 as in their published paper. However, when we ran their code on the other benchmarks the accuracies were much lower.

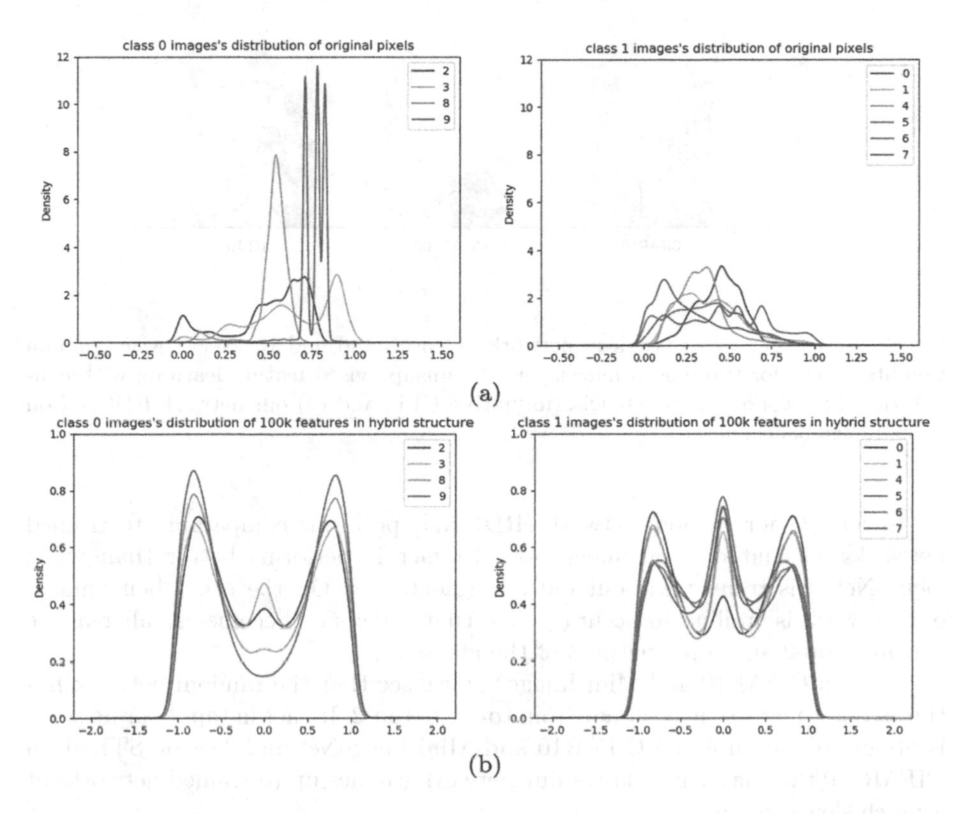


**Fig. 4.** Pixel distribution of images from two classes in the original and new representations

The accuracies of random networks on CIFAR10 that we report here are lower than those of previously reported by Saxes et al. [9] (53.2%) and Gilbert et al. [12] (74.8%). It's likely they may have trained additional parameters besides just the final layer that we do. Jarrett et al. [10] report 99.46% on MNIST with random weights and we perform comparably by reaching 99.4%.

### 3.5 Comparison to Trained Convolutional Networks

In Fig. 6 we show the test accuracies of the top-1, top-2, and top-3 outputs from all networks on four benchmarks. The top-k outputs are obtained by considering images with the top $k$ highest outputs in the final layer. In our case we use the SVM discriminant to rank the outputs.

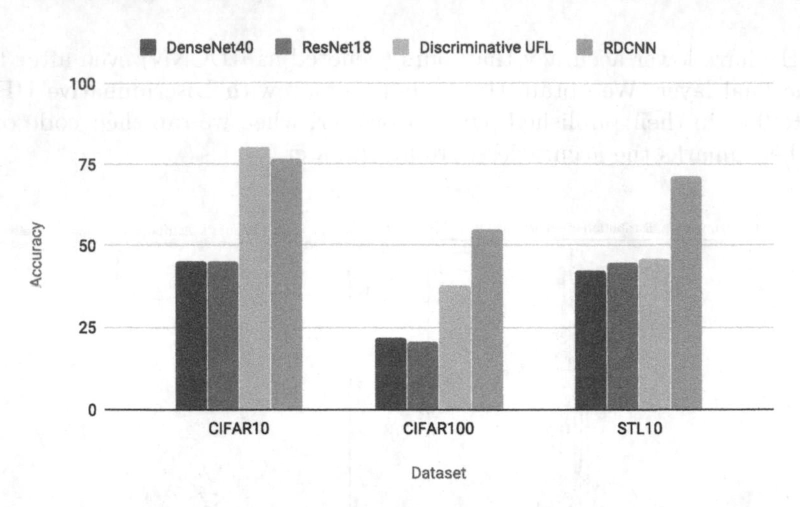


**Fig. 5.** Test accuracy of (1) deep networks DenseNet40 and ResNet18 with random weights except for training in final layer, (b) unsupervised feature learning with convolutional networks denoted as Discriminative UFL, and (3) our network RDCNN on three image benchmarks

In STL10 our random network (RDCNN) performs comparably to trained networks without data augmentation. In fact it performs better than when DenseNet40 is trained without data augmentation. On the other benchmarks our network is trailing in accuracy but that difference becomes smaller as we consider top-2 and top-3 outputs of the classifier.

In both CIFAR10 and Mini ImageNet we see that the random network has the steepest increase in accuracy from top-1 to top-2. In fact in top-2 our network is about 90% accurate on CIFAR10 and Mini ImageNet and 85% on STL10. In CIFAR100 that has a 100 classes our network catches up to trained networks at a much slower pace.

*Data Augmentation.* For our method we first separately generated augmented images with flips and random rotations from STL10 (10 augments per input image). We then combined these into the original training set and used them as input to our network. Our final test accuracy, however, was no better than training on just the original data. To understand this we compare the cosine similarity of each augmented image to its original and plot the cosine similarity value distribution across all images. We perform the cosine similarity in the feature space given by the final layer of our network. In Fig. 7 we see that the flips and rotations produce images that are highly similar to the original in our feature representation. However, if we perform the cutout augmentation [6] the images are relatively more different. Perhaps this augmentation or something even stronger may boost the accuracy of our network.

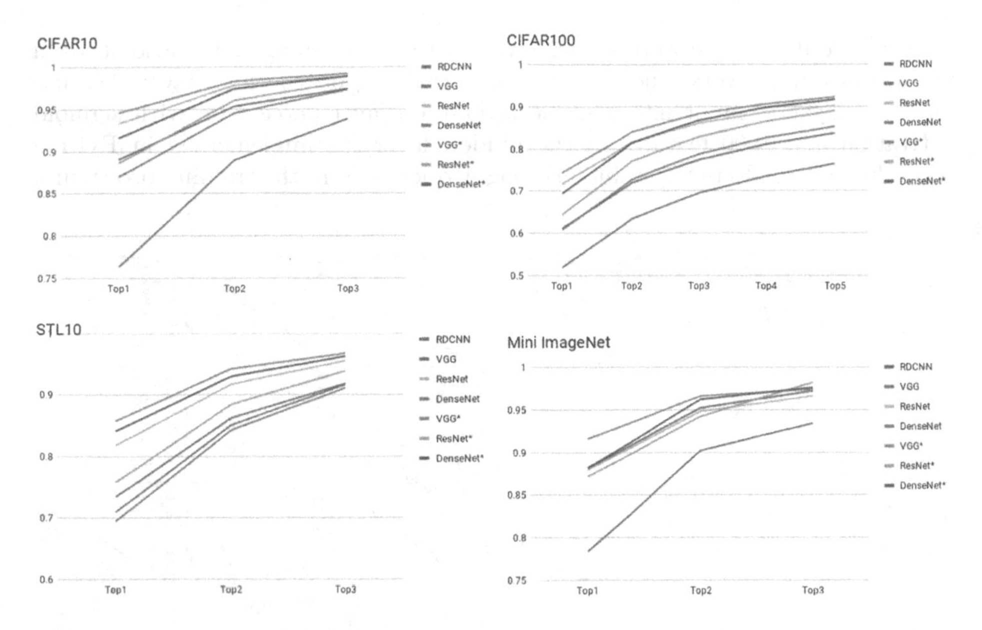


**Fig. 6.** Top-k accuracy of different networks on benchmarks. Methods with asterisk denote data augmentation was enabled.

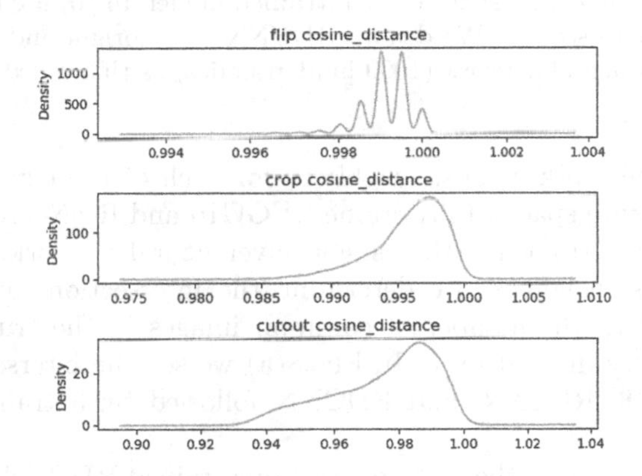


**Fig. 7.** Cosine similarity distribution across all STL10 images. Each cosine value is between an image and its augmentation in our network's final layer feature space.

### 3.6 Similar Image Retrieval

We train the VGG16 and ResNet18 networks on all 10,000 images in the Corel Princeton Image Similarity dataset [14] except for the eight queries. In this benchmark the most similar images are ranked and given a score by human observers. We use the last hidden layer for representing all images. Our choice for this comes from previous studies that have shown the final hidden layer (that is usually a dense layer) works best for image retrieval with deep convolutional neural networks [25–27].

Here we use our network's final layer to represent images. We also study a version of our network where we perform training on the final layer. We use our final layer as input into a single layer multilayer perceptron with sigmoid activation and 1000 features in the hidden layer (as implemented in Python scikit-learn [28]). Here we exclude the eight queries from the training procedure.

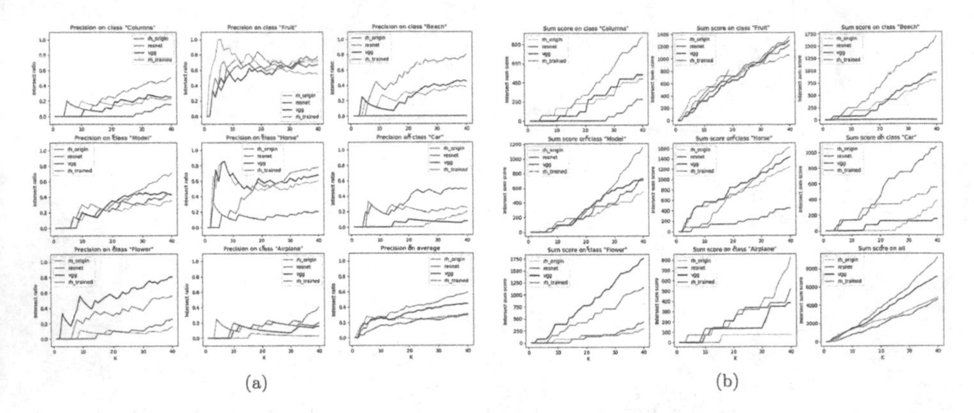


**Fig. 8.** Shown here in (a) is the intersection ratio (also known as precision) of top-k images returns in the feature space of each trained model. In (b) are the sum of scores of images in the intersection. We denote RDCNN as rh_origin and RDCNN trained with a single layer neural network (1000 hidden nodes) as rh_trained.

We obtain the top-k nearest neighbors for each of the eight queries in the four different feature spaces: fully trained VGG16 and ResNet18, RDCNN, and RDCNN final layer trained with a single layer neural network. For each value of $k$ (in $k$-nearest neighbor) we determine the intersection ratio (also known as the precision) as the number of common images in the true top $k$ images for the given query divided by $k$. In Fig. 8(a) we see the intersection ratios for VGG16, Renset18, RDCNN, and RDCNN followed by a trained single layer neural network.

On the average across the eight queries our trained RDCNN has the highest precision values as we cross values of $k$ above 10. If we sum the score of all images in the intersection there too we see our trained network in the lead (Fig. 8(b)). Our random one (without training) is behind VGG but better than ResNet.

While our method performs well for image similarity on this benchmark we see that the fully trained networks produce a better classification of the eight queries. We consider the correct classification of the query to be the category it belongs to in the database images. Both VGG16 and ResNet18 classify 7 out of 8 correctly whereas our RDCNN (final trained layer) does 6 out of 8 correctly. On the training VGG and ResNet achieve 98% accuracy whereas our trained RDCNN gets to 76% with the multilayer perceptron. Thus it appears that RDCNN captures image similarity better than image content.

### 3.7 Sensitivity to Adversarial Attacks

We explore whether the sign activation and lack of gradient in our model offers a defense to adversarial attacks. We perform a practical black box attack [29] on our model and the other trained ones. For each model including ours we first input and output examples from the STL10 dataset. We then use these examples to learn a two layer multi-layer perceptron from which we produce 5,000 adversarial examples. In Table 3 we see that all models including ours fail on the adversarial examples. This despite the sign activation function and lack of gradient our model can be attacked.

**Table 3.** Accuracy of our model and others on STL10 adversarial examples generated with a black box attack

| | RDCNN | VGG16 | DenseNet40 | ResNet18 |
|---|---|---|---|---|
| STL10 | 51% | 58% | 46% | 56% |

## 4 Discussion

We found that our network is somewhat more sensitive to background than trained networks. From STL10 we pick a bird image and show the top similar images to it as given by VGG16 and our network. Here we measure cosine similarity in final layer of VGG and our network. In Fig. 9(a) we see that VGG16 reports almost all birds as the most similar image to the original bird. In our network shown in Fig. 9(b) we see that images similar to the bird are also similar in color and background and are mostly not birds.

Thus our random network is sensitive to background and color which possibly accounts for its trailing accuracy in classification tasks. This can also be seen in the image retrieval experiments where RDCNN performs better in retrieval but worse when we classify the queries. While this obviously is a pitfall it may also be advantageous in problems where similarity plays an important role such as medical imaging [30, 31]. If we train our network's final layer we begin to see birds in top similar images as shown in Fig. 9(c). Even in the image retrieval by similarity task RDCNN performs better with some training.

To determine the sensitivity of our method to training set size we reversed our Mini ImageNet training and test: we use the much smaller test set of 500 images as training and 12,370 for test. In this experiment we found our method to give 58% accuracy while a trained ResNet18 gives 48% and 63.7% without and with data augmentation respectively.

One challenge in RDCNN is the high dimensional feature space. This can be a challenge for large datasets like ImageNet [21]. We propose to solve this by compressing the final layer with a simple bit-wise encoding method. We plan to study this and applications on medical images as part of future work.

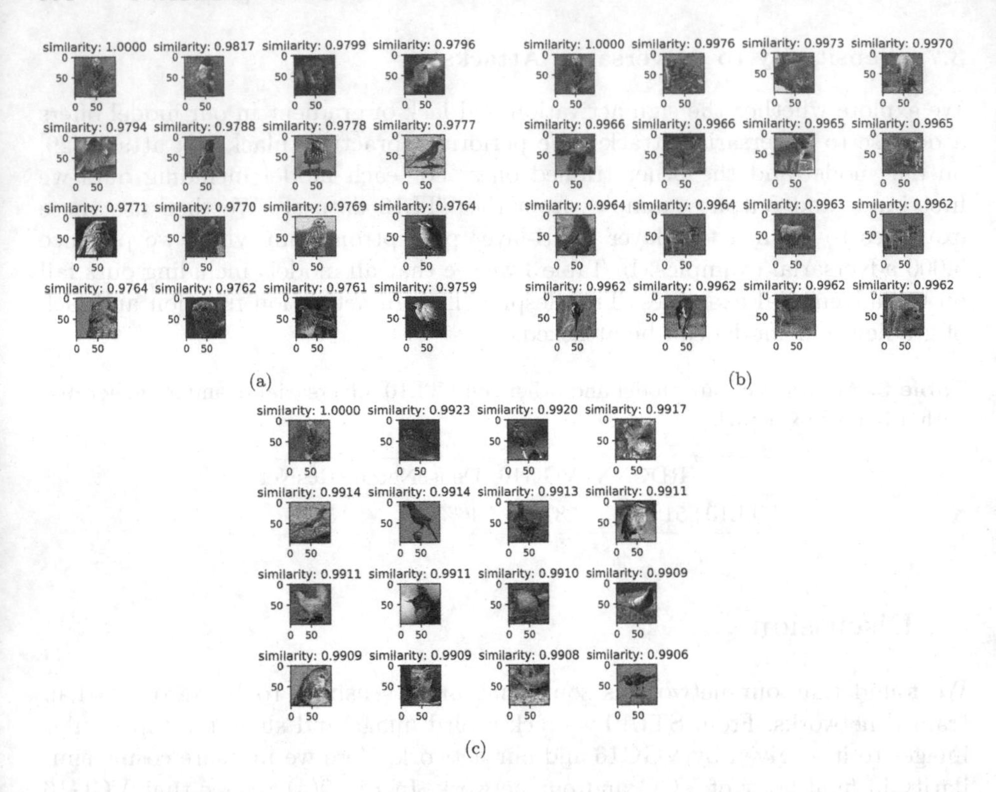


**Fig. 9.** Shown here are cosine similarity values of the top 16 images similar to the bird in the upper left. In (a) are images from a trained VGG and in (b) are images from RDCNN's untrained final layer. In (c) are images similar to the bird after training the final layer of RDCNN.

# 5 Conclusion

We propose a random depthwise convolutional neural network with thousands of convolutional kernels per layer. We show that our network can achieve better accuracies than other random networks and unsupervised feature learning methods, and competitive accuracies in image classification and retrieval compared to trained networks. We highlight sensitivity to background which is a pitfall but also a potential advantage that remains to be explored.

# References

1. Krizhevsky, A., Sutskever, I., Hinton, G.E.: Imagenet classification with deep convolutional neural networks. In: Pereira, F., Burges, C.J.C., Bottou, L., Weinberger, K.Q. (eds.) Advances in Neural Information Processing Systems, vol. 25, pp. 1097–1105. Curran Associates Inc. (2012)
2. Bottou, L.: Large-scale machine learning with stochastic gradient descent. In: Proceedings of COMPSTAT'2010, pp. 177–186. Springer (2010)

3. Salamon, J., Bello, J.P.: Deep convolutional neural networks and data augmentation for environmental sound classification. IEEE Signal Process. Lett. **24**(3), 279–283 (2017)
4. Le, Q.V.: Building high-level features using large scale unsupervised learning. In: 2013 IEEE International Conference on Acoustics, Speech and Signal Processing (ICASSP), pp. 8595–8598. IEEE (2013)
5. Srivastava, N., Hinton, G., Krizhevsky, A., Sutskever, I., Salakhutdinov, R.: Dropout: a simple way to prevent neural networks from overfitting. J. Mach. Learn. Res. **15**(1), 1929–1958 (2014)
6. DeVries, T., Taylor, G.W.: Improved regularization of convolutional neural networks with cutout. arXiv preprint arXiv:1708.04552 (2017)
7. Krizhevsky, A., Sutskever, I., Hinton, G.E.: Imagenet classification with deep convolutional neural networks. In: Advances in Neural Information Processing Systems, pp. 1097–1105 (2012)
8. Murata, N.: A statistical study of on-line learning. In: Murata, N. (ed.) Online Learning and Neural Networks, pp. 63–92. Cambridge University Press, Cambridge (1998)
9. Saxe, A.M., Koh, P.W., Chen, Z., Bhand, M., Suresh, B., Ng, A.Y.: On random weights and unsupervised feature learning. In: ICML, pp. 1089–1096 (2011)
10. Jarrett, K., Kavukcuoglu, K., LeCun, Y., et al.: What is the best multi-stage architecture for object recognition? In: 2009 IEEE 12th International Conference on Computer Vision, pp. 2146–2153. IEEE (2009)
11. Pinto, N., Doukhan, D., DiCarlo, J.J., Cox, D.D.: A high-throughput screening approach to discovering good forms of biologically inspired visual representation. PLoS Comput. Biol. **5**(11), e1000579 (2009)
12. Gilbert, A.C., Zhang, Y., Lee, K., Zhang, Y., Lee, H.: Towards understanding the invertibility of convolutional neural networks. arXiv preprint arXiv:1705.08664 (2017)
13. He, K., Wang, Y., Hopcroft, J.: A powerful generative model using random weights for the deep image representation. In: Advances in Neural Information Processing Systems, pp. 631–639 (2016)
14. Corel-princeton image similarity benchmark. http://www.cs.princeton.edu/cass/benchmark/
15. He, K., Zhang, X., Ren, S., Sun, J.: Deep residual learning for image recognition. In: Proceedings of the IEEE Conference on Computer Vision and Pattern Recognition, pp. 770–778 (2016)
16. Huang, G., Liu, Z., Weinberger, K.Q., van der Maaten, L.: Densely connected convolutional networks. In: Proceedings of the IEEE Conference on Computer Vision and Pattern Recognition, vol. 1, p. 32 (2017)
17. Simonyan, K., Zisserman, A.: Very deep convolutional networks for large-scale image recognition. arXiv preprint arXiv:1409.1556 (2014)
18. LeCun, Y., Bottou, L., Bengio, Y., Haffner, P.: Gradient-based learning applied to document recognition. Proc. IEEE **86**(11), 2278–2324 (1998)
19. Krizhevsky, A.: Learning multiple layers of features from tiny images (2009)
20. Coates, A., Ng, A., Lee, H.: An analysis of single-layer networks in unsupervised feature learning. In: Proceedings of the Fourteenth International Conference on Artificial Intelligence and Statistics, pp. 215–223 (2011)
21. Russakovsky, O., et al.: ImageNet large scale visual recognition challenge. Int. J. Comput. Vis. (IJCV) **115**(3), 211–252 (2015)
22. Fan, R.-E., Chang, K.-W., Hsieh, C.-J., Wang, X.-R., Lin, C.-J.: Liblinear: a library for large linear classification. J. Mach. Learn. Res. **9**, 1871–1874 (2008)

23. Fuglede, B., Topsoe, F.: Jensen-Shannon divergence and hilbert space embedding. In: International Symposium on Information Theory, ISIT 2004, Proceedings, p. 31. IEEE (2004)
24. Wu, Z., Xiong, Y., Yu, S.X., Lin, D.: Unsupervised feature learning via non-parametric instance discrimination. In: Proceedings of the IEEE Conference on Computer Vision and Pattern Recognition, pp. 3733–3742 (2018)
25. Wang, H., Cai, Y., Zhang, Y., Pan, H., Lv, W., Han, H.: Deep learning for image retrieval: what works and what doesn't. In: 2015 IEEE International Conference on Data Mining Workshop (ICDMW), pp. 1576–1583. IEEE (2015)
26. Razavian, A.S., Azizpour, H., Sullivan, J., Carlsson, S.: CNN features off-the-shelf: an astounding baseline for recognition. In: Proceedings of the IEEE Conference on Computer Vision and Pattern Recognition Workshops, pp. 806–813 (2014)
27. Wan, J., et al.: Deep learning for content-based image retrieval: a comprehensive study. In: Proceedings of the 22nd ACM International Conference on Multimedia, pp. 157–166. ACM (2014)
28. Pedregosa, F., et al.: Scikit-learn: machine learning in Python. J. Mach. Learn. Res. **12**, 2825–2830 (2011)
29. Papernot, N., McDaniel, P., Goodfellow, I., Jha, S., Celik, Z.B., Swami, A.: Practical black-box attacks against machine learning. In: Proceedings of the 2017 ACM on Asia Conference on Computer and Communications Security, pp. 506–519. ACM (2017)
30. Shapiro, L.G., Atmosukarto, I., Cho, H., Lin, H.J., Ruiz-Correa, S., Yuen, J.: Similarity-based retrieval for biomedical applications. In: Perner, P. (ed.) Case-Based Reasoning on Images and Signals. SCI, vol. 73, pp. 355–387. Springer, Heidelberg (2008). https://doi.org/10.1007/978-3-540-73180-1_12
31. Town, C.: Content-based and similarity-based querying for broad-usage medical image retrieval. In: Sidhu, A., Dhillon, S. (eds.) Advances in Biomedical Infrastructure 2013. SCI, vol. 477, pp. 63–76. Springer, Heidelberg (2013). https://doi.org/10.1007/978-3-642-37137-0_8

# Exploring Classification, Clustering, and Its Limits in a Compressed Hidden Space of a Single Layer Neural Network with Random Weights

Meiyan Xie and Usman Roshan(✉)

Department of Computer Science,
New Jersey Institute of Technology, Newark, NJ 07090, USA
{mx42,usman}@njit.edu

**Abstract.** Classification in the hidden layer of a single layer neural network with random weights has shown high accuracy in recent experimental studies. We further explore its classification and clustering performance in a compressed hidden space on a large cohort of datasets from the UCI machine learning archive. We compress the hidden layer with a simple bit-encoding that yields a comparable error to the original hidden layer thus reducing memory requirements and allowing to study up to a million random nodes. In comparison to the uncompressed hidden space we find classification error with the linear support vector machine to be statistically indistinguishable from that of the network's compressed layer. We see that test error of the linear support vector machine in the compressed hidden layer improves marginally after 10,000 nodes and even rises when we reach one million nodes. We show that k-means clustering has an improved adjusted rand index and purity in the compressed hidden space compared to the original input space but only the latter by a statistically significant margin. We also see that semi-supervised $k$-nearest neighbor improves by a statistically significant margin when only 10% of labels are available. Finally we show that different classifiers have statistically significant lower error in the compressed hidden layer than the original space with the linear support vector machine reaching the lowest error. Overall our experiments show that while classification in our compressed hidden layer can achieve a low error competitive to the original space there is a saturation point beyond which the error does not improve, and that clustering and semi-supervised is better in the compressed hidden layer by a small yet statistically significant margin.

## 1 Introduction

Single layer neural networks are known to approximate any non-linear function within a threshold of error [1,2]. We also know from Cover's theorems [3] that there exists a linearly separable feature space given sufficient nodes in the single layer. Their applicability to classification is limited though because we usually do


I. Rojas et al. (Eds.): IWANN 2019, LNCS 11506, pp. 507–516, 2019.
https://doi.org/10.1007/978-3-030-20521-8_42

not know the non-linear function underlying the data and Cover's work doesn't tell us exactly how to obtain node weights that give a linearly separable space. As a result we initialize with random weights and use the back propagation algorithm to fit the weights on training data [4].

It is well known that optimizing a single layer network can overfit on the training data and give poor results on test data [5]. To relieve this overfitting we use methods like dropout [6] and stochastic gradient descent [7] while training the model. These methods introduce randomness into the training process with the intent to relieve overfitting.

On the extreme end we can use entirely random weights in the single hidden layer. In fact random weights have been studied as early as Rosenblatt's original paper on perceptrons [8]. Recent studies have shown that random weights in the hidden single layer followed by trained weights in the output layer give highly accurate classification results [9–12]. Of these the random bit regression and random bit forest give the highest accuracies [11,12]. This work is different from random projections [13,14] because here we apply a non-linear activation function on the hidden layer output whereas random projections work on the actual output.

In this paper we further explore classification in the feature space given by the hidden layer of a single layer random network. Specifically we compress the hidden layer with bit-encoding to reduce memory requirements and thus allowing for up to a million random nodes in the network. We compare the test error of the linear support vector machine and other classifiers in the compressed hidden layer compared to the original input space as well as k-means clustering and k-nearest neighbor semi-supervised learning.

In conclusion we see that adding random nodes reaches a saturation point and the error begins to increase when we are at a million random nodes. We see that the linear support vector machine in the hidden layer achieves classification errors comparable to state of the art methods in the original and compressed hidden space. We also see that both clustering and k-nearest neighbor semi-supervised learning are better in the compressed hidden layer by statistically significant margins (except for the adjusted rand index in k-means clustering). Overall our study shows that a single hidden layer with random weights improves classification and clustering by moderate statistically significant margins.

## 2 Methods

### 2.1 Single Layer Random Weights Network

In Fig. 1 we show a basic single layer neural network. Each node in the middle layer represents a linear classifier (or peceptron). The weights of each node are usually determined by the back propagation algorithm [4]. Briefly, this algorithm starts with initial random weights and optimizes each node at a time with gradient descent (here the gradient is given by the chain rule [4]). It iterates over the entire network until we have convergence.

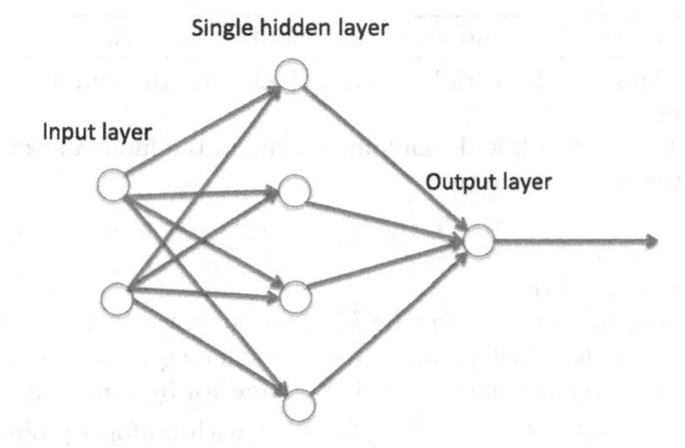


**Fig. 1.** Toy example of a single layer network. Here we see that the original input data is two dimensional. There are four nodes in the hidden layer which corresponds to a feature space of four dimensions. From this feature space we learn a classifier in the output layer.

In our work we use random weights in the hidden single layer followed by a trained linear support vector machine in the output layer. We describe our single layer network in detail in Algorithm 1. We see that our algorithm is similar to the random bit regression method [11] except for two things: the output layer and selection of offset. Our output layer is a support vector machine whereas they use logistic regression, and our offset is chosen randomly between projected points while theirs is an actual projected point (also randomly chosen).

Since the hidden layer has random weights the only optimization required is in the final layer. The outputs of the hidden layer represent a new feature space on which we optimize a linear support vector machine with the liblinear program [15].

# 3 Results

## 3.1 Experimental Performance Study

In order to evaluate the performance of the random weights network we study it on several dataset from the UCI repository at https://archive.ics.uci.edu/ml/.

**Datasets:** We obtained 52 datasets from the UCI repository. The datasets include data from different sources such as biological, medical, robotics, and business. Some of the datasets are multi-class and since we are studying only binary classification in this paper we convert them to binary. We label the largest class to be $-1$ and remaining as $+1$. We trim down excessively large datasets and ignore instances with missing values across the datasets. Thus, the number of instances in some of our datasets are different from that given in the UCI website https://archive.ics.uci.edu/ml/. For example the SUSY dataset originally has 5 million entries randomly ordered from which we choose the first 5000 for our study. We provide our data on the website http://web.njit.edu/~usman/randomnet.

**Algorithm 1.** Single layer random weight neural network

**Input:** Training data $x_i \in R^d$ with labels $y_i \in \{+1, -1\}$, the number of nodes $m \in N$ in the hidden layer

**Output:** Single layer network with random weights in the hidden layer and optimized SVM weights in the final output layer

**Procedure:**

**Single hidden layer:**

**for** node $k = 0$ to $m - 1$ **do**

1. Create a random vector $w_k$ for the $k^{th}$ node such that $w_{ki} \sim Uniform[-1, 1]$
2. In order to set the offset parameter $w_0$ first let $w_k^T x_i, \forall i = 0, ..., n-1$ be the projection of the training data on $w$. Determine $w_0$ by randomly picking it from $\{\frac{w^T x_0 + w^T x_1}{2}, \frac{w^T x_1 + w^T x_2}{2}, ..., \frac{w^T x_{n-2} + w^T x_{n-1}}{2}\}$ with uniform probability.
3. For each training point $x_i$ the output given by node $k$ is $z_{ki} = sign(w^T x_i + w_0)$

**end for**

**Final output layer:** Learn a linear cross-validated SVM model (which we do with the liblinear program [15]) on the outputs given by the hidden layer. This classifier represents a single node in the output layer.

**Software:** We use the Python scikit-learn library version 0.19.1. to study other classification and clustering programs in the original and our hidden space, except for linear support vector machine. For that we use the fast liblinear version 2.20 software.

- XGboost ver 0.81: a Python implementation of gradient boosting [16] for decision trees. We use the scikit-learn API called XGBClassifier.
- Multi-layer perceptron: Python sci-kit [17] implementation with a single layer of 100 hidden nodes.
- Liblinear ver 2.20: a fast linear support vector machine program [15] with cross-validated value of $C$ from the set $\{0.000001, 0.00001, 0.0001, 0.001, 0.01, 0.1, 1, 10, 100\}$.
- $K$-nearest neighbor: Python sci-kit [17] implementation with cross-validated value of number of neighbors $k$ from the set $\{1, 2, 5, 7, 10, 20, 50\}$ for the semi-supervised study. For the classification we use $k = 5$ neighbors.
- $K$-means clustering: Python sci-kit [17] implementation with 1000 random restarts. We cross-validate the number of clusters $k$ from the set $\{2, 3, 4, 5, 6, 7, 8, 16, 20, 24, 28\}$ for the respective accuracy measure.

**Experimental Platform:** We run our experiments on a cluster of computing nodes equipped with Intel Xeon E5-2660v2 2.27 GHz processors with one method and dataset exclusively on a processor core.

**Train and Test Splits:** For each dataset we create a single random partition into training and test datasets in the ratio of 90% to 10%. We run all programs on each training dataset and predict the labels in the corresponding test set.

**Measure of Accuracy:** We measure error as the number of misclassifications divided by the number of test datapoints. We use two measures for clustering accuracy: the adjusted Rand index [18] (ARI) and purity [19]. Briefly, the Rand index is a pairwise similarity measure defined by $RI = \frac{TP+TN}{TP+FP+FN+TN}$ whereas purity is the classification accuracy given by labeling clusters to maximize the accuracy. We use the adjusted Rand index (ARI) that accounts for pairs occurring due to chance.

### 3.2 Compressed Hidden Layer with Bit-Encoding

Since the output of each hidden layer is a 0 or 1 we perform a simple bit encoding. We divide the hidden nodes into sets of $k$ and replace each set of $k$ outputs with their bit encoded single value. For example for a set of all 0's except for 1's in position 0 and 2 we would give it the number $2^0 + 2^2 = 5$. A 64 bit compression on 100,000 features would thus give us 1563 features in the end.

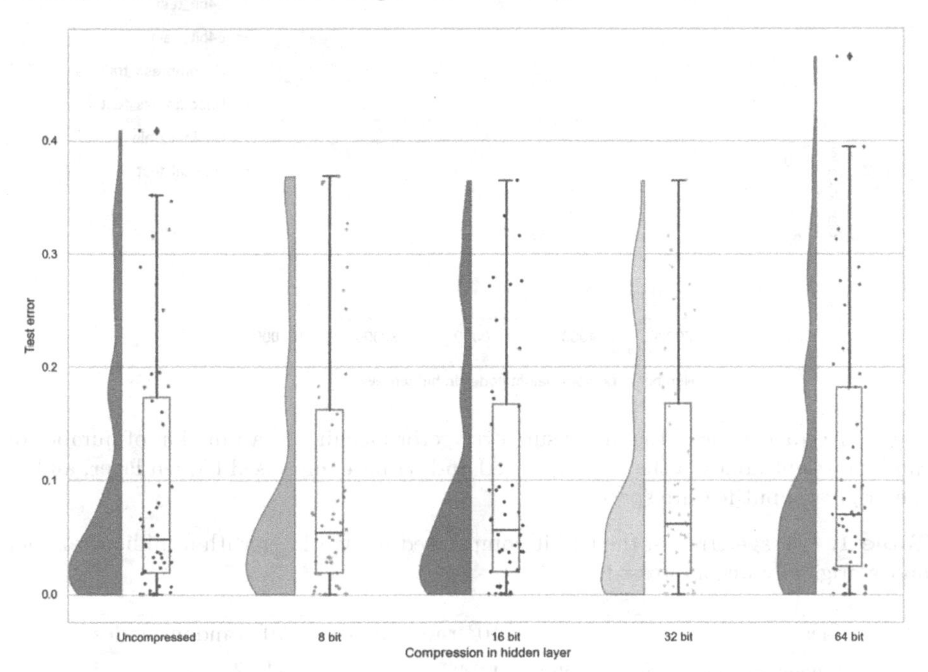


**Fig. 2.** Raincloud plot showing distribution of errors across the 52 datasets as well as the five summary statistics of min, max, first and third quartiles, and median. The graph shows errors of the linear support vector machine in the uncompressed hidden space vs. 8, 16, 32, and 64 bit compressions.

In Fig. 2 we see Raincloud plots showing the five summary statistics of median, min, max, first, and third quartile as well as the distribution of errors across the 52 datasets for different $k$-bit encodings. The uncompressed hidden layer has 100,000 random nodes. We see that the median error increases somewhat marginally as we increase the number of bits in the encoding. At 64 bits the median error is statistically indistinguishable by the t-test from the uncompressed version even though it is a little higher.

### 3.3 Effect of Number of Nodes on the Accuracy of the Linear Support Vector Machine

In Fig. 3 we see that as we increase the number of nodes both the train and test error in the uncompressed and 64-bit encoded compressed layers decrease but become flat after 10,000 nodes. With the compressed features we can now study classification accuracy in the hidden layer up to a million random nodes compressed with 64 bits. In Table 1 we see that the error of both the linear support vector machine and $k$-nearest neighbor increases as we go from 100,000 to a million random nodes.

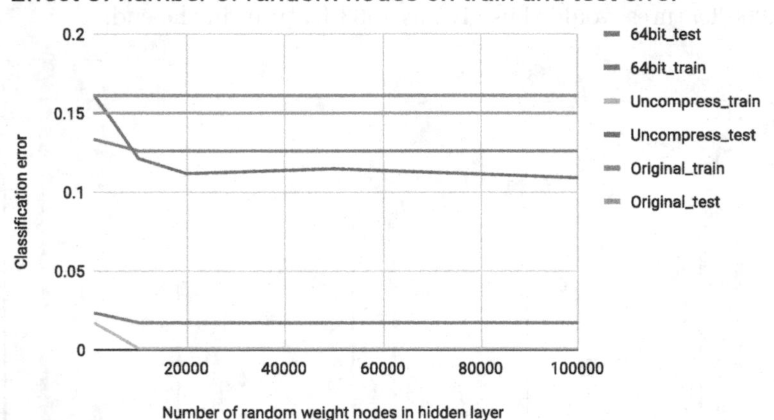


**Fig. 3.** Average error of the linear support vector machine as a function of number of random weight nodes in the uncompressed and 64-bit compressed hidden layer, and in the original input feature space

**Table 1.** Average error in the 64-bit compressed hidden layer with a million random nodes originally uncompressed

| Classifier | $10^5$ random nodes | $10^6$ random nodes |
|---|---|---|
| Linear support vector machine | 11.4% | 11.8% |
| K nearest neighbor | 13.4% | 13.9% |

### 3.4 Performance of k-Means Clustering in Original and 64 Bit Compressed Hidden Layer

In Fig. 4 we see the Raincloud plot of adjusted rand index (ARI) and the purity of k-means in the original and 64 bit encoded hidden layers. In the original uncompressed layer we have 100,000 random nodes. In both cases the improvement in median ARI and purity over the original space is small (about 2%) but statistically significant for purity under the t-test (p-value $< 0.01$).

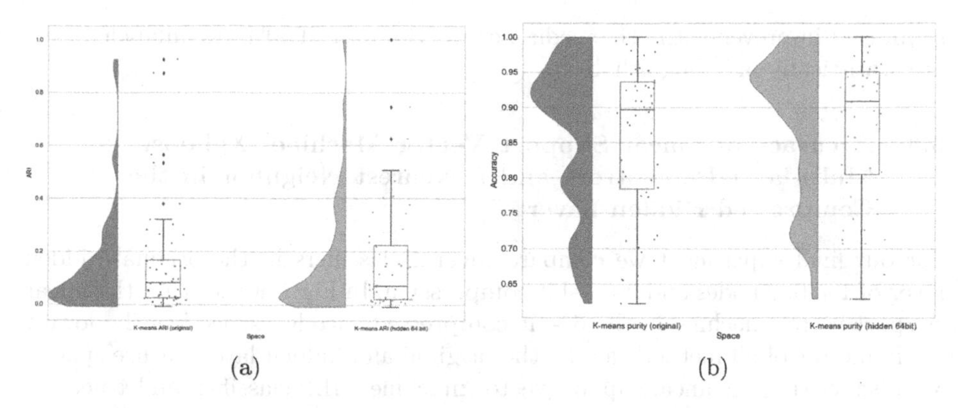


**Fig. 4.** Raincloud plots of k-means adjusted rand index and purity across the 52 datasets in the original space and the 64 bit compressed hidden layer

### 3.5 Performance of Semi-supervised k-Nearest Neighbor in Original and 64 Bit Compressed Hidden Layer

In this setup we randomly pick 90% (and 50% separately) of the training data and set it aside as unlabeled. We use the label propagation method [20] based on k-nearest neighbor to determine labels of the remainder of the training dataset. We then predict the test dataset with k-nearest neighbor.

In Fig. 5 we see the Raincloud plot of test error across the 52 datasets in the original hidden space of 100,000 random nodes and the compressed one. We see that when 90% of the training data is unlabeled there is a statistically

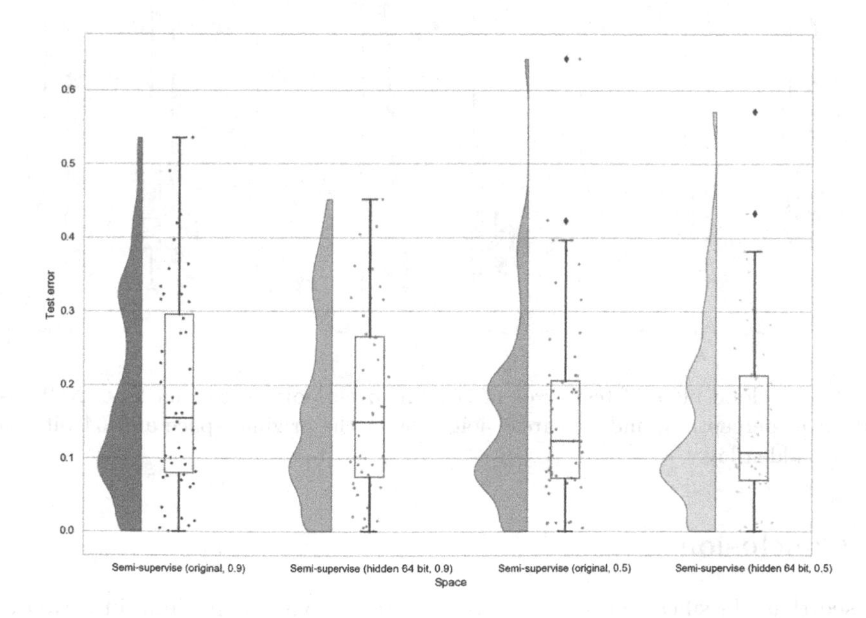


**Fig. 5.** Raincloud plots of k-nearest neighbor test error across the 52 datasets in the original space and the 64 bit compressed hidden layer

significant improvement in the median error. However, if 50% are unlabeled they are statistically indistinguishable.

### 3.6 Accuracy of Linear Support Vector Machine, Xgboost, Multi-layer Perceptron, and $k$-Nearest Neighbor in the Compressed Hidden Layer

For our final experiment we compare several classifiers in the original hidden layer of 100,000 nodes and its 64-bit compression. In Fig. 6 we see that the linear support vector machine in the 64-bit compressed encoded space has the lowest median error of all methods across the original and hidden layer feature spaces. We also see that the linear support vector machine, MLP classifier, and $k$-nearest neighbor all improve in the new feature space by statistically significant margins under the t-test (p-value $< 0.01$). However xgboost performs slightly better in the original space and is statistically indistinguishable in the hidden space under the t-test.

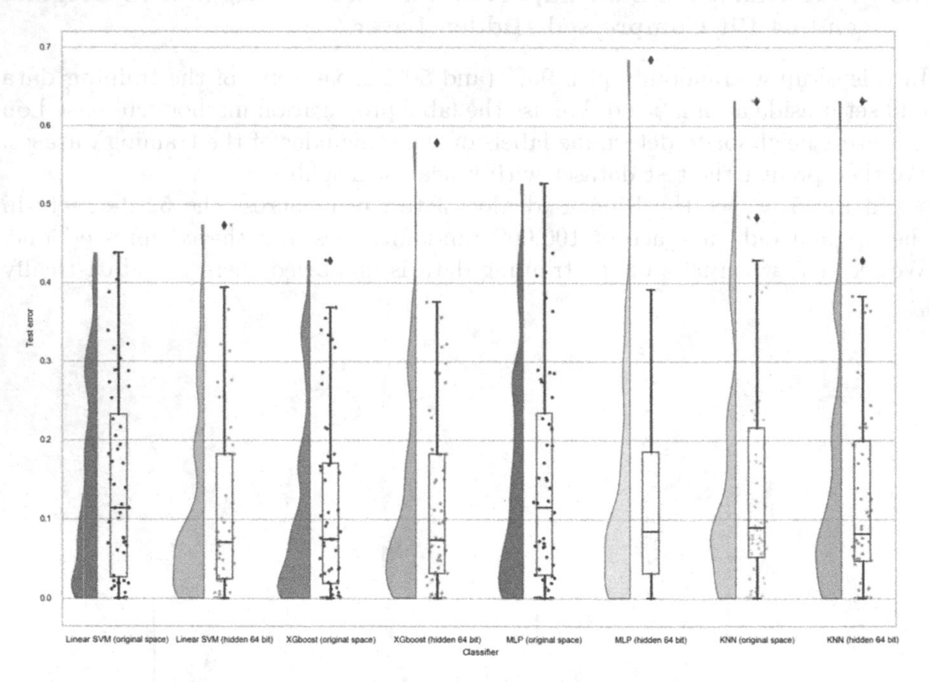


**Fig. 6.** Raincloud plots of test error of the linear support vector machine, XGboost, multi-layer perceptron, and $k$-nearest neighbor in the original space and 64 bit compressed hidden layer

## 4 Conclusion

We see that classification with the linear support vector machine in a random network's compressed hidden layer achieves low errors competitive with state of the art classifiers but does not decrease considerably beyond 10,000 random nodes. Both semi-supervised and clustering are better in the network's compressed hidden layer by statistically significant margins except for the case of adjusted rand index in clustering.

## References


1. Hornik, K.: Approximation capabilities of multilayer feedforward networks. Neural Netw. **4**(2), 251–257 (1991)
2. Cybenko, G.: Approximation by superpositions of a sigmoidal function. Math. Control Sign. Syst. **2**(4), 303–314 (1989)
3. Cover, T.M.: Geometrical and statistical properties of systems of linear inequalities with applications in pattern recognition. IEEE Trans. Electron. Comput. **14**(3), 326–334 (1965)
4. Rumelhart, D.E., Hinton, G.E., Williams, R.J., et al.: Learning representations by back-propagating errors. Cogn. Model. **5**(3), 1 (1988)
5. Caruana, R., Lawrence, S., Lee Giles, C.: Overfitting in neural nets: backpropagation, conjugate gradient, and early stopping. In: Advances in Neural Information Processing Systems, pp. 402–408 (2001)
6. Srivastava, N., Hinton, G., Krizhevsky, A., Sutskever, I., Salakhutdinov, R.: Dropout: a simple way to prevent neural networks from overfitting. J. Mach. Learn. Res. **15**(1), 1929–1958 (2014)
7. Bottou, L.: Large-scale machine learning with stochastic gradient descent. In: Lechevallier, Y., Saporta, G. (eds.) Proceedings of COMPSTAT 2010, pp. 177–186. Springer, Heidelberg (2010). https://doi.org/10.1007/978-3-7908-2604-3_16
8. Rosenblatt, F.: The perceptron: a probabilistic model for information storage and organization in the brain. Psychol. Rev., 65–386 (1958)
9. Schmidt, W.F., Kraaijveld, M.A., Duin, R.P.W.: Feedforward neural networks with random weights. In: 11th IAPR International Conference on Pattern Recognition, vol. II. Conference B: Pattern Recognition Methodology and Systems, pp. 1–4. IEEE
10. Huang, G.-B., Zhu, Q.-Y., Siew, C.-K.: Extreme learning machine: theory and applications. Neurocomputing **70**(1–3), 489–501 (2006)
11. Wang, Y., Li, Y., Xiong, M., Shugart, Y.Y., Jin, L.: Random bits regression: a strong general predictor for big data. Big Data Analytics **1**(1), 12 (2016)
12. Wang, Y., et al.: Random bits forest: a strong classifier/regressor for big data. Sci. Rep. **6** (2016)
13. Bingham, E., Mannila, H.: Random projection in dimensionality reduction: applications to image and text data. In: Proceedings of the Seventh ACM SIGKDD International Conference on Knowledge Discovery and Data Mining, pp. 245–250. ACM (2001)
14. Johnson, W.B., Lindenstrauss, J.: Extensions of lipschitz mappings into a hilbert space. Contemp. Math. **26**, 189–206 (1984)
15. Fan, R.-E., Chang, K.-W., Hsieh, C.-J., Wang, X.-R., Lin, C.-J.: LIBLINEAR: a library for large linear classification. J. Mach. Learn. Res. **9**, 1871–1874 (2008)
16. Friedman, J.H.: Greedy function approximation: a gradient boosting machine. Ann. Stat., 1189–1232 (2001)
17. Pedregosa, F., et al.: Scikit-learn: machine learning in Python. J. Mach. Learn. Res. **12**, 2825–2830 (2011)

18. Rand, W.M.: Objective criteria for the evaluation of clustering methods. J. Am. Stat. Assoc. **66**(336), 846–850 (1971)
19. Manning, C., Raghavan, P., Schütze, H.: Introduction to information retrieval. Nat. Lang. Eng. **16**(1), 100–103 (2010)
20. Zhu, X., Ghahramani, Z.: Learning from labeled and unlabeled data with label propagation. Technical report. Citeseer (2002)

# Improving Randomized Learning of Feedforward Neural Networks by Appropriate Generation of Random Parameters

Grzegorz Dudek(✉)

Electrical Engineering Department, Częstochowa University of Technology, Częstochowa, Poland
dudek@el.pcz.czest.pl

**Abstract.** In this work, a method of random parameters generation for randomized learning of a single-hidden-layer feedforward neural network is proposed. The method firstly, randomly selects the slope angles of the hidden neurons activation functions from an interval adjusted to the target function, then randomly rotates the activation functions, and finally distributes them across the input space. For complex target functions the proposed method gives better results than the approach commonly used in practice, where the random parameters are selected from the fixed interval. This is because it introduces the steepest fragments of the activation functions into the input hypercube, avoiding their saturation fragments.

**Keywords:** Function approximation · Feedforward Neural Networks · Neural networks with random hidden nodes · Randomized learning algorithms

## 1 Introduction

Feedforward neural networks (FNNs) learn from data by iteratively tuning their parameters, weights and biases, using some form of gradient descent method. Due to the layered structure of the network, the learning process is complicated, inefficient and time consuming. The network converges to the local optima, and the final result is very sensitive to the initial values of parameters.

Some of these drawbacks can be avoided by using a randomized learning approach. In this case the weights and biases of the hidden nodes are randomly selected from given intervals according to any continuous sampling distribution and remain fixed. The only parameters that are learned are the weights between the hidden layer and the output layer. The resulting optimization task becomes convex and can be formulated as a linear least-squares problem [1]. So, the problem can be considered as a linear one and the gradient descent method is not

Supported by Grant 2017/27/B/ST6/01804 from the National Science Centre, Poland.


I. Rojas et al. (Eds.): IWANN 2019, LNCS 11506, pp. 517–530, 2019.
https://doi.org/10.1007/978-3-030-20521-8_43

needed to solve it. The output weights can be analytically determined through a simple generalized inverse operation of the hidden layer output matrices. Due to the non-iterative nature, the randomized learning of FNNs can be much faster than classical gradient descent-based learning. In addition, it is easy to implement in any computing environment.

It was theoretically proven that FNN is a universal approximator for a continuous function on a bounded finite dimensional set, when the random parameters are selected from a uniform distribution within a proper range [2]. Husmeier showed that the universal approximation property also holds for a symmetric interval for random parameters if the function to be approximated meets the Lipschitz condition [3]. However, how to select the range for the random parameters remains an open question. It is well known that this range has a significant impact on the performance of the network. This has already been noted in early works on randomized learning algorithms, e.g. authors of [4] and [3] recommended optimizing the interval for a specified task. In [5] a series of simulations were carried out to illustrate the significance of the range of random parameters on modeling performance. The authors empirically showed that a widely used setting for this range, usually $[-1, 1]$, is misleading because the network is unable to model nonlinear maps, no matter how many training samples are provided or what sized networks are used. Although, they observed that for some specific ranges the network performs better in both learning and generalization than for other ranges, they do not give any tips on how to select an appropriate range.

In [6] the problem of the random parameters range is investigated by introducing a scaling factor to control the ranges of the randomization: $[-s, s]$ is used for weights and $[0, s]$ for biases. The authors observed that the commonly adopted approach where $s = 1$ may not lead to optimal performance. The network performs poorly when the range of the random parameters becomes either too large or too small. Setting small $s$ to increase the discrimination power of the features in the hidden neurons may cause more neurons to saturate. On the other hand, setting large $s$ to reduce the possibility of neuronal saturation may reduce the discrimination power of the features in the hidden neurons. To find the optimal symmetric interval for weights and biases in [7] a stochastic configuration algorithm is used. Random parameters are generated with an inequality constraint adaptively selecting the range for them, ensuring the universal approximation property of the model.

In [8] it was noted that if the network nodes are chosen at random and not subsequently trained, they are usually not located in accordance with the density of the input data. Consequently, the training of linear parameters becomes ineffective at reducing errors in the nonlinear part of the network. Moreover, the number of nodes needed to approximate a nonlinear map grows exponentially, and the model is very sensitive to the random parameters. To improve the effectiveness of the network, the unsupervised placement of network nodes according to the input data density could be combined with the subsequent supervised or reinforcement learning values of the linear parameters of the approximator.

Despite intensive research in recent years into randomized learning of FNNs, there are still several open problems which need to be addressed, such as how to generate random parameters. This issue remains untouched in the literature and is considered to be one of the most important research gaps in the field of randomized algorithms for training NNs [9,10]. In many practical applications in classification or regression problems the ranges are selected as fixed without scientific justification, typically $[-1, 1]$, regardless of the data and activation function type.

Recently, in [11] a new method of random parameters generation was proposed. The formulas for weights and biases were derived assuming that the steepest fragments of the activation functions are located in the input space region and their slopes are adjusted to the target function complexity. This method of generating random parameters allows us to control the generalization degree of the model and leads to an improvement in the approximation performance of the network.

This work follows on from [11]. Here we propose an alternative way of initially setting the activation functions for single-hidden-layer FNN, where we randomly determine their slope angles and rotations in space. Instead of selecting the weights and biases from the assumed interval, we randomly select the slope angles from the interval adjusted to the target function, and then randomly rotating the activation functions and shifting them into the input hypercube, we calculate weights and biases. As shown in Sect. 3, the proposed method gives much better results than the standard approach with fixed intervals for random parameters. It is also more intuitive than the method proposed in [11] as it has parameters which are easily interpreted.

## 2 Generating Random Parameters of Hidden Nodes

For brevity, we use the following acronyms:

CSs - constructional sigmoids, i.e. the set of sigmoid activation functions of hidden nodes whose linear combination forms the function fitting data,
II - input interval,
FF - fitted function (curve or surface) constructed by FNN,
TF - target function.

In this section we analyze how the random parameters affect the approximation abilities of FNN. The standard approach for generating random parameters is analyzed and a new approach is proposed. We consider a single-hidden-layer FNN with one output, $m$ hidden neurons and $n$ inputs. A sigmoid is used as an activation function. The output weights are calculated using the Moore-Penrose pseudo-inverse operation.

To illustrate the results, let us use a single-variable TF in the form:

$$g(x) = \sin(20 \cdot \exp(x)) \cdot x^2 \quad (1)$$

where $x \in [0, 1]$.

This function is shown by the dashed line in the upper chart of Fig. 1. As can be seen in this figure, variation of TF (1) increases along the II = $[0, 1]$. At the left bound of the II the TF is flat, while towards the right bound its fluctuations increase.

For NN learning we generate a training set containing 5000 points $(x_l, y_l)$, where $x_l$ are uniformly randomly distributed on $[0, 1]$ and $y_l$ are calculated from (1) and then distorted by adding the uniform noise distributed in $[-0.2, 0.2]$. A test set of the same size is created similarly but without noise. The outputs are normalized in the range $[-1, 1]$.

We consider the sigmoid activation function for hidden nodes:

$$h(x) = \frac{1}{1 + \exp(-(a \cdot x + b))} \tag{2}$$

where $a$ is a weight deciding about the slope of the sigmoid ($\mathrm{d}h/\mathrm{d}x$) and $b$ is a bias shifting it along the x-axis.

The set of sigmoids included in the hidden neurons are the basis functions whose linear combination forms the function fitting data. For nonlinear TF this set should deliver the nonlinear parts of sigmoids (avoiding their saturated fragments) to model the TF with required accuracy. The sigmoids should also be distributed properly in the II so that their steep fragments correspond to the steep fragments of the TF. Are these requirements met when the sigmoids weights and biases are both generated randomly from the typical interval $[-1, 1]$?

The left panel of Fig. 1 shows results of function (1) fitting when using FNN with 100 hidden nodes whose parameters $a$ and $b$ are both selected randomly from $[-1, 1]$. The upper chart shows the training points and the FF (solid line). The CSs are shown in the middle chart and CSs multiplied by the output weights are shown in the bottom chart. Note that CSs are flat in the II, which is shown as a gray field, and many of them have their steepest parts, which are around their inflection points (sigmoid value for the inflection point is 0.5), outside the II. Thus, the CSs slopes and their distribution in II do not correspond to TF fluctuations. This results in a very weak fitting. This simple example leads to the conclusion that random parameters of hidden nodes cannot be generated from the fixed interval $[-1, 1]$. The intervals for them should, instead, be estimated taking into account the II and the TF features, such as fluctuations.

### 2.1 The Idea of the Proposed Method

Instead of determining the interval for weights $a$, we determine the interval for the slope angles of CSs. By the slope angle $\alpha$ we mean the angle between a tangent line to the sigmoid at its inflection point and the x-axis. Using the slope angles instead of the weights is more intuitive because we can imagine without any effort a sigmoid having slope angle $\alpha$, but it is hard to imagine a sigmoid having the weight $a$. So, it is clear to us what CSs for $\alpha \in [30°, 60°]$ look like, and we have no idea what CSs for $a \in [2.31, 6.93]$ look like. In fact, they look similar because these are corresponding intervals.

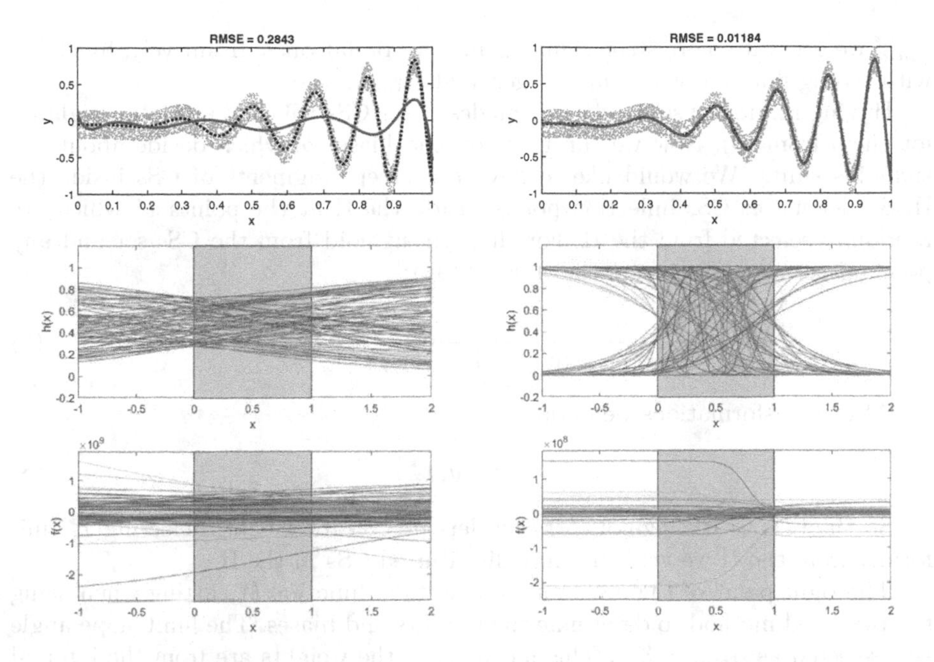


**Fig. 1.** Fitted curve (upper panel), CSs (middle panel) and weighted CSs (bottom panel) for the standard method, $a, b \in [-1, 1]$ (left panel), and the proposed method, $\alpha_{min} = 30°$ (right panel).

The weight $a$ translates nonlinearly into the slope angle. To find the relationship between them, it should be noted that the derivative of a sigmoid at the inflection point is equal to $\tan \alpha$:

$$ah(x)(1 - h(x)) = \tan \alpha \tag{3}$$

Setting this sigmoid in such a way that its inflection point is in $x = 0$ we get:

$$a \frac{1}{1 + \exp(-(a \cdot 0 + 0))} \left(1 - \frac{1}{1 + \exp(-(a \cdot 0 + 0))}\right) = \tan \alpha \tag{4}$$

From (4) we achieve:

$$a = 4 \tan \alpha \tag{5}$$

Note that for a typical interval for $a$, i.e. $[-1, 1]$ we get $\alpha \in [-14°, 14°]$, so rather flat CSs. Note also that selecting $a$ uniformly from a certain interval (typical case) we get nonlinearly distributed angles $\alpha$, according to the tangent function. We propose in this work instead of selecting weights $a$ for CSs, randomly selecting the slope angles $\alpha$ for them from the interval:

$$\Gamma = (-90°, -\alpha_{min}) \cup (\alpha_{min}, 90°) \tag{6}$$

where $\alpha_{min} \in [0°, 90°]$ is a limit angle adjusted to the TF. For lower value of $\alpha_{min}$ we have both flat and steep sigmoids in the set of CSs. When increasing

$\alpha_{min}$ we generate more steep sigmoids instead of flat ones. Intuitively, functions with strong fluctuations require larger angle $\alpha_{min}$.

Having randomly selected slope angles $\alpha$ for CSs we can calculate weights $a$ for them from (5). Now we can focus on the biases $b$, which decide about the sigmoids shifts. We would like to have the steep fragments of CSs inside the II. So, let us set CSs inflection points inside the II at the points $x^*$ which are randomly selected from the II. For the $i$-th sigmoid from the CSs set and any point $x_i^*$ selected for it from the II we obtain:

$$\frac{1}{1+\exp(-(a_i x_i^* + b_i))} = 0.5 \tag{7}$$

After transformations we obtain:

$$b_i = -a_i x_i^* \tag{8}$$

So the bias of the sigmoid strictly depends on its weight. Selecting $x_i^*$ uniformly from the II we get uniformly distributed CSs in the II.

The right panel of Fig. 1 shows the results of function (1) fitting when using the proposed method to determine the weights and biases. The limit slope angle was selected as $\alpha_{min} = 30°$. This means that the weights are from the interval $(-\infty, -2.31) \cup (2.31, \infty)$. But note that we do not uniformly select weights from this interval. If that were the case, then most of them would have high values corresponding to angles close to 90°. Instead, we select the angles randomly from the uniform distribution on (6), and then calculate weights. Compare the right panel of Fig. 1 with the left one and note the completely different CSs distribution and slopes resulting in a pretty good fit.

### 2.2 Multidimensional Case

The idea of random parameters generation is now expanded for the multi-dimensional case. Let us consider a multi-dimensional sigmoid $S$ of the form:

$$h(\mathbf{x}) = \frac{1}{1+\exp\left(-\left(\mathbf{a}^T\mathbf{x} + b\right)\right)} \tag{9}$$

which has one of its inflection points, point $P$, in the origin of the Cartesian coordinate system, i.e. in $O = (0, 0, ..., 0)$. (The sigmoid inflection points are all points for which $h(\mathbf{x}) = 0.5$). Let $T$ be the tangent hyperplane to the sigmoid at point $P$. This hyperplane takes the form:

$$a_1' x_1 + ... + a_n' x_n + a_0' y + b' = 0 \tag{10}$$

where $x_i$ are $n$ independent variables and $y$ is a dependent variable. A normal vector to this hyperplane is $\mathbf{n} = [a_1', ..., a_n', a_0']^T$. Let $\alpha \in (0°, 90°)$ be an angle between the normal vector $\mathbf{n}$ and the unit vector in the direction of the y-axis: $\mathbf{u} = [0, ..., 0, 1]^T$ (see Fig. 2). The vector $\mathbf{u}$ is normal for a hyperplane containing all x-axes. So, $\alpha$ is also an angle between this hyperplane and $T$, and decides

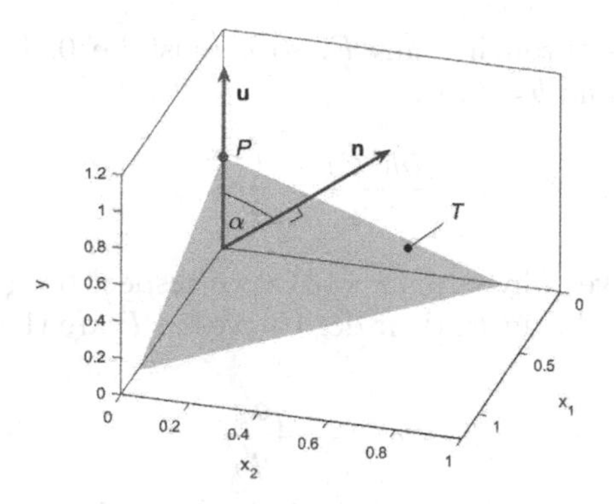


**Fig. 2.** Slope angle $\alpha$ as an angle between the normal vector $\mathbf{n}$ and the unit vector $\mathbf{u}$.

about the slope of the sigmoid. When $\alpha$ is near 0° we get flat sigmoid; when $\alpha$ is near 90° we get very steep sigmoid.

The cosine of this angle is expressed as:

$$\cos\alpha = \frac{\mathbf{u}\cdot\mathbf{n}}{\|\mathbf{u}\|\|\mathbf{n}\|} = \frac{0\cdot a'_1 + ... + 0\cdot a'_n + 1\cdot a'_0}{1\cdot\sqrt{(a'_1)^2 + ... + (a'_n)^2 + (a'_0)^2}} = \frac{a'_0}{\sqrt{(a'_0)^2 + ... + (a'_n)^2}} \tag{11}$$

From this equation we obtain:

$$a'_0 = \pm\sqrt{\frac{\cos^2\alpha}{1-\cos^2\alpha}\left((a'_1)^2 + ... + (a'_n)^2\right)} = \pm\frac{\sqrt{(a'_1)^2 + ... + (a'_n)^2}}{\tan\alpha} \tag{12}$$

where $\cos\alpha \neq 1$, $\tan\alpha \neq 0$, i.e. $\alpha \neq 0$.

Now we can construct the hyperplane $T$ inclined to the hyperplane containing all x-axes at an angle of $\alpha$, passing through point $P$ and randomly rotated around the y-axis. To do so, first we assume $\alpha \in (0°, 90°)$. Then, we generate randomly the first $n$ components of the normal vector $\mathbf{n}$: $a'_1, ..., a'_n$. Finally, we calculate its last component from (12). Note that this hyperplane defines the random rotation of the sigmoid around the y-axis and also its slope angle, which is $\alpha$. It is convenient to express hyperplane $T$ in the form:

$$y = -\frac{a'_1}{a'_0}x_1 - ... - \frac{a'_n}{a'_0}x_n + 0.5 \tag{13}$$

We achieve this equation from (10) assuming that $T$ passes through $P = (0, ..., 0, 0.5)$, so the intercept term must be 0.5.

Having randomly rotated a tangent hyperplane to sigmoid $S$, we are looking for weights $a_k$. A partial derivative of $S$ with respect to $x_k$ is:

$$\begin{aligned}\frac{\partial h(\mathbf{x})}{\partial x_k} &= a_k h(\mathbf{x})(1-h(\mathbf{x})) \\ &= a_k\frac{1}{1+\exp(-(\mathbf{a}^T\mathbf{x}+b))}\left(1-\frac{1}{1+\exp(-(\mathbf{a}^T\mathbf{x}+b))}\right)\end{aligned} \tag{14}$$

Sigmoid $S$ passes through point $P$, so $b$ must be 0. Derivative (14) in $P$, where $x_1, ..., x_n = 0$ and $b = 0$, is:

$$\frac{\partial h(P)}{\partial x_k} = \frac{1}{4} a_k \tag{15}$$

A partial derivative of hyperplane (13) with respect to $x_k$ is $-a'_k/a'_0$. Because $T$ is tangent to sigmoid $S$ in $P$, their derivatives in $P$ are the same, so: $1/4 \cdot a_k = -a'_k/a'_0$, and finally:

$$a_k = -4\frac{a'_k}{a'_0} \tag{16}$$

This equation expresses the relationship between the sigmoid weights and the normal vector to the tangent hyperplane defining the sigmoid slope and rotation. We propose to select the slope angles for CSs from the interval:

$$\Delta = (\alpha_{min}, 90°) \tag{17}$$

where $\alpha_{min} \in [0°, 90°]$ is a limit angle adjusted to the TF.

Having $\alpha$ for the sigmoid, we determine its random orientation (rotation) selecting randomly from any symmetric interval $n$ first components of the normal vector $\mathbf{n}$, and calculating the last one from (12). Repeating this for each sigmoid from the CSs set, we achieve a set of CSs randomly rotated around the y-axis, having a random slope angle greater than $\alpha_{min}$. Finally, we distribute the CSs into the input space. To do so, for each $i$-th sigmoid we randomly select point $\mathbf{x}_i^*$ from the input space and shift the sigmoid in such a way that its inflection point is in $\mathbf{x}_i^*$. Thus:

$$\frac{1}{1 + \exp(-(a_{i,1}x_{i,1}^* + ... + a_{i,n}x_{i,n}^* + b_i))} = 0.5 \tag{18}$$

Hence:

$$b_i = -a_{i,1}x_{i,1}^* - ... - a_{i,n}x_{i,n}^* \tag{19}$$

Note that the biases of the hidden nodes are not selected from a certain interval, as they are in the typical approach, but they are dependent on the node weights. Equation (19) ensures the distribution of CSs in the input space according to the data distribution.

### 2.3 Discussion

The proposed method of generating random parameters of hidden nodes ensures that the steep fragments of CSs are put inside the input space. The input space is an $n$-dimensional hypercube $H = [x_{\min 1}, x_{\max 1}] \times [x_{\min 2}, x_{\max 2}] \times ... \times [x_{\min n}, x_{\max n}]$, where $x_{\min k}$ and $x_{\max k}$ are the lower and upper bounds, respectively, for data in the dimension $k$. In most cases it is convenient to normalize all input variables into the range $[0, 1]$, so the input hypercube is $H = [0, 1]^n \subset \mathbb{R}^n$.

The steepness of CSs is controlled by the slope angle $\alpha_{min}$, i.e. the CSs generated have a slope angle from $\alpha_{min}$ to 90°. If we want to limit the maximum value of the slope angle, we can use $\alpha_{max} > \alpha_{min}$ as an upper bound for $\alpha$:

$$\Delta = (\alpha_{min}, \alpha_{max}) \tag{20}$$

Both parameters $\alpha_{min}$ and $\alpha_{max}$ should be adjusted to TF, e.g. in the cross-validation, such as the number of hidden nodes $m$. The number of hidden nodes depends on the TF as well. More complex TFs need more neurons to get sufficient approximation accuracy. To minimize the computational effort in the hyperparameters selection phase, we can consider fixed $\alpha_{min} = 0°$ and $\alpha_{max} = 90°$ and only search for $m$. If this does not bring satisfactory results we can also search for $\alpha_{min}$ assuming fixed $\alpha_{max} = 90°$. Finally, when this is still unsatisfactory we can also search for $\alpha_{max}$.

The CS rotation is determined by the normal vector to the tangent hyperplane $T$. Its first $n$ components are selected randomly, independently and uniformly from the same interval, which should be symmetrical, i.e. $[-d, d]$. In such a case each rotation is just as likely. The interval limit value $d$ is not important. We recommend the interval for $a'_k$ selection as $[-1, 1]$. The last component of the normal vector, $a'_0$, is calculated from (12) to ensure the slope of the CSs at an angle of $\alpha$. The sign for $a'_0$ is selected randomly.

In the bias determination step we select points $\mathbf{x}^*$ from hypercube $H$ and then shift the sigmoids to these points. To avoid shifting CSs to the empty region without data, it is reasonable to shift them to the points from the training set. In such a case, points $\mathbf{x}^*$ are the randomly selected training points. This ensures that all CSs have their steep fragments in the regions containing data. Another way of choosing points $\mathbf{x}^*$ is to group training points into $m$ clusters and take the prototypes $\mathbf{p}$ of these clusters (centroids) as $\mathbf{x}^*$.

In this work, sigmoids are considered as activation functions of the hidden nodes. But similar considerations can be made for other types of activation functions, such as Gaussian, softplus, sine and others (see [11], where an alternative method of random parameters generation was proposed and different activation functions were considered).

The procedure of generating the random parameters of FNNs described above is summarized in Algorithm 1.

## 3 Simulation Study

The proposed method of selecting random parameters for FNN is illustrated by several examples. The first one concerns a single-variable function approximation, where TF is in the form:

$$g(x) = 0.2e^{-(10x-4)^2} + 0.5e^{-(80x-40)^2} + 0.3e^{-(80x-20)^2} \tag{21}$$

**Algorithm 1.** Generating Random Parameters of FNNs

**Input:**

Number of hidden nodes $m$
Number of inputs $n$
Input hypercube $H$
Training set $\Phi$ (optionally)
Set of prototypes $\{\mathbf{p}_i\}_{i=1,\dots,m}$ (optionally)

**Output:**

Weights $\mathbf{A} = \begin{bmatrix} a_{1,1} & \dots & a_{m,1} \\ \vdots & \ddots & \vdots \\ a_{1,n} & \dots & a_{m,n} \end{bmatrix}$
Biases $\mathbf{b} = [b_1, \dots, b_m]$

**Procedure:**

Set $\alpha_{min} = 0°$ or choose $\alpha_{min} \in [0°, 90°]$
Set $\alpha_{max} = 90°$ or choose $\alpha_{max} \in [\alpha_{min}, 90°]$
**for** $i = 1$ **to** $m$ **do**
  Choose randomly $\alpha \sim U(\alpha_{min}, \alpha_{max})$
  Choose randomly i.i.d. $a'_1, \dots, a'_n \sim U(-1, 1)$
  Calculate

$$a'_0 = (-1)^c \frac{\sqrt{(a'_1)^2 + \dots + (a'_n)^2}}{\tan \alpha}$$

    where $c \sim U\{0,1\}$
  Calculate

$$a_{i,k} = -4\frac{a'_k}{a'_0} \quad \text{for } k = 1, 2, \dots, n$$

  Choose randomly $\mathbf{x}^* = [x^*_1, \dots, x^*_n]$ from $H$
    or set $\mathbf{x}^* = \mathbf{x}_j \in \Phi$, where $j \sim U\{1, 2, \dots, N\}$
    or set $\mathbf{x}^* = \mathbf{p}_i$, where $\mathbf{p}_i$ is a prototype of the $i$-th
      cluster of $\mathbf{x} \in \Phi$
  Calculate

$$b_i = -\sum_{k=1}^{n} a_{i,k} x^*_k$$

**end for**

The training set contains 1000 points $(x_l, y_l)$, where $x_l$ are uniformly randomly distributed on $[0, 1]$ and $y_l$ are calculated from (21). A test set of size 300 is generated from a regularly spaced grid on $[0, 1]$.

The left panel of Fig. 3 shows root-mean-square error (RMSE) while searching for the number of hidden nodes $m$ and $\alpha_{min}$ in 10-fold cross-validation procedure ($\alpha_{max}$ was set as 90°). The lowest RMSE ($1.34 \cdot 10^{-6}$) was achieved for $\alpha_{min} = 85$ and $m = 320$. The RMSE on the test set was $9.35 \cdot 10^{-7}$ at the optimal values of the hyperparameters. When using the standard method of random parameters generation, i.e. the weights and biases selected from the interval $[-1, 1]$, the test RMSE was above 0.1 regardless of the number of neurons. FFs for the proposed and standard methods of random parameters generation are shown in the right panel of Fig. 3 (500 nodes were used for the standard method). Note that the standard method is not able to fit the TF. This is because CSs are too flat in the II.

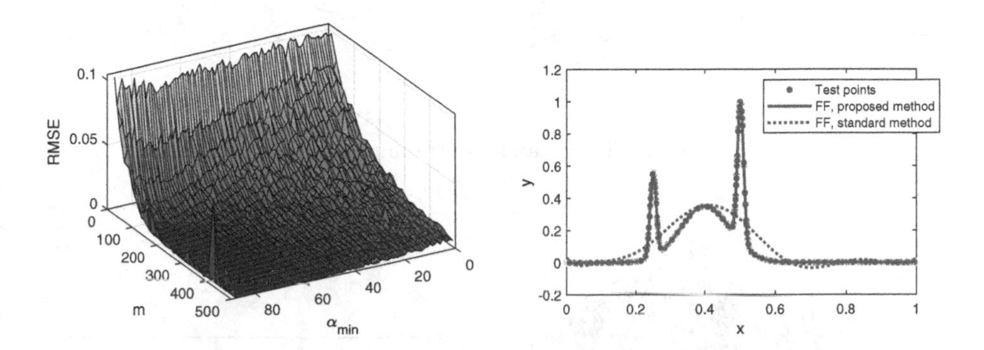


**Fig. 3.** TF (21) fitting using the proposed method: RMSE in the grid search (left panel) and fitted curves (right panel).

In the second example we use two-variable TF in the form:

$$g(\mathbf{x}) = \sin\left(20e^{x_1}\right) \cdot x_1^2 + \sin\left(20e^{x_2}\right) \cdot x_2^2 \tag{22}$$

The training set contains 5000 points $(\mathbf{x}_l, y_l)$, where both components of $\mathbf{x}_l$ are independently uniformly randomly distributed on $[0, 1]$ and $y_l$ are distorted by adding the uniform noise distributed in $[-0.2, 0.2]$. The test set containing 100000 points is distributed in the input space on a regularly spaced grid and is not disturbed by noise. The outputs are normalized in the range $[-1, 1]$. The TF and training points are shown in Fig. 4. Note that variation of the TF is the lowest in the corner $(0, 0)$ and increases towards the corner $(1, 1)$.

The left panel of Fig. 5 shows RMSE while searching for the hyperparameters values in a 10-fold cross-validation procedure. In this figure, we can observe a large plateau region for $0° < \alpha_{min} < 70°$ and $m > 500$. The RMSE in this region is less than 0.0850 at the lowest value of 0.0690 for $\alpha_{min} = 29°$ and $m = 700$.

The right panel of Fig. 5 shows RMSE when changing the upper bound slope angle $\alpha_{max}$ from $\alpha_{min}$ to 90° at the optimal values for $m$ and $\alpha_{min}$. As we can see from this figure, the lowest error is when $\alpha_{max}$ is above 70°.

Figure 6 shows the FF when using the proposed method with the optimal values of hyperparameters and the standard method when 700 neurons are used (increasing the neuron number did not improve the results). As we can see from these figures, the proposed method maps the TF quite well (RMSE = 0.0287) while the standard method fails (RMSE = 0.2326).

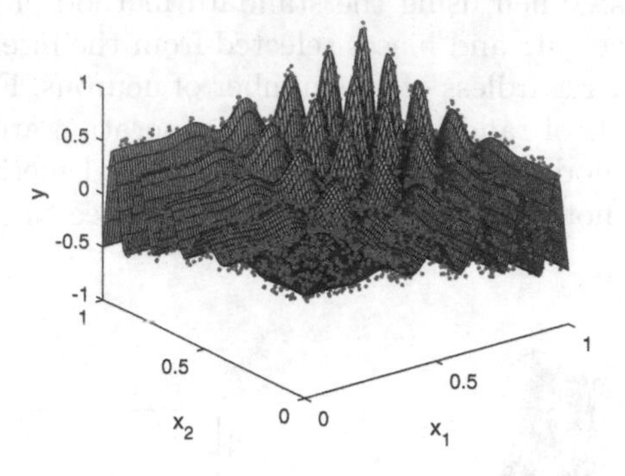


**Fig. 4.** TF (22) and the training points.

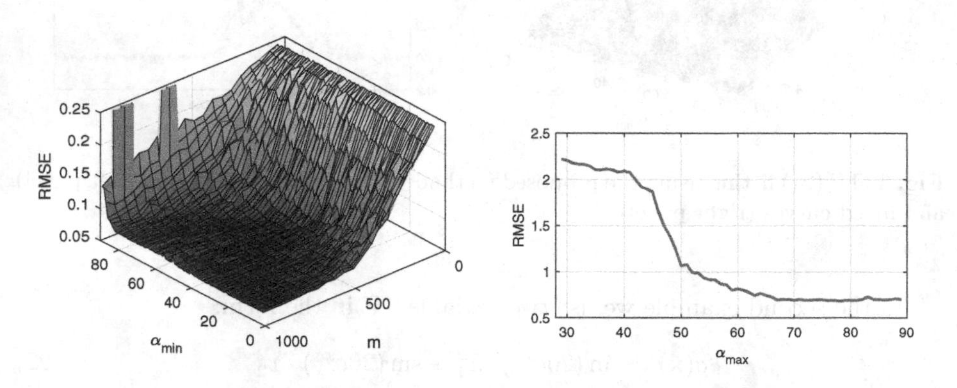


**Fig. 5.** TF (22) fitting using the proposed method: RMSE in the grid search (left panel) and impact of $\alpha_{max}$ on RMSE at $\alpha_{min} = 29°$ and $m = 700$ (right panel).

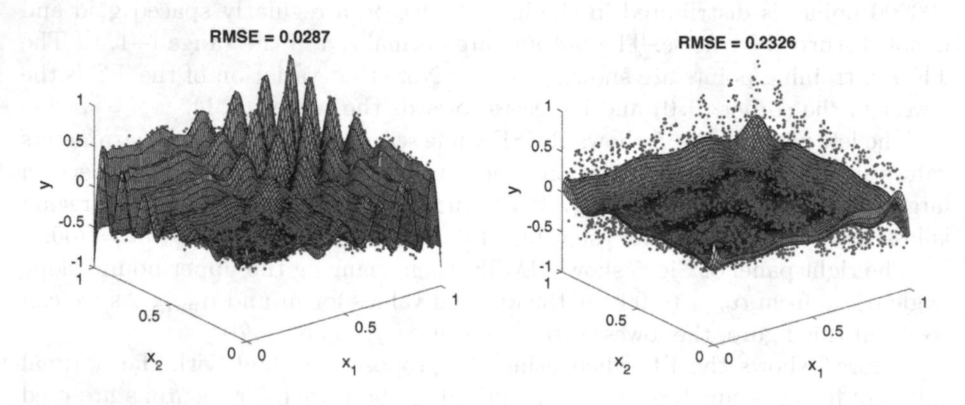


**Fig. 6.** Fitted surface for TF (22), the proposed (left panel) and standard method (right panel).

The last example concerns a 21-dimensional modeling problem: Compactiv - the Computer Activity dataset which is a collection of computer systems activity measures. The data was collected from a Sun Sparcstation 20/712 with 128 Mbytes of memory running in a multi-user university department. The task is to predict the portion of time that CPUs run in user mode. There are 8192 samples composed of 21 input variables (activity measures) and one output variable. The whole data set was divided into a training set containing 75% of samples selected randomly, and a test set containing the remaining samples. The dataset was downloaded from KEEL (Knowledge Extraction based on Evolutionary Learning) dataset repository (http://www.keel.es). The input and output variables are normalized into [0, 1].

The RMSE in the grid search procedure using a 10-fold cross-validation in Fig. 7 is shown. The lowest value of RMSE was 0.0309 for $\alpha_{min} = 45°$ and $m = 600$. The mean value of the test error for 100 trials of the learning sessions carried out at the optimal values of hyperparameters was 0.0335. For the standard method it was 0.0358. The difference between RMSE for the proposed and standard method is not as high as for the TFs (21) and (22). This is probably because the TF in this case has no strong fluctuations and can be modeled using flat neurons.

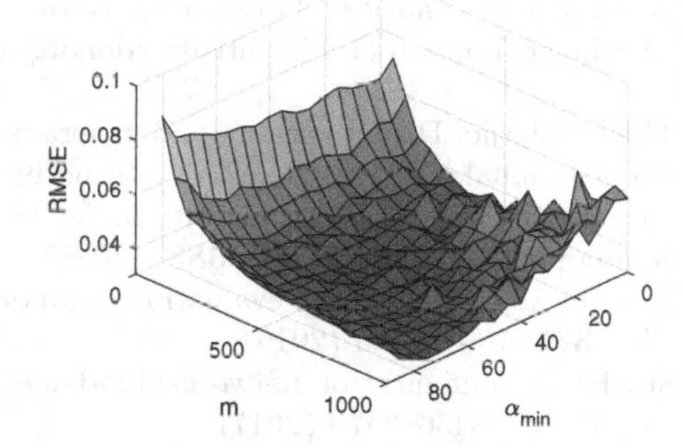


**Fig. 7.** Compactive data fitting using the proposed method: RMSE in the grid search.

## 4 Conclusions

One of the most important issues in the randomized learning of FNNs is the selection of random parameters: weights and biases of hidden nodes. In the existing methods the random parameters are selected uniformly from the fixed interval, such as [−1, 1] or another symmetrical interval whose bounds are adjusted to the problem being solved. We have shown that the weights and biases of the hidden nodes have a different meaning and should not be selected from the same interval.

In this work we recommend generating random parameters in single-hidden-layer FNNs taking into account the input space location and size, target function complexity, and activation function type. We propose a method which randomly selects the slope angles for the activation functions from the interval adjusted to the target function. Then, after rotating randomly the activation functions and shifting them into the input hypercube, we calculate weights and biases. The proposed approach turned out to be much more accurate than the existing one in regression problems. In the simulation study generating random parameters from the fixed interval $[-1, 1]$ brought very poor fitting, while the proposed method performed very well on target functions with strong fluctuations.

Future work will focus on the better adjustment of the hidden nodes to data. This should translate into a more compact network structure without redundant nodes. Additionally, the adaptation of the method to classification problems is planned.

## References


1. Principe, J., Chen, B.: Universal approximation with convex optimization: gimmick or reality? IEEE Comput. Intell. Mag. **10**, 68–77 (2015)
2. Igelnik, B., Pao, Y.-H.: Stochastic choice of basis functions in adaptive function approximation and the functional-link net. IEEE Trans. Neural Netw. **6**(6), 1320–1329 (1995)
3. Husmeier, D.: Random vector functional link (RVFL) networks. Neural Networks for Conditional Probability Estimation: Forecasting Beyond Point Predictions, chap. 6, pp. 87–97. Springer, London (1999). https://doi.org/10.1007/978-1-4471-0847-4_6
4. Pao, Y.-H., Park, G.H., Sobajic, D.J.: Learning and generalization characteristics of the random vector functional-link net. Neurocomputing **6**(2), 163–180 (1994)
5. Li, M., Wang, D.: Insights into randomized algorithms for neural networks: practical issues and common pitfalls. Inf. Sci. **382–383**, 170–178 (2017)
6. Zhang, L., Suganthan, P.: A comprehensive evaluation of random vector functional link networks. Inf. Sci. **367**, 1094–1105 (2016)
7. Wang, D., Li, M.: Stochastic configuration networks: fundamentals and algorithms. IEEE Trans. Cybern. **47**(10), 3466–3479 (2017)
8. Gorban, A.N., Tyukin, I.Y., Prokhorov, D.V., Sofeikov, K.I.: Approximation with random bases: pro- et contra. Inf. Sci. **364**, 146–155 (2016)
9. Zhang, L., Suganthan, P.: A Survey of randomized algorithms for training neural networks. Inf. Sci. **364**, 146–155 (2016)
10. Weipeng, C., Wang, X., Ming, Z., Gao, J.: A review on neural networks with random weights. Neurocomputing **275**, 278–287 (2018)
11. Dudek, G.: Generating random weights and biases in feedforward neural networks with random hidden nodes. Inf. Sci. **481**, 33–56 (2019)
12. Scardapane, S., Wang, D.: Randomness in neural networks: an overview. WIREs Data Min. Knowl. Discov. **7**(2), e1200 (2017)

# Pattern Recognition

# Detector of Small Objects with Application to the License Plate Symbols

Alexey Alexeev[1(✉)], Yuriy Matveev[1], Anton Matveev[1], Georgy Kukharev[2], and Sattam Almatarneh[3]

[1] ITMO University, Saint-Petersburg, Russia
aaalexeev@corp.ifmo.ru, matveev@speechpro.com, aush.tx@gmail.com
[2] LETI University, Saint-Petersburg, Russia
Kuga41@mail.ru
[3] Universidad de Santiago de Compostela, Santiago, Spain
sattam.almatarneh@usc.es

**Abstract.** In the article we look at an architecture of a detector of groups of small objects in close proximity to each other with distances between them as short as couples of pixels. In modern days the issue with detection of such small objects using a neural network is often the pooling based architecture leading to spatial information loss. We suggest a model of a convolutional network based on a fully connected convolutional network such as Network in Network (NiN). Accuracy of the detector is measured in a license plate recognition problem when images of license plates are produced by roads and highways video surveillance systems. Our aim is to present a solution to a specific problem without regards to use case specifics such as license plate edge detection, segmentation, and binarization of symbols. We focus on symbol detection and we process raw grayscale data. Furthermore we avoid license plate pattern detection and matching. In spite of narrow conditions we put on the problem the result we achieve is useful since it can be universally applied to many kinds of real world problems due to it being invariant to orientation in space and having low requirements to quality of an image. There are no particular requirements to size of an image being processed, but scaling might require to be executed in order to fit symbols in a predefined range, which in most commonly used systems is achievable due to positions of cameras and surveilled objects being known in advance. In our benchmarking we achieved mean Average Precision (mAP) of 90.25% which is on the level with modern automatic recognition systems for license plates.

**Keywords:** Object detection · Region proposal · CNN · NiN · License plates

This work was financially supported by the Government of the Russian Federation (Grant 08-08).


I. Rojas et al. (Eds.): IWANN 2019, LNCS 11506, pp. 533–544, 2019.
https://doi.org/10.1007/978-3-030-20521-8_44

## 1 Introduction

In recent years, the composition of image processing methods has changed significantly and now fast and scalable convolutional networks (CNN) take the major role [1,2]. With their arrival, it became possible to solve many problems of image processing, including object detection tasks [3–8]. For successive reduction of dimensionality of networks from the initial to the last levels, pooling is used. But as we know, when we use pooling, most of spatial information is lost. Therefore, standard detectors show good results mostly when determining objects of medium or large sizes. So in the description of the YOLO algorithm [4] it is stated that an error in determining locations of objects does not depend on their size. This leads to the fact that a relative positioning error increases with decrease of size of an object. The main purpose of detection is localization and recognition of objects with the highest mAP and the minimum probability of false positives. In this paper, we present our detector of small objects, which we test on license plates symbols detection task. Symbols on a license plate are viewed as small objects, centers of which sometimes are located within 4 pixels of each other and can represent the same class. Detection of small objects is a difficult task to standard convolutional networks. This is primarily due to the use of pooling, which leads to a loss of spatial information about objects. Working on the license plates symbols detection task we are only interested in universal neural network detection methods, which potentially might be applied for detection of many other types of objects.

## 2 Related Work

Modern license plate detectors often split the task in a number of steps: plate localization, segmentation and recognition of symbol, and also intermediate steps, such as pattern matching. This technique provides robust, though not quite agile, solutions. Reviewing related works we try to avoid methods heavily tuned to a specific environment and focus on neural network based methods, which can be applied universally for detection of small objects, independent of specifics of a particular task.

Among CNN based works we reviewed, most do not solve the problem of license plate symbol detection on a raw image and without preparational work [9–11]. Absence of such solutions comes from high error rate of localization of small objects: symbols on a license plate, in this particular case. [12] demonstrates a use of the classical YOLO detector, but employs it only after preparational work to locate a license plate is done; furthermore, it uses information about timespans between keyframes. [13] also suggests to employ the dimension of time alongside spatial dimensions. Since presence of this kind of information is rather use case specific, and it yet has not been shown that it can provide significant improvement, we left it out of the scope of our research. [7] shows how a model can be trained on a set of rescaled input images. Again, this puts additional restrictions on an environment, for which aforementioned methods could be used. In [14] a license number is sought for targeting symbols on it, but symbols themselves are not recognized and are to be precisely located in later stages of the algorithm. It is worth mentioning that most of the well-known systems based on neural networks are not fit for pixel precision localization task [3,4,7,8]. We also have not found a work, which would target detection of small objects in groups and in close proximity to each other.

# 3 Detector

Some well-known detection algorithms use a concept, known as "anchors", and requiring a number of classifier outputs with the size $N_{anchors} * M_{classes}$. Presence of such a large amount of output data may complicate work of a detector. For our detector, we have introduced a limited set of reference coordinates of objects (we name them boxes) in the image like anchors, which are uniformly spread across a search window and for which only the binary classification task is performed. Multi-classification is performed at a later stage, where coordinates of candidate objects are used as inputs. This subsequent classifier will have a single complete set of non-repeatable classes, unlike standard algorithms that use anchors. Thus, the total amount of detector outputs is defined proportionally $N_{boxess} * 2 + M_{classes}$ for two classes: presence or absence of an object, and subsequent multi-classification. In this paper, we describe all stages of detection and focus on an algorithm for determining locations of symbols on an automobile license plate, regardless of its type. The detector does not perform any specific to license plates image enhancing operations. The developed detection mechanism makes it possible to recognize symbols, regardless of a license plate type. To achieve that, we have to train it on different types of license plates.

The scheme of processing an image is shown in Fig. 1. The scheme differs from the scheme from [15], primarily due to splitting the detector into localizer and classifier, as well as the use of two-level clustering. Image processing includes image normalization and conversion from a full color to a grayscale image. Network block solves the problem of classification and regression of coordinates of symbols. Immediately after obtaining groups of coordinates of objects, the Density-Based Spatial Clustering of Applications with Noise (DBSCAN) can be used, optionally, which allows to calculate an average size of an object; this value can be used to more precisely configure the subsequent clustering of the centers of the objects at a later stage, which is performed using the mean-shift clusterizer. After the clustering, objects are classified.

## 3.1 Training Data

In order to overcome the problem of overfitting, as well as to obtain translation-invariant representations, the best solution is to use a large dataset of training data [16]. Also, a model trained on a larger set has better generalizing abilities. We decided to follow this route by choosing patches with random coordinates during training from whole images making up the dataset (Fig. 2). When choosing a size of patches, we wanted to provide detection on images of arbitrary size, as well as to take into account the context area [17], in this case, part of a license plate. We chose a detection area of $48 \times 48$ pixels with an average symbol size of $8 \times 12$ pixels. To ensure uniformity of processing of image areas and eliminate redundancy, we narrowed the viewing area to a square with a side of 24 pixels. The control points, which we name boxes, are in the center of the detection area ($24 \times 24$ inside and $48 \times 48$ in the center) set in chess order (Fig. 3). The distance of the symbol's search area for each box were set equal to the root of the distance between boxes. The configuration of boxes and the chosen distance ensure that significant part of symbols and a license plate surface will be inside detection area in case symbols are activated by any boxes.

In Fig. 3 a few of possible grid boxes configurations are represented. Blue dots show the grids of the boxes themselves, and the green dots mark the centers of the symbols. The aim of the detector is to activate the boxes that are closest to the centers of the symbols. The best results in terms of mAP in training were got using a $9 \times 9$ grid of boxes. In the rightmost column, images of symbols that have gone beyond the grid area are shown.

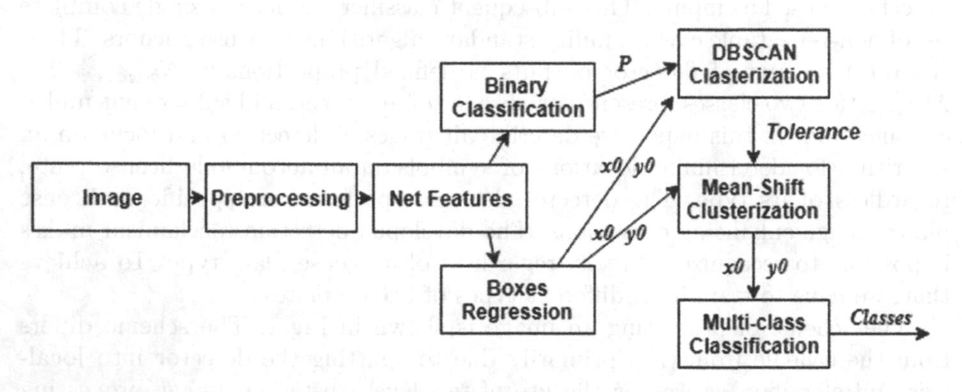


**Fig. 1.** Block diagram of image detection and subsequent clustering of data of object classes

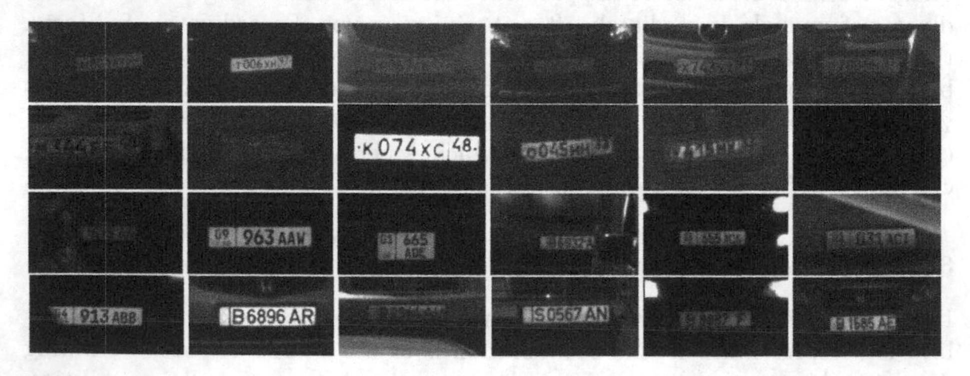


**Fig. 2.** Examples of license plates after coarse detection. The upper two rows are Russian license plates and the rest are Kyrgyz ones.

### 3.2 Algorithms

**Localization of License Plates.** The main purpose of localization of license plates is to narrow down the area, processed by the detector of symbols, and to reduce likelihood of false detections of symbols from an image background during training. Localization requirements for license plate are not that high. It is possible to use any simple localization algorithm like [18], which produces a license plate area with the minimum size.

**Binary Classifier and Regression.** Training is done on images with a fixed size of $48 \times 48$ pixels with a total stride of 24 pixels and an internal grid of reference boxes in the center of an image with a size of $24 \times 24$ pixels. Each activated reference box calculates parameters $\{x0, y0, p\}$, where the parameter p is calculated by the binary classifier and represents the probability of presence of an object, and the remaining parameters {x0, y0} are calculated by the neural network regressor and represent the normalized coordinate offsets to objects from boxes, active for them.

Loss function (the Eq. 1) is calculated for all boxes for the classifier and only for active boxes for the regressor.

$$L(\{p_{all}\}, \{coord_{all}\}) = \sum_{i=0}^{N_{boxes}} (\lambda_p L_{cls}(p_i, \overset{*}{c}_{i0,1}) + \overset{*}{c}_{i0,1}\ \lambda_x L_{reg_x}(x_i - \overset{*}{x}_i)^2 + \overset{*}{c}_{i0,1}\ \lambda_y L_{reg_y}(y_i - \overset{*}{y}_i)^2) \quad (1)$$

where i-index of a box, $p_i$-predicted probability that a box is associated with an object (symbol of license plate), $c_i^*$-label that a box is positive (1) or negative (0), $L_{cls}$-classification error, $L_{reg}$-regression error.

For a square plate, values of $\lambda$ across x and y should be about the same. We conducted our experiments with $\lambda_p = 0.1$, $\lambda_x = 1.0$, and $\lambda_y = 0.3$. Different values of the $\lambda$ coefficients for x and y axes are due to the rectangular shape and spatial orientation of the license plate. So, the $y$ coordinates of the license plate typically have similar values, and the $x$ coordinates are different.

In the process of recognition due to the convolutional architecture, the detector seems to be moving with a sliding window with the corresponding stride along

**Fig. 3.** An example of image areas being processed (Color figure online)

the x and y coordinates (Fig. 4). Each shift of the window provides a uniform overlay of the grid boxes on an image plane. The image in Fig. 4 is indicative for the fact that it has extraneous symbols. It also shows the difference between the size of the shift window and the inner area of the grid boxes.

**Fig. 4.** A sliding window of the convolutional network in the image and the direction of its movement and the box grid area.

**Calculation of Distance Between Symbols.** For clustering groups of points, describing positions of symbols, we use mean-shift [19] clustering. Because size of symbols can vary, for accurate clusterization it has to be provided with the average size of a symbol, which can be indirectly obtained from the averaged values of the $x$ coordinates of symbols from a license plate. To find out which points belong to the selected license plate, it is sufficient to cluster the points around the license plate area. We solved this problem using the DBSCAN [20] clustering algorithm with calculation of algorithm's tolerance parameter for next symbol's clusterization algorithm.

**Clustering of Symbols of License Plate.** As mentioned above, as a result of the detection and activation of boxes, distributions of X and Y coordinate points are formed around the centers of symbols relative to the active box (upper raw in Fig. 6). To calculate the center points, it is necessary to combine the coordinate points of the distributions to identify the centers of symbols. This task is complicated by markup errors during training, as well as in some cases by a very small distance between symbols, sometimes up to 4 pixels. A clustering algorithm is needed, which allows computing clusters with respect to distribution centers; this algorithm also does not need to know in advance the number of classes. The optimal clustering algorithm for solving this problem is the mean-shift [19] algorithm. It allows you to determine the location of the maxima of the probability density of the distribution of coordinate points on the plane. This algorithm is iterative, and its running time is defined as $O(Tn^2)$, where T is the number of iterations, and n is the number of points. For a fixed T and a large number of points, the algorithm is computationally expensive. But since the number of points of license plate symbols in clusters usually does not exceed 5, the costs become insignificant against the background of the operation of a neuro-network detector. The minimum number of cluster points are 2 points on the plane. This allows you to filter out single incorrectly formed data. After the clusters are calculated, their center coordinates are calculated as averaging over all coordinates of the cluster points (lower raw in Fig. 6).

**Recognition of Symbols.** The multi-class classification was carried out by a network with the same structure as the detector network (Table 1, except for the absence of symbol coordinate outputs and the use of multi-class classification $p \in R^{N_{classes}}$. Despite the fact that the network works only on the selected coordinates, it does not take advantage of the large stride on large images, as it is applied on each candidate symbol separately, therefore its running time is comparable to the operation time of the binary detector. For the training of the classifier, the same set of images was used as for the binary classifier with the regressor, with the only exception that the training took place symbol-by-symbol. To account for markup errors and coordinate calculation errors, training for each symbol occurred within the specified offsets along the x and y coordinates. As these offsets, half the average distance between the symbols in the license plate with respect to the coordinate was given x $shift_x = \frac{0.5}{N-1}\sum_{i=0}^{N-1}(Sym_{i+1} - Sym_i)$. The size of the images was fixed ($23 \times 23$) and the same for all symbols (Fig. 5). We did not combine the symbols 0 and O, so the total number of classes for symbols $N_{classes}$ is 36 (10 numbers and 26 letters).

**Network Settings.** The issue with large batch values is poor convergence [21]. On the other hand, low batch values result in low learning rates. A common approach for training a network is to gradually reduce learning rate at a constant batch size. This contributes to the smoothness of the cost function, but does not provide the same rate of convergence, the speed drops in the learning process. An approach to iterative (by analogy with a decrease in learning rate) increase in batch size was proposed in [22]. With an increase in batch, with a constant learning rate, it becomes possible to obtain approximately the same learning curve network as with a decrease in learning rate, but with a faster passing of epochs, since increasing batch proportionally decreases the number of iterations in it. We used batch from 8 to 128 with a smooth increase in the value of batch in the learning process for all time. The high parameter of learning speed learning_rate = 0.001 remained unchanged throughout the whole network training time.

**Fig. 5.** Images for symbols recognition. For example, from left to right, the central symbols are given: 9, 9, 6, 7, Y, 3, M

### 3.3 Selecting the Size of the Layers

In the selection of the size of the layers of network, we proceeded from the problem to be solved - the problem of calculating the coordinates of symbols. It is necessary to choose such size of layers (the number of neurons in fully connected networks) in a given network architecture (the number of layers is set like as the stride), which will provide the minimum average error of determining first of all the X coordinates of small objects, so providing the minimum $\sigma$ (Eq. 2). Weak networks, as well as Overfitted networks, demonstrate high value of $\sigma$. The first is due to the fact that it is not capable of averaging the coordinates of the symbols, the second is due to overfitting. Thus, through the search for an extremum, the values of $\sigma$ selected the dimensions of its layers for a given network architecture and thus, as it were, ensured its focusing. As a result, the layers of the network presented in the Table 1 were formed.

$$\sigma^2 = \frac{1}{N_{classes}} \sum_{i=0}^{N_{classes}} \frac{1}{N_{symbols_i}} \sum_{j=0}^{N_{symbols_j}} (X_{i_i} - \mu_i)^2 \tag{2}$$

**Table 1.** Network composition.

| Network composition | | | | |
|---|---|---|---|---|
| | Conv3D 1, 4 × 4, stride 4 | Conv3D 2, 3 × 3, stride 3 | Conv3D 3, 2 × 2, stride 2 | Conv3D 4, 2 × 2, stride 1 |
| 1 | FC 1, 16, lrelu | FC 1, 64, lrelu | FC 1, 128, lrelu | FC 1, 512, lrelu |
| 2 | FC 2, 16, lrelu | FC 2, 64, lrelu | FC 2, 128, lrelu | FC 1, 243, lrelu |

# 4 Experiments

Two threads were used on CPU - one for the neural network operation, the other thread provided data uploading from the dataset, image decompression and selection of the next active area for training. We tested the algorithm performance on three datasets, two of which are open and one is ours. All datasets are described below.

### 4.1 Detection Quality Evaluation Method

[23] presents a method for calculating mean average precision (mAP). It comes from the presence of ground truth and real areas of the object. The regions are represented by bounding boxes, between which the Intersection over union (IoU) parameter is calculated. It is usually considered true positive (TP) when IoU > 0.5. For all other values, the result is considered as false positive (FP). In our case, we do not have information on the ground truth of license plates, but we have only their coordinates on the plane, so to calculate the mAP, we need

to introduce a new metric replacing IoU. You can set the offset from the center in pixels, but then there are two problems - the error of the markup itself, which can have a strong effect on the difference in deflection, as well as the degree of influence of the error on symbols of different sizes. The effect of the offset from the true center is more significant for small objects. Therefore, in order to solve the correspondence of the calculated offset from the ground truth offset, we decided to compute through symbol recognition in a cascade that solves this problem. If the symbol on the license plate (separately) is recognized correctly, then this is TP and if vice versa, then FP. So, mAP was calculated for only one threshold, and can be perceived as the equivalent of $mAP_{50}$. Otherwise, the mAP calculation corresponds to [23]. By the way, a supporter of the old $mAP_{50}$ metric and the author of YOLOv3 cites his arguments in [5]. As a result of the detector operation on our dataset, mAP = 0.9025 % averaged over all symbols was achieved.

### 4.2 Background Accounting

Due to the fact that license plate detection training was taking place on images of limited size, as well as a desire to reduce the likelihood of false positives, we had to take into account background images. We conducted this by setting 50,000 images randomly selected from the dataset [24] into the dataset of license plates. These images were taken by car recorders at different times of the day and year and have all the properties of a road context.

### 4.3 Joint Dataset of Russian and Kyrgyz License Plates

The dataset was obtained from the video traffic violation video surveillance cameras and includes 34,000 and 35,000 Russian and Kyrgyz license plates, respectively. The frames presented in them were received at different times of the year, time of day. At night, license plates were illuminated. Only some of them are represented by a series of images. This dataset is closed, but upon request it can be demonstrated via remote access means.

### 4.4 SSIG Dataset

Dataset [25] consists of approximately 800 training images taken during the daytime. Images are a sequential series of a specific small set of scenes.

### 4.5 UFPR-ALPR Dataset

Dataset [12] includes 1800 training images taken during the daytime. Images are also a sequential series of a specific small set of scenes.

### 4.6 Detection Results

The results of the Recall are listed in the Table 2. The same table contains the results of the work of third-party detectors for other open datasets and the results of our detector's work on these datasets, when training on all of the existing datasets, and not separately on each of them. Though we calculated the mAP separately for symbols, Recall parameter (or the percentage of correct recognitions) were calculated for the whole license plates.

**Table 2.** List of results in terms of Recall for the third-party detectors and our detector.

| | Algorithm name | SSIG, Recall % | UFPR-ALPR, Recall % | Our, Recall % | Time localization, ms | Time recognition, ms |
|---|---|---|---|---|---|---|
| 1 | Sighthound | 73.13 | 47.39 | - | - | - |
| 2 | OpenALPR | 87.44 | 50.94 | - | - | - |
| 3 | UFPR-ALPR | 85.45 | 64.89 | - | 1.6555 (GPU) on LP area only | 11.5164 (GPU) |
| 4 | **Our** | 91.0 | 93.0 | 89.0 | 1.5 (CPU) on image $180 \times 80$ | 2 (CPU) |

Figure 6 shows the results of the detection of license plates symbols together for Russian and Kyrgyz datasets. Detection for each license plate is presented in three rows. In the first row, points are marked immediately after the symbol localization algorithm. One can observe a high degree of clustering of points. In the second row, the result of the DBSCAN clustering algorithm. The last third row shows the result of the operation of the cluster of initial symbol localization points by the mean-shift algorithm.

**Fig. 6.** An example of image areas of license plates after the detection of symbols on two types of license plates Russian and Kyrgyz.

## 5 Conclusion

In our work we demonstrated the possibility of detection of small objects based of a neural network with a fully connected convolution core NiN. We define a small object as an object for which the distance to its neighboring objects or groups of objects can be as short as couples of pixels. The fact that neighbors in a group can be of the same type complicates the problem even further. This led us to a decision to abandon the idea of an end-to-end detector, since the localization problem with simultaneous classification seems to us unattainable with the selected network architecture and the requirement to achieve an acceptable performance. Therefore, we have proposed a cascade method of alternately coarse localization of the area of a group of objects, then the subsequent binary classification of each object in the group, as well as the last stage of the multi-class classification of objects detected by the binary classifier. As a test of the accuracy of the algorithms, we chose the task of detection symbols on car license plates, images of which were obtained from video surveillance systems. It was important for us to take into account a context of an image of a license plate, so we suggested choosing a rectangular area of the boxes in the center of the image with inactive edges that allow us to capture a sufficient area providing us with enough information to recognize symbols on the license plate and not on another basis. The boxes represent a predetermined lattice structure of the reference points located in the center of the image and relative to which the regression of the coordinates of the closest symbol to each of them is calculated within the specified range. We have shown that the use of a network based on a NiN-based convolution is capable of storing and conveying spatial information about the location of objects to its output. The detector is able to work on images of arbitrary size and at relatively high speed on a standard CPU. In addition to solving the problem of detecting small objects that we are interested in, we nevertheless made our contribution to the area of symbols detection of car license plates. For example, in our work, we simultaneously successfully detected the symbols on license plates of Russian and Kyrgyz types.

## References


1. Agarwal, S., Terrail, J.O.D., Jurie, F.: Recent advances in object detection in the age of deep convolutional neural networks. CoRR, abs/1809.03193 (2018)
2. Zhao, Z., Zheng, P., Xu, S., Wu, X.: Object detection with deep learning: a review. CoRR, abs/1807.05511 (2018)
3. Ren, S., He, K., Girshick, R., Sun, J.: Faster R-CNN: towards real-time object detection with region proposal networks. ArXiv e-prints, June 2015
4. Redmon, J., Divvala, S.K., Girshick, R.B., Farhadi, A.: You only look once: unified, real-time object detection. CoRR, abs/1506.02640 (2015)
5. Redmon, J., Farhadi, A.: Yolov3: an incremental improvement. arXiv (2018)
6. Lin, T., Goyal, P., Girshick, R.B., He, K., Dollár, P.: Focal loss for dense object detection. CoRR, abs/1708.02002 (2017)
7. Liu, W., et al.: SSD: single shot multibox detector. CoRR, abs/1512.02325 (2015)

8. Wong, A., Shafiee, M.J., Li, F., Chwyl, B.: Tiny SSD: a tiny single-shot detection deep convolutional neural network for real-time embedded object detection. CoRR, abs/1802.06488 (2018)
9. Szegedy, C., Toshev, A., Erhan, D.: Deep neural networks for object detection. In: Advances in Neural Information Processing Systems 26: 27th Annual Conference on Neural Information Processing Systems 2013. Proceedings of a meeting held 5–8 December 2013, Lake Tahoe, Nevada, United States, pp. 2553–2561 (2013)
10. Li, H., Wang, P., Shen, C.: Towards end-to-end car license plates detection and recognition with deep neural networks. CoRR, abs/1709.08828 (2017)
11. Masood, S.Z., Shu, G., Dehghan, A., Ortiz, E.G.: License plate detection and recognition using deeply learned convolutional neural networks. CoRR, abs/1703.07330 (2017)
12. Laroca, R., et al.: A robust real-time automatic license plate recognition based on the YOLO detector. In: 2018 International Joint Conference on Neural Networks (IJCNN), pp. 1–10, July 2018
13. LaLonde, R., Zhang, D., Shah, M.: Fully convolutional deep neural networks for persistent multi-frame multi-object detection in wide area aerial videos. CoRR, abs/1704.02694 (2017)
14. Li, H., Shen, C.: Reading car license plates using deep convolutional neural networks and LSTMS. CoRR, abs/1601.05610 (2016)
15. Alexeev, A., Matveev, Y., Kukharev, G.: Using a fully connected convolutional network to detect objects in images, pp. 141–146, October 2018
16. Kauderer-Abrams, E.: Quantifying translation-invariance in convolutional neural networks. CoRR, abs/1801.01450 (2018)
17. Ren, Y., Zhu, C., Xiao, S.: Small object detection in optical remote sensing images via modified faster R-CNN. Appl. Sci. **8**(5), 813 (2018)
18. Kukharev, G., Kamenskaya, E., Matveev, Y., Shchegoleva, N.: Metody obrabotki i raspoznavanija izobrazhenij lic v zadachah biometrii. SPb, Politechnika (2013)
19. Cheng, Y.: Mean shift, mode seeking, and clustering. IEEE Trans. Pattern Anal. Mach. Intell. **17**, 790–799 (1995)
20. Ester, M., Kriegel, H.-P., Sander, J., Xu, X.: A density-based algorithm for discovering clusters a density-based algorithm for discovering clusters in large spatial databases with noise. In: Proceedings of the Second International Conference on Knowledge Discovery and Data Mining, KDD 1996, pp. 226–231. AAAI Press (1996)
21. Keskar, N.S., Mudigere, D., Nocedal, J., Smelyanskiy, M., Tang, P.T.P.: On large-batch training for deep learning: generalization gap and sharp minima. CoRR, abs/1609.04836 (2016)
22. Smith, S.L., Kindermans, P., Le, Q.V.: Don't decay the learning rate, increase the batch size. CoRR, abs/1711.00489 (2017)
23. Everingham, M., Van Gool, L., Williams, C.K.I., Winn, J., Zisserman, A.: The pascal visual object classes (VOC) challenge. Int. J. Comput. Vis. **88**, 303–338 (2010)
24. Gu, J., et al.: Recent advances in convolutional neural networks. CoRR, abs/1512.07108 (2015)
25. Goncalves, G.R., da Silva, S.P.G., Menotti, D., Schwartz, W.R.: Benchmark for license plate character segmentation. J. Electron. Imaging **25**(5), 1–5 (2016)

# Failure Diagnosis of Wind Turbine Bearing Using Feature Extraction and a Neuro-Fuzzy Inference System (ANFIS)

Mojtaba Kordestani[1], Milad Rezamand[2], Rupp Carriveau[2], David S. K. Ting[2], and Mehrdad Saif[1](✉)

[1] Electrical Department, University of Windsor, Windsor, ON, Canada
msaif@uwindsor.ca

[2] Turbulence and Energy Laboratory, University of Windsor, Windsor, ON, Canada

**Abstract.** Bearing failures are the most common type of malfunction in wind turbines. As such, isolating these defects enables maintenance scheduling in advance; hence, preventing further damage to turbines. This paper introduces a new fault detection and diagnosis (FDD) method to isolate two types of bearing failures in Wind turbines (WTs). The proposed FDD method consists of a feature extraction/feature selection and an adaptive neuro-fuzzy inference system (ANFIS) method. The feature extraction and selection phase identifies proper features to capture the nonlinear dynamics of the failure. Then, the ANFIS classifier diagnoses the failure type using the extracted features. Several experimental test studies with the historical data of wind farms in South-western Ontario are performed to evaluate the performance of the FDD system. Test results indicate that the proposed monitoring system is accurate and effective.

**Keywords:** Failure diagnosis · Feature extraction · Neuro fuzzy

## 1 Introduction

Wind turbines play an essential role in power generation as they are a viable source of renewable energy generation in the world. However, working under harsh operating conditions, they suffer from many reliability issues [1]. Maintenance comprises a high proportion of the total operational cost of wind turbines. It is estimated that the maintenance comprises 10–15% of power generation cost [2]. A significant percentage of the maintenance cost often relates to unexpended failures. These failures often lead to an extended downtime as the main components are required to move to a maintenance site which may be far from the wind farm [3].

Condition monitoring can assist to reduce the maintenance cost by monitoring the critical component through real-time measurement of data [4]. Various types


I. Rojas et al. (Eds.): IWANN 2019, LNCS 11506, pp. 545–556, 2019.
https://doi.org/10.1007/978-3-030-20521-8_45

of measurements, such as vibration, temperature, electrical quantities (voltage, current, power), and blade direction can be utilized for monitoring purposes.

Fault detection and diagnosis (FDD) are applied as an important step in condition monitoring systems to detect and isolate the type of failures. FDD approaches are divided into three categories: (1) model-based methods; (2) data-driven methods; (3) knowledge-based methods. Model-based methods employ an accurate mathematical model of systems to identify a fault. This group includes Kalman filter [5], extended Kalman filter [6], unscented Kalman Filter [7], and observer-based methods [8]. However, complex engineering systems often contain nonlinear dynamics with uncertainty or noise which make it hard to obtain a confidential model [9]. Data-driven methods, which are also known as model-free models, only need historical data of the system to construct fault diagnosis systems. Neural networks (NN) [10], adaptive neuro-fuzzy inference system (ANFIS) [11], support vector machine (SVM) [12] are some well-known data-driven methods. Nevertheless, Data-driven methods require complete data from all operating conditions [13]. Knowledge-based FDD methods are regularly developed based on expert knowledge on a specific process. The most common methods in this group are fuzzy logic [14], signal processing methods [15,16]. However, the knowledge of an expert is not always available or may not be precise enough to build an FDD system.

Bearings are critical mechanical parts in wind turbines and are prone to faults due to harsh operating conditions such as high-speed winds or high temperature. Bearing failures in wind turbines can lead to low production or even a turbine shutdown [17]. Therefore, fault detection and diagnosis of the bearings is essential to identify the type of abnormality and avoid further damage to the turbines through maintenance. A combination of feature selection and learning vector quantization (LVQ) neural network method is introduced in [18] to diagnose wind turbine bearing fault. Empirical Mode Decomposition (EMD) is utilized to extract the proper features. Then, the LVQ neural network is implemented to classify various failures. The experimental test results indicate high accuracy of the proposed fault diagnosis method. Hu et al. [12] develop a fault diagnosis method to isolate rotor bearing defects in wind turbines. They apply ensemble empirical mode decomposition to extract features of the bearing fault. Then, the SVM method is considered to classify the failures. Numerical results show that the proposed method can efficiently detect the bearing faults.

The work presented in this research project is based on data from a wind farm in Southwestern Ontario, Canada. The data acquisition protocols and settings are summarized below:

- Integrated Circuit-Piezoelectric (ICP) accelerometer sensors are utilized to measure vibration signal.
- "M-system" device is considered to record and process the data in the nacelle of the turbine. Then, the collected data is transmitted to the Turbine Condition Monitoring (TCM) site server in the wind farm station to perform real-time condition monitoring.

The objective of this paper is to develop an FDD system to detect and isolate various types of bearing failures based on the vibration signal measurements. First, four features are extracted from the vibration signals. Then, the ANFIS classifier is implemented to classify the failure type. The main contributions of this study are illustrated as follows:

- The proposed feature extraction and selection techniques capture failure dynamics and supply suitable inputs for classification phase.
- The ANFIS classifier not only detects and isolates the types of the failures, but it can also detect unknown failures.

The paper is organized as follows: Sect. 2 provides the description of the wind turbine bearings. The primary theory of the proposed FDD method is illustrated in Sect. 3. Section 4 explain the design implementation and experimental test results. A conclusion and future direction are summarised in Sect. 5.

## 2 Wind Turbine Bearing

A bearing is a component that is utilized to facilitate rotational or linear movement while lessening friction and handling stress. Bearings typically must deal with two classes of loading: radial and thrust. Bearings are designed to manage a specific kind of load and different amounts of weight so they can be employed in several applications. It is worth noting that the differences between types of bearings concern load type and strength to handle the weight. One of these applications is the use of bearings in wind turbines. In this paper, two main bearing failures including outer raceway and inner raceway failures are investigated.

## 3 Primary Theory of Fault Detection Method

In this section, the primary theory of the FDD method is introduced. Figure 1 demonstrates the proposed FDD system.

The proposed FDD system consists of vibration data acquisition, feature extraction, feature selection, and ANFIS classifier. In the following, first, the feature extraction and selection are demonstrated. Then, the ANFIS classifier is introduced.

### 3.1 Feature Extraction Techniques

One of the most common avenues for fault diagnosis of bearings is based on vibration signal analysis. However, the vibration signal cannot directly identify the types of failures. Therefore, various feature extraction techniques are often applied to the raw vibration signal to identify failure dynamics. Time domain statistical features have proven their high efficiency in bearing failure detection. Some time-domain statistical features are presented as follows [19]:

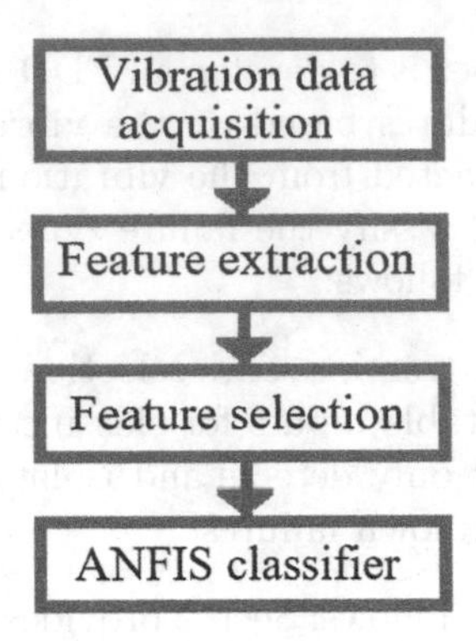


**Fig. 1.** The proposed FDD system

1. Root mean square (RMS): It is also known as quadratic mean and used to identify differences between vibration signals.
2. Variance: It measures the distribution of a signal from its average value and can characterize how far a set of numbers (signals) are spread out.
3. Peak to peak (P2P): It shows the difference between the maximum and minimum amplitude of a signal.
4. Band power: It indicates the average power of the signal.

### 3.2 Feature Selection Techniques

After identifying all features, a mechanism is required to select the most appropriate features. It should be noted that all the features need not be utilized in the FDD system as it may increase the computation complexity of the classifier or even reduce the accuracy of the FDD system due to inappropriate features. Therefore, a few most relevant feature must be chosen.

There are two general feature selection methods in the literature, namely, filter based and wrapper methods. The wrapper method tries to select a few features as research set and look for optimal or suboptimal features between all the features by choosing all available combination of the features. However, the wrapper is often computationally expensive due to the repeated learning steps [20]. The filter-based method applies a metric to select the most appropriate feature based on that metric. Then, in the end, it selects the highest ranked features. Correlation coefficient analysis is a commonly used approach in filter-based categories [21].

### 3.3 ANFIS Method

ANFIS applies an integrated framework by combining neural network and fuzzy logic, and therefore, could achieve higher accuracy in comparison with Neural networks [22]. The adaptive structure of the ANFIS uses a supervised learning algorithm constructed by an inference system like Takagi-Sugeno which is known as neuro-fuzzy system. A set of rules similar to the fuzzy system are driven as follows:

**Rule 1:** If $x_1$ is $A_1$ and $x_2$ is $B_1$, then $f_1 = p_1 x_1 + q_1 x_2 + r_1$

**Rule 2:** If $x_1$ is $A_2$ and $x_2$ is $B_2$, then $f_2 = p_2 x_1 + q_2 x_2 + r_2$

where $A_i$ and $B_i$ represent membership functions (MFs) for input $x_1$ and $x_2$, respectively, and $p_i$, $q_i$ and $r_i$ indicates adaptive parameters. Figure 2 presents ANFIS network structure. It is noted from Fig. 2 that the ANFIS structure consists of five feed forward layers. In the following, a mathematical model of each layer is formulated [23]:

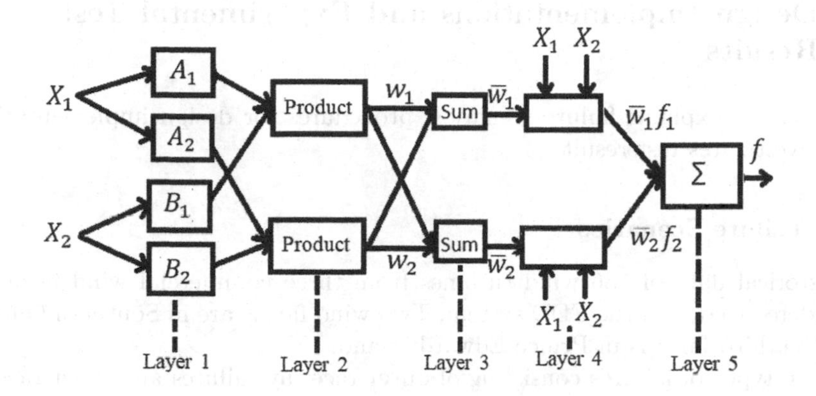


**Fig. 2.** ANFIS network structure.

**Layer 1:** Each node in this layer is formulated by a fuzzy MF and signs with a linguistic label:

$$\begin{aligned} O_{1,i} &= \mu_{A_i}(x_1), \quad i = 1, 2. \\ O_{1,i} &= \mu_{B_{i-2}}(x_2), \quad i = 3, 4. \end{aligned} \tag{1}$$

**Layer 2:** This layer consists of fixed nodes with a firing strength ($w_i$) assigned for the rules. The output of the nodes are constructed by product T-norm operator as follows:

$$O_{2,i} = w_i = \mu_{A_i}(x_1) \times \mu_{B_i}(x_2) \quad i = 1, 2. \tag{2}$$

**Layer 3:** In this layer, the nodes are also fixed. Normalization is performed to form relative firing strengths ($\bar{w}_i$) as follows:

$$O_{3,i} = \bar{w}_i = \frac{w_i}{\sum_{j=1}^{2} w_i} \quad i = 1, 2. \tag{3}$$

**Layer 4:** An adaptive structure is provided for this layer. The mathematical model of the nodes is computed by multiplication of the adaptive parameters and the relative firing strengths:

$$O_{4,i} = \bar{w}_i f_i = \bar{w}_i(p_i x_1 + q_i x_2 + r_i) \quad i = 1, 2. \tag{4}$$

**Layer 5:** In this layer, a summation of all signals is computed as follows:

$$O_{5,1} = \sum_{i=1}^{2} \bar{w}_i f_i = \frac{\sum_i^2 w_i f_i}{\sum_i^2 w_i} \tag{5}$$

To train the ANFIS network, backpropagation algorithm or hybrid learning using a combination of the least-squares and the gradient descent methods can be applied. The adaptive parameters in layers 1 and 4 are optimized in the training procedure to minimize the errors of the ANFIS network.

## 4 Design Implementations and Experimental Test Results

This section explains failure scenarios, procedures for design implementations, and investigates test results.

### 4.1 Failure Scenarios

A historical data of 136 wind turbines from three commercial wind farms are considered to design the FDD system. Two wind farms are in Southern Ontario, and the third farm is in Prince Edward Island.

Two types of failures consisting of outer raceway failures and inner raceway failures are investigated. Figure 3 depicts an outer raceway failure (A) of a main-shaft bearing and an inner raceway failure (B) of a generator DE bearing.

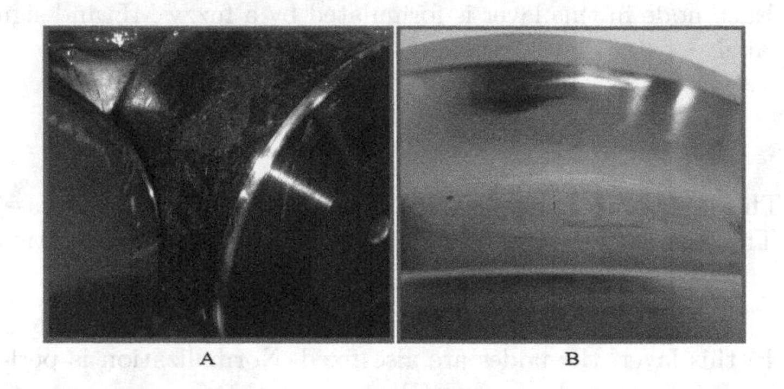


**Fig. 3.** Outer raceway failure (A) and inner raceway failure (B)

In this research work, vibration data of healthy and faulty bearings have been studied for ten years. The vibration signal of the main-shaft bearing is recorded twice per month for 90 s with a sampling rate of 6401 s. Regarding the generator DE bearing, the vibration signal duration is 4 s with a sampling rate of 153601 s.

### 4.2 Design Implementation of the Proposed Feature Extraction and Selection Methods

Time domain statistical techniques are utilized to extract the best features from the vibration signals. For this aim, RMS, variance, peak to peak and Band power features are extracted for healthy and faulty vibration signals.

In the feature selection step, the filter-based method is applied using correlation analysis to correlate the features with the fault labels. Table 1 presents the features and their correlation coefficients.

It is noted that the RMS and the peak to peak (P2P) have the highest correlations with the fault labels.

**Table 1.** The features and their correlation coefficients.

| Measurement | Correlation coefficient |
|---|---|
| RMS | 0.5053 |
| Peak to peak (P2P) | 0.6233 |
| Variance | 0.4556 |
| Band power | 0.4108 |

### 4.3 Design Implementation of the ANFIS Classifier

The RMS and the peak to peak (P2P) features obtained from feature selection techniques are considered as inputs of ANFIS classifier. Figure 4 shows the structure of the proposed ANFIS classifier. It is noted from Fig. 4 that the one delay features are also included to the inputs to add dynamics to the classifier. Therefore, it enhances the accuracy of the classifier.

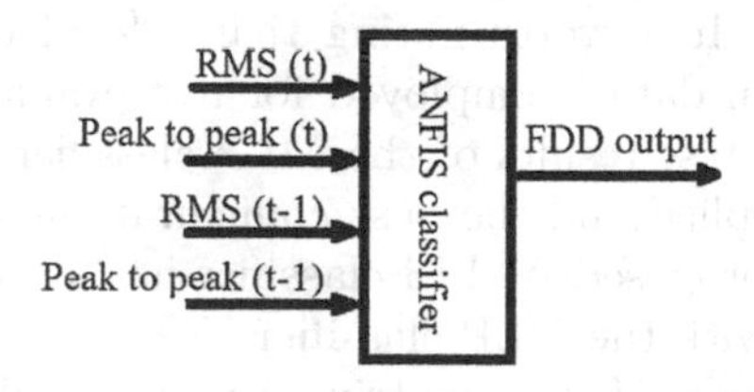


**Fig. 4.** The structure of the proposed ANFIS classifier.

To train the FDD system, the output of the ANFIS classifier is set to 0 for the healthy bearing data. For the faulty data, the output is labelled to 1 and 2 for outer raceway and inner raceway failures, respectively. The Hybrid learning algorithm is used to train the network. Table 2 illustrates the design criteria and test results of the ANFIS classifier.

Under a real-time testing scenario, the output of the system takes any value between $-0.5$ and $+0.5$, in which case it is concluded that the bearings are

**Table 2.** The design criteria and test results of the ANFIS and MLP classifiers.

| Networks | Parameters | Description or values |
|---|---|---|
| ANFIS | Membership functions for inputs | Gaussian |
| ANFIS | Number of membership functions | 5 for each input |
| ANFIS | Membership functions for output | Linear |
| ANFIS | Training method | Hybrid |
| ANFIS | Test Error (MSE) | 0.0295 |
| ANFIS | Maximum number of epochs | 58 |
| MLP | Inputs nodes type | Buffer |
| MLP | Hidden layer node type | Log-sigmoid function |
| MLP | Number of nodes in Hidden layer | 7 |
| MLP | Output Node type | Linear function |
| MLP | Training method | Back propagation |
| MLP | Test Error (MSE) | 0.114 |
| MLP | Maximum number of epochs | 86 |

healthy. If the FDD system takes a value between +0.5 and +1.5 or +1.5 and +2.5, it is concluded that the outer raceway or inner raceway failures have occurred, respectively. If the output takes any other values, the FDD assumes that an unknown failure is detected in the bearings. Under such a condition, the output of the FDD system is set to −1. Moreover, the FDD system must see two repeated failures to isolate the type of failure. We impose this condition to reduce the chance of false-alarm rate due to noise or disturbance in the system.

**Remark 1:** To illustrate the advantage and superior performance of the ANFIS classifier, a multilayer perceptron (MLP) network is constructed with the same input-output structure. It is worth noting that 70% of the dataset is used for training and the remain data is employed for test evaluation. Table 2 explains the design criteria and test results of the MLP classifier. Monte Carlo method with 100 run test is applied and the best configurations are chosen. It is noted from Table 2 that the proposed ANFIS classifier has a lower mean square error (MSE) in comparison with the MLP classifier.

Table 3 discusses the confusion matrix of various failure types. It is noted from Table 3 that ANFIS classifier correctly isolates 84% of failure type 2 with only 12% of false classification to failure type 1 and 4% false classification as unknown failures. Regarding failure type 1, the accuracy of the ANFIS classifier is 78%. The accuracy of MLP classifier is lower with 80% and 72% of accuracy for failures type 2 and 1, respectively.

### 4.4 Experimental Test Results (Two Case Studies)

In this subsection, experimental test results are investigated to evaluate the time response of the proposed FDD system.

**Table 3.** The confusion matrix of various failure types

| Networks | Failures | Failure type 1 | Failure type 2 |
|---|---|---|---|
| ANFIS | Failure type 1 | 78% | 12% |
| ANFIS | Failure type 2 | 16% | 84% |
| ANFIS | Unknown | 6% | 4% |
| MLP | Failure type 1 | 72% | 14% |
| MLP | Failure type 2 | 18% | 80% |
| MLP | Unknown | 10% | 6% |

**4-4-1 Outer Raceway Failure:** An outer raceway failure in the main shaft bearing of the wind turbine is utilized. This bearing started a failure on Oct 17, 2017. Figure 5 shows the RMS and the peak to peak (P2P) features, ANFIS and MLP classifier results for outer raceway failure.

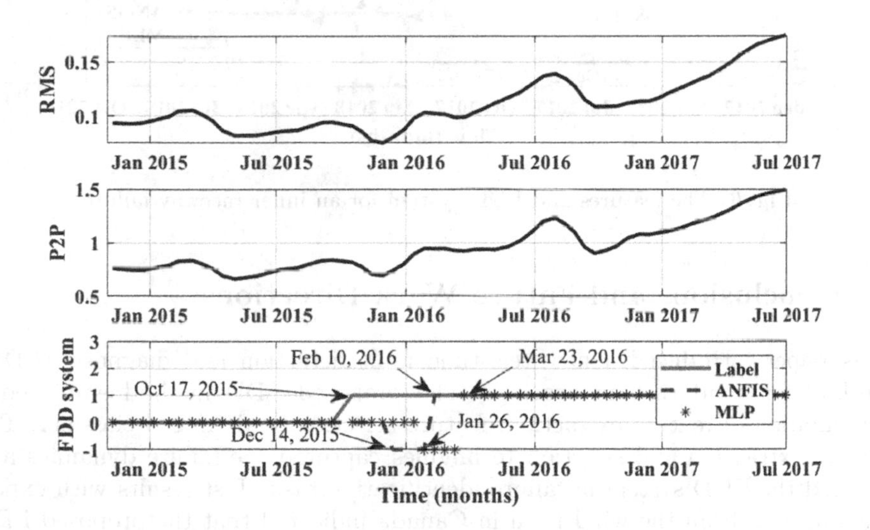


**Fig. 5.** The features and FDD system for an outer raceway failure.

It is noted from Fig. 5 that the ANFIS classifier detected the failure on Dec 14, 2015, and isolated the type of failure on Feb 10, 2016. The ANFIS classifier has a delay of 46 day between detection and isolation which is due to slow dynamics of the failure degradation. The performance of the MLP classifier is lower with detection on Jan 26, 2016, and isolation on Mar 23, 2016.

**4-4-2 Inner Raceway Failure:** An inner raceway failure in the generator DE bearing occurred on Dec 08, 2017. Figure 6 shows the features and FDD system for the inner raceway failure.

It is noted from Fig. 6 that both ANFIS and MLP classifier detected the failure on Dec 24, 2017. Moreover, ANFIS classifier identified the type of failure on Jan 22, 2018, with less delay in comparison with MLP classifier.

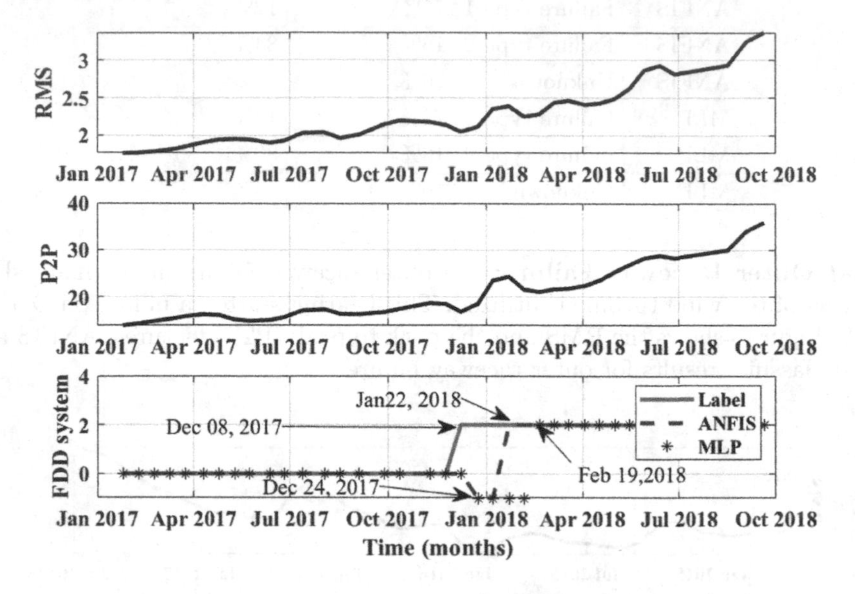


**Fig. 6.** The features and FDD system for an inner raceway failure.

## 5 Conclusions and Future Work Directions

This paper introduced a new real-time fault detection and diagnosis (FDD) method for wind turbine bearings. The proposed FDD method employed a combination of feature extraction, feature selection, and ANFIS classifier. The feature extraction and selection techniques captured the failure dynamics and assisted the FDD system in failure identification task. Test results with experimental data from the wind farm in Canada indicated that the proposed FDD system using ANFIS classifier had higher accuracy in isolating failures in comparison with the FDD system based on MLP classifier.

Finally, some areas for further research are:

1. Applying other approaches such as image detection to detect the type of damage such as crack and erosions.
2. Performing probabilistic methods to predict remaining useful life of the bearings.

**Acknowledgment.** The authors would like to acknowledge that this work is part of the YR21 Investment Decision Support Program supported by progressive industrial partners, Natural Sciences and Engineering Research Council (NSERC) of Canada, and the Ontario Centres of Excellence (OCE).

## References


1. Carroll, J., McDonald, A., McMillan, D.: Failure rate, repair time and unscheduled O&M cost analysis of offshore wind turbines. Wind Energy **19**(6), 1107–1119 (2016)
2. Musial, W., Ram, B.: Large-Scale Offshore Wind Power in the United States: Assessment of Opportunities and Barriers. National Renewable Energy Lab, Golden (2010)
3. Lu, D., Qiao, W., Gong, X., Qu, L.: Current-based fault detection for wind turbine systems via Hilbert-Huang transform. In: 2013 IEEE Power and Energy Society General Meeting, pp. 1–5. IEEE (2013)
4. Gong, X., Qiao, W.: Simulation investigation of wind turbine imbalance faults. In: 2010 International Conference on Power System Technology, pp. 1–7. IEEE (2010)
5. Wang, J., Peng, Y., Qiao, W., Hudgins, J.L.: Bearing fault diagnosis of direct-drive wind turbines using multiscale filtering spectrum. IEEE Trans. Ind. Appl. **53**(3), 3029–3038 (2017)
6. Shi, Y., Hou, Y., Qiao, S., Liu, W., Li, Z., Sun, D., Wen, C.: Research on predictive control and fault diagnosis of wind turbine based on MLD. In: 32nd Chinese Control Conference, pp. 6166–6173. IEEE (2013)
7. Yu, S., Emami, K., Fernando, T., Iu, H.H., Wong, K.P.: State estimation of doubly fed induction generator wind turbine in complex power systems. IEEE Trans. Power Syst. **31**(6), 4935–4944 (2016)
8. Wu, C.: Multiplicative fault estimation using sliding mode observer with application. In: 2015 International Conference on Control, Automation and Robotics, pp. 163–167. IEEE (2015)
9. Kordestani, M., Samadi, M.F., Saif, M., Khorasani, K.: A new fault prognosis of MFS system using integrated extended Kalman filter and Bayesian method. IEEE Trans. Ind. Inform. **99**, 1–11 (2018)
10. Rahimilarki, R., Gao, Z., Zhang, A., Binns, R.J.: Robust neural network fault estimation approach for nonlinear dynamic systems with applications to wind turbine systems. IEEE Trans. Ind. Inform. (2019)
11. Attoui, I., Boudiaf, A., Fergani, N., Oudjani, B., Boutasseta, N., Deliou, A.: Vibration-based gearbox fault diagnosis by DWPT and PCA approaches and an adaptive neuro-fuzzy inference system. In: 2015 16th International Conference on Sciences and Techniques of Automatic Control and Computer Engineering, pp. 234–239. IEEE (2015)
12. Hu, C.-z., Huang, M.-y., Yang, Q., Yan, W.-j.: On the use of EEMD and SVM based approach for bearing fault diagnosis of wind turbine gearbox. In: Control and Decision Conference, pp. 3472–3477. IEEE (2016)
13. Kordestani, M., Samadi, M.F., Saif, M., Khorasani, K.: A new fault diagnosis of multifunctional spoiler system using integrated artificial neural network and discrete wavelet transform methods. IEEE Sensors J. **18**(12), 4990–5001 (2018)
14. Bae, K.-H., Choi, B.-O., Park, J.-W., Kim, B.-K.: A study on crack fault diagnosis of wind turbine simulation system. In: 2014 International Conference on Reliability, Maintainability and Safety, pp. 53–57. IEEE (2014)
15. Balan, H., Cozorici, I., Buzdugan, M., Karaisas, P.: Signal processing software techniques for the monitoring and the diagnosis of the wind turbines. In: 4th International Symposium on Electrical and Electronics Engineering, pp. 1–6. IEEE (2013)
16. Kordestani, M., Saif, M.: Data fusion for fault diagnosis in smart grid power systems. In: 2017 IEEE 30th Canadian Conference on Electrical and Computer Engineering (CCECE), pp. 1–6. IEEE (2017)

17. Cheng, F., Qu, L., Qiao, W., Hao, L.: Enhanced particle filtering for bearing remaining useful life prediction of wind turbine drive train gearboxes. IEEE Trans. Ind. Electron. **66**, 4738 (2018)
18. Shi, X., Li, W., Gao, Q., Guo, H.: Research on fault classification of wind turbine based on IMF kurtosis and PSO-SOM-LVQ. In: IEEE 2nd Information Technology, Networking, Electronic and Automation Control Conference, pp. 191–196. IEEE (2017)
19. Caesarendra, W., Tjahjowidodo, T.: A review of feature extraction methods in vibration-based condition monitoring and its application for degradation trend estimation of low-speed slew bearing. Machines **5**(4), 21 (2017)
20. Das, S.: Filters, wrappers and a boosting-based hybrid for feature selection. In: ICML, vol. 1, pp. 74–81 (2001)
21. Siegel, D., Ly, C., Lee, J.: Methodology and framework for predicting helicopter rolling element bearing failure. IEEE Trans. Reliabil. **61**(4), 846–857 (2012)
22. Kordestani, M., Zanj, A., Orchard, M.E., Saif, M.: A modular fault diagnosis and prognosis method for hydro-control valve system based on redundancy in multi-sensor data information. IEEE Trans. Reliab. **99**, 1–12 (2018)
23. Morshedizadeh, M., Kordestani, M., Carriveau, R., Ting, D.S.-K., Saif, M.: Power production prediction of wind turbines using a fusion of MLP and ANFIS networks. IET Renew. Power Gener. **12**(9), 1025–1033 (2018)

# OnMLM: An Online Formulation for the Minimal Learning Machine

Alan L. S. Matias[1], César L. C. Mattos[1], Tommi Kärkkäinen[2], João P. P. Gomes[1(✉)], and Ajalmar R. da Rocha Neto[3]

[1] Federal University of Ceará, Fortaleza, CE, Brazil
matiasalsm@gmail.com, cesarlincoln@dc.ufc.br, jpaulo@lia.ufc.br
[2] University of Jyvaskyla, Jyvaskyla, Finland
tommi.karkkainen@jyu.fi
[3] Federal Institute of Ceará, Fortaleza, CE, Brazil
ajalmar@ifce.edu.br

**Abstract.** Minimal Learning Machine (MLM) is a nonlinear learning algorithm designed to work on both classification and regression tasks. In its original formulation, MLM builds a linear mapping between distance matrices in the input and output spaces using the Ordinary Least Squares (OLS) algorithm. Although the OLS algorithm is a very efficient choice, when it comes to applications in big data and streams of data, online learning is more scalable and thus applicable. In that regard, our objective of this work is to propose an online version of the MLM. The Online Minimal Learning Machine (OnMLM), a new MLM-based formulation capable of online and incremental learning. The achievements of OnMLM in our experiments, in both classification and regression scenarios, indicate its feasibility for applications that require an online learning framework.

**Keywords:** Online learning · Incremental learning · Stochastic optimization · Minimal Learning Machine

## 1 Introduction

The learning algorithm of many popular machine learning methods, such as the Least Squares Support Vector Machine (LSSVM) [14] and the Extreme Learning Machine (ELM) [3], are based on a batch method, such as the Ordinary Least Squares (OLS). Even though batch approaches perform well in some cases, they present some major drawbacks: (i) they lack scalability for larger datasets; (ii) they cannot be used in scenarios where data are sequentially made available [7].

Both aforementioned issues are usually tackled by online learning techniques. For instance, the classical backpropagation algorithm [8] has been fundamental to enable the use of machine learning models in real world applications with large and/or sequential data. Thus, some authors have directed their efforts to propose alternative formulations, both algorithmically and mathematically,


I. Rojas et al. (Eds.): IWANN 2019, LNCS 11506, pp. 557–568, 2019.
https://doi.org/10.1007/978-3-030-20521-8_46

for standard batch learning-based methods in order to perform online learning. Santos and Barreto [12] proposed an approach to train the LSSVR following an online method which is able to obtain sparse solutions incrementally from the data. Jian *et al.* [4] proposed the Budget Online LSSVM, which significantly scales LSSVM through an online learning approach. Online variants of the ELM can also be found in the literature, such as the Online Sequential ELM (OS-ELM) [2] and its recent improvement by Matias *et al.* [10]. Efforts towards online learning can also be found for ensemble methods such as Gradient Boosting [1] and Random Forests [6].

The recently proposed Minimal Learning Machine (MLM) [13] aims to tackle the general supervised learning task also from a batch learning perspective. The MLM considers a linear mapping between input and output distance matrices, computed from the available data points and $M$ reference points randomly chosen from the dataset. In the generalization step, the learned distance map is used to provide an estimate of the distance from the output reference points to the unknown target output value [11]. This mapping assumes that a multi-response linear regression method is applied between the distance matrices. In the original MLM formulation, such solution is obtained via the batch OLS algorithm. Thus, the MLM inherits the lack of scalability and the inability to handle streaming data of other batch learning techniques.

In this context, our objective is to introduce a feasible online and incremental learning formulation for the MLM. Our proposal, named Online MLM (OnMLM), is able to handle both regression and classification tasks in scenarios where batch approaches are not feasible. The OnMLM starts with one input and one output reference points and, as the training is performed, it incrementally increases the multi-response linear system by adding new reference points based on a given criterion. The new mechanisms provided by the OnMLM requires the addition of three new hyperparameters. However, such hyperparameters are very intuitive and can be easily set.

The properties offered by OnMLM makes it suitable and scalable to applications involving large datasets, and since the training is performed online, the OnMLM is also applicable to datastreams scenarios. Also, we must emphasize that OnMLM reaches always sparser solutions in comparison to MLM, as a consequence of its ability to incrementally add new reference points. Furthermore, our formulation for OnMLM can be optimized with any variant of the Stochastic Gradient Descent (SGD) algorithm. Thus, one can use OnMLM with more efficient SGD variants, boosting its generalization performance and speeding the training phase. To summarize, our contribution with this work is an online formulation based on the SGD algorithm for the MLM model which incrementally includes reference points in the online learning step, with the additional benefit of reaching much sparser solutions.

We have organized our paper by presenting a full description of the MLM and the OnMLM models in Sect. 2 – in the OnMLM formulation, we show a special focus in providing an overview about the newly included three hyperparameters. We then follow with the experiments performed in both regression and classification scenarios in Sect. 3. Finally, the conclusions we have obtained with this work are presented in Sect. 4.

# 2 Methods

## 2.1 Minimal Learning Machine

Consider a set of input points $\mathcal{X} = \{\boldsymbol{x}_i \in \mathcal{R}^p : i = 1, \ldots, N\}$ and their corresponding outputs $\mathcal{Y} = \{\boldsymbol{y}_i \in \mathcal{R}^q : i = 1, \ldots, N\}$. We can define the following distance matrices $\mathbf{D}$, whose elements are defined by $d_{ij} = \|\boldsymbol{x}_i - \boldsymbol{s}_j\|_2$, and $\boldsymbol{\Delta}$ whose elements are given by $\delta_{ij} = \|\boldsymbol{y}_i - \boldsymbol{t}_j\|_2$. The point denoted by $\boldsymbol{r}_j$ and $\boldsymbol{t}_j$, named reference points, are randomly selected vectors from $\mathcal{X}$ and $\mathcal{Y}$, respectively. Their corresponding sets are defined as $\mathcal{S} = \{\boldsymbol{s}_j \in \mathcal{X} : j = 1, \ldots, M\}$ and $\mathcal{T} = \{\boldsymbol{t}_j \in \mathcal{Y} : j = 1, \ldots, M\}$, where $M$ is the number of reference points. Under the assumption of a linear mapping between $\mathbf{D}$ and $\boldsymbol{\Delta}$, we can obtain a multiresponse regression model

$$\boldsymbol{\Delta} = \mathbf{D}\mathbf{B} + \mathbf{E}, \tag{1}$$

where $\mathbf{E}$ is the residual matrix.

Thus, the objective of the MLM is to obtain the optimal parameters $\hat{\mathbf{B}}$ which minimizes the multivariate residual sum of squares:

$$\underset{\mathbf{B}}{\arg\min}\ \mathcal{J}(\mathbf{B}) = \mathrm{tr}[(\boldsymbol{\Delta} - \mathbf{D}\mathbf{B})^T(\boldsymbol{\Delta} - \mathbf{D}\mathbf{B})]. \tag{2}$$

It is noteworthy that the reference points are randomly chosen before optimizing the objective function in Eq. (2). Furthermore, the optimization problem above can be solved in two different ways: (*i*) with the usual optimal least squares when $M < N$, and (*ii*) uniquely solvable with the direct inverse of $\mathbf{D}$ when $M = N$. To better understand the training process of the MLM, step by step, we describe its training algorithm in the Algorithm 1.

**Algorithm 1.** Training procedure for MLM.

**Input:** Training sets $\mathcal{X}$ and $\mathcal{Y}$, and the number of reference points $M$.
**Output:** The optimal weights $\hat{\mathbf{B}}$, and the sets $\mathcal{S}$ and $\mathcal{T}$.

1: Randomly selects $M$ reference points from $\mathcal{X}$ and their correspond output from $\mathcal{Y}$, in order to obtain the sets $\mathcal{S}$ and $\mathcal{T}$, respectively.
2: Compute the input distance matrix $\mathbf{D}$ from $\mathcal{X}$ and $\mathcal{S}$.
3: Compute the output distance matrix $\boldsymbol{\Delta}$ from $\mathcal{Y}$ and $\mathcal{T}$.
4: Calculate $\hat{\mathbf{B}} = (\mathbf{D}^T\mathbf{D})^{-1}\mathbf{D}^T\boldsymbol{\Delta}$.

Supposing that $\mathbf{B}$ was either estimated ($M < N$) or uniquely solved ($M = N$), for an arbitrary input test point $\boldsymbol{x}_i \in \mathcal{R}^p$, and its representation in the distance space $\boldsymbol{d}_i = [\|\boldsymbol{x}_i - \boldsymbol{s}_1\|_2, \ldots, \|\boldsymbol{x}_i - \boldsymbol{s}_M\|_2]^T$, the MLM will predict the output related to the distance space

$$\hat{\boldsymbol{\delta}}_i = \boldsymbol{d}_i^T \mathbf{B}. \tag{3}$$

The vector $\hat{\boldsymbol{\delta}}_i$ provides an estimate of the geometrical configuration of $\boldsymbol{y}_i$ related to the input test $\boldsymbol{x}_i$ and the reference set $\mathcal{T}$. To recover the estimation of $\boldsymbol{y}_i$ from $\hat{\boldsymbol{\delta}}_i$, MLM solves a multilateration problem. From a geometric view point, locating $\boldsymbol{y}_i \in \mathcal{R}^q$ is equivalent to solve the overdetermined set of $M$ nonlinear equations corresponding to $q$-dimensional hyper-spheres centered in $\boldsymbol{t}_j$ which passes through $\boldsymbol{y}_i$. Figure 1 graphically depicts the problem for $q = 2$.

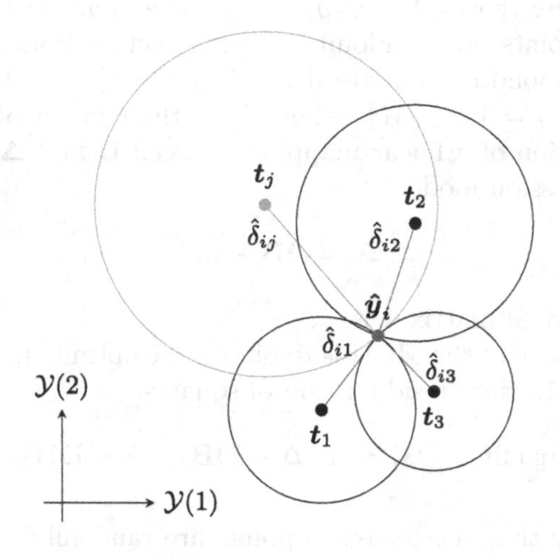


**Fig. 1.** Output estimation.

Thus, given the set of $j = 1, \ldots, M$ hyper-spheres, each with radius equal to $\hat{\boldsymbol{\delta}}_i$, the location of $\boldsymbol{y}_i$ can be estimated by finding the optimal solution of the following objective function:

$$\underset{\hat{\boldsymbol{y}}_i}{\arg\min}\ \mathcal{J}(\boldsymbol{y}_i) = \sum_{j=1}^{M} [(\boldsymbol{y}_i - \boldsymbol{t}_j)^T (\boldsymbol{y}_i - \boldsymbol{t}_j) - \hat{\delta}_{ij}^2]^2. \tag{4}$$

## 2.2 Online Minimal Learning Machine (Proposal)

To initiate the OnMLM formulation, we must highlight that, unlike MLM, which have the same number of reference points $M$ in the input set $\mathcal{S}$ and in the output set $\mathcal{T}$, the OnMLM may have different number of reference points $M_s$ and $M_t$, respectively. The original formulation of MLM adds pairs of points as reference points $(\boldsymbol{x}_i \in \mathcal{X}, \boldsymbol{y}_i \in \mathcal{Y})$ for each $i$-th indexation chosen to represent a reference point. On the other hand, the OnMLM adds new reference points in $\mathcal{S}$ and

$\mathcal{T}$ independently – in fact, the new reference point criterion, which we will be introduced later, determines if an input $\boldsymbol{x}_i$ or its related output $\boldsymbol{y}_i$ will become a new reference input point or a new reference output point, respectively, in a supervised way. To proceed with the OnMLM formulation let us redefine sets $\mathcal{S}$ and $\mathcal{T}$ as $\mathcal{S} = \{\boldsymbol{s}_j \in \mathcal{X} : j = 1, \dots, M_s\}$ and $\mathcal{T} = \{\boldsymbol{t}_k \in \mathcal{Y} : k = 1, \dots, M_t\}$.

Let $\mathcal{X}_l$ and $\mathcal{Y}_l$ be the sets of the input patterns and the expected outputs of the $l$-th mini batch. As in the standard MLM learning algorithm, we must compute the distance input matrix $\mathbf{D}_l = [d_{ij}] = [\|\boldsymbol{x}_i - \boldsymbol{s}_j\|_2]$ and the distance output matrix $\boldsymbol{\Delta}_l = [\delta_{ik}] = [\|\boldsymbol{y}_i - \boldsymbol{t}_k\|_2]$, where $i = 1, \dots, N_l$, $j = 1, \dots, M_s$, and $k = 1, \dots, M_t$. Under the assumption of a linear mapping between $\mathbf{D}_l$ and $\boldsymbol{\Delta}_l$ we have that:

$$\boldsymbol{\Delta}_l = \mathbf{D}_l \mathbf{W} + \mathbf{E}_l, \tag{5}$$

where $\mathbf{W} \in \mathcal{R}^{M_s \times M_t}$ is the weight matrix and $\mathbf{E}_l$ is the $l$-th residual matrix.

Our objective is to find the optimal $\mathbf{W}$ that minimizes the mean squared sum of residuals for each $\mathbf{D}_l$ and $\boldsymbol{\Delta}_l$, so that

$$\mathcal{J}(\mathbf{W}) = \frac{1}{2N_l} \sum_{i=1}^{N_l} \sum_{k=1}^{M_t} (\delta_{ik} - \boldsymbol{d}_i^T \boldsymbol{w}_{:k})^2, \tag{6}$$

where $\boldsymbol{w}_{:k}$ is the $k$-th column of $\mathbf{W}$ and $\boldsymbol{d}_i^T$ is the $i$-th row of $\mathbf{D}_l$.

Considering the SGD algorithm, the matrix $\mathbf{W}$ can be updated by

$$\boldsymbol{w}_{:k}^{new} = \boldsymbol{w}_{:k} - \eta \frac{\partial \mathcal{J}(\mathbf{W})}{\partial \boldsymbol{w}_{:k}}, \ \forall k = 1, \dots, M_t, \tag{7}$$

where $\eta$ is the learning rate parameter.

The formulation presented above composes the weight updating rule of OnMLM, however it does not explains how the reference points are included during the training step. We must emphasize that OnMLM initiates the training process with only one input and one output reference points $(\boldsymbol{s}_1, \boldsymbol{t}_1)$ that were randomly selected from $\mathcal{X}$ and $\mathcal{Y}$, respectively. After that, at several iterations, one point is evaluated and included in the set reference points if the error reduction is higher than a pre-defined threshold. The criterion for adding new reference points is based on the hyperparameters $\Gamma > 0$, $\gamma \geq 0$, and $0 \leq \beta \leq 1$. The value $\Gamma$ determines the iterations in which the OnMLM will verify if a new set of reference points can be added, while $\gamma$ is the error threshold in the new reference point criterion. The $\beta$ parameter is used to update the value of $\gamma$, in order to adjust the error threshold along with the learning process of the OnMLM. Algorithm 2 summarizes how new reference points are included, being triggered after the weight matrix updating.

It is important to highlight that the weight matrix $\mathbf{W}$ increases as the number of input and output reference points increases – remember that $\mathbf{W} \in \mathcal{R}^{M_s \times M_t}$, which implies that for each new input reference point the number of lines in $\mathbf{W}$ increases by one. A similar consequence is observed for each new output reference points, however, in this last case, the number of columns in $\mathbf{W}$ is increased. Furthermore, from Algorithm 2, we can see that the $\gamma$ value is not

**Algorithm 2.** New reference point criterion.

**Input:** The hyperparameters $\Gamma$ and $\beta$, the initial value for $\gamma$, the current iteration $l$, the updated weight matrix $\mathbf{W}$, the distance matrices $\mathbf{D}_l$ and $\mathbf{\Delta}_l$, and the sets $\mathcal{X}_l$ and $\mathcal{Y}_l$.
**Output:** The new sets of input and output reference points $\mathcal{S}^{new}$ and $\mathcal{T}^{new}$.
1: Verify if $l \pmod{\Gamma} = 0$. If yes, then go to step 2; otherwise, stop the algorithm.
2: Compute the expected output distance matrix: $\hat{\mathbf{\Delta}}_l = \mathbf{D}_l\mathbf{W}$.
3: Compute the mean squared error obtained by each input vector $\boldsymbol{d}_i$:
$e^2 = \frac{1}{M_t} diag[(\mathbf{\Delta}_l - \hat{\mathbf{\Delta}}_l)(\mathbf{\Delta}_l - \hat{\mathbf{\Delta}}_l)^T]$.
4: Update $\gamma$:
$\gamma = \beta\gamma + (1-\beta)\overline{\gamma}, \ \overline{\gamma} = \max\{e_i^2 : i = 1\dots, N_l\}$.
5: Update the sets $\mathcal{S}$ and $\mathcal{T}$:
$\mathcal{S}^{new} = \mathcal{S} \cup \{\boldsymbol{x}_i \in \mathcal{X}_l : e_i^2 > \gamma, i = 1, \dots, N_l\}$,
$\mathcal{T}^{new} = \mathcal{T} \cup \{\boldsymbol{y}_i \in \mathcal{Y}_l : e_i^2 > \gamma, i = 1, \dots, N_l\}$.

constant (step 4), but rather updated each time the new reference point criterion algorithm is executed. In summary, the criterion enables the model to decide if a new input pattern or expected output will become a reference point looking at the related error of each input pattern at the current $l$-th mini-batch.

As stated before, the OnMLM model includes 3 new parameters ($\Gamma$, $\gamma$ and $\beta$) and the task of adjusting such parameters may seem challenging. However, as we will argue, the adjustment intuitive. First, consider $\Gamma > 0$. Such parameter controls the frequency in which the model executes Algorithm 2, checking for new input and output reference points in the current $l$-th mini-batch. To choose a coherent value for $\Gamma$, one must consider the following: (*i*) each line and column of $\mathbf{W}$ is randomly initialized with positive and negative values; and (*ii*) the OnMLM training is based on SGD. From (*i*) and (*ii*), one can realize that, for each new input or output reference point, a line or a column with random values will be included in $\mathbf{W}$, and the model will require a number of iterations large enough to reach stability before adding new reference points, since the model is trained via SGD. This implies that $\Gamma$ must not be a small value, but large enough to reach some stability in the current matrix $\mathbf{W}$. Also, a large value for $\Gamma$ can lead the training process of the model to be slow. Taking this in consideration, and after some experiments with Adam algorithm [5], we verified that $\Gamma = \lceil \frac{N}{N_l} \rceil \times 50$ is a good value – note that $N$ is the number of training samples and $N_l$ the number of training samples in the $l$-th mini-batch.

Let us now focus on the $\gamma \geq 0$ parameter, which delimits if a input/output pattern will become a reference point. As shown in Algorithm 2, the parameter $\gamma$ is updated considering the maximum quadratic error $\overline{\gamma}$ among the errors in the $l$-th mini-batch. For regression problems we recommend to use as initial value $\gamma = 0$ because $\gamma$ will be adjusted during the training process. For instance, assume $\beta > 0$, then the $\gamma$ provides the lower bound for adding new reference points. If $\overline{\gamma} > \gamma$, then the new value of $\gamma$ increases, but under the convex combination between $\overline{\gamma}$ and $\gamma$, it remains lower than $\overline{\gamma}$, so that at least the pair $(\boldsymbol{x}_i, \boldsymbol{y}_i)$ related to the maximum quadratic error $\overline{\gamma}$ of the current mini-batch will become

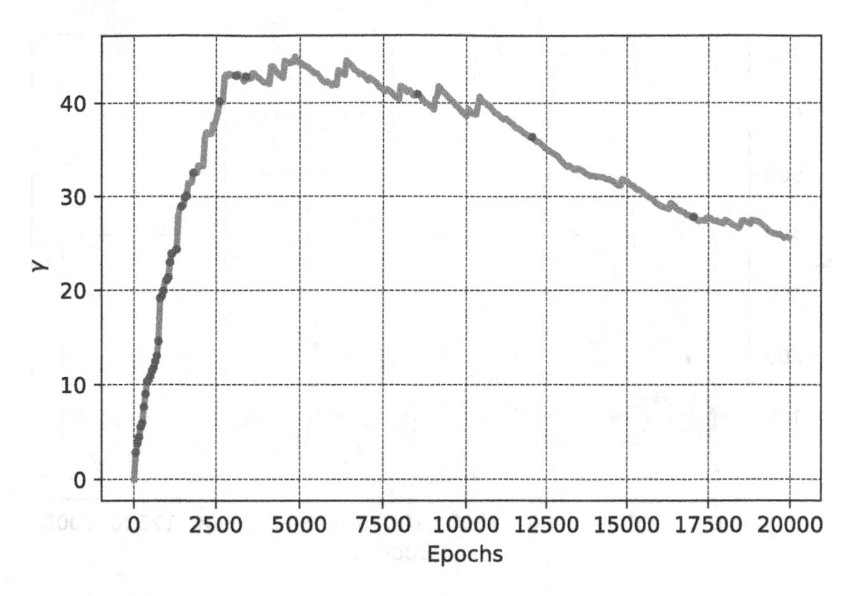


**Fig. 2.** $\gamma$ value fluctuation along 20000 epochs with $\beta = 0.99$ applied to the BHG dataset. (Color figure online)

a reference point. On the other hand, if $\overline{\gamma} < \gamma$, then the new value of $\gamma$ decreases, but remains greater than $\overline{\gamma}$ and no reference point will be added. This implies that as the error is increasing, new reference points are being added; unlike, when the error is decreasing, no reference points are being added. Also, we must emphasize that at some point of the training process, the model can experience a continuous error reduction, which can lead $\gamma$ to be very small. As a consequence of a small $\gamma$, a new maximum quadratic error $\overline{\gamma} > \gamma$ can arise, adding new reference points. This dynamic adds new reference points in a supervised way, so that, when the model is experiencing excessive error values (instability) during the training process, the number of reference points tends to increase. On the other hand, when the model has the necessary information provided by the current reference points and performed a large enough number of iterations to stabilize $\mathbf{W}$, the training error tends to stabilizes, making it difficult to include new reference points. The same dynamic applies for classification scenario, however, in such a scenarios, we recommend the initial value to be $\gamma = 1$ to speed up the training process (this only applies for 1-hot coded expected outputs).

In Fig. 2 we show the $\gamma$ value fluctuation for 20000 epochs and $\beta = 0.99$ applied to the Boston Housing (BHG) dataset from UCI Machine Learning [9] – some information about such a dataset can be seen in Table 1. The model was trained via Adam algorithm. In this Figure, one can see the blue line, which corresponds to the values of $\gamma$ through the epochs and the red spots, showing the time at which at least one reference point was added. Taking a closer look on this Figure, one can see that at the beginning of the training process, where the model is experiencing large errors, a bigger number of reference points is added

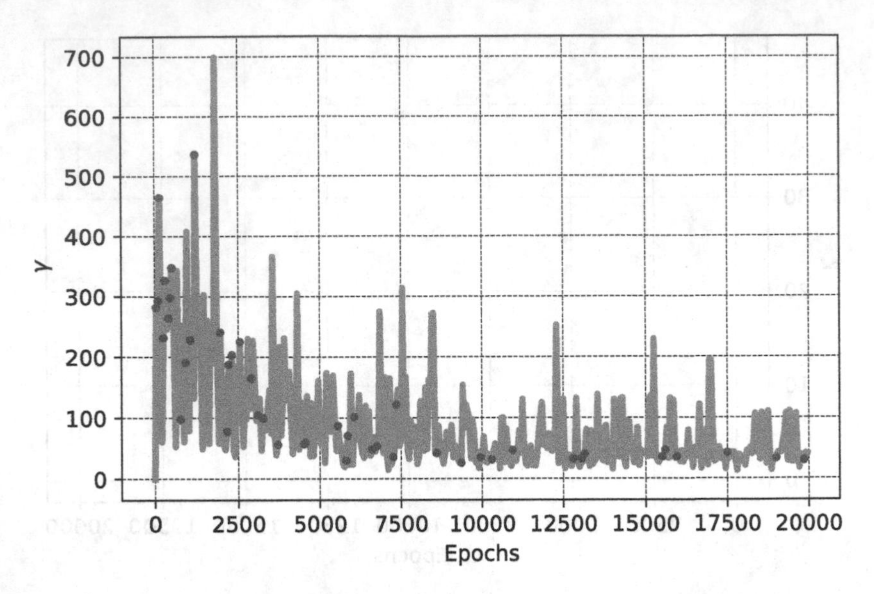


**Fig. 3.** $\gamma$ value fluctuation through 2$e$+4 epochs with $\beta = 0.01$ applied to BHG dataset.

(epochs 0–5000). However, after the epoch 5000, the algorithm stabilizes and a reduced number of reference points are added. It is also noticeable the occurrence of some peaks in $\gamma$ between the epochs 5000–12500 where no reference point is being added. This behavior occurs because this peaks of $\gamma$ are being provided by patterns that are already reference points, and since the operation for adding new reference points is based in the union operation, a pattern that is already a reference point which provides a maximum quadratic error $\overline{\gamma} > \gamma$ will not be added as a reference point.

At last, we can make some considerations about the $\beta \in [0, 1]$ parameter. This parameter is used in the convex combination between $\overline{\gamma}$ and $\gamma$. For $\gamma \rightarrow 0$, the value of gamma will tend to reach the current value of $\overline{\gamma}$. In such a scenario, fewer reference points will be added, which can lead to poor generalization, and the value of $\gamma$ will suffer some deep fluctuations. This scenario is depicted in Fig. 3. The only difference concerning that one from Fig. 2 is that $\beta = 0.01$, and the number of input and output reference points reached is 46 and 35, respectively. From Figs. 2 and 3, one can see that a value for $\beta \rightarrow 1$ provides a more stable learning process.

Despite the three new OnMLM parameters, a main difference between OnMLM and MLM relies on the fact that OnMLM do not use duplicated output reference points. For instance, to make it clear, let us consider a binary classification problem with $M > 2$. In this scenario, the MLM will use $M$ reference points, so that there will be $M > 2$ output reference points and some of than will be duplicated. On the other hand, the union operation used by OnMLM in the Algorithm 2 avoid such a duplication, providing much sparse solutions at the output reference point space.

## 3 Experiments

To provide an overview of OnMLM capabilities, we have performed experiments considering both classification and regression tasks. We compared the OnMLM with the standard MLM and the Multilayer Perceptron (MLP). For each task we have used 3 datasets obtained from UCI Machine Learning repository [9], totalizing 6 datasets: Abalone with 3 classes (ABA3C), Adult (ADU), Bank Marketing (BKM), Boston Housing (BHG), Abalone (ABA), and Wine Quality (WIN). Table 1 summarizes the information datasets information: The column *task* identifies the task proposed by each dataset, where we have $C$ for classification and $R$ for regression. The columns $p$ and $q$ specify the dimension of the input and output vectors, respectively – notice that for classification tasks, $q$ represents the number of classes. The column *test size* says the proportion used to evaluate the model in each scenario.

**Table 1.** Summary of datasets informations.

| *Dataset* | *Task* | *#samples* | *p* | *q* | *Test size* (%) |
|---|---|---|---|---|---|
| ABA3C | C | 4177 | 8 | 3 | 33% |
| ADU | C | 45222 | 14 | 2 | 33% |
| BKM | C | 45211 | 16 | 2 | 33% |
| BHG | R | 506 | 13 | 1 | 20% |
| ABA | R | 4177 | 8 | 1 | 33% |
| WIN | R | 6497 | 11 | 1 | 33% |

The training of both OnMLM and MLP (with one hidden layer) was performed using the Adam algorithm with $\beta_1 = 0.9$, $\beta_2 = 0.999$, and $\epsilon = 1e-8$ [5]. Concerning the learning rate, we have used $\eta = 1e{-}3$ for MLP, while for OnMLM we have set $\eta = 1e-2$ in regression tasks and $\eta = 1e-3$ in classification tasks. We have performed a 5-fold cross validation to optimize the number of reference points of the standard MLM and the size of the MLP hidden layer. For both OnMLM and MLP, we performed the simulations with mini batches of sizes $N_l = 256$ (ABA3C, ABA, and WIN), $N_l = 64$ (BHG), and $N_l = 512$ (BKS, ADU, and BKM). The maximum number of epochs was 2000. The parameters for OnMLM were fixed: we set $\Gamma = \lceil \frac{N}{N_l} \rceil \times 50$ for all datasets, $\gamma = 1.0$ for classification and $\gamma = 0$ for regression. We used $\beta = 0.95$ for all datasets.

The results for classification and regression can be viewed in Table 2. We must emphasize that, for ADU and BKM datasets, we have trained MLM with 6000 training samples in order to ensure computational efficiency. The first comment we should make about OnMLM is related to its superiority concerning sparsity degree compared to standard MLM. For all applications, the OnMLM achieves fewer number of input and output reference points, with special attention to BKM and WIN datasets – even when OnMLM performance was inferior to MLM (WIN dataset), the standard MLM required almost the entire training dataset to be used as reference points.

**Table 2.** Classification accuracy and regression MSE, number of input/output reference points, and hidden layer size (# Neurons) averaged over 10 runs.

| *Dataset* | *Model* | *Accuracy* ($\sigma$) | *# Neurons* | $M_s$ | $M_t$ |
|---|---|---|---|---|---|
| ABA3C | MLP | **67.5** (0.9) | 40.1 | | |
| | MLM | 65.6 (1.6) | – | 238.0 | 238.0 |
| | OnMLM | 65.6 (1.0) | | **153.3** | **3.0** |
| ADU | MLP | 80.2 (0.4) | 78.1 | | |
| | MLM | 78.3 (0.5) | | 540.0 | 540.0 |
| | OnMLM | **82.5** (0.6) | | **317.3** | **2.0** |
| BKM | MLP | **89.0** (0.1) | 119.1 | | |
| | MLM | 88.3 (0.5) | | 900.0 | 900.0 |
| | OnMLM | 88.9 (0.3) | | **150.7** | **2** |
| *MSE* ($\sigma$) | | | | | |
| BHG | MLP | $1.4e+1$ (3.7) | 40.1 | – | – |
| | MLM | $\mathbf{1.2e+1}$ (4.4) | – | 396.0 | 396.0 |
| | OnMLM | $1.5e+1$ (4.6) | – | **110.4** | **84.3** |
| ABA | MLP | 4.6 ($1.7e-1$) | 5.0 | – | – |
| | MLM | **4.5** ($2.2e-1$) | – | 210.0 | 210.0 |
| | OnMLM | 4.6 ($2.4e-1$) | – | **172.6** | **22.6** |
| WIN | MLP | $5.5e-1$ ($1.1e-2$) | 29.3 | – | – |
| | MLM | $\mathbf{3.9e-1}$ ($2.4e-2$) | | 3764.7 | 3764.7 |
| | OnMLM | $5.0e-1$ ($2.3e-2$) | – | **352.7** | **7.0** |

Concerning the accuracy of the methods, we can say that OnMLM in general outperforms MLM, with an exception for ABA3C dataset. Also, although the OnMLM reached lower accuracy than MLP in the ABA3C dataset and equivalent accuracy in the BKM dataset, the OnMLM performed well when compared to its counterparts in the ADU dataset. Finally, taking a close look into the MSE measures for the regression tasks, one can see that the MLM is always superior, although the OnMLM still obtains much sparser solutions.

## 4 Conclusions

We have presented the OnMLM, a feasible formulation based on MLM for performing online and incremental learning. Thus, OnMLM greatly increases the applicability of MLM, since it is able to tackle large datasets and data streams. Furthermore, the proposed OnMLM is able to reduce the number of input and output reference points when its hyperparameters are carefully chosen. The new reference points criterion is of central importance because of the incremental nature of the training procedure and the sparseness of the generated model. Those particularities turn OnMLM a promising online learning method. Future works intend to provide more detailed analysis of the OnMLM incremental behavior and investigate alternatives for enabling it to remove reference points in the training step.

**Acknowledgements.** The work was supported by the Academy of Finland from grants 311877 and 315550.

## References

1. Beygelzimer, A., Hazan, E., Kale, S., Luo, H.: Online gradient boosting. In: Cortes, C., Lawrence, N.D., Lee, D.D., Sugiyama, M., Garnett, R. (eds.) Advances in Neural Information Processing Systems 28, pp. 2458–2466. Curran Associates Inc., New York (2015)
2. Huang, G., Liang, N., Rong, H., Saratchandran, P., Sundararajan, N.: On-line sequential extreme learning machine. In: the IASTED International Conference on Computational Intelligence (CI 2005), Calgary, Canada, 4–6 July 2005
3. Huang, G.B., Zhu, Q.Y., Siew, C.K.: Extreme learning machine: theory and applications. Neurocomputing **70**(1), 489–501 (2006). Neural Networks
4. Jian, L., Shen, S., Li, J., Liang, X., Li, L.: Budget online learning algorithm for least squares SVM. IEEE Trans. Neural Netw. Learn. Syst. **28**(9), 2076–2087 (2017)
5. Kingma, D.P., Ba, L.J.: Adam: a method for stochastic optimization. In: International Conference on Learning Representations, ICLR (2015)
6. Lakshminarayanan, B., Roy, D.M., Teh, Y.W.: Mondrian forests: efficient online random forests. In: Proceedings of the 27th International Conference on Neural Information Processing Systems, NIPS 2014, vol. 2. pp. 3140–3148. MIT Press, Cambridge (2014)
7. Langford, J., Li, L., Zhang, T.: Sparse online learning via truncated gradient. J. Mach. Learn. Res. **10**, 777–801 (2009)
8. LeCun, Y., Bottou, L., Orr, G.B., Müller, K.-R.: Efficient BackProp. In: Orr, G.B., Müller, K.-R. (eds.) Neural Networks: Tricks of the Trade. LNCS, vol. 1524, pp. 9–50. Springer, Heidelberg (1998). https://doi.org/10.1007/3-540-49430-8_2
9. Lichman, M.: UCI machine learning repository (2013). http://archive.ics.uci.edu/ml
10. Matias, T., Souza, F., Araújo, R., Gonçalves, N., Barreto, J.P.: On-line sequential extreme learning machine based on recursive partial least squares. J. Process. Control. **27**, 15–21 (2015)
11. Mesquita, D.P.P., Gomes, J.P.P., Souza Junior, A.H.: Ensemble of efficient minimal learning machines for classification and regression. Neural. Process. Lett. **46**(3), 751–766 (2017). https://doi.org/10.1007/s11063-017-9587-5

12. Santos, J.D.A., Barreto, G.A.: A regularized estimation framework for online sparse LSSVR models. Neurocomputing **238**, 114–125 (2017)
13. de Souza Júnior, A.H., Corona, F., Barreto, G.A., Miche, Y., Lendasse, A.: Minimal learning machine: a novel supervised distance-based approach for regression and classification. Neurocomputing **164**, 34–44 (2015)
14. Suykens, J., Vandewalle, J.: Least squares support vector machine classifiers. Neural Process. Lett. **9**(3), 293–300 (1999)

# Adversarial Examples are a Manifestation of the Fitting-Generalization Trade-off

Oscar Deniz(✉), Noelia Vallez, and Gloria Bueno

VISILAB, E.T.S. Ingenieros Industriales, Universidad de Castilla-La Mancha, Avda. Camilo Jose Cela s/n, 13071 Ciudad Real, Spain
Oscar.Deniz@uclm.es

**Abstract.** In recent scientific literature, some studies have been published where recognition rates obtained with Deep Learning (DL) surpass those obtained by humans on the same task. In contrast to this, other studies have shown that DL networks have a somewhat strange behavior which is very different from human responses when confronted with the same task. The case of the so-called "adversarial examples" is perhaps the best example in this regard. Despite the biological plausibility of neural networks, the fact that they can produce such implausible misclassifications still points to a fundamental difference between human and machine learning. This paper delves into the possible causes of this intriguing phenomenon. We first contend that, if adversarial examples are pointing to an implausibility it is because our perception of them relies on our capability to recognise the classes of the images. For this reason we focus on what we call *cognitively adversarial examples*, which are those obtained from samples that the classifier can in fact recognise correctly. Additionally, in this paper we argue that the phenomenon of adversarial examples is rooted in the inescapable trade-off that exists in machine learning (including DL) between fitting and generalization. This hypothesis is supported by experiments carried out in which the robustness to adversarial examples is measured with respect to the degree of fitting to the training samples.

**Keywords:** Adversarial examples · Deep learning · Bioinspired learning

## 1 Introduction

While advances in Deep Learning (DL) have been significant and have propelled AI into the spotlight, many researchers know that the abilities of this methodology are being at times overestimated [1]. In recent scientific literature, some studies have been published where recognition rates obtained with DL surpass those obtained by humans on the same task. In contrast to this, some studies have shown that DL networks have a somewhat strange behavior which is very


I. Rojas et al. (Eds.): IWANN 2019, LNCS 11506, pp. 569–580, 2019.
https://doi.org/10.1007/978-3-030-20521-8_47

different from human responses when confronted with the same task. Perhaps the best example to describe it is the case of the so-called "adversarial examples" [2], see the following figure. Adversarial examples are apparently identical to the original example versions except for a very small change in pixels of the image. Despite being perceived by humans as completely equal to the originals, DL techniques fail miserably at classifying them (Fig. 1).

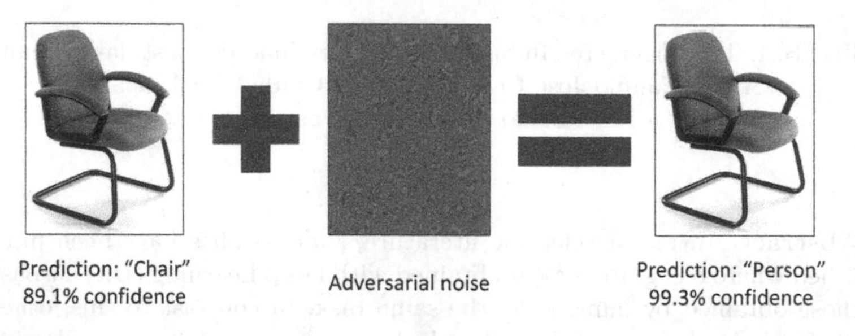


**Fig. 1.** Adversarial example. "Person" is the so-called *target* class (AE generation or "'attack"' methods aim at making the original class be confused with a target class).

Thus, while apparently having *superhuman* capabilities, DL also seems to have weaknesses that are not coherent with human performance. Not only that, from the structure of DL (essentially an interconnected network of neurons with numerical weights), it is unclear what gives rise to that behavior. The problem is not also in maliciously-selected noise, since some transformations involving rescaling, translation, and rotation produce the same results [3].

The contributions of this paper are as follows. We first contend that, if adversarial examples are pointing to an implausibility it is because our perception of them relies on our capability to recognise the classes involved. For this reason we focus on what we call *cognitively adversarial examples*, which are those obtained as variations of samples that the classifier can recognise correctly. On the other hand, while other underlying reasons have been proposed in the literature for the existence of adversarial examples, in this paper we argue that the phenomenon of adversarial examples is rooted in the inescapable trade-off that exists in machine learning (including DL) between fitting and generalization. This hypothesis is supported by experiments carried out in which the robustness to adversarial examples is measured with respect to the degree of fitting to the training samples. The goal of this paper is not to introduce a novel method, but to advance our knowledge about the phenomenon, its root causes and implications.

This paper is organised as follows. In Sect. 2 we summarize previous work that focused on the nature of adversarial examples. Section 3 describes the methods and datasets used. Section 4 shows the experimental results. The paper concludes with a broader discussion (Sect. 5) and the main conclusions.

## 2 Previous Work

Since adversarial examples (henceforth AEs) first drawn attention of researchers, the two major lines of associated research have been: (1) generating AEs and (2) defending against AEs. In this work we will not cover either, and the reader is referred to recent surveys. In parallel to those two lines, however, a significant body of work has been carried out to delve into the root causes of AEs and their implications.

In early work, the high nonlinearity of deep neural networks was suspected as a possible reason explaining the existence of adversarial examples [2]. On the other hand, later in [4] it is argued that high-dimensional linearities cause the adversarial pockets in the classification space. This suggests that generalization (as implied by the less complex linear discrimination boundaries) have a detrimental effect that produces AEs. In the same line, in [5] it is stated: "*Unlike the initial belief that adversarial examples are caused by the high non-linearity of neural networks, our results suggest instead that this phenomenon is due to the low flexibility of classifiers*".

In [2] the authors had suggested a preliminary explanation for the phenomenon, arguing that low-probability adversarial "pockets" are densely distributed in input space. In later work [6] the authors probed the space of adversarial images using noise of varying intensity and distribution. They showed that adversarial images appear in large regions in the pixel space instead.

In [7] the authors review the existing literature on the topic and conclude that up to 8 different explanations have been given for AEs. The prevailing trend, however, seems to focus on the linear/non-linear and in general in the overfitting problems of the classifier. Under two interpretations (the boundary tilting hypothesis [8] and in [9]) the authors argue that the phenomenon of AEs is essentially due to overfitting and can be alleviated through regularisation or smoothing of the classification boundary.

In the recent work [10] the authors perform experiments on a contrived synthetic dataset and conclude that low (test) error classification and AEs are intrinsically associated. They contend that this does not imply that defending against adversarial examples is impossible, only that success in doing so would require improved model generalization. Thus, they argue that the only way to defend against AEs is to massively reduce that error. However, we note that this would be in apparent contradiction with the main finding in that paper (that AEs appear with low classification error). Thus, while generalization may help reduce error in general, without additional considerations it would not necessarily remove AEs.

In [11] the authors point out that an implicit assumption underlying most of the related work is that the same training dataset that enables good *standard* accuracy also suffices to train a robust model. The authors argue that the assumption may be invalid and suggest that, for high-dimensional problems, adversarial robustness can require a significantly larger number of samples. Similar conclusions are drawn in [12], where it is stated that adversarial vulnerability increases with input dimension.

One interesting work that indirectly relates to ours is [13], where a *Deep K-Nearest Neighbors* (DkNN) method is proposed. DkNN inspects the internals of a deep neural network (DNN) at test time to provide 3 outputs: prediction, confidence and credibility. The DkNN algorithm uses a standard already-trained deep neural network. During inference, for a test input it compares each layer's outputs with those of the nearest neighbors used to train the model (including the network's output layer). The authors show that for adversarial examples the number of different nearest-neighbor labels obtained from the layers is often high. While there are practical issues with the method (specifically: (a) it is not clear how one would use the three outputs from DkNN in practice and (b) the algorithm requires of a calibration set -a holdout set that does not overlap with the training or test sets-), the connection made with nearest neighbors is intriguing.

In the related literature some findings have transpired that appear to be particularly interesting:

- The phenomenon does not seem to be tied to specific architectures or particular subsets of the training data [14,15]. In particular, it is possible to transfer adversarial examples from models with known parameters and architecture to other models with unknown parameters and architecture
- The phenomenon is not exclusive of deep learning [5,16]
- Despite the fact that methods have been proposed to increase robustness to AEs, the phenomenon appears to be essentially unavoidable (see for example [17–19])

In summary, despite substantial research, the exact cause of the phenomenon is still poorly understood and remains unsolved. It is not clear, for example, whether the phenomenon is due to overfitting or, on the contrary, to a lack of expressive power. Neither it is clear if the phenomenon is due to the (limited) amount of training samples that are available or due to the large input dimension (or the relationship between these two). On the other hand, even though there appears to be consensus that AEs are unavoidable, it is not clear why this has to be so.

## 3 Datasets and Methods

In existing research work, AEs are typically generated as variations from the test samples, irrespective of the classifier's decisions on those. This means that some AEs may be generated from test samples that are themselves not correctly classified. In contrast to this, as humans the cognitive dissonance we experience with AEs occurs because we can perceive the classes involved. Thus, for a classifier it should not come as a surprise that AEs are missclassified when they are tiny variations of samples that are missclassified. For these reasons in the following we focus on AEs generated from samples from the test set that are classified correctly (i.e. *cognitively adversarial examples*).

In an adversarial attack, an imperceptible change on an image from an *original* class makes the classifier confuse it with the so-called *target* class. When we think of the actual cause of error in this adversarial example we may ask, which class manifold is not being correctly modelled? that of original class? that of the target class? (see Fig. 2). The fact is that the AE actually implies a bad modelling of both. AEs are consistent with an incorrect modeling of the classes whereby manifolds span excess or defect space.

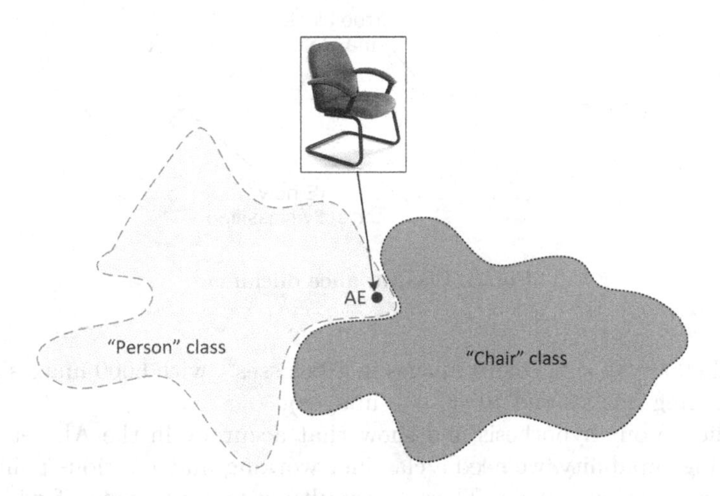


**Fig. 2.** Adversarial example *vs* trivial example. Is the error shown here due to a bad modelling of the Person class manifold or a bad modelling of the Chair class manifold?

In general, we can always analyze our approximations to the class manifolds in terms of bias and variance. Bias measures how far off the estimated class sample positions are from the true positions, while variance refers to the spread of the estimated positions. Because of the so-called bias-variance dilemma, reducing one quantity automatically increases the other. On the other hand, AEs are practically co-located with (i.e. extremely close to) samples that are correctly handled by the classifier, i.e. they are practically situated on true class positions. This means that reducing model bias (and therefore increasing model variance) should reduce error in those samples (albeit error may increase for other samples), see Fig. 3. That is exactly the hypothesis that we address below in the experiments. The bias and variance errors are in general controlled by the classifier's trade-off between fitting and generalization. Our aim is to test if such change in the fitting-generalization trade-off point reflects in the robustness to AEs.

In the experiments we used the MNIST [20] and CIFAR-10 [21] datasets, arguably the two most common datasets used in research on the nature of adversarial examples (used in 39 of the 48 papers reviewed in [7]). MNIST is a dataset of handwritten grayscale $28 \times 28$ images representing the digits 0 to 9. Typically, 60000 images are used for training and 10000 for testing. The CIFAR-10 dataset

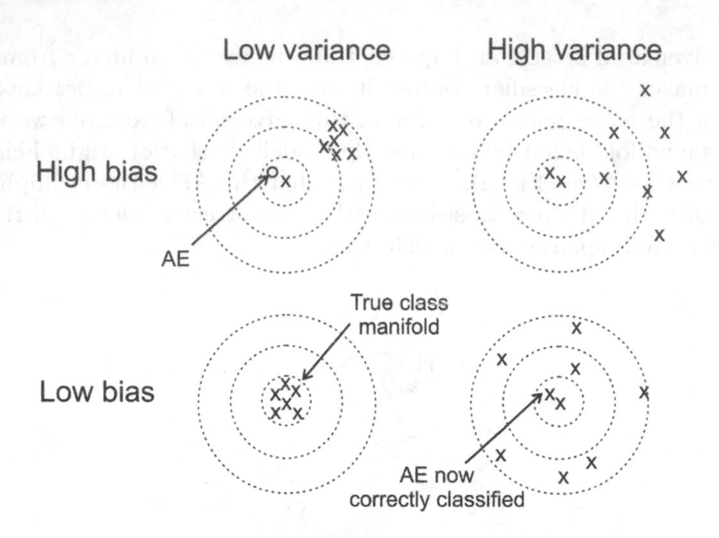


**Fig. 3.** Bias-variance dilemma.

consists of 60000 $32 \times 32$ colour images in 10 classes[1], with 6000 images per class (50000 training images and 10000 test images).

To validate our hypothesis and show that accuracy in the AE set is linked to the fitting capability, we need a classifier working under various points of the fitting-generalization regime. There are multiple factors, some of which inter-related, that affect the fitting-generalization trade-off point in a deep network: dimensionality of the input space, number of samples available, number of layers, number of epochs, etc. In order to have a direct control of the trade-off point based on a single parameter we decided to use a K-Nearest Neighbor Classifier instead, for K values equal and greater than 1. The K-NN classifier is a natural choice here. It is widely known that large values of K are used to have better generalization, while lower values (down to K = 1) may produce overfitting.

Given the dimensionalities involved, an efficient KD-tree-based implementation was used for the K-NN classifier.

The AEs were obtained using two methods: the Fast Gradient Sign Method (a so-called white-box targeted attack, introduced in [4]) and DeepFool [22], targeting all classes and keeping the maximum perturbation threshold at $\epsilon = 0.1$. For both methods we used the LENET-5 network architecture. Figure 4 shows some examples of the AEs generated.

## 4 Experimental Results

As mentioned above, the AEs were obtained from test samples that were correctly classified by the K-NN, for a $K = Z$ (for a fixed $Z > 1$, $Z$ odd). The attack method (FGSM or DeepFool) was then used to generate a set of AEs

[1] Airplane, automobile, bird, cat, deer, dog, frog, horse, ship and truck.

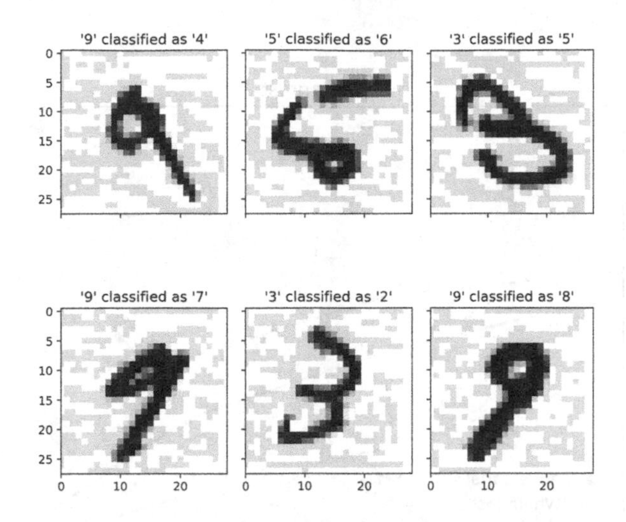


**Fig. 4.** Sample AEs generated with FGSM for the MNIST dataset.

from those. This set was filtered to discard samples that were correctly classified by the Z-NN. This leaves a set of samples that was generated from test samples correctly classified by Z-NN but which were not themselves correctly classified by Z-NN (thus being adversarial examples). Therefore Z-NN accuracy in this set is 0%.

Then we measured the K-NN classifier accuracy on this set, for values of K smaller than Z, down to K = 1. Again, our hypothesis is that accuracy in this set should increase as K gets smaller. In order to discard the possibility of this being a general trend with lower values of K, we also obtained the accuracy for the whole test set and for the subset of test samples in which the classifier gave a wrong decision. The latter is therefore the set of samples for which the classifier gave a wrong decision for K = Z. Note that, by definition, this set and the adversarial set have both 0% accuracy for K = Z.

We repeated the experiment a number of times, each run performing a stratified shuffling of the dataset between the training and test sets (always leaving 60000 samples for training and 10000 samples for test for MNIST, and 50000 samples for training with 10000 samples for test in CIFAR-10). The results are shown in Figs. 5 and 6.

The results show that accuracy in the adversarial set has the highest increase rate as K gets smaller. The accuracy values obtained in the whole test set are always very stable (and very close 100% in the case of MNIST) which makes it difficult to establish a trend in that case. For the other two sets, in order to check if there was a statistically significant difference of trends we applied hypothesis testing in the following way. Let $A_i$ be the accuracy obtained using $K = i$. We calculated the *slope* between accuracies for successive values of K in the following way: $S_j = A_j/A_{j+2}$, for $j = \{1, 3, 5\}$. Then we used the slopes as

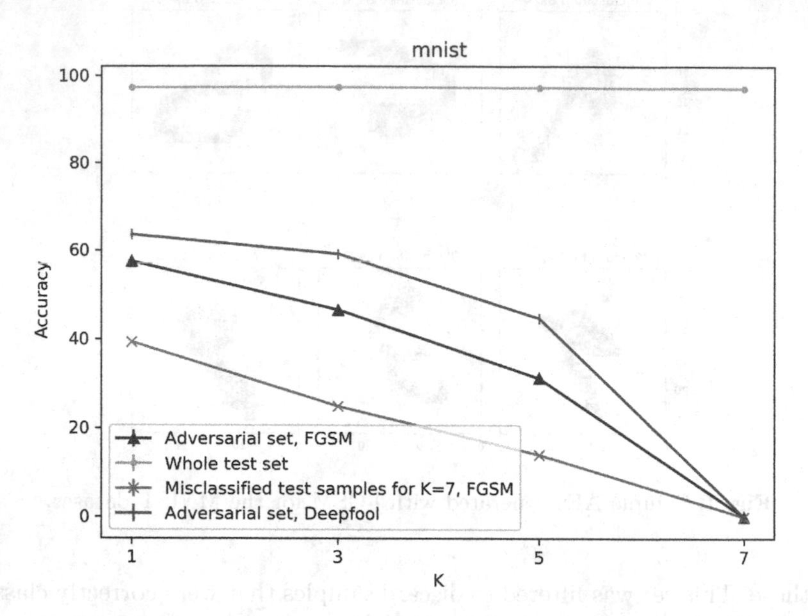


**Fig. 5.** Left: Accuracy values for the MNIST dataset, using Z = 7. Averages of 12 runs.

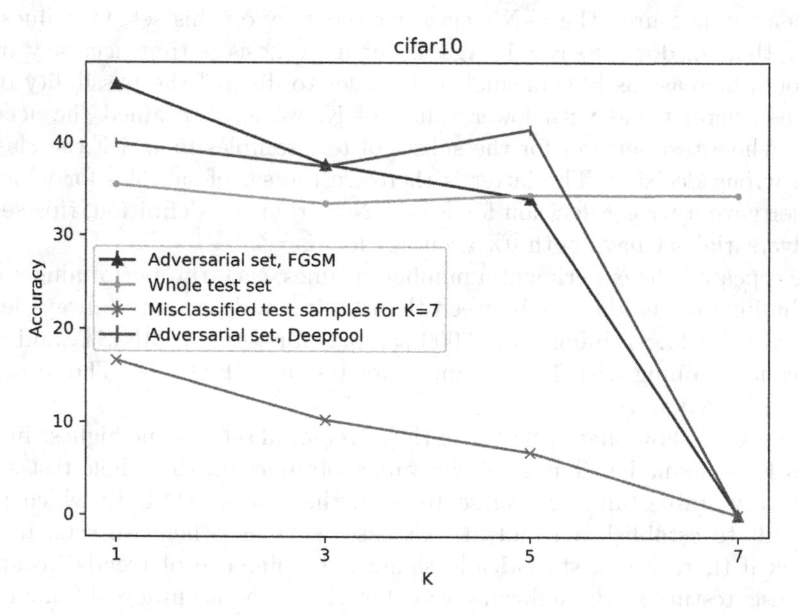


**Fig. 6.** Left: Accuracy values for the CIFAR-10 dataset, using Z = 7. Averages of 8 runs.

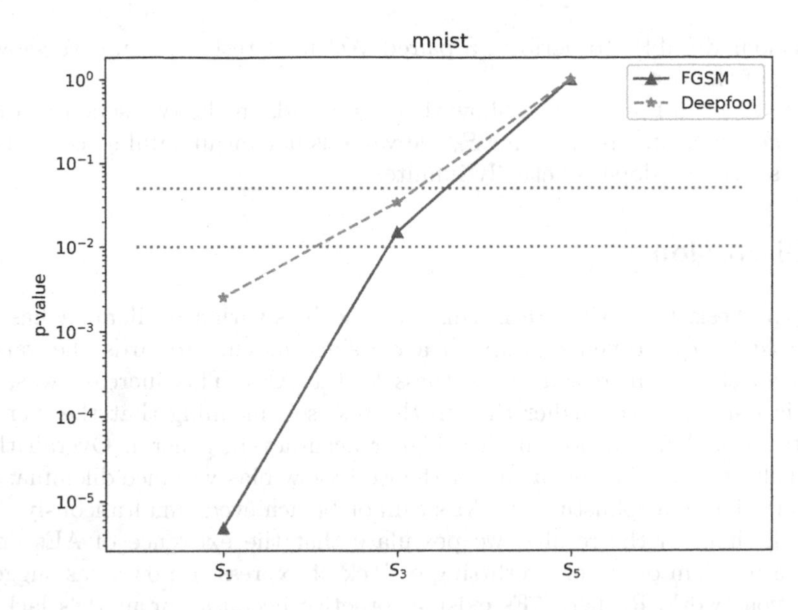


**Fig. 7.** p-values (represented in logarithmic scale) obtained by the paired Welch's t-test between results in the Adversarial set and those in the misclassified test samples. The dashed horizontal lines represent the 95% and 99% confidence thresholds.

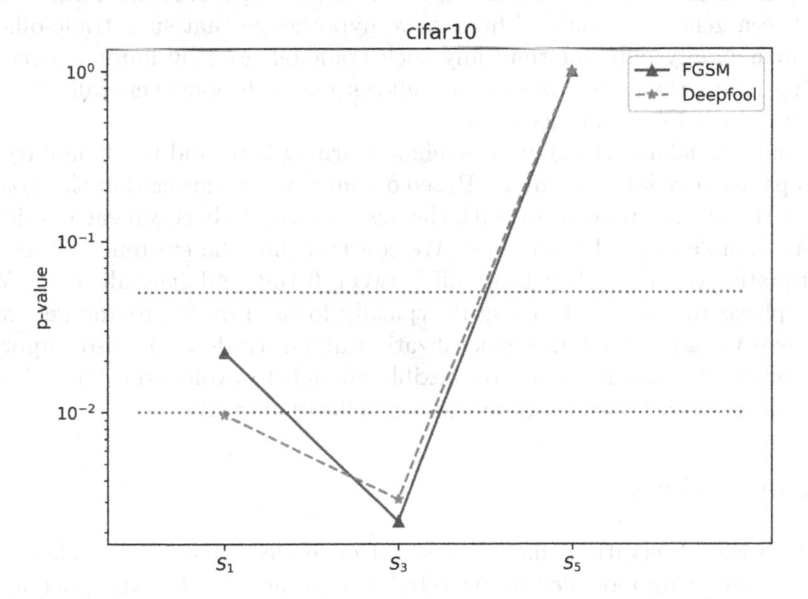


**Fig. 8.** p-values (represented in logarithmic scale) obtained by the paired Welch's t-test between results in the Adversarial set and those in the misclassified test samples. The dashed horizontal lines represent the 95% and 99% confidence thresholds.

the random variables to perform a paired Welch's t-test. In Fig. 7 we show the p-values of the test.

The values in Figs. 7 and 8 show that the trends in the two sets are statistically different. Note also that for $S_5$ the value is not meaningful since $A_7$ is 0 in both cases so the slope is actually infinite.

## 5 Discussion

Our hypothesis that AEs are intrinsic to the bias-variance dilemma has been supported by experiments in which a classifier moving towards the variance extremum showed increasing robustness to the AEs. This increase was, with statistical significance, higher that in the test set, meaning that the increased robustness to AEs did not entail a higher accuracy in general. Overall this is essentially the expected behavior for the well-know bias-variance dilemma: good generalization and robustness to AEs cannot be achieved simultaneously.

In the light of the results, we postulate that the existence of AEs do not reflect a problem of either overfitting or lack of expressive power, as suggested by previous work. Rather, AEs exist in practice because our models lack *both* aspects simultaneously. Rather than being an impossibility statement, this actually calls for methods that have more flexibility to reflect both aspects. While practically all machine learning methods already incorporate some form of trade-off between generalization and fitting, we hypothesize that such trade-offs may be fundamentally different than any such trade-off used by human perceptual learning (since the latter presumably allows for both good generalization and robustness to AEs simultaneously).

Broadly speaking, the goal of machine learning is to find the boundary that best separates two sets of samples. Based on our work, we argue that this goal can be achieved but not necessarily with the best separation between each individual training sample and other samples. We contend that the existence of AEs is a manifestation of the implicit trade-off between fitting and generalization. While the emphasis in machine learning is typically focused on improving generalization, here we argue that the generalization-fitting trade-off is also important. Thus, while the classifier must be flexible enough to avoid overfitting, it must be *flexible* enough to accomodate the *good* effects of overfitting.

## 6 Conclusions

Based on the observation that adversarial examples indicate that class manifolds are not being modelled accurately, we have argued that the phenomenon is rooted in the inescapable trade-off that exists in machine learning (including DL) between fitting and generalization. This hypothesis is supported by experiments carried out in which the robustness to adversarial examples is measured with respect to the degree of fitting to the training samples, as measured by the K value of a nearest neighbor classifier. As far as the authors know, this is the first time that such reason is proposed as the underlying cause for AEs. The hypothesis should in any case receive additional support through future work in which deep networks are used instead of a K-NN classifier.

While the bias-variance dilemma is posited as the root cause, that should not be considered an impossibility statement. Rather, this would actually call for methods that have more flexibility to reflect both aspects. Current trade-offs between bias and variance or equivalently between fitting and generalization would seem to be themselves biased towards generalization. The cost of that is precisely a lack of robustness to cases such as AEs. If human learning uses (or can be modelled with) any such trade-off, then it should be fundamentally different because presumably it allows for both good generalization and robustness to AEs simultaneously.

**Acknowledgments.** This work was partially funded by projects TIN2017-82113-C2-2-R by the Spanish Ministry of Economy and Business and SBPLY/17/180501/000543 by the Autonomous Government of Castilla-La Mancha and the ERDF.

## References

1. Yuille, A.L., Liu, C.: Deep nets: What have they ever done for vision? CoRR abs/1805.04025 (2018). http://arXiv.org/abs/1805.04025
2. Szegedy, C., et al.: Intriguing properties of neural networks, CoRR abs/1312.6199 (2013). http://dblp.uni-trier.de/db/journals/corr/corr1312.html#SzegedyZSBEGF13
3. Athalye, A., Engstrom, L., Ilyas, A., Kwok, K.: Synthesizing robust adversarial examples, CoRR abs/1707.07397 (2017). arXiv:1707.07397
4. Goodfellow, I.J., Shlens, J., Szegedy, C.: Explaining and harnessing adversarial examples, arXiv preprint arXiv:1412.6572 (2014)
5. Fawzi, A., Fawzi, O., Frossard, P.: Fundamental limits on adversarial robustness. In: Proceedings of ICML, Workshop on Deep Learning (2015). http://infoscience.epfl.ch/record/214923
6. Tabacof, P., Valle, E.: Exploring the space of adversarial images. In: 2016 International Joint Conference on Neural Networks (IJCNN), pp. 426–433 (2016)
7. Serban, A.C., Poll, E.: Adversarial examples: a complete characterisation of the phenomenon, CoRR abs/1810.01185 (2018). arXiv:1810.01185
8. Tanay, T., Griffin, L.D.: A boundary tilting persepective on the phenomenon of adversarial examples, CoRR abs/1608.07690 (2016). arXiv:1608.07690
9. Fawzi, A., Moosavi-Dezfooli, S., Frossard, P.: Robustness of classifiers: from adversarial to random noise, CoRR abs/1608.08967 (2016). arXiv:1608.08967
10. Gilmer, J., et al.: Adversarial spheres, CoRR abs/1801.02774 (2018). arXiv:1801.02774
11. Schmidt, L., Santurkar, S., Tsipras, D., Talwar, K., Madry, A.: Adversarially robust generalization requires more data, CoRR abs/1804.11285 (2018). arXiv:1804.11285
12. Simon-Gabriel, C.-J., Ollivier, Y., Schölkopf, B., Bottou, L., Lopez-Paz, D.: Adversarial vulnerability of neural networks increases with input dimension, CoRR abs/1802.01421 (2018)
13. Papernot, N., McDaniel, P.D.: Deep k-nearest neighbors: towards confident, interpretable and robust deep learning, CoRR abs/1803.04765 (2018). arXiv:1803.04765

14. Papernot, N., McDaniel, P., Goodfellow, I.: Transferability in machine learning: from phenomena to black-box attacks using adversarial samples, arXiv preprint arXiv:1605.07277 (2016)
15. Charles, Z.B., Rosenberg, H., Papailiopoulos, D.S.: A geometric perspective on the transferability of adversarial directions, CoRR abs/1811.03531 (2018)
16. Wang, Y., Jha, S., Chaudhuri, K.: Analyzing the robustness of nearest neighbors to adversarial examples. In: ICML (2018)
17. Bortolussi, L., Sanguinetti, L.: Intrinsic geometric vulnerability of high-dimensional artificial intelligence, CoRR abs/1811.03571 (2018). arXiv:1811.03571
18. Tsipras, D., Santurkar, S., Engstrom, L., Turner, A., Madry, A.: Robustness may be at odds with accuracy. In: International Conference on Learning Representations (2019). https://openreview.net/forum?id=SyxAb30cY7
19. Shamir, A., Safran, I., Ronen, I., Dunkelman, O.: A simple explanation for the existence of adversarial examples with small hamming distance, CoRR abs/1901.10861 (2019). arXiv:1901.10861
20. LeCun, Y., Cortes, C.: MNIST handwritten digit database (2010). http://yann.lecun.com/exdb/mnist/. (cited 2016-01-14 14:24:11)
21. Krizhevsky, A., Nair, V., Hinton, G.: CIFAR-10 (Canadian Institute for Advanced Research). http://www.cs.toronto.edu/~kriz/cifar.html
22. Moosavi-Dezfooli, S., Fawzi, A., Frossard, P.: Deepfool: a simple and accurate method to fool deep neural networks, CoRR abs/1511.04599 (2015). arXiv:1511.04599

# Deep Learning and Natural Language Processing

# Some Insights and Observations on Depth Issues in Deep Learning Networks

Arindam Chaudhuri(✉)

Samsung R&D Institute, Noida 201304, Delhi, India
arindam_chau@yahoo.co.in,
arindamphdthesis@gmail.com

**Abstract.** In deep neural networks the depth of network is specified by number of layers and neurons in each layer. These parameters are basically set through trial and error methods. During past few years deep networks have provided successful results for various categories of constrained optimization problems at the cost of high memory and computation. With this motivation, here we propose some valuable insights and observations towards depth aspects in deep learning networks. As the number of paramcters in these networks are redundant in nature they are often replaced through subtle architectures. The number of neurons in each layer of a deep network are obtained automatically through various complex functions. The different parameters of network are obtained various regularizers by certain operations on network neurons. This provides a single coherent framework which optimizes memory and computation time thereby generalizing network architectures. The process reduces number of parameters upto an appreciable amount while improving network accuracy. This has provided superior results for several regression and optimization computational scenarios.

**Keywords:** Shallow networks · Deep networks · Network architectures · Network layers · Neurons per layer

## 1 Introduction

With growth of large-scale datasets and computation power, deep networks have evolved as most preferred tool for unstructured data. Inspite of this growth sketching a deep network still is a very complex activity and done using trial and error. Some important aspects like number of layers and the neurons each layer are set through brute force approach. Well questions have also been raised on shallow and depth aspects of neural networks. The question still remains that whether we really need a deep network.

We consider a situation with 1B training points as labeled data points. On training shallow network with fully connected feed forward hidden layer with the said data an accuracy of about 90% is achieved for test data. While training deep network with convolutional, pooling and three fully connected feed forward layers for similar data an accuracy of around 95% is obtained. How can we improve these accuracy figures? This accuracy increase happens because deep network has more parameters, can learn more complex functions with identical parameters number, has better inductive bias, better learning with convolution as well as learning algorithms and regularization methods [1].


I. Rojas et al. (Eds.): IWANN 2019, LNCS 11506, pp. 583–595, 2019.
https://doi.org/10.1007/978-3-030-20521-8_48

Several attempts have been made to address these questions. The deep networks with unsupervised layer wise pre-training [2, 3] holds good. The network depth combined with pre-training provides better prior considering weights and improves generalization. A network with large single hidden layer for sigmoid units approximates any decision boundary [4]. But training shallow networks as accurately as deep networks is not an easy task. Some references are available in [5, 6]. The trend goes forward towards deep networks [7, 8] that are more expressive. This is achieved at memory and speed cost that prevents such network's deployment for constrained platforms. This also complicates process of learning because of gradients which vanish rapidly. The automatic model selection has been studied with considerable success. Beginning from shallow networks additional parameters can be placed incrementally or to network layers [7]. The shallow networks are not as expressive as deep networks and do not provide good initialization with newer layers. Again, deep networks have significant redundant parameters [9, 10]. Thus, given an initial deep network the objective is to reduce parameters while preserving power of representation. This is achieved initially by removing parameters [11, 12] or neurons [13–15] which influences output minimally. This requires analyzing every parameter independently and it does not scale well. The recent trend goes towards performing network reduction where focus is on training shallow networks in order to mimic large and deep network's behavior [16, 17]. This method is post-processing and looks towards successful training of deep network initially.

In this article, we investigate certain aspects related to deep neural networks' depth which tries to address the network's overall architecture. The network's parameters are basically defined using trial and error approach. The pointers revolve around considering optimum number of layers and neurons in deep network architecture. The compactness in the deep network happens when the learning process is performed using group sparsity regularizer. The estimation of number of neurons is done in order to produce the steadiness in network architecture. This leads to higher memory and computation advantages. The network architecture is also evaluated with shallow networks training against the deep networks. The redundancy in deep network architecture is also reduced. The experimental analysis with several large-scale datasets validates the benefits of the present approach. This also addresses the generalization ability of the present approach with respect to different architectures. This analysis provides useful results with the regression and optimization computational scenarios. This paper is structured through following sections. The Sect. 2 places work related on deep networks towards number of layers and neurons selection. This is followed by selection of deep network architecture in Sect. 3. The Sect. 4 presents the shallow networks training in terms of deep networks. In Sect. 5 results for the experiments are highlighted. Finally, in Sect. 6 conclusion is presented.

## 2 Related Work

The best model selection for deep networks viz number of layers and neurons each layer needs fair amount of exploration. This is achieved by manually tuning hyper-parameters through validation data or deep networks [7, 8]. These large networks come at high memory footprint cost and lower test time speed. Also, most network parameters are always repeated [9, 10]. An initial approach on model selection for deep learning is available in [18]. The shallow networks do not handle non-linearities as deeper networks [19]. Thus, initial shallow architectures get easily trapped at bad optima and give poor initialization. There are networks for which the aim is to reduce it while keeping the behavior unchanged [11–15, 20]. The recent networks consist of learning a shallower or thinner network that mimics an initial deep network [16, 17]. In designing of compact networks, convolutional neural networks have proposed to decompose filters of pre-trained network into low-rank filters reducing number of parameters [21–24]. Another approach by [25] uses regularizers that eliminates some network parameters. The regularizers are minimized simultaneously as network is learned with no pre-training. Some other parameter regularization techniques are available in [18].

# 3 Selection of the Deep Network Architecture

The pointer now moves towards automatically determining neurons for every layer of deep network while performing learning on network parameters. We present here the framework for general deep network. Some specific deep network architectures are highlighted in experimental results. A deep neural network is presented as $A$ layers successively which performing linear and non-linear operations. This comes along with several non-linear aspects and potential operations on pooling. Every layer $a$ has $N_a$ neurons, with each neuron being encoded through parameters $\varphi_a^v = \left[\mathbf{wp}_a^v, bs_a^v\right]$. Here $\mathbf{wp}_a^v$ and $bs_a^v$ represents linear or non-linear operator acting on input from layer and bias respectively. All these parameters constitute parameters $\emptyset = \{\varphi_a\}_{1 \le a \le A}$ with $\varphi_a = \{\varphi_a^v\}_{1 \le v \le N_a}$. Assuming input $\mathbf{y}$ which may be a video frame, network's output is $\hat{z} = h(\mathbf{y}, \emptyset)$ with $h(\cdot)$ transforming linear, non-linear and pooling operations successively. We have training set with $N$ pairs $\{(\mathbf{y}_i, z_i)\}_{1 \le i \le N}$ where parameters of network are learned and are obtained by solving the optimization problem:

$$\min_{\emptyset} \frac{1}{N} \sum_{i-1}^{N} loss(z_i, h(\mathbf{y}_i, \emptyset)) + regularizer(\emptyset) \tag{1}$$

In Eq. (1) $loss(\cdot)$ and $regularizer(\cdot)$ presents loss which compares predicted with actual output and regularizer on network parameters respectively. The various choices of $loss(\cdot)$ and $regularizer(\cdot)$ are available in [18]. Here the objective is to find number of neurons in network's each layer. This is initiated from an overcomplete network and effect of some neurons are ruled out. It is to be noted that none of the regularizers in [18] reach the stated objective. The initial one supports small parameters and next one eliminates out individual parameters. The neuron is transformed through parameters' group and it translates the entire groups towards zero. This is achieved through group sparsity [18]. The potential regularizer we have here:

$$regularizer(\emptyset) = \sum_{a=1}^{A} p\left(I_a (V_a)^{\frac{1}{p}}\right) \sum_{v=1}^{N_a} \left\|\varphi_a^v\right\|_2 \tag{2}$$

In Eq. (2) it is assumed that each neuron in layer $a$ parameters are placed together in vector size $V_a$ and $I_a$ sets tone for penalty. It is to be noted that weight is different towards each layer $a$. It has been observed that it is better to have two different weights. The smaller weight is attributed towards first few layers and larger weight is for rest layers. This allows finishing many neurons in first few layers and restores adequate information for the rest. The sparsity in group allows effective removal of some neurons and exploits standard regularizers on individual parameters towards generalization [18]. The Lasso's sparse group from [26] provides road towards automatic model selection approach. Incorporating this idea in regularizer we have:

$$regularizer(\emptyset) = \sum\nolimits_{a=1}^{A} \left((1-\alpha)p\left(I_a(V_a)^{\frac{1}{p}}\right)\sum\nolimits_{v=1}^{N_a} \|\varphi_a^v\|_2 + \alpha I_a \|\varphi_a\|_1\right) \quad (3)$$

In Eq. (3) $\alpha \in [0,1]$ places relative influence towards both terms. Several experiments are performed with $\alpha = 0, 0.25, 0.50, 0.75$. In order to solve optimization problem in Eq. (1) alongwith following regularizer, proximal gradient descent approach is adopted [18]. The proximal gradient descent takes gradient step of size $st$ iteratively for loss function $\ell oss(z_i, h(\mathbf{y}_i, \emptyset))$. Then on resulting solution regularizer's proximal operator is applied. Here as groups are non-overlapping, proximal operator is applied independently towards each group. For single group parameters are updated as:

$$\overline{\varphi}_a^v = \mathrm{argmin}_{\varphi_a^v}\left(\frac{1}{2(st)}\|\varphi_a^v - \widehat{\varphi}\|_2^2\right) + regularizer(\emptyset) \quad (4)$$

In Eq. (4) $\widehat{\varphi}_a^v$ represents solution from gradient step-based loss. Adopting thoughts from [18] and considering regularizers stated in Eqs. (2) and (3) there is a closed-form solution which is presented as follows:

$$\overline{\varphi}_a^v = \left(1 - \frac{st(1-\alpha)p\left(I_a(V_a)^{\frac{1}{p}}\right)}{\left\|Soft\left(\widehat{\varphi}_a^v, (st)\alpha I_a\right)\right\|_2}\right)_* \quad (5)$$

In Eq. (5) $*$ takes the maximum between zero and argument. The $soft(\cdot)$ represents soft thresholding operator which is presented elementwise as follows:

$$(Soft(\mathbf{s}, \tau))_i = \mathrm{sign}(\mathbf{s}_i)(|\mathbf{s}_i| - \tau)_* \quad (6)$$

The learning algorithm moves forward by iteratively taking gradient step for loss function. Then it updates variables of all groups as per Eq. (5). The stochastic gradient descent approach is adopted here considering mini batches. The proximal operator is applied at each epoch end. The algorithm then runs for fixed epochs. As learning terminates, some neurons parameters go towards zero. These neurons are removed completely as they do not impact the output. When taking fully connected layers, neurons on zeroed-out neurons from previous layers become unnecessary and are removed. On removing all these neurons, we have compact network architecture. In

Sect. 5, to have a better comparative analysis results on model selection through cross-validation with grid-search and random search are highlighted.

## 4 Performing Shallow Networks Training as Deep Networks

After presenting some thoughts towards number of neurons in each deep network's layer, the next issue is shallow and depth aspects in neural networks. This goes along with number of neurons which makes the network shallow or deep. The training of shallow networks in terms of deep networks is achieved through model compression using regression logits learning with *L*2 loss and growing learning process through introduction of linear or non-linear layers.

The prima face behind model compression [18] is to train a compact model to approximate function through them. A single neural network is trained to mimic larger ensembles. The smaller networks have less parameters with same accuracy as ensembles. The model compression works by passing unlabeled data through large accurate model to collect model's scores. This synthetically labeled data trains smaller model which is not trained through original labels. Rather it is trained to learn larger model function. If compressed model mimics large one accurately, similar predictions are achieved as complex model. Training a small network on original training data accurately as complex model is not always possible. The network compression highlights that a small network learns more accurate function. However, current learning algorithms are not able to train a model with accuracy of original data. Rather, intermediate model must be trained initially and then network should be trained to mimic it. As and when it is possible to mimic function learned by complex model with smaller network, function learned by complex model is not actually complex to be learned by smaller network. Considering TIMIT and CIFAR-10 datasets [18], model compression is used for training shallow mimic networks using labeled data. The shallow mimic models are trained with cross-entropy on 186 $p_k$ output values. Here $p_k$ presents output from deep network's softmax layer. The training is done directly on 186 log probability logit values before activation of softmax. Training logits makes learning easier for student by placing equal emphasis on relationships learned by teacher across all targets. When teacher predicts targets with some probability and those values are used as prediction targets, cross entropy is minimized. The student focuses on target with maximum probability value. A student trained on logits for these targets will better mimic teacher's model behavior. Again, taking a situation where teacher predicts different logit sets. After softmax, these logits yield same predicted probabilities as actual one, but teachers are different. By training student on logits, he better learns internal model which teacher learns. We now have learning objective function which is a regression problem with training data $\left\{\left(y^{(1)}, z^{(1)}\right), \ldots\ldots\ldots\ldots\ldots, \left(y^{(N)}, z^{(N)}\right)\right\}$:

$$\mathcal{L}eaning(Wt, \alpha) = \frac{1}{2N}\sum_n \left\| h\left(y^{(n)}; Wt, \alpha\right) - z^{(n)} \right\|_2^2 \quad (7)$$

In Eq. (7) $Wt$ represents weight matrix between input features $y$ and hidden layer, $\alpha$ weights from hidden to output units and $h(y^{(n)}; Wt, \alpha) = \alpha v(Wty^{(n)})$ is model's prediction on $n^{th}$ training data point and $v(\cdot)$ is non-linear hidden units' activation. The parameters $Wt$ and $\alpha$ are updated using standard error back-propagation algorithm with stochastic gradient descent through momentum. The experiments have been performed with other mimic loss functions and are available in [18]. To match deep network's parameters, a shallow network has more non-linear hidden units in a single layer to produce a large $Wt$. When training a large shallow neural network with many hidden units, it is very slow to learn large number of parameters in $Wt$ between input and hidden layers. Because of many highly correlated parameters in large $Wt$, gradient descent converges slowly. On introducing a bottleneck linear layer with $h$ linear hidden units between input and non-linear hidden layer, learning is speeded up. The matrix $Wt$ is factorized into low-rank product matrices $A$ and $B$. The new cost function is:

$$\mathcal{Leaning}(A, B, \alpha) = \frac{1}{2N} \sum_n \left\| \alpha v\left(ABy^{(n)}\right) - z^{(n)} \right\|_2^2 \tag{8}$$

The weights $A$ and $B$ are learnt through back-propagation of linear layer. This reparameterization of $Wt$ not only increases convergence of shallow mimic networks but also minimizes memory space appreciably. The factorizing of $Wt$ matrices are available in [18]. It is applied between the input and hidden layers to improve training convergence.

# 5 Experiments and Analysis

The experimental results are presented in this section. The prima face is to figure out number of neurons towards large scale classification task. In order to achieve this, five different architectures are studied and behavior analysis is performed on five variant datasets considering parameter reduction. The experiments are done on four large scale image classification and one-character recognition datasets. Some benefits are also achieved at test time which are available in [18]. These datasets include ImageNet, MS-COCO, Street View House Numbers (SVHN), Places2-401 and MNIST [27] from which variant datasets are created. The ImageNet has over 15 million labeled images split into 22,000 categories. Here a variant of ILSVRC-2014 [18] image subset with 5000 categories with 1.6 million training images and 250,000 validation images are used. The MS-COCO dataset has over 250,000 labeled images split into 80,000 categories. Here a variant of MS-COCO [18] image subset with 8000 categories with 50,000 training images and 25,000 validation images are used. The SVHN dataset has over 600,000 labeled images which are basically digits split into 10 categories one per digit. Here a variant of this dataset [18] is used with 75,000 image digits for training, 30,000 image digits for testing and over 500,000 additional digits as extra training data. The Places2-401 [18] dataset consists of more than 10 million images with 401 unique scene categories. The variant dataset consists of training set which ranges between

10,000 and 50,000 images per category and validation set between 10,000 and 20,000 images per category. The MNIST [18] dataset consists of 60,000 training and 10,000 test samples split into 10 categories. The variant dataset consists of training samples depict handwritten digits collected various source datasets. The test set is derived from training set after pre-processing. For the sake of convenience, the results on the image classification and character recognition datasets are placed in two different subsections.

Architectures considered: With ImageNet, MS-COCO, SVHN and Places2-401, variant architectures are based on VGG-B network (BNet) [7] and DecomposeMe$_8$ (Dec$_8$) [36]. BNet consists of 14 convolutional layers followed by 5 fully connected layers. In order to reach better network accuracy some of the fully connected layers are removed randomly. This results in the reduction of network's parameters. The Dec$_8$ has 20 convolutional layers with 1D kernels. With MNIST, variant architecture based on [18] is used. Here, 12 1D convolutional layers are placed along with max-pooling layers.

Implementation details: All the models alongwith baseline networks are trained through same random seed and framework. The ImageNet, MS-COCO, SVHN and Places2-401 used torch-7 multi-GPU [18]. All models are trained considering 100 epochs with 14,000 batches per epoch. The batch size for BNet and Dec$_8$ is 60 and 240 respectively. The initial learning rate is 0.01 and then updated to 0.1. The data augmentation is done using random crops and horizontal flips with probability 0.5. For MNIST each network is trained considering single Tesla K20 m GPU for 100 epochs with 512 batch size of and 1000 iterations per epoch. The initial learning rate is 0.1 which is updated with 1.0 momentum. The hyper parameters are adjusted regularly and considered as $\lambda_{layer} = 0.119$ towards first 6 layers and $\lambda_{layer} = 0.286$ for rest of the layers.

Evaluation aspects: The classification performance is measured through Top-1 accuracy considering center crop. The results are compared considering similar standard architectures. Since the present approach determines number of neurons per layer, results are also placed considering variant number of neurons for over complete network. Due care is also given considering convolutional layers, neurons percentage, zero valued parameters percentage and total percentage of zero parameters which also adheres to parameters set to zero in non-completely zeroed-out neurons. Attention is given towards total percentage of zero-valued parameters induced through zeroed-out neurons.

### 5.1 ImageNet, MS-COCO, SVHN and Places2-401 Results

The experiments on ImageNet, MS-COCO, SVHN and Places2-401 datasets are performed through BNet and Dec$_8$ architectures. These architectures used group sparsity (GS) regularizer in Sect. 3. In Dec$_8$ evaluation was performed considering five versions with 640, 768, 800, 886 and 996 neurons per layer. For 640, 800 and 886 neurons per layer, evaluation is done by GS and sparse group lasso (SGL) regularizers in Sect. 3 with $\alpha = 0.5$. The Tables 1, 2 and 3 (for ImageNet) 4, 5, 6 (for MS-COCO) present a comparative accuracy representation considering actual and baseline architectures. It is to be noted that except $Dec_8$ versions of 768 and 996, all methods have improvements over original network with appreciable performance differences for BNet and $Dec_8$ versions of 640, 800 and 886. As additional baseline evaluation is also performed through naive approach which consists of reducing model's each layer by factor 30%. The results for SVHN and Places2-401 datasets are available in [18]. As obvious from the Tables 1, 2, 3, 4, 5 and 6 there is a relative saving obtained using present approach in terms of percentage of zeroed-out neurons/parameters for BNet and $Dec_8$ respectively. For BNet the present approach reduces number of neurons by over 16% while improving its generalization ability as presented by the accuracy gaps. The number of neurons in subsequent fully connected layer is considerably reduced which leads to 35% reduction in total number of parameters. For $Dec_8$ it is observed that when considering actual architecture with 512 neurons per layer, present approach gives a small reduction in parameters with least performance gains. However, as actual number of neurons in each layer is increased present approach has higher benefits. For 640 neurons per layer with GS regularizer, a reduction in number of parameters is observed with higher generalization ability. With SGL regularizer, reduction is higher. For 768, 800, 886, 996 neurons per layer, 30%, 35%, 40%, 45% of neurons are removed that maps to about 55%, 60%, 65%, 70% of parameters. During learning notable reduction in training validation accuracy gaps are observed when applying present regularization approach. This is always achieved through experiments on 768, 800, 886 and 996 versions of $Dec_8$. This indicates that networks trained using present approach have better generalization ability even when they have lesser parameters. A similar observation is achieved for other architectures. The present approach's sensitivity is analyzed by $\lambda_{layer}$ considering equations in Sect. 3. We considered $768_{GS}$, $800_{GS}$, $886_{GS}$ and $996_{GS}$ versions of $Dec_8$ with different values in $\lambda_{layer} = [0.050, \ldots\ldots, 0.50]$. We also considered 100 variant value pairs $(\lambda^a, \lambda^b)$ where former is used towards first six layers and rest on remaining layers. The further details are available in [18].

**Table 1.** Top-1 accuracy results for several architectures vs present approach (ImageNet)

| Model | Top-1 accuracy (in percentage) |
|---|---|
| BNet | 62.5 |
| BNet (cross validation – grid search) | 62.5 |
| BNet (cross validation – random search) | 62.5 |
| BNet (variant) | 62.6 |
| ResNet50 [18] | 67.3 |
| $Dec_8$ | 68.8 |
| $Dec_8$-640 | 70.9 |
| $Dec_8$-768 | 72.5 |
| $Dec_8$-800 | 75.5 |

(*continued*)

**Table 1.** *(continued)*

| Model | Top-1 accuracy (in percentage) |
|---|---|
| $Dec_8$-886 | 78.5 |
| $Dec_8$-996 | 80.5 |
| **Model** | **Top-1 accuracy (in percentage)** |
| **Present-$BNet_{GS}$ (variant)** | **65.8** |
| **Present-$Dec_{8(GS)}$** | **66.8** |
| **Present-$Dec_8$-$640_{(SGL)}$** | **68.5** |
| **Present-$Dec_8$-$640_{(GS)}$** | **69.6** |
| **Present-$Dec_8$-$768_{(GS)}$** | **69.0** |
| **Present-$Dec_8$-$800_{(SGL)}$** | **69.6** |
| **Present-$Dec_8$-$800_{(GS)}$** | **70.0** |
| **Present-$Dec_8$-$886_{(SGL)}$** | **70.6** |
| **Present-$Dec_8$-$886_{(GS)}$** | **72.0** |
| **Present-$Dec_8$-$996_{(GS)}$** | **75.0** |

**Table 2.** The reduction of parameters on ImageNet for BNet (variant)

| BNet (variant) on ImageNet (in percentage) | |
|---|---|
| | GS |
| Neurons | 18.70 |
| Group parameters | 19.60 |
| Total parameters | 19.60 |
| Total induced | 30.39 |
| Accuracy gap | 1.2 |

**Table 3.** The reduction of parameters on ImageNet for $Dec_8$

| $Dec_8$ on ImageNet (in percentage) | | | | | | | | | |
|---|---|---|---|---|---|---|---|---|---|
| | $Dec_8$ | $Dec_8$-640 | | $Dec_8$-768 | $Dec_8$-800 | | $Dec_8$-886 | | $Dec_8$-996 |
| | GS | GS | SGL | GS | GS | SGL | GS | SGL | GS |
| Neurons | 4.39 | 12.09 | 14.46 | 13.69 | 14.09 | 16.5 | 16 | 18.5 | 18.86 |
| Group parameters | 4.46 | 14.48 | 14.69 | 14.96 | 16.48 | 16.6 | 18 | 18.6 | 20.69 |
| Total parameters | 4.46 | 14.48 | 24.75 | 14.96 | 16.48 | 25.5 | 18 | 27.5 | 20.69 |
| Total induced | 4.86 | 21.12 | 24.46 | 22.50 | 23.12 | 25.5 | 25 | 27.5 | 27.69 |
| Accuracy gap | 0.008 | 2.00 | 0.86 | 1.90 | 1.86 | 0.69 | 1.6 | 0.65 | 1.4 |

**Table 4.** Top-1 accuracy results for several architectures vs present approach (MS-COCO)

| Model | Top-1 accuracy (in percentage) |
|---|---|
| BNet | 64.5 |
| BNet (cross validation – grid search) | 64.5 |
| BNet (cross validation – random search) | 64.5 |
| BNet (variant) | 64.6 |
| ResNet50 [18] | 69.3 |
| $Dec_8$ | 70.8 |
| $Dec_8$-640 | 72.9 |
| $Dec_8$-768 | 74.5 |
| $Dec_8$-800 | 77.5 |
| $Dec_8$-886 | 79.5 |
| $Dec_8$-996 | 80.5 |
| **Model** | **Top-1 accuracy (in percentage)** |
| **Present-$BNet_{GS}$ (variant)** | **66.8** |
| **Present-$Dec_{8(GS)}$** | **67.8** |
| **Present-$Dec_8$-$640_{(SGL)}$** | **69.5** |
| **Present-$Dec_8$-$640_{(GS)}$** | **70.6** |
| **Present-$Dec_8$-$768_{(GS)}$** | **70.0** |
| **Present-$Dec_8$-$800_{(SGL)}$** | **70.6** |
| **Present-$Dec_8$-$800_{(GS)}$** | **72.0** |
| **Present-$Dec_8$-$886_{(SGL)}$** | **72.6** |
| **Present-$Dec_8$-$886_{(GS)}$** | **74.0** |
| **Present-$Dec_8$-$996_{(GS)}$** | **76.0** |

**Table 5.** The reduction of parameters on MS-COCO for BNet (variant)

| BNet (variant) on MS-COCO (in percentage) | |
|---|---|
| | GS |
| Neurons | 19.70 |
| Group parameters | 20.60 |
| Total parameters | 20.60 |
| Total induced | 31.39 |
| Accuracy gap | 1.0 |

**Table 6.** The reduction of parameters on MS-COCO for $Dec_8$

| $Dec_8$ on MS-COCO (in percentage) | | | | | | | | | |
|---|---|---|---|---|---|---|---|---|---|
| | $Dec_8$ | $Dec_8$-640 | | $Dec_8$-768 | $Dec_8$-800 | | $Dec_8$-886 | | $Dec_8$-996 |
| | GS | GS | SGL | GS | GS | SGL | GS | SGL | GS |
| Neurons | 4.86 | 12.19 | 14.69 | 13.86 | 14.50 | 16.9 | 17 | 18.9 | 19.86 |
| Group parameters | 4.69 | 14.69 | 14.86 | 14.99 | 16.69 | 16.9 | 19 | 19.0 | 21.69 |
| Total parameters | 4.69 | 14.69 | 24.86 | 14.99 | 16.69 | 25.9 | 19 | 27.9 | 21.69 |
| Total induced | 4.96 | 21.19 | 24.69 | 22.69 | 23.19 | 25.9 | 27 | 27.9 | 28.69 |
| Accuracy gap | 0.007 | 1.98 | 0.96 | 1.86 | 1.69 | 0.86 | 1.4 | 0.65 | 1.2 |

### 5.2 MNIST Results

While considering MNIST [18] dataset, $Dec_3$ architecture is used with last four layers initially have 512 neurons. The objective is towards reaching an optimal architecture. The Table 7 places results considering GS and SGL regularization with respect to baseline method. From the comparison between $MaxPool_{2Dneurons}$ and $Dec_3$, it is observed that better performance is achieved through learning 1D filters than their 2D kernels. The present algorithm brings number of parameters down by about 86% and improves network's performance appreciably.

**Table 7.** The analysis results on MNIST with $Dec_3$

| $Dec_3$ performance on MNIST (in percentage) | | |
|---|---|---|
| | GS | SGL |
| Neurons | 60.12 | 40.69 |
| Group parameters | 70.69 | 35.75 |
| Total parameters | 70.69 | 80.46 |
| Total induced | 86.46 | 80.40 |
| Accuracy gap | 1.86 | 1.37 |
| **Top-1 accuracy on MNIST (in percentage)** | | |
| $MaxOut_{Dec}$ [18] | 91.30 | |
| MaxOut [18] | 89.80 | |
| $MaxPool_{2Dneurons}$ | 83.80 | |
| $Dec_3$ (baseline) | 89.30 | |
| **Present-$Dec_{3(GS)}$** | **95.50** | |
| **Present-$Dec_{3(SGL)}$** | **94.86** | |

# 6 Conclusion

In this research work, some insights and observations are placed on the depth issues of deep neural networks. These parameters are generally fixed through trial and error methods. The investigation considers selecting optimum number of layers and neurons selection in the deep network architecture. The group sparsity regularizer allows the learning process to happen considering one compact computational framework. The number of neurons is effectively estimated and also produces a steady architecture. This gives better test time memory as well as computation benefits. The deep network architecture is further evaluated through training the shallow networks in terms of deep networks. This also reduces redundancy in these deep network architectures. The experimental analysis with various computer vision and character recognition datasets has highlighted the inherent advantages of the present approach. It also goes forward towards the generalization ability of this approach considering variant architectures. This analysis has provided appreciable results towards several regression and optimization computational scenarios. The future work involves performing the experiments with other large-scale datasets such that the optimality of the deep architecture in terms of the network parameters is further improved.

## References


1. Erhan, D., Bengio, Y., Courville, A., Manzagol, P.A., Vincent, P., Bengio, S.: Why does unsupervised pre-training help deep learning? J. Mach. Learn. Res. **11**, 625–660 (2010)
2. Hinton, G.E., Salakhutdinov, R.R.: Reducing the dimensionality of data with neural networks. Science **313**(5786), 504–507 (2006)
3. Vincent, P., Larochelle, H., Lajoie, I., Bengio, Y., Manzagol, P.A.: Stacked denoising autoencoders: learning useful representations in a deep network with a local denoising criterion. J. Mach. Learn. Res. **11**, 3371–3408 (2010)
4. Cybenko, G.: Approximation by superpositions of a sigmoidal function. Math. Control Sig. Syst. **2**(4), 303–314 (1989)
5. Dauphin, Y.N., Bengio, Y.: Big neural networks waste capacity. arXiv:1301.3583 (2013)
6. Eigen, D., Rolfe, J., Fergus, R., LeCun, Y.: Understanding deep architectures using a recursive convolutional network. arXiv:1312.1847 (2013)
7. Simonyan, K., Zisserman, A.: Very deep convolutional networks for large-scale image recognition. arXiv:1409.1556 (2014)
8. He, K., Zhang, X., Ren, S., Sun, J.: Deep residual learning for image recognition. arXiv: 1512.03385 (2015)
9. Denil, M., Shakibi, B., Dinh, L., Ranzato, M.A., De Freitas, N.: Predicting parameters in deep learning. arXiv:1306.0543 (2013)
10. Cheng, Y., Yu, F.X., Feris, R.S., Kumar, S., Choudhary, A.N., Chang, S.F.: An exploration of parameter redundancy in deep networks with circulant projections. In: ICCV (2015)
11. LeCun, Y., Denker, J.S., Solla, S.A.: Optimal brain damage. In: NIPS (1990)
12. Hassibi, B., Stork, D.G., Wolff, G.J.: Optimal brain surgeon and general network pruning. In: ICNN (1993)
13. Mozer, M., Smolensky, P.: Skeletonization: A technique for trimming the fat from a network via relevance assessment. In: NIPS (1988)
14. Ji, C., Snapp, R.R., Psaltis, D.: Generalizing smoothness constraints from discrete samples. Neural Comput. **2**(2), 188–197 (1990)
15. Reed, R.: Pruning algorithms – a survey. IEEE Trans. Neural Netw. **4**(5), 740–747 (1993)
16. Hinton, G.E., Vinyals, O., Dean, J.: Distilling the knowledge in a neural network. arXiv (2014)
17. Romero, A., Ballas, N., Kahou, S.E., Chassang, A., Gatta, C., Bengio, Y.: Fitnets: hints for thin deep nets. In: ICLR (2015)
18. Chaudhuri, A.: Some investigations in deep neural networks for image and text datasets. Technical report TR–1896, Samsung R & D Institute Delhi India (2018)
19. Montufar, G.F., Pascanu, R., Cho, K., Bengio, Y.: On the number of linear regions of deep neural networks. In: NIPS (2014)
20. Liu, B., Wang, M., Foroosh, H., Tappen, M., Penksy, M.: Sparse convolutional neural networks. In: CVPR (2015)
21. Jaderberg, M., Vedaldi, A., Zisserman, A.: Speeding up convolutional neural networks with low rank expansions. In: BMVC (2014)
22. Denton, E.L., Zaremba, W., Bruna, J., LeCun, Y., Fergus., R.: Exploiting linear structure within convolutional networks for efficient evaluation. In: NIPS (2014)

23. Gong, Y., Liu, L., Yang, M., Bourdev, L.D.: Compressing deep convolutional networks using vector quantization. arXiv:1412.6115 (2014)
24. Srivastava, R.K., Greff, K., Schmidhuber, J.: Training very deep networks. In: NIPS (2015)
25. Collins, M.D., Kohli, P.: Memory bounded deep convolutional networks. arXiv:1412.1442 (2014)
26. Simon, N., Friedman, J., Hastie, T., Tibshirani, R.: A sparse-group lasso. J. Comput. Graph. Stat. **22**(2), 231–245 (2013)
27. Experimental datasets. https://www.analyticsvidhya.com/blog/2018/03/comprehensive-collection-deep-learning-datasets/

# Multi-input CNN for Text Classification in Commercial Scenarios

Zuzanna Parcheta[1(✉)], Germán Sanchis-Trilles[1], Francisco Casacuberta[2], and Robin Redahl[3]

[1] Sciling S.L., Carrer del Riu 321, 46012 Pinedo, Spain
{zparcheta,gsanchis}@sciling.com
[2] PRHLT Research Center, Camino de Vera s/n, 46022 Valencia, Spain
fcn@prhlt.upv.es
[3] Northfork, Regeringsgatan 65, 11156 Stockholm, Sweden
robin@northfork.ai

**Abstract.** In this work we describe a multi-input Convolutional Neural Network for text classification which allows for combining text preprocessed at word level, byte pair encoding level and character level. We conduct experiments on different datasets and we compare the results obtained with other classifiers. We apply the developed model to two different practical use cases: (1) classifying ingredients into their corresponding classes by means of a corpus provided by *Northfork*; and (2) classifying texts according to the English level of their corresponding writers by means of a corpus provided by *ProvenWord*. Additionally, we perform experiments on a standard classification task using *Yahoo! Answers* and *GermEval2017 task A* datasets. We show that the developed architecture obtains satisfactory results with these corpora, and we compare results obtained for each dataset with different state-of-the-art approaches, obtaining very promising results.

**Keywords:** Text classification · Document classification · CNN · Multi-input network · Gastrofy · ProvenWord · Use case · Northfork · GermEval2017 · Agglutinative language · Swedish · German

## 1 Introduction

Text mining is one of the fundamental tasks in Natural Language Processing (NLP) which has steadily gained importance in recent years. Enormous amounts of data produced by a wide range of sources such as social networks, blogs or forums increase the need for developing an automatic system for processing and organising the information. Text data in particular is an example of unstructured information which is very easy to understand for humans but much harder to understand for machines. For that reason, efficient and effective techniques and algorithms are required to discover useful patterns.

One of the applications for text mining is the classification of sentences or documents into pre-determined classes with the assumption that each class consists of similar texts. Different techniques are used to classify texts, e.g. a simple


I. Rojas et al. (Eds.): IWANN 2019, LNCS 11506, pp. 596–608, 2019.
https://doi.org/10.1007/978-3-030-20521-8_49

naive Bayes classifier [4], Support Vector Machines [12], Nearest Neighbours [18], and Neural Networks [5]. Recently, models based on neural networks have become increasingly popular [7], achieving very good performance [22].

The need for automatic text classification can be a great business opportunity, with many companies applying automatic text classification to develop their services. As a company dedicated to machine learning solutions, at Sciling we work closely with companies dealing with text classification problems. For that reason, the main motivation to develop this work was to create a high quality model which brings satisfactory results applied to their specific use cases.

We present a multi-input Convolutional Neural Network (CNN) which combines embeddings at the word, byte pair encoding (BPE) [17] and character levels. We analyse the performance of the developed network on four classification tasks.

This paper is organised as follows: Sect. 2 reports on the previous work on text classification. Section 3 describes the implementation of the developed model. In Sect. 4 we describe the experimental setup, the datasets used for our experiments, and the results obtained. In Sect. 5 we analyse and compare results obtained in four classification tasks. Finally, in Sect. 6 we describe conclusions derived from the present work.

## 2 Related Work

Text classification is a very popular task in which different techniques are applied to improve the classification accuracy. Due to increasing interest in neural networks, many recent works involve them in order to obtain promising results. In [21], the authors propose a deep architecture which can extract high-level word features to perform text classification. Authors use different temporal convolution filters, which vary in size, to capture different contextual features. In [10], the authors implement recurrent networks for text classification. In their model they apply a recurrent structure to capture contextual information. Also, there are works which combine different input types. In [11], word-level and character-level information is combined in a similar way as in the current work, but without including the BPE level. As will be shown, the combination of different token types helps the model learn better representations for informal text.

Due to the long training time derived from neural network training, the creation of a time efficient, high quality model becomes a challenge. In [6], authors show that linear models with a rank constraint and a fast loss approximation can train on a billion words within ten minutes, while achieving performance on par with the state-of-the-art. As a result, the authors created *fastText*, an open-source, free, lightweight library that allows users to train text representations and text classifiers with the option of reducing model size. *fastText* combines techniques such as bag of words (BoW), which is a representation of words as vectors of linear space, and also linear classification. In the present work, we used *fastText* as a baseline to assess the quality of the results obtained, given the fast turnaround of experiments in *fastText*. In [22] the authors describe a similar architecture to our network applying ConvNets. Authors apply their model to various large-scale datasets, in a similar way as we train the network described throughout this article. They show that temporal ConvNets can achieve astonishing performance without the knowledge of words, phrases, sentences or any other syntactic or semantic structures with regards to a human language. In their work, the authors only use character level information to classify sentences. In [19] we can find a different system description presented to the *GermEval2017 A* classification task. Here, the best-ranked teams used bidirectional LSTMs, and also classification libraries such as fastText [6] and xgboost [1].

In this work, we developed a multi-input Convolutional Neural Network (CNN) which combines three different levels of preprocessed data: word-level, BPE-level and character level. The model developed was applied to two different practical use-cases: (1) the *Gastrofy* task, provided by *Northfork*, which consists on classifying ingredients into different classes and (2) *ProvenWord*, which consists on classifying sentences into three different English levels. In addition, we conducted experiments on two standard classification tasks: *Yahoo! Answers* and *GermEval2017* (task A), for comparison purposes. We mainly compared the proposed model against [6] and [22], and also against the winners from *GermEval2017*, achieving promising results.

## 3 Multi-input CNN Model

As mentioned earlier, we use three types of input layers where basic units can be words, BPE segments or characters. Depending on the input layer used, the texts are tokenised and represented as a sequence of one-hot vectors. A one-hot encoding is a representation of categorical variables as binary vectors. The categorical values, e.g. words or characters, are mapped to integer values using a previously created dictionary based on the vocabulary $V_k$ used, where $k$ is the specific tokenisation used. In this paper we used $k = \{\text{word, BPE, char}\}$. Each integer value is represented as a one-hot encoding. Therefore, each text $\mathbf{X} = (\mathbf{X}_1, \ldots, \mathbf{X}_k, \ldots, \mathbf{X}_K)$ is, for every tokenisation $k \in K$, a sequence of one-hot vectors $\mathbf{X}_k = (\mathbf{x}_{k,1}, \ldots, \mathbf{x}_{k,t}, \ldots, \mathbf{x}_{k,T})$, where $x_{k,t}$ is representing each specific tokenisation in $k$, and $T$ is the length of the longest input text from the training set. Note that $T$ in fact depends on $k$, and should be denoted as $T_k$, and $t$ as $t_k$, but we omit the subindex to avoid clogging the notation. The size of vector $\mathbf{x}_{k,t}$ is $|V_k|$. Formally, to project the sequence of one-hot vectors into its embedding vector $\mathbf{E}_k$ it is necessary to multiply the one-hot vector with a weight matrix $\mathbf{W}_k \in \mathbb{R}^{d \times |V_k|}$ where $d$ is the size of word embedding vectors, $\mathbf{e}_{k,t} = \mathbf{W}_k \mathbf{x}_{k,t}$.

After the embedding layer, the input sequence of one-hot vectors becomes a sequence of dense, real valued vectors $\mathbf{E}_k = (\mathbf{e}_{k,1}, \ldots, \mathbf{e}_{k,t}, \ldots, \mathbf{e}_{k,T})$. In the convolutional layer, a filter $\mathbf{Q}$ of size $r \times d$, $\mathbf{Q} \in \mathbb{R}^{r \times d}$, is applied to the input sequence of embedding vectors

$$\mathbf{f}_{k,t} = \phi(\mathbf{Q}[\mathbf{e}_{k,t-\lfloor\frac{r-1}{2}\rfloor}; \ldots; \mathbf{e}_{k,t}; \ldots, \mathbf{e}_{k,\lceil\frac{r-1}{2}\rceil}]) \tag{1}$$

where $\phi$ is an activation function. This process is done for every time step of the input sequence, giving as result sequence $\mathbf{F}_k = (\mathbf{f}_{k,1}, \ldots, \mathbf{f}_{k,T-r+1})$. Then, sequence $\mathbf{F}_k$ is max-pooled with size $m$ resulting in $\mathbf{f}'_{k,t}$

$$\mathbf{f}'_{k,t} = \max[\mathbf{f}_{k,t-m/2+1}, \ldots, \mathbf{f}_{k,t}, \ldots, \mathbf{f}_{k,t+m/2}] \tag{2}$$

where the max operation is applied for each element of the vectors, resulting in a sequence $\mathbf{F}'_k = (\mathbf{f}'_{k,1}, \ldots, \mathbf{f}'_{k,(T-r+1)/m})$. Finally, $\mathbf{F}'_k$ from different levels (word $w$, BPE $b$, character $r$) are concatenated into $\mathbf{F}' = [\mathbf{F}'_w; \mathbf{F}'_b; \mathbf{F}'_r]$. The output layer transforms $\mathbf{F}'$ sequences from the convolutional layer and transforms them by applying an activation function [15] to compute the predictive probabilities for all the categories using the following equation:

$$p(y = c \mid \mathbf{X}) = \frac{\exp(\mathbf{w}_c^\top \mathbf{F}' + b_c)}{\sum_{c'=1}^{C} \exp(\mathbf{w}_{c'}^\top \mathbf{F}' + b_{c'})} \tag{3}$$

where $\mathbf{w}_c$ is a weight vector and $b_c$ is a bias vector. $C$ is the number of classes, and $\mathbf{X} = [\mathbf{X}_w; \mathbf{X}_b; \mathbf{X}_r]$.

## 4 Experimental Setup

In our experiments, the embedding layer size was set to 128 units, the convolutional layer size was set to 512 units, with filter sizes of 3, 4, and 5. The convolutional layers use ReLU as an activation function [20]. Dropout of 0.5 probability is added to the output layer. The activation function of output layer is Softmax. We use a batch size of 50 samples and Adam as the optimisation technique [8]. The network was developed in Keras [2]. To train the model, we dedicated 10% of the data available for development purposes. The configuration of the model developed is shown in Fig. 1.

In this article we deal with different datasets, some of them containing large documents. Due to their size, two different versions of classifier, were implemented where *(1)* the first one loads all data into RAM and the data is moved from RAM into GPU memory directly; and *(2)* the second one stores huge amounts of numerical data on disk, and trains the model by loading into memory one batch at a time using h5py[1]. The second implementation allowed us to train with very large amounts of data, while still using a limited amount of RAM.

In all experiments conducted in this work, we compare the obtained results with *fastText*. *fastText* is a linear model for text classification whose algorithm is similar to the Continuous Bag of Words algorithm [13], where the middle word is replaced by the label. For experiment we used the standard values of *fastText*, as implemented per default in the toolkit, where embedding vector size is 100 neurons, softmax as loss function, and learning rate of 0.1 which updates after 100 updates. The best model was selected according to development accuracy.

All experiments were conducted on a Titan Xp 16 G GPU device with A10-7700K 3.4 GHz Quad-Core FM2+ Processor with 15 GB of RAM.

[1] https://www.h5py.org/.

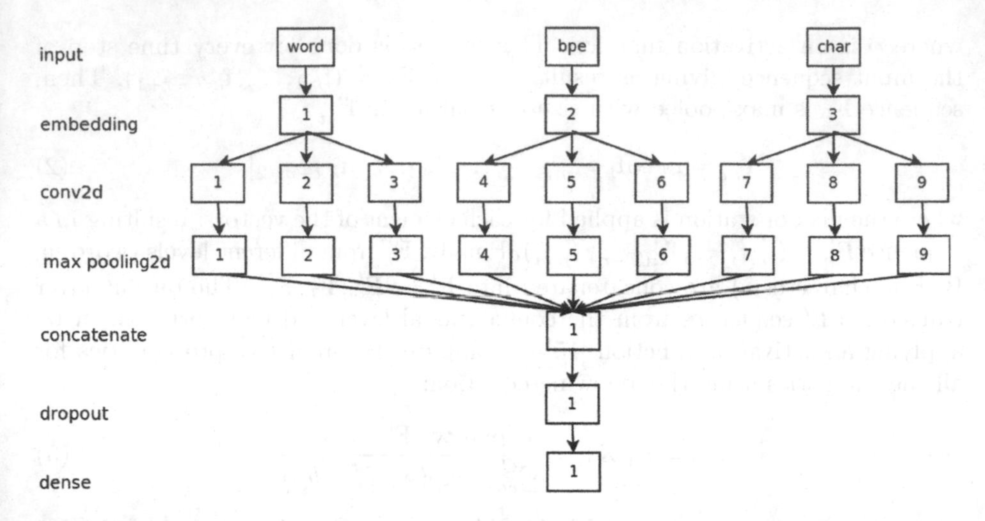


**Fig. 1.** Multi-input CNN model with three input layers for word, BPE and character levels. Note that the length of the word, BPE and character inputs will be different.

### 4.1 Datasets and Results

In this section we describe the different tasks where the model developed was tested. First we elaborate on the distinctive characteristics of each dataset, and then we show the results achieved. We compare the results obtained with *fastText*. To obtain test accuracy and confidence intervals we conducted 1000 repetitions of bootstrap resampling [9], and show the 95% confidence level.

**Yahoo! Answers** is a website where users can post questions and answers, where all content is publicly visible. We downloaded version 1.0 of the dataset[2] which contains category, subcategory, title of question, question text, best answer text and the rest of answers. In total, the corpus contains 4.4 M English samples.

We replicated the experimental setup described in [22] where state-of-the-art results are reported. From the data available, we selected 140k examples from each of the 10 main categories for training purposes, and 6k examples from each category for test. Table 1 shows main figures of *Yahoo! Answers* dataset.

From all fields given, we only used the title, best answer content and the main category information. Also, we processed text as described in [22] by fixing the length of each sample to 1014 characters. We compared our results with the results reported in [22]. We performed experiments with three different types of

[2] https://bit.ly/2DwXyME L6 - Yahoo! Answers Comprehensive Questions and Answers version 1.0 (multi part).

**Table 1.** *Yahoo! Answers* main figures. k denotes thousands of elements, $|S|$ stands for number of documents, $|W_{word}|$ for number of running words, $|V_{word}|$ for vocabulary size, $\overline{S}$ for mean length of documents (in words), and C for number of classes. In the case of the *Gastrofy* dataset, the specific number of classes cannot be revealed for confidentiality reasons.

| Dataset | Language | Subset | $\|S\|$ | $\|W_{word}\|$ | $\|V_{word}\|$ | $\overline{S}$ | C |
|---|---|---|---|---|---|---|---|
| Yahoo! Answers | English | Train | 1.4 M | 121.8 M | 1.6 M | 87 | 10 |
| | | Test | 60 k | 5.2 M | 65 k | | |
| Gastrofy | Swedish | Train | 37.8 k | 150 k | 13.8 k | 4 | >1k |
| | | Test | 4.2 k | 14.3 k | 1.2 k | | |
| GermEval2017 | German | Train | 20.9 k | 1.8 M | 117 k | 84 | 2 |
| | | Test 1 | 2.6 k | 4 M | 21 k | | |
| | | Test 2 | 1.8 k | 3.8 M | 16 k | | |
| ProvenWord | English | Train | 3.8 k | 4.2 M | 53.8 k | 151 | 3 |
| | | Test | 0.4 k | 66 k | 4.7 k | | |

input data: word, BPE and character embeddings. The maximum length of the inputs were 502 words, 1014 BPE segments, and 1014 characters[3].

Due to the fact that in this task we have a large amount of data available (1.4 M sentences with 2530 variables in word, BPE and character levels jointly) we use the version of the classifier which stores data on disk using h5py to train the model. We trained 7 different models. In addition, we compared the classification results obtained with *fastText*, where the best model is selected according to development set accuracy. We also compared the results obtained with the accuracy reported in [22]. All results are shown in Table 2.

Given that the training set is very large (1.4 M examples with up to 2.5 k variables), we evaluated the model every 5 k updates. We stopped the training when development accuracy did not improve in the following 5 updates. We selected the best model according to the best development accuracy. We obtained the best accuracy of 74.5% in the development set using only one input layer of character processing. The accuracy obtained with this model in the test set was 72.9%. The time per epoch in all models is approximately 3 h, with the fastest model using only word level information and spending 2.5 h per epoch, and the slowest one using three types of input and spending 2.8 h per epoch. At this point, it should be noted that difference in time is so small because reading data from disk takes up most of the time required.

As we can see in Table 2, in each combination of input granularity, we obtained the best model very quickly, usually in the 3rd or 4th epoch.

[3] Note that maximum length of BPE and character sequences coincide. However, this does not mean that BPE splits sequences into characters, given that the BPE implementation used adds extra tokens for later recovering the original words.

**Table 2.** Accuracy of Yahoo! Answers classification. Best model in bold selected according to best development accuracy. Differences are not statistically significant.

| System | Best epoch | Best update | Train | Dev | Test |
|---|---|---|---|---|---|
| fastText [6] | 7 | N/A | 73.0 | 70.0 | 72.0±0.2 |
| Zhang et al. [22] | N/A | N/A | 75.6 | N/A | 71.1 |
| char-bpe-word | 3 | 16 | 72.3 | 74.0 | 73.0±0.2 |
| bpe-word | 3 | 15 | 71.9 | 74.4 | 72.8±0.2 |
| bpe-char | 2 | 11 | 71.6 | 73.6 | 71.3±0.1 |
| char-word | 4 | 17 | 75.7 | 74.3 | 72.5±0.2 |
| char | 4 | 17 | 75.3 | **74.5** | **72.9±0.1** |
| bpe | 3 | 14 | 73.7 | 74.1 | 73.2±0.1 |
| word | 4 | 17 | 75.6 | 73.8 | 72.7±0.3 |

Note that the dataset used in this section is reproduced from [22], thus results are not strictly analogous, since there can be small differences in the training, development and test sets. As in [22], the best results obtained in this task were obtained using character level information. However, the differences observed using different input types are not significant, which means that it can not be concluded that other tasks of moderate length should be dealt with using only character information. When comparing model the complexity of Zhang et al., we conclude that the model described in the present work is much simpler in the number of layers used. Specifically, the authors in [22] used 6 convolutional layers and 3 fully-connected layers. Hence, the training time in Zhang et al. is much higher, spending 24 h per epoch which is almost 10 times slower than the training time of the model implemented in this work.

In the case of *fastText*, the experiments were conducted on the exactly same dataset than the one use for evaluating our model, and we can observe that our model is able to improve their results by about 0.9%. However, *fastText* is able to train the model within minutes.

We can conclude that the model developed is a robust network that is suitable to classify documents of moderate length: the model was able to achieve a good score which is able to improve over the two baselines in the state of the art.
***Gastrofy*** is turning meal planning, grocery shopping, and recipe creation into an incredibly simple, healthy, and personalised 1-min process. Their goal is to simplify the process from inspiration to food on the table - whether it is following a diet, trying out new dishes or throwing a great dinner party.

The *Gastrofy* data is an interesting application of text classification. The task consists of classifying ingredients into different classes. The particularity of this dataset is that input sentences are very short: in fact, the average length of the input sentences is of 4 words. Another interesting aspect is that the language of this corpus is Swedish, which implies that words are compounds. Table 1 shows the main figures of the *Gastrofy* dataset. We performed experiments using the model described, with different types of inputs. Due to the small size of the dataset provided, we used the implementation which loads all data into GPU memory. For comparison purposes, experiments using *fastText* were performed. All results obtained with this task are shown in Table 3, where confidence intervals are computed at the 95% confidence level via bootstrap resampling.

**Table 3.** Accuracy with the *Gastrofy* dataset. Best model in bold selected according to best development accuracy. Differences are statistically significant.

| System | Best epoch | Train | Dev | Test |
|---|---|---|---|---|
| fastText | 80 | 99.1 | 77.2 | 76.0± 0.2 |
| bpe_word_char | 23 | 99.4 | **85.5** | **84.2±0.1** |
| bpe_word | 39 | 98.6 | 85.0 | 81.9±0.2 |
| word_char | 43 | 98.6 | 85.2 | 83.2±0.1 |
| bpe_char | 88 | 98.7 | 85.4 | 88.0±0.1 |
| bpe | 80 | 97.6 | 85.6 | 88.4±0.1 |
| word | 72 | 99.3 | 82.5 | 65.4±0.2 |
| char | 40 | 95.0 | 84.9 | 88.3±0.1 |

As observed, the model developed is able to improve over the accuracy obtained by *fastText*. We obtained 84.2% accuracy comparing with 76.7% accuracy using *fastText* model. Total training time in this case was about 30 min, compared with just a few minutes of training time of *fastText*.

At this point, it is important to point out that the results obtained were satisfactory not only according to accuracy, but also in terms of being able for their business purpose, as analysed by *Northfork*.

**GermEval2017** [19] is a set of tasks on aspect-based sentiment in social media customer feedback in German language. Specifically, we focused on task A, where the goal is to determine whether a review is relevant for a specific topic. In this case, reviews relevant to the German train service (Deutsche Bahn) should be filtered and processed further. Note that, as Swedish, German is an agglutinative language, where the term "Bahn" can refer to many different things in German: the rails, the train, any track or anything that can be laid in straight lines. Therefore, it is important to remove documents about e.g. the "Autobahn" (highway), which are not relevant to the Deutsche Bahn service. This is similar for other query terms that are used to monitor web sites and microblogging services. In task A, the documents are labelled as relevant (true) or irrelevant (false). Below is a relevant document about bad behaviour in a Deutsche Bahn train, and an irrelevant document about stock exchange developments:

True Ehrlich die männer in Der Bahn haben keine manieren?
(Seriously, the men in those trains have no manners!)
False Japanische S-Bahn wird mit Spiegelwaggons "unsichtbar"
(Japanese urban railway becomes "invisible" thanks to reflecting wagons)

Table 1 shows the main figures of the *GermEval2017 task A* dataset. We performed experiments using the model described in Sect. 3, with different types of inputs. Due to the fact that the dataset contains a large amount of data, we used the version of the classifier which stores data on disk using h5py. In addition, we limited the number tokens to 2000 words/BPE segments/characters in each document. We compared the results obtained in this task with the winners of the competition in [19], who used an SVM and a random forest classifier with XGBoost[4] [16]. The evaluation was done using micro-average F1-measure using an executable jar file provided by the organization. In Table 5 we shows the best results obtained in *GermEval2017* task A, along with the scores obtained by multi-input CNN developed, where model selection was done according to development F1 score. In this task we used the F1 score for comparison purposes, since the shared task was evaluated with this score. As in the previous tasks, 95% level confidence intervals. As shown, the multi-input CNN is able to improve the best result in the *GermEval2017* task A by about 0.017 in test set 1, and 0.013 in test set 2, by using only BPE input tokens. It should be noted that the results are statistically significant, according to the confidence intervals computed. Finally, training time per epoch was approximately 6 min (Table 4).

**Table 4.** F1-measure for the *GermEval 2017 task A* dataset. Best model in bold selected according to best development accuracy. Differences are statistically significant.

| System | Best epoch | Train | Dev | Test 1 | Test 2 |
|---|---|---|---|---|---|
| Sayyed et al. [16] | N/A | N/A | N/A | .903 | .906 |
| Hövelmann and Friedrich [3] | N/A | N/A | N/A | .899 | .897 |
| Mishra et al. [14] | N/A | N/A | N/A | .879 | .870 |
| char-bpe-word | 4 | .962 | .910 | .920±.008 | .928±.006 |
| bpe-word | 5 | .998 | .910 | .910±.005 | .918±.007 |
| bpe-char | 13 | .934 | .918 | .925±.005 | .931±.006 |
| char-word | 11 | .997 | .916 | .906±.006 | .912±.007 |
| char | 15 | .985 | .918 | .906±.006 | .919±.006 |
| bpe | 21 | .994 | **.935** | **.920±.005** | **.919±.006** |
| word | 16 | .999 | .908 | .891±.007 | .902±.006 |

[4] https://github.com/dmlc/xgboost.

**ProvenWord** was founded by a group of educators who are enthusiastically committed to helping English learners develop their writing skills. The members of the *ProvenWord* team are firmly dedicated to language improvement and the refinement of academic writing through the application of cutting-edge technology and innovation. The task proposed by *ProvenWord* consists of classifying English sentences into three language levels, according to the English level of the corresponding writer. Table 1 shows the main figures of the *ProvenWord* dataset. Note that in this case the dataset provided is much smaller than the other datasets, with only 4.2 k samples. We divided the dataset provided into a training set with 90% of the total sentences and a test set with the remaining 10%. As in the previous cases, we conducted experiments combining different types of input, and the results obtained are shown in Table 5, along with the results obtained with *fastText*. The best model obtained for this dataset includes all types of input data, achieving 61.8% accuracy. The best score obtained with *fastText* was 60.2%, slightly below the results obtained by our model.

## 5 Results Analysis

After seeing the results obtained in the previous Section, we pursued to analyse the results obtained, with the purpose of understanding why the classifier developed delivers better results in certain tasks.

One interesting fact that can be derived from the results shown above is that different input types lead to varying results in each task. This can be explained by the language in each task: in the cases of *Gastrofy* and *GermEval2017*, it becomes evident that BPE information provides important improvements when the language of the task is agglutinative, i.e., Swedish and German, respectively. Given that BPE is able to split words into simpler units, the resulting vocabulary size is smaller, leading to less sparse input features, and ultimately to better accuracy when training the model.

**Table 5.** Accuracy of *ProvenWord* dataset classification. Best model in bold selected according to best development accuracy. Differences are not statistically significant.

| System | Best epoch | Train | Dev | Test |
|---|---|---|---|---|
| fastText | 25 | 84.2 | 57.4 | 60.2±2.6 |
| char-bpe-word | 31 | 98.0 | **62.9** | **61.8±2.5** |
| bpe-word | 17 | 85.0 | 56.9 | 57.8 ±2.3 |
| bpe-char | 40 | 70.2 | 60.4 | 59.8 ±2.2 |
| char-word | 38 | 99.3 | 56.9 | 61.8 ±2.3 |
| char | 35 | 70.0 | 59.7 | 57.8 ±2.5 |
| bpe | 39 | 71.7 | 59.7 | 59.2 ±2.6 |
| word | 31 | 98.0 | 61.0 | 59.5 ±2.6 |

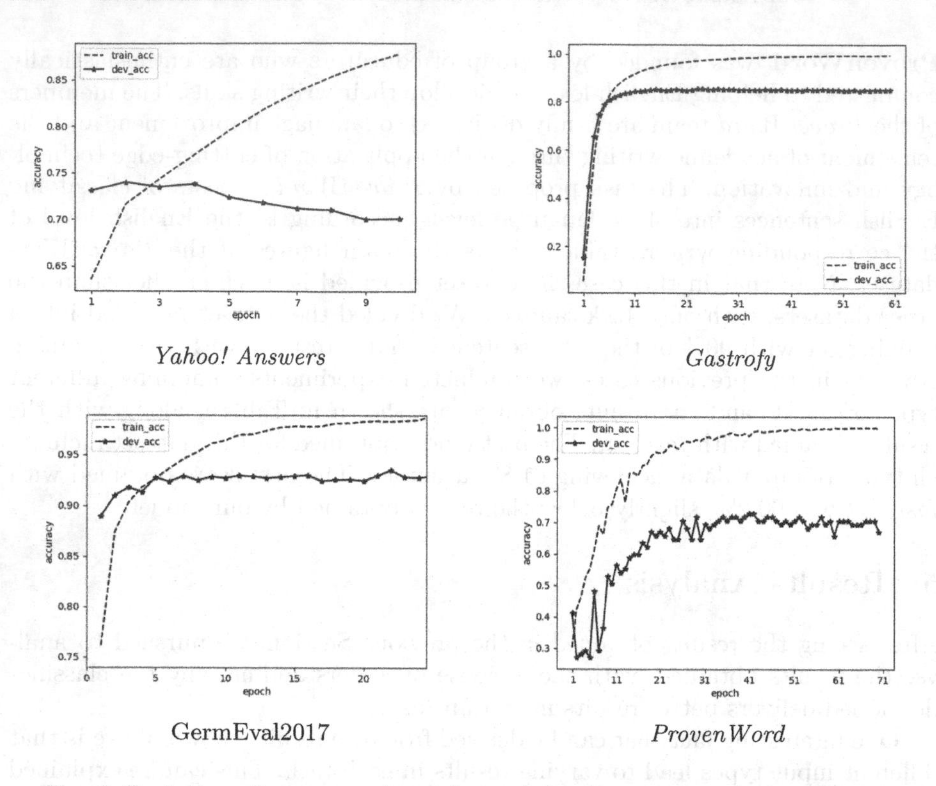


*Yahoo! Answers* *Gastrofy*

GermEval2017 *ProvenWord*

**Fig. 2.** Evolution of accuracy of training and development sets during training.

In contrast, in the case of English language datasets (i.e., *Yahoo! Answers* and *ProvenWord*) the advantage of including BPE information is less obvious. In the case of the *Yahoo! Answers* dataset, the results obtained point to a small difference of 0.9% accuracy in training, and 1.9% in test. In the case of the *ProvenWord* dataset, the results obtained lead to less than 6% difference between the best and worst models for the development set, and 4% for the test set. In both English tasks, the combination of three input levels is the best option, even though differences are smaller than in the *Gastrofy* and *GermEval2017* tasks.

Another interesting aspect of the *Gastrofy* dataset is the number of classes. In this task there are thousands different classes, whereas typical classification tasks seldom have more than a couple of tens of classes. To understand the difference in performance achieved across different tasks, it is interesting to see the accuracy evolution for each task during the training process (Fig. 2). For the *Yahoo! Answers* dataset, accuracy in development starts dropping after the 4th epoch, which seems to point to possible overfitting. In contrast, the case of *Gastrofy* is an typical example of training process where development and training accuracy tend to stabilise after a certain number of epochs, in this case after 15th epoch approximately. Regarding the *Germeval2017* plot, development accuracy is very high since the very beginning, leading to only a small improvement of 2.5% during the full training process. Finally, the *ProvenWord* plot shows an oscillating behaviour in between consecutive epochs, which seems to point towards data scarcity problems, even though the direction of the curves seem to indicate that training converged reasonably well.

## 6 Conclusions

In this work we implemented a multi-input Convolutional Neural Network which combines word-level, BPE-level and character-level sentence representations to classify text. We combined the different input types in each experiment, and demonstrate that our network makes good quality predictions within a reasonable amount of time. We measured the performance of the developed network in four very different scenarios, with varying language and training data size, and on commercial and standard use cases, obtaining competitive results in every task. Furthermore, we analysed the impact leveraging BPE-level information, which ended up being especially important in the case of agglutinative languages such as Swedish and German.

**Acknowledgments.** Work partially supported by MINECO under grant DI-15-08169 and by Sciling under its R+D programme. The authors would like to thank NVIDIA for their donation of Titan Xp GPU that allowed to conduct this research.

## References

1. Chen, T., Guestrin, C.: XGBoost: a scalable tree boosting system. In: Proceedings of the 22nd KDD - ACM SIGKDD, pp. 785–794. ACM (2016)
2. Chollet, F., et al.: Keras: deep learning library for Theano and TensorFlow, vol. 7(8) (2015). https://keras.io/k
3. Hövelmann, L., Allee, S., Friedrich, C.M.: Fasttext and gradient boosted trees at GermEval-2017 on relevance classification and document-level polarity. In: Shared Task on Aspect-Based Sentiment in Social Media Customer Feedback, pp. 30–35 (2017)
4. Jiang, L., Li, C., Wang, S., Zhang, L.: Deep feature weighting for Naive Bayes and its application to text classification. Eng. Appl. AI **52**, 26–39 (2016)
5. Jiang, M., et al.: Text classification based on deep belief network and softmax regression. Neural Comput. Appl. **29**(1), 61–70 (2018)
6. Joulin, A., Grave, E., Bojanowski, P., Mikolov, T.: Bag of tricks for efficient text classification. In: Proceedings of EACL 2017, vol. 2, pp. 427–431. ACL (2017)
7. Kim, Y.: Convolutional neural networks for sentence classification. arXiv preprint arXiv:1408.5882 (2014)
8. Kingma, D.P., Ba, J.: Adam: a method for stochastic optimization. CoRR abs/1412.6980 (2014)
9. Koehn, P.: Statistical significance tests for machine translation evaluation. In: Proceedings of the EMNLP 2004, pp. 388–395 (2004)
10. Lai, S., Xu, L., Liu, K., Zhao, J.: Recurrent convolutional neural networks for text classification. In: AAAI, vol. 333, pp. 2267–2273 (2015)

11. Liang, D., Xu, W., Zhao, Y.: Combining word-level and character-level representations for relation classification of informal text. In: Proceedings of the 2nd Workshop on Representation Learning for NLP, pp. 43–47. ACL (2017)
12. Lilleberg, J., Zhu, Y., Zhang, Y.: Support vector machines and Word2vec for text classification with semantic features. In: ICCI*CC, pp. 136–140. IEEE (2015)
13. Mikolov, T., Chen, K., Corrado, G., Dean, J.: Efficient estimation of word representations in vector space. In: Proceedings of Workshop at ICLR 2013 (2013)
14. Mishra, P., Mujadia, V., Lanka, S.: Germeval 2017: sequence based models for customer feedback analysis. In: Shared Task on Aspect-Based Sentiment in Social Media Customer Feedback, pp. 36–42 (2017)
15. Bridle, J.S.: Probabilistic interpretation of feedforward classification network outputs, with relationships to statistical pattern recognition. In: Soulié, F.F., Hérault, J. (eds.) Neurocomputing. NATO ASI Series (Series F: Computer and Systems Sciences), vol. 68, pp. 227–236. Springer, Heidelberg (1990). https://doi.org/10.1007/978-3-642-76153-9_28
16. Sayyed, Z.A., Dakota, D., Kübler, S.: IDS IUCL: investigating feature selection and oversampling for GermEval2017. In: Shared Task on Aspect-based Sentiment in Social Media Customer Feedback, pp. 43–48 (2017)
17. Sennrich, R., Haddow, B., Birch, A.: Neural machine translation of rare words with subword units. In: Proceedings of ACL 2016, vol. 1, pp. 1715–1725. ACL (2016)
18. Tan, Y.: An improved KNN text classification algorithm based on K-medoids and rough set. In: 2018 10th International Conference on IHMSC, vol. 1, pp. 109–113. IEEE (2018)
19. Wojatzki, M., Ruppert, E., Holschneider, S., Zesch, T., Biemann, C.: GermEval 2017: shared task on aspect-based sentiment in social media customer feedback
20. Xu, B., Wang, N., Chen, T., Li, M.: Empirical evaluation of rectified activations in convolutional network. CoRR abs/1505.00853 (2015)
21. Xu, J., Zhang, C., Zhang, P., Song, D.: Text classification with enriched word features. In: Geng, X., Kang, B.-H. (eds.) PRICAI 2018. LNCS (LNAI), vol. 11013, pp. 274–281. Springer, Cham (2018). https://doi.org/10.1007/978-3-319-97310-4_31
22. Zhang, X., LeCun, Y.: Text understanding from scratch. arXiv preprint arXiv:1502.01710 (2015)

# Applying Sentiment Analysis with Cross-Domain Models to Evaluate User eXperience in Virtual Learning Environments

Rosario Sanchis-Font, Maria Jose Castro-Bleda(✉), and José-Ángel González

Universitat Politècnica de València, Valencia, Spain
rosanfon@doctor.upv.es, mcastro@dsic.upv.es, jogonba2@inf.upv.es

**Abstract.** Virtual Learning Environments are growing in importance as fast as e-learning is becoming highly demanded by universities and students all over the world. This paper investigates how to automatically evaluate User eXperience in this domain. Two Learning Management Systems have been evaluated, one system is an ad-hoc system called "Conecto" (in Spanish and English languages), and the other one is an open-source Moodle personalized system (in Spanish). We have applied machine learning tools to all the comments given by a total of 133 users (37 English speakers and 96 Spanish speakers) to obtain their polarity (positive, negative, or neutral) using cross-domain models trained with a corpus of a different domain (tweets for each language) and general models for the language. The obtained results are very promising and they give an insight to keep going the research of applying sentiment analysis tools on User eXperience evaluation. This is a pioneering idea to provide a better and accurate understanding on human needs in the interaction with Virtual Learning Environments. The ultimate goal is to develop further tools of automatic feed-back of user perception for designing Virtual Learning Environments centered in user's emotions, beliefs, preferences, perceptions, responses, behaviors and accomplishments that occur before, during and after the interaction.

**Keywords:** Machine learning · Sentiment analysis · Polarity · User eXperience · Virtual Learning Environments · Learning Management Systems

## 1 Introduction

Human Computer Interaction (HCI) tools developers, agents and industry require to focus their interactive systems on end-users in order to design and provide quality systems upon the international standards requirements ISO.

Partially supported by the Spanish MINECO and FEDER founds under project TIN2017-85854-C4-2-R. Work of J. A. González is financed under grant PAID-01-17.


I. Rojas et al. (Eds.): IWANN 2019, LNCS 11506, pp. 609–620, 2019.
https://doi.org/10.1007/978-3-030-20521-8_50

These interactive systems are the "combination of hardware, software and/or services that receives input from, and communicates output to, users" (ISO 9241-20: 2010) [1]. This international standard is related to ergonomics of human system-interaction and human-centered design for interactive systems. It provides requirements and recommendations for human-centered design principles and activities throughout the life cycle of computer-based interactive systems. It is intended to be used by those managing design processes, and is concerned with ways in which both hardware and software components of interactive systems can enhance human-system interaction.

Therefore "User eXperience" (UX) enhances human interaction within the hardware or software components, being the UX concept multidimensional and centered in human needs. This UX concept goes beyond usability, interaction experience and design by involving two main qualities: traditional HCI usability and accessibility balanced with hedonic and affective design [19]. In this perspective, in [6], UX is described as a consequence of a user's internal state (predispositions, expectations, needs, motivation, mood, etc.), the characteristics of the designed system (e.g. complexity, purpose, usability, functionality, etc.) and the context (or the environment) within which the interaction occurs (e.g. organizational/social setting, meaningfulness of the activity, voluntariness of use, etc.). Therefore, these authors conclude that UX is considering three perspectives: emotion and affect of the user, technology and the hedonic instrument and the experiential aspect. As a result, UX includes a multidimensional concept and focuses in human needs and the aspects of beauty, fun, pleasure, and personal growth rather than the value of the product or instrument used [6], which improves or worsens along the time of use [9].

UX has to be considered when designing and redesigning hardware and software applications. In this way, in the last years, UX has been taken into account when designing Virtual Learning Environments (VLEs) [19]. VLEs includes a wide range of technology-enabled learning environments, such as Learning Management Systems (LMSs), computer games or Virtual Worlds.

In order to evaluate UX in VLEs, we have used the validated User Experience Questionnaire (UEQ) [16], addressed to 559 users of biomedical postgraduate studies. Two LMSs have been evaluated using this adapted UEQ: one LMS is an ad-hoc system called "Conecto" (in Spanish and English languages), and the other one is an open-source Moodle personalized system (in Spanish).

We have applied machine learning tools to all user's comments using a cross-domain model trained with tweets for each language [3] and a general system for text analytics (MeaningCloud [11]). The application of sentiment analysis tools on UX comments will provide a better and accurate understanding on human needs in the interaction with VLE for postgraduate and biomedical online learning. The ultimate goal is to develop further tools of automatic feed-back of user perception for designing user-centered VLE valued by users for its usability, quality and pleasure of use.

This paper is organized as follows. Next Section gives a brief view about the state of the art of the work presented here. Section 3 describes the data collection and preprocessing from the questionnaires. The used models for sentiment analysis are described in Sect. 5. Section 6 presents the experimental results and their analysis. Finally, the conclusions are drawn in the last Section.

## 2 State of the Art

Sentiment analysis is one of the most active areas in Natural Language Processing since the early 2000s. Concretely, since Pang et al. [14], who addressed the importance of "sentiment classification" for a large number of tasks such as *message filtering*, *recommender systems* or *business intelligence applications*. A decade after, until our days, the popularity of the sentiment analysis has been increasing and Deep Learning has consolidated as a well-established alternative to the previous machine learning systems. Thus, Deep Learning is the state of the art in sentiment analysis [2,4,8,18]. Our approach uses state-of-the-art models, neural networks trained with tweets in English and Spanish, as described in Sect. 5.1.

In addition, a large number of commercial products and frameworks have also proliferated to facilitate the development and deployment of sentiment analysis systems based on machine learning, such as Google Cloud [5], IBM Watson [7], Microsoft Text Analytics [13] and MeaningCloud [12]. This kind of products allow us to perform text analytics such as sentiment analysis, in a broad variety of domains and languages in an easy way, obtaining also competitive results. For this reason, besides our neural network models, MeaningCloud models will be used in our work as explained in Sect. 5.2.

But, though the promising results of natural language processing and, in particular, of sentiment analysis, generally speaking, UX evaluation is immature in most applications and, especially, in VLEs. Some work has been done in eCommerce, using natural language processing to improve their UX, for instance, to search products in a more intelligent way, using sentiment analysis to extract insights from the reviews made by the customers on the product or identifying trends and trying to answer best to the customers' concerns. Several new conferences have recently been launched around these ideas (see, for example, https://julielab.de/econlp/2019/ or https://www.aclweb.org/portal/content/first-international-workshop-e-commerce-and-nlp).

Another research line covered in this paper is the use of cross-domain polarity classification approach, that is, the texts to be classified belong to a different domain from those used in the training phase. Most work have been done within the classic approach, the so-called single-domain polarity classification, which classifies texts in the same domain to which the texts used in the training phase belong to. Due to the lack of training data (only comments from 133 users), we used cross-domain models, those trained with another domain (tweets) and those trained with general data.

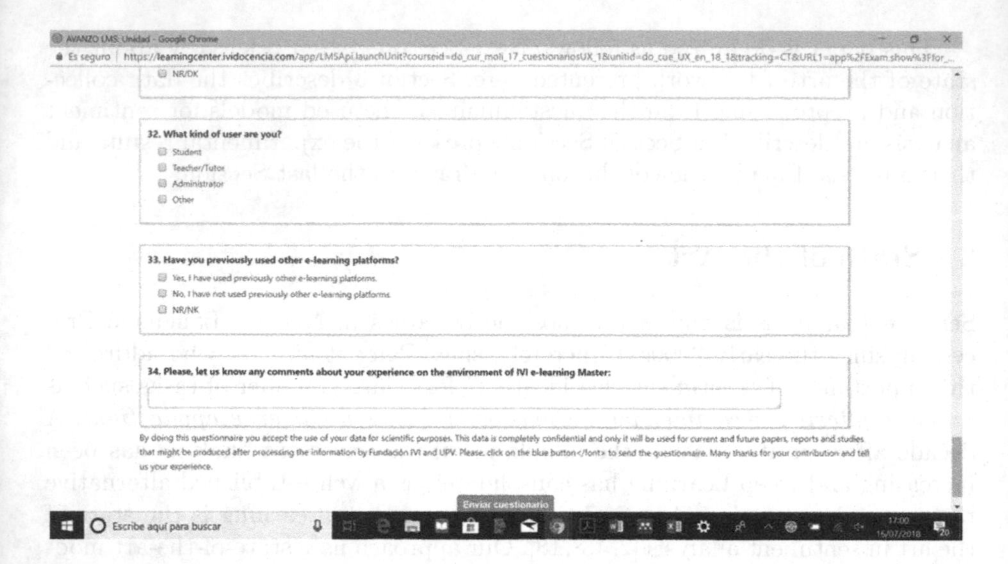


**Fig. 1.** "Other comments" box from UX questionnaire delivered to English speaker users on Conecto LMS of IVI Foundation Biomedical International Master (edition 2017–18).

## 3 Experimental Data

The validated User Experience Questionnaire (UEQ) [16] was used in order to automatically evaluate UX in our VLEs. This questionnaire is a list of close-ended questions, but we added questions concerning to sociodemographics data (age, sex, etc.) and an open field "Other comments" (see Fig. 1 for a screenshot of the questionnaire in one VLE). It is a text entry box to express any comments related to UX in the course, which is an opportunity to get new and more precise information about their experience, not only by close-ended questions.

Two LMSs have been evaluated: one system is an ad-hoc system called "Conecto" (in Spanish and English language)[1], and the other one is an open-source Moodle personalized system (in Spanish)[2]. We have collected data in different editions: 2016–17, 2017–18, January 2018, and April 2018, and at the middle and final term of each course. The UEQ was addressed to 559 users. Only 133 users (37 English speakers and 96 Spanish speakers) filled the "Other comments" box.

We have performed experiments at three different semantic levels of decreasing complexity:

1. *Observation.* We measured the polarity of the whole observation. Each entry is composed by one or more sentences. There were 96 Spanish comments and 37 English observations.

[1] https://postgrado.adeituv.es/es/cursos/salud-7/assisted-reproduction/datos_generales.htm.

[2] https://medicinagenomica.com/eugmygo/.

2. *Sentence.* As an observation from one user can be composed by one or more sentences, we automatically split each observation into sentences, being one sentence the text between points. We got 151 Spanish sentences and 61 English sentences.
3. *Meaningful unit.* Also, we parsed complex statements in a semi-supervised way in order to obtain meaningful units, that is, one sentence or part of one sentence which has own meaning. We obtained in this way 227 Spanish units and 70 English units.

The observation can be composed of more than one sentence, and it is very usual to mix positive and negative comments about different concepts in different sentences, so many comments are tagged as neutral (see Table 1 and some examples in Table 2 to illustrate this idea). This fact hides the intention of the user, which is tagged as neutral when she or he is not, that is the reason we automatically split the original observations into sentences and measuring the polarity of each sentence. Finally, we desired in this study to explore the idea of detecting polarity with cross-domain models in less complex structures, splitting the sentences in meaningful units which are usually only positive or negative.

After this process, all of these units (whole observations, automatic sentences and meaningful units) were manually tagged according its polarity (positive, negative, or neutral). Positive and negative sentiment units were annotated, being tagged as neutral those units without presence of any emotion or feelings (i.e, "No applicable.") or when the unit provided the same amount of positive and negative feelings (i.e.,"Some of the modules were very interesting and valuable but some of them confusing as too genetic details involved."). Two human taggers did both the parsing of complex statements in meaningful units and the annotation of each unit as positive, negative or neutral. See the total number of units and the class distributions in Table 1. As it can be observed, there are more positive than negative samples. The neutral category decreased from the whole comment (a complex statement) to the meaning unit (usually, with polarity or, less frequently, with lack of sentiment). All these samples were used as test set, and they were automatically labeled by using the proposed models (neural networks and MeaningCloud models). Following, Table 2 gives some examples of tagged observations, sentences and meaningful units.

## 4 Evaluation Metrics

In order to evaluate the systems with the gold standard, different evaluation metrics were used. Concretely, as defined below, we used Accuracy (*Acc*, Eq. 1) and macro $F_1$ ($MF_1$, Eq. 3) to reduce the impact of corpus imbalance in the evaluation. Moreover, the $F_1$ per class (being $c$ the *positive*, *negative*, and *neutral* class in $F_1^c$, Eq. 2) is shown to observe the behavior of our systems at class level.

$$Acc = \frac{\sum_{c \in C} \sum_{x \in \Omega_c} [f(x) = c]}{|\Omega|} \tag{1}$$

**Table 1.** Different units extracted from the "Other comments" box, from 133 users (37 English speakers and 96 Spanish speakers).

| Type of unit | Language | VLE | Total | Positive | Negative | Neutral |
|---|---|---|---|---|---|---|
| Observation | Spanish | Conecto | 31 | 12 (39%) | 8 (26%) | 11 (35%) |
| | | Moodle | 65 | 35 (54%) | 13 (20%) | 17 (26%) |
| | | Total | 96 | 47 (49%) | 21 (22%) | 28 (29%) |
| | English | Conecto | 37 | 18 (49%) | 8 (22%) | 11 (30%) |
| Sentence | Spanish | Conecto | 51 | 21 (41%) | 23 (45%) | 7 (14%) |
| | | Moodle | 100 | 49 (49%) | 32 (32%) | 19 (19%) |
| | | Total | 151 | 70 (46%) | 55 (37%) | 26 (17%) |
| | English | Conecto | 61 | 31 (51%) | 18 (29%) | 12 (20%) |
| Meaningful unit | Spanish | Conecto | 63 | 31 (49%) | 31 (49%) | 1 (2%) |
| | | Moodle | 164 | 97 (59%) | 59 (36%) | 8 (5%) |
| | | Total | 227 | 128 (56%) | 90 (40%) | 9 (4%) |
| | English | Conecto | 70 | 37 (53%) | 25 (36%) | 8 (11%) |

**Table 2.** Examples of tagged observations, sentences and meaningful units, with its polarity.

| Unit | Example | Polarity |
|---|---|---|
| Observation | *Excellent opportunity to learn with our busy routine concerns regarding very low volume of speakers as a very quiet room required even a fan disturbs the volume.* | Neutral |
| Sentence | *Excellent opportunity to learn with our busy routine concerns regarding very low volume of speakers as a very quiet room required even a fan disturbs the volume.* | Neutral |
| Meaningful unit | *Excellent opportunity to learn with our busy routine* | Positive |
| Meaningful unit | *concerns regarding very low volume of speakers as a very quiet room required even a fan disturbs the volume.* | Negative |
| Observation | *Well-organized and structured course. Great study material (articles) but not enough time to read them all. Keep up the good work.* | Neutral |
| Sentence | *Well-organized and structured course.* | Positive |
| Sentence | *Great study material (articles) but not enough time to read them all.* | Neutral |
| Sentence | *Keep up the good work.* | Positive |
| Meaningful unit | *Well-organized and structured course.* | Positive |
| Meaningful unit | *Great study material (articles)* | Positive |
| Meaningful unit | *but not enough time to read them all.* | Negative |
| Meaningful unit | *Keep up the good work.* | Positive |

$$F_1^c = \frac{2 \cdot P_c \cdot R_c}{P_c + R_c} \tag{2}$$

$$MF_1 = \frac{1}{|C|} \sum_{c \in C} F_1^c \tag{3}$$

where $\Omega$ is the set of samples, $\Omega_c$ are the samples of class $c$ in $\Omega$, $y(x)$ is the prediction of the model $f$ for a given sample $x$, $C$ is the set of classes, $[\cdot]$ denotes the Iverson bracket, and $P_c$ and $R_c$ are the prediction and recall measure of each class:

$$P_c = \frac{\sum_{x \in \Omega_c} [f(x) = c]}{\sum_{x \in \Omega} [y(x) = c]} \qquad R_c = \frac{\sum_{x \in \Omega_c} [f(x) = c]}{|\Omega_c|} \tag{4}$$

## 5 Polarity Models

Cross-domain models for both Spanish and English are used to address the problem of sentiment analysis on VLEs. On the one hand, Convolutional Neural Networks (CNN) were used to train models for sentiment analysis tasks on Twitter, both in Spanish and English, proposed in national and international competitions [10,17]. On the other hand, we used the sentiment analysis module provided by the product "Software as a Service" MeaningCloud [12], which acts as a general domain polarity classifier both for English and Spanish.

### 5.1 Convolutional Neural Networks Models

To determine the polarity of the students' opinions, we used polarity models based on the use of word embeddings and deep learning. Unfortunately, due to the lack of training data, it was not possible to learn robust models specifically for the task described in this paper. Instead, we used models trained, by our research group, for similar tasks related to the social network Twitter [10,17] both for English and Spanish. Table 3 shows some details of the corpora used to train the models.

Specifically, our trained models are based on the use of Convolutional Neural Networks. This architecture is inspired by the work described in [8], which has obtained competitive results in text classification tasks such as sentiment analysis or irony detection. Each opinion is represented as a $50 \times 300$ matrix where each word of the opinion - up to a maximum of 50 - is represented as a 300-dimensional embedding. Zero padding at the start of the matrix was used for opinions with less than 50 words. We applied several one-dimensional (the width of the filter is constant and equal to the dimension of the embeddings) convolutions with different height filters in order to extract the sequential structure of the text. Subsequently, we applied Global Max Pooling to the feature maps in order to extract the most salient features for each region size. The final decision is carried out by a softmax fully-connected layer. Table 4 shows the performance of the models for the test set of the two tasks [3,15], along with their 95% confidence

**Table 3.** Characteristics of the corpora used to train the CNN models (both for English and Spanish).

| Task | Set | Total | Positive | Negative | Neutral | None |
|---|---|---|---|---|---|---|
| SemEval 17 (English) | **Train** | 39656 | 15705 (40%) | 6203 (15%) | 17748 (45%) | N/A |
| | **Test** | 12284 | 2375 (19%) | 3972 (32%) | 5937 (49%) | N/A |
| | **Total** | 51940 | 18080 (35%) | 10175 (19%) | 23685 (46%) | N/A |
| TASS 17 (Spanish) | **Train** | 1008 | 318 (32%) | 418 (41%) | 133 (13%) | 139 (14%) |
| | **Test** | 1899 | 642 (34%) | 767 (40%) | 216 (11%) | 274 (15%) |
| | **Total** | 2907 | 960 (33%) | 1185 (41%) | 349 (12%) | 413 (14%) |

**Table 4.** Performance of the CNN (in grey) and MeaningCloud (in white) systems on SemEval 2017 Task 4 (English) and TASS 17 (Spanish) for the test set.

| **Task** | *Acc* | | $F_1$ | | $F_1^{pos}$ | | $F_1^{neg}$ | | $F_1^{neu}$ | | $F_1^{none}$ | |
|---|---|---|---|---|---|---|---|---|---|---|---|---|
| **English** | 0.72±0.00 | 0.52±0.01 | 0.73 | 0.52 | 0.73 | 0.47 | 0.75 | 0.57 | 0.70 | 0.53 | N/A | |
| **Spanish** | 0.62±0.02 | 0.54±0.02 | 0.47 | 0.44 | 0.67 | 0.64 | 0.71 | 0.60 | 0.08 | 0.14 | 0.42 | 0.37 |

intervals. Note that for TASS 2017, the distinction between the classes Neutral (*with both positive and negative feelings*) and None (*lack of sentiment*) is made during training and test. However, when the model trained for TASS 2017 is applied to our UX evaluation task, both classes are considered equal. That is, given a test opinion $x$, $\text{argmax}_y p(y|x) \in \{Neutral, None\} \rightarrow y = Neutral$.

### 5.2 MeaningCloud Models

MeaningCloud is a Software As A Service product [12] that provides a large number of tools, easy to use and to deploy, for text processing, analytics and text/audio mining, with the aim of facilitating the resolution of natural language processing problems to developers. It includes tools for summarization, topic extraction, language identification and sentiment analysis.

We have used the sentiment analysis module. This module allowed us to use a classifier, trained in a general domain with texts in multiple languages, to determine the global polarity of user opinions on VLEs. Concretely, we use the field *score tag* in the response of the MeaningCloud API, that indicates the global polarity of the text in 6 different levels: strong positive, positive, neutral, negative, strong negative and without sentiment (None). To carry out our experiments, we collapsed the strong sentiments, i.e., strong positive and strong negative are considered as Positive and Negative, respectively. Moreover, the neutral/none classes are fused in only one class (Neutral). Table 4 shows the performance of the MeaningCloud system for the test set of the two tasks used to train the CNN models in order to compare both systems. As it could be expected, performance is much better with the specific trained CNN models than when using general language models provided by MeaningCloud for a specific task.

In addition, the module is also capable of performing sentiment analysis at segment and aspect level. Thus, it is possible to detect words that express polarity and relate them with the objects of such polarity. That is interesting to capture relevant aspects that influence on the students and were not considered during the questionnaire, or to analyze which aspects of a course tend to be more negative or positive for the students.

On the other hand, the module has a series of additional capabilities such as the detection of subjectivity, the "agreement" and the irony, as well as disambiguation utilities of entities to enrich the sentiment analysis at the level of aspects. However, all these extended features are out of the scope of this work and we have planned to approach them in future work.

## 6 Experimental Results

We applied these two systems to the proposed task which consists of determining the polarity (positive, negative, or neutral) of each unit. One unit, as explained in Sect. 3, can be one observation, one sentence or one meaning unit.

The results obtained with the two systems for each type of segmentation can be seen in Tables 5, 6 and 7, along with their confidence intervals. It is possible to observe that in all cases, the *neutral* class is the worst detected. That means that it is more difficult to detect the absence of feelings (or the same amount of mixed positive and negative feelings) than detect isolated positive or negative feelings in this kind of comments. Also, *pos* is the best classified class almost in all cases, i.e., positive feelings are the easiest to detect.

It is possible to compare the behavior of the systems in Spanish and English (only on the Conecto system). It is striking that $F_1^{pos}$ is better (or equal) in English than in Spanish, while $F_1^{neg}$ is always worse in English than in Spanish. It seems more difficult to classify the negative class in English than in Spanish, whereas it is easier to classify positive and neutral classes.

In general, every metric is higher when detecting feelings for manually extracted semantic units. This is especially noticeable in the English language. The difference is not so marked in the case of whole comments and automatically extracted sentences.

CNN models have a slightly better behavior on the English language than MeaningCloud models (only in absolute values, their 95% confidence intervals overlap). This may be due to the fact that there are much more training samples for English than for Spanish (see Table 3) and therefore the trained model is able to generalize better. On the contrary, for Spanish, the results in terms of *Acc* and $F_1$ are better with MeaningCloud models. In this case, CNN models have only been trained with 1000 samples (versus 40k samples for English) and the generalization is much worse than using the general models trained for Spanish in the MeaningCloud system.

**Table 5.** Experiments with the whole comments. The unit is the whole comment, composed by one or more sentences. First column of each evaluation metric (in grey) is obtained with our approach based on CNN, the second column (in white) is obtained with MeaningCloud models.

| Lang. | VLE | Tot. | *Acc* | | $MF_1$ | | $F_1^{pos}$ | | $F_1^{neg}$ | | $F_1^{neu}$ | |
|---|---|---|---|---|---|---|---|---|---|---|---|---|
| Spanish | Conecto | 65 | 0.58±0.12 | 0.65±0.12 | 0.44 | 0.53 | 0.71 | 0.80 | 0.61 | 0.57 | 0.00 | 0.23 |
| | Moodle | 31 | 0.48±0.18 | 0.58±0.17 | 0.43 | 0.55 | 0.69 | 0.69 | 0.59 | 0.71 | 0.00 | 0.25 |
| | Total | 96 | 0.55±0.10 | 0.62±0.10 | 0.43 | 0.55 | 0.70 | 0.77 | 0.60 | 0.63 | 0.00 | 0.24 |
| English | Conecto | 37 | 0.62±0.16 | 0.57±0.16 | 0.52 | 0.44 | 0.77 | 0.74 | 0.50 | 0.36 | 0.29 | 0.24 |

**Table 6.** Experiments with automatically extracted sentences from the whole comments. The unit is one sentence, automatically extracted from the whole comment. First column of each evaluation metric (in grey) is obtained with our approach based on CNN, the second column (in white) is obtained with MeaningCloud models.

| Lang. | VLE | Tot. | *Acc* | | $MF_1$ | | $F_1^{pos}$ | | $F_1^{neg}$ | | $F_1^{neu}$ | |
|---|---|---|---|---|---|---|---|---|---|---|---|---|
| Spanish | Conecto | 100 | 0.56±0.10 | 0.62±0.10 | 0.45 | 0.54 | 0.66 | 0.76 | 0.62 | 0.58 | 0.07 | 0.29 |
| | Moodle | 51 | 0.57±0.14 | 0.61±0.13 | 0.45 | 0.51 | 0.75 | 0.75 | 0.59 | 0.65 | 0.00 | 0.12 |
| | Total | 151 | 0.56±0.08 | 0.62±0.08 | 0.45 | 0.53 | 0.69 | 0.76 | 0.61 | 0.60 | 0.04 | 0.24 |
| English | Conecto | 61 | 0.61±0.12 | 0.57±0.12 | 0.49 | 0.45 | 0.79 | 0.78 | 0.50 | 0.33 | 0.18 | 0.23 |

**Table 7.** Experiments with semi-supervised extracted sentences from the whole comments. The unit is one sentence which has own meaning. First column of each evaluation metric (in grey) is obtained with our approach based on CNN, the second column (in white) is obtained with MeaningCloud models.

| Lang. | VLE | Tot. | *Acc* | | $MF_1$ | | $F_1^{pos}$ | | $F_1^{neg}$ | | $F_1^{neu}$ | |
|---|---|---|---|---|---|---|---|---|---|---|---|---|
| Spanish | Conecto | 164 | 0.65±0.07 | 0.66±0.07 | 0.50 | 0.53 | 0.76 | 0.82 | 0.63 | 0.60 | 0.12 | 0.16 |
| | Moodle | 63 | 0.67±0.12 | 0.65±0.12 | 0.49 | 0.52 | 0.81 | 0.77 | 0.67 | 0.65 | 0.00 | 0.13 |
| | Total | 227 | 0.65±0.06 | 0.66±0.06 | 0.50 | 0.53 | 0.77 | 0.81 | 0.64 | 0.62 | 0.09 | 0.16 |
| English | Conecto | 70 | 0.66±0.11 | 0.66±0.11 | 0.53 | 0.53 | 0.81 | 0.83 | 0.59 | 0.51 | 0.20 | 0.26 |

## 7 Conclusions and Future Work

In this paper, we have presented a sentiment analysis task to observations written in natural language extracted from questionnaires of postgraduate biomedical online learning students. As stated in the introduction, the application of sentiment analysis tools on UX comments will provide a better and accurate understanding on human needs in the interaction with VLEs. Two Learning Management Systems have been evaluated, both in Spanish and English, applying cross-domain polarity models trained with a corpus of a different domain (tweets for each language) and general models for the language. The obtained results are very promising and they give an insight to keep going the research of applying sentiment analysis tools on User eXperience evaluation.

The ultimate goal is to develop further tools of automatic feed-back of user perception for designing virtual learning environments valued by users for its usability, quality and pleasure of use. For this, as a future work we will address automatic aspect detection (*pleasure of use, pleasure of learning, learning platforms, video, slides, usability, etc.*) and we will analyze the aspect polarity to capture relevant aspects that influence on the students and, possibly, were not considered during the questionnaire, or to analyze which aspects of a course tend to be more negative or positive for the students. Finally, we are now working with questionnaires on Massive Open Online Course (MOOC) to collect large amounts of data in order to train models for the task of sentiment analysis at global and aspect level on VLEs. A transfer learning approach from models trained with data of other domains could also be applied in order to have more robust models for the task.

## References

1. Ergonomics of Human System Interaction-Part 210: Human-centred design for interactive systems. International Standardization Organization (ISO), Switzerland (2009)
2. Baziotis, C., Pelekis, N., Doulkeridis, C.: Datastories at SemEval-2017 task 4: deep LSTM with attention for message-level and topic-based sentiment analysis. In: Proceedings of the 11th International Workshop on Semantic Evaluation (SemEval-2017), pp. 747–754 (2017)
3. González, J.Á., Plà, F., Hurtado, L.F.: ELiRF-UPV at SemEval-2017 task 4: sentiment analysis using deep learning. In: SemEval@ACL (2017)
4. González, J.A., Hurtado, L.F., Pla, F.: Análisis de sentimientos en Twitter Basado en Aprendizaje Profundo. In: TASS 2018: Workshop on Semantic Analysis at SEPLN (TASS 2018). CEUR Workshop Proceedings, vol. 2172 (2018)
5. GoogleCloud: Cloud Natural Language API (2019). https://cloud.google.com/natural-language/
6. Hassenzahl, M., Tractinsky, N.: User experience - a research agenda. Behav. Inf. Technol. **25**(2), 91–97 (2006). https://doi.org/10.1080/01449290500330331
7. IBM: Natural Language Understanding (2019). https://www.ibm.com/watson/services/natural-language-understanding/
8. Kim, Y.: Convolutional neural networks for sentence classification. arXiv preprint arXiv:1408.5882 (2014)
9. Kujala, S., Roto, V., Väänänen-Vainio-Mattila, K., Karapanos, E., Sinnelä, A.: UX Curve: a method for evaluating long-term user experience. Interact. Comput. **23**(5), 473–483 (2011)
10. Martínez-Cámara, E., Díaz-Galiano, M., García-Cumbreras, M., García-Vega, M., Villena-Román, J.: Overview of TASS 2017. In: Proceedings of TASS 2017: Workshop on Semantic Analysis at SEPLN (TASS 2017). CEUR Workshop Proceedings, vol. 1896 (2017)
11. MeaningCloud: Demo de Analítica de Textos (2019). https://www.meaningcloud.com/es/demos/demo-analitica-textos

12. MeaningCloud: MeaningCloud: Servicios web de analítica y minería de textos (2019). https://www.meaningcloud.com/
13. MicrosoftAzure: Text Analytics API (2019). https://azure.microsoft.com/es-es/services/cognitive-services/text-analytics/
14. Pang, B., Lee, L., Vaithyanathan, S.: Thumbs up? Sentiment classification using machine learning techniques. In: Proceedings of the ACL-02 Conference on Empirical Methods in Natural Language Processing, vol. 10, pp. 79–86 (2002)
15. Pla, F., Hurtado, L.F.: Spanish sentiment analysis in Twitter at the TASS workshop. Lang. Resour. Eval. **52**(2), 645–672 (2018). https://doi.org/10.1007/s10579-017-9394-7
16. Rauschenberger, M., Schrepp, M., Cota, M.P., Olschner, S., Thomaschewski, J.: Efficient measurement of the user experience of interactive products. How to use the User Experience Questionnaire (UEQ). Example: Spanish language version. Int. J. Interact. Multimedia Artif. Intell. **2**(1), 39–45 (2013). https://doi.org/10.9781/ijimai.2013.215
17. Rosenthal, S., Farra, N., Nakov, P.: Semeval-2017 task 4: sentiment analysis in Twitter. In: Proceedings of the 11th International Workshop on Semantic Evaluation (SemEval-2017), pp. 502–518 (2017)
18. Socher, R., et al.: Recursive deep models for semantic compositionality over a sentiment treebank. In: Proceedings of the 2013 Conference on Empirical Methods in Natural Language Processing, pp. 1631–1642 (2013)
19. Zaharias, P., Mehlenbacher, B.: Editorial: Exploring User Experience (UX) in virtual learning environments. Int. J. Hum.-Comput. Stud. **70**(7), 475–477 (2012). https://doi.org/10.1016/j.ijhcs.2012.05.001

# Document Model with Attention Bidirectional Recurrent Network for Gender Identification

Bassem Bsir(✉) and Mounir Zrigui(✉)

Research Laboratory in Algebra, Numbers Theory and Intelligent Systems, University of Monastir, Monastir, Tunisia
Bsir.bassem@yahoo.fr, mounir.zrigui@fsm.rnu.tn

**Abstract.** Author profiling is an important statistical and semantic processing task Author profiling is an important statistical and semantic processing task in the field of natural language processing (NLP). It refers to the extraction of information from author's texts such as gender, age and other kinds of personality traits. Author profiling can be applied in various fields like marketing, security and forensics. In this work, we explore how bi-directional deep learning architectures can be used to learn the abstract and higher-level features of the document, which could be employed to identify the author's gender. To deal with this, we extend Bidirectional Long Short-Term Memory Networks Language Models with an attention mechanism. The originality of our approach lays in its ability to capture the most important semantic information in a sentence. The experimental results on Facebook and twitter corpus show that our method outperformed the majority of the existing methods.

**Keywords:** Attention · GRU · LSTM neural network · Author profiling · Gender identification · Deep learning

## 1 Introduction

The rise of social media is so pervasive in communication and social relationships (more than 2 billion monthly users, for instance, access to the platform Facebook[1]). Among the major tasks of social media analysis, we can mention author profiling (AP) which aims at detecting information about the background of the author of an anonymous text based on the language of the text. It can be applied in a variety of applications, including security and forensics, when, for example, identifying the gender in which a given threat is written can help limit the search space of the author of this threat [4, 8, 18, 47].

DNN have recently achieved state-of-the-art performance in several NLP tasks such as machine translation, paraphrase identification, question answering and text summarization [9, 14, 36, 41, 46, 48]. Indeed, AP has benefited greatly from the resurgence of deep neural networks (DNNs), due to their high performance with less need of engineered features [6, 22].

[1] https://techcrunch.com/2017/06/27/facebook-2-billion-users/.


I. Rojas et al. (Eds.): IWANN 2019, LNCS 11506, pp. 621–631, 2019.
https://doi.org/10.1007/978-3-030-20521-8_51

In this paper, we propose an attention mechanism to enforce the DNN model to attend to the important part of a sentence for the detection of author's profile.

In order to capture important information in response to a given aspect, we design an attention model based LSTM that can concentrate by focusing on the key part of a sentence. To evaluate the efficacy of our model, we conducted experiments on a dataset consisting of twitter texts presented by the PAN Lab at CLEF 2018 [31]. We notice that our proposed model achieved significant improvement over the state-of-the-art results in terms of accuracy.

The rest of our paper is structured as follows: Sect. 2 discusses related works. Section 3 gives a detailed description of our attention-based proposals. Section 4 presents extensive experiments to justify the effectiveness of our proposals, and Sect. 5 summarizes this work and our future direction. We analyze its effectiveness on a dataset consisting of twitter texts presented by the PAN Lab at CLEF 2018 [31] in terms of accuracy. Furthermore, we investigate techniques for decreased training times and compare the different neural network architectures.

## 2 State of the Art

Authorship profiling is a well-studied task in natural language processing. [8, 12, 17] presented the various ways machine learning has been applied to this task. In each of the methods, the models rely on hand-coded vocabulary and stylometric features. Deep learning moves beyond this hand coded features and allows for a more flexible model.

Gender is perhaps the most studied social factor in many disciplines that examine the link between language use and the social world. Indeed, Sap et al. [34] derived a predictive lexica (words and weights) for age and gender identification using regression and classification models based on word usage in Facebook, blog and twitter data with associated demographic labels.

Werlen [24] categorized motion, anger and religion based frequency of words that are helpful while classifying the age and gender in hotel reviews. The writing style, word choice and grammar rule is solely depend on the topic of interest and the differences were found with topic variations. It is observed that the gender specific topic will have an impact in their writing styles. It is observed that the female tend to write more about wedding styles and fashions and where as male bloggers stress more on technology and politics.

Clauset et al. [10] used LIWC to analyze 46 million words produced by 11,609 participants. The studied texts include written texts composed as part of psychology experiments carried out by universities in the US, New Zealand and England. Full texts of fictional novels, essays written for university evaluation and transcribed free conversations from research interviews were investigated in these studies. The authors concluded that some of their LIWC variables were statistically significant and showed a gender effect.

We can also mention, for instance, the work of Wang et al. [36] who examined 10,000 short blog messages; each of which contains 15 words. They got 72.1% accuracy for gender prediction. Peersman et al. [28] presented an exploratory study in which they applied a text categorization approach for the prediction of age and gender on a corpus of chat texts collected from the Belgian social networking site Net log. They tried to identify the types of features that are most informative for a reliable prediction of age and gender on these difficult text types.

Bamman et al. [5] examined the most frequent words in a corpus of short messages written by more than 14,000 users of Twitter for gender patterns to predict the gender of anonymous text. The results they obtained confirmed that social network homophile is correlated to the use of same-gender language markers.

In [13], the authors worked on the automatic classification of emails; they got a rate of 81.5% of well classified documents for the gender dimension and 72% for the age dimension [13].

Poulston et al. [1] used the genism Python library for LDA topic extraction with SVM classifiers. Their results proved that the topic models are useful in developing author's profiling systems.

Argamon et al. [3] analyzed an analogous sample taken from the BNC consisting of fiction and non-fiction documents. Their corpus includes 604 texts equally divided by genre and controlled for authorial origin for a total size of 25 million words. Their analysis consists in a frequency count of basic and most frequent function words, part-of-speech tags and part-of-speech two-grams and three-grams. The counts were processed by a machine-learning algorithm used to classify the texts according to the author's gender. They obtained an accuracy of 80%.

Martinc et al. [26] based on the corpus collected from Twitter text written by four different languages (Arabic, English, Portuguese and Spanish), [23] obtained 70.02 by using logistic regression by combining character, word POS n-grams, emoji's, sentiments, character flood in gland lists of words per variety in PAN 2017 competition [30].

González-Gallardo et al. [7] predicted the gender, age and personality traits of Twitter users. They accounted stylistic features represented by character N grams and POS N-grams to classify tweets. They applied Support Vector Machine (SVM) with a linear kernel called LinearSVC and obtained 83.46% for gender detection.

Recently, few architectures and models have been introduced for authorship attribution employing Deep learning frameworks including LSTM, CNN and Recursive Neural Network [6, 22, 41]. Models, in this paradigm, can take the advantage of the general learning procedures relying on back-propagation, 'deep learning', a variety of efficient algorithms and some other tricks to further improve the training procedure.

For instance, Savoy et al., in 2018, evaluated two neural models for gender profiling on the PAN@CLEF 2018 tweet collection [31]. The first model is a character-based Convolutional Neural Network (CNN), while the second is an Echo State Network-based (ESN) recurrent neural network with various features. A lot of features based on words, characters and grammar were suggested to identify the characteristics of authors. There has been a recent revival of interest in using deep learning methods to solve various machine learning problems and NLP issues for the ultimate purpose of learning more robust features using easily available unlabelled data.

Kodiyan et al. [21] in their latest research in 2017 implemented a bi-directional Recurrent Neural Network with a Gated Recurrent Unit (GRU) combined with an Attention Mechanism for author profiling detection. In this work, researchers achieved an average accuracy of 75.31%, in gender classification, and 85.22% in language variety classification. Based on the same corpus, Miura et al. [24] presented model integrating word and character information with multiple conventional neural network layers. Their models marked joint accuracies of 64, 86% in the gender identification and the language classification, respectively in PAN 2017.

In 2018, Bsir and Mounir implemented a bi-directional Recurrent Neural Network with a Gated Recurrent Unit (GRU) combined with stylometric features for author profiling detection. In this work, researchers achieved an accuracy of 79%, in gender classification [6].

Luka Stout et al., identified the gender of authors based on written texts and shared images. They proposed a way to combine multiple predictions on shared content into a single prediction on user level. Their systems compare Naive Bayes model and a RNN. Authors got accuracy scores varying between 62.3% and 78.8% depending on the language and whether we used models that classify based on text or on images [31].

The approach of Sebastian Sierra et al., consisted of evaluating gender by using multi-modal information (texts and images). They learned their multimodal representation by employing GMUs. They obtained 0.80, 0.74 and 0.81 of accuracy rate in the multi-modal scenario for the test partition for English, Spanish and Arabic respectively [31].

Another way to solve the long-term dependency problem is to use an attention mechanism. They were recently demonstrated to have success in a wide range of NLP tasks [27, 37, 39, 43].

In 2016, Yang et al. proposed a hierarchical attention network for document classification. Our model has two distinctive characteristics: (i) it has a hierarchical structure that mirrors the hierarchical structure of documents; (ii) it has two levels of attention mechanisms applied at the word and sentence-level. Visualization of the attention layers illustrates that the model selects qualitatively informative words and sentences [39].

For sentiment classification, Wang et al. revealed that the sentiment polarity of a sentence is not only determined by the content but is also highly related to the concerned aspect. they proposed an Attention-based Long Short-Term Memory Network for aspect-level sentiment classification. They experiment on the SemEval 2014 dataset and results show that their model achieves state-of the-art performance on aspect-level sentiment classification [39].

In 2018, [44] proposed a novel attention mechanism in which the attention between elements from input sequence(s) is directional and multi-dimensional (i.e., feature-wise). A light-weight neural net, "Directional Self-Attention Network (DiSAN)," is then proposed to learn sentence embedding, based solely on the proposed attention without any RNN/CNN structure. DiSAN is only composed of a directional self-attention with temporal order encoded, followed by a multi-dimensional attention that compresses the sequence into a vector representation. It achieves the best test accuracy among all sentence encoding methods and improves the most recent best result by 1.02% on the Stanford Natural Language Inference (SNLI) dataset.

# 3 The Proposed Approach

In this section we propose an Attention based LSTM ATT-BLSTM model in detail. The standard LSTM cannot detect which is the important part for aspect-level sentiment classification. In order to address this issue, we propose to design an attention mechanism that can capture the key part of sentence in response to a given aspect.

## 3.1 Long Short-Term Memory (LSTM)

Standard RNN has the gradient vanishing or exploding problems. In order to overcome the issues, Long Short-term Memory network (LSTM) was developed and achieved superior performance [16].

LSTM networks introduce a new structure called a memory cell where each memory cell is made of two memory blocks and an output layer.

Each LSTM cell computes its internal state by applying the following iterative process and for multiple blocks; the calculations were randomly repeated for each block.

$$i_t = \sigma(W_{hi}h_{t-1} + W_{xi}xt + W_{ci}c_{t-1} + b_i) \quad (1)$$

$$f_t = \sigma(W_{hf}h_{t-1} + W_{xf}xt + W_{cf}c_{t-1} + b_f) \quad (2)$$

$$o_t = \sigma(Woh_{t-1} + W_{xo}xt + Woc_t + b_o) \quad (3)$$

$$c_t = f_t\Theta c_{t-1} + i_t\Theta \tanh(W_{xc}x_t + W_{hc}h_{t-1} + b_c) \quad (4)$$

$$h_t = o_t\Theta \tanh(c_t) \quad (5)$$

Where

$i_t$: Input gate. It shows the amount of new information that will be transmitted through the memory cell.
$f_t$: Forget gate is responsible for throwing way information from memory cell.
$o_t$: Output gate show how much information will be passed to expose to the next step.
$c_t$: Self-recurrent, which is equal to the standard RNN.
$\sigma$: The sigmoid function.
$h_t$: Final output.
$\Theta$: denotes the element-wise vector product.
W: matrices with different subscripts are parameter matrices.
b: the bias vector.

## 3.2 Attention-Based LSTM (ATT-LSTM)

Attention mechanisms have recently attracted enormous interest due to their highly parallelizable computation, significantly less training time, and flexibility in modeling dependencies [44]. As shown in Fig. 1 the model proposed in this paper contains four components: (1) Embedding layer: map each word into a low dimension vector; (2) LSTM layer: utilize BLSTM to get high level features from step (1); (3) Attention layer: produce a weight vector, and merge word-level features from each time step into a sentence-level feature vector, by multiplying the weight vector; (4) Output layer: the tweet-level feature vector is finally used for gender identification.

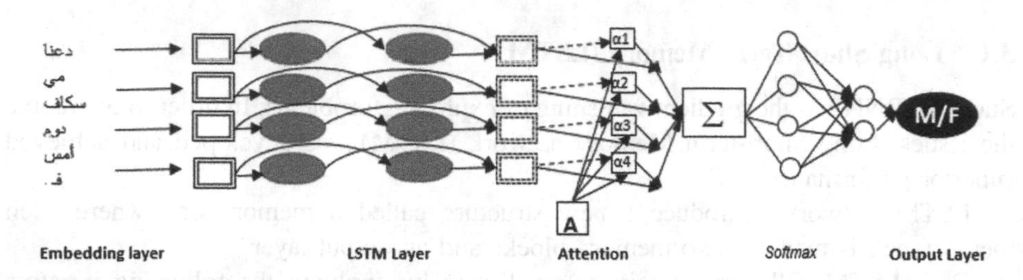


**Fig. 1.** Representation of the bi-LSTM+Attention model. We used n_ 4 and u _ 3 for visualization purposes.

The attention parameters weight the information to be taken into account in the sequence and allow focusing on the discriminating elements. The general idea is either a sequence of elements W = $\{w_1, \ldots, w_n\}$.

In fact, the sum of its elements if weighted by coefficients $\alpha_i$ (Eq. 6):

$$e_s = \sum_{i=1}^{n} \alpha_i w_i \tag{6}$$

In fact, there are different ways to obtain these coefficients $\alpha_i$. In fact, we rely on a vector of attention used by Yang et al. [39].

This coefficient $\alpha_i$ is then interpreted as the word or sentence score of importance for the considered task. The learning phase allows identifying the attention vector used to obtain the optimal weighting of the involved terms.

$$\alpha_i = \frac{\exp(a^T s_i)}{\sum_i \exp(a^T s_i)} \tag{7}$$

### 3.3 Data

The dataset is part of author profiling task of PAN@CLEF 2018 [31], as shown in Table 1. It was collected from Twitter. For each tweet collections, Arabic texts are composed of tweets written by 2400 authors; 100 tweets per authors. For the Arabic language, four varieties were used in this corpus: Egypt, Gulf, Levantine, and Maghrebi.

# 4 Experimentation

## 4.1 Performance Metric

In this section, we will evaluate the prediction accuracy of our method using the aforementioned corpora. We used the best result obtained in PAN@CLEF2018 as a baseline method to assess our technique and show its efficiency as shown in Table 2 [31].

**Table 1.** The best accuracy result of classifying gender for Arabic language on PAN 2018 testing.

| Authors | Methods | Accuracy |
|---|---|---|
| Takahashi et al. [31] | SVM+tf-idf | 81.70% |

## 4.2 Training Setup

Word2Vec model pre-trained word vectors are used since Word2Vec captures the word context information (such as word similarity). We mainly focus on Recurrent Neural Network (RNN) on Word2Vec word vectors since we believe that RNN can also capture the word/ sentence sequence information which can help us to better classifiy the authorship of tweets. The experiment results confirm this assumption. The Word2Vec model was formed by the corpus of Arabic Wikipedia[2] with 4 million tweets extracted in order to enrich the vocabulary list with words that do not exist in Wikipedia. For training, we used the skip-gram neural network model with a window of size 5 (1 center word + 2 words before and 2 words after), a minimum frequency of 15 and a dimension equal to 300.

For training our model, we used 10 fold cross-validation. The dataset was divided at the note level. We separated out 10% of the training set to form the validation set. We used the ADAM optimizer for parameter updates and Binary Cross Entropy Loss as our loss function. During training we based on the Keras library using a TensorFlow backend.

## 4.3 Evaluation

As mentioned previously, we use parts of the PAN 2018 author profiling data set. The results during the development phase were achieved on the provided training corpus with cross validation. We also utilized a maximum of 100 epochs to train our model on an Intel core i7 machine with 16 GB memory. The accuracy result obtained by applying our method for gender identification is 80.23% for test data, as show Fig. 2.

[2] Wikipedia, "WikimediaDownloads." https://dumps.wikimedia.org/arwiki/20170401/, 2017. [Online; accessed 10-April-2017].

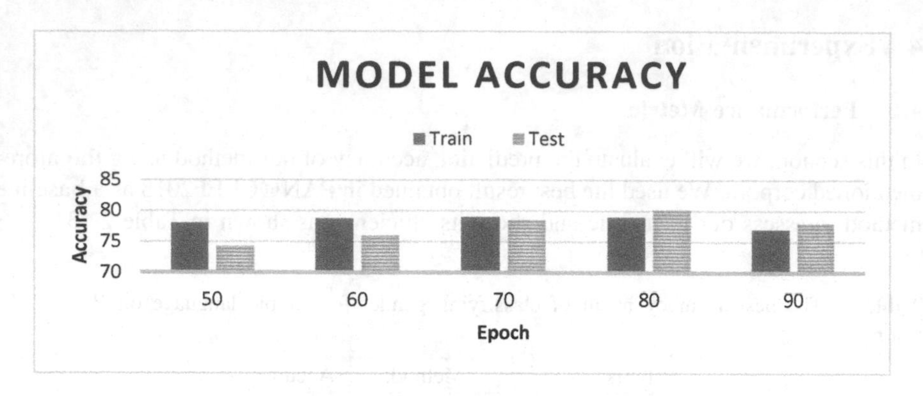


**Fig. 2.** Gender accuracy for train and test corpus.

## 5 Conclusion

In this paper, we extend Bidirectional Long Short-Term Memory Networks Language Models with an attention mechanism to predict the authors' gender of Twitter texts. We compared the average accuracy of our models with a previously developed SVM model. Our ATT-LSTM model showed a good performance for gender identification. The obtained results were encouraging, since neural network models were efficiently applied to solve natural language processing problems (sentiment analysis, text classification, etc.).

As future works, we plan to extend to other attributes for Arabic author's identification such as language variety and personality features.

## References

1. Poulston, A., Stevenson, M., Bontcheva, K.: Topic models and n–gram language models for author profiling. In: Proceedings of CLEF 2015 Evaluation Labs (2015)
2. Alvarez-Carmona, M.A., et al.: INAOE's participation at PAN'15: Author profiling task. Working Notes Papers of the CLEF (2015)
3. Argamon, S., Koppel, M., Fine, J., Shimoni, A.R.: Gender, genre, and writing style in formal written texts. Text-The Hague Then Amsterdam Then Berlin **23**(3), 321–346 (2003)
4. Aslam, T., Krsul, I., Spafford, E.H.: Use of a taxonomy of security faults (1996)
5. Bamman, D., Eisenstein, J., Schnoebelen, T.: Gender identity and lexical variation in social media. J. Sociolinguistics **18**(2), 135–160 (2014)
6. Bassem, B., Zrigui, M.: Enhancing deep learning gender identification with gated recurrent units architecture in social text. Computación y Sistemas **22**(3), 757–766 (2018)
7. González-Gallardo, C.E., et al.: Tweets classification using corpus dependent tags, character and POS N-grams. In: Proceedings of CLEF 2015 Evaluation Labs (2015)
8. Chaski, C.E.: Who wrote it? Steps toward a science of authorship identification. Natl. Inst. Justice J. **233**, 15–22 (1997)

9. Cho, K., et al.: Learning phrase representations using RNN encoder-decoder for statistical machine translation. arXiv preprint arXiv:1406.1078 (2014)
10. Clauset, A., Moore, C., Newman, M.E.: Hierarchical structure and the prediction of missing links in networks. Nature **453**(7191), 98 (2008)
11. Collobert, R., et al.: Natural language processing (almost) from scratch. J. Mach. Learn. Res. **12**, 2493–2537 (2011)
12. Ding, H., Samadzadeh, M.H.: Extraction of Java program fingerprints for software authorship identification. J. Syst. Softw. **72**(1), 49–57 (2004)
13. Estival, D., et al.: Author Profiling for English and Arabic Emails (2008)
14. Gehring, W.J., et al.: A neural system for error detection and compensation. Psychol. Sci. **4**(6), 385–390 (1993)
15. Gokturk, S.B., et al.: System and method for providing objectified image renderings using recognition information from images. U.S. Patent No. 9,430,719, 30 August 2016
16. Hochreiter, S., Schmidhuber, J.: LSTM can solve hard long time LAG problems. In: Advances in Neural Information Processing Systems, pp. 473–479 (1997)
17. Hochreiter, S., Schmidhuber, J.: Long short-term memory. Neural Comput. **9**(8), 1735–1780 (1997)
18. Inches, G., Crestani, F.: Overview of the International Sexual Predator Identification Competition at PAN-2012. CLEF (Online working notes/labs/workshop), vol. 30 (2012)
19. Joachims, T.: Text categorization with support vector machines: learning with many relevant features. In: Nédellec, C., Rouveirol, C. (eds.) ECML 1998. LNCS, vol. 1398, pp. 137–142. Springer, Heidelberg (1998). https://doi.org/10.1007/BFb0026683
20. Kalchbrenner, N., Blunsom, P.: Recurrent continuous translation models. In: Proceedings of the 2013 Conference on Empirical Methods in Natural Language Processing (2013)
21. Kalchbrenner, N., Grefenstette, E., Blunsom, P.: A convolutional neural network for modelling sentences. arXiv preprint arXiv:1404.2188 (2014)
22. Kodiyan, D., et al.: Author profiling with bidirectional RNNs using attention with GRUs: notebook for PAN at CLEF 2017. In: CLEF 2017 Evaluation Labs and Workshop–Working Notes Papers, Dublin, Ireland, 11–14 September 2017 (2017)
23. LeCun, L.B., Bengio, Y., Haffner, P.: Gradient-based learning applied to document recognition. Proc. IEEE **86**(11), 2278–2324 (1998)
24. Werlen, L.M.: Statistical learning methods for profiling analysis. In: Proceedings of CLEF 2015 Evaluation Labs (2015)
25. Maraoui, M., Terbeh, N., Zrigui, M.: Arabic discourse analysis based on acoustic, prosodic and phonetic modeling: elocution evaluation, speech classification and pathological speech correction. Int. J. Speech Technol. **21**(4), 1071–1090 (2018)
26. Martinc, M., Škrjanec, I., Zupan, K., Pollak, S.: Pan 2017: Author Profiling, gender and Language Variety Prediction. CLEF (Working Notes) 2017. CEUR Workshop Proceedings 1866, CEUR-WS.org (2017)
27. Miura, Y. et al.: Author Profiling with Word+Character Neural Attention Network. CLEF (Working Notes) 2017. CEUR Workshop Proceedings 1866, CEUR-WS.org (2017)
28. Peersman, C., Daelemans, W., Van Vaerenbergh, L.: Predicting age and gender in online social networks. In: Proceedings of the 3rd International Workshop on Search and Mining User-Generated Contents, pp. 37–44. ACM (2011)
29. Pham, D.D., Tran, G.B., Pham, S.B.: Author profiling for Vietnamese blogs. In: International Conference on Asian Language Processing, IALP 2009, pp. 190–194. IEEE (2009)

30. Rangel, F., et al.: Overview of the 5th author profiling task at PAN 2017: gender and language variety identification in Twitter. Working Notes Papers of the CLEF (2017)
31. Rangel, F., Rosso, P., Montes-y-Gómez, M., et al.: Overview of the 6th author profiling task at PAN 2018: multimodal gender identification in Twitter. Working Notes Papers of the CLEF (2018)
32. Säily, T.: Variation in morphological productivity in the BNC: Sociolinguistic and methodological considerations. Corpus Linguist. Linguist. Theory **7**(1), 119–141 (2011)
33. Sallis, P.J., et al.: Identified: software authorship analysis with case-based reasoning (1998)
34. Sap, M., et al.: Developing age and gender predictive lexica over social media. In: Proceedings of the 2014 Conference on Empirical Methods in Natural Language Processing (EMNLP), pp. 1146–1151 (2014)
35. Socher, R., et al.: Recursive deep models for semantic compositionality over a sentiment treebank. In: Proceedings of the 2013 Conference on Empirical Methods in Natural Language Processing (2013)
36. Wang, P., et al.: Semantic clustering and convolutional neural network for short text categorization. In: Proceedings of the 53rd Annual Meeting of the Association for Computational Linguistics and the 7th International Joint Conference on Natural Language Processing (Volume 2: Short Papers), vol. 2, pp. 352–357 (2015)
37. Wang, Y., Huang, M., Zhao, L.: Attention-based LSTM for aspect-level sentiment classification. In: Proceedings of the 2016 Conference on Empirical Methods in Natural Language Processing, pp. 606–615 (2016)
38. Williams, J.D., Zweig, G.: End-to-end LSTM-based dialog control optimized with supervised and reinforcement learning. arXiv preprint arXiv:1606.01269 (2016)
39. Yang, Z., Yang, D., Dyer, C., He, X., Smola, A., Hovy, E.: Hierarchical attention networks for document classification. In: Proceedings of the 2016 Conference of the North American Chapter of the Association for Computational Linguistics: Human Language Technologies, pp. 1480–1489 (2016)
40. Yih, X.H., Meek, C.: Semantic parsing for single-relation question answering. In: Proceedings of ACL 2014 (2014)
41. Yin, W., et al.: Comparative study of CNN and RNN for natural language processing. arXiv preprint arXiv:1702.01923 (2017)
42. Zhou, C., Sun, C., Liu, Z., Lau, F.: A C-LSTM neural network for text classification. arXiv preprint arXiv:1511.08630 (2015)
43. Zhou, P., et al.: Attention-based bidirectional long short-term memory networks for relation classification. In: Proceedings of the 54th Annual Meeting of the Association for Computational Linguistics (Volume 2: Short Papers), vol. 2, pp. 207–212 (2016)
44. Zhou, T., Shen, T., Long, G., Jiang, J., Pan, S., Zhang, C.: DiSAN: directional self-attention network for RNN/CNN-free language understanding. In: Thirty-Second AAAI Conference on Artificial Intelligence, April 2018
45. Zrigui, M., Ayadi, R., Mars, M., Maraoui, M.: Arabic text classification framework based on latent Dirichlet allocation. J. Comput. Inform. Technol. **20**(2), 125–140 (2012)
46. Zrigui, M., Charhad, M., Zouaghi, A.: A framework of indexation and document video retrieval based on the conceptual graphs. J. Comput. Inform. Technol. **18**(3), 245–256 (2010)
47. Zouaghi, A., Zrigui, M., Antoniadis, G.: Compréhension automatique de la parole arabe spontanée. Traitement Automatique des Langues **49**(1), 141–166 (2008)

48. Zouaghi, A., Merhbene, L., Zrigui, M.: A hybrid approach for Arabic word sense disambiguation. Int. J. Comput. Process. Lang. **24**(02), 133–151 (2012)
49. Zouaghi, A., Zrigui, M., Antoniadis, G., Merhbene, L.: Contribution to semantic analysis of Arabic language. In: Adv. Artif. Intell. (2012)
50. Zouaghi, A., Merhbene, L., Zrigui, M.: Combination of information retrieval methods with LESK algorithm for Arabic word sense disambiguation. Artif. Intell. Rev. **38**(4), 257–269 (2012)
51. Zouaghi, A., Zrigui, M., Ahmed, M.B., Riadi, L.: Évaluation des performances d'un modèle de langage stochastique pour la compréhension de la parole arabe spontanée. In: Proceedings of the TALN (2007)

# Visual Disambiguation of Prepositional Phrase Attachments: Multimodal Machine Learning for Syntactic Analysis Correction

Sebastien Delecraz[1(✉)], Leonor Becerra-Bonache[2], Alexis Nasr[1], Frederic Bechet[1], and Benoit Favre[1]

[1] Aix-Marseille Univ, Université de Toulon, CNRS, LIS, UMR 7020, Marseille, France
{sebastien.delecraz,alexis.nasr,frederic.bechet,benoit.favre}@univ-amu.fr
[2] Univ Lyon, UJM-Saint-Etienne, CNRS, Laboratoire Hubert Curien, UMR 5516, Saint-Étienne, France
leonor.becerra@univ-st-etienne.fr

**Abstract.** Prepositional phrase attachments are known to be an important source of errors in parsing natural language. In some cases, pure syntactic features cannot be used for prepositional phrase attachment disambiguation while visual features could help. In this work, we are interested in the impact of the integration of such features in a parsing system. We propose a correction strategy pipeline for prepositional attachments using visual information, trained on a multimodal corpus of images and captions. The evaluation of the system shows us that using visual features allows, in certain cases, to correct the errors of a parser. It also helps to identify the most difficult aspects of such integration.

**Keywords:** Prepositional phrase attachments · Multimodality · Correction strategy

## 1 Introduction

Natural languages are intrinsically ambiguous. Some of these ambiguities can be solved by using only syntactic information, but many others require access to the context in which they have been produced. For instance, the ambiguity involved in the famous example of "John saw a man with a telescope" could be easily solved if we could have access to that scene.

Prepositional phrase (PP) attachments are a common source of ambiguity. They constitute one of the most difficult constructions to deal with [20], representing around 20% of errors in parsing natural languages [18]. The main difficulty lies in the fact that this kind of ambiguity can often not be solved using

The work of Leonor Becerra-Bonache has been performed during her teaching leave granted by the CNRS (French National Center for Scientific Research) in Laboratoire d'Informatique et Systèmes of Aix-Marseille University.


I. Rojas et al. (Eds.): IWANN 2019, LNCS 11506, pp. 632–643, 2019.
https://doi.org/10.1007/978-3-030-20521-8_52

only linguistic cues (as shown in the previous example). Even if it can be overcame, a parser often does not have semantic constraints to rule out incorrect attachments. However, adding information from a visual source to the analysis of a sentence is a difficult task because it is necessary to extract relevant information from that source and to be able to relate this information to the sentence.

The work that we present in this article has a twofold objective. First, to propose a method for resolving PP-attachments based on the use of visual cues. Second, to analyze the impact of adding visual information on the task of syntactic analysis. To do this, we use a corpus made up of pairs of an image and a caption describing it. This corpus has been annotated manually at different levels. At the image level, rectangles (which we will call boxes) have been identified and a semantic category has been associated with each of them. At the text level, some noun phrases have been identified, as well as some PP-attachments for a subset of frequent prepositions. In addition, boxes corresponding to noun phrases were aligned to the latter. The fact of having simultaneously the analysis of the image (via boxes) and the text (through certain PP-attachments), as well as the alignment between boxes and noun phrases makes it possible to establish a link between the two modalities and to use information from the image to process the text.

The system we propose is based on an attachment error detector, that offers an alternative attachment if it detects an error. The originality of this detector is that it allows to take lexical, but also visual and conceptual clues as input. For example, in the noun phrase *a ball in_front_of a dog with a red collar*, the decision to attach *with* to *dog* rather than *ball* may be based on obvious lexical evidence, but could also be based on visual clues by studying, for example, the relative positions of the boxes corresponding to the words *ball*, *dog* and *collar*.

The paper is organized as follows. Related work is presented in Sect. 2. We describe our model in Sect. 3, which is composed of two different modules: automatic multimodal alignment and PP-attachment detection/correction. Finally, Sect. 4 presents our experiments and a discussion of the results obtained.

## 2 Related Work

The problem of finding the correct PP-attachment has attracted the attention of many researchers in the field of Natural Language Processing. It constitutes an important and challenging problem in parsing natural languages. Many different methods and sources of information have been proposed for the resolution of PP-attachment ambiguities.

Two kind of resources have mainly been used in the literature to solve the PP-attachment problem: semantic knowledge bases [1,7], and corpora [2,19,23]. To our knowledge, there are not too many works using *multimodal information* to deal with this problem. The most relevant work to us is [4]; their approach consists in simultaneously perform object segmentation and PP-attachment resolution for captioned images. In order to do that, they produce a set of possible hypothesis for both tasks, and then they jointly rerank them to select the most consistent pair. The main difference between their work and ours is that we produce a unique syntactic analysis and it is corrected according to visual informations. Moreover, we perform experiments with a much bigger number of images/caption pairs (22,800 vs. 1,822).

The disambiguation of PP-attachments by using visual information is also related to *visual relation learning*. The most related work to us is [25], in which the authors developed new visual descriptors for representing object relations in an image. Their model relies on a multimodal representation of object configurations for each relation, and it is able to learn classifiers for object relations from image-level supervision only (i.e., from image-level annotations such as "person on bike", without annotating the objects involved in the relation). While we could use their spatial relation classifiers, the focus of our work is different. We deal with the problem of disambiguating PP-attachments. We use similar visual features for representing the spatial configuration of objects, but objects are detected and represented in a different way.

Our system also aligns fragments of sentences (more concretely, noun phrases) with boxes in the image in order to be able to use multimodal information, without this being our main goal. A first step in this task is to detect objects in an image [13,14,27]. This requires two things: (i) to find the position of the object in the image, often by calculating the coordinates of the rectangle that surrounds the object; (ii) to predict a semantic class to the object. Many works have tried to learn correspondences between a part of a sentence and a part of an image, with different kind of applications in mind, such as caption generation [11,16,33] and image retrieval [3,6].

Many researchers in psycholinguistics and cognitive psychology have also studied the interaction between vision and ambiguous language in human sentence processing, such as [5,32]. These works provide evidence of the relevance of visual information for humans to solve linguistic ambiguity. This information is also of great relevance during the first stages of children's language acquisition, since much of the sentences received by children are linked to their immediate visual environment [30,31]. Our work is inspired by these ideas and address the problem of disambiguating PP-attachments by an artificial system that uses, among other cues, the visual information linked to a concrete ambiguous sentence.

## 3 Our Model

The model that we propose in this paper takes as input an image/sentence pair, and provides a syntactic analysis of the sentence by performing two different tasks. First, an automatic alignment of boxes detected in the image and noun phrases in the corresponding captions. Second, detection and correction of incorrect PP-attachments. We explain them in detail in the next sections.

### 3.1 Automatic Multimodal Alignment

This task is divided into three steps: detection of boxes in the image, detection of noun phrases in a sentence, and, finally, their alignment.

**Detection of Boxes.** The task of detecting boxes in an image consists of predicting the presence or absence of an object in an image given a list of objects that the system is able to recognize. When an object is recognized, the coordinates of the box containing it are produced. We have used here the real-time object detection model based on neural networks called YOLOv2 [28], which produces, for a given image, a list of boxes associated with semantic labels.

This system is broken down as follows: it takes an image as input and then cuts it into a grid. For each cell of the grid the system predicts a fixed number of bounding boxes, a confidence score for each box, and a probability for each semantic category. The final predictions are made by multiplying the confidence scores by the probability of the semantic categories.

**Detection of Noun Phrases.** Although the detection of noun phrases is a widely studied task, the target phrases in our work correspond to visual objects and may differ in nature from the typical noun phrases resulting from syntactic analysis.

It consists of a simple detector of the beginning and the end of noun phrases, which associates to any word in the sentence a label in the form $B$ (*begin*), $I$ (*inside*) and $O$ (*outside*) depending on whether the word starts a noun phrase, is within a noun phrase without being the first word or is outside a noun phrase. The prediction is made using an average perceptron based on the words of the sentence and their parts-of-speech. An evaluation by using our test corpus indicates an error rate of 2.2% per word.

**Alignment.** The alignment problem consists of determining for each noun phrase which is its corresponding object detected in the image. For example, given the caption *Someone is holding out a punctured ball in front of a brown dog with a red collar*, it is necessary to find among the boxes corresponding to the objects detected in the corresponding image (e.g., the balloon, the arm, the dog, the collar) which noun phrases they correspond to (*someone*, *a punctured ball*, *a brown dog*, *a red collar*). It is a difficult artificial vision problem because of the very different nature of the aligned objects: on the one hand, the pixels of the image and, on the other hand, a sequence of words. The problem become even more difficult when some noun phrases may correspond to several objects in the image (e.g. *children playing soccer*), some objects are only partially represented in the image (*people standing in a train*), and the object detector may have detected objects not represented in the caption.

To tackle this problem, we divide this task into two sub-tasks: the first is to calculate an association score between each visual object and each noun phrase, the second is to decide which of these potential associations will be retained for the future. We are not addressing the problem of multiple associations.

The association score between a visual object and a sequence of words is calculated by projecting the pixels of the image and the words of the caption towards the same representation space. Each visual object and each textual sequence is represented by vectors in this common space, which makes it possible to calculate a similarity between the vectors to obtain an association score. This projection in a common space is carried out using neural networks. The parameters of this network can be driven from known image/text pairs using a method based on *visual semantic embeddings* [10], which take advantage of a convolutional neural network to create image representations and recurrent neural network to create word sequence representations.

Once the alignment score is obtained for each image/text pair, it is necessary to determine a global association taking into account that it is neither injective nor surjective (some elements are not associated, others have multiple associations). This association is achieved using the following heuristic: pairs with the highest score are iteratively selected, each box can be assigned to no more than one noun phrase. Only pairs of scores greater than 0.3 are considered (threshold determined on a development corpus).

### 3.2 Detection and Correction of PP-attachments

The automatic alignment between the image and its caption allows the integration of visual features for the task of correcting PP-attachments produced by a parser. This section presents the correction module, which is divided into two steps. A first step that detects the attachment errors produced by the parser using a classifier. Then the correction strategy in which candidate governors at the target preposition are selected and then evaluated again with a classifier. The governor with the highest score is selected.

**Detection of Errors in the Attachment.** The detection of attachment errors is carried out using the AdaBoost algorithm [12]. To train this classifier we used two types of features: lexical and visual. These features concern to the $p$ preposition, its governor $G$ and its object $O$. When the governor is a verb, the subject of the verb serves as $G$. So, in the sentence "Jean eats with gloves on.", we get $G = Jean$, $p = with$ and $O = gloves$.

For the *lexical features*, starting from the dependent tree produced by a parser, we use: (a) the lemma and the grammatical category of the governor and the object; (b) the distance between the preposition and its governor. A detailed description of this feature is presented in previous works [8].

*Visual features* are calculated from the bounding boxes that the alignment system has associated to the governor and the object of the preposition. We distinguish two types of visual features:

- Conceptual features: person, body part, animal, clothing, instrument, vehicle and other. They are predicted when objects are detected in the image. If the alignment module has not selected any boxes for one of the two selected noun phrases (governor or object), the $UNK$ value is used to represent the concept of this noun phrase and none of the spatial features are calculated.
- Spatial features: given the governor's box $b_G = [x_g, y_g, w_g, w_g, h_g]$ and the object box $b_O = [x_d, y_d, w_d, h_d, h_d]$ of the preposition, where $(x, y)$ are the coordinates of the box center, and $(w, h)$ are the box height and width, we use the features proposed by [25]:

$$V_{S1} = \frac{x_d - x_g}{\sqrt{w_g h_g}} \; V_{S3} = \sqrt{\frac{w_d h_d}{w_g h_g}} \; V_{S2} = \frac{y_d - y_g}{\sqrt{w_g h_g}} \; V_{S4} = \frac{b_g \cap b_d}{b_g \cup b_d}$$

Features $V_{S1}$ and $V_{S2}$ represent the horizontal and vertical relative positions between the two boxes, respectively. $V_{S3}$ is the ratio of box sizes, $V_{S4}$ the overlap between boxes, and $V_{S5} = \frac{w_g}{h_g}$, $V_{S6} = \frac{w_d}{h_d}$ the aspect ratio of each box, respectively.

Based on all these features, the classifier checks whether the alignment proposed by the parser is correct or not.

**Correction Strategy.** In order to increase the accuracy of the parser, we use a correction strategy that consists in changing the attachment proposed by the parser using an error corrector [8]. When a connection is detected as incorrect by the classifier, we apply rules to the syntax tree generated by the analyzer to obtain a set of alternative connections. These possible new attachments are given to the classifier to make a final decision by selecting the one with the highest probability of attachment.

# 4 Experiments

## 4.1 Dataset

There is a lack of datasets that provide not only paired sentences and images, but detailed information about the correspondence between regions in images and phrases in captions. In this paper we focus in a multimodal corpus called *Flickr30K Entities* (F30kE) that provides such type of annotations [26]. It constitutes an extension of the original Flickr30k dataset [34], which is a well-known benchmark for sentence-based image description.

F30kE is composed of almost 32K images with five captions per image. Their annotation consists of identifying which mentions among the captions of the same image refer to the same set of entities (a total of 244k co-reference chains were annotated), and associating them with bounding boxes localizing those entities (a total of 276K bounding boxes were manually annotated). Each mention in the captions is categorized, using manually constructed dictionaries, into the following eight types: people, body parts, animals, clothing, instruments, vehicles, scene, and other.

In order to use this corpus for our task, we enriched it with a manual attachment of 29,068 prepositions to their governor. The attachment correction was made by a single annotator, who had only access to the caption, the target preposition and the corresponding image. More details can be found in [9]. For our experiments, we subdivided this corpus into three sets: learning (23,254 prepositions), development (2,907 prepositions) and test (2,907 prepositions).

### 4.2 Setup

**Alignment Module**

*Image processing:* We re-trained the YOLOv2 model on the F30kE corpus using as initialization the weights provided by the authors and limiting the number of categories to the eight semantic categories of the F30kE corpus. Only predictions with a confidence score above 0.1 were retained. The system detected 7,110 of the 14,229 boxes of images from our test corpus. An object is considered detected if the ratio of the area of the intersection over union between its ground truth box and the predicted box is greater than 0.5.

The detector achieves a recall of 0.49 and an accuracy of 0.29 on the test set. If we take into account the semantic categories, these performances fall down to 0.25 and 0.15 respectively. These results show us that this is a difficult task and that automatic image processing in this detection task represents a first barrier to the use of visual information.

Afterwards, the content of each box is resized to 224 by 224 pixels, then passed to the input of a ResNet-152 [15] network pre-trained for the image classification task in thousand scene categories from the ImageNet Large Scale Visual Recognition Challenge [29]. The last layer of the network is replaced by a dense layer (i.e., a linear transformation) that projects the representations to a vector size 1,024.

*Text processing:* The words of the noun phrases are first projected into a 300 size representation space that provides inputs to a GRU (*gated recurrent unit*) recurrent layer whose hidden representation is of size 1,024. The hidden representation of the recurrent network, after reading the words of a noun phrase, is used as a representation for the textual modality.

*Alignment:* A $\ell_2$-normalization is applied to neural networks activations of both modalities in order to be compared using the scalar product; this is equivalent to calculating the cosine similarity between the two vectors. Learning is performed on batches (*batches*) of size 48 for 30 periods using the Adam [17] optimization method. This is a model of *triplet ranking* whose learning is performed by calculating the similarity between an image/text pair existing in the learning data and a random pair with one of the two members in common and modifying the model so that the score of the valid pair is higher than that of the invalid pair.

Table 1 shows the performance of the alignment system between boxes and noun phrases. The error rate is calculated as follows: for each noun phrase, the association is considered correct if the box with the highest similarity to this noun phrase is the one corresponding to it in the ground truth data. The results are calculated according to two methods, VSE and VSE+++, which differ by the cost function used for learning [10]. According to the model used, those provided with the VSEpp tool were trained on the Flickr30k and Microsoft Common Objects in Context [21] corpus, or the model was re-trained on the data of our task (F30kE). The models available with the VSEpp implementation were trained on complete images and complete description sentences. Their performance falls on boxes containing only one object in our corpus, doubling the number of association errors, compared to the same model re-trained on the target data (from 21% to 37%) which demonstrates the importance of re-training the model under the same conditions as the test.

**Table 1.** Alignment error rate on our test set by comparing ground truth boxes and ground truth noun phrases, depending on the model (VSE, VSE+++) and training corpus (Flickr30k and Microsoft COCO are the models provided with the tool, trained on complete sentence/image pairs rather than noun phrases and box contents)

| Approaches | Training corpus | Error rate |
|---|---|---|
| VSE++ | Flickr30k | 42.07% |
| VSE++ | MS-COCO | 38.90% |
| VSE | MS-COCO | 37.17% |
| VSE | Fine-tuning | 21.47% |

**PP-attachment Detector Module.** In order to identify PP-attachments in captions we used a standard transition parser [24] trained on the Penn Treebank corpus [22]. The PP-attachment error detector is trained on our learning corpus. The classifier parameters were adjusted according to its performance on the development corpus.

We also used the development corpus to evaluate the performance of the rules we used to find potential governors $G_p$. At the output of the parser the rules allow us to retrieve the correct manually annotated governor for the preposition in the set $G_p$ in 92.28% of cases. This score therefore represents an upper bound for PP-attachments.

### 4.3 Results

The experiments presented here assume a scenario in which lexical information is not available (the semantics of words are not known by the system), in order to highlight the information that can be used in the visual part of the task. The case in which lexical features are used is then seen as an upper bound. In this context, we are interested in two questions: what is the impact of semantic categories in the visual modality, and what is the impact of spatial information in the same modality?

Table 2 shows the good attachment rate for the 10 most common prepositions of the test corpus. The F30kE corpus being mainly composed of captions that can be understood by a human without seeing the associated image (semantically unambiguous), real ambiguous cases are therefore rare.

**Table 2.** Correct attachment rate to the test: the number of occurrences is given for each preposition; the *baseline* is produced by the parser without correction; $V_C$ is obtained after correction by using only visual concepts, $V_S$ only spatial features, $V$ the combination of conceptual and spatial features, $L$ lexical features and $V + L$ is the combination of all features.

| Prepositions | Occurrences | Baseline | $V_C$ | $V_S$ | $V$ | $L$ | $V + L$ |
|---|---|---|---|---|---|---|---|
| *in* | 369 | 0.76 | 0.76 | 0.77 | 0.76 | 0.85 | 0.84 |
| *with* | 310 | 0.65 | 0.68 | 0.68 | 0.70 | 0.78 | 0.78 |
| *for* | 168 | 0.73 | 0.73 | 0.73 | 0.72 | 0.82 | 0.83 |
| *near* | 159 | 0.33 | 0.53 | 0.50 | 0.59 | 0.84 | 0.84 |
| *through* | 145 | 0.95 | 0.95 | 0.95 | 0.95 | 0.96 | 0.96 |
| *on* | 143 | 0.85 | 0.85 | 0.85 | 0.85 | 0.89 | 0.87 |
| *from* | 140 | 0.76 | 0.76 | 0.76 | 0.76 | 0.86 | 0.85 |
| *next to* | 137 | 0.89 | 0.89 | 0.89 | 0.89 | 0.90 | 0.89 |
| *into* | 116 | 0.89 | 0.89 | 0.89 | 0.89 | 0.92 | 0.95 |
| *over* | 111 | 0.66 | 0.64 | 0.66 | 0.68 | 0.85 | 0.84 |

As we can see in Table 2, the overall accuracy of the parser (without correction) is 75%. Note that the results vary a lot depending on the prepositions, ranging from 95% for the preposition *through*, to 33% for the preposition *near*.

Visual concepts ($V_C$) and spatial information ($V_S$) provide different improvements depending on the prepositions, but there seems to be mainly a gain with locative prepositions such as *near* (17% points). Since our visual features are focused on spatial information, it is logical that they have an impact on this kind of prepositions. One might think that conceptual categories are sufficient to solve the problem, but it should be noted that the task of visual categorization is difficult (and therefore the classifier is often wrong), and that categories are rather rough and may not remove all ambiguities. The combination of $V_C$ and $V_S$ gives the best results and corrects about 3% of the errors.

Lexical information ($L$) has a drastic effect since performance increases for most of the prepositions, underlining our hypothesis that the text is unambiguous in the absence of an image. These results are not surprising because it is well known that some ambiguities can be resolved by using only syntactic information. The problem is that this type of information is not always available. The fact that the gain is higher for lexical information than for visual information can also be explained by the unreliability of predictions in the visual modality. There is no difference on average between the use of lexical information only and the use of all features ($V + L$). However, this result highlights the fact that the fusion of features from text and image modalities takes advantage of the strongest modality without suffering from a modality with lower performance.

We present here some examples of image/text pairs for which the image has allowed, or not, to perform a correct attachment using sets of different features. In all examples the target preposition is in bold, the governor chosen by the analyzer is underlined and the new governor after correction is in square brackets.

Figure 1a shows a sentence for which the parser has made a bad attachment but the classifier has allowed to correct it by using only visual information. Concretely, the preposition *near* is incorrectly attached to *area* by the parser, and the classifier corrects the attachment by selecting *are* as a governor. This example is the justification for this study: to correct poor connections thanks to visual information.

Figure 1b shows a sentence for which the use of only visual features did not help to correct a wrong attachment. Concretely, the preposition *on* is incorrectly attached to the word *building*. The alignment system did not find the bounding box for at least one of the two noun phrases. This example shows one of the limitations of this study: the difficulty of the detection and alignment phase between boxes and noun phrases, limits the impact of visual features in correcting erroneous analyses.

Even if lexical features are the most effective, if the learning corpus does not contain enough examples for some entities, visual features may be more effective. Thus Fig. 1c shows a sentence for which the use of only visual features allows to obtain the correct alignment, while the use of only linguistic features produces an error. Concretely, the preposition *with* is incorrectly attached to the word *jeans* instead of the word *wearing*.

a – Two children [are] in a grassy <u>area</u> **near** two horses.

b – Two people sitting in front of an older <u>building</u> **on** a bench.

c – A dog is wearing [<u>jeans</u>] and a blue and yellow shirt **with** a black vehicle in the background.

**Fig. 1.** Examples of image/text pairs for which have been found, or not, a correct PP-attachment, by using different kind of features.

## 5 Conclusion

This work explores the possibility of using images to disambiguate prepositional phrases attachments in sentences that describe them. Visual features improve the performance by three points on average depending on the prepositions, and sometimes drastically, as in the case of the preposition *near*. However, the difficulty of the problem lies in the detection and categorization of objects, as well as in the alignment between text and images. Indeed, the errors and lack of information resulting from the automation of this step inevitably reduces the overall performance of the attachment corrector.

However, the gain obtained between the output of the parser and the output of the corrector, even when using only visual features, proves two things. First, that information was found at the image level and that an alignment, even partial, was produced. And secondly, that this information could be properly used despite the use of relatively simple descriptors.

A better use of the information from the image is a main direction to explore in order to improve the system with, in particular, the integration of information directly from the pixels, such as the use of the space representation for the image or directly at the level of the bounding boxes.

## References


1. Agirre, E., Baldwin, T., Martinez, D.: Improving parsing and PP attachment performance with sense information. In: ACL HLT, pp. 317–325 (2008)
2. Belinkov, Y., Lei, T., Barzilay, R., Globerson, A.: Exploring compositional architectures and word vector representations for prepositional phrase attachment. TACL **2**, 561–572 (2014)
3. Chang, A.X., Monroe, W., Savva, M., Potts, C., Manning, C.D.: Text to 3D scene generation with rich lexical grounding. In: ACL-IJCNLP:2015, pp. 53–62 (2015)
4. Christie, G., Laddha, A., Agrawal, A., et al.: Resolving language and vision ambiguities together: joint segmentation & prepositional attachment resolution in captioned scenes. In: EMNLP, pp. 1493–1503 (2016)
5. Coco, M.I., Keller, F.: The interaction of visual and linguistic saliency during syntactic ambiguity resolution. QJEP **68**(1), 46–74 (2015)
6. Coyne, R., Sproat, R.: WordsEye: an automatic text-to-scene conversion system. In: SIGGRAPH, pp. 487–496 (2001)
7. Dasigi, P., Ammar, W., Dyer, C., Hovy, E.: Ontology-aware token embeddings for prepositional phrase attachment. In: ACL, vol. 1, pp. 2089–2098 (2017)
8. Delecraz, S., Nasr, A., Bechet, F., Favre, B.: Correcting prepositional phrase attachments using multimodal corpora. In: IWPT, pp. 72–77 (2017)
9. Delecraz, S., Nasr, A., Béchet, F., Favre, B.: Adding syntactic annotations to flickr30k entities corpus for multimodal ambiguous prepositional-phrase attachment resolution. In: LREC (2018)
10. Faghri, F., Fleet, D.J., Kiros, J.R., Fidler, S.: Vse++: Improved visual-semantic embeddings. arXiv preprint arXiv:1707.05612 (2017)
11. Fang, H., Gupta, S., Iandola, F.N., et al.: From captions to visual concepts and back. In: CVPR, pp. 1473–1482 (2015)

12. Freund, Y., Schapire, R., Abe, N.: A short introduction to boosting. JSAI **14**(771–780), 1612 (1999)
13. Girshick, R.: Fast R-CNN. In: ICCV, pp. 1440–1448 (2015)
14. Girshick, R., Donahue, J., Darrell, T., Malik, J.: Rich feature hierarchies for accurate object detection and semantic segmentation. In: CVPR, pp. 580–587 (2014)
15. He, K., Zhang, X., Ren, S., Sun, J.: Deep residual learning for image recognition. In: CVPR, pp. 770–778 (2016)
16. Karpathy, A., Fei-Fei, L.: Deep visual-semantic alignments for generating image descriptions. In: CVPR, pp. 3128–3137 (2015)
17. Kingma, D.P., Ba, J.: Adam: a method for stochastic optimization. arXiv preprint arXiv:1412.6980 (2014)
18. de Kok, D., Hinrichs, E.W.: Transition-based dependency parsing with topological fields. In: ACL, vol. 2: short paper (2016)
19. de Kok, D., Ma, J., Dima, C., Hinrichs, E.: PP attachment: Where do we stand? In: EACL, vol. 2, pp. 311–317 (2017)
20. Kummerfeld, J.K., Hall, D.L.W., Curran, J.R., Klein, D.: Parser showdown at the wall street corral: an empirical investigation of error types in parser output. In: EMNLP-CoNLL, pp. 1048–1059 (2012)
21. Lin, T.-Y., et al.: Microsoft COCO: common objects in context. In: Fleet, D., Pajdla, T., Schiele, B., Tuytelaars, T. (eds.) ECCV 2014. LNCS, vol. 8693, pp. 740–755. Springer, Cham (2014). https://doi.org/10.1007/978-3-319-10602-1_48
22. Marcus, M.P., Marcinkiewicz, M.A., Santorini, B.: Building a large annotated corpus of English: the penn treebank. Comput. Linguist. **19**(2), 313–330 (1993)
23. Mirroshandel, S.A., Nasr, A.: Integrating selectional constraints and subcategorization frames in a dependency parser. Comput. Linguist. **42**, 55–90 (2016)
24. Nasr, A., Béchet, F., Rey, J.F., Favre, B., Le Roux, J.: MACAON: an NLP tool suite for processing word lattices. In: ACL HLT, pp. 86–91 (2011)
25. Peyre, J., Laptev, I., Schmid, C., Sivic, J.: Weakly-supervised learning of visual relations. In: ICCV, pp. 5189–5198 (2017)
26. Plummer, B.A., Wang, L., Cervantes, C.M., Caicedo, J.C., Hockenmaier, J., Lazebnik, S.: Flickr30k entities: collecting region-to-phrase correspondences for richer image-to-sentence models. ICCV **123**(1), 74–93 (2017)
27. Redmon, J., Divvala, S., Girshick, R., Farhadi, A.: You only look once: unified, real-time object detection. In: CVPR, pp. 779–788 (2016)
28. Redmon, J., Farhadi, A.: Yolo9000: better, faster, stronger. In: CVPR, pp. 6517–6525. IEEE (2017)
29. Russakovsky, O., Deng, J., Su, H., et al.: ImageNet large scale visual recognition challenge. IJCV **115**(3), 211–252 (2015)
30. Shaerlaekens, A.: The Two-Word Sentence in Child Language Development: A Study Based on Evidence Provided by Dutch-Speaking Triplets. Mouton, The Hague (1973)
31. Snow, C.E.: Mothers' speech to children learning language. Child Dev. **43**(2), 549–565 (1972)
32. Spivey, M.J., Tanenhaus, M.K., Eberhard, K.M., Sedivy, J.C.: Eye movements and spoken language comprehension: effects of visual context on syntactic ambiguity resolution. Cogn. Psychol. **45**(4), 447–481 (2002)
33. Vinyals, O., Toshev, A., Bengio, S., Erhan, D.: Show and tell: a neural image caption generator. In: CVPR, pp. 3156–3164 (2015)
34. Young, P., Lai, A., Hodosh, M., Hockenmaier, J.: From image descriptions to visual denotations: new similarity metrics for semantic inference over event descriptions. TACL **2**, 67–78 (2014)

# Meeting Summarization, A Challenge for Deep Learning

Francois Jacquenet(✉), Marc Bernard, and Christine Largeron

Univ Lyon, UJM-Saint-Etienne, CNRS, Institut d'Optique Graduate School, Laboratoire Hubert Curien UMR 5516, 42023 Saint-Etienne, France
{Francois.Jacquenet,Marc.Bernard,Christine.Largeron}@univ-st-etienne.fr

**Abstract.** Text summarization is one of the challenges of Natural Language Processing. Given the volume of texts produced daily on the Internet, managers can no longer have an exhaustive reading of current events, or progress reports from their employees, etc. They urgently need tools to automatically produce a summary of this flow of information. As a first approach, extractive summarization tools have been produced and there are now commercial tools available. However, this family of systems is not well suited to certain types of texts such as written transcriptions of dialogues or meetings. In that case, abstractive summarization tools are needed. Research in that field is very old but has been particularly stimulated since the mid-2010s by the recent successes of deep learning. This paper presents a short survey of deep learning approaches to abstractive text summarization and then highlights the various challenges that will have to be solved in the coming years to deal with meeting summaries in order to be able to provide a text summarization tool that generates good quality summaries.

**Keywords:** Text summarization · Deep learning · Meeting summarization · System evaluation

## 1 Introduction

Several studies have shown that a large part of employees' working time is spent attending meetings and that 50% of meeting time is unproductive with up to 25% spent discussing irrelevant issues. 69% of participants felt that these meetings were unproductive and unnecessary, and worst, results indicated that 9 out of 10 people daydreamed in meetings. Romano and Nunamaker proposed an interesting analysis of meetings [24]. Their conclusion were that meetings are costly in terms of money and time and are in general unproductive and wasteful. In fact many meetings end without any clear summary of what has been said during the meetings and moreover what has been decided. However, it is

This work is being carried out as part of the REUS project funded under the FUI 22 by BPI France, the Auvergne Rhône-Alpes Region and the Grenoble metropolitan area, with the support of the competitiveness clusters Minalogic, Cap Digital and TES.


I. Rojas et al. (Eds.): IWANN 2019, LNCS 11506, pp. 644–655, 2019.
https://doi.org/10.1007/978-3-030-20521-8_53

not possible to prohibit any meeting within an organisation. It means we need to design techniques to automate the supervision of meetings. This has been proposed for example in [25] that suggested to use machine learning and data mining techniques to help manage meetings. A huge research effort has also been deployed through various European projects to work around the concept of smart meeting rooms. The IM2 project[1], the AMI project[2] and it's sequel, the AMIDA project aim to design, among other things, tools to assist people during meetings such as finding information from a recorded meeting, finding documents generated from a previous meeting, generating minutes, etc.

The REUS project, which supports this work, aims to provide tools to help people manage meetings and particularly automatically generate summaries at the end of meetings. Compared to those huge european projects that aim to equip meeting rooms with sophisticated and costly sensors, the REUS project aims to automatically gather information from meetings with lightweight material, that is just the microphone of each meeting's participant laptop or smartphone. In this paper we will only focus on the generation of minutes of meetings, based on text summarization techniques and the difficulties attached to that part of the project.

In the domain of text summarization [23,34], we can first consider two main approaches: *mono-document* (respectively *multi-document*) *summarization* where the goal is to produce a summary of one (respectively several) document(s). In this paper we only consider mono-document summarization as we want to summarize one meeting at a time. Now, there is again two main approaches. The first and oldest one, which is called *Abstractive Text Summarization*, aims to understand the input text and build a structured and/or semantic representation, then extract the most essential components of that representation to finally generate an output text from this condensed representation. The second approach, called *Extractive Text Summarization*, aims to discover the most relevant top-k sentences from a text, k being a parameter of the system. Hence, an extractive summary of a text is just the set of those top-k sentences. The difference between those two approaches may be illustrated by the following example. Consider the following input text:

> *"The raven seized a piece of cheese and carried his spoils up to his perch high in a tree. A fox came up and walked in circles around the raven, planning a trick. 'What is this?' cried the fox. 'O raven, the elegant proportions of your body are remarkable, and you have a complexion that is worthy of the king of the birds! If only you had a voice to match, then you would be first among the fowl!' The fox said these things to trick the raven and the raven fell for it: he let out a great squawk and dropped his cheese. By thus showing off his voice, the raven let go of his spoils. The fox then grabbed the cheese and said, 'O raven, you do have a voice, but no brains to go with it!'*
> *If you follow your enemies' advice, you will get hurt. "*

[1] http://www.im2.ch/.
[2] http://www.amiproject.org/.

An extractive summarizer will just extract from the text some sentences it considers as crucial. For example:

> *"A fox came up and walked in circles around the raven, planning a trick. He let out a great squawk and dropped his cheese. If you follow your enemies' advice, you will get hurt."*

An ideal abstractive summarizer would understand the text and generate a new shorter one:

> *"A raven carrying a cheese met a fox. The fox complimented the crow who dropped his cheese. The fox grabbed it and left."*

As we can see, an abstractive summary is more informative than an extractive one. Moreover, in some cases, texts generated by an extractive approach may be inconsistent due to the lack of anaphora resolution (for example in the extractive summary above, in the sentence: 'He let out ...' we don't know if 'He' refers to the fox or to the raven).

In the 70's, a major research effort has been dedicated to knowledge representation, text understanding, inference, and generation, that is, various domains related to semantics. Some concepts were born from this research at that time such as schemes, frames, conceptual dependencies, scripts [39,50,51]. So at that time, people working on text summarization tried to use such concepts to build a semantic representation of the text to be summarized, then work on this representation and finally generate a new text, that is, they focused on abstractive approaches. Nevertheless, given the state of the art in computer hardware at that time, this approach was very inefficient.

In the 90's, with the advent of digital libraries, machine learning, and statistical natural language processing [37], extractive text summarization has been put forward and now we can even find commercial tools based on that technique.

Nevertheless, in the context of meeting summarization, extractive summarization is not a good approach. Indeed, given the fact that the text of a meeting transcription is a dialogue between people attending this meeting, extractive summarization can lead to misinterpretations. Consider for example the following extract of a meeting transcription:

> *Bart: I think that John should write a small report by next Friday.*
> *Mary: No reason to limit John to writing a small report.*
> *Bart: ok, so John should write a report by next Friday.*
> *John: I already have to prepare a presentation for Friday could we postpone this to Monday 12th?*
> *Bart: ok no problem.*
> *Bart: By the way, John, could you fix the computer for Thursday?*
> *John: I can do this for Thursday without any problem.*

An extractive summarizer could for example generate the following summary:

> *Bart: I think that John should write a small report by next Friday.*
> *John: I can do this for Thursday without any problem.*

We can see that this summary overlooks the fact that John has to fix the computer and also makes it look as if John has agreed to write a report by Thursday when he's planning to do it by Monday 12th.

In fact, Vodolazova et al. have shown [62] that in the domain of extractive text summarization, we cannot use the same technique for any text. Indeed extractive sumarization can be well suited to summarize news to get a compact summary of information of the day, but as we have seen in the example above, it is absolutely not suited to summarize texts made up of dialogues. For that kind of texts we need to consider abstractive text summarization techniques.

In the next section we first present the datasets that are commonly used to develop research in the domain of text summarization. We then focus on deep learning techniques designed nowadays to build abstractive text summarizer. Finally, we present the challenges faced by deep learning to address text summarization in the context of meeting summarization.

## 2 Datasets for Text Summarization

Most approaches in machine learning for text summarization are supervised approaches. As we only consider the mono-document summarization approaches, the datasets considered for that task are made up of pairs of texts with their associated human-generated summaries. Historically, the first main dataset that has been used to advance research in text summarization was the DUC dataset [44] that contains 500 news articles from the New York Times and Associated Press. For each article, 4 summaries have been produced by some humans. Nevertheless, this dataset is too small to train deep learning prototypes so researchers have then used the annotated GigaWord dataset [42] which contains approximately 4 million texts, each of them averaging 31.4 words and the associated abstract 8.2 words. This dataset has been used to design summarizers that produced very short abstracts (one sentence) from one long sentence. More recently, researchers have used the CNN/Daily Mail corpus made up of news from CNN and the Daily Mail journal and for each news, a human generated bullet list of the highlights of the news. This dataset has been built by Hermann et al. [20] and adapted to the text summarization task by Nallapati et al. [41]. This corpus contains longer texts and abstracts than the GigaWord dataset as it is made up of approximately 300,000 news and the average length of each news is 775 words and that of the abstracts is 48 words.

Less popular at the moment is the dataset built by Cohan et al. [11] that has been developed to consider summarization techniques on even longer texts than news. It has been built using the arxiv and pubmed repositories and contains approximately 350,000 scientific texts with an average length of 4200 words. It is almost certain that this dataset will become popular in the coming years because summarizing news is certainly interesting but summarizing long texts is even more interesting as we have less time to read them and therefore a summary is even more useful.

In the context of meeting summarization, only two datasets are available, these are the AMI and ICSI Meeting corpora. The ICSI Meeting Corpus [22] consists of 75 meetings recorded in a conference room at the International Computer Science Institute in Berkeley between the years 2000–2002. This corpus has been developped in the context of the Meeting Project at ICSI [40].

The AMI corpus [6] has been design in the context of the European AMI project. It consists of 142 meetings recorded in a smart meeting room while people artificially met to design new objects.

Those corpora are not only dedicated to text summarization, they contain many other textual data but also videos and speech recordings. Concerning text summarization, AMI and ICSI both contain human-generated transcriptions of meetings (which ensure a good quality of the texts) and human-generated extractive and abstractive summaries. Obviously, what makes those corpora original and interesting to handle is that they contain dialogues between several people attending the same meetings. In fact, the corpora contain manually annotated dialogue acts of each meeting.

Technically, it's important to note that those two corpora are not made up of raw texts but unfortunately of XML Nite format files which makes them difficult to handle rapidly.

## 3 Deep Learning Approaches for Abstractive Text Summarization

So, in fact, although less commercially mediatized, research in abstractive text summarization has never stopped as we can read in [19]. Nevertheless, we are far from being able to produce commercial tools based on that technique.

From 2015, with the strong development of deep learning, abstractive text summarization techniques have seen a renewed interest and high expectations of progress. The first research using deep learning to build an abstractive text summarizer is the one of Rush et al. in [49]. They used the GigaWord dataset which means they focused on the generation of headlines from short texts. Their system uses a convolutional neural network to encode the input texts and a context-sensitive attentional feed-forward neural network to generate the associated headlines. Also based on GigaWord, Nallapati et al. proposed in [41] to use the Seq2Seq model presented in [57] and generally used in the context of machine translation. They proposed an encoder-decoder RNN with attention mechanisms [3] and introduced the CNN/Daily Mail dataset in that context. A lot of other research has been done using the GigaWord dataset such as [9,17,58,65], nevertheless we do not detail them here because generating headlines from small texts is not very useful in the context of meeting summarization.

More interestingly, many people began to use the CNN/Daily Mail dataset when it has been produced by Nallapati et al. in 2016. For example, See et al. [52] introduced a copy-mechanism based on [61] to avoid repeated and unknown words in summaries. Gehrmann et al. [16] extended their work to improve the content selection process. Kryscinski et al. [26] proposed to improve the abstractiveness of the summaries by combining, in the decoder, a contextual network that retrieve the relevant parts of the input text and a language model that incorporates prior knowledge about language generation. Incorporating prior knowledge is also what has been done in [1] that integrated entity commonsense representation in the decoding step of the Seq2Seq model. Li et al. considered structural compression and structural coverage to incorporate some structural properties of texts and summaries in their prototype [30]. Tan et al. [60] tried to overcome the weaknesses of the encoding-decoding approaches to deal with long sequences. To do so, they proposed to integrate a graph-based attentional mechanism in the Seq2Seq framework. Liu et al. used for the first time a Generative Adversarial Network (GAN) approach [33] and get results approaching the state of the art in deep learning for text summarization. Fernandes et al. proposed in [15] to introduce some linguistic structure such as named entities and entity coreferences and combine a sequential encoder with a graph neural network.

In many studies, the general strategy is to combine extractive and abstractive approaches. They use an extractive approach to first select the most interesting sentences and then apply an abstractive approach to generate an abstractive summary from those filtered sentences. This is the case, for example, of the works presented in [5,21,31,54].

Deep reinforcement learning has also been recently used in the context of abstractive summarization. Paulus et al. focused on avoiding the problem of repeating sentences [47] by introducing rewards from policy gradient reinforcement learning. Pasunuru and Bansal used reinforcement learning in [46] with the objective of ensuring that the salient information of the input text is selected, that an abstract is logically entailed from the input text and no redundancy is generated. To do this, they designed two rewards functions: "ROUGESal" and "Entail". In a similar way to this work, Chen and Bansal [8] designed a two step approach: they first extracted the salient sentences and then used a policy based reinforcement learning approach to rewrite them abstractively.

With the availability of the AMI/ICSI corpora in the early 2000s, work on meeting summarization first focused on extractive summarization. Since the early 2010s, researchers are more interested in abstractive summarization such as in [4,38,45,63] nevertheless no one has yet used a deep learning approach to generate abstractive meeting summaries based on the AMI/ICSI corpora so this remains an open research domain.

To evaluate the performance of a text summarization tool, the most popular measure is the ROUGE measure proposed by Lin and Hovy in [32] that measures the quality of an abstract with respect to a given gold standard abstract. Even if other measures have been designed, see [35] for an exhaustive presentation of the measures designed until now, ROUGE remains today the measure used in all the papers in the field. ROUGE is made up of several measures but the three main measures are: ROUGE-1, ROUGE-2 and ROUGE-L whose values are percentages. Given an abstract to be evaluated and a gold standard abstract, ROUGE-1 (respectively ROUGE-2) measures the overlap of words (respectively bigrams) between those two abstracts. ROUGE-L is based on the Longest Common Subsequence measure between the abstract to be evaluated and the gold standard abstract.

At the moment, for the best abstractive summarizers we have ROUGE-1 $\approx$ 42%, ROUGE-2 $\approx$ 19% and ROUGE-L $\approx$ 38%. Despite the significant amount of work done in the last three years in this area, the various prototypes developped in recent years do not have performances that exceed the ROUGE measure values given above. If we consider most of the papers whose experiments are based on the CNN/Daily Mail dataset, the improvement remains quite low and it seems difficult to make a significant step forward today. In fact, perhaps the ROUGE measure is not well suited to assess the quality of an abstract. We talk about that, among others, in the next section.

# 4 Challenges for Meeting Summarization

## 4.1 Issues with Datasets

As previously mentioned, little work has been done in the context of summarizing long texts. However, meeting transcriptions are quite often (very) long texts. Thus, techniques that work well on news may no more work on very long texts due to the fact that the Seq2Seq/LSTM/Attention frameworks have some difficulties to take into account long dependencies. People wishing to work on meeting summarization should consider datasets of long texts to design their prototypes. The effort of Cohan et al. that built the dataset based on the arxiv and pubmed repositories [11] is a first step to develop research in long text summarization.

In fact, few large adequate datasets are available, which is a bottleneck for the development of text summarization techniques based on deep learning. As explained in [13], at the moment each dataset has its own format and using a dataset requires an investment for preprocessing the data that cannot be mutualized over the various datasets. Moreover, in the domain of meeting summarization, the only datasets available (AMI and ICSI) are too small to be used in a deep learning context. A first solution would be to manually enlarge those datasets but this is not a realistic solution because the human effort is far too important. In addition, AMI and ICSI are in English and developing new corpora for other languages would be far too expensive.

A first solution can be to *generate synthetic datasets* with GAN approaches in a similar way they have been used for text generation for example in [18,55] and more specifically for dialogue generation in [29]. A big challenge is that the generated texts must be very long consistent dialogues which is not studied at the moment.

A second solution, which may be complementary to the first, is to use *unsupervised summarization techniques*. That eliminates the problem of having to provide human-written abstracts associated with each text in the training set. [10] proposed such a solution in the case of multi-document summarization and [53] for the specific case of meeting summarization. We can also have a look at recent work on *unsupervised machine translation* [2,27] that provides similar results to supervised neural machine translation without the need of parallel corpora.

To overcome the variety of natural languages, as well as the different fields covered by the texts in the datasets, we can work on *transfer learning techniques* [59]. Having learned a summarizer of english spoken meetings (respectively meetings about chemistry for example), we can use transfer learning to adapt the model to french speaking meetings (respectively meetings about marketing for example). This is a difficult task and a significant effort have to be made in this area.

### 4.2 Issues with Noise

Nowadays, speech processing techniques are very efficient in a quiet environment, without simultaneous speakers, which is the case, for example, in the context of TV news. However, this is not the case in the context of meeting transcriptions where the quality of texts by a speech-to-text tool can be very poor. Of course, in the REUS project, one task is dedicated to improving the speech-to-text techniques in the context of meetings. However, it may take time before speech recognition techniques achieve the same performance on meetings as on TV news. This means that we will still have to deal with noisy transcriptions, that is texts with missing words, syntactic and semantic inconsistencies, words out of the scope of the topic of the meeting. This is a new situation for current deep learning techniques that are presently working on relatively clean datasets.

### 4.3 Issues with Evaluation

Until now the performance of prototypes has been measured using the popular ROUGE measure [32]. Nevertheless it has been recognized that ROUGE is not a really good measure for abstractive text summarization [12,43]. Indeed, instead of processing the percentage of common sub-sequences, using a semantic similarity between texts seems much more appropriate. We could explore, for example, the use of sentence embeddings [28,56], universal sentence encoder [7], etc.

### 4.4 Issues with Content Understanding

We finally believe that in order to go one step further in the performance of abstractive text summarization techniques, we need to add more prior knowledge into the deep learning architectures. We can identify two possible approaches to address this problem. First, we can work on a semantic representation of the texts instead of just working on raw texts. In fact, integrating word embeddings already embed semantics, but we think we need to go further considering, for example, semantic structures such as semantic graphs or conceptual graphs. Looking at knowledge graph embedding techniques [64] will be useful for that purpose.

Second, we think it's important to be able to do some reasoning on what we understand from the text to be summarized. In that context, looking at the work that tries to combine for example logic programming and deep learning may lead to improve abstractivness in abstractive text summarization. For example, end-to-end differentiable proving [48], deep neural machines [14], neural probabilistic logic programming [36] are techniques that certainly deserve attention.

## 5 Conclusion

As we have seen in this paper, to summarize meeting transcriptions, it is essential to use an abstractive approach. Although deep learning has contributed to the renewal of abstractive text summarization, there are still many problems to be solved. We listed four main issues in the specific context of meeting summarization that concern the availability of datasets, the presence of noise in the meeting transcriptions, the inappropriateness of the ROUGE measure with the abstractive approach and the need for a deep understanding of the text to be summarized, which requires reasoning skills during the generation steps of the summaries. We believe that no significant improvement of abstractive summarization techniques based on deep learning can be achieved without addressing those issues.

## References


1. Amplayo, R.K., Lim, S., Hwang, S.-w.: Entity commonsense representation for neural abstractive summarization. In: Proceedings of ACL, pp. 697–707. ACL (2018)
2. Artetxe, M., Labaka, G., Agirre, E.: Unsupervised statistical machine translation. In: Proceedings of EMNLP, pp. 3632–3642. ACL (2018)
3. Bahdanau, D., Cho, K., Bengio, Y.: Neural machine translation by jointly learning to align and translate. CoRR, abs/1409.0473 (2014)
4. Banerjee, S., Mitra, P., Sugiyama, K.: Generating abstractive summaries from meeting transcripts. In: Proceedings of the 2015 ACM Symposium on Document Engineering, pp. 51–60. ACM (2015)
5. Cao, Z., Li, W., Li, S., Wei, F.: Retrieve, rerank and rewrite: soft template based neural summarization. In: Proceedings of ACL, pp. 152–161. ACL (2018)
6. Carletta, J., et al.: The AMI meeting corpus: a pre-announcement. In: Renals, S., Bengio, S. (eds.) MLMI 2005. LNCS, vol. 3869, pp. 28–39. Springer, Heidelberg (2006). https://doi.org/10.1007/11677482_3
7. Cer, D., et al.: Universal sentence encoder. CoRR, abs/1803.11175 (2018)
8. Chen, Y.-C., Bansal, M.: Fast abstractive summarization with reinforce-selected sentence rewriting. In: Proceedings of ACL, pp. 675–686. ACL (2018)
9. Chopra, S., Auli, M., Rush, A.M.: Abstractive sentence summarization with attentive recurrent neural networks. In: The 2016 Conference of the North American Chapter of the Association for Computational Linguistics: Human Language Technologies, pp. 93–98. ACL (2016)
10. Chu, E., Liu, P.J.: MeanSum: a neural model for unsupervised multi-document abstractive summarization. CoRR, abs/1810.05739 (2018)
11. Cohan, A., et al.: A discourse-aware attention model for abstractive summarization of long documents. In: Proceedings of ACL, pp. 615–621. ACL (2018)
12. Cohan, A., Goharian, N.: Revisiting summarization evaluation for scientific articles. In: Proceedings of LREC (2016)

13. Dernoncourt, F., Ghassemi, M., Chang, W.: A repository of corpora for summarization. In: Proceedings of LREC (2018)
14. Dong, H., Mao, J., Lin, T., Wang, C., Li, L., Zhou, D.: Neural logic machines. In: Proceedings of ICLR (2019)
15. Fernandes, P., Allamanis, P., Brockschmidt, M.: Structured neural summarization. CoRR, abs/1811.01824 (2018)
16. Gehrmann, S., Deng, Y., Rush, A.M.: Bottom-up abstractive summarization. In: Proceedings of EMNLP, pp. 4098–4109. ACL (2018)
17. Guo, H., Pasunuru, R., Bansal, M.: Soft layer-specific multi-task summarization with entailment and question generation. In: Proceedings of ACL, pp. 687–697. ACL (2018)
18. Guo, J., Lu, S., Cai, H., Zhang, W., Yu, Y., Wang, J.: Long text generation via adversarial training with leaked information. In: Proceedings of AAAI, pp. 5141–5148. AAAI Press (2018)
19. Gupta, S., Gupta, S.K.: Abstractive summarization: an overview of the state of the art. Expert Syst. Appl. **121**, 49–65 (2019)
20. Hermann, K.M., et al.: Teaching machines to read and comprehend. In: Proceedings of NIPS, pp. 1693–1701 (2015)
21. Hsu, W.T., Lin, C.-K., Lee, M.-Y., Min, K., Tang, J., Sun, M.: A unified model for extractive and abstractive summarization using inconsistency loss. In: Proceedings of ACL, pp. 132–141. ACL (2018)
22. Janin, A., et al.: The ICSI meeting corpus. In: International Conference on Acoustics, Speech, and Signal Processing, pp. 364–367. IEEE (2003)
23. Jones, K.S.: Automatic summarising: the state of the art. Inf. Process. Manage. **43**(6), 1449–1481 (2007)
24. Romano Jr., N.C., Nunamaker Jr., J.F.: Meeting analysis: findings from research and practice. In: 34th Annual Hawaii International Conference on System Sciences. IEEE Computer Society (2001)
25. Kim, B., Rudin, C.: Learning about meetings. Data Min. Knowl. Disc. **28**(5–6), 1134–1157 (2014)
26. Kryscinski, W., Paulus, R., Xiong, C., Socher, R.: Improving abstraction in text summarization. In: Proceedings of EMNLP, pp. 1808–1817. ACL (2018)
27. Lample, G., Ott, M., Conneau, A., Denoyer, L., Ranzato, M.: Phrase-based & neural unsupervised machine translation. In: Proceedings of EMNLP, pp. 5039–5049. ACL (2018)
28. Le, Q.V., Mikolov, T.: Distributed representations of sentences and documents. In: Proceedings of ICML, pp. 1188–1196 (2014)
29. Li, J., Monroe, W., Shi, T., Jean, S., Ritter, A., Jurafsky, D.: Adversarial learning for neural dialogue generation. In: Proceedings of EMNLP, pp. 2157–2169. ACL (2017)
30. Li, W., Xiao, X., Lyu, Y., Wang, Y.: Improving neural abstractive document summarization with explicit information selection modeling. In: Proceedings of EMNLP, pp. 1787–1796. ACL (2018)
31. Li, W., Xiao, X., Lyu, Y., Wang, Y.: Improving neural abstractive document summarization with structural regularization. In: Proceedings of EMNLP, pp. 4078–4087. ACL (2018)
32. Lin, C.-Y., Hovy, E.H.: Automatic evaluation of summaries using n-gram co-occurrence statistics. In: Proceedings of the Human Language Technology Conference of the North American Chapter of the Association for Computational Linguistics. ACL (2003)

33. Liu, L., Lu, Y., Yang, M., Qu, Q., Zhu, J., Li, H.: Generative adversarial network for abstractive text summarization. In: Proceedings of AAAI, pp. 8109–8110. AAAI Press (2018)
34. Lloret, E., Palomar, M.: Text summarisation in progress: a literature review. Artif. Intell. Rev. **37**(1), 1–41 (2012)
35. Lloret, E., Plaza, L., Aker, A.: The challenging task of summary evaluation: an overview. Lang. Resour. Eval. **52**(1), 101–148 (2018)
36. Manhaeve, R., Dumancic, S., Kimmig, A., Demeester, T., De Raedt, L.: DeepProbLog: neural probabilistic logic programming. In: Proceedings of NIPS, pp. 3753–3763 (2018)
37. Manning, C.D., Schütze, H.: Foundations of statistical natural language processing. MIT Press, Cambridge (2001)
38. Mehdad, Y., Carenini, G., Tompa, F.W., Ng, R.T.: Abstractive meeting summarization with entailment and fusion. In: Proceedings of the European Workshop on Natural Language Generation, pp. 136–146. ACL (2013)
39. Minsky, M.: A framework for representing knowledge. In: The Psychology of Computer Vision. McGraw-Hill, New York (1975)
40. Morgan, N., et al.: The meeting project at ICSI. In: Proceedings of the First International Conference on Human Language Technology Research. Morgan Kaufmann (2001)
41. Nallapati, R., Zhou, B., dos Santos, C.N., Gülçehre, Ç., Xiang, B.: Abstractive text summarization using sequence-to-sequence RNNs and beyond. In: Proceedings of the 20th SIGNLL Conference on Computational Natural Language Learning, pp. 280–290. ACL (2016)
42. Napoles, C., Gormley, M.R., Van Durme, B.: Annotated gigaword. In: Proceedings of the Joint Workshop on Automatic Knowledge Base Construction and Web-scale Knowledge Extraction, pp. 95–100. ACL (2012)
43. Ng, J.-P., Abrecht, V.: Better summarization evaluation with word embeddings for ROUGE. In: Proceedings of EMNLP, pp. 1925–1930. ACL (2015)
44. Over, P., Dang, H., Harman, D.: DUC in context. Inf. Process. Manage. **43**(6), 1506–1520 (2007)
45. Oya, T., Mehdad, Y., Carenini, G., Ng, R.T.: A template-based abstractive meeting summarization: leveraging summary and source text relationships. In: Proceedings of the Eighth International Natural Language Generation Conference, pp. 45–53. ACL (2014)
46. Pasunuru, R., Bansal, M.: Multi-reward reinforced summarization with saliency and entailment. In: Proceedings of ACL, pp. 646–653. ACL (2018)
47. Paulus, R., Xiong, C., Socher, R.: A deep reinforced model for abstractive summarization. CoRR, abs/1705.04304 (2017)
48. Rocktäschel, T., Riedel, S.: End-to-end differentiable proving. In: Proceedings of NIPS, pp. 3791–3803 (2017)
49. Rush, A.M., Chopra, S., Weston, J.: A neural attention model for abstractive sentence summarization. In: Proceedings of EMNLP, pp. 379–389. ACL (2015)
50. Schank, R.C.: Conceptual dependency: a theory of natural language understanding. Cogn. Psychol. **3**(4), 552–631 (1972)
51. Schank, R.C., Abelson, R.P.: Scripts, Plans, Goals, and Understanding: An Inquiry into Human Knowledge Structures. Lawrence Erlbaum Associates, Hillsdale (1977)
52. See, A., Liu, P.J., Manning, C.D.: Get to the point: summarization with pointer-generator networks. In: Proceedings of ACL, pp. 1073–1083. ACL (2017)

53. Shang, G., et al.: Unsupervised abstractive meeting summarization with multi-sentence compression and budgeted submodular maximization. In: Proceedings of ACL, pp. 664–674. ACL (2018)
54. Song, S., Huang, H., Ruan, T.: Abstractive text summarization using LSTM-CNN based deep learning. Multimedia Tools Appl. **78**(1), 857–875 (2019)
55. Subramanian, S., Rajeswar, S., Sordoni, A., Courville, A.C., Trischler, A., Pal, C.: Towards text generation with adversarially learned neural outlines. In: Proceedings of NIPS, pp. 7562–7574 (2018)
56. Subramanian, S., Trischler, A., Bengio, Y., Pal, C.J.: Learning general purpose distributed sentence representations via large scale multi-task learning. In: Proceedings of ICLR (2018)
57. Sutskever, I., Vinyals, O., Le, Q.V.: Sequence to sequence learning with neural networks. In: Proceedings of NIPS, pp. 3104–3112 (2014)
58. Suzuki, J., Nagata, M.: Cutting-off redundant repeating generations for neural abstractive summarization. In: Proceedings of EACL, pp. 291–297 (2017)
59. Tan, C., Sun, F., Kong, T., Zhang, W., Yang, C., Liu, C.: A survey on deep transfer learning. In: Kůrková, V., Manolopoulos, Y., Hammer, B., Iliadis, L., Maglogiannis, I. (eds.) ICANN 2018. LNCS, vol. 11141, pp. 270–279. Springer, Cham (2018). https://doi.org/10.1007/978-3-030-01424-7_27
60. Tan, J., Wan, X., Xiao, J.: Abstractive document summarization with a graph-based attentional neural model. In: Proceedings of ACL, pp. 1171–1181. ACL (2017)
61. Vinyals, O., Fortunato, M., Jaitly, N.: Pointer networks. In: Proceedings of NIPS, pp. 2692–2700 (2015)
62. Vodolazova, T., Lloret, E., Muñoz, R., Palomar, M.: Extractive text summarization: can we use the same techniques for any text? In: Métais, E., Meziane, F., Saraee, M., Sugumaran, V., Vadera, S. (eds.) NLDB 2013. LNCS, vol. 7934, pp. 164–175. Springer, Heidelberg (2013). https://doi.org/10.1007/978-3-642-38824-8_14
63. Wang, L., Cardie, C.: Domain-independent abstract generation for focused meeting summarization. In: Proceedings of ACL, pp. 1395–1405. ACL (2013)
64. Wang, Q., Mao, Z., Wang, B., Guo, L.: Knowledge graph embedding: a survey of approaches and applications. IEEE Trans. Knowl. Data Eng. **29**(12), 2724–2743 (2017)
65. Zhou, Q., Yang, N., Wei, F., Zhou, M.: Selective encoding for abstractive sentence summarization. In: Proceedings of ACL, pp. 1095–1104. ACL (2017)

# Semantic Fake News Detection: A Machine Learning Perspective

Adrian M. P. Braşoveanu[1,2(✉)] and Răzvan Andonie[3]

[1] Electronics and Computers Department, Transilvania University of Braşov, Braşov, Romania
[2] MODUL Technology GmbH, Vienna, Austria
adrian.brasoveanu@modul.ac.at
[3] Computer Science Department, Central Washington University, Ellensburg, WA, USA
andonie@cwu.edu

**Abstract.** Fake news detection is a difficult problem due to the nuances of language. Understanding the reasoning behind certain fake items implies inferring a lot of details about the various actors involved. We believe that the solution to this problem should be a hybrid one, combining machine learning, semantics and natural language processing. We introduce a new semantic fake news detection method built around relational features like sentiment, entities or facts extracted directly from text. Our experiments show that by adding semantic features the accuracy of fake news classification improves significantly.

**Keywords:** NLP · Semantics · Relation Extraction · Deep learning

## 1 Introduction

Teaching an automated system to recognize fake news is a challenging task, especially due to its interdisciplinary nature. At a superficial level it is important to distinguish between satire and political weapons (or any other kind of weapons built on top of deceptive news) [4], but when examining a news item, it might help to deploy a varied Natural Language Processing (NLP) arsenal that includes sentiment analysis, Named Entity Recognition Linking and Classification (NERLC [12]), n-grams, topic detection, part-of-speech (POS) taggers, query expansion or relation extraction [34]. Quite often such tools are supported by large Knowledge Bases (KBs) like DBpedia [16], which collects data about entities and concepts extracted from Wikipedia. The extracted named entities and relations will be linked to such KBs whenever possible, whereas various sentiment aspects, polarity or subjectivity might be computed according to the detected entities. Features like sentiment, named entities or relations render a set of shallow meaning representations, and are typically called *semantic features.* In contrast, POS or dependency trees render *syntactic features.*


I. Rojas et al. (Eds.): IWANN 2019, LNCS 11506, pp. 656–667, 2019.
https://doi.org/10.1007/978-3-030-20521-8_54

The underlying assumption made by most models used for detecting fake news is that the title and style of an article are sufficient to identify it as fake news. This is mostly true for news that originate from verifiable bad sources, which is rarely the case anymore. Therefore, we think that taking a holistic approach, that includes a machine generated Knowledge Graph (KG) [20] of all the stakeholders involved in the various events we are interested in is absolutely needed. Such a holistic approach includes methods which can generate and learn graphs of entities associated to fake news.

Our contribution is a method used to integrate semantic features in the training of fake news classifiers. The goal is to show how to use semantic features to improve fake news detection. For this, we compute semantic features (sentiment analysis, named entities and relations) which will be added to a set of syntactic features (POS - part-of-speech and NPs - Noun Phrases) and to the features of the original input dataset. On the resulted augmented dataset we apply various classifiers, including Deep Learning (DL) models: Long-Short Term Memory (LSTM), Convolutional Neural Network (CNN), and Capsule Networks. For the Liar data set [32], using semantic features improves the fake news recognition accuracy on average by 5–10%.

The paper is organized as follows. Section 2 presents the most recent results in fake news recognition. Section 3 introduces our approach for building machine generated KGs for semantic fake news detection. Section 4 describes the experimental results. The paper is concluded in Sect. 5.

## 2 Related Work

An exploration of the fake news phenomena during more than a decade (2006–2017) was built around Twitter rumor cascade by a series of social scientists [31]. Multiple surveys (e.g., [26,35]) were focused on building various fake news classifications. Rubin [22] defined a set of criteria for creating a good text corpora for fake news detection, namely that (i) it should only contain verifiable facts, (ii) happened in a certain interval, and (iii) were reported using similar style though with various degrees of cultural influences. Any such corpora should only focus on text-only item, as they would be easier to process.

Most of the time, simply analyzing the text will not get us very far. Recent models incorporate some data about the networks (e.g., social media, organizations) through which the news was spread. Ruchansky [23] proposed a CSI model which stands for Capture, Score and Integrate, therefore combining information on the temporal activity of the users, their behavior, and a classifier. The 3HAN network [28] is a Hierarchical Attention Network (HAN) network with three layers used to examine different parts of articles.

A model for early detection of fake news based on news propagation paths is described in [17] and is based on a hybrid time-series classifier that contains both Recurrent Neural Networks (RNNs) and CNNs. Wu [33] assumed that intentional fake news are typically manipulated to look like real news. He built a classifier based on social media propagation pathways using LSTM-RNN and embeddings. Vo and Lee [30] took a different approach, focusing on the story told by fake news URLs and the co-occurence of various entities through such links.

A set of LSTMs was used for performing a multi-source multi-class fake news detection (or MMFD) in [14]. The advantage of this method is the multi-source fusion of the MMFD framework, since it can determine various degrees of fake news. The accuracy of the approach is not very high, but given the fact that it combines three large components (automated feature extraction, multi-source fusion and fakeness discrimination) it is promising. Aghakhani [2] showed that a Generative Adversarial Network (GAN) [8] can perform relatively well for detecting deceptive reviews.

A good review of the state-of-the-art DL applications in NLP, that also includes details about sentiment analysis or named entities extraction/classification, is [34].

# 3 Our Approach

In this section, we introduce our approach for semantic information extraction and then describe how we use the extracted information to classify fake news. We present techniques related to: metadata collection, extraction of relations, and inclusion of embeddings to neural classifiers.

Our main research question is: *what are the most useful semantic features for improving fake news detection?* Ideally, such features should be integrated into the neural models, whenever possible. Today, due to the cost of developing good semantic systems, some of these features might come from various external tools. The semantic features need to be selected according to the task and dataset at hand. If the task refers to the detection of fake news as spread by people via their statements, then the main entities we will be interested in might include people, organizations, locations and events.

In order to fully exploit the relations between the entities mentioned in a news statement, our procedure includes the following steps:

- **Metadata collection.** The first step is to simply collect the sentiment, entities and additional metadata available from third party tools.
- **Relation Extraction.** A second pass will collect both (i) the general relations found in a KG, and (ii) those computed from the current texts.
- **Embeddings.** Last step refers to the adaptation of various neural models (e.g., by adding a layer of embeddings) for improving fake news detection.

The features included in the last step will be only internal, whereas the features included on the other steps can also be external. The entire process is illustrated in Fig. 1.

The intuition behind the current data modeling that lead to the additional semantic features is that by adding extracted entities and making a clear distinction between direct and indirect speech, we can create the premises for more sophisticated analysis that may pinpoint the personal history of a speaker with

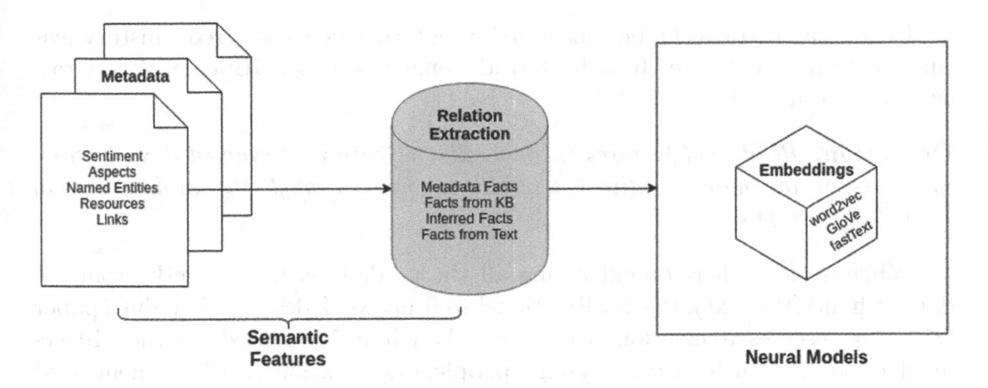


**Fig. 1.** External and internal semantic features for neural network models.

both the issue at hand (or subject), as well as with all the parties involved in the respective issue. If such an analysis is extended, down the road, it should also be possible to identify more obscure details about a speaker, for example if (s)he follows the party line or not. In other words, it opens up the possibility of using the graphs to peak behind the scenes of various declarations.

### 3.1 Fake News Detection and Knowledge Graphs

There are various definitions of fake news. Most of them refer to Alcott and Gentzkow's paper that examines the impact of fake news on the 2016 US Election [3].

**Definition (based on** [3]**).** *A news item or a part of a news item will be considered fake if it can be verified that its content is false.*

In order to perform semantic fake news detection, some additional statements like the past truth history of a speaker or the relations between speakers and publishers should be considered if possible. The idea of using past inaccuracies for each speaker was introduced with the Liar data set [32] and named credit history, but it is rarely used in practice.

**Definition (based on** [32]**).** *Credit History (CH) is the historical count of false (or provably untrue) statements for an actor.*

A credit history score can also be replaced by a single aggregated count of all the untrue values. Such credit scores allows us to understand diverse perspectives when analyzing news and helps determine which person or group might benefit from spreading certain news. An earlier iteration of this idea was also explored in the context of social media networks: credibility propagation [13].

**Definition.** *A credit history graph is a graph that contains all the entities, their credit histories and links between them as they are available from a Knowledge Graph (KG) or generated from a collection of texts.*

Relational features can be considered an alternative to the credit history features and can be extracted from both traditional KGs (e.g., DBpedia, Wikidata), as well as from text.

**Definition.** *Relational features include all the features extracted directly from the texts or the named entities detected in them through the exploitation of Knowledge Graphs.*

While we focus here on extracting all the needed features directly from the data at hand (the text), the Tri-Relationship framework described in Shu's paper [27] also deserves a mention here, even though it is focused on the objects involved in distributing the news (e.g., people, organizations). All the mentioned approaches share the idea of enriching the fake news text with a set of annotations, in order to provide some context.

### 3.2 Metadata Collection Pipeline

Our pipeline for generating metadata has following components:

- *Sentiment Analysis (SA).* Sentiment annotations can exist on multiple levels: (i) document; (ii) sentence; (iii) aspect-based [34]. Current state-of-the-art systems are typically aspect-based, therefore all the aspect of the entity features can get an estimate of the sentiment value. Since our data set (the Liar data) contains short statements, we use aggregated sentence level sentiment polarity and subjectivity values.
- *Named Entities (NE).* Since the results for NE extractions are typically good enough [12], almost any modern NLP library can be used for this task.
- *Named Entity Links (NEL).* Generally NERLC (NER+linking and classification) tasks are considered more complicated and typically require dedicated NEL engines [12]. Any good NEL engine can be used for this task. We use a wrapper built on top of DBpedia Spotlight [7].

### 3.3 Relation Extraction

Instead of using existing solutions, we develop a simple *Relation Extraction (REL)* component that queries DBpedia. Where possible, the existing entities are enriched with additional data obtained via a SPARQL query from DBpedia. This is particularly important in order to discover more relations between a speaker (which we will call *source entity*) and his/her subject (which we will call *target entity*). We consider two types of relations:

- (i) extracted directly from the provided news statements by defining the types of relations we are interested in via POS tags (for example, for extracting relations between two entities we will generally be interested in NP - V - NP chains - a verb between two proper nouns, whereas additional relations for an entity can be added by extracting S - V - O triplets);

- (ii) extracted from the DBpedia Knowledge Base (e.g., if dbr:Donald_Trump mentions dbr:Barack_Obama in a document, all the triples that belong to these entities are extracted from DBpedia and a subset of common links like dbo:orderInOffice or dbo:President is identified).

The machine generated KG includes all the DBpedia triples that belong to the entities collected from the data set. The relations extracted from text are schemaless, whereas the relations extracted from KG are grounded to a schema (e.g., DBpedia ontology). This component is implemented with the Python libraries RDFLib, SPARQLWrapper and Spacy[1].

### 3.4 Embeddings

Shallow neural architectures that learn word embeddings from distributional semantics (e.g., continuous bag of words architectures like Word2Vec, GloVe or fastText [19]) have been successfully applied to classic NLP problems [34], and should be an integral part of any NLP architecture. Such architectures generally provide fast computation times and lead to good results due to the fact that they capture relational similarities.

If the used corpora is clean and large enough (several tens of thousands of examples [19]), embeddings can be an ideal solution for building baselines. Only the most used (word2vec, GloVe, fastText) pre-computed embeddings were included for the top 60k English words. The component that loads them uses negative sampling and a fixed size of 300. The Keras API offers the possibility to add an embeddings layer to a neural network. This layer can be used for: *(i)* learning and saving the embeddings together with the word vectors; *(ii)* loading pre-trained embeddings. In all our DL models, we place such a layer after the inputs and use it for loading embeddings. Such a layer is effective especially when the number of training examples is relatively small [21].

## 4 Experiments

The success of our approach depends on a series of components for extracting sentiment scores, named entities, or relations. Therefore, if those components do not perform well, the whole approach will be flawed. First, we would like to find out if such an approach is valid. Therefore, missing a named entity from a statement might not be extremely important at this stage. If the approach proves to be valid, then further work needs to include additional evaluations for all the components in the pipeline, or at least some of their performance scores (when available).

We use the Liar data set [32] for our experiments. It contains politics-related articles classified based on the degree of truth, while also offering credit histories that tracks the accuracy of the speaker statements. The data set is split into three partitions (**train**, **test** and **validation**) and includes six classes that need

[1] https://spacy.io/.

to be predicted: **False**, **Barely-true**, **Half-true**, **Mostly-true**, **True**, **Pants on fire**. The initial paper about the Liar data set [32] identified SVMs as best classical models and CNNs as the best Deep Learning classifiers. A follow-up paper [18] indicates that LSTMs would be even better. Since our focus is not on credit history (five counts for all the classes that are not True including the score for the current statement) but on the impact of the relational features, we do not reproduce those results and do not compare with them.

**Table 1.** Accuracy for the test set runs on the Liar dataset. The best results are presented in **bold**. **T** stands for text, **A** for attributes and **R** for relations.

| Model | T | T+A | T+R | ALL |
|---|---|---|---|---|
| *Classic ML* | | | | |
| Multinomial Naive Bayes | 0.230 | 0.234 | 0.244 | **0.269** |
| SGDClassifier | 0.229 | 0.216 | 0.225 | **0.246** |
| Logistic regression (OneVsRest) | 0.226 | 0.240 | 0.249 | **0.262** |
| Random forest | **0.249** | 0.219 | 0.231 | 0.238 |
| Decision trees | 0.226 | 0.240 | 0.249 | **0.262** |
| SVM | 0.240 | 0.250 | 0.259 | **0.284** |
| *Deep Learning* | | | | |
| CNN | 0.260 | 0.274 | 0.280 | **0.290** |
| BasicLSTM | 0.225 | 0.255 | 0.265 | **0.324** |
| BiLSTM attention | 0.408 | 0.435 | 0.444 | **0.496** |
| GRU attention | 0.460 | 0.496 | 0.506 | **0.549** |
| CapsNet | 0.555 | 0.583 | 0.593 | **0.644** |

We consider four cases, as depicted in Table 1. The texts themselves (named **text (T)**) are simply statements that are taken out of their original context. The features included in the original data set (**text+attributes (T+A)**) contain information about the subject, speaker (including his job title, state and party affiliation), as well as credit history, and the context (the speech's location). The set **text+relations (T+R)** has semantic features (sentiment polarity, sentiment subjectivity, entities, links, and relations), syntactic features (NP), and the aggregated score of the credit history counts. The features included in the **T+R** data set are all extracted directly from the statements - there is no need to use the full text of the articles to compute them. This is an important detail, since this operation can always be performed if we have a good set of tools for metadata generation, even when the full articles are no available. The last set of features (identified as **all (ALL)**) includes all the previous features.

The classes are balanced and the split between train and test is 4:1. In Tables 1 and 2 we report the test set accuracy scores for all considered models and additional features.

**Table 2.** Accuracy for the test set runs using different combinations of semantic profile attributes (**T+R**). The best results are presented in **bold**.

| Feature | Acc |
|---|---|
| BasicLSTM (T+A) | 0.255 |
| Sentiment Polarity | 0.260 |
| Sentiment Subjectivity | 0.267 |
| NP | 0.273 |
| Entities | 0.284 |
| Links | 0.285 |
| Relations | 0.303 |
| CH | 0.324 |
| All | **0.324** |

We start by testing several "classic" models [10] that were built with scikit-learn (Table 1). For these models, using the relational features (**T+R**) shows some improvements, typically 2–3% above the original features (**T+A**) of the data set. However, the best score are far from optimal. Logistic regression and decision trees scores prove to be quite similar for all the three runs, while simultaneously being the worst scores. We notice a single case (the random forest classifier) in which the added relational features do not yield improvements over a run with only the original text. The best "classic" ML classifier proves to be the SVM, confirming the results from [18].

In the second phase, we test several DL models. The DL models are built with Keras [6] and TensorFlow [1], and use hot encoding of the class labels. For the DL models, the reported evaluation metric is accuracy with Adam optimizer [15].

The following DL classifiers are used:

- CNN - based on the model described in [18].
- BasicLSTM - a simple LSTM with a GlobalMaxPool layer, dropout set at 0.1 and dense layers;
- BiLSTM [5] - a bidirectional LSTM with attention, dropout and recurring dropout set at 0.1, which also includes an embeddings layer and the rest of the layers from the BasicLSTM;
- GRU [11] - a GRU with attention, otherwise similar to the previous BiLSTM model;
- CapsNetLSTM [24] - uses a Capsule layer instead of the GlobalMaxPool layer used in the other models.

All the DL models, besides CNN and BasicLSTM, use embeddings. We did not perform additional tuning of the DL models. We noticed that the embeddings for the most used 60k words from the English language have almost no effect on the results. The input vectors were loaded using Keras's embeddings layers which is defined as the first hidden layer of a network. For the DL experiments, we used whenever possible pre-trained models. Of course, fine-tuning the architectures may improve these results.

In all cases, relational features (**T+R**) perform better than the original features of the data set (**T+A**), which suggests that in some cases it might be enough to simply collect texts and build the rest of the features from metadata.

We note that all the DL models obtain better scores than the classic models with the same features. While the current literature is mostly focused on CNNs and basic LSTMs, we observe that attention models and CapsNet models performed best. For all DL models, adding our features results in an accuracy increase of up to 5–6%. This could be caused by the fact that the embeddings represent internal features of our DL models.

We have not repeated all feature combinations presented in Wang [32] and Long [18], but rather took the best feature combinations found in those papers and added new combinations based on the relational features proposed by us. The scores obtained obtained by us for SVMs, basic CNNs and LSTMs confirm their results. Using relational features (sentiment, recognized named entities, named entities links, relations) together with syntactic features (NP), it is already possible to beat the baselines at a comfortable distance, even without using advanced architectures. It is even possible to use only these semantic and syntactic features, instead of the original ones, and the scores will still be better than the baselines.

We tried to minimize the number of input features. Depending on the length of the text and number of entities involved, the number of additional features can be increased - which may lead to some increase in the overall performance. The most important thing when using our technique is to select the appropriate additional features that can lead to performance improvements. According to the results (Table 2), a good choice is to select relations, sentiments and entities.

## 5 Conclusions

While the literature on fake news detection is increasing at fast pace, the accuracy of the various models greatly varies depending on the data sets and the number of classes involved. In our view, good models should be adaptive and should not require a lot of fine-tuning on data sets. According to our results, by also considering relational features like sentiment, named entities or facts extracted from both structured (e.g., Knowledge Graphs) and unstructured data (e.g., text), we generally obtain better scores on most classifiers.

Currently, most models are based on word embeddings, even though phrases and multi-words expressions perform better for longer texts. This is due to the fact that the language used in a fake news article may differ from the language used in a normal article, as it is often needed to reinforce certain claims. Some future investigation areas include exploiting these relational features together with graph neural networks, like the recently developed R-GCN [25] or using a single multi-head attention architecture [29] to generate all the semantic features. Another interesting direction is to use semantic features for detecting fake reviews. While this is somewhat similar to the fake news detection, the goal here is to detect fake accounts on websites like TripAdvisor or fake authorships.

## References

1. Abadi, M., et al.: Tensorflow: a system for large-scale machine learning. CoRR abs/1605.08695 (2016). http://arxiv.org/abs/1605.08695
2. Aghakhani, H., Machiry, A., Nilizadeh, S., Kruegel, C., Vigna, G.: Detecting deceptive reviews using generative adversarial networks. CoRR abs/1805.10364 (2018). http://arxiv.org/abs/1805.10364
3. Allcott, H., Gentzkow, M.: Social media and fake news in the 2016 election. J. Econ. Perspect. **31**(2), 211–236 (2017)
4. Berghel, H.: Lies, damn lies, and fake news. IEEE Comput. **50**(2), 80–85 (2017). https://doi.org/10.1109/MC.2017.56
5. Chiu, J.P.C., Nichols, E.: Named entity recognition with bidirectional LSTM-CNNs. TACL **4**, 357–370 (2016). https://transacl.org/ojs/index.php/tacl/article/view/792
6. Chollet, F.: Deep Learning with Python. Manning Publications Co., Shelter Island (2017)
7. Daiber, J., Jakob, M., Hokamp, C., Mendes, P.N.: Improving efficiency and accuracy in multilingual entity extraction. In: Sabou, M., Blomqvist, E., Noia, T.D., Sack, H., Pellegrini, T. (eds.) I-SEMANTICS 2013–9th International Conference on Semantic Systems, ISEM 2013, Graz, Austria, 4–6 September 2013, pp. 121–124. ACM (2013). https://doi.org/10.1145/2506182.2506198
8. Goodfellow, I.J., et al.: Generative adversarial nets. In: Ghahramani, Z., Welling, M., Cortes, C., Lawrence, N.D., Weinberger, K.Q. (eds.) Advances in Neural Information Processing Systems 27: Annual Conference on Neural Information Processing Systems 2014, 8–13 December 2014, Montreal, Quebec, Canada, pp. 2672–2680 (2014). http://papers.nips.cc/paper/5423-generative-adversarial-nets
9. Guyon, I., et al. (eds.): Advances in Neural Information Processing Systems 30: Annual Conference on Neural Information Processing Systems 2017, 4–9 December 2017, Long Beach, CA, USA (2017)
10. Hastie, T., Tibshirani, R., Friedman, J.: The Elements of Statistical Learning. SSS. Springer, New York (2009). https://doi.org/10.1007/978-0-387-84858-7. http://www.worldcat.org/oclc/300478243
11. Irie, K., Tüske, Z., Alkhouli, T., Schlüter, R., Ney, H.: LSTM, GRU, highway and a bit of attention: an empirical overview for language modeling in speech recognition. In: Morgan, N. (ed.) Interspeech 2016, 17th Annual Conference of the International Speech Communication Association, San Francisco, CA, USA, 8–12 September 2016, pp. 3519–3523. ISCA (2016). https://doi.org/10.21437/Interspeech.2016-491
12. Ji, H., Nothman, J.: Overview of TAC-KBP2016 tri-lingual EDL and its impact on end-to-end KBP. In: Eighth Text Analysis Conference (TAC). NIST (2016). https://tac.nist.gov/publications/2016/additional.papers/
13. Jin, Z., Cao, J., Zhang, Y., Luo, J.: News verification by exploiting conflicting social viewpoints in microblogs. In: Schuurmans, D., Wellman, M.P. (eds.) Proceedings of the Thirtieth AAAI Conference on Artificial Intelligence, 12–17 February 2016, Phoenix, Arizona, USA, pp. 2972–2978. AAAI Press (2016). http://www.aaai.org/ocs/index.php/AAAI/AAAI16/paper/view/12128

14. Karimi, H., Roy, P., Saba-Sadiya, S., Tang, J.: Multi-source multi-class fake news detection. In: Bender, E.M., Derczynski, L., Isabelle, P. (eds.) Proceedings of the 27th International Conference on Computational Linguistics, COLING 2018, Santa Fe, New Mexico, USA, 20–26 August 2018, pp. 1546–1557. Association for Computational Linguistics (2018). https://aclanthology.info/papers/C18-1131/c18-1131
15. Kingma, D.P., Ba, J.: Adam: a method for stochastic optimization. CoRR abs/1412.6980 (2014). http://arxiv.org/abs/1412.6980
16. Lehmann, J., et al.: DBpedia - a large-scale, multilingual knowledge base extracted from wikipedia. Semantic Web **6**(2), 167–195 (2015). https://doi.org/10.3233/SW-140134
17. Liu, Y., Wu, Y.B.: Early detection of fake news on social media through propagation path classification with recurrent and convolutional networks. In: McIlraith, S.A., Weinberger, K.Q. (eds.) Proceedings of the Thirty-Second AAAI Conference on Artificial Intelligence, New Orleans, Louisiana, USA, 2–7 February 2018. AAAI Press (2018). https://www.aaai.org/ocs/index.php/AAAI/AAAI18/paper/view/16826
18. Long, Y., Lu, Q., Xiang, R., Li, M., Huang, C.: Fake news detection through multi-perspective speaker profiles. In: Kondrak, G., Watanabe, T. (eds.) Proceedings of the Eighth International Joint Conference on Natural Language Processing, IJCNLP 2017, Taipei, Taiwan, 27 November–1 December 2017, Volume 2: Short Papers, pp. 252–256. Asian Federation of Natural Language Processing (2017). https://aclanthology.info/papers/I17-2043/i17-2043
19. Mikolov, T., Grave, E., Bojanowski, P., Puhrsch, C., Joulin, A.: Advances in pre-training distributed word representations. In: Calzolari, N., et al. (eds.) Proceedings of the Eleventh International Conference on Language Resources and Evaluation, LREC 2018, Miyazaki, Japan, 7–12 May 2018. European Language Resources Association (ELRA) (2018). http://www.lrec-conf.org/lrec2018
20. Nickel, M., Murphy, K., Tresp, V., Gabrilovich, E.: A review of relational machine learning for knowledge graphs. Proc. IEEE **104**(1), 11–33 (2016). https://doi.org/10.1109/JPROC.2015.2483592
21. Qi, Y., Sachan, D.S., Felix, M., Padmanabhan, S., Neubig, G.: When and why are pre-trained word embeddings useful for neural machine translation? In: Walker, M.A., Ji, H., Stent, A. (eds.) Proceedings of the 2018 Conference of the North American Chapter of the Association for Computational Linguistics: Human Language Technologies, NAACL-HLT, New Orleans, Louisiana, USA, 1–6 June 2018, Volume 2 (Short Papers), pp. 529–535. Association for Computational Linguistics (2018). https://aclanthology.info/papers/N18-2084/n18-2084
22. Rubin, V., Conroy, N., Chen, Y., Cornwell, S.: Fake news or truth? using satirical cues to detect potentially misleading news. In: Proceedings of the Second Workshop on Computational Approaches to Deception Detection, pp. 7–17 (2016)
23. Ruchansky, N., Seo, S., Liu, Y.: CSI: a hybrid deep model for fake news detection. In: Lim, E., et al. (eds.) Proceedings of the 2017 ACM on Conference on Information and Knowledge Management, CIKM 2017, Singapore, 06–10 November 2017, pp. 797–806. ACM (2017). https://doi.org/10.1145/3132847.3132877
24. Sabour, S., Frosst, N., Hinton, G.E.: Dynamic routing between capsules. In: Guyon et al. [9], pp. 3859–3869. http://papers.nips.cc/paper/6975-dynamic-routing-between-capsules
25. Schlichtkrull, M., Kipf, T.N., Bloem, P., van den Berg, R., Titov, I., Welling, M.: Modeling relational data with graph convolutional networks. In: Gangemi, A., et al. (eds.) ESWC 2018. LNCS, vol. 10843, pp. 593–607. Springer, Cham (2018). https://doi.org/10.1007/978-3-319-93417-4_38

26. Shu, K., Sliva, A., Wang, S., Tang, J., Liu, H.: Fake news detection on social media: a data mining perspective. SIGKDD Explor. **19**(1), 22–36 (2017). https://doi.org/10.1145/3137597.3137600
27. Shu, K., Wang, S., Liu, H.: Exploiting tri-relationship for fake news detection. CoRR abs/1712.07709 (2017). http://arxiv.org/abs/1712.07709
28. Singhania, S., Fernandez, N., Rao, S.: 3HAN: a deep neural network for fake news detection. In: Liu, D., Xie, S., Li, Y., Zhao, D., El-Alfy, E.S. (eds.) ICONIP 2017. LNCS, vol. 10635, pp. 572–581. Springer, Cham (2017)
29. Vaswani, A., et al.: Attention is all you need. In: Guyon et al. [9], pp. 6000–6010. http://papers.nips.cc/paper/7181-attention-is-all-you-need
30. Vo, N., Lee, K.: The rise of guardians: fact-checking url recommendation to combat fake news. In: Collins-Thompson, K., Mei, Q., Davison, B.D., Liu, Y., Yilmaz, E. (eds.) The 41st International ACM SIGIR Conference on Research & Development in Information Retrieval, SIGIR 2018, Ann Arbor, MI, USA, 08–12 July 2018, pp. 275–284. ACM (2018). https://doi.org/10.1145/3209978.3210037
31. Vosoughi, S., Roy, D., Aral, S.: The spread of true and false news online. Science **359**(6380), 1146–1151 (2018)
32. Wang, W.Y.: Liar, Liar Pants on Fire: A New Benchmark Dataset for Fake News Detection. CoRR abs/1705.00648 (2017). http://arxiv.org/abs/1705.00648
33. Wu, L., Liu, H.: Tracing fake-news footprints: characterizing social media messages by how they propagate. In: Chang, Y., Zhai, C., Liu, Y., Maarek, Y. (eds.) Proceedings of the Eleventh ACM International Conference on Web Search and Data Mining, WSDM 2018, Marina Del Rey, CA, USA, 5–9 February 2018, pp. 637–645. ACM (2018). https://doi.org/10.1145/3159652.3159677
34. Young, T., Hazarika, D., Poria, S., Cambria, E.: Recent trends in deep learning based natural language processing [review article]. IEEE Comp. Int. Mag. **13**(3), 55–75 (2018). https://doi.org/10.1109/MCI.2018.2840738
35. Zannettou, S., Sirivianos, M., Blackburn, J., Kourtellis, N.: The web of false information: rumors, fake news, hoaxes, clickbait, and various other shenanigans. CoRR abs/1804.03461 (2018). http://arxiv.org/abs/1804.03461

# Unsupervised Inflection Generation Using Neural Language Modelling

Octavia-Maria Şulea[1,2,3](✉) and Steve Young[4]

[1] Faculty of Mathematics and Computer Science, University of Bucharest, Bucharest, Romania
[2] Human Language Technologies Research Center, University of Bucharest, Bucharest, Romania
[3] GumGum, Santa Monica, CA, USA
mary.octavia@gmail.com
[4] Pronto.AI, San Francisco, CA, USA
steve.young@pronto.ai

**Abstract.** The use of Deep Neural Network architectures for Language Modelling has recently seen a tremendous increase in interest in the field of NLP with the advent of transfer learning and the shift in focus from rule-based and predictive models (supervised learning) to generative or *unsupervised* models to solve the long-standing problems in NLP like Information Extraction or Question Answering. While this shift has worked greatly for langauges lacking in inflectional morphology, such as English, challenges still arise when trying to build similar systems for morphologically-rich langauges, since their individual words shift forms in context more often [8]. In this paper we investigate the extent to which these new *unsupervised* or *generative* techniques can serve to alleviate the type-token ratio disparity in morphologically rich languages. We apply an off-the-shelf neural language modelling library [20] to the newly introduced [9] task of unsupervised inflection generation in the nominal domain of three morphologically rich languages: Romanian, German, and Finnish. We show that this neural language model architecture can successfully generate the full inflection table of nouns without needing any pre-training on large, wikipedia-sized corpora, as long as the model is shown enough inflection examples. In fact, our experiments show that pre-training hinders the generation performance.

## 1 Introduction

Modelling variability, or the certain pattern a variable follows in context, is one of the major goals of Statistical Machine Learning and Data Science. In Artificial Intelligence, the notion whose variability researchers attempt to account for or simulate is the assumed concept of human intelligence. Within Natural Language Processing (NLP), where the variable scrutinized is language, this goal can surface into many different tasks, depending at which level of the language the analysis is being carried out. When we focus our attention at the word level,


I. Rojas et al. (Eds.): IWANN 2019, LNCS 11506, pp. 668–678, 2019.
https://doi.org/10.1007/978-3-030-20521-8_55

as researchers within Computational Morphology have done, we notice a clear distinction between langauges that have a high *stem variance*, meaning that the same word (stem) will shift its form often in the context of different sentences, or low *stem variance*, where a word will appear in the same form in most contexts. The ones with high stem variance, like Romanian, French, or German, have been dubbed langauges with *rich* inflectional morphology [10–12], while languages with low variance, like English, are considered to have a *poor* inflectional morphology. This distinction has made applying machine learning algorithms to NLP tasks like Named Entity Recognition for languages like English much easier than for languages like French or German [19], where a base form (uninflected, dictionary form) of a word can go through many phonological transformations (apophonies) when they merge with affixes (sub-word character clusters) denoting gender, person, number, case, definiteness to reflect the context (inflection) they are being used in. While great progress in text classification, information extraction, or language understanding tasks has been made through language modelling with deep neural networks and transfer learning [3,4], these models still suffer when the background information needed for the task is out of vocabulary or rarely occuring and require fine-tuning on labeled datasets [17].

Tables 1 and 2 show the stem alternations occuring in Romanian and German nouns respectively as opposed to English nouns which incur no shifts in the stem. The ending of the stem and beginning of the contextual marker (inflectional affix) is marked with a hyphen. Note that although Romanian and German both are rich in inflectional morphology they do not share the same contexts for which alternation occurs. Specifically, Romanian nouns do not sift from Nominative to Accusative or from Genitive to Dative, unlike German, but they do shift depending on whether they are being marked for definiteness, a notion that in English is marked by the presence of determiners (i.e. *the*).

**Table 1.** Stem alternations in the nominal domain for Romanian vs. English

| Tag | English | Romanian |
|---|---|---|
| N-Acc.sg.indef | (a) door- | p*oa*r*t*-ă |
| N-Acc.pl.indef | (a few) door-s | por*ț*-i |
| N-Acc.sg.def | (the) door- | p*oa*r*t*-a |
| N-Acc.pl.def | (the) door- | por*ț*-ii |
| G-D.sg.indef | (a) door-'s | por*ț*-i |
| G-D.pl.indef | (a few) door-s' | por*ț*-i |
| G-D.sg.def | (the) door-'s | por*ț*-ile |
| G-D.pl.def | (the) door-s' | por*ț*-ilor |

Over the last 10 years, there have been extensive efforts within the field of NLP to apply supervised learning to predict which inflectional class (morphological paradigm) words fall into: [2,8,10–13]. However, attempting to generate

**Table 2.** Stem alternations in the nominal domain for German vs. English

| Tag | English | German |
|---|---|---|
| N.sg | power- | m*a*cht- |
| N.pl | power-s | m*ä*cht-e |
| G.sg | power-'s | m*a*cht- |
| G.pl | power-s' | m*ä*cht-e |
| D.sg | power | m*a*cht- |
| D.pl | power-s | m*ä*cht-en |
| Acc.sg. | power- | m*a*cht- |
| Acc.pl | power-s | m*ä*cht-e |

all inflected forms of a word from its base form in a fully unsupervised manner, given no explicit information of the existing paradigms of a language, has only very recently been studied for the verbal domain of a few morphologically rich languages [9]. We would like to pursue this investigation further and see to what extent it works in the more variable nominal domain. Our interest in the task of unsupervised inflection generation introduced in [9] was sparked by the application of neural generative models to the related task of morphological reinflection [1,14,21] and by the recent availability of morphological datasets coming from Wikitionary [13] and the SIGMORPHON shared tasks [7].

We make use of the same attention-based recurrent neural architecture [20] applied in [9], which has become a popular plug-and-play text generation tool outside of the academic community[1] and investigate the best practices and limitations to generating the paradigm of a noun given only its uninflected (dictionary) form and being trained in a fully unsupervised fashion. While [9] report state-of-the-art results for Romanian verb generation and show their models to require no fine-tuning, we show that in fact fine-tuning hinders the ability of the model to generate any inflection list. This comes as an argument against the pre-trained and fine-tuned strategy advocated in the past few years by the transfer learning community within NLP [3,4,17]. We invite the NLP community to further test other neural models with this training scheme.

## 2 Datasets

For our experiments, we used a subset of the wikitionary corpus [13] and the dataset subset of the corpus introduced in [6] and used in [8], focusing on nominal inflection generation in German, Romanian, and Finnish. We chose the first two languages because they are known to display the morpho-phonlogical phenomenon of apophony, or stem alternations, and also represent two different language families: the latin and germanic. We also decided to look at Finnish

[1] airweirdness.com has used it extensively.

since it is a Fino-Ugric language highly unrelated to the other two and, while it does have a rich inflectional morphology, it displays a characteristic that is arguably the opposite of apophony - agglutination (the stem does not alter during inflectional affixation).

While the German and Finnish datasets contained no direct (labeled) information about inflectional class, for Romanian we were able to attain the labeled dataset introduced in [8] which associates the dictionary form of a noun (nominative-accusative singular indefinite form) with a number from 0 to 20 representing the inflectional class label which they identified based on various linguistic works [5,16] and which was successfully predicted by their proposed model using only character ngrams of the base form. This label is supposed to encode the pattern of inflectional endings the noun receives as well as the pattern of alternations the stem goes through during inflection. The experiments in [9] show that making use of this label (by conditioning the language model) leads to lower generation performance so we did not carry this line of research over to the nominal domain.

## 3 TextGenRNN Architecture

We adapt the character-level language model `textgenrnn` developed in [15,20], whose architecture is reviewed here for completeness.

Input sequences to the model are strings of up to $T$ characters. Each character in an input sequence is first translated into a 100-D embedding-vector. These are then fed through two bi-directional 128-unit LSTM layers. Next, the outputs of the embedding and both LSTM layers are concatenated and fed into an attention layer which weights the most important temporal features and averages them together. Note that this skip-connects the embedding and first LSTM layer to the attention layer, which helps alleviate vanishing gradients. Finally, the output of the attention layer is routed through a fully-connected MLP layer with output dimension equal to the number of possible characters.

## 4 Generation Experiments

The first set of experiments investigated the generation of full inflection given the dictionary form of the noun. Specifically, we wanted to see if the model could generate all forms of a noun in one go, given only the uninflected (dictionary) form. For Romanian this would be the nominative-accusative singular indefinite form (N-Acc.sg.indef in Table 1), whereas for German and Finnish this would be the nominative singular. To this end, the training datasets were arranged such that each row contained this baseform of the noun, which is the first form to appear in dictionaries, followed by all the inflected forms resulted from combining case (nominative, genitive, accusative, and dative for German; nominative-accusative, genitive-dative for Romanian; 14 for Finnish), number (i.e. singular, plural) and definiteness.

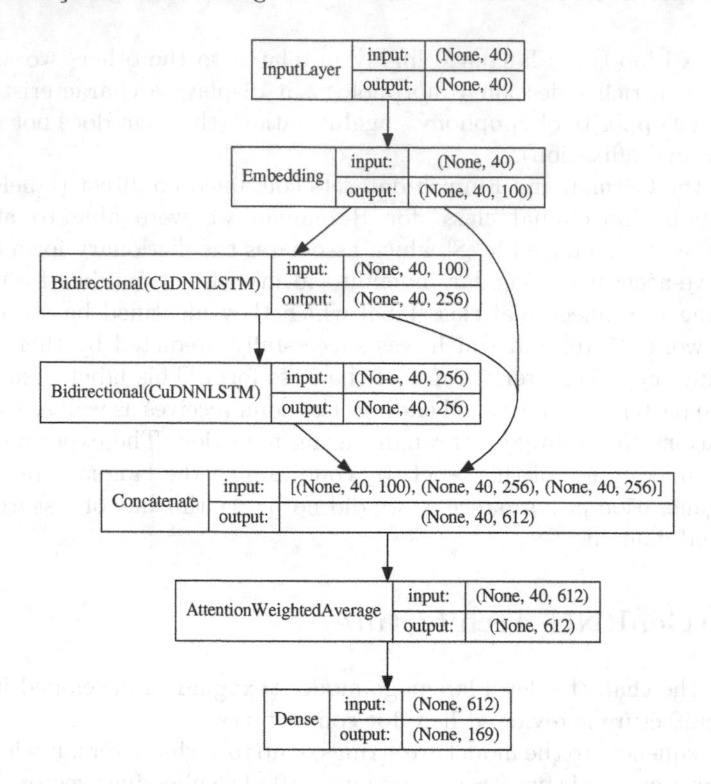


**Fig. 1.** Structure of the model. We use a max-length $T = 40$, a 100-D embedding matrix, 2 128-unit LSTMs, and an attention weighting with 169 outputs.

Each form in the training set was separated by a comma and a space and written on the same line. The test sets were formatted the same. For German and Finnish, we consolidated the training and development sets together and used the test set for generation. For the Romanian set, since we also had access to the inflectional class for each base form, we split the 30k examples in train and test sets, making sure that both sets maintained the same distribution based on this inflectional class (Fig. 1).

During the generation phase, the trained language model is given the uninflected form from the test set as the value for the *prefix* parameter in the *generate* function of the textgenrnn model. If the forms generated by the model match exactly with the test set forms, then the corresponding entry in the test set is counted as having been correctly generated.

The experiments were carried out with different training parameters. We trained the three language models for 14 epochs and different values for the *max_length* parameter which determines how many tokens (characters) are taken as context during training. Since the Romanian models performed less accurately than the German and Finish ones in the nominal domain but also than Spanish and Finnish in the verbal domain according to our previous study [9], we investigated if pre-training of the model with the Romanian EuroParl corpus [18] and fine-tuning it over the morphological corpus helped boost generation accuracy for the verbal model. We also tested the influence of the number of tokens to consider before predicting the next one by changing *max_length* from 40 characters (default value) to 90, chosen based on the distribution of lengths in the training corpora for Romanian Fig. 2 and German Fig. 3. Interestingly, the Finnish corpus had a much larger character length average than the other two languages since it has considerably more cases (14 vs. 4). We decided to set the max_length parameter to 100 for Finnish.

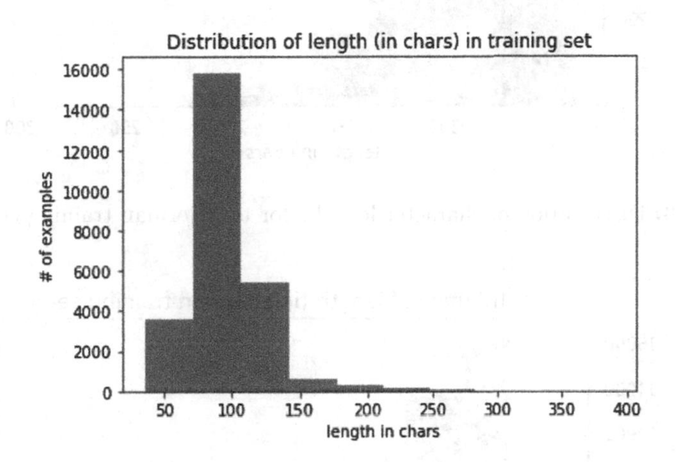


**Fig. 2.** Distribution of character lengths for the Romanian training corpus

A final variable we investigated in our ablation experiments was the number of forms in each training example. Specifically, we wanted to see if, beside the number of characters we told the language model to look at when learning to predict the next character sequence, we also indirectly *conditioned* the model to "understand" that it needs to generate a different amount of forms than the norm for certain words. Indeed, unlike the German and the Finnish datasets, whose examples consistently had 8 and 30 forms respectively, the Romanian noun dataset contained certain nouns which had both male and female variants and therefore the same baseform would lead to more than 8 inflected forms. We tested the extent to which this architecture is able to learn, without explicit labeling of the dataset, so still in a fully unsupervised manner, the distinction between those nouns who get more than 8 forms by training two separate version of the model: one on the available Romanian nouns, another just on the part of the dataset containing nouns with only 8 forms.

## 5 Results

Unlike the verbal experiments [9], we first see that our language models for the Finnish as well as the German noun corpus are not able to reach performance

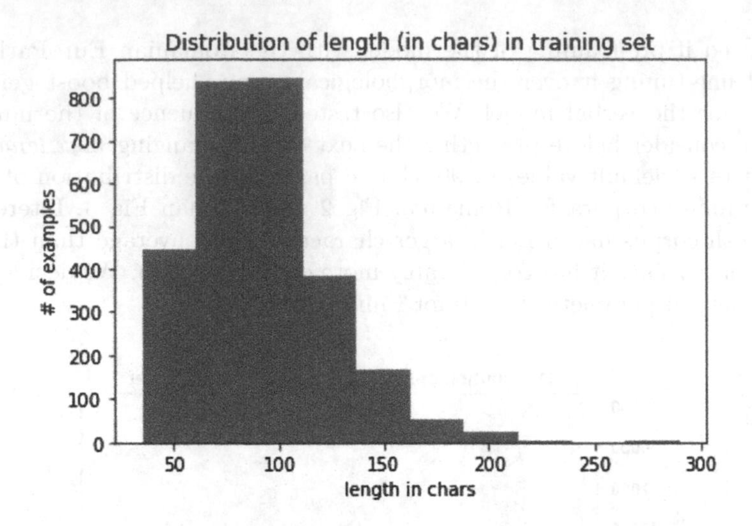


**Fig. 3.** Distribution of character lengths for the German training corpus

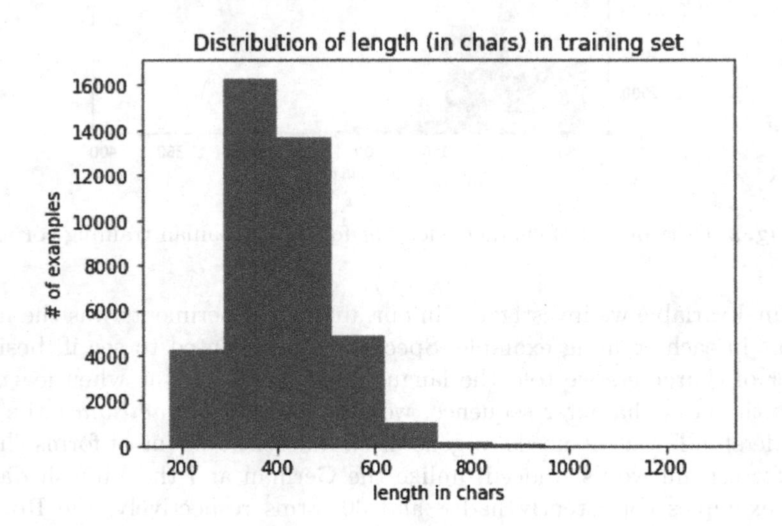


**Fig. 4.** Distribution of character lengths for the Finnish training corpus

anywhere close to the state-of-the-art [1]. In fact, the Finnish is only able to produce the entire list of 30 inflected forms for only 5 nouns in the test corpus and the German one for 18. We identify two reasons for this. The first one is related to the order of magnitude greater stem variance in comparison to the size of the datasets. Specifically, the Finnish noun train set contains around 6k examples, which is close to the 7k Finish verb train set, but every noun has 30 inflected forms, unlike the 6 forms for Finnish verbs. For German, the noun dataset is even smaller than the Spanish verbal dataset (2k for 8 noun forms versus 3.5k for 6 verbal forms). Clearly, the variance is too high for the language model to learn anything about the affixation patterns. Another thing it fails to learn is that it needs to generate exactly 30 forms for Finnish and 8 for German. The second reason for the failure of the model has to do with Finnish being an agglutinative language, which means that the type of stem alternations triggered during inflection (affixation) does not lead to a character cluster in the stem to be replaced with another, close in size, rather it grows in size (Fig. 4).

For Romanian, we report the accuracy of generating the full sequence of inflected forms for the test set containinng exclusively nouns with 8 forms and the set containing all nouns by traininng two separate models. As we can see, indirectly giving the language model the information that all generated examples should have 8 forms by never showing it examples with less or more forms (the Nominal_8 models) leads to just a percentage point improvement. This is in line with the German and Finnish models either predicting too many or too little forms. The deviation is however greatly improved on the Romanian dataset, which makes us conclude that it is an issue related with size of the training since the Romanian one is around 30k. We also see that increasing the max_length from 40 to 90 does not seem to help the system significantly and that training it for longer (from 14 to 28 epochs) leads the model to overfit to regular nouns and therefore generate poorer quality inflection tables (generation accuracy for the 8 form model drops 2% points from 14 to 28 epochs).

We also include the results from previous work for reference, although these models are either fully supervised or semi-supervised and use quite a lot of extensive pretraining. In contrast, our models use no supervision such as tags related to the context in which each generated form needs to appear (person, number) (Table 3).

**Table 3.** Results for full inflection generation and previous work

| Model | Supervision | Forms | Epochs | Language | max_len | Accuracy % |
|---|---|---|---|---|---|---|
| Verbal [9] | Unsupervised | 6 | 28 | Romanian | 40 | **84.86** |
| Nominal_all | Unsupervised | varies, 8 on avg. | 14 | Romanian | 40 | 71.5 |
| Nominal_all | Unsupervised | varies, 8 on avg. | 14 | Romanian | 90 | 71.8 |
| Nominal_8 | Unsupervised | 8 | 14 | Romanian | 40 | 72.6 |
| Nominal_8 | Unsupervised | 8 | 14 | Romanian | 90 | 72.2 |
| Nominal_8 | Unsupervised | 8 | 28 | Romanian | 90 | 70.6 |
| [21] | Semi-Supervised | - | - | Romanian | - | 78.6 |
| Nominal [1] | Seq2Seq | 6 | - | Finnish | - | 95.75 |
| Nominal [1] | Seq2Seq | 6 | - | German | - | 88.87 |

In another set of experiments, we pre-trained textgenrnn on the Romanian EuroParl corpus for 14 epochs and then fine-tuned the model for another 14 epochs using our Romanian inflectional training set described above. We then proceeded to generate inflected forms by calling the *generate* function with the prefix parameter set to the uninflected infinitive forms from the test set. We observed that no inflected form was generated in this setting. Instead, the model would generate chunks of text similar in style and word choice to the pre-training EuroParl corpus, suggesting that the fine-tuning corpus was too different in style and too small to impact the pre-trained model in any way. Below we reproduce one such example, when the value for the *prefix* parameter was inputed as *abandona* (to abandon).

abandon, aderarea lucrătorilor în cadrul negocierilor comunitare care au fost adoptate în mod eficient și la noul acord care constituie o soluţie pentru produsele alimentare și în același timp este deosebit de important să se prezinte o politică externă şi de aplicare a programelor de

# 6 Conclusions

In this paper, we've shown that the method introduced in [9] of generating in a fully unsupervised manner full inflections using neural language models implemented in an off-the-shelf artificial deep recurrent neural network architecture works well in the nominal domain of morphologically rich languages. which for Romanian and Finnish is larger. in order to generate full inflections for verbs in the morphologically-rich Romanian, Spanish, and Finnish, introducing the task of *unsupervised inflection generation.* We showed that even when the training dataset is small for deep learning standards and without any supervision, we can achieve accuracy close to the state of the art for Finish and Spanish and we surpass previous state-of-the-art for Romanian.

Our generation results with the pre-trained EuroParl models also suggest that if the fine-tuning corpus is substantially different (in number of examples and structure) from the pre-training corpus (EuroParl vs. inflection lists), the generated text will maintain the structure of the larger corpus, used in initial training. This goes against the current assumption in transfer learning which maintains that language models are unsupervised multi-task learners which can be successfully used for a new task with minimal need for fine-tuning [4].

**Acknowledgements.** SY would like to thank Noisebridge Hackerspace in San Francisco for use of their computing facilities.

## References

1. Aharoni, R., Goldberg, Y.: Morphological inflection generation with hard monotonic attention. In: Proceedings of the 55th Annual Meeting of the Association for Computational Linguistics, ACL 2017, Vancouver, Canada, 30 July–4 August, Long Papers, vol. 1, pp. 2004–2015 (2017). https://doi.org/10.18653/v1/P17-1183
2. Ahlberg, M., Forsberg, M., Hulden, M.: Paradigm classification in supervised learning of morphology. In: Mihalcea, R., Chai, J.Y., Sarkar, A. (eds.) The 2015 Conference of the North American Chapter of the Association for Computational Linguistics: Human Language Technologies, NAACL HLT 2015, Denver, Colorado, USA, 31 May–5 June, 2015, pp. 1024–1029. The Association for Computational Linguistics (2015). http://aclweb.org/anthology/N/N15/N15-1107.pdf
3. Radford, A., Narasimhan, K., Salimans, T., Sutskever, I.: Improving language understanding by generative pre-training (2018)
4. Radford, A., Wu, J., Child, R., Luan, D., Amodei, D., Sutskever, I.: Language models are unsupervised multitask learners (2019)
5. Barbu, A.M.: Conjugarea verbelor româneşti. Dicţionar: 7500 de verbe româneşti grupate pe clase de conjugare, 4th edition, revised, 263 pp. Coresi, Bucharest (2007). (in Romanian)
6. Barbu, A.M.: Romanian lexical databases: inflected and syllabic forms dictionaries (2008)
7. Cotterell, R., Kirov, C., Sylak-Glassman, J., Yarowsky, D., Eisner, J., Hulden, M.: The SIGMORPHON 2016 shared task–morphological reinflection. In: Proceedings of the 2016 Meeting of SIGMORPHON. Association for Computational Linguistics, Berlin, Germany, August 2016
8. Şulea, O.M.: Semi-supervised approach to Romanian noun declension. In: Knowledge-Based and Intelligent Information Engineering Systems: Proceedings of the 20th International Conference KES-2016, York, UK, 5–7 September 2016, pp. 664–671 (2016)
9. Şulea, O.M., Young, S., Dinu, L.P.: Morphogen: full inflection generation using recurrent neural networks. In: CICLing 2019 (to appear)
10. Dinu, L.P., Şulea, O.M., Niculae, V.: Sequence tagging for verb conjugation in Romanian. In: RANLP, pp. 215–220 (2013)
11. Dinu, L.P., Ionescu, E., Niculae, V., Şulea, O.M.: Can alternations be learned? A machine learning approach to verb alternations. In: Recent Advances in Natural Language Processing 2011, pp. 539–544 (2011)
12. Dinu, L.P., Niculae, V., Şulea, O.M.: Learning how to conjugate the Romanian verb. Rules for regular and partially irregular verbs. In: European Chapter of the Association for Computational Linguistics 2012, pp. 524–528, April 2012
13. Durrett, G., DeNero, J.: Supervised learning of complete morphological paradigms. In: Vanderwende, L., Daumé III, H., Kirchhoff, K. (eds.) Human Language Technologies: Conference of the North American Chapter of the Association of Computational Linguistics, Proceedings, 9–14 June 2013, Westin Peachtree Plaza Hotel, Atlanta, Georgia, USA, pp. 1185–1195. The Association for Computational Linguistics (2013). http://aclweb.org/anthology/N/N13/N13-1138.pdf
14. Faruqui, M., Tsvetkov, Y., Neubig, G., Dyer, C.: Morphological inflection generation using character sequence to sequence learning. CoRR abs/1512.06110 (2015). http://arxiv.org/abs/1512.06110

15. Felbo, B., Mislove, A., Søgaard, A., Rahwan, I., Lehmann, S.: Using millions of emoji occurrences to learn any-domain representations for detecting sentiment, emotion and sarcasm. In: Conference on Empirical Methods in Natural Language Processing (EMNLP) (2017)
16. Guţu-Romalo, V.: Morfologie Structurală a limbii române. Editura Academiei Republicii Socialiste România (1968). (in Romanian)
17. Howard, J., Ruder, S.: Universal language model fine-tuning for text classification. In: ACL. Association for Computational Linguistics (2018). http://arxiv.org/abs/1801.06146
18. Koehn, P.: Europarl: a parallel corpus for statistical machine translation (2005)
19. Şulea, O.M., Nisioi, S., Dinu, L.P.: Using word embeddings to translate named entities. In: Proceedings of the Tenth International Conference on Language Resources and Evaluation, LREC 2016, Portorož, Slovenia, 23–28 May 2016 (2016). http://www.lrec-conf.org/proceedings/lrec2016/summaries/1167.html
20. Woolf, M.: textgenrnn. http://github.com/minimaxir/textgenrnn
21. Zhou, C., Neubig, G.: Morphological inflection generation with multi-space variational encoder-decoders. In: Proceedings of the CoNLL SIGMORPHON 2017 Shared Task: Universal Morphological Reinflection, Vancouver, BC, Canada, 3–4 August 2017, pp. 58–65 (2017). https://doi.org/10.18653/v1/K17-2005

# AL4LA: Active Learning for Text Labeling Based on Paragraph Vectors

Damián Nimo-Járquez(✉), Margarita Narvaez-Rios, Mario Rivas, Andrés Yáñez, Guillermo Bárcena-González, M. Paz Guerrero-Lebrero, Elisa Guerrero, and Pedro L. Galindo

Grupo Sistemas Inteligentes de Computación, Universidad de Cádiz, Cádiz, Spain
damian.nimojarquez@gm.uca.es

**Abstract.** Nowadays, despite the huge amount of digitized information, the biggest drawback to use machine learning in text mining is the lack of availability of a set of tagged data due to mainly, that it requires a great user effort that it is not always viable. In this paper, with the aim of reducing the great workload required to manually processing the contents of large volumes of documents, we present a methodology based on probabilistic inference and active learning to label documents in Spanish using a semi-supervised approach. First, a vector representation of the documents is generated, and then an interactive learning process to apply both, automatic and manual labeling is proposed. To evaluate the accuracy of the predictions and the efficiency of the methodology, different configurations regarding the automatic and manual labeling processes have been studied. The proposed methodology reduces the need for a large corpus of manually labeled texts by introducing a self-labeling process during training. We have shown that both tagging approaches can be combined maintaining accuracy and reducing user intervention.

**Keywords:** Paragraph vectors · Text labeling · Active learning · Text categorization

## 1 Introduction

With the rapid advance of information technologies, we can access large amounts of data that can be structured (for example, databases) or unstructured (such as web content). This fact has given rise to the development of tools that allow processing and extracting underlying information of interest. In the case of unstructured data, one of the tasks in which companies and organizations have shown great interest in recent years and, therefore, has become a relevant research topic in the field of Natural Language Processing (NLP), is text categorization [1]. This field aims at the automatic assignment of texts to predefined categories according to their content, and constitutes a support technology in various information processing tasks, including opinion mining, sentiment analysis, controlled vocabulary indexing, routing and packaging of news and other text transmissions, content filtering (spam, etc.), information security and others.


I. Rojas et al. (Eds.): IWANN 2019, LNCS 11506, pp. 679–687, 2019.
https://doi.org/10.1007/978-3-030-20521-8_56

Automatic text classification is, in general, a supervised process, where a set of documents labeled by human experts are prepared to train the system, before obtaining the correct inference over new documents of unknown categories [2]. The main drawback lies on the available large amount of unlabeled texts against the few volume of categorized texts, since the manual labeling process is extremely slow and expensive. A completely unsupervised process could be fast, but given the particularities in NLP, where other factors such as semantic double meanings, ambiguities, etc., are involved, there is no guarantee of obtaining accurate results [3].

As an alternative, Active Learning (AL) can be considered as a supervised strategy which takes into account human intervention during the learning process in order to manually label some samples of the training set, with this manual setting a better performance is expected to be achieved, even with less data for training than with traditional methods [4, 5]. AL has inspired different interactive machine learning systems mixing supervised and semi-supervised paradigms, such as DUALIST [6], designed to build classifiers for text processing tasks, by soliciting and learning from labels on both features (e.g., words) and instances (e.g., documents), the annotation tool based on active learning, known as PRODIGY [7], or Infer.net [8] that includes graphical information and lets users incorporate domain knowledge into their model, building an algorithm tailor-made for a specific purpose.

Prior to categorization it is necessary to accomplish an adequate text representation, in order to capture the underlying semantics and relationships between the words in the document. Bag Of Words (BOW) and distributed text representations were the most common methods [9, 10] until distributed word embedding was popularized [11–13]. Word embedding represents words in a continuous vector space in which words with similar meanings are mapped closer to each other. Thus, a word vector consists of a set of real numbers where each value represents a dimension of the word's meaning and where semantically similar words have similar vectors. New words in application texts that were missing in training texts can still be classified through similar words. Mikolov et al. [14] introduced Paragraph Vectors (PV). PV learns vector representations from texts of variable length belonging to both, sentences and documents. The paragraph vectors are obtained by training a neural network on the task of predicting a probability distribution of words in a paragraph given a randomly-sampled word from the paragraph [15]. Different works have demonstrated the applicability of PV to different NLP tasks [16–20]. In [21] the showed that PV can perform very robustly when using models trained on large external corpora, and can be further improved by using pre-trained word embeddings.

In this work we propose a complete methodology based on active and semi-supervised learning for Spanish document tagging, using PV for the numerical representation of documents. The proposed methodology reduces the need for a large corpus of manually labeled texts by introducing a self-labeling process during training. This work explores different settings in order to optimize the performance of the overall process.

## 2 AL4LA: Active Learning Methodology for Text Labeling

### 2.1 Pre-trained Word Embeddings

The first step is to create a vector that represents the meaning of a document, which can then be used as input to a supervised machine learning algorithm to associate documents with labels.

Wikipedia is a rich source of well-organized textual data that can be used to pre-train our model. In recent years Wikipedia has attracted many NLP researchers due to the great variety of topics and the open nature and free access that differentiates it from other data sources [23].

From the pre-trained model, we use documents from the Spanish Official State Gazette (Boletines Oficiales del Estado, BOE) for which we focus on a specific topic, those documents related to this particular topic will be labeled as positive while other non-related documents will be labeled as negative. The samples have three attributes: id, text, class (positive or negative).

Figure 1 shows an example of a BOE labeled as positive, and an example labeled as negative is shown in Fig. 2.

V. Anuncios B. Otros anuncios oficiales MINISTERIO DE FOMENTO 9375 Anuncio de la Demarcación de Carreteras del Estado en Galicia sobre Resolución del Ministerio de Fomento de Aprobación Provisional, del Proyecto de Trazado: "Supresión del ramal de acceso al polígono de O Pino en la autovía Lugo-Santiago (A-54). Tramo: Lavacolla-Arzúa oeste". Clave: 12-LC-5720. Provincia de A Coruña. El Director General de Carreteras, con fecha 25 de enero de 2017 ha resuelto: 1. "Aprobar provisionalmente el proyecto de trazado sobre supresión del ramal de acceso al polígono de O Pino en la autovía Lugo-Santiago (A-54). Tramo: Lavacolla-Arzúa oeste, con la siguiente prescripción: o Antes de la aprobación definitiva del proyecto de trazado se procederá a la realización de la correspondiente auditoría de seguridad viaria, en cumplimiento del Real Decreto 345/2011, de 11 de marzo, sobre gestión de la seguridad de las infraestructuras viarias de la Red de Carreteras del Estado.

**Fig. 1.** BOE text labeled as positive

<PAG:1>BOJA Boletín Oficial de la Junta de Andalucía Número 240 - Viernes, 16 de diciembre de 2016 página 204 5. Anuncios 5.1. Licitaciones públicas y adjudicaciones Consejería de Fomento y Vivienda Resolución de 12 de diciembre de 2016, de la Delegación Territorial de Fomento y Vivienda en Sevilla, por la que se anuncia la contratación del suministro que se indica por el procedimiento abierto y un único criterio de adjudicación. (PD. 3082/2016). La Consejería de Fomento y Vivienda de la Junta de Andalucía ha resuelto anunciar por el procedimiento abierto y un único criterio de adjudicación el siguiente suministro: 1. Entidad adjudicadora. a) Organismo: Consejería de Fomento y Vivienda. b) Dependencia que tramita el expediente: Delegación Territorial de Fomento y Vivienda de Sevilla. c) Número de expediente: 2016/010-SUM. 2. Objeto del contrato. a) Descripción del objeto: «Suministro de vestuario para el personal del servicio de carretera.

**Fig. 2.** BOE text labeled as negative

### 2.2 Learning Process

The proposed methodology is based on an iterative and semi-supervised probabilistic text categorization. The training data set, $T$, composed by the feature vectors obtained by applying doc2vec to the set of BOE texts, is successively growing with new labeled (manually or automatically) samples.

Table 1 summarizes the proposed algorithm. In each iteration, the process consists of feeding the labeled data to inputs, and obtain a probability distribution over a set of classes, in this work, two different categories. The fitted model is used to predict the responses for the unlabeled observations (dataset $U$). This dataset acts as a validation set during training, but its size decreases as the learning evolves. Some samples from $U$ are selected to be manually labeled, while other samples will be automatically tagged, assuming the predicted class given by the fitted model.

**Table 1.** Algorithm of AL4LA methodology.

1. Obtain the PV representation of $N$ documents
2. Generate the Training Set, $T$, labeling $N_t$ documents
3. The rest of documents ($N_u = N\text{-}N_t$ ) will form the unlabeled set, $U$
4. While $U$ is not empty:
    4.1 Train the classifier using $T$
    4.2 Predict the categories of $U$
    4.3 If a predetermined iteration, $I$, has been reached, activate the self-labeling process:
        Remove $N_s$ ($N_s+N_m<=N_u$) samples from $U$ and add to $T$, using the predicted categories as labels of these texts
    4.4 Select $N_m$ samples ($N_m<=N_u$) to be manually labeled

We aim at maximising the performance of the tagging process, thus the proposed methodology combines manual and automatic labeling. Since we use a probabilistic classifier, the automatic labeling is performed by taking those samples whose predictions are very likely to belong to the predicted category. Manual labeling must correct or confirm the prediction made by the classifier or the self-tagging process in some selected samples. In this work we study what should be such selected samples to learn from in the next iteration, we consider different selection strategies that have been described in the experimental setup.

# 3 Experimental Setup

A dataset of 3977 documents from the Spanish Wikipedia, have been used to pre-train the PV model, and then, we obtain the vector representation of 86 BOE documents. The paragraph vectors are obtained by training a neural network on the task of predicting a probability distribution of words in a paragraph given a randomly-sampled word from the paragraph. From the Gensim implementation [22] different parameters have been setup:

- Number of features to be represented = 20
- Window size to capture the context = 10

The resulting PV dataset of 86 feature vectors has been divided into 44 samples to train (and validate during training) while the remaining 42 samples were used for testing, that is, to obtain an unbiased evaluation of the final model fit on the training dataset. We start with Nt = 4, an initial training set of 4 labeled samples (2 positive and 2 negative) and 40 unlabeled samples will form dataset U. These 40 samples will become successively part of the labeled training set, as they are either manually or automatically labeled.

Naive Bayes classifier has been applied due to its simplicity, efficiency and efficacy [24] in text categorization. This algorithm calculates the posterior probability that the document belongs to different categories and assigns the document to the class with the highest posterior probability. The self-labeling process is based on the output provided by this classifier to assign the most likely class and starts at first iteration ($I = 1$), while the manual labeling requires the user interaction to correct or confirm the predictions made by the model and is carried out in every iteration.

In order to analyse the methodology, we posed the following questions:

- Is it worth activating the automatic labeling process? Our proposal aims to reduce the manual effort of labeling as long as the tagging error rate does not increase.
- How many samples should be manually labeled? We consider two different values, Nm = 2, 4 or 12 samples in each iteration. A higher value of Nm does not make sense in our approach, since it would be similar to traditional tagging processes.
- Which should be the upper limit of the number of samples to be automatic labeled in each iteration? Three different values: $Ns$ = 4, 12, 20, 30 or a maximum of 40 have been considered.

As the training evolves, the number of categorized samples increases and the algorithm provides the possibility of reviewing some samples manually. Additionally, in this work we study what should be these samples, two options have been considered:

(a) Strategy 1: Select up to $Nm$ samples randomly.
(b) Strategy 2: Select up to $Nm$ samples that have been incorrectly categorized.

Experiments have been repeated 1000 times in order to obtain unbiased estimates of the test error, when the fitted model is evaluated with samples that have never been used before.

## 4 Results and Discussion

In order to answer the first question: is it worth activating the automatic labeling process? We have compared test error rates when all the training samples are manually tagged, against a combination of both, manual and automatic labeling processes. When all the samples are manually labeled, an average test error of 0.2919 is reached with a standard deviation around 0.05 (we will take this test error as reference), while combining both labeling approaches can yield to very similar values as we will see in next figures.

Figure 3 shows error rates in function of Ns (x axis) and Nm (line plots), when the samples to be manually labeled are chosen randomly (Fig. 3a) and from the incorrectly labeled samples (Fig. 3b), meaning that in each iteration Ns (4 12 20 30 or 40) samples are automatically labeled and Nm (2, 4 or 12) samples are manually labeled.

As we can observe in both figures, the random selection of the samples that will be reviewed manually (Fig. 3a) provides lower error rates than the selection of those that are incorrectly labeled (Fig. 3b). Therefore, we corroborate the objective of this work of diminishing the intervention and the effort that the user must devote to the selection of the samples to be reviewed.

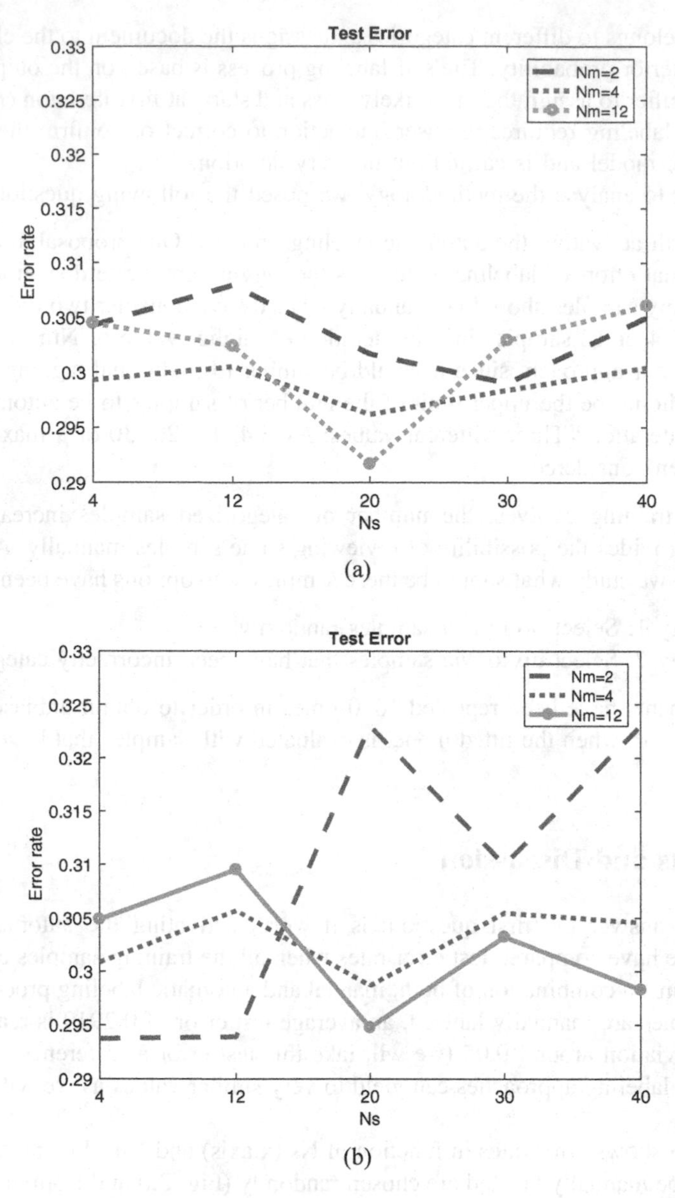


**Fig. 3.** Test error in function of Ns, the upper limit of samples to be automatically labeled. Nm refers to the upper limit of samples to be manually labeled in each iteration. (a) Strategy 1 for selecting the samples to be manually labeled. (b) Strategy 2 for selecting the samples to be manually labeled (Color figure online)

In both cases, best values are obtained when 12 samples are manually tagged in each iteration of the learning process (green dotted line), as expected, the more training samples are labeled manually, the better. Surprisingly, this configuration yields to the lowest error rates when up to 20 samples can be automatically tagged in the same iteration (test error = 0.292 and std = 0.06).

Focusing on Fig. 3(a) and still regarding Nm, we can observe that there exists one more configuration where the error rate is very similar to the above reference, namely Nm = 4 and Ns = 20. Figure 4(a) and (b) show both test error distributions. Although an manual labeling of 12 samples provides better results, the alternative of limiting this value to 4 samples in each iteration does not yield to significant differences on average.

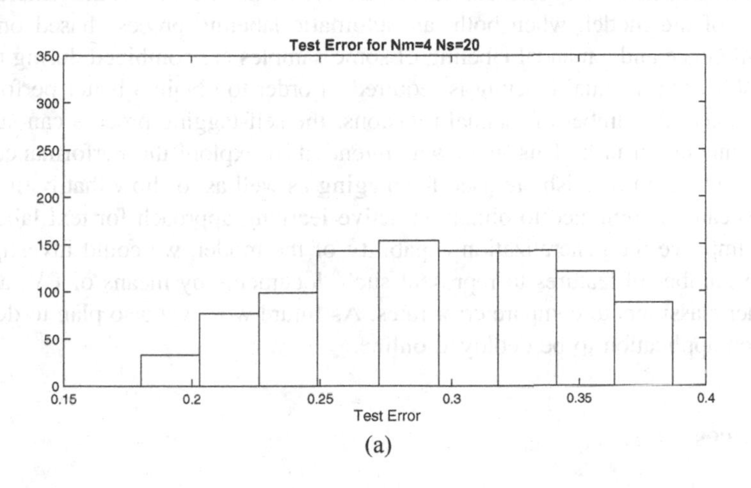


(a)

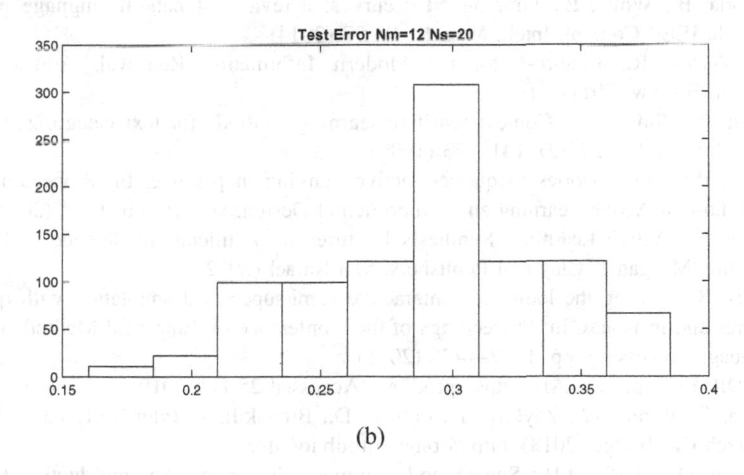


(b)

**Fig. 4.** Test error distributions for an upper limit of automatic labeled samples of Ns = 20 and selecting randomly Nm samples to be manually labeled. (a) Nm = 4 (b) Nm = 12

Thus, the combination of both, manual and self-labeling process can yield to results of comparable accuracy, as long as the model is trained with a certain number of manually labeled samples.

## 5 Conclusions

A semi-supervised approach to text categorization has been proposed. Paragraph vectors have been applied in order to get word embedding representations and generate a pre-trained model from the Spanish Wikipedia, and then, the vector representation of BOE documents has been obtained. Subsequently, we have explored the generalization capability of the model, when both, an automatic labeling process based on probabilistic inference and a manual labeling of some samples are combined during training. Results show that manual labeling is required in order to obtain a better performance, but with a certain number of manual revisions, the self-tagging process can substitute the user annotation task. This work was intended to explore the performance of PV when documents in Spanish are used for tagging as well as to show that both labeling processes can be combined to obtain an active learning approach for text labeling. In order to improve the generalization capability of the model, we could investigate the optimum number of features to represent such documents by means of PV, and consider other classifiers to compare error rates. As future work we also plan to develop a web-based application to be deployed online.

## References

1. Cambria, B., White, E.: Jumping NLP curves: a review of natural language processing research. IEEE Comput. Intell. Mag. **9**, 48–57 (2014)
2. Baeza-Yates, R., Ribeiro-Neto, B.: Modern Information Retrieval. Addison-Wesley, Pearson, Harlow (2011)
3. Cohen, W., Singer, Y.: Context-sensitive learning methods for text categorization. ACM Trans. Inform. Syst. **17**(2), 141–173 (1999)
4. Settles, B.: From theories to queries: active learning in practice. In: Proceedings of the Workshop on Active Learning and Experimental Design, vol. 16, pp. 1–18 (2011)
5. Settles, B.: Active Learning. Synthesis Lectures on Artificial Intelligence and Machine Learning. Morgan & Claypool Publishers, San Rafael (2012)
6. Settles, B.: Closing the loop: fast, interactive semi-supervised annotation with queries on features and instances. In: Proceedings of the Conference on Empirical Methods in Natural Language Processing, pp. 1467–1478 (2011)
7. PRODIGY (Explosion AI). https://prodi.gy. Accessed 25 Feb 2019
8. Minka, T., Winn, G.J., Zaykov, Y., Fabian, D., Bronskill, J.: Infer.NET 0.3, J. Microsoft Research Cambridge (2018). http://dotnet.github.io/infer
9. Jurafsky, D., Martin, J.H.: Speech and Language Processing An Introduction to Natural Language Processing, Computational Linguistics, and Speech Recognition. Pearson, Upper Saddle River (2018)
10. Melamud, M., McClosky, O., Patwardhan, D., Bansal, S.: The role of context types and dimensionality in learning word embeddings. In: Proceedings of NAACL-HLT, pp. 1030–1040 (2016)
11. Al-Rfou, R., Perozzi, B., Skiena, S.: Polyglot: distributed word representations for multilingual NLP. In: Proceedings of the Seventeenth Conference on Computational Natural Language Learning, pp. 183–192. Association for Computational Linguistics, Sofia (2013)

12. Ma, C., Xu, W., Li, P., Yan, Y.: Distributional representations of words for short text classification. In: Proceedings of the 1st Workshop on Vector Space Modeling for Natural Language Processing, no. 21, pp. 33–38 (2015)
13. Rudkowsky, E., Haselmayer, M., Wastian, M., Jenny, M., Emrich, S., Sedlmair, M.: More than bags of words: sentiment analysis with word embeddings. Commun. Methods Measures **12**(2–3), 140–157 (2018)
14. Mikolov, T., Chen, K., Corrado, G., Dean, J.: Efficient estimation of word representations in vector space. arXiv preprint arXiv:1301.3781 (2013)
15. Le, Q., Mikolov, T.: Distributed representations of sentences and documents. In: International Conference on Machine Learning (ICML), Beijing, China, pp. 1188–1196 (2014)
16. Maslova, N., Potapov, V.: Neural network doc2vec in automated sentiment analysis for short informal texts. In: Karpov, A., Potapova, R., Mporas, I. (eds.) SPECOM 2017. LNCS (LNAI), vol. 10458, pp. 546–554. Springer, Cham (2017). https://doi.org/10.1007/978-3-319-66429-3_54
17. Markov, I., Gómez-Adorno, H., Posadas-Durán, J.-P., Sidorov, G., Gelbukh, A.: Author profiling with doc2vec neural network-based document embeddings. In: Pichardo-Lagunas, O., Miranda-Jiménez, S. (eds.) MICAI 2016. LNCS (LNAI), vol. 10062, pp. 117–131. Springer, Cham (2017). https://doi.org/10.1007/978-3-319-62428-0_9
18. Bilgin, M., Şentürk, İ.F.: Sentiment analysis on Twitter data with semi-supervised doc2Vec. In: 2017 International Conference on Computer Science and Engineering (UBMK), Antalya, pp. 661–666 (2017)
19. Kottur, D., Vedantam, S., Moura, R., Parikh, J.M.: Visualword2vec(vis-w2v): learning visually grounded word embeddings using abstract scenes. In: Proceedings of the IEEE Conference on Computer Vision and Pattern Recognition, pp. 4985–4993 (2016)
20. Dai, A.M., Olah, C., Le, Q.V.: Document embedding with paragraph vectors. In: Proceedings of the NIPS Deep Learning Workshop, pp. 1–8 (2015)
21. Lau, J.H., Baldwin, T.: An empirical evaluation of doc2Vec with practical insights into document embedding generation. arXiv preprint arXiv:1607.05368 (2016)
22. GENSIM. https://radimrehurek.com/gensim/. Accessed 25 Feb 2019
23. Dang, Q.V., Ignat, C.L.: Quality assessment of Wikipedia articles: a deep learning approach. ACM SIGWEB Newslett. **3**, 1–6 (2016)
24. Dai, W., Xue, G., Yang, Q., Yu, Y.: Transferring Naive Bayes classifiers for text classification. In: Proceedings of the 22nd Association for the Advancement of Artificial Intelligence (AAAI) Conference on Artificial Intelligence, pp. 540–545 (2007)

# On Transfer Learning for Detecting Abusive Language Online

Ana-Sabina Uban[1,2](✉) and Liviu P. Dinu[1,2]

[1] Faculty of Mathematics and Computer Science, University of Bucharest, Bucharest, Romania
ana.uban@gmail.com, liviu.p.dinu@gmail.com

[2] Human Language Technologies Research Center, University of Bucharest, Bucharest, Romania

**Abstract.** Abusive language online has become a growing social issue in our age of social media. Given the massive amounts of data being generated daily on social platforms, manually detecting and regulating such behavior has become unfeasible, so automatic solutions are necessary, and tasks related to identifying abusive language, in its various forms, from hate speech to bullying, have come into the focus of the natural language processing research community. In this paper, we focus on two subtypes of abusive language: aggressive language and offensive language, for which we implement a deep learning model based on convolutional neural networks. We further propose a new approach using transfer learning to boost performance of abusive language detection by leveraging data annotated with a different type of label, related to sentiment. We show how transferring knowledge between these tasks affects performance of detecting abusive language, offering insights into how these tasks are related, and how the more traditional task of sentiment analysis can be leveraged to help with solving the newer and less data rich task of abusive language detection.

**Keywords:** Abusive language · Deep learning · Tweet · Sentiment analysis · Transfer learning · Convolutional neural network

## 1 Introduction

Along with the accelerated rise of social media activity, and online communication, online abuse and aggression have become growing social issues in recent years. Online abuse is a social problem from different perspectives: the risk of being harassed on a social platform can be generally unpleasant and harmful for the online community; but more particularly, aggression can be targeted toward specific social groups, in which case it can become an issue of social justice or discrimination, such as sexism or xenophobia. Social media services such as Facebook and Twitter have been criticized for not having done enough to prohibit the use of their services for attacking people belonging to some specific race,


I. Rojas et al. (Eds.): IWANN 2019, LNCS 11506, pp. 688–700, 2019.
https://doi.org/10.1007/978-3-030-20521-8_57

minority etc. Regulating abusive behavior online, by identifying and restricting it, has thus become a real necessity, and with the overwhelming amount of data flowing constantly on social platforms, manually detecting it is unfeasible, so automatic solutions are necessary. As a consequence, identification of abusive language has become a concern in natural language processing research, and it has been identified as a task in itself, to be studied and found its particular intelligent solutions. Abusive language online can manifest in different ways, and experts studying the phenomenon (which is still fairly recent) have identified various types of abusive language, such as: hate speech, bullying, trolling, aggressive or offensive language etc. These subtypes have things in common, but also their own specific features, and have been treated as relatively separate tasks.

As an indication of this topic's popularity in the NLP community, as well as its relative freshness, in recent years, many workshops and shared tasks dedicated to some form or another of online abuse have appeared. Some of the first workshops dedicated to analyzing online trolling have appeared as early as 2015. In 2017, the first workshop on hate-speech was held alongside the ACL conference. That same year, the first workshop on abusive language online was organized. In 2018, the First Workshop on Trolling, Aggression and Cyberbullying has organized the First Shared Task on Aggression Identification [8]. In 2019, the conference on semantic evaluation SemEval organized a special shared task dedicated to offensive language detection [20]. Despite the high popularity of these tasks, and the many solutions that have been attempted only over the course of the past few years, detecting abusive language, in its various forms, proves to be a relatively challenging task, and there is still potential for improving the performance over existing solutions.

In this paper, we perform experiments using deep learning to predict abusive language online, using two separate datasets, one related to aggressive language and one to offensive language. We propose a fairly general model and show that it can perform reasonably well for predicting the various types of abusive language. The main contributions of our paper are around studying the effect of transfer learning on the task of abusive language detection. To this effect, we focus on a general deep learning model which is generally successful across NLP tasks, and show how training it on a larger dataset from a similar but distinct domain, that of sentiment analysis, can affect the way it learns our main task. We show that indeed pretraining is useful for improving performance on the main task, proving both the validity of transfer learning for this task, and possibly pointing to a relatedness between the two tasks that we use to share knowledge: sentiment and abuse. We look at what transfer strategy is most effective, and at what the results might entail for the directions to follow in further research related to this topic.

### 1.1 Previous Work

The task of detecting abusive speech, in its various forms, has recently become a very popular one in NLP research, and given the appearance of many workshops and shared tasks dedicated to the topic in the past few years, many solutions have been proposed, from simple SVMs or Naive Bayes models, to complex neural networks. Most of the work has been done on social media data, especially Facebook posts and Twitter data.

In this section we aim to show a general overview of the work that has been done in the field, and mention some of the most noteworthy articles published in literature on this topic, either from the point of view of performance or original approach. [17] are among the first researchers to study abusive language online. Among some of the first approaches to use embeddings, are [2], who propose an embedding-based solution to hate speech detection, where in they learn to project comments in social media in lower dimensional space.

In [18], the authors introduce a dataset of tweets labelled as hate-speech (or non hate-speech), and divide it into two subcategories: sexism and racism. They attempt to predict hate-speech using character n-grams along with various user-level features such as geographical location and gender. On the same dataset, [14] identify hate-speech using a language-agnostic model in the form of a recurrent neural network, incorporating both word and user features, such as tendency for racism. [13] perform the classification in a two step process, where abusive text first is distinguished from the non-abusive, and then the class of abuse (sexism or racism) is determined. [4] employed pretrained CNN vectors in an effort to predict four classes, achieving slightly higher F-score than character n-grams. [15] present a more exhaustive survey of hate-speech detection literature up to 2017.

In an experiment similar to ours, in the sense of a meta-analysis of how abusive language detection can be generalized and extended across datasets and domains, [6] analyze how easy it is to perform cross-domain detection of abusive language, and find that abusive language detection models fail to generalize, and that at least some in-domain data is needed for performance to be reasonable.

In [3], the authors use a deep learning model that is designed to detect various types of abuse with a unified solution. The authors also experiment with using a fine-tuning approach instead for combining different types of features, and suggest that this transfer learning approach may be superior to training a unified model in a single step. In a multi-task learning approach, [19] show that using data from various datasets and subtasks can help improve performance of detecting hate speech.

Among other recent approaches using deep neural networks for identifying aggressive language we mention [1,9,11,16], where a combination of CNNs, RNNs and ensembles are used. In [12], the authors try to use word embeddings and sentiment features to classify aggressive language.

The deep learning architectures proposed in literature as solutions to abusive language detection are not unique to this task. Deep learning models are state-of-the-art in most text classification tasks, with recurrent neural networks and convolutional neural networks being among the most popular, and widely successful across NLP applications. RNNs have been proven very useful at modelling longer-term dependencies between words and are very successful on classifying longer input texts, as well as in sequence-to-sequence tasks, such as machine translation. CNNs are also very successful in text classification, especially when applied on shorter-length texts, and are very useful at capturing relevant subsequences of the text: both in their word-level version, as proposed in [7], and the character-level version [21]. In this paper we propose using a word-level CNN for classifying texts into abusive and non-abusive according to our task.

## 2 Datasets

As part of our experiments, we use two datasets of social media posts annotated with abusive language labels, and one dataset of posts annotated with sentiment labels. In turn, our experiments will consist of three parts, where we train our network on each of the three datasets, in various configurations corresponding to the knowledge transfer strategy between these stages.

**Aggressive Language Dataset.** For the first of the abusive language related tasks, we use a dataset consisting of examples of aggressive language, made up of Facebook and Twitter posts labelled as "aggressive"/"non-aggressive". The dataset was proposed within the Shared Task on Aggression Identification organized as part of the First Workshop on Trolling, Aggression and Cyberbullying [8].

Aside from the simple "aggressive"/"non-aggressive" labels, the dataset provides three high-level tagsets, namely: covertly aggressive, non-aggressive, and overtly aggressive. The dataset covers both English and Hindi posts, but we focused only on the English data for the sake of our experiments. The English dataset consists of 12,000 training examples from Facebook, and 1,257 examples from an undisclosed social media platform. The number of examples per class is unbalanced, with more examples belonging to the non-aggressive class.

For the purpose of our experiments, we only consider the English text, and ignore the source of the text. We also only consider whether a text is aggressive or not, and disregard the details on the type of aggression: overt or covert. Thus, we merge the two subtypes of aggressive texts into one set, and label them all as "aggressive", converting the problem into a binary one.

**Offensive Language Dataset.** As a second dataset related to abusive language, we use the Offensive Language Identification Dataset (OLID), proposed in the recent shared task on offensive language identification at SemEval 2019 [20]. The dataset consists of a total of 14,100 tweets, labelled as either offensive and non-offensive. Further, the offensive posts are categorized hierarchically into subtypes of offenses (and labelled accordingly): targeted or untargeted, and targeting an individual or group. In this case also, we disregard the more fine-grained labels, and only focus on the binary problem of whether a post is "offensive"/"not offensive", merging all of the offensive texts into one group. This dataset is also imbalanced, with approximately two thirds of the posts belonging to the "non-offensive" group.

The definition of "offensive language" is slightly different than the definition of "aggressive language" in the dataset described in the previous section. This is also reflected in the data collection process. The authors of the offensive language dataset, for example, mention that for finding offensive posts, keyword searches were performed on Twitter, and that many of them needed to be political keywords, for enough offensive posts to be found. A large part of the data contains texts with a political message: 50% of the tweets come from political keywords and 50% come from non-political keywords.

**Sentiment Dataset.** The Stanford Twitter sentiment corpus[1], introduced by [5], consists 1.6 million tweets automatically labelled as positive or negative based on emotions. The dataset was not manually, but automatically built, by assuming that any tweet with positive emoticons, such as ":)", are positive, and tweets with negative emoticons, such as ":(", are negative. Examples are labelled with either "0", denoting negative emotion, or "4", denoting positive emotion. The dataset is balanced in positive and negative tweets.

The fact that the texts in this dataset are automatically labelled creates the risk of adding noise to the data. Nevertheless, as a precondition of our experiments, we assume that there is some connection between negative emotion and aggressive or offensive language, and hypothesize that it can be exploited through transfer learning. As an advantage, this sentiment dataset is two orders of magnitude larger than our abusive language datasets, making it suitable, despite the possible noise, for learning more general information, such as training word embeddings, which are difficult to train on small datasets.

## 3 Methodology

Since our main goal is to study the effect of transferring knowledge across domains from a model trained on a sentiment analysis task to one for aggressive/offensive language detection, we choose a relatively simple, general model, that we use across our experiments. We choose a deep convolutional network taking words as input, which has proven successful across a range of NLP tasks, including hate-speech detection. At the same time, we do not try to engineer the features or the details of the model to suit this particular task or dataset (for example we don't include separate hashtag features, which are specific to our dataset), in the hope that our results can also generalize, as much as possible, to other types of data and tasks.

We formulate the problem as a supervised binary classification task, and as a precursory step make sure all of our data is labelled with binary labels. In this way, we make sure the tasks are all similar enough in structure so that knowledge transfer is facilitated.

[1] http://help.sentiment140.com/.

### 3.1 Preprocessing and Features

As features for our model, we use the words in the text, seen as ordered sequences. This type of encoding is very common throughout natural language processing tasks. Extracting these features entails preprocessing the text according to regular practice in natural language processing applications: lowercasing and tokenizing the text into words, and excluding the very rare words. We discard any non-letter characters (including emoticons and the hashtag symbol). We also discard usernames where they appear, since we only want to learn from word data, but make sure to include the text in the hashtags, which are usually made up of simple words in the English language. We treat the hashtag words as usual word features, and don't leverage them as separate features, trying to keep our network architecture more general and suited to any dataset. We construct a vocabulary of the most frequent 50,000 words using the larger sentiment dataset, and encode all out of vocabulary words with the same "unknown" token.

We further transform the texts into same-length sequences. We choose as a sequence length the number of 35 words, since this is a reasonably long sentence length for our dataset, a large part of which is comprised of tweets, and many of the texts will fall below this length. In case a text is longer than 35 words, we truncate it; in case it is shorter, we pad it with a special "empty" token.

### 3.2 Model

As our model we use a convolutional neural network, that uses word embeddings as input. Together with recurrent neural networks, convolutional neural networks are among the most successful models for machine learning on text data, across a wide range of NLP tasks. Although CNNs were not traditionally used on texts, [7] has shown that they can very successfully perform text classification tasks as well. As opposed to when they are applied on images, in text processing the convolutions are 1-dimensional, or "temporal", with convolutions essentially passing through the word sequence and finding relevant word n-grams in the input sequence.

We use an architecture similar to the one proposed in [7]. The input to the network is composed of our word sequences, encoded as one-hot vectors of words, which are then passed through an embedding layer. The embedding layer has the role of transforming one-hot sparse representations into shorter dense representations, in a vector space where points corresponding to words that are similar (according to the task the model is trained for) are close together, and words that are not similar are further apart. We use word embeddings of dimensionality 300, and for the convolutional layers, we use 100 filters of width 3. We finally pass the data through a fully connected layer of 100 hidden units, and train the model parameters using binary cross entropy loss. The network's architecture is illustrated in Fig. 1.

*word2vec.* Word embeddings have been proven very successful all across NLP tasks since the introduction of the *word2vec* model [10]. Since then, many improved variants have been proposed, but *word2vec* remains widely used and can help with performance on most NLP tasks. *Word2vec* is a model that generates word embeddings, which are lower-dimensional continuous vectorial representations of words in an embedding space such that semantically similar words are also close together in this space (geometrically). This is accomplished by training the model to either predict a target word given the words found in its surrounding context in the training set (in the CBOW model), or predict the context given the target word (in the skip-gram model). A popular corpus of such representations is the set of the 300-dimensional word2vec embeddings trained on the large (3 billion words) dataset of Google News.

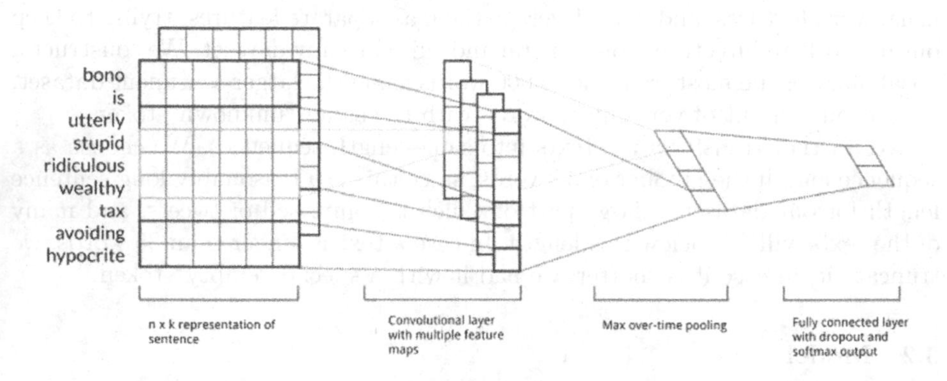


**Fig. 1.** Convolutional neural network for sentence classification

We also use these pretrained word embeddings in some of our experiments, by initializing the weights of our embedding layer with the pretrained weights of the word2vec model. Since they are pretrained on a different task than ours (predicting context words), using these can also be seen as a form of transfer learning, and we experiment with this as described in the following chapter.

### 3.3 Experiments

We aim to study the effect of transferring knowledge from one task to another, namely from the task of predicting sentiment of a text to the task of predicting abusive language. To this effect, we design several experiments to help us gain insight into the problem: we will in turn compare how a model trained for predicting abusive language can perform without transferring knowledge from the separate sentiment analysis task, and how it performs in the context of transfer learning, using various strategies. To make sure the effect we observe is robust and generalizable, we also use two different datasets for the abusive language detection phase, which correspond to slightly different tasks within the issue of online abuse: aggressive language and offensive language.

For our first set of experiments, we train our convolutional network to predict whether a text contains abuse or not, using no pretraining. All weights are initialized randomly.

In the second stage, we perform our transfer learning experiments. First, we train our network on the sentiment analysis task. This is the pretraining phase, which will be followed by the fine-tuning phase on the abusive language task. For pretraining, we train for only 2 epochs, so as not to overfit the model to the pretraining task, considering also the sentiment dataset is much larger than the abusive language datasets. For the training or fine-tuning phase, we train for 200 epochs in each of our experiments. For all experiments we use an initial learning rate of 0.0001 and gradually decrease it using a cosine annealing scheme. Specifically in the experiments performed on the offensive language data, we use a weighted loss function, where more weight is given to the examples in the negative class (the offensive texts), so as to compensate for the imbalance in the data. We included a dropout layer with a probability of 0.6 and additionally used *l2* regularization, with a weight decay parameter set to 0.001 in the pretraining experiments and 0.0001 when training on the main task. Below we describe the various setups for transferring knowledge from one task to another.

**Knowledge Transfer from Word2vec Embeddings.** In a first setting, we only look at transferring knowledge from word2vec embeddings, so we initialize the embedding layer weights of the network to the trained embedding weights of word2vec.

**Knowledge Transfer from Sentiment Task**

- **Sharing all weights**, and allowing them to be learned in the fine-tuning phase. In this setting, before starting training on our main task, we initialize the weights of the network with the weights learned in our pretraining task (on the sentiment analysis dataset), for all parameters of the network. The architecture of the network, including the number of classes, is the same between the two tasks, so sharing all parameters is possible. We then allow all parameters to be learned while training on our main task.
- **Sharing all weights**, but **freezing** the embedding weights. In this second setting, we freeze the embedding weights while training on the main task, so they can no longer be updated. This is so that the smaller model doesn't overfit, since the amount of data on the second tasks (related to language abuse) is much smaller, and the number of parameters corresponding to embeddings is very large (of the order of the vocabulary size multiplied by the embedding size, namely $50,000 \times 300$).
- **Sharing only word embedding weights**. As in the experiment with knowledge transferred from word2vec embeddings, here we try to reuse embeddings, this time trained on the larger sentiment task, and initialize the network for predicting abusive language with these, then allow them to be further learned by the network. The rest of the parameters are randomly initialized.
- **Sharing only word embedding weights, frozen**. In this setting, as in the previous one, we only share the embedding weights with the network trained on the sentiment analysis task, but this time we freeze them so that they are not learnable by the network while training on the main task.

### 3.4 Metrics

We use various metrics to measure the performance of our models in the different setups. Since we model the problem as a binary classification problem, measures such as precision and recall are suitable, aside from the general accuracy measure. Given some datasets are imbalanced, we also report the F1-score.

- **Accuracy** is measured as the number of correctly classified examples

$$Acc = \frac{TP + TN}{TP + TN + FP + FN} \tag{1}$$

  where $TP$ = true positives, $TN$ = true negatives, $FP$ = false positives, $FN$ = false negatives.
- **Precision** and **recall**: defined as the number of correctly classified examples from a class among the examples classified as that class for precision, and from among the examples of that class in the case of recall.

$$Prec = \frac{TP}{TP + FP}; Rec = \frac{TP}{TP + FN} \tag{2}$$

- **F1-score**: is the harmonic mean of precision and recall and is used as a more fair metric than simple precision and recall when datasets are imbalanced.

$$F1 = \frac{2 \times Prec \times Rec}{Prec + Rec} \tag{3}$$

## 4 Results

The results of the pretraining experiments, performed on the sentiment analysis dataset, are shown in Table 1. These are relevant only as a starting point for our following experiments, showing that the network already has reasonable performance on the sentiment analysis task, thus we can expect it to have gained some knowledge about the data, that could help with our main task.

**Table 1.** Results on sentiment dataset (pretraining)

| Experimental setting | Accuracy | Precision | Recall | F1 |
|---|---|---|---|---|
| Network trained on sentiment dataset | 71.63 | 83.43 | 52.96 | 64.79 |

Tables 2 and 3 show our results on the main task, in the various transfer learning experiments, on the two datasets related to abusive language. The higher performance obtained in almost all scenarios where the network was first pretrained on the sentiment analysis task show that knowledge transfer does occur between the two tasks. In the case of detecting abusive language, pretraining all the weights of the network helps most with performance; whereas for aggressive

**Table 2.** Results on offensive language dataset

| Experimental setting | Acc | Prec | Rec | F1 |
|---|---|---|---|---|
| 1. No pretraining | 66.90 | 47.45 | 48.49 | 47.96 |
| 2. Embeddings pretrained word2vec | 69.9 | 55.2 | 55.94 | 55.57 |
| 3. Embeddings pretrained on sentiment, learnable | 67.27 | 49.7 | 47.76 | 48.71 |
| 4. Embeddings pretrained on sentiment, frozen | 63.72 | 49.30 | 44.92 | 47.01 |
| 5. Model pretrained on sentiment, frozen embeddings | 67.63 | 51.67 | 62.26 | 56.47 |
| 6. Model pretrained on sentiment | 68.09 | 51.87 | 63.85 | **57.24** |

**Table 3.** Results on aggressive language dataset

| Experimental setting | Acc | Prec | Rec | F1 |
|---|---|---|---|---|
| 1. No pretraining | 61.89 | 76.57 | 48.84 | 59.64 |
| 2. Embeddings pretrained word2vec | 62.63 | 77.27 | 50.54 | 61.11 |
| 3. Embeddings pretrained on sentiment, learnable | 64.54 | 75.35 | 57.52 | **65.24** |
| 4. Embeddings pretrained on sentiment, frozen | 57.18 | 72.58 | 43.16 | 54.13 |
| 5. Model pretrained on sentiment, frozen embeddings | 60.81 | 71.86 | 54.86 | 62.26 |
| 6. Model pretrained on sentiment | 61.54 | 70.67 | 58.47 | 64 |

language detection, using only the pretrained embedding weights performs best, suggesting that in this case word embedding representations capture the most useful information in a way that is transferrable between tasks.

Word2vec embeddings help to a lesser extent, showing perhaps that the sentiment analysis task we pretrained our network on has more in common with our main task than the more general task that word2vec was trained on.

While results are generally better on the aggressive language dataset than on the offensive language dataset, showing that offensive language is possibly more challenging to detect; transfer learning seems to improve results on offensive language more, significantly increasing recall of offensive tweets (from 48% to 63%), suggesting offensive tweets may have more in common with negative emotion language.

Table 4 presents some examples of previously misclassified offensive/aggressive tweets, which the trained classifier was able to correctly classify when first using pretraining (more specifically, in experimental setting 6, when all weights are pretrained, as opposed to experimental setting 1, where none are).

**Table 4.** Abusive tweets and predicted labels with and without pretraining

| Tweet | Without Pretr(1) | With Pretr(6) | True label |
|---|---|---|---|
| *Ramos sounds like a shitty ump* | non-off | off | off |
| *coming from guy who should be arrested and charged as an accessory. I hat a deranged individual.* | non-off | off | off |
| *Two prominent backwards conservatives removed from Texas education. Looks like they're as blind deaf and dumb as Helen Keller once was.* | non-off | off | off |
| *i love how active you are on twitter. Forever shining on my TL - Dislike: dont event know you to dislike anything tbh* | off | non-off | non-off |
| *Needs another one tight slap.. SRK anyone?* | non-agg | agg | agg |
| *since when loudspeakers were used for centuries* | non-agg | agg | agg |
| *None of the media shown us such index..... Thank you Sourav Ghosh for your information* | agg | non-agg | non-agg |

## 5 Conclusions and Future Work

Our experiments have shown that transfer learning can be successful in the task of abusive language detection. We chose a general model, so it may be possible to extend our conclusions to other types of datasets and domains. The drawback of our method, and a possible direction for future work is around using word embeddings, which can help, but can also hinder in experiments using social media text, where there are many variations in spelling that our model disregards. Especially in the case of offensive language, spelling variations can be a problem, since users may intentionally change characters in the offensive words they use (while keeping them intelligible to human readers) in order to obfuscate them. A character-based method, such as character-level CNNs, could help overcome these weaknesses.

The insights into how knowledge gained in the sentiment-related task as well as leveraged from the general pretrained word2vec embeddings for each of the abusive language detection tasks have interesting implications from different points of view. They show that sentiment analysis, which is an old subfield of NLP with a long tradition, and relatively robust results, can be leveraged to jumpstart results on newer tasks, such as abusive language detection. This suggests that the two tasks may have things in common - and that abusive language shares features with negative sentiment language - but can also serve as an encouragement to research into other tasks between which knowledge transfer may be feasible and possibly rely on such methods for advancing new subfields of research in natural language processing, as a continuously growing research area.

## References

1. Aroyehun, S.T., Gelbukh, A.: Aggression detection in social media: Using deep neural networks, data augmentation, and pseudo labeling. In: Proceedings of the First Workshop on Trolling, Aggression and Cyberbullying, TRAC 2018, pp. 90–97 (2018)
2. Djuric, N., Zhou, J., Morris, R., Grbovic, M., Radosavljevic, V., Bhamidipati, N.: Hate speech detection with comment embeddings. In: Proceedings of the 24th international conference on world wide web, pp. 29–30. ACM (2015)
3. Founta, A.M., Chatzakou, D., Kourtellis, N., Blackburn, J., Vakali, A., Leontiadis, I.: A unified deep learning architecture for abuse detection. arXiv preprint arXiv:1802.00385 (2018)
4. Gambäck, B., Sikdar, U.K.: Using convolutional neural networks to classify hate-speech. In: Proceedings of the First Workshop on Abusive Language Online, pp. 85–90 (2017)
5. Go, A., Bhayani, R., Huang, L.: Twitter sentiment classification using distant supervision. Processing **150**, 1–6 (2009)
6. Karan, M., Šnajder, J.: Cross-domain detection of abusive language online. In: Proceedings of the 2nd Workshop on Abusive Language Online (ALW2), pp. 132–137 (2018)
7. Kim, Y.: Convolutional neural networks for sentence classification. arXiv preprint arXiv:1408.5882 (2014)
8. Kumar, R., Ojha, A.K., Malmasi, S., Zampieri, M.: Benchmarking aggression identification in social media. In: Proceedings of the First Workshop on Trolling, Aggression and Cyberbullying, TRAC 2018, pp. 1–11 (2018)
9. Madisetty, S., Desarkar, M.S.: Aggression detection in social media using deep neural networks. In: Proceedings of the First Workshop on Trolling, Aggression and Cyberbullying, TRAC 2018, pp. 120–127 (2018)
10. Mikolov, T., Sutskever, I., Chen, K., Corrado, G.S., Dean, J.: Distributed representations of words and phrases and their compositionality. In: Advances in neural information processing systems, pp. 3111–3119 (2013)
11. Orabi, A.H., Orabi, M.H., Huang, Q., Inkpen, D., Van Bruwaene, D.: Cyber-aggression detection using cross segment-and-concatenate multi-task learning from text. In: Proceedings of the First Workshop on Trolling, Aggression and Cyberbullying, TRAC 2018, pp. 159–165 (2018)
12. Orasan, C.: Aggressive language identification using word embeddings and sentiment features. Association for Computational Linguistics (2018)
13. Park, J.H., Fung, P.: One-step and two-step classification for abusive language detection on twitter. arXiv preprint arXiv:1706.01206 (2017)
14. Pitsilis, G.K., Ramampiaro, H., Langseth, H.: Detecting offensive language in tweets using deep learning. arXiv preprint arXiv:1801.04433 (2018)
15. Schmidt, A., Wiegand, M.: A survey on hate speech detection using natural language processing. In: Proceedings of the Fifth International Workshop on Natural Language Processing for Social Media, pp. 1–10 (2017)
16. Singh, V., Varshney, A., Akhtar, S.S., Vijay, D., Shrivastava, M.: Aggression detection on social media text using deep neural networks. In: Proceedings of the 2nd Workshop on Abusive Language Online (ALW2), pp. 43–50 (2018)
17. Warner, W., Hirschberg, J.: Detecting hate speech on the world wide web. In: Proceedings of the Second Workshop on Language in Social Media, pp. 19–26. Association for Computational Linguistics (2012)

18. Waseem, Z., Hovy, D.: Hateful symbols or hateful people? Predictive features for hate speech detection on twitter. In: Proceedings of the NAACL Student Research Workshop, pp. 88–93 (2016)
19. Waseem, Z., Thorne, J., Bingel, J.: Bridging the gaps: multi task learning for domain transfer of hate speech detection. In: Golbeck, J. (ed.) Online Harassment. HIS, pp. 29–55. Springer, Cham (2018). https://doi.org/10.1007/978-3-319-78583-7_3
20. Zampieri, M., Malmasi, S., Nakov, P., Rosenthal, S., Farra, N., Kumar, R.: Predicting the type and target of offensive posts in social media (2019)
21. Zhang, X., Zhao, J., LeCun, Y.: Character-level convolutional networks for text classification. In: Advances in Neural Information Processing Systems, pp. 649–657 (2015)

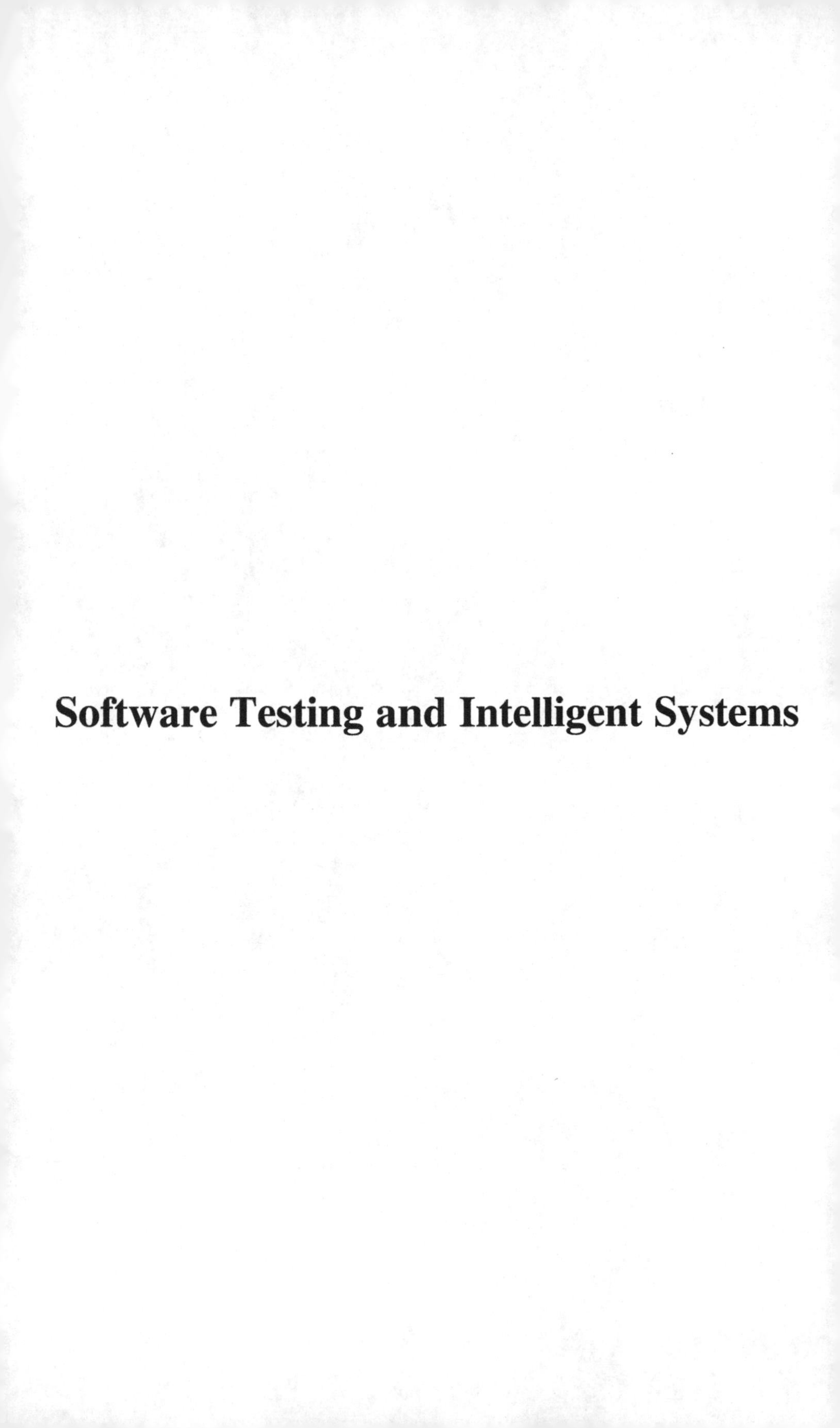

# Software Testing and Intelligent Systems

# Security Testing for Multi-Agent Systems

Damas P. Gruska[1] and M. Carmen Ruiz[2(✉)]

[1] Comenius University, Bratislava, Slovakia
[2] Universidad de Castilla-La Mancha, Albacete, Spain
MCarmen.Ruiz@uclm.es

**Abstract.** A combination of formal methods and security testing for communication layer of multi-agent systems is proposed and studied. We start with security property called bisimulation process opacity. Unfortunately, this property is undecidable in general so we propose its more realistic variant based on simulation and tests and testing. A test represents an attacker's (i.e. possibly one of the agents) scenario to obtain some confidential information on systems. Here we consider system to be secure if it cannot be compromised by a given test or set of tests. By test we can express also capabilities of an attacker related to time properties such as time measurement accuracy, duration of tests (attacks), complete lack of attacker's time information and so on. At the end we state a decidability result for testing.

**Keywords:** Multi-Agent systems · Timed process algebras · Information flow · Security · Testing

## 1 Introduction

Formal methods which include formal specifications and formal verifications play an important role to guarantee software quality. On the other side their application is frequently either rather expensive or practically impossible in case of complex systems. On top of these there are properties which cannot be formally verified in general due to their undecidability. For such cases tests and testing offer more realistic approach.

In this paper we address security properties of systems, particularly multi-agent (MAS), and means how to test these properties. In a sense, we combine formal verification with testing. We propose a formalism for a security analysis and testing of systems specified by timed process algebras which could be seen as a model for communication layer of MAS. On this layer all MAS's actions are considered to be just communications among agents but they still could exhibit some vulnerable behaviour.

Work supported by the grant VEGA 1/0778/18, the Spanish Ministry of Science and Innovation and the European Union FEDER Funds under grant CAS18/00106 and TIN2015-65845-C3-2-R and by the JCCM regional project SBPLY/17/180501/000276.


I. Rojas et al. (Eds.): IWANN 2019, LNCS 11506, pp. 703–715, 2019.
https://doi.org/10.1007/978-3-030-20521-8_58

The proposed approach allows us to formalize security properties based on an absence of information flow between private and public system's activities or states. The presented approach combines several ideas emerged from the security theory. We exploit an idea (of an absence) of information flow between public and private system's behaviour (see [GM82]). This concept has been exploited many times in various formalisms. For example, the security property called Bisimulation Strong Nondeterministic Non-Interference requires that it cannot be distinguished (by means of bisimulation) between forbidding and hiding of private actions. In [Gru13] we have exploited this idea but we weaken it by requiring that forbidding and hiding of the private actions cannot be distinguished by a given test, i.e. we exploit a kind of testing equivalence (see also [NH84, SL95]).

Here we start with security concept called opacity. To explain the opacity principle, let's suppose that we have some process's property. This might be an execution of one or more classified actions, an execution of actions in a particular classified order which should be kept hidden, etc. We suppose that this property is expressed by predicate $\phi$ over process' traces. We would like to know whether an observer can deduce the validity of the property $\phi$ just by (partially) observing sequences of actions (traces) performed by the given process. The observer cannot deduce the validity of $\phi$ if there are two traces $w, w'$ such that $\phi(w) \wedge \neg\phi(w')$ holds and the traces cannot be distinguished by the observer.

Many security properties could be viewed as special cases of opacity (see, for example, [Gru12, Gru17, Gru18]). In [Gru15] opacity is modified (the result is called process opacity) in such a way that instead of a process' traces we focus on properties of reachable states. Hence we assume an intruder who is not primarily interested in whether some sequence of actions performed by a given process has some given property but we consider an intruder who wants to discover whether this process reaches a state which satisfied some given (classified) predicate. It turned out that in this way we could capture many new security flaws. On the other hand some security flaws, particularly important for multi-agent systems, are not covered by this state-based security property neither by its variant called an initial state opacity, studied in [Gru17].

In this paper we extend process opacity to reflect attackers which can interact with systems and not only observe their traces. This approach is particularly appealing for multi-agent systems where one of the agents could be an attacker which can not only observe but also interact with other agents. The resulting property called bisimulation process opacity is even more general then process opacity.

Qualitative security properties are often criticized for being either too restrictive or too benevolent. For example, a standard access control process should be considered insecure even if there always exists some (even very small) information flow which could help an attacker who tries to learn a password. By every attempt an attacker can learn, at least, what is not the correct one. There are several ways to overcome these disadvantages i.e. either quantify information flow or put some restrictions attackers capabilities. An amount of leaked information could be expressed by means of the Shannon's information theory as it was done, for example, in [CHM07, CMS09] for simple imperative languages and in [Gru08] for process algebras. Another possibility is to exploit the probability theory as it was done for process algebras in [Gru09]. In this way we can obtain quantification of information flow either as a number of bits of private information which could leak or as a probability that an intruder can learn some secrete property.

Here we exploit a different approach. We define tests and testing of security properties. Hence instead of requiring general security we require security with respect to given set of tests. Each test represents a possible scenario of an attacker as well his or her capabilities. For example, an access control system with strong password policy should be considered secure with respect to a "small" attackers (tests), which can try only few passwords. The presented testing approach is strictly stronger then that of [Gru11], which is based on simple process's observations. On the other side, bisimulation based approach, which is stronger, is often too strong and does not correspond to real(istic) possible intruders. Moreover, testing allow us, besides other advantages, to express security of a system with respect to size of the test which could jeopardize its security. Hence the resulting level of security gives us relevant information on real (practical) system security.

The paper is organized as follows. Our working formalism, i.e. the timed process algebra, is introduced in Sect. 2. In Sect. 3 we describe information flow security properties of interest. In Sect. 4 we define tests and testing and in Sect. 5 we relate testing and security.

## 2 Timed Process Algebra

In this section we define Timed Process Algebra, TPA for short. TPA is based on Milner's CCS but the special time action $t$ which expresses elapsing of (discrete) time is added. The presented language is a slight simplification of Timed Security Process Algebra introduced in [FGM00]. We omit an explicit idling operator $\iota$ used in tSPA and instead of this we allow implicit idling of processes. Hence processes can perform either "enforced idling" by performing $t$ actions which are explicitly expressed in their descriptions or "voluntary idling" (i.e. for example, the process $a.Nil$ can perform $t$ action since it is not contained the process specification). But in both cases internal communications have priority to action $t$ in the parallel composition. Moreover we do not divide actions into private and public ones as it is in tSPA. TPA differs also from the tCryptoSPA (see [GM04]). TPA does not use value passing and strictly preserves *time determinancy* in case of choice operator $+$ what is not the case of tCryptoSPA.

To define the language TPA, we first assume a set of atomic action symbols $A$ not containing symbols $\tau$ and $t$, and such that for every $a \in A$ there exists $\overline{a} \in A$ and $\overline{\overline{a}} = a$. We define $Act = A \cup \{\tau\}, At = A \cup \{t\}, Actt = Act \cup \{t\}$. We assume that $a, b, \ldots$ range over $A$, $u, v, \ldots$ range over $Act$, and $x, y \ldots$ range over $Actt$. Assume the signature $\Sigma = \bigcup_{n \in \{0,1,2\}} \Sigma_n$, where

$$\Sigma_0 = \{Nil\}$$
$$\Sigma_1 = \{x. \mid x \in A \cup \{t\}\} \cup \{[S] \mid S \text{ is a relabeling function}\}$$
$$\cup\{\backslash M \mid M \subseteq A\}$$
$$\Sigma_2 = \{|, +\}$$

with the agreement to write unary action operators in prefix form, the unary operators $[S], \backslash M$ in postfix form, and the rest of operators in infix form. Relabeling functions, $S : Actt \rightarrow Actt$ are such that $\overline{S(a)} = S(\bar{a})$ for $a \in A, S(\tau) = \tau$ and $S(t) = t$.

The set of TPA terms over the signature $\Sigma$ is defined by the following BNF notation:

$$P ::= X \mid op(P_1, P_2, \ldots P_n) \mid \mu XP$$

where $X \in Var$, $Var$ is a set of process variables, $P, P_1, \ldots P_n$ are TPA terms, $\mu X-$ is the binding construct, $op \in \Sigma$.

The set of CCS terms consists of TPA terms without $t$ action. We will use an usual definition of opened and closed terms where $\mu X$ is the only binding operator. Closed terms which are t-guarded (each occurrence of $X$ is within some subterm $t.A$ i.e. between any two $t$ actions only finitely many non timed actions can be performed) are called TPA processes.

We give a structural operational semantics of terms by means of labeled transition systems. The set of terms represents a set of states, labels are actions from $Actt$. The transition relation $\rightarrow$ is a subset of TPA $\times$ $Actt$ $\times$ TPA. We write $P \xrightarrow{x} P'$ instead of $(P, x, P') \in \rightarrow$ and $P \not\xrightarrow{x}$ if there is no $P'$ such that $P \xrightarrow{x} P'$. The meaning of the expression $P \xrightarrow{x} P'$ is that the term $P$ can evolve to $P'$ by performing action $x$, by $P \xrightarrow{x}$ we will denote that there exists a term $P'$ such that $P \xrightarrow{x} P'$. We define the transition relation as the least relation satisfying the inference rules for CCS plus the following inference rules:

$$\frac{}{Nil \xrightarrow{t} Nil} A1 \qquad \frac{}{u.P \xrightarrow{t} u.P} A2$$

$$\frac{P \xrightarrow{t} P', Q \xrightarrow{t} Q', P \mid Q \not\xrightarrow{\tau}}{P \mid Q \xrightarrow{t} P' \mid Q'} Pa \qquad \frac{P \xrightarrow{t} P', Q \xrightarrow{t} Q'}{P + Q \xrightarrow{t} P' + Q'} S$$

Here we mention the rules that are new with respect to CCS. Axioms $A1, A2$ allow arbitrary idling. Concurrent processes can idle only if there is no possibility of an internal communication ($Pa$). A run of time is deterministic ($S$) i.e. performing of $t$ action does not lead to the choice between summands of $+$. In the definition of the labeled transition system we have used negative premises (see $Pa$). In general this may lead to problems, for example with consistency of the defined system. We avoid these dangers by making derivations of $\tau$ independent of derivations of $t$.

For $s = x_1.x_2.\ldots.x_n, x_i \in Actt$ we write $P \xrightarrow{s}$ instead of $P \xrightarrow{x_1}\xrightarrow{x_2} \ldots \xrightarrow{x_n}$ and we say that $s$ is a trace of $P$. The set of all traces of $P$ will be denoted by $Tr(P)$. By $\epsilon$ we will denote the empty sequence of actions, by $Succ(P)$ we will denote the set of all successors of $P$ i.e. $Succ(P) = \{P' | P \xrightarrow{s} P', s \in Actt^*\}$. If the set $Succ(P)$ is finite we say that $P$ is a finite state process. We define modified transitions $\stackrel{x}{\Rightarrow}_M$ which "hide" actions from $M$. Formally, we will write $P \stackrel{x}{\Rightarrow}_M P'$ for $M \subseteq Actt$ iff $P \xrightarrow{s_1}\xrightarrow{x}\xrightarrow{s_2} P'$ for $s_1, s_2 \in M^\star$ and $P \stackrel{s}{\Rightarrow}_M$ instead of $P \stackrel{x_1}{\Rightarrow}_M \stackrel{x_2}{\Rightarrow}_M \cdots \stackrel{x_n}{\Rightarrow}_M$. We will write $P \stackrel{x}{\Rightarrow}_M$ if there exists $P'$ such that $P \stackrel{x}{\Rightarrow}_M P'$. We will write $P \stackrel{\widehat{x}}{\Rightarrow}_M P'$ instead of $P \stackrel{\epsilon}{\Rightarrow}_M P'$ if $x \in M$. Note that $\stackrel{x}{\Rightarrow}_M$ is defined for arbitrary action $x$ but in definitions of security properties we will use it for actions (or sequence of actions) not belonging to $M$. We can extend the definition of $\Rightarrow_M$ for sequences of actions similarly to $\xrightarrow{s}$. Let $s \in Actt^\star$. By $|s|$ we will denote the length of $s$ i.e. a number of action contained in $s$. By $s|_B$ we will denote the sequence obtained from $s$ by removing all actions not belonging to $B$. For example, $|s|_{\{t\}}|$ denote a number of occurrences of $t$ in $s$, i.e. time length of $s$. By $Sort(P)$ we will denote the set of actions from $A$ which can be performed by $P$. By $Succ(P)$ we denote the set of all successors of $P$ i.e. $Succ(P) = \{P' | P \xrightarrow{s} P'$, for $s \in Actt^*\}$. The set of traces of process $P$ is defined as $L(P) = \{s \in Actt^\star | \exists P'.P \stackrel{s}{\Rightarrow} P'\}$. The set of weak timed traces of process $P$ is defined as $L_w(P) = \{s \in (A \cup \{t\})^\star | \exists P'.P \stackrel{s}{\Rightarrow}_{\{\tau\}} P'\}$. Two processes $P$ and $Q$ are weakly timed trace equivalent ($P \simeq_w Q$) iff $L_w(P) = L_w(Q)$. We conclude this section with definitions of M-bisimulation and weak timed trace equivalence.

**Definition 1.** *Let $(TPA, Actt, \rightarrow)$ be a labelled transition system (LTS). A relation $\Re \subseteq TPA \times TPA$ is called a* M-bisimulation *if it is symmetric and it satisfies the following condition: if $(P,Q) \in \Re$ and $P \xrightarrow{x} P', x \in Actt$ then there exists a process $Q'$ such that $Q \stackrel{\widehat{x}}{\Rightarrow}_M Q'$ and $(P',Q') \in \Re$. Two processes $P,Q$ are M-bisimilar, abbreviated $P \approx_M Q$, if there exists a M-bisimulation relating $P$ and $Q$.*

We conclude this section with definitions of trace equivalence, weak simulation and weak bisimulation.

**Definition 2.** *The set of weak traces of process $P$ with respect to the set $M, M \subseteq A$ is defined as $Tr_{wM}(P) = \{s \in A^\star | \exists P'.P \stackrel{s}{\Rightarrow}_M P'\}$. Instead of $Tr_{w\emptyset}(P)$ we will write $Tr_w(P)$.*

*Two processes $P$ and $Q$ are weakly trace equivalent with respect to $M$ ($P \approx_{wM} Q$) iff $Tr_{wM}(P) = Tr_{wM}(Q)$. We will write $\approx_w$ instead of $\approx_{w\emptyset}$.*

**Definition 3.** Let $(CCS, Act, \rightarrow)$ be a labeled transition system (LTS). A relation $\Re \subseteq CCS \times CCS$ is called a *weak bisimulation* if it is symmetric and it satisfies the following condition: if $(P,Q) \in \Re$ and $P \xrightarrow{x} P', x \in Act$, then there exists a process $Q'$ such that $Q \stackrel{\widehat{x}}{\Rightarrow} Q'$ and $(P',Q') \in \Re$. Two processes $P,Q$ are *weakly bisimilar*, abbreviated $P \approx Q$, if there exists a weak bisimulation relating $P$ and $Q$. If it is not required that relation $\Re$ is symmetric we call it simulation and we say that process $P$ simulates process $Q$, abbreviated $P \prec Q$, if there exists a simulation relating $P$ and $Q$.

## 3 Information Flow Security

In this section we will present our working security concepts. First we define the absence-of-information-flow property - Strong Nondeterministic Non-Interference (SNNI, for short, see [FGM00]). Suppose that all actions are divided into two groups, namely public (low level) actions $L$ and private (high level) actions $H$. It is assumed that $L \cup H = A$. SNNI property assumes an intruder who tries to learn whether a private action was performed by a given process while (s)he can observe only public ones. If this cannot be done then the process has SNNI property. Namely, process $P$ has SNNI property (we will write $P \in SNNI$) if $P \setminus H$ behaves like $P$ for which all high level actions are hidden (namely, replaced by action $\tau$) for an observer. To express this hiding we introduce the hiding operator $P/M, M \subseteq A$, for which it holds that if $P \xrightarrow{a} P'$ then $P/M \xrightarrow{a} P'/M$ whenever $a \notin M \cup \overline{M}$ and $P/M \xrightarrow{\tau} P'/M$ whenever $a \in M \cup \overline{M}$. Formally, we say that $P$ has SNNI property, and we write $P \in SNNI$ iff $P \setminus H \simeq_w P/H$. A generalization of this concept is given by opacity (this concept was exploited in [BKR04] and [BKMR06] in a framework of Petri Nets, transition systems and process algebras, respectively). Actions are not divided into public and private ones at the system description level but a more general concept of observations and predicates are exploited. A predicate is opaque if for any trace of a system for which it holds, there exists another trace for which it does not hold and the both traces are indistinguishable for an observer (which is expressed by an observation function). This means that the observer (intruder) cannot say whether a trace for which the predicate holds has been performed or not. Now let us assume a different scenario, namely that an intruder is not interested in traces and their properties but he or she tries to discover whether a given process always reaches a state with some given property which is expressed by a (total) predicate. This property might be process deadlock, capability to execute only traces with time length less then $n$ time unites, capability to perform at the same time actions form a given set, incapacity to idle (to perform $t$ action) etc. We do not put any restriction on such predicates but we only assume that they are consistent with some suitable behavioral equivalence. The formal definition follows.

**Definition 4.** *We say that the predicate $\phi$ over processes is consistent with respect to relation $\cong$ if whenever $P \cong P'$ then $\phi(P) \Leftrightarrow \phi(P')$.*

As the consistency relation $\cong$ we could take bisimulation ($\approx_\emptyset$), weak bisimulation ($\approx_{\{\tau\}}$) or any other suitable equivalence. A special class of such predicates are such ones (denoted as $\phi^Q_{\cong}$) which are defined by a given process $Q$ and equivalence relation $\cong$ i.e. $\phi^Q_{\cong}(P)$ holds iff $P \cong Q$.

We suppose that the intruder can observe only some activities performed by the process. Hence we suppose that there is a set of public actions which can be observed and a set of hidden (not necessarily private) actions. To model such observations we exploit the relation $\stackrel{s}{\Rightarrow}_M$ where actions from $M$ are those ones which could not be seen by the observer. The formal definition of process opacity (see [Gru15]) is the following.

**Definition 5 (Process Opacity).** *Given process $P$, a predicate $\phi$ over processes is process opaque w.r.t. the set $M$ if whenever $P \stackrel{s}{\Rightarrow}_M P'$ for $s \in (Actt \setminus M)^*$ and $\phi(P')$ holds then there exists $P''$ such that $P \stackrel{s}{\Rightarrow}_M P''$ and $\neg\phi(P'')$ holds. The set of processes for which the predicate $\phi$ is process opaque w.r.t. to the $M$ will be denoted by $POp_M^{\phi}$.*

Note that if $P \cong P'$ then $P \in POp_M^{\phi} \Leftrightarrow P' \in POp_M^{\phi}$ whenever $\phi$ is consistent with respect to $\cong$ and $\cong$ is such that it is a subset of the trace equivalence (defined as $\simeq_w$ but instead of $\stackrel{s}{\Rightarrow}_{\{\tau\}}$ we use $\stackrel{s}{\Rightarrow}_{\emptyset}$).

$$P \stackrel{s}{\Longrightarrow}_M \phi(P')$$

$$P \stackrel{s}{\Longrightarrow}_M \neg\phi(P'')$$

**Fig. 1.** Process opacity

The security property process opacity expects an attacker who can just observe process' traces but cannot interact with the process. In many cases this does not cover real attacks and attackers, particularly in the case of multi-agent systems where a possible attacker could be one of the agents. Hence we generalize process opacity in such a way that also every successor of a process has to be process opaque as well. The formal definition follows.

**Definition 6 (Bisimulation Process Opacity).** *Given a predicate $\phi$ over processes and set of actions $M$. We say $\Im$ is the set of processes which are bisimulation process opaque w.r.t. $\phi$ and $M$ if for every $P \in \Im$ it holds $Succ(P) \subseteq \Im$ and if $P \stackrel{x}{\Rightarrow}_M P'$ for $x \in Actt \setminus M$ and $\phi(P')$ holds then there exists $P''$ such that $P \stackrel{s}{\Rightarrow}_M P''$ and $\neg\phi(P'')$ holds. The maximal bisimulation process opaque set w.r.t. to $\Phi$ and $M$ will be denoted by $BPOp_M^{\phi}$. Process is said to be bisimulation process opaque iff $P \in BPOp_M^{\phi}$.*

It is easy to prove, directly form Definitions 5 and 6 that bisimulation process opacity is the stronger property than process opacity.

**Proposition 1.** $POp_M^{\phi} \subset BPOp_M^{\phi}$.

The previous definition mimics process opacity in such a sense that whenever a reached process satisfies property $\phi$ than there such exists another process, reached by the same observable way, which does not satisfy $\phi$. In this case an attacker cannot learn whether $\phi$ is satisfied but its negation can still be learned. Hence we define stronger variant of bisimulation process opacity.

**Definition 7 (Strong Bisimulation Process Opacity).** *Given a predicate $\phi$ over processes and set of actions $M$. We say $\Im$ is the set of processes which are bisimulation process opaque w.r.t. $\phi$ and $M$ if for every $P \in \Im$ it holds $Succ(P) \subseteq \Im$ and if $P \stackrel{x}{\Rightarrow}_M P'$ for $x \in Actt \setminus M$ and $\phi(P')$ holds then there exists $P''$ such that $P \stackrel{s}{\Rightarrow}_M P''$ and $\neg\phi(P'')$ holds. Moreover if $P \stackrel{x}{\Rightarrow}_M P'''$ for $x \in Actt \setminus M$ and $\neg\phi(P''')$ holds then there exists $P''''$ such that $P \stackrel{s}{\Rightarrow}_M P''''$ and $\phi(P'''')$ holds. The maximal strong bisimulation process opaque set w.r.t. to $\Phi$ and $M$ will be denoted by $sBPOp^{\phi}_M$. Process is said to be bisimulation process opaque iff $P \in sBPOp^{\phi}_M$.*

**Proposition 2.** *In general it holds $sBPOp^{\phi}_M \subset BPOp^{\phi}_M$.*

*Proof.* Sketch. The inclusion follows directly from Definitions 6 and 7. Clearly, a process with all its successors for which it hold $\neg\phi$, belongs to $BPOp^{\phi}_M$ but not to $sBPOp^{\phi}_M$, hence the inclusion is proper.

**Proposition 3.** *Property bisimulation process opacity is undecidable in general.*

*Proof.* The main idea. Let property $\phi$ express that corresponding process "halts". The rest follows from undecidability of halting problem for Turing machine due to Turing machine power of TPA processes.

Note that a similar property holds also for strong bisimulation process opacity. In the subsequent sections we propose a realistic way how to obtain decidability of bisimulation process opacity as well as its more realistic variant by means of testing.

# 4 Testing

In this section we define basic testing scenario which will be applied in the subsequent section. First we start with asymmetric variant of M-bisimulation, called M-simulation. Basically, it requires, that every visible action (i.e. not belonging to $M$) performed by one process has to be emulated by another process which, moreover, can perform also some invisible actions (i.e. belonging to $M$). Then we define passing of a test which puts some requirements (i.e. properties expressed by some given relations $\Re, \Re'$) on processes resulting from M-simulation.

**Definition 8.** *Let $(TPA, Actt, \rightarrow)$ be a labelled transition system (LTS). A relation $\Re \subseteq TPA \times TPA$ is called a* M-simulation *if it satisfies the following condition: if $(P, Q) \in \Re$ and $P \stackrel{x}{\rightarrow} P', x \in A \setminus M$ then there exists a process $Q'$ such that $Q \stackrel{x}{\Rightarrow}_M Q'$ and $(P', Q') \in \Re$. We will denote the union of all $M$-simulations as $\prec_M$. If $M = \emptyset$ we will write $\prec$ instead of $\prec_{\emptyset}$.*

**Definition 9.** *Let $M \subset A$. We say that process $P$ passes test $T$ with respect to $M$ and relations $\Re, \Re'$ such that $\emptyset \times Succ(P) \subseteq \Re \cup \Re'$ (denoted by $\preceq_{M,\Re,\Re'}$) iff whenever $(T, P) \in \Re \cup \Re'$ then for every $a \in A$ if $T \stackrel{a}{\rightarrow} T'$ and there exists a process $P'$ such that $P \stackrel{\bar{a}}{\Rightarrow}_M P'$ and $(T', P') \in \Re$ than there exists also a process $P''$ such that $P \stackrel{\bar{a}}{\Rightarrow}_M P''$ and $(T', P') \in \Re'$.*

Now we will present some useful properties of testing. The first one claims that whenever a process passes a stronger test (in a sense of simulation) then it has to pass also a weaker test. By tests $T$ $(T_1, T_2, \dots)$ we will use processes which does not contain $\tau$ and $M$ actions i.e. such that $Sort(T) \cap (M \cup \{\tau\}) = \emptyset$. Now we will formulate some properties, including compositional ones, of M-simulation and testing.

**Lemma 1.** *Let $T_1 \prec T_2$. Then $T_2 \preceq_{M,\Re,\Re'} P$ implies $T_1 \preceq_{M,\Re,\Re'} P$.*

*Proof.* Let $T_2 \preceq_{M,\Re,\Re'} P$ but $T_1 \not\preceq_{M,\Re,\Re'} P$. Since $T_1 \prec T_2$ we know that every "computation" (execution) of $T_1$ could be simulated by $T_2$. This is true also for a computation due to which process $P$ does not pass test $T_1$ and hence we have a contradiction with $T_2 \preceq_{M,\Re,\Re'} P$.

The following Lemma is mostly a consequence of the previous one. It states some simple compositional results.

**Lemma 2.** *Let $T_1 + T_2 \preceq_{M,\Re,\Re'} P$ or $T_1|T_2 \preceq_{M,\Re,\Re'} P$. Then $T_1 \preceq_{M,\Re,\Re'} P$. Let $T \preceq_{M,\Re,\Re'} P$. Then $T \setminus M' \preceq_{M,\Re,\Re'} P$ for every $M', M' \subseteq A$. Let $T \preceq_{M,\Re,\Re'} P$. Then $x.T \preceq_{M,\Re,\Re'} x.P$.*

*Proof.* Except the last implication, all properties are direct consequences of the previous lemma since $T_1 \prec T_1 + T_2$, $T_1 \prec T_1|T_2$, $T \setminus M' \prec T$. The last implication follows directly from Definition 9.

A kind of an inverse Lemma to Lemma 1 holds: if a weaker process passes a test then also a stronger one has to pass the test.

**Lemma 3.** *Let $P_1 \prec P_2$. Then $T \preceq_{M,\Re,\Re'} P_1$ implies $T \preceq_{M,\Re,\Re'} P_2$.*

*Proof.* Again let $T \not\preceq_{M,\Re,\Re'} P_2$. But since every computation of $P_1$ could be simulated by $P_2$ and $P_1$ passes the test we have a contradiction.

We present some compositional properties for tested processes.

**Lemma 4.** *Let $T \preceq_{M,\Re,\Re'} P$. Then $T \preceq_{M,\Re,\Re'} P + Q$, $T \preceq_{M,\Re,\Re'} P|Q$ and $x.T \preceq_{M,\Re,\Re'} x.P$ for $x \notin M$.*

*Proof.* The proof follows directly from the previous lemma and the fact that $P \prec P + Q, P|Q$.

Now we will define the test equivalence with respect to the given test $T$. Two process will be test equivalent if they cannot be distinguished by test $T$. In fact we should prove that this relation is really equivalence relation but this follows from properties of logical equivalence ($\leftrightarrow$).

**Definition 10.** *We define the test equivalence with respect to test $T$, relations $\Re, \Re'$ and set $M, M \subset A$ (we will denote it as $\approx_{T,M,\Re,\Re'}$) as $P \approx_{T,M,\Re,\Re'} Q$ iff $T \preceq_{M,\Re,\Re'} P \leftrightarrow T \preceq_{M,\Re,\Re'} Q$.*

The test equivalence expresses that security of processes could not be distinguished by test $T$. Now will formulate a stronger variant of testing which would correspond, as we will see later, to the stronger variant of bisimulation process opacity.

**Definition 11.** *Let $M \subset A$. We say that process $P$ passes test $T$ with respect to $M$ and relations $\Re, \Re'$ such that $\emptyset \times Succ(P) \subseteq \Re \cup \Re'$ (denoted by $\preceq_{sM,\Re,\Re'}$) iff whenever $(T, P) \in \Re \cup \Re'$ then for every $a \in A$ if $T \xrightarrow{a} T'$ then there exist processes $P', P''$ such that $P \stackrel{\bar{a}}{\Rightarrow}_M P'$, $(T', P') \in \Re$ and $P \stackrel{\bar{a}}{\Rightarrow}_M P''$, $(T', P') \in \Re'$.*

Clearly, strong testing is really stronger than testing. The following two Lemmas, which are counterparts of Lemmas 1 and 2 hold also for the strong testing and also their proofs are similar. Note that the other two Lemmas (3 and 4) which hold for the testing do not hold for the strong testing. But a weaker variant of Lemma 3 still holds (Lemma 7).

**Lemma 5.** *Let $T_1 \prec T_2$. Then $T_2 \preceq_{sM,\Re,\Re'} P$ implies $T_1 \preceq_{sM,\Re,\Re'} P$.*

**Lemma 6.** *Let $T_1 + T_2 \preceq_{sM,\Re,\Re'} P$ or $T_1|T_2 \preceq_{sM,\Re,\Re'} P$. Then $T_1 \preceq_{sM,\Re,\Re'} P$.*

**Lemma 7.** *Let $T \preceq P$ and $T \preceq Q$. Then $T \preceq_{sM,\Re,\Re'} P + Q$.*

*Proof.* The proof follows from the fact that every successor of $P+Q$ is a successor either of $P$ or $Q$. Note that without the assumption $T \preceq Q$ (as it is in Lemma 7) this lemma does not hold.

Now we can define the strong test equivalence similarly to Definition 10.

**Definition 12.** *We define strong test equivalence with respect to test $T$ and set $M, M \subset A$ (we will denote it as $\approx_{s,T,M}$) as $P \approx_{s,T,M\Re,\Re} Q$ iff $T \preceq_{sM,\Re,\Re'} P \leftrightarrow T \preceq_{sM,\Re,\Re'} Q$.*

## 5 Bisimulation Process Opacity Testing

In this section we will study a relation between testing and bisimulation process opacity. In general, this property is not only undecidable but for many applications too strong or too restrictive. In many cases it is more realistic to require system's security only with respect to a given set of tests, which represents attackers (as an agent of MAS) capabilities. First we need some notations. Let set $\phi(TPA)$ denotes processes for which $\phi$ holds, that means $\phi(TPA) = \{P|$ such that $\phi(P)$ holds$\}$. Let $\Re_\phi = TPA \times \phi(TPA)$ and $\Re'_{\neg\phi} = TPA \times \neg\phi(TPA)$. If a process is secure withe respect to bisimulation process opacity then it should pass any test as it is stated by the following proposition.

**Proposition 4.** *$P \in BPOp^{\phi}_M$ iff for every process $T, Sort(T) \subset (A \cup \{t\}) \setminus M$ it holds $T \preceq_{M,\Re_\phi,Re'_{\neg\phi}} P$.*

*Proof.* The main idea. Suppose that $P \notin BPOp_M^\phi$ i.e. $P$ does not satisfy Definition 6. Hence we can construct a test $T$ which mimic why $P$ does not satisfy the property. It easy to see that for such process $T \preceq_{M,\Re_\phi,Re'_{\neg\phi}} P$ does not hold.

The similar property holds also for strong bisimulation process opacity as it is stated by the following theorem.

**Proposition 5.** *$P \in sBPOp_M^\phi$ iff for every process $T, Sort(T) \subset (A \cup \{t\}) \setminus M$ it holds $T \preceq_{sM,\Re_\phi,Re'_{\neg\phi}} P$.*

On the other site we might require security of systems only with respect to a given set of specific tests. For example, let us suppose that $T$ is such that $Sort(T) \subseteq A$ i.e. the test represents an attacker who has no notion about elapsing of time. Then passing such test represents robustness with respect to timing attacks. Let $\mathcal{T}$ be a set of tests. We can define passing test with respect the set of test as $\mathcal{T} \preceq_{M,\Re_\phi,Re'_{\neg\phi}} P$ iff $T \preceq_{M,\Re_\phi,Re'_{\neg\phi}} P$ for every $T \in \mathcal{T}$.

**Corollary 1.** *Let $P \approx_{T,M\Re,\Re} Q$ ($P \approx_{s,T,M\Re,\Re} Q$) for every test $T$. Then $P \in BPOp_M^\phi$ ($P \in sBPOp_M^\phi$) iff $Q \in BPOp_M^\phi$ ($Q \in sBPOp_M^\phi$), respectively.*

As regards complexity of testing, sets $\Re, \Re'$ represent the crucial problem. If these sets are easy to compute the testing becomes feasible. Suppose that there exists process $F$ such that for every $P$, $P \sim F$ iff $\phi(P)$ holds. That means that property $\phi$ can by defined by bisimulation with $F$. In this case we say that $\phi$ is defined by $F$.

**Proposition 6.** *Let $\mathcal{T}$ is finite set of finite state tests. Let $\phi$ is defined by finite state process $F$. Then property $\mathcal{T} \preceq_{M,\Re_\phi,Re'_{\neg\phi}} P$ is decidable.*

*Proof.* The main idea. Bisimulation is decidable for finite state processes. Similarly it can be shown that also simulation and testing and its variant are decidable.

## 6 Conclusions

We have presented the new security concept called bisimulation process opacity and its stronger variant. We have proposed a way how to (partially) verify these properties by means of tests and testings. Each test represents a scenario for an attacker as well his or her capabilities. Instead of verifying bisimulation process opacity we define system security with respect to a given set of tests. We have shown, that under some restriction, this testing is feasible or at least decidable.

As regards future work, we plan to define and study a minimal set of tests which are necessary to be passed to guarantee some security property. In this way we would simplify overall testing.

## References


[BKR04] Bryans J., Koutny, M., Ryan, P.: Modelling non-deducibility using Petri Nets. In: Proceedings of the 2nd International Workshop on Security Issues with Petri Nets and other Computational Models (2004)

[BKMR06] Bryans, J.W., Koutny, M., Mazaré, L., Ryan, P.Y.A.: Opacity generalised to transition systems. In: Dimitrakos, T., Martinelli, F., Ryan, P.Y.A., Schneider, S. (eds.) FAST 2005. LNCS, vol. 3866, pp. 81–95. Springer, Heidelberg (2006). https://doi.org/10.1007/11679219_7

[CHM07] Clark, D., Hunt, S., Malacaria, P.: A static analysis for quantifying the information flow in a simple imperative programming language. J. Comput. Secur. **15**(3), 321–371 (2007)

[CMS09] Clarkson, M.R., Myers, A.C., Schneider, F.B.: Quantifying information flow with beliefs. J. Comput. Secur., (2009, to appear)

[NH84] De Nicola, R., Hennessy, M.C.B.: Testing equivalences for processes. Theoret. Comput. Sci. **34**, 83–133 (1984)

[FGM00] Focardi, R., Gorrieri, R., Martinelli, F.: Information flow analysis in a discrete-time process algebra. In: Proceedings of 13th Computer Security Foundation Workshop. IEEE Computer Society Press (2000)

[GSS95] van Glabbeek, R.J., Smolka, S.A., Steffen, B.: Reactive, generative and stratified models of probabilistic processes. Inf. Comput. **121**, 59–80 (1995)

[GM04] Gorrieri, R., Martinelli, F.: A simple framework for real-time cryptographic protocol analysis with compositional proof rules. Sci. Comput. Program. **50**(1–3), 23–49 (2004)

[GM82] Goguen, J.A., Meseguer, J.: Security policies and security models. In: Proceedings of IEEE Symposium on Security and Privacy (1982)

[Gru18] Gruska, D.P., Ruiz, M.C.: Opacity-enforcing for process algebras. In: CS&P 2018 (2018)

[Gru17] Gruska, D.P., Ruiz, M.C.: Initial process security. In: Specification and Verification CS&P 2017 (2017)

[Gru15] Gruska, D.P.: Process opacity for timed process algebra. In: Voronkov, A., Virbitskaite, I. (eds.) PSI 2014. LNCS, vol. 8974, pp. 151–160. Springer, Heidelberg (2015). https://doi.org/10.1007/978-3-662-46823-4_13

[Gru13] Gruska, D.P.: Information flow testing. Fundam. Inf. **128**(1–2), 81–95 (2013)

[Gru12] Gruska, D.P.: Informational analysis of security and integrity. Fundam. Inf. **120**(3–4), 295–309 (2012)

[Gru11] Gruska, D.P.: Gained and excluded private actions by process observations. Fundam. Inf. **109**(3), 281–295 (2011)

[Gru09] Gruska, D.P.: Quantifying security for timed process algebras. Fundam. Inf. **93**(1–3), 155–169 (2009)

[Gru08] Gruska, D.P.: Probabilistic information flow security. Fundam. Inf. **85**(1–4), 173–187 (2008)

[HJ90] Hansson, H., Jonsson, B.: A calculus for communicating systems with time and probabilities. In: Proceedings of 11th IEEE Real - Time Systems Symposium, Orlando (1990)

[LN04] López, N., Núñez, M.: An overview of probabilistic process algebras and their equivalences. In: Baier, C., Haverkort, B.R., Hermanns, H., Katoen, J.-P., Siegle, M. (eds.) Validation of Stochastic Systems. LNCS, vol. 2925, pp. 89–123. Springer, Heidelberg (2004). https://doi.org/10.1007/978-3-540-24611-4_3

[SL95] Segala, R., Lynch, N.: Probabilistic simulations for probabilistic processes. Nord. J. Comput. **2**(2), 250–273 (1995)

[YL92] Yi, W., Larsen, K.G.: Testing probabilistic and nondeterministic processes. In: Proceeding Proceedings of the IFIP TC6 - WG6.1 (1992)

# GPTSG: A Genetic Programming Test Suite Generator Using Information Theory Measures

Alfredo Ibias[1], David Griñán[2], and Manuel Núñez[1](✉)

[1] Complutense University of Madrid, 28040 Madrid, Spain
{aibias,manuelnu}@ucm.es
[2] Polytechnic University of Madrid, 28223 Madrid, Spain
david.grinanm@alumnos.upm.es

**Abstract.** The automatic generation of test suites that get the best score with respect to a given measure is costly in terms of computational power. In this paper we present a genetic programming approach for generating test suites that get a good enough score for a given measure. We consider a black-box scenario and include different Information Theory measures. Our approach is supported by a tool that will actually generate test suites according to different parameters. We present the results of a small experiment where we used our tool to compare the goodness of different measures.

**Keywords:** Testing · Genetic programming · Test generation · Information Theory

## 1 Introduction

Software testing [2,20] is an important research topic because it helps to ensure that the systems work as expected. Specially important have been the efforts to define testing in a formal way, which is still an active research area [3]. There are two orthogonal approaches to test a system: Consider that the system is a white-box or consider that it is a black-box. In this paper, we will focus on the latter as it poses a higher challenge because we have to rely on a model of the system, but we cannot see how the implementation of the system works. One of the main components of black-box testing consists in generating test suites that find the maximum number of faults. Actually, this is a critical part when we are talking about systems that require a lot of time to process an input or that have a small time window where you can test them. As it is a critical task, a lot of work around generating and selecting test suites and constructing tools supporting the theoretical frameworks has been performed [23].

Research partially supported by the Spanish project DArDOS (TIN2015-65845-C3-1-R) and the Comunidad de Madrid project FORTE-CM (S2018/TCS-4314).


I. Rojas et al. (Eds.): IWANN 2019, LNCS 11506, pp. 716–728, 2019.
https://doi.org/10.1007/978-3-030-20521-8_59

Genetic programming has been a successful technique to find *good* enough solutions to NP-hard problems. Specifically, genetic programming was developed to solve the restrictions of genetic algorithms [15]. Instead of representing each solution as a vector, each element of the solutions space is coded as a tree, with no size limitations. In order to ensure that these trees always represent feasible solutions, the *grammar-based genetic programming* paradigm was proposed [18]. Trees according to this paradigm are produced as derivations from a grammar that is specifically designed so that every solution is feasible. This approach has been used to search for neural net structures [4], Bayesian network structures [21] and rule-based systems [17]. Another field that has been applied to find *good* test suites is Information Theory [6]. There are several approaches proposing different measures to select the *better* test suites from a given set and comparing them [9,11] and the ones that show better performance are measures based on Information Theory.

In order to find the best test suite, up to a given size, to test a certain system we confront the classical combinatorial explosion: we have to check how good are all the subsets (up to a given size) of the set of available test cases. Previous work has automatically generated test suites (or test cases) using genetic algorithms [7,8,16,22]. Most of these approaches usually consider the basic conception of a genetic algorithm, with the risk of losing the correctness of the test suite, or needing to introduce some special items (as a *do not care element* [10]) in order to preserve it. Finally, although there are some work using genetic programming, we are not aware of any work that uses genetic programming ensuring at the same time the correctness and length of the test suite.

In this paper we propose a genetic programming algorithm to generate test suites, guided by a grammar that ensures their correctness. The algorithm generates test suites that get a *good* score for a given measure, and with a fixed length, in order to avoid the generation of extremely long and computationally heavy (or short and useless) test suites. This algorithm is supported by a tool that implements it, using already proved Information Theory based measures. This tool also allows users to compare the test suites generated by the algorithm using two different measures and see how well each of them performs. Finally, the tool allows to include measures defined by the user.

The rest of the paper is organized as follows. In Sect. 2 we introduce the main concepts that we will use along the paper. In Sect. 3 we introduce the core of our genetic programming algorithm. In Sect. 4 we present the main features of our tool and the results of an experiment. Finally, in Sect. 5 we give the conclusions of our work.

## 2 Preliminaries

In this section we present the main definitions and concepts that we use throughout this paper. Most of the concepts are based on the classical notions while some notation is adapted to facilitate the formulation of subsequent definitions.

Given a set $A$, we let:
- $A^*$ denote the set of finite sequences of elements of $A$.
- $\epsilon \in A^*$ denote the empty sequence.
- $A^+$ denote the set of non-empty sequences of elements of $A$.
- $|A|$ denote the cardinal of set $A$.

Given a sequence $\sigma \in A^*$, we have that $|\sigma|$ denotes its length. Given a sequence $\sigma \in A^*$ and $a \in A$, we have that $\sigma a$ denotes the sequence $\sigma$ followed by $a$ and $a\sigma$ denotes the sequence $\sigma$ preceded by $a$.

Throughout this paper we let $I$ be the set of input actions and $O$ be the set of output actions. In our context an input of a system will be a non-empty sequence of input actions, that is, an element of $I^+$ (similarly for outputs and output actions).

A *Finite State Machine* is a (finite) labelled transition system in which transitions are labelled by an input/output pair. We use this formalism to define specifications.

**Definition 1.** *We say that* $M = (Q, q_{in}, I, O, T)$ *is a* Finite State Machine *(FSM), where* $Q$ *is a finite set of states,* $q_{in} \in Q$ *is the initial state,* $I$ *is a finite set of inputs,* $O$ *is a finite set of outputs, and* $T \subseteq Q \times (I \times O) \times Q$ *is the transition relation. A transition* $(q, (i, o), q') \in T$*, also denoted by* $q \xrightarrow{i/o} q'$ *or by* $(q, i/o, q')$*, means that from state* $q$ *after receiving input* $i$ *it is possible to move to state* $q'$ *and produce output* $o$*.*

*We say that* $M$ *is* deterministic *if for all* $q \in Q$ *and* $i \in I$ *there exists at most one pair* $(q', o) \in Q \times O$ *such that* $(q, i/o, q') \in T$*.*

*We say that* $M$ *is* input-enabled *if for all* $q \in Q$ *and* $i \in I$ *there exists* $(q', o) \in Q \times O$ *such that* $(q, i/o, q') \in T$*.*

*We let* $\mathtt{FSM}(I, O)$ *denote the set of finite state machines with input set* $I$ *and output set* $O$*.*

In this paper we assume that FSMs are deterministic. We make this assumption because most Information Theory measures are applied to code and code is usually deterministic. We do not impose that FSMs are input-enabled. We will assume the *test hypothesis* [14]: the *System Under Test* (SUT) can be modelled as an object described in the same formalism as the specification (in our case, an FSM). Note that we do not need to have access to this description; we are indeed in a black-box testing framework because we only assume the existence of such FSM. Actually, it would be enough to assume that each time that the SUT receives a sequence of input actions, it returns a sequence of output actions. As usual, we do need access to the specification.

Our main goal while testing is to decide whether the behaviour of an SUT conforms to the specification of the system that we would like to build. In order to detect differences between specifications and SUTs, we need to compare the behaviours of specifications and SUTs and the main notion to define such behaviours is given by the concept of *trace*.

**Definition 2.** *Let $M = (Q, q_{in}, I, O, T)$ be an FSM, $q \in Q$ be a state and $\sigma = (i_1, o_1) \dots (i_k, o_k) \in (I \times O)^*$ be a sequence of pairs. We say that $M$ can perform $\sigma$ from $q$ if there exist states $q_1 \dots q_k \in Q$ such that for all $1 \leq j \leq k$ we have $(q_{j-1}, i_j/o_j, q_j) \in T$, where $q_0 = q$. We denote this by either $q \stackrel{\sigma}{\Longrightarrow} q_k$ or $q \stackrel{\sigma}{\Longrightarrow}$. If $q = q_{in}$ then we say that $\sigma$ is a* trace *of $M$. We denote by $\mathtt{traces}(M)$ the set of traces of $M$. Note that for every state $q$ we have that $q \stackrel{\epsilon}{\Longrightarrow} q$ holds. Therefore, $\epsilon \in \mathtt{traces}(M)$ for every FSM $M$.*

Using the notion of trace, we can introduce the notion of test: a test is a sequence of (input action, output action) pairs. A test suite will be a set of tests.

**Definition 3.** *Let $M = (Q, q_{in}, I, O, T)$ be an FSM. We say that a sequence $t = (i_1, o_1) \dots (i_k, o_k) \in (I \times O)^+$ is a* test *for $M$ if $t \in \mathtt{traces}(M)$. We define the* length *of $t$ as the length of the sequence, that is, $|t| = k$. We define the sequence of inputs of $t$ as $\alpha = i_1 \dots i_k$ and the sequence of outputs of $t$ as $\beta = o_1 \dots o_k$ (we will sometimes use the notation $t = (\alpha, \beta) \in (I^+ \times O^+)$). A* test suite *for $M$ is a set of tests for $M$. Given a test suite $\mathcal{T} = \{t_1, \dots, t_n\}$, we define the* length *of the test suite as the sum of the lengths of its tests, that is, $|\mathcal{T}| = \sum_{i=1,\dots,n} |t_i|$.*

*Let $t = (\alpha, \beta)$ be a test for $M$. We say that the application of $t$ to an FSM $M'$ fails if there exists $\beta'$ such that $(\alpha, \beta') \in \mathtt{traces}(M')$ and $\beta \neq \beta'$. Similarly, let $\mathcal{T}$ be a test suite for $M$. We say that the application of $\mathcal{T}$ to an FSM $M'$ fails if there exists $t \in \mathcal{T}$ such that the application of $t$ to $M'$ fails.*

Intuitively, a test $(\alpha, \beta)$ for $M$ denotes that the application of the sequence of input actions $\alpha$ to a correct system (with respect to $M$) should show the sequence of output actions $\beta$. Note that if we would allow non-determinism, then the previous inequality must be appropriately replaced to express that the behaviours of the SUT must be a subset of those of the specification. For now, we will assume the determinism of the FSMs.

In order to select the tests that can detect the higher amount of fails in the program, it is useful to have a *measure* on the goodness of a test suite. Let us emphasize that measures will be, in general, heuristics to find good solutions and that each measure should be validated with experiments. Usually, higher values of a measure will be associated with better solutions, but this relation need not be monotonic. The measures that we use in this paper have been introduced in previous work and it has been shown that they are useful to find good test suites. We introduce a general notion of measure.

**Definition 4.** *A* measure *is a function*

$$f : \mathit{FSM}(I, O) \times \mathcal{P}(I^+ \times O^+) \rightarrow \mathbb{R}^+ \cup \{0\}$$

Intuitively, a measure is a function that receives an FSM and a test suite and returns a real number representing how good the measure considers that this test suite is to detect fails in an SUT. This notion of measure allows us to use information both from the specification and the test suite that we are evaluating, although it not necessarily has to use information from both, that is, a measure could work only with the information from the test suite and not use the specification at all. Finding the best test suite according to a measure (that is, the test suite that gets the best score) is usually an NP-hard problem (due to the combinatorial explosion). Therefore, we decided to rely on genetic programming in order to obtain *relatively good* test suites. A genetic algorithm is composed by:

```
Initialize population;
Evaluate population;
while termination criterion not reached do
    Select next population;
    Perform crossover;
    Perform mutation;
    Evaluate population;
end
```

**Algorithm 1.** Genetic algorithm: general scheme

- An encoding of the population in *genes*.
- An *initial population*, that is, randomly generated individuals expressed in the selected codification.
- A *fitness function* to evaluate the population.
- A *stopping criterion.*
- A *next population selection method*, which usually keeps the best individuals and discards the worst ones (with respect to the fitness function values).
- A *crossover method* that generates new individuals from the mixture of the genes of the existing ones.
- A *mutation method* that can modify some individuals in order to obtain new genes that might have not been present before.

The structure of a genetic algorithm is given in Algorithm 1. A basic genetic programming algorithm is a genetic algorithm where the codification of the population in genes does not use a linear structure (as a vector) but a tree-like structure [15]. Most of the work using genetic algorithms to generate test suites rely on a linear structure to represent the test suite. Specially, they use to rely on a vector of the inputs of the test suite [7,8,16,22]. This encoding of a test suite presents a problem: if the FSM is not input-enabled, then the algorithm could generate invalid tests that will always fail when applied to the SUT, even if this is totally equivalent to the FSM. As we are working with deterministic but not necessarily input-enabled FSMs, we have to face this problem and using a grammar-guided genetic programming algorithm allows us to ensure the correctness of the generated test suites. This approach also allows us to use the information from the output that each input generates in each state of the FSM (as the inputs do not have to generate the same output in all the states).

# 3 The Genetic Programming Algorithm

In this section, we will present all the components of our genetic algorithm.

## 3.1 Encoding

The first and most important choice of a genetic approach is to select a good encoding. As we are working with test suites generated from an `FSM`, we need to preserve the structure of the `FSM` in order to generate correct tests for it. Therefore, we decided to use a tree structure as an encoding of our tests and we use a genetic programming algorithm. Specifically, we decided to use a grammar-guided genetic programming approach, which solves the correctness issues from just using genetic programming. This implies that the first step of our genetic programming algorithm will be to generate the grammar that the `FSM` produces. We have the following components:

- A start non-terminal symbol $S$ that starts the grammar.
- A non-terminal symbol $T$ that introduces each test of the test suite.
- A non-terminal symbol $N$ for each state, where $N \in \mathbb{N}$ is the state number.
- A terminal symbol $'a/b'$ for each input/output pair present on the `FSM`, where $a$ is the input and $b$ is the output.
- A terminal symbol $'null'$ to represent the end of a test.
- A production rule $S \longrightarrow T$ to generate the initial test.
- A production rule $T \longrightarrow T + T$ to introduce a new test.
- A production rule $T \longrightarrow$ 0 to start each test in the `FSM` initial state.
- A production rule $N \longrightarrow$ $'a/b' + M$ for each transition from the state $N$ to a state $M$ with input/output pair $(a, b)$.
- A production rule $N \longrightarrow$ $'null'$ for each state $N$ to a terminal to represent the end of the test.

Given an `FSM`, the generation of the associated grammar is automatic (and it has been implemented as part of our tool).

## 3.2 Initial Population

As an initial population we randomly generate 100 test suites of the length given by the user using the grammar previously derived from the `FSM`. Each rule in the grammar has the same probability of being triggered. This allows a uniform random initialization.

## 3.3 Fitness Function

The fitness function of our genetic programming algorithm will be the available measures. As previously defined, they will receive the test suite and the `FSM` and will return a real value that represents how *good* is this test suite according to the measure. An important remark about fitness functions is that they should be easy to compute, as they will be invoked many times during the execution of the algorithm. Therefore, fitness functions with high computational cost will lead to a higher computational cost of the algorithm.

We decided to give the users the capability to select the fitness function that better suites their problem, along with the decision on whether the score should be maximized or minimized. As we explained before, fitness functions should have similar performances and this is the case for the measures based on Information Theory that we include in our tool. Among them, we can mention, due to the big improvement with respect to previous measures, the *Test Set Diameter* (TSDm) based measures [9]. We implemented the Input-TSDm, the Output-TSDm and the InputOutput-TSDm. Also, we implemented a measure that we have developed in our research group and that it is called *Biased Mutual Information*. Note that users of our tool can add their own measures. So, our tool can be use to evaluate the usefulness of new proposals because they can be compared with existing ones.

### 3.4 Stopping Criterion

The algorithm performs at most 100 epochs and at least 20 epochs. Once we have passed the 20 epochs, the stop criterion will be fulfilled if the best test suite is the same along $0.2 \times \mathit{NumberOfPassedEpochs}$ epochs.

### 3.5 Selection Method

We use a variant of elitist reduction [19]. First, the test suites that got a fitness score over the mean (or under the mean if we want to minimize) go directly to the next epoch. In addition, the ones that are under the mean can pass to the next epoch if their score is higher than the mean minus a random number modulo the distance between the mean and the best score.

### 3.6 Crossover Method

The choice of crossover method depends on our encoding and the characteristics we want the produced test suites to have. As we use a grammatical encoding, we need to use a grammatical crossover. We have considered a mixture between the Whigham crossover [18] and the standard grammatical crossover [5]. Also, as we want all our test suites to have the same length (as previously defined), we need to slightly modify crossovers in order to achieve a crossover that keeps the length fixed. Algorithm 2 shows how crossover is performed.

Finally, we need to set the probability of producing the crossover. In our case, giving how hard is to perform a crossover, we decided to set this probability to 90%, so that we favour the mixture between test suites.

**Data:** $TS1$, $TS2$ test suites
**Result:** Crossover of $TS1$ and $TS2$
*match* = *false*;
**while** !*match* **do**
  Select a random node $t1$ from $TS1$;
  **for** *each node* $t2$ *of* $TS2$ **do**
    **if** $t2$ *non-terminal* == $t1$ *non-terminal and* $t2$ *length* == $t1$ *length* **then**
      Set $t2$ as valid node.
    **end**
  **end**
  **if** *valid nodes* > 0 **then**
    *match* = *true*;
  **end**
**end**
Select a random valid node $t2$;
Get parent $p1$ of $t1$;
Get parent $p2$ of $t2$;
Set $t2$ as child of $p1$;
Set $t1$ as child of $p2$;

**Algorithm 2.** Crossover algorithm

### 3.7 Mutation Method

A mutation consists in generating a new test with the same length. The probability of performing a mutation will be, as usual [19], equal to 5% for each test of each test suite of the population.

## 4 GPTSG

We have implemented a tool[1] supporting our framework. The tool has two main uses: generate a test suite with a giving length according to a selected measure and compare different measures. In order to develop the tool, we looked for libraries dealing with FSMs and we decided to use the OpenFST library [1]. Therefore, input files must be in OpenFST format, with the .fst extension. The tool will have two kind of calls *generate* and *compare*. The syntax of the two calls is:

```
gptsg generate inputFile length {max|min} fitness
gptsg compare length {max|min} fitness {max|min} fitness
```

[1] The tool can be downloaded from https://github.com/Colosu/gptsg.

and two examples of calls are:

```
gptsg generate ./test/binary.fst 50 max ITSDm
gptsg compare 50 max ITSDm min OTSDm
```

Currently, our tool supports the following fitness functions:

- BMI: Biased Mutual Information.
- ITSDm: Input Test Set Diameter.
- OTSDm: Output Test Set Diameter.
- IOTSDm: Input-Output Test Set Diameter.
- Own: For your own developed measure.
- random: generates a totally random test suite.

Let us emphasize that an important feature of our tool is that it is possible to define new measures, so that they can be compared with the already existing ones. The user only needs to open the `src/Measures.cpp` file and modify the `OwnFunction` method. Once the code is compiled, the inserted measure can be called as the *Own* fitness function.

### 4.1 Test Suite Generation

In order to generate a *good* enough test suite, we implemented the genetic programming algorithm explained in the previous section, giving some configuration to the user. The tool needs that the input `FSM` is in OpenFST format (in a .fst file). This format is easy to use and can be learned quickly. Also, the tool needs to know the length of the expected test suite, in terms of input actions, and the measure to use as a fitness function. Then, the user will receive a .txt file with the generated test suite, with each test conformed by a succession of input/output pairs.

### 4.2 Test Suite Comparison

The tool allows users to compare two measures. It needs to know the length of the desired test suite, the two measures to be compared and if it should maximize or minimize each measure. Essentially, the tool takes the set of 100 `FSMs` that are shipped with the tool, representing different and diverse scenarios and characteristics, and for each one of them it generates two test suites according to the corresponding measures. Then, the tool produces 1000 mutants of the corresponding `FSM` and checks which test suite kills more mutants. With the results for each `FSM`, the tool produces an output telling the percentage of cases where each test suite has killed more mutants, along with a percentage of how many mutants where killed by each test suite. This process is repeated 50 times, getting 50 results, and at the end, the program gives a mean of all the results obtained for the 50 repetitions. This process is given in Algorithm 3.

```
Data: length, measure1, measure2
Result: .txt file with the values
REP = 50;
FSM = 100;
for each REP do
    Set control values to 0;
    for each FSM F do
        Generate TS1 genetic test suite using measure measure1;
        Generate TS2 genetic test suite using measure measure2;
        Generate 1000 mutants of F;
        Check which test suite kills more mutants;
    end
    Output the percentage of runs TS1 killed more mutants;
    Output the percentage of runs TS2 killed more mutants;
    Output the percentage of mutants killed by TS1;
    Output the percentage of mutants killed by TS2;
end
Output the average percentage of runs TS1 killed more mutants;
Output the average percentage of runs TS2 killed more mutants;
Output the average percentage of mutants killed by TS1;
Output the average percentage of mutants killed by TS2;
```

**Algorithm 3.** Test suite comparison algorithm

### 4.3 Experiment

Next we show the results of a small experiment to evaluate our genetic programming algorithm and tool. First, we compared the Input Test Set Diameter measure, used as fitness function, and a random test suite generation. We observed that the genetically generated test suite killed more mutants than the randomly generated test suite in a 75.3% of the cases, killing an average of 47.1% of the mutants, while the randomly generated test suite killed more mutants the 24.7% of the cases, killing an average of 43.9% of the mutants. We can see the full comparison in Fig. 1 (left), where each of the first 50 rows shows the result of an iteration of the experiment. In order to see how two measures are compared, we rerun the comparison algorithm to compare the maximization of the Input Test Set Diameter and the maximization of the Output Test Set Diameter. The results can be seen in Fig. 1 (right). As expected, they obtain similar results, getting better results the Output TSDm due to the randomization involved in the genetic algorithm. On average, the Input TSDm killed more mutants the 49% of the cases, killing 47.3% of the mutants, while Output TSDm killed more mutants the 51% of the cases, killing 47.5% of the mutants.

| Iteration Number | % wins ITSDm | % wins random | % mutants killed by ITSDm | % mutants killed by random |
|---|---|---|---|---|
| 1 | 0.757576 | 0.242424 | 0.475081 | 0.439909 |
| 2 | 0.744898 | 0.255102 | 0.476306 | 0.443786 |
| 3 | 0.75 | 0.25 | 0.46496 | 0.43275 |
| 4 | 0.8 | 0.2 | 0.47095 | 0.43229 |
| 5 | 0.75 | 0.25 | 0.46347 | 0.43407 |
| 6 | 0.69 | 0.31 | 0.46947 | 0.4419 |
| 7 | 0.72449 | 0.27551 | 0.475051 | 0.447755 |
| 8 | 0.72449 | 0.27551 | 0.475051 | 0.447857 |
| 9 | 0.74 | 0.26 | 0.46035 | 0.43142 |
| 10 | 0.767677 | 0.232323 | 0.471525 | 0.441646 |
| 11 | 0.72449 | 0.27551 | 0.473204 | 0.443663 |
| 12 | 0.767677 | 0.232323 | 0.470525 | 0.441 |
| 13 | 0.646465 | 0.353535 | 0.465929 | 0.440162 |
| 14 | 0.717172 | 0.282828 | 0.468394 | 0.437384 |
| 15 | 0.81 | 0.19 | 0.47143 | 0.43202 |
| 16 | 0.686869 | 0.313131 | 0.469606 | 0.441101 |
| 17 | 0.76 | 0.24 | 0.46867 | 0.43234 |
| 18 | 0.81 | 0.19 | 0.47033 | 0.43285 |
| 19 | 0.79798 | 0.20202 | 0.468172 | 0.438222 |
| 20 | 0.83 | 0.17 | 0.46732 | 0.43214 |
| 21 | 0.77 | 0.23 | 0.46903 | 0.43535 |
| 22 | 0.757576 | 0.242424 | 0.473949 | 0.440232 |
| 23 | 0.78 | 0.22 | 0.46716 | 0.43258 |
| 24 | 0.744898 | 0.255102 | 0.475673 | 0.442582 |
| 25 | 0.79798 | 0.20202 | 0.475343 | 0.446455 |
| 26 | 0.79 | 0.21 | 0.4678 | 0.43391 |
| 27 | 0.68 | 0.32 | 0.46184 | 0.43662 |
| 28 | 0.79 | 0.21 | 0.46923 | 0.43317 |
| 29 | 0.75 | 0.25 | 0.4663 | 0.43457 |
| 30 | 0.79 | 0.21 | 0.46861 | 0.43607 |
| 31 | 0.747475 | 0.252525 | 0.474848 | 0.436535 |
| 32 | 0.663265 | 0.336735 | 0.469214 | 0.445235 |
| 33 | 0.73 | 0.27 | 0.46731 | 0.43186 |
| 34 | 0.721649 | 0.278351 | 0.478278 | 0.449763 |
| 35 | 0.72 | 0.28 | 0.46624 | 0.43619 |
| 36 | 0.79798 | 0.20202 | 0.472141 | 0.439152 |
| 37 | 0.783505 | 0.216495 | 0.48668 | 0.450495 |
| 38 | 0.75 | 0.25 | 0.46744 | 0.43613 |
| 39 | 0.767677 | 0.232323 | 0.465687 | 0.432909 |
| 40 | 0.676768 | 0.323232 | 0.469 | 0.443343 |
| 41 | 0.742268 | 0.257732 | 0.484216 | 0.450979 |
| 42 | 0.75 | 0.25 | 0.46626 | 0.43473 |
| 43 | 0.77551 | 0.22449 | 0.477082 | 0.443643 |
| 44 | 0.77 | 0.23 | 0.46771 | 0.43251 |
| 45 | 0.76 | 0.24 | 0.47459 | 0.43538 |
| 46 | 0.757576 | 0.242424 | 0.471333 | 0.439051 |
| 47 | 0.744898 | 0.255102 | 0.479857 | 0.444378 |
| 48 | 0.806122 | 0.193878 | 0.478143 | 0.437531 |
| 49 | 0.777778 | 0.222222 | 0.473505 | 0.440222 |
| 50 | 0.74 | 0.26 | 0.46422 | 0.43529 |
| Mean | 0.752723 | 0.247277 | 0.470851 | 0.438578 |

| Iteration Number | % wins ITSDm | % wins OTSDm | % mutants killed by ITSDm | % mutants killed by OTSDm |
|---|---|---|---|---|
| 1 | 0.494845 | 0.505155 | 0.484216 | 0.48366 |
| 2 | 0.546392 | 0.453608 | 0.481206 | 0.479309 |
| 3 | 0.44 | 0.56 | 0.46812 | 0.47277 |
| 4 | 0.525253 | 0.474747 | 0.473485 | 0.468192 |
| 5 | 0.443299 | 0.556701 | 0.481979 | 0.489619 |
| 6 | 0.515152 | 0.484848 | 0.474323 | 0.473404 |
| 7 | 0.525253 | 0.474747 | 0.477051 | 0.474596 |
| 8 | 0.56 | 0.44 | 0.4725 | 0.4662 |
| 9 | 0.48 | 0.52 | 0.46631 | 0.47339 |
| 10 | 0.45 | 0.55 | 0.46801 | 0.47551 |
| 11 | 0.515152 | 0.484848 | 0.469535 | 0.472646 |
| 12 | 0.489796 | 0.510204 | 0.476612 | 0.478694 |
| 13 | 0.51 | 0.49 | 0.46966 | 0.46947 |
| 14 | 0.40404 | 0.59596 | 0.468949 | 0.478323 |
| 15 | 0.47 | 0.53 | 0.46272 | 0.46797 |
| 16 | 0.52 | 0.48 | 0.46723 | 0.46533 |
| 17 | 0.367347 | 0.632653 | 0.468765 | 0.480704 |
| 18 | 0.535354 | 0.464646 | 0.471131 | 0.471232 |
| 19 | 0.428571 | 0.571429 | 0.47351 | 0.48201 |
| 20 | 0.51 | 0.49 | 0.46843 | 0.47081 |
| 21 | 0.561224 | 0.438776 | 0.483551 | 0.479612 |
| 22 | 0.43 | 0.57 | 0.46509 | 0.46947 |
| 23 | 0.414141 | 0.585859 | 0.470253 | 0.475808 |
| 24 | 0.555556 | 0.444444 | 0.473596 | 0.469566 |
| 25 | 0.494949 | 0.505051 | 0.474384 | 0.475111 |
| 26 | 0.52 | 0.48 | 0.46671 | 0.46436 |
| 27 | 0.505155 | 0.494845 | 0.480866 | 0.481639 |
| 28 | 0.515152 | 0.484848 | 0.472889 | 0.473586 |
| 29 | 0.469388 | 0.530612 | 0.474306 | 0.476204 |
| 30 | 0.545455 | 0.454545 | 0.477 | 0.473242 |
| 31 | 0.510204 | 0.489796 | 0.476571 | 0.477184 |
| 32 | 0.51 | 0.49 | 0.46423 | 0.46837 |
| 33 | 0.469388 | 0.530612 | 0.477316 | 0.479204 |
| 34 | 0.408163 | 0.591837 | 0.475296 | 0.484643 |
| 35 | 0.545455 | 0.454545 | 0.468455 | 0.471949 |
| 36 | 0.494949 | 0.505051 | 0.475091 | 0.475495 |
| 37 | 0.5 | 0.5 | 0.46762 | 0.46903 |
| 38 | 0.46 | 0.54 | 0.46477 | 0.47045 |
| 39 | 0.474227 | 0.525773 | 0.48034 | 0.484113 |
| 40 | 0.45 | 0.55 | 0.46602 | 0.47369 |
| 41 | 0.489796 | 0.510204 | 0.480306 | 0.480286 |
| 42 | 0.546392 | 0.453608 | 0.484423 | 0.484804 |
| 43 | 0.51 | 0.49 | 0.46951 | 0.46428 |
| 44 | 0.47 | 0.53 | 0.46927 | 0.47261 |
| 45 | 0.469388 | 0.530612 | 0.478684 | 0.477918 |
| 46 | 0.43 | 0.57 | 0.46678 | 0.47126 |
| 47 | 0.438776 | 0.561224 | 0.479082 | 0.480398 |
| 48 | 0.515464 | 0.484536 | 0.479969 | 0.483546 |
| 49 | 0.56 | 0.44 | 0.4709 | 0.46138 |
| 50 | 0.484848 | 0.515152 | 0.472162 | 0.471455 |
| Mean | 0.489581 | 0.510419 | 0.47293 | 0.474633 |

**Fig. 1.** Results of ITSDm vs random (left) and ITSDm vs OTSDm (right).

## 5 Conclusions

The automatic generation of good test suites is a fundamental task when limitations in testing complex systems come into play. In this paper we have presented a genetic programming algorithm to generate these test suites. We have implemented a tool to support our algorithm so that any potential user can apply it. The tool allows users to compare genetically generated test suites that outperform different measures, so that we can compare the performance of each measure. We have relied on Information Theory to define the measures that will work as fitness functions of our genetic programming algorithm. Finally, we have performed several experiments with our tool and report on two of them. There are some lines for future work.

We are considering several lines of future work. First, it would be interesting to find and define new measures so that we can extend the catalogue of our tool. Second, we are working on extending our algorithm to deal with the generation of test suites to test systems with distributed interfaces [12,13].

## References


1. Allauzen, C., Riley, M., Schalkwyk, J., Skut, W., Mohri, M.: OpenFst: a general and efficient weighted finite-state transducer library. In: Holub, J., Žd'árek, J. (eds.) CIAA 2007. LNCS, vol. 4783, pp. 11–23. Springer, Heidelberg (2007). https://doi.org/10.1007/978-3-540-76336-9_3
2. Ammann, P., Offutt, J.: Introduction to Software Testing, 2nd edn. Cambridge University Press, Cambridge (2017)
3. Cavalli, A.R., Higashino, T., Núñez, M.: A survey on formal active and passive testing with applications to the cloud. Ann. Telecommun. **70**(3–4), 85–93 (2015)
4. Couchet, J., Manrique, D., Porras, L.: Grammar-guided neural architecture evolution. In: Mira, J., Álvarez, J.R. (eds.) IWINAC 2007. LNCS, vol. 4527, pp. 437–446. Springer, Heidelberg (2007). https://doi.org/10.1007/978-3-540-73053-8_44
5. Couchet, J., Manrique, D., Rios, J., Rodríguez-Patón, A.: Crossover and mutation operators for grammar-guided genetic programming. Soft Comput. **11**(10), 943–955 (2007)
6. Cover, T.M., Thomas, J.A.: Elements of Information Theory, 2nd edn. Wiley, Hoboken (2006)
7. Derderian, K., Merayo, M.G., Hierons, R.M., Núñez, M.: Aiding test case generation in temporally constrained state based systems using genetic algorithms. In: Cabestany, J., Sandoval, F., Prieto, A., Corchado, J.M. (eds.) IWANN 2009. LNCS, vol. 5517, pp. 327–334. Springer, Heidelberg (2009). https://doi.org/10.1007/978-3-642-02478-8_41
8. Derderian, K., Merayo, M.G., Hierons, R.M., Núñez, M.: A case study on the use of genetic algorithms to generate test cases for temporal systems. In: Cabestany, J., Rojas, I., Joya, G. (eds.) IWANN 2011. LNCS, vol. 6692, pp. 396–403. Springer, Heidelberg (2011). https://doi.org/10.1007/978-3-642-21498-1_50
9. Feldt, R., Poulding, S.M., Clark, D., Yoo, S.: Test set diameter: quantifying the diversity of sets of test cases. In: 9th IEEE International Conference on Software Testing, Verification and Validation, ICST 2016, pp. 223–233. IEEE Computer Society (2016)
10. Guo, Q., Hierons, R.M., Harman, M., Derderian, K.: Computing unique input/output sequences using genetic algorithms. In: Petrenko, A., Ulrich, A. (eds.) FATES 2003. LNCS, vol. 2931, pp. 164–177. Springer, Heidelberg (2004). https://doi.org/10.1007/978-3-540-24617-6_12
11. Henard, C., Papadakis, M., Harman, M., Jia, Y., Traon, Y.L.: Comparing white-box and black-box test prioritization. In: 38th International Conference on Software Engineering, ICSE 2014, pp. 523–534. ACM Press (2016)
12. Hierons, R.M., Merayo, M.G., Núñez, M.: Bounded reordering in the distributed test architecture. IEEE Trans. Reliab. **67**(2), 522–537 (2018)
13. Hierons, R.M., Núñez, M.: Implementation relations and probabilistic schedulers in the distributed test architecture. J. Syst. Softw. **132**, 319–335 (2017)

14. ISO/IEC JTCI/SC21/WG7, ITU-T SG 10/Q.8: Information Retrieval, Transfer and Management for OSI; Framework: Formal Methods in Conformance Testing. Committee Draft CD 13245–1, ITU-T proposed recommendation Z.500. ISO - ITU-T (1996)
15. Koza, J.R.: Genetic Programming. MIT Press, Cambridge (1993)
16. Lefticaru, R., Ipate, F.: Automatic state-based test generation using genetic algorithms. In: 9th International Symposium on Symbolic and Numeric Algorithms for Scientific Computing, SYNASC 2007, pp. 188–195. IEEE Computer Society (2007)
17. Luna, J.M., Romero, J.R., Ventura, S.: Design and behavior study of a grammar-guided genetic programming algorithm for mining association rules. Knowl. Inf. Syst. **32**(1), 53–76 (2012)
18. McKay, R.I., Hoai, N.X., Whigham, P.A., Shan, Y., O'Neill, M.: Grammar-based genetic programming: a survey. Genet. Program. Evolvable Mach. **11**(3–4), 365–396 (2010)
19. Mitchell, M.: An Introduction to Genetic Algorithms. MIT Press, Cambridge (1998)
20. Myers, G.J., Sandler, C., Badgett, T.: The Art of Software Testing, 3rd edn. Wiley, Hoboken (2011)
21. Regolin, E.N., Pozo, A.T.R.: Bayesian automatic programming. In: Keijzer, M., Tettamanzi, A., Collet, P., van Hemert, J., Tomassini, M. (eds.) EuroGP 2005. LNCS, vol. 3447, pp. 38–49. Springer, Heidelberg (2005). https://doi.org/10.1007/978-3-540-31989-4_4
22. Samarah, A., Habibi, A., Tahar, S., Kharma, N.N.: Automated coverage directed test generation using a cell-based genetic algorithm. In: 11th Annual IEEE International High-Level Design Validation and Test Workshop, pp. 19–26. IEEE Computer Society (2006)
23. Shafique, M., Labiche, Y.: A systematic review of state-based test tools. Int. J. Softw. Tools Technol. Transf. **17**(1), 59–76 (2015)

# An Intelligent System Integrating CEP and Colored Petri Nets for Helping in Decision Making About Pollution Scenarios

Gregorio Díaz[1(✉)], Enrique Brazález[1], Hermenegilda Macià[1], Juan Boubeta-Puig[2], and Valentín Valero[1]

[1] School of Computer Science, University of Castilla-La Mancha, 02071 Albacete, Spain
{gregorio.diaz,enrique.brazalez,hermenegilda.macia, valentin.valero}@uclm.es

[2] Department of Computer Science and Engineering, University of Cádiz, Avda. de la Universidad de Cádiz 10, 11519 Puerto Real, Cádiz, Spain
juan.boubeta@uca.es

**Abstract.** Air pollution is currently a great concern especially in large cities. To reduce pollution levels, governments are imposing traffic restrictions. However, the decision about which grade of traffic restriction must be applied in a particular city zone is a cumbersome task. This decision depends on the pollution scenario occurred at a time period. To face this issue, we propose an analyzable and flexible intelligent system integrating Complex Event Processing (CEP) technology and Colored Petri Net (CPN) formalism to help domain experts to conduct such a decision-making process. This system uses a CEP engine to automatically analyze and correlate real air sensing data to detect pollutant averages at a particular sensor station. This produced information is then consumed in runtime by a CPN model in charge of obtaining the pollution scenarios, which are the basis to make decisions on the traffic regulations.

**Keywords:** Intelligent system · Complex Event Processing · Formal method · Simulation · Colored Petri Net · Decision making

## 1 Introduction

Air pollution is a great concern in cities from small to large. All of them share the same problem about air pollutants mainly produced by road traffic. This situation is not only a concern for the global warming, but also for citizens' health.

This study was funded in part by the Spanish Ministry of Science and Innovation and the European Union FEDER Funds under Grants TIN2015-65845-C3-2-R, TIN2015-65845-C3-3-R and TIN2016-81978-REDT, and also by the JCCM regional projects SBPLY/17/180501/000276/1 and SBPLY/17/180501/000276/2, both of them co-financed by the European Union FEDER Funds.


I. Rojas et al. (Eds.): IWANN 2019, LNCS 11506, pp. 729–740, 2019.
https://doi.org/10.1007/978-3-030-20521-8_60

Several solutions have been proposed with different time scopes, for instance, the substitution of the combustion engines by electric ones. This solution has a medium time scope that may be reached in several decades. In the meantime, cities try to deal with this problem taking some other more immediate measures. In this sense, during the last years, road traffic regulations have been enacted in cities all around the world. These regulations describe the protocols that their citizens are obliged to follow when certain conditions regarding air pollutants occur.

International organizations, such as World Health Organization (WHO) [15], have proposed several guidelines to be followed by governments. Additionally, other national agencies have enacted regulations, such as EPA and EU [8]. All of these agencies describe their own Air Quality Index (AQI). In the case of EU, the environment commission establishes that more than 400, 000 people die prematurely in the EU every year as a result of poor air quality and that millions more suffered respiratory and cardiovascular diseases. Under EU law, member states have to implement air quality plans to bring the levels back down, when air pollution limits are breached. In this context, Madrid city has its action plan for $NO_2$ pollution episodes [11].

In this work, we propose an intelligent system to help domain experts to conduct the decision-making process about when these protocols must be applied. The main features of this system are: (a) **analyzable**, easy to be studied and validated either in an hypothetical or a real situation, (b) **adaptable**, able to be easily adapted to a wide range of air pollution protocols and (c) **agile**, able to process huge amounts of data. To achieve this goal, we use the Colored Petri Net (CPN) formalism [10], and the Complex Event Processing (CEP) technology [9]. By using CPNs, we can simulate both real and hypothetical situations to validate the proposed protocols. In this way, the first feature is covered. Due to the graphical description that CPNs provide, it is easy to adapt the provided CPN model to sustain other proposed protocols, so covering the second feature. Finally, CEP is able to correlate huge amounts of data in real time detecting situations of interest used to feed the CPNs simultaneously, fulfilling the third feature.

The proposed intelligent system consists of a CPN model able to process the data provided by the CEP technology and produce information of interest to make a decision. To accomplish it, a CEP engine is integrated with an Enterprise Service Bus (ESB), which gathers the information from the environment using the data provided from a sensor network. This information (simple events) will be processed by the CEP engine by using patterns, thus producing new information (complex events) that the ESB gathers and sends to the CPN model.

CPNs have been widely used in several domains and in the related bibliography we can find several works combining CEP systems with this formalism. Thus, formal methods as Petri nets can help in the design and implementation of CEP systems which are underdeveloped as stated by Offel et al. [12]. They apply this point of view in the development process of CEP systems using envisioned verification. A model of event processing networks using CPNs with priorities and time (PTCPNs) is proposed by Weidlich et al. [14]. As an illustration, they present an example implemented in the ETALIS framework [3]. They present the EPN (Event Processing Networks) architecture as the overall system and propose a general translation of this concept.

The structure of the paper is as follows. Section 2 presents a brief background about Colored PNs and CEP technology. The protocol and the specific model of PNs we use are described in Sect. 3. Section 4 presents the event patterns. The interconnection between CEP and the protocol model via an ESB is explained in Sect. 5. Finally, Sect. 6 highlights our conclusions and lines of future work.

## 2 Background

In this section, we introduce the main features of both technologies, CPNs and CEP.

### 2.1 Colored Petri Nets

CPNs are a formalism which provides mathematical rigor and a graphical representation of the model, which allows us to have a better comprehension of the system behavior. In addition, the analysis capabilities provided by the model allow us to obtain important results about the system behavior. Furthermore, CPNs are supported by tools, which allow us the simulation and analysis of the behavior of a given system in a suitable way.

A PN is a directed bi-partite graph with nodes of two types: places, which represent system conditions (drawn as circles), and transitions, which represent actions that change the system state (drawn as rectangles). An arc can connect either a place with a transition (pt-arc) or a transition with a place (tp-arc). Places linked with a transition $t$ by input arcs are called the precondition places of $t$, while the places linked with output arcs are called postconditions of $t$, and the same notation applies to transitions linked with places.

In CPNs, places have an associated *color set* (a data type). Thus, a place can be marked with a natural number of tokens, all of them with some attached information belonging to the associated place color set. There is an empty color set *(UNIT)*, which corresponds to no information attached to tokens. We can define some other color sets, by using the basic data types or by combinations of them. For instance, we have the set of integer numbers $INT$, a Cartesian product of two or more color sets, as $INT2 = INT \times INT$, a string (*STRING*), etc. CPNs are supported by a widely used tool, CPN Tools [1], which allows us to create, edit, simulate and analyze CPNs. The notation described below is the one used in this tool.

Furthermore, we can use the timed features of CPNs. In this type of nets, a discrete global clock is used to represent the total time elapsed in the system model, and places can be either timed or untimed. In the case of timed places, their tokens have an associated timestamp, which indicates the time at which they will be available and thus usable to fire a transition.

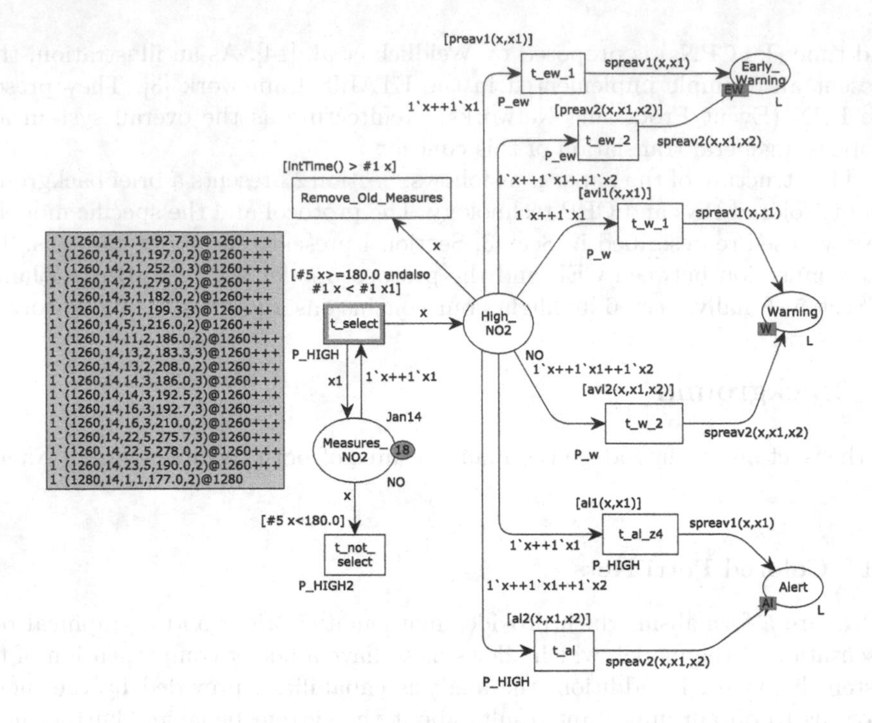


**Fig. 1.** CPN model for Madrid air pollution protocol (part I).

In CPN Tools, the current number of tokens on every place is drawn on green at the right-hand side of the place circle, and the specific colors of these tokens are indicated by the notation $n$‘$v$, meaning that there are $n$ instances of color $v$. The symbol ‘++’ (respectively ‘+++’ for timed tokens) is used to represent the union of colors in CPN Tools. Thus, a timed integer place (color set *int timed*) with a marking $2‘3@5 + + + 1‘9@10$ has 2 tokens with value 3 and timestamp 5 and 1 token with value 9 and timestamp 10.

Figure 1 shows a CPN, in which we can see the color sets annotated beside the places, as *NO* for the *Measures_NO2* place, which is a timed color set defined as $INT \times INT \times INT \times INT \times REAL \times INT$, capturing $NO_2$ measures with the following information: timestamp, day, station number, zone, average and time window, i.e., the time window used to calculate the average.

Arcs have color set expressions as inscriptions, which are constructed using variables, constants, operators and functions. Arc expressions must evaluate to a color or multiset of colors in the *color set* of the attached place. Tp-arcs[1] can have a delay associated, with the syntax $n$‘$v@x$ to denote that $n$ tokens with value $v$ are produced with a timestamp equal to the current time increased by $x$ time units. Furthermore, transitions can have guards that can restrict their firing[2], as well as priorities. Guards are boolean expressions constructed by using

[1] Pt-arcs can also have delays associated, but this feature is not used in this paper.

[2] Indicated above the transition, by enclosing them in brackets.

the variables, constants, operators and functions of the model, and they must evaluate to true for the transition to be *fireable*. For instance, [#5 $x < 180.0$] is a guard that will allow the firing of transition *t_not_select* by using one token on the *Measaures_NO2* place with its fifth field smaller than 180.0. Transitions can also have an associated priority, so in the event of a conflict between two transitions that can be fired at a given time, the transition with the highest level of priority is fired first. Notice the priority levels below the transitions. The priority levels we use are *P_LOW*, *P_NORMAL* (by default), *P_ew*, *P_w*, *P_HIGH2* and *P_HIGH*, following this increasing order of priority.

For any transition $t$ with variables $x1, x2, \ldots$ on its input and output arc expressions, we call a *binding* of $t$ to an assignment of concrete values to each of these variables. A binding of a transition $t$ is then *enabled* if there are tokens on its precondition places matching with the values of the corresponding inscriptions. Thus, arc expressions are evaluated by assigning values to the variables, and these values are then used to select the tokens that must be removed or added when firing the corresponding transition. When a transition is fired, for each precondition place we remove the tokens matching with the corresponding arc inscription, and we add new tokens on each postcondition place, according to the associated arc inscriptions. Moreover, when no transition can be fired at the current time, time elapsing occur, but only up to a time in which some transition can be fired.

The CPN in Fig. 1 is an excerpt of the CPN that models the whole system (see Fig. 3), which will be described in detail in Sect. 3.1.

### 2.2 Complex Event Processing

CEP provides users with facilities for analyzing and correlating large volumes of data in the form of events with the aim of detecting relevant or critical situations for a particular domain in real time. To fulfill this objective, the conditions describing the situations of interest to be detected must be specified as event patterns. Patterns are implemented by using the languages provided by CEP engines, the so-called Event Processing Languages (EPLs) [6], and once the patterns are defined they can be deployed in the CEP engine in question.

To spare domain experts this implementation using an EPL, the MEdit4CEP approach [5] provides them with a graphical modeling editor for specifying the CEP domain, event patterns and action definitions. From the graphical models thus produced, the editor automatically generates a corresponding Esper EPL code [2]. Furthermore, MEdit4CEP was extended by using the Prioritized Colored Petri Net (PCPN) [13] formalism, resulting in the MEdit4CEP-CPN approach [4], in which event pattern models can be automatically transformed into PCPN models, and then into the corresponding PCPN code executable by CPN Tools.

## 3 Modeling the Air Pollution Protocol with Petri Nets

WHO recommends that cities do not exceed 40 µg/m$^3$ of $NO_2$ (nitrogen dioxide) annually or 200 µg/m$^3$ over one hour. In this sense, and due to the high concentration of $NO_2$ in Madrid city, the Council was obliged to implement the high-pollution protocol, imposing restrictions on road traffic in the city. To implement this protocol, the city has deployed a sensor network to measure different types of pollutants. Figure 2(a) depicts the Madrid map with its different stations. This protocol consists of different phases that imply greater restrictions depending on the pollution levels, their persistence over time and the weather forecast taking into account the different zones in which Madrid city is divided (see Fig. 2(b)). The different air pollution levels considered in this protocol are as follows:

- **Early Warning:** either 180 µg/m$^3$ are simultaneously exceeded in two stations of the same zone during two consecutive hours, or 180 µg/m$^3$ are simultaneously exceeded in three stations during three consecutive hours.
- **Warning:** either 200 µg/m$^3$ are simultaneously exceeded in two stations of the same zone during two consecutive hours, or 200 µg/m$^3$ are simultaneously exceeded in three stations during three consecutive hours.
- **Alert:** 400 µg/m$^3$ are simultaneously exceeded in three stations of the same zone (two stations for Zone 4) during three consecutive hours.

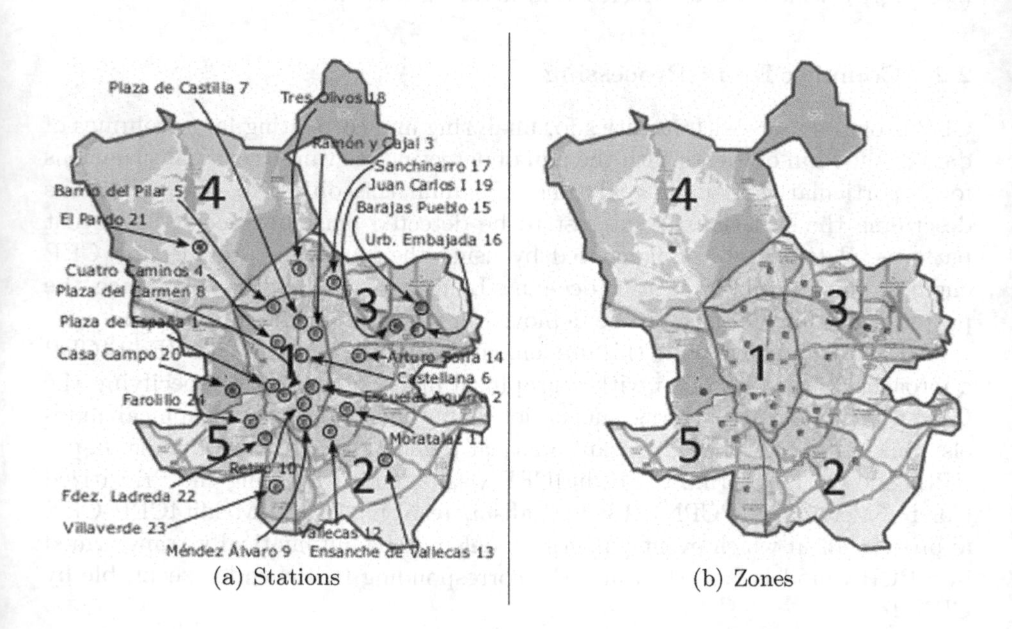


(a) Stations (b) Zones

**Fig. 2.** Map of Madrid.

An *air pollution episode* will be considered if any of the above levels is reached. Depending on its time duration and weather forecasts, an episode entails different degrees of restriction for road traffic according to different scenarios:

- **Scenario 1:** one day exceeding the Early Warning level.
- **Scenario 2:** two consecutive days exceeding the Early Warning level or one day exceeding the Warning level.
- **Scenario 3:** three consecutive days exceeding the Early Warning level or two consecutive days exceeding the Warning level.
- **Scenario 4:** four consecutive days exceeding the Warning level.
- **Scenario 5:** one day exceeding the Alert level.

### 3.1 The Colored Petri Net Model

The CPN depicting the protocol is shown by pieces, using the fusion places of CPN Tools to split a CPN model. Fusion places are identified by a blue label below the place, and they are functionally a same place, used in different parts (pages) of a CPN. The CPN model is then shown in Figs. 1 and 3. Figure 1 is the part of the model that processes the $NO_2$ levels from the input data to recognize the air pollution levels, while Fig. 3 is used to establish the scenarios to be applied. The starting point of this CPN is the *Measures_NO2* place, which is connected with two transitions, which are used to classify the measures into two types: those higher than or equal to 180.0 (*t_select*) and those smaller (*t_not_select*), which will be discarded. Higher measures will be transferred to place *High_NO2*, as a batch of measures gathered at the same time. As we can see in Fig. 1, transition *t_select* is enabled at the shown marking. This transition will fire several times, until all the tokens in *Measures_NO2* with the same current time (@1260) are consumed and the corresponding tokens are written in the *High_NO_2* place.

Transitions *t_ew_i*, *t_w_i* and *t_al_i* ($i = 1, 2$) will then be enabled, depending on the tokens in *High_NO_2*. With the marking indicated in the figure, transition *t_w_1* will be fired, since the warning preconditions are fulfilled by the following two tokens:

$$1`(1260, 14, 5, 1, 216.0, 2)@1280 + + + 1`(1260, 14, 2, 1, 279.0, 2)@1280$$

representing the first condition for a **warning** situation. Specifically, stations 5 and 2 in the same zone exceed 200 $\mu$g/m$^3$ (216.0 and 279.0) during the last two consecutive hours (at minute 1260, day 14). By firing this transition, a new token $1`(1260, 14, [2, 5], 1260)@1280$ is produced in place *Warning.* Here we indicate the minute, day, list of stations (2 and 5) where the warning was detected, and the time at which this warning situation was detected (1260). After firing this transition, transitions *t_ew_i* ($i = 1, 2$) become enabled, since they have a smaller priority. Their firings produce the corresponding tokens in the *Early_Warning* place. Finally, transition *Remove_Old_Measures* becomes enabled and it removes all the remaining tokens with time stamp 1260, so a new batch of measures can be processed.

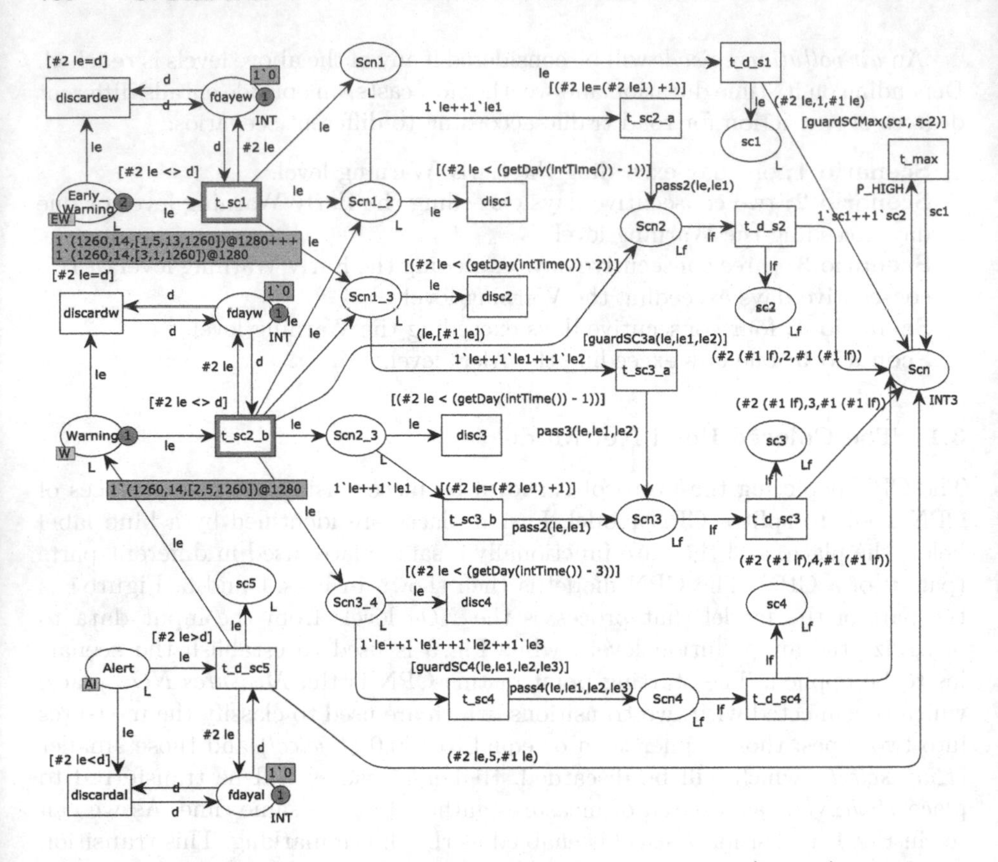


**Fig. 3.** CPN model for Madrid air pollution protocol (part II).

The tokens produced at places *Warning* and *Early_Warning* can be observed in Fig. 3. There are now two enabled transitions, *t_sc1* and *t_scd2_b*. By firing *t_scd2_b*, places *Scn1_2*, *Scn1_3*, *Scn2*, *Scn2_3* and *Scn3_4* become marked again with the same previous token `1‘(1260,14,[2,5,1260])@1280` and only transition *t_d_s2* is now enabled. This token will be stored in the place *sc2* and the token `1‘(14,2,1260)@1280` is produced in *Scn*, which represents that a scenario 2 occurs at minute 1260, day 14. In a similar way, tokens at *Early_Warning* enable the firing of transition *t_sc1*, after which *t_d_s1* and *discardew* are also fired. Transition *t_sc1* stores the token at place sc1 and produces a new token `1‘(14,1,1260)@1280` at place *Scn*, which will be discarded by transition *t_max*, since the previous token at this place corresponds to a higher scenario (2).

Finally the last token at *Early_Warning* will be discarded by transition *discardew*. The implementation of this feature of discarding tokens allows us to obtain a faster simulation, because memory usage is reduced significantly. Notice that similar situations can be generated for the rest of scenarios 3, 4 and 5.

### 3.2 Model Validation

The validation of the model has been performed using different techniques. An untimed version of the model has been tested using the state space generation tool that CPN Tools provides. Starting with an initial marking we can conclude that all the resulting scenarios can be obtained from this initial set up. The second technique used to validate the model is the simulation technique available at CPN Tools via the **Simulation** tool set, which provides functions to execute a predefined number of transitions automatically. The simulations performed can be classified into two types: randomly generated or statically generated. To generate random situations, we have introduced different types of probability distributions, and static data have been obtained from real data provided by the Air Quality Site of Madrid city council[3]. The main objective has been to check whether the CPN model has been able to produce a correct result in relation to the scenarios identified. A second objective was to analyze and predict possible scenarios in the following days. For instance, three tokens at *Scn3_4* can produce a fourth type episode in the next day.

## 4 Modeling the Event Patterns with MEdit4CEP-CPN

The *AirMeasurement* domain and two patterns have been graphically modeled by using MEdit4CEP-CPN [4], to detect the last 2-hour and 3-hour $NO_2$ averages at a particular station. We assume data are received in this station according to such a domain, which has been modeled and transformed into EPL code:

```
create schema NO2Measurement(
  timestamp long, station integer, zone integer, no2 double);
```

This EPL schema defines the event information required for the $NO_2$ measurements. It contains the time at which the measurement is taken, the station and zone identifiers and the $NO_2$ pollutant. Once the *AirMeasurement* domain is designed, the event pattern editor is automatically reconfigured for this domain. Figure 4(a) shows the design of a pattern that computes the average value for $NO_2$ at every station based on the $NO_2$ measurements received during the last 2 h. Therefore, from all the simple events of *NO2Measurement* for a same station the average value for $NO_2$ is obtained, and a new complex event with the current timestamp, the station, the computed average value and the window size is created and inserted into the *Last2h_NO2_Avg* flow, so as to have all *Last2h_NO2_Avg* average values obtained over the time period. These average values are computed as they are received by using time sliding data windows. Once the pattern is modeled and syntactically validated, the EPL code automatically generated for the *Last2h_NO2_Avg* pattern is shown in Fig. 4(b). As shown in Fig. 5(a), the *Last3h_NO2_Avg* pattern is modeled analogously to the *Last2h_NO2_Avg* pattern, but replacing both the 2-h sliding data window by a 3-h sliding data window, and value 2 for the window size complex event property by value 3. The EPL code generated for this pattern is shown in Fig. 5(b).

[3] Available at: http://www.mambiente.madrid.es/sica/scripts/index.php.

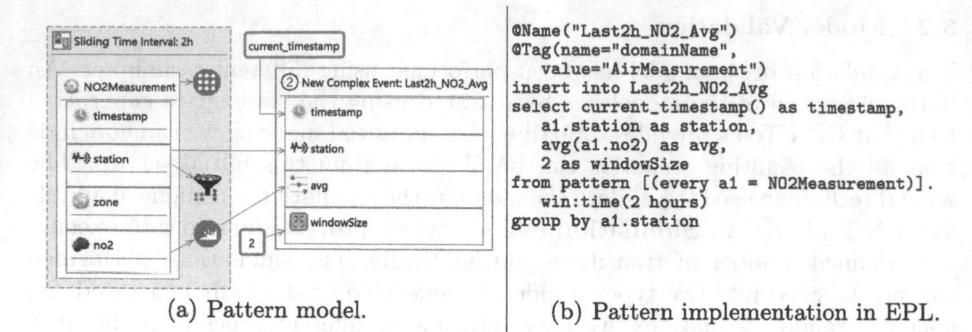
Sliding Time Interval: 2h
current_timestamp
NO2Measurement
timestamp
station
zone
no2
2
New Complex Event: Last2h_NO2_Avg
timestamp
station
avg
windowSize
@Name("Last2h_NO2_Avg")
@Tag(name="domainName",
value="AirMeasurement")
insert into Last2h_NO2_Avg
select current_timestamp() as timestamp,
a1.station as station,
avg(a1.no2) as avg,
2 as windowSize
from pattern [(every a1 = NO2Measurement)].
win:time(2 hours)
group by a1.station

(a) Pattern model. (b) Pattern implementation in EPL.

**Fig. 4.** *Last2h_NO2_Avg* pattern.

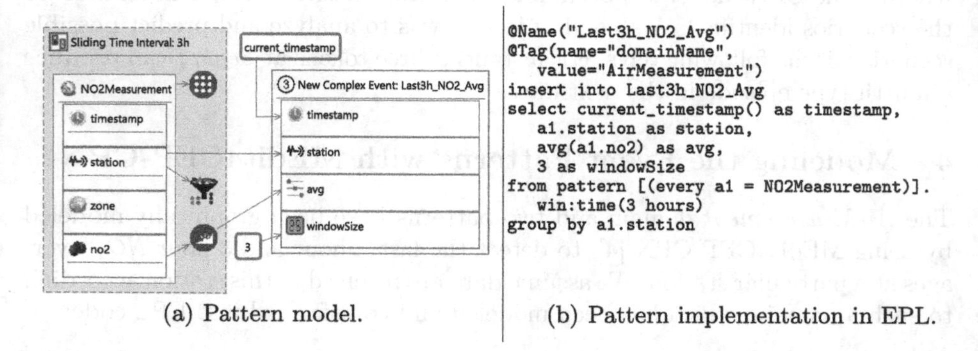
Sliding Time Interval: 3h
current_timestamp
NO2Measurement
timestamp
station
zone
no2
3
New Complex Event: Last3h_NO2_Avg
timestamp
station
avg
windowSize
@Name("Last3h_NO2_Avg")
@Tag(name="domainName",
value="AirMeasurement")
insert into Last3h_NO2_Avg
select current_timestamp() as timestamp,
a1.station as station,
avg(a1.no2) as avg,
3 as windowSize
from pattern [(every a1 = NO2Measurement)].
win:time(3 hours)
group by a1.station

(a) Pattern model. (b) Pattern implementation in EPL.

**Fig. 5.** *Last3h_NO2_Avg* pattern.

# 5 Integrating a CEP Engine with the Petri Net Model

This section describes our novel ESB-based solution that makes it possible the interconnection between the Esper CEP engine [2] and the CPN model presented in Sect. 3.

More specifically, the ESB allows us to integrate the sensing data producers with the CEP engine, and also the CEP engine with the CPN model.

As a data producer, we have created a Python simulator that retrieves the real sensing air data from the Madrid city council (http://www.mambiente.madrid.es/sica/scripts/index.php) and sends it to the ESB.

The ESB is responsible for receiving this real data, transforming it to an event format according to the *NO2Measurement* schema (see Sect. 4), as well as sending the transformed events to the connected Esper engine.

The Esper engine is in charge of analyzing and correlating the received simple events, as well as generating complex events when the conditions defined in the *Last2h_NO2_Avg* and *Last3h_NO2_Avg* event patterns, described in Sect. 4 and previously deployed in the engine, are satisfied.

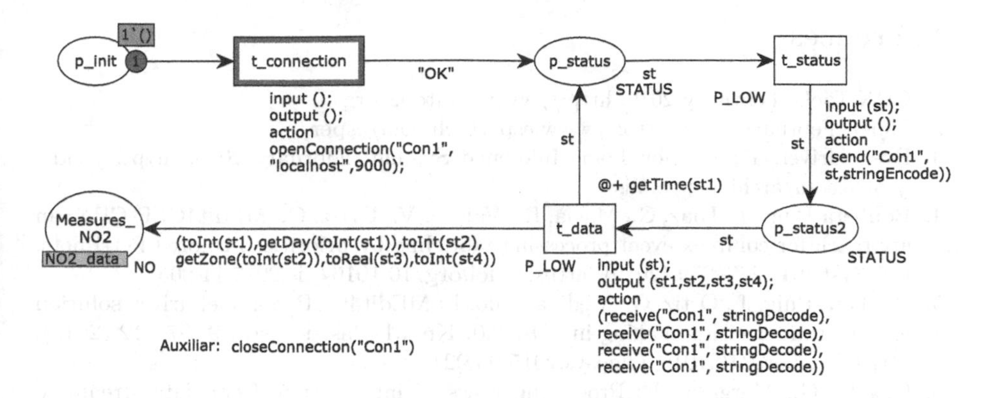


**Fig. 6.** Petri net connection.

The detected complex events are sent to the ESB, where we have implemented a new client/server component responsible for sending these events to the page of the CPN model shown in Fig. 6. This component uses an intermediary TCP/IP socket implemented in Java at both sides.

In particular, this CPN page starts the communication via socket through port 9000 by firing the *t_connection* transition. Then, a cycle continues. The *t_status* transition communicates the *OK* status to the ESB, indicating "*ready to receive data*". Next, the ESB sends a $NO_2$ measurement that the page receives via transition *t_data* and the cycle starts again. Although the communication established by the socket is asynchronous, this cycle allows the ESB to send only one measurement since it needs to receive an *OK* status to send a new one. Note that this cycle ensures the traceability of the information sent.

As a result, the events produced by the Esper engine can be consumed at runtime by the CPN model in charge of making simulations with CPN Tools to determine whether a pollution scenario occurs (see Fig. 3).

# 6 Conclusions and Future Work

In this work, we have proposed an easily analyzable, flexible and agile intelligent system capable of conducting the decision-making process about when air pollution protocols must be applied. To achieve this goal, we combine the CPN formalism and the CEP technology by using an ESB. More specifically, the proposed intelligent system consists of an unprecedented CPN model able to process the data provided by a CEP engine and assists stakeholders to make a decision.

As future work, we pretend to include new protocols used by other cities in this intelligent system, implement our own sensor network as well as considering other data, like weather forecast, to study the implementation of these protocols. Additionally, this approach could be integrated with our intelligent transportation system proposed in [7].

## References

1. CPN Tools, February 2019. http://www.cpntools.org/
2. Esper, February 2019. http://www.espertech.com/esper/
3. Event-driven Transaction Logic Inference System, February 2019. https://code.google.com/archive/p/etalis/
4. Boubeta-Puig, J., Díaz, G., Macià, H., Valero, V., Ortiz, G.: MEdit4CEP-CPN: an approach for complex event processing modeling by prioritized colored Petri nets. Inf. Syst. **81**, 267–289 (2019). https://doi.org/10.1016/j.is.2017.11.005
5. Boubeta-Puig, J., Ortiz, G., Medina-Bulo, I.: MEdit4CEP: a model-driven solution for real-time decision making in SOA 2.0. Knowl.-Based Syst. **89**, 97–112 (2015). https://doi.org/10.1016/j.knosys.2015.06.021
6. Cugola, G., Margara, A.: Processing flows of information: from data stream to complex event processing. ACM Comput. Surv. **44**(3), 15:1–15:62 (2012)
7. Díaz, G., Macià, H., Valero, V., Boubeta-Puig, J., Cuartero, F.: An intelligent transportation system to control air pollution and road traffic in cities integrating CEP and Colored Petri Nets. Neural Comput. Appl. (in press). https://doi.org/10.1007/s00521-018-3850-1
8. EU Environment: Air quality standards, February 2019. http://ec.europa.eu/environment/air/quality/standards.htm
9. Etzion, O., Niblett, P.: Event Processing in Action, 1st edn. Manning Publications Co., Greenwich (2010)
10. Jensen, K., Kristensen, L.M.: Coloured Petri Nets: Modelling and Validation of Concurrent Systems, 1st edn. Springer, Heidelberg (2009). https://doi.org/10.1007/b95112
11. City Council of Madrid: Madrid protocol for $NO_2$ pollution episodes, February 2019. https://www.madrid.es/UnidadesDescentralizadas/Sostenibilidad/CalidadAire/Ficheros/ProtocoloNO2AprobFinal_201809.pdf
12. Offel, M., van der Aa, H., Weidlich, M.: Towards net-based formal methods for complex event processing. In: Proceedings of the Conference "Lernen, Wissen, Daten, Analysen", LWDA 2018, Mannheim, Germany, 22–24 August 2018, pp. 281–284 (2018)
13. van der Aalst, W.M., Stahl, C.: Modeling Business Processes - A Petri Net-Oriented Approach. Cooperative Information Systems Series. MIT Press, Cambridge (2011)
14. Weidlich, M., Mendling, J., Gal, A.: Net-based analysis of event processing networks – the fast flower delivery case. In: Colom, J.-M., Desel, J. (eds.) PETRI NETS 2013. LNCS, vol. 7927, pp. 270–290. Springer, Heidelberg (2013). https://doi.org/10.1007/978-3-642-38697-8_15
15. WHO: Health risk assessment of air pollution. General principles, February 2019. http://www.euro.who.int/en/health-topics/environment-and-health/air-quality/publications/2016/health-risk-assessment-of-air-pollution.-general-principles-2016

# Using Genetic Algorithms to Generate Test Suites for FSMs

Miguel Benito-Parejo[1], Inmaculada Medina-Bulo[2], Mercedes G. Merayo[1], and Manuel Núñez[1](✉)

[1] Departamento Sistemas Informáticos y Computación, Universidad Complutense de Madrid, Madrid, Spain
{mibeni01,mgmerayo,manuelnu}@ucm.es

[2] Departmento de Ingeniería Informática, Escuela de Ingeniería, Universidad de Cádiz, Cádiz, Spain
inmaculada.medina@uca.es

**Abstract.** It is unaffordable to apply all the possible tests to an implementation in order to assess its correctness. Therefore, it is necessary to select relatively small subsets of tests that can detect many errors. In this paper we use different approaches to select these test suites. In order to decide how good a test suite is, we confront it with a set of *mutants*, that is, small variations of the specification of the system to be developed. The goal is that our algorithms build test suites that *kill* as many mutants as possible. We compare the different approaches (consider all the possible subsets up to a given number of inputs, intelligent greedy algorithm and different genetic algorithms) and discuss the obtained results. The whole framework has been fully implemented and the tool is freely available.

**Keywords:** Genetic algorithms · Testing from FSMs · Mutation testing

## 1 Introduction

Testing is one of the main techniques to validate the correctness of software systems [1]. Quite often we find ourselves with a group of properties that should be satisfied by the system under development and we want to reassure that it does. In testing, these properties are encoded as tests and we have to check that the system, usually called System Under Test (SUT), successfully passes them. In practice, this approach is unfeasible because the number of tests may be astronomical (one property will give rise to many tests). In addition, we may have a bound on the number of tests that we can apply (e.g. due to budget or temporal constraints). Therefore, it is important to *wisely* choose among these tests a subset that is able to detect most faults. Clearly, the method to select

Research partially supported by the Spanish MINECO/FEDER project DArDOS (TIN2015-65845-C3-1-R and TIN2015-65845-C3-3-R) and the Comunidad de Madrid project FORTE-CM (S2018/TCS-4314).


I. Rojas et al. (Eds.): IWANN 2019, LNCS 11506, pp. 741–752, 2019.
https://doi.org/10.1007/978-3-030-20521-8_61

these tests should rely on a measure of how good a test is. In this line, mutation testing [3,8,11] is a useful tool. The idea behind mutation testing is that if a test suite distinguishes the SUT from other similar, but faulty, systems then it is probably good at discovering faults. The technique introduces small changes in the SUT, one at a time by applying *mutation operators*, to generate a set of *mutants*. Intuitively, good test suites are the ones *killing* most of the mutants.

In this paper we analyze different strategies to select *good* sets of tests. We assume that we have a formal representation of the SUT (its specification) and that we are provided with a set of mutants (possibly constructed from the specification and representing most representative faults) and a set of tests (usually huge) that we might apply to the SUT. Our goal is to select a subset of tests up to a certain *complexity* (we will measure the complexity of a test suite in terms of the number of inputs included in it) that kills as many mutants as possible. If $T$ is the whole set of tests and $I$ is the bound on the number of inputs, then the obvious solution is to compute all the subsets of $T$ with up to $I$ inputs, apply them to the set of mutants and choose the subset killing more mutants. Unfortunately, in this case we have a combinatorial explosion that disallows us to use this approach. A second possibility, based on previous work [2], considers a greedy algorithm where we select the best tests individually (according to the number of mutants that they kill) until we reach the limit of $I$ inputs. This technique will generally provide good results, both in cost and in faults detected, but it may not always yield the best result. For instance, there could be a combination of individually worse elements that are able to cover more faults. In order to solve this problem, and this is the main contribution of this paper, we developed a genetic algorithm to find better solutions than the greedy algorithm. The algorithm is versatile and allows users to exercise different variants. We have developed a tool that fully implements all the algorithms, with its variants, presented in the paper. Finally, we have performed several experiments to compare the different methods.

The rest of the paper is structured as follows. In Sect. 2 we introduce the main concepts used in the paper. In Sect. 3 we enumerate the considered methods to select test suites. In Sect. 4 we report on our experiments. Finally, in Sect. 5 we present our conclusions and some lines for future work.

## 2 Preliminaries

In this section, we introduce the main background and concepts required for the paper.

**Definition 1.** *A* Finite State Machine, *in the following FSM, is a tuple* $M = (S, I, O, Tr, s_{in})$ *where* $S$ *is a finite set of* states, $I$ *is the set of* input actions, $O$ *is the set of* output actions, $Tr$ *is the set of* transitions *and* $s_{in} \in S$ *is the* initial state. *A transition belonging to* $Tr$ *is a tuple* $(s, s', i, o)$ *where* $s, s' \in S$ *are the initial and final states of the transition,* $i \in I$ *is the input action and* $o \in O$ *is the output action. We say that* $M$ *is* input-enabled *if for each* $s \in S$ *and input*

*$i \in I$, there exist $s' \in S$ and $o \in O$ such that $(s, s', i, o) \in Tr$. We say that $M$ is* deterministic *if for each $s \in S$ and $i \in I$, there exists at most one transition $(s, s', i, o)$ belonging to $Tr$.*

In this paper we will restrict ourselves to input-enabled deterministic FSMs, that is, from each state of the machine, it is possible to perform all the inputs and there will be only one possible evolution. This restriction mimics testing of programs: programs are (usually) deterministic and should react to any received input. FSMs will be used to represent specifications and mutants. Although the FSM formalism looks simple, it has a common base with more complex frameworks to represent black-box systems. Therefore, the framework presented in this paper can be extended as long as we consider systems where we apply a sequence of inputs and decide whether the returned output sequence is expected or nor. Next we introduce the notion of mutant.

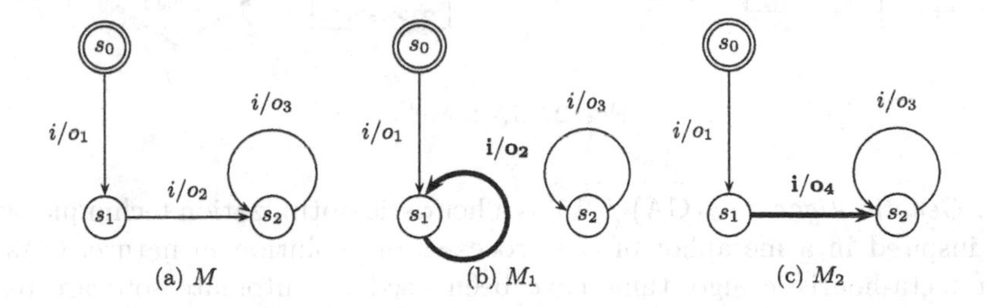


**Fig. 1.** An FSM and two of its mutants

**Definition 2.** *Let $M = (S, I, O, Tr, s_{in})$ be an FSM. We say that an FSM $M' = (S, I, O, Tr', s_{in})$ is a* mutant *of $M$ if $Tr'$ differs from $Tr$ in only one transition. This* mutation *can be produced by choosing one transition $(s, s', i, o) \in Tr$ and replacing it by either $(s, s', i, o') \in Tr'$, with $o \neq o'$, or by $(s, s'', i, o) \in Tr'$, with $s' \neq s''$.*

Note that mutants are still deterministic and input-enabled. For example, consider the FSM given in Fig. 1a, being $s_0$ the initial state. Two possible mutants are shown in Figs. 1b and c: the first one represents the change of final state of a transition while the second one represents a change in the output. The next concept that we need is the notion of test.

**Definition 3.** *Let $M = (S, I, O, Tr, s_{in})$ be an FSM. A* test *for $M$ is a pair $\sigma = (\sigma_{in}, \sigma_{out})$ where $\sigma_{in} \in I^*$ is a sequence of inputs and $\sigma_{out} \in O^*$ is the sequence of outputs that $M$ produces when applying $\sigma_{in}$.*

*Let $t = (\sigma_{in}, \sigma_{out})$ be a test for $M$. We say that a system $M'$ passes $t$ if the application of $\sigma_{in}$ produces $\sigma_{out}$; otherwise, we say that the $M'$ fails $t$.*

Consider again $M$, $M_1$ and $M_2$. For example, we have that $t_1 = (i, o_1)$, $t_2 = (ii, o_1o_2)$ and $t_3 = (iii, o_1o_2o_3)$ are tests for $M$. Moreover, $M_1$ passes $t_1$

and $t_2$ and fails $t_3$ while $M_2$ passes $t_1$ and fails $t_2$ and $t_3$. The next concept allows us to represent in a matrix whether and when a test kills a mutant. It will be important to record the first input showing a faulty behavior because we might be able to discard the rest of the test.

**Definition 4.** *Let $M = (S, I, O, Tr, s_{in})$ be an FSM, $\{t_i\}_{i=1}^n$ be a set of tests for $M$ and $\{M_j\}_{j=1}^m$ be a set of mutants of $M$. We define a* results table *as a matrix $(a_{ij})_{i=1,j=1}^{n,m}$, where $a_{ij}$ is the length of the shortest prefix of test $t_i$ that kills mutant $M_j$. In the case that such mutant passes the test, this distance will be equal to infinity.*

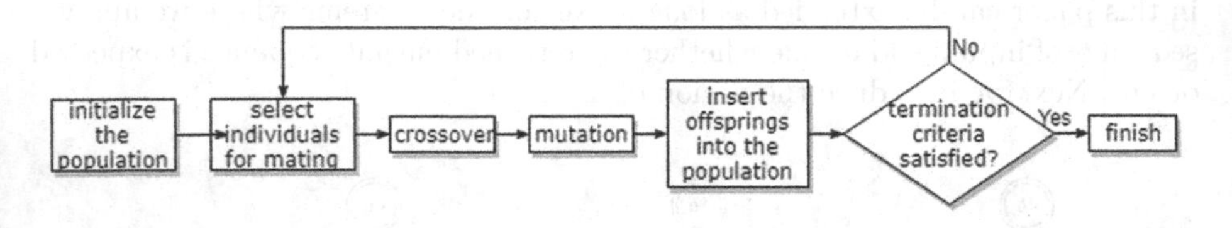


**Fig. 2.** GA flowchart

A *Genetic Algorithm* (GA) [5,15] is a heuristic optimization technique, which it is inspired in a metaphor of the processes of evolution in nature. GAs and other meta-heuristic algorithms have been used to automate software testing [4,6,12,13]. Generally, a GA consists of a group of individuals (population of genomes), each representing a potential solution to the problem in hand. An initial population with such individuals is usually selected at random. Then, a parent selection process is used to pick a few of these individuals. New offspring individuals are produced using *crossover*, keeping some of the characteristics of their parents, and *mutation*, which introduces some new genetic material. The quality of each individual is measured by a fitness function, defined for the particular search problem. Crossover exchanges information between two or more individuals. The mutation process randomly modifies offspring individuals. The population is iteratively recombined and mutated to evolve successive populations, known as generations. When the termination criterion specified is satisfied, the algorithm terminates. A flowchart for a simple GA is presented in Fig. 2.

## 3 Selection Methods

In this section we present each of the considered approaches to solve the problem of finding good sets of tests. Each of them will be based on how good a test is, which is represented by the number of mutants that it kills and its length (number of inputs). This shows that two tests killing the same mutants might not be equally good, as one detecting them sooner will have a better score[1] because it will save resources.

[1] The score is computed by dividing the fitness of an individual between the sum of all fitness.

### 3.1 Global Search

The global search approach looks through all the possible combinations of the initial set of tests having less inputs than the given bound. This approach always provides the best solution because it explores all the possible subsets. Therefore, it is useful because it helps to compare the results with the forthcoming algorithms. The negative part is that for non-trivial systems, it gets impossible to apply it because it suffers of a combinatorial explosion. In fact, we were able to compute it only for the smallest of our experiments.

$$\begin{pmatrix} 4 & 7 & 5 & \infty & 12 & \infty \\ 13 & \infty & \infty & 6 & 1 & \infty \\ \infty & 9 & \infty & 8 & \infty & \infty \\ 5 & \infty & \infty & \infty & \infty & \infty \\ \infty & \infty & \infty & \infty & \infty & 7 \\ \infty & \infty & \infty & \infty & \infty & \infty \end{pmatrix}$$

(a) Original matrix

$$\begin{pmatrix} 6 & \infty \\ 8 & \infty \\ \infty & \infty \\ \infty & 7 \\ \infty & \infty \end{pmatrix}$$

(b) Reduced matrix

**Fig. 3.** Matrix simplification

### 3.2 Greedy Algorithm

Our greedy algorithm is based on the matrix including information about tests and mutants (see Definition 4). Essentially, the algorithm sorts the matrix in a way such that the first test is the best, the second is the second best and so on. Then, we add the first test to the subset that we are constructing. Afterwards, we remove that test from the matrix, as well as all the mutants that it has killed. After reducing the matrix, we iterate the process until either we have killed all the mutants or until we reach the upper bound on the number of inputs. We can see an example in Fig. 3a: a matrix where rows represent tests and columns represent mutants. As such, the first row is the best test, which will be selected, and will kill the mutants 1, 2, 3 and 5. As a result, we obtain the reduced matrix given in Fig. 3b, where one test and four mutants have being removed. The resulting matrix needs to be resorted, and equivalent rows do not necessarily represent the same test, since several entries (comparing with the dead mutants) have been omitted. In addition, we have to take into account that the selected set of tests has length 12 (the length needed to kill all the mutants).

A good property of this algorithm is that it works in low polynomial order over an space, that is the matrix, that reduces its size after each iteration. This is the less costly, in terms of time needed to compute the solution, algorithm that we present in this paper. Our greedy method shows great results and will also help bounding the number of generations and the global cost of our next algorithm.

### 3.3 Genetic Algorithm

As we know, GAs shine when we seek for a good enough approximation of the solution of problems whose optimal solution needs a combinatorial approach to compute all the potential candidates. This is the case of our problem and its solution, as discussed in Sect. 3.1. Therefore, a genetic algorithm is a sensible approach to compete with our greedy algorithm, in particular, taking into account that our greedy algorithm computes relatively good solutions using a short period of time.

In our population, an individual only has one chromosome. Our GA has a population of initial test subsets that will evolve to generate better subsets until either we reach the optimum or we stop the execution because we have reached the time limit. Even though we will spend more time than the greedy algorithm in the computation of the solution, we would like to have a reasonable bound.

We have implemented our algorithms and the tool is freely available at https://github.com/miguelbpsg/IWANN19. Our tool needs different parameters that can be received in a dynamic way. We have an interface where we can provide how many inputs we expect in the solution, the type of selection of the population, the type and rate of crossover, the rate of mutation (of the genetic algorithm, not to be confused with the mutants of the specification), and more concrete fields in specific cases. Next, we briefly describe these parameters.

**Initialization Method.** The type of initialization that we have designed is *incremental initialization*. It provides a variety of chromosomes, with a different amount of test and inputs each, which generates more diversity. Such initialization follows the idea of minimizing the number of inputs to apply. As some chromosomes may have too few inputs and others too many, the execution of the algorithm will mix them at some point and improve the general result.

**Selection of Population Methods.** As some individuals might be better than others, the transition from one generation to the next one has to ensure that the representatives of the foremost have to be selected. The idea is to reward the best ones with more appearances and the worst ones with even no appearances at all. We allow the user of our tool to use alternative selection models: *Linear rank*, *remains*, *roulette wheel*, *tournament*, *truncation* and *stochastic universal*. Since these selection models are quite common [5,14], we do not explain the details about how each of them is applied to our populations.

**Crossover Methods.** Concerning crossover, combining elements of the population to produce a new generation after an iteration of the algorithm, we have designed two methods. First, *standard crossover* takes two chromosomes of the population and chooses a random point at each of their respective list of tests. Then, the left part of each list is preserved and the right part is exchanged. If such modification generates a test suite with more inputs than the fixed bound, then the last tests are discarded until the bound is reached. Second, *continuous* *crossover* takes two chromosomes and randomly exchanges the test at each position for both lists. Therefore, more possible combinations of swaps can happen.

**Mutation of Population Methods.** As the initial seeds for the population might not be complete, it is sensible to refresh the population with some slight changes that could renew some stale state. In our case, we designed two different techniques. The first one considers *adding mutation*, that is, it is oriented to not-complete subsets. In these cases it is possible to introduce an extra test to an individual without exceeding the bound of inputs. Despite increasing the number of tests on the whole population, using the appropriate crossover method, it is possible to generate an exchange of tests where some of them have to be discarded. As a consequence, this method complements the possible loss of these tests. The second method is based on *replacing mutation*. This method allows an individual to change one of its test by another one coming from the initial set of all provided tests. This technique will include some slight changes to specific individuals that might either increase or decrease the relevance of a subset of tests into the population.

The last step of each generation on the genetic algorithm is to replace the population by the new one. Again, we have two possibilities. On one hand, the trivial option would be to simply substitute the last population by the new one, even if it could be worse. In this way, less operations are performed at this stage and, as a result, the execution will be faster. On the other hand, we could replace a high percentage of the new generation on the previous one, where the best individuals could still stay. This approach will always allow the population to keep the best partial solution until it is improved. As a counterpart, more calculations have to be made and that cost might decelerate the program so that the GA will perform less iterations in the same amount of time.

Finally, the heuristics that we use to define the fitness function enhances the individuals that locally improve the expected results of testing because it takes into account how many mutants are killed. Specifically, the fitness is calibrated by adding the minimum number of inputs required to kill a mutant considering the subset of tests that the chromosome has at a moment, through all the mutants. This value does punish leaving mutants alive, as a penalty is added to the final value for each living mutant. Therefore, the more mutants a subset kills, the lower the score will be, and a lesser number of inputs required to kill more mutants will also reduce the value, leading us to a minimization problem (a lower value of fitness denotes a better population).

In Fig. 4 we show the results of one experiment concerning how fitness varies along generations. The results are as expected. In short, there is a relatively big variance concerning the worse individual of each generation (that is, the highest value of fitness), this variance is smaller for average fitness (and the value notoriously improves after only 10 generations), generational best stabilizes (although there are small variations), while absolute best quickly converges.

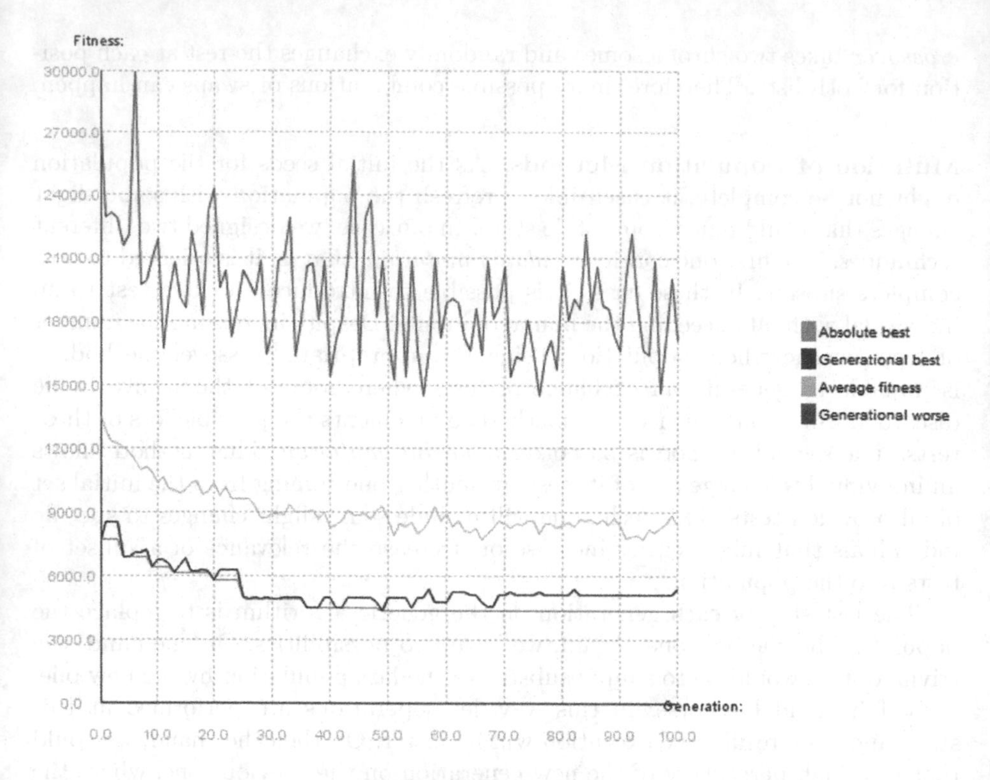


**Fig. 4.** Genetic evolution

# 4 Experiments

In this section we comment on the experiments that we have performed, on why we chose to make them, the results obtained in each of them provided, and the main differences between experiments. We also discuss the results of trying our three previous methods into a specification, with different bounds on the number of inputs that a solution could have. We have analyzed the time that the different approaches need to compute their solutions.

## 4.1 Description of the Experiments

Our experiments consisted in the execution of the three described algorithms over the same specification, mutants, initial tests and maximum number of allowed inputs, for several combinations of them. Afterwards, we compare the results both in time needed to compute the solution and in the goodness of the solution. Note that the initial set of tests should not be confused with the set of tests produced in the initialization of the GA. The former is provided as a precondition of the problem. This is the set of tests that we should aim to exercise but if our resources do not allow us to try all of them (for example, because they have more inputs than the considered bound), then we should apply a *good* subset of them (computing this final subset is the goal of our approaches). The latter is computed as the first step of the solution of the problem.

It is clear that the smaller the FSM is, the lesser number of mutants we will have. We have considered a fixed specification with 10 states, 3 different inputs and 5 different outputs generating near 300 mutants. We also considered three possible bounds on the number of inputs, and two possible initial sets of tests, resulting in six representative cases. Next we give the details of each of them.

1. The first experiment consisted of allowing a maximum of 30 inputs in the solution and starting with a set of 99 tests. All algorithms yield a very good (if not the best) solution on a reasonable amount of time as there are few possible results. This experiment is a good baseline that all three algorithms provide good solutions.
2. Next, we increased the bound on the number of inputs, from 30 to 80, maintaining the initial set of tests. This experiment should show us the evolution of the algorithms depending on the number of inputs.
3. To conclude the 99 initial tests experiments, we increased the bound on the number of inputs up to 150 inputs.
4. The next variation considered the smallest bound on the number of inputs (that is, 30 inputs) on a bigger set of tests (we consider a set of 957 tests).
5. The next experiment considers the intermediate bound (that is, 80) and the big set of tests.
6. Finally, the last experiment consider the biggest bound on the number of inputs and the bigger set of tests (150 inputs and 957 tests, respectively).

These experiments were performed with our default options, that is, tournament selection over 3 participants with a ratio of 80%, continuous crossover with a probability of 0.6, the extra test mutation option with a 0.02 coefficient and a direct replacement. It is worth to mention that most of the combinations of our GAs gave very similar results. It should be also noted that the elitist replacement always produced better fitness results at the cost of a longer execution time while roulette wheel and stochastic universal selection methods produced slightly worse results.

We also tested what bounds and initial sets of tests the full search was able to compute. We were not able to exceed a bound of 60 inputs over 99 initial tests, taking over a week to compute the solution.

## 4.2 Evaluation

As expected, as soon as the systems were sizable enough and we had a nontrivial set of tests to choose start with, generating all the possible subsets was unfeasible due to the combinatory explosion on the number of subsets. For example, if we had 40 initial tests with an average of 10 inputs, and we could choose tests up to 150 inputs to finally execute them, we would have more than 40 billion possible subsets, with 150 inputs, of the original set of tests. Nevertheless, whenever we find ourselves with small bounds, it is possible to compute an optimal result, guaranteeing that we obtain the best subset of tests to use on the SUT.

In terms of relative cost, we have that the greedy algorithm was always the fastest and we can see this in Table 1. The time needed to compute the solution mainly depended on the size of the given set of tests but it also had a small dependence on the number of inputs allowed, since this determines how many times the matrix has to be sorted. Considering its efficiency, our GA was able to provide very good results, almost equivalent to the optimal in the only experiment where we were able to compute all the combinations, as Table 2 shows (11.761 fitness of the full search vs. 11.887 of the GA). However, the solution was computed in a much faster time, showing evidence of the usefulness of this technique.

Focusing now on Tables 1 and 2, we have that a higher bound on the number of inputs always increases the execution time, as more possibilities are available. In contrast, the fitness of the obtained solution improves. Also, comparing all the experiments, we observe that a bigger initial set of tests induces a higher execution time, but better results are obtained for the same bound of inputs.

**Table 1.** Time results (in milliseconds)

| | Time Exp. 1 | Time Exp. 2 | Time Exp. 3 | Time Exp. 4 | Time Exp. 5 | Time Exp. 6 |
|---|---|---|---|---|---|---|
| Genetic | 91 | 190 | 220 | 208 | 296 | 462 |
| Greedy | 22 | 23 | 26 | 147 | 178 | 263 |
| Full search | 1.355 | – | – | – | – | – |

**Table 2.** Fitness results (higher values denote worse results)

| | Fitness Exp. 1 | Fitness Exp. 2 | Fitness Exp. 3 | Fitness Exp. 4 | Fitness Exp. 5 | Fitness Exp. 6 |
|---|---|---|---|---|---|---|
| Genetic | 11.887 | 5.866 | 2.877 | 12.225 | 3.212 | 1.988 |
| Greedy | 16.166 | 7.687 | 5.415 | 22.746 | 13.543 | 7.429 |
| Full search | 11.716 | – | – | – | – | – |

These experiments show that the GA can adequately compete, depending on the resources, and complement the results of the greedy algorithm. It is true that the GA requires some more time to evaluate, but considering the obtained results we find it worth to use this extra computing power. Finally, let us note that the whole implementation and some additional examples are freely available at https://github.com/miguelbpsg/IWANN19. In Fig. 5 we give a screenshot of our tool.

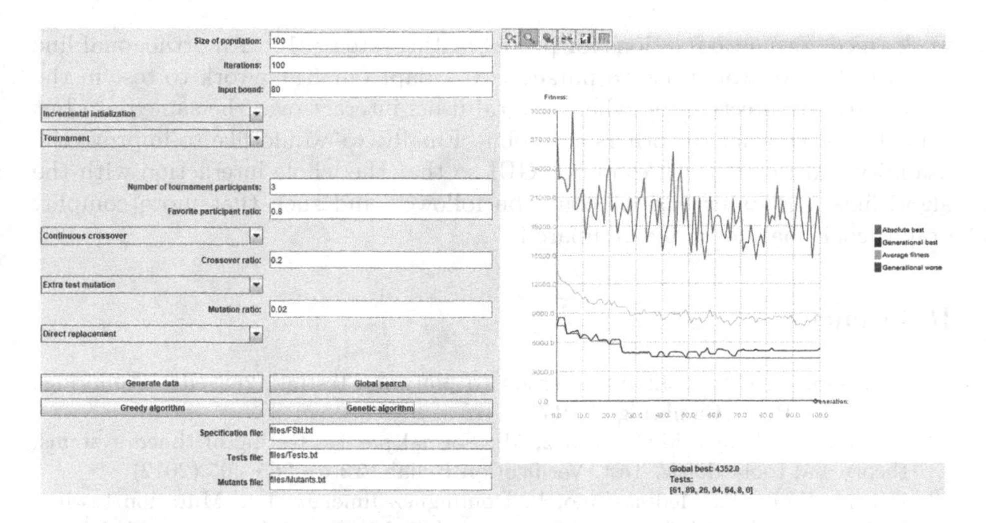


**Fig. 5.** GUI of our tool

# 5 Conclusions and Future Work

In this paper we present different solutions to the problem of obtaining good sets of tests out of big sets. Ideally, if a tester is provided with a whole set of tests then the tester should apply all of them to the SUT. However, the time and resources devoted to testing are usually limited and the tester can apply only a subset of these tests. If we are working within a framework where the tester applies inputs and receive outputs, then this bound is given by the number of inputs that the tester can apply. This is an important problem in testing and in addition to provide a sound theoretical framework, it is a must to develop tools supporting the frameworks. We have a tool implementing all the algorithms presented in the paper, that is, our tool is able to, given an initial set of tests and the maximum number of inputs that we can really apply, compute a subset of the initial set with any of the algorithm, and the different variants of the GA, discussed in this paper. In addition, the tool supports the process of generating mutants from a specification of the SUT.

The results show that our GA usually finds an excellent solution. In general, the GA beats the greedy algorithm needing a slightly higher amount of time to compute the result. For smaller experiments, where full search could be effectively computed, the differences between the best solution and the one obtained through the GA were very small: the fitness of the full search approach was around 1,5% better but it needed 14 times longer to compute it. Therefore, we can be satisfied with the results considering the complexity of the problem.

As future work, we plan to extend the framework to deal with other FSM-like formalisms. A first line of work is to consider probabilistic FSMs, where nondeterminism is probabilistically quantified. We will take as initial step our previous work on mutation testing of probabilistic FSMs [7] complemented with our recent work on conformance relations for probabilistic systems [10]. An orthogonal line or work that we would like to pursue is to adapt our framework to test in the distributed architecture [9], where several users interact over the same data but cannot observe what the others are doing. Finally, we would like to improve the usability and report features of our GUI so that the whole interaction with the algorithms and its extensions could be followed and such that more complex graphs could be shown and compared.

## References

1. Ammann, P., Offutt, J.: Introduction to Software Testing, 2nd edn. Cambridge University Press, Cambridge (2017)
2. Andrés, C., Merayo, M.G., Núñez, M.: Formal passive testing of timed systems: theory and tools. Softw. Test. Verification Reliab. **22**(6), 365–405 (2012)
3. Delgado-Pérez, P., Medina-Bulo, I., Domínguez-Jiménez, J.J.: Mutation testing. In: Encyclopedia of Information Science and Technology, 3rd edn., pp. 7212–7221. IGI Global (2014)
4. Derderian, K., Merayo, M.G., Hierons, R.M., Núñez, M.: A case study on the use of genetic algorithms to generate test cases for temporal systems. In: Cabestany, J., Rojas, I., Joya, G. (eds.) IWANN 2011. LNCS, vol. 6692, pp. 396–403. Springer, Heidelberg (2011). https://doi.org/10.1007/978-3-642-21498-1_50
5. Goldberg, D.: Genetic Algorithms in Search Optimisation and Machine Learning. Addison-Wesley, Boston (1989)
6. Harman, M., McMinn, P.: A theoretical and empirical study of search-based testing: local, global, and hybrid search. IEEE Trans. Softw. Eng. **36**(2), 226–247 (2010)
7. Hierons, R.M., Merayo, M.G.: Mutation testing from probabilistic and stochastic finite state machines. J. Syst. Softw. **82**(11), 1804–1818 (2009)
8. Hierons, R.M., Merayo, M.G., Núñez, M.: Mutation testing. In: Laplante, P.A. (ed.) Encyclopedia of Software Engineering, pp. 594–602. Taylor & Francis (2010)
9. Hierons, R.M., Merayo, M.G., Núñez, M.: Bounded reordering in the distributed test architecture. IEEE Trans. Reliab. **67**(2), 522–537 (2018)
10. Hierons, R.M., Núñez, M.: Implementation relations and probabilistic schedulers in the distributed test architecture. J. Syst. Softw. **132**, 319–335 (2017)
11. Jia, Y., Harman, M.: An analysis and survey of the development of mutation testing. IEEE Trans. Softw. Eng. **37**(5), 649–678 (2011)
12. Jones, B.F., Eyres, D.E., Sthamer, H.H.: A strategy for using genetic algorithms to automate branch and fault-based testing. Comput. J. **41**(2), 98–107 (1998)
13. Michael, C.C., McGraw, G., Schatz, M.A.: Generating software test data by evolution. IEEE Trans. Softw. Eng. **27**(12), 1085–1110 (2001)
14. Michalewicz, Z.: Genetic Algorithms + Data Structures = Evolution Programs, 3rd edn. Springer, Heidelberg (1996). revised and extended
15. Srinivas, M., Patnaik, L.M.: Genetic algorithms: a survey. IEEE Comput. **27**, 17–27 (1994)

# Conformance Relations for Fuzzy Automata

Iván Calvo[1], Mercedes G. Merayo[1], Manuel Núñez[1(✉)], and Francisco Palomo-Lozano[2]

[1] Departamento Sistemas Informáticos y Computación, Universidad Complutense de Madrid, Madrid, Spain
{ivcalvo,mgmerayo,manuelnu}@ucm.es

[2] Departmento de Ingeniería Informática, Escuela de Ingeniería, Universidad de Cádiz, Cádiz, Spain
francisco.palomo@uca.es

**Abstract.** The use of formal methods improves the reliability of computer systems. In this context, fuzzy logic provides a tool to formally specify systems where uncertainty and imprecision play an important role. In this paper, we propose an extension of the fuzzy automata formalism and establish different conformance relations. The main goal of these relations is to formally capture the idea of a system behaving as specified by a specification. We sketch how our conformance relations can be alternatively characterized as a testing process by producing sound and complete sets of tests.

**Keywords:** Fuzzy automata · Conformance relations · Formal approaches to testing

## 1 Introduction

The use of formal methods can notoriously improve the reliability of systems: a formal development, with a mathematical basis, will help to skip faults. In particular, the existence of a formal specification will guide the development of the systems and will help to assess the correctness of the developed system. Unfortunately, Computer Science is a *strange* Engineering discipline because it is not usually understood, specially in industrial environments, the importance of a formal plan to build complex systems [9]. The second mentioned use of a specification, validate the developed system, has been combined with testing [1] to provide a powerful methodology: formal approaches to testing [6,8]. Essentially, formal testing takes a specification (the formal description of the system that we are developing) and a system under test (SUT, the implementation of

Research partially supported by the Spanish MINECO/FEDER project DArDOS (TIN2015-65845-C3-1-R and TIN2015-65845-C3-3-R) and the Comunidad de Madrid project FORTE-CM (S2018/TCS-4314).


I. Rojas et al. (Eds.): IWANN 2019, LNCS 11506, pp. 753–765, 2019.
https://doi.org/10.1007/978-3-030-20521-8_62

the system) and checks, by applying tests, whether the SUT shows a behavior contradicting what the specification states. There are many alternative formal testing frameworks but the ones considering state-based systems are more frequent and are usually supported by tools [12].

The main problem with mainstream formalisms is that they are not suited to specify systems with specific features. These formalisms are appropriate to specify properties such as "if the system receives input $i$ then it should produce output $o$", that is, causality relations that should always hold. However, there are systems where this information can be more imprecise. For example, there are situations where we would like to specify a property such as "if the system receives an input $i$ then it should produce output $o$ most of the times, while it should produce output $o'$ sometimes". In this case, we need a *tailored* formalism. In this paper we will consider fuzzy logic [15,16] as the theory underlying our *fuzzy automata*. There are many proposals to include fuzzy logic into automata [7,11,14] and our research group has been particularly active in this area [2–5]. The main goal of this paper is to present a formal framework to assess the correctness of an SUT with respect to a specification expressed as a fuzzy automata. The first step was to take our last formalism [4] and slightly modify it to incorporate the improvements that we have detected since its creation. Next, we had to provide an operational semantics to formally define how a system evolves. This is a fundamental step because conformance relations rely on comparing the behavior of the specification and the SUT. The study of an appropriate conformance relation revealed that in a *fuzzy framework* there is not a unique way to define it. Therefore, we propose alternative notions and show how they are related. A final step is to characterize conformance relations via testing. Due to space limitations we cannot present the complete testing framework and we only sketch how tests are defined and how test derivation is implemented.

The rest of the paper is structured as follows. In Sect. 2 we review some concepts related to fuzzy logic. In Sect. 3 we introduce our latest proposal of fuzzy automata by formally defining its syntax and semantics. In this section we also introduce some relevant concepts about the operational behavior of these automata. In Sect. 4 we define our conformance relations and include some interesting observations and results, which will be needed, in Sect. 5, to define our approach to testing. Finally, in Sect. 6 we present our conclusions and some lines for future work.

## 2 Preliminaries

In this section we review the basic definitions underlying our formalism. These definitions represent an extension of the concepts presented in our previous work [4] but revised and refined in order to fit the needs of the current version of our formalism. In usual *crisp* logic we have that truth values can be either *True* or *False*, represented respectively by 1 and 0. In contrast, *fuzzy* logic allows truth values to be anywhere in between these bounds. In order to combine these truth values, we need to define the concept of *triangular norms* or $t$-norms.

**Definition 1.** *A function* $\triangle : [0,1] \times [0,1] \longrightarrow [0,1]$ *is said to be a t-norm when it satisfies the following properties:*

- $\triangle$ *is commutative:* $x \triangle y = y \triangle x$ *for all* $x, y \in [0,1]$.
- $\triangle$ *is associative:* $(x \triangle y) \triangle z = x \triangle (y \triangle z)$ *for all* $x, y, z \in [0,1]$.
- $\triangle$ *is monotonic: if* $x_1 \leq y_1$ *and* $x_2 \leq y_2$ *then* $x_1 \triangle x_2 \leq y_1 \triangle y_2$ *for all* $x_1, x_2, y_1, y_2 \in [0,1]$.
- $\triangle$ *has an identity element:* $1 \triangle a = a$ *for all* $a \in [0,1]$.

*In this paper we will only consider* strictly monotonic *t-norms, that is, t-norms such that for all* $0 \leq x, y, z \leq 1$ *we have that* $y < z$ *and* $x > 0$ *implies* $(x \triangle y) < (x \triangle z)$.

Some examples of strictly monotonic $t$-norms are the Hamacher $t$-norm

$$(x, y) \mapsto \frac{x \cdot y}{x + y - x \cdot y}$$

and the product $t$-norm

$$(x, y) \mapsto x \cdot y$$

Not strictly monotonic $t$-norms do not have the properties that we need and are not considered in this paper. For example, the minimum $t$-norm

$$(x, y) \mapsto \min(x, y)$$

is not strictly monotonic. Our proposed formalism is based on finite automata equipped with a finite set of real valued variables. These variables are used to track and process relevant data. A precise definition of the syntax and semantics of arithmetic expressions over the set of variables needs to be defined in order to specify how data is meant to be processed.

**Definition 2.** Expressions *are formed by variable evaluations, usual arithmetic operators* $(+, -, \cdot, /)$*, clamped t-norms (functions that are t-norms on the interval* $[0,1]$ *and identical to* $0$ *outside the interval), and the reserved variable* $\mu$.

The alphabet labelling automata transitions includes *input* and *output actions.*

**Definition 3.** *An* input action *consists of its name, followed by a tuple of variable names taking real values. The name is a string whose first character is* ?. *The set of all input actions is denoted by* $I$. *An* output action *consists of its name, followed by a tuple of real-valued expressions. The name is a string whose first character is* !. *The set of all output actions is denoted by* $O$.

*We assume that all the actions have different names, so that there is no pair of actions with the same name and different associated tuples. The set of all defined actions is denoted by* $\text{Acts} = I \bigcup O$.

The actions received and produced by a system are parameterized with actual values instead of variable names or expressions.

**Definition 4.** *Let* Acts *be a set of actions. We denote by* IActs *the set of actions from* Acts *parametrized with all possible tuples of values of the same arity as the corresponding tuple of variable names or expressions.*

For example, consider an input $?i(x) \in$ Acts. Then we have that $?i(r)$ belongs to IActs for all $r \in \mathbb{R}$. Similarly, if an output $!o(x+1) \in$ Acts then $!o(r) \in$ IActs for all $r \in \mathbb{R}$.

A test case is determined by a sequence of input actions and output names. The idea is that we will apply inputs with specific values to an SUT and receive outputs from it. The elements forming these sequences will be called *test actions*.

**Definition 5.** *Let* Acts $= I \cup O$ *be a set of actions. We define an input action for each input in $I$ and an output action for each output in $O$.* Test inputs *consist of their name and a tuple of values.* Test outputs *consist only of their name. The set of all test inputs and outputs is called* TActs.

We will also define a mapping from IActs to TActs which removes output parameters. This will be used in the definition of our conformance relations.

**Definition 6.** *Let* Acts *be a set of actions,* IActs *be its associated set of parameterized actions and* TActs *be the set of test actions from* Acts*. The mapping* $\mathcal{D}$ : IActs $\longrightarrow$ TActs *is defined as:*

$$\mathcal{D}(a) = \begin{cases} ?i(\chi) & \text{if } a = ?i(\chi) \\ !o & \text{if } a = !o(\zeta) \end{cases}$$

*Given a sequence* $a_1 \cdots a_n \in \text{IActs}^*$ *we define* $\mathcal{D}(a_1, \cdots, a_n)$ *in the natural way, that is as* $\mathcal{D}(a_1) \cdots \mathcal{D}(a_n) \in \text{TActs}^*$.

Variables are updated when a transition of an automaton is triggered. In order to include this idea, each transition is labelled with a *variable transformation.*

**Definition 7.** *Let* $x_1, \ldots, x_m$ *be variable names and* $t_1, \ldots, t_m$ *be real valued expressions. A* variables transformation, *denoted by* $[t_1/x_1, \ldots, t_m/x_m]$*, assigns to each variable* $x_i$ *the value obtained after evaluating the term* $t_i$*. The evaluations of all the terms are computed before any assignment is performed. The set of variable transformations is denoted by* $\mathcal{VT}$.

We use *fuzzy relations*, instead of the usual *crisp* relations (e.g. $\leq, \geq, =$), to compare input values. The fuzzy counterparts are functions from $\mathbb{R}^2$ to $[0, 1]$, taking the value 1 when the corresponding crisp relation holds. For example, the fuzzy relation $\overline{x \leq y}^{\delta}$, defined as

$$\overline{x \leq y}^{\delta} \equiv \begin{cases} 1 & \text{if } x < y \\ \frac{\delta+y-x}{\delta} & \text{if } y \leq x \leq y+\delta \\ 0 & \text{if } y+\delta < x \end{cases}$$

represents an extension of the crisp relation $x \leq y$. An extended discussion on fuzzy relations can be found in previous work [3].

Transitions may be triggered depending on the parameters of the received input action. In addition, a transition will be triggered with a certain *Grade of Confidence* belonging to the interval $[0,1]$. More specifically, a transition will be triggered with a Grade of Confidence equal to the *Satisfaction Degree* of its associated fuzzy constraint.

**Definition 8.** *A fuzzy constraint is a formula consisting of fuzzy relations, possibly combined with t-norms, which may have free variables. The set of all fuzzy constraints is denoted by $\mathcal{FC}$. When the set of free variables of a constraint, $C$, is contained in the tuple of variable names of an input action, $?i(x_1 \cdots x_n)$, then we can define the function $\mu_{C,?i(x_1 \cdots x_n)}(v_1 \cdots v_n)$ as the value obtained by evaluating the constraint $C$ after performing the substitution $x_1 \mapsto v_1, \ldots, x_n \mapsto v_n$. This value represents the Satisfaction Degree of the constraint with respect to the input values $v_1, \ldots, v_n$.*

## 3 Fuzzy Automata: Syntax and Semantics

In this section we review the basic syntax and semantics of our formalism and extend it with alternative semantics. These alternative semantics will be used in the definition of the proposed conformance relations. First, we introduce our notion of fuzzy automata.

**Definition 9.** *A* fuzzy automaton *is a tuple*

$$(S, \text{Acts}, L, X, X_0, s_0, Tr)$$

*where:*

- *$S$ is a finite set of states.*
- Acts *is a finite set of actions, partitioned into a set of inputs $I$ and a set of outputs $O$.*
- *$L \subseteq O$ is a distinguished subset of outputs. We refer to these actions as localized outputs.*
- *$X$ is a set of variable names.*
- *$X_0 : X \longrightarrow \mathbb{R}$ is the initial mapping from variable names to real values. We call these mappings variable states.*
- *$s_0 \in S$ is the initial state.*
- *$Tr \subseteq S \times \text{Acts} \times \mathcal{FC} \times \mathcal{VT} \times S$ is a set of transitions satisfying the following conditions:*
  - *Output transitions have the constant True as constraint.*
  - *Input transitions have a constraint whose free variables are a subset of the variables of the input action.*
  - *For each localized output, $l \in L$, there exists a state, $s \in S$, such that, for all transitions of the form $(s_1, l, \text{True}, \text{vt}, s_2) \in Tr$, we have $s_2 = s$.*

Now, we will review the basic semantics of our version of *fuzzy automata*.

**Definition 10.** *Let $A = (S, \text{Acts}, L, X, X_0, s_0, Tr)$ be a fuzzy automaton and $\triangle$ be a strictly monotonic t-norm. Let $s_1, s_2 \in S$ be states and $X_1, X_2 : X \longrightarrow \mathbb{R}$ be variable states. We have a* transition *from $(s_1, X_1)$ to $(s_2, X_2)$, after performing the* input action *$?i(\chi) \in I$ for the real valued tuple $\alpha$ with confidence $\epsilon$, denoted by*

$$(s_1, X_1) \xrightarrow{?i(\alpha)}_{\epsilon} (s_2, X_2)$$

*if the following conditions hold:*

- *There exists a fuzzy constraint $C \in \mathcal{FC}$ such that $(s_1, ?i(\chi), C, VT, s_2) \in Tr$.*
- *The grade of confidence of $C$, with the corresponding variables mapped to the values in $\alpha$, is higher than 0. We denote this grade of confidence by $\epsilon := \mu_{C,?i(\chi)}(\alpha)$.*
- *$X_2$ is the result of applying $VT$ to $X_1$ considering:*
  - *The reserved variable $\mu$ takes the value $\epsilon$.*
  - *The values of the variables in the expressions of $VT$ are taken from $\alpha$ when the corresponding variable belongs to the input $?i(\chi)$. The values for the rest of variables in the expressions of $VT$ are taken from $X_1$.*

*We have a* transition *from $(s_1, X_1)$ to $(s_2, X_2)$, after performing the* output action *$!o(\zeta) \in O$, denoted by*

$$(s_1, X_1) \xrightarrow{!o(\alpha)}_{1} (s_2, X_2)$$

*if the following conditions hold:*

- *There exists $(s_1, !o(\zeta), True, VT, s_2) \in Tr$.*
- *$X_2$ is the result of applying $VT$ to $X_1$ and $\alpha$ is the tuple of expressions from the output $o$, considering:*
  - *The reserved variable $\mu$ takes the value 1.*
  - *The values of the variables in the expressions, $\zeta$, are taken from $X_1$.*

*We say that a sequence*

$$(s_0, X_0) \xrightarrow{a_1(\alpha_1)}_{\epsilon_1} (s_1, X_1) \cdots \xrightarrow{a_n(\alpha_n)}_{\epsilon_n} (s_n, X_n)$$

*of consecutive transitions starting in the initial state of the automaton is a $\triangle$-*trace *of $A$ if $\epsilon = \triangle\{\epsilon_1, \ldots, \epsilon_n\}$ is greater than zero. In this case, we write*

$$(s_0, X_0) \xRightarrow{a_1(\alpha_1)\cdots a_n(\alpha_n)}_{\epsilon} (s_n, X_n)$$

Next we will define two transition concatenation rules that will be later used to define the proposed conformance relations. The first one is based upon the idea of imposing a minimum confidence level.

**Definition 11.** *Let $A = (S, \text{Acts}, L, X, X_0, s_0, Tr)$ be a fuzzy automaton, $\alpha \in [0, 1]$ and $a_1 \cdots a_n \in \text{IActs}^+$ be a sequence of actions. We write*

$$(s_0, X_0) \xRightarrow[\epsilon]{a_1 \cdots a_n}{}^{\text{conf}^\alpha} (s, X)$$

*if $\epsilon > \alpha$ and we have $(s_0, X_0) \xRightarrow{a_1 \cdots a_n}_{\epsilon} (s, X)$.*

The difference with the basic transition semantics is that $\text{conf}^\alpha$ does not consider transitions below a certain confidence threshold. The following result means that the higher $\alpha$ is, the more restrictive $\text{conf}^\alpha$ becomes (the proof is immediate).

**Proposition 1.** *Let $A = (S, \text{Acts}, L, X, X_0, s_0, Tr)$ be a fuzzy automaton, $a_1 \cdots a_n \in \text{IActs}^+$ be a sequence of actions and $\alpha_1, \alpha_2 \in [0,1]$ such that $\alpha_1 < \alpha_2$. We have that*

$$(s_0, X_0) \xRightarrow{a_1 \cdots a_n}_{\epsilon}^{\text{conf}^{\alpha_2}} (s, X) \text{ implies } (s_0, X_0) \xRightarrow{a_1 \cdots a_n}_{\epsilon}^{\text{conf}^{\alpha_1}} (s, X)$$

The second transition concatenation rule is a little more complex. Its purpose is to accept only the most likely executions.

**Definition 12.** *Let $A = (S, \text{Acts}, L, X, X_0, s_0, Tr)$ be a fuzzy automaton and $a_1 \cdots a_n \in \text{IActs}^+$ be a sequence of actions. We write*

$$(s_0, X_0) \xRightarrow{a_1 \cdots a_n}_{\epsilon}^{\text{confm}} (s, X)$$

*if we have that*

$$(s_0, X_0) \xRightarrow{a_1 \cdots a_n}_{\epsilon} (s, X)$$

*and one of the following four conditions holds.*

- *$n = 1$ and $\mathcal{D}(a_1) \notin L$.*
- *$n = 1$, $\mathcal{D}(a_1) \in L$ and for all $a'_1$ such that $\mathcal{D}(a_1) = \mathcal{D}(a'_1)$ and such that $(s_0, X_0) \xRightarrow{a'_1}_{\epsilon'} (s, X)$ we have $\epsilon \geq \epsilon'$.*
- *$n > 1$, $\mathcal{D}(a_n) \notin L$ and there exist $\epsilon', \epsilon'', s'$ and $X'$ such that $\epsilon = \epsilon' \triangle \epsilon''$, $(s_0, X_0) \xRightarrow{a_1 \cdots a_{n-1}}_{\epsilon'}^{\text{confm}} (s', X')$ and $(s', X') \xrightarrow{a_n}_{\epsilon''} (s, X)$.*
- *$n > 1$, $\mathcal{D}(a_n) \in L$ and for all $a'_1 \cdots a'_n \in \text{IActs}^*$ such that $\mathcal{D}(a_1 \cdots a_n) = \mathcal{D}(a'_1 \cdots a'_n)$ and $(S_0, X_0) \xRightarrow{a'_1 \cdots a'_n}_{\epsilon'} (S, X')$ we have that $\epsilon \geq \epsilon'$.*

The idea behind this definition is that if an action in $L$ is received, then the system must reach a specific state: the state corresponding to that particular action in $L$. Having this property in mind, we know that the higher the Grade of Confidence is at that point, the higher it will be after successive actions. Therefore, the previous definition translates into imposing that output actions must be parameterized with values that maximize the Grade of Confidence of that execution.

## 4 Conformance Relations

In this section we define our conformance relations. The general framework considers a *System Under Test (SUT)*, which is receiving input actions and producing output actions. We will say that an SUT conforms to a specification if every sequence of actions observed in the SUT does not show a mismatch with respect to what the specification says. In addition, we need to parameterize the conformance relations with respect to a specific $t$-norm. Formally, our two relations should have a $\triangle$ parameter, that is, a strictly monotonic $t$-norm. However, we will omit this relation to not overload the notation. The definition of our conformance relations is based on the operational semantics previously defined in Sect. 3. We first define a conformance relation based on accepting the sequences with the highest Grade of Confidence.

**Definition 13.** *Let $Spec = (S, \text{Acts}, L, X, X_0, s_0, Tr)$ be a fuzzy automaton. We say that an SUT maximally conforms to Spec, and we write*

$$\text{SUT } \texttt{confmax } Spec$$

*if for all sequence of actions $a_1 \cdots a_n \in \text{IActs}^*$ observed in the SUT, we have that there exist $\epsilon$, $s$ and $X$ such that $(s_0, X_0) \xRightarrow{a_1 \cdots a_n}{}_{\epsilon}^{\text{confm}} (s, X)$.*

The `confmax` relation only holds when every observed sequence of actions is *accepted* by the specification. In particular, sequences that have not terminated yet should also be *accepted* if the complete sequence is going to be *accepted.* This property is satisfied because of the following property of our operational semantics. The proof is based on the strict monotonicity of the $t$-norms that we consider in our framework.

**Proposition 2.** *Let $A = (S, \text{Acts}, L, X, X_0, s_0, Tr)$ be a fuzzy automaton and $a_1 \cdots a_n \in \text{IActs}^*$ be a sequence of actions. If we have*

$$(s_0, X_0) \xRightarrow{a_1, \cdots, a_n}{}_{\epsilon}^{\text{confm}} (s, X)$$

*for some state $s$, confidence $\epsilon$ and variable state $X$, then we also have that there exist $\epsilon'$, $S'$ and $X'$ such that*

$$(s_0, X_0) \xRightarrow{a_1, \cdots, a_{n-1}}{}_{\epsilon'}^{\text{confm}} (s', X')$$

Next, we define a conformance relation based on accepting only sequences whose *Grade of Confidence* is higher than a certain bound.

**Definition 14.** *Let $Spec = (S, \text{Acts}, L, X, X_0, s_0, Tr)$ be a fuzzy automaton and $\alpha \in [0, 1]$. We say that an SUT conforms to Spec with respect to $\alpha$, and we write*

$$\text{SUT } \texttt{conf}^{\alpha} \ Spec$$

*if for all sequence of actions $a_1 \cdots a_n \in \text{IActs}^*$ observed in the SUT, we have that there exist $\epsilon$, $S$ and $X$ such that*

$$(s_0, X_0) \xRightarrow{a_1, \cdots, a_n}{}_{\epsilon}^{\text{conf}^{\alpha}} (s, X)$$

Similarly to the previous case, given an SUT that conforms to a specification with respect to the $\texttt{conf}^{\alpha}$ relation, if a sequence of actions is observed in the SUT then every prefix of the sequence has been observed too. The proof of this result relies again in the monotonicity of $t$-norms.

**Proposition 3.** *Let $A = (S, \text{Acts}, L, X, X_0, s_0, Tr)$ be a fuzzy automaton and $a_1 \cdots a_n \in \text{IActs}^*$ be a sequence of actions. If we have*

$$(s_0, X_0) \xRightarrow[\epsilon]{a_1 \cdots a_n} {}^{\text{conf}^\alpha} (s, X)$$

*for some state $s$, confidence $\epsilon$ and variable state $X$, then we also have that there exist $\epsilon'$, $S'$ and $X'$ such that*

$$(s_0, X_0) \xRightarrow[\epsilon']{a_1 \cdots a_{n-1}} {}^{\text{conf}^\alpha} (s', X')$$

Finally, we define a conformance relation that imposes both a certain bound and the maximal *Grade of Confidence* condition.

**Definition 15.** *Let $Spec = (S, \text{Acts}, L, X, X_0, s_0, Tr)$ be a fuzzy automaton and $\alpha \in [0, 1]$ . We say that an SUT maximally conforms to Spec with respect to $\alpha$, and we write*

$$\text{SUT } \mathtt{confmax}^\alpha \; Spec$$

*if for all sequence of actions $a_1 \cdots a_n \in \text{IActs}^*$ observed in the SUT we have that there exist $\epsilon$, $S$, and $X$ such that*

$$(s_0, X_0) \xRightarrow[\epsilon]{a_1 \cdots a_n} {}^{\text{conf}^\alpha} (s, X) \text{ and } (s_0, X_0) \xRightarrow[\epsilon]{a_1 \cdots a_n} {}^{\text{confm}} (s, X)$$

This relation represents the conjunction of the former two relations, as stated in the following result.

**Proposition 4.** *If we have* SUT $\mathtt{confmax}^\alpha$ *Spec then we also have* SUT $\mathtt{confmax}$ *Spec and* SUT $\mathtt{conf}^{\alpha_1}$ *Spec, for all $\alpha_1 \leq \alpha$.*

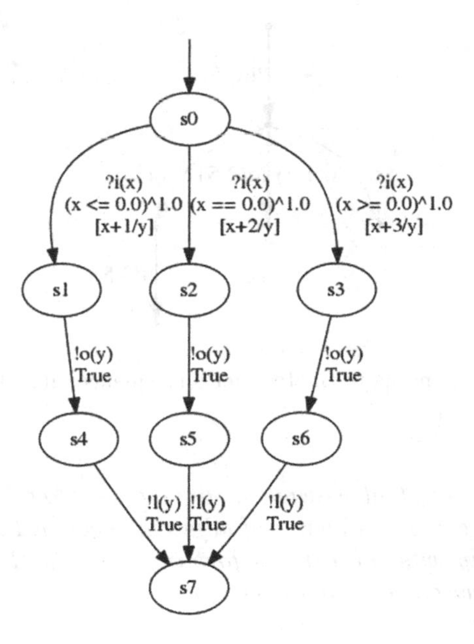


**Fig. 1.** Example of a fuzzy automaton.

## 5 Test Definition and Generation

Although conformance relations give a precise notion to define when an SUT conforms to a specification, in other words, when we have no evidence to claim that the system under development is incorrect, it is important to have an alternative characterization of these relations. Testing provides such a tool. The idea is to define a testing framework such that the SUT successfully passes a set of tests (extracted from the specification) if and only if the conformance of the SUT with respect to the specification holds. Due to space restrictions we cannot present all the details. We will follow the classical approach where conformance was characterized as a testing framework [13] and we will focus on the non-standard intricacies that we had to confront. During the rest of this section we will briefly describe the data structure that will represent each test and we will define what does it mean for an SUT to pass a test. We will also sketch a test derivation algorithm that produces sound and complete test suites with respect to a given specification and a conformance relation.

We begin by defining the trees that represent the possible executions of a particular test case.

**Definition 16.** *A test tree is defined to be a rooted tree whose edges are labelled with actions from* IActs. *The set of all test trees is denoted by* TTrees.

An example of a *test tree* is shown in Fig. 2. This test tree corresponds to a particular sequence of test actions. Next, we have to define what it means for a sequence of actions to satisfy a given test tree.

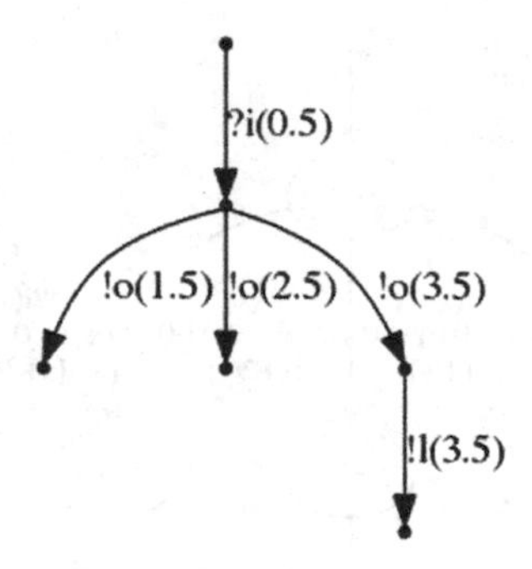


**Fig. 2.** Test tree corresponding to the test action sequence $?i(0.5), !o, !l$ for the automaton given in Fig. 1.

**Definition 17.** *We say that a sequence* $a_1 \cdots a_n \in \text{IActs}^*$ *satisfies a test tree* $T \in \text{TTrees}$ *if there exists a sequence of adjacent edges in* $T$*, starting in its root, labelled consecutively with the actions forming* $a_1 \cdots a_n$*. In this case we write* $a_1 \cdots a_n \models T$*; otherwise, we write* $a_1 \cdots a_n \not\models T$

The idea is that a test tree provides a compact representation of the set of accepted actions sequences. We can now define what does it mean to pass or fail a given test case.

**Definition 18.** *A test case is defined to be a pair* $(t, T)$ *with* $t \in \text{TActs}^*$ *and* $T \in \text{TTrees}$. *A sequence of actions* $a_1 \cdots a_n \in \text{IActs}^*$ *observed in the SUT fails the test* $(t, T)$, *and we write* $a_1 \cdots a_n \nvDash (t, T)$, *if* $\mathcal{D}(a_1 \cdots a_n)$ *is a prefix of* $t$ *and* $a_1 \cdots a_n \nvDash T$; *otherwise we say that the test is passed and we write* $a_1 \cdots a_n \vDash (t, T)$

Intuitively, sequences of test actions are associated with test trees in the sense that if a sequence of actions is a fragment of the sequence of test actions, then it must be a fragment of the corresponding test tree. The most important property that a test must have is to be *sound*. This means that the test only fails when the sequence of actions is not valid with respect to a specification. In order to determine whether a sequence of actions is valid, both a specification and a conformance relation must be considered.

**Definition 19.** *Let* $Spec = (S, \text{Acts}, L, X, X_0, s_0, Tr)$ *be a fuzzy automaton. A test case* $(t, T) \in \text{TActs}^* \times \text{TTrees}$ *is said to be sound with respect to Spec under the conformance relation* `rel` *if for each sequence of actions* $a_1 \cdots a_n$ *we have that*

$$(s_0, X_0) \overset{a_1 \cdots a_n}{\Longrightarrow}_{\epsilon}^{\text{rel}} (s, X)$$

*implies that* $a_1 \cdots a_n \vDash (t, T)$.

For example, in Fig. 2 we show a sound test case with respect to the specification given in Fig. 1 and the conformance relation `confmax`. We can see it is sound since after receiving $?i(0.5)$ the only possible outputs are $!o(1.5)$, $!o(2.5)$ and $!o(3.5)$ and, after that, the localized action $!l$ must take the value $!l(3.5)$ since it is the only value which maximices the Grade of Confidence in the sequence of actions.

The correctness of an SUT is tested by applying a *test suite*.

**Definition 20.** *A test suite is defined to be a set of tests. An SUT is said to pass a test suite if for each sequence of actions observed during the application of all the tests to the SUT we have that the tests are passed.*

The most important property that a test suite may have, once we know that all the tests are sound, is to be *complete*. This means that if the SUT shows a sequence of actions that it is not valid with respect to a specification, then at least one of the tests will fail. In order for a test suite to be actually complete it would need to contain both an infinite number of parameters for each action and possibly also an infinite number of actions. In practice, samples from random variables modelling the distribution of the parameters may be used in order to produce a realistic pool of parameterized input actions. A longer explanation about this (classical) problem to achieve real completeness can be found in previous work [10].

The last step of our framework was to define a *test derivation algorithm* (the actual algorithm cannot be included in the paper due, as we have already said, to space limitations). Essentially, we traverse the specification and generate tests that appropriately reflect the structure of the specification (the interested reader can see similar algorithms, although for simpler formalisms, in previous work [10,13]).

## 6 Conclusions and Future Work

Designing and building software with the aid of formal methods and models is certainly one of the most promising ways to improve the reliability of critical systems. Fuzzy logic has proven to offer a powerful and expressive language to model and reason about the implicit uncertainty present in many fields.

In this paper we have successfully extended our previous framework based on fuzzy automata, making it more suitable to specify a wide variety of software systems. A family of conformance relations has been defined, taking advantage of the monotonicity properties of $t$-norms, which makes them consistent with respect to the partial execution of an implementation. This translates into a better suitability to model systems meant to run over long periods of time, whose execution at some point before completion may need to be proved correct. A testing methodology has been presented in order to apply these formal notions to actual software.

Our next goal consists in applying this extension to real use cases in order to show the potential of the proposed framework. Besides developing models relying on the framework, software tools are planned to be constructed in order to automatize as much as possible the generation of test cases. This task implies both the need to formally prove the soundness of the different test generation algorithms and to empirically show the adequacy of different strategies meant to choose a rich enough subset of the sound test cases in order to obtain complete test suites.

## References

1. Ammann, P., Offutt, J.: Introduction to Software Testing, 2nd edn. Cambridge University Press, Cambridge (2017)
2. Andrés, C., Llana, L., Núñez, M.: Self-adaptive fuzzy-timed systems. In: 13th IEEE Congress on Evolutionary Computation, CEC 2011, pp. 115–122. IEEE Computer Society (2011)
3. Boubeta-Puig, J., Camacho, A., Llana, L., Núñez, M.: A formal framework to specify and test systems with fuzzy-time information. In: Rojas, I., Joya, G., Catala, A. (eds.) IWANN 2017. LNCS, vol. 10306, pp. 403–414. Springer, Cham (2017). https://doi.org/10.1007/978-3-319-59147-6_35
4. Calvo, I., Merayo, M.G., Núñez, M.: An improved and tool-supported fuzzy automata framework to analyze heart data. In: Nguyen, N.T., Hoang, D.H., Hong, T.-P., Pham, H., Trawiński, B. (eds.) ACIIDS 2018. LNCS (LNAI), vol. 10751, pp. 694–704. Springer, Cham (2018). https://doi.org/10.1007/978-3-319-75417-8_65

5. Camacho, A., Merayo, M.G., Núñez, M.: Using fuzzy automata to diagnose and predict heart problems. In 19th IEEE Congress on Evolutionary Computation, CEC 2017, pp. 846–853. IEEE Computer Society (2017)
6. Cavalli, A.R., Higashino, T., Núñez, M.: A survey on formal active and passive testing with applications to the cloud. Ann. Telecommun. **70**(3–4), 85–93 (2015)
7. Doostfatemeh, M., Kremer, S.C.: New directions in fuzzy automata. Int. J. Approximate Reasoning **38**(2), 175–214 (2005)
8. Hierons, R.M., et al.: Using formal specifications to support testing. ACM Comput. Surv. **41**(2), 9:1–9:76 (2009)
9. Lamport, L.: Who builds a house without drawing blueprints? Commun. ACM **58**(4), 38–41 (2015)
10. Merayo, M.G., Núñez, M., Rodríguez, I.: Formal testing from timed finite state machines. Comput. Netw. **52**(2), 432–460 (2008)
11. Mordeson, J.N., Malik, D.S.: Fuzzy Automata and Languages: Theory and Applications. Chapman & Hall/CRC (2002)
12. Shafique, M., Labiche, Y.: A systematic review of state-based test tools. Int. J. Softw. Tools Technol. Transf. **17**(1), 59–76 (2015)
13. Tretmans, J.: Model based testing with labelled transition systems. In: Hierons, R.M., Bowen, J.P., Harman, M. (eds.) Formal Methods and Testing. LNCS, vol. 4949, pp. 1–38. Springer, Heidelberg (2008). https://doi.org/10.1007/978-3-540-78917-8_1
14. Wee, W.G., Fu, K.S.: A formulation of fuzzy automata and its application as a model of learning systems. IEEE Trans. Syst. Sci. Cybern. **5**(3), 215–223 (1969)
15. Zadeh, L.A.: Fuzzy sets. Inf. Control **8**(3), 338–353 (1965)
16. Zadeh, L.A.: Fuzzy sets, fuzzy logic, and fuzzy systems. In: Advances in Fuzzy Systems - Applications and Theory, vol. 6. World Scientific Press (1996)

# Investigating the Effectiveness of Mutation Testing Tools in the Context of Deep Neural Networks

Nour Chetouane, Lorenz Klampfl, and Franz Wotawa(✉)

CD Lab for Quality Assurance Methodologies for Cyber-Physical Systems, Institute for Software Technology, Graz University of Technology, Inffeldgasse 16b/2, 8010 Graz, Austria
{nour.chetouane,lklampfl,wotawa}@ist.tugraz.at

**Abstract.** Verifying the correctness of the implementation of machine learning algorithms like neural networks has become a major topic because – for example – its increasing use in the context of safety critical systems like automated or autonomous vehicles. In contrast to evaluating the learning capabilities of such machine learning algorithms, in verification, and particularly in testing we are interested in finding critical scenarios and in giving some sort of guarantees with respect to the underlying used tests. In this paper, we contribute to the area of testing machine learning algorithms and investigate the effectiveness of traditional mutation tools in the context of Deep Neural Networks testing. In particular, we try to answer the question whether mutated neural networks can be identified considering their learning capabilities when compared to the original network. To answer this question, we performed an empirical study using Java code implementations of such networks and a mutation tool to create mutated neural networks models. As an outcome, we are able to identify some mutations to be more likely to be detected than others.

**Keywords:** Testing neural networks · Mutation testing · Mutation score for neural networks

## 1 Introduction

The use of machine learning for practical applications even in case of safety critical systems like autonomous vehicle has become more and more important. In particular, deep neural networks are often used for classifying images during the operation of an autonomous car for detecting boundaries of lanes or obstacles (see e.g., [5,16]). Such classification results are important for the autonomous car to perceive its environment and are the basis for smart and safe reactions on environmental changes. Hence, it is of crucial importance to thoroughly test the correctness and predictability of such deep neural networks in order to meet the requirements of relevant safety standards. Evaluating and testing neural


I. Rojas et al. (Eds.): IWANN 2019, LNCS 11506, pp. 766–777, 2019.
https://doi.org/10.1007/978-3-030-20521-8_63

networks, therefore, have become major research directions, see e.g., [6] or most recently [18]. Testing neural networks in the case of autonomous driving has many facets ranging from analyzing the consequences of disturbances in the inputs, e.g., the images, [3], to faults occurring in the network structure [10].

In this paper, we contribute to previous research in the area of testing neural networks but considering a more software testing oriented approach where questions like how to assure the quality of programs or test suites is of importance. Here often coverage criteria including the mutation score are used to justify that the underlying test suite is somehow "good enough" for a particular program. In case of safety-critical systems, for example, the test suite has to fulfill the Modified Condition/Decision Coverage (MC/DC) criterion. When testing neural networks, the question whether coverage or mutation score is a valuable measure immediately becomes apparent especially in the case where neural networks are used in a safety-critical context like in the autonomous driving domain. In particular, we carry out experiments where we analyze whether it is possible to distinguish small changes, i.e., mutations, induced in the program implementing a neural network. In the experiments, we specifically focus on the differences in the learning capabilities of the original network and its mutation and in case of substantial deviations are able to classify the mutant as being detected, which is called "killed" in the mutation testing terminology.

What we expect from the experimental evaluation is to find out whether there exist mutations that can never be killed and thus maybe part of neural network implementations, and whether there are mutations that can be more easily identified when evaluating neural networks. In both cases, we gain information about the resilience of neural networks regarding their learning capabilities in case of software bugs, when assuming that the mutations themselves can be considered as a certain class of bugs. In addition, we are able to find out whether mutation testing in principle can be applied for assuring the quality of the neural network implementation.

This paper is organized as follows: First, we introduce the background information comprising deep learning systems, convolutional neural networks, and mutation testing. Afterwards, we discuss the evaluation comprising information regarding the used experimental process, and the experimental results obtained. Finally, we give an overview of related research and conclude the paper.

## 2 Background

In this section, we briefly introduce deep learning systems including convolutional neural networks and mutation testing.

### 2.1 Deep Learning Systems

Deep learning uses massive amounts of data and computing power to simulate Deep Neural Networks (DNNs). These networks imitate the human brain's connectivity, they can be used for different machine learning tasks such as classifying data and recognizing patterns and finding correlations between them. Such systems learn progressively by training on examples in order to improve their ability to do a specific task.

A DNN consists of multiple interconnected neurons organized on layers: the input layer, the output layer, and one or multiple hidden layers. Each layer consists of nodes which are called neurons. Each neuron is a computing unit that computes its output by applying an activation function to the input. In classic DNNs, each neuron is fully connected with all neurons on the next layer, and each edge has a coefficient or a weight, which indicates the strength of the connections between neurons. These weights are mainly used to assign significance to inputs for the task the algorithm is trying to learn (Fig. 1).

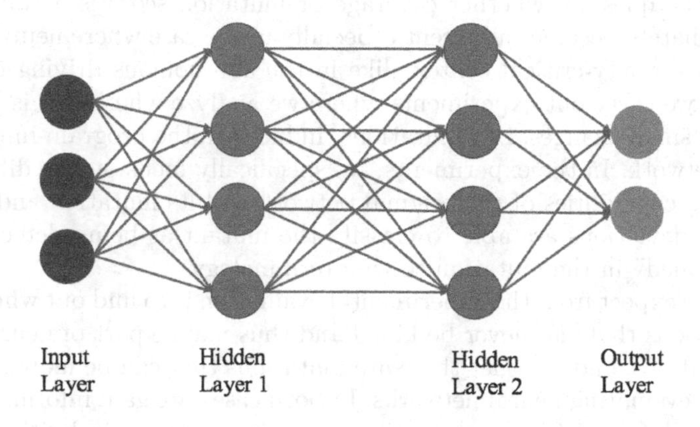


**Fig. 1.** A regular 4-layer fully connected Neural Network (NN)

## 2.2 Convolutional Neural Networks

Convolutional Neural Network (CNN, or ConvNet) is a class of deep artificial neural networks that has successfully been applied for analyzing visual data. They can be used to classify images, cluster them by similarity and perform object recognition within scenes. The effectiveness of convolutional nets in image recognition is one of the main reasons why the world has woken up to the efficacy of deep learning. They have shown capabilities in identifying faces, individuals, street signs and many other aspects of visual data.

Moreover, images are known to be high-dimensional which makes them hard to deal with. They require a lot of time and computing power to process. Convolutional networks are quite employed in image analysis since they are intended to reduce the dimensionality of images in a variety of ways. A ConvNet is composed of an input, an output layer, and multiple hidden layers. The hidden layers of a CNN usually consist of convolutional layers, pooling layers and fully connected layers.

The convolutional layer is the essential element of a CNN. The layer's parameters consist of a set of learnable filters (or kernels). During the forward pass, the convolution layer basically slides each filter through the image and computes the scalar product between the entries of the filter and the input. A scalar product value indicates whether the pixel pattern in the underlying image matches the pixel pattern expressed by the filter. The output of the convolutional layer is a 2-dimensional activation map which is then fed into a pooling layer, this layer serves to progressively reduce the spatial size of the representation in order to decrease the number of parameters and amount of computation in the network. There are several ways to implement pooling. The most common one is max pooling. It partitions the input image into a set of non-overlapping rectangles and takes the maximum value for each sub-region. After applying several convolutional and max pooling layers, the high-level reasoning in the neural network is done via fully connected layers which are basically composed of regular back-propagation neural networks that learn on the resulting data from the convolution and pooling layers (Fig. 2).

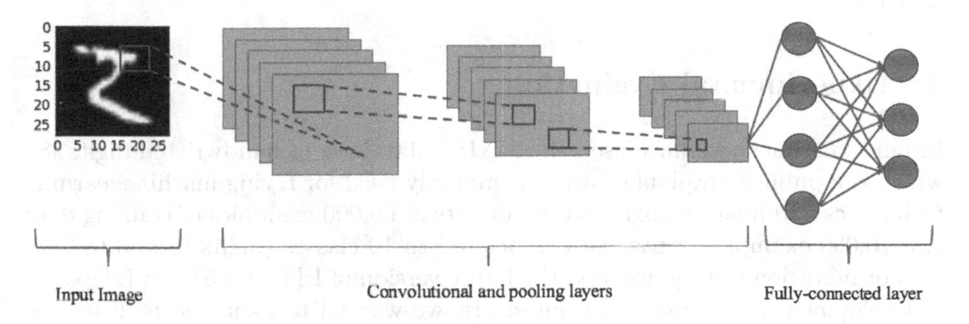


**Fig. 2.** A regular convolutional neural network

## 2.3 Mutation Testing

Mutation testing is a fault-based testing technique [2] where syntactic modifications are applied on the model or program (SUT) and lead to faulty versions of the SUT. These faulty versions are called mutants. A mutant is detected if the execution leads to different results than expected, here the mutant is said to be killed if it is distinguished by the test suite from the original program. Mutation testing has been used for evaluating test suites where a test suite is considered a better one if it kills more mutants. Mutation testing is based on the competent programmer hypothesis and the coupling effect. The former [2] states that the introduced faults in a program are due to small syntactical errors. The latter [14] assumes that even small faults may lead to forming other emergent faults.

In order to evaluate test suites in case of mutation testing, the mutation score $\mu$ is used, which is defined as the ratio of killed mutants $r$ to the number of existing mutants $s$:

$$\mu = r/s \tag{1}$$

It is well known that there might be mutants that are equivalent to the original program with respect to its behavior. In this case, we speak about equivalent mutants that can never be killed. Hence, it is important either to detect equivalent mutants or to avoid constructing them for mutation analysis. See [13] for a survey on overcoming the equivalent mutant problem.

There are several code-level based mutation tools for traditional software testing such as [7] and [12]. In our study we choose to work with PIT [1]. It is a practical open source mutation testing tool. It is considered fast since it works on byte code instead of source code and minimizes mutant executions by exercising tests which execute the instruction where the mutant is located. Some mutators provided by PIT may generate equivalent mutants, yet the default activated mutators are quite stable and tend to minimize the number of equivalent mutations. PIT is well integrated with development environments (e.g. Eclipse) and affords many build tools (e.g. Maven). Besides it is quite robust and actively updated.

## 3 Experimental Evaluation

In our experiment we have used the MNIST database of handwritten digits [8], which is a publicly available dataset commonly used for trying machine learning techniques and image recognition. It consists of 60,000 examples of training data and 10,000 examples of test data associated to 10 classes (digits form 0 to 9).

For mutation testing, we used the latest version of PIT (1.4.5) and Eclipse as a development environment. In our study, we wanted to focus more on testing different Neural Network code implementations. Therefore, we selected two NN java implementation examples; a feedforward Multi-Layer Perceptron network and a convolutional neural network, which are available in [4]. Besides, we have added a third convolutional neural network which is a modified version of the first one. Table 1 presents the structure and configuration of the studied Neural Networks and their prediction accuracy.

The MLP network has 784 nodes in the input layer corresponding to the pixels of the image. The first hidden layer in the network has 500 nodes and 100 nodes in the second hidden layer which passes 100 outputs to the final output layer which has 10 nodes corresponding to the 10 classes. After training, The MLP Network achieves a test accuracy of 96.53% on the MNIST Dataset.

The first convolutional NN is composed of 2 convolutional layers both with a kernel size of [5, 5] and two max pooling layers with a [2, 2] kernel. It also contains a dense layer (fully-connected) which has 500 nodes. It predicts the images of the test set with an accuracy of 98.87%. The second convolutional NN has a larger structure than the first one. It is composed of four convolutional layers with a kernel size of [3, 3], each two are followed by a max pooling layer with a [2, 2] kernel. Besides, it has two fully connected layers consecutively composed of 100 and 200 nodes. This CNN achieves 98.91% test accuracy.

**Table 1.** Model structures and configurations.

| MNIST | | |
|---|---|---|
| MLP | CNN 1 | CNN 2 |
| Dense (784, 500) + ReLU | Conv (20, 5, 5) + ReLU | Conv (16, 3, 3) + ReLU |
| Dense (500, 100) + ReLU | MaxPooling (2, 2) | Conv (32, 3, 3) + ReLU |
| OutputLayer (100, 10) + Softmax | Conv (50, 5, 5) + ReLU | MaxPooling(2, 2) |
| | MaxPooling (2, 2) | Conv (64, 3, 3) + ReLU |
| | Dense (500) + ReLU | Conv (64, 3, 3) + ReLU |
| | OutputLayer(10) + Softmax | MaxPooling (2, 2) |
| | | Dense (100) + ReLU |
| | | Dense (200) + ReLU |
| | | OutputLayer (10) + Softmax |
| **Total parameters:** 443,610 | 431,080 | 184,934 |
| **Training accuracy:** 97.24 % | 99.02 % | 98.96 % |
| **Testing accuracy:** 96.53% | 98.87 % | 98.91 % |

The MLP, CNN 1 and CNN 2 source code implementations have consecutively 40, 79, 91 lines of code. Each NN implementation includes one main java class that implements four main phases, i.e.: data preprocessing (i.e. data loading, data splitting, data normalization), setting model configurations (i.e. layers, activation functions, kernel sizes, strides), training model and model evaluation (i.e. measuring test accuracy).

Algorithm 1 describes our experimental process including how to identify mutants as being able to be killed. The algorithm starts with computing the average accuracy of the original neural network in Line 5. This is done via running the original network 10 times and computing the average prediction accuracy. For this case study we use the PIT tool to generate mutants (Line 6) where in each iteration the tool applies different types of mutations (see Table 2). For each mutant in lines 7 to 16, we do the following: We train the mutant on the same dataset used for the original network. We test them by comparing their prediction accuracy with the original network. For comparing the trained mutants with the original model we allow a deviation of 1% of the average prediction accuracy. A mutant is killed if its accuracy is not in the defined margin of accuracy. Let us say for example we have an average accuracy of the original programs equal to 98%, a mutant is killed if it has less than 97% (i.e. 0.98−0.01) prediction accuracy on the test set. Therefore, mutants which are not able to learn due to the mutation operator applied get killed because either they have no prediction capability at all or a very poor one. Note that we set a 1% deviation boundary for the prediction accuracy, because all the obtained prediction accuracies of the original network, also were within this boundary. When applying this test oracle to other networks, we may have to change the deviation boundary.

**Algorithm 1. Test Oracle**

**Input:** OriginalModels
**Output:** MutationScore $\mu$.

1: Let $AverageAcc_{orig}$ be the average test accuracy of the trained original models.
2: Let $MutantTestAcc$ be the test accuracy of one mutant.
3: Let $s$ be the total number of created mutants by PIT
4: Let $r$ be the number of killed mutants
5: Let $AverageAcc_{orig} = ComputeAverageAccuracy(OriginalModels)$
6: Create mutants by running PIT on the main class
7: **while** New mutant **do**
8: $TrainMutant()$
9: Let $MutantTestAcc$ be $EvaluateAccuracy()$
10: **if** $MutantTestAcc \geq AverageAcc_{orig} - 0.01$ **then**
11: Mutant survives.
12: **else**
13: Mutant is killed.
14: Let $r$ be $r + 1$
15: **end if**
16: **end while**
17: $\mu = r/s$
18: **return** $\mu$

### 3.1 Experimental Results

As shown in Table 2 PIT provides several examples of mutation operators. We try to limit the number of equivalent mutants by selecting the default mutation operators, however some still might be created since we have activated the inline constant mutator. This mutator has a significant impact on the NN configuration parameters which are generally defined as a set of constants. Tables 3 and 4 present the results after applying mutations on the studied models. Note that mutations which timed out are also considered killed.

The number of input channels for convolutional networks is set according to the type of images in the data set. The MNIST dataset for instance contains greyscale images which have only one channel. Inline constant mutation that changes the number of channels from 1 to 0 gets killed because in this case the first layer of the network will not receive a valid input. Besides, mutating the number of output classes causes a major change in the structure of the network. Increasing or decreasing the number of nodes composing the output layer of the network stops it from learning since we have a mismatch with the number of classes in the training set.

Furthermore, a learning rate is defined in order to minimize the loss function of the network. When the PIT tool changes the learning rate value to a larger one (e.g. from 0.0015 to 1) the network fails to converge or even diverge and so it can not minimize the loss function, the mutant is then killed because its prediction accuracy gets really low. In case of convolutional neural networks, a scheduler with a set of starting point iterations and corresponding learning rates is defined. Mutations on the learning rate values get also killed but mutations which change the iteration number in the scheduler survive.

**Table 2.** Some examples of the available mutation operators provided by PIT.

| Mutator type | Description |
|---|---|
| Conditionals Boundary Mutator | Relational operators are changed to their boundary counterpart (e.g. $<$ to $\leq$, $>$ to $\geq$). |
| Increments Mutator | Replaces increments with decrements of local variables and vice versa. |
| Inline Constant Mutator | Changes inline constants (e.g. *true* to *false*, 1 to 0, $-1$ to 1, increment by 1). |
| Invert Negatives Mutator | Inverts negation of integer and floating point numbers. |
| Math Mutator | Replaces binary arithmetic operations for either integers or floats with another operation (e.g. $+$ to $-$, $*$ to $/$, $\&$ to $\vert$). |
| Negate Conditionals Mutator | Mutates all found conditionals (e.g. $==$ to $!=$, $\leq$ to $>$, $\geq$ to $<$, $<$ to $\geq$, $>$ to $\leq$). |
| Return Values Mutator | Mutates the return values of method calls depending on the return type of the method (e.g. boolean: *false* to *true*, Object: *non* $-$ *null* to *null*). |
| Void Method Call Mutator | Removes method calls to void methods |

**Table 3.** Results after applying mutations on the Multi-Layer Perceptron Neural Network.

| Mutation operators | Generated | Killed | Survived | Mutation score |
|---|---|---|---|---|
| **Inline Constant Mutator** | 15 | 6 | 9 | 40% |
| Change number of output classes | | × | | |
| Change learning rate | | × | | |
| Change number of epochs | | | × | |
| Change batch size | | | × | |
| **Void Method Call Mutator** | 3 | 1 | 2 | 33% |
| **Increments Mutator** | 1 | 1 | 0 | 100% |
| **Math Mutator** | 1 | 1 | 0 | 100% |
| **Conditionals Boundary Mutator** | 1 | 0 | 1 | 0% |
| **Negate Conditionals Mutator** | 1 | 1 | 0 | 100% |
| Total mutations | 22 | 10 | 12 | 45% |

Moreover, in convolutional networks, the sliding movement of the filters across the input image space is controlled by the stride. If this later is mutated to 0 the filters get disabled, otherwise if it is incremented by 1, no significant impact on the learning process of the network is observed. Other constant mutations such as increasing or decreasing the batch size or the number of epochs by one do not really affect the training of the network and hence this mutants survive. However, mutants get killed in case the number of epochs is set to 0 which means that there was no training.

**Table 4.** Results after applying mutations on the two convolutional neural networks.

| Mutation operators | Generated | | Killed | | Survived | | Mutation score | |
|---|---|---|---|---|---|---|---|---|
| | CNN 1 | CNN 2 | CNN 1 | CNN 2 | CNN 1 | CNN 2 | CNN 1 | CNN 2 |
| **Inline Constant Mutator** | 77 | 103 | 48 | 73 | 29 | 30 | 62% | 71% |
| Change no. of input channels | | | × | × | | | | |
| Change no. of output classes | | | × | × | | | | |
| Change learning rate | | | × | × | | | | |
| Change no. of epochs | | | | | × | × | | |
| Change batch size | | | | | × | × | | |
| Change stride to 0 | | | × | × | | | | |
| **Void Method Call Mutator** | 10 | 10 | 4 | 4 | 6 | 6 | 40% | 40% |
| **Increments Mutator** | 1 | 1 | 1 | 1 | 0 | 0 | 100% | 100% |
| **Cond. Boundary Mutator** | 1 | 1 | 1 | 1 | 0 | 0 | 100% | 100% |
| **Negate Cond. Mutator** | 3 | 3 | 1 | 1 | 2 | 2 | 33% | 33% |
| Total mutations | 92 | 118 | 55 | 80 | 37 | 38 | 60% | 68% |

For the Void Method Call Mutator, some mutants get killed because the method calls that were removed are used for the training of the model (e.g. the evaluation method which computes the accuracy of the network on the test set) otherwise, they survive if the removed method has no direct influence on the prediction accuracy of the network.

## 4 Related Work

Usually testing a machine learning system mainly consists of measuring their accuracy on given test inputs which are randomly drawn from manually labeled datasets and ad hoc simulations [19]. Considering the limited access to high quality test data, good accuracy performance on test data is not always sufficient enough to provide confidence to the testing adequacy and generality of DL systems [9]. Recently testing Deep Neural Networks has been tackled in many research studies. The first white-box framework for NN testing was introduced in [15]. In this paper the authors present a neuron coverage metric which systematically measures the parts of the DL system exercised by the test input. However, this metric has been criticized later in [17]. Experiments made in the latest study have shown that in particular, a test suite with high neuron coverage is not sufficient to increase the confidence of neural networks in safety-critical domains. They have proved that 100% neuron coverage can easily be achieved by a simple test suite comprised of few input vectors from the training dataset.

In this study they have also introduced a test criterion inspired by the traditional MC/DC coverage. They propose a set of four test criteria that are personalized to the distinct features of DNNs. These metrics include coverage, adversarial ratio and adversarial quality to measure the safety/robustness of a DNN. Besides, they provide an algorithm for each criterion for generating test cases based on linear programming (LP). They validate their method on a set of networks trained on the

MNIST data set, they obtained promising results, indicating the effectiveness of generated test suites in detecting safety bugs of the DNN (i.e., adversarial examples). In another study [9], the authors propose, a set of multi-granularity testing criteria based on multi-level and multi-granularity coverage for DL systems and measure the testing quality which aims to render a multi-faceted portrayal of the testbed.

In [11] they adapt the concept of combinatorial testing and propose a set of combinatorial testing (CT) criteria based on the neuron input interaction for each layer of DNNs. Their main objective is to guide test generation towards achieving a balance between the detection ability and a reasonable number of tests. They study the feasibility and usefulness of CT to testing massive runtime states of DL systems. They show that generating a sufficient test set of reasonable size and test effectiveness can be achieved in parallel by leveraging CT.

The majority of these research studies have focused on creating specific coverage criteria to test DNN, yet to the best of our knowledge, until now, the first attempt to create mutation testing techniques specialized for testing deep neural networks is introduced in [10]. In this study they design a mutation tool which provides two types of mutation operators specific to deep learning network architecture; data mutation such as data repetition, label errors, missing data, noise perturbation and program mutation operators like layer removal, layer addition and activation function removal. After injecting faults into the training data or the program, the training process of NN is re-executed and mutated models are generated. These model mutants are then executed and analyzed against the test set T to be compared later with the original DL model. The more behavior differences detected the higher the quality of the test set is indicated.

## 5 Conclusion

In this paper, we tackled the question whether it is possible to use available program mutation tools in the context of Deep Neural Networks. In particular, we have been interested in evaluating whether mutants that do not crash can be identified when being used for learning and classification. In this paper, we outline a test oracle for this purpose that compares the average prediction accuracy of the original neural network with the one of a mutant. In case of deviations larger than a given threshold, e.g., 1% in our case, we are able to classify mutants as being killed. In addition, we described a case study applying mutation testing based on the PIT tool to 3 different neural networks. In all cases, we were able to identify at least some mutants to be killed. The obtained mutation score ranges between 40 and 65%. Some mutation operators, and in particular those that have an effect on the learning behavior, have been more effective than others. In future research, we plan to extend the case study using different neural networks and learning tasks. In addition, we want to clarify whether certain mutants can be detected when using different input data for learning and evaluation and thus gaining experience on the robustness of neural networks with respect to small faults introduced in the source code.

**Acknowledgement.** The financial support by the Austrian Federal Ministry for Digital and Economic Affairs and the National Foundation for Research, Technology and Development is gratefully acknowledged.

## References

1. Coles, H., Laurent, T., Henard, C., Papadakis, M., Ventresque, A.: PIT: a practical mutation testing tool for Java. In: Proceedings of the 25th International Symposium on Software Testing and Analysis, pp. 449–452. ACM (2016)
2. DeMillo, R.A., Lipton, R.J., Sayward, F.G.: Hints on test data selection: help for the practicing programmer. IEEE Comput. **11**(4), 34–41 (1978)
3. Eykholt, K., et al.: Robust physical-world attacks on deep learning visual classification. In: Proceedings CVPR (2018). arXiv: 1707.08945v5
4. Gibson, A., et al.: Deeplearning4j: distributed, open-source deep learning for Java and Scala on Hadoop and Spark (2016)
5. Hadsell, R., Erkan, A., Sermanet, P., Scoffier, M., Muller, U., LeCun, Y.: Deep belief net learning in a long-range vision system for autonomous off-road driving. In: IEEE/RSJ International Conference on Intelligent Robots and Systems, Nice, France, September 2008
6. Huval, B., et al.: An empirical evaluation of deep learning on highway driving (2015). arXiv: 1504.01716v3
7. Just, R.: The major mutation framework: efficient and scalable mutation analysis for Java. In: Proceedings of the 2014 International Symposium on Software Testing and Analysis, pp. 433–436. ACM (2014)
8. LeCun, Y.: The MNIST database of handwritten digits (1998). http://yann.lecun.com/exdb/mnist/
9. Ma, L., et al.: DeepGauge: multi-granularity testing criteria for deep learning systems. In: Proceedings of the 33rd ACM/IEEE International Conference on Automated Software Engineering, pp. 120–131. ACM (2018)
10. Ma, L., et al.: DeepMutation: mutation testing of deep learning systems. In: 2018 IEEE 29th International Symposium on Software Reliability Engineering (ISSRE), pp. 100–111. IEEE (2018)
11. Ma, L., et al.: Combinatorial testing for deep learning systems. arXiv preprint arXiv:1806.07723 (2018)
12. Ma, Y.S., Offutt, J., Kwon, Y.R.: MuJava: a mutation system for Java. In: Proceedings of the 28th International Conference on Software Engineering, pp. 827–830. ACM (2006)
13. Madeyski, L., Orzeszyna, W., Torkar, R., Jòzala, M.: Overcoming the equivalent mutant problem: a systematic literature review and a comparative experiment of second order mutation. IEEE Trans. Softw. Eng. **40**(1), 23–42 (2014)
14. Offutt, A.J.: Investigations of the software testing coupling effect. ACM Trans. Softw. Eng. Method. **1**(1), 5–20 (1992)
15. Pei, K., Cao, Y., Yang, J., Jana, S.: DeepXplore: automated whitebox testing of deep learning systems. In: Proceedings of the 26th Symposium on Operating Systems Principles, pp. 1–18. ACM (2017)
16. Sallab, A.E., Abdou, M., Perot, E., Yogamani, S.: Deep reinforcement learning framework for autonomous driving. In: Proceedings IS&T International Symposium on Electronic Imaging, Autonomous Vehicles and Machines. Society for Imaging Science and Technology (2017). https://doi.org/10.2352/ISSN.2470-1173.2017.19.AVM-023

17. Sun, Y., Huang, X., Kroening, D.: Testing deep neural networks. arXiv preprint arXiv:1803.04792 (2018)
18. Tian, Y., Pei, K., Jana, S., Ray, B.: DeepTest: automated testing of deep-neural-network-driven autonomous cars. In: Proceedings of the ACM/IEEE 40th International Conference on Software Engineering. ACM, New York, Gothenburg, Sweden, May–June 2018. https://doi.org/10.1145/3180155.3180220
19. Witten, I.H., Frank, E., Hall, M.A., Pal, C.J.: Data Mining: Practical Machine Learning Tools and Techniques. Morgan Kaufmann, Burlington (2016)

# Data-Driven Intelligent Transportation Systems

# SGD-Based Wiener Polynomial Approximation for Missing Data Recovery in Air Pollution Monitoring Dataset

Ivan Izonin[1(✉)], Michal Greguš ml.[2], Roman Tkachenko[1], Mykola Logoyda[1], Oleksandra Mishchuk[1], and Yurii Kynash[1]

[1] Lviv Polytechnic National University, Lviv, Ukraine
ivanizonin@gmail.com, roman.tkachenko@gmail.com, mykola.m.lohoida@lpnu.ua, oleksandra.myroniuk@gmail.com, kynash@ukr.net

[2] Comenius University in Bratislava, Bratislava, Slovakia
Michal.Gregusml@fm.uniba.sk

**Abstract.** This paper describes the developed SGD-based Wiener polynomial approximation method for the missing data recovery of air pollution monitoring tasks. The main steps of algorithmic implementation of the method have been described and the necessity of a combination of both of these tools is substantiated. The basic parameters of the method (the degree of the polynomial, the loss function of the SGD algorithm) for design an optimal variant of it are experimentally investigated. One out of four studied loss functions was chosen for the practical implementation of the method for the design of the future applied air pollution monitoring system. It is founded that high degrees of the Wiener polynomial significantly increase the training time with a slight increase in accuracy. That's why a second-degree polynomial was chosen. The simulation of the method showed high as accuracy (based on MAPE, RMSE, MAE) and low computation time. Comparison of the developed method's results with the existing regression analysis methods (Adaptive Boosting, GRNN, SVR with different kernels) confirmed the high efficiency of its work. The proposed combination of the method allows obtaining an effective result from the point of view of accuracy-speed for the large volumes of data processing. The developed method will be useful when solving different tasks, for example, for a smart home or a smart city, medicine, economics, etc. That is, for those tasks where the problem of missing data does not allow conducting further effective intellectual analysis.

**Keywords:** Air pollution · Missing data recovery · Regression approach · Wiener polynomial · Stochastic gradient descent · Supervised learning method

## 1 Introduction

Air pollutants can cause various problems, including corrosion, erosion [1], visually impaired, unpleasant smells, damage to plants and grain crops, and negative effects on the animals and humans health [2]. Air monitoring is important both at the micro level,


I. Rojas et al. (Eds.): IWANN 2019, LNCS 11506, pp. 781–793, 2019.
https://doi.org/10.1007/978-3-030-20521-8_64

in particular in smart home systems [3], for example, for identification and notification of carbon monoxide emissions, fire signs, etc.; and at macro level, in particular for smart city systems [4] for example, for the purpose of monitoring and control over sources of pollution, emissions of harmful substances within the city, etc.

The modern development of the Internet of things [5] allows building both large and very small monitoring devices based on a set of sensors capable of gathering a huge amount of necessary information. Intelligent analysis of this information, especially in the online mode [6] for the data, representing the object of attention, remains an urgent task. The problem is complicated in case of the impossibility of processing the part of the data due to missing values in them. An analysis which are based on such data may be distorted which, in the case of air monitoring and control, can lead to very high losses [7]. This situation arises for a variety of reasons: a malfunction or complete failure of the sensor for collecting information, an imperfect system for transmitting information or a problem with its storage [8]. One of the methods for solving this task is to reduce data, that is, to remove rows or columns containing missing values or if the data is not informative in terms of the group of used criteria [9]. However, it should be noted that the using of this method can lead to the execution of useful information, which in turn will increase the error in the environmental monitoring task.

The selection of the correct method for solving this task plays a very important role. The unreliable values of the recovered data and their untimely receipt may be more harmful than the missing ones. Such situations may arise where monitoring, control or reporting subsystems based on the analysis of such data may provide incorrect or non-timely advice to the operator resulting in significant losses [10, 11]. More damage can be caused by fully automated control systems that decision solely on the analysis of received information [12, 13]. Therefore the missing data recovery with high accuracy and speed is an important task.

## 2 Review and Problem Statement

This paper considers a regression approach, which is very common to the missing data recovery in the air pollution monitoring dataset. In [14] the missing data recovery task based on GRNN ensemble is considered. The method shows satisfactory results for accuracy but large time delays for the training procedure implementation by all neural networks from the ensemble. In [15], authors consider boosting methods, in particular, based on the AdaBoost algorithm to recover missing values in the data sample. Due to a large number of iterations by the proposed modified approach, this method imposes a number of limitations on the application of the method in online monitoring environment systems.

In [16] a number of SVR modifications were developed using fuzzy logic and optimization methods. The hybrid method proposed by the authors ensures high accuracy of the solution of the regression task, but again, the working time of such a method at the expense of using particle swarm optimization is quite large. Paper [17] represents the use of linear regression to recover missing values. The regression model looks like this:

$$y = \theta_0 + \theta_1 x_1 +, \ldots, \theta_n x_n, \quad (1)$$

where $x_1, \ldots, x_n$ - is the vector of $n$ features, $\theta_0, \theta_1, \ldots, \theta_n$ are coefficients.

Authors in [18] propose to search coefficients of the model (1) using the Stochastic Gradient Descent (SGD) algorithm instead of the least squares method when working with large data samples. In [19] it is proposed to carry out the approximation of linear models using the SGD algorithm to solve the stated task. The authors prove the convergence of the algorithm when working with the large volumes of data.

In this paper, we propose to use a Wiener polynomial (WP) instead of (1) to improve the accuracy of the method for missing data recovery. For the efficient, and most important, for the fast search of the polynomial's coefficients under conditions of large samples of air pollution monitoring data processing, it is proposed to use the Stochastic Gradient Descent (SGD) algorithm. It is a very simple but effectiveness approach.

# 3 Proposed SGD-Based Wiener Polynomial Approximation

## 3.1 Wiener Polynomial Approximation

Approximation of multiparametric dependencies in conditions of large volumes of data processing is one of the important tasks of the present. In [20], the authors use the model of approximation of complex multiparametric dependencies based on the Wiener polynomial. The advantages of using this tool stem from the effects of Stone-Weierstrass theorem [21] on increasing the accuracy of approximation using a polynomial functions. The general form of the Wiener polynomial can be written as follows [22]:

$$\begin{aligned} Y(x_1, \ldots, x_n) = \theta_i + \sum_{i=1}^{n} \theta_i x_i + \sum_{i=1}^{n} \sum_{j=i}^{n} \theta_{i,j} x_i x_j + \sum_{i=1}^{n} \sum_{j=i}^{n} \sum_{l=j}^{n} \theta_{i,j,l} x_i x_j x_l + \ldots + \\ + \sum_{i=1}^{n} \sum_{j=i}^{n} \sum_{l=j}^{n} \ldots \sum_{z=k-1}^{n} \theta_{i,j,l,\ldots,z} x_i x_j x_l \ldots x_z \end{aligned}, \quad (2)$$

where $k = 1, 2, 3, \ldots$ is the degree of the Wiener polynomial, $y$ is output attribute, $\theta_j$ are regression parameters, $n$ is the feature's number in each vector $x^{(i)}$, $i = \overline{1,n},\ j = \overline{i,n},\ l = \overline{j,n}\ , \cdots, z = \overline{k-1,n}$ [22]. However, as has been noted in [18, 22] the search of $\theta_j$ for (2) in the case of large volumes of data processing on the basis of the least squares method is not expedient. That is why the authors propose to use the SGD algorithm for search $\theta_j$ of the model (2).

## 3.2 Stochastic Gradient Descent Algorithm

Stochastic Gradient Descent algorithm is an optimization algorithm that provides the ability to carry out an online machine learning [23]. Its main advantage over the

Gradient Descent (GD) algorithm is the ability to quickly work with large data samples. This is because each SGD iteration decreases the error of not the entire sample (for which GD spends a lot of time), but only a certain selected object. In addition, it does not require the memory of the entire training sample, which significantly reduces the computing resources needed to work the method. Due to this statement, SGD allows investigated linear models in large dimensional samples that do not fit into the computer's memory.

# 4 Simulation Results

## 4.1 Air Pollution Monitoring Dataset Characteristics

Air pollution monitoring dataset [24] has been used for the method modelling. It contains collected data about the hourly chemical air composition near an Italian town [25]. The device, based on a set of chemical sensors, collected information on 12 different chemical air components during some parts of years 2004–2005.

Visualization of a data set with all parameters is shown in Fig. 1. Detailed review and analysis of each dataset parameter are given in [25]. The size of the data set is 9,358 observations, 2,407 of which contain omissions in the data.

In order to simulate the work of the method, the observation of the missing data was removed from the data, and the data sample (6,950 observations) was randomly divided into a training and test samples (80% and 20%).

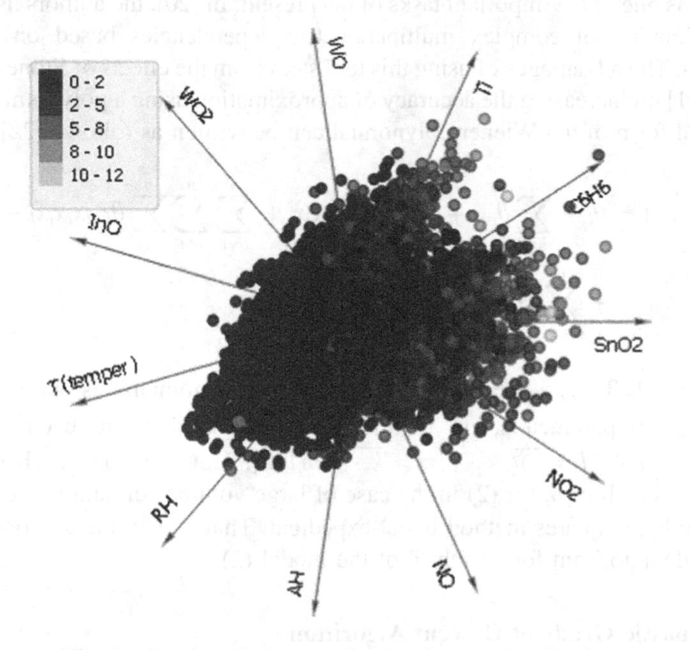


**Fig. 1.** An air pollution monitoring dataset visualization.

The solution to the missing data recovery task was focused on missing value imputation of carbon monoxide (CO) due to the high number of passes and its high toxicity.

### 4.2 SGD Simulation Results for Different Loss Functions

The four different loss functions of the SGD algorithm are used in this work, which are described in detail in [18]:

- *"squared loss"*;
- *"huber"*;
- *"epsilon insensitive"*;
- *"squared epsilon insensitive"*.

Table 1 shows the results of the existing SGD method based on (1) for all four loss functions in training and testing modes. As can be seen from Table 1, the best results of an existing algorithm are shown when using the two last loss functions.

**Table 1.** Evaluation results for existing SGD algorithm with different loss functions for training and test modes based on MAPE, RMSE, MAE and training time.

| Loss function's type | MAPE, % | RMSE | MAE | Training time |
|---|---|---|---|---|
| Training mode | | | | |
| *"squared_loss"* | 30,164739 | 0,6218984 | 0,370407 | 0,011968136 |
| *"huber"* | 65,783303 | 1,1890067 | 0,845971 | 0,023936272 |
| *"epsilon insensitive* | 28,908624 | 0,6273006 | 0,367068 | 0,012964964 |
| *"squared epsilon insensitive"* | 28,814964 | 0,5929891 | 0,35128 | 0,006980658 |
| Test mode | | | | |
| *"squared_loss"* | 26,89860468 | 0,5839834 | 0,3698654 | |
| *"huber"* | 62,02685178 | 1,2035563 | 0,8641756 | |
| *"epsilon insensitive* | 25,88601225 | 0,5914623 | 0,3691373 | |
| *"squared epsilon insensitive"* | 25,85183687 | 0,5528597 | 0,3494847 | |

### 4.3 Proposed Method Results for Different Wiener Polynomial Degrees

In this paper, authors have investigated the application of high WP's degrees based on the SGD algorithm to solve the missing data recovery task for the air environmental monitoring dataset. Appendix A, Table A.1 shows the experimental results of such studies for the training and test modes. We will highlight some important results on the charts for greater clarity. Figure 2 shows the proposed method simulation results (four types of loss functions) in the test mode with the application of Wiener Polynomials of different degree (from 2-nd to 6-th) based on MAPE. Other indicators from Appendix A Table A.1 show the same dependencies.

Since the time of the training procedure of the chosen supervised learning method for solving the missing data recovery task in case of the large volumes of data processing plays an important role, the results of such simulation are given. The time of the training procedure for the proposed method in the application of various polynomial's degrees is given in Appendix A in Table A.1.

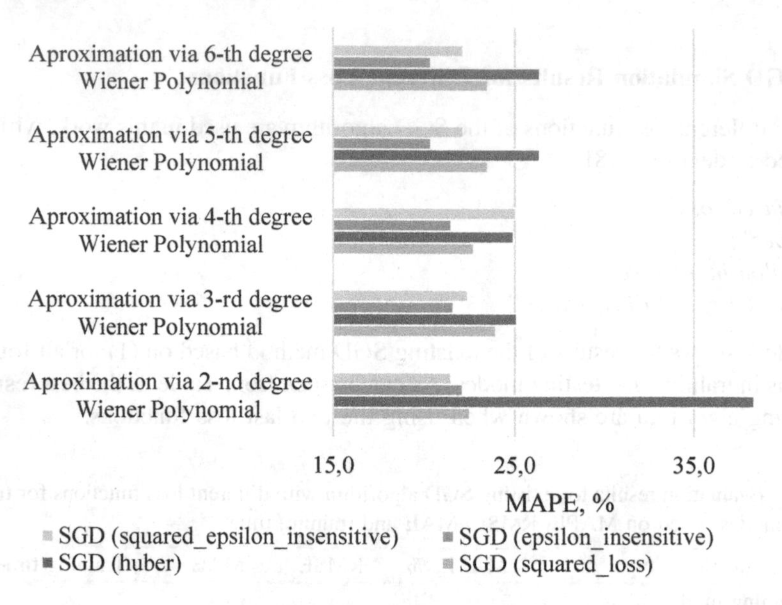


**Fig. 2.** The dependence of the approximation accuracy on the degree of the Wiener polynomial in the test mode (MAPE, %).

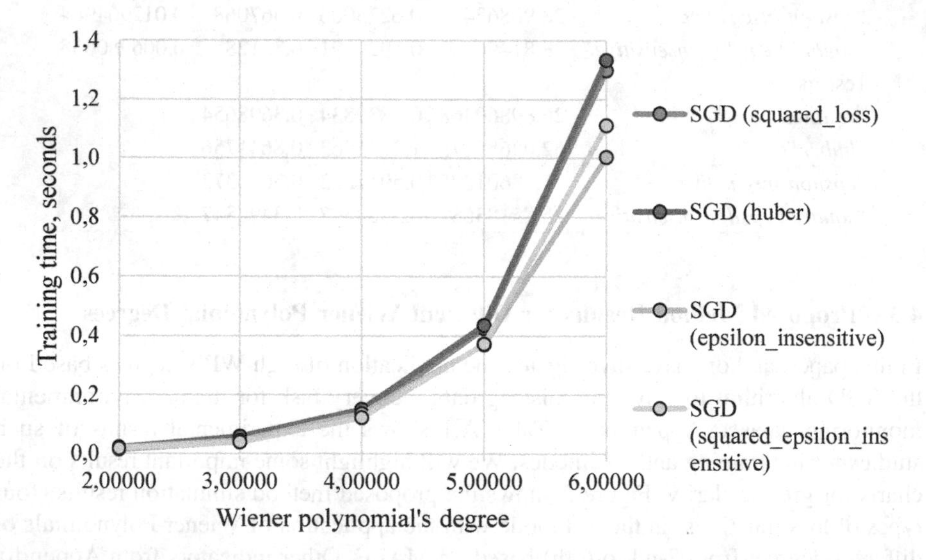


**Fig. 3.** The dependence of the training time on the degree of the Wiener polynomial (seconds).

In Fig. 3 the dependence of the training procedure's duration (in seconds) on the selected Wiener polynomial's degree is visualized for all four loss functions. It should be noted that the simulation of the proposed method was carried out on the author's software (console application) with the use of some Python's libraries [26].

# 5 Comparisons and Discussions

## 5.1 Discussion of the Obtained Results

The application of the regression approach to the solution of the missing data recovery task in the air pollution datasets is due to both a large amount of input data and the

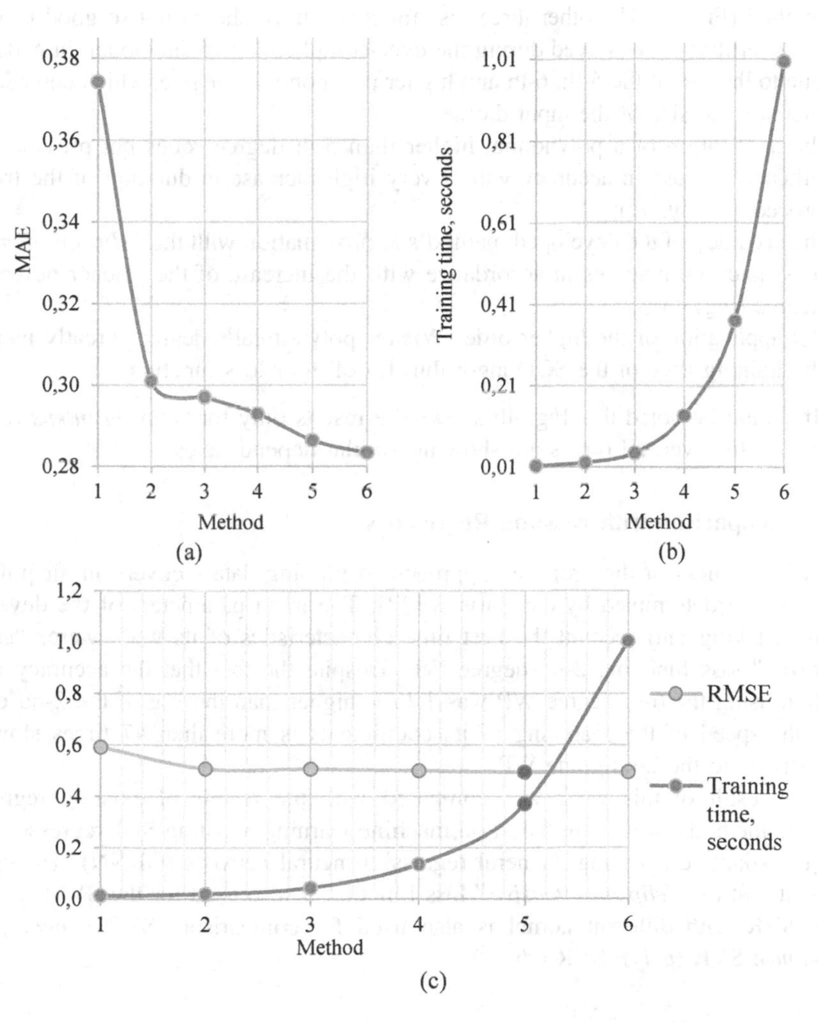


**Fig. 4.** Visualization of the results: (a) the dependence of the accuracy on the polynomial degree (test mode); (b) the dependence of the training procedure duration on the polynomial degree; (c) comparison of increasing as the method's accuracy (test mode) and the time of its training. The *x*-axis represents SGD-based methods: 1. existing regressor using (1); 2. with 2-nd degree WP; 3. with 3-rd degree WP; 4. with 4-th degree WP; 5. with 5-th degree WP; 6. with 6-th degree WP.

complex interconnections between the individual parameters of each individual observation (Fig. 1). The using of the Wiener polynomial as an approximation model, as expected, shows an increase of the solution accuracy (Appendix A, Table A.1, or Fig. 2). The SGD algorithm is one of the few existing ones, which allows finding the coefficients of the Wiener polynomial with high speed.

However, based on the analysis of the obtained results, it is should note the following:

- the SGD-based method using the *"elipson intensive"* loss function demonstrates the monotony of the precision increase when increasing the polynomial degree (Fig. 2). Moreover, it provides the best results with respect to the time characteristics of the method (Fig. 3). The other three loss functions have shown not so good results;
- the overfitting is observed during the over-complication of the model, in particular, due to the use of the 5-th, 6-th and higher polynomial's degree, which considerably increase the size of the input dataset;
- the application of a polynomial higher than 5-th degrees does not provide a significant increase in accuracy with a very high increase in duration of the training procedure (Fig. 4c);
- the accuracy of the developed method's approximation with the *"elipson intensive"* loss function increases in accordance with the increase of the Wiener polynomial degree (Fig. 4a);
- the application of the higher order Wiener polynomial's degrees greatly increases the training time of the SGD algorithm for all four loss functions.

It should be noted that Fig. 4b shows the results only for "*elipson intensive*" loss function. However, all others are showing similar dependencies.

### 5.2 Comparison with Existing Regressors

The effectiveness of the proposed approach to missing data recovery in air pollution datasets was determined by comparison [27]. The main parameters of the developed method, taking into account the best time characteristics of its work were: *"elipson intensive"* loss function, 2-nd degree WP. Despite the fact that the accuracy of the method using the 6-th degree WP was 1,17% higher than the use of the 2-nd degree WP, the speed of the searching of its coefficients is more than 47 times slower in comparison to the 2-nd grade WP.

The result of this work was compared with the results of existing regression analysis methods, which are based on machine learning: existing SGDregressor based on (1); Adaptive Boosting; General regression neural network (GRNN). Taking into account that the *"elipson intensive"* loss function is also used for the SVM regressor [22], SVR with different kernel is also used for comparison: SVR (*linear*); SVR (*sigmoid*); SVR (*poly*); SVR (*rbf*).

The comparison was made on the basis of MAPE, RMSE, MAE and training time both for the training and test modes. The implementation of known methods is taken from the Python library. All parameters of the known methods are described in [26].

The simulation results (test mode) for the accuracy (based on MAPE) and the speed of the developed method (SGD + WP) are shown in Figs. 5 and 6 respectively. As can be seen from Fig. 5 the proposed approximation method using the SGD machine learning algorithm provides the highest accuracy among all considered methods.

Figure 6 represents the training time for all methods, where the fastest is the linear kernel SVR method. However, this method according to Fig. 5 demonstrates low accuracy. The accuracy of the GRNN method shows only 1% less accuracy (according to MAPE) compared to the developed method, and this method works more than 200 times slower than developed. Despite the increase in the dimension of the input data space using the second degree Wiener polynomial, the developed method shows only 1.75 times less training time compared with the common SGD method based on (1). However, it is more accurate than the common SGD based on (1) of more than 3% (according to MAPE).

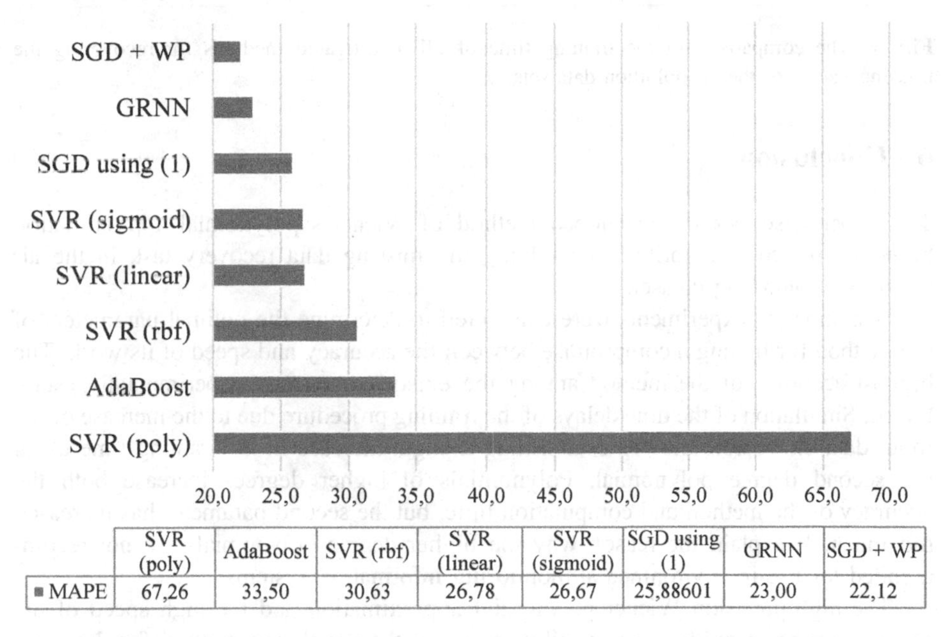


| | SVR (poly) | AdaBoost | SVR (rbf) | SVR (linear) | SVR (sigmoid) | SGD using (1) | GRNN | SGD + WP |
|---|---|---|---|---|---|---|---|---|
| ■ MAPE | 67,26 | 33,50 | 30,63 | 26,78 | 26,67 | 25,88601 | 23,00 | 22,12 |

**Fig. 5.** The comparison of the accuracy of all investigated methods for recovering the missing values of the air pollution data sets.

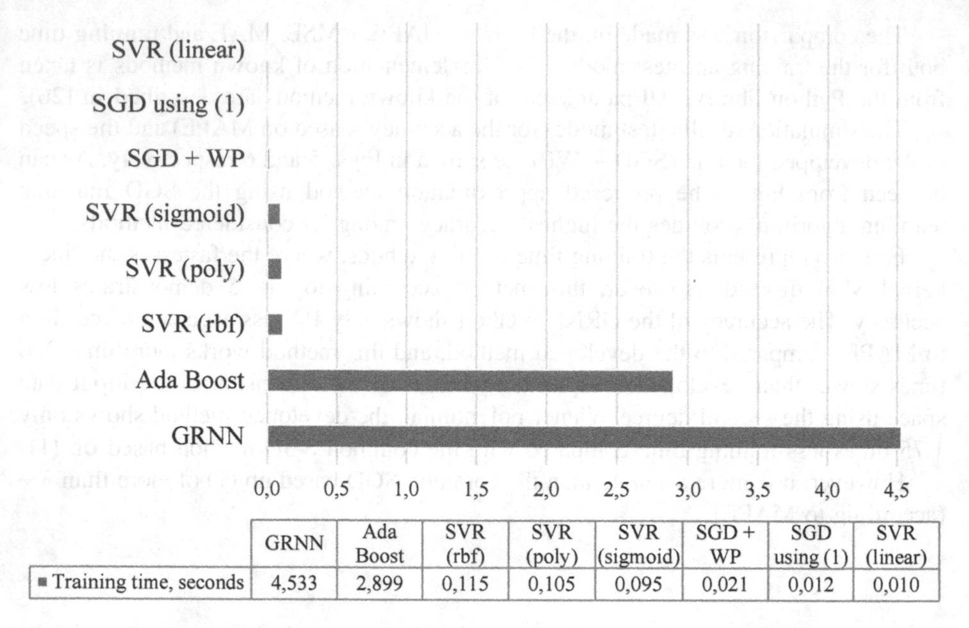


**Fig. 6.** The comparison of the training time of all investigated methods for recovering the missing values of the air pollution data sets.

## 6 Conclusion

The paper describes the developed method of Wiener's polynomial approximation based on the SGD algorithm for solving the missing data recovery task in the air pollution monitoring dataset.

A number of experiments were conducted to determine the optimal parameters of the method for finding a compromise between the accuracy and speed of its work. The highest accuracy of the method among the existing ones was experimentally established. Simulation of the time delays of the training procedure due to the increase of the input data dimension allowed establishing a slight increase in the training time using the second degree polynomial. Polynomials of higher degrees increase both the accuracy of the method and computation time, but the second parameter has increased too much. It explain the reason why the higher degree polynomials are not recommended for use in environmental monitoring information systems.

The high-precision Wiener polynomial approximation and the high speed of the SGD algorithm provide the possibility of using the developed method for the large volumes of data processing. The proposed method can be used in various areas, especially: smart home, medicine, materials science, economics, service science, etc.

# Appendix A

(See Table A.1)

**Table A.1.** Obtained results for the developed method with different Wiener Polynomial degree

| | Training mode | | | | Test mode | | |
|---|---|---|---|---|---|---|---|
| Loss function's type | MAPE, % | RMSE | MAE | Training time, sec. | MAPE, % | RMSE | MAE |
| Wiener polynomial of the second degree | | | | | | | |
| "squared loss" | 25,48957 | 38,85126 | 0,07075 | 0,01396 | 22,52820 | 0,51299 | 0,30527 |
| "huber" | 41,78869 | 53,90374 | 0,12017 | 0,02195 | 38,37121 | 0,73514 | 0,51135 |
| "epsilon insensitive" | 25,01448 | 38,40137 | 0,06821 | 0,02094 | 22,11927 | 0,50706 | 0,29595 |
| "squared epsilon insensitive" | 24,22143 | 38,50167 | 0,07055 | 0,01691 | 21,48941 | 0,50745 | 0,29961 |
| Wiener polynomial of the third degree | | | | | | | |
| "squared loss" | 26,92991 | 38,80563 | 0,07118 | 0,04089 | 23,95975 | 0,51603 | 0,30853 |
| "huber" | 27,97489 | 40,26512 | 0,07789 | 0,05884 | 25,10743 | 0,53654 | 0,33495 |
| "epsilon insensitive" | 24,53636 | 38,18539 | 0,06765 | 0,04488 | 21,61379 | 0,50609 | 0,29199 |
| "squared epsilon insensitive" | 25,25040 | 38,53493 | 0,07063 | 0,03990 | 22,41027 | 0,50805 | 0,30036 |
| Wiener polynomial of the fourth degree | | | | | | | |
| "squared loss" | 25,70013 | 38,54858 | 0,07041 | 0,54516 | 22,74281 | 0,50796 | 0,30125 |
| "huber" | 27,73087 | 41,30131 | 0,07964 | 0,58409 | 24,93140 | 0,56800 | 0,34461 |
| "epsilon insensitive" | 24,43369 | 37,90109 | 0,06630 | 0,53600 | 21,49122 | 0,49885 | 0,28797 |
| "squared epsilon insensitive" | 28,12322 | 38,40460 | 0,07017 | 0,54312 | 25,07391 | 0,50364 | 0,30832 |
| Wiener polynomial of the fifth degree | | | | | | | |
| "squared loss" | 26,46229 | 38,84947 | 0,07010 | 0,42390 | 23,50990 | 0,51709 | 0,31011 |
| "huber" | 29,17308 | 42,17964 | 0,08161 | 0,43484 | 26,38790 | 0,58821 | 0,35750 |
| "epsilon insensitive" | 23,13723 | 37,64923 | 0,06551 | 0,37001 | 20,36566 | 0,49423 | 0,28142 |
| "squared epsilon insensitive" | 24,97069 | 37,87880 | 0,06753 | 0,37201 | 22,14904 | 0,49605 | 0,29547 |
| Wiener polynomial of the sixth degree | | | | | | | |
| "squared loss" | 24,68688 | 38,06546 | 0,06866 | 1,29753 | 21,79333 | 0,49950 | 0,29547 |
| "huber" | 29,39637 | 41,94738 | 0,08229 | 1,33245 | 26,59210 | 0,57815 | 0,35789 |
| "epsilon insensitive" | 23,75924 | 37,50750 | 0,06424 | 1,00331 | 20,93933 | 0,49463 | 0,27837 |
| "squared epsilon insensitive" | 29,47514 | 39,03136 | 0,07269 | 1,11103 | 26,39987 | 0,51317 | 0,32470 |

## References

1. Duryahina, Z.A., Kovbasyuk, T.M., Bespalov, S.A., et al.: Micromechanical and electrophysical properties of Al2O3 nanostructured dielectric coatings on plane heating elements. Mater. Sci. **52**, 50 (2016)
2. Atmospheric chemistry. https://www.tankonyvtar.hu/hu/tartalom/tamop412A/2011-0073_atmospheric_chemistry/adatok.html. Accessed 09 Feb 2019
3. Artem, K., Ivan, T., Vasyl, T.: Intelligent house as a service and its practical usage for home energy efficiency. In 2017 12th International Scientific and Technical Conference on Computer Sciences and Information Technologies (CSIT), vol. 1, pp. 220–223 (2017)
4. Boreiko, O., Teslyuk, V.: Developing a controller for registering passenger flow of public transport for the 'smart' city system. East. Eur. J. Enterp. Technol. **6**(3(84)), 40–46 (2016)
5. Vynokurova, O., Peleshko, D., Oskerko, S., Lutsan, V., Peleshko, M.: Multidimensional wavelet neuron for pattern recognition tasks in the internet of things applications. In: Hu, Z., Petoukhov, S., Dychka, I., He, M. (eds.) ICCSEEA 2018. AISC, vol. 754, pp. 64–73. Springer, Cham (2019). https://doi.org/10.1007/978-3-319-91008-6_7
6. Shakhovska, N., Shamuratov, O.: The structure of information systems for environmental monitoring. In: 2016 XIth International Scientific and Technical Conference Computer Sciences and Information Technologies (CSIT), pp. 102–107 (2016)
7. Lytvyn, V., Vysotska, V., Veres, O., et al.: The risk management modelling in multi project environment. In: 2017 12th International Scientific and Technical Conference on Computer Sciences and Information Technologies (CSIT), vol. 1, pp. 32–35 (2017)
8. Riznyk, O., Yurchak, I., Povshuk, O.: Synthesis of optimal recovery systems in distributed computing using ideal ring bundles. In: 2016 XII International Conference on Perspective Technologies and Methods in MEMS Design (MEMSTECH), pp. 220–222 (2016)
9. Babichev, S., Lytvynenko, V., Gozhyj, A., Korobchynskyi, M., Voronenko, M.: A fuzzy model for gene expression profiles reducing based on the complex use of statistical criteria and shannon entropy. In: Hu, Z., Petoukhov, S., Dychka, I., He, M. (eds.) ICCSEEA 2018. AISC, vol. 754, pp. 545–554. Springer, Cham (2019). https://doi.org/10.1007/978-3-319-91008-6_55
10. Molner, E., Molner, R., Kryvinska, N., et al.: Web intelligence in practice. J. Serv. Sci. Res. **6**, 149 (2014). https://doi.org/10.1007/s12927-014-0006-4
11. Kaczor, S., Kryvinska, N.: It is all about services - fundamentals, drivers, and business models. J. Serv. Sci. Res. **5**(2), 125–154 (2013)
12. Gregus, M., Kryvinska, N.: Service orientation of enterprises - aspects, dimensions, technologies. Comenius University in Bratislava (2015)
13. Kryvinska, N., Gregus, M.: SOA and its business value in requirements, features, practices and methodologies. Comenius University in Bratislava (2014)
14. Gheyas, I.A., Smith, L.S.: A neural network-based framework for the reconstruction of incomplete data sets. Neurocomputing **73**(16–18), 3039–3065 (2010)
15. Wang, C.Y., Feng, Z.: Boosting with missing predictors. Biostat. (Oxford, England) **11**(2), 195 (2010)
16. An Imputation Method for Missing Traffic Data Based on FCM Optimized by PSO-SVR. https://www.hindawi.com/journals/jat/2018/2935248/. Accessed 09 Feb 2019
17. Missing data imputation: focusing on single imputation. https://www.ncbi.nlm.nih.gov/pmc/articles/PMC4716933/. Accessed 09 Feb 2019
18. Zhang, T.: Solving large scale linear prediction problems using stochastic gradient descent algorithms. In: Proceedings of the Twenty-First International Conference on Machine learning, ICML 2004, Banff, Alberta, Canada, pp. 116–120 (2004)

19. Stochastic Gradient Descent for Linear Systems with Missing Data. https://arxiv.org/abs/1702.07098. Accessed 09 Feb 2019
20. Ivakhnenko, A.G.: Polynomial theory of complex systems. IEEE Trans. Syst. Man Cybern. **SMC-1**(4), 364–378 (1971)
21. Stone-Weierstrass theorem. https://en.wikipedia.org/wiki/Stone%E2%80%93Weierstrass_theorem. Accessed 09 Feb 2019
22. Vitynskyi, P., Tkachenko, R., Izonin, I., Kutucu, H.: Hybridization of the SGTM neural-like structure through inputs polynomial extension. In: 2018 IEEE Second International Conference on Data Stream Mining and Processing (DSMP), Lviv, Ukraine, pp. 386–391 (2018)
23. Stochastic Optimization for Machine Learning. http://www.cse.ust.hk/~szhengac/papers/pqe.pdf. Accessed 09 Feb 2019
24. UCI Machine Learning Repository: Air Quality Data Set. http://archive.ics.uci.edu/ml/datasets/air+quality. Accessed 09 Feb 2019
25. De Vito, S., Vito, S.D., Massera, E., et al.: On field calibration of an electronic nose for benzene estimation in an urban pollution monitoring scenario. Sens. Actuators B: Chem. **129**(2), 750–757 (2008)
26. sklearn.linear_model.SGDRegressor - scikit-learn 0.20.2 documentation. https://scikit-learn.org/stable/modules/generated/sklearn.linear_model.SGDRegressor.html. Accessed 09 Feb 2019
27. Kryvinska, N.: Building Consistent Formal Specification for the Service Enterprise Agility Foundation. J. Serv. Sci. Res. **4**(2), 235–269 (2012)

# Heavy Duty Vehicle Fuel Consumption Modelling Based on Exploitation Data by Using Artificial Neural Networks

Oskar Wysocki[1(✉)], Lipika Deka[2], David Elizondo[2], Jacek Kropiwnicki[1], and Jacek Czyżewicz[1]

[1] Gdańsk University of Technology, Narutowicza 11/12, 80-233 Gdańsk, Poland
oskwys@gmail.com
[2] De Montfort University, Gateway House, Leicester LE1 9BH, UK

**Abstract.** One of the ways to improve the fuel economy of heavy duty trucks is to operate the combustion engine in its most efficient operating points. To do that, a mathematical model of the engine is required, which shows the relations between engine speed, torque and fuel consumption in transient states. In this paper, easy accessible exploitation data collected via CAN bus of the heavy duty truck were used to obtain a model of a diesel engine. Various polynomial regression, K-Nearest Neighbor and Artificial Neural Network models were evaluated, and based on RMSE the most relevant sets of parameters for the given algorithm were selected. Finally, the models were compared by using RMSE and Absolute Relative Error scores for 5 test samples. These represent the whole engine's operating area. Apart from goodness of fit, the models were analyzed in terms of sensitivity to the size of the training samples. ANN and KNN proved to be accurate algorithms for modeling fuel consumption by using exploitation data. The ANN model was ranked best, as it required less observations to be trained in order to achieve an absolute relative error which was lower than 5%. A conventional method, i.e. polynomial regression, performed significantly worse than either the ANN or the KNN models. The approach presented in this study shows the potential for using easy accessible exploitation data to modeling fuel consumption of heavy duty trucks. This leads to the reduction of fuel consumption having a clear positive impact on the environment.

**Keywords:** Neural Networks · Combustion engine · Heavy duty truck · Fuel economy · Fuel consumption · Refuse Collection Vehicle

## 1 Introduction

Trucks, whether used for freight transportation or as utility vehicles, play an important role in a countries economy and improving their fuel efficiency can undoubtable prove highly beneficial. One of the ways to improve the fuel economy is to operate the combustion engine at its most efficient operating points. It is particularly significant when it comes to performing duty cycles by the vehicle. For instance, when the engine runs the hydraulic or electric power receiver in the body. However, the information about the engine provided by the manufacturers does not cover the full engine


I. Rojas et al. (Eds.): IWANN 2019, LNCS 11506, pp. 794–805, 2019.
https://doi.org/10.1007/978-3-030-20521-8_65

characteristics and it is impossible to determine an optimal duty cycle in terms of the engine's speed and load. Conventional methods for obtaining general engine characteristics use engine or chassis dynamometers. This is associated with high costs. What is more, they provide steady state characteristics, which can lead to relatively high errors when used to model fuel consumption in transient states. This study presents methods for using exploitation data collected from a utility truck in order to obtain a mathematical model of the engine. Such model should provide an accurate fuel consumption prediction for any simulated duty cycle of the vehicle. Then, it can be used to compare the duty cycles and lead to an optimal design of the drivetrain, which will set the engine into its more efficient operating points during its work. In this study three models are presented: polynomial regression, K-Nearest Neighbor and Artificial Neural Network. The performance levels of the models are compared based on the RMSE and Absolute Relative Error over 5 test samples. Each of the samples has various observation distributions representing various operating areas of the engine. The models are also analyzed in terms of the influence of the trainset size on the model accuracy. As the result of this study, the KNN and the ANN models show high accuracy in fuel consumption prediction for even relatively small trainsets. Thus, by using easily accessible exploitation data it is possible to obtain an accurate model of the engine and use it to improve fuel consumption.

This paper is organized as follows. Section 2 describes the collected data and the principles of the truck's operation. Section 3 explains the selection of test samples. Polynomial regression, KNN and ANN models are described in Sects. 4, 5 and 6 respectively. Section 7 explains the validation of the models by using test samples, followed by discussion in Sect. 8. The conclusions derived from the results are presented in the last section.

## 2 Exploitation Data Description

The analysis presented in the paper is based on data collected using the vehicle's CAN bus. This was done using the FMS-interface [1]. This is an open standard, which gives third parties access to vehicle data, containing among others: engine speed **n**, torque **T** and fuel consumption $\mathbf{G_e}$. The examined truck was a Scania P320, working as a Refuse Collection Vehicle (RCV). The chassis was equipped with a diesel engine (9.3 $dm^3$ of displacement, max power = 235 kW, max torque = 1600 Nm) [2]. The engine parameters were recorded for 8 h, with the frequency of 10 Hz. This resulted in 290000 observations. This corresponds to a representative day of RCV operation, which is described further in the text. Although the FMS-interface is an optional interface in a truck, a large number of RCV (and other heavy duty vehicles) are equipped with it when purchased. Therefore, data collection does not require any modification of the vehicle's systems nor does it require additional measuring devices, except for the data logger. Similarly, the tests conducted on the trucks using FMS interface are described in [3, 4].

A regular day of the RCV operation consists of short distance travels and frequent stops. The truck stops close to a bin or set of bins, collects the waste and moves to another bin. The distances between stops and time spend collecting waste is dependent on the number of bins to be emptied, their size and distribution in the operating area. Approximately twice a day, the RCV needs to be discharged at the garbage dump. However, this has a minor effect on the presented analysis.

Most of rear loaders RCVs are equipped with a hydraulic system powered by a twin-flow, fixed displacement pump. The pump supplies the power to two separate circuits: the compactor circuit and the lifter circuit. The former drives a compaction mechanism, the ejector plate and the cylinders which lift the tailgate. The latter is used to operate the lifting mechanism. The principles of RCV operations are discussed in detail in [5–7]. In terms of truck's engine operation, it is important that when the RCV stops, the engine is idling i.e. its speed is close to 600 rpm. Then the hydraulic pump is turned on by the driver, and it is possible to operate the lifting mechanism. The engine speed remains the same, due to low power demand and sufficient oil flow provided by the pomp. When the compactor is activated, the engine speed is raised to 900 rpm. This value can be arbitrarily set by the RCV manufacturer, and usually ranges from 850 to 1050 rpm. The lifting mechanism can be operated regardless of the engine speed, and is not related to the compactor.

The above description explains measuring points distribution in an n-T domain of the data collected during a regular operation day. This is presented as a 2D histogram in Fig. 1. Approximately 50% of the time the engine speed was around 600 rpm, which corresponds to vehicle's stops and idling. This time periods are related to the stops due to traffic, the stops pending pump activation, and the lifter operation (without compacting). In general, the low values of torque for n = 600 rpm on the histogram are related to the former, and higher values to the latter. The second concentration of the measuring points is related to waste compaction, when the engine speed is set to 900 rpm. It represents about 15% of the daily operation time.

## 3 Selection of Test Samples

All models presented in further sections, are evaluated using a cost function. In most cases, the cost function is related to the difference between the real response value and the one calculated by the model. The final model is obtained by minimizing this error using e.g. Least Squares Method, Gradient Descend [8] etc. A widely used approach is to split the dataset into a training and a testing set, typically in a percentage proportion of 70:30, 80:20 or 85:15. Then the model is trained using the train set, to obtain the lowest product of the cost function. Then, a response value is predicted using the test set, and real values are compared with predicted ones. Based on this, a model accuracy score can be calculated, such as Root Mean Squared Error (RMSE). The non-uniform distribution of the observations leads to a good fit of the model in the areas of high observation density, but a relatively poorer fit elsewhere. Splitting the dataset randomly does not solve this problem, because distribution in the test set is analogous to the one in the train set. In order to assess accuracy of the reconstructed performance map of the engine, test samples from different areas in n-T domain are needed. Thus, 5 test samples were selected, each representing 20 succeeding seconds of the engine's operation. The paths in n-T domain of each sample are presented in Fig. 2. Test sample 1 covers the range of 1200–1450 rpm and 200–750 Nm. Test sample 2 may be assigned to the RCV body operation when the engine speed is raised from idling at 600 rpm to 900 rpm while the compacting mechanism is activated. Test sample 3 covers 1100–1250 rpm and a larger spectrum of torque: 0–1000 Nm. Test samples 4 and 5 represent the engine operation during driving, from idle engine speed to higher values of up to 1600 rpm. They also cover a wide range of torque, up to 900 and 1400 Nm per sample 4 and 5 respectively. In the following sections, the models are validated and compared by using RMSE and Absolute Relative Error for these 5 test samples.

## 4 Polynomial Regression Model

The data collected can be used to reconstruct the engine's performance map, as they contain information about the engine speed, the torque and the fuel consumption. Conventional methods, used during tests on chassis dynamometers or in laboratories, assume that the engine operates in a quasi–steady state i.e. the changes on its parameters do not exceed previously specified values at a given time [9]. The engine's performance map is also termed as a general engine characteristic. It is a vector function given by equation: $Y = f(n, T); (n, T) \in L$. Where L is a range of engine operating points. A polynomial surface of $G_e$ values is obtained by an approximation of the measured values of $G_e$ in the n-T domain using the least squares method [10, 11].

However, as described in [12], during regular daily operation the engine works mostly in transient states. Thus, applying the conventional method to the data presented in this paper, may lead to inaccuracy in the fuel prediction when using polynomial regression model. This is due to the time delay $\Delta t$ between the change of input variables: engine rotational speed and torque, and the engine's response (Ge). This is presented in Fig. 3, where the $\Delta t$ can reach up to 1.5 s. Because of that, a large number of similar observations with similar n and T can correspond to significantly different Ge.

A common practice is to use the engine's performance map obtained in quasi-steady states in order to model fuel consumption in transient states [10, 11, 13]. In this paper, transient state observations are used to evaluate the performance map, and then the model is used for fuel consumption modelling. This is the only possible approach for applying the conventional method using exploitation data, due to the very limited number of observations present in quasi-steady states [12].

A polynomial regression model was used, as one of the three algorithms in this study, to reconstruct the engine performance map. Later on is referred to as the Polynomial Regression Model (PRM). The polynomial function may be of arbitrary degree, however higher degrees are associated with higher risk of overfitting. In this paper polynomial surfaces of $2^{nd}$, $3^{rd}$, $4^{th}$ and $5^{th}$ degree are exanimated.

Each of the models was trained 30 times on the given train set. The train set consisted each time of a randomly selected sample from the dataset. The influence of the sample size was also considered. The models were trained on 22 sample sizes,

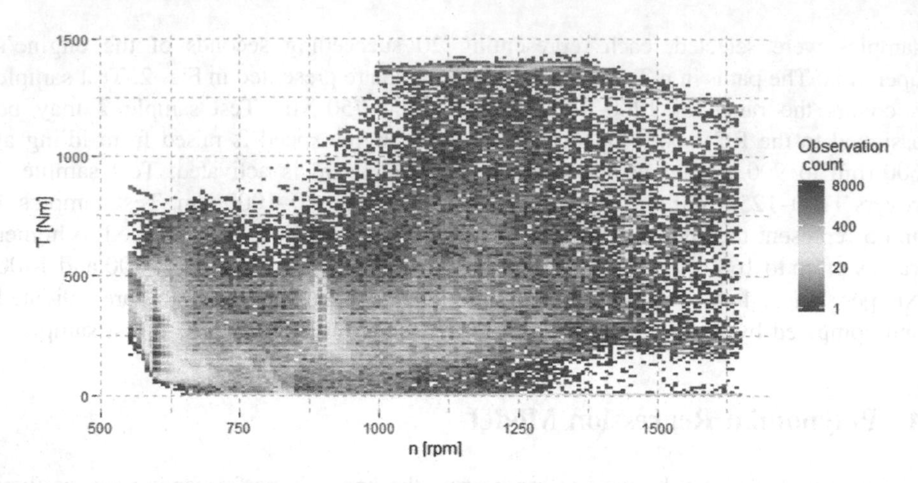


**Fig. 1.** Histogram of the observations. Engine data collected during a regular working day of the RCV.

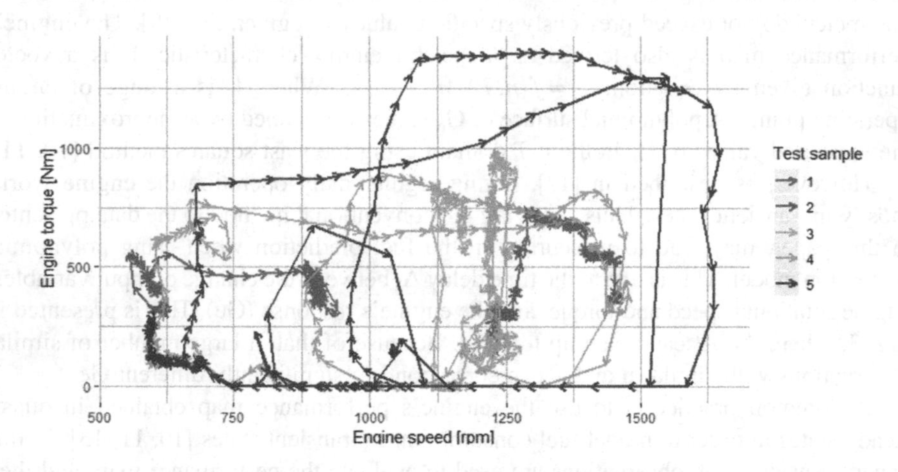


**Fig. 2.** Test samples measuring points presented as paths in n-T domain.

which logarithmically increase in the range between 100–250000 observations. As a result, for each model 660 RMSE scores were calculated, 30 for each sample size. These scores were calculated both for the training set and for the 5 test samples and are presented in Fig. 4. For sample sizes larger than 1000 observations, the RMSE become similar for both the training and the testing data sets. However, for small number of observations (<1000) the overfitting in 4th and 5th degree models is apparent. Thus, for the comparison with alternative algorithms in following sections, the 3rd degree polynomial regression model was selected.

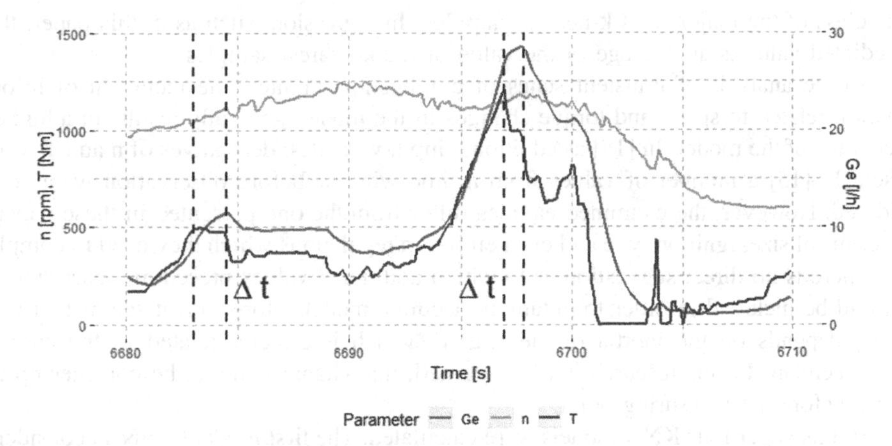


**Fig. 3.** Engine operation in a transient state. Time delay Δt between the input values (n and T) and the response (Ge)

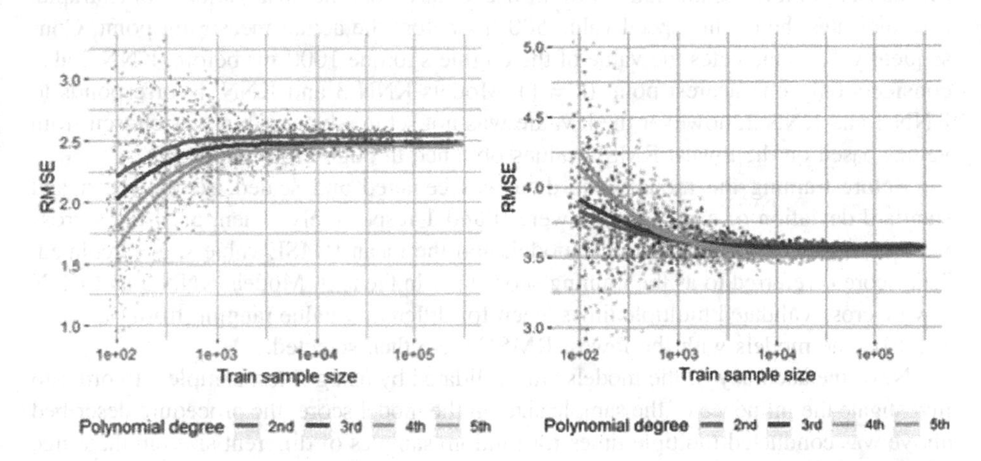


**Fig. 4.** Polynomial regression models scores: (left) RMSE train; (right) mean RMSE for all 5 test samples. Scores smoothed by using LOESS.

# 5 K-Nearest Neighbor Model

PRM uses a surface to approximate the engine's performance map. However, given enough observations, it is possible to find similar observations to the one used to predict engine fuel consumption. Moreover, this method is also useful with relatively large datasets with a uniform distribution of observations. The accuracy of the model is expected to correlated with the size of the dataset size and/or its uniform distribution. To find similar observations, a K-Nearest Neighbors (KNN) algorithm can be applied [8]. The principle of KNN is that the algorithm calculates the Euclidean distances of all of the observations in the dataset for a given input value. Then the algorithm chooses k-number of the closest observations. In classification problems, the predicted class is the class of the majority of k-nearest samples. In regression, such as in this paper, the predicted value is an average of the value of the k-nearest samples.

In the analysis of transient states of combustion engines, the inclusion of information related to speed and torque changes in the model, generally results in a higher accuracy of the model. In [14] as additional input variables: derivatives of n and T were used. In [15] a number of values from a time window before observation were considered. However, the examined engines differ from the one presented in these papers in terms of size, ignition type and even emission regulations which they need to comply to. There is no direct suggestion as to which additional values preceding observations should be included in order to obtain an accurate model. Moreover, it seems that this issue depends on the inertia of the engine, which is directly related to the engine displacement. In this research it was assumed, that changes should be consider up to 1.5 s before the measuring point.

In this paper four KNN models were calculated. The first model (KNN 1) considers n and T as input variables, and assign $G_e$ to the one nearest value (k = 1). The second model (KNN 2) consists of 8 input variables: n, T, $n_{500}$, $T_{500}$, $n_{1000}$, $T_{1000}$, $n_{1500}$, $T_{1500}$. The subscript indicates the number of milliseconds before the observation. For example, $n_{500}$ indicates the engine speed value 500 ms before the actual measuring point. Consequently $T_{1000}$ indicates the value of the engine's torque 1000 ms before. KNN 2 also considers only one nearest point (k = 1). Models KNN 3 and KNN 4 corresponds to KNN 1 and KNN 2, however the k value was not set in advance, but was chosen from values based on the lowest RMSE values obtained through cross validation.

Before training the models, the data was centered and scaled, so the mean and standard deviation of each variable were 0 and 1 respectively. Then, a 10-folds cross validation was performed for each model, and the mean RMSE value was calculated. This score is referred to as the training score later in the text. Models KNN 3 and KNN 4 were cross-validated multiple times, each for different k value ranging from (3, 5, …, 17, 19). The models with the lowest RMSE were then selected.

Next, the accuracy of the models was validated by using 5 test samples. In order to investigate the influence of the sample size on the model score, the procedure described above was conducted multiple times for random samples of different sizes in the range of 100 to 250000 observations, similarly to PRM, 30 times per sample size. Results for train RMSE scores and mean RMSE for 5 test samples are presented in Fig. 5. KNN models with input 8 variables clearly outperform the ones with 2 variables. Moreover, lower RMSE values for KNN 4 than KNN 2 suggest, that for k-values greater than 1 the accuracy of the model improves. The best K-Nearest Neighbors model was selected for further comparison. This corresponds to the KNN 4.

## 6 Artificial Neural Network Model

The artificial neural network model used in this study is the ANN. This model is based on a multi-layer feedforward artificial neural network, that is trained with stochastic gradient descent using back-propagation. A rectifier was used as activation function [16]. In order to produce an optimal ANN, a combination of 3 sets of predictors and 7 sets of hidden layers were analyzed, resulting in 21 models. The models are summarized in Table 1. A subscript in the predictors notation indicates the time in milliseconds before the observation, similarly as for KNN models. Additionally, Δ indicates that the variable corresponds to the difference between the actual value and the value of x milliseconds before; e.g. $\Delta n_{1500}$ corresponds to difference between actual n and $n_{1500}$. One hidden layer or two hidden layers with equal number of neurons were used.

**Table 1.** ANN model possible configuration: 3 sets of predictors and 7 sets of hidden layer resulted in combination of 21 models

| | Set of predictors | Neurons in hidden layers |
|---|---|---|
| Set 1 | n, T | 5, 10, 20, 50 |
| Set 2 | n, T, $\Delta n_{500}$, $\Delta n_{1000}$, $\Delta n_{1500}$, $\Delta T_{500}$, $\Delta T_{1000}$, $\Delta T_{1500}$ | 5-5, 10-10, 20-20 |
| Set 3 | n, T, $n_{500}$, $T_{500}$, $n_{1000}$, $T_{1000}$, $n_{1500}$, $T_{1500}$ | |

Before training the models, the data was centered and scaled, so that the mean and standard deviation of each variable were 0 and 1 respectively. Each neural network model was computed 30 times with a 10-fold cross-validation for each size of training sample in the range of 100–250000; the epochs were set to a 100. To measure the performance of the model RMSE was used. To select most accurate ANN model from 21 possible, the mean cross-validation RMSE values from the training process were compared. Results are shown in Fig. 6. On the left graph, the scores are grouped by the set of predictors. Set 1 is clearly underperforming regardless of sample size, with a constant RMSE around 2.5. Sets 2 and 3 are convergent for large sample sizes to RMSE = 0.6, but the latter is much better for sample sizes smaller than 5000 observations. On the right graph, only the scores from Set 3 were shown and grouped by the set of hidden layers. The RMSE decreases along with the increase of sample size, and the largest differences between the models can be observed for small training sample size. Based on that, the best ANN model was selected for comparison with PRM and KNN. The best model has 8 input variables (Set 3) and 50 neurons in 1 hidden layer.

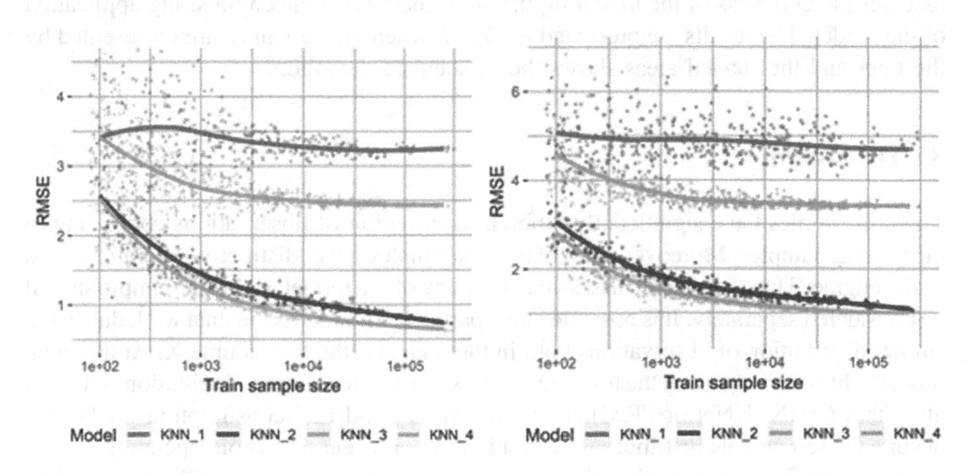


**Fig. 5.** KNN models comparison for increasing training sample size: (left) mean cross-validation RMSE; (right) mean RMSE for all 5 test samples. Scores smoothed by using LOESS.

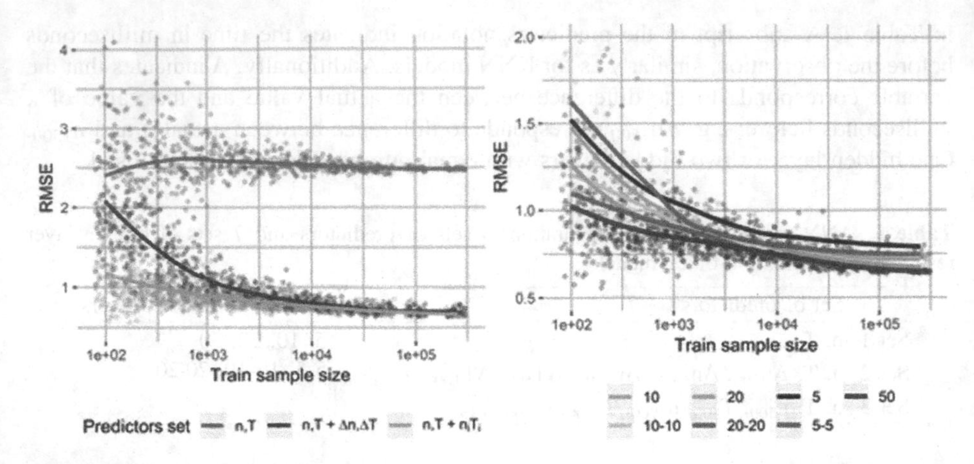


**Fig. 6.** ANN models comparison: (left) mean cross-validation RMSE for 3 sets of predictors; (right) mean cross-validation RMSE for best set of predictors (Set 3) for different configuration of hidden layers. Scores smoothed by using LOESS.

## 7 Models' Validation by Using 5 Test Samples

In previous sections three algorithms for diesel engine fuel consumption model were described. They were polynomial regression, K-Nearest Neighbors and Artificial Neural Networks. For each of the algorithms a model with optimal parameters was selected, according to the lowest RMSE values. In this section a validation of these three models is presented by using 5 test samples described in Sect. 2. Each model was trained 30 times on a random sample of the given size and for each test sample the fuel consumption was predicted. Two measures of model accuracy were used: RMSE and the Absolute Relative Error (ARE). The ARE is a percentage difference between real total fuel used and predicted total fuel used in a given test sample. While RMSE assesses the goodness of the model fit, the ARE focuses on the engineering application of the model. The results are presented in Fig. 7, where mean values are represented by the lines and the shaded areas shows the ± standard deviation.

## 8 Discussion

The test samples have significantly different distribution of observations compering to the training samples. Moreover, each of the test samples cover distinctive operating areas of the engine. Thus, in order to assess the accuracy of the model, each test sample should be considered separately. It is possible that a particular model fits the data well, due to the similar distribution of observations both in the train and the test sample X. At the same time, the fit is rather poor in the test sample Y, which has a different distribution. The best algorithm (ANN, KNN or PRM) should deliver a good fuel consumption prediction, despite the test sample distribution. It should also represent the whole operating area of the engine (n-T domain) with satisfactory accuracy. Considering RMSE values, ANN and KNN clearly outperform the more traditional method PRM in every test sample. Although the scores are similar for both methods ANN and KNN for large sample sizes, the KNN performs significantly worse than the ANN for smaller sample sizes. The ANN model appears to be relatively insensitive to trainset size, e.g. in test sample 1 the RMSE slightly decreased from 1.1 to 0.8, while for the KNN model it started of with a RMSE value of 3.2 for sample size equal 100 and decreased to 0.9. There is no improvement for the PRM model when the trainset size exceeds 1000 observations and the RMSE is several times higher than in other models, however the standard deviation decreases. Although the RMSE values vary in every test sample, the ANN model appears to be the best model in each of them.

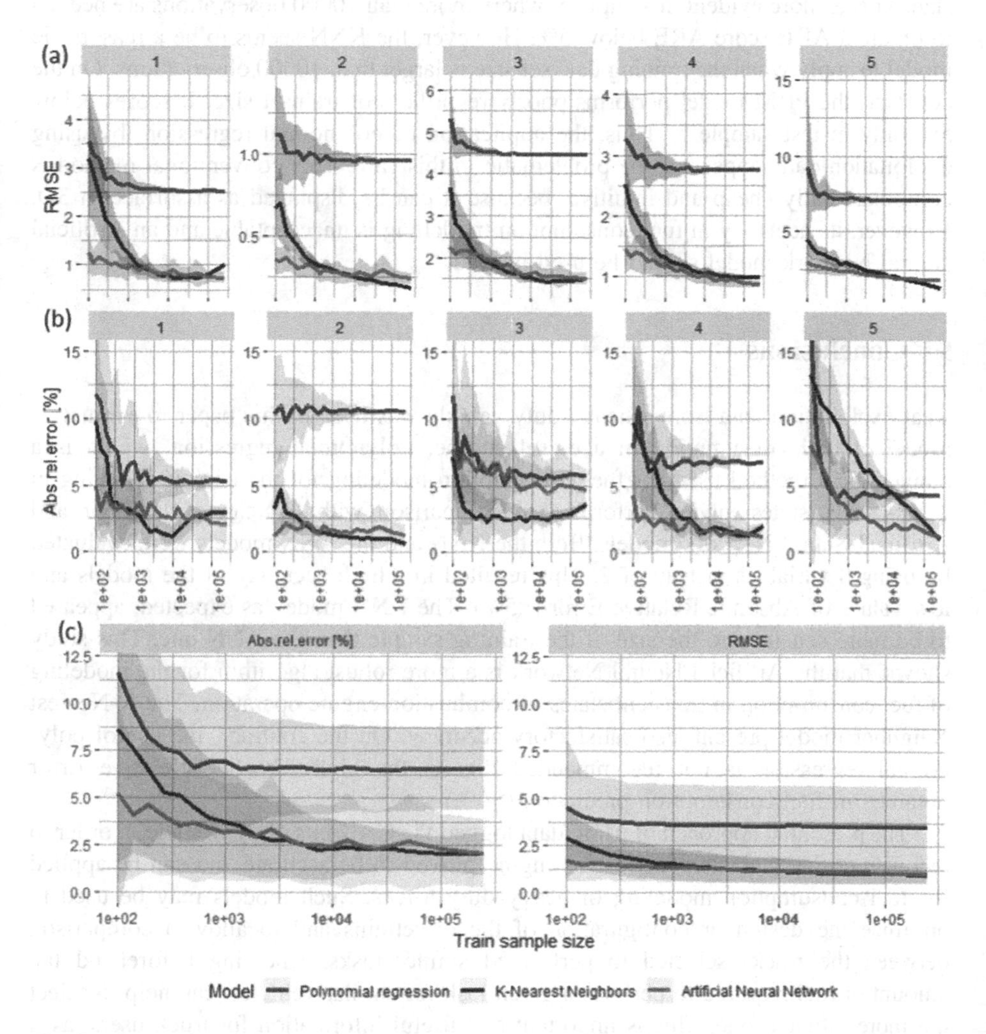


**Fig. 7.** Comparison between the models: ANN (red), KNN (black) and polynomial regression (blue): (a) Mean RMSE for each test sample; (b) Absolute Relative Error for each test sample; (c) mean values of RMSE and Absolute Relative Error for all test samples. Shaded areas represent ± standard deviation (Color figure online)

The Absolute Relative Error allows comparing models in terms of the final result of fuel consumption prediction. Even if a particular model does not fit well to each of the observation (high RMSE), the total amount of fuel predicted in the test sample may be close to the real value. In other words, for some points the model predicts higher $G_e$, and for other points lower values of $G_e$. Then, the differences are neutralized when the total fuel consumption is calculated. Values of ARE below 5% are considered as satisfactory from the perspective of engineering application.

In test sample 1, 2 and 4 the ANN model fulfills the above criterion regardless of trainset size. In samples 3 and 5 more than 5000 and 500 data samples are needed, respectively. The KNN model performs worse and is highly sensitive to the trainset size. This is more evident in sample 5, where more than 10000 observations are needed to obtain a ARE score ARE below 5%. However, the KNN seems to be a reasonable model to apply when the training data set size is larger than 10000 observations. On the contrary, the PRM model performs poorly regardless of trainset size. It scores below 5% only in test sample 5. Thus, the application of polynomial regression, by using exploitation data, appears to be problematic in this case. This conventional method is computationally cheap and intuitive, because it can be displayed as a surface in 3D. However, its accuracy in fuel consumption modelling is unacceptable and an Artificial Neural Network model should be used instead.

## 9 Conclusions

Real exploitation data from a heavy duty vehicle was used in this paper to evaluate a model of fuel consumption in a diesel engine. Polynomial regression, which is a conventional method used for fuel consumption modeling for the engine operating in quasi-steady states, under performed in comparison with K-Nearest Neighbor and Artificial Neural Network models. Both the KNN and the ANN models were evaluated by using 8 variables instead of 2. This resulted in a high accuracy of the models and low values of Absolute Relative Error (<5%). The KNN model, as expected, appeared to be more sensitive to the size of the training sample than the ANN one. This study shows, that the Artificial Neural Network is a more robust algorithm for the modeling of fuel consumption in transient states of combustion engine operation. The K-Nearest Neighbor model present also satisfactory accuracy. On the contrary, the use of polynomial regression is not recommended due to the high Absolute Relative Error obtained in fuel consumption prediction (>6%).

The presented approach of using data logged via CAN bus of the vehicle in order to reconstruct the characteristics of the engine proved to be accurate and can be applied for fuel consumption modeling of heavy duty trucks. Such models may be used to optimize the design or configuration of the drivetrains and to allow a comparison between the trucks selected to performed similar tasks. Knowing beforehand the amount of fuel required to perform a given task for similar vehicles, can help to select the more efficient one. This is important and useful information for truck users, as it will allow them to make fuel savings while at the same time reducing the impact to the environment.

**Acknowledgements.** The authors would like to thank the company Ekocel and Zoeller Tech for the assistance and collaboration provided in preparing this work, as well as for the financial support needed to perform the tests.

## References

1. FMS-Standard Startpage. http://www.fms-standard.com/. Accessed 1 Dec 2018
2. Scania Trucks website: www.scania.com. Accessed 1 Dec 2018
3. Wysocki, O., Kropiwnicki, J., Zajdziński, T., Czyżewicz, J.: Experimental determination of general characteristic of internal combustion engine using mobile test bench connected via power take-off unit. In: IOP Conference, Materials Science and Engineering, vol. 421 (2018)
4. Wysocki, O., Czyżewicz, J., Kropiwnicki, J.: Design of a test bench for determining the general characteristics of an internal combustion engine using a hydraulic power take-off system. In: XXIII International Symposium, Research-Education-Technology, Stralsund, 12–13 October 2017, pp. 2–7 (2017)
5. Wysocki, O., Zajdziński, T., Czyżewicz, J., Kropiwnicki, J.: Evaluation of the efficiency of the duty cycle of refuse collection vehicle based on real-world data. In: IOP Conference, Materials Science and Engineering, vol. 421 (2018)
6. Czyżewicz, J., Kropiwnicki, J., Wysocki, O.: Model of the hydraulic pump powertrain of refuse collection vehicle compaction mechanism. Combust. Engines **162**, 626–630 (2015)
7. Zajdziński, T., Wysocki, O., Czyżewicz, J., Kropiwnicki, J.: Energetic model of hydraulic system of refuse collection vehicle based on simulation and experimental data. In: IOP Conference, Materials Science and Engineering, vol. 421 (2018)
8. James, G., Witten, D., Hastie, T., Tibshirani, R.: An Introduction to Statistical Learning. STS, vol. 103. Springer, New York (2013). https://doi.org/10.1007/978-1-4614-7138-7
9. Chłopek, Z.: Some remarks on engine testing in dynamic states. Combust. Engines **49**, 61–71 (2010)
10. Kropiwnicki, J.: Modelowanie układów napędowych pojazdów z silnikami spalinowymi. Gdańsk (2016)
11. Kropiwnicki, J.: The possibilities of using of the engine multidimensional characteristic in fuel consumption prediction. J. Kones Intern. Combust. Engines **9**, 127–134 (2002)
12. Wysocki, O., Kropiwnicki, J., Czyżewicz, J.: Analysis of the possibility of determining the general characteristics using the operational data of a vehicle engine. Combust. Engines **171**, 33–38 (2017)
13. Steckelberg, D.B., Pacifico, A.L.: A methodology for measuring an internal combustion engine performance map using on-board acquisition. In: 23rd ABCM International Congress of Mechanical Engineering (2015)
14. Bera, P.: Fuel consumption analysis in dynamic states of the engine with use of artificial neural network. Combust. Engines **52**(4), 16–25 (2013)
15. Thompson, G.J., Atkinson, C.M., Clark, N.N., Long, T.W., Hanzevack, E.: Neural network modelling of the emissions and performance of a heavy-duty diesel engine. Proc. Inst. Mech. Eng. Part D J. Automob. Eng. **214**, 111–126 (2000)
16. H2O documentation: docs.h2o.ai/h2o/latest-stable/h2o-docs/data-science/deep-learning.html. Accessed 1 Dec 2018

# A Deep Ensemble Neural Network Approach to Improve Predictions of Container Inspection Volume

Daniel Urda Muñoz[1](✉), Juan Jesus Ruiz-Aguilar[2], Javier González-Enrique[1], and Ignacio J. Turias Domínguez[1]

[1] Departamento de Ingeniería Informática, Universidad de Cádiz, EPS de Algeciras, Algeciras, Spain
daniel.urda@uca.es

[2] Departamento de Ingeniería Civil, Universidad de Cádiz, EPS de Algeciras, Algeciras, Spain

**Abstract.** The use of predictive models at the border inspection posts in a port may help to manage and plan operations processes in such a way that time delays and congestion issues are minimized. In this paper, an enriched time series database containing records of the number of inspections carried out in the Port of Algeciras Bay between 2010 and 2018 is analyzed using two well-known statistical and computational intelligence methods such as linear regression (baseline model) and deep-fully connected neural networks. Additionally, a deep ensemble neural network approach is proposed in order to try to boost predictive performance even further. The results of the analysis show how deep fully-connected neural networks outperform a simple linear regression model, in particular the ensemble approach obtains performances of $\sigma = 0.813$ and $MSE = 330.160$ in contrast to $\sigma = 0.804$ and $MSE = 342.721$ achieved by linear regression. A visual comparison of the original and predicted time series shows how the ensemble approach is able to model better high and low peaks than the time series predicted by linear regression.

**Keywords:** Deep learning · Neural networks · Ensembles · Time series analysis

## 1 Introduction

With ever increasing shipping transport the number of seaport terminals and competition among them have become quite remarkable. Operations are nowadays unthinkable without effective and efficient use of information technology. The need for forecasting in port terminal operation has become more and more important in recent years. This is because the logistics especially of large container terminals has already reached a degree of complexity that further improvements require scientific methods. The characteristics of a port operation demands


I. Rojas et al. (Eds.): IWANN 2019, LNCS 11506, pp. 806–817, 2019.
https://doi.org/10.1007/978-3-030-20521-8_66

forecasting and decision. However, most of the processes occurring at ports cannot be foreseen for a longer time span and, in general, the planning horizon is very short [16].

Border Inspection Posts (BIPs) constitute a critical subsystem. BIPs of ports could lead to significant time delays and congestion problems decreasing port efficiency. The use of predictive models provides information that may be helpful as a decision-making tool in the management and planning of operations processes in the whole system. The BIPs process is becoming more and more important since the global trade and security of both goods and passengers have been increased in recent years. Thus, a short-term prediction of daily inspections at BIPs may be a powerful tool to avoid congestion or delays. In this sense, authors of this work have previously used Artificial Neural Networks (ANNs) to forecast several flows and traffics at the Port of Algeciras Bay: Moscoso-López et al. [8] predicted Ro-Ro traffic using ANNs comparing with other well-known forecasting techniques; Ruiz-Aguilar et al. [13,14] predicted flow freight congestion using different forecasting methods including ANNs; And a two-stage approach using an aggregation-disaggregation procedure together with Bayesian regularization was proposed in [9] to predict the number of inspections at the BIP in the Port of Algeciras Bay.

This paper aims to continue our forecasting studies regarding the inspection process at BIPs by using deep learning models on a new database provided by the authorities of the Port of Algeciras containing records from 2010 to 2018. The original time series given was enriched by adding three new time series which takes into account possible relationships between the number of inspections in the BIP with the day of the week, day of the year or season. Linear regression, as baseline model, and deep fully-connected neural networks (deep-FCNNs) were used to forecast container inspection volume based on the records of the last four weeks. Furthermore, this work proposes the use of a deep ensemble neural network (DeNN) approach which allows to use a given number of deep-FCNNs within an ensemble, each one with a specific network architecture and hyper-parameters setting learned through Bayesian optimization from data, aiming at boosting predictive performance even further.

The rest of the paper is organized as follows. Section 2 describes the dataset used to carry out the experiments, the ML models that will be fitted to data, the DeNN approach proposed in this paper and the experimental design of this study. Results of the analysis are presented and discussed in Sect. 3 and some conclusion of this work are provided in Sect. 4.

## 2 Materials and Methods

This section describes in detail the time series under study at the BIP in the Port of Algeciras Bay and the ML models used to carry out the analysis. Moreover, it includes a description of the DeNN approach proposed in this study as well as the validation strategy chosen to assess the performance of the ML models used.

### 2.1 Dataset

The authorities in the Port of Algeciras provided a database consisting of daily records regarding the number of inspections carried out at the BIP of this port. In concrete, a time series of 3103 points (starting in 1st January, 2010, and ending in 30th June, 2018) is initially available with an average of 70.78 daily inspections. Usually, the inspection process at the BIP may be influenced by several factors such as seasonality or peaks at certain days of the week and/or year, among others. However, this additional and valuable information is not a priory included in the original database provided. This database containing the time series of the inspections carried out at the BIP was enriched by adding three more time series corresponding to (i) the day of the week of each time point ($Mon = 1, \ldots, Sun = 7$), (ii) the day of the year of each time point ($1st\ January = 1, \ldots, 31st\ December = 365$), and (iii) the season associated to each time point ($Winter = 1$, $Spring = 2$, $Summer = 3$, $Fall = 4$). In this sense, statistical and computational intelligence methods may be able to capture existing relationships between the number of inspections and the day of the week, the day of the year and the current season of the year, respectively, and use this information to forecast the number of inspections at the BIP of the port in the future. Figure 1 shows the original time series (first sub-figure) and the three added ones within the first two years, 2010 to 2012.

Given this time series problem, partial autocorrelation plots [3] are typically used in data analysis to check randomness in a dataset and, particularly, to identify the order of an autoregressive model. However, in this work a fixed size of $W = 28$ days was chosen for the size of the autoregressive window since studying the optimal size of the window is out of the scope of this paper. On the other hand, the goal of this study is to forecast the number of inspections at the BIP of the port on the next day $(t+1)$. Therefore, the four time series were pre-processed using a 28-day autoregressive window to build a dataset $\mathcal{D} = \{\boldsymbol{X}_i, y_i\}$ consisting of $N = 3075$ samples and $P = 112$ input features which will allow the use of statistical and computational intelligence methods to forecast the number of inspections on the next day. In this sense, one sample is represented by 112 input features (28 days in the past for each of the four time series) and the expected outcome on the following day as depicted in Eqs. 1 and 2 ($t$ corresponds to a specific time point of the time series):

$$\begin{aligned} \forall i \in [1, N],\ \boldsymbol{X}_i = BIP(t),\ & BIP(t-1),\ \ldots,\ BIP(t-27), \\ & DayWeek(t),\ DayWeek(t-1),\ \ldots,\ DayWeek(t-27), \\ & DayYear(t),\ DayYear(t-1),\ \ldots,\ DayYear(t-27), \\ & Season(t),\ Season(t-1),\ \ldots,\ Season(t-27) \end{aligned} \tag{1}$$

$$\forall i \in [1, N],\ y_i = BIP(t+1) \tag{2}$$

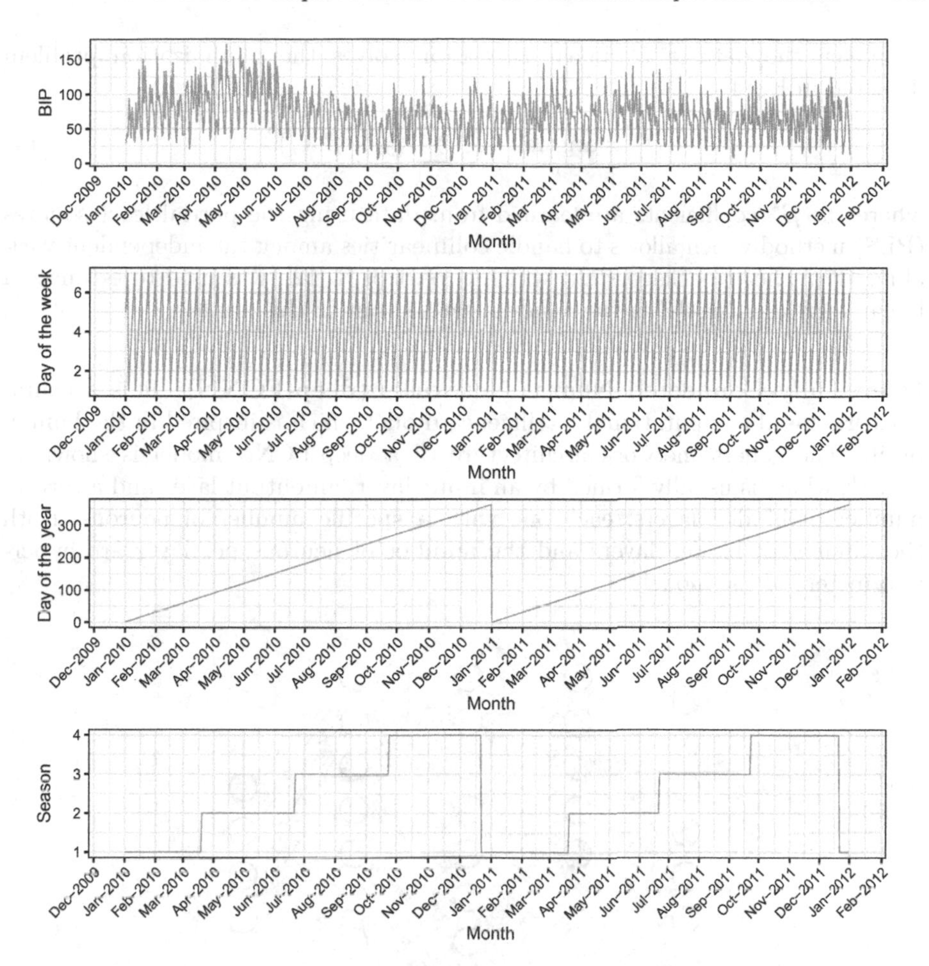


**Fig. 1.** Overview of the first two years for the raw time series of daily inspections at the BIP and the three corresponding time series added (day of the week, day of the year and season) for each time point.

## 2.2 Methods

Two well-known ML models were used in this study to forecast the number of inspections at the BIP of the port given the measures on the previous 28 days. Next, a description of both models is provided.

**Linear Regression (linreg).** It is used as baseline model in order to compare the performance results of the proposed DeNN approach. Linear regression [11] is the simplest statistical method which aims to model the dependent variable $y_i$ as a linear combination of the independent variables $\boldsymbol{X_i}$ (see Eq. 3).

$$y_i = \beta_0 + \beta_1 X_{i1} + \beta_2 X_{i2} + ... + \beta_P X_{iP} \tag{3}$$

Given the dataset $\mathcal{D}$, linear regression solves the minimization problem depicted in Eq. 4:

$$\boldsymbol{\beta}^* = \arg\min_{\boldsymbol{\beta}} \frac{1}{N} \sum_{i=1}^{N} (y_i - \boldsymbol{\beta} \boldsymbol{X}_i)^2 \tag{4}$$

where the $\boldsymbol{\beta}^*$ coefficients are learned from data using the partial least squares (PLS) method which allows to handle collinearities among the independent variables [17]. The R package *stats*, which is part of R [12], was used to estimate a linear regression model in the analysis carried out in this study.

**Deep Fully-Connected Neural Networks (deep-FCNN).** It is a neural network-based computational intelligence model which is inspired in the human brain. The general network architecture of a deep-FCNN model is shown in Fig. 2[1], which is usually formed by an input layer, an output layer and a certain number of hidden layers, each one with a specific number of neurons. Both the number of hidden layers and the number of neurons per layer are hyper-parameters of the model.

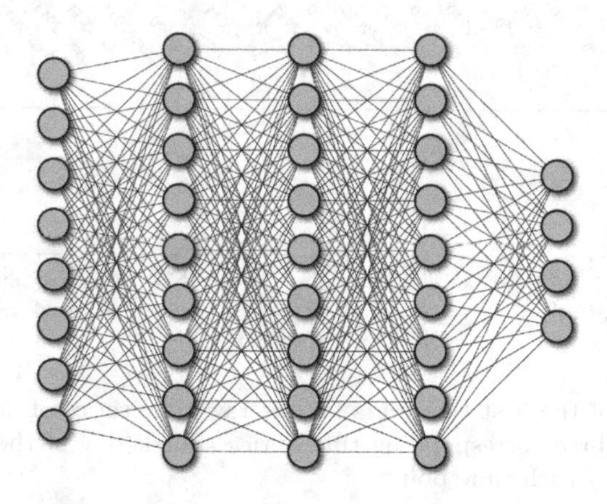

**Fig. 2.** General deep-FCNN architecture with one input layer, one output layer and as many hidden layers as needed (3 in this example), each of them composed by a specific number of neurons.

In contrast to linear regression, deep-FCNN models allow to learn existing non-linearities between the input and output spaces. Among the different activation functions available, Rectified Linear Units (ReLUs) are usually preferred due to its better handling of the vanishing gradient problem and for allowing a faster model estimation [10]. Regarding the training speed, batch normalization was introduced in [5] to accelerate deep networks training process by reducing

[1] Image source: O'Reilly Media (https://www.oreilly.com/library/view/tensorflow-for-deep/9781491980446/ch04.html).

internal covariate shift. Additionally, the dropout strategy [15] allows to overcome over-fitting issues which may appear as soon as the size of the network architecture increases by "disconnecting" some of the neurons of the hidden layers during the training process, thus decreasing the number of synaptic weights that are learned.

The R package *keras* [4] was used to model and train a deep-FCNN. With respect to the hyper-parameters of the model, the number of hidden layers, the number of neurons per hidden layer, the use of batch normalization layers and the dropout rate per hidden layer are learned by performing Bayesian optimization using the R package *mlrMBO* [2]. The remaining hyper-parameters were fixed as follow: (i) *activation function*: ReLU, (ii) *loss function*: mean squared error (commonly used in regression tasks), (iii) *optimizer*: adam [6] (fast and efficient optimization of stochastic objective functions), (iv) *epochs*: 500, and (v) *early stopping*: 20 (training process is stopped after 20 epochs with no improvement in the validation loss).

### 2.3 Deep Ensemble Neural Network Approach

This work proposes the use of a deep ensemble neural network (DeNN) approach in order to improve predictions of a simple linear regression or a deep-FCNN model. The ensemble consists of several deep-FCNN trained over the same data, each one with its own network architecture and hyper-parameters setting, and for which their individual predictions are combined by computing the average to produce the final DeNN prediction. Figure 3 shows an overview of this DeNN architecture.

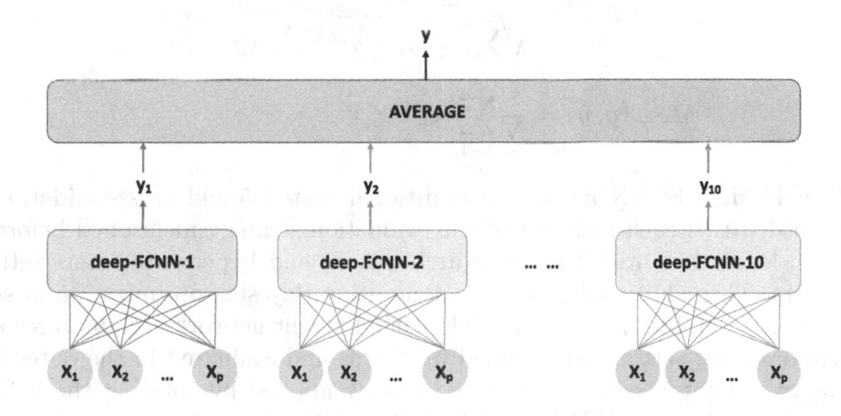


**Fig. 3.** DeNN approach consisting of 10 deep-FCNN, each one with a given architecture and hyper-parameters setting (different colour coding), and for which their individual outputs are averaged to produce the final ensemble output. (Color figure online)

The DeNN approach proposed considers a maximum of 10 deep-FCNN as part of the ensemble. In this sense, this study will evaluate the impact of adding one by one, up to 10, a new deep-FCNN model to the ensemble in terms of predictive accuracy. The procedure of choosing 10 network architectures and hyper-parameters setting to fit the deep-FCNN models to data is described in more details in Sect. 2.4. However, it basically consists of learning the bests networks configurations that provide higher average performance from data instead of randomly choosing 10 networks configurations that may lead to poor results. Moreover, the 10 networks configurations learned are sorted according to their predictive performance in such a way that the best setting is added first to the DeNN approach and the remaining ones are subsequently added in a similar way.

## 2.4 Experimental Design

In order to estimate the performance of each model, a 10-fold cross-validation [7] evaluation strategy was performed partitioning the complete dataset in 10 folds of equal sizes. Both linear regression and the deep-FCNN models were fitted in 9 folds (train set) and evaluated in the unseen test fold left apart (test set) within an iterative procedure that rotates the train and test folds used. Moreover, this procedure was repeated 20 times in order to guarantee the randomness of the partitioning process and the absence of any bias introduced due to the train/test partitions used. Two performance measures were used to test the goodness of each model: the Pearson's correlation coefficient ($\sigma$) and the mean squared error ($MSE$). Equations show how both performance measures are calculated given the observed ($\boldsymbol{y}$) and predicted ($\hat{\boldsymbol{y}}$) outcome. Higher values of $\sigma$ shows better performance while lower values of $MSE$ indicates more accurate predictions.

$$\sigma(\boldsymbol{y}, \hat{\boldsymbol{y}}) = \frac{\sum_{i=1}^{N}(y_i - \bar{y})(\hat{y}_i - \bar{\hat{y}})}{\sqrt{\sum_{i=1}^{N}(y_i - \bar{y})^2 \sum_{i=1}^{N}(\hat{y}_i - \bar{\hat{y}})^2}} \tag{5}$$

$$MSE(\boldsymbol{y}, \hat{\boldsymbol{y}}) = \frac{1}{N}\sum_{i=1}^{N}(y_i - \hat{y}_i)^2 \tag{6}$$

For the deep-FCNN model, an additional nested 5-fold cross-validation is performed within each train set of the evaluation strategy mentioned before in order to learn the optimal network architecture and hyper-parameters setting from data. This additional process is done using Bayesian optimization in such a way that several deep-FCNN models with different network architectures and hyper-parameter settings are trained in 4 folds and evaluated in the outer fold left apart. This procedure is again iteratively applied by rotating the 4 folds used to train the deep-FCNN models with the different settings and the outer fold left apart to evaluate their performance. Since the proposed DeNN approach considers 10 deep-FCNN models, the 10 settings that achieve lower average $MSE$ in the outer folds are finally chosen to be part of the ensemble and trained over the specific train set.

# 3 Results

Table 1 shows the quantitative performance results for the analysis carried out in this work. In concrete, the average Pearson's correlation coefficient ($\sigma$) and mean squared error (MSE) obtained by both linear regression and the DeNN approach, which takes into account a different number of deep-FCNNs within the ensemble (from 1 to 10), are shown. In the best of the cases, a DeNN approach which uses 10 deep-FCNNs within the ensemble is able to outperform a simple linear regression model, achieving a predictive performance of $\sigma = 0.813$ and $MSE = 330.160$ in contrast to $\sigma = 0.804$ and $MSE = 342.721$ obtained by linear regression. Additionally, the average time required (in minutes) to fit each model tested to a single train set is also included, and it is worth mentioning that the improvement achieved comes at an extra cost in terms of computational efficiency. In general, the DeNN will require approximately 2 h to learn the optimal network architecture and hyper-parameters settings and fit the model to data compared to a linear regression model which is trained in the order of milliseconds.

Furthermore, it is possible to see in Table 1 a dependency between the performance measures and the number of deep-FCNNs used within the DeNN approach. In this sense, Fig. 4 shows a visual representation of the $MSE$ and $Time$ with respect to the number of deep-FCNNs used in the ensemble. The optimal number of deep-FCNNs within the DeNN approach was selected according to the elbow method [1], which basically consists of visually inspecting a line chart and choosing the point in the graph from where the event of interest is not significantly improved, usually identified as an elbow shape. In this particular case, the $MSE$ was the measure used to choose the optimal cut-off point in the line chart (the time needed by the DeNN is linear and increases similarly with every deep-FCNN added), thus resulting in an optimal cut-off of 5 deep-FCNNs within the DeNN approach.

Once the number of deep-FCNNs within the DeNN approach was fixed 5, the model was fit to a train set consisting of time points between 2010 and 2017. In this sense, Table 2 shows the 5 bests network architectures and hyper-parameters settings learned through Bayesian optimization. Regarding network architectures, 4 out of 5 settings with better $MSE$ on the validation set consist of a one hidden layer network with approximately 60 neurons and only one setting adds a second hidden layer. With respect to the hyper-parameters of the model, the dropout rates learned for each hidden layer on the 5 settings are shown and, as concluded in [5], the use of batch normalization layers helps to obtain better performance results. Moreover, this global DeNN approach of 5 deep-FCNNs trained on data between years 2010 and 2017 was finally evaluated on a test set consisting of time points between January, 2018 and June, 2018. Figure 5 shows the original number of inspections at the BIP of the port (gray colour), the predictions made by a simple linear regression model (blue colour)

**Table 1.** Average performance measures (Pearson's correlation coefficient, $\sigma$, and mean squared error, MSE) obtained by the different models over the test set after performing 20 repetitions of 10-fold cross-validation. The time in minutes required for a single training step is also shown under the *Time* column.

| Model | $\sigma$ | MSE | Time (mins) |
|---|---|---|---|
| linreg | 0.804 | 342.721 | 0.003 |
| deep-FCNN-1 | 0.809 | 336.974 | 96.954 |
| deep-FCNN-2 | 0.811 | 332.904 | 100.185 |
| deep-FCNN-3 | 0.812 | 331.998 | 103.408 |
| deep-FCNN-4 | 0.812 | 331.313 | 106.644 |
| deep-FCNN-5 | 0.813 | 330.866 | 109.860 |
| deep-FCNN-6 | 0.813 | 330.515 | 113.247 |
| deep-FCNN-7 | 0.813 | 330.418 | 116.594 |
| deep-FCNN-8 | 0.813 | 330.236 | 119.971 |
| deep-FCNN-9 | 0.813 | 330.189 | 123.429 |
| deep-FCNN-10 | 0.813 | 330.160 | 126.920 |

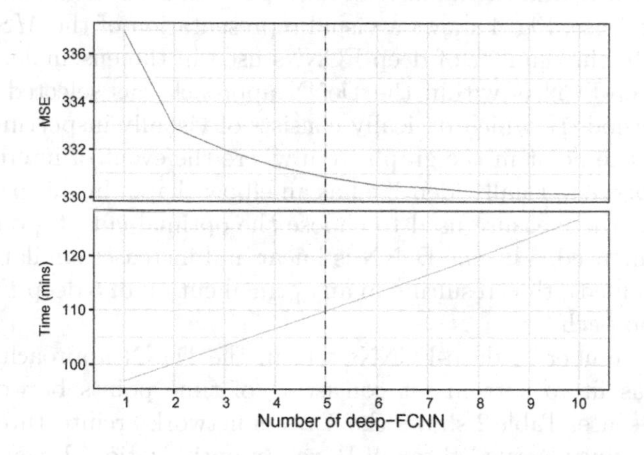


**Fig. 4.** Optimal cut-off point for the number of deep-FCNN to consider in the ensemble taking into account both the mean squared error (MSE) and the time required to estimate the model.

and predictions made by the DeNN approach proposed (red colour) on both train and test sets. Although predictions of linear regression and the DeNN approach seem to be similar in medium values of the original time series, one can appreciate in the figure that the DeNN approach tends to model better high and lower peaks of it, thus performing better than linear regression.

**Table 2.** Best network architectures and hyper-parameters settings learned through Bayesian optimization for the *deep-FCNN-5* model trained over the time points between year 2010 and 2017, both included. The *MSE* column indicates the average mean squared error on the validation set obtained by performing 5-fold cross-validation within the training set.

| Layers | Units | Dropouts | Batch normalization | MSE |
|---|---|---|---|---|
| 1 | (58) | (0.4675) | TRUE | 329.900 |
| 1 | (60) | (0.1790) | TRUE | 330.063 |
| 1 | (44) | (0.4739) | TRUE | 331.339 |
| 1 | (61) | (0.1695) | TRUE | 331.522 |
| 2 | (55, 25) | (0.3255, 0.2573) | TRUE | 331.692 |

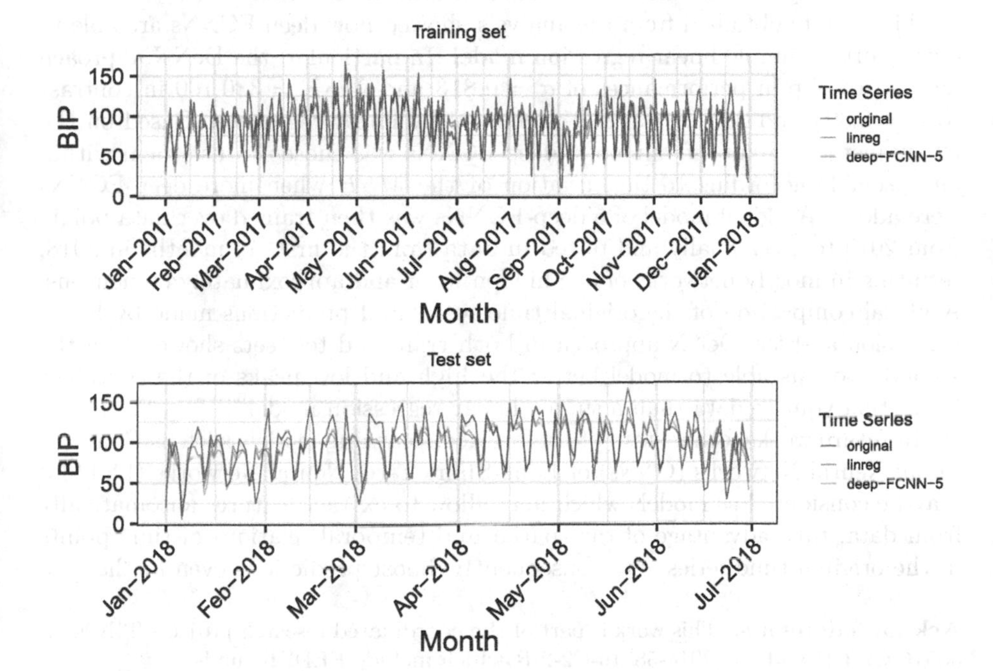


**Fig. 5.** Original PIB time series and the corresponding prediction made by the *linreg* and *deep-FCNN-5* models in a subset of the training set used to estimate the models (which included in total time points between 2010 and 2017) and in the complete test set left aside (only the first six months of 2018) which was not used at any moment during the training process. (Color figure online)

## 4 Conclusions

In this work a DeNN approach has been presented to forecast freight inspection volume at the BIP in the Port of Algeciras Bay. The database provided by the authorities of this port consisted of daily records regarding the number of inspections carried out at the BIP since 2010 until the 30th June, 2018. Three additional time series were attached to the original one provided taking into account possible relationships between the number of inspections and the day of the week, day of the year or season. Two well-know statistical and computational intelligence models were used to predict container inspection volume at the BIP: linear regression as a baseline model and deep-FCNNs as a more sophisticated one which allows to capture non-linearities between input and output spaces. For the deep learning model, the optimal network architecture and hyper-parameters settings were learned from data through Bayesian optimization. Moreover, a DeNN approach which uses several deep-FCNNs within an ensemble was presented in order to improve predictions. A 10-fold cross-validation procedure, which was repeated 20 times varying the data partitioning, was chosen as validation strategy to test their performance in unseen test data.

The results obtained from the analysis showed how deep-FCNNs are able to outperform a simple linear regression model. In particular, the DeNN approach allowed to obtain performances of $\sigma = 0.813$ and $MSE = 330.160$ in contrast to $\sigma = 0.804$ and $MSE = 342.721$ achieved by linear regression. Based on the elbow method, a cut-off value of 5 deep-FCNNs was chosen to use them within an ensemble according to the variation of the $MSE$ when more deep-FCNNs were added. A global model of 5 deep-FCNNs was then trained over data points from 2010 to 2017 finally and tested in data from the first six months in 2018, resulting in mostly networks of one hidden layer and approximately 60 neurons. A visual comparison of the original time series and predictions made by linear regression and the DeNN approach in both train and test sets showed how the second one was able to model better the high and low peaks of the container inspection volume data than a simple linear regression model.

In future works, other state-of-the-art deep learning models such as Convolutional Neural Networks (CNNs) or Long-Short Term Neural Networks (LSTMs) may be considered as models which may allow to extract features automatically from data, take advantage of the spatial and temporal relations of time points in the original time series and, consequently, boost predictions even further.

**Acknowledgments.** This work is part of the coordinated research projects TIN2014-58516-C2-1-R and TIN2014-58516-C2-2-R which include FEDER funds.

## References

1. Bholowalia, P., Kumar, A.: EBK-means: a clustering technique based on elbow method and k-means in WSN. Int. J. Comput. Appl. **105**(9), 17–24 (2014). Full text available
2. Bischl, B., Richter, J., Bossek, J., Horn, D., Thomas, J., Lang, M.: mlrMBO: A Modular Framework for Model-Based Optimization of Expensive Black-Box Functions. http://arxiv.org/abs/1703.03373
3. Box, G.E.P., Jenkins, G.: Time Series Analysis, Forecasting and Control. Holden-Day Inc, San Francisco (1990)
4. Chollet, F., Allaire, J., et al.: R interface to keras (2017). https://github.com/rstudio/keras

5. Ioffe, S., Szegedy, C.: Batch normalization: Accelerating deep network training by reducing internal covariate shift. CoRR abs/1502.03167 (2015). http://arxiv.org/abs/1502.03167
6. Kingma, D.P., Ba, J.: Adam: A method for stochastic optimization. CoRR abs/1412.6980 (2014). http://arxiv.org/abs/1412.6980
7. Kohavi, R.: A study of cross-validation and bootstrap for accuracy estimation and model selection. In: Proceedings of the 14th International Joint Conference on Artificial Intelligence, IJCAI 1995, vol. 2, pp. 1137–1143 (1995)
8. Moscoso-López, J., Turias, I., Come, M., Ruiz-Aguilar, J., Cerbán, M.: Short-term forecasting of intermodal freight using ANNs and SVR: Case of the Port of Algeciras Bay. Transp. Res. Procedia **18**, 108–114 (2016). https://doi.org/10.1016/j.trpro.2016.12.015, http://www.sciencedirect.com/science/article/pii/S2352146516307700. Efficient, Safe and Intelligent Transport. Selected papers from the XIIConference on Transport Engineering, Valencia (Spain) 7-9 June
9. Moscoso-López, J.A., Turias, I., Jiménez-Come, M.J., Ruiz-Aguilar, J.J., Cerbán, M.D.M.: A two-stage forecasting approach for short-term intermodal freight prediction. Int. Trans. Oper. Res. **26**(2), 642–666 (2019). https://doi.org/10.1111/itor.12337
10. Nair, V., Hinton, G.E.: Rectified linear units improve restricted Boltzmann machines. In: Proceedings of the 27th International Conference on International Conference on Machine Learning, ICML 2010, pp. 807–814 (2010)
11. Neter, J., Kutner, M.H., Nachtsheim, C.J., Wasserman, W.: Applied Linear Statistical Models. Irwin (1996)
12. R Core Team: R: A Language and Environment for Statistical Computing. R Foundation for Statistical Computing, Vienna, Austria (2017). https://www.R-project.org/
13. Ruiz-Aguilar, J.J., Turias, I., Moscoso-López, J.A., Jiménez-Come, M.J., Cerbán-Jiménez, M.: Efficient goods inspection demand at ports: a comparative forecasting approach. Int. Trans. Oper. Res. https://doi.org/10.1111/itor.12397
14. Ruiz Aguilar, J.J., Turias, I., Moscoso López, J.A., Jiménez, M., Cerábn, M.D.M.: Forecasting of short-term flow freight congestion: a study case of Algeciras Bay Port (Spain). DYNA **83**(195), 163–172 (2016). https://doi.org/10.15446/dyna.v83n195.47027
15. Srivastava, N., Hinton, G., Krizhevsky, A., Sutskever, I., Salakhutdinov, R.: Dropout: a simple way to prevent neural networks from overfitting. J. Mach. Learn. Res. **15**(1), 1929–1958 (2014)
16. Steenken, D.: Optimised vehicle routing at a seaport container terminal. Orbit **4**, 8–14 (2003)
17. Wold, S., Ruhe, A., Wold, H., Dunn III, W.: The collinearity problem in linear regression. the partial least squares (PLS) approach to generalized inverses. SIAM J. Sci. Stat. Comput. **5**(3), 735–743 (1984)

# Ro-Ro Freight Forecasting Based on an ANN-SVR Hybrid Approach. Case of the Strait of Gibraltar

José Antonio Moscoso-López(✉), Juan Jesús Ruiz-Aguilar, Daniel Urda, Javier González-Enrique, and Ignacio José Turias

Intelligent Modelling of Systems Research Group, Polytechnic School of Engineering (Algeciras), University of Cádiz, Avda. Ramón Puyol s/n, 11202 Algeciras, Cádiz, Spain
{joseantonio.moscoso, juanjesus.ruiz, daniel.urda, javier.gonzalezenrique, ignacio.turias}@uca.es

**Abstract.** The Ro-Ro (Roll-on Roll-off) freight forecasting plays an important role in ports management in the logistic node of the Strait of Gibraltar. International freight trips are subject to variable schedule or calendar. The use of the prediction in seven days in advance may be helpful as a decision-making tool in ports operations. This work addresses the forecasting problem on a daily time series by a novel ANN-SVR hybrid approach for 7 days ahead. The study compares the performance in the framework of several autoregressive windows and the improve of the performance through the stages of the hybrid model. The hybrid approach is based on Artificial Neural Networks, Support Vector Machines for regression and an ensemble approach in order to obtain an accurate forecasting. The results show that the presented models are an auspicious tool to predict Ro-Ro freight.

**Keywords:** Freight forecasting · Intelligent transport · Ro-Ro · Hybrid approach · ANNs · SVRs

## 1 Introduction

Ports play an important role of the entire supply chain where the information knowledge of freight flow is a key factor in ports operations and management. An accuracy freight forecasting is a valuable decision making tool and can help to the different actors in the planning and management of the ports operations. The capacity to accurately forecast demand in freight could minimize the operation cost and improve the service quality in the ports services, and thereby increase competitiveness.

In European Union (EU), all the maritime import freight from third countries must be checked in ports facilities. The complexity of the import process during the port phase and the variation of the scheduled arrival of the cargo causes problems as delays and congestions that provoke a critical point in the supply chain. These problems imply continuous pre-planning modifications and generate high variability in ports environment [1–3]. The agility of the import processes depends on the short-term future behavior of the freight flow. Hence, is very important the use of decision-making tools


I. Rojas et al. (Eds.): IWANN 2019, LNCS 11506, pp. 818–831, 2019.
https://doi.org/10.1007/978-3-030-20521-8_67

to anticipate the planning of the transport system [4, 5]. This fact is more important in the case of perishable freight, because delays or bottlenecks could be supposed losses of the cargo value.

During last few decades, numerous forecasting approaches have been used for the problem of traffic flows prediction. These problems have been addressed by times series modelling approaches such statistical methods and artificial neural networks (ANN). Generally, these methods develop their predictions on historical data analysis and present good performance when the time series vary temporally. Nowadays, statistical methods have been used for comparison purposes whenever a new forecasting approach is proposed [6–9]. Due that, freight flows are usually non-linear functions of exogenous variables ANN-based methods have commonly used to solve freight-forecasting problems [4, 10–12].

In recent times, Support Vector Machines (SVMs) have been introduced in the context of traffic flow forecasting and have been extensively adopted in pattern recognition. SVM for regression (SVR) was developed afterwards and applied successfully to solve short-term transport forecasting problems [13, 14].

Many authors employ the combinations of intelligence techniques with statistical methods to improve their forecasting. These hybrid or combined models based on intelligence techniques always outperform simple statistical models and single pattern models [15]. ANN and SVM are commonly combined with ARIMA, MLR or other statistical approaches. The use of combined models with ANNs are frequently focused to solve problems in transport engineering as transport mode choice [16] or passenger flows short-term forecasting [17]. Van der Voort et al. [18] combined Kohonen self-organizing maps (SOMs) with ARIMA. Ruiz-Aguilar et al. forecasted freight traffic in ports facilities combining statistical techniques in two different approaches: a hybrid approaches based in SARIMA and ANN [19] and in the second case, a three-step procedure with the combination of statistical (ARIMA+SOMs) and ANN models [20].

Peng [21] presented two statistical combination models among six univariate forecasting models proposed with the purpose to predict container throughput in the three major ports in Taiwan. Others works used SVR in hybrid approaches to solve container throughput forecasting. The employment of SVR with statistical methods achieves better forecasting performance than individual approaches [22]. SVR was also employed in combination with SOM to forecast maritime freight traffic at (port) Border Inspection Posts (BIPs) [23].

The methodologies namely above are very extended in forecasting time series in transportation and others fields. Despite of the advances in the improvement of the forecasting-development performance, this issue awakens a large interest in the transport research (or between transport researchers).

In this paper, we propose a three-stage hybrid approach for short-term Ro-Ro freight forecasting in the case of the Strait of Gibraltar. This paper seeks to show that forecasting could be a making decision tool in ports planning and management.

The remainder of the paper is organized as follow, in Sect. 2, the basic concepts of the ANN and SVR are briefly reviewed, Sect. 3 presents the database and the experiment procedure. Section 4 showed the results and the comparison of the performance with simple models. Finally, the concluding is exposed.

## 2 Methodology

### 2.1 Support Vector Regression (SVR)

This method is proposed by Vapnik [24] which is based on recent advances in statistical learning theory. SVM established the structural risk minimization principle (minimizing the fitting error and reducing the upper bound of the generalization error). Support Vector Regression (SVR) has been developed by incorporating Vapnik's *ε-intensive* loss function into SVMs to solve non-linear regression estimation problems. SVR approximates an unknown function in order to map the input data into a high-dimensional feature space through a nonlinear mapping function, a then a linear regression problem is constructed in this feature space. Given a set of data points $(x_1, y_1), (x_2, y_2), \ldots, (x_m, y_m)$ where $x_m, y_m \in \mathbb{R}$ are the input and output vector correspondingly and *m* is the total number of training samples. In the high-dimensional feature space, there theoretically exists a linear function, *f*, to formulate the non-linear relationship between input and output data. Such a linear regression function, namely SVR function, is as Eq.:

$$y = f(x, w) = w^T \varphi(x) + b \tag{1}$$

Where $f(x, w)$ is the forecasting values; $\varphi(x)$ denotes the Kernel function which maps the input $(x)$ to a vector in a feature space; the coefficients *w* (weight vector) and *b* (bias term) are adjustable and they are estimated by the minimizing the regularized risk function *R* (2):

$$R(w, \mathrm{b}) = \frac{1}{2}\|w\|^2 + \frac{C}{N}\sum_{i=1}^{N} |y_i - f(x_i, w)|_\varepsilon \tag{2}$$

where *C* is the regularization parameter and determines the trade-off between the flatness of the $f(x)$ and the amount up to which deviations larger than ε are tolerated, $y - f(x, w)$ is the ε-insensitive loss function. Where ε denotes the maximum deviation allowing during the training, $C > 0$ decides the trade-off of generalization ability and training error. This problem can be solved by making use of the Karush-Kuhn-Tucker's conditions [25]. The performance of SVR is determined by the type of kernel function and the setting of kernel parameters. In this work three types of kernel function, namely linear, polynomial and Gaussian Kernels, are employed: Artificial Neural Networks.

Shallow Artificial Neural Networks (ANNs) are universal approximators [26] that can approximate a large class of functions with high accuracy. The capacity of learning examples based on historic data is probably the most important property of neural networks. ANNs do not require imposing any assumptions regarding the model building process. Instead, the network model is largely determined by the features of the database.

The configuration of the model depends of the number of the units in each layer. The input layer depends of the number of the past samples (lags) selected as inputs, the hidden layer is composed of different number of neurons and the output layer contains only one output, the forecasted value for the prediction horizon considered. All the neurons are connected between layers by weighted links and the relationship between the output ($y_t$) and the inputs ($y_{t-1}$, ..., $y_{t-1}$) could be expressed by equation:

$$y_t = w_0 + \sum_{j=1}^{M} w_j g\left(w_{0j} + \sum_{i=1}^{D} w_{ij} y_{t-i}\right) + e_t \quad (3)$$

Where $w_{ij}$ ($i = 0, 1, 2, ..., D; j = 1, 2, ..., M$) and $w_j$ ($j = 1, 2, ..., M$) are the matrix of connection weight. $D$ is the total number of the inputs unit and $M$ is the number of the neurons in the hidden layer. The selection of the number of hidden nodes ($M$) establishes the complexity of the network and could determine the best generalization error ($e_t$). In order to select the optimal number of $D$ and $M$ during learning phase, an experimental procedure is carried out. During the training phase, the training inputs can be well adjusted but the performance during the learning phase can result insufficient. This is called overfitting. In this work, the models were trained using Levenberg-Marquadt and an early stopping procedure has been applied to avoid overfitting.

### 2.2 Database

We had used a database provided by the Port Authority of Algeciras Bay where the daily traffic of Ro-Ro transport in the Strait of Gibraltar is recorded. The strait of Gibraltar is the main gate between North Africa and Europe for the Ro-Ro transport. The original database is composed of over three millions of records during eight full years We use in this work the fresh vegetable freight in the period mentioned above. Vegetable freight is the main kind of good in the Strait of Gibraltar. In the import of fresh freight, the full process in port may be the most efficient possible due to the value of the good decrease along the days.

We assess some features of the database by statistical analysis. An analysis of the stationarity of the time series was developed. The period with higher freight volume is during September and March and the rest of the year the volume is lowest during all the study period. Hence, the time series have seasonality behaviour.

## 3 Experimental Procedure

We propose a hybrid approach, which combined the strengths of ANN and SVR models. This hybrid approach was developed in three stages in order to improve the prediction of the daily vegetable freight in the logistic node of the Strait of Gibraltar. At the first stage (Stage-I), we developed a procedure to obtain a prediction for 7 days ahead using a simple forecasting approach based on ANN or SVR separately. At the second stage (Stage-II), we obtained the best forecasting values of each prediction approach from Stage-I (ANN and SVR) and estimated the error with the real data. Afterwards, we selected the best models configuration applying several multicomparison test and finally an ensemble approach determined the best output and the error. At the third stage (Stage-III), we developed a model in order to correct the estimated error obtained in the Stage-II. ANN and SVR are used in the stages I and III alternatively in a way that in this work is determined the hybrid approach ANN-SVR. When in the first stage is used ANN, SVR is used to correct the error. Conversely, the hybrid approach SVR-ANN uses in the first stage SVR and ANN in the second stage. Finally, the results of both hybrid approaches (ANN-SVR and SVR-ANN) are compared to establish the best approach (Fig. 2).

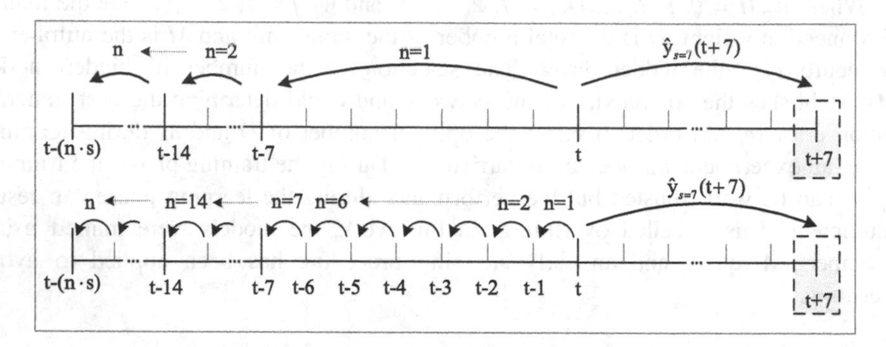


**Fig. 1.** Lag selection schemes

**Stage-I: Simple Forecasting Approach (ANN-SVR)**

We have used two methods (ANN and SVR) to forecast the future values seven days ahead. The aim of this stage is to obtain the best prediction values of the whole time series for each simple forecasting methods described above (ANN and SVR). These predicted values are used with two different purposes: on the one hand, as input of the Stage-II and, on the other hand, as benchmark of the forecast model proposed in this work. In order to select the best configuration of the hyperparameters for each method, we used an iterative grid search procedure. The ranges used in this work are collected in Table 1.

**Table 1.** Hyperparameters used in ANNs and SVR models.

| | Stage | ANN | SVR |
|---|---|---|---|
| Time horizon | I/III | 7 | 7 |
| Number of lags | I | 1, 7 | 1, 7 |
| | III | 1, 7 | 1, 7 |
| Hidden layer *(nhhiden)* | | 1:5, 10, 15, 20 | |
| Size of autoregressive window | | 1:4, 7, 14, 21, 28, 52 | 1:4, 7, 14, 21, 28, 52 |
| Kernel function | | | Polynomial; Linear; Gaussian* |
| C | | | 0.05,0.5,1:5,10:10:50 |
| Epsilon (ε) | | | 2–8, 2–7, …, 2–2 |
| *Gamma (γ) | | | 2–8, 2–7,…, 2–2 |

In the case of the ANN-SVR hybrid approach, ANNs was used as basic forecasting model and the number of hidden neurons was selected using a resampling procedure using the same subsets of training (20 times) and two-fold crossvalidation (2-CV).

In the case of SVR-ANN where SVR is the model applied at the Stage-I, the performance depends on the combination of several parameters: $C$, $\varepsilon$ and $\gamma$ (in the case of Gaussian kernel) the values of these parameters are shown in the Table 1.

**Stage-II: Multicomparison and Ensemble Approach**

Stage-II determines the configuration of the model (and the models without significantly differences) which obtain the best performance using multicomparison tests. Afterwards, in Stage-II an ensemble using results of the best configuration and the results of the models not significantly different have been applied.

In order to select the configuration of the best performance model, a multicomparison approach is developed. A Friedman/LSD Fisher test [28] compare the accuracy of the models analysed in order to resolve if the models have significant differences between them [29]. Friedman test is the non-parametric alternative to the ANOVA test. Friedman test assess the mean and variance of the performance indexes, with a probability lower than 0.05, and determines the existence of significant difference when the null hypothesis is rejected. This test is carried out when the assumptions of the ANOVA test (normality and homogeneity of variance) are not achieved.

Once the null hypothesis is rejected, the post-hoc LSD Fisher test determines which models are not equivalent. LSD Fisher test is a post-hoc method, which computed the error of the difference between two means and compared it with LSD. The LSD test is expressed as:

$$\left|\bar{y}_i - \bar{y}_j\right| \geq t_{\alpha/2}\sqrt{S_w^2\left(\frac{2}{n}\right)} \Rightarrow \text{Reject } \mathrm{H}_0 \tag{4}$$

Where $t_{\alpha/2}$ is the t value for a significance level of $\alpha/2$, in this case $\alpha = 0.05$, $n$ is the number of observation for each model and $S_w^2$ is the variance of the residual error.

Each selected ANN model has been repeated 30 times in order to calculate average measurements.

Finally, in Stage-II, the outputs of the best models selected by the multicomparison test, are improved by an ensemble approach. The accuracy of the results of the best model and the models without significantly differences generally is similar, therefore, determination of the best overall model is not an easy task which model is the best overall. Basically, the ensemble is the combination of individual outputs from several models in order to improve the results of each model obtained individually. The ensemble has the advantage of providing better generalization error than the mean of the generalization error of the single members.

The outputs obtained in the proposed ensemble are based in the follow six operators: Mean, Median, Kernel Density Function, Multiple Linear Regression, Least Square Boosting and Bagging.

**Stage-III: Correction Error and Hybrid Forecasting**

The hybrid approach assumes that the time series can be decomposed as the sum of the prediction values obtained in Stage-II and the estimation of their residual $e_t$ at time $t$ obtained in Stage-II too. Then, the experimental procedure used in Stage-III is applied to the prediction of the residual $e_t$. The use of ANNs or SVR depend on the method used in the forecasting in Stage-I. As consequence, the residuals can be predicted as

$$\hat{e}_{t+nh} = f(e_t + e_{t-s}, \ldots, e_{t-n \cdot s}) + \varepsilon_t \quad (5)$$

Where $f$ is the nonlinear function determined by SVR or ANNs, respectively, $nh$ is the prediction horizon (in this case is 7 days), $s$ is the size of the step used to select the past values in a past window of size $n$. The values of these parameters are shown in Table 1.

The challenge of this stage is to correct the error (residual) between the values obtained in the second stage and the target values. Thus, the two forecasting method employed in this stage are applied over the error in order to capture the nonlinear relationship of the time series,

Finally, the prediction is obtained by the sum of the terms obtained in the Stage-II (Lt) and the predicted error in the third stage. That is showed in Fig. 2. The proposed hybrid approach for Ro-Ro freight forecasting.

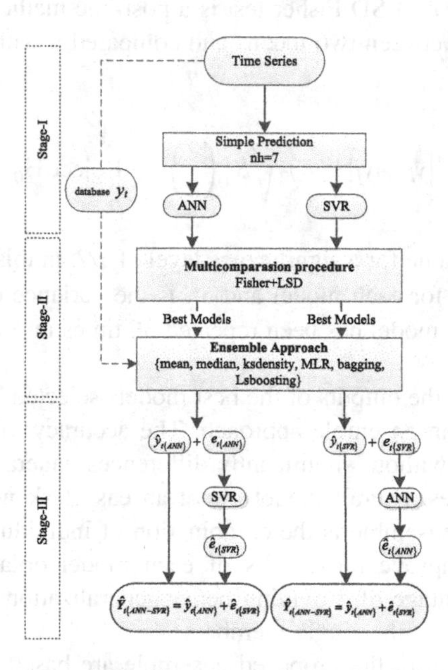


**Fig. 2.** The proposed hybrid approach for Ro-Ro freight forecasting.

# 4 Results

In this section, the prediction performance of the proposed hybrid approaches is presented, compared and interpreted. The results represent the forecasting values of the daily vegetable freight by Ro-Ro transport which are imported in the Port of Algeciras Bay. In this work, we estimated 7 days of prediction horizon. For a clear discussion, the results achieved are presented and discussed for each stage. In order to evaluate the prediction performance of the proposed models, standard correlation coefficient (*R*), agreement index (*d*), root mean squared error (*RMSE*) and symmetric mean square error (*sMAPE*) have been the quality indexes employed. The equations of these performance criteria are shown as equations in Table 2, where *O* is the observed value, *F* the forecasted value and *N* the total number of data observations.

**Table 2.** Performance criteria

| | |
|---|---|
| $R = \frac{\sum_{t=1}^{N}(O_t - \bar{O})(F_t - \bar{F})}{\sqrt{\sum_{t=1}^{N}(O_t - \bar{O})^2} \cdot \sqrt{\sum_{t=1}^{N}(F_t - \bar{F})^2}}$ | $d = 1 - \frac{\sum_{t=1}^{N}(F_t - O_t)^2}{\sum_{t=1}^{N}(\lvert F_t - \bar{O}\rvert - \lvert O_t - \bar{O}\rvert)^2}$ |
| $MAPE = \frac{1}{N}\sum_{t=1}^{N}\left\lvert\frac{O_t - F_t}{O_t}\right\rvert \times 100\%$ | $RMSE = \sqrt{\frac{1}{N}\sum_{t=1}^{N}(O_t - F_t)^2}$ |

Besides the prediction and the performance assessment, the performance indexes were taken into account to validate the results and to analyze the improvement of the three-stage process.

Besides the prediction and the performance assessment, these performance indexes were taken into account to validate the results and to analyze the improvement of the three-stage process. The final performance indexes have been computed to unseen samples (test sets) in order to check out the generalization capabilities of the different approaches.

Finally, in order to select the best forecasting values, the accuracy of the models is assessed by Diebold Mariano test [27]. The results achieved are assessed throughout the three stages in order to analysed the improvement of the results.

## 4.1 Stage-I. Simple Prediction

In the first stage, we analyzed the prediction of a simple forecasting model in two approaches. Both prediction approaches were applied over the whole database to obtained the best prediction models, using the last year of the database to test the models. The range of the parameters used in the prediction in the Stage-I are collected in the Table 1. This table defines the different parameters in Stage-I, both for ANN and for SVR. In both cases, the time series inputs are defined by an iterative process. The inputs vary in function of the steps number or temporal leaps. In this work, we define two temporal leaps for one day and for seven days Fig. 2. The proposed hybrid approach for Ro-Ro freight forecasting. explains the architecture of the inputs.

The performance indexes d and R at Stage-I suggest slightly better results in ANNs regarding SVR in steps of 1 day and 7 days. If we analyze the different kernel used in SVR polynomic kernel offers better performance indices than lineal or Radial Basic Function. The values of the performance indices obtained in Stage-I are shown in Table 3.

**Table 3.** Performance index in Stage-I in the ANN and SVR approaches.

| Steps | Index | ANN | SVR(Lineal) | SVR(Polynomic) | SVR(RBF) |
|---|---|---|---|---|---|
| s = 1 day | d | **0,9275** | 0,7704 | 0,9220 | 0,9192 |
| | R | **0,8750** | 0,6211 | 0,8593 | 0,8512 |
| | sMAPE | 0,3664 | 0,6505 | **0,3535** | 0,3644 |
| s = 7 days | d | **0,9380** | 0,7708 | 0,9216 | 0,9188 |
| | R | **0,8898** | 0,6216 | 0,8533 | 0,8504 |
| | sMAPE | 0,3676 | 0,6404 | **0,3538** | 0,3643 |

### 4.2 Stage-II. Multicomparison and Ensemble Procedure

In order to select the best configuration of the Stage-I a multicomparison approach was carried out for each proposed models (ANN and SVR). This multicomparison approach consists in the application of two statistical test (Friedman+LSD) to the models and resulting in the models with no significant differences with the optimal one and therefore resulting in a group of best models. The statistical tests were applied on the performance values obtained in the Stage-I for each configuration proposed. We have selected the best ten configurations for each performance index.

In the case that there are no significant differences, Occam's razor's criterion should be used. Occam's razor is the principle that states a preference for simple theories: "Accept the simplest explanation". We use the predictions outputs of the best configurations as inputs of the ensemble procedure. Finally, an ensemble of the best configurations selected after the multicomparison improves the performance in comparison of the simple model forecasting (benchmark models) in all analyzed cases. In the same way as Stage-I, ANNs outperform SVR in the performance of the stage-II. The ensemble approach is divided in two phases, where for each configuration selected has been applied 30 repetitions. In order to obtain optimal values, an iterative process (mean, median, ksDensity and RML) is applied for the repetitions. The optimal values obtained for the repetitions were used as inputs in the ensemble second phase. In the second phase we add two techniques to the iterative process, Bagging and LSBoosting.

In the case of the ANNs with s = 1, the best configuration of the ensemble is (mean, bagging). And for ANN with s = 7 the optimal configuration is (RML, bagging). In this sense, Bagging offer the best performance in the ensemble approach when we use ANNs.

On the other hand, the proposed model based on SVR obtained the best values when MLR is applied for all kernels (lineal, polynomic and RBF) and all steps (s = 1 and s = 7). For the second phase of the ensemble the SVR models with s = 1 obtained theirs best performance with Bagging, KsDensity and MLR for Linear, polynomic and

RBF kernels respectively. For SVR models with s = 7 the models obtained the best performance for bagging with linear and polynomic kernels and MLR in RBF. The performance of the ensemble is quite good in a general way. In all cases improve the performance in comparison of the simple models. The results of the ensemble indicate that ANNs models have slightly better performance than SVR in all of the performance indexes analyzed. The values of the first and the second Stage are collected in Table 4. The difference between the results of each stage denotes that the ensemble approach improves the performance of the prediction in all performance indexes and in both steps tested ($s = 1$ and $s = 7$). The best models in Stage-II are ANN in comparison of the SVR models. The best performance for SVR are produced for the lineal kernel for $s = 1$ and $s = 7$.

**Table 4.** Performance index after ensemble procedure in the two proposed approaches ANN and SVR

| Steps | Index | ANN | SVR(Lineal) | SVR(Polynomic) | SVR(RBF) |
|---|---|---|---|---|---|
| s = 1 day | d | 0,9479 | 0,9357 | 0,9248 | 0,9251 |
| | R | 0,9043 | 0,88 | 0,8642 | 0,87 |
| | sMAPE | 0,2833 | 0,3320 | 0,3554 | 0,3573 |
| s = 7 days | d | 0,9475 | 0,9304 | 0,9253 | 0,9238 |
| | R | 0,9038 | 0,8751 | 0,8653 | 0,8636 |
| | sMAPE | 0,2858 | 0,3412 | 0,3568 | 0,3601 |

## 4.3 Stage-III. Error Prediction

In order to determine the best results, a third stage is proposed. A hybrid model approach where the residual (error) is calculated with the difference between the real values and the predicted values of the best configuration for ANN and the best configuration for SVR in the first stage. Therefore, we can obtain an error for ANN ($e_{ann}$) and for each kernel of SVR ($e_{svr\text{-}lin}$; $e_{svr\text{-}pol}$; $e_{svr\text{-}pol}$). The hybrid models used a different technique to obtain the error prediction, in such a way, the forecasting of the $e_{\mathrm{ann}}$ is carried out with SVR and we call ANN-SVR (kernel) hybrid model and the forecasting of each $e_{svr}$ is calculated with ANNs call SVR(kernel)-ANN hybrid model. In the application of the SVR in both hybrid models is considered the three kernels explained above.

The procedure to forecast the error is the same as the original time series. Firstly, the error is predicted using a simple forecast model, secondly the best models are selected by a multicomparison procedure, and those models selected are the inputs to the final ensemble procedure and then, the best accurate error is added to the best forecasting value obtained in the Stage-II. Therefore, at the third stage obtain the best prediction of the error (in each case, using ANNs and SVR). Tables 5 and 6 show the values of the performance indexes.

The analysis of the results reveals that the overall best model is ANN7-SVR$_{(\mathrm{pol})7}$. This approach obtains the best performance in the three indexes analysed. The use of steps of seven days offers better results than the best models with steps of one-day.

**Table 5.** Performance indexes values for Stage III in models with step of one day at the Stage-I.

| Model | Model No. | Steps stage-III | D | R | sMAPE | RMSE |
|---|---|---|---|---|---|---|
| ANN-SVR(lin) | 1 | s = 1 | 0,9411 | 0,8931 | 0,3209 | 460.104,09 |
| | 2 | s = 7 | 0,9416 | 0,8935 | 0,3194 | 463.537,73 |
| ANN-SVR(pol) | 3 | s = 1 | 0,9416 | 0,8939 | 0,3233 | **458.920,45** |
| | 4 | s = 7 | 0,9418 | 0,8935 | 0,3196 | 463.224,57 |
| ANN-SVR(RBF) | 5 | s = 1 | 0,9418 | 0,8936 | 0,3207 | 459.382,94 |
| | 6 | s = 7 | **0,9418** | **0,8939** | **0,3195** | 462.364,72 |
| SVR(lin)-ANN | 7 | s = 1 | 0,9380 | 0,8880 | 0,3300 | 471.266,05 |
| | 8 | s = 7 | 0,9404 | 0,8899 | 0,3290 | 470.126,23 |
| SVR(pol)-ANN | 9 | s = 1 | 0,9293 | 0,8700 | 0,3544 | 502.775,51 |
| | 10 | s = 7 | 0,9354 | 0,8792 | 0,3462 | 495.268,80 |
| SVR(RBF)-ANN | 11 | s = 1 | 0,9314 | 0,8773 | 0,3658 | 490.422,94 |
| | 12 | s = 7 | 0,9371 | 0,8830 | 0,3645 | 483.322,58 |

**Table 6.** Performance indexes values for Stage-III in models with step of seven days at the Stage-I.

| Model | Model No. | Steps Stage-III | d | R | sMAPE | RMSE |
|---|---|---|---|---|---|---|
| ANN-SVR(lin) | 13 | s = 1 | 0,9477 | 0,9036 | 0,2869 | 443.121,68 |
| | 14 | s = 7 | 0,9489 | 0,9058 | 0,2890 | 442.728,99 |
| ANN-SVR(pol) | 15 | s = 1 | 0,9477 | 0,9037 | **0,2859** | 442.617,45 |
| | 16 | s = 7 | **0,9490** | **0,9059** | 0,2890 | **442.457,68** |
| ANN-SVR(RBF) | 17 | s = 1 | 0,9419 | 0,8924 | 0,3238 | 462.228,76 |
| | 18 | s = 7 | 0,9421 | 0,8927 | 0,3218 | 465.493,52 |
| SVR(lin)-ANN | 19 | s = 1 | 0,9358 | 0,8807 | 0,3294 | 492.659,78 |
| | 20 | s = 7 | 0,9339 | 0,8770 | 0,3310 | 499.160,57 |
| SVR(pol)-ANN | 21 | s = 1 | 0,9353 | 0,8800 | 0,3499 | 492.274,34 |
| | 22 | s = 7 | 0,9335 | 0,8776 | 0,3538 | 495.852,96 |
| SVR(RBF)-ANN | 23 | s = 1 | 0,9343 | 0,8782 | 0,3500 | 495.464,41 |
| | 24 | s = 7 | 0,9343 | 0,8777 | 0,3509 | 497.412,00 |

On the other hand, the SVR models obtained better performance in the prediction of the error than ANNs. The hybridization of ANNs to predict values of the daily freight and the SVR to predict the error of the previous forecasting works better than the other approach where SVR predicts the freight and ANN the residuals.

Figure 3 shows the evolution of each performance index for the approaches proposed in this work. The values of each bar correspond to the best performance of each model in each steps. The figure shows the improvement throughout the stages of the proposed process, where the Stage-I, which is considered as benchmark model, obtains worse performance than the other stages. On the other hand, Stage-III obtains the best performance result in comparison with the other stages. The comparison between the two hybrid approaches indicates that ANN-SVR hybrid model obtains the best performance in all the indexes proposed.

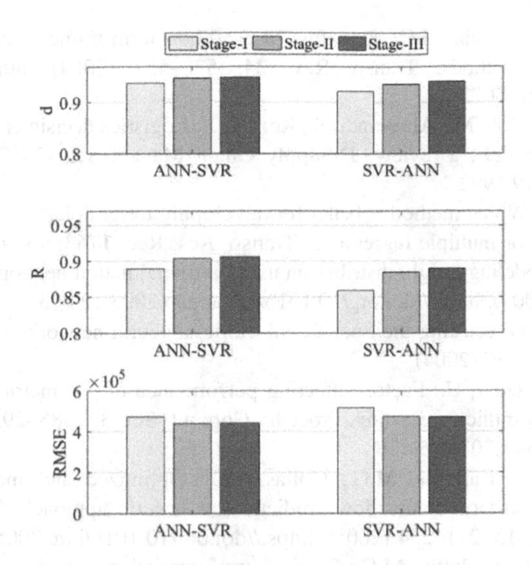


**Fig. 3.** Performance evolution and comparison of the best hybrid-models throughout the three stages.

## 5 Conclusions

An accurate prediction of Ro-Ro traffic in the Strait of Gibraltar is an important challenge in order to improve the operations management. This accurate prediction is achieved with the proposal ANN-SVR hybrid approach. This knowledge is an important making decision tool in ports environment and therefore in the Strait of Gibraltar This work reveals that the ANN-SVR hybrid approach is able to outperform results obtained by either of the models used separately.

The comparison is carried out with *R*, *d* and *sMAPE* as performance index. The results suggest that ANN-SVR hybrid approach is a promising methodology to forecast Ro-Ro freight in ports environment. The contribution of this work is to provide a useful knowledge for improve ports planning and management.

## References

1. Bilegan, C.I., Crainic, T.G., Gendreau, M.: Forecasting freight demand at intermodal terminals using neural networks - an integrated framework. Eur. J. Oper. Res. **13**, 22–36 (2008)
2. Romero, G., Durán, G., Marenco, J., Weintraub, A.: An approach for efficient ship routing. Int. Trans. Oper. Res. **20**, 767–794 (2013). https://doi.org/10.1111/itor.12021
3. Farhan, J., Ong, G.P.: Forecasting seasonal container throughput at international ports using SARIMA models. Marit. Econ. Logist. **20**, 131–148 (2018). https://doi.org/10.1057/mel.2016.13

4. Vlahogianni, E.I., Golias, J.C., Karlaftis, M.G.: Short-term traffic forecasting: overview of objectives and methods. Transp. Rev. **24**, 533–557 (2004). https://doi.org/10.1080/0144164042000195072
5. Tavasszy, L., de Bok, M., Alimoradi, Z., Rezaei, J.: Logistics decisions in descriptive freight transportation models: a review. J. Supply Chain Manag. Sci. (2019). https://doi.org/10.18757/jscms.2019.1992
6. Al-Deek, H.M.: Which method is better for developing freight planning models at seaports - neural networks or multiple regression? Transp. Res. Rec. **1763**, 90–97 (2001)
7. Celik, H.M.: Modeling freight distribution using artificial neural networks. J. Transp. Geogr. **12**, 141–148 (2004). https://doi.org/10.1016/j.jtrangeo.2003.12.003
8. Mostafa, M.M.: Forecasting the Suez Canal traffic: a neural network analysis. Marit. Policy Manag. **31**, 139–156 (2004)
9. Ratrout, N.T., Gazder, U.: Factors affecting performance of parametric and non-parametric models for daily traffic forecasting. Procedia Comput. Sci. **32**, 285–292 (2014). https://doi.org/10.1016/j.procs.2014.05.426
10. Vlahogianni, E.I., Karlaftis, M.G., Golias, J.C.: Optimized and meta-optimized neural networks for short-term traffic flow prediction: a genetic approach. Transp. Res. Part C Emerg. Technol. **13**, 211–234 (2005). https://doi.org/10.1016/j.trc.2005.04.007
11. Vlahogianni, E.I., Karlaftis, M.G.: Testing and comparing neural network and statistical approaches for predicting transportation time series. Transp. Res. Rec. J. Transp. Res. Board. **2399**, 9–22 (2013). https://doi.org/10.3141/2399-02
12. Sun, S., Huang, R., Gao, Y.: Network-scale traffic modeling and forecasting with graphical lasso and neural networks. J. Transp. Eng. **138**, 1358–1367 (2012). https://doi.org/10.1061/(ASCE)TE.1943-5436.0000435
13. Hwang, W.-Y., Lee, J.-S.: A new forecasting scheme for evaluating long-term prediction performances in supply chain management. Int. Trans. Oper. Res. **21**, 1045–1060 (2014). https://doi.org/10.1111/itor.12098
14. Marković, N., Milinković, S., Tikhonov, K.S., Schonfeld, P.: Analyzing passenger train arrival delays with support vector regression. Transp. Res. Part C Emerg. Technol. **56**, 251–262 (2015). https://doi.org/10.1016/j.trc.2015.04.004
15. Karlaftis, M.G., Vlahogianni, E.I.: Statistical methods versus neural networks in transportation research: differences, similarities and some insights. Transp. Res. Part C Emerg. Technol. **19**, 387–399 (2011). https://doi.org/10.1016/j.trc.2010.10.004
16. Andrade, K., Uchida, K., Kagaya, S.: Development of transport mode choice model by using adaptive neuro-fuzzy inference system. Transp. Res. Rec. **1977**(1), 8–16 (2006)
17. Ma, Z., Xing, J., Mesbah, M., Ferreira, L.: Predicting short-term bus passenger demand using a pattern hybrid approach. Transp. Res. Part C Emerg. Technol. **39**, 148–163 (2014). https://doi.org/10.1016/j.trc.2013.12.008
18. Van Der Voort, M., Dougherty, M.S., Watson, S.: Combining Kohonen maps with ARIMA time series models to forecast traffic flow. Transp. Res. Part C Emerg. Technol. **4**, 307–318 (1996). https://doi.org/10.1016/S0968-090X(97)82903-8
19. Ruiz-Aguilar, J.J., Turias, I.J., Jiménez-Come, M.J.: Hybrid approaches based on SARIMA and artificial neural networks for inspection time series forecasting. Transp. Res. Part E Logist. Transp. Rev. **67**, 1–13 (2014). http://dx.doi.org/10.1016/j.tre.2014.03.009
20. Ruiz-Aguilar, J.J., Turias, I.J., Jiménez-Come, M.J.: A novel three-step procedure to forecast the inspection volume. Transp. Res. Part C Emerg. Technol. **56**, 393–414 (2015). https://doi.org/10.1016/j.trc.2015.04.024
21. Peng, W.Y., Chu, C.W.: A comparison of univariate methods for forecasting container throughput volumes. Math. Comput. Model. **50**, 1045–1057 (2009)

22. Geng, J., Li, M.-W., Dong, Z.-H., Liao, Y.-S.: Port throughput forecasting by MARS-RSVR with chaotic simulated annealing particle swarm optimization algorithm. Neurocomputing. **147**, 239–250 (2015). https://doi.org/10.1016/j.neucom.2014.06.070
23. Ruiz-Aguilar, J.J., Turias, I.J., Jiménez-Come, M.J.: A two-stage procedure for forecasting freight inspections at Border Inspection Posts using SOMs and support vector regression. Int. J. Prod. Res. **53**, 2119–2130 (2015). https://doi.org/10.1080/00207543.2014.965852
24. Vapnik, V.: Statistical Learning Theory. Wiley, New York (1998)
25. Schölkopf, B., Smola, A.J.: Learning with Kernels: Support Vector Machines, Regularization, Optimization, and Beyond. MIT Press, Cambridge (2002)
26. Hornik, K., Stinchcombe, M., White, H.: Multilayer feedforward networks are universal approximators. Neural Netw. **2**, 359–366 (1989)
27. Diebold, F.X., Mariano, R.S.: Comparing predictive accuracy. J. Bus. Econ. Stat. **13**, 253–263 (1995)
28. Fisher, R.A.: Statistical Methods and Scientific Inference. Hafner Publishing Co, Oxford (1956)
29. Kourentzes, N., Barrow, D.K., Crone, S.F.: Neural network ensemble operators for time series forecasting. Expert Syst. Appl. **41**, 4235–4244 (2014). https://doi.org/10.1016/j.eswa.2013.12.011

# Infering Air Quality from Traffic Data Using Transferable Neural Network Models

Miguel A. Molina-Cabello[1(✉)], Benjamin N. Passow[2], Enrique Dominguez[1], David Elizondo[2], and Jolanta Obszynska[3]

[1] Department of Computer Science, ETSI Informatica, University of Malaga, Málaga, Spain
{miguelangel,enriqued}@lcc.uma.es

[2] De Montfort University's Interdisciplinary Group in Intelligent Transport Systems (DIGITS), De Montfort University, Leicester LE1 9BH, UK
benpassow@ieee.org, elizondo@dmu.ac.uk

[3] Leicester City Council's Pollution Team, Leicester LE1 6ZG, UK
jolanta.obszynska@leicester.gov.uk

**Abstract.** This work presents a neural network based model for inferring air quality from traffic measurements. It is important to obtain information on air quality in urban environments in order to meet legislative and policy requirements. Measurement equipment tends to be expensive to purchase and maintain. Therefore, a model based approach capable of accurate determination of pollution levels is highly beneficial. The objective of this study was to develop a neural network model to accurately infer pollution levels from existing data sources in Leicester, UK. Neural Networks are models made of several highly interconnected processing elements. These elements process information by their dynamic state response to inputs. Problems which were not solvable by traditional algorithmic approaches frequently can be solved using neural networks. This paper shows that using a simple neural network with traffic and meteorological data as inputs, the air quality can be estimated with a good level of generalisation and in near real-time. By applying these models to links rather than nodes, this methodology can directly be used to inform traffic engineers and direct traffic management decisions towards enhancing local air quality and traffic management simultaneously.

**Keywords:** Neural network · Inferring pollution concentration · Air quality · Traffic management

## 1 Introduction

Detailed air quality simulation models have been developed for many cities. These mechanistic and deterministic models are used as tools to help understand the basic air quality and traffic management. They support legislative and


I. Rojas et al. (Eds.): IWANN 2019, LNCS 11506, pp. 832–843, 2019.
https://doi.org/10.1007/978-3-030-20521-8_68

policy requirements which depend on air quality in urban environments. Hardware equipment to measure air quality is normally expensive to purchase and maintain. Therefore, a model based approach capable of accurate prediction of pollution levels is highly beneficial.

This paper is organized as follows. First of all, the background is presented in Sect. 2. After that, the neural network model is reported in Sect. 3, while Sect. 4 exhibits the data sources. Then, the results are depicted in Sect. 5. Finally some conclusions are presented in the last section.

## 2 Background

City councils and local authorities across the world are managing and controlling urban traffic. This involves monitoring traffic conditions and designing, maintaining and controlling traffic signals. It also involves dealing with faults, roadworks, air quality, noise and traffic. Across Europe, a large number of research projects are related to air pollution and its impact on the quality of people's health.

Emissions from motor vehicles are a major source of air pollution. Despite the fact that air quality has improved significantly in the last decades, there is string scientific evidence suggesting that current levels of air pollution are a high risk for the environment and to human health [1,2].

Therefore, there is a need for air quality information in urban environments to meet legislative and policy requirements. Recent research in this topic using Computational Intelligence (CI) techniques can be categorized into two areas: Modelling and Forecasting.

### 2.1 Air Quality Forecasting

The ability to forecast air quality is becoming increasingly important as links are made between poor air quality and adverse health effects. With accurate forewarning of low air quality levels, it becomes possible to take preventive action thereby reducing the severity of the pollution incident.

There is a myriad of air quality forecasting methods. In particular, Artificial Neural Networks (ANNs) have been employed for this purpose. For example, an ANN with weather parameters and traffic conditions developed by Viotti et al. [17] was used to forecast pollutant concentrations (benzene, carbon monoxide, $NO_2$, $NO_x$, Ozone) at both 1 h and 24 h intervals into the future. They point out that different models are required for each pollutant type and forecasting interval.

A good illustration of this can be found in a comparison study, where five different ANN models, one linear statistical model and one deterministic modelling system were assessed in terms of their capability to forecast urban $NO_2$ and $PM_{10}$ concentrations in Helsinki [7]. The ANN models were generally found to perform better than the linear and deterministic models, especially for forecasting $PM_{10}$ which is considered more difficult to predict than $NO_2$. However, it is pointed out that ANNs are limited to the forecast window and location for which they were trained, and that they require retraining if either of these parameters change.

Another example is presented in Ibarra-Berastegi et al. [6], where the use of neural networks is investigated for providing air quality forecasts up to 8 hours in advance for five different pollutants ($SO_2$, $CO$, $NO_2$, $NO$ and $O_3$) at six different locations in Bilbao, Spain. They report a best case for $NO_2$ with $R^2 = 0.88$.

Other option could be the use of a Genetic Algorithm to optimise the design of an ANN. Niska et al. [11] apply a model of this kind for predicting $NO_2$ levels in Helsinki. It was found that any improvements offered by the Genetic Algorithm were negligible when compared with a reference model. Moreover, the use of neural networks optimised by genetic algorithms for predicting $PM_{10}$ concentrations at 4 locations in the Greater Athens area is presented in [4]. When compared with linear regression methods, the neural networks were found to be superior, with $R^2$ in the range 0.50–0.67 compared to 0.29–0.35 for linear regression.

Different neural network topologies are also considered. The performance of three different neural network topologies (MLP, RBF and Modular-ANN) for predicting roadside concentrations of $CO$ and $NO_2$ in Melton Mowbray, UK, are compared in [18]. The transferability of these models is then considered by applying them to data from another site in Leicester, UK, concluding that whilst performance is degraded, it is still able to provide a useful prediction capability.

### 2.2 Air Quality Modelling

The behaviour of pollution dispersion, especially in urban areas is complex and difficult to model. Dispersion is influenced by many factors such as temperature, wind speed, wind direction, wind strength and terrain, all of which exhibit significant levels of uncertainty on any mathematical models used. The application of CI methods to such problems, especially those that do not require an underlying model (such as neural networks and fuzzy sets) is therefore becoming the focus of much ongoing research. Some examples of recent work are summarised here.

Pérez-Roa et al. [14] use a neural network to predict the eddy diffusivity ($K_v$), an important parameter used in the study of pollution dispersion, for which values supplied in literature are often inaccurate. Based on the neural network predicted values of $K_v$, a dispersion model was able to forecast peak carbon monoxide concentrations with improved accuracy.

Neural networks as a filter for the concentration levels produced by an air pollution model were applied by Pelliccioni and Tirabassi [13] to account for disagreement between the measured and predicted values. Their results demonstrated the efficiency of Neural Networks.

The modelling of the nitrogen dioxide ($NO_2$) dispersion phenomena using the ANN technique was also assessed by Nagendra and Khare [10]. A satisfactory performance of the ANN-based $NO_2$ models on the evaluation data set is shown.

Another example of the successful application of ANN to calculate the average spatial distribution of air pollutants based on diffusive sampling measurements in Cyprus was demonstrated in Pfeiffera et al. [15]. Their results illustrate the application of ANN resulting in realistic maps of the annual average distribution of $NO_2$ in Cyprus.

In addition to ANN, fuzzy sets provide another tool in dealing with uncertainty in dispersion modelling. Fisher [3] reviewed the various uncertainties existing in dispersion modelling, and highlighted the feasibility of fuzzy approaches to environmental decisions. Another fuzzy based system for predicting modelling air quality was proposed by [16], where the use of a trapezoidal membership function was proposed. The model is based on data collected over a year using 5 locations in Tehran.

Furthermore, the feasibility of using ANNs in predicting air quality based on the SCOOT data (stops, flows, congestions and delays) per link-intersection and the weather conditions (temperature, wind speed and direction) was performed in a preliminary study at the University of New Castle by Bell et al. [18]. The results indicate that this approach might be a promising option.

In this research we propose to expand this work by providing a global against a local approach for modelling air quality which can be easily adapted to new locations. Our model uses a combination of all the SCOOT data (combining all the links) and the Climate data as inputs. Another novelty of this study consists on the use of a Self-Organizing Map (SOM) neural network to filter the dataset.

## 3 Neural Network Model

Of the neural network algorithms available, two of the most common ones are the multilayer perceptron (MLP) for supervised learning and the Kohonen self organising maps for unsupervised learning. In supervised learning, the inputs are presented to the neural network and the output produced y the neural network is compared with the desired output. After this, the weights of the neural network are adjusted taking into account any error in the output pattern. In unsupervised learning, the weight adjustments are not made based on comparison with some target output. Instead, the neural network self organises its weights.

The proposed neural model to infer the air quality is based on a MLP of feedforwad network, since this kind of neural networks is widely used for the function fitting problem.

The proposed MLP uses the back propagation algorithm and it consists on one single hidden layer of simply interconnected neurons. Both the input and output layers are determined by the problem. In this case, the output layer is composed by an only neuron, which is used for the air quality. The input layer is composed by nine neurons: five for the climate data (wind speed, wind direction, rain, radiation and air pressure) and four for SCOOT data (stops, delays, flows and congestions). The output of a neuron is given by the following expression:

$$y_i = f(\sum_{j=0}^{N_i} w_{ji} x_j + b_i) \tag{1}$$

where $N_i$ is the number of inputs of the neuron $i$, $w_{ji}$ is the synaptic weight between the $j$-th input and the neuron $i$, $b_i$ is the bias and $f$ is the transfer function. In the hidden layer, a tangential sigmoid function:

$$f(x) = \frac{sinhx}{coshx} = \frac{e^x - e^{-x}}{e^x + e^{-x}} = \frac{e^{2x} - 1}{e^{2x} + 1} \tag{2}$$

is used as transfer function, although a pure linear function is used in the output layer.

Generally, only one hidden layer is required to approximate any measurable function [5,8]. The optimum number of neurons in the hidden layer depends on the desired accuracy and a trial and error approach is normally used to determine it. In our experiments, a small hidden layer composed by only ten neurons was sufficient to achieve good results.

The learning of the proposed MLP is supervised, hence a training data set is required. This training data set consists of a set of pairs, that is, an input vector (climate and scoot data) and an associated target (the air quality). During the training, the synaptic weights in the neural network are adjusted until the output produced by the network matches the target within a certain error, which it is used for the weights adaptation according to the learning algorithm. In this case, the backpropagation algorithm is proposed to minimise the mean squared error.

Overtraining of the neural network is an undesirable characteristic, which may occur when the neural model learns too much the details of the training data. Therefore, the results produced by network are poor generalization capability when new data are presented. A bad choice of the training data set or an inappropriate learning algorithm typically impair the network's generalization capability and lead to model overfitting. In this sense, a validation data set was used during the training to avoid the overtraining of the neural network and to check the generalization performance. Training is stopped when the performance on the validation data starts to decline.

## 4 Data Sources

In a long term collaboration with Leicester City Council's Traffic and Air Quality Teams a large amount of data from near real time feeds has been collected over the past years. This data includes:

- Traffic detection via inductive road loops
- Air quality measurements from road-side monitors
- Meteorological measurements from a mast/measurement station

The data collected from street loops is commonly noisy and often corrupted. The level of performance achieved with an ANN model is directly linked with the quality of the data sets. This data is often surrounded by errors which can be caused by noise and equipment malfunction. These errors are referred to as outliers, which need to be identified and eliminated from the data sets in order to improve the models performance. In this work a Self-Organizing Map (SOM) is proposed for filtering and handling outliers [12].

### 4.1 Road Traffic Condition Data

Road traffic related data has been collected on an ongoing basis from Leicester City Council's traffic management systems. Specifically, a large number of SCOOT M02 messages have been collected via FTP. SCOOT uses raw induction loop data to measure and estimate four traffic flow and congestion related parameters [9]. These parameters are reported every 15 min via M02 system messages.

The four parameters collected by every link are flow, delay, stops and congestion. *Flow* is estimated as vehicles per hour. *Delay* is estimated as total delay in vehicles per hour. *Stops* is estimated as the number of vehicles that have stopped at least once on the link in the given period. *Congestion* is measured as percentage of 4 second intervals when a detector is occupied by traffic. This data has been accumulated and used as hourly means in order to match the time-base of the air quality and meteorological measurements. Two possibilities have been addressed in this work:

- Aggregated links. Road traffic data from all links is previously aggregated and then accumulated for every site. Therefore, a fixed number of four traffic inputs is established. Note that this option is needed to build transferable models.
- Separated links. Road traffic data is separately accumulated for every link and site. Consequently, the number of traffic inputs is four times the number of links composing every site.

### 4.2 Air Quality Data

A number of pollutants are continuously being monitored and recorded by Leicester City Council's Air Quality team. The ratified data for $NO_2$ and PM10 has been used for this study from the following sites:

- Abbey Lane
- Uppingham Road
- Glenhills Lane
- AURN, New Walk Centre
- Imperial Av/Narborough Rd
- London Road

Figure 1 shows a map with the locations of these monitors, which are commonly placed in critical sites (e.g. crossing roads) composed by several link roads.

### 4.3 Meteorological Data

A meteorological mast in Leicester (location, see 1) automatically provided weather data as an hourly average. A NRT stream of meteorological data (met data) was used as part of the model to determine and forecast the emission level at a given location. There were 6 indicators for weather condition in the model (see Table 1).

**Table 1.** Meteorological measurements available and used in this work.

| Parameter | Unit of measurement |
|---|---|
| Cloud coverage | Amount of sunshine reaching the surface [0 → 1] |
| Air pressure | millibar |
| Temperature | degrees Celsius |
| Rain (amount of) | millimetres per hour |
| Wind direction | degrees from North |
| Wind speed | metres per second |

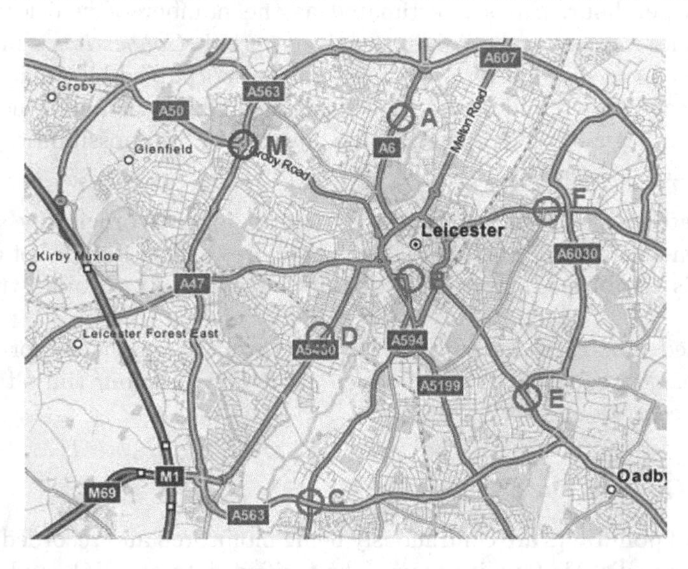


**Fig. 1.** Map of Leicester City, UK, with locations of air quality measurement stations (A to F) and meteorological mast (M). The labels represent the following measurement stations: A = Abbey Lane, B = AURN, New Walk, C = Glenhills Way, D = Imperial Avenue, E = London Road, F = Uppingham Road.

## 5 Results

Several experiments have been carried out. Three different proposals have been considered. The first one presents a single ANN model with two outputs ($NO_2$ and PM10), and with SOM based filtering. The second proposal is composed by two separate ANN models, each one with one output ($NO_2$ or PM10), and without SOM based filtering. The third approach consists of two separate ANN models, each one with one output ($NO_2$ or PM10), and with SOM based filtering.

Some well-known measures have been selected in order to compare the selected proposals from a quantitative point of view. The traditional R and $R^2$ measures are the main ones. If the R performance is positive that means a positive correlation exists, where higher is better. On the other hand, $R^2$ is a positive value, where higher is better. The root mean square error (RMSE), the mean square error (MSE) and the mean absolute error (MAE) are also taken into account, where lower is better.

Tables 2, 3 and 4 exhibit the performance of the three different considered models, respectively. As it can be observed, the third considered proposal (two separate ANN models with one output and without SOM based filtering) offers the best performance of the three approaches. It must be highlighted that each proposal works with a similar performance in each studied site.

**Table 2.** Results for a single ANN model with 2 outputs, $NO_2$ and PM10, and with SOM based filtering

| | $NO_2$ | | | | | | PM10 | | | | | |
|---|---|---|---|---|---|---|---|---|---|---|---|---|
| | Separated links | | | Aggregated links | | | Separated links | | | Aggregated links | | |
| Site | R | RMSE | MAE | R | RMSE | MAE | R | RMSE | MAE | R | RMSE | MAE |
| Abbey Lane | 0.65 | 5.05 | 3.42 | 0.60 | 5.27 | 3.64 | 0.61 | 7.04 | 4.84 | 0.39 | 9.13 | 5.59 |
| AURN, New Walk | 0.36 | 7.81 | 5.49 | 0.42 | 7.36 | 5.38 | 0.47 | 5.93 | 4.41 | 0.45 | 6.53 | 4.71 |
| Glenhills Way | 0.81 | 6.17 | 4.69 | 0.74 | 6.82 | 4.75 | 0.65 | 8.12 | 5.62 | 0.53 | 11.05 | 6.74 |
| Imperial Avenue | 0.77 | 2.84 | 2.12 | 0.79 | 2.70 | 1.97 | 0.60 | 6.39 | 4.34 | 0.50 | 8.20 | 4.67 |
| London Road | 0.78 | 4.12 | 2.97 | 0.78 | 4.22 | 3.18 | 0.70 | 6.22 | 4.34 | 0.62 | 7.53 | 4.61 |
| Uppingham Road | 0.68 | 4.55 | 3.33 | 0.73 | 4.22 | 3.19 | | | | | | |
| Average | 0.68 | 5.09 | 3.67 | 0.68 | 5.10 | 3.69 | 0.61 | 6.74 | 4.71 | 0.50 | 8.49 | 5.26 |

**Table 3.** Results for two separate ANN models each with 1 output, $NO_2$ or PM10, and without SOM based filtering.

| | $NO_2$ | | | | | | PM10 | | | | | |
|---|---|---|---|---|---|---|---|---|---|---|---|---|
| | Separated links | | | Aggregated links | | | Separated links | | | Aggregated links | | |
| Site | R | RMSE | MAE | R | RMSE | MAE | R | RMSE | MAE | R | RMSE | MAE |
| Abbey Lane | 0.65 | 8.33 | 6.28 | 0.64 | 8.38 | 6.35 | 0.54 | 11.55 | 8.28 | 0.57 | 11.23 | 7.82 |
| AURN, New Walk | 0.33 | 13.77 | 10.65 | 0.34 | 13.71 | 10.59 | 0.51 | 10.97 | 7.38 | 0.53 | 10.79 | 7.14 |
| Glenhills Way | 0.73 | 9.88 | 7.51 | 0.66 | 10.79 | 8.25 | 0.57 | 12.97 | 9.43 | 0.52 | 13.42 | 9.83 |
| Imperial Avenue | 0.67 | 5.20 | 3.99 | 0.66 | 5.21 | 3.99 | 0.55 | 10.59 | 7.37 | 0.55 | 10.61 | 7.42 |
| London Road | 0.68 | 7.47 | 5.45 | 0.68 | 7.51 | 5.44 | 0.58 | 10.62 | 7.39 | 0.58 | 10.63 | 7.42 |
| Uppingham Road | 0.68 | 7.13 | 5.46 | 0.63 | 7.25 | 5.55 | | | | | | |
| Average | 0.62 | 8.63 | 6.56 | 0.60 | 8.81 | 6.69 | 0.55 | 11.34 | 7.97 | 0.55 | 11.34 | 7.92 |

Nevertheless, the site *AURN, New Walk Centre* is the most problematic. The $NO_2$ MSE of the different studied places are reported in Fig. 2. It is shown that *AURN, New Walk Centre* presents the highest MSE. In addition, Fig. 3 depicts the $NO_2$ regressions. It can be noticed that the $R^2$ performance is extremely low.

After the training of the proposed neural models and the evaluation of their performances, the input variables were examined in order to figure out the most important input to estimate the roadside concentration. A simple sensitivity analysis was carried out by increasing a certain percentage (in our case 5%) each

**Table 4.** Results for two separate ANN models each with 1 output, $NO_2$ or PM10, and with SOM based filtering.

| | $NO_2$ | | | | | | PM10 | | | | | |
|---|---|---|---|---|---|---|---|---|---|---|---|---|
| | Separated links | | | Aggregated links | | | Separated Links | | | Aggregated links | | |
| Site | R | RMSE | MAE | R | RMSE | MAE | R | RMSE | MAE | R | RMSE | MAE |
| Abbey Lane | 0.72 | 4.47 | 3.08 | 0.68 | 4.67 | 3.29 | 0.62 | 6.32 | 4.68 | 0.77 | 5.30 | 3.75 |
| AURN, New Walk | 0.50 | 7.21 | 4.97 | 0.53 | 6.46 | 4.72 | 0.69 | 4.53 | 3.21 | 0.61 | 5.29 | 3.79 |
| Glenhills Way | 0.86 | 5.18 | 3.85 | 0.81 | 5.72 | 4.26 | 0.63 | 8.38 | 5.67 | 0.50 | 10.88 | 6.31 |
| Imperial Avenue | 0.81 | 2.53 | 1.87 | 0.81 | 2.50 | 1.82 | 0.58 | 7.26 | 4.71 | 0.64 | 6.74 | 4.02 |
| London Road | 0.83 | 3.65 | 2.66 | 0.78 | 4.06 | 2.93 | 0.66 | 6.98 | 4.66 | 0.75 | 5.56 | 3.84 |
| Uppingham Road | 0.76 | 3.88 | 2.94 | 0.74 | 4.07 | 5.46 | | | | | | |
| Average | 0.75 | 4.49 | 3.23 | 0.73 | 4.58 | 3.75 | 0.64 | 6.69 | 4.59 | 0.65 | 6.75 | 4.34 |

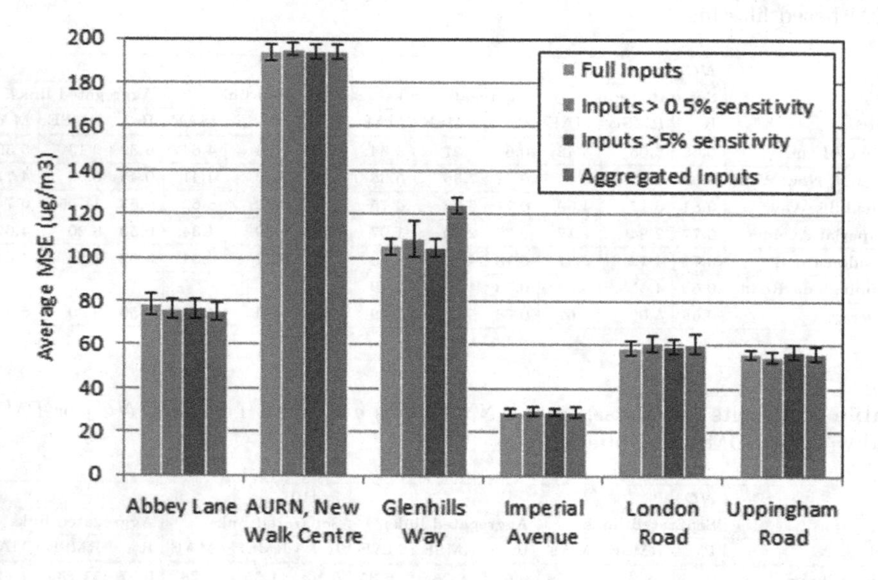


**Fig. 2.** $NO_2$ mean squared error.

of the input variables, and then calculating the change caused in the output. The sensitivity of each input is defined by $S(x) = 100\frac{\Delta V_{output}}{\Delta V_{input}}$ where $x$ is an input variable. Figure 4 shows the sensitivity of each input variable for each monitored location. Note that the wind direction (H) is not important in the estimation, since it is always below 5% for all stations. However, the air pressure (A) is very significant, except for the station placed at Imperial Avenue.

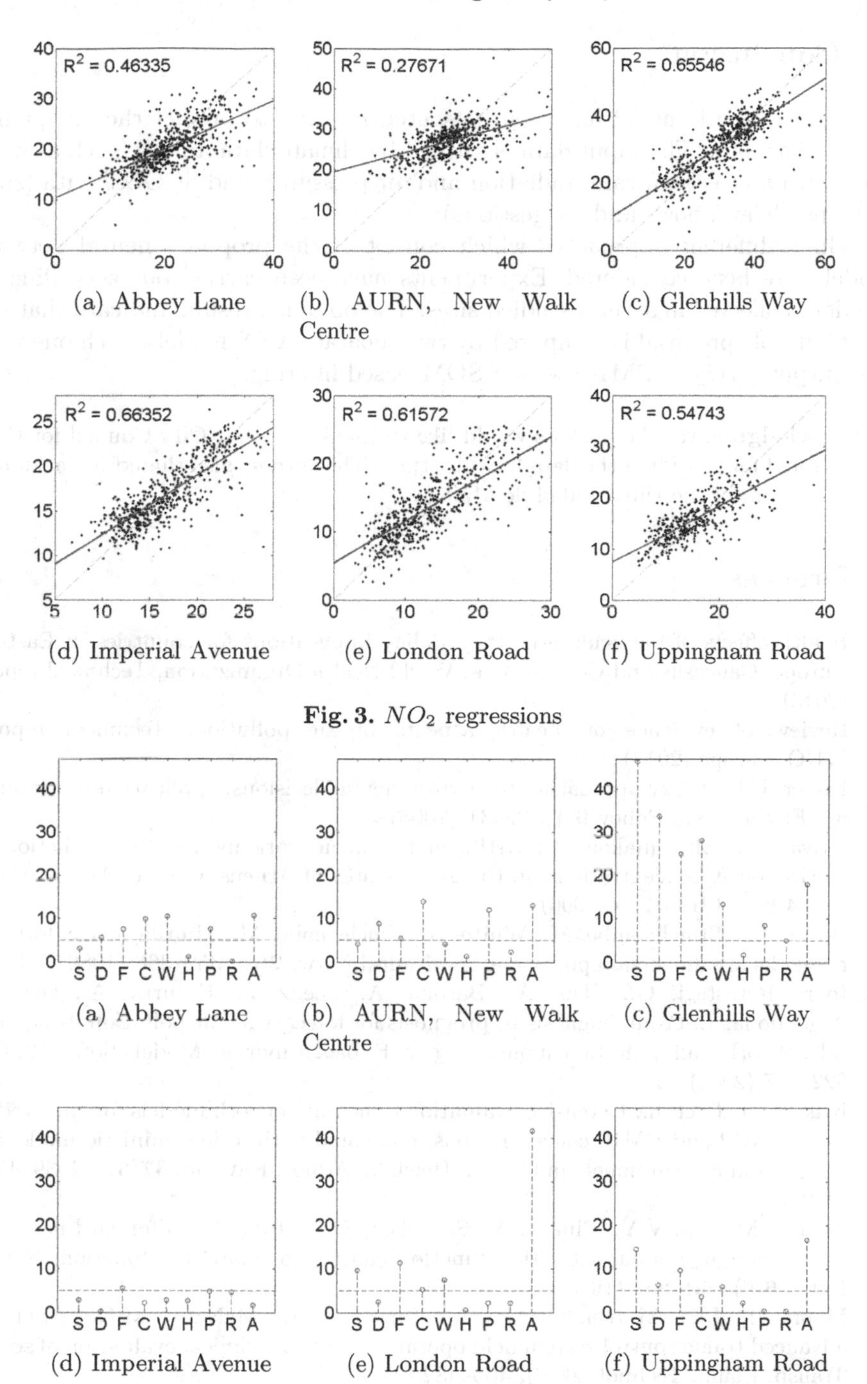


**Fig. 3.** $NO_2$ regressions



**Fig. 4.** Sensitivity analysis (%) of inputs by increasing 5%. The different inputs of the network are shown in horizontal axis: Stops, Delays, Flows, Congestions, Wind speed, Wind direction (H), Precipitations (Rain), Radiation and Air pressure.

## 6 Conclusion

A neural network model has been presented in order to predict the air quality from traffic data. The input data is formed by climate data inputs (such as wind speed, wind direction, rain, radiation and air pressure), and SCOOT data (such as stops, delays, flows and congestions).

Three different approaches which consist of the proposed neural network model have been considered. Experiments have been carried out according to the data collected in several studied sites. The obtained results indicate that the most suitable proposal is composed by two separate ANN models, each one with one output ($NO_2$ or PM10) with a SOM based filtering.

**Acknowledgment.** The authors would like to thank Leicester City Council for their support and for providing the data for this study. The authors also thankfully acknowledge the grant of the Universidad de Málaga.

## References

1. Health effects of particulate matter: Policy implications for countries in Eastern Europe, Caucasus and Central Asia. World Health Organization, Technical report (2013)
2. Review of evidence on health aspects of air pollution. Technical report, WHO/Europe (2013)
3. Fisher, B.E.: Fuzzy approaches to environmental decisions: application to air quality. Environ. Sci. Policy **9**(1), 22–31 (2005)
4. Grivas, G., Chaloulakou, A.: Artificial neural network models for prediction of PM10 hourly concentrations, in the greater area of Athens, Greece. Atmos. Environ. **40**(7), 1216–1229 (2006)
5. Hornik, K., Stinchcombe, M., White, H., Tinchcombe, M.: Multilayer feedforward networks are universal approximators. Neural Netw. **2**(5), 359–366 (1989)
6. Ibarra-Berastegi, G., Elias, A., Barona, A., Saenz, J., Ezcurra, A., Diaz de Argandoña, J.: From diagnosis to prognosis for forecasting air pollution using neural networks: air pollution monitoring in Bilbao. Environ. Model. Softw. **23**(5), 622–637 (2008)
7. Kukkonen, J., et al.: Extensive evaluation of neural network models for the prediction of NO2 and PM10 concentrations, compared with a deterministic modelling system and measurements in Central Helsinki. Atmos. Environ. **37**(32), 4539–4550 (2003)
8. Leshno, M., Lin, V.Y., Pinkus, A., Schocken, S.: Multilayer feedforward networks with a nonpolynomial activation function can approximate any function. Neural Netw. **6**(6), 861–867 (1993)
9. Moore II, J.E., Mattingly, S.P., MacCarley, C.A., McNally, M.G.: Anaheim advanced traffic control system field operations test: a technical evaluation of scoot. Transp. Plann. Technol. **28**(6), 465–482 (2005)
10. Nagendra, S.S., Khare, M.: Artificial neural network approach for modelling nitrogen dioxide dispersion from vehicular exhaust emissions. Ecol. Model. **190**(1–2), 99–115 (2006)

11. Niska, H., Hiltunen, T., Karppinen, A., Ruuskanen, J., Kolehmainen, M.: Evolving the neural network model for forecasting air pollution time series. Eng. Appl. Artif. Intell. **17**(2), 159–167 (2004)
12. Passow, B.N., Elizondo, D., Chiclana, F., Witheridge, S., Goodyer, E.: Adapting traffic simulation for traffic management: a neural network approach. In: 16th International IEEE Conference on Intelligent Transportation Systems (ITSC 2013), pp. 1402–1407. No. ITSC/IEEE, October 2013. http://ieeexplore.ieee.org/document/6728427/
13. Pelliccioni, A., Tirabassi, T.: Air dispersion model and neural network: a new perspective for integrated models in the simulation of complex situations. Environ. Model. Softw. **21**(4), 539–546 (2006)
14. Pérez-Roa, R., Castro, J., Jorquera, H., Pérez-Correa, J., Vesovic, V.: Air-pollution modelling in an urban area: correlating turbulent diffusion coefficients by means of an artificial neural network approach. Atmos. Environ. **40**(1), 109–125 (2006)
15. Pfeiffera, H., Baumbacha, G., Sarachaga-Ruiza, L., Kleanthousb, S., Poulidab, O., Beyaz, E.: Neural modelling of the spatial distribution of air pollutants. Atmos. Environ. **43**(20), 3289–3297 (2009)
16. Sowlat, M.H., Gharibi, H., Yunesian, M., Mahmoudi, M.T., Lotfi, S.: A novel, fuzzy-based air quality index (FAQI) for air quality assessment. Atmos. Environ. **45**(12), 2050–2059 (2011)
17. Viotti, P., Liuti, G., Di Genova, P.: Atmospheric urban pollution: applications of an artificial neural network (ANN) to the city of Perugia. Ecol. Model. **148**(1), 27–46 (2002)
18. Zito, P., Chen, H., Bell, M.: Predicting real-time roadside CO and $NO_2$ concentrations using neural networks. IEEE Trans. Intell. Transp. Syst. **9**(3), 514–522 (2008)

# Deep Learning Based Ship Movement Prediction System Architecture

Alberto Alvarellos[1(✉)], Andrés Figuero[2], José Sande[2], Enrique Peña[2], and Juan Rabuñal[3]

[1] RNASA Group, Computer Science Department, Research Center on Information and Communication Technologies, University of A Coruña, Elviña, 15071 A Coruña, Spain
alberto.alvarellos@udc.es

[2] Water and Environmental Engineering Group (GEAMA), Center of Technological Innovations in Construction and Civil Engineering, University of A Coruña, Elviña, 15071 A Coruña, Spain

[3] RNASA Group, Computer Science Department, Center of Technological Innovations in Construction and Civil Engineering, University of A Coruña, Elviña, 15071 A Coruña, Spain

**Abstract.** In this work we present the software architecture used to implement a ship movement prediction system based on a deep learning model. In previous works of the group we recorded the movement of several cargo vessels in the Outer Port of Punta Langosteira (Spain) and created a deep neural network that classifies the vessel movement given the vessel dimensions, the sea state and weather conditions. In this work we present the architectural design of a software system that allows to deploy machine learning models and publish the results it provides in a web application. We later use this architecture to deploy the deep neural network we have mentioned, creating a tool that is able to predict the behavior of a moored vessel 72 h in advance. Monitoring the movement of a moored vessel is a difficult and expensive task and port operators do not have a tool that predicts whether a moored vessel is going to exceed the recommended movements limits. With this work we provide that tool, believing that it could help to coordinate the vessel operations, minimizing the economic impact that waves, tides and wind have when cargo vessels are unable to operate or suffer damages. Although we use the proposed system architecture for solving a particular problem, it is general enough that it could be used for solving other problems by deploying any machine learning model compatible with the system.

**Keywords:** Vessel movement prediction · System architecture · Deep learning · Node-RED · Anaconda · Tensorflow · Python

## 1 Introduction

The Spanish state-run port system activity accounts for 1.1% of the Spanish GDP and nearly 20% of the transport sector GDP, handling nearly 60% of exports and 85% of imports. This accounts for 53% of Spanish foreign trade with the European Union and 96% with third countries. It also employs more than 35000 workers directly and around


I. Rojas et al. (Eds.): IWANN 2019, LNCS 11506, pp. 844–855, 2019.
https://doi.org/10.1007/978-3-030-20521-8_69

110000 indirectly [1]. These characteristics, the fact that Spain is the European Union country with the longest coastline (8000 km) and the closest European country to the axis of one of the world's major maritime routes, reveal the importance of ports in the Spanish and European economy.

There are several sea states and meteorological conditions that can disrupt the regular port operations. These conditions can have a significant economic impact due to the importance of the port system in the Spanish economy. The events that can produce such disruptions are waves, wind, currents etc. They are classified in extreme and mean conditions, depending on their severity. The only way to reduce the impact of extreme conditions is to reinforce the existing infrastructures or to build new ones and regular port operations are rarely possible in these conditions. In mean meteorological conditions the port must ensure that the regular vessel operations can be carried out safely.

A port operability is its ability to be kept in a safe and reliable functioning condition, according to predefined operational requirements. The Spanish port authorities are subject to several European and international regulations that establish these requirements [2, 3]. These regulations are summarized in a document [4], hereinafter referred to as ROM, published by the Spanish State Port System. This document sets the mandatory maritime-port safety requirements, seeking to guarantee minimum vessel navigation and maneuverability conditions in port waters.

A usual way to quantify a port operability is to measure the movements of vessels moored in it. These movements are affected by the sea state and meteorological conditions in the port, therefore the lower the effect the meteorological conditions have on vessel movements during operations inside the port, the greater the operability of the port. There are many variables that could potentially influence the movement of a vessel, such as the sea level, the waves height and period, the wind speed and direction, the vessel type, the vessel orientation, the mooring system, the distribution of cargo, etc. These variables have complex interactions and the relationship between all of them is not known.

We can measure a vessel movement following two approaches: using physical modeling or measuring it in a real environment. The first approach has several limitations [5, 6]. In a previous work, using the second and more precise approach, we measured several vessels at the facilities of the Outer Port of Punta Langosteira (A Coruña), Spain, located 10 km southeast of the city of A Coruña [5, 6].

In another work (currently in revision) we created a model, based on a deep neural network, that classifies a vessel movement given several sea state and meteorological variables and vessel characteristics. This classification is given in the form of an output that indicates if the movement of the ship is going to exceed the limits established by ROM. The model achieved a 91% accuracy on the test set.

The model input variables not related to vessel characteristic were chosen considering that they would have to be available as sea state and weather forecasts. This was done in order to be able to achieve what we are presenting in this work: we have built a tool based on a deep neural network that allows to predict, up to 72 h in advance, if the movement of a vessel is going to exceed the limits that the ROM regulation establishes. The purpose of this tool is to contribute to optimize the operating system of the port by helping to decide, before a vessel is operating, if the port operators should take special measures or reschedule the vessel loading and unloading of cargo in order to optimize the port operability.

## 2 System Architecture

The aim of this work was to create a production-ready deep learning based tool for predicting a moored vessel movement up to 72 h in advance. We also designed the system architecture having several aspects in consideration: the tool should provide the prediction in a timely manner, it should be simple to use, and it should be based on open source software.

People that work in the civil engineering field, and more precisely in maritime and port engineering field, are accustomed to using computational modeling for solving civil engineering problems. This type of modeling has been used for modeling the movement of a vessel, but the results obtained with this approach are not satisfactory. Also, a moored vessel is a complex system, so computational modeling requires extensive computational resources: each time we want to predict how a vessel is going to behave in the future (using a forecast for the sea state and weather conditions) we would have to run a resource and time-consuming simulation. We think that using a deep learning approach trained with real data of vessel movements is going to provide better results than numerical modeling and in a timely manner, much faster than those provided by numerical modeling.

We also think that creating an easy to use tool would further increase its acceptance by the maritime and port engineering community. Computational modeling often requires that many parameters must be tuned and used when doing a simulation. Our tool only requires the user to provide some vessel characteristic in order to provide the prediction: the vessel length and its breadth.

We designed the system to be a web tool. In software engineering such tools can be separated by concerns between the presentation layer (frontend), and the data access layer (backend). The whole system architecture can be seen in Fig. 1.

There are three main tasks that need to be considered when designing a system that uses a machine learning model in a production environment, i.e. a machine learning model that is used in a real environment once trained:

- The interaction with the user. In our system this is done via the frontend.
- Accessing the data that the model needs to make predictions or classifications. In our system this is done mainly in the backend by the *data collection microservice* and partially in the frontend for the data that needs to be imputed by the user.
- Handling all the operations related to the model: model instantiation, preparing the data for the model, retrieving the model outputs, etc. In our system this is done in the backend by the *prediction microservice*.

The functionality of each layer and microservice is explained in the following sections.

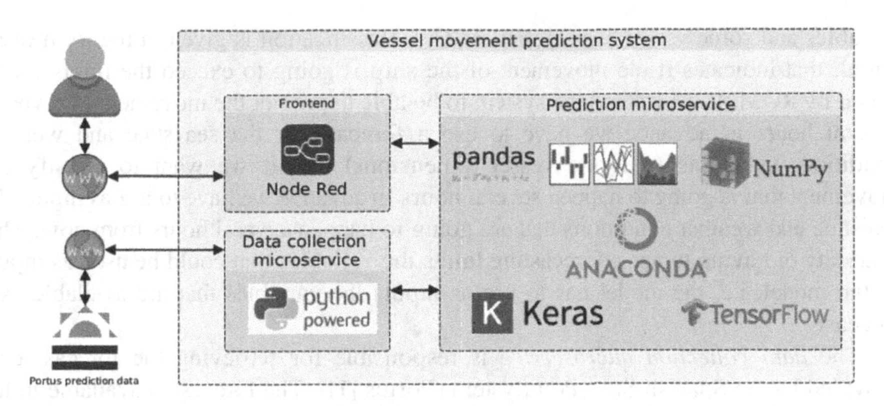


**Fig. 1.** Vessel movement prediction system architecture.

## 2.1 Backend

We have used a microservice architecture [7] for developing the backend functionality. This approach allows to separate the backend concerns into different microservices. This separation allows us to make changes or improve each microservice independently, making the system easier to maintain. It also allows us to reuse the microservices in other projects that need the functionality of any of the microservices.

The backend is composed by two microservices (see Fig. 1): one in charge of collecting and preparing some of the data needed by the deep neural network to make predictions (the sea state and weather forecasts) named *data collection microservice* and another, named *prediction microservice*, responsible for handling the deep neural network and using the data provided by the *data collection microservice* and the data provided by the user, as inputs for the model in order to obtain a prediction. The functionality of each microservice is explained in the following sections.

**Data Collection Microservice**

This microservice, provided in the form of a Python package, is responsible for collecting and preparing some of the data needed as input of the deep neural network. All the input data the network needs can be classified in three categories: the vessel dimensions, the sea state and the weather conditions. The *data collection microservice* is responsible for providing the sea state and weather conditions. The information regarding the vessel dimensions (vessel length (**L**) and breadth (**B**)) has to be provided by the user via the frontend (see Sect. 2.2). The ship geometric characteristics have a high relevance in its dynamic behavior, so using certain vessel characteristics allows the deep neural network to create different internal representations for different vessel types, making the prediction model more precise. These two variables were chosen from all possible vessel characteristic because it is public information that can be easily obtained when we want to use the system for making a prediction. Other vessel characteristics that govern the vessel behavior, such as cargo weight, cargo distribution, mooring configuration, etc., are more difficult or practically impossible to obtain.

As explained before, the deep neural network we are using in our system was created to classify a vessel movement given several sea state and meteorological variables and some vessel characteristics. This classification is given in the form of an output that indicates if the movement of the ship is going to exceed the limits established by ROM. If we wanted the system to be able to predict the movement of a vessel several hours in advance we have to use a forecast for the sea state and weather conditions as inputs (plus the vessel dimensions), i.e. if we want to classify the movement that is going to happen several hours in advance we have to use as inputs the sea state and weather conditions that are going to happen several hours from now. This necessity of having to use a forecasting limits the variables that could be used as inputs of the model, i.e. the model has to use as inputs the variables that are available as a forecast.

The *data collection microservice* is responsible for retrieving the forecast data provided by the Spanish State Port System (Portus [1]). The forecast is available in the SIMAR data set [8], which comprises several time series of wind and wave parameters from numerical modeling. From this data, we used the following sea state and meteorological variables as inputs of the deep neural network:

- $\mathbf{H_0}$: sea level, with respect to the zero of the port.
- $\mathbf{H_s}$: significant wave height, i.e. the mean of the highest third of the waves in a time-series of waves representing a certain sea state.
- $\mathbf{T_p}$: peak wave period, i.e. the period of the waves with the highest energy (extracted from the spectral analysis of the wave energy).
- $\boldsymbol{\theta_m}$: mean wave propagation direction, i.e. the mean of all the individual wave directions in a time-series representing a certain sea state.
- $\mathbf{W_s}$: mean wind speed.
- $\mathbf{W_d}$: mean wind direction.

The deep neural network has been trained using historical real data for the previous variables. The sea state data is provided by the REDCOS [9] and REMPOR [10] data sets and the weather condition is provided by the REDMAR [11] data set. The REDCOS data set is made up of the measures coming from the Coastal Buoy Network of the Spanish Port System. The REDMAR data source provides measures taken from the Spanish Port System Network of Tide Gauges. The REMPOR data source provides measures taken from the Spanish Port System Meteorological Stations Network. Each device has a unique identification number in the previous networks.

The model was trained using historical data of three measurement devices from these networks: a weather station, a directional wave buoy and a tide gauge. The weather station is located at the end of the port main breakwater (outer port of a Coruña, Spain). The directional wave buoy is located 1.8 km off the main breakwater of the port (43º 21′ 00″ N 8º 33′ 36″ W). This buoy has the code 1239 in the REDCOS network. The tide gauge has the code 3214 on the REDMAR network.

As explained before, the data the deep neural network uses for making predictions is provided by the SIMAR data source. Being this data source the outputs of a computational model, it provides the forecast for coordinates belonging to points in a mesh. We were careful in choosing the points for the SIMAR forecast to coincide with the coordinates of the devices the deep neural network was trained on.

The *data collection microservice* accesses the SIMAR data source, downloads the appropriate data, parses it and makes the necessary transformations (such as cleaning errors, eliminating NaN, etc.)[1]. It returns a Pandas [12] Data Frame as an output.

**Prediction Microservice**

The prediction microservice has been implemented as a RESTful web service, i.e. it conforms to the REST architectural style. This allows to easily increase the service functionalities as new REST operations and also allows the service to be used by any frontend technology that can consume REST services. Its functionality was implemented as a Flask application [13] that is used in conjunction with the Apache HTTP Server and the mod_wsgi [14] module. This architecture allows to create a web service that is able to use the Conda package and environment management system [15] and the Anaconda scientific distribution [16]. Using Conda allows to create separate python environments for different microservices, where in each environment we can install only the libraries that are needed by the microservice, avoiding conflicts with others. The libraries installation and dependencies resolution are also managed by Conda. Using Anaconda gives access to thousands of Python Scientific libraries that makes the development and maintenance of the system easier.

The prediction microservice uses the data collection microservice to obtain the sea state and weather forecasts. Then it scales the data, to match the distribution of the data used during the deep neural network training, using the stored data of a scikit-learn [17] Standard Scaler created with the dataset used for training the deep neural network. Once the data is prepared, the microservice uses Keras [18] with a Tensorflow [19] backend for loading the previously trained deep neural network that classifies the vessel movements. The data is then used as input for the model and the result is provided in form of a json object. This object contains both the sea state and weather forecast as well as the deep neural network output with the vessel movements predictions.

Regarding the network training, there were some hyperparameters that were fixed during all the training process. Some of them were imposed by the problem nature and other were chosen because the literature suggests certain optimal values. In this manner, all the hidden units use the ReLU activation function [20] and the output unit uses a sigmoid activation function (imposed because we have a binary classification problem). Another fixed hyperparameter is the batch size. We had 759 training examples, so we did not use minibatches in the training process.

We used an iterative approach for tuning the network hyperparameters. We used this approach because the hyperparameter space for our model is large, and even a grid search in a small subset of hyperparameters results in a training time of several months.

The iterative approach was implemented through two techniques. We tested several hyperparameters values in a grid search and chose the ones that had the best results. In the next iteration, we searched for other hyperparameters values in a similar manner. When searching for a given hyperparameter, the rest were kept with the default values

---

[1] Although the data is publicly available, as required by Spanish law, it cannot be exploited easily: it is provided in the form of an HTML table inside a frame. We had to use a sophisticated HTML scraper that uses a WebDriver (a virtual web browser) in order to process the HTML and AJAX of the SIMAR web.

or the best values if we already searched for that hyperparameter. In each iteration, the search was done using a parameter grid and cross validation. The other technique is searching for hyperparameters in a coarse grid first and then, in the next iteration, search in a finer grid focusing on the range of values that gave the best results in the coarse search. The grid search was done using the Keras wrappers for the Scikit-Learn API. All the grids in each iteration were created using cross validation with a 10-fold cross validation set.

For those hyperparameters that were not fixed, the values tested using this iterative approach were the network architecture, the weights (kernel) initialization, the optimizer, the learning rate, the training iterations and the weights regularization.

In Table 1 we can see the results for the final iteration of this iterative approach. The mean values in each row correspond to the averaged values over the 10 folds of the cross validation set. The results are ordered in descending order of cross validation test score (accuracy).

**Table 1.** Final iteration model hyperparameter tuning results

| Rank | Mean fit time (s) | Mean val. score | Mean train score | Dropout ratio | Kernel constraint |
|---|---|---|---|---|---|
| 1 | **333** | **0.9144** | **0.9652** | **0.56** | **None** |
| 2 | 413 | 0.9051 | 0.9564 | 0.60 | 10 |
| 3 | 463 | 0.8986 | 0.9501 | 0.66 | 5 |
| 3 | 482 | 0.8986 | 0.9470 | 0.66 | 10 |
| 5 | 384 | 0.8959 | 0.9587 | 0.60 | None |
| 6 | 398 | 0.8946 | 0.9611 | 0.60 | 5 |
| 7 | 446 | 0.8933 | 0.9488 | 0.66 | None |
| 8 | 262 | 0.8920 | 0.9816 | 0.46 | 5 |
| 8 | 369 | 0.8920 | 0.9584 | 0.56 | 15 |
| 10 | 287 | 0.8906 | 0.9731 | 0.50 | None |
| 11 | 503 | 0.8893 | 0.9534 | 0.66 | 15 |
| 12 | 271 | 0.8880 | 0.9791 | 0.46 | 10 |
| 12 | 323 | 0.8880 | 0.9713 | 0.50 | 15 |
| 14 | 311 | 0.8841 | 0.9750 | 0.50 | 10 |
| 15 | 236 | 0.8827 | 0.9880 | 0.40 | 10 |
| 15 | 356 | 0.8827 | 0.9640 | 0.56 | 10 |
| 17 | 251 | 0.8814 | 0.9805 | 0.46 | None |
| 17 | 344 | 0.8814 | 0.9656 | 0.56 | 5 |
| 19 | 244 | 0.8788 | 0.9886 | 0.40 | 15 |
| 19 | 429 | 0.8788 | 0.9586 | 0.60 | 15 |
| 21 | 277 | 0.8775 | 0.9794 | 0.46 | 15 |
| 22 | 221 | 0.8762 | 0.9908 | 0.40 | None |
| 23 | 160 | 0.8722 | 0.9987 | 0.00 | 5 |
| 24 | 300 | 0.8696 | 0.9744 | 0.50 | 5 |
| 25 | 229 | 0.8643 | 0.9908 | 0.40 | 5 |
| 26 | 169 | 0.8577 | 0.9996 | 0.00 | 10 |
| 27 | 178 | 0.8551 | 0.9981 | 0.00 | 15 |
| 28 | 151 | 0.8472 | 0.9990 | 0.00 | None |

We can see that the best model achieved a 96.52% training accuracy and a 91.44% validation accuracy in the cross validation set. This model was selected as the final model. The selected hyperparameters values of the final model for all the iterations are the following:

– **Network architecture**: 8 input units, 50 elements in the first hidden layer, 50 elements in the second hidden layer, 50 elements in the third hidden layer and 1 output unit. All layers are feed forward and fully connected.
– **Activation function**: ReLU in all units in hidden layers, sigmoid in the output unit.
– **Optimizer**: Adam optimizer with $\beta\_1 = 0.9$ and $\beta\_2 = 0.999$.
– **Learning rate**: 0.001.
– **Kernel initialization**: He Uniform in the hidden layer and random uniform in the output layer.
– **Regularization**: Dropout regularization in all hidden layers (except output), with dropout rates per layer [0.56, 0.56, 0.56] and no kernel constraints.

A representation of the best model can be seen in the Fig. 2, where the input variables, hidden, dropout and output layers are represented. Each hidden layer, in grey color, has 50 elements.

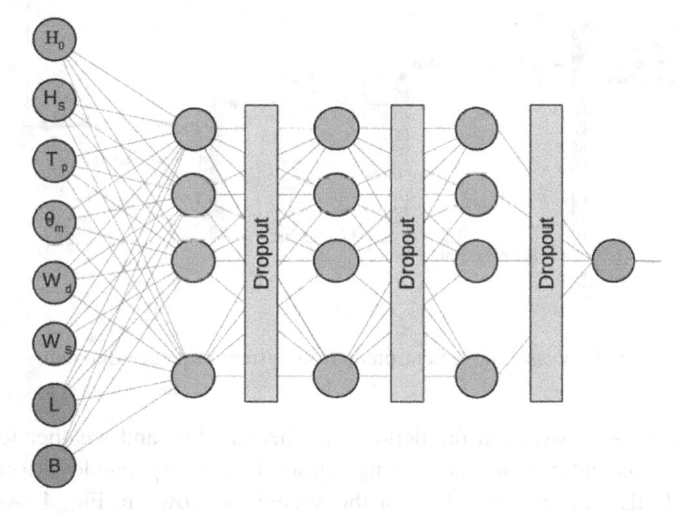


**Fig. 2.** The deep learning model architecture used to predict the ship movements.

## 2.2 Frontend

In a system that uses a machine learning model in a real environment (*in production*) the frontend (presentation layer) is responsible for asking the user for necessary data and presenting back the results of the model. The user of the system could be a human or another system in the production environment. In our system the frontend was implemented using the open source solution Node-RED [20]. This flow-based programming tool allows us to describe the frontend behavior as a network of *nodes*, where each node has a well-defined purpose: given some input data the node processes it and then outputs the results. The network is responsible for the flow of data between the nodes.

In our case, the user (human) needs to interact with the system through the frontend in order to provide the vessel dimensions (**L**ength and **B**readth). This data input in the frontend is the event that triggers the flow of information through all the microservices that provides an output to the user. A Node-RED node handles this data input. Its output is then processed by another node that calls the *prediction microservice*, using these values as arguments. The *prediction microservice* uses the *data collection microservice* to retrieve the sea state and weather forecasts. Once obtained, it joins the forecasts with the vessel dimensions and uses this data as input of the deep neural network to obtain a prediction. The prediction and the forecasts are then returned to the frontend. This communication between the frontend and microservices is represented in the sequence diagram shown in Fig. 3.

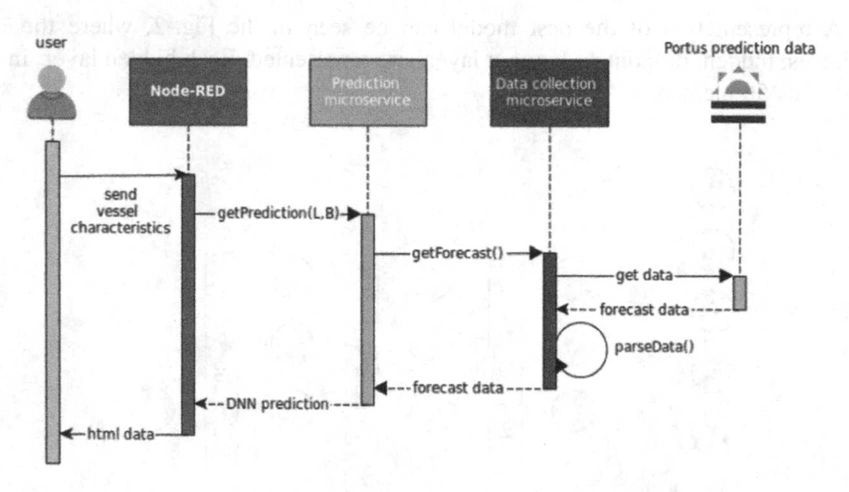


**Fig. 3.** Vessel movement prediction system sequence diagram.

Once the vessel movement prediction and the sea state and weather forecasts are obtained, the frontend exposes them using a node that is responsible of formatting the results as HTML. The *user interface* of the system is shown in Fig. 4. As explained before, the prediction of the deep neural network is given in the form of an output that indicates if the movement of the ship is going to exceed the limits established by ROM. This output is provided by a sigmoid function, with values between 0 and 1. The system shows this value to the user and prints the table cell in red if the limit is greater than 0.5, indicating that the movement is going to exceed the ROM limit, and green otherwise.

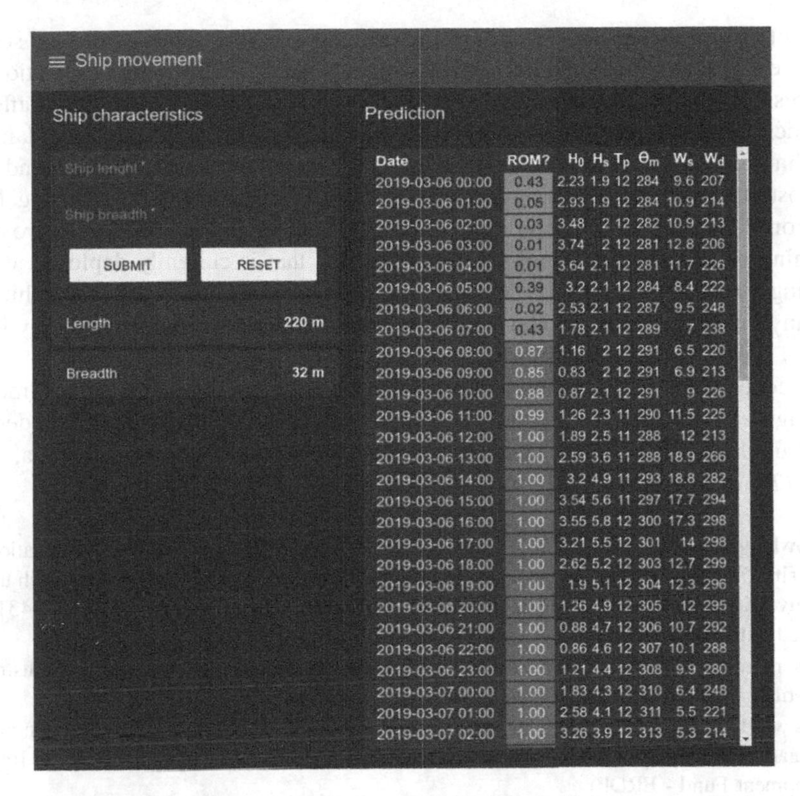


**Fig. 4.** The user interface of the system frontend. (Color figure online)

## 3 Results and Conclusions

The aim of this work was to design the architecture and create a production-ready deep learning based web tool for predicting a moored vessel movement up to 72 h in advance. We also designed the tool having several aspects in consideration: the tool should provide the prediction in a timely manner, it should be simple to use, and it should be based on open source software.

The results of this work provide an architectural design for any system that uses a machine learning model for solving any problem. We used this architecture to implement a system that uses a deep neural network for predicting the movements of moored vessels in the outer port of A Coruña, Spain. The tool can be accessed using the following url: http://ojo.citic.udc.es:1881/ui/.

The system was designed to be a web application and to use the microservices architectural design. Its frontend was implemented using Node-RED and its backend is divided in two microservices: one responsible for providing some of the data needed for the deep neural network model and one responsible for handling the model and providing the outputs of the model in form of a REST service. We chose a REST approach because we plan to implement more microservices in the future that will constitute a port management tool. This architectural design ensures that new services can be easily incorporated into the *prediction microservice* as a REST operation.

This separation allows to improve any microservice individually without affecting the others or substituting one microservice with another that uses a different language or technology but provides the same service interface. For instance, the frontend could be substituted with another frontend using a web framework such as Spring MVC, Ruby on Rails, Django, etc. The deep neural network model can also be improved by retraining it with more data and substitute the one that is currently deployed without affecting the other microservices. This separation also allows us to use this architecture with any machine learning model available in the Python ecosystem trough any library such as scikit-learn, Keras, Tensorflow, etc.

In conclusion, we have obtained a robust system that allows to deploy machine learning models and show its results in a web application. We have used it to deploy a model used for predicting the vessel movements on the outer port of A coruña, Spain, up to 72 h in advance.

**Acknowledgments.** This work has been funded by the Ministry of Culture Education and University Organization, aid for the consolidation and structuring of competitive research units of the University System of Galicia of the Xunta de Galicia and Singular Centers (ED431G/01) endowed with FEDER funds of the EU.

This research has also been funded by the Spanish Ministry of Economy, Industry and Competitiveness, R&D National Plan, within the project BIA2017-86738-R.

This work has received financial support from the Xunta de Galicia (Centro singular de investigación de Galicia accreditation 2016–2019) and the European Union (European Regional Development Fund - ERDF).

This work has been partially financed by the GERIA-TIC Project, a project co-funded by the Galician Innovation Agency (GAIN) through the PEME Connect Program (3rd edition) (IN852A 2016/10) and EU FEDER funds, the Collaborative Integration Project of Genomic data (CICLOGEN). Data mining techniques and molecular docking for analysis of integrative data in colon cancer. "Funded by the Ministry of Economy, Industry and Competitiveness. Galician Network of Colorectal Cancer Research (REGICC) ED431D 2017/23, Galician Network of Medicines (REGID) ED431D 2017/16".

## References

1. Spanish Port System Homepage. http://www.puertos.es/en-us/nosotrospuertos/Pages/Nosotros.aspx. Accessed 10 Mar 2018
2. MarCom Working Group 115: Criteria for the (Un)loading of Container Vessels. http://www.pianc.org/edits/articleshop.php?id=2012115
3. Llorca, J., Gonzalez Herrero, J.M., Ametller, S.: Rom 2.0-11: recomendaciones para el proyecto y ejecución en obras de atraque y amarre. Puertos del Estado, Madrid (2012)
4. Figuero, A., Rodriguez, A., Sande, J., Peña, E., Rabuñal, J.R.: Dynamical study of a moored vessel using computer vision. J. Mar. Sci. Technol. **26**, 240–250 (2018)
5. Rabuñal, J.R., Rodriguez, A., Figuero, A., Sande, J., Peña, E.: Field measurements of angular motions of a vessel at berth: Inertial device application. J. Control Eng. Appl. Inform. **20**, 79–88 (2018)

6. Vural, H., Koyuncu, M., Guney, S.: A systematic literature review on microservices. In: Gervasi, O., et al. (eds.) ICCSA 2017. LNCS, vol. 10409, pp. 203–217. Springer, Cham (2017). https://doi.org/10.1007/978-3-319-62407-5_14
7. Spanish State Port System. SIMAR data set. http://calipso.puertos.es//BD/informes/INT_8.pdf. Accessed 10 Mar 2018
8. Spanish State Port System. REDCOS data set. http://calipso.puertos.es//BD/informes/INT_1.pdf. Accessed 10 Mar 2018
9. Spanish State Port System. REMPOR data set. http://calipso.puertos.es//BD/informes/INT_4.pdf. Accessed 10 Mar 2018
10. Spanish State Port System. REDMAR data set. http://calipso.puertos.es//BD/informes/INT_3.pdf. Accessed 10 Mar 2018
11. Mckinney, W.: Pandas: a foundational python library for data analysis and statistics. Python High Perform. Sci. Comput. (2011)
12. Flask Homepage. https://palletsprojects.com/p/flask/. Accessed 10 Mar 2018
13. mod_wsgi documentation. https://modwsgi.readthedocs.io/en/develop/. Accessed 10 Mar 2018
14. Conda documentation. https://conda.io/en/latest/. Accessed 10 Mar 2018
15. Anaconda Software Distribution. https://www.anaconda.com/. Accessed 10 Mar 2018
16. Pedregosa, F., et al.: Scikit-learn: machine learning in Python. J. Mach. Learn. Res. **12**, 2825–2830 (2011)
17. Chollet, François: Home - Keras Documentation. https://keras.io/. Accessed 10 Mar 2018
18. Abadi, M., et al.: TensorFlow: large-scale machine learning on heterogeneous distributed systems. arXiv:1603.04467 (2016)
19. Node-RED : About. https://nodered.org/about/. Accessed 10 Mar 2018
20. Nair, V., Hinton, G.E.: Rectified linear units improve restricted boltzmann machines. In: Proceedings of ICML, pp. 807–814 (2010)

# A Genetic Algorithm and Neural Network Stacking Ensemble Approach to Improve $NO_2$ Level Estimations

Javier González-Enrique[1(✉)], Juan Jesús Ruiz-Aguilar[2], José Antonio Moscoso-López[2], Steffanie Van Roode[1], Daniel Urda[1], and Ignacio J. Turias[1]

[1] Department of Computer Science Engineering, Polytechnic School of Engineering, University of Cádiz, Avda. Ramón Puyol, s/n, 11202 Algecira, Cádiz, Spain
{javier.gonzalezenrique, steffanie.vanroode, daniel.urda, ignacio.turias}@uca.es

[2] Department of Industrial and Civil Engineering, Polytechnic School of Engineering, University of Cádiz, Avda. Ramón Puyol, s/n, 11202 Algeciras, Cádiz, Spain
{juanjesus.ruiz, joseantonio.moscoso}@uca.es

**Abstract.** This work investigates the possible improvements that a stacked ensemble can provide to $NO_2$ estimations in a monitoring network located in the Bay of Algeciras (Spain). In the proposed ensemble, ANNs, linear and nonlinear genetic algorithms models have been used as the individual learners in the first stage. The non-linear GA models produce better results than linear GA models as they are able to detect useful relationships between variables that are ignored in the linear case. The outputs of the individual learners have been employed as the inputs of the ANN models of the second stage. The most accurate of these models produced the final $NO_2$ estimation. The obtained results are promising as this final stage-2 model is able to outperform all the other estimation models considered in this work. This can be explained due to its ability to exploit the advantages offered by each individual model from stage-1 and then find an optimal combination of their outputs in order to increase the global estimation performance. The improvement of these $NO_2$ estimations can be very useful to improve the autonomous capacities for monitoring networks.

**Keywords:** Artificial neural networks · Genetic algorithms · Air pollution · $NO_2$ · Ensembles

## 1 Introduction

Ensemble methods are machine learning algorithms where the performance or classification accuracy is improved as a result of the combination of individual models. Different variants and approaches can be found in the scientific literature to create these ensembles. A first approach employs the same learner but changes the training datasets. Between this type of ensembles Boosting [1], Bagging [2], Random forest [3] and


I. Rojas et al. (Eds.): IWANN 2019, LNCS 11506, pp. 856–867, 2019.
https://doi.org/10.1007/978-3-030-20521-8_70

AdaBoost [4] can be cited. Another possible approach relies on the use of different learning methods. In this case, majority voting, weighted voting and averaging are the most common techniques. Finally, stacking ensembles [5] are based on the use of the outputs of individual models as inputs of a second stage algorithm as a way to improve the performance of the models.

Air pollution is one of the most important environmental problems that must be faced in order to preserve the quality of living of the population. Nitrogen dioxide ($NO_2$) is one of the main pollutants. Its origins are manifold, but it is very related to combustion processes [6] and the reactions between nitrogen oxides and ozone [7]. It has harmful effects on human health [8] and is considered to be the main reason for air quality loss in urban areas [9].

The main objective of this paper is to improve $NO_2$ estimations in a monitoring network located in the Bay of Algeciras (Spain). To achieve this goal, a stacking ensemble is proposed. Artificial neural networks (ANNs), linear and nonlinear genetic algorithms (GAs) are employed as individual learners. Besides, ANNs are used as the second stage algorithm. This ensemble produces promising results outperforming all the individual models and other stacking ensembles that are also calculated. The importance of improving the $NO_2$ estimations is related to their ability to give monitoring networks autonomous capabilities, such as missing data imputation or detection of decalibration situations.

The rest of this paper is organized as follows. Section 2 describes the area of study and the database. Section 3 presents the methods used in this work. Section 4 describes the experimental design. Results are discussed in Sect. 5. Finally, the conclusions are shown in Sect. 6.

## 2 Data and Area Description

The Bay of Algeciras area is a heavily industrialized region which is located in the south of Spain and includes a population of nearly 300,000 inhabitants. The sources of $NO_2$ are numerous including not only the mentioned industries but very heavy traffic in the urban areas. Additionally, the Port of Algeciras Bay is one of the most prominent ship-trading ports in Europe. Thus, vessels constitute another important source of gaseous air pollution in this area.

All the aforementioned facts highlight the importance that an adequate pollution control strategy has to preserve the wellbeing of the population. With this purpose, a pollution monitoring network is located in this area. It is composed of 14 stations and records hourly measures of $NO_2$. Figure 1 shows the location of the Bay of Algeciras and the situation of the monitoring stations (depicted using their codes). Table 1 shows the correspondence between stations and their codes.

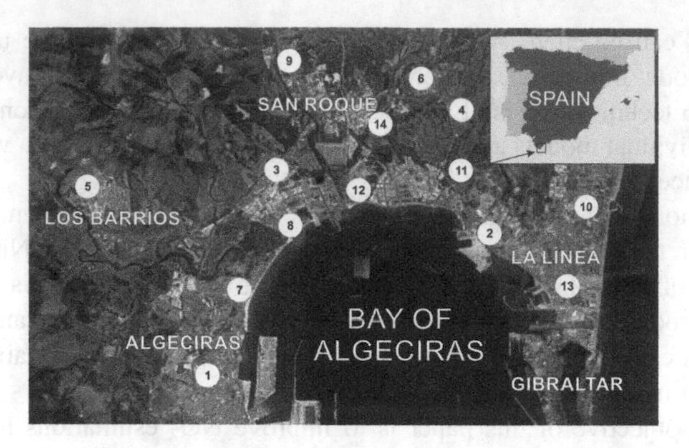


**Fig. 1.** Area of study

**Table 1.** Location of the $NO_2$ monitoring stations

| Monitoring station | Code | Lat. | Long. |
|---|---|---|---|
| EPSA Algeciras | 1 | 36°8′11.70″ N | 5°27′11.44″ W |
| Campamento | 2 | 36°10′45.96″ N | 5°22′37.9″ W |
| Los Cortillijos | 3 | 36°11′25.74″ N | 5°26′8.73″ W |
| Esc. Hostelería | 4 | 36°12′13.97″ N | 5°23′1.33″ W |
| Col. Los Barrios | 5 | 36°11′5.154″ N | 5°29′18.362″ W |
| Col. Carteya | 6 | 36°12′34.798″ N | 5°23′29.803″ W |
| El Rinconcillo | 7 | 36°09′42.95″ N | 5°26′31.765″ W |
| Palmones | 8 | 36°10′35.274″ N | 5°26′24.648″ W |
| Est. San Roque | 9 | 36°12′47.218″ N | 5°25′55.945″ W |
| El Zabal | 10 | 36°10′21.47″ N | 5°20′30.192″ W |
| Economato | 11 | 36°11′23.032″ N | 5°22′51.85″ W |
| Guadarranque | 12 | 36°10′55.55″ N | 5°24′41.6″ W |
| La Línea | 13 | 36°09′22.252″ N | 5°20′22.823″ W |
| Madrevieja | 14 | 36°12′6.66″ N | 5°24′19.8″ W |

The database used in this work contains hourly $NO_2$ concentration measures that were obtained by the aforementioned monitoring stations during a period of 6 years (2010–2015). This database was normalized as a previous step. Then, it was split into two different datasets. A first one including records from 2010 to 2014, which was used to select the best parameters of the models and train them. The second one includes only measures taken in 2015 and was used as the test set. The results are provided using only the test set in order to determine the performance of the models with unseen data.

# 3 Methods

This section presents a brief description of the methods and techniques used in this work.

## 3.1 Artificial Neural Networks

Backpropagation feedforward multilayer perceptron [10], which includes at least one hidden layer different from the input and output layers, is the most widely used design for ANNs. According to [11], ANNs with enough neurons and a single hidden layer can be considered as universal approximators of any nonlinear function.

In this work, backpropagation neural networks (BPNNs) with a single hidden layer have been used to create hourly $NO_2$ estimation models. The Levenberg–Marquardt algorithm [12] has been employed for optimization purposes. Additionally, the early stopping technique [13] has been applied to the training process with the aim of avoiding overfitting and ensuring good generalization capabilities in the models.

So as to determine the optimal number of hidden neurons, authors have used a 5-fold cross validation resampling procedure, which has been used previously with good results [14–18].

## 3.2 Genetic Algorithms

Genetic algorithms [19] are search methods inspired by the natural selection processes. The decision variables for a particular problem are encoded into strings of a certain alphabet. This strings are known as chromosomes and act as candidate solutions of the problem (which are known as individuals). The set of all the individuals is known as the population. In order to determine the goodness of each possible solution, a fitness value is calculated for each individual of the population.

The general process starts with the generation of a random initial population. This population evolves from one generation to another through the application of genetic operators. It moves towards a global optimum solution of the problem according to the fitness values obtained. Selection, crossover and mutation can be found among the genetic operators. The process continues and new generations are created until the stopping criteria are met. The interested reader can find a more detailed explication of this process in the work of [20].

In this work, four different genetic algorithms models have been developed in order to estimate the hourly $NO_2$ concentrations at the EPSA monitoring station (see Table 1). In all these cases, the fitness function that must be minimized is the mean squared error (MSE) between the dependent variable and the estimation produced as the output of a function which is specific for each case. This is shown in Eq. 1.

$$err = MSE(y, \widehat{y}) \tag{1}$$

where $y$ is the dependent variable and $\widehat{y}$ is the estimation produced by the GA model. The main differences between these models lie on the specific function that is used to produce the estimations. Equations 2, 3, 4 and 5 show the estimation functions corresponding to GA model 1 (GA-1), GA model 2 (GA-2), GA model 3 (GA-3) and GA model 4 (GA-4) respectively.

$$\hat{y} = \sum\nolimits_{i=1}^{n} \left(w_{1_i} \cdot \left(S(w_{2_i} \cdot x_i) + w_{3_i} \cdot x_i + w_{4_i} \cdot x_i^{w_{5_i}} + e^{w_{6_i}} \cdot x_i + w_{7_i}\right)\right) + k \tag{2}$$

$$\hat{y} = \sum\nolimits_{i=1}^{n} \left(S(w_{1_i}) \cdot \left(S(w_{2_i} \cdot x_i) + w_{3_i} \cdot x_i + w_{4_i} \cdot x_i^{w_{5_i}} + e^{w_{6_i}} \cdot x_i + w_{7_i}\right)\right) + k \tag{3}$$

$$\hat{y} = \sum\nolimits_{i=1}^{n} \left(w_{1_i} \cdot x_i + S(w_{2_i} \cdot x_i) + w_{3_i} \cdot x_i^{w_{4_i}} + e^{(w_{5_i} \cdot x_i)}\right) + k \tag{4}$$

$$\hat{y} = \sum\nolimits_{i=1}^{n} (w_i \cdot x_i) + k \tag{5}$$

where $y$ is the dependent variable, $x_i$ are the independent variables (predictors), $n$ is the total number of predictors, $w_{1_i}$, $w_{2_i}$, $w_{3_i}$, $w_{4_i}$, $w_{5_i}$, $w_{6_i}$, $w_{7_i}$, $w_i$, and $k$ are the weights determined by the GA and $S$ is the sigmoid function, which is expressed in Eq. 6.

$$S(n) = \frac{1}{1 + e^{(-n)}} \tag{6}$$

It is important to note that $w_{6_i}$ in Eqs. 2, 3, and 4 has been constrained within the $[10^{-12}, +\infty)$ interval. The genetic algorithm function provided by MATLAB R2016b has been used to develop the GA models. In this software, the codification of the variables in chromosomes is done internally without any intervention by the user. As can be seen in Eqs. 2–5, GA-1, GA-2, GA-3 present a non-linear behaviour whereas GA-4 fitness function is linear.

### 3.3 Stacked Ensembles

Stacked ensembles [21] are techniques which are intended to supply an overall prediction or estimation value based on the combination of the outputs of individual models. This type of ensemble can be beneficiated from the different perspectives offered by the individual models and usually improve their results. A brief description of the ensembles used in this work is presented next:

- Average (*avg*): The final estimation is calculated as the average of the individual models' estimations.
- Weighted average (*wavg*): In this case, each individual model has a different contribution to the final estimation according to the goodness of its estimation power, as is shown in Eq. 7.

$$E_{final} = \sum\nolimits_{i=1}^{j} (w_i \cdot E_i) \tag{7}$$

where $w_i \in [0, 1]$ for each $i \in [1, \ldots, j]$, $\sum_{i=1}^{j} w_i = 1$ and $E_i$ represent an individual estimation model.

- ANN weighted ensemble (*ANNwe*): Inspired in the *wavg* ensemble, this work proposes a type of ensemble which uses the individual models as inputs of a BPNN. The obtained model represents the best possible combination of the inputs in order to produce the aggregated output.

## 4 Experimental Procedure

The objective of this study is to determine the possible improvements in $NO_2$ estimation models performance when a proposed stacked ensemble is applied. In this approach, GAs are used in conjunction with ANNs. The proposed fitness functions (see Eqs. 2–6) let the GA models capture linear and nonlinear relations between variables and increase their estimation performance.

For estimation purposes, the hourly $NO_2$ values measured at the EPSA monitoring station (see Table 1) was considered as the dependent variable. In contrast, the hourly $NO_2$ values corresponding to the rest of the monitoring stations were used as predictor variables. As an initial step, the original database was normalized and divided into two disjoint groups. The first one included hourly $NO_2$ records going from 2010 to 2014 and was used as the training set. The second one included records belonging to 2015 and acted as the test set.

The experimental process was divided into two different stages. In the first stage, five estimation models were developed, one using ANNs and the rest using different genetic algorithms approaches. In the case of the ANN models, the BPNNs used a single hidden layer and a different number of hidden neurons (hns) (1 to 50). The Levenberg–Marquardt was selected as the optimization algorithm and the early stopping technique was employed to improve the generalization capabilities of the models. Starting with the training set, a random resampling procedure using 5-fold cross-validation was used for each number of hns and the average performance measures were calculated. This process was repeated 20 times to avoid the effect of randomness in the ANN weights initialization, and the average results were also calculated. Additionally, the individual results for each repetition were also stored so that a multicomparison procedure could determine meaningful differences within the models in a later step. Regarding performance measures, the Pearson correlation coefficient (R), the mean squared error (MSE), the index of agreement (d) and the mean absolute error (MAE) [22] were calculated. These performance indexes are defined in Eqs. (8–11).

$$R = \frac{\sum_{i=1}^{N} (O_i - \overline{O}) \cdot (P_i - \overline{P})}{\sqrt{\sum_{i=1}^{N} (O_i - \overline{O})^2 \cdot \sum_{i=1}^{N} (P_i - \overline{P})^2}} \tag{8}$$

$$MSE = \frac{1}{N} \sum_{i=1}^{N} (P_i - O_i)^2 \tag{9}$$

$$d = 1 - \frac{\sum_{i=1}^{N} (P_i - O_i)^2}{\sum_{i=1}^{N} \left( \left| P_i - \overline{O} \right| + \left| O_i - \overline{O} \right| \right)^2} \tag{10}$$

$$MAE = \frac{1}{N}\sum\nolimits_{i=1}^{N} |P_i - O_i| \quad (11)$$

where P indicates predicted values and O indicates observed values.

Finally, the best model was obtained using the Friedman test [23] and the Bonferroni method [24], jointly with the mentioned performance measures. The Friedman test let us determine if meaningful differences were present between the models. The Bonferroni method evaluated which of the models were not statistically equivalent. Following the Occam's razor principle, the model with fewer hns was selected among those showing no significant differences with the model that produced best performance indexes.

After the model selection, a new BPNN model was trained using the whole training dataset and the number of hns of the most accurate model. Then, this model was fed with the inputs of the test set in order to obtain the final $NO_2$ estimation for the year 2015. Finally, performance measures were calculated through the comparison of observed vs. estimated values.

In the case of the GA models, the fitness functions presented in Sect. 3.2 (Eqs. 1 to 6) were minimized using the training data set. Regarding the parameters that control the genetic algorithms, different tests were carried out in order to select the best possible parameter combination and each combination was repeated 20 times. Table 2 shows the possible values that were tested for each parameter. A detailed description of each parameter can be found in Matlab's Genetic Algorithm Options web page [25]. Table 3 shows the final combination selected per each GA model.

**Table 2.** Parameters tested in the GA models

| Population size | Crossover function | Crossover fraction (%) | Selection function | Mutation rate (%) | Max. generations |
|---|---|---|---|---|---|
| 200, 250, 350, 450, 550, 650 | Scattered<br>Single point<br>Two point<br>Intermediate<br>Heuristic<br>Arithmetic | 75, 80, 85, 90 | Stochastic uniform<br>Remainder<br>Uniform<br>Roulette<br>Tournament | 2, 4, 6, 8, 10 | 2000 |

**Table 3.** Selected parameters for each AG model

| Model | Population size | Crossover function | Crossover fraction (%) | Selection function | Mutation rate (%) |
|---|---|---|---|---|---|
| GA-1 | 450 | Heuristic | 80 | Roulette | 4 |
| GA-2 | 450 | Heuristic | 80 | Roulette | 8 |
| GA-3 | 650 | Arithmetic | 80 | Tournament | 6 |
| GA-4 | 650 | Arithmetic | 80 | Stochastic uniform | 8 |

Once the stopping criteria were met, the corresponding weights were stored for each specific GA model. As the last step, the final $NO_2$ estimations for the year 2015 were obtained using Eqs. 2, 3, 4, 6 and the corresponding weights and test sets for each case. Finally, values for R, MSE, d and MAE were calculated after comparing the observed $NO_2$ values against the estimated ones obtained with each model.

In the second stage, *avg*, *wavg* and *ANNwe* ensembles (see Sect. 3.3) were calculated. In the case of the *avg* ensemble, the calculation is straightforward as it only averages the obtained estimations obtained with the individual models. For the *wavg* ensemble (see Eq. 7), each of the estimations from stage-1 was weighted according to its MSE value following an inversely proportional distribution. To calculate the *ANNwe* ensemble, ANN models were trained using the outputs from stage-1 as their inputs and the 2015 $NO_2$ measured values as their targets. In this case, the same network configuration as stage-1 ANN models was applied. The final output was obtained as the one which produced a lesser MSE value after 20 repetitions.

## 5 Results and Discussion

The results of the experimental procedure are presented in this section. In the first stage, different models have been developed to estimate the hourly $NO_2$ concentration values at the EPSA monitoring station (station 1). $NO_2$ values measured at the other stations have been used as inputs of the models (see Table 1 and Fig. 1). The initial data set has been split into two disjoint datasets and the results are obtained through the comparison of observed vs. estimated values for the test set (2015). This lets us evaluate the performance of the models with unseen data. For comparative purposes, a Lasso model using the same datasets and 5-fold cross validation has also been included. Table 4 shows the performance measures corresponding to stage-1 models.

**Table 4.** Performance indexes for stage-1 estimation models

| Model | nh | $\overline{R}$ | $\overline{MSE}$ | $\overline{MAE}$ | $\overline{d}$ |
|---|---|---|---|---|---|
| ANN | 6 | 0.761 | 238.769 | 10.988 | 0.845 |
| GA-1 | - | 0.716 | 276.931 | 11.904 | 0.814 |
| GA-2 | - | 0.713 | 277.556 | 11.956 | 0.811 |
| GA-3 | - | 0.716 | 276.093 | 11.980 | 0.815 |
| GA-4 | - | 0.650 | 326.148 | 13.512 | 0.757 |
| Lasso | - | 0.650 | 326.155 | 13.520 | 0.760 |

As can be expected, the best ANN model outperforms all the GA models. However, it can be noted that non-linear GA models (GA-1, GA-2 and GA3) beat easily the performance offered by the GA-4 linear model. This indicates that Eqs. 2, 3 and 4 are able to capture linear and also an important amount of non-linear relations between input and output variables. However, the proposed fitness functions cannot compete with ANNs' ability to act as universal approximator of any nonlinear function, as was mentioned in Sect. 3.1.

**Table 5.** Performance indexes for stage 2 ensembles

| Model | nh | $\overline{R}$ | $\overline{MSE}$ | $\overline{MAE}$ | $\overline{d}$ |
|---|---|---|---|---|---|
| *avg* | - | 0.726 | 268.741 | 11.755 | 0.815 |
| *wavg* | - | 0.729 | 266.139 | 11.679 | 0.8175 |
| *ANNwe* | 40 | 0.784 | 215.608 | 10.503 | 0.8690 |

Table 5 shows the results obtained by the proposed ensembles in the second stage. These methods combine the outputs of stage-1 models with the aim of improving the estimation results.

Results show how *avg* and *wavg* ensembles constitute an improvement over GA models but do not reach the estimation goodness offered by the stage-1 ANN model. This can be explained by the fact that the average and the weighted average operations (to a lesser extent) are highly influenced by extreme values that are far from the mean of the individual learners considering a specific instant of time. In our case, this influence comes primarily from GA4 output. As an example, if AG4 output is removed from the ensemble, the value of MSE for *avg* drops to 260.837 and its R-value rises to 0.735.

In the case of the *ANNwe* ensemble, its performance indexes are far superior when compared to those belonging to all the proposed models in the first and second stages. As can be seen, the second stage ANN can take advantage of the different linear and non-linear relations captured by the GA and ANN of the first stage. Some of them are already present in the ANN model of the first stage, but other ones are provided by the GA models. Considering the results, the proposed two-stage approach guarantees a better estimation performance of the $NO_2$ concentration values at the EPSA monitoring station.

A comparison between the best models of the first and second stages is presented in Figs. 2 and 3 where estimated versus measured $NO_2$ hourly values are depicted for January 2015. As can be seen, the fit and adjustment to the observed values are superior in the case of *ANNwe* when compared to the ANN model of the first stage. This confirms what was stated before about the improvement of the estimation models provided by the proposed approach (Fig. 3).

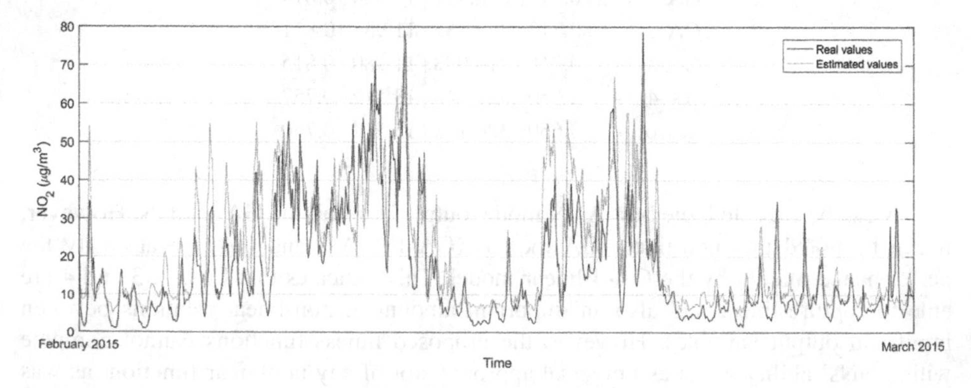


**Fig. 2.** Estimated vs. real values for January 2015 using the stage-1 ANN most accurate model

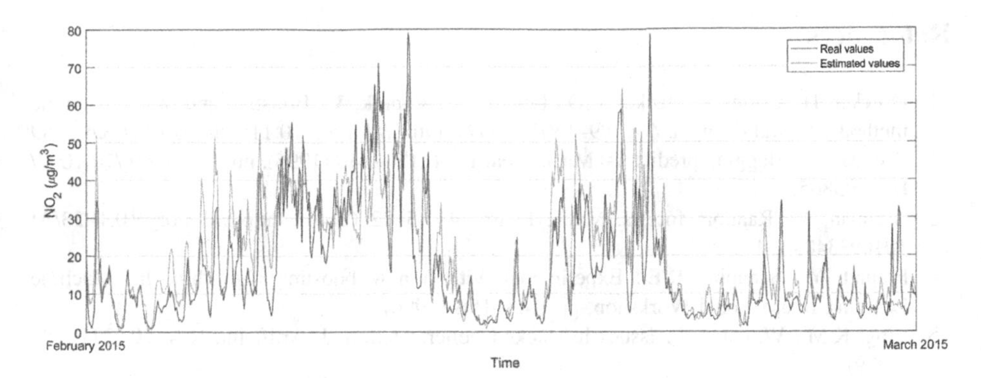


**Fig. 3.** Estimated vs. real values for January 2015 using the stage-2 *ANNwe* ensemble

## 6 Conclusions

The aim of this paper is to verify the possible improvements that a stacked ensemble approach can provide to $NO_2$ estimations over compared to other individual models. This approach uses artificial neural networks, linear and nonlinear genetic algorithms as individual learners. Then, their outputs are used as inputs of the ANN models of the second stage.

Regarding the first stage results, the proposed GA models that use non-linear functions produce much better results if they are compared to GA using a linear fitness function. This indicates that their fitness functions can detect useful relationships between variables that are ignored in the linear approaches. However, ANNs outperform them due to their ability to act as universal approximators (see Sect. 3.1).

The results of both stages show how the *ANNwe* approach outperforms all the other proposed approaches, ensuring a better estimation performance of $NO_2$ in the monitoring network. The main reason can be found in the fact that all stage-1 models capture different linear and nonlinear relations between the inputs and the targets. Therefore, the *ANNwe* approach is able to exploit the advantages offered by each individual model. Additionally, it is able to find an optimal combination of their outputs in order to increase the global estimation performance.

The use of the proposed model can provide better and more reliable $NO_2$ estimations if it is compared to the other proposed models. This can be very useful as these estimations can provide robustness and autonomous capabilities to the monitoring network. They also can be helpful in missing values or detection of decalibration situations.

**Acknowledgements.** This work is part of the coordinated research projects TIN2014-58516-C2-1-R and TIN2014-58516-C2-2-R supported by MICINN (Ministerio de Economía y Competitividad-Spain). Monitoring data have been kindly provided by the Environmental Agency of the Andalusian Government.

## References


1. Drucker, H., Cortes, C., Jackel, L.D., LeCun, Y., Vapnik, V.: Boosting and other ensemble methods. Neural Comput. **6**, 1289–1301 (1994). https://doi.org/10.1162/neco.1994.6.6.1289
2. Breiman, L.: Bagging predictors. Mach. Learn. **24**, 123–140 (1996). https://doi.org/10.1007/BF00058655
3. Breiman, L.: Random forests. Mach. Learn. **45**, 5–32 (2001). https://doi.org/10.1023/A:1010933404324
4. Freund, Y., Schapire, R.E.: Experiments with a new boosting algorithm. In: Machine Learning International Workshop, pp. 148–156 (1996)
5. Ting, K.M., Witten, I.H.: Issues in stacked generalization. J. Artif. Int. Res. **10**, 271–289 (1999)
6. Rivera, C., et al.: Spatial distribution and transport patterns of $NO_2$ in the Tijuana - San Diego area. Atmos. Pollut. Res. **6**, 230–238 (2015)
7. Finlayson-Pitts, B.J., Pitts, J.N.J.: The atmospheric system. In: Finlayson-Pitts, B.J., Pitts, J.N.J. (eds.) Chemistry of the Upper and Lower Atmosphere: Theory, Experiments, and Applications, pp. 15–42. Academic Press, San Diego (2000)
8. Faustini, A., Rapp, R., Forastiere, F.: Nitrogen dioxide and mortality: review and meta-analysis of long-term studies. Eur. Respir. J. **44**, 744–753 (2014)
9. Westmoreland, E.J., Carslaw, N., Carslaw, D.C., Gillah, A., Bates, E.: Analysis of air quality within a street canyon using statistical and dispersion modelling techniques. Atmos. Environ. **41**, 9195–9205 (2007)
10. Rumelhart, D.E., Hinton, G.E., Williams, R.J.: Learning internal representations by error propagation. In: Rumelhart, D.E., McClelland, J.L., PDP Research Group (eds.) Parallel Distributed Processing: Explorations in the Microstructure of Cognition. Foundations, vol. 1, pp. 318–362. MIT Press, Cambridge (1986)
11. Hornik, K., Stinchcombe, M., White, H.: Multilayer feedforward networks are universal approximators. Neural Netw. **2**, 359–366 (1989). https://doi.org/10.1016/0893-6080(89)90020-8
12. Marquardt, D.W.: An algorithm for least-squares estimation of nonlinear parameters. J. Soc. Ind. Appl. Math. **11**, 431–441 (1963)
13. Gardner, M.W., Dorling, S.R.: Artificial neural networks (the multilayer perceptron)—a review of applications in the atmospheric sciences. Atmos. Environ. **32**, 2627–2636 (1998). https://doi.org/10.1016/S1352-2310(97)00447-0
14. Turias, I.J., González, F.J., Martin, M.L., Galindo, P.L.: Prediction models of CO, SPM and $SO_2$ concentrations in the Campo de Gibraltar Region, Spain: a multiple comparison strategy. Environ. Monit. Assess. **143**, 131–146 (2008). https://doi.org/10.1007/s10661-007-9963-0
15. Muñoz, E., Martín, M.L., Turias, I.J., Jimenez-Come, M.J., Trujillo, F.J.: Prediction of $PM_{10}$ and $SO_2$ exceedances to control air pollution in the Bay of Algeciras, Spain. Stoch. Environ. Res. Risk Assess. **28**, 1409–1420 (2014). https://doi.org/10.1007/s00477-013-0827-6
16. Turias, I.J., et al.: Prediction of carbon monoxide (CO) atmospheric pollution concentrations using meteorological variables. WIT Trans. Ecol. Environ. **211**, 137–145 (2017). https://doi.org/10.2495/AIR170141
17. González-Enrique, J., Turias, I.J., Ruiz-Aguilar, J.J., Moscoso-López, J.A., Jerez-Aragonés, J., Franco, L.: Estimation of $NO_2$ concentration values in a monitoring sensor network using a fusion approach. Fresen. Environ. Bull. **28**, 681–686 (2019)

18. González-Enrique, J., Turias, I.J., Ruiz-Aguilar, J.J., Moscoso-López, J.A., Franco, L.: Spatial and meteorological relevance in $NO_2$ estimations. A case study in the Bay of Algeciras (Spain). Stoch. Environ. Res. Risk Assess. **33**, 801–815 (2019). https://doi.org/10.1007/s00477-018-01644-0
19. Holland, J.H.: Adaptation in Natural and Artificial Systems. University of Michigan Press, Ann Arbor (1975)
20. Goldberg, D.E.: Genetic Algorithms in Search, Optimization and Machine Learning. Addison-Wesley Longman Publishing Co., Inc., Reading (1989)
21. Polikar, R.: Ensemble based systems in decision making. Circuits Syst. Mag. IEEE. **6**, 21–45 (2006). https://doi.org/10.1109/MCAS.2006.1688199
22. Willmott, C.J.: Some comments on the evaluation of model performance. Am. Meteorol. Soc. **63**, 1309–1313 (1982)
23. Friedman, M.: The use of ranks to avoid the assumption of normality implicit in the analysis of variance. J. Am. Stat. Assoc. **32**, 675–701 (1937). https://doi.org/10.1080/01621459.1937.10503522
24. Hochberg, Y., Tamhane, A.C.: Multiple Comparison Procedures. Wiley, New York (1987)
25. The Mathworks Inc.: Genetic Algorithm Options. https://es.mathworks.com/help/gads/genetic-algorithm-options.html

# Deep Learning Models in Healthcare and Biomedicine

# Convolutional Neural Network Learning Versus Traditional Segmentation for the Approximation of the Degree of Defective Surface in Titanium for Implantable Medical Devices

Ruxandra Stoean[1(✉)], Catalin Stoean[1], Adriana Samide[2], and Gonzalo Joya[3]

[1] Department of Computer Science, Faculty of Sciences, University of Craiova, Craiova, Romania
{ruxandra.stoean,catalin.stoean}@inf.ucv.ro

[2] Department of Chemistry, Faculty of Sciences, University of Craiova, Craiova, Romania
samide_adriana@yahoo.com

[3] School of Telecommunication Engineering, University of Malaga, Málaga, Spain
gjoya@uma.es

**Abstract.** One prevalent option used in the manufacturing of dental and orthopedic medical implants is titanium, since it is a strong, yet light, biocompatible metal. Nevertheless, possible micro-defects due to earlier chemical treatment can alter its surface morphology and lead to less resistance of the material for implantation. The scope of the present paper is to give an estimate of the defectuous area in titanium laminas by analysing microscopic images of the surface. This is done comparatively between traditional segmentation with thresholding and a sliding window classifier based on convolutional neural networks. The results show the supportive role of the proposed means towards a timely recognition of defective titanium sheets in the fabrication process of medical implants.

**Keywords:** Medical implant · Titanium · Defect demarcation · Surface estimation · Convolutional neural networks

## 1 Introduction

When manufacturing a medical implantable device, attention must be primarily given to the decision regarding the type of material that will be used. There are several common as well as new materials that can be used for medical implants [5]. A good candidate is the one that provides bio-compatibility, light weight and resistance [1,8].

Titanium is one of the frequent choices for medical implants and is mainly used in dentistry and orthopedics. It exhibits all the required features for a material suitable for implantation. However, some micro-defects as an effect of


I. Rojas et al. (Eds.): IWANN 2019, LNCS 11506, pp. 871–882, 2019.
https://doi.org/10.1007/978-3-030-20521-8_71

chemical treatment may exist on the surface of titanium plates, yet observable only under the microscope and as such invisible to the human eye.

The aim of the current paper then becomes to put forward means that would efficiently give an estimate on the defective area of a titanium lamina. This value offers decisive support information for manufacturers of medical implants when choosing between several titanium sheets provided. The straightforward possibility lies in a traditional automated segmentation of the images, based on pixel differences between clean and defective areas. The threshold under which a pixel should be considered as defect or non-defect can be established based on a set of microscopy images where the pitting is manually annotated by the chemical engineer.

On the other hand, a machine learning approach can also be used to conduct learning on the same annotated examples. To this scope, small windows are randomly cropped from both the defective as well as the faultless regions of each available large microscopy image and labelled in a binary manner as defect/non-defect. The collection of small samples is then given to a convolutional neural network (CNN) in order to learn the difference. Once the image features corresponding to the two classes have been learnt, a window slides over the large titanium samples from the test set and labels each such square as either defective or non-defective.

The precision and recall of the prediction is computed and the results of the two approaches are compared. In the end, the estimated total defective area of the titanium sample is provided.

The paper is structured in the following manner. The problem presentation and data preparation from the chemical engineering perspective is provided in Sect. 2. The recent existing entries for conceptually different defect detection problems are described in Sect. 3. The details of the combination between the CNN classification of cropped tiles and a sliding labelling window over the original images are given in Sect. 4. The methodology part ends with exhibiting an alternative common approach for segmentation, with the support of an established threshold, in Sect. 5. The results related both to the CNN classification accuracy and the performance of the CNN window in estimating the area of defect, as opposed to the traditional segmentation method, are found in Sect. 6. The conclusions and prospects of continuing the study are given in Sect. 7.

## 2 Problem Definition and Data Collection

Titanium is used as a dental and orthopedic bioimplant, being considered an inert and biocompatible material with a high capacity of osteointegration [3]. In contact with the tissue environment, titanium is inactivated due to the formation of a stable oxide layer that sits at the tissue/material interface, allowing the implant integration into the bone and restricting the release of potentially toxic titanium ions into the body [2]. The implant shape and dimensions as well as the surface chemistry and topography are essential features for the osteoacceptation of the material [4].

Chemical treatments with acidic solutions could alter the titanium surface morphology due to the formation of some micro-fissures, which reduce the implant fatigue resistance. In this context, the implant surface optimal characterization leads to the detection of possible micro-defects, thus preventing the occurrence of subsequent implantation inconveniences.

The bioimplant micro and macro-structural characterization was performed using a metallographic microscope, Euromex type, with the Canon camera and included software, having the magnification power (x80). Before the microscopic examination, the titanium surface was mechanically finished with sand paper of various sizes, ultrasonically cleaned, degreased with ethyl alcohol and dried in warm air.

A number of 128 microscopy images were acquired from the same surface, with a dimension of $3888 \times 2592$ pixels.

As the annotation procedure is laborious and time consuming for the human expert, the image collection was split into a training and a validation set where the manual delineation of defective areas is done only roughly, i.e. only the most prominent zones are marked. The test data were however charted up to the finest parts, in order to test the demarcation ability of the segmentation and deep learning approach.

## 3 Related Literature Review

The state of the art in automatic surface defect detection in metallic sheets includes several recent techniques. There are direct classification problems where images must be categorized into several types of surface defects. One such example uses CNN for this discrimination in the case of steel sheets [15].

There are however some literature entries that are more closely related to the current task of defect delineation in entire images. The first one [11] targets micro-structure defect detection of Ti-6Al-4V titanium alloy from images. Material grains are segmented and the defects are found by region-based graphs.

A second approach targets defect images from the production line of a flat metal part visualized under an industrial microscope [14]. A cascaded autoencoder performs defect localization, through a pixel-wise prediction mask from semantic segmentation. The found regions are then classified into several known categories with the help of a CNN.

Images of metal boxes containing or not manufacturing defects undergo preprocessing and are then binarily categorized through an autoencoder that learns a reduced set of features, whose output is fed to a Gaussian process classifier [6]. The procedure is further on applied deeper to patches of $32 \times 32$ pixels from the initial images, where some carry defects and others do not, where the method finds those that do.

The tandem window cropping - crop classification - window sliding - location heat map that is proposed in this study for defect area approximation is known as an effective option in object localization tasks. Two papers [7,9] are thus connected to our study as regards this methodological approach. They however employ it for the task for which it has been initially designed, i.e. object localization from X-ray images, where this concerns aluminium casting defects in the automotive industry. The first difference is therefore that this type of defects due to metal casting are objects, with several specific patterns that distinguish them from regular areas. Some such examples are cavities, air bubbles, human-induced drilling, out projections, foreign particles [7,9], several of which also have a certain probable range in millimeters. Conversely, chemical treatment induces non-homogeneous, irregular shapes for defects at the surface level of the titanium lamina, which are invisible to the human eye.

The study [9] proposes cropping small images of 32 × 32 pixels from the initial X-ray data and employs several classifiers for the discrimination between defective and non-defective pictures. They are either combinations between traditional segmentation approaches and shallow machine learners or transfer learning from GoogleNet, AlexNet, several VGG models and a personally designed CNN. The subsequent sliding window strategy is moved one pixel at a time and the heat map is obtained through a Gaussian mask superimposition, with defect localization resulting from thresholding the map. The best results were obtained for the combination between local binary patterns and support vector machines. The paper [7] tests the same CNN patch classification in combination with a sliding window heat map. A selective search algorithm is used for appointing bounding boxes from the resulting heat map. Those bounding boxes with an average value higher than 0.5 for the heat map are considered defects. The approach is however deemed inferior to CNN methods (R-CNN, R-CFN, SSD) tailored specifically for their studied task of object localization.

In this paper, we also employ a version of the window cropping - CNN patch classification - area heat map flow, which is expected to be more appropriate for the current, more indefinite surface defect area assessment task.

## 4 Sliding Window Classifier Based on CNN

The sliding CNN window methodology is tailored for the problem at hand as shown in the following steps.

Defective and non-defective tile samples are randomly taken from the original images. Contrary to the defect localization problems of the state of the art presented in Sect. 3, titanium micro-defective areas have irregular, sizeable shapes. Hence, a bad tile does not contain a defect against a normal background, but it is defective surface in its entirety. Conversely, for this problem there are no defects of a certain size that could be chosen as the dimension of the tile, but entire areas of various shapes.

The CNN subsequently learns the distinction between defective areas and normal zones from the training data tiles. Once an accurate model is chosen, based on the validation data, the original test images are traversed by consecutive sliding CNN windows, where each gives a prediction upon the crossed area as to whether it is a defective area or not. The precision (recall) and sensitivity of the prediction are evaluated against the manual annotation on the test samples and the total defective area is calculated. A schematic representation of the flow of the considered approach is given in Fig. 1.

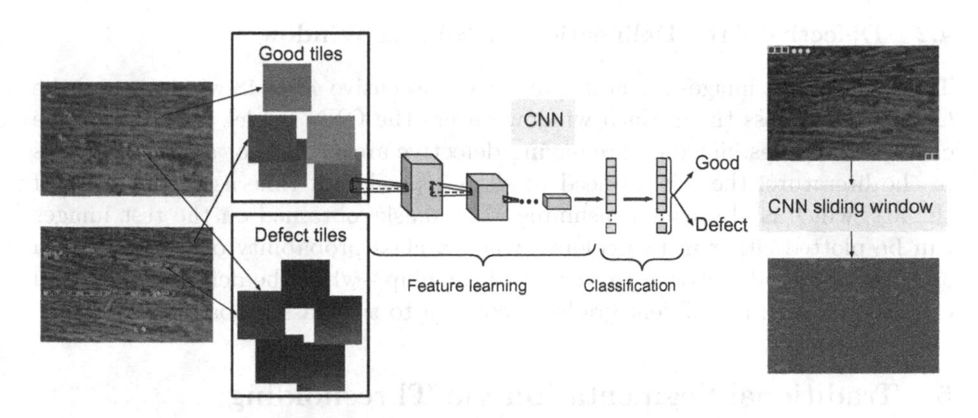


**Fig. 1.** Overview of the sliding window CNN methodology for detecting good and defective tiles in titanium images. Defective areas are manually annotated first, then tiles of fixed sizes are randomly taken from the inside of the defective areas and from the outside, the CNN is trained on these patches, and finally the learned model classifies new images by considering sliding windows. In the final image (bottom-right) the blue represents the good areas, while the red represents the defective ones. (Color figure online)

### 4.1 CNN with Transfer Learning for Defective/Non-defective Tile Recognition

From each training, validation and test set, a number of tiles of size 48 × 48 are randomly excerpted such as to have a balanced number of examples of both completely defective and normal zones. The defective tiles are extracted from the manually annotated regions denoting defects, while the non-defective examples are taken from the remaining parts of the slides. Once the training, validation and test collections are established, a CNN architecture is appointed to discover and learn the features that differentiate between defective and non-defective areas.

Due to the small number of original images in the collection that poses a limit on the number of non-overlapping tiles that could be randomly extracted for pattern learning, a pre-trained CNN model is the best choice for this task. Among the several possibilities, it was opted for VGG-16, due to its capability of producing reasonable results even when having as input pictures of the given size. A number of its first layers were pre-trained on the ImageNet collection and the remaining are trained on the current titanium tile collection. Good tiles and defective tiles enter the convolutional layers, then into the fully connected ones and eventually their probability of belonging to each of the two classes is given. Once the CNN model has been trained and the best weights as found from the validation data have been saved, it is applied to the test tiles in order to get its performance of prediction.

### 4.2 Defective Area Delineation by Sliding Window

The original test images are now split into consecutive $48 \times 48$ windows and the CNN slides across them. Each window enters the CNN model and obtains the class probabilities. Since we are having defective areas, and not single defects as in the literature, there is no need to slide 1 pixel at a time as in the state of the art, which is also time consuming. The masks obtained on the test images can be plotted either in two colors - when a class probability over 0.5 is taken as pointing to a defective area - or as a heat map - when the defective potential can be measured in different grades according to the class probabilities.

## 5 Traditional Segmentation via Thresholding

A legitimate question would be why try a complex paradigm like the CNN instead of a straightforward classical image segmentation approach. We investigate this simpler option for our task by means of thresholding.

The segmentation of an image through binary thresholding is obtained by converting the input image to gray scale and then considering a degree of comparison against all image pixels. If the current pixel is smaller than this specified threshold, it is changed to 1, otherwise to 0. Accordingly, a binary image is obtained, where 1 corresponds to white and 0 to black. In this particular case, 0 encodes a normal pixel and 1 a defective one. This procedure leads however to many isolated pixels (black on white spots and white on the black areas) and, for eliminating these, morphological transformations like closing and opening are applied. The amount of noise that gets eliminated depends on the kernel size (not to be mistaken with the homonymous notion of filter dimension from CNN) of the structuring element. When these kernel sizes are larger, more noise is eliminated, but the number of detected objects decreases as well. To conclude, the efficiency of the thresholding procedure depends on the choice of two key parameters, the threshold parameter and the size of the structuring element.

This classical approach should work well because it examines the image at the lowest level, that of the pixel. It does not encompass however relations between neighbouring pixels. It might as well recognize as defect illuminations of the image that are actually noise generated at data preparation with the microscope. Additionally, there is a low contrast between some defective zones and the regular color of titanium. The method will also need a certain number of minimally annotated training images in order to set the threshold level. As running time is concerned, it is a fast option.

The CNN, on the other hand, captures the information from the small cropped windows as a whole. It might also extract other features besides shades of colors, such as shapes of the defect. The running time poses the same issue as with all applications of CNN. However, transfer learning is appointed in this case, which leaves a significantly smaller number of parameters to be trained, hence leading to a shorter training time just for the last layers. Finally, the CNN approach was planned also as a result of its recent successful use in the same area of surface science, namely on the differential analysis of formed coatings as a consequence of corrosion [12,13].

# 6 Experimental Results

The CNN training and the parameter setting of the segmentation based on thresholding were conducted during experimentation. The results of the two approaches - deep and traditional - are measured against each other in terms of over- and under- estimation of the defective area.

## 6.1 Performance of CNN for Tile Classification

We take random crops from each set (training, validation, test) holding the original microscope images with a size of 48 × 48. This dimension had been chosen as to capture enough detail but still be quick in its slide over the initial picture.

The number of defective tiles is proportional to the size of the annotated region and that is given precisely by the division between the area of the annotation and the double of the area of a cropped tile. The number of the good tiles is taken equal to the number of the defective ones. However, if an image does not contain any defect, 100 "good" tiles are taken from it. The training set contains 2456 defective tiles and 2656 good ones. The validation collection holds 624 defective patches and 724 clean ones. The test set has 1325 defective mini-images and 1325 good tiles. Data augmentation was performed to the tile collection in the form of rotation (range = 20°), shifting (width and height ranges equal to 0.2), and horizontal flipping.

As mentioned before in Sect. 4, the method for transfer learning was chosen to be VGG-16, since it is a model that allows images of the chosen size. Eight layers are left as they were pre-trained on the ImageNet collection and the other half were trained on the current set of entirely defective and completely clean tiles. The output of the last convolutional layer is flattened and enters a fully-connected layer with 1024 neurons. In the end, a last fully-connected layer with a sigmoid activation function gives the binary prediction. The number of epochs was set to 60 and the batch size to 32.

The accuracy results on the training and validation sets were of 93.98% and 95.40%, respectively. The obtained test accuracy in discriminating between defective and non-defective tiles is of 96.71%. These results are obtained on the tiles from these corresponding sets. Next, the model is applied for entire images using sliding windows.

### 6.2 Estimation of Defective Surface

The resulting model was applied to predict the label (defective vs. non-defective) for every consecutive window with the same 48 × 48 size from the complete images in the test collection. The resulting masks denoting the found defective areas against the normal surrounding zone are depicted in Fig. 2, side by side with the original image under the microscope. The predicted defective portions can be either roughly (binary) estimated or in shades of defectiveness.

**Fig. 2.** Initial image and the obtained masks after the application of the CNN: the middle image shows the binary classification (defective areas in red, i.e. where probabilities were below 0.5), while the third illustrates the class probabilities for each sliding window in turn. (Color figure online)

### 6.3 Traditional Segmentation

The training and validation sets that consist of the initial large (i.e. complete, not cropped) images were used in searching for the optimal settings for the thresholding procedure. In the current study, the shape of the structuring element is kept as the implicit square. Recall, precision and F1 score are used for measuring the efficiency of the methodology using the different parameter values selected. The precision is computed as the ratio between the true positives and the sum between true positives and false positives. The recall is given by the ratio between the true positives and the sum between true positives and false negatives. The F1 score is the harmonic mean between precision and recall.

Figure 3 shows the efficiency of the values for the threshold and the size involved in the parametrization of this image segmentation procedure in terms of the precision, recall and F1 measures. The lighter the color, the better the configuration.

The inputs that lead to the highest F1 score on the training and validation sets (for this method their separation is not needed) are used for the application of the segmentation procedure on the test images.

### 6.4 Comparison Between Deep and Thresholding Segmentation

An example of how the two techniques find defects in a test sample is shown in Fig. 4.

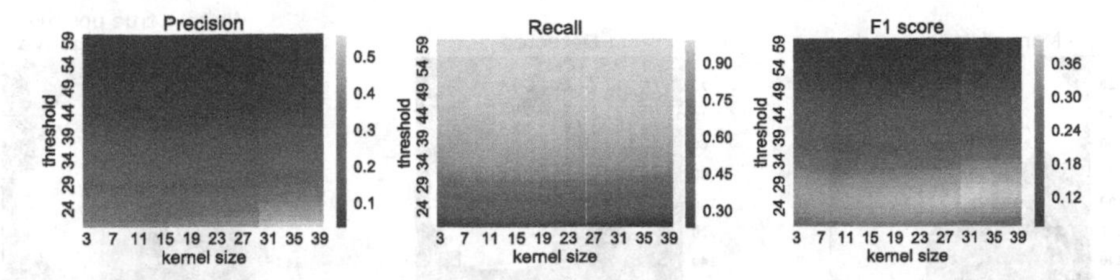


**Fig. 3.** Results for the thresholding parameter tuning on the training and validation images taken together. The three plots correspond to the precision, recall and F1 score and they are computed against the manual annotations. A lighter color corresponds to better parametrization. (Color figure online)

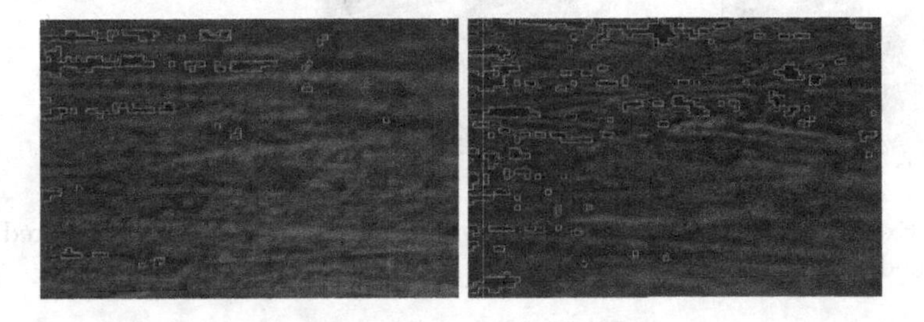

**Fig. 4.** Two test image samples having manual annotations in blue, thresholding ones in red and CNN in green. (Color figure online)

Figure 5 illustrates how the true positives, true negatives, false positives and false negatives are calculated for one annotation. This verification is made at the pixel level. In order to compare the two techniques - deep and traditional - precision, recall and F1 score are redefined to measure the match between the predicted and the actual defective areas, as described in Subsect. 6.3. A methodology must achieve a balance between predicting those areas that are indeed defective and not mistaking other zones which are in fact not defective.

**Table 1.** Comparison between the results obtained by the two methods for the test set.

| Percentage measures | CNN | Thresholding | Manual |
|---|---|---|---|
| Precision | 68.04 | 68.43 | - |
| Recall | 52.02 | 48.96 | - |
| F1 score | 58.49 | 55.49 | - |
| Overall mean defective area | 2.58 | 2.66 | 3.35 |

Table 1 shows results in the form of precision, recall, F1 score and even average over the defective areas for the test images. The last measure is provided also for the manual annotations. The two methods have a surprisingly similar

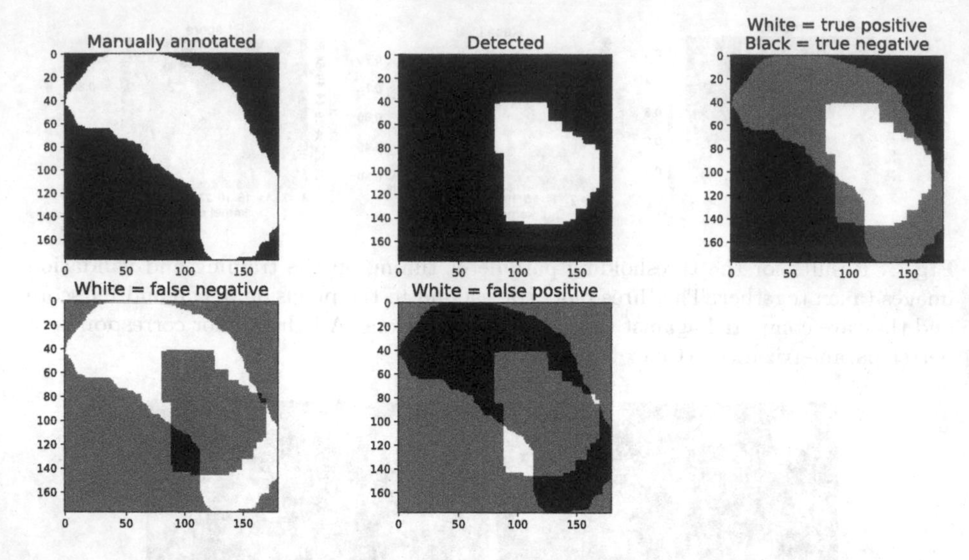


**Fig. 5.** Exemplification of the manner in which the detected object is checked against the ground truth.

precision value, meaning that there might be a percent of defective shapes easier to delineate. A difference between the two in favor of the CNN occurs for recall, indicating that the thresholding procedure finds more false negatives. This difference naturally means that the F1 score will also favor CNN. Nevertheless, the two methodologies lead overall to very similar results and make a relatively good approximation of the defective area.

For CNN the decision for good/defective crop is taken for a confidence threshold of 0.5, i.e. as illustrated in the middle image from Fig. 2. However, the confidence threshold could be tuned in the (0, 1) interval in search for an optimal value. Some preliminary tuning on the validation set (i.e. using steps of 0.05 between 0 and 1 for the confidence level), although having led to small gains on the validation images, did not brought accompanying improvement of the results on the test set. Moreover, the best result was very close to the usual confidence threshold of 0.5. Perhaps further parameter tuning, like using an evolutionary approach to direct the search in the promising areas [10] could lead to improved results.

The degree of defective area coverage for each image in the test set was quantified and is shown in Fig. 6. Each triplet of bars corresponds to a test image and the vertical measurements indicate the percentage of the good and defective areas. The detected areas are very similar in size, although there are also some false negatives present, i.e. those shown in the results of Table 1. A large difference is observed for the 7th image, where the amount of the manual defective area is significantly larger than the detected ones.

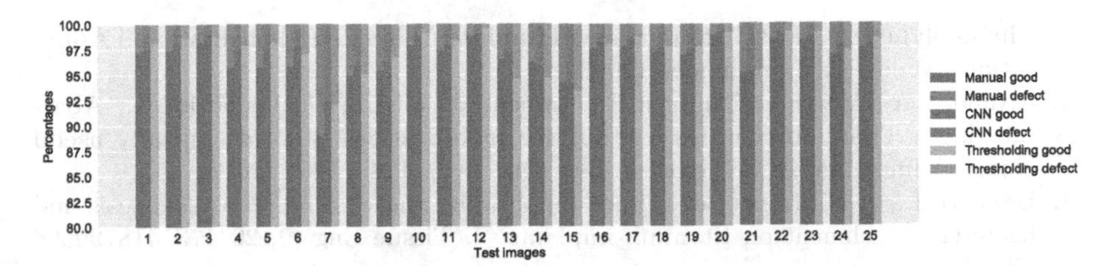


**Fig. 6.** The percentages of good and defective parts in each test image as manually annotated, as detected by CNN and by thresholding, respectively.

# 7 Conclusions and Future Work

Although the F1 scores may not seem high, one has to take into account that for every image having an area of over $10^7$ pixels, only a fraction of 3.35% in average needs to be identified for every one of the 25 test images. Moreover, as Fig. 6 indicates, the amount of defective areas for each test image is very close to the size of the manual annotations as achieved by both methodologies. CNN achieves a slightly better localization of the defective areas as opposed to the thresholding, but at the expense of a higher running time and more computations, respectively. However, if the user is only interested in the amount of the defective area and not necessarily in its precise localization, thresholding would provide a rapid and satisfying assessment.

In order to further improve the accuracy, a sliding window smaller than the current 48 by 48 pixels would make a clear fine tuning as concerns the detected shapes for CNN. The approach makes only squared slices, while the shapes of the defective areas are irregular. A smaller sliding window would allow the little portions to belong to various squares and further compute for each pixel an average over the probabilities provided by the multiple applications of the CNN.

Another direction lies in checking the amount of defective areas in titanium when it is further treated with a protective coating layer against the extension of the defects. This increase in the number of instances will also allow for the additional check of the eventuality of an abrupt difficulty gap in defect formation in the sample leading to such close precision for the two methods. Finally, a network pre-trained on similar material data might improve the results.

**Acknowledgements.** This work has been partially supported by the Spanish Ministry of Science, Innovation and Universities, through the Plan Estatal de Investigación Científica y Técnica y de Innovación, Project TIN2017-88728-C2-1-R.

# References

1. Ahmed, W., Elhissi, A., Jackson, M., Ahmed, E.: Precision machining of medical devices. In: Davim, J.P. (ed.) The Design and Manufacture of Medical Devices, pp. 59–113. Woodhead Publishing Reviews, Mechanical Engineering Series. Woodhead Publishing (2012)

2. Christoph Leyens, M.P.: Titanium and Titanium Alloys - Fundamentals and Applications. Wiley-VCH, Weinheim (2003)
3. Civantos, A., Martínez-Campos, E., Ramos, V., Elvira, C., Gallardo, A., Abarrategi, A.: Titanium coatings and surface modifications: toward clinically useful bioactive implants. ACS Biomater. Sci. Eng. **3**(7), 1245–1261 (2017)
4. Damiati, L., et al.: Impact of surface topography and coating on osteogenesis and bacterial attachment on titanium implants. J. Tissue Eng. **9**, 2041731418790694 (2018)
5. Edwards, C.: Materials used in medical implants: how is the industry breaking the mould? Verdict Medical Devices (2018). https://www.medicaldevice-network.com/features/materials-used-medical-implants-industry
6. Essid, O., Laga, H., Samir, C.: Automatic detection and classification of manufacturing defects in metal boxes using deep neural networks. PLoS One **13**(11), e0203192 (2018)
7. Ferguson, M., Ak, R., Lee, Y.T., Law, K.H.: Automatic localization of casting defects with convolutional neural networks. In: 2017 IEEE International Conference on Big Data (Big Data), pp. 1726–1735 (2017)
8. Li, J., Stachowski, M., Zhang, Z.: Application of responsive polymers in implantable medical devices and biosensors. In: Zhang, Z. (ed.) Switchable and Responsive Surfaces and Materials for Biomedical Applications, pp. 259–298. Woodhead Publishing, Oxford (2015)
9. Mery, D., Arteta, C.: Automatic defect recognition in x-ray testing using computer vision. In: 2017 IEEE Winter Conference on Applications of Computer Vision (WACV), pp. 1026–1035 (2017)
10. Preuss, M., Stoean, C., Stoean, R.: Niching foundations: basin identification on fixed-property generated landscapes. In: Krasnogor, N., Lanzi, P.L. (eds.) 13th Annual Conference on Genetic and Evolutionary Computation (GECCO-2011), pp. 837–844. ACM (2011)
11. Ren, R., Hung, T., Tan, K.C.: Automatic microstructure defect detection of Ti-6AL-4V titanium alloy by regions-based graph. IEEE Trans. Emerg. Top. Comput. Intell. **1**(2), 87–96 (2017)
12. Samide, A., Stoean, C., Stoean, R.: Surface study of inhibitor films formed by polyvinyl alcohol and silver nanoparticles on stainless steel in hydrochloric acid solution using convolutional neural networks. Appl. Surf. Sci. **475**, 1–5 (2019)
13. Samide, A., Stoean, R., Stoean, C., Tutunaru, B., Grecu, R.: Investigation of polymer coatings formed by polyvinyl alcohol and silver nanoparticles on copper surface in acid medium by means of deep convolutional neural networks. Coatings **9**, 105 (2019)
14. Tao, X., Zhang, D., Ma, W., Liu, X., Xu, D.: Automatic metallic surface defect detection and recognition with convolutional neural networks. Appl. Sci. **8**(9), 1575 (2018)
15. Zhou, S., Chen, Y., Zhang, D., Xie, J., Zhou, Y.: Classification of surface defects on steel sheet using convolutional neural networks. Mat. Technol. **51**(1), 123–131 (2017)

# Convolutional Neural Networks and Feature Selection for BCI with Multiresolution Analysis

Javier León[1](✉), Julio Ortega[1], and Andrés Ortiz[2]

[1] Department of Computer Architecture and Technology, University of Granada, Granada, Spain
jaleon@correo.ugr.es, jortega@ugr.es

[2] Department of Communications Engineering, University of Málaga, Málaga, Spain
aortiz@ic.uma.es

**Abstract.** Classification in high-dimensional feature spaces is a difficult task, often hindered by the curse of dimensionality. This is the case of Motor Imagery tasks involving Brain-Computer Interfaces through electroencephalography, where the number of available patterns is limited, making more noticeable the effect of the high dimensionality on the generalization capabilities of the models. This paper tackles classification in that particular setting, drawing a comparison between an explicit feature selection procedure using evolutionary computation and an implicit feature selection using Convolutional Neural Networks. These two alternatives are also compared to a Support Vector Machine approach that serves as a baseline quality threshold. According to the experiments performed in this paper, Convolutional Neural Networks are able to produce promising results when compared to the Support Vector Machine models and, after a partial hyperparameter optimization stage, also to previous work on the same dataset. Furthermore, this raises the issue of the trade-off between computational cost and classification accuracy, which is briefly discussed when assessing the quality of the results in relation to existing work.

**Keywords:** Brain-Computer Interfaces · Convolutional Neural Networks · Support Vector Machines · Feature selection · Genetic algorithms · Multiresolution Analysis

## 1 Introduction

In many applications related to bioinformatics it is usual to find data samples composed of a large number of features. In addition, the number of observations in such samples tends to be very small in comparison, since the obtention of numerous patterns is unfeasible or at least cumbersome for the user. The problem at hand, Electroencephalography (EEG) classification for Brain-Computer Interface (BCI) tasks, is a good example of this: in order to obtain new data,


I. Rojas et al. (Eds.): IWANN 2019, LNCS 11506, pp. 883–894, 2019.
https://doi.org/10.1007/978-3-030-20521-8_72

test subjects are required to repeat the process of imagining limb movements. Understandably, the cost of repeating the recording sessions several times seems unreasonable, both for researchers and for participants.

Consequently, it is unsurprising to face issues like the curse of dimensionality problem [1], for which a solution often involves feature selection. Some of the benefits of this procedure that should be brought into focus are: a decrease in overall computational costs; the elimination of noise and redundancy; and, in cases where the number of final features is low enough, real interpretability of the classifiers. The most important of them, though, is the potential increase in classification accuracy.

Concerning the BCI application considered here, the EEG readings to be classified correspond to Motor Imagery (MI), a paradigm that uses a series of brief amplifications and attenuations conditioned by limb movement imagination called Event-Related Desynchronization (ERD) and Event-Related Synchronization (ERS), respectively. The patterns were built using a type of Multiresolution Analysis (MRA) [2], specifically the discrete wavelet transform, as shown in [3].

Existing literature on the topic of BCI [4] points to feature selection as a key step, highlighting that it helps to achieve real-time performance, as well as to better understand the inner workings of the brain. Feature selection is an NP-hard problem, meaning that for considerable input sizes a brute-force search is out of the question because the search space is exponential in size. From the three main approaches (i.e. filter, wrapper, and embedded methods) the choice is genetic algorithms, a wrapper method. The algorithm uses a population of individuals that encode different feature subsets, which will evolve throughout a fixed number of generations in order to find an optimal configuration.

This paper discusses, for the case of MI with EEG, the role of the aforementioned feature selection process followed by a model optimization step that will depend on the classifier. The considered classifiers are Convolutional Neural Networks (CNNs) and Support Vector Machines (SVMs).

Regarding the contents, Sect. 2 explains the structure of the data; Sect. 3 briefly presents the measures used to quantify the success of the process; Sect. 4 describes the feature selection approach; afterwards, Sect. 5 details each classifier included in this paper and its corresponding optimization procedure; Sect. 6 shows the most relevant experimental results and statistical comparisons. Finally, Sect. 7 contains the corresponding conclusions.

## 2 Data Description

Going into more detail about the structure of the data, an EEG pattern is a time series built by feature extraction: the signal provided by each electrode is comprised of several segments to which a set of wavelet details and approximation coefficients are assigned. In total, the pattern will have $2 \times S \times E \times L$ sets of coefficients, with $S$ being the number of segments, $E$ the number of electrodes and $L$ the number of wavelet levels. In the particular case of the dataset we have used, recorded at the BCI Laboratory of the University of Essex, $S = 20$, $E = 15$, and $L = 6$, therefore resulting in 3,600 sets of coefficients with sizes ranging from 4 to 128 and a total of 151,200 coefficients. As seen in [3], it is possible to reduce this number to a more tractable 3,600 by computing the variance of each set (adding a normalization step at the end). However, given that the amount of available training patterns is about half a tenth of that (precisely, 178 for training and 178 for testing), further reductions are still needed. Finally, the classification step deals with three balanced classes, namely: imagined left and right hand movement, and imagined feet movement; this step will be carried out for three BCI subjects (104, 107, and 110).

## 3 Performance Measurement and Estimation

When deciding the means through which the quality of the different stages of the process will be assessed, it is important to remember the supervised nature of the learning problem. This means that the labels of our dataset are known, and that they are used to train machine learning models. Accordingly, our measures take advantage of this fact to directly estimate quality. The two metrics used throughout this paper are the following:

- Cohen's Kappa error [5]: similar to the test set error, but also taking into account the possibility of classifying correctly by chance. It is computed as:

$$\kappa = \frac{p_0 - p_c}{1 - p_c},$$

  where $p_0$ is equivalent to the test set error, and $p_c$ is the sum of the probabilities of random agreement for all possible classes. Its values lie within the range $[-1, 1]$.
- Cross-validation error: splits the training set in $n$ sections, using $n-1$ of them for model training and leaving the remaining one out for testing. The final value is the arithmetic mean of the $n$ testing errors. Its range of possible values is $[0, 1]$.

The first measure provides the accuracy on actual data not seen by the classifier, while the second one gives an estimate of the generalization capabilities of the model.

## 4 Feature Selection Using Evolutionary Computing

As was previously stated, the chosen feature selection approach is a genetic algorithm whose typical structure is shown in Fig. 1.

The individuals or chromosomes represent feature subsets. Because the memory consumption is manageable, they are binary-encoded: a chromosome contains as many binary genes as there are features. In this context, a value of 1 in the $i$-th position means the $i$-th feature is active, and a value of 0 means the opposite.

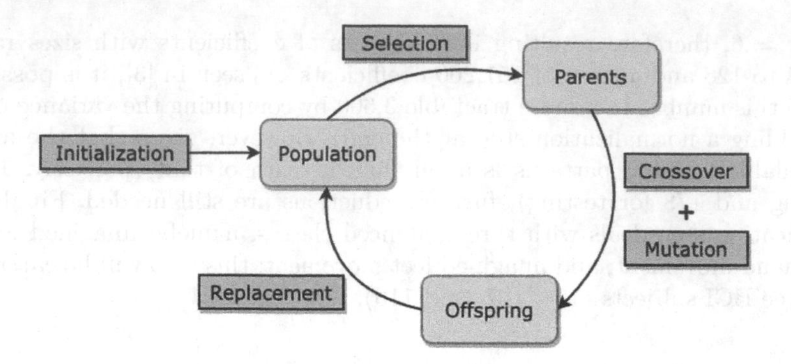


**Fig. 1.** General cyclical overview in a genetic algorithm. Blue boxes represent operations. Violet boxes represent sets of individuals. (Color figure online)

Different crossover operators have been considered at this point, named single-point, two-point, and uniform crossover, from which the last one has been selected. The mutation operator consists in randomly flipping one bit out of the 3,600 that make up an individual.

Machine learning models come into play during fitness evaluation, which is usually applied to rank potential parents from the current generation, or to know which individuals will be passed on to the next generation.

In principle, one or more functions can be used for fitness evaluation, which is known as multi-objective optimization. We can define the general case formally in its minimization version as:

$$\min_{x \in X}(f_1(x), f_2(x), ..., f_n(x)), \forall n = 1..N, \tag{1}$$

where $X$ represents the space of possible solutions, and $N$ is the number of objective functions.

The procedure in this paper only requires one fitness function. This is mainly due to the additional cost of training more models, but it also simplifies and reduces the stage of solution sorting to a simple ranking of the fitness values of the individuals in ascending order.

Now that the choice to solve the optimization problem has been explained, the next section describes the different machine learning algorithms that take part in the experimentation.

## 5 The Classifiers: CNN and SVM

### 5.1 Support Vector Machines

Popular in the field of BCI [6,7], the Support Vector Machine [8] is a linear model (the type used here), but it can be extended to non-linear classification boundaries if necessary by means of the kernel trick. It attempts to find the maximum-margin hyperplane, which is the separating hyperplane that is most distant to any training point located in the boundary of any class, and is based on the notion of larger margins being associated with better generalization.

However, in real-world applications data points are seldom linearly separable. For this reason, a soft-margin policy is needed to allow the model to accept a certain degree of misclassification. The hyperparameter that controls this leniency, $C$, will take higher values to produce less error tolerance.

## 5.2 Convolutional Neural Networks

Neural networks are machine learning models loosely based on the structure of the human brain. In their most basic form, they are made of simple processing units (*neurons*) arranged in three general types of layers:

- Input layer: With as many units as features describing a pattern. Each unit can receive the original value of its corresponding feature or, alternatively, a preprocessed version.
- Output layer: The number of possible outputs (in this instance, three) dictates its number of units, from which only one, indicating the answer to the classification problem, can be active at a time.
- Hidden layers: Inserted between the previous two layers, their interactions yield the predictive capability of the network, with more layers and bigger sizes corresponding to potentially more complex decision boundaries.

The units contained in these layers can be thought of and linked together in a variety of shapes. The simplest one involves forward, one-way connections (hence the term *feed-forward*, with no cycles in the graph). To each connection a *weight* is assigned by the learning algorithm.

In turn, CNNs are driven by the convolution operator, which takes its most intuitive form when dealing with images. The concept behind it is to overlay a region of the input data with a filter and compute their dot product. This process is applied successively in order to cover the whole input.

Unlike Feed-Forward Neural Networks, CNNs make use of several types of layers in their architectures, although the input and output layers remain the same. Using images as the example data type, a rough summary of available choices is the following:

- Convolutional layer: This kind of layer can be composed of many filters, whose weights will be learned through the training process. When a filter is applied throughout an image (assuming a single color channel), a two-dimensional output is produced and thus the global output of a convolutional layer contains an additional depth dimension in which all filter outputs are stacked. Tunable hyperparameters here include the number of filters or the stride that separates each application of the same filter along the image.
- Pooling layer: Essentially a downsampling process. It can be non-linear, usually in the form of a *max pooling* function that is again applied to the whole image by non-overlapping regions. It serves a variety of purposes, including

the progressive reduction of the input representation, along with a consequent reduction in the number of learnable parameters. The underlying reasoning for this step is that the relative position of a feature with respect to the others is more important than its exact location. For these layers, the downsampling function and the filter size are hyperparameters to tune.
- Activation layer: Often a ReLU layer for its benefits in training speed, it introduces non-linearity with the function $f(x) = max(0, x)$ whose main practical effect is the elimination of negative values. It is frequently used just after every convolutional layer.
- Fully-connected layer: Placed at the end of the neural network, one or more of these layers are used to process the features from the previous layers and to output the verdict of the classification. Here, the structure can be tuned.

Looking at the whole network, there are still two relevant hyperparameters related to the training stage whose values we have to choose:

- Number of training epochs: An epoch corresponds to one forward and one backward pass of the training data through the network, in order to calculate its current error rate and act on it by adjusting weights.
- Learning rate: The proportion of the classification error that the algorithm uses to correct the weights of the network at each training epoch.

Since the goal of this paper is to discuss the potential of CNNs compared to other alternatives, only these two learning hyperparameters will be optimized and therefore the structure of the network will be fixed.

# 6 Experimental Results

## 6.1 Experimental Setup

The genetic algorithm has been implemented from scratch in *Python 3.4.9*, with custom operators in the case of neural networks. *Scikit-Learn 0.19.2* has been used for its SVM implementation, as well as for model evaluation with the Kappa and cross-validation measures. Neural networks have been built using *Keras 2.2.2* with *Tensorflow 1.10.1* as its backend. Data operations have been handled with *NumPy 1.14.5*.

The experiments have been run in a server node equipped with an Intel® Xeon® E5-2620 v2 @ 2.10 GHz, 32 GB DDR3 RAM memory and an NVIDIA® Tesla K20c, making use of either CPU (evaluation of feature subsets) or GPU (CNN training) parallelism. Relevant experiments have been repeated at least five times in order to perform a statistical test to find out the significance of the observed differences.

### 6.2 Results

The first decision to make is whether feature selection can be a meaningful component of the final process. With this in mind, a comparison will be drawn between the use of feature selection and the use of the full feature sets for the two classifiers considered in this paper (CNN and SVM). Then, there should be enough information to choose which alternative to optimize for each classifier.

Both types of models use default hyperparameters in order to ensure minimal external influence. For SVM, the only hyperparameter is $C$, whose value is fixed to 1. For the 15 runs of CNN, the hyperparameters and their values are the following:

- Structure: One convolutional layer composed of 130 one-dimensional kernels with size 5 and ReLU activation, followed by a fully-connected, softmax layer for outputting classification labels.
- Learning algorithm: The Adam [9] optimizer with the recommended parameter values by the authors.
- Training epochs: 60.

The genetic algorithm for feature selection has been run with a population of 1,000 individuals evaluated through 100 generations using 5-fold cross-validation, yielding 25 features for subject 104 and 28 features for subjects 107 and 110. After that, the outcomes have been used to train SVM and CNN models. Table 1 contains the experimental results for SVM (run only once since the training is deterministic). Analogously, Table 2 contains the results for CNN but averaged 15 times in order to account for the dependence on weight initialization.

As can be observed in Table 1, the efficacy of feature selection for the case of SVM is mixed: it can produce similar or even better performance, but it can also be significantly worse. For this reason, the model will be optimized later for both alternatives so that their differences can be better studied.

**Table 1.** Comparison of Kappa values in the test set for SVM with feature selection (SVM-fs) and no feature selection (SVM-nfs).

| Subject | SVM-nfs | SVM-fs |
|---|---|---|
| **104** | 0.66950 | 0.65304 |
| **107** | 0.66345 | 0.54652 |
| **110** | 0.52790 | 0.56166 |

Table 2, on the contrary, displays a marked contrast in favor of not carrying out the explicit feature selection. After performing one Kruskal-Wallis test for each subject to assess differences between medians, we obtain $p$-values of $2.97 \cdot 10^{-6}$, $3.03 \cdot 10^{-6}$, and $4.42 \cdot 10^{-6}$, respectively, which all confirm that this is true for a significance level of 95%. In view of this, only the CNN alternative using all 3,600 features will be considered for the hyperparameter optimization phase.

**Table 2.** Average and best Kappa values in the test set obtained by CNN with a previous feature selection step (CNN-fs) versus no feature selection (CNN-nfs).

| | CNN-nfs | | CNN-fs | |
|---|---|---|---|---|
| **Subject** | **Average** | **Best** | **Average** | **Best** |
| **104** | $0.70958 \pm 0.0130$ | 0.72946 | $0.62102 \pm 0.0098$ | 0.63528 |
| **107** | $0.70992 \pm 0.0125$ | 0.73061 | $0.54142 \pm 0.0114$ | 0.56315 |
| **110** | $0.61850 \pm 0.0170$ | 0.63756 | $0.56056 \pm 0.0180$ | 0.58714 |

This latter result could be attributed to the often not necessary feature selection when using CNNs, since their internal way of functioning serves as an implicit extraction of the most useful features for detecting patterns in the data. Our procedure is likely interfering with the attempts of the network at finding locally-related occurrences of values; this could stem from the selection of individual predictors by measuring the quality of a model that does not exploit local association (as CNNs do).

The next step is, as was anticipated, an optimization procedure for the most promising models to make a final comparison. On account of the differences in complexity between both models, a separate tuning method is employed for each one:

- SVM: Given the simpler nature of this model, random search [10] is our choice. Unlike grid search, random search consists in a number of unstructured combinations which do not use fixed intervals between tried values. 100 trials will be used to find a value for the hyperparameter $C$.
- CNN: The same genetic algorithm that was used for feature selection can be adapted for the optimization of different aspects in a neural network. In this instance, it will help to tune the two aforementioned learning hyperparameters: the learning rate and the number of training epochs. Due to time constraints, the experiments are run with 40 individuals and 10 generations.

For the sake of statistically significant comparisons, the SVM optimization experiments have been repeated 15 times, while the CNN ones have been repeated 5 times (results are shown in Table 3). As this difference is related to the running times, its implications will be discussed later. The procedure for selecting a solution from each run consists in picking the individual with the best cross-validation error for the case of random search, and in evaluating the performance of the final generation in the test set and choosing the best for the case of the genetic algorithm.

According to the average values in Table 3, there appears to be a clear prevalence of the classification power of the CNN models over that of the SVM ones. Between the two SVM alternatives, nevertheless, the results are mixed again: the feature selection step is able to approximately match or even surpass the use of the whole set, but in one test subject it significantly decreases the accuracy.

Table 4 provides the $p$-values for the multiple comparison tests performed with a significance level of 95% for all three subjects.

It is noticeable that every alternative is labeled as different from the others in all three datasets. This means that the CNN method performs notably better in general, while the other two seem to be more case-dependent.

**Table 3.** Average and best Kappa values in the test set obtained by the optimized models: SVM with feature selection (SVM-fs-opt), SVM with no feature selection (SVM-nfs-opt) and CNN with no feature selection (CNN-nfs-opt).

| | SVM-fs-opt | | SVM-nfs-opt | | CNN-nfs-opt | |
|---|---|---|---|---|---|---|
| **Subject** | **Average** | **Best** | **Average** | **Best** | **Average** | **Best** |
| **104** | $0.68785 \pm 0.0029$ | 0.69525 | $0.69514 \pm 0.0038$ | 0.70336 | $0.73589 \pm 0.0062$ | 0.74576 |
| **107** | $0.52123 \pm 0$ | 0.52123 | $0.62320 \pm 0.0028$ | 0.63042 | $0.73406 \pm 0.0041$ | 0.73917 |
| **110** | $0.60356 \pm 0.0174$ | 0.62110 | $0.58360 \pm 0.0068$ | 0.58699 | $0.65104 \pm 0.0126$ | 0.66284 |

**Table 4.** Comparison of $p$-values for the three optimized alternatives. Non-significant differences ($p > 0.05$) in **bold**.

| **104** | **SVM-fs-opt** | **SVM-nfs-opt** |
|---|---|---|
| **CNN-nfs-opt** | $p = 0.000464$ | $p = 0.000805$ |
| **SVM-fs-opt** | | $p = 0.001754$ |
| **107** | **SVM-fs-opt** | **SVM-nfs-opt** |
| **CNN-nfs-opt** | $p = 0.000017$ | $p = 0.000121$ |
| **SVM-fs-opt** | | $p = 1.64 \cdot 10^{-7}$ |
| **110** | **SVM-fs-opt** | **SVM-nfs-opt** |
| **CNN-nfs-opt** | $p = 0.000952$ | $p = 0.000329$ |
| **SVM-fs-opt** | | $p = 0.000264$ |

Addressing the topic of running times, there is a clear advantage for the SVM-based approaches. The complexity of neural networks, both in the number of learnable parameters and of hyperparameters (fixed before training), makes optimization much costlier: every CNN optimization run took from 10 to 14 hours per subject, while the equivalent process for SVM needed 15 to 20 min when using 3,600 features and about 30 seconds when using the reduced feature set (25 to 28 features). This raises the issue of whether the observed increases in predictive power are worth the additional computational cost.

**Table 5.** Kappa values for CNN (CNN-opt) and DBN (DBN-opt) in the test set.

| | CNN-opt | | DBN-opt | |
|---|---|---|---|---|
| **Subject** | **Average** | **Best** | **Average** | **Best** |
| **104** | $0.736 \pm 0.006$ | 0.746 | $0.733 \pm 0.011$ | 0.750 |
| **107** | $0.734 \pm 0.004$ | 0.739 | $0.723 \pm 0.007$ | 0.733 |
| **110** | $0.651 \pm 0.013$ | 0.663 | $0.672 \pm 0.008$ | 0.683 |

Finally, given the existence of previous work on this specific dataset, a comparison is made below. In particular, in Table 5 and Fig. 2 the performances of Deep Belief Networks (DBNs) [11] and the proposed CNNs are compared.

As can be seen, the 2-layer CNN models produce results that are comparable to those of 6-layer DBNs (except for subject 110, where the DBNs score slightly higher), which highlights the cost-saving benefits of the former alternative.

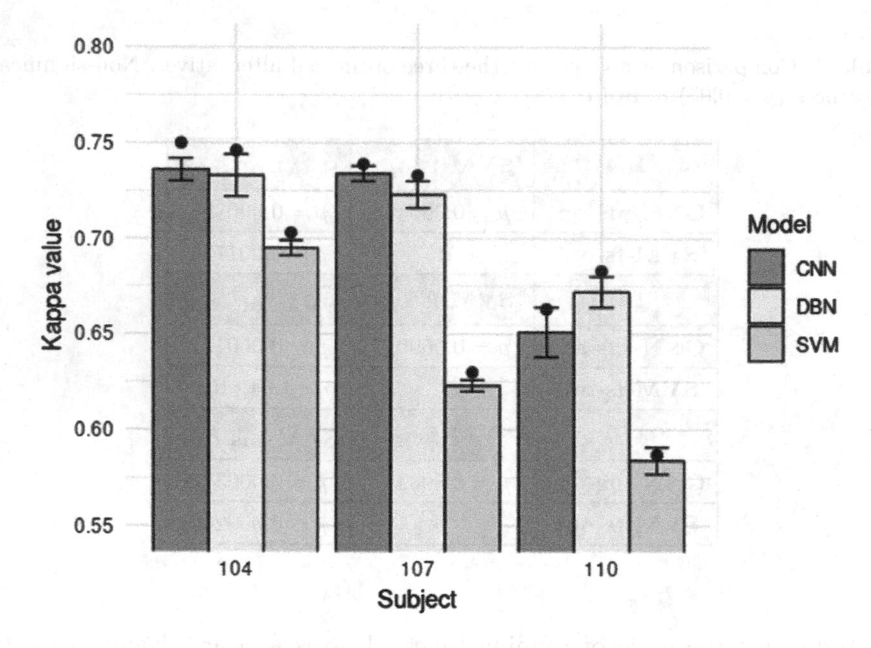


**Fig. 2.** Comparison of test-set performances between CNN and DBN, with bars representing average Kappa values, points meaning peak performance, and lines delimiting the range of the standard deviation. SVM-nfs-opt added for a baseline reference.

## 7 Conclusions and Future Work

When faced with a dataset of high-dimensional observations and its associated classification problem, feature selection becomes essential in many cases, not only for quality reasons and but also for the computational resources needed to build competitive models. Electroencephalogram-based MI is an example of this, as the datasets of measured patterns that characterize a movement are often a combination of complex electrode readings. Previous work on this topic using ten original test subjects supports this claim and also confirms the widely-documented variability of brain patterns that tends to make each BCI user unique (thus requiring individualized configuration).

At the moment of carrying out a feature selection process, it is common to evaluate the quality of the candidate feature sets by training models and testing them on a specifically dedicated subset of the data. However, this is not feasible under all circumstances, as the training time of powerful classifiers would place a limit on the algorithm. For this reason, lighter models such as Logistic Regression, Support Vector Machines, or Linear Discriminant Analysis are popular choices on the condition that a certain bias of the final outcome towards the chosen model is acknowledged. On the other hand, alternatives such as Convolutional Neural Networks allow implicit feature selection when fed the original feature set. This way, there is no longer a need for an external procedure. The results shown in this paper provide evidence that an explicit feature selection stage can interfere with the internal operations of CNNs, due to their leveraging of local structure being disrupted by individual picking of features according to different criteria.

Regarding the use of CNNs for the problem and dataset at hand, the accuracy of the proposed configuration appears to agree with existing literature in that shallower architectures tend to produce more promising results. This is mainly due to the small size of BCI datasets and the computational costs of model training. However, for this particular dataset a further look into this aspect is still needed in order to be able to confidently back up said claim. What is clear, nonetheless, is that CNNs are capable of handling a high-dimensional feature space with reasonable success.

The experimental results achieved in this paper point to a promising direction in classification for EEG-based MI tasks using CNN models. Future work could delve into other tunable parts of this type of neural network, such as overall structure, parameter configuration for each layer or a combination of these with what we propose here. Additional potential alternatives or improvements include the use of other models, such as Random Forests, and especially the creation of an adaptive classifier which can be incrementally re-trained as new data become available.

**Acknowledgements.** This work was partly funded by the Spanish MINECO and FEDER funds under TIN2015-67020-P, PSI2015-65848-R and PGC2018-098813-B-C32 projects.

## References

1. Raudys, S.J., Jain, A.K.: Small sample size effects in statistical pattern recognition: recommendations for practitioners. IEEE Trans. Pattern Anal. Mach. Intell. **3**, 252–264 (1991)
2. Daubechies, I.: Ten Lectures on Wavelets, vol. 61. Siam, Philadelphia (1992)
3. Asensio-Cubero, J., Gan, J., Palaniappan, R.: Multiresolution analysis over simple graphs for brain computer interfaces. J. Neural Eng. **10**(4), 046014 (2013)
4. Lotte, F., et al.: A review of classification algorithms for EEG-based brain-computer interfaces: a 10 year update. J. Neural Eng. **15**(3), 031005 (2018)
5. Cohen, J.: A coefficient of agreement for nominal scales. Educ. Psychol. Measur. **20**(1), 37–46 (1960)
6. Yi, W., Qiu, S., Qi, H., Zhang, L., Wan, B., Ming, D.: EEG feature comparison and classification of simple and compound limb motor imagery. J. Neuroengineering Rehabilitation **10**(1), 106 (2013)
7. Li, J., Zhang, L.: Bilateral adaptation and neurofeedback for brain computer interface system. J. Neurosci. Methods **193**(2), 373–379 (2010)
8. Cortes, C., Vapnik, V.: Support-vector networks. Mach. Learn. **20**(3), 273–297 (1995)
9. Kingma, D.P., Ba, J.: Adam: a method for stochastic optimization, arXiv preprint arXiv:1412.6980 (2014)
10. Bergstra, J., Bengio, Y.: Random search for hyper-parameter optimization. J. Mach. Learn. Res. **13**, 281–305 (2012)
11. Ortega, J., Ortiz, A., Martín-Smith, P., Gan, J.Q., González-Peñalver, J.: Deep belief networks and multiobjective feature selection for BCI with multiresolution analysis. In: Rojas, I., Joya, G., Catala, A. (eds.) IWANN 2017. LNCS, vol. 10305, pp. 28–39. Springer, Cham (2017). https://doi.org/10.1007/978-3-319-59153-7_3

# Attention-Based Recurrent Neural Networks (RNNs) for Short Text Classification: An Application in Public Health Monitoring

Oduwa Edo-Osagie[1](✉), Iain Lake[1], Obaghe Edeghere[2], and Beatriz De La Iglesia[1]

[1] University of East Anglia, Norwich, UK
{o.edo-osagie,i.lake,b.iglesia}@uea.ac.uk
[2] Public Health England, Birmingham, UK
obaghe.edeghere@phe.gov.uk

**Abstract.** In this paper, we propose an attention-based approach to short text classification, which we have created for the practical application of Twitter mining for public health monitoring. Our goal is to automatically filter Tweets which are relevant to the syndrome of asthma/difficulty breathing. We describe a bi-directional Recurrent Neural Network architecture with an attention layer (termed ABRNN) which allows the network to weigh words in a Tweet differently based on their perceived importance. We further distinguish between two variants of the ABRNN based on the Long Short Term Memory and Gated Recurrent Unit architectures respectively, termed the ABLSTM and ABGRU. We apply the ABLSTM and ABGRU, along with popular deep learning text classification models, to a Tweet relevance classification problem and compare their performances. We find that the ABLSTM outperforms the other models, achieving an accuracy of **0.906** and an $F1$-score of **0.710**. The attention vectors computed as a by-product of our models were also found to be meaningful representations of the input Tweets. As such, the described models have the added utility of computing document embeddings which could be used for other tasks besides classification. To further validate the approach, we demonstrate the ABLSTM's performance in the real world application of public health surveillance and compare the results with real-world syndromic surveillance data provided by Public Health England (PHE). A strong positive correlation was observed between the ABLSTM surveillance signal and the real-world asthma/difficulty breathing syndromic surveillance data. The ABLSTM is a useful tool for the task of public health surveillance.

**Keywords:** Syndromic surveillance · Sequence modelling · Deep learning · Natural Language Processing

Supported by Health Protection Research Unit, Public Health England.


I. Rojas et al. (Eds.): IWANN 2019, LNCS 11506, pp. 895–911, 2019.
https://doi.org/10.1007/978-3-030-20521-8_73

## 1 Introduction

Text classification is a well established field related to Natural Language Processing (NLP) and data mining which has seen a lot of activity. Usually, literature published in this domain studies medium to large bodies of text such as film and internet reviews as well as news articles. However, with the proliferation of social media as a viable source of data for data mining, the issue of Tweet classification has become more prominent. Tweet classification is a natural yet specific extension of the text classification problem. Tweets are very short pieces of text, each limited to 280 characters only. Forms of expression vary when they are constrained in this way. This means that although we can apply existing text classification techniques, we have to pay special attention to the concise nature of Tweets so that it does not negatively impact the workings of these techniques.

We are motivated by a real world problem. This is the analysis of ***Tweets*** for the purpose of public health surveillance. Specifically, we have investigated the use of Tweets to obtain a signal for a given syndrome [6], that is asthma and/or difficulty breathing. For this, we collected Tweets related to our syndrome of interest - asthma and/or difficulty breathing - using keywords. Unfortunately, as explained previously [6], many Tweets contain terms like "*asthma*" or "*can't breathe*" but are not actually related to individuals expressing concern over asthma or difficulty breathing. Hence the classification of relevant/irrelevant Tweets for this particular syndrome is our problem. For some context, examples of Tweets that contain the keyword "*asthma*" include "*oh I used to have asthma but I managed to control it with will power*" or "*Does your asthma get worse when you exercise?*". However, we do not consider these Tweets as relevant for our purposes. On the other hand, Tweets such as "*why is my asthma so bad today?*" express a person currently affected and will be considered as relevant.

Text classification using neural networks has been widely investigated and found to yield positive results [6,14,15]. These neural network models look at a document as a whole, examining the interrelations of words and word vectors in the document without giving any words special treatment. However, we believe that texts usually contain a number of *keywords* that inform the meaning and sentiment of the whole text. Such keywords should be used to inform the classification process. To this end, we propose to apply an attentive approach, which makes use of an encoder-decoder architecture, to short text classification, and we demonstrate its value specifically in the context of Tweet classification for public health monitoring.

Attentive neural networks pioneered for machine translation [9] have recently seen success in a range of tasks ranging from question answering, speech recognition to image captioning [1,3,32]. We propose adapting the attention mechanism for short text classification tasks such as Tweet analysis. We apply our attention mechanism using two popular RNN setups - Long Short Term Memory (LSTM) [10] and Gated Recurrent Unit (GRU) [2] networks in order to derive attention-based variants for comparisons. We call our attention-based LSTM, ***ABLSTM*** and our attention-based GRU, ***ABGRU***. After we employ our attention-based RNN classifiers to Tweet classification to generate an 'activity' signal over time, we compare our results to the activity recorded by syndromic surveillance systems maintained by Public Health England (PHE).

Our proposed approaches combine the characteristics of both deep learning and traditional classification algorithms. We combine the self-learning and intrinsic pattern recognition capabilities of deep learning with the use of keywords in classification employed by traditional classification methods. Through our experiments, we find that the *ABLSTM* and *ABGRU* are able to identify keywords in a Tweet relevant to its meaning and improve classification accuracy. As an example, Fig. 1 shows a Tweet heatmap of perceived word importance generated with our *ABLSTM* network. The darker areas/words represent words which the model deems key to the message of the Tweet. We can see that the model does a good job of recognizing that *swelling*, *throat* and *difficulty*, *breathing* are important for determining whether the Tweet is relevant to our health context. We also found that using this attention-based approach to syndromic Tweet classification yields a signal that correlates well with the signal recorded by syndromic surveillance systems put in place by PHE for England.

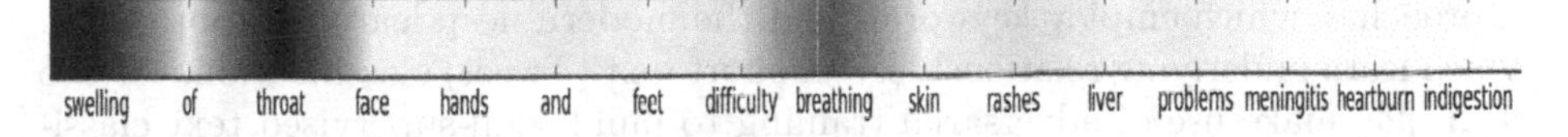


**Fig. 1.** Heatmap showing weights placed on words in a Tweet by our attentive RNN model

## 2 Related Work

The problem of text classification has a long history. In the 1960s, it was often referred to as *text categorization*, and was approached by employing a set of handcrafted logical rules based on the specific language and its grammar and idiosyncracies [24]. In the 1990s, the study of automatic text categorization became more prominent. The approaches used in these studies involved the use of pre-marked data to automatically learn discriminatory rules for classification with which new samples could then be classified [12,18,30,33,34]. This was the precursor to the approaches used today. A number of learning algorithms have been applied to text which had been vectorised using a *tf*idf* weighting method, including Support Vector Machines [12], regression models [34], nearest neighbour classification [33], Bayesian models [30], and inductive learning [18]. These algorithms assume that independent key words or phrases are important to the text category and extract vector features representing those key words or phrases using statistical methods [29]. These methods generally yield successful results but the assumption is an oversimplification that brings some shortcomings. While independent keywords and phrases are important, there are other linking words which also give meaning to a text. The way words relate can provide context and disambiguation and without this, we potentially lose some information.

Recently, deep-learning-based methods have seen a lot of success for text classification. This is mostly due to the fact that such methods can automatically and effectively learn underlying features and interrelationships in data. Some authors [14,16] have adapted Convolutional Neural Networks (CNNs), which are normally used for images, to the task of text classification. They propose a semi-supervised approach by first learning word or region embeddings from a large unstructured corpus to be used as inputs to the CNN. The work was also expanded to use RNNs for the generation of region embeddings and text classification [15]. Lee and Dernoncourt [17] make use of an RNN for short-text classification. We previously employed CNNs and RNNs for the task of Tweet classification for syndromic surveillance and compared their performance [6]. Similar work was carried out by Hankin et al. [7] in the USA. Both our work and that of Hankin et al. attempt to use deep learning models for Tweet and news article classification in order to detect symptoms reported on these platforms. The identified reported cases can then be aggregated to create an estimate of the prevalence of the syndrome(s) under investigation.

While deep learning models have seen widespread success, they treat all the words as a block of input without giving any words or phrases special treatment. We would like to leverage the advantages of both the classical text categorization approaches, which employ keywords, and the modern deep learning approaches, which learn underlying relationships, for short text (Tweet) classification. Miyato et al. [23] make use of adversarial training to build semi-supervised text classification models. They make use of LSTMs and BLSTMs, making small changes or perturbations to the word embeddings during training. This approach is also similar to the semi-supervised transductive SVM approach [13] in that both families of methods push the decision boundary far from training examples. Zhou et al. [36] make use of attention BLSTMs for entity relation classification, which is the task of finding relations between pairs of nominal values. That is useful for applications such as information extraction and question answering. Yang et al. [35] proposed an attention network with a hierarchical architecture for document classification. The hierarchical structure of the attention mechanisms is intended to mirror the hierarchical nature of documents. As such, it involves two levels of attention applied at the sentence level and the word level. It is better suited to large-scale text classification tasks that short text classification problems such as Tweet classification.

Our work is also related to Du and Huang [5] who used a BLSTM with attention for text classification. However, in their work, they compute the attention or weight of a word as the similarity between the embedding for that word and the hidden state of the BLSTM at the time step for that word. Rather than computing the attention from the hidden state, we opt for learning a function to approximate the values of the attention vector through back-propagation. Hence, the attention vector is a parameter to be learned directly. Furthermore, for the input to the classifier, Du and Huang [5] concatenate the attention vector and BLSTM output states. We propose the computation of a Tweet (or document) representation from the attention weights and hidden state. Such a representation could then also be used in a similar manner as a document embedding.

## 3 Model

In this section, we describe the proposed attention-based RNN. The attention RNN can be broken down into four parts:

1. ***Word Embedding***: This step vectorises the Tweet. It involves mapping each word in the Tweet to a fixed-dimension word embedding. In our work, we make use of GloVe embeddings which we build from a large unlabelled corpus of Tweets.
2. ***RNN***: Takes the output of the previous step as input. The RNN learns high level features from the given input.
3. ***Attention Layer***: Produces a weight vector which it uses in conjunction with the output states of the RNN to form a new Tweet representation.
4. ***Classification***: The attention-powered vector representation of the Tweet is fed into a classifier to obtain a prediction

Figure 2 shows a simple illustration of the workflow of the attentive RNN model. Each component of the process will subsequently be explored in more detail below.

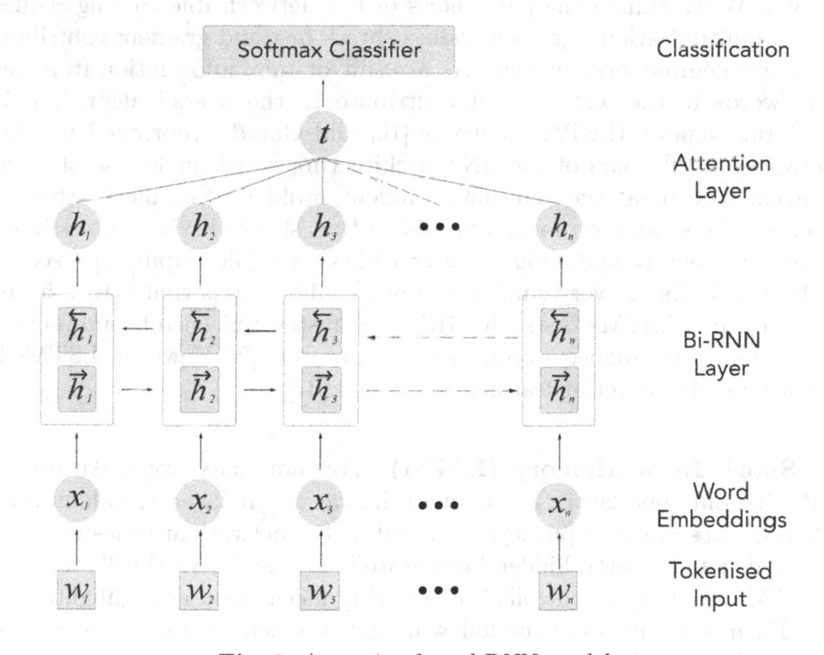


**Fig. 2.** Attention-based RNN model

### 3.1 Word Embeddings

Word embeddings (sometimes referred to as word vectors) are a powerful distributed representation of text learned using neural networks that have been shown to perform well in similarity tasks [11]. They encode semantic information of words in dense low-dimensional vectors. There are many different ways to learn word embeddings [19,22,26]. After learning, an embedding matrix $X$ of size $|V| \times d$ is produced where $V$ is the set of all the words in our vocabulary and $d$ is the dimension of each word embedding. Given a Tweet $T$ consisting of $n$ words, $T = \{w_1, w_2, ..., w_n\}$, each word $w_i$ is converted to a real-valued vector $x_i$ by performing a lookup from the embedding matrix $X$. For this work, we built GloVe embeddings [26] from a set of 5 million unlabelled Tweets.

### 3.2 RNNs

RNNs are a category of neural networks that incorporate sequential information. That is to say, while in a traditional neural network, inputs are independent, in RNNs each node depends on the output of the previous node. This is particularly useful for sequential data such as text where each word depends on the previous one. While in theory, RNNs can make use of information in arbitrarily long lengths of text, in practice, they are limited to looking back only a few steps due to the vanishing gradient problem which occurs during the back-propagation algorithm. When tuning the parameters of the network due to long sequences of matrix multiplications, gradient values shrink fast and gradient contributions from earlier neurons become zero. As a result of this, information from earlier inputs (words in the text) do not contribute to the overall algorithm. Long Short Term Memory (LSTM) networks [10] and Gated Recurrent Unit (GRU) [2] networks are flavours of the RNN architecture which make use of a gating mechanism to combat the vanishing gradient problem. Succinctly, they are a solution for the short-term memory problem that simple RNNs possess in which they cannot properly update and learn weights for earlier inputs in a sequence. LSTMs and GRUs are very similar, the main difference is that GRUs have less parameters than LSTMs. As such, GRUs are faster and have been observed to exhibit better performance on some smaller datasets [2]. However, LSTMs have been shown to be better at learning in general [31].

**Long Short Term Memory (LSTM).** For simplicity, we make use of an LSTM with only one layer. The network has an input layer $x$, hidden layer $h$, LSTM cell state $c$ and output layer $y$. Input to the network at timestep $t$ is $x(t)$, output is denoted as $y(t)$, hidden layer state is $h(t)$ and LSTM cell state is $c(t)$. The LSTM cell state is controlled by a gating mechanism as highlighted above briefly. Each cell consists of the following gates which interact with each other to dictate the overall cell state:

- input gate ($i$)
- forget gate ($f$)
- write gate ($g$)
- output gate ($o$)

Each of these gates has its own weights and biases and is a function of the previous time step's hidden state $h(t-1)$. The hidden state of a layer can then be computed as a function of the cell state as shown below:

$$c(t) = f(t) \cdot c(t-1) + i(t) \cdot g(t) \tag{1}$$

$$h(t) = o(t) \cdot tanh(c(t)) \tag{2}$$

For the sake of brevity and simplicity of our equations, let us assume that there is only one hidden layer $l$ so that we do not have to specify different equations for the different edge cases that would come with multiple layers, such as when execution is in the first layer and has no previous layer or when it is in a middle layer or the final layer. In the real world scenario, this is not the case as each hidden layer state is influenced by the hidden state in the previous time step as well as the state of the previous hidden layer. To adapt this, one may simply add the product of the weights and input of the previous layer to each activation function. The activation functions for the gates are computed as:

$$f(t) = sigmoid(W_{xf} \cdot x_t + W_{hf} \cdot h_{t-1} + b_f) \tag{3}$$

$$g(t) = tanh(W_{xg} \cdot x_t + W_{hg} \cdot h_{t-1} + b_g) \tag{4}$$

$$i(t) = sigmoid(W_{xi} \cdot x_t + W_{hi} \cdot h_{t-1} + b_i) \tag{5}$$

$$o(t) = sigmoid(W_{xo} \cdot x_t + W_{ho} \cdot h_{t-1} + b_o) \tag{6}$$

where $W_{pq}$ are the weights that map $p$ to $q$ and $b_p$ refers to the bias vector of $p$. For example, if we look at Eq. 3, $W_{xf}$ refers to the weights going from input $x$ to the forget gate $f$ and so on while $b_f$ refers to the bias of the forget gate $f$.

**Gated Recurrent Unit (GRU).** Again, for the sake of brevity and simplicity of our equations, let us assume that there is only one hidden layer $l$. The GRU cell state is controlled by a gating mechanism similar to the LSTM. Each cell consists of the following gates which interact with each other to dictate the overall cell state:

- update gate ($z$)
- reset gate ($r$)

The gates can be formalised as follows:

$$z(t) = sigmoid(W_{xz} \cdot x_t + W_z \cdot h_{t-1} + b_z) \tag{7}$$

$$r(t) = sigmoid(W_{xr} \cdot x_t + W_r \cdot h_{t-1} + b_r) \tag{8}$$

The hidden state of a layer is computed as a function of the input and gates as shown below:

$$h(t) = z(t) \cdot h(t-1) + (1 - z(t-1)) \cdot tanh(W_x + r(t) \cdot W_h \cdot h(t-1)) \tag{9}$$

where $W_{pq}$ are the weights that map $p$ to $q$ and $b_p$ refers to the bias vector of $p$. For example, if we look at Eq. 7, $W_{xz}$ refers to the weights going from input $x$ to the update gate $z$ and so on, while $b_z$ refers to the bias of the update gate $z$ and $W_z$ refers to the weights for the update gate itself.

**Bi-directional Networks.** The above RNNs process sequences in time steps with subsequent time steps taking in information from the hidden state of the previous time steps. This means that they ignore future context. Bi-directional RNNs (Bi-RNNs) extend this by adding a second layer where execution flows in reverse order [28]. Hence, each layer in a Bi-RNN has two sub-layers: one moving forward in time steps and one moving backwards in time steps. To compute the hidden state $h(t)$ of a Bi-RNN layer, we perform an element-wise sum of the hidden states computed from both its sublayers:

$$h(t) = \overrightarrow{h(t)} \bigoplus \overleftarrow{h(t)} \tag{10}$$

where $\overrightarrow{h(t)}$ and $\overleftarrow{h(t)}$ are the hidden states of the forward and backward traversals of the bi-directional RNN.

### 3.3 Attention

In this section, we describe the attention mechanism used. The Bi-RNN layer takes in a sequence of vectors for each of the words in an $n$-worded Tweet $\{x_1, x_2, ..., x_n\}$, resulting in hidden states $\{h_1, h_2, ..., h_n\}$ where $h_i$ is a vector derived from Eq. 10. That is, the hidden state of the Bi-RNN for the word $w_i$ is $h_i$. Let $H$ be a matrix containing these vectors such that $H \in \mathbb{R}^{k \times n}$ where $k$ is the number of neurons in the hidden layer. The Tweet representation $t$ is derived by taking a weighted sum of the hidden vectors with the attention weight for the relevant words. The attention weights, $\alpha$ such that $\alpha_i$ represents the attention weight for $w_i$, are obtained from trainable parameters and so are adjusted as the optimization algorithm trains the network:

$$M = tanh(H) \tag{11}$$

$$\alpha = softmax(w^T M) \tag{12}$$

$$t = M\alpha^T \tag{13}$$

where $w$ is a trainable parameter in the network and $w^T$ is its transpose. $w$, $\alpha$ and $t$ have the dimensions $k$, $n$ and $k$ respectively. Finally, the hyperbolic tangent function (tanh) is applied to $t$, the Tweet attention vector, in order to squash it between the range $[-1, 1]$ and make it easier to train with the network:

$$t^* = tanh(t) \tag{14}$$

### 3.4 Softmax Classifier

Once, the new attention-based representation for the Tweet has been obtained, it is passed to a softmax classifier to make the class prediction. The softmax classifier predicts a class $y$ from a discrete set of $m$ classes $Y$ by calculating the probability that the observed Tweet belongs to each class, $P(y|T)$, and assigning the Tweet the class with the highest probability:

$$P(y|T) = softmax(W_s t^* + b_s) \tag{15}$$

$$y = argmax_y P(y|T) \quad (16)$$

where $W_s$ represents the softmax classifier network weight and $b_s$ represents its bias term. The loss function used to train the entire network is the cross entropy loss function [4]:

$$L = -\frac{1}{m}\sum_{i}^{m} e_i log(o_i) \quad (17)$$

where $L$ estimates loss between the observed and expected values. $e$ is a one-hot encoded vector of the ground truth for $t$ and $o$ is the probability of each class being the target according to the softmax classifier.

## 4 Experiments and Results

We evaluate the performance of our attention-based RNN for Tweet classification. First, we evaluate our proposed approach's ability to automatically classify Tweets as **"relevant"** or **"irrelevant"** based on whether they associate with an individual expressing concern or discomfort over asthma/difficulty breathing or its symptoms. In these experiments, we compare the classification ability of our proposed approach to that of popular existing approaches. Next, we apply our

**Table 1.** Performance of different classifiers on Tweet relevance classification tast.

| Classifier | Metric | |
|---|---|---|
| *ABGRU* | *Accuracy* | 0.900 |
| | *Precision* | 0.734 |
| | *Recall* | 0.656 |
| | *F1* | 0.682 |
| | *F2* | 0.666 |
| *ABLSTM* | *Accuracy* | **0.906** |
| | *Precision* | 0.752 |
| | *Recall* | **0.672** |
| | *F1* | **0.710** |
| | *F2* | **0.687** |
| Convolutional Neural Network (CNN) | *Accuracy* | 0.850 |
| | *Precision* | 0.507 |
| | *Recall* | 0.562 |
| | *F1* | 0.533 |
| | *F2* | 0.550 |
| Recurrent Neural Network (LSTM) | *Accuracy* | 0.889 |
| | *Precision* | **0.762** |
| | *Recall* | 0.557 |
| | *F1* | 0.644 |
| | *F2* | 0.589 |

attention-based RNN classifier to a continuous period of collected unlabelled Twitter data in order to generate a public health signal representing Twitter activity for asthma/difficulty breathing. We then compare this signal to data from real-world syndromic surveillance systems for evaluation.

### 4.1 Tweet Relevance Classification

Tweets were collected using the official Twitter streaming Application Programmer's Interface (API). The streaming API contains parameters which can be used to restrict the Tweets obtained (e.g. keyword search, where only Tweets containing the given keywords are returned). In conjunction with epidemiologists from Public Health England (PHE), we built a set of keywords likely to be connected to the symptoms for asthma/difficulty breathing syndrome. We then expanded on this initial set using synonyms from regular thesauri as well as from the urban dictionary in order to capture some of the more colloquial language used on Twitter. This set of keywords was used to restrict our Tweet collection. We also only collected Tweets we found to be geolocated to the UK, marked as originating from a place in the UK or marked as originating from a profile with its time zone set as the UK as our syndromic surveillance problem is in fact restricted to the UK. The collected Tweets had to be cleaned with the removal of duplicates and retweets and replacing URLs and user mentions with the tokens "<URL>" and "<MENTION>" respectively. We considered implementing measures to prevent the false amplification of signals from users tweeting multiple times, potentially about the same thing. After further inspection however, we found that this was not necessary as it is discouraged by Twitter [8]. A similar concern existed for a single user posting Tweets across multiple accounts but this is also handled by Twitter's anti-spam efforts [27].

Five million Tweets were collected in total. 8000 of these Tweets were randomly selected and labelled to be used for development and experimentation. Tweets were labelled as relevant if they declared or hinted at an individual displaying symptoms pertaining to respiratory difficulties or asthma. The labelling was done by three volunteers. A first volunteer initially labelled the Tweets. A second volunteer checked the labels and flagged up any Tweets with labels that they did not agree with. These flagged Tweets were then sent to the third volunteer who then decided on which label to use. 23% of the labelled Tweets were labelled as relevant while 77% were labelled as irrelevant. This labelled dataset was then partitioned into a 70:30 training-test split. The 5 million Tweets were used to construct GloVe word embeddings while the labelled Tweets were used for experimentation. To assess the models under evaluation, accuracy can be a misleading metric as it may only be reflecting the prevalence of the majority class which is especially problematic in this application, as our dataset is quite unbalanced. Our aim is to detect Tweets which might suggest cases of a syndrome under surveillance (which for the purposes of this study was symptoms of asthma/difficulty breathing). As this is a health surveillance application, we need to prioritise that relevant Tweets are kept. We would like to reduce the number of irrelevant Tweets but not at the expense of losing the relevant Tweets in the signal. In essence, errors are not of equal cost for our application. Relevant

Tweets that are classified as irrelevant (False Negative (FN) errors) should have a higher cost and hence be minimised; we can have more tolerance of irrelevant Tweets classified as relevant (False Positive (FP) errors). These subtleties are well captured by additional measures of model performance such as *Recall*, which can be interpreted as the probability that a relevant Tweet is identified by the model and *Precision*, which is the probability that a Tweet predicted as relevant is indeed relevant. The *F-measure* (sometimes referred to as *F-score*) combines precision and recall together in a meaningful way. The formula for positive real $\beta$ is defined as:

$$F_\beta = (1 + \beta^2) \times \frac{Precision \times Recall}{(\beta^2 \times Precision) + Recall}. \tag{18}$$

The traditional $F$-measure or balanced $F1$-score [21] uses a value of $\beta = 1$. A variation of this, the $F_2$ measure, which uses $\beta = 2$, is more suited to our purpose as it weighs recall higher than precision. For this reason, in addition to accuracy, we also examine the $F1$-score for an insight into classification power and the $F2$-score for its utility in the context of syndrome detection. We implemented and applied our *ABLSTM* and *ABGRU* networks to the Tweet relevance classification task. The hyperparameters of the attention networks were selected using grid search. The dimension of our word vectors $d$ was 200. The hidden layer size $k$ was also 200. The learning rate of the optimization algorithm was 0.001. The dropout rate was set to 0.3 and the networks were trained for 50 epochs. The other parameters such as weights and biases were initialised randomly. We compare the results of applying both flavours of our proposed model and we also compare them to established deep learning text classification methods as a baseline. For this, we implemented the text classification CNN by Kim [16] and the short-text classification RNN by Nowak et al. [25]. The results of these comparisons are shown in Table 1. Note that all our results were computed from the test partition.

We found that the attentive RNNs outperformed the other architectures, with the ABLSTM being the stronger attentive RNN. As shown in Sect. 3.2, the gating mechanism used by the GRU is smaller and less complex than that of the LSTM. This means that ABGRU is faster but not quite as accurate as the ABLSTM. The LSTM RNN was seen to achieve a higher precision than the ABLSTM and ABGRU but it fell behind in terms of recall. Its recall was quite low and negatively impacted its overall performance. In effect, this translates to it being more likely to find negative class examples which were the majority class in the dataset and it suggests that it may be more suited to balanced datasets. However, our task of syndrome monitoring using social media deals with highly unbalanced data as most social media posts are not about health reporting. We also observed that the text CNN scored the worst in every metric so it performed quite badly at the Tweet relevance classification, even though it had perform well at other text classification tasks [16]. CNNs are good at extracting position-invariant features in space. They represent text as a 2D matrix made up of the word vectors of the constituent words from which the CNN learns which regions are important. However when applied to short Tweets, CNNs do not have a lot of salient spatial information to work with and so do not perform nearly as well as they would when applied to larger texts.

## 4.2 Document Embedding Capabilities

As was mentioned in Sect. 3, the output of the attention layer is a Tweet attention vector, $t$. This vector summarizes the input word vectors while putting emphasis on important words. $t$ is subsequently used as a vector representation for the Tweet in the classification part of the model. As such, the described model could also be applied to documents in other problems to create meaningful embeddings for them. We collected a random sample of Tweets, computed their attention vectors and performed t-distributed stochastic neighbour embedding (t-SNE) [20] dimensionality reduction to reduce their dimensions to 2. We then plotted these 2D attention vectors, shown in Fig. 3 in order to spatially visualize them. We found that Tweets with similar meanings and words appeared to be clustered together and away from irrelevant Tweets. In fact, it could be possible from 3 to draw a decision boundary line that roughly separates both classes. Below the red line, we see Tweets which are symptomatic of the asthma/difficulty breathing syndrome. Above the line, we see Tweets which may contain keywords related to asthma/difficulty breathing but are not expressing concern or suffering. It is also worth noting that "*wheezing*" is often used as slang to exaggerate laughter. Social media contains a lot of slang. The Tweet attention vectors capture the semantics of the different contexts of slang words, such as "*wheezing*", and this boosts its discriminatory ability. The attention vectors give us a semantic and discriminatory vector representation for our Tweets. In addition to its utility for short text classification, the attentive model we have described has the added ability to create useful document embeddings.

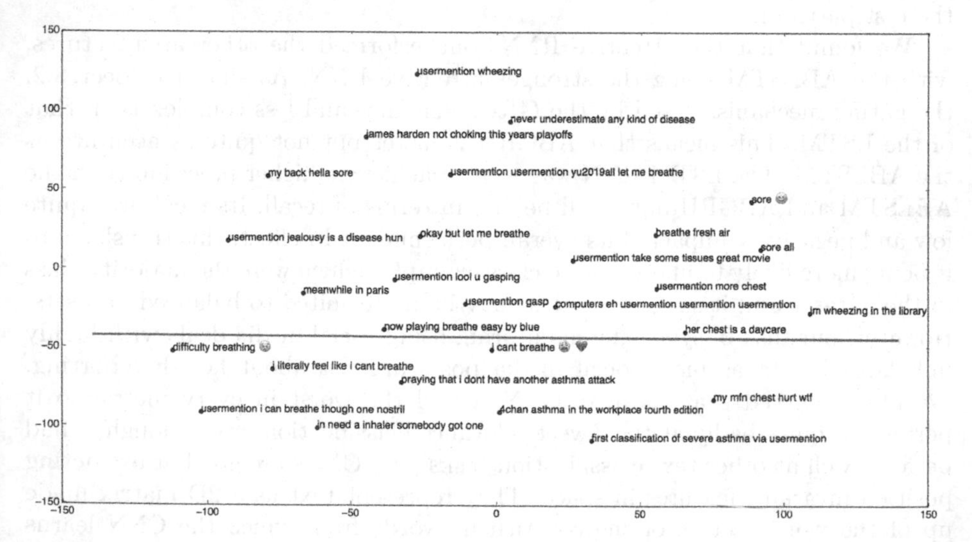


**Fig. 3.** Plot of Tweets representative of distances in embedding space. The axes represent t-SNE dimensional values.

**Table 2.** Pearson correlations and P-Values for extracted Twitter signals with syndromic surveillance signals.

| | Twitter with ABLSTM filtering | Twitter with LSTM filtering | Twitter without filtering |
|---|---|---|---|
| GPOOH Asthma/Wheeze/Difficulty Breathing | 0.792($p < 0.001$) | 0.637($p < 0.001$) | 0.555($p < 0.001$) |
| NHS 111 Difficulty breathing | 0.830($p < 0.001$) | 0.586($p < 0.001$) | 0.361($p < 0.001$) |
| NHS 111 Diarrhoea | 0.207($p = 0.09$) | 0.125($p = 0.3$) | 0.027($p = 0.8$) |

### 4.3 Syndromic Surveillance Evaluation

While we have shown in the previous section that the ABLSTM performs well at the task of Tweet relevance classification, we would like to demonstrate its utility to generate a signal for syndromic surveillance. To do this, we employ our ABLSTM to mine relevant Tweets for asthma/difficulty breathing in the UK. We then compare the results of our ABLSTM with recorded public health data. PHE runs a number of syndromic surveillance systems across England. For this experiment, we used Tweets outside of the labelled dataset used to build the classifier. We used unlabelled Tweets collected continuously between June 21, 2016 and August 30, 2016. We performed comparisons with relevant anonymised data from PHE's syndromic surveillance systems for this time period. PHE systems use primary care (general practitioner in hours and out of hours) consultations, emergency department (ED) attendances and tele-health (NHS 111) calls We performed retrospective analyses comparing the signals generated by some of these systems to that generated by our ABLSTM. For this analysis, a number of 'syndromic indicators' monitored by PHE's syndromic surveillance systems were selected based upon their availability, quality and potential association to asthma/difficulty breathing. These indicators were ***"difficulty breathing"*** and ***"asthma/wheeze/difficulty breathing"***. We also made use of ***"diarrhoea"*** as a control indicator. *Difficulty breathing* and *diarrhoea* are generated from NHS 111 calls while *asthma/wheeze/difficulty breathing* are generated from GP Out-of-hours (GPOOH) consultations. For all indicators, daily counts of consultations for relevant syndromic indicators, together with daily counts of the consultations overall were used to compute daily proportions of consultations related to the indicators. Similarly, for ABLSTM we computed daily proportions of Tweets that were relevant to the syndrome of asthma/difficulty breathing relative to the number of Tweets collected each day. We used these daily proportions to plot comparative time series shown in Fig. 4. We also included the LSTM from our comparisons in Sect. 4.1 in this experiment. We included it because it performed the best at the Tweet relevance classification task after our attentive RNNs and we wanted to observe how it measured against our attentive RNN in the real world and not just in the classification task with the limited test-partition data.

We smoothed the time series signals using a 7-day average to minimise the irregularities caused by the differences between weekend and weekday activities for GP out-of-hours services. Figure 4 shows that the signals

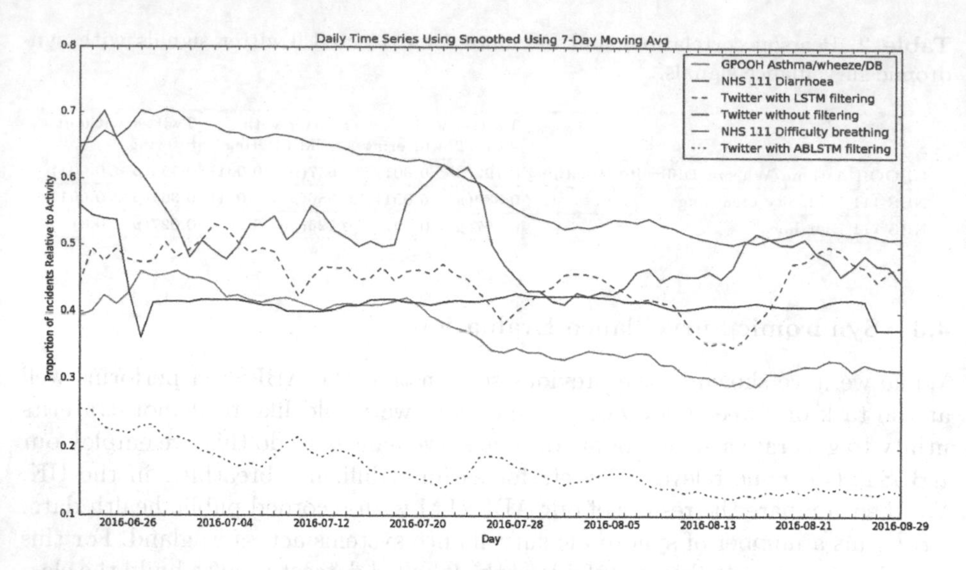


**Fig. 4.** Comparison of PHE syndromic surveillance indicators with Twitter signals.

for *asthma/wheeze/difficulty breathing*, *difficulty breathing* and Twitter with ABLSTM filtering follow very similar trends and have similar shapes. The signal for *diarrhoea* on the other hand, does not appear to be related to any others as we may expect. We also show a time series for the Twitter system without filtering. For this, we used the daily counts of collected Tweets and normalised each day's count by the average Tweet count for that week. We see in Fig. 4 that this raw Twitter signal does not match well with the *asthma/wheeze/difficulty breathing* signal. However, it still seems to match better than that of *diarrhoea.* To gain a clearer picture of how well the signals matched, we calculated the Pearson correlations between them. The results of this are shown in Table 2. Table 2 confirms that the Attentive RNN (ABLSTM) does indeed perform well at Twitter mining for syndromic surveillance for this specific syndrome and displays a strong positive correlation ($r = 0.830$) with the recorded public health signal for asthma/ difficulty breathing. The Twitter signal with LSTM demonstrated a lower correlation with what may be considered the 'ground truth' ($r = 0.586$), and was not that far off from the correlation between the ground truth and Twitter without any filtering classifier applied to it.

## 5 Conclusion

We describe an attention-based RNN architecture for short text classification. We find from the literature that most Neural Network models used to classify Tweets treat all words as equal while focusing on making use of semantic relationships between words to get the overall meaning. Our proposed approach takes this a step further by not only trying to employ these semantic relationships, but also acknowledging the presence of key words and capitalizing on them. We demonstrate the utility of the described model for Tweet classification in a syndromic surveillance context. We monitor Twitter and employ our text classifiers to detect Tweets relevant to asthma/difficulty breathing. After learning and converting the words in Tweets to vectors, the Attentive bi-directional RNN derives a vector representation for the Tweet, which places emphasis on important words in the Tweet. We experimented with LSTM and GRU units for the cells in our attentive bi-directional RNN. The attentive bi-directional LSTM (ABLSTM) approach was found to outperform the popular text-CNN and LSTM at the task of Tweet relevance classification.

The also show that the attentive model has strong understanding capabilities that can not only be used for accurate short text classification, but could also be taken advantage of for building informative document embeddings.

We then evaluate the ABLSTM performance on the real-world task of syndromic surveillance by using it to generate a public health signal from Twitter and comparing it to the signal detected by PHE syndromic surveillance systems. We found that the signal generated using the ABLSTM had a strong correlation with the 'ground truth' signal generated by PHE.

While we found strong correlations between the ABLSTM and the syndromic surveillance data, we are yet to fully assess the syndromic surveillance utility of its application to Twitter as there were no real-world major incidents during our investigated periods and we only have Twitter data from these periods. We intend to repeat this analysis prospectively over a longer time period, where incidents may occur. Another limitation is that our syndromic surveillance data was collected with the geographical scope of England. However, as described in Sect. 4.1, our location filtering was not accurate. Our Tweet filtering system focused on Tweets geolocated to the UK or marked loosely as originating from a place in the UK (e.g. by time zone). This makes our geographical filtering larger in scope (UK-level) than that of the syndromic surveillance data (England-level) and possibly inaccurate. Better Twitter location filtering needs to be carried out in order to further fine-tune our syndromic surveillance framework. Despite that, we show that the ABLSTM performs better than popular neural network architectures for short text classification, i.e. Tweet classification. We also show that the described attention model can be used for creating meaningful document embeddings which not only summarize and encode the semantics of the document, but also automatically encodes emphasis on keywords in the document.

## References

1. Bahdanau, D., Cho, K., Bengio, Y.: Neural machine translation by jointly learning to align and translate. arXiv preprint arXiv:1409.0473 (2014)
2. Cho, K., et al.: Learning phrase representations using RNN encoder-decoder for statistical machine translation. arXiv preprint arXiv:1406.1078 (2014)
3. Chorowski, J.K., Bahdanau, D., Serdyuk, D., Cho, K., Bengio, Y.: Attention-based models for speech recognition. In: Advances in Neural Information Processing Systems, pp. 577–585 (2015)

4. De Boer, P.T., Kroese, D.P., Mannor, S., Rubinstein, R.Y.: A tutorial on the cross-entropy method. Ann. Oper. Res. **134**(1), 19–67 (2005)
5. Du, C., Huang, L.: Text classification research with attention-based recurrent neural networks. Int. J. Comput. Commun. Control **13**(1), 50–61 (2018)
6. Edo-Osagie, O., De La Iglesia, B., Lake, I., Edeghere, O.: Deep learning for relevance filtering in syndromic surveillance: a case study in asthma/difficulty breathing. In: International Conference on Pattern Recognition Applications and Methods 2019, no. 8 (2019)
7. Serban, O., Thapen, N., Maginnis, B., Hankin, C., Foot, V.: Real-time processing of social media with SENTINEL: a syndromic surveillance system incorporating deep learning for health classification. Inf. Process. Manag. **56**(3), 1166–1184 (2019). https://doi.org/10.1016/j.ipm.2018.04.011
8. Fennell, K.: Everything you need to know about repeating social media posts, March 2017. https://mavsocial.com/repeating-social-media-posts/. Accessed 12 Mar 2017
9. Hermann, K.M., et al.: Teaching machines to read and comprehend. In: Advances in Neural Information Processing Systems, pp. 1693–1701 (2015)
10. Hochreiter, S., Schmidhuber, J.: Long short-term memory. Neural Comput. **9**(8), 1735–1780 (1997)
11. Jin, L., Schuler, W.: A comparison of word similarity performance using explanatory and non-explanatory texts. In: Proceedings of the 2015 Conference of the North American Chapter of the Association for Computational Linguistics: Human Language Technologies, pp. 990–994 (2015)
12. Joachims, T.: Text categorization with support vector machines: learning with many relevant features. In: Nédellec, C., Rouveirol, C. (eds.) ECML 1998. LNCS, vol. 1398, pp. 137–142. Springer, Heidelberg (1998). https://doi.org/10.1007/BFb0026683
13. Joachims, T.: Transductive inference for text classification using support vector machines. In: ICML, vol. 99, pp. 200–209 (1999)
14. Johnson, R., Zhang, T.: Semi-supervised convolutional neural networks for text categorization via region embedding. In: Advances in Neural Information Processing Systems, pp. 919–927 (2015)
15. Johnson, R., Zhang, T.: Supervised and semi-supervised text categorization using LSTM for region embeddings. arXiv preprint arXiv:1602.02373 (2016)
16. Kim, Y.: Convolutional neural networks for sentence classification. arXiv preprint arXiv:1408.5882 (2014)
17. Lee, J.Y., Dernoncourt, F.: Sequential short-text classification with recurrent and convolutional neural networks. arXiv preprint arXiv:1603.03827 (2016)
18. Lewis, D.D., Ringuette, M.: A comparison of two learning algorithms for text categorization. In: Third Annual Symposium on Document Analysis and Information Retrieval, vol. 33, pp. 81–93 (1994)
19. Luong, T., Socher, R., Manning, C.: Better word representations with recursive neural networks for morphology. In: Proceedings of the Seventeenth Conference on Computational Natural Language Learning, pp. 104–113 (2013)
20. van der Maaten, L., Hinton, G.: Visualizing data using t-SNE. J. Mach. Learn. Res. **9**(Nov), 2579–2605 (2008)
21. Maynard, D., Bontcheva, K., Rout, D.: Challenges in developing opinion mining tools for social media (2012)
22. Mikolov, T., Sutskever, I., Chen, K., Corrado, G.S., Dean, J.: Distributed representations of words and phrases and their compositionality. In: Advances in Neural Information Processing Systems, pp. 3111–3119 (2013)

23. Miyato, T., Dai, A.M., Goodfellow, I.: Adversarial training methods for semi-supervised text classification. arXiv preprint arXiv:1605.07725 (2016)
24. Nosofsky, R.M., Gluck, M.A., Palmeri, T.J., McKinley, S.C., Glauthier, P.: Comparing modes of rule-based classification learning: a replication and extension of Shepard, Hovland, and Jenkins (1961). Mem. Cogn. **22**(3), 352–369 (1994)
25. Nowak, J., Taspinar, A., Scherer, R.: LSTM recurrent neural networks for short text and sentiment classification. In: Rutkowski, L., Korytkowski, M., Scherer, R., Tadeusiewicz, R., Zadeh, L.A., Zurada, J.M. (eds.) ICAISC 2017. LNCS (LNAI), vol. 10246, pp. 553–562. Springer, Cham (2017). https://doi.org/10.1007/978-3-319-59060-8_50
26. Pennington, J., Socher, R., Manning, C.: Glove: global vectors for word representation. In: Proceedings of the 2014 Conference on Empirical Methods in Natural Language Processing (EMNLP), pp. 1532–1543 (2014)
27. Roeder, L.: What Twitter's new rules mean for social media scheduling (March 2018). https://meetedgar.com/blog/what-twitters-new-rules-mean-for-social-media-scheduling/. Accessed 13 Mar 2018
28. Schuster, M., Paliwal, K.K.: Bidirectional recurrent neural networks. IEEE Trans. Signal Process. **45**(11), 2673–2681 (1997)
29. Sebastiani, F.: Machine learning in automated text categorization. ACM Comput. Surv. (CSUR) **34**(1), 1–47 (2002)
30. Tzeras, K., Hartmann, S.: Automatic indexing based on Bayesian inference networks. In: Proceedings of the 16th Annual International ACM SIGIR Conference on Research and Development in Information Retrieval, pp. 22–35. ACM (1993)
31. Weiss, G., Goldberg, Y., Yahav, E.: On the practical computational power of finite precision RNNs for language recognition. arXiv preprint arXiv:1805.04908 (2018)
32. Xu, K., et al.: Show, attend and tell: neural image caption generation with visual attention. In: International Conference on Machine Learning, pp. 2048–2057 (2015)
33. Yang, Y.: Expert network: effective and efficient learning from human decisions in text categorization and retrieval. In: Croft, B.W., van Rijsbergen, C.J. (eds.) SIGIR 1994, pp. 13–22. Springer, London (1994). https://doi.org/10.1007/978-1-4471-2099-5_2
34. Yang, Y., Chute, C.G.: An example-based mapping method for text categorization and retrieval. ACM Trans. Inf. Syst. (TOIS) **12**(3), 252–277 (1994)
35. Yang, Z., Yang, D., Dyer, C., He, X., Smola, A., Hovy, E.: Hierarchical attention networks for document classification. In: Proceedings of the 2016 Conference of the North American Chapter of the Association for Computational Linguistics: Human Language Technologies, pp. 1480–1489 (2016)
36. Zhou, P., et al.: Attention-based bidirectional long short-term memory networks for relation classification. In: Proceedings of the 54th Annual Meeting of the Association for Computational Linguistics (Volume 2: Short Papers), vol. 2, pp. 207–212 (2016)

# A Transfer-Learning Approach to Feature Extraction from Cancer Transcriptomes with Deep Autoencoders

Guillermo López-García(✉), José M. Jerez, Leonardo Franco, and Francisco J. Veredas

Departamento de Lenguajes y Ciencias de la Computación, ETSI Informática, Universidad de Málaga, Málaga, Spain
guilopgar@uma.es

**Abstract.** The diagnosis and prognosis of cancer are among the more challenging tasks that oncology medicine deals with. With the main aim of fitting the more appropriate treatments, current personalized medicine focuses on using data from heterogeneous sources to estimate the evolution of a given disease for the particular case of a certain patient. In recent years, next-generation sequencing data have boosted cancer prediction by supplying gene-expression information that has allowed diverse machine learning algorithms to supply valuable solutions to the problem of cancer subtype classification, which has surely contributed to better estimation of patient's response to diverse treatments. However, the efficacy of these models is seriously affected by the existing imbalance between the high dimensionality of the gene expression feature sets and the number of samples available for a particular cancer type. To counteract what is known as the curse of dimensionality, feature selection and extraction methods have been traditionally applied to reduce the number of input variables present in gene expression datasets. Although these techniques work by scaling down the input feature space, the prediction performance of traditional machine learning pipelines using these feature reduction strategies remains moderate. In this work, we propose the use of the Pan-Cancer dataset to pre-train deep autoencoder architectures on a subset composed of thousands of gene expression samples of very diverse tumor types. The resulting architectures are subsequently fine-tuned on a collection of specific breast cancer samples. This transfer-learning approach aims at combining supervised and unsupervised deep learning models with traditional machine learning classification algorithms to tackle the problem of breast tumor intrinsic-subtype classification. Our main goal is to investigate whether leveraging the information extracted from a large collection of gene expression data of diverse tumor types contributes to the extraction of useful latent features that ease solving a complex prediction task on a specific neoplasia.

**Keywords:** Next-generation sequencing · Deep learning · Autocoders · Machine learning · Transfer-learning · Predictive modelling


I. Rojas et al. (Eds.): IWANN 2019, LNCS 11506, pp. 912–924, 2019.
https://doi.org/10.1007/978-3-030-20521-8_74

## 1 Introduction

Over the last decade, Next Generation Sequencing (NGS) techniques have transformed fields such as biochemistry, biology or medicine, generating an unprecedented vast amount of data that is analyzed by the omics disciplines: genomics, transcriptomics, proteomics, metabolomics and epigenomics [1]. In particular, gene expression data analysis (transcriptomics) plays an increasingly important role in *P4 medicine*–which stands for predictive, preventive, personalized and participatory–, due to the advent of the high-throughput sequencing technology called RNA-Seq [2]. In areas such as oncology, gene expression data offers a new way of describing the molecular state of a patient. As cancer is considered to be a genetic disease, a gene expression sample from a patient–which describes the genetic changes responsible for the progression of the disease, such as the over-activity or the repression of genes–contains information of paramount importance for the prevention, diagnosis and treatment of this malignant disease.

Enormous potential exists for machine learning (ML) methods to analyze these data in order to solve many different cancer prediction tasks. In fact, numerous ML studies have been proposed to tackle the problem of cancer diagnosis and prediction using gene expression data [3,4]. However, in clinical tasks such as cancer detection, the number ($M$) of available samples to solve a concrete problem is usually scarce (300–1$K$), while the number ($N$) of input features (genes or transcripts) is extremely large (10$K$–60$K$). This existing imbalance between both figures, seriously diminishes the performance of ML approaches when applied to gene expression data. To counteract the effects of what is known as the curse of dimensionality ($N \gg M$) [5], various traditional ML dimensionality reduction techniques, such as feature selection and extraction methods [6], have been applied to reduce the number of input variables. Although these techniques scale down the input feature space, the prediction performance of traditional ML methods remains moderate, as the features-samples imbalance problem is only partly solved. In this way, the reduced number of labeled samples used to train the ML models does not allow them to extract from the data the hidden patterns that contribute most to improve the performance of the predictive models.

With the aim of solving the problematic effects derived from the curse of dimensionality in a more effective way, a deep learning (DL) approach can be adopted. Nowadays, DL is the state-of-the-art technology in fields such as image recognition and natural language processing [7]. In particular, deep autoencoders (AEs) are specifically designed to exploit unlabeled data and learn high-level features, being widely employed as a feature extraction procedure [8]. In this work, we will apply different deep AE models to perform feature extraction on gene expression data, hence reducing the high number of initial features.

On the other hand, when having such a reduced number of samples, training a DL architecture from scratch would lead the model to serious over-fitting issues. Diverse strategies, such as data augmentation or transfer learning (TL) approaches are commonly used to prevent these issues. Namely, in a typical TL approach an initial DL model is pre-trained on a *base* dataset aimed at solving a *base* task. The pre-trained model is subsequently fine-tuned on a *target* dataset used to solve a *target* task, i.e. the final task (notice that *base* and *target* refer to different datasets and tasks). For the TL approach to work properly, the *base* dataset must contain a much greater number of samples than the *target* dataset. This technique has been successfully applied to many different domains, such as text classification, image processing or software error detection [9]. Here, we apply a TL approach to pre-train several deep AE models in an unsupervised manner using a large collection of unlabeled tumor samples. These pre-trained AE models are further fine-tuned on a smaller collection of labeled samples to solve a concrete supervised task for breast cancer (BRCA) subtype classification.

In fact, although the contributions of DL to cancer prediction using gene expression data are just starting to emerge and there are not yet numerous studies [10,11], a few recent works have already successfully applied a TL strategy using AEs to solve different cancer classification tasks. In [12], the authors pre-trained a stacked sparse autoencoder (SSAE) using unlabeled samples from two different tumors, and then fine-tuned the architecture using labeled samples from a third tumor type to differentiate between normal and tumor samples. In [13], traditional ML classifiers were applied using the high-level features extracted by a SSAE model in order to separate samples from two distinct tumor types. However, the cancer prediction tasks tackled by these preliminary studies are very general and relatively simple, as they aimed at classifying gene expression samples into tumor or normal classes, or distinguishing between different tumor types, which are manageable task successfully tackled by traditional feature selection and ML methods. Furthermore, the number of samples used in these studies to pre-train the deep models could still be considered as scarce ($\sim$400–1.5$K$).

In this work, we use the Pan-Cancer dataset to pre-train deep AE architectures on 9$K$ samples obtained from 32 different tumor types. The resulting architectures are then fine-tuned on a collection composed of $\sim$900 BRCA samples, aimed at solving a very specific cancer prediction task: breast tumor intrinsic subtypes classification. Our main goal is to investigate whether pre-training these DL models, by using a large collection of heterogeneous gene expression data from 32 distinct tumor types, contributes to the extraction of useful latent features that ease solving a complex cancer prediction task, such as the classification of BRCA subtypes. By means of a TL strategy, in this study we train and fine-tune different AE models and architectures to work as feature extractors, and use the extracted latent features as the input of three different ML classification algorithms that are analyzed in a comparative manner: logistic regression (LR), support vector machines (SVM) and shallow artificial neural networks (ANN). To evaluate the efficacy of the proposed TL approach, we compare the results obtained using the AE models with the performance achieved by those same three ML models when using four traditional dimensionality reduction methods: analysis of variance (ANOVA) feature selection, mutual information feature selection, chi-squared feature selection and principal component analysis (PCA).

The rest of the paper is organized as follows. Section 2 describes the gene expression datasets used within the analysis, as well as the different AE models, the transfer-learning strategy and the feature selection/extraction techniques used in combination with ML classifiers, which deal with the cancer prediction task being tackled in this study. The cross-validation strategy followed to assess the performance of the different approaches compared in this paper is also presented in that section. The results obtained from the analysis are given in Sect. 3 and, finally, some conclusions are provided in Sect. 4.

## 2 Materials and Methods

The work-flow of our TL approach is shown in Fig. 1, and the details of our method are discussed in the next subsections.

### 2.1 Gene Expression Data and Feature Pre-selection

The Pan-Cancer dataset from The Cancer Genome Atlas (TCGA) project was used in this study [14], accessed from the UCSC Xena data browser [15]. This dataset consists of $\sim 11K$ RNA-Seq gene expression samples from 33 different tumor types, which have been previously pre-processed to take into account batch effects, using $log_2(TPM + 0.001)$ transformed RSEM values. The initial number of features (transcripts) was 60498, which is an intractable number for any ML model. In order to perform an initial reduction of the feature space, we applied an unsupervised feature selection strategy. Firstly, using the standard deviation (SD) measure, the variables with constant expression values across all the samples were removed. Then, according to the median absolute deviation (MAD), the $\sim 9K$ most variably expressed genes across the samples were selected, having a final dataset of 10535 samples and 9076 features.

### 2.2 Dataset Split

Rather than using the whole Pan-Cancer dataset, we split the data into two distinct subsets: one of the subsets contains only the BRCA tumor samples (BRCA dataset, 1212 samples), whereas the other one includes the remaining samples from the rest of the 32 tumor types (non-BRCA dataset, 9323 samples). The rationale for this split is that the labeled information relating to the cancer prediction task tackled in this work is only contained in the BRCA tumor samples (see Sect. 2.4). For that reason, following a TL approach, as it is described in Fig. 1, the non-BRCA dataset is used during the unsupervised pre-training phase, while BRCA data containing the available labeled information is used to perform the supervised fine-tuning of the models.

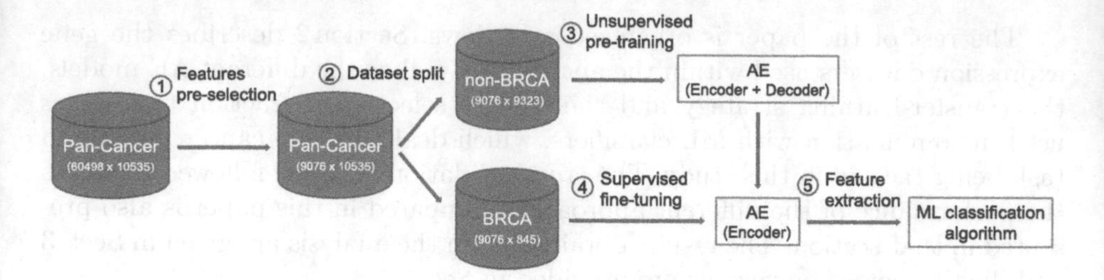


**Fig. 1.** A general overview of the TL strategy used in this work to perform BRCA intrinsic-subtypes classification.

### 2.3 Unsupervised Pre-training of Deep AE Models

An AE, in its simplest (i.e. shallowest) form, is a feed-forward neural network with only three layers: an input, a hidden and an output layer (Fig. 2A). It is an unsupervised learning method for which the main aim is to reconstruct, at the output layer, a pattern given to the input layer, so that the reconstructed output pattern is as closely similar as possible to the original input pattern. This is done by training the network using the back-propagation algorithm to minimize the reconstruction error, a function that computes the difference between the input and the output vectors.

Given an input $x = \{x_1, x_2, ..., x_n\}$, an AE tries to learn a function $\hat{x} \approx x$ (Fig. 2A). The function that transforms the input into a hidden representation is called the encoder, and can be expressed as $h(x) = f(Wx + b)$, where $f$ is the hidden activation function, $W$ is the hidden weight matrix and $b$ is the bias vector of the hidden layer. Given $n$ the number of units in the input layer and $k$ the number of hidden units, the matrix $W$ is of dimensions $n \times k$. On the other hand, the function that takes the hidden representation and transforms it into the reconstructed input representation is called the decoder, and can be expressed as $\hat{x}(h) = g(W'h + b')$, where $g$ is the output activation function, $W'$ is the output weight matrix and $b'$ is the bias vector of the output layer.

Having a hidden layer with fewer units than the input layer (i.e. $k < n$), forces the AE to compress the input vector into a lower dimensional representation, which can be reconstructed to its initial representation. In this case, the AE can be used as a dimensionality reduction method, in particular as a feature extraction procedure.

Constraining the network, such as using a small number of hidden units, has demonstrated to force the AE to extract more abstract and meaningful features in the hidden representations. In addition to reducing the number of hidden units, another popular way of constraining the model is using what is called a sparsity penalty [16]. This penalty creates sparse representations, in which hidden units tend to be inactive most of the time, favoring the units specialization. The sparsity constraint can be implemented using L1-regularization in the hidden layers, which is added to the reconstruction error function. In this way, for an input vector $x \in \Re^n$, if the mean squared reconstruction error as well as positive hidden activation functions are used, the overall loss function minimized during the learning procedure can be expressed as:

$$J_{sparse}(W,b) = \frac{1}{n}\sum_{i}^{n}(x_i - \hat{x}_i)^2 + \lambda \sum_{j}^{k} |h_j|$$

where $n$ is the number of input and output units, $k$ is the number of hidden units, $h_j$ is the activation value of the $j$-th hidden unit and $\lambda$ is the L1-regularizer penalty. The first term corresponds to the input reconstruction error, whereas the second term represents the L1-regularization, which tends to decrease the absolute values of the hidden activations towards zero, acting as a sparsity constraint.

Finally, another widely used approach to constrain the network is known as denoising AE [17]. During training, noise is added to the input data, and the difference between the input reconstruction and the original noiseless data is minimized using back-propagation. Thus, the goal of the network is to obtain a hidden representation robust to the introduction of noise at the input layer. In order to be able to reconstruct the input correctly, the corruption of the input data forces the network to extract more abstract and meaningful features in the hidden representation. This can be easily implemented using dropout in the input layer of the AE network [18].

In this work, with the purpose of extracting complex non-linear patterns from the high-dimensional gene-expression data, two different deep AE approaches have been implemented and analyzed: a deep sparse model with 3 hidden layers (see Fig. 2B), and a deep sparse denoising AE with 5 hidden layers. In both cases, the sparsity constraint has been implemented using an L1-regularization penalty for all the hidden layers in the encoder sub-networks. On the other hand, dropout has been used to introduce the noise necessary for the deep sparse denoising AE to work as expected. In addition, with the aim of reducing the initial number of features (9076) in an incremental way, the deep sparse AE uses $5K$ units in hidden layer 1 and 500 unit in hidden layer 2. For its part, the deep sparse denoising network counts on $5K$ nodes in hidden layer 1, $2K$ nodes in hidden layer 2 and 500 nodes in hidden layer 3. In terms of the number of units of each layer, both architectures are symmetric with respect to the central hidden layer, i.e. hidden layer 2 in case of the deep sparse AE and hidden layer 3 in the deep sparse denoising model. Additionally, for comparison reasons, we have also implemented a shallow sparse architecture (see Fig. 2A), which uses a single hidden layer of 500 units in which L1-regularization is employed as the sparsity constraint.

Finally, with the purpose of training the AEs using a large collection of unlabeled gene expression samples from 32 different tumor types, the models are pre-trained using the non-BRCA dataset in an unsupervised way. Before pre-training the AEs, the non-BRCA gene expression dataset is normalized using zero-one scaling. The activation function of the output layer of all three AE models is a sigmoid.

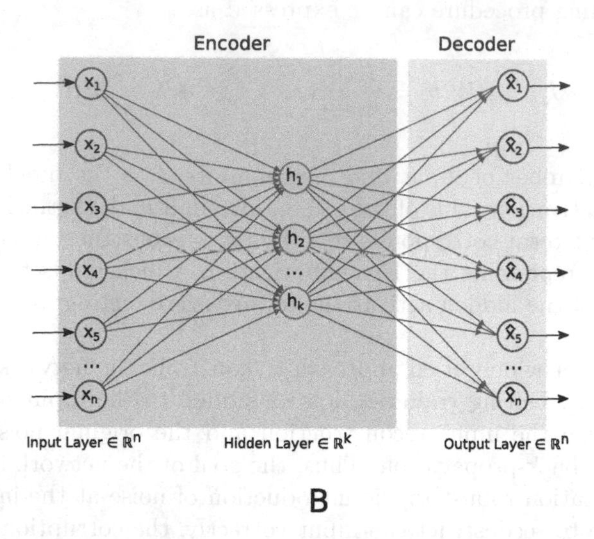


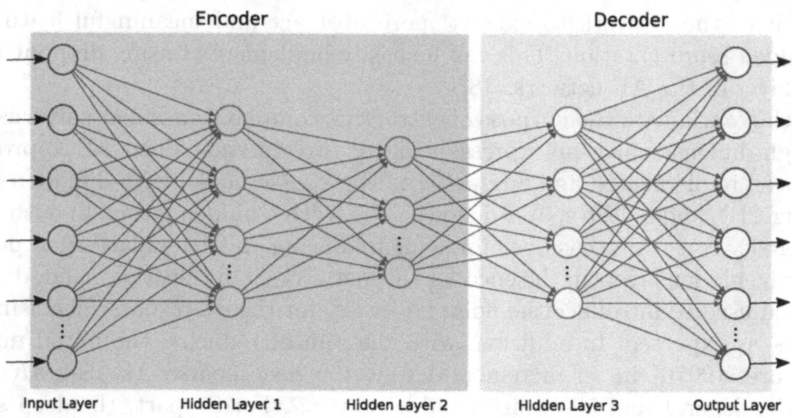


**Fig. 2.** Different AE architectures. **A** The architecture of a basic AE, where $\{x_1, x_2, ..., x_n\}$ are the units of the input layer, $\{h_1, h_2, ..., h_k\}$ are the hidden units and $\{\hat{x}_1, \hat{x}_2, ..., \hat{x}_n\}$ represent the output neurons. **B** The architecture of a deep AE with 3 hidden layers.

## 2.4 Supervised Fine-Tuning

Once pre-trained, the resulting AEs are fine-tuned using the BRCA dataset. The cancer prediction task tackled in this work is breast tumor intrinsic-subtypes classification. Hence, the variable to be predicted is the PAM50 intrinsic subtype, included among the clinical output variables contained in the BRCA samples from the Pan-Cancer dataset. PAM50 is a widely used 50-gene BRCA intrinsic subtype predictor [19], which groups the samples into four main subtypes: Luminal A,

Luminal B, Basal-like and Her-2 enriched. From the 1212 samples contained in the BRCA dataset (see Sect. 2.2), we only select the samples for which PAM50 subtypes information is known, giving a final BRCA dataset composed of 845 labeled samples (see Fig. 1). Since each sample in this dataset is labeled with one of the 4 possible PAM50 subtypes, the classification task becomes a multi-class prediction problem with 4 different classes.

To perform the fine-tuning of the AEs using the BRCA samples labeled with the PAM50 subtype labels, the unsupervised AE models have to be transformed into supervised models. To do so, the decoder part of the networks (see Fig. 2) is replaced by a softmax output layer with 4 units (one for each BRCA intrinsic subtype). Finally, the resulting network architectures are fine-tuned in an supervised manner, using back-propagation to minimize the categorical cross-entropy loss function.

### 2.5 Autoencoders for Feature Extraction

After fine-tuning the models, we eliminate the softmax output layer from the AEs, so that only the encoder part of the networks remains available. In this way, the resulting fine-tuned encoders are used as feature extraction mechanisms. Thus the encoders work by propagating forward the high-dimensional patterns given as their input, so that they are transformed, layer by layer, to get a final latent representation with fewer number of variables than the original gene-expression data. These extracted features are then used as inputs that are fed into three different ML classifiers, namely LR, SVM and ANN, which are both trained in a supervised manner and evaluated using the PAM50 subtypes information.

### 2.6 Comparison to Other Dimensionality Reduction Methods

With the aim of evaluating the efficacy of the deep AEs as feature extraction methods, we compare them to other classical feature extraction and selection algorithms when they are used in combination with the same three ML supervised models (i.e. LR, SVM and shallow NN) to tackle the PAM50-subtypes prediction task. Namely, we compare the AE feature-extraction networks to three different feature selection methods, ANOVA, mutual information and chi-squared feature selection, as well as a feature extraction procedure, PCA. Like the encoders obtained from the pre-trained and fine-tuned AEs, these algorithms are also applied to reduce the number of features contained in the labeled BRCA dataset. Again, the selected/extracted variables given by these methods are used as the input for three ML classifiers (LR, SVM, ANN), which are trained and evaluated using the PAM50 intrinsic subtypes labels.

Note that, on the one hand, the main difference between the TL approach (feature extraction via AE + ML classifier) and the traditional ML pipeline (classical feature selection/extraction algorithm + ML classifier) used in this study is the strategy employed to reduce the dimensionality of the gene expression data, as the same classification algorithms are used by both methods to perform the PAM50 subtypes prediction task. On the other hand, while the TL strategy makes use of both the non-BRCA—for unsupervised learning—and the BRCA dataset—for supervised learning—, only the BRCA dataset is used in a supervised manner in the traditional ML pipeline followed in this work for comparison purposes.

### 2.7 Validation Scheme

In this work, a 10-fold cross-validation (CV) scheme is used to estimate the predictive performance of each model using the labeled BRCA dataset. The average accuracy (ACC) calculated across the 10 test folds is used as the evaluation measure. Regarding the optimization of models' hyper-parameters, Random Search [20] with 20 iterations was performed using 5-fold CV within each of the 10 train folds, thus carrying out a nested CV procedure, using the inner 5-fold CV for model selection and the outer 10-fold CV for model evaluation. In case of the TL approach, both the fine-tuning hyper-parameters of the deep AE models (such as dropout, learning rate, momentum and number of epochs) and the hyper-parameters of the ML classifiers (such as kernel function, C and gamma for SVM and the hidden layer size, learning rate and momentum for ANN) are tuned, whereas in case of the traditional ML pipeline, only the hyper-parameters of the ML supervised models are optimized.

## 3 Results

Table 1 shows the average accuracy (ACC) and standard deviation from the 10-fold cross-validation obtained by each combination of feature selection/extraction method and ML model, when predicting PAM50 intrinsic subtypes. The rows in the table represent the different methods used to reduce the dimensionality of the gene-expression data, whereas the columns stand for the classification algorithms used to perform the PAM50-subtype prediction task. While the first four rows in Table 1 correspond to the classical feature selection/extraction procedures analyzed in this study, the last three represent the distinct AE architectures used within the TL approach for feature extraction proposed in this paper. Additionally, Fig. 3 contains a box-plot that depicts the 10-fold CV ACC test values distribution obtained by each ML classifier when using the selected/extracted features given by the different dimensionality reduction procedures.

In terms of the average test ACC, if we compare only the results obtained by the approaches using the different AE models as feature extraction methods, for all ML classifiers (i.e. LR, SVM and ANN), the deep sparse architecture performs better (88.31, 88.68 and 89.91, respectively) than the shallow sparse network (87.10, 88.08 and 88.21). Moreover, the deep sparse denoising AE obtains better performance rates for the three ML classifiers (89.29, 89.88 and 90.26) than the deep sparse model. Thus, we can conclude that the deeper the AE architecture is, the better results are achieved, showing the great potential of this sort of DL models to extract complex patterns from high-dimensional data useful for

**Table 1.** Average and standard deviation from 10-fold CV ACC test results.

| Feature selection/extraction method | ML classifier | | |
|---|---|---|---|
| | LR | SVM | ANN |
| ANOVA | $90.76 \pm 3.03$ | $90.99 \pm 2.87$ | $91.24 \pm 3.37$ |
| Mutual Information | $90.99 \pm 1.94$ | $91.35 \pm 1.90$ | $90.75 \pm 1.74$ |
| Chi-Squared | $88.07 \pm 3.04$ | $86.35 \pm 3.86$ | $86.96 \pm 3.62$ |
| PCA | $90.62 \pm 2.71$ | $90.50 \pm 3.72$ | $90.62 \pm 3.37$ |
| Sparse AE | $87.10 \pm 2.84$ | $88.08 \pm 3.83$ | $88.21 \pm 4.53$ |
| Deep Sparse AE | $88.31 \pm 2.95$ | $88.68 \pm 3.10$ | $89.91 \pm 3.97$ |
| Deep Sparse Denoising AE | $89.29 \pm 4.13$ | $89.88 \pm 2.77$ | $90.26 \pm 2.85$ |

classification purposes. On the other hand, if we focus separately on the analysis of the predictive capacity of each of the three ML classification algorithms, when using the features extracted by the AEs (Sparse, Deep Sparse and Deep Sparse Denoising), SVM (88.08, 88.68 and 89.88) performs better than LR (87.10, 88.31 and 89.29), whereas shallow ANN (88.21, 89.91 and 90.26) obtains better results than SVM. Since the shallow ANN is a connectionist model, it takes more advantage of the features extracted by the AEs—which are also feed-forward NNs—to perform the final classification task.

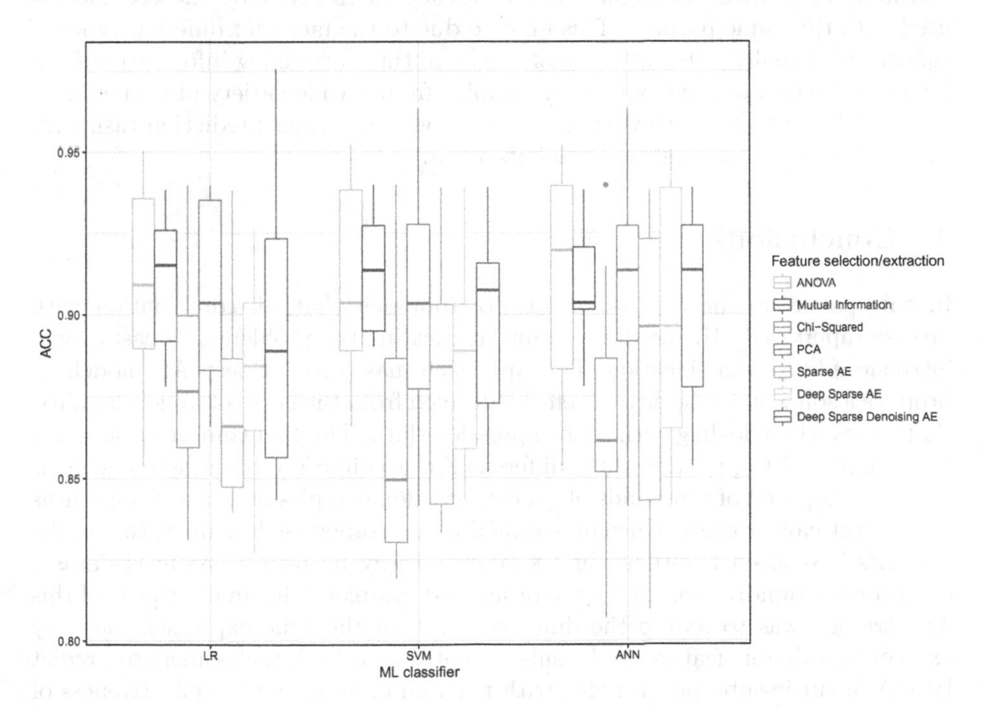


**Fig. 3.** Box-plot describing the 10-fold CV ACC test values distribution obtained by each combination of feature selection/extraction method and ML classification algorithm.

However, when comparing the traditional ML pipeline with the TL approach, the classical feature selection/extraction algorithms contribute to undoubtedly better performance than the AE models do. In terms of ACC, ANOVA and Mutual Information feature selection, as well as PCA feature extraction, outperform any of the AE models, and only the Chi-Squared feature selection procedure is surpassed by our TL approach with AEs. Among the traditional dimensionality reduction algorithms, ANOVA and Mutual Information contribute to the best predictive performances of the ML classifiers, and the highest average ACC (91.35) among all models is obtained when combining Mutual Information with SVM classifier.

The TL approach proposed in this work aims to apply deep AE models in combination with ML classification algorithms to tackle the problem of breast tumor intrinsic-subtypes classification using a scarce (845 samples) gene expression dataset. By pre-training the model with a large collection of unlabeled samples—from 32 tumors different from BRCA—and fine-tuning the resulting architecture using the reduced collection of BRCA samples, the deep AEs are able to make use of the knowledge extracted from data of other tumors to solve a particular cancer prediction task. However, the efficacy of this strategy has been shown to be limited. Thus, in terms of accuracy, the ML classifiers analyzed in this work achieve better predictive performance rates when they are preceded by classical dimensionality reduction algorithms as feature selection/extraction methods, in contrast to slightly lower efficacy rates given by the AE models used with this same purpose. This may be due to the fact that different types of cancer are actually different kinds of diseases, thus leveraging information from a large collection of gene expression samples from a wide variety of tumors does not contribute to a great extent to solve a complex cancer prediction task such as the BRCA intrinsic-subtypes classification.

## 4 Conclusions

In this paper, we have presented a TL approach that, in combination with diverse supervised ML algorithms, aims at tackling the problem of breast cancer intrinsic-subtype classification. This approach makes use of deep AE models to propose a solution to the adverse effects derived from the curse of dimensionality, that arises when dealing with gene expression data. The Pan-Cancer dataset has been employed to pre-train three different AE architectures on a heterogeneous dataset composed of thousands of gene expression samples obtained from tenths of different cancer types. Once pre-trained in a unsupervised manner, the resulting AEs have been fine-tuned in a supervised way by using a reduced dataset composed of hundreds of breast tumor labeled samples. The final purpose of this TL strategy was to reduce the dimensionality of the gene expression data by extracting valuable features to be subsequently used by ML classifiers to predict BRCA intrinsic-subtypes. Finally, with the aim of assessing the effectiveness of AE models as feature extraction mechanisms, we have analysed the contribution of three different AE architectures to the performance of several ML classifiers, and compared it to the efficacy achieved by these same ML models when preceded by four different traditional feature selection/extraction algorithms in a classical ML pipeline.

The results of the analysis showed that, on the one hand, the deep AE architectures extracted more useful features for classification purposes than the shallow AE model. On the other hand, the features selected/extracted by the traditional methods, led the ML classifiers to achieve slightly better predictive performance rates than the AE models. Hence, leveraging information from many cancer types does not seems to contribute to solve a more complex and specific cancer classification task such as prediction of breast tumor intrinsic subtypes. This findings support the hypothesis stating that different types of cancer are merely different types of diseases, all of them called cancer.

In future work, authors will continue to explore the adaptation of different DL approaches to be applied to the biomedical/bioinformatics domain, in particular to gene expression data. Special attention will be paid to model interpretability, as though many efforts have already been made in this particular field, most of DL models are still considered as "black-boxes". In areas such as oncology, if these algorithms aim to become a benchmark, interpretability must not be a lacking quality, but a main characteristic.

**Acknowledgments.** This work was partially supported by the project TIN2017-88728-C2-1-R, MINECO, Plan Nacional de I+D+I, and I Plan Propio de Investigación y Transferencia of the Universidad de Málaga.

## References

1. Schuster, S.C.: Next-generation sequencing transforms today's biology. Nat. Methods **5**(1), 16 (2007)
2. Wang, Z., Gerstein, M., Snyder, M.: RNA-seq: a revolutionary tool for transcriptomics. Nat. Rev. Genet. **10**(1), 57 (2009)
3. Kourou, K., Exarchos, T.P., Exarchos, K.P., Karamouzis, M.V., Fotiadis, D.I.: Machine learning applications in cancer prognosis and prediction. Comput. Struct. Biotechnol. J. **13**, 8–17 (2015)
4. Bashiri, A., Ghazisaeedi, M., Safdari, R., Shahmoradi, L., Ehtesham, H.: Improving the prediction of survival in cancer patients by using machine learning techniques: experience of gene expression data: a narrative review. Iran. J. Public Health **46**(2), 165–172 (2017)
5. Guyon, I.: An introduction to variable and feature selection. J. Mach. Learn. Res. **3**, 1157–1182 (2003)
6. Saeys, Y., Inza, I., Larrañaga, P.: A review of feature selection techniques in bioinformatics. Bioinformatics **23**(19), 2507–2517 (2007)
7. LeCun, Y., Bengio, Y., Hinton, G.: Deep learning. Nature **521**(7553), 436 (2015)
8. Hinton, G.E., Salakhutdinov, R.R.: Reducing the dimensionality of data with neural networks. Science **313**(5786), 504–507 (2006)
9. Pan, S.J., Yang, Q., et al.: A survey on transfer learning. IEEE Trans. Knowl. Data Eng. **22**(10), 1345–1359 (2010)

10. Fakoor, R., Ladhak, F., Nazi, A., Huber, M.: Using deep learning to enhance cancer diagnosis and classification. In: Proceedings of the International Conference on Machine Learning, vol. 28. ACM, New York (2013)
11. Danaee, P., Ghaeini, R., Hendrix, D.A.: A deep learning approach for cancer detection and relevant gene identification. In: Pacific Symposium on Biocomputing, World Scientific, pp. 219–229 (2017)
12. Xiao, Y., Wu, J., Lin, Z., Zhao, X.: A semi-supervised deep learning method based on stacked sparse auto-encoder for cancer prediction using RNA-seq data. Comput. Methods Programs Biomed. **166**, 99–105 (2018)
13. Sevakula, R.K., Singh, V., Verma, N.K., Kumar, C., Cui, Y.: Transfer learning for molecular cancer classification using deepneural networks. IEEE/ACM Trans. Comput. Biol. Bioinform. 1–1 (2018). https://doi.org/10.1109/TCBB.2018.2822803
14. Weinstein, J.N., et al.: The cancer genome atlas pan-cancer analysis project. Nat. Genet. **45**(10), 1113–1120 (2013)
15. Goldman, M., Craft, B., Kamath, A., Brooks, A.N., Zhu, J., Haussler, D.: The UCSC xena platform for cancer genomics data visualization and interpretation. bioRxiv (2018)
16. Poultney, C., Chopra, S., Cun, Y.L., et al.: Efficient learning of sparse representations with an energy-based model. In: Advances in Neural Information Processing Systems, pp. 1137–1144 (2007)
17. Vincent, P., Larochelle, H., Bengio, Y., Manzagol, P.A.: Extracting and composing robust features with denoising autoencoders. In: Proceedings of the 25th International Conference on Machine learning, pp. 1096–1103. ACM (2008)
18. Srivastava, N., Hinton, G., Krizhevsky, A., Sutskever, I., Salakhutdinov, R.: Dropout: a simple way to prevent neural networks from overfitting. J. Mach. Learn. Res. **15**(1), 1929–1958 (2014)
19. Parker, J.S., et al.: Supervised risk predictor of breast cancer based on intrinsic subtypes. J. Clin. Oncol. **27**(8), 1160 (2009)
20. Bergstra, J., Bengio, Y.: Random search for hyper-parameter optimization. J. Mach. Learn. Res. **13**, 281–305 (2012)

# Dementia Detection and Classification from MRI Images Using Deep Neural Networks and Transfer Learning

Amen Bidani[1], Mohamed Salah Gouider[1(✉)], and Carlos M. Travieso-González[2(✉)]

[1] SMART Lab, High Institute of Management of Tunis, University of Tunis, Tunis, Tunisia
amen_bidani@yahoo.fr, ms.gouider@yahoo.fr
[2] Signals and Communications Department - IDeTIC, Universidad de Las Palmas de Gran Canaria, Las Palmas de Gran Canaria, Spain
carlos.travieso@ulpgc.es

**Abstract.** In this paper, we present a new approach in the field of Deep Machine Learning, that comprises both DCNN (Deep Convolutional Neural Network) model and Transfer Learning model to detect and classify the dementia disease. This neurodegenerative disease which is described as a decline in memory, language, and other problems of cognitive skills to make daily activities, is identified in this study by using MRI (Magnetic Resonance Imaging) brain scans from OASIS dataset. These MRI brain scans are normalized before the image extraction with Bag of the features and the Learning classification methods into no-demented, very mild demented, and mild demented. Results showed that the DCNN model achieved significant accuracy for better Dementia diagnosis.

**Keywords:** Dementia · MRI · Bag of feature · K-means · Deep Machine Learning · DCNN · Transfer Learning

## 1 Introduction

Dementia disease is a neurological disorder that is characterized by a decline in memory, language, problem-solving and other cognitive skills. It involves a person's facility to perform everyday activities. The gradual loss and damage of the brain's nerve cells is considered as one of the major causes of Dementia [6]. It is responsible for more than 50% of dependency situations of the elderly people. In recent years, some studies of this disorder have estimated that by the year 2050, the number of 36 million new cases is expected to reach 100 trillion [1]. Neurological MRI is an imaging technique that provides doctors and radiologists with better tools to research a specific dementia diagnosis [6]. Therefore, it particularly helps doctors and specialists to make the best judgment about the reasons of symptoms based on the achieved tests. Classification methods in Data Mining and Machine Learning fields are used for early prediction and prevention of the disease. They are used to process large amounts of data compared to Regression method [3]. Recent research studies in neural classification have indicated


I. Rojas et al. (Eds.): IWANN 2019, LNCS 11506, pp. 925–933, 2019.
https://doi.org/10.1007/978-3-030-20521-8_75

that neural networks are an alternative to traditional classification methods. In this work, we apply Deep Machine Learning techniques for dementia detection based on brain MRI dataset. Our goal is to provide a technique which aims to classify those MRI scans that display the most potential for the diagnosis of this neurodegenerative disease. The problem can be solved with minimal error rate by using the CNN that is a hybrid technique of Kernel convolutions and neural networks [16]. The rest of this paper structured as follows: First, in the Sect. 2, we present the OASIS dataset that contains both healthy and diseased brains. Second, Sect. 3 is about the related works of dementia detection and classification using Machine and Deep Learning algorithms. Third, Sect. 4 describes the methodology of this paper. Finally, Sect. 5, the results of Learning Classification models DCNN and Transfer Learning are presented.

## 2 MRI Data Collection

For our research work, we have utilized the MRI images from the open source, OASIS (Open Access Series of Imaging Studies) where the free project of this open source provides these MRI neuroimaging datasets of the brain. In particular, we have considered the cross-sectional MRI datasets (Oasis 1) for demented and non-demented subjects (Fig. 1). The subjects are right-handed and include both men and women, and the datasets includes young, middle aged and older adults. This set consists of a cross-sectional collection of about 416 subjects, who are aged between 18 to 96 years and in total of 436 imaging sessions [3, 4, 11].

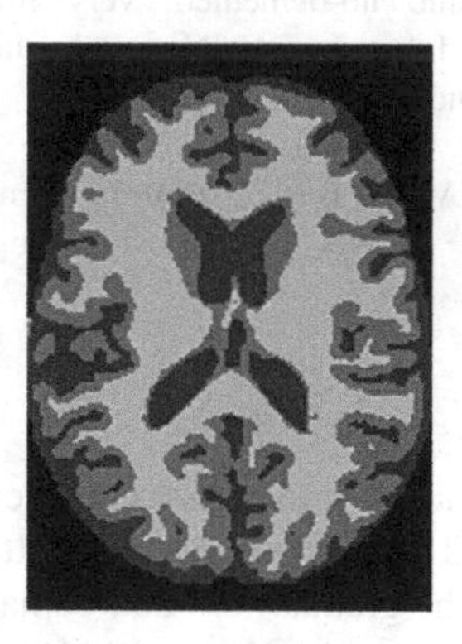

**Fig. 1.** Typical axial scan of a patient who does not have dementia from OASIS data collection.

In each case of those subjects, we can find the three or four individual T1-weighted MRI scans. Similarly, all right-handed obtained in single scan sessions are also included. 100 subjects have mild to moderate dementia and 198 subjects are all over the age of 60 as shown in Table 1. The existing data contain factors like CDR and MMSE, which are important in neurodegenerative diseases diagnosis such as Alzheimer's disease and Dementia.

*MMSE* is a mini mental state examination and CDR is clinical dementia rating which is an efficient scale to measure the severity of dementia.

*CDR takes into* consideration factors like impairments in memory orientation, judgement and problem solving.

**Table 1.** Type of OASIS dataset.

| Dataset's name | Total of subjects aged 18 to 96 | Subjects over the age of 60 | Reliability dataset |
|---|---|---|---|
| OASIS 1 | 416 Subjects | 100 Subjects, have been clinically diagnosed with very Mild to Moderate AD | 20 Subjects as non-demented on subsequent visit within 90 days of their initial session |

The dataset includes 218 no-demented subjects from the age group of 18 to 59 years and includes demented subjects from the age of 60 to 96 years with CDR scores shown in Table 2. Besides, it has 98 elderly subjects over the age of 60 with no-demented and CDR = 0. In our research study, we will use images within OASIS 1 dataset. In addition, we principally based our work on three cases of subjects; healthy with CDR = 0, Mild demented with CDR = 0.5 and Very Mild demented with CDR = 1, as found in Sect. 5 for results and discussion.

**Table 2.** Demographics of dementia status.

| Age | Dataset's name | | | |
|---|---|---|---|---|
| | Number of subjects via age | Very mild dementia CDR = 0.5 | Mild dementia CDR = 1 | Moderate dementia CDR = 2 |
| 60–69 | 15 | 12 | 3 | - |
| 70–79 | 48 | 32 | 15 | 1 |
| 80–89 | 32 | 22 | 9 | 1 |
| 90–96 | 5 | 4 | 1 | - |

## 3 Related Works

In recent years, the existing approaches in health informatics and medical imaging using machine learning and Deep Learning have been subject to extensive research [4, 5, 13, 14]. Concerning neuroimaging studies, we have reviewed the literature from 2015 to late 2018, for machine and Deep Learning approaches for early dementia detection, and classification, using Brain MRI datasets such as OASIS dataset [2–4, 9, 10, 12, 13, 15–17].

*Machine learning approach for identifying Dementia from MRI images*, proposed by [3] presents a framework for classifying MRI images for dementia from OASIS dataset. Initially, they used the Gabor filters for extracted features with 0, 30, 60, 90 orientations and the GLCM (Gray Level Co-occurrence Matrix) to normalize and fuse the features. Then, they used ICA (Independent Component Analysis) for the features selection. Lastly, this study evaluated the dementia detection. Results showed that the proposed feature fusion classifier achieves higher classification accuracy.

*Identifying Dementia in MRI scans using ANN (Artificial Neural Network) and KNN (K-Nearest Neighbor)* is proposed by [10]. They used two models for dementia detection in MRI scans: (1) Artificial Neural Network (ANN) and (2) K-Nearest Neighbor (KNN). The first model is used in segmentation and the second model is used in the normalization of the images. Machine learning and artificial intelligence can generate adequate accuracy in the classification of demented and non-demented MRI scans. Based on the result, the KNN model has performed better than ANN model in all statistical measures of performance of binary classification models. The ANN implementation resulted in 69.81% accuracy and the KNN implementation resulted in 81.13% significant accuracy.

*A brief review of automated identification of dementia using medical imaging: a survey from a pattern classification perspective*, is provided by [17]. It is divided into two parts: feature extraction and classification. They reviewed in the first, the voxel-based, vertex-based, and ROI-based feature extraction methods and LDA-based, Bayesian, SVM-based, and in the second, ANN-based pattern classification methods used in various dementia identification algorithms in brain images of MCI patients. Furthermore, they compared the performance of some of those algorithms.

*Review article* presented by [16] using deep learning methods and applications to investigate the neuroimaging correlates of psychiatric and neurologic disease. They introduced the primary concepts of deep learning approach and reviewed the neuroimaging studies that have used this approach to classify brain-based disorders. The results of these studies indicate that Deep Learning could be a powerful tool in the current search.

*A method to classify Dementia from MRI images with Feature extraction and classification using ANN,* is proposed by [2]. Primarily, they used techniques of image processing for the statistical extraction features from MRI images of brain. Afterwards, they used discrete wavelet transform.

*Deep machine learning application to the detection of preclinical neurodegenerative diseases of aging*, presented by [15]. They explore an alternate model for artificial intelligence deployment. Correspondingly, they proposed that AI might provide highly accurate and reliable detection of preclinical disease states associated with neurodegenerative diseases of aging. In addition, they investigated an approach to developing AI platforms for individual monitoring and preclinical disease detection and examined its potential benefits. One of the major challenges facing clinical detection of preclinical phases of diseases such as dementia is the high degree of inter-individual variability in aging-related changes to cognitive function.

*Machine learning of neuroimaging to diagnose cognitive impairment and dementia: a systematic review and comparative analysis* is presented by [13]. They indicated that the development of more machine learning methods in neuroimaging requires the prediction of risks of dementia, which are not yet ready for routine use. They focused also on relevant outcomes to ensure that the resulting machine learning methods are robust and reliable before testing in clinical trials.

*Detection of Alzheimer's disease from MRI using Convolutional Neural Network with Tensorflow* is proposed by [4]. They have presented a new system with the latest dataset and the latest technology such as Tensorflow. They can scale a model to production instantly. Thus, they observed in testing part that the CNN has been successful and the problem is solved with minimal error rate. They propose to use the libraries (CUDA, CuDNN) for building and the use of GPU to achieve high performance with multiple cores/parallel computing to train models.

## 4 Methodology

In this project, our proposed methodology is a Deep Machine-Learning model as shown in Fig. 2. It includes the following steps: (1) Image pre-processing, (2) feature extraction, and (3) Learning classification model.

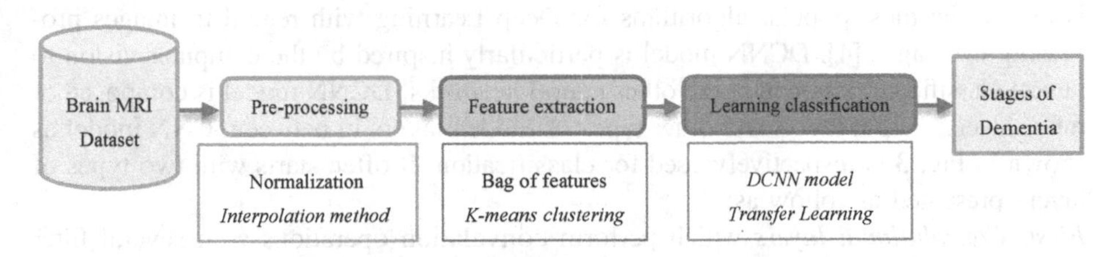


**Fig. 2.** Diagram of proposed methodology.

### 4.1 Image Pre-Processing

**Image Pre-processing.** The pre-processing is one of the major phases that needs to be considered in dementia detection using Brain MRI images from Oasis dataset. These images files contain raw information about each voxel. In fact, this type of image consists of a large number of features and a complexity in the classification model. Therefore, the images files cannot be directly used in detection and classification models. In our work, the pre-processing phase is based on normalization of the images before the extraction of the features for the deep machine-learning model.

**Normalized Image Data.** 23 layers from the top and 25 layers from the bottom of these images were loading by AlexNet, which is pre-trained for neural classification model. The size of the MRI images is reduced from 176 * 208 * 176 to 227 * 227 * 1 to facilitate our work. This process is repeated three times using the cubic interpolation method in order to reduce all images included in each folder of Brain MRI dataset.

### 4.2 Feature Extraction

From the above-mentioned Image Pre-processing, we mostly use Bag of features approach in Feature extraction phase. This phase is created and divided into three categories: one category for healthy patients who do not have dementia and two categories for patients who have dementia (mild and very mid demented). These extracted categories that measured from normalized brain image, use a custom feature extraction function from 84 images and 192192 features. Thus, bag of features for the deep machine-learning model kept 80% of the strongest feature from each category.

**K-means Clustering.** That is a vector quantization method and probably the most common way of constructing the visual vocabulary. It aims to partition N descriptors into k clusters in which each descriptor belongs to the cluster with the nearest mean, serving as a prototype of the cluster [7]. In our work, K-means clustering method used to create a 500 words visual vocabulary where the number of the features 153753, 100% for the initialization clusters centers, and the number of clusters was 500. This clustering are completed 38% where 100 iterations converged in 38 iterations.

## 4.3 Learning Classification Models

**DCNN Training Model.** Deep Neural Network method is used for training. The CNN is one of the most popular algorithms for Deep Learning with regard to images processing of images [4]. DCNN model is particularly inspired by the computer vision to neural classification models. Like other neural networks, DCNN model is composed of input layers, output layers and other types of hidden layers in between. CNN model as shown in Fig. 3 is respectively used for classification. It often starts with two types of layers presented as follow as:
*First, Convolutional layers* which perform convolution operations with several filter maps of equal size. Second, *Sub-sampling layers* which reduce the sizes of proceeding layers by averaging pixels within a small neighbor. Accordingly, DCNN model is designed in feature detection phase with several layers performing one of three types of operations on the data: convolution, pooling, or rectified linear unit (ReLU). Each layer is connected to every other layer. These three operations are repeated over tens or hundreds of layers, with each layer learning to detect different features.

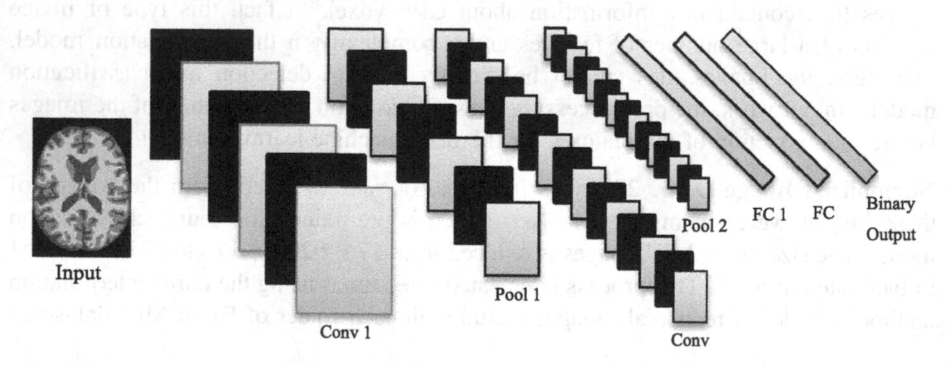


**Fig. 3.** Standard CNN model [4].

*Convolution layers* display the input images through a set of convolutional filters, each of which activates certain features from the images.

Additionally, *Pooling layers* simplify the output by performing nonlinear down sampling, and by reducing the number of parameters.

*Rectified linear unit (ReLU)* on the other hand allows for faster and more effective training by mapping negative values to zero and maintaining positive values. After the feature detection phase, the architecture of a CNN shifts to classification. The next-to-last layer is a fully connected layer (FC) that outputs a vector of K dimensions where K is the number of classes that the network will be able to predict. This vector contains the probabilities for each class of any image being subject to classification. The final layer of the CNN architecture uses a soft-max function to provide the classification output. However, with a small training dataset, this may result in the cost function that is being stuck in local minima, which may lead to overfitting/underfitting. A better alternative in such case is Transfer Learning, where pre-trained weights from the same architecture, but on a different larger dataset from the same/different domain is used to initialize the network layers. Only the last fully connected layer is pre-trained with the training data. This not only provides us with a robust set of pre-trained weights to work with, but also it gives us the opportunity to employ proven network architectures in our problem.

**Transfer Learning for Testing Model.** *VGG16* is a 16-layer network constructed by Oxfords Visual Geometry Group (VGG) [8]. It contributed in the ImageNet competition in ILSVRC-2014. One of the main reasons that VGG16 won the competition, is that it is one of the first architectures to discover network depth by pushing up to 16–19 layers and using very small (2 × 2) convolution filters.

## 5 Results and Discussion

In this paper, MRI images from the OASIS 1 dataset are being used for the detection of dementia and its classification into three stages; No-demented, very mild demented, and mild demented. Each class contains 14 subjects were used to train the DCNN training model. In addition, three classes are used for DCNN model such as for testing model where each class contains 14 subjects and it had different subjects than in the training model as shown in Table 3.

**Table 3.** OASIS 1 dataset for training and testing.

| Dementia status | No dementia | Very mild dementia | Mild dementia |
|---|---|---|---|
| CDR | 0 | 0.5 | 1 |
| Number of subjects for training | 14 | 14 | 14 |
| Number of subjects for testing | 14 | 14 | 14 |

The features extracted from the Brain MRI dataset for training and testing model are trained all through three stages; no demented, very mild demented, and mild demented as illustrated in Fig. 4 and displayed by MATLAB command.

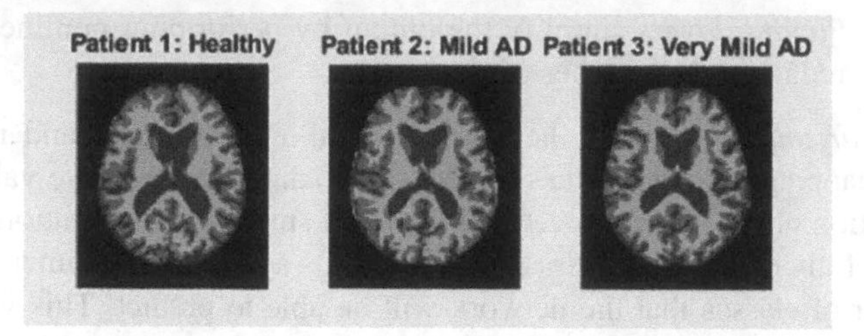


**Fig. 4.** The different stages of Dementia disease.

In our work, the performance of the Learning classification model is measured with reference to the accuracy as follows:

$$\text{Accuracy} = \frac{(\mathrm{T_p} + \mathrm{T_n})}{(\mathrm{T_p} + \mathrm{T_n} + \mathrm{F_p} + \mathrm{F_n})} \tag{2}$$

$T_p$: True positive,
$T_n$: True negative,
$F_p$: False positive,
$F_n$: False negative.

The dementia detection and classification approach from brain MRI images using DCNN model produced an important classification accuracy of 81.94% when the features are extracted. In opposition, the Transfer Learning model resulted an accuracy of 68.13%.

## 6 Conclusion and Future Works

Deep Machine Learning and computer vision domain can produce adequate accuracy in classification methods of no-demented, very mild demented and mild demented on the basis of brain MRI scans. From the above-mentioned results, we can see that the T-DCNN model has outperformed. Therefore, the level of diagnosis accuracy (>80%) for the dataset used is acceptable and produced by the DCNN model which has given maximum accuracy. In the future research work, we will use other neurodegenerative diseases such as dementia and we will use a large brain dataset is a wide range of subjects which could improve the accuracies of these Learning models by yielding better results.

## References

1. Alzheimer's disease International: World Alzheimer Report 2018
2. Akhila, J.A., Markose, C.: Feature extraction from MRI images and classification of dementia using ANN. Int. J. Adv. Res. Trends Eng. Technol. (IJARTET) **4**(6), 94–97 (2017)
3. Aruna, S.K., Chitra, S.: Machine learning approach for identifying dementia from MRI images. World academy of science, engineering and technology. Int. J. Comput. Inf. Eng. **9**(3), 881–888 (2015)

4. Awate, G.J., et al.: Detection of Alzheimer's disease from MRI using Convolutional Neural Network with Tensorflow (2018)
5. Ching, T., et al.: Opportunities and obstacles for deep learning in biology and medicine. J. R. Soc. Interface **15**, 20170387 (2018). https://doi.org/10.1098/rsif.2017.0387
6. Dubois, B., et al.: Perspective preclinical Alzheimer's disease: definition, natural history, and diagnostic criteria. Alzheimer's Dement. **12**, 292–323 (2016)
7. Chougrad, H., Zouaki, H.: Bag of features model using the new approaches: a comprehensive study. (IJACSA). Int. J. Adv. Comput. Sci. Appl. **7**(1), 226–234 (2016)
8. Hon, M., Khan, N.M.: Towards Alzheimer's Disease Classification through Transfer Learning, 29 November 2017
9. Jyoti, I., Zhang, Y.: Early diagnosis of Alzheimer's disease: a neuroimaging study with deep learning architectures. In: CVF, CVPR Workshop Paper. IEEE Explore (2017)
10. Manandhar, A., et al.: Identifying dementia in MRI scans using artificial neural network and k-nearest neighbor. Zerone Scholar **1**(1), 22–25 (2016)
11. Marcus, D.S., Wang, T.H., Parker, J., Csernansky, J.G., Morris, J.C., Buckner, R.L.: OASIS fact sheet: cross-sectional data across the adult lifespan. J. Cogn. Neurosci. **19**(9), 1498–1507 (2007)
12. Mathotaarachchi, S., et al.: Identifying incipient dementia individuals using machine learning and amyloid imaging. Neurobiol. Aging **59**, 80–90 (2017)
13. Pellegrini, E., et al.: Machine learning of neuroimaging for assisted diagnosis of cognitive impairment and dementia: a systematic review. Alzheimer's Dement. Diagn. Assess. Dis. Monit. **10**, 519–535 (2018)
14. Ravi, D., et al.: Deep learning for health informatics. IEEE J. Biomed. Health Inform. **21**(1), 4–21 (2017)
15. Summers, M.J., et al.: Deep machine learning application to the detection of preclinical neurodegenerative diseases of aging. DigitCult Sci. J. Digit. Cult. **2**(2), 9–24 (2017)
16. Vieira, S., Pinaya, W.H.L., Mechelli, A.: Using deep learning to investigate the neuroimaging correlates of psychiatric and neurological disorders: methods and applications. Neurosci. Biobehav. Rev. **74**, 58–75 (2017)
17. Zheng, C., Xia, Y., Pan, Y., Chen, J.: Automated identification of dementia using medical imaging: a survey from a pattern classification perspective. Brain Inform. **3**(1), 17–27 (2016)

# Author Index

Printed in the United States
By Bookmasters